Oceans and Human Health

Oceans and Human Health

Risks and Remedies From the Seas

Edited by
Patrick J. Walsh
Sharon L. Smith
Lora E. Fleming
Helena M. Solo-Gabriele
William H. Gerwick

ELSEVIER

AMSTERDAM • BOSTON • HEIDELBERG • LONDON
NEW YORK • OXFORD • PARIS • SAN DIEGO
SAN FRANCISCO • SINGAPORE • SYDNEY • TOKYO
Academic Press is an imprint of Elsevier

Academic Press is an imprint of Elsevier
30 Corporate Drive, Suite 400, Burlington, MA 01803, USA
525 B Street, Suite 1900, San Diego, California 92101-4495, USA
84 Theobald's Road, London WC1X 8RR, UK

This book is printed on acid-free paper. ∞

Cover design: Joanne Blank

Library of Congress Cataloging-in-Publication Data
Application submitted

British Library Cataloguing-in-Publication Data
A catalogue record for this book is available from the British Library.

ISBN: 978-0-12-372584-4

For information on all Academic Press publications
visit our Web site at www.books.elsevier.com

Printed in the United States of America
Transferred to Digital Printing, 2012

To our families and to future oceans and human health scientists;
may you be thrilled by the challenges and opportunities.

Contents

List of Contributors

Amir M. Abdelzaher University of Miami NSF NIEHS Oceans & Human Health Center, University of Miami, College of Engineering, Department of Civil, Arch., and Environ. Engineering, 1251 Memorial Drive, McArthur Building, Room 325, Coral Gables, FL 33146 USA

Lorraine C. Backer Center for Disease Control, NCEH, 4770 Buford Highway NE, MS F-46, Chamblee, Georgia 30341 USA

Roberta Balstad CIESIN, Center for Research on Environmental Decisions, Columbia University, 406 Schermerhorn Hall, MC 5501, New York, NY 10027 USA

George S. Bailey The Linus Pauling Institute, Marine and Freshwater Biomedical Sciences Center, Department of Environmental and Molecular Toxicology, Oregon State University, Corvallis, OR 97331 USA

Stephen S. Bates Fisheries and Oceans Canada, Gulf Fisheries Centre, P.O. Box 5030, 343 Université Ave., Moncton, New Brunswick, E1C 9B6 Canada

Abby D. Benninghoff Marine and Freshwater Biomedical Sciences Center, Department of Environmental and Molecular Toxicology, Oregon State University, Corvallis, OR 97331 USA

RR Bidigare University of Hawaii NSF NIEHS Center for Oceans and Human Health, Pacific Research Center for Marine Biomedicine, School of Ocean and Earth Science & Technology, University of Hawaii, Manoa, Honolulu, HI 86822 USA

Paul K. Bienfang University of Hawaii NSF NIEHS Oceans & Human Health Center, Department of Oceanography, School of Ocean & Earth Science & Technology, University of Hawaii at Manoa, 1000 Pope Road, MSB 608, Honolulu, HI 96822 USA

Jessica Bolson Division of Marine Affairs and Policy, Rosenstiel School of Marine and Atmospheric Science, University of Miami, 4600 Rickenbacker Causeway, Miami, FL 33149 USA

Sílvia Bofill-Mas Department of Microbiology, Faculty of Biology, University of Barcelona, Av. Diagonal 645, 08028 Barcelona Spain

J. Alfredo Bonilla University of Miami NSF NIEHS Oceans & Human Health Center, University of Florida, Department of Infectious Diseases and Pathology, P.O. Box 110880, Gainesville, FL 32611 USA

Tonya D. Bonilla University of Miami NSF NIEHS Oceans & Human Health Center, University of Florida, Department of Infectious Diseases and Pathology, P.O. Box 110880, Gainesville, FL 32611 USA

Robert E. Bowen Department of Environmental, Coastal & Ocean Sciences, University of Massachusetts, 100 Morrissey Blvd., Boston, MA 02125 USA

Kenneth Broad Division of Marine Affairs and Policy, Rosenstiel School of Marine and Atmospheric Science, Abbes Center for Ecosystem Science and Policy, University of Miami, 4600 Rickenbacker Causeway, Miami, FL 33149 USA

Joanna Burger Division of Life Sciences, 604 Allison Road, Piscataway, NJ 08854 USA

R. Andrew Cameron Center for Computational Regulatory Genomics, Beckman Institute 139-74, California Institute of Technology, 1200 East California Blvd., Pasadena, CA 91125 USA

Guy T. Carter Wyeth Research, 401 N Middletown Road, Pearl River NY 10965 USA

Amy Clement Division of Meteorology and Physical Oceanography, Rosenstiel School of Marine and Atmospheric Science, University of Miami, 4600 Rickenbacker Causeway, Miami, FL 33149 USA

Mary Culver NOAA/Coastal Services Center, 2234 South Hobson Ave, Charleston, SC 29405 USA

Eric Dewailly Public Health Research Unit, CHUL-CHUQ, Laval University, 945, avenue Wolfe, Ste-Foy, Québec, G1V 5B3 Canada

Tim Dixon Divison of Marine Geology and Geophysics, Rosenstiel School of Marine and Atmospheric Science, University of Miami, 4600 Rickenbacker Causeway, Miami, FL 33149 USA

Samir M. Elmir University of Miami NSF NIEHS Oceans & Human Health Center, Environmental Health and Engineering, Miami-Dade County Health Department, 1725 NW 167th Street, Miami, Florida 33056 USA

Ian R. Falconer Pharmacology, Medical Sciences, University of Adelaide, Adelaide, South Australia 5005, Cooperative Research Centre for Water Quality and Treatment, Salisbury, South Australia 5108, Australia

Lynne A. Fieber Division of Marine Biology and Fisheries, Rosenstiel School of Marine and Atmospheric Science, University of Miami, 4600 Rickenbacker Causeway, Miami, FL 33149 USA

Lora E. Fleming University of Miami NSF NIEHS Oceans & Human Health Center, Depts of Epidemiology & Public Health, University of Miami School of Medicine, Division of Marine Biology & Fisheries, Rosenstiel School of Marine and Atmospheric Sciences, 1801 NW 9th Ave Suite 200 (R-669), Miami, FL 33136 USA

Leanne Flewelling Florida Marine Research Institute, 100 Eighth Avenue SE, St. Petersburg, FL 33701 USA

M. Dolores Furones IRTA-Sant Carles de la Ràpita, Crta. Poble Nou s/n, 43540 Sant Carles de la Ràpita, Tarragona Spain

Jennifer Galvin Harvard School of Public Health, Boston, 02138 USA

William Gerwick Center for Marine Biotechnology and Biomedicine, Scripps Institution of Oceanography, University of California at San Diego, 9500 Gilman Drive, MC 0212, La Jolla, California 92093-0212 USA

Rosina Girones Dep. Microbiology, Faculty of Biology, University of Barcelona, Diagonal, 645, 08028-Barcelona, Spain

Michael Gochfeld Environmental & Occupational Health Sciences Institute, 170 Frelinghuysen Road, Piscataway, NJ 08854 USA

Kelly D. Goodwin National Oceanographic and Atmospheric Administration, Atlantic Oceanographic and Meteorological Laboratories, Stationed at NOAA/SWFC, 8604 La Jolla Shore Drive, La Jolla, CA 92037 USA

Mary Hagedorn Smithsonian Institution National Zoological Park and, The Hawaii Institute of Marine Biology, P. O. Box 1346, Kane'ohe, Hawaii 96744 USA

Mark E. Hahn Woods Hole NSF/NIEHS Center for Oceans and Human Health, Biology Department, Woods Hole Oceanographic Institution, Woods Hole, MA 02543 USA

Steven Hand Biological Sciences, 202 Life Sciences, Louisiana State University, Baton Rouge, LA 70803 USA

Joel Hansel USEPA REGION 4, 61 Forsyth Street, S.W., Atlanta, GA 30303 USA

Barbara Hickey University of Washington NSF NIEHS Oceans & Human Health Center, Box 355351, School of Oceanography, University of Washington, Seattle, WA 98125 USA

Stephen M. Highstein Washington University School of Medicine, Box 8115, 4566 Scott Avenue, St. Louis, MO 63110 USA

Christer Hogstrand King's College London, School of Biomedical and Life Sciences, Nutritional Sciences Division of Franklin-Wilkins Building 3.35, 150 Stamford Street, SE1 9NH, London, UK

John A. Kalaitzis Center for Marine Biotechnology and Biomedicine, Scripps Institution of Oceanography, University of California at San Diego, 9500 Gilman Drive, MC 0212, La Jolla, CA 92093 USA

Barry D. Keim Louisiana State Climatologist, Louisiana State University, Baton Rouge, LA 70803 USA

Christopher J. Kennedy Dept. of Biological Sciences, Simon Fraser University, Burnaby, BC, V5A 1S6 Canada

Barbara A. Kirkpatrick Mote Marine Laboratory, 1600 Ken Thompson Parkway, Sarasota, FL 34236 USA

Anthony Knap Bermuda Biological Station for Research, St-George's, GE 01 Bermuda

Jan Landsberg Florida Marine Research Institute, 100 Eighth Avenue SE, St. Petersburg, FL 33701 USA

Edward Laws University of Hawaii NSF NIEHS Oceans & Human Health Center, Louisiana State University, School of the Coast and Environment, 1002R Energy, Coast and Environment Building, Baton Rouge, LA 70803 USA

Jocelyn LeBlanc Department of Neurobiology, Harvard Medical School, Boston, MA 02115 USA

David Letson Division of Marine Affairs and Policy, Rosenstiel School of Marine and Atmospheric Science, University of Miami, 4600 Rickenbacker Causeway, Miami, FL 33149 USA

R. Wayne Litaker NOAA, National Ocean Service, 101 Pivers Island Road, Beaufort, North Carolina 28516 USA

Vincent J. Lovko Dept. of Environmental and Aquatic Animal Health, Virginia Institute of Marine Science,, The College of William and Mary, Rt. 1208, Gloucester Point, Virginia 23062 USA

Tom Malone Horn Point Laboratory, University of Maryland Center for Environmental Science, P.O. Box 775, Cambridge, MD 21613 USA

Sabine Marx Center for Research on Environmental Decisions, Columbia University, 406 Schermerhorn Hall, MC 5501, New York, NY 10027 USA

M. Danielle McDonald Division of Marine Biology and Fisheries, Rosenstiel School of Marine and Atmospheric Science, University of Miami, 4600 Rickenbacker Causeway, Miami, FL 33149 USA

Allen F. Mensinger Biology Dept., University of Minnesota, 10 University Drive, Duluth, MN 55812 USA

P. D. R. Moeller Special Projects Program, NOS/NOAA, 331 Hollings Marine Laboratory, Charleston, SC 29412 USA

Bradley S. Moore Center for Marine Biotechnology and Biomedicine, Scripps Institution of Oceanography, University of California at San Diego, 9500 Gilman Drive, MC 0212, La Jolla, California 92093 USA

Robert A. Muller Department of Geography and Anthropology, Louisiana State University, Baton Rouge, LA 70803 USA

Emile A. Okal Department of Earth & Planetary Sciences, Locy Hall, 1850 Campus Drive, Northwestern University, Evanston, IL 60208 USA

Baldomero M. Olivera Biology Department, University of Utah, Room 115 South Biology, 257 South 1400 East, Salt Lake City, UT 84112 USA

Gayle A. Orner The Linus Pauling Institute, Marine and Freshwater Biomedical Sciences Center, Oregon State University, Corvallis, OR 97331 USA

Richard Owen Environment Agency, BSIO 6BF Bristol, UK

Carol J. Palmer University of Miami NSF NIEHS Oceans & Human Health Center, Univ. of Florida, Dept. of Infectious Disease and Pathology, 2015 SW 16th Avenue, Gainesville, FL 32611 USA

M. L. Parsons Department of Marine and Ecological Sciences, Florida Gulf Coast University, 10501 FGCU Boulevard South, Fort Myers, FL 33965 USA

Daria Pereg Public Health Research Unit, CHUL-CHUQ, Laval University, 945, avenue Wolfe, Ste-Foy, Québec, G1V 5B3 Canada

Nicole Peterson Center for Research on Environmental Decisions, Columbia University, 406 Schermerhorn Hall, MC 5501, New York, NY 10027 USA

Ruben Rabinsky Philosophy Department, University of Miami, P.O. Box 248054, Coral Gables, FL 33124 USA

Ivan J. Ramirez Department of Geography, Michigan State University, 118 Geography Building, East Lansing, MI 48824 USA

Christopher M. Reddy Department of Marine Chemistry and Geochemistry, Woods Hole Oceanographic Institution, Woods Hole, MA 02543 USA

Kimberly S. Reece Dept. of Environmental and Aquatic Animal Health, Virginia Institute of Marine Science, The College of William and Mary, Rt. 1208, Gloucester Point, VA 23062 USA

Daniel Riemer Division of Marine and Atmospheric Chemistry, Rosenstiel School of Marine and Atmospheric Science, University of Miami, 4600 Rickenbacker Causeway, Miami, FL 33149 USA

Chris Rodgers IRTA-Sant Carles de la Ràpita, Crta. Poble Nou s/n, 43540 Sant Carles de la Ràpita, Tarragona Spain

Philippe Rouja PAHO/WHO Collaborating Center on Environmental and Occupational Health., CHUQ Québec, Canada

Jorge W. Santo Domingo US Environmental Protection Agency, NRMRL/WSWRD/MCCB, 26 W. Martin Luther King Dr., MS 387, Cincinnati, OH 45268 USA

Michael C. Schmale Division of Marine Biology and Fisheries, Rosenstiel School of Marine and Atmospheric Science, University of Miami, 4600 Rickenbacker Causeway, Miami, FL 33149 USA

T. Luke Simmons Center for Marine Biotechnology and Biomedicine, Scripps Institution of Oceanography, University of California at San Diego, 9500 Gilman Drive, MC 0212, La Jolla, California 92093 USA

Helena M. Solo-Gabriele University of Miami NSF NIEHS Oceans & Human Health Center, University of Miami, College of Engineering, Department of Civil, Arch., and

Environ. Engineering, 1251 Memorial Drive, McArthur Building, Room 325, Coral Gables, FL 33146 USA

John Stegeman Woods Hole NSF NIEHS Center for Oceans and Human Health Center, Biology Department, Woods Hole Oceanographic Institution, Woods Hole, MA 02543 USA

Karen A. Steidinger Florida Fish and Wildlife Conservation Commission, Fish and Wildlife Research Institute, 100 Eighth Avenue SE, St. Petersburg, FL 33701 USA

Ian Stewart School of Public Health, Griffith University, Queensland Health Forensic and Scientific Services, 39 Kessels Road, Coopers Plains, QLD 4108, Australia

Russell W. Teichert Biology Department, University of Utah, Room 115 South Biology, 257 South 1400 East, Salt Lake City, UT 84112 USA

Keith B. Tierney Dept. of Biological Sciences, Simon Fraser University, Burnaby, BC, V5A 1S6 Canada

Susan C. Tilton Environmental and Occupational Health Sciences, University of Washington, Seattle, WA USA

Graham S. Timmins University of New Mexico Health Science Center, College of Pharmacy, Albuquerque, NM 87131 USA

Vera L. Trainer NOAA OHH Center, Marine Biotoxins Program, Northwest Fisheries Science Center, Environmental Conservation Division, 2725 Montlake Boulevard E., Seattle, WA 98112 USA

Daniel W. Udwary Center for Marine Biotechnology and Biomedicine, Scripps Institution of Oceanography, University of California at San Diego, 9500 Gilman Drive, MC 0212, La Jolla, California 92093 USA

Wolfgang K. Vogelbein Virginia Institute of Marine Science, The College of William and Mary, Rt. 1208, Gloucester Point, VA 23062 USA

Patrick J. Walsh Department of Biology, Centre for Advanced Research in Environmental Genomics, University of Ottawa, 30 Marie Curie, Ottawa, ON, K1N 6N5 Canada

Ronald B. Walter Department of Chemistry & Biochemistry, Texas State University, 601 University Drive, San Marcos, TX 78666 USA

Jörg Wiedenmann University of Southampton, National Oceanography Centre, European Way, Southampton, SOI4 3ZH, UK

David E. Williams The Linus Pauling Institute, Marine and Freshwater Biomedical Sciences Center, Department of Environmental and Molecular Toxicology, 435 Weniger Hall, Oregon State University, Corvallis, OR 97331 USA

Robert W. Zeller Department of Biology, San Diego State University, 5500 Campanile Drive, San Diego, CA 92182 USA

Leonard I. Zon HHMI/Children's Hospital, Karp Family Research Laboratories, Rm 7211, 300 Longwood Avenue, Boston, MA 02115 USA

Foreword

The impetus for this book comes from a growing sense among the editors, authors and others that a new "meta-discipline" is emerging, namely the integrated study of how the oceans affect human health (and vice versa) in both positive and negative ways. Clearly, the individual disciplines that contribute to this newer area of research (e.g., oceanography, toxicology, natural products chemistry, environmental microbiology, comparative animal physiology, epidemiology and public health, social sciences, engineering, etc.) have existed as established areas of study for decades, even centuries. Interestingly, the list above reflects already highly inter-disciplinary areas, and we believe that these disciplines are now taking on yet another layer of inter- or multi-disciplinary interaction to form the new discipline of "Oceans and Human Health". Indeed, environmental scientists and physicians are beginning to cooperate and collaborate on an unprecedented level, and it would seem to be timely that this discipline now merits "textbook" status.

How did this new set of collaborations begin? It is difficult to say exactly. Certainly within the framework of the chapters in this book, the specific scientific landmarks emerge. But, convenient benchmarks in scientific histories can also come when society, government and science combine to fund and accelerate new initiatives. In the U.S., one such initiative began in 1978 and continued through the early 1980s when the National Institute of Environmental Health Science (NIEHS) of the National Institutes of Health (NIH), under the directorship of Dr. David Rall, founded its Marine and Freshwater Biomedical Sciences Center program. At NIEHS, Dr. Christopher Schonwalder oversaw a program that began with five Centers at Duke University, Mount Desert Island Biological Laboratory, Oregon State University, the University of Milwaukee, Wisconsin, and the University of Washington (and in 1991 at the University of Miami). While the program was initiated largely to recognize and support the growing need for alternative (aquatic) test systems and model systems for toxicology and carcinogenesis, clearly these Centers brought to the forefront the need and opportunities for research in topics such as seafood safety, harmful algal blooms and carcinogenesis as related to human exposures. During the same time period, the National Center for Research Resources (NIH) also began to fund Resources that focused on rearing and supplying aquatic species for research. Several of these National Resources continue today (see Section on Aquatic Animal Models of Human Health).

A second important benchmark came in the form of a National Research Council (drawing on members from the National Academy of Sciences/Institute of Medicine) report, entitled "From Monsoons to Microbes: The Role of Oceans in Human Health" (Fenical *et al.*, 1999). This report was stimulated by several scientific findings in the late 1980s through the mid 1990s that suggested a strong link between climate, microbes, and human health (see e.g., Colwell, 1996). The Consortium for Oceanographic Research and Education (CORE), its Founding President (Adm. James Watkins) and several members of the Board of Governors (including Drs. D. Jay Grimes, Paul Sandifer, Donald Boesch *et al.*) worked hard: 1) to encourage Congress to establish the National Oceans Partnership Act (1996) to enact the cooperation of nine governmental agencies concerned with the Oceans; and 2) with the NRC, to commission the 1999 report. The Staff of the NRC's Ocean Studies Board at the time, including Drs. Susan Roberts (Study Director), M. Elizabeth Clarke, Kennneth Brink and Shari Maguire and Morgan Gopnik, and the Report's authors (see Fenical *et al.*, 1999) should be commended for their role in generating this report. Notably, 1998 was also a "Year of the Ocean," and the U.S. appropriately sponsored an OHH-themed pavilion at the 1998 World's Fair in Lisbon, Portugal. Exhibits were sponsored by NIEHS and designed by University of Miami scientists, including Drs. Daniel Baden, Michael Schmale,

Lynne Fieber, Lora Fleming, and Eric Speyer, and Tom Capo.

Both the NOP Act and the NRC report, and the impact they had on scientific and lay audiences alike, created a highly collaborative climate for scientific program personnel at agencies like NIEHS, the National Science Foundation (NSF), and the National Oceanic and Atmospheric Administration (NOAA). The interagency linkages that were established, and the scientific workshops that took place, coalesced the scientific community and initiated one set of centers, the NSF-NIEHS Oceans and Human Health Centers program, with founding centers established at the University of Hawaii, the University of Miami, the University of Washington, and the Woods Hole Oceanographic Institution. Many people at NSF (including Drs. Rita Colwell, Margaret Leinen, Donald Rice, and Lawrence Clark) and NIEHS (including Drs. Kenneth Olden, Samuel Wilson, Allen Dearry, Fredrick Tyson and David Schwartz) deserve credit for this cooperation. In parallel, Congress also established (through Sen. Hollings' sponsorship and the assistance of Margaret Spring and Lila Helms from his office, and Penny Dalton of CORE) the OHH Act (2004), enabling NOAA to establish a second set of parallel Centers at the Northwest Fisheries Center in Seattle, the Great Lakes Environmental Research Laboratory in Milwaukee, and the Hollings Marine Laboratory in Charleston. There is a growing literature describing this process (Knap *et al.*, 2002; Dewailly *et al.*, 2002; Sandifer *et al.*, 2004; Tyson *et al.*, 2004; Rice *et al.*, 2004; Bowen *et al.*, 2006; Fleming and Laws, 2006; Fleming *et al.*, 2006), and these articles will include the many key players in this process, some of whom we surely have inadvertently forgotten to include in our kudos.

In picking the topics and authors for this text, we had some difficult choices to make. First, what audience should we aim it at? Rather than produce a text in which scientists were simply writing for the benefit of other established colleagues (i.e., "preaching to the choir"), our goal was to produce a text for use in both advanced undergraduate and introductory post-graduate courses. In this way, we wished to try to have the greatest effect on students seeking that "aha" moment when they choose a career/study path. In so doing, we hope to help recruit the next generation of young scientists and physicians interested in Oceans and Human Health. We believe that students at this level will have most of the right background in terms of the basic biology and physical science courses to comprehend the subjects covered.

With a multi-authored text, there was bound to be some heterogeneity in the breadth vs. depth that authors chose for coverage of their subjects. However, we think that this diversity is beneficial as it makes at least parts of the book and subject material accessible to a broader range of students. We also recognize that most instructors developing courses in this area will also not be well versed in all of the subject matter (as evidence, it took the diverse and complementary expertise areas of five co-editors and 90 authors to produce this book!).We think that this diversity of coverage depth, and of subject areas, allows for instructors to choose which chapters will be the most appropriate for the background and interests of their particular group of students. So, new courses in "Oceans and Human Health" are likely to use some, but perhaps not all, of the chapters in this text as lecture material, likely using some chapters for the particularly advanced student or for group discussion sections that explore the primary literature from the many citations given in the references sections. We also hope that the 'Study Questions' after most chapters will also be useful in generating discussion and beyond-class exploration.

A second important choice we had to make was "how saline" should the subjects be? Given the title, and the "buzz" about this emerging discipline, it was tempting to limit our authors to truly "oceanic" subjects. However, clearly, all aquatic systems are incredibly interlinked, and as the authors in our Preface and the Section on 'Effects of the Physical Environment' adeptly point out, so are the land, atmosphere/climate and society. So, early on in our planning, we made the decision to include freshwater bodies and freshwater species, and even hypersaline species (evidence the brine shrimp model in the final chapter), and more generally to try to treat the topic as an interlinked "planetary" subject.

Finally, we had to decide how to group our chapters/sections. We followed the Risks and Remedies concept, largely because there was a similar framework in the 1999 NRC report. However, we caution both students and instructors to not take these divisions too literally. Clearly, there are two sides to many of these coins, and just, as for example, a cyanobacterial toxin can be a risk (e.g., Chapter 15), so can a cyanobacterial product be a potential remedy (e.g., Chapter 22), and that even within a 'toxic' red tide dinoflagellate bloom, potentially beneficial compounds might be produced (see Chapter 13; Potera, 2007). Also, recognizing that each of the five subsections would represent a relatively large change of gears in an academic course, we attempted to provide some introductory material to the sections. Thus, for each section, there is either an overview of the section to introduce the topic, written by one of the editors, or a disciplinary/methodological chapter to give students certain tools/concepts that they might not have (yet) experienced in a standard undergraduate science curriculum (e.g., see Chapters 1, 6, and 11).

The Editors wish to acknowledge support from the National Science Foundation and National Institute of Environmental Health Sciences Oceans and Human Health Center at the University of Miami Rosenstiel School (NSF 0CE0432368; NIEHS 1 P50 ES12736), the former National Institute of Environmental Health Sciences Marine and Freshwater Biomedical Sciences Center at the University of

Miami Rosenstiel School (NIEHS P30ES05705), the National Institute of Environmental Health Sciences Red Tide POI (P01 ES 10594), NSF grant OCE 0554402, the Florida Dept of Health, the National Center for Environmental Health of the Centers for Disease Control and Prevention (CDC), the Florida Harmful Algal Bloom Taskforce, the Natural Sciences and Engineering Council of Canada, and the Canada Research Chairs Program. These and many other grants fostered the scientific discussions leading to the development of this textbook. We also wish to acknowledge the work of Ms Julie Hollenbeck (Administrator of the University of Miami NSF NIEHS Oceans and Human Health Center) in coordinating the communication and correspondence among editors and authors.

As with any text, there are bound to be errors and omissions, and for these we apologize in advance. We had a great deal of fun in editing this text, and as diverse as our interests are, reading and editing these chapters have clearly broadened each of our own horizons. We hope that you will enjoy learning from it as well, and for our more junior colleagues, that you will consider careers in this emerging area. We certainly encourage you to approach and share your ideas with the hundreds of scientists and physicians, who are authors of either these chapters or the materials referenced within, and who are actively contributing to the emerging discipline of Oceans and Human Health.

References

Bowen, R.E., Halvorson, H., Depledge, M., 2006. Editorial: the Ocean and Human Health. Marine Pollution Bulletin 52, 541–544.

Colwell, R.R., 1996. Global climate and infectious disease: the cholera paradigm. Science 274, 2025–2031.

Dewailly E, Furgal C, Knap A, Galvin J, Baden D, Bowen B, Depledge M, Duguy L, Fleming LE, Ford T, Moser F, Owen R, Suk W, Unluata U. Indicators of ocean and human health. Canadian Journal of Public Health. Revue Canadienne de Sante Publique. 2002;93 Suppl 1: S34–8.

Fenical, W., Baden, D., Burg, M., De Ville De Goyet, C., Grimes, D.J., Katz, M., Marcus, N., Pomponi, S., Rhines, P., Tester, P., Vena, J. 1999. From Monsoons to Microbes: Understandint the Ocean's Role in Human Health. National Academy Press, Washington, DC, 132pp.

Fleming, L.E., Broad, K., Clement, A., Dewailly, E., Elmir, S., Knap, A., Pomponi, S.A., Smith, S., Solo Gabriele, H., Walsh, P., 2006. Oceans and Human Health: Emerging Public Health Risks in the Marine Environment. Marine Pollution Bulletin. 53, 545–560.

Fleming, L.E., Laws, E., 2006. Overview of the oceans and human health special issue. Oceanography 19, 18–23.

Knap, A., Dewailly, É., Furgal, C., Galvin, J., Baden, D., Bowen, R.E., Depledge, M., Duguy, L., Fleming, L.E., Ford, T. Moser, F., Owen, R., Suk, W., Unluata, U., 2002. Indicators of ocean health and human health: A research framework. Environmental Health Perspectives 110, 839–845.

Potera, C., 2007. Florida red tide brews up drug lead for cystic fibrosis. Science 316, 1561–1562.

Rice, D., Dearry, A., Garrison, D., 2004. Pioneering research initiatives for Oceans and Human Health. Ecohealth. 1, 220–225.

Sandifer, P.A., Holland, A.F., Rowles, T.K., Scott, G.I., 2004. The oceans and human health. Environmental Health Perspectives. 112, A454–455.

Tyson, F.L., Rice, D.L., Dearry, A., 2004. Connecting the oceans and human health. Environmental Health Perspectives. 112, A455–456.

Patrick J. Walsh
Sharon L. Smith
Lora E. Fleming
Helena M. Solo-Gabriele
William H. Gerwick

November, 2007

Globalization and Global Ocean Change: An Overview of Influences on Human Health

ROBERT E. BOWEN

INTRODUCTION

The concept of global environmental dynamics is hardly a novel concept. The atmospheric science and oceanographic communities have long acknowledged integrated global-wide environmental systems. Climate change and the ENSO (El Ñino/Southern Oscillation) phenomenon represent well this view of global systems. However, more recently other equally vital and challenging views of global change are gaining resonance.

One, assessed in many of the pages of this textbook on Oceans and Human Health, asserts that local changes in coastal and watershed ecologies can take on a global context when imposed pressures on those systems reach a point where the cumulative impacts are globally significant. Yet another view of global change, and focus of this brief overview, acknowledges the global integration of socio-economic dynamics. Taken as a whole it is clear that traditional views of global interdependencies are quickly being altered. Humans now live in a world that is more integrated, more interdependent, and more dynamic than at any point in history. It has been argued that those born within a decade of the new millennium will see more global change in their lifetimes than any generation that has ever lived.

The pace and scale of that change is driven by myriad themes, however, this overview has a constrained scope. First, it will illustrate this new global dynamic by use of a limited number of essential socio-economic drivers. Second, it endeavors to link these changes to the core themes of the text. It can be easily argued that the most important impact of global change lies within its influence on our health, the health of our children, and, hopefully, a compassionate concern for those whom we have yet to touch. Four attributes of global socio-economics will focus this treatment: human population dynamics; changes in the patterns of global economic growth; the terms and direction of international trade; and, international touristic development.

GLOBALIZATION: HUMAN POPULATION DYNAMICS

The most common indicator of human population change conveys estimates of total global population. As noted elsewhere in this text, the United Nations (UN) currently places the earth's human population at around six billion – a 300% increase since World War II. And, current projections argue that within 20 years that number will increase by an additional 50%. These are essential numbers, but they lack the richness afforded by a more regional focus (UN, 2006). Not all countries will grow at the same rate; indeed, the population in most developing countries will increase by several times the global average, while other countries (e.g. many in Western and Eastern Europe) will actually lose population. It is the subtleties of population change that best convey essential understanding of environmental dynamics and associated impacts on human health.

Perhaps the best way to illustrate the theme is to examine the extremes. While global population will, indeed, increase, the growth rate for individual countries and regions span a startlingly broad range. For example, over the next fifty years the population of Uganda is estimated to grow by 375%, while Ukraine may lose half its present population (UN, 2006). The importance of understanding regional differences in population and environmental changes are well illustrated by a cursory view of two areas: the Adriatic and Central Africa.

The Adriatic

The countries surrounding the Adriatic provide a fascinating illustration of why a singular emphasis on total population can be analytically limited and misleading. Of the six countries bordering the Adriatic, four are projected to lose more than 10% of their population by mid-century (Italy –11%; Slovenia –20%; Croatia –16%; Bosnia –24%), while the other two will either lose population throughout the period (Montenegro) or will be in negative growth by 2050 (Albania).

However, an assessment of changes in regional population density results in a quite different, even counter-intuitive conclusion. The Defense Meteorological Satellite Program (DMSP) measures the amount of human-generated light (nightlight) emerging from the surface of the globe. One algorithm used by the Program has assessed the average annual amount and intensity of nightlight within the Adriatic region for two years, 1993 and 2000 (FAO, 2005; Bowen, et al., 2006a). During the seven years included in these data, it can be easily argued that between 15–20% of the areas where there was little or no human habitation in 1993, had been developed to the point where measurable, if not substantial, nightlight could be discerned by 2000. The simple conclusion is that even where total population is lost, the *sprawl* of new human development can be extensive and important.

Two patterns in these data are notable. First, much of this new development is focused on the coast (particularly in eastern and southern Italy, and along the west coast of the Adriatic). Second, industrial development in northern Italy (particularly in areas surrounding the Po River) hold challenges to the riverine/watershed system supporting both nearby riparian population, as well as downstream impacts on the Adriatic. Simply, the environmental footprint of a diminishing population is expanding and extending onto new environments that had until recently seen little population pressure. This kind of developed state sprawl can impose several pressures on the state of environmental condition and human health:

- New human development can substantially destroy and fragment critical habitat reducing the ecological service value formally contributed by those areas. The destruction of spawning habitat for both commercial and recreational fisheries, and of the hazard buffer of coastal wetlands are just two examples;
- More people residing along the coast means a greater number of people exposed to natural hazards like tsunami's and intense coastal storms;
- Changes in nutrient dynamics can lead to eutrophication, and to changes to the frequency, distribution and intensity of harmful algal activity;
- Industrialization within coastal areas and watershed can increase the level of anthropogenic chemical pollutants in coastal waters and other aquatic systems. Bioaccumulation in fish stocks can transmit this increased risk to humans locally and more broadly if that product enters international trading markets.

In short, the footfall of new human habitation can leave a deep and wide environmental imprint.

Central Africa

The other extreme is Central Africa where the average population growth is estimated to be several times the global average. If one were to illuminate a broad brush sweeping across central Africa from Liberia on the west to Ethiopia in the east, virtually all the highlighted countries are estimated to, at least, double their populations by 2050 (UN, 2006). While few of these countries are coastal, all are reliant on aquatic systems tied to human health. This case study can convey a different set of challenges:

- This region is already among the poorest in the world, and while estimates suggest economic improvements for the region (mostly driven by historically high prices for metals and other natural resources), those gains will be more than offset by stunning population growth rates;
- If one were to overlay this brushstroke with recent estimates of the Palmer Drought Severity Index provided by the Intergovernmental Panel on Climate Change (IPCC), severe intensification of regional drought is the forecast (IPCC, 2007a);
- If true, one should expect increased famine, increased incidence of waterborne diseases, and an increase in civil unrest.
- One consequence of greater civil unrest will almost certainly lead to an increase of number of refugees seeking asylum elsewhere in the region. Africa already accounts for about a quarter of displaced peoples worldwide and that number, given the challenges above, will almost certainly increase substantially (UNHCR, 2006).

These points underlie an essential argument within this text. Water, whether saline or fresh, defines one of the most essential connections between humans and the environment.

These two extremes illustrate quite different examples of how environmental dynamics influence human health. In the Adriatic, expanding development pressures are leading to the destruction and fragmentation of critical habitat and to the introduction of excessive pollutant loads. In Africa, acute poverty and the near absence of wastewater treatment are driving increases in disease exposure through enteric pathogens sourced in the water supply. These two regions are connected to the health of aquatic systems. They are tied by an acute need to refine our understanding of population dynamics, the environment, and the water we need.

GLOBALIZATION: PATTERNS OF GLOBAL ECONOMIC GROWTH

Humans often define themselves in economic terms. The study of economics is, by definition, an anthropocentric (i.e. human-centered) enterprise. And, the dominant forces in today's economics are driven by economic globalization. As a global people, we are more economically connected than at any point in history and those growing economic independencies may well be viewed historically as the most important social force of the first half of this century.

Global economic integration has been a theme of note since the post World War II meeting at Bretton Woods (New Hampshire) which, among other agreements, created the World Bank, the International Monetary Fund (IMF), and the General Agreement on Tariffs and Trade (GATT). However, until quite recently, global economics had been dominated by the developed nations of Europe and North America. With the opening decade of the new century, that dominance has been mitigated, if not ended, by emerging economies within the so-called "developing world." Of particular note is the emerging importance of the "BRIC" nations: Brazil, Russia, India, and China. While the BRIC nations are estimated to account for about 40% of new economic growth (a number generally proportional to their population), the fact that four countries can contribute so substantially is obviously noteworthy.

Existing data and estimates argue that emerging economies of the developing world will grow at a substantially greater rate than will the currently developed states. The International Monetary Fund argues that over the first decade of the century, emerging economies will reach annual economic growth rates of around 7%, while those in the traditional western democracies will be limited to rates averaging about 2.5% (IMF, 2004; including 2006 data updates). That is nearly a three-fold difference in the relative rate of economic growth.

Admittedly, growth rates tell only a partial story. However, other data indicate the importance, present and future, of emerging economies in both global economic and environmental terms. Not surprisingly, economic data suggest that China will be a, or even the, dominant global economic actor in coming years. It is likely that China will surpass the United States as the world's leading economic actor well before 2030. And, it is equally likely that India will soon surpass Japan in total contribution to global economic production (IMF, 2004).

The inference to the current text of economic globalization has less to do with relative rates of economic growth than does the sector source of that growth and the impact on global trading patterns. For example, in most of the developed world, future economic growth will focus in expansions in the service sector. Alternatively, much of the growth in emerging economies lies in expansion of manufacturing and resource development (aquaculture is a useful example). How that emerging economic expansion is conducted will influence greatly its environmental impact. Environmentally sensitive manufacturing and resource use practices are broadly available; however, the degree to which emerging economies will embrace them is unclear.

Greenhouse gas emissions provide an example of the challenge. At present, on a per-capita basis, both India and China emit about one-tenth the amount of anthropogenic carbon into the atmosphere than does the U.S, while Japan and Italy introduce about half the U.S. per capita average. How India and China will respond to economic growth – whether the average will be at the level of the U.S. or that of Japan – will be one of the defining environmental attributes in the decades to come. For example, the British journal, *The Economist*, estimates that both India and China will have more cars on the road than the U.S. by 2035; indeed, 2006 marked the year in which China surpassed the United States in the number of cars produced annually (*Economist*, 14 April 2007). Increase in car ownership is linked to another forecast reported in The Economist:

> "Over the next decade, almost a billion new consumers will enter the global marketplace as household incomes rise above the threshold at which people generally begin to spend on non-essential goods." (Economist, 14 September 2006)

That a billion new consumers will enter the global economy driving changes in product supply and demand in a single decade is unquestionably unprecedented.

The myriad economic and environmental impacts one could expect are well beyond the scope of this overview, however, two attributes are particularly essential in understanding the global importance of the relationship between aquatic systems and human health. Changes in the terms and directions of international commodity trade and international tourism are both a part of the globalization story and essential to the understanding of the integration of the environment and human health.

GLOBALIZATION: TERMS AND DIRECTION OF INTERNATIONAL TRADE

The intensity, breadth and direction of global trade are all essential to understanding the scale of global interaction. As late as 1970, international trade accounted for about a tenth of global economic output. Today, it accounts for more than a quarter of gross global product. If trade is important understanding the direction of trade is essential. Emerging economies are now approaching half of total export trade (FAO, 2007a).

The importance of trade is, perhaps, best represented in the question of international seafood trade. The United States, for example, imports as much as 80% of seafood destined for human consumption. Seafood has become the most broadly and intensely traded commodity in the global market. For example, in 1980, the value of coffee exports entering the U.S. was twice that of seafood. By 1990, seafood had supplanted coffee as the dominant export commodity and by 2000 seafood imports were twice the value of coffee imports (FAO, 2007b).

More than virtually any other commodity, the safety of seafood is linked to the quality of the environment from which it is harvested. The risks detailed within the pages of this text take on a central interest if the harvest of the environments herein assessed finds its way into global markets and onto the tables of people with little understanding of the source of the food they eat. This is not an argument for closing international commodity markets – far from it. However, it is an illustration of the need for a greater and broader understanding of aquatic systems and human health. The challenge in that understanding lies both in the paucity of relational studies (this textbook is a fundamental contribution to that challenge), and in the lack of information on the quality of aquatic systems within the jurisdictional borders of emerging economies. The point is not that commodities imported from emerging economies are, in fact, less safe than those from developed states; rather, the point is that relatively little is known about either the quality of environmental systems at an effective level of detail, or about the so-called "chain-of-custody" that brings product from the environment to the plates of consumers (Yasuda and Bowen, 2006).

GLOBALIZATION: INTERNATIONAL TOURISTIC DEVELOPMENT

A further underscore to the link between intensifying globalization and human health is tourism. While trade provides a central driver to economic integration, a more human form of global dynamics resides in the increasingly common activity of international travel. In the time before commercial jet aircraft, international travel for tourism was restricted to an elite few. In the decades since, the number of individuals traveling abroad has grown at a steady and steep pace. As late as 1995, the number of international arrivals had just broken the 500 million threshold. It took nearly half a century to reach that point. The estimates for 2006 (UNWTO, 2007) assesses a 65–70% increase over the 1995 levels. While the rate of increase in international arrivals is important, the absolute number of travelers is equally impressive. The 2006 UN World Tourism Organization (UNWTO) estimates argue that nearly 850 million people traveled to another country during that year. This means that worldwide about one in seven people are traveling internationally annually.

There are three essential points connecting touristic travel and the themes of this textbook. First, while North America and Europe still dominate the absolute numbers of touristic arrivals, recent trends suggest that South Asia, South-East Asia and Sub-Saharan Africa are increasingly important destinations. Indeed, those trends suggest that arrivals in those regions are growing at a rate nearly double those in the developed west (UNWTO, 2007). Second, given the arguments surrounding the challenges of understanding aquatic-sourced human health risks in emerging economies, tourism can be viewed as an additional risk vector exposing travelers to pathogenic organisms, tropical toxins and other anthropogenic compounds of risk. Arguably, travelers are at greater risk to environmental risk than the resident population because they may lack essential antibodies common in the resident population, but absent in visitors. Third, one of the drivers to coastal development in many places is the capture of more of this tourism market. Tourism expansion is often viewed solely as an economic benefit. Often, it is. Development supporting new touristic destinations can spill over into improvements in public infrastructure – such as airport/seaport improvement, utilities and roadways/railways. The successful entrainment of tourists can provide new and stable jobs, and general economic improvement for populations starving for opportunity (there are, of course numerous conditions and constraints to this optimistic view of local touristic value).

However, while new touristic development can be conducted with an eye to environmental sustainability and equitable distribution of benefit, it can also be conducted in a quite different way. Without an eye toward sustainability, these developments can, and do, destroy and fragment the very environments they are marketing. Furthermore, touristic development can mitigate or eliminate environmental system values and critical habitats whose total social benefits are poorly understood or, more often, simply ignored.

GLOBALIZATION: THE NEED TO INTEGRATE INFORMATION AND GOVERNMENT ACTION

This overview merely touches the surface of issues of great depth and breadth. Even the most cursory touch engages the obvious need to better integrate information and government policy. An important constraint in any governance effort is the need for, and the common lack of, essential empirical studies. The complexity and pace of change relating to aquatic systems and human health highlights the need for more directed study and active attention within the policy community (at all levels of government). This point has been specifically addressed by two recent efforts.

The recently released IPCC Working Group II, "Summary for Policymakers," argues that our understanding of the inevitability of climate change must lead the policy community to substantially enhance efforts to create measures that allow society to adapt.

> "Adaptation will be necessary to address results from the warming which is already unavoidable due to past emissions . . . A wide array of adaptation options is available, but more extensive adaptation than is currently occurring is required to reduce vulnerability to future climate change. There are barriers, limits and costs, but these are not fully understood." (IPCC, 2007b)

On the specific questions at the core of this textbook, the need for action was similarly raised in the "Oristano Declaration on the Ocean and Human Health" (Bowen et al, 2006).

> ". . . the global coastal environment is under threats through intensified natural resource utilization brought about by higher densities of settlements, increasing shipping, rapidly growing aquaculture production, expanding tourism activities, massive resource exploitation and other activities. All of these have shown to contribute individually, but more importantly cumulatively, to higher risks for public health and the global burden of disease."
>
> ". . . added human pressures to inhabit, develop and exploit the coast and its resources have brought a pace and scale of change deserving of acute attention and response."

We are clearly beyond the point where a more active and expansive sense of partnership between the natural science, social science and policy communities is needed.

SUMMARY

The pages of this overview, quite admittedly, tell too neat a story and are analytically coarse. However, its goals are less to assess the impact of globalization on human health than to illustrate and support the essential importance of the arguments presented in this textbook on oceans and human health. The omissive and selected use of information on the growing role of globalization in our collective daily lives simply extends the need to integrate the pages of this volume into the socio-economic dimension.

If, as the New York Times columnist Thomas Friedman puts it, the "world is flat" (his enticing metaphor for globalization), it remains a world where texture and contour are both uncertain and in process. Globalization is neither inherently good nor bad. While the reality of globalization is unquestioned, the future influence of globalization on environmental and social sustainability is uncertain, at best. If sustainability is to be achieved, however, a greater and broader sense of social and environmental linkage is essential. Herein resides both motivation and an essential value of this textbook. It elegantly presents persuasive arguments that a better and more integrated understanding of aquatic systems is essential to a better and more integrated understanding of human health.

The changing shape of our social world both mirrors and drives the state of our natural environment. The argument here is that refined understanding is needed if we are to mitigate the risks and enhance the opportunities that most directly influence the health of our global community. It is hard to imagine a more important impact of scientific understanding.

References

Bowen, R.E., Halvorson, H., Depledge, M., (Eds.), 2006. The Ocean and Human Health. Marine Pollution Bulletin 52, 539–540.

Bowen, R.E., Davis, M., Frankic, A., 2006a. Human Development and Resource Use in the Coastal Zone: Influences on Public Health. Oceanography 19, 62–71.

Food and Agricultural Organization of the United Nations (FAO), 2005. Coastal GTOS Draft Strategic Design and Phase I Implementation Plan. By Christian, R., Baird, D., Bowen, RE., Clark, D., DiGiacomo, P., de Mora, S., Jimenez, J., Kineman, J., Mazzilli, S., Servin, G., Talaue-McManus, L., Viaroli, P., Yap, H., GTOS Report No. 36. Rome: FAO, 93 pp.

Food and Agricultural Organization of the United Nations (FAO), 2007a. FAOSTAT. http://faostat.fao.org/site/342/default.aspx.

Food and Agricultural Organization of the United Nations (FAO), 2007b. FishSTAT. www.fao.org/fi/statist/FISOFT/FISHPLUS.asp

Intergovernmental Panel on Climate Change (IPCC), 2007a. Climate Change 2007: The Physical Science Basis. Working Group I Contribution to the IPCC Fourth Assessment Report. Niarobi, 6 February 2007 http://www.ipcc.ch/present/WMEF_FINAL.ppt

Intergovernmental Panel on Climate Change (IPCC), 2007b. Working Group II Contribution to the Intergovernmental Panel on Climate Change, Fourth Assessment Report. Climate Change 2007: Climate Change Impacts, Adaptation and Vulnerability. http://www.ipcc.ch/SPM13apr07.pdf

International Monetary Fund (IMF), 2004. World Economic Outlook: The Global Demographic Transition. http://www.imf.org/external/pubs/ft/weo/2004/02/

United Nations Department of Social and Economic Affairs (UN), 2006. World Population Prospects: The 2006 Revision Population Database. http://esa.un.org/unpp/index.asp?panel=2.

United Nations High Commission on Refugees (UNHCR), 2006. Refugees by Numbers, 2006 Edition. http://www.unhcr.org/basics/BASICS/4523b0bb2.pdf

United Nations World Tourism Organization (UNWTO), 2007. World Tourism Barometer, 5, 1, January 2007.http://www.unwto.org

Yasuda, T., Bowen, R.E., 2006, Chain of Custody as an Organizing Framework in Seafood Risk Reduction. Marine Pollution Bulletin 52, 640–649.

SECTION I

RISKS

A. Effects of the Physical Environment

CHAPTER

1

Background Oceanography

EDWARD LAWS

INTRODUCTION

Climate is generally considered to be the long-term average of weather. One might say somewhat flippantly that climate is what you expect, and weather is what you get. Factors typically taken into consideration when characterizing climate include average temperature, the range of temperatures, and average precipitation. One may also consider factors such as humidity, wind speeds, snow and ice, photoperiod, and so forth. Broadly speaking, one can divide the Earth's climate into three zones based on latitude: polar, temperate, and tropical. However, climatic regimes can also be characterized in many other ways based on a variety of factors, such as maritime (influenced by the ocean), continental (typical of the interior of large land masses and far from the influence of the ocean), alpine (high altitude—above the tree line), and arid (dry).

Most scientists now agree that human activities are causing the climate of the Earth to change and that the changes, now subtle, will become more apparent during the next several centuries. The effects of projected climate changes on the human population are likely to be profound (Patz *et al.*, 2005). Impacts will include *inter alia*, changes in temperature and precipitation and associated effects on agricultural productivity, a sea level rise, and a spread toward higher latitudes of the prevalence of tropical diseases such as yellow fever and malaria (Laws, 2007). By far the most important cause of anthropogenic effects on climate has been the release of carbon dioxide (CO_2) into the atmosphere as a result (primarily) of fossil fuel burning and deforestation. Because CO_2 is a greenhouse gas (i.e., it effectively traps infrared radiation that would otherwise escape to outer space), its presence in the atmosphere helps to warm the Earth. If human beings burn most of the remaining fossil fuels (coal, oil, and natural gas) over the course of the next 100 to 200 years, the concentration of CO_2 in the atmosphere will likely rise to ~1900 parts per million by volume (ppmv)[1] from its current value of 380 ppmv (Caldeira and Wicket, 2003), enough to raise global temperatures by ~10°C (Berner, 1994). The ocean has the potential to absorb virtually all of this anthropogenic CO_2, but the response time of the ocean is very slow, on the order of 10,000 years, because the air-sea boundary is a considerable limiting factor to gaseous exchange. More efficient use of fossil fuels will not change this picture. Because the response time of the ocean is so long, it makes little difference whether the fossil fuels are burned over the course of the next 100 years or the next 300 years. Either way, the CO_2 concentration in the atmosphere would rise to ~1900 ppmv. In this chapter, we review some of the basic information needed to understand the climate of the Earth, the variations of climate from one region of the globe to another, and the impact of the ocean on climate and climate change, and hence its potential for impacting human health.

THE CLIMATE OF THE EARTH OVER GEOLOGICAL TIME

To put our discussion in context, it is important to realize that over geological time the climate of the Earth has in fact changed dramatically. Despite geological evidence for oxygen-producing photosynthesis as early as 3.5 billion years ago (e.g., widespread deposits of oxidized iron called banded iron formations) the Earth's atmosphere appears to

[1]One ppmv is one liter of CO_2 in 1 million liters of air. Because air behaves much like an ideal gas, 1900 ppmv is equivalent to 1900 molecules of CO_2 for every million molecules of N_2 plus O_2, the principal components of the Earth's atmosphere.

have remained devoid of oxygen for roughly another 1.5 billion years. Most of the oxygen produced by photosynthetic processes was apparently consumed by reactions with (primarily) ferrous iron and (secondarily) sulfide in seawater (Schlesinger, 1997). Following this so-called rusting of the oceans, it was possible for oxygen to diffuse into the atmosphere, but atmospheric O_2 concentrations comparable to present values (21%) were probably not reached until the Silurian, roughly 430 million years ago. Initially much of the oxygen released to the atmosphere was apparently consumed by reactions with reduced minerals such as pyrite (FeS_2), resulting in fluvial transfer of Fe_2O_3 to the ocean. This process of terrestrial weathering is evidenced by the accumulation of the so-called red beds, deposits of Fe_2O_3 alternating with layers of other lithogenous ocean sediments. Consistent with this scenario is the fact that the earliest occurrence of red beds roughly coincides with the latest banded iron formation deposits (Schlesinger, 1997).

There is good reason to believe that atmospheric O_2 levels have not fluctuated outside the 15% to 35% range since the Silurian (Berner and Canfield, 1989). At O_2 concentrations less than 15%, fires would not burn (Lovelock, 1979), and at concentrations greater than about 25%, even wet organic matter would burn freely (Watson *et al.*, 1978). The principal mechanism responsible for the stability of atmospheric O_2 concentrations appears to be the negative feedback between O_2 concentrations and the long-term burial of organic matter in sedimentary rocks (Schlesinger, 1997).

Particularly noteworthy from the standpoint of current global climate change issues is the fact that atmospheric CO_2 concentrations during Phanerozoic time (approximately the past 570 million years) have generally been higher than current values, perhaps by as much as a factor of 20 to 25 during the Cambrian (Berner and Kothavala, 2001). The impact of these elevated CO_2 concentrations on the climate of the Earth has been profound (Fig. 1-1). Since the formation of the solar system, the luminosity of the Sun has increased by about 43%, a result of the Sun's slow expansion associated with the conversion of hydrogen to helium in its core (Sagan and Chyba, 1997). In the absence of greenhouse gases to trap infrared radiation, the Earth would have been fully glaciated until roughly 1 billion years ago, but geological evidence indicates that there has been abundant liquid water on the Earth's surface for more than 3 billion years (Sagan and Chyba, 1997). Ammonia may have accounted for much of the greenhouse effect in the reducing atmosphere of the early Earth (Sagan and Mullen, 1972; Sagan, 1977), but once atmospheric O_2 levels rose to ~21%, ammonia concentrations were probably far too low to provide much of a greenhouse effect. At the present time, water vapor accounts for about 95% of the total greenhouse effect, CO_2 for 3.6%, N_2O for about 1%, and CH_4 for 0.4%. In the absence of an atmosphere, the Earth's surface

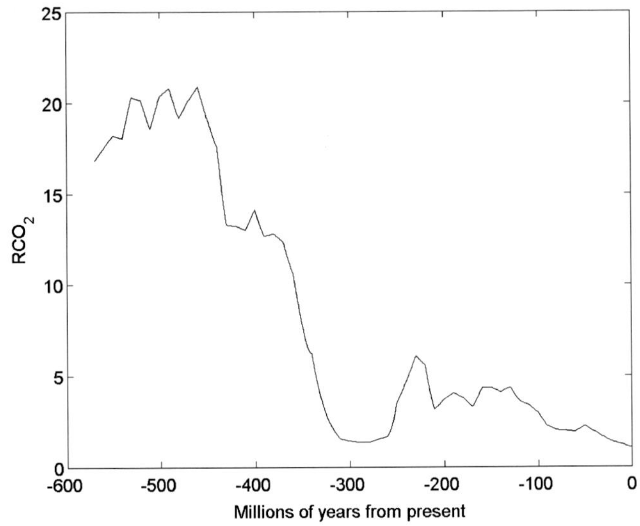

FIGURE 1-1. Ratio of atmospheric CO_2 in times past to the present concentration (RCO_2) as determined from the Geocarb II model (Berner, 1994).

temperature would average about 255°K or −18°C. The fact that the Earth's surface temperature averages about 288°K or 15°C is largely attributable to the fact that greenhouse gases are rather opaque to infrared radiation. At the beginning of the Phanerozoic eon, the solar constant was about 5% less than it is today. Had atmospheric CO_2 concentrations been the same then as now, the Earth's surface temperature would have averaged about 2°C (Berner, 1994).

In addition to climatic effects associated with variations in atmospheric CO_2 concentrations, the Earth has experienced dramatic climatic changes manifested by the advance and retreat of continental ice sheets and polar ice caps. Continental drift is certainly one factor that has influenced the ice age cycle; the movement of Antarctica to the South Pole is a case in point. The most recent ice age began roughly 40 million years ago with the accumulation of ice on Antarctica, but it intensified during the Pleistocene with the development of continental ice sheets in the Northern Hemisphere. During the Pleistocene ice age there was a cyclical advance and retreat of the Northern Hemisphere ice sheets that is most commonly attributed to variations in the eccentricity, axial tilt, and precession of the Earth's orbit around the Sun. This explanation of glacial/interglacial periodicity was initially advanced by the Serbian geophysicist Milutin Milanković, but it did not gain widespread acceptance until studies of deep-sea sediments during the 1960s and 1970s produced evidence consistent with so-called Milankovitch cycles (Hays *et al.*, 1976). These cycles are clearly apparent in the record of atmospheric CO_2 in the Vostok ice core (Fig. 1-2). Evident in this figure is a systematic pattern of atmospheric CO_2 variation from roughly 180 to 280 ppmv. Low CO_2 concentrations are associated

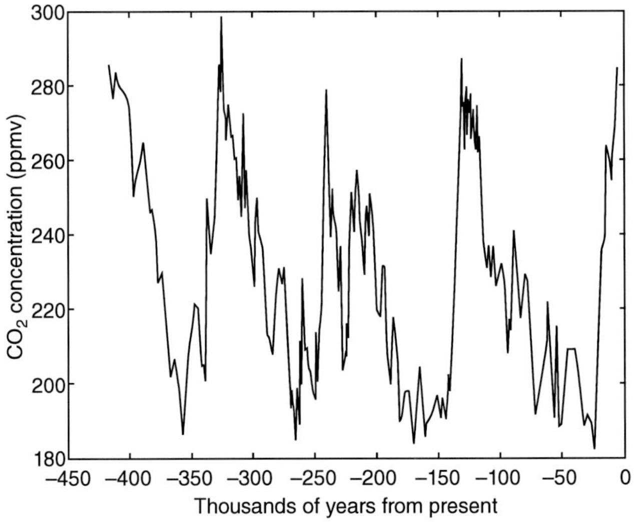

FIGURE 1-2. Atmospheric CO_2 concentrations during the past 420,000 years based on the composition of air entrapped in the Vostok ice core (Barnola *et al.*, 1999).

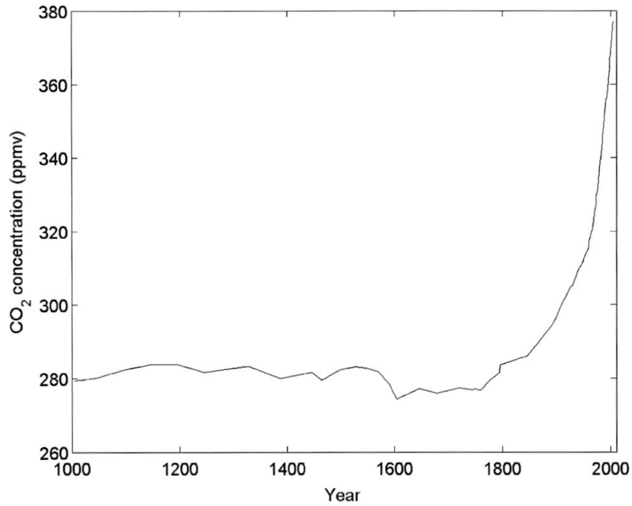

FIGURE 1-3. Atmospheric CO_2 concentrations since 1000 A.D. estimated from ice core data and monitoring of CO_2 at Mauna Loa (Etheridge *et al.*, 2006; Keeling *et al.*, 2006).

with glacial periods, the most recent of which have been the Wisconsinan (~15 to 70 thousand years ago) and Illinoian (~125 to 200 thousand years ago). High CO_2 concentrations are associated with interglacial periods, the most recent of which have been the Eemian (~115 to 130 thousand years ago) and Holocene (~11,500 years ago to the present). The record clearly implicates CO_2 as an amplifier of the effect of orbital forcing on the glacial/interglacial cycle.

As noted, climate change at the present time is largely associated with the accumulation of CO_2 in the atmosphere resulting from fossil fuel burning and deforestation. Fossil fuel burning, which currently releases about seven billion tons of carbon to the atmosphere each year, is generally blamed for roughly 70% of anthropogenic CO_2 emissions. Much of the rest is attributed to deforestation, because of the decrease in the uptake of CO_2 by plants (Raven and Falkowski, 1999). Although the oceans and continental vegetation absorb roughly half of the anthropogenic CO2 released to the atmosphere, the rest accumulates in the atmosphere. The result is clearly apparent in Figure 1-3, which documents the rise in atmospheric CO_2 concentrations by roughly 100 ppmv during the past two centuries.

CONTROLS ON THE CLIMATE OF THE EARTH

Understanding the general characteristics of the Earth's climate requires a modest amount of information and an understanding of a few important concepts. The first important piece of information is the fact that the radiant energy from the Sun is not equally distributed over the surface of the Earth. Equatorial latitudes receive much more energy than polar latitudes, and as a result the atmosphere near the surface of the Earth is much warmer near the equator than near the poles. Heating air causes it to expand, become less dense, and rise (a phenomenon routinely used by hot air balloon enthusiasts). Cooling air causes it to sink. Because equatorial latitudes receive more solar energy than the poles, the differential heating of the Earth-atmosphere system causes air to rise near the equator and to descend near the poles. One might imagine that the atmosphere would therefore move directly north and south, rising at the poles and sinking at the equator, as shown in Figure 1-4.

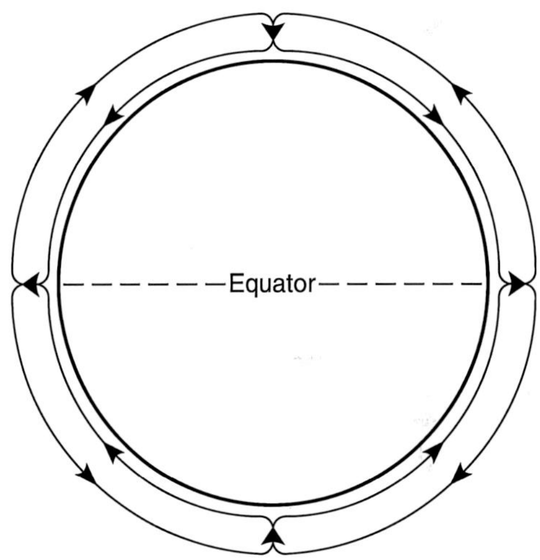

FIGURE 1-4. Cross section of the Earth showing the pattern of circulation of the lower atmosphere that might be expected from differential heating of the Earth-atmosphere system by the Sun.

In fact, atmospheric circulation is not so simple. Although air tends to rise near the equator, as it moves poleward it

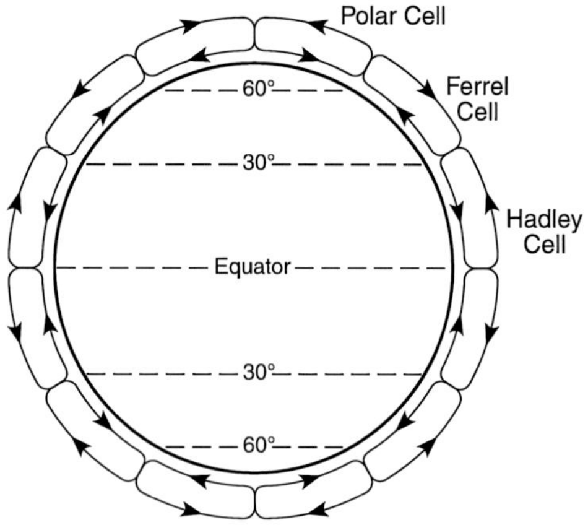

FIGURE 1-5. Meridional circulation that results from differential heating of the Earth-atmosphere system by the Sun. Note that the vertical scale of circulation cells is greatly exaggerated. The vertical extent of the cells is approximately 10 km.

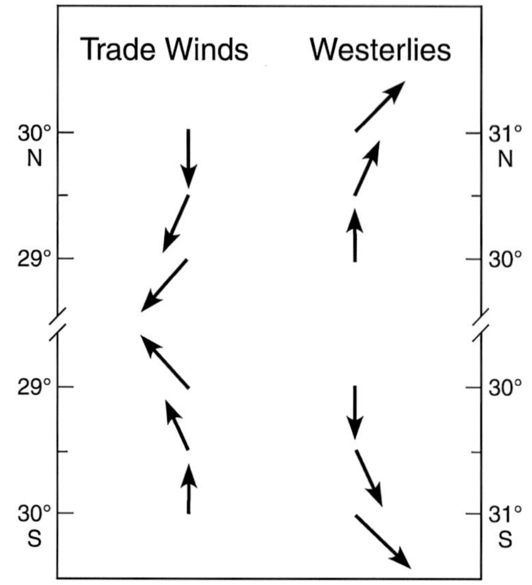

FIGURE 1-6. The effect of the rotation of the Earth on a parcel of air initially at a latitude of 30° and moving at a speed of 8 m s^{-1} directly toward the equator (trade winds) or directly away from the equator (westerlies). No east-west forces are assumed to act on the parcel of air. By the time the air has moved 1°, its direction has changed by about 45°. In the trade wind zone, the parcel of air acquires a westerly component, whereas in the region of the westerlies it acquires an easterly component. The effect of the Earth's rotation is always to divert the air to the right of its direction of motion in the northern hemisphere and to the left in the southern hemisphere.

radiates heat into outer space and eventually cools and sinks at about 30° latitude. Similarly, cold air that sinks at the poles tends to be warmed as it flows along the surface of the Earth toward the equator and to rise near 60° latitude. The vertical circulation of the atmosphere, in simplified terms, consists of three circulation cells as shown in Figure 1-5. The subtropical and temperate-latitude circulation cells are referred to as Hadley cells and Ferrel cells, respectively, after the scientists who discovered them. The high-latitude cells are called polar cells.

The Effect of the Earth's Rotation

In most respects Figure 1-5 is an accurate characterization of the overall meridional (north-south) circulation of the atmosphere, but it is an oversimplification. The real circulation pattern is neither as uniform nor as continuous as Figure 1-5 implies. The figure suggests, for example, that surface winds would blow directly toward the equator in tropical and subtropical latitudes and directly toward the poles in temperate latitudes. This is only partly true.

If we were to slice up the Earth along its latitude lines, we would get a series of rings, the largest at the equator and diminishing in size toward the poles. Because the Earth is rotating as a solid body, a point on a large ring moves faster than a point on a small ring. At 30° latitude, for example, the circumference of our latitudinal ring would be about 34,600 kilometers. A point on the Earth's surface at that latitude is moving toward the east at a rate of 34,600 kilometers per day, or 1442 kilometers per hour. At 29° latitude, the surface of the Earth is moving faster, at 1458 kilometers per hour, because the circumference of a cross section there is 35,000 kilometers.

If there are no other zonal (east-west) forces acting on it, a mass of air flowing toward the equator across the surface of the Earth will appear to be deflected toward the west, because the underlying Earth is moving faster toward the east the closer to the equator the air travels (Fig. 1-6). The surface winds that blow from about 30° toward the equator are referred to as the trade winds. Because winds are customarily named on the basis of the direction *from* which (rather than *to* which) they are flowing, these winds are known as the northeast trades in the northern hemisphere and the southeast trades in the southern hemisphere.

Now consider the air that sinks at 30° and flows toward the poles. At higher latitudes the surface of the Earth is moving to the east more slowly than at 30°, so this air will acquire an apparent eastward motion. The surface winds between 30° and 60° are more complex and unstable than the trade winds, but they consistently have a west-to-east component and hence are known as the westerlies. Because surface winds between the poles and 60° are moving toward the equator, they are affected by the Earth's rotation in the same way as the trade winds, blowing out of the northeast in the northern hemisphere and the southeast in the southern hemisphere (Fig. 1-7).

Once again though, the situation is more complicated. The continental landmasses influence the flow of the wind, and because the land is unevenly distributed between the northern and southern hemispheres, the winds do not blow in an entirely symmetrical manner with respect to the equator. In fact, the entire wind system shown in Figure 1-7 is shifted about 5 to 10° to the north. In addition, in temper-

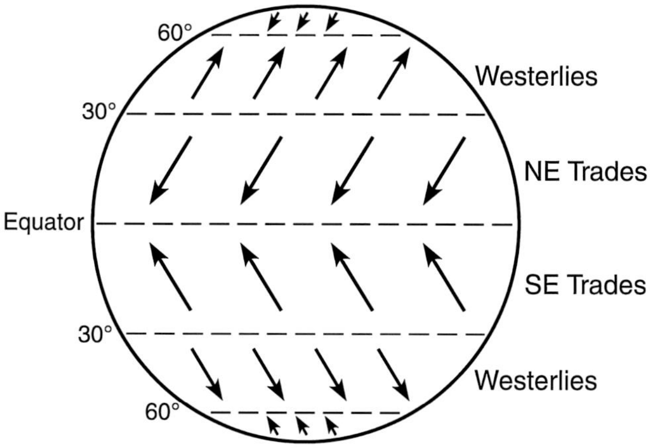

FIGURE 1-7. Direction of surface winds resulting from the combined effects of the Coriolis force and meridional cell circulation.

ate latitudes surface winds tend to circulate about high-pressure ridges and low-pressure troughs, and shifts in the positions of these ridges and troughs can produce important climatological effects.

Finally, the difference in the heat capacity of the continents and oceans causes seasonal temperature differentials to develop between them. Because it takes a great deal of heat to warm a mass of water, and because the upper mixed layer of the ocean is large (typically it extends to tens of meters in the summer and perhaps hundreds of meters in the winter), the temperature of the ocean remains relatively constant compared to the temperature of the continents. During the summer the continents are warmer than the ocean, and during the winter they are cooler. The exchange of heat between the Earth and atmosphere therefore causes the air over the continents to be warmer and less dense than the air over the surrounding oceans during the summer. During the winter, the conditions are reversed. As the continental air warms and rises during the summer, air overlying the surrounding ocean is drawn in to replace it. In the winter, the cool, dense air over the continents tends to sink and flow toward the surrounding ocean. The winds associated with this seasonal circulation pattern are referred to as monsoon winds and are best developed over India, Southeast Asia, and Australia.

The Effect of Surface Winds and the Coriolis Force on Ocean Currents

Because the Earth is a rotating sphere, it appears to an observer on Earth that a force is always pushing the wind to the right of the direction of motion in the northern hemisphere and to the left in the southern hemisphere (e.g., Fig. 1-6). This force is called the Coriolis force, and it affects the oceans as well as the atmosphere. The Coriolis force is directly proportional to the speed of motion and to the sine

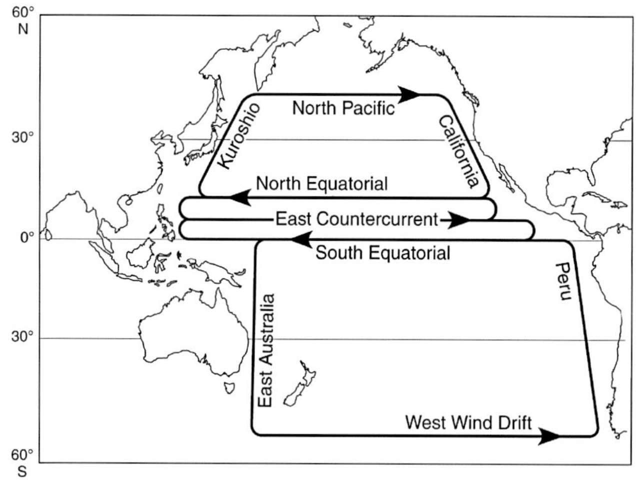

FIGURE 1-8. The Pacific Ocean subtropical gyre current systems. Note that the current gyres are not symmetric with respect to the equator. The equatorial countercurrent actually flows between about 4° and 10° N latitude.

of the latitude. The force is zero at the equator and a maximum at the poles (see References at the end of this chapter).

One would expect that ocean currents would flow in the same direction as the surface winds, but they rarely do. Just as landmasses affect the flow of winds, they impose some constraints on the direction in which ocean currents can flow. Virtually all coastal current systems flow parallel to the coast, regardless of the direction in which the wind is blowing. But even in the open ocean, surface currents do not tend to move in the same direction as the wind. Again, this is due to the Coriolis force, which causes those currents to flow at an angle to the right of the wind in the northern hemisphere and to the left of the wind in the southern hemisphere. The transport of currents at an angle to the wind is referred to as Ekman transport, after the Scandinavian oceanographer who explained the phenomenon theoretically.

The combination of the Coriolis force and Ekman transport causes ocean surface currents in the region of the trade winds to flow almost exactly due west across the ocean basins, whereas in the vicinity of the westerlies the flow is due east. When these transoceanic surface currents encounter continental landmasses, they may either turn and flow parallel to the coastline or completely reverse direction and flow back across the ocean basin. In the former case, they are called boundary currents; in the latter case, they are called countercurrents. The major current systems driven by the trade winds and westerlies in the Pacific Ocean are shown in Figure 1-8. The transoceanic currents to the north of the equator are the North Pacific Current and the North Equatorial Current, and the corresponding boundary currents are the California and Kuroshio currents. The analogous current systems in the South Pacific are the West Wind Drift, the South Equatorial Current, the Peru Current, and the East Australia Current, respectively. The South Equatorial Current actually extends to about 4°N, and much of the flow in the West Wind Drift is actually circumpolar, as there are no continental landmasses to impede it between roughly 55° and 65°S. The Equatorial Countercurrent flows from west to east across the Pacific between approximately 4° and 10°N. Another eastward-flowing countercurrent, called the Equatorial Undercurrent, is at the equator at depths of approximately 100 to 200 meters. Obviously neither the Equatorial Countercurrent nor the Equatorial Undercurrent is driven directly by the wind. The Equatorial Countercurrent, in particular, would seem to be flowing into the teeth of the prevailing trade winds, but it flows through a region of light and variable winds called the Doldrums, which offers little resistance. The more-or-less continuous current system consisting of the California, North Equatorial, Kuroshio, and North Pacific currents is called the North Pacific subtropical gyre, and its counterpart in the South Pacific is the South Pacific subtropical gyre.

Table 1-1 compares the major boundary currents in the Atlantic and Pacific oceans. The poleward flowing boundary currents (Gulf Stream, Kuroshio, Brazil, East Australia, North Atlantic Drift, and Alaska) are particularly important from the standpoint of climate because they transport large amounts of heat from low latitudes to high latitudes. The impact of the heat transported by the combined Gulf Stream/North Atlantic Drift current system, for example, warms northwestern Europe by an annual average of as much as 5 to 10°C (Manabe and Stouffer, 1988; Rahmstorf and Ganopolski, 1999). There is no subpolar gyre current system in the southern hemisphere, because there are no continental landmasses to block the West Wind Drift, a circumpolar current system that forms the southern boundary of the subtropical gyres in both the Atlantic, Pacific, and Indian ocean basins.

An important point about the subtropical and subpolar gyres is the fact that Coriolis forces tend to push water toward their interior and exterior, respectively. This fact is apparent from an examination of Figure 1-8, taking into account the fact that the Coriolis force pushes to the right of the direction of motion in the northern hemisphere and to the left in the southern hemisphere. The result is that the sea surface is actually somewhat higher to the right of a current system flowing in the northern hemisphere and to the left of a current system flowing in the southern hemisphere. In a steady state situation, the force of gravity acting on the tilted sea surface exactly balances the Coriolis force. When this happens, the current is said to be in geostrophic balance, and the current is characterized as a geostrophic current. The difference in sea surface height (SSH) across the Gulf Stream, for example, is about one meter, with SSH being higher to the interior of the North Atlantic subtropical gyre (Kelly *et al.*, 1999).

Similar considerations influence the circulation of the atmosphere, but with the caveat that the analogs of high and low SSH are high and low atmospheric pressure, respectively. Thus, in the northern hemisphere winds tend to blow in a clockwise direction around a region of high pressure and in a counterclockwise direction around a region of low pressure. In each case, the pressure gradient force is in the opposite direction of the Coriolis force. In the southern hemisphere, the circulation is in the opposite sense because the Coriolis force pushes to the left of the direction of motion. Thus, a satellite image of a cyclone or hurricane (extreme low pressure system) in the northern hemisphere always reveals a pattern of counterclockwise circulation (Fig. 1-9). In the southern hemisphere, cyclonic winds blow in a clockwise sense. Appropriately enough, the circulation of winds or currents around any region of low pressure or low SSH is characterized as cyclonic circulation (i.e., counterclockwise in the northern hemisphere and clockwise in the southern hemisphere). The circulation of winds or currents around any region of high pressure or high SSH is characterized as anticyclonic circulation.

With this introduction, it is straightforward to understand some of the major patterns of the climate of the Earth. As the trade winds blow across the tropical ocean, they pick up both heat and water vapor. Because warm, moist air is less

TABLE 1-1. Comparison of major boundary current systems in the Atlantic and Pacific oceans.

	North Atlantic	North Pacific	South Atlantic	South Pacific
	Subtropical Gyre Current Systems			
Western boundary current	Gulf Stream and North Atlantic current	Kuroshio	Brazil	East Australia
Eastern boundary current	Canary	California	Benguela	Peru
	Subpolar Gyre Current Systems			
Western boundary current	Labrador	Oyashio		
Eastern boundary current	North Atlantic Drift	Alaska		

FIGURE 1-9. Hurricane Katrina in the Gulf of Mexico.

dense than cold, dry air,[2] this air tends to rise where the northeast and southeast trade winds converge. This region is known as the intertropical convergence zone, or ITCZ. As the air rises, the water vapor condenses and falls as rain. The ITCZ is therefore characterized by excess precipitation over evaporation. Once the air has risen to an altitude of roughly 3 km it is transported to higher latitudes by Hadley cell circulation (Fig. 1-5). Having lost most of its water vapor to condensation, the air is now dry, and as it moves poleward, it radiates heat into outer space. As the air approaches a latitude of roughly 30°, it becomes sufficiently dense (i.e., cold and dry) that it begins to sink. The climate near 30° is therefore characterized by very low humidity and an excess of evaporation over precipitation. Most of the major desert areas of the world (the Sahara Desert in northern Africa, the Namib and Kalahari deserts in southern Africa, the Great Victoria Desert in Australia, the Arabian Desert, and the Great Desert of the southwestern United States and northern Mexico) are all found near 30° latitude.[3]

In the polar gyre systems air moving over the ocean toward the equator picks up heat and water vapor as do the Trade Winds in the tropics. The combination of increased temperature and humidity causes the air to rise at roughly 60° latitude. Like the ITCZ, the region near 60° latitude is also characterized by an excess of precipitation over evaporation. When the air rises to an altitude of roughly 3 km it moves either toward the poles (polar cell circulation) or toward the equator (Ferrel cell circulation). Having lost most of its water vapor, it now loses heat to outer space via radiation and eventually sinks near the poles or near 30° latitude. We can now understand why the climate of the Earth is wet near the equator at 60° and dry near 30° at the poles. It is no accident, for example, that rain forests are found in the tropics. Superimposed on this pattern precipitation and evaporation is a meridional[4] temperature gradient, warm at the equator and cold at the poles.

This analysis can also account for some of the general features of atmospheric pressure at the surface of the Earth. Keeping in mind that cold, dry air is more dense than warm, moist air, we can easily see that sea level pressure will be relatively low near the equator and 60° latitude and relatively high near 30° at the poles. The lowest sea level pressure tends to be found near the equator (warm, moist air) and the highest near the poles (cold, dry air).

In the tropics an important east-west asymmetry in both precipitation and sea level pressure is also apparent across the major ocean basins. The explanation is apparent from an examination of Figure 1-10. The trade winds blow both toward the equator and toward the west. For reasons already noted, they become warm and moisture laden as they move from east to west over the tropical ocean. The result is an east-west asymmetry in sea level pressure and precipitation near the equator, with the lowest pressure and greatest precipitation at the western edge of the ocean basin. At the western edge of the ocean basin, part of the rising air mass moves back toward the east. As it moves, it radiates heat into the surrounding atmosphere and eventually cools and sinks near the eastern edge of the ocean basin. This circulation pattern is called a Walker cell, after British mathematician Sir Gilbert Walker, who made major contributions to

[2]Air behaves very much like an ideal gas, for which PV = nRT. The number of moles of air (n) per unit volume (V) therefore equals P/(RT). At constant pressure (P), n/V is inversely proportional to the absolute temperature (T). Water (H_2O) has a molecular weight of 18. N_2 and O_2, the principal gases in air, have molecular weights of 28 and 32, respectively. When water displaces nitrogen and oxygen, the average molecular weight of the gases in the air decreases. Therefore, warm, moist air is less dense than cold, dry air because there are fewer molecules per unit volume in warm air and because the average molecular weight of the molecules is lower in moist air.

[3]One major desert that does not fit this pattern is the Gobi Desert at approximately 40 to 45°N latitude. It cannot be attributed to the sinking of cool, dry air in the subtropics. However, it does lie in the region of the westerlies, one manifestation of Ferrel cell circulation, and its location places it in the rain shadow of some very high mountain ranges.

[4]Along a meridian or line of constant longitude.

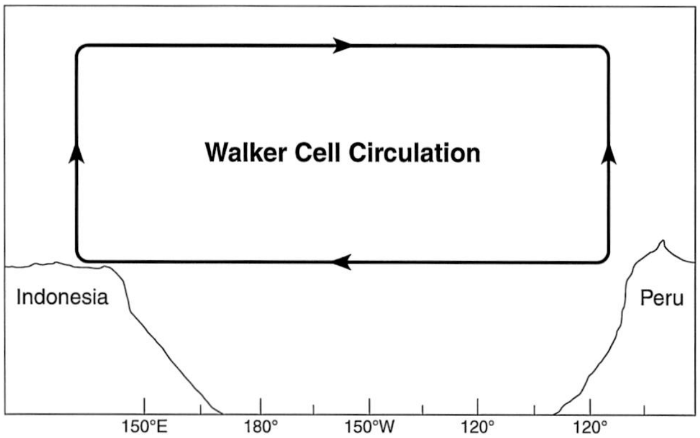

FIGURE 1-10. The Walker cell circulation cycle over the Pacific Ocean. The vertical scale is exaggerated, the height of the circulation cell being about 15 km. This atmospheric circulation pattern tends to produce low atmospheric pressure and a warm, moist climate over Indonesia. Atmospheric pressure is relatively high and the climate cool and dry along the coast of northern Peru.

our understanding of tropical meteorology in the first half of the 20th century.

Because the air that sinks near the equator near the eastern edge of the ocean basin has lost heat as well as water vapor, it tends to be denser than the air that rises along the equator in the west. Consequently there is a small east-west difference in sea level pressure between the eastern and western sides of ocean basins in the Trade Wind zone.

The pressure differentials associated with Walker cell and Hadley cell circulation are both manifestations of the impact of the trade winds on climate. Within the Trade Wind zone, the pressure will be highest near the eastern side of ocean basins at ~30° latitude and lowest at the equator near the western side of ocean basins. In the Pacific Ocean this pressure differential is known as the Southern Oscillation Index (SOI). One common measure of the SOI is the sea level pressure difference between Easter Island (27°S) and Darwin, Australia (12°S).

THE OCEAN AND CLIMATE CHANGE

Now that we have a basic understanding of how the oceans influence climate, let's consider the issue of climate change. We will consider two kinds of climate change, one with a relatively short-term periodicity, the El Niño–Southern Oscillation (ENSO) cycle, and the other with a much longer time constant, the thermohaline circulation of the ocean. We will begin with the ENSO cycle.

El Niño was originally the name given to a dramatic shift in weather and sea conditions off the coast of Peru. Because of the tendency of the change to begin near Christmas, it was given the name El Niño, literally "the child" in Spanish. The changes observed included a warming of the ocean and, in extreme cases, torrential rains in a region normally characterized by very dry conditions.[5] At one time El Niño was regarded as an abnormal event. However, scientists currently view El Niño as simply one phase of a natural cycle, the El Niño–Southern Oscillation, or ENSO cycle, that occurs every several years and is no more usual or unusual than the conditions during any other phase of the cycle. Furthermore, they now recognize that the changes in climate observed during El Niño years along the coast of Peru are simply a local manifestation of a much larger phenomenon that is driven by interactions between the ocean and atmosphere in the subtropics.

The history of El Niños has been reconstructed from as early as 1525 using proxy information, and the record indicates that they occur about every 4 years, with strong events separated by an average of 10 years. Unfortunately for purposes of prediction, the interval between El Niños is irregular. It is not uncommonly 6 or 7 years, but some events have been separated by as little as 1 year. The most recent El Niños occurred in 1957–1958 (strong), 1965 (moderate), 1969 (weak), 1972–1973 (strong), 1976 (moderate), 1982–1983 (very strong), 1986–1987 (strong), 1991–1992 (very strong), 1993 (weak), 1994 (weak), 1997–1998 (very strong), and 2002–2003 (weak).

El Niño conditions are triggered by a movement of warm water from the western Pacific to the eastern Pacific via the Equatorial Countercurrent and Undercurrent. The water is transported largely in the form of so-called Kelvin waves. Kelvin waves and similar waves known as Rossby waves are internal waves (they have their maximum amplitude below the surface of the ocean) whose dynamics are affected by the Coriolis force. Their wavelengths are on the order of thousands of kilometers, and their effects can be felt across an entire ocean basin. Kelvin waves cross the Pacific in 2 to 3 months. As their warm water reaches the coast of South America, it flows over the cooler water of the Peru Current system. The result is an elevation of sea level (Fig. 1-11) and an increase in sea-surface temperature. Some of the warm water flows north along the coast. Some flows south and causes El Niño conditions off the coasts of Ecuador and Peru. As sea level rises and warm water accumulates in the eastern equatorial Pacific, air-sea interactions generate Rossby waves that move westward across the Pacific. The time they take to cross the ocean is strongly dependent on latitude; it is about 9 months near the equator and 4 years at a latitude of 12°. When the Rossby waves reach the western Pacific, they travel toward the equator in the form of coastal

[5]The normally dry weather reflects the fact Peru lies in the rain shadow of the Andes Mountains and that the sea surface temperature is cool for the latitude (e.g., 12°S for Lima), a reflection of the cold water transported by the Peru current (Fig. 1-8) and the fact that the Southeast Trade Winds and Ekman transport induce upwelling of cold water along the coast.

Kelvin waves. Upon reaching the equator, they turn east and begin another crossing of the Pacific. When this second set of Kelvin waves reaches the eastern Pacific, sea level is lowered, the sea-surface temperature declines, and conditions along the coast of Peru return to "normal." Since roughly 1985, these "normal" conditions have come to be known as La Niña (literally "the girl" in Spanish). However, the air-sea interactions associated with the lowered sea-surface temperatures intensify the trade einds, and this shift in the winds sends Rossby waves westward across the Pacific. Upon reaching the western Pacific, these waves travel toward the equator as coastal Kelvin waves and then return to the east along the equator. This final set of equatorial Kelvin waves raises the sea level in the eastern Pacific and completes the El Niño cycle. The entire process is illustrated in Figure 1-12.

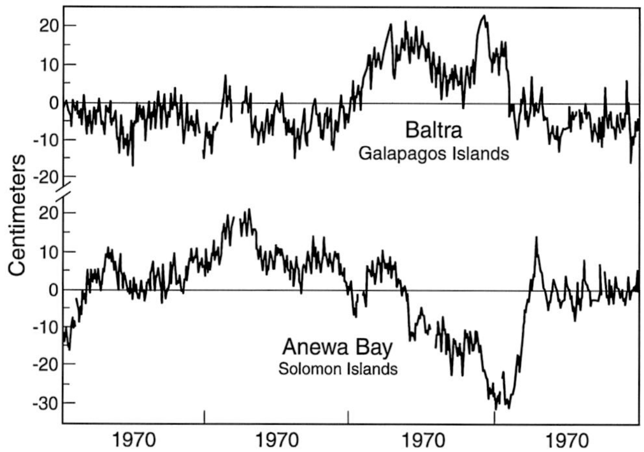

FIGURE 1-11. The response of sea level in the equatorial Pacific Ocean to the 1972 El Niño. Note that sea level was high in the western Pacific (Solomon Islands) preceding El Niño but dropped dramatically by the end of 1972 as water flowed toward the east along the Equatorial Countercurrent and Undercurrent. Sea level was relatively low in the eastern Pacific (Galapagos Islands) preceding El Niño but rose by almost 30 cm as water arrived from the western Pacific. Redrawn from Wyrtki (1979).

Air-Sea Interactions

Because of the exchange of heat between the atmosphere and ocean, changes in sea-surface temperature in the eastern Pacific can have a significant effect on the intensity of the trade wind system. When the eastern Pacific warms during an El Niño year, the Walker cell circulation is slowed because the temperature difference between the eastern and western Pacific is reduced. Thus, the speed of the equatorial trade winds, and consequently the speed of both the South Equatorial and North Equatorial Currents, decreases. A decline in the strength of the equatorial trades allows more warm water to flow from the western to the eastern Pacific, further reducing the temperature differential between the eastern and western Pacific. On the other hand, when the eastern Pacific is cool, the Walker cell circulation increases, because there is a greater temperature differential between the eastern and

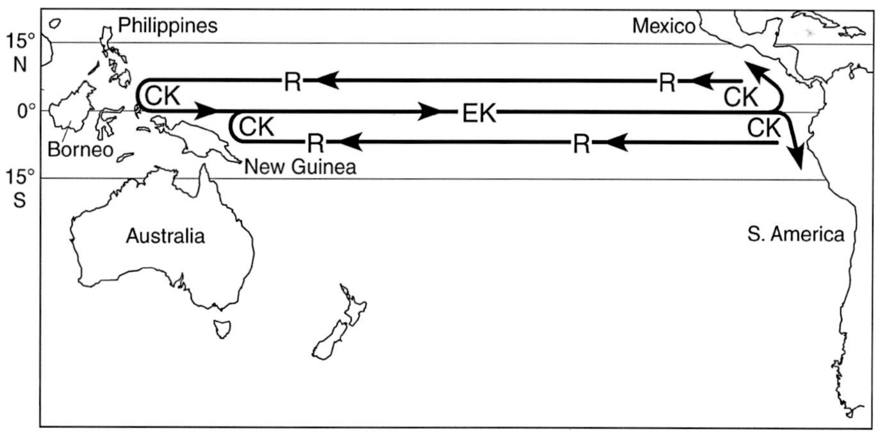

FIGURE 1-12. The wave system that constitutes the negative feedback mechanism in the El Niño cycle. Equatorial Kelvin waves (EK) travel west to east across the Pacific Ocean raising sea levels. When they reach the coastline of South America, they propagate poleward and are clearly identifiable as coastal Kelvin waves (CK) at latitudes higher than 5°. Air-sea interactions associated with the arrival of warm water in the eastern equatorial Pacific cause the Trade Winds to slacken. This shift in the winds sends a series of off-equatorial Rossby waves (R) that lower sea levels back across the Pacific. These Rossby waves reach the western Pacific and propagate toward the equator in the form of coastal Kelvin waves (CK) that also lower sea levels. The Kelvin waves reach the equator, turn east, and move back across the Pacific as sea-level-lowering equatorial Kelvin waves. The equatorial Kelvin waves require about 2 to 3 months to cross the Pacific, but the off-equatorial Rossby waves require anywhere from a few months to a few years. A complete El Niño cycle requires that the Pacific be crossed by two sets of Rossby waves and Kelvin waves, one set raising sea levels in the direction they are moving and the other lowering them. Hence, a complete El Niño cycle typically requires 3 to 5 years.

western Pacific. The trade winds become stronger, and the North and South Equatorial Currents intensify. The strengthening of the trade winds opposes the transport of warm water *via* the Equatorial Countercurrent and Undercurrent, further increasing the temperature difference between the eastern and western Pacific. Those air-sea interactions are an example of what is known as a positive feedback loop. They tend to reinforce El Niño or La Niña conditions, whichever condition prevails.

The reason there is an oscillation between El Niño and La Niña conditions is the negative feedback loop created by the movement of the Kelvin and Rossby waves across the Pacific. During El Niño conditions, the eastern equatorial Pacific warms and the trade winds slacken. The change in trade wind intensity generates off-equatorial Rossby waves that lower sea levels in the western Pacific. Ultimately these lower sea levels generate Kelvin waves that travel back east and lower sea levels in the eastern Pacific.

One implication of this analysis of air-sea interactions is that the Southern Oscillation Index may provide a useful predictor of forthcoming El Niños. The index is high (the pressure differential is large) when the trade winds are strong (La Niña conditions). The index is low (the pressure differential is small) when the trade winds are weak (El Niño conditions). Figure 1-13 shows the behavior of the Southern Oscillation Index and sea-surface temperatures off the coast of Peru for the period from 1968 to 1985. The El Niños of 1972–1973, 1976, and 1982–1983 are all apparent as increases in sea-surface temperature of at least 2°C above long-term monthly averages over a period of several months, and each El Niño is associated with a drop in the Southern Oscillation Index of at least 8 millibars (mb). A drop of greater than 4 mb is usually a sign that an El Niño is approaching.

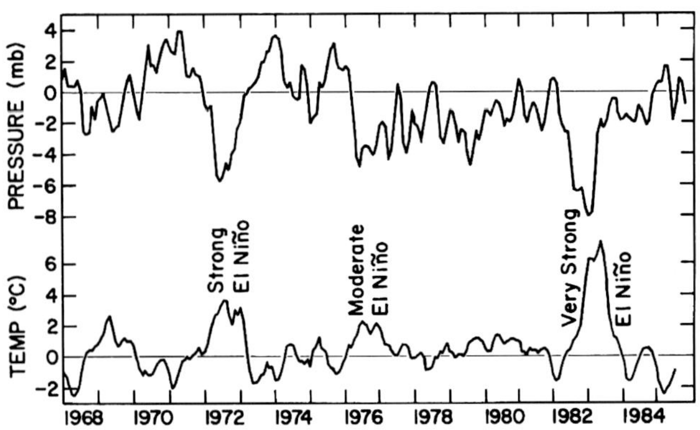

FIGURE 1-13. Three-month running mean variations in the Southern Oscillation Index (top) and sea-surface temperature (SST) off the coast of Chimbote, Peru (bottom) from 1968 to 1985. Monthly variations are the difference between the value for a given month and the long-term average value for that month. During this period, El Niños occurred in 1972–1973 (strong), 1976 (moderate), and 1982–1983 (very strong). The El Niños of 1972–1973, 1976, and 1982–1983 are all apparent as increases in temperature of at least 2°C over a period of several months, and each El Niño is associated with a drop in the Southern Oscillation Index of at least 8 millibars (mb).

Recognition of the connection between the Southern Oscillation Index and El Niño has given rise to the acronym ENSO, which, as noted earlier, stands for El Niño–Southern Oscillation. The ENSO cycle is understood to consist of an irregular meteorological oscillation characterized by two extreme conditions, a warm phase (El Niño) and a cool phase (La Niña), driven by exchanges of heat and water between the ocean and atmosphere in the tropical Pacific.

Shutdown of the North Atlantic Conveyer Belt

Not all of the water transported to the North Atlantic by the North Atlantic Current and North Atlantic Drift is returned *via* the Labrador Current (Table 1-1). Instead, evaporation of water vapor from these warm currents causes the salinity of their surface waters to increase and the temperature to decrease. Sea ice formation is not a factor, but during the winter the combined effect of increased salinity and decreased temperature causes some of the water transported by these currents to sink to depths of 2 to 4 km in the Greenland Sea and Labrador Sea off Greenland.

In the southern hemisphere bottom waters are formed along Antarctic ice shelves during the time of sea ice formation in the winter. The fact that surface waters sink to depths of several kilometers results from the surface waters' being very cold and saline, but the mechanism responsible for creating these conditions differs somewhat in the North Atlantic and Southern Ocean. In the Southern Ocean surface waters sink to the bottom because of an increase in salinity associated with the formation of sea ice.[6]

Because the formation and movement of water masses at intermediate and bottom depths in the ocean are driven by temperature and salinity effects, the deep water current system is referred to as the ocean's thermohaline circulation. Once formed, bottom waters remain submerged for roughly 1000 years, but they eventually return to the surface. From there, surface currents transport them back to the regions of deep and bottom water formation in the North Atlantic and Southern Ocean, respectively. The grand pattern of surface and bottom water circulation in the ocean is referred to as the ocean's conveyer belt.

The analogs of the Gulf Stream and the North Atlantic Drift in the North Pacific Ocean are the Kuroshio Current and North Pacific Current, respectively, but there is no analogous formation of bottom water. Why does bottom water form in the North Atlantic but not in the North Pacific? The answer is that of the major ocean basins, the North Atlantic

[6]Sea ice contains very little salt compared to the water from which it was formed. The liquid brines that remain after sea ice forms are literally at the freezing point of seawater and are hypersaline because of the exclusion of salt from the ice.

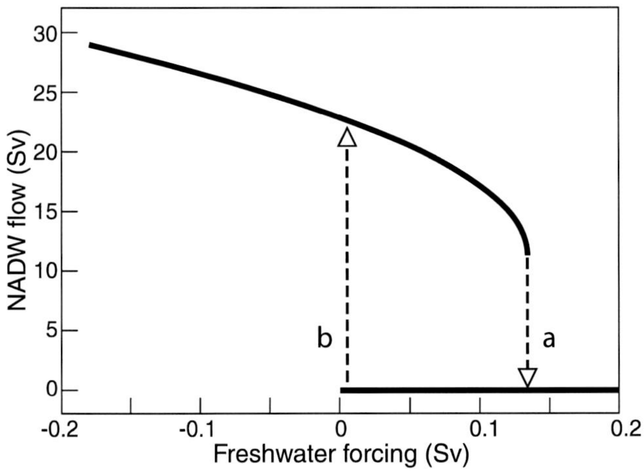

FIGURE 1-14. Relationship between freshwater forcing in the North Atlantic and the rate of formation of North Atlantic Deep Water. One Sverdrup (Sv) = 10^6 m^3 s^{-1} (Rahmstorf, 2000).

has the highest salinity and the North Pacific the lowest. The low salinity of the North Pacific relative to the North Atlantic is primarily the result of differences in rainfall. Precipitation on the Pacific and Atlantic Ocean averages about 120 and 80 cm per year, respectively (Gross, 1982). The result is that surface waters at high latitudes in the North Pacific are less saline than underlying waters, and cooling of surface waters during the winter is insufficient to make them denser than the more saline waters beneath them. In the North Atlantic, on the other hand, the salinity gradient is small, and cooling during the winter is sufficient to cause surface waters to sink to depths of several kilometers.

This comparison underscores the importance of freshwater inputs in determining whether bottom water is formed. In the Southern Ocean bottom waters are formed because freshwater is effectively removed by the formation of sea ice during the winter. In the North Atlantic, deep waters are formed in the winter because freshwater and heat are removed by evaporation. In the North Pacific, freshwater and heat are also removed by evaporation, but the effect of evaporation on the density of the surface waters is more than offset by the input of freshwater from rainfall. Because global warming will warm the ocean's surface waters and accelerate the hydrologic cycle, it is reasonable to ask what impact global warming may have on the thermohaline circulation.

Figure 1-14 illustrates the nature of the problem. Freshwater forcing is here defined to be the net effect of surface exchange, wind-driven ocean currents, and thermohaline circulation. When freshwater forcing is in the range zero to roughly 0.13 Sverdrup (Sv)[7], two very different but stable modes of the Atlantic thermohaline circulation are possible, one in which there is no deep water formation and the other in which North Atlantic Deep Water (NADW) is formed at rates ranging between roughly 11 and 22 Sv. Although the Atlantic is a net evaporative basin (i.e., net surface exchange of freshwater is negative), the overall freshwater forcing is believed to be positive at the present time but almost certainly less than 0.05 Sv (Rahmstorf, 2000). Hence, either of two modes of NADW formation is compatible with the present rate of freshwater forcing, and an increase on the order of 0.1 Sv in freshwater forcing could cause the system to undergo the transition indicated by the (a) arrow. Once the system settles into that mode, it will remain there until freshwater forcing drops below zero, at which point the system transitions back to the current mode as indicated by the (b) arrow.

The ocean contains about 1.3×10^9 km^3 of water. Under current conditions, deep and bottom water is formed in the Southern Ocean and North Atlantic at a combined rate equal to about 0.1% of this volume per year or about 43 Sv (Broecker, 1997). About 47% of this deep water formation occurs in the North Atlantic—that is, the NADW flow is about 20 Sv (Broecker, 1997). Based on Figure 1-14, this would imply that freshwater forcing is roughly 0.02 Sv, and an increase of about 0.1 Sv in freshwater forcing would indeed be necessary to shut down the North Atlantic component of the conveyer belt. Is there any evidence that this has happened in the past?

The short answer to this question is yes. During the most recent glacial period (Wisconsinan), there was a series of brief warm periods known as Dansgaard-Oeschger events and extreme cold periods known as Heinrich events. The best known of the Heinrich events is the Younger Dryas cold event, which lasted from roughly 12,700 to 11,500 years ago and immediately preceded the transition to the present Holocene interglacial. Many paleoclimatologists believe that the Younger Dryas was triggered by the draining of about 9.5×10^3 km^3 of water from Lake Agassiz[8] through the St. Lawrence River into the Atlantic Ocean (Perkins, 2002). Similar emptying of large lakes formed along the edge of northern hemisphere ice sheets[9] may have triggered other Heinrich events. The resultant influx of freshwater was presumably sufficient to shut down the North Atlantic Drift and NADW formation (Fig. 1-14). The associated drop in heat transport

[7]One Sverdrup = 10^6 m^3s^{-1} or 3.2×10^4 km^3y^{-1}.

[8]Lake Agassiz was an immense lake, larger than the area of the present-day Great Lakes combined, and covered much of Manitoba, Ontario, Saskatchewan, and northern Minnesota and North Dakota. It appears to have formed ~13,000 years ago and was fed by glacial runoff. At various times it discharged to the south through the Mississippi River system or to the northwest through the Mackenzie River. The event that triggered drainage of about 85% of Lake Agassiz's volume through the St. Lawrence River about 12,700 years ago was apparently the failure of an ice dam. Modern remnants of Lake Agassiz include *inter alia*, Lake Winnipeg, Lake Winnipegosis, Lake Manitoba, and Lake of the Woods.

[9]For example, large ice-dammed lakes that are known to have formed in the Siberian Altai Mountains.

to the North Atlantic and Europe would have produced a dramatic transition to frigid conditions in Europe and the accumulation of sea ice in the North Atlantic. Eventually, however, conditions along the ice edge during winter months may have led to the formation of bottom water by the same mechanism currently operative in the Southern Ocean (see the earlier discussion). With the formation of NADW thus renewed, the transport of heat by the North Atlantic Drift would have returned, eventually leading to the next Dansgaard-Oeschger event. Thus, during glacial periods such as the Wisconsinan, a plausible mechanism exists to explain alternating Dansgaard-Oeschger and Heinrich events.

One might naively assume that abrupt drainages of ice-dammed lakes would not be a factor during interglacial periods, but this is not entirely true. During the Younger Dryas, the Laurentide ice sheet moved south again, eventually blocking the outflow of Lake Agassiz through the St. Lawrence River. Lake Agassiz refilled with glacial meltwater and eventually merged with another meltwater lake, Lake Ojibway. During the early years of the Holocene interglacial the combined volume of the two lakes is estimated to have been about 2×10^5 km^3, about 60% more than the combined volume of all the world's lakes today (Barber *et al.*, 1999). As the Holocene climate warmed, the ice dam again failed, this time over the Hudson Bay. Geological studies indicate that most of the enormous volume of the combined meltwater lakes drained into the Labrador Sea within 1 year, a flux of roughly 6 Sv (Barber *et al.*, 1999). It is likely that this influx of freshwater completely blocked formation of deep water in the Labrador Sea and may have significantly reduced formation of NADW in the Greenland Sea as well. The result, once again, was a dramatic reduction in the transport of heat to the North Atlantic and Europe. The failure of the Hudson Bay ice dam occurred about 8470 years ago and led to a cold event that lasted roughly 400 years.

The cold event of ~8200 years ago is the most recent climate change attributed to large influxes of freshwater to the North Atlantic, but it is by no means the most recent Holocene climate change. Both Bond *et al.* and deMenocal *et al.* have argued persuasively that climate during both glacial and interglacial periods is modulated by a cycle with a period of 1500 ± 500 years (Bond *et al.*, 1997; deMenocal *et al.*, 2000). Although the ultimate mechanism responsible for producing this modulation is unknown, the process appears to be independent of high-latitude ice sheets and involves "large-scale ocean and atmosphere reorganizations that were completed within decades or centuries, perhaps less" (deMenocal *et al.*, 2000, p. 2201). The most recent manifestation of this climate cycle was the Little Ice Age, which lasted for a period of several hundred years following the so-called Medieval Warm Period (Fig. 1-15) and was associated with bitterly cold winters in North America and Europe (Fig. 1-16). The fact that such climatic changes can occur by mechanisms we do not currently understand raises serious concerns about our ability to predict the impact of global warming on the dynamics of ocean/atmospheric interactions and future climate.

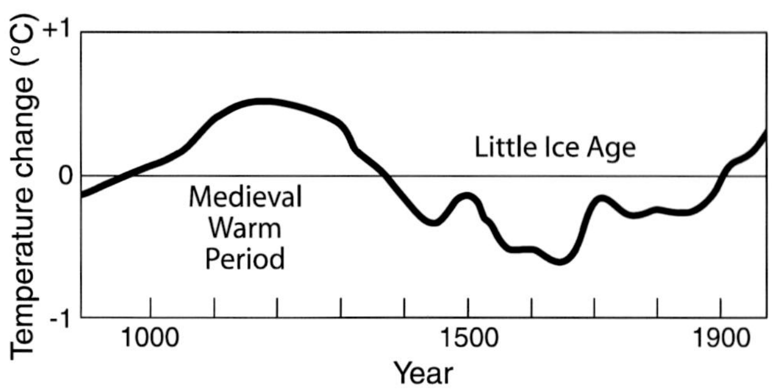

FIGURE 1-15. Reconstruction of global temperature anomalies during the past 1000 years. From http://en.wikipedia.org/wiki/Image:1000_Year_Temperature_Comparison.png.

One obvious concern is whether global warming could shut down the formation of bottom water in the North Atlantic and thereby trigger a prolonged period of cooling. Based on computer simulations, Rahmstorf has argued that a shutdown of the North Atlantic conveyer is unlikely to occur through temperature effects alone (Rahmstorf, 2000). A large influx of freshwater is a much more likely trigger, and as Rahmstorf noted, "The location of the freshwater perturbation is also important—a rule of thumb is: the closer to the deep water formation regions, the more effective it is" (Rahmstorf, 2000, p. 251). Gregory *et al.* have argued that the Greenland icecap will begin to melt if air temperatures rise more than 2.7°C and that a temperature increase of 8°C would cause most of the Greenland icecap to melt within 1000 years (Gregory *et al.*, 2004). Is this likely to happen, and if so, would the influx of freshwater be sufficient to shut down the North Atlantic conveyer?

If the entire Greenland icecap were to melt, sea level would rise by about 7 meters (Gregory *et al.*, 2004). Because the surface area of the ocean is 3.6×10^{14} m^2, the volume of water added to the ocean by melting the Greenland icecap would be 2.5×10^{15} m^3. If this amount of freshwater were added to the ocean over a period of 1000 years, the average flux would be 0.08 Sv. Based on Figure 1-14 and the foregoing discussion, this might be insufficient to literally shut down the formation of NADW, but it would certainly reduce the rate of formation, perhaps by as much as 30% to 40%. An important caveat to this argument is that melting of the Greenland icecap would almost certainly not result in a steady flux of freshwater into the North Atlantic Ocean for 1000 years. The flux might be substantially less than 0.08 Sv for extended periods of time and substantially greater than 0.08 Sv during other times.

Is there any reason to believe that the temperature over Greenland will increase by as much as 8°C? The Intergovernmental Panel on Climate Change (IPCC) projections

FIGURE 1-16. *A Scene on the Ice* by Hendrick Avercamp was inspired by the harsh winter of 1608 in Europe. Source: http://en.wikipedia.org/wiki/Image:SCENEONICE.jpg.

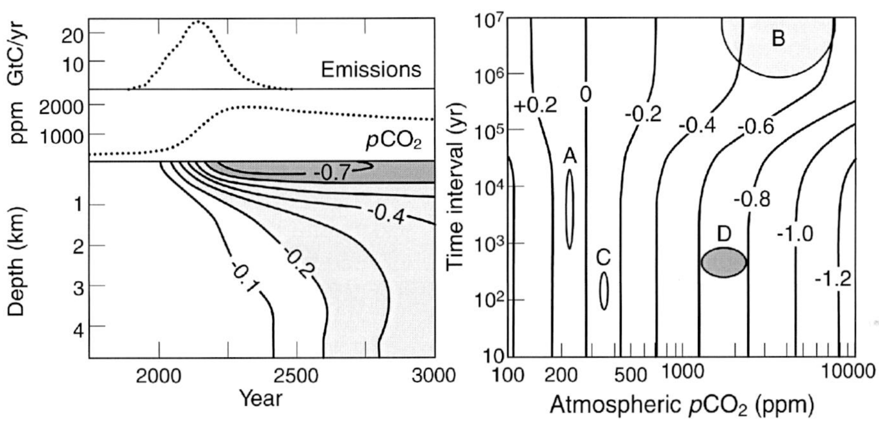

FIGURE 1-17. (a) Atmospheric CO_2 emissions, historical atmospheric CO_2 levels and predicted CO_2 concentrations, together with changes in ocean pH based on horizontally averaged chemistry. (b) Estimated maximum change in surface ocean pH as a function of final atmospheric CO_2 pressure, and the transition time over which this CO_2 pressure is linearly approached from 280 p.p.m. A, glacial–interglacial CO_2 changes; B, slow changes over the past 300 Myr; C, historical changes in ocean surface waters; D, unabated fossil-fuel burning over the next few centuries. Reprinted by permission from Macmillan Publishers Ltd: [*Nature*]; Caldeira and Wickett (2003), copyright 2003.

indicate that by the end of the 21st century, atmospheric CO_2 concentrations will have increased to 710 ppmv and temperatures will have risen by 1.4–5.8°C.[10] What happens after that? Caldeira and Wickett have addressed this question with the use of a computer simulation model in which they assume that we continue to burn fossil fuels until there is literally nothing left (Fig. 1-17) (Caldeira and Wickett, 2003). Their model says that atmospheric CO_2 concentrations will rise to a peak of ~1900 ppmv around the year 2300 and then slowly decline. Based on Berner's GEOCARB II model, an increase in atmospheric CO_2 from 380 to 1900 ppmv would increase average global temperatures by about 9.7°C (Berner, 1994). The temperature rise would be substantially greater at high northern latitudes, because the melting of Arctic sea ice would substantially reduce the albedo of the Arctic Ocean. So there is a distinct possibility that burning fossil fuels until there is literally nothing left will melt the Greenland icecap and raise sea level by 7 meters. What then?

There are several issues to consider. First, the icecap will require roughly 1000 years to melt. The rise in sea level will therefore average about 7 mm per year. Second, a complete shutdown of NADW formation will require several centuries (Rahmstorf, 2000). Although most of the anthropogenic CO_2

[10]The IPCC Web site is http://www.ipcc.ch.

added to the atmosphere will eventually be taken up by the ocean, the process of air/sea exchange will require thousands of years to effect a significant drawdown of atmospheric CO_2 concentrations. Caldeira and Wickett's model, for example, indicates that atmospheric CO_2 concentrations will decline from 1900 ppmv in the year 2300 to ~1500 ppmv by the year 3000 (Caldeira and Wickett, 2003). Thus, the global warming caused by the rise in atmospheric CO_2 concentrations will remain in effect for centuries. As Rahmstorf noted, "A serious cooling of the North Atlantic region (including northwestern Europe) results only in the longer term, when greenhouse gases decline again and the circulation remains in the 'off' mode" (Rahmstorf, 2000, p. 253). One major uncertainty in the long-term climate change forecasts concerns the role of the ENSO cycle. Currently, freshwater export from the Atlantic increases by about 0.1 Sv during El Niño versus La Niña years, and "in one model, increased El Niño frequency resulting from global warming draws enough water vapor from the subtropical Atlantic across into the Pacific to cancel out the weakening effects on the thermohaline circulation" (Rahmstorf, 2000, p. 252). It is therefore possible that after melting of the Greenland icecap the increased frequency of El Niño events associated with global warming would drive freshwater forcing of the North Atlantic to the left of transition (b) in Figure 1-14 and turn on the North Atlantic conveyer, if indeed it had been turned off.

ADVANCED READING: WHY IS THERE A CORIOLIS FORCE?

Let's begin with a simple example. Suppose that we get on an airplane in London and fly to New York City. We board the airplane at noon. The flight takes 8 hours. When we arrive in New York City, the local time is 3 p.m. Given the fact that the flight took 8 hours, why is the local time not 8 p.m.?

The answer, of course, is that we have flown across five time zones. The time zones reflect the differences in longitude between one location and another. The longitude of London is 0°W, and the longitude of New York City is 74°W. As we go forward to the west in longitude we go backward in time. There are a total of 360° of longitude (180° to the west and another 180° to the east from Greenwich, England), and those 360° of longitude represent 24 hours of time. Hence 74° of longitude is the equivalent of (74°/360°)(24 hours) = 4.93 hours. Time zones in effect round this difference to the nearest hour.

Now let's suppose that we are mathematically inclined and decide to write a differential equation to describe the effect of time (t) and longitude (ϕ) on local time, which we will call T. The rate of change of T is given by the equation

$$\frac{dT}{dt} = \frac{\partial T}{\partial t} + \frac{\partial T}{\partial \phi}\frac{\partial \phi}{\partial t} \qquad (1)$$

The first term on the right-hand side of Equation (1) is the partial derivative of local time with respect to time when there is no change in longitude. It should be obvious that $\frac{\partial T}{\partial t} = 1$. The second term describes the effect on local time when the longitude changes. When longitude is increasing toward the west, $\frac{\partial T}{\partial \phi}$ = −(24 hours)/360° = −(1/15) hour per degree longitude. So Equation (1) becomes

$$\frac{dT}{dt} = 1 - \frac{1}{15}\frac{\partial \phi}{\partial t} \qquad (2)$$

Now let's go ahead and integrate this equation. We conclude that

$$\Delta T = \Delta t - \frac{1}{15}\Delta \phi \qquad (3)$$

If Δt = 8 hours and $\Delta \phi$ = 74°, we conclude that ΔT = 8 − 74/15 = 3 hours, which is exactly the change in local time we observed on our flight from London to New York. Note that what we subtract from Δt to correct for our change in longitude has nothing to do with how long it takes us to fly from London to New York City. The correction depends only on the change in longitude. Assuming that we can ignore the general theory of relativity, time (t) in this example can be taken to be absolute time. Local time, T, is clearly relative. It is relative to longitude on the Earth.

From this example, we can see why it is important to take into account the rotation of the Earth when we are trying to understand something that is a function of local time. Now let's consider another example, this time involving a vector (local position) rather than a scalar (local time). At the risk of seeming Earth-centric, let's consider the center of the Earth to be the center of the universe. This will be the center of our coordinate system, which based on our Earth-centric viewpoint, we assume to be fixed (i.e., not moving). We will set up a right-handed coordinate system at the center of the Earth, and we will draw a vector (**R**) from the center of this coordinate system to a reference point on the surface of the Earth (Fig. 1-18). The place we choose on the surface of the Earth is arbitrary. Let's assume that it is Baton Rouge, Louisiana. Note that while the latitude and longitude of Baton Rouge are constant in the Earth's rotating coordinate system,[11] the longitude of Baton Rouge is not constant if the center of the Earth is the origin of our coordinate system. If the longitude of Baton Rouge is measured relative to the latter (fixed) coordinate system, it will change by 360° (2π radians) over the course of 24 hours. An analogy would be the fact that the local time in Greenwich, England, is always

[11]For purposes of this discussion, we will ignore plate tectonics.

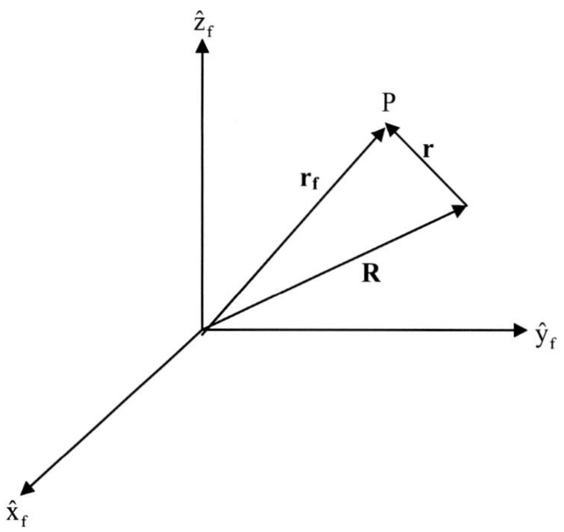

FIGURE 1-18. Coordinate system and vectors used to analyze effect of Earth's rotation on equations of motion.

6 hours ahead of the local time in Baton Rouge, even though the absolute time in both locations is constantly changing.

Now let's imagine that we find ourselves at point P on the surface of the Earth. The vector from the center of the Earth to point P is $\mathbf{r_f}$. This is our location in the fixed coordinate system. We can also locate point P with respect to Baton Rouge. The vector from Baton Rouge to point P is $\mathbf{r}$. All of this is illustrated in Figure 1-18, where $\hat{x}_f$, $\hat{y}_f$, and $\hat{z}_f$ are unit vectors in the x, y, and z directions, respectively, in the fixed coordinate system. We will assume that $\hat{z}_f$ points due north. From simple vector addition, it follows that $\mathbf{r_f} = \mathbf{R} + \mathbf{r}$. Now if ϕ is our longitude and θ is our latitude relative to the center of the Earth, it follows that

$$\begin{aligned} x_f &= r_f \cos\theta \cos\phi \\ y_f &= r_f \cos\theta \sin\phi \\ z_f &= r_f \sin\theta \end{aligned} \tag{4}$$

Here we have implicitly assumed that the surface of the Earth is a perfect sphere. Later we will see that this assumption is not entirely true, but for now it will suffice. Now let's assume that our position P changes only because of the rotation of the Earth. If that be the case,

$$\begin{aligned} \frac{dx_f}{dt} &= -r_f \cos\theta \sin\phi \frac{d\phi}{dt} = -y_f \frac{d\phi}{dt} = -y_f \omega \\ \frac{dy_f}{dt} &= r_f \cos\theta \cos\phi \frac{d\phi}{dt} = x_f \frac{d\phi}{dt} = x_f \omega \\ \frac{dz_f}{dt} &= 0 \end{aligned} \tag{5}$$

where we have defined $\omega \equiv \frac{d\phi}{dt}$. Now consider the determinant $\begin{vmatrix} \hat{x}_f & \hat{y}_f & \hat{z}_f \\ 0 & 0 & \omega \\ x_f & y_f & z_f \end{vmatrix}$. When we expand this determinant, we see that it equals $-y_f\omega\hat{x}_f + x_f\omega\hat{y}_f$. In other words, the determinant equals $d\mathbf{r_f}/dt$. Furthermore, if we remember the definition of a vector cross product in Cartesian coordinates, we see that the determinant equals $\boldsymbol{\omega}\mathrm{x}\mathbf{r_f}$, (i.e., the vector cross product of $\boldsymbol{\omega}$ and $\mathbf{r_f}$) where $\boldsymbol{\omega}$ is a vector pointing in the $\hat{z}_f$ direction (i.e., due north). So if our position in the fixed coordinate system changes only because the Earth is rotating, it follows that

$$\frac{d\mathbf{r_f}}{dt} = \boldsymbol{\omega}\mathrm{x}\mathbf{r_f} \tag{6}$$

This is the case when $\mathbf{r}$ is constant in the spherical Earth's rotating reference frame (i.e., when the latitude and longitude of point P are constant relative to the latitude and longitude of Baton Rouge). Now let's assume that the latitude and longitude of point P change relative to the latitude and longitude of Baton Rouge. In that case, $\mathbf{r}$ is not constant in the spherical Earth's rotating frame of reference and

$$\frac{d\mathbf{r_f}}{dt} = \frac{d\mathbf{r}}{dt} + \boldsymbol{\omega}\mathrm{x}\mathbf{r_f} \tag{7}$$

In Equation (7) $d\mathbf{r_f}/dt = \mathbf{v_f}$, and $d\mathbf{r}/dt = \mathbf{v_r}$, where $\mathbf{v_f}$ is the velocity of point P relative to the center of the Earth and $\mathbf{v_r}$ is the velocity of point P relative to the location of Baton Rouge. Note that while $\mathbf{v_f}$ and $\mathbf{v_r}$ are measured with respect to different reference points, both are reported in the center-of-the-Earth fixed coordinate system (e.g., the three components of $\mathbf{v_r}$ are the projections of $\mathbf{v_r}$ on $\hat{x}_f$, $\hat{y}_f$, and $\hat{z}_f$). Equation (7) can be written as

$$\mathbf{v_f} = \mathbf{v_r} + \boldsymbol{\omega}\mathrm{x}\mathbf{r_f} \tag{8}$$

If we differentiate Equation (8) with respect to time, we get

$$\frac{d\mathbf{v_f}}{dt} = \frac{d\mathbf{v_r}}{dt} + \boldsymbol{\omega}\mathrm{x}\frac{d\mathbf{r_f}}{dt} \tag{9}$$

where all derivatives are to be evaluated in the fixed coordinate system. Here we have assumed that $d\omega/dt = 0$ (i.e., the Earth's rotation rate is not changing). Clearly $d\mathbf{v_f}/dt$ is the acceleration of point P relative to the center of the Earth, i.e., $\mathbf{a_f}$. From Equation (7) it follows that

$$\left(\frac{d\mathbf{v_r}}{dt}\right)_{\text{fixed}} = \left(\frac{d\mathbf{v_r}}{dt}\right)_{\text{rotating}} + \boldsymbol{\omega}\mathrm{x}\mathbf{v_r} \tag{10}$$

Clearly $\left(\frac{d\mathbf{v_r}}{dt}\right)_{\text{rotating}}$ is the acceleration of point P relative to the location of Baton Rouge (i.e., $\mathbf{a_r}$). So Equation (9) becomes

$$\begin{aligned} \mathbf{a_f} &= \mathbf{a_r} + \boldsymbol{\omega}\mathrm{x}\mathbf{v_r} + \boldsymbol{\omega}\mathrm{x}\frac{d\mathbf{r_f}}{dt} = \mathbf{a_r} + \boldsymbol{\omega}\mathrm{x}\mathbf{v_r} + \boldsymbol{\omega}\mathrm{x}\left(\frac{d\mathbf{r}}{dt} + \boldsymbol{\omega}\mathrm{x}\mathbf{r_f}\right) \\ &= \mathbf{a_r} + 2\boldsymbol{\omega}\mathrm{x}\mathbf{v_r} + \boldsymbol{\omega}\mathrm{x}(\boldsymbol{\omega}\mathrm{x}\mathbf{r_f}) \end{aligned} \tag{11}$$

According to Newton's second law, in an inertial frame of reference, $\mathbf{F} = m\mathbf{a}$, where $\mathbf{F}$ is force, $\mathbf{a}$ is acceleration, and m is the mass of the object to which the force is applied. An inertial frame of reference is a frame of reference that is not accelerating. In our Earth-centric model, the coordinate system with its origin at the center of the Earth is an inertial frame of reference. Multiplying both sides of Equation (11) by mass m and rearranging gives

$$m\mathbf{a}_r = m\mathbf{a}_f - 2m\boldsymbol{\omega}\mathrm{x}\mathbf{v}_r - m\boldsymbol{\omega}\mathrm{x}(\boldsymbol{\omega}\mathrm{x}\mathbf{r}_f) \qquad (12)$$

If we use some fixed point on the surface of the Earth (e.g., Baton Rouge) as our frame of reference, then mass times acceleration equals the right-hand side of Equation (12). The true force, in the Newtonian sense, is $m\mathbf{a}_f$. However, the right-hand side of Equation (12) contains two additional terms. The first term, $-2m\boldsymbol{\omega}\mathrm{x}\mathbf{v}_r$, is the Coriolis force. It is named after G. G. Coriolis, who first published an equation equivalent to Equation (12) in 1835. The second term, $-m\boldsymbol{\omega}\mathrm{x}(\boldsymbol{\omega}\mathrm{x}\mathbf{r}_f)$, is the centrifugal force. Physicists generally regard these as pseudo-forces; they appear in Equation (12) only because the acceleration $\mathbf{a}_r$ is not relative to an inertial frame of reference.

Now let's spend a little time thinking about the direction and magnitude of these pseudo-forces. From physics we know that the magnitude of the cross product of two vectors is equal to the product of the magnitudes of the two vectors times the sine of the angle between them: $\omega = 2\pi/(24\text{ hours}) = 2\pi/(86{,}400\text{ seconds}) = 7.27 \times 10^{-5}$ radians per second. The magnitude of $\boldsymbol{\omega}\mathrm{x}\mathbf{r}_f$ is $\omega r_f \cos\theta$, where θ is the latitude. The cosine of θ appears here because the angle between $\boldsymbol{\omega}$ and $\mathbf{r}_f$ is the complement of the latitude (i.e., the sine of the angle between them is the cosine of the latitude). $\boldsymbol{\omega}\mathrm{x}\mathbf{r}_f$ is perpendicular to both $\boldsymbol{\omega}$ and $\mathbf{r}_f$, so the magnitude of $-\boldsymbol{\omega}\mathrm{x}(\boldsymbol{\omega}\mathrm{x}\mathbf{r}_f)$ is $\omega^2 r_f \cos\theta$. The vector $-\boldsymbol{\omega}\mathrm{x}(\boldsymbol{\omega}\mathrm{x}\mathbf{r}_f)$ lies in a plane at right angles to the axis of rotation, and it points away from the axis of rotation, as illustrated in Figure 1-19.

Since the radius of the Earth is 6.38×10^6 meters, the magnitude of $\omega^2 r_f$ is $(7.27 \times 10^{-5})^2(6.38 \times 10^6) = 3.37 \times 10^{-2}\text{ m s}^{-2}$. This is about 291 times smaller than the acceleration of gravity, which is 9.8 m s^{-2}. The centrifugal acceleration can be expressed as the sum of two acceleration vectors, one being tangent to the surface of the spherical Earth in the direction of the equator and the other perpendicular to the surface of the spherical Earth. Our centrifugal acceleration will point perpendicular to the surface of the spherical Earth if we are standing on the equator, and it will point parallel to the surface of the spherical Earth as we approach the North or South Pole. In between, the magnitude of the centrifugal acceleration tangent to the surface of the spherical Earth will equal $\omega^2 r_f \cos\theta$ times the sine of the latitude. In other words, the magnitude of the centrifugal acceleration tangent to the surface of the spherical Earth will equal $\omega^2 r_f \cos\theta \sin\theta = \frac{1}{2}\omega^2 r_f \sin 2\theta$. This will be a maximum when

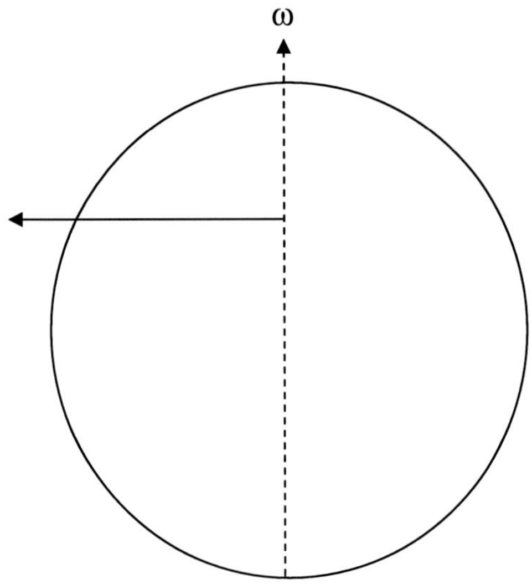

FIGURE 1-19. The direction of the centrifugal force is perpendicular to the axis of rotation.

$\theta = \pi/4$ radians or 45°, and at that angle it will equal $\dfrac{3.37 \times 10^{-2}}{2\sqrt{2}} = 1.19 \times 10^{-2}\,\text{m s}^{-2}$.

Now let's consider the Coriolis acceleration, $-2\boldsymbol{\omega}\mathrm{x}\mathbf{v}_r$. We can use Figure 1-19 to analyze the situation when something is moving directly east or west in the northern hemisphere. Let's assume that in Figure 1-19 we are looking directly west and that an object is moving toward us (i.e., toward the east). In this case, $\boldsymbol{\omega}$ and $\mathbf{v}_r$ will in fact be perpendicular to each other, and the direction of $-2\boldsymbol{\omega}\mathrm{x}\mathbf{v}_r$ will be as indicated by the arrow (i.e., in the same direction as the centrifugal acceleration). Following the same logic as before, this acceleration includes a component perpendicular to the surface of the Earth and a component parallel to the surface of the Earth. The magnitude of the parallel component will be $2\omega v_r \sin\theta$, where θ is the latitude.

Now let's consider a second situation where we are moving directly north in the northern hemisphere, as illustrated in Figure 1-20. In this case, $\mathbf{v}_r$ and $\boldsymbol{\omega}$ are clearly not perpendicular. The angle between them is in fact our latitude. The vector $-2\boldsymbol{\omega}\mathrm{x}\mathbf{v}_r$ will point to the east, and its magnitude will again be $2\omega v_r \sin\theta$. Note that in this case $\sin\theta$ appears because $\mathbf{v}_r$ and $\boldsymbol{\omega}$ are not perpendicular, whereas for an object moving east or west, $\sin\theta$ appears because we care only about the component of the acceleration tangent to the surface of the Earth. The net result, however, is that the relevant component of the Coriolis acceleration is $2\omega v_r \sin\theta$, regardless of which direction an object is moving, and directional analysis shows that the direction of the Coriolis acceleration is to the right of the direction of motion in the northern hemisphere and to the left of the direction of motion in the southern hemisphere.

Now how big is the Coriolis acceleration? The answer, of course, depends on how fast an object is moving. Let's

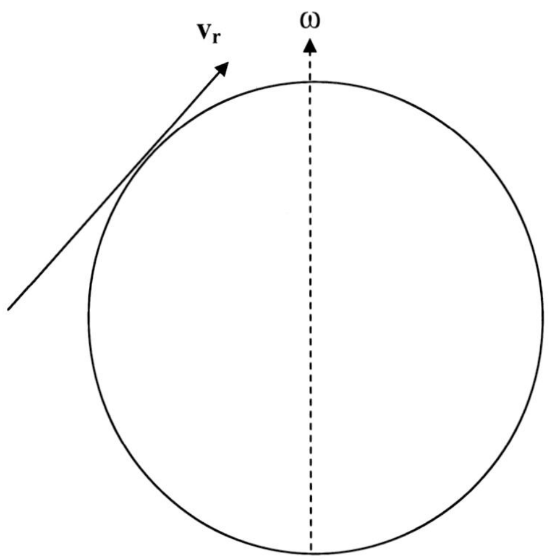

FIGURE 1-20. Relationship between v_r and ω for an object moving along the surface of the Earth toward the north in the northern hemisphere.

assume that we are concerned with an ocean current with a speed of 1.0 knot or 0.514 meters per second. In that case, $2\omega v_r = (2)(7.27 \times 10^{-5})(0.514) = 7.47 \times 10^{-5}$ m s^{-2}. The acceleration of gravity is 9.8 m s^{-2}, about 1.3×10^5 times larger. Even if we were talking about hurricane winds with speeds of 150 knots, the Coriolis acceleration will be very small compared to the acceleration of gravity, so we can ignore any component of the Coriolis acceleration perpendicular to the surface of the Earth. However, it appears that we need to pay attention to the components of the centrifugal and Coriolis accelerations that are tangent to the surface of the Earth. These are

Centrifugal acceleration
$= \frac{1}{2}\omega^2 r_f \sin 2\theta = 1.69 \times 10^{-2} \mathrm{Sin} 2\theta$ m s^{-2} toward the equator

Coriolis acceleration $= 2\omega v_r \sin\theta = 1.45 \times 10^{-4} v_r \sin\theta$ m s^{-2} to the right of the direction of motion in the northern hemisphere and to the left of the direction of motion in the southern hemisphere. In this equation, v_r has units of meters per second.

Now comes a caveat. Up to this point we have considered the Earth to be a perfect sphere, and we have talked about radial and tangential forces in the same context. Now we need to be a bit more sophisticated. The reason is that *inter alia* is the centrifugal acceleration. To understand why, let's consider a parcel of seawater at a latitude of 45° that is initially going nowhere relative to the surface of a perfectly spherical Earth. The centrifugal acceleration at this latitude is 1.19×10^{-2} m s^{-2}. After 1000 seconds and in the absence of any other forces, the parcel of seawater will have acquired a velocity of $(1.19 \times 10^{-2})(1000) = 11.9$ meters per second or 23 knots toward the equator.

In fact, this sort of thing does not happen, but why not? The explanation lies in the fact that the force of gravity and the centrifugal force are both so-called conservative forces, which means that they are a function only of position. And it is the combination of these two forces that we experience as gravity. So the effective acceleration of gravity is actually the combination of the true gravitational acceleration and the centrifugal acceleration. What we mean by vertical is the direction defined by the vector sum of the acceleration of gravity and centrifugal acceleration. Because the centrifugal acceleration is small compared to the true acceleration of gravity, vertical is nearly coincident with the direction of the true gravitational acceleration, but not exactly. And a horizontal surface is a surface that is normal (perpendicular) to the vector sum of the centrifugal and true gravitational accelerations. Such a surface is called a geopotential surface, and the corresponding surface of the ocean in the absence of winds, currents, and horizontal density gradients is called the geoid.

So on a geopotential surface there is, by definition, no net tangential force associated with the combined gravitational and centrifugal forces. However, if objects are moving over a geopotential surface, they will experience the Coriolis force, and hence it must be taken into account when we study ocean and atmospheric circulation.

References

Barber, D.C., Dyke, A., Hillaire-Marcel, C., Jennings, A.E., Andrews, J.T., Kerwin, M.W., Bilodeau, G., McNeely, R., Southon, J., Morehead, M.D., Gagnon, J.-M., 1999. Forcing of the cold event of 8,200 years ago by catastrophic drainage of Laurentide lakes. Nature 400, 344–348.

Barnola, J.M., Raynaud, D., Lorius, C., Barkov, N.I., 1999. Historical CO_2 record from the Vostok ice core. In Trends: A compendium of data on global change, Carbon Dioxide Information Analysis Center, Oak Ridge National Laboratory, US Department of Energy: Oak Ridge. (Available online at http://cdiac.esd.ornl.gov/trends/co2/vostok.htm.)

Berner, R.A., 1994. Geocarb II: A revised model of atmospheric CO_2 over Phanerozoic time. Am. J. Sci. 294, 56–91.

Berner, R.A., Canfield, D.E., 1989. A new model for atmospheric oxygen over Phanerozoic time. Am. J. Sci. 289, 333–361.

Berner, R.A., Kothavala, Z., 2001. Geocarb III: A revised model of atmospheric CO2 over Phanerozoic time. Am. J. Sci. 301, 182–204.

Bond, G., Showers, W.J., Cheseby, M., Lotti, R., Almasi, P., deMenocal, P., Priore, P., Cullen, H., Hajdas, I., Bonani, G., 1997. A pervasive millennial-scale cycle in North Atlantic Holocene and glacial climates. Science 278, 1257–1266.

Broecker, W.S., 1997. Thermohaline circulation, the Achilles heel of our climate system: Will man-made CO_2 upset the current balance? Science 278, 1582–1588.

Caldeira, K., Wickett, M.E., 2003. Anthropogenic carbon and ocean pH. Nature 425, 365.

deMenocal, P., Ortiz, J., Guilderson, T., Sarnthein, M., 2000. Coherent high- and low-latitude climate variability during the Holocene warm period. Science 288, 2198–2202.

Etheridge, D.M., Steele, L.P., Langenfelds, R.L., Francey, R.J., Barnola, J.M., Morgan, V.I., 2006. Historical CO_2 records from the Law Dome DE08, DE08–2, and DSS ice cores. In Division of Atmospheric Research, CSIRO, Aspendale, Victoria, Australia. (Available online athttp://cdiac.ornl.gov/trends/co2/lawdome.html.)

Gregory, J.M., Huybrechts, P., Raper, S.C.B., 2004. Threatened loss of the Greenland ice-sheet. Nature 428, 616.

Gross, M.G., 1982. Oceanography: A View of the Earth. New York, Prentice-Hall.

Hays, J.D., Imbrie, J., Shackleton, N.J., 1976. Variations in the earth's orbit: Pacemaker of the ice ages. Science 194, 1121–1132.

Keeling, C.D., Whorf, T.P., 2006. Atmospheric CO_2 concentrations (ppmv) derived from in situ air samples collected at Mauna Loa Observatory, Hawaii. In Carbon Dioxide Research Group, Scripps Institution of Oceanography. (Available online at http://cdiac.ornl.gov/ftp/trends/co2/maunaloa.co2.)

Kelly, K.A., Singh, S., Huang, R.X., 1999. Seasonal variations of sea surface height in the Gulf Stream region. Journal of Physical Oceanography 29, 313–327.

Laws, E.A., 2007. Climate change, oceans, and human health. Ocean Yearbook 21.

Lovelock, J.E., 1979. Gaia: A New Look at Life on Earth. Oxford, Oxford University Press.

Manabe, S., Stouffer, R.J., 1988. Two stable equilibria of a coupled ocean-atmosphere model. Journal of Climate 1, 841–866.

Patz, J.A., Campbell-Lendrum, D., Holloway, T., Foley, J.A., 2005. Impact of regional climate change on human health. Nature 438, 310–317.

Perkins, S., 2002. Once upon a lake. Science News 162, 283.

Rahmstorf, S., 2000. The thermohaline ocean circulation: A system with dangerous thresholds? Climate Change 46, 247–256.

Rahmstorf, S., Ganopolski, A., 1999. Long-term global warming scenarios computed with an efficient coupled climate model. Climatic Change 43, 353–367.

Raven, J.A., Falkowski, P.G., 1999. Oceanic sinks for atmospheric CO_2. Plant, Cell & Environ. 741–755.

Sagan, C., 1977. Reducing greenhouses and the temperature of Earth and Mars. Nature 269, 224–226.

Sagan, C., Chyba, C., 1997. The early faint sun paradox: Organic shielding of ultraviolet-labile greenhouse gases. Science 276, 1217–1221.

Sagan, C., Mullen, G., 1972. Earth and Mars: Evolution of atmospheres and surface temperature. Science 177, 52–56.

Schlesinger, W.H., 1997. Biogeochemistry: An Analysis of Global Change. San Diego, CA, Academic Press.

Watson, A.J., Lovelock, J.E., Margulis, L., 1978. Methanogenesis, fires, and the regulation of atmospheric oxygen. Biosystems 10, 293–298.

Wyrtki, K., 1979. An abnormal event in the ocean atmosphere system. La Recherche 10, 1212–1220.

STUDY QUESTIONS

1. Explain why the Coriolis force tends to focus waves traveling from west to east along the equator. Why would it be difficult for equatorial Kelvin waves to travel from east to west?
2. Explain why the atmospheric circulation cells tend to produce cool, dry air at latitudes near 30°.
3. Explain how air-sea interactions constitute a short-term positive feedback loop but are a component of a long-term negative feedback loop in the control of the El Niño cycle.
4. How would you account for the fact that many of the world's important rain forests are found in tropical latitudes?
5. The coastal region of Peru is normally very dry. In some places, no rain may fall for years. However, during El Niños it is not uncommon for torrential rains to fall in these areas. How would you account for this fact based on what you know about air-sea interactions during the El Niño cycle?
6. Show that if the circumference of the Earth is 40×10^3 km, then the speed of the Earth's surface toward the east (V) is given by the equation $V = 40 \times 10^3 \cos(\theta)$ km d^{-1}, where θ is the latitude.
7. Suppose that the Earth rotated in the opposite direction. Refer to Figures 1–5 through 1–8 and then draw the surface winds and major ocean surface currents that you would expect to find on such an Earth. How would the climate on the Earth be affected?
8. Assume that a satellite in polar orbit passes directly over the South Pole at midnight. Immediately afterward the satellite is headed due north along the 90°W meridian. The orbit of the satellite is circular, and its speed is such that it takes 12 hours to make one complete orbit around the Earth. What are the latitude and longitude of the satellite each hour over a 24–hour period? Plot the position of the satellite on a globe of the Earth. Explain why the orbit does not appear to be circular. The Coriolis force on a moving object depends on both the speed and the latitude of the object. In this case, the speed of the satellite is constant, but its latitude is constantly changing. By examining the orbit you have plotted on a globe of the Earth, at which latitudes would you say the Coriolis force is a maximum and a minimum?
9. Seaman Sanford is assigned to a tour of duty aboard a military submarine in the Arctic Ocean beneath the North Pole, ostensibly to make scientific observations related to climate warming and the thinning of the Arctic icecap. During long weeks below the surface of the ocean Sanford becomes bored and spends much of his time reading and re-reading a physics textbook, the only reading material available in the submarine's library. Sanford also eats too much military food, and due to the confined quarters aboard the submarine gets little exercise. As a result his weight balloons to 291 pounds. In order to lose weight, Sanford decides to request a transfer to a naval facility at the equator. His reasoning is that weight is the product of mass and the acceleration of gravity, and based on what he has read in the physics textbook, he knows that centrifugal acceleration at the equator is in direct opposition to the acceleration of gravity. How much weight will Sanford lose if he is transferred to the equator?

C A S E   S T U D Y

1

Managing Public Health Risks: Role of Integrated Ocean Observing Systems (IOOS)

TOM MALONE AND MARY CULVER

INTRODUCTION

This is an exciting era in Earth observations in which local, regional, and global observations of terrestrial, atmospheric, and oceanic systems are being linked into a Global Earth Observing System of Systems (GEOSS). With the involvement of more than 60 countries and 40 international organizations, the overall aim of these types of systems is to leverage data collections from satellites, ocean buoys, weather stations, and other observing instruments to maximize the amount of environmental data that can be integrated into new information products. These systems will build on and improve data collection systems to develop products that will to enable powerful predictions and early warning systems that are relevant to the oceans and their impact on human health.

To address U.S. needs for ocean information, the U.S. Commission on Ocean Policy has called for the development of an Integrated Ocean Observing System (IOOS)[1] to provide data and information needed to address seven societal goals (U.S. Commission on Ocean Policy, 2004):

1. Improve predictions of climate variability and change.
2. Improve the safety and efficiency of maritime operations.
3. Improve national and homeland security.
4. More effectively mitigate the effects of natural hazards.
5. Reduce public health risks.
6. More effectively protect and restore healthy coastal ecosystems.
7. Enable the sustained use of ocean and coastal resources.

Achieving these goals depends on establishing an integrated, multidisciplinary system of systems that routinely, reliably, and continuously provides data and information on oceans and coasts, in forms and at rates specified by groups that use, depend on, manage, and study marine systems (Malone *et al.*, 2005). Provisional products for each of the seven goals have been identified by groups of experts from government, academia, industry, and nongovernmental organizations (Table CS1-1). Although each societal goal and associated product have unique requirements for data and information, together they have many common data and information needs that can be more effectively met by sharing and integrating data collected by a broad spectrum of state and federal agencies, research programs, industries, and nongovernmental organizations (NGOs). Likewise, many requirements for data management are similar across all seven societal goals, and, as will be illustrated here for the public health goal, all seven goals require analyses and models of physical states and processes. Thus, an integrated approach to developing a multiuse, multidisciplinary observing system that transcends institutional and programmatic boundaries is feasible, sensible, and cost-effective (Ocean. US, 2002, 2006a).

THE INTEGRATED OCEAN OBSERVING SYSTEM (IOOS)

Effectively linking societal needs for environmental information to measurements requires a managed, efficient, two-way flow of data and information among three essential

[1] IOOS is the U.S. contribution to the Global Ocean Observing System (GOOS), which is the oceans and coastal component of the Global Earth Observing System of Systems (GEOSS); see www.earthobservations.org/index.html.

TABLE CS1-1. IOOS is being established to provide the data and information required to address seven societal goals (see text in the Introduction). Examples of products and services are given that will be improved by integrating data across government agencies and programs (Ocean.US, 2006a, 2006b).

Societal Benefit Areas	Examples of Products and Services
(1) Climate Prediction	Estimates of global distributions of surface fluxes of heat and freshwater on monthly to decadal time scales Estimates of the state of ocean circulation and transports of heat, fresh water, and carbon on annual to decadal time scales Annual estimates of regional sea level change Provide improved global climatologies of key ocean variables (e.g., temperature, salinity, carbon)
(2 and 3) Marine Operations and Security	Maintain up-to-date, high-resolution bathymetry of shipping channels and near-shore shipping lanes Improve nowcasts and forecasts of sea surface current velocity, directional waves, and vector wind fields Improve real-time vessel tracking in coastal waters
(4) Natural Hazards	Improve forecasts of tsunamis on local to ocean basin scales Improve forecasts of time-space extent of coastal inundation caused by tropical storms, extra-tropical storms, nor'easters, and tsunamis Maintain up-to-date maps of changes in near-shore bathymetry-topography and the extent and condition of near-shore coastal habitats that affect resiliency and vulnerability to coastal inundation
(5) Public Health	Waterborne pathogens, and HAB organisms and their toxins: increase the accuracy and timeliness of nowcasts and forecasts of exposure risk through proximity (e.g., inhaling aerosols), direct contact (e.g., swimming), and seafood consumption Provide data and information needed to quantify relationships between changes in land use and land-based inputs to coastal waters and changes in public health risks
(6) Ecosystem Health and Water Quality	More rapid detection and accurate-timely predictions of the impacts of land-based sources of nutrients on phytoplankton biomass, water clarity, and dissolved oxygen fields More rapid detection and timely prediction of the transport, dispersion, and fate of suspended sediments, water-borne contaminants, and toxic harmful algal blooms Up-to-date GIS-based maps of the extent and condition of subtidal and intertidal habitats including barrier islands, sea grass beds, kelp beds, coral reefs, tidal wetlands, beaches, and dunes Repeated surveys of biodiversity and invasive species Models in support of ecosystem-based management of water quality
(7) Living Marine Resources	Annual documentation of the frequency and magnitude of mass mortalities (fish, mammals, birds) Improve (more accurate and timely) predictions of annual fluctuations in spawning stock size, distribution, recruitment, and sustainable yield for exploitable fish stocks Improve detection and prediction of the effects (commercial and recreational) of human uses (fishing, boating, commercial shipping, etc.) on habitats and biodiversity Increase the number of marine protected areas and monitor their effectiveness in terms of the sustainability of habitats, biodiversity, and fisheries Develop models in support of ecosystem-based management of water quality and living marine resources

"subsystems":[2] (an "end-to-end" system) for (1) measurements (remote and *in situ* observations) and data telemetry, (2) data management and communications (DMAC), and (3) data analysis and modeling (Fig. CS1-1a). The observing subsystem incorporates two interdependent components: a global ocean component with an emphasis on ocean-basin scale observations and a coastal component that focuses on the U.S.'s Exclusive Economic Zone (EEZ), Territorial Waters, Great Lakes, and estuaries. The coastal component can be further broken down into a National Backbone with Regional Coastal Ocean Observing Systems (RCOOSs) nested within it.

The global ocean component and the National Backbone monitor a set of core variables that are required for all seven IOOS goals (Table CS1-2). Recognizing that user groups and priorities for data and information vary from region to region, RCOOS collects observations specific to the issues for that region and provides data at greater resolution for these and other core variables. The development and design of the RCOOS is coordinated by IOOS regional associations, which are responsible for coordinating with stakeholders in the region. The regional component of IOOS will leverage observation data collected by state and local moni-

[2]The term *subsystem* is used here to indicate necessary functions of the IOOS, not to identify actual organizational entities or programs per se.

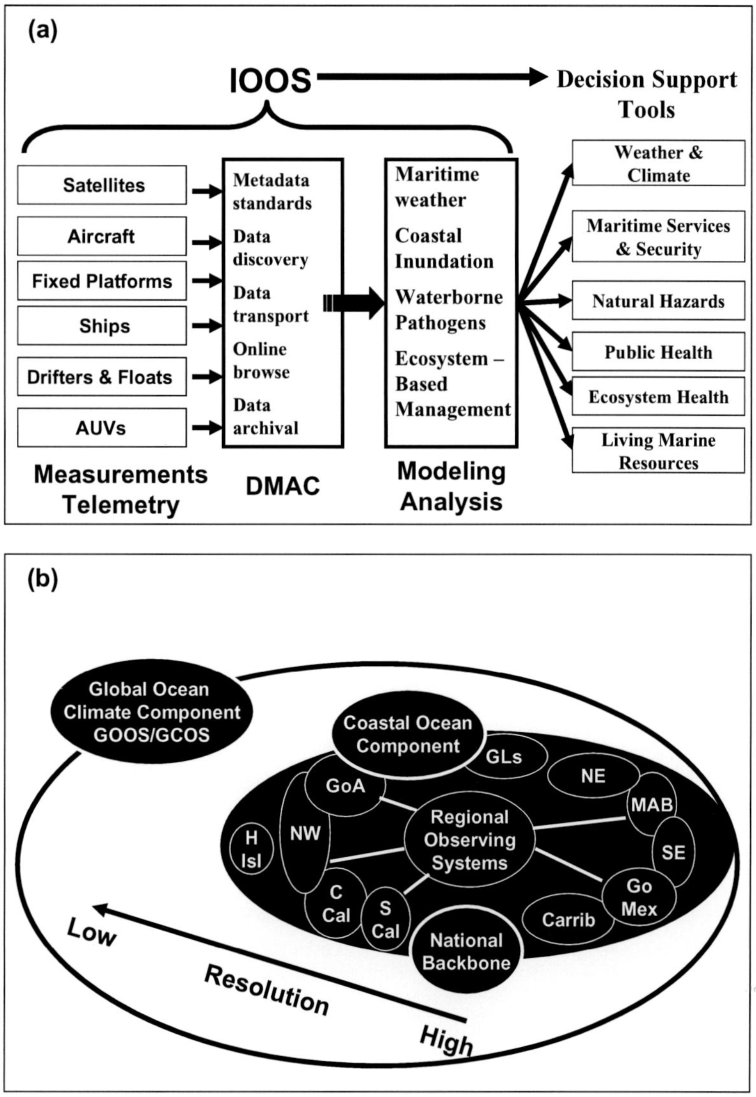

FIGURE CS1-1. (a) The IOOS is an "end-to-end" system of systems consisting of three efficiently linked subsystems for (1) observations and data telemetry, (2) data management and communications (DMAC), and (3) data analysis and modeling. The integrating engines are the DMAC and modeling subsystems. The atmospheric observing system of the National Weather Service is the kind of operational, end-to-end system that IOOS is envisioned to be. In contrast to the NWS observing system, the IOOS will serve a much greater diversity of multidisciplinary data and information for multiple applications. AUV = autonomous underwater vehicle. (b) The observing subsystem is multiscale consisting of global and coastal components. The latter can be further broken down into a National Backbone with Regional Coastal Ocean Observing Systems (RCOOSs) nested in it. The National Backbone measures core variables (Table CS1–2) required by federal agencies and IOOS Regional Associations (RAs) as a group. Abbreviations: GLs (Great Lakes), NE (Northeast), MAB (Mid-Atlantic Bight), SE (Southeast), GoMex (Gulf of Mexico), Carrib (Caribbean), SCal (Southern California), CCal (Central California), NW (Northwest), GoA (Gulf of Alaska), HIsl (Hawaiian Islands).

toring programs through integration with other relevant data sources and distribution to a broader community.

Together, the global ocean and coastal components of the IOOS constitute a hierarchy of observations (Fig. CS1-1b) required to detect, assess, and predict the effects of large-scale changes in the oceans, atmosphere, and land-based inputs on coastal ecosystems, resources, and human populations. Data management and communication provide rapid access to diverse data from many sources, and they are the primary means of integration. Models are the primary tools of synthesis required for rapid and timely detection and prediction of changes.

Successful development of the IOOS depends on making more effective use of the collective resources of U.S. institu-

TABLE CS1-2. Provisional IOOS core variables for the global component and National Backbone and their relevance to the seven societal goals of the IOOS (indicated by "X"). Physical variables are ranked high because they are required to achieve all seven societal goals. Note that natural hazards such as oxygen depletion and harmful algal blooms are addressed in the ecosystem health category. This list of variables is augmented by data on atmospheric, land-based, and anthropogenic forcings (Ocean.US, 2006a).

Core Variables	Weather and Climate	Marine Operations and Security	Natural Hazards	Public Health	Healthy Ecosystems	Sustained Resources
Salinity	X	X	X	X	X	X
Temperature	X	X		X	X	X
Bathymetry	X	X	X	X	X	X
Sea level	X	X	X		X	X
Surface waves	X	X	X	X	X	X
Surface currents	X	X	X	X	X	X
Ice distribution	X	X	X			
Contaminants		X		X	X	X
Dissolved nutrients				X	X	X
Fish species					X	X
Fish abundance					X	X
Zooplankton species				X	X	X
Optical properties		X		X	X	X
Heat flux	X				X	X
Ocean color	X	X		X	X	X
Bottom character	X	X			X	X
Pathogens		X		X	X	X
pCO_2	X					
Dissolved O_2					X	X
Phytoplankton species	X	X		X	X	X
Zooplankton abundance				X	X	X

tions and leveraging them to improve operational[3] capabilities for all seven societal goals. To these ends, IOOS development is guided by the following design principles:

1. Begin with the integration of existing observing assets that will improve the nation's ability to achieve the seven societal goals and regional priorities.
2. Enable data users and providers to achieve their missions and goals more effectively and efficiently.
3. Implement a scientifically sound system with guidance from users and providers from both public and private sectors.
4. Improve operational capabilities of the IOOS by enhancing and supplementing the initial system over time based on user needs and advances in technology and scientific understanding.
5. Routinely, reliably, and continuously provide rapid access to and disseminate data and information for multiple applications.
6. Share data and information produced at the public expense in a timely manner.
7. Ensure data quality and interoperability by meeting federally approved standards and protocols for observations, data discovery and transport, and modeling.
8. Establish procedures to ensure reliable and sustained data streams, routinely evaluate the performance of the IOOS, assess the value of the information produced, and improve operational elements of the system as new capabilities become available and user requirements evolve.
9. Improve the capacity of states and regions to contribute to and benefit from the IOOS through training and infrastructure development nationwide.

[3]*Operational* is used here to mean (1) the provision reliable, quality controlled assessments, and predictions (hind-, now- or forecasts) used by decision makers responsible for one or more of the seven societal goals; (2) the provisions of these assessments and predictions in forms and at rates specified by the users (on a schedule or on demand); and (3) activities that are performed by a responsible body and meet performance standards agreed to by both operators and users.

10. Demonstrate that observing systems, or elements thereof, that are incorporated into the operational system either benefit from being a part of an integrated system or contribute to improving the integrated system in terms of the delivery of new or improved products that serve the needs of user groups.

THE OCEANS AND HUMAN HEALTH

Climate, People, and Coastal Ecosystems

The cumulative effects of natural hazards, human activities, and climate change are and will continue to be most pronounced in the coastal zone where people and ecosystem goods and services are most concentrated, exposure to natural hazards is greatest, and inputs of energy and matter from land, sea, and air converge (Costanza *et al.*, 1993; McKay and Mulvaney, 2001; Nicholls and Small, 2002; Small and Nicholls, 2003). Changes occurring in coastal waters affect public health and well-being, the safety and efficiency of marine operations, and the capacity of ecosystems to support goods and services (including the sustainability of living marine resources and biodiversity). Although these changes tend to be local in scale, they are occurring in coastal ecosystems worldwide and are often local expressions of larger scale variability and change, including both natural and anthropogenic drivers or "forcings":

- Natural hazards (Epstein, 1999; Flather, 2000; Michaels *et al.*, 1997)
- Global warming and sea level rise (Barry *et al.*, 1995; Levitus *et al.*, 2000; Najjar *et al.*, 1999)
- Basin scale changes in ocean-atmosphere interactions (El Niño Southern Oscillation, North Atlantic Oscillation, and Pacific Decadal Oscillation) (Barber and Chavez, 1986; Beaugrand *et al.*, 2003; Koblinsky and Smith, 2001; Wilkinson *et al.*, 1999)
- Human alterations of the environment (Group of Experts on the Scientific Aspects of Marine Pollution [GESAMP], 2001; Heinz Center, 2002; Peierls *et al.*, 1991; Vitousek *et al.*, 1997),
- Exploitation of living resources (Jackson *et al.*, 2001; Myers and Worm, 2003)
- Introductions of nonnative species (Carlton, 1996; Hallegraeff, 1998)

Each of these drivers of change has been shown to influence human health risks, from exposure to waterborne human pathogens to the toxins produced by harmful algae bloom (HAB) organisms (affecting people through direct contact, inhalation of aerosols, and seafood consumption). The clearest and most direct impacts on the oceans and human health occur in coastal areas that are subject to intense human use (sewage discharge, agriculture and aquaculture practices, human habitation and recreation, fishing, etc.) and are susceptible to flooding from tsunamis, storm surges, and excessive rainfall associated with tropical storms and monsoons (National Research Council [NRC], 1999). There is also increasing evidence that global scale changes in the abundance and distribution of both waterborne and vector-borne diseases are occurring in response to global warming and changes in the hydrological cycle (Colwell, 1996; Epstein, 1999; Haines and Parry, 1993; Rogers and Packer, 1993).

Ecosystem-Based Approaches to Managing Health Risks

The oceans and Great Lakes are conduits for many pathogenic microorganisms and their toxins (Table CS1-3). Their distributions and exposure risks in aquatic systems are governed by their sources; their behavior once introduced into the aquatic environment (e.g., rates of growth, mortality, migration, buoyancy, etc.); their place in the food web; and by water motions that transport, disperse, or concentrate them. The most effective ways to reduce the immediate cost of lives and human suffering from exposure to waterborne pathogens and harmful algal blooms is to detect changes in risk more rapidly, provide timely accurate predictions of changes in risk in both time and space, and control the sources (e.g., reduce inputs of untreated sewage wastes that transport pathogens, reduce land-based inputs of anthropogenic nutrients that stimulate some HAB organisms, and reduce the temporal and spatial extent of coastal flooding that can promote events such as cholera epidemics and the growth of HAB organisms).

Increases in risk to levels that lead to beach and shellfish bed closures are typically localized, episodic, and dynamic. Consequently, rapid, timely, and accurate assessments of risk are difficult if not impossible based on traditional sampling regimes (e.g., monthly or biweekly monitoring of sewage outfalls and daily shoreline sampling at a limited number of beach sites). Remote sensing and the development of species-specific *in situ* sensors for waterborne pathogens and HABs thus have great potential for providing the means to address these challenges.

For example, satellite-based synthetic aperture radar (SAR) provides high resolution (<100 m) active microwave observations of sea surface roughness that are independent of cloud cover and time of day. At surface wind speeds between 2 m sec^{-1} and 7 m sec^{-1}, areas with biogenic or anthropogenic surfactant films that dampen small waves are detected by SAR as patterns of low backscatter return. Studies in the Southern California Bight illustrate the ability of SAR to detect and track the fate of storm-water runoff and sewage discharge (DiGiacomo *et al.*, 2004; Svejkovsky and Jones, 2001). In combination with field surveys, land-based high-frequency radar, and numerical models, these studies demonstrate the potential for rapid detection and

TABLE CS1-3. Selected lists of agents of waterborne disease, their pathologies, and routes of transmission (modified from NRC, 1999).

Agent	Pathology	Transmission Route
Pathogens		
Viruses		
Hepatitis A	Infectious hepatitis	Seafood,[1] Water ingestion
Hepatitis B	Hepatitis	Water ingestion
Caliciviruses		
Rotaviruses	Gastroenteritis	Water ingestion
Astroviruses	Gastroenteritis	Seafood,[1] Water ingestion
Enteroviruses	Gastroenteritis	Water ingestion
Autochthonous Bacteria		
Mycobacterium marinum	Granuloma	Water contact
Vibrio alginolyticus	Wound infections	Water contact
V. cholerae	Cholera	Seafood,[1] Water ingestion
V. parahaemolyticus	Gastroenteritis, Wound infections	Seafood,[1] Water contact
Allochthonous Bacteria		
Escherichia coli	Dysentery, Gastroenteritis	Water contact
Leptospira interrogans	Leptospirosis	Water contact
Listeria monocytogenes	Listeriosis	Seafood[1]
Morganella morganii	Scomboid food poisoning	Seafood[1]
Salmonella spp.	Typhoid, Gastroenteritis	Water ingestion
Shigella spp.	Dysentery	Water ingestion
Harmful Algae		
Dinoflagellates		
Alexandrium catenella	Paralytic shellfish poisoning	Seafood[2]
A. tamaense	Paralytic shellfish poisoning	Seafood[2]
Gymnodinium catenatum	Paralytic shellfish poisoning	Seafood[2]
Pyrodinium bahamense	Paralytic shellfish poisoning	Seafood[2]
Karenia brevis	Neurotoxic shellfish poisoning	Seafood[2,3]
Gambierdiscus toxicus	Ciguatera fish poisoning	Seafood[2]
Prorocentrum spp.	Ciguatera fish poisoning	Seafood[2]
Prorocentrum spp.	Diarrheic shellfish poisoning	Seafood[2]
Dinophysis spp.	Diarrheic shellfish poisoning	Seafood[2]
Diatoms		
Pseudo-nitzschia multiseries	Amnesic shellfish poisoning	Seafood[2]
P. australis	Amnesic shellfish poisoning	Seafood[2]
Cyanobacteria		
Anabena spp.	Diarrhea	Water ingestion
Microcystis spp.	Diarrhea	Water ingestion
Oscillatoria spp.	Diarrhea	Water ingestion

[1]Consumption of raw or undercooked seafood.
[2]Cooked or uncooked.
[3]Inhalation of aerosol particles coated with the toxin can also cause respiratory problems (tightness of breath, coughing and sneezing, tearing eyes, burning sensation in mucous membranes, and excess mucous discharge).

timely predictions that can be used to inform management and mitigation decisions that reduce public health risks and increase the economic and social value of beaches and living resources. The IOOS provides a platform for achieving these objectives.

Two case studies, adapted from Malone and Rockwell (in press), are described next to illustrate the common need for hydrodynamic models (and meteorological and physical oceanographic observations) and the challenges and potential of developing early warning observing systems for more cost-effective management and mitigation of their impacts (e.g., reducing health risks and realizing economic benefits through fewer closures of shellfish beds and beaches).

Harmful Algal Bloom Observing Systems

Concerns about harmful algal blooms (HABs) have increased since the mid-1990s largely because of the perceived increase in the number and duration of events. The toxins produced by these species cause finfish and shellfish poisoning, a variety of human illnesses that can lead to death, and mass mortalities of marine organisms including fish, mammals, and birds. The case study described here is the Harmful Algal Bloom Forecasting System for the Gulf of Mexico (www.csc.noaa.gov/crs/habf). The Gulf of Mexico has many features that made this region suitable for a pilot project that can rapidly develop into an operational system if successful. These include the following:

- It is an interested public and active management community with a long history of monitoring HABs and associated shellfish bed closures.
- Although home to 30 or more toxic species, areas of the Gulf are most impacted by *Karenia brevis* (formerly *Gymnodinium breve*), which causes respiratory distress and neurotoxic shellfish poisoning (NSP) in humans and is responsible for mass mortalities of marine organisms, including fish, sea turtles, and dolphins, and shellfish beds closures (Kirkpatrick *et al.*, 2004; Steidinger *et al.*, 1997).
- *K. brevis* can be detected through a combination of remote and autonomous *in situ* sensing (Hu *et al.*, 2003; Schofield *et al.*, 1999; Stumpf *et al.*, 2003). It is pigmented with concentrations of chlorophyll *a* and cells correlated over the dynamic range of concentrations typically found in the Gulf (<1–20 μg Chl liter^{-1}); it has a surface expression that can be detected by satellite-based remote sensing (Fig. CS1-2); and it has a specific signature of inherent optical properties that allows recognition by *in situ* optical instrumentation (Stumpf *et al.*, 2003).
- *K. brevis* blooms are a regional problem (i.e., they are not confined to one state but are often transported across state jurisdictions) (Stumpf *et al.*, 2003).
- *K. brevis* blooms occur most frequently during late summer and fall when diatom biomass tends to be low, therefore *K. brevis* often dominates the chlorophyll *a* field (Kusek *et al.*, 1999).
- Blooms of *Trichodesmium* spp., a genus of cyanobacteria with an optical signature (phycobilins) different from that of *K. brevis* (Subramaniam and Carpenter, 1994), often precede *K. brevis* blooms (Walsh and Steidinger, 2001).
- The frequency and timing of blooms may be a useful indicator of global warming (Tester and Steidinger, 1997).

A comprehensive review of the occurrence of *K. brevis* blooms, the biology and ecology of *K. brevis* and its impacts (real and perceived) are given by Kusek *et al.* (1999) and Kirkpatrick *et al.* (2004) (also see Chapter 13) (Fig. CS1-2).

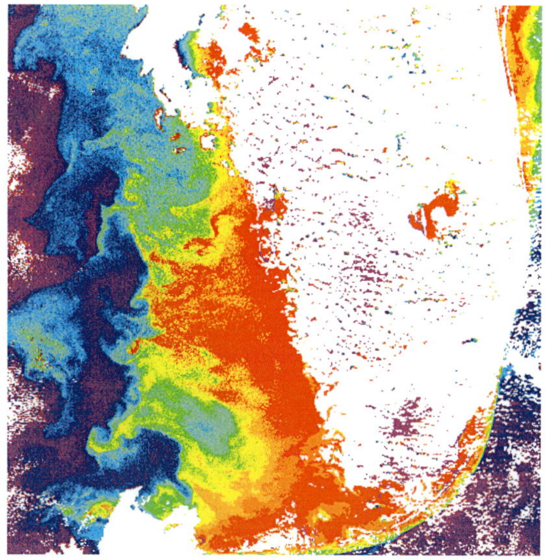

FIGURE CS1-2. *Karenia brevis* surface chlorophyll signature detected by the NASA satellite, Coastal Zone Color Scanner, along the west coast of Florida.

Stumpf (2001) developed a climatological approach to detecting areas of anomalous chlorophyll *a* that are most likely caused by *K. brevis* (75% of the time during 1999 through 2001), and it has been used to provide early warnings of new blooms, to track the movements of existing blooms, and to forecast trajectories (Stumpf *et al.*, 2003). An anomaly is identified when differences between the current chlorophyll field (from satellite-based remote sensing of ocean color) and a mean field exceeds 1 μg liter-1 (equivalent to 100 cells ml^{-1}), where the mean is a 2-month running mean ending 2 weeks before the current image. Two months provides a sufficient number of images over a long enough period to be representative of the season. A 2-week lag minimizes the possibility that a persistent and stationary bloom will bias the mean. This information is communicated to user groups responsible for monitoring or managing HAB events in near real time via a Web-based HAB bulletin.

The ultimate goal is to predict when and where a bloom is likely to impact people and fisheries. This requires knowledge of the physical environment and the organism's physiology to understand the initiation and transport of blooms. *K. brevis* has been reported to be both positively and negatively phototactic (Kamykowski *et al.*, 1998; Steidinger, 1975). Blooms along the west coast of Florida often appear during upwelling-favorable winds when preceded by a period of calm winds (Stumpf *et al.*, 1998). Although it is believed that blooms are initiated offshore, blooms often appear along the shoreline with no indication of offshore development based on satellite imagery (Stumpf *et al.*,

2003). Together, these observations suggest that blooms are initiated and develop offshore at depth under calm conditions. With the development of wind-driven upwelling circulation, blooms are transported onshore with bottom water and transported to the surface by coastal upwelling. The set of observations that will be needed to analyze the conditions for bloom initiation and transport are as follows:

1. *Forcings.* Real-time wind fields and precipitation updated three to four times daily; river and stream flows and associated transports of sediments and nutrients updated weekly
2. *Ecosystem properties.* Surface current, temperature, salinity, and chlorophyll fields updated daily; vertical profiles of temperature, salinity, dissolved oxygen, inorganic nutrients, chlorophyll, colored dissolved organic matter, and *K. brevis* cell densities at weekly (hot spots that have a history of frequent HABs) to monthly intervals (larger area, more stations)

Using observations from satellite and buoy platforms, alerts of near-shore blooms have been communicated to government agencies responsible for public health, management of living marine resources, and environmental protection. These alerts allow the agencies to effectively plan their monitoring efforts to target areas where a bloom may impact public health or seafood safety (http://coastwatch.noaa.gov/hab/bulletins_ns.htm).

The information on bloom location gathered through ocean measurements can be combined with wind speed and direction information to estimate the health impact of a coastal bloom on the beach (Kirkpatrick *et al.*, 2004). Stronger winds blowing across a bloom area onto the beach will be more likely to impact air quality than calmer winds blowing away from the beach. The public, particularly those living and recreating on the beach where *K. brevis* blooms affect air quality, is very interested in receiving daily, even hourly, information on the presence of *K. brevis* at the shore. This need for information dictates frequent measurements of *K. brevis* and its toxins, as well as human symptoms and other health information, from the surf-zone, beaches, and coastal areas where persons are exposed with input of these data into the predictive model (Reich *et al.*, 2006).

Developing a comprehensive monitoring design that collects data at the time and space scales relevant to the dynamics of the bloom and connecting that environmental information to the health impacts is a critical step to providing useful information from IOOS for assessing risk from a *K. brevis* event.

Beach Closures in the Great Lakes

Improved estimates of land-based inputs of human pathogens and models of coastal circulation and water quality will reduce risks of human exposure, provide the data and information needed for more effective control of anthropogenic inputs, and maximize recreational income (Cheves, 2003).

Land-based sources of pathogens and bacterial indicators of pathogens include both point source discharges of untreated wastewater (e.g., combined sewer overflows, SSO, boat sanitary discharges, and septic systems) and nonpoint (diffuse) surface runoff of fecal contaminants from wildlife and livestock. Both inputs are exacerbated by excessive rainfall (e.g., associated with tropical storms and El Niño–Southern Oscillation [ENSO] events) and coastal inundation (caused by tropical storms, extratropical storms, northeasters, and tsunamis). Once introduced into aquatic systems, determination of health risks and subsequent beach notifications or swim closures depends on timely and accurate estimates of concentrations of waterborne pathogens, their distribution, and changes in distribution over time.

Because of uncertainties associated with both estimates of pathogen concentrations and distributions, public officials face the probability that the decision to close a beach will be in error because little correlation has been found between measurements from one day to the next day and distributions change with time because of circulation, mixing, and mortality (Olyphant *et al.*, 2003). The waters at the beach may be safe for swimming, but the official closes the waters; or the waters at the beach may be unsafe for swimming, yet they are kept open in error. Given that the official is likely to use the precautionary principle and keep the beaches closed, more swimming at beaches is likely to be prevented when they could be open if uncertainty were reduced.

Because of the high diversity of waterborne microbial pathogens and the cost of direct measurements of their concentration, *E. coli* and enterococci (known as fecal indicator organisms) are used in the United States to indicate the risk of exposure to pathogens in freshwaters (Dufour and Wymer, 2006). Current cost-effective tests using culture techniques require 12 to 24 hours to determine concentrations of *E. coli* and enterococci. Thus, decisions to close a beach or take no action (keep the beach open and permit swimming) are based on information that does not become available until 18 to 24 hours after the fact. Errors are also introduced because concentrations of fecal indicators such as *E. coli* and enterococci are often not correlated with concentrations of the actual microbial pathogens themselves (NRC, 1999). This is exacerbated by the fact that some strains of *E. coli* produce human toxins and can survive in a viable but unculturable state for days to weeks (NRC, 1999).

Pathogens of human origin (e.g., allochthonous bacteria, Table CS1-3) enter coastal waters with surface runoff and sewage discharges. Once in coastal waters, changes in the distribution and concentration of human pathogens are determined primarily by the "half life" of pathogens and by their rates of transport, concentration, and dispersion. Pathogens introduced via storm water runoff and point source discharges are typically concentrated in buoyant plumes that can be observed and modeled to provide "nowcasts" of location and forecasts of transport and dilution. As discussed previously, an optimal mix of *in situ* observations and remote sensing can be used to provide the required observations. Predictive models have been developed to estimate the dilution, dispersion, and decline of fecal indicators (Intergovernmental Oceanographic Commission [IOC], 2003). These coupled physical-biological models incorporate the effects of (1) dilution resulting from turbulent mixing, (2) advective transport, and (3) the rate of decay in toxicity or mortality (including the dormancy rate).

In the Great Lakes, direct measurement of surface winds and currents have been used in an experimental mode to provide nowcasts and forecasts of *E. coli* distributions that can be used by beach managers to advise the swimming public more effectively. Model predictions of *E. coli* distribution use wind speed and direction, wave height, water turbidity, suspended sediment concentration, rainfall in the preceding 24 hours, and water temperature to make such forecasts (Francy and Darner, 1998; Francy *et al.*, 2003; Whitman *et al.*, 1999). Great Lake beaches have numerous and varying conditions that affect bacterial conditions (Nevers and Whitman, 2004), but a regional nowcast model for southern Lake Michigan is being developed that shows clusters of beaches responding in similar ways. Monitoring systems provide daily metrological measurements that are assimilated into a nowcast model with prediction accuracy ranging from 54% to 91% depending on beach zone and high or low predictions (Whitman, 2005). The data and information provided by an integrated and sustained Great Lakes observing system coupled with beach-specific models will reduce these uncertainties, leading to fewer beach closures and lower health risks.

Current approaches for managing public use of beaches and near-shore waters for swimming suffer from three sources of uncertainty: (1) the time required to detect *E. coli* indicators of pathogens (24 to 48 hours); (2) the need to measure indicator concentrations in the context of environmental parameters that determine their distribution and concentration; and (3) need for operational, data-assimilating models that can provide nowcasts and forecasts of the distribution of pathogens using coupled hydrodynamic-biological models. Reducing these sources of uncertainty requires (1) the development of species-specific, inexpensive methods of rapidly measuring the concentration of waterborne pathogens (rapid detection) and (2) observations and models of the aquatic environment that can be used to nowcast and forecast their distributions (timely prediction).

To address these problems, the U.S. Environmental Protection Agency (EPA), the U.S. Geological Survey (USGS), and the National Oceanographic and Atmospheric Administration (NOAA) are collaborating to support an integrated effort to develop an observing system for more timely and accurate forecasts of the distribution and abundance of waterborne pathogens in beach environments. An important part of this effort is the development of a beach forecasting tool (BFT) (Frick *et al.*, 2004). The BFT combines data on bacteria sources and concentrations with hydrodynamic and statistical models to generate 3-day forecasts of beach concentrations. In this interagency collaboration, the NOAA provides a wind-driven, three-dimensional hydrodynamic model (Beletsky *et al.*, 2003); the EPA provides point source plume models (Frick *et al.*, 2003); and the EPA and USGS have compiled data and developed source strength and mortality models to enable initial bacterial concentrations and decay rates to be estimated (Frick *et al.*, 2004; Myers *et al.*, 1998; Olyphant *et al.*, 2003).

The BFT is the basis of "visual beach," which will make forecasts available via the World Wide Web and disseminate forecasts to registered users (Frick *et al.*, 2004). Monitoring results will be used to initialize model runs and to verify model results. The goal is to overcome the long time lag between changes in environmental concentrations of *E. coli* and their detection (24 to 48 hours), help guide adaptive sampling, and inform the public and decision makers responsible for beach closures. The effectiveness of this approach was illustrated during high and low flows of the Milwaukee River in May and June of 2004 (Fig. CS1-3). The integration of three-dimensional hydrodynamic models using wind direction modeling, wind direction and speed measurements, and geosynchronous orbiting satellites can provide the means to predict and verify the growth, movement, and dis-

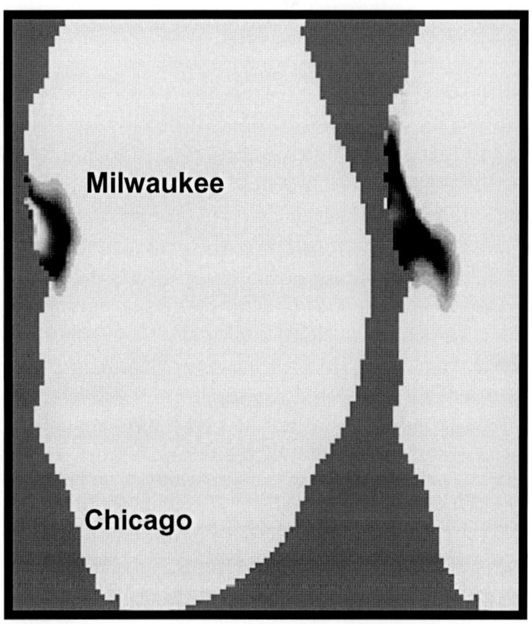

FIGURE CS1-3. Milwaukee River plume simulations using the NOAA hydrodynamic model for May 2004 (left, low flow) and June 2004 (right, high flow) correspond well with observations (Frick *et al.*, 2004). Degree of darkness indicates relative concentration of river water.

sipation of buoyant plumes and possibly their potential impact on coastal beaches. Linkage of these hydrodynamic, forecast, and nowcast models to computer Web sites offer the promise of providing monitoring and modeling results so the swimming community can plan their weekend (Frick *et al.*, 2004).

DATA REQUIREMENTS FOR MANAGING PUBLIC HEALTH RISKS AS A DRIVER FOR IOOS DEVELOPMENT

Design and implementation of the IOOS is guided by the data and information needs of those that use, manage, or depend on the oceans and its resources. The public health goal (Table CS1-1) encompasses risks associated with the consumption of seafood and contact with waterborne toxins (including aerosols) and microbial pathogens (viruses, bacteria, and harmful algae). This contribution focuses on the latter to illustrate the need for an IOOS that provides data and information needed for more rapid detection, assessments, and predictions of change in marine systems that affect the health and well-being of people.

The current and potential IOOS contributions to achieving the societal goal of reducing public health risks were addressed in detail at a workshop, *Public Health Risks: Coastal Observations for Decision Making and Management*, held in January 2006 (Ocean.US, 2006b). Observations can contribute to public health management by providing the underlying data to inform several types of decisions, including the following:

- Short-term decisions to provide health advisories and warnings for the opening and closing of beaches and shellfish management areas (e.g., decisions regarding food-borne and environmental exposure risks)
- Longer-term decisions on remediation strategies for coastal areas (e.g., improving water quality through shoreline revitalization or nonpoint source pollution control)

Timely and credible short- and long-term decisions require rapid identification of sources and timely nowcasts and forecasts of their transport and fate. Improving the operational capabilities of IOOS to achieve these objectives depends on major advances on at least three fronts:

1. Development of integrated data communications and management systems that leverage existing monitoring data and provide rapid access to data on key environmental parameters (e.g., current, temperature, and salinity fields) and exposure risk (e.g., distribution of indicators of waterborne pathogens, the pathogens themselves, or toxicity)
2. Development of *in situ* and remote sensing sensors for near real-time provision of data on these environmental parameters and exposure risks and placement of these sensors in the near-shore areas where public health impacts occur
3. Development of data assimilation techniques and coupled physical-ecological models that can be run in an operational mode to provide nowcasts and forecasts of the distribution of exposure risk

Harmful algal blooms and waterborne pathogens are important cases in point. Ecosystem-based management strategies are aimed at preventing HABs (e.g., reduce nutrient loading) and pathogen contamination (e.g., sewage treatment) before they occur, mitigating their effects (e.g., close shellfish beds, close beaches, move net pens of cultured salmon) and controlling them once they occur (e.g., reducing their magnitude, containing their distribution). Achieving these objectives requires the development of four related capabilities for both implementing adaptive management and assessing their efficacy:

- More rapid detection of waterborne pathogens, and HAB organisms and their toxins
- Timely predictions of where and when public health risks are likely to be unacceptably high
- Timely forecasts of trajectories and contaminated water masses in time and space
- Products developed that provide relevant information at the appropriate time and space scales and in the format needed for the user community to implement prevention, mitigation, and control strategies

Rapid detection is a high priority for both waterborne pathogens and HABs, and molecular techniques offer a way forward. For example, species-specific DNA probes from ribosomal sequences have the potential to provide accurate and rapid diagnostic tools for the evaluation of environmental samples (NRC, 1999). When combined with polymerase chain reaction (PCR), these probes allow detection of an increasing number of pathogens and indicators. The recent application of real-time PCR to field diagnostics of microbial pathogens reveals the potential of this approach for rapid and reliable detection of pathogens in aquatic systems. The IOOS will provide a platform for testing and deploying sensors that directly measure biological and chemical variables, such as pathogens and toxins, in near real-time to verify the location and intensity of events.

To the extent that allochthonous waterborne pathogens behave as passive particles with known half-lives and their sources are well documented and quantified, the development of operational nowcasting and forecasting capabilities depends primarily on more rapid detection of pathogens and increases in the spatial resolution of hydrodynamic models.

The HAB challenge is more complex. It will not be possible to develop operational models for HAB prediction based on environmental conditions until the combination of environmental factors that promote the growth and accumulation of one species over others are quantified and physical-biological interactions are parameterized (e.g., how environmental factors such as turbulence, advective transport, light, nutrient, grazing, and inherent biological attributes interact with each other to favor the development of a given species). Developing this capacity will require significant advances in our understanding of the processes of species succession; in the development of coupled, data assimilating physical-ecological models; and in the capacity to estimate the distribution and abundance of HAB species and toxins rapidly with time-space estimates of physical and chemical fields using a combination of *in situ* and remote sensing techniques. These include (1) more accurate estimates of sea surface chlorophyll *a* and accessory pigment fields on space scales of < 1 km based on ocean leaving radiance measurements from satellites and aircraft (improve the skill of coastal algorithms and increase spatial resolution); (2) long-term, high resolution time series by instrumenting moorings and fixed platforms with sensors to measure apparent optical properties and nutrient concentrations (N, P, Si) synoptically with temperature, salinity, currents, and waves; (3) techniques for rapid, species-specific identification and enumeration, including near real-time measurement and telemetry of HAB cell densities; (4) techniques for more rapid measurement of HAB toxins, including *in situ* detection and near real-time telemetry; and (5) rapid access to data from both *in situ* and satellite-based observations.

Until then, statistical models will be used to predict where and when HABs are likely to occur based on historical records of the location, frequency, and magnitude of HABs or on correlations of HABs with environmental variables or indices. This not only places a high priority on research (predictive models, real-time sensing technologies), it places a high priority on detection: initially, IOOS must focus on the development of the capacity to detect HAB organisms and toxins routinely and rapidly in the context of changes in the distribution of key environmental factors. To these ends, priority should be placed on the establishment of a global network of sentinel (early warning) and reference (to develop climatologies of HABs and associated environmental conditions) stations for long-term time series observations and the development of an integrated data communications and management system for rapid access to and dissemination of data on HAB organisms abundance, toxin concentrations, and key environmental variables (temperature, salinity, surface waves and currents, concentrations of inorganic and organic forms of N, P, and Si). This information from an IOOS system integrated with epidemiological data on symptoms and diseases in human and other animal populations will provide a comprehensive data source to assess the risks and health impacts associated with HABs.

Programs such as those described in the Gulf of Mexico and Great Lakes and others on the U.S. West Coast, in Washington state, southern California, and central California, provide early warning systems for HABs, but they will have a limited lifetime unless they are made operational as part of a multipurpose, integrated observing system for the oceans, Great Lakes, and coastal ecosystems. The incorporation of well-tested detection and forecasting systems for HABs and pathogens into the IOOS bridges the gap between advances in science and the application of these advances to develop information products for the public good.

References

Barber, R.T., Chavez, F.P., 1986. Ocean variability in relation to living resources during the 1982–83 El Niño. Nature 319, 279–285.

Barry, J.P., Baxter, C.H., Sagarin, R.D., Gilman, S.E., 1995. Climate-related, long-term faunal changes in a California rocky intertidal community. Science 267, 672–675.

Beaugrand, G., Brander, K.M., Lindley, J.A., Souissi, S., Reid, P.C., 2003. Plankton effect on cod recruitment in the North Sea. Nature 426, 661–664.

Beletsky, D., Schwab, D.J., Roebber, R.P., McCormick, M.J., Miller, G.S., Saylor, J.H., 2003. Modeling wind-driven circulation during the March 1998 sediment resuspension even in Lake Michigan. J. Geo. Res. 108(C2), 20–1 to 20–13.

Carlton, J.T., 1996. Marine bioinvasions: The alteration of marine ecosystems by nonindigenous species. Oceanography 9, 36–45.

Cheves, M., 2003. Monitoring the heartbeat of mother earth. EOM 12, 6–10.

Colwell, R., 1996. Global climate and infectious disease: The cholera paradigm. Science 274, 2025–2031.

Costanza, R., Kemp, W.M., Boynton, W.R., 1993. Predictability, scale, and biodiversity in coastal and estuarine ecosystems: Implications for management. Ambio 22, 88–96.

DiGiacomo, P.M., Washburn, L., Holt, B., Jones, B.H., 2004. Coastal pollution hazards in southern California observed by SAR imagery: Stormwater plumes, wastewater plumes, and natural hydrocarbon seeps. Mar. Pollution Bull. 49, 1013–1024.

Dufour, A., Wymer, L.J., 2006. Microbes, Monitoring and Human Health. Oceanography 19,72–80.

Epstein, P.R. (ed.), 1999. Extreme Weather Events: The Health and Economic Consequences of the 1997/98 El Niño and La Niña. The Center for Health and the Global Environment, Boston, Harvard Medical School.

Flather, R.A., 2000. Existing operational oceanography. Coastal Eng. 41, 13–40.

Francy, D.S., Darner, R.A., 1998, Factors affecting *Escherichia coli* concentrations at Lake Erie public bathing beaches. U.S. Geological Survey, Columbus Ohio, Water Resources Investigations Report 98–4241.

Francy, D.S., Gifford, A.M., Darner, R.A., 2003. *Escherichia coli* at Ohio bathing beaches: Distribution, sources, wastewater indicators, and predictive modeling. U.S. Geological Survey, Columbus Ohio, Water-Resources Investigations Report 02–4285. Full report available at http://oh.water.usgs.gov/beaches/index.html.

Frick, W., Francy D.S., Rockwell, D.C., Schwab, D., Beletsky, D, Lunetta, R., 2004. A Vision for a Beach Forecasting Tool, presented at MWWD

2004, 3rd International Conference on Marine Waste Water Disposal and Marine Environment, and IEMES 2004, 1st International Exhibition on Material Equipment and Services for Coastal WWTP Outfalls and Sealines, Catania (I).

Group of Experts on the Scientific Aspects of Marine Pollution (GESAMP), 2001. Protecting the Oceans from Land-based Activities: Land-based Sources and Activities Affecting the Quality and Uses of the Marine, Coastal and Associated Freshwater Environments. GESAMP Reports and Studies No. 71.

Haines, A., Parry, M., 1993. Climate and human health. J. Roy. Soc. Med. 86, 707–711.

Hallegraeff, G.M., 1998. Transport of toxic dinoflagellates via ship's ballast water: Bioeconomic risk assessment and efficacy of possible ballast water management strategies. Mar. Ecol. Prog. Ser. 168, 297–309.

Heinz Center, 2002. Human Links to Coastal Disasters. Washington, D.C., The H. John Heinz III Center for Science, Economics and the Environment.

Hu, C., Hackett, K.E., Callahan, M.K., Andréfouët, S., Wheaton, J.L., Porter, J.W., Muller-Karger, F.E., 2003. The 2002 ocean color anomaly in the Florida Bight, a cause of local coral reef decline? Geophys. Res. Letters 30.

IOC (Intergovernmental Oceanographic Commission), 2003. The Integrated Strategic Design Plan for the Coastal Ocean Observations Module of the Global Ocean Observing System. GOOS Report No. 124, UNESCO (http://ioc.unesco.org/goos/docs/doclist.htm).

Jackson, J.B.C., Kirby, M.X., Berger, W.H., Bjorndal, K.A., Botsford, L. W., Bourque, B.J., Bradbury, R.H., Cooke, ,R., Erlandson, J., Estes, J. A., Hughes, T.P., Kidwell, S., Lange, H.S., 2001. Historical overfishing and the recent collapse of coastal ecosystems. Science 293, 629–643.

Kamykowski, D., Milligan, E.J., Reed, R.E., 1998. Biochemical relationship with the orientation of the autotrophic dinoflagellate *Gymnodinium breve* under nutrient replete conditions. Mar. Ecol. Prog. Ser. 167, 105–117.

Kirkpatrick, B., Fleming, L.E., Squicciarini, D., Backer, L.C., Clark, R., Abraham, W., Benson, J., Cheng, Y.S., Johnson, D., Pierce, R., Zaias, J.,Bossart, G.D., Baden, D.G., 2004. Literature review of Florida red tide: Implications for human health effects. *Harmful Algae* 3, 99–115.

Koblinsky, C., Smith, N.R., 2001. Observing the Oceans in the 21st Century. Melbourne, Australia, GODAE Project Office, Bureau of Meteorology.

Kusek, K.M., Vargo, V., Steidinger, S., 1999. *Gymnodinium breve* in the field, in the lab, and in the newspaper: A scientific and journalistic analysis of Florida red tides. In Villarcal, T.A. (cd.), Contributions in Marine Science, 34, 1–229.

Levitus, S., Antonov, J.L., Boyer, T.P., Stephens, C., 2000. Warming of the world ocean. Science 287, 2225–2229.

Malone, T.C., Knap, C., Fogarty, M., 2005. Overview of science requirements. In Robinson, A.R., and Brink, K. (eds.), The Global Coastal Ocean: Multiscale Interdisciplinary Processes, The Sea, vol. 13, pp. 757–784, Cambridge, MA, Harvard University Press.

Malone, T.C., Rockwell, D.C., in press. Ocean observation networks: Their role in monitoring water quality. In Rose, J.B., and Dreelin, E.A. (eds.), Effective Cross-Border Monitoring Systems for Waterborne Microbial Pathogens: A Plan for Action. London, IWA.

McKay, B., Mulvaney, K., 2001. A review of marine major ecological disturbances. Endangered Species Update 18, 14–24.

Michaels, A., Malmquis T.D., Knap, A., Close, A., 1997. Climate science and insurance risk. Nature 389,225–227.

Myers, R.A., Worm, B., 2003. Rapid worldwide depletion of predatory fish communities. Nature 423, 280–283.

Najjar, R.G., Walker, H.A., Anderson, P.J., Barro, E.J., Bord, R.J., Gibso, J.R., Kennedy, V.S., Knight, C.G., Megonigal, J.P., O'Connor, R.E., Polsky, C.D., Psuty, N.P., Richards, B.A., Sorenson, L.G., Steele, E.M., Swanson, R.S., 1999. The potential impacts of climate change on the Mid-Atlantic coastal region. Climate Res. 14, 219–233.

National Research Council (NRC), 1999. From Monsoons to Microbes: Understanding the Ocean's Role in Human Health. Washington, D.C., National Academy Press.

Nevers M.B., Whitman R.L., 2004, Protecting visitor health in beach waters of Lake Michigan: Problems and opportunities in the state of Lake Michigan. In Edsall, T., and Munawar, M., Ecology, Health and Management, Ecovision World Monograph Series, Aquatic Ecosystem Health and Management Society, 18.

Nicholls, R.J., Small, C., 2002. Improved estimates of coastal population and exposure to hazards. Eos Transactions 83, 301–305.

Ocean.US., 2002. An integrated and sustained ocean observing system (IOOS) for the United States: Design and Implementation. Ocean.US Report No. 2 (www.ocean.us/oceanus_publications).

Ocean.US., 2006a. The First Annual Integrated Ocean Observing System (IOOS) Development Plan. Ocean.US Report No. 9 (www.ocean.us/oceanus_publications).

Ocean.US., 2006b. Public Health Risks: Coastal Observations for Decision-Making. Ocean.US Report No. 15 (www.ocean.us/oceanus_publications).

Olyphant, G.A., Thomas J., Whitman R.L., Harper, D., 2003, Characterization and statistical modeling of bacterial (Escherichia coli) outflows from watersheds that discharge into southern Lake Michigan. Environ. Monitoring Assessment 81, 289–300.

Peierls, B.L., Caraco, N.F., Pace, M.L., Cole, J.J., 1991. Human influences on river nitrogen. Nature 350, 386–387.

Reich, A., Backer, L.C., Kirkpatrick, B., Stumpf, R., Fleming, L.E., Stephan, W.B., Weisman, R., Jerez, E., Heil, C., Steidinger, K., Landsberg, J., Connor, J., DeThomasis, J., Baden, D.G., 2006. Public health and Florida red tide: From remote sensing to poison information (published abstract). Am. Public Health Association Annual Meeting, Boston, MA.

Rogers, D.J., Packer, M.J., 1993. Vector-borne diseases, models and global change. Lancet 342, 1282–1284.

Schofield, O., Grzymski, J., Bisset, W.P., Kirkpatrick, G.J., Millie, D.F., Moline, M., Roesler, C.S., 1999. Optical monitoring and forecasting systems for harmful algal blooms: Possibility or pipe dream? J. Phycol. 35, 1477–1496.

Small, C., Nicholls, R.J., 2003. A global analysis of human settlement in coastal zones. J. Coastal Res. 19, 584–599.

Steidinger, K.A., 1975. Implications of dinoflagellate life cycles on initiation of *Gymnodinium breve* red tides. Environ. Lett. 9, 129–139.

Steidinger, K.A., Vargo, G.A., Tester, P.A., Tomas, C.R., 1997. Bloom dynamics and physiology of Gymnodinium breve. In Anderson, D.M., Cembrella, A.E., and Hallegraeff, G.M. (eds.), The Physiological Ecology of Harmful Algal Blooms, Amsterdam, Elsevier.

Stumpf, R.P., 2001. Applications of satellite ocean color sensors for monitoring and predicting harmful algal blooms. J. Hum. Ecol. Risk Assess 7, 1363–1368.

Stumpf, R.P., Culver, M.E., Tester, P.A., Tomlinson, M., Kirkpatrick, G.J., Pederson, B.A., Truby, E., Ransibrahmanakul, V., and Soracco, M., 2003. Monitoring *Karenia brevis* blooms in the Gulf of Mexico using satellite ocean color imagery and other data. Harmful Algae 25, 1–14.

Stumpf, R.P., Ransibrahmanakul, V., Steidinger, K.A., Tester, P.A., 1998. Observations of sea surface temperature and winds in association with Florida, USA red tides (*Gymnodinium breve*). In Reguera, B., Blanco, J., Fenandez, M.L., and Wyatt, T. (eds.), Harmful Algae, pp. 145–148. Paris, Xunta de Galacia and Intergovernmental Oceanographic Commission, UNESCO.

Subramaniam, A., Carpenter, E.J., 1994. An empirically derived protocol for the detection of blooms of the marine cyanobacterium Trichodesmium using CZCS imagery. Int. J. Remote Sensing 5, 1559–1569.

Svejkovsky, J., Jones, B., 2001. Satellite imagery detects coastal stormwater and sewage runoff. Eos 82, 621, 624–625, 630.

Tester, P.A., Steidinger, K.A., 1997. *Gymnodinium breve* red tide blooms: Initiation, transport, and consequences of surface circulation. Limnol. Oceanogr. 42, 1039–1051.

U.S. Commission on Ocean Policy, 2004. An Ocean Blueprint for the 21st Century, www.oceancommission.gov/documents/welcome.html.

Vitousek, P.M., Aber, J.D., Howarth, R.W., Likens, G.E., Matson, P.A., Schindler, Schlesinger, W.H., Tilman, G.D., 1997. Human alterations of the global nitrogen cycle: Causes and consequences. Ecol. Appl. 7, 737–750.

Walsh, J.J., Steidinger, K.A., 2001. Saharan dust and Florida red tides: The cyanophyte connection. J. Geophys. Res. 106, 11597–11612.

Whitman, R.L., 2005, Progress Report Grant Project Number GL98500001, City of Chicago Department of Environment, David Rockwell, Project Officer.

Whitman, R.L., Nevers, M.B., Gerovac, P.J., 1999. Interaction of ambient conditions and fecal coliform bacteria in southern Lake Michigan waters: Monitoring program implications. Nat. Area J. 19, 166–171.

Wilkinson, C., Linden, O., Cesar, H., Hodgson, G., Rubens, J., Strong, A.E., 1999. Ecological and socioeconomic impacts of the 1998 coral mortality in the Indian Ocean: An ENSO impact and a warning of future change? Ambio 28, 188–196.

STUDY QUESTIONS

1. The IOOS will initially provide data primarily on the physical environment (such as winds, currents, and temperatures) preceding, during, and after an event. Describe how knowledge of environmental conditions preceding an event may affect the health community's ability to detect and respond to an event.
2. Choose a public health event (such as a harmful algal bloom or toxin contamination event that has occurred in your area), and describe the environmental parameters that would need to be measured to predict or detect that event. Discuss the sampling location, frequency, and types of information that would need to be collected.
3. Choose a public health event that has occurred in your area, and describe the ideal sensor and sampling pattern for detecting that contaminant or effector. Consider needs for real-time information, geographic scale, acceptable error, type of technology, duration of event, and platform.

CHAPTER

2

Climate and Human Health

Physics, Policy, and Possibilities

KENNETH BROAD, JESSICA BOLSON, AMY CLEMENT, ROBERTA BALSTAD, SABINE MARX, NICOLE PETERSON, AND IVAN J. RAMIREZ

INTRODUCTION

There always have been and will continue to be climate surprises (Glantz *et al.,* 1998). The past 10,000 years have been relatively stable, arguably providing the baseline conditions for rapid societal development through advances in agriculture, allowing for more permanent human settlements and the accompanying cultural and sociopolitical structures that remain with us in varying forms to the present. Even during this "stable" climate regime, humans continue to experience devastating famines, disease outbreaks, and water shortages. These crises are often a combination of unfavorable climate patterns and extreme weather events in conjunction with sociopolitical and cultural changes (such as civil strife, urbanization patterns, and increased global travel). But what if climate is changing much more rapidly than humans have ever experienced before, as a result of not only natural variability but also the anthropogenic (human) influence through emissions of greenhouse gasses? Are we prepared for what perhaps could be a new set of conditions—that is, emerging infectious diseases, changes in storm patterns and intensity, and changes in the characteristics of the seasons that we have not experienced?

That climate affects human health is perhaps so obvious that we often take it for granted, and we assume that societies have reached some sort of equilibrium with climate. Humans are, however, in a perpetual state of adjusting behavior to the elements through innovations such as fire, air conditioning, drought resistant crops, and the Weather Channel, to mention just a few diverse examples. The more we understand about climate and the way disease functions, the more we realize that even subtle shifts in climate can affect human health in less noticeable ways. A warming climate in some regions, for example, may result in a lack of freezes or predator-prey controls that limit certain pest and rodent populations associated with infectious diseases. Warming oceans can alter the coastal ecosystems, increasing the likelihood for harmful algal outbreaks. Reduced water availability because of changes in the hydrologic cycles or salt water intrusion into freshwater aquifers can directly affect our ability to grow crops. The examples linking infectious diseases and climate change are numerous (Wyman, 1991). Even our psychological well-being is tied to the climate, evident in the syndrome known as seasonal affect disorder (SAD) (Rosenthal *et al.,* 1984).

In this chapter, we outline the important physical and societal factors that link climate and human health, and then we illustrate how these factors play out in case studies from around the world. We first present a brief overview of the observed connections between climate and human health. This is followed by a section in which we describe the basic physical phenomena that cause the regional climate changes that occur from year-to-year and also those resulting in longer-term global trends. Next we lay out the fundamental societal factors that influence how human health can be impacted by climate changes, which together are determinants of what is often referred to as "vulnerability." We then present four case studies. The first study examines the possible connection between year-to-year variations in climate and the incidence of Hantavirus cases in New Mexico. The second case study looks at the connection between climate variability and malaria epidemics. The third study examines the effects of urban heat waves on mortality rates. A theme throughout each of these three studies is that the ability to cope with climate changes to reduce the effect of these changes on human health is strongly mediated by societal constraints. Thus, a final study examines a case in Malawi in which proactive policy measures have been taken to limit the effect of year-to-year rainfall changes on drought. Finally, some key points of the connection between climate physics, social policies, and possibilities for reducing vulnerability and improving human health are highlighted in a concluding section.

LINKING HUMAN HEALTH AND CLIMATE FACTORS

Research scientists have been studying the links between climate and human health since the early 20th century. Efforts in the 1920s sought to understand how seasonal climate conditions affected incidences of malaria and cholera in India and Bangladesh (Epstein *et al.,* 1998; Gill, 1920, 1921) and the connections between shorter timescale weather conditions and diseases such as smallpox and tuberculosis (Rogers, 1926; Swaroop, 1949). However, it was not until the late 20th century, with the increasing availability of climate data, the advent of satellite technologies, and the development of standardized and computerized health diagnostic systems, that statistical relationships between human health and climate parameters have become more readily quantifiable (Kunkel *et al.,* 1999; Pascual *et al.,* 2002; World Health Organization [WHO], 2004b). Quantifying relationships between climate data and disease data increases the potential for monitoring meteorological parameters and using them to predict future outbreaks (Davies *et al.,* 1991; Haines *et al.,* 2006; WHO, 2004b).

There has been an increase in the scope of research efforts and methods to include studies of potential impacts of future global climate change upon human health (Intergovernmental Panel on Climate Change [IPCC], 2001; Martens *et al.,* 1995; McMichael *et al.,* 1996). Some of these studies use global circulation models (GCMs) in combination with health impact assessments (HIAs) to evaluate potential health impacts of climate change, such as temperature-related mortality, health effects of extreme weather events, air-pollution related health effects, water and food borne diseases, and vector and rodent-borne diseases (Casimiro *et al.,* 2006; Colwell, 1996; Patz *et al.,* 2001). In addition, studies have begun to focus on the indirect impacts of climate on human health, including those resulting from shifts in agricultural production areas and the associated changes in migratory patterns that occur as a result (Epstein, 1997; Parry *et al.,* 2004; Rosenzweig and Parry, 1994).

As our scientific understanding of the relationship between climate and human health increases, so have efforts to develop early warning systems in disease management (Davies *et al.,* 1991; Travis, 1997). This work is largely focused upon understanding the role of climate variability in seasonal variation of disease epidemics (Franke *et al.,* 2002; Gubler *et al.,* 2001) and famine early warning systems (Broad and Agrawala, 2000; Dilley, 2000; Hansen *et al.,* 2004). Although both of these avenues of research hold promise for improved future disease management, the effective implementation of applied work in the field has been met by analytical and sociological challenges, including data limitations of both morbidity and mortality statistics, financial constraints on implementing measures even when early warnings are distributed, actions that may discount information that may be politically disadvantageous, or conditions of civil strife that may prevent action from being taken (Ebi *et al.,* 2006; Patz *et al.,* 2001, 2005).

The scientific community has made impressive strides in inventorying *potential* human health impacts of global climate change in different regions of the world (IPCC, 2001; WHO, 2004a). However, the human dimensions that mediate societies' vulnerability to climate change impacts involve such a range of complex, and often uncontrollable, factors that challenge the application of this scientific knowledge in a *proactive* manner. Thus, our efforts are in need of a great deal more interdisciplinary collaboration among the physical, social, and political science and health educators and practitioners (Kunkel *et al.,* 1999).

CLIMATE AND WEATHER BASICS

"Climate is what you expect, weather is what you get."—
—Anonymous

The concepts of "what you expect" and "what you get" are particularly useful for understanding the basics of climate and weather, and how they may impact human health. Let us begin with what to expect, which are the main climate zones and seasonal variations of those zones. These are caused by the geometry of the planet and its orbit. The equatorial region receives the highest amount of solar radiation on a year-round basis because it is closest to the Sun. As such, it is not only the warmest place on the planet, but it is also the wettest place because the solar radiation drives strong evaporation over the tropical oceans resulting in large amounts of rainfall. Moving to higher latitudes, one expects the temperature and precipitation to decrease because there is less solar radiation. In the most general sense, this is indeed the pattern we find in the climate.

There are, however, some exceptions that arise primarily because of the atmospheric circulation. Consider, for example, the European climate, which is considerably warmer than comparable latitudes on the east coast of North America (say Newfoundland). This is because the atmosphere brings air from the southwest to the European continent, which is relatively warm and moist (Seager, 2006). Another example is the contrast in climate between the Sahara and the Florida Everglades, again which are at similar latitudes. It is now thought that the dryness of Africa is caused by dry air coming from the north (Hoskins *et al.,* 1999).

These climate zones undergo large changes from season to season, which is expected on the basis of the tilt of the axis of rotation of the Earth. For example, during the month of July, the northern hemisphere is tilted toward the Sun and hence absorbs more solar radiation than in January when it is tilted away. Generally speaking, we expect the warmest temperatures and highest precipitation to move north and south with the Sun. At high latitudes, however, the wettest seasons can be the winter season. This is because the precipitation (whether snow or rain) is brought to those regions by storms, which are more active in the winter.

Now let us contrast these ideas of what we expect on the basis of the seasonal and latitudinal distribution of solar radiation with what we get. Although one may expect cold weather and snowfall in the winter in high latitudes, any given day may differ significantly from that expectation. What happens on a day-to-day basis is the result of weather. We will define *weather* as daily or weekly variations in temperature, humidity, winds, rainfall, and so on, and the phenomena that cause these variations (i.e., storms). *Climate*, on the other hand, we will define as variations in these same quantities that occur on timescales of seasons and longer, which include *interannual* (year-to-year variations), *decadal* (decade-to-decade), and *century* (changes that persisted throughout the 20th century). A useful way to conceptualize climate and weather together is with statistics. We can define a mean quantity, and then variations about this mean can be represented as a spread. This is illustrated in Figure 2-1a with a probability distribution function (PDF) of mean temperature at Chicago O'Hare airport in July. The mean value of 23°C is at the midpoint of the horizontal axis, and the probability that this mean temperature will occur can be read off the y-axis, which is about 12% of the time. (This value of mean temperature may seem low, but recall that it is an average of day and nighttime values.) As you go to extreme high and low temperatures, they become less probable, so that a daily temperature of 30°C only occurs 2% of the time. In this PDF, the mean represents the climate (what we expect), and the spread about the mean represents the weather (what we get).

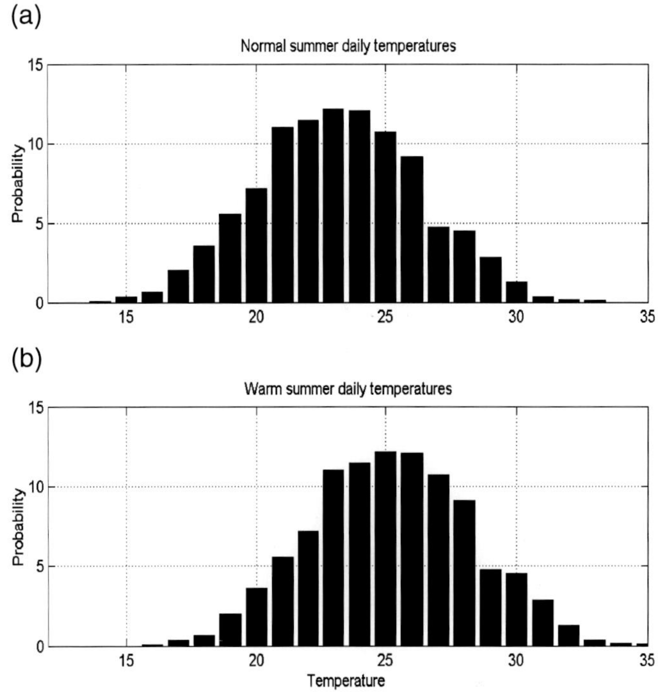

Figure 2-1. (a) The probability distribution function (PDF) of daily mean temperatures (in degrees C) in July recorded for Chicago since the 1950s. The vertical axis shows the probability (as a percentage) for a given temperature (horizontal axis). (b) PDF if the mean temperature is raised by 2°C. This bottom panel is meant to illustrate how a shift in the mean temperatures can alter the likelihoods of extreme temperatures.

When "What You Expect" Changes: Interannual and Longer-Term Changes

Interannual Variability

A given season can look quite different from year to year. A winter season, for example, can be particularly mild or have a low seasonal total of snowfall. That does not mean that every day in that season was warm or dry. Let us reconsider the Chicago PDF (Fig. 2-1b) in order to illustrate this with statistics. We could characterize an anomalously warm Chicago summer, for example, with a shift in the PDF. In this case, the mean temperature gets warmer by 2°C, but there is still spread about that mean such that there is a possibility that a particularly cold day can still occur. Note that this shift does result in a change in the probability of high temperatures, so that in a "normal" summer (i.e., Fig. 2-1a), the probability of a mean temperature of 30°C is 2%, whereas in the warm summer, this probability goes up to almost 5%.

Such year-to-year variations are commonly referred to as interannual variability. Much of this variability is essentially random, but there is a component that is influenced by a well-known physical process and also has some predictability. That process is the El Niño/Southern Oscillation, or ENSO. Over the past several hundred years, Peruvian fishermen noted an episodic warming of the coastal waters, which tended to occur around the end of the calendar year and was hence referred to as El Niño. Meanwhile, in the early 20th century, British scientists were studying variability in the atmosphere in hopes of predicting monsoon rains in Asia. Sir Gilbert Walker made the important observation that when sea level pressure over Darwin, Australia, was anomalously high, it was anomalously low over Tahiti. He referred to this phenomenon as the "Southern Oscillation," essentially a shift of air from west to east. Together, this oceanic El Niño and atmospheric Southern Oscillation are the main indicators of what we now know of as an episodic reorganization of the tropical Pacific climate (for a comprehensive overview or the history and scientific aspects of El Niño, see Glantz, 2001).

To understand what happens during an El Niño event, let us first consider the "normal" conditions in the tropical Pacific, shown schematically in Figure 2-2a. The western part of the Pacific is the wettest place on Earth. Southeast Asian, Northern Australia, and the islands of Indonesia are tropical humid climates with extensive rain forests. In contrast, the eastern Pacific climate is extremely dry. The Atacama Desert at latitudes of about 20° is the driest place on Earth, and closer to the equator, the coasts of northern Chile and Peru are essentially deserts. The ocean and atmospheric circulations together are responsible for these differences. Rising motion in the atmosphere over the western Pacific is associated with rainfall and low sea level pressure, whereas air descends over the eastern tropical Pacific causing low rainfall amounts and high sea level pressure. This circuit of air with rising in the west and descent in the east is connected at the surface by the trade winds blowing from east to west. This circulation is called the Walker cell, in honor of Sir Gilbert Walker. Trade winds blowing over the ocean have the effect of essentially piling up water in the western Pacific, where the sea level is actually about half a meter higher than in the east. The water that is pushed to the west is replaced by water from below through a phenomenon called *upwelling*. This subsurface water is much colder, so that upwelling causes a significant reduction in the ocean surface temperatures in the east. Temperatures in the eastern Pacific are typically around 20°C in contrast with the warm surface temperatures in the west, which are around 30°C. The upwelled water is also rich in nutrients so that the coastal and equatorial regions have high primary productivity, which maintains a large commercial fishery.

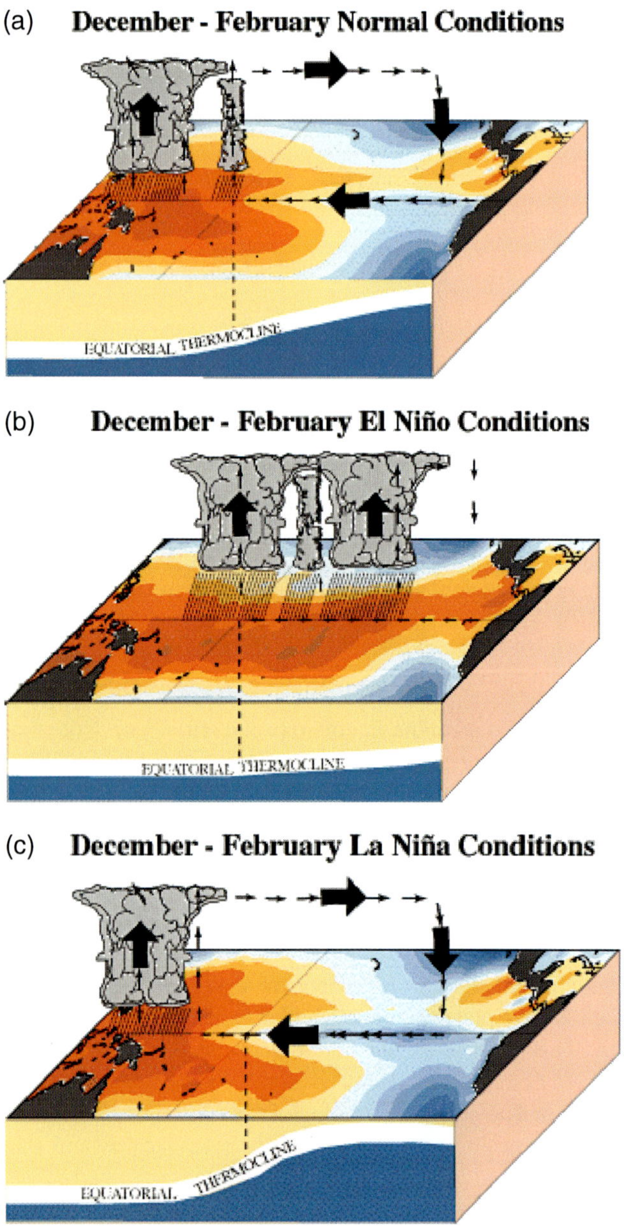

FIGURE 2-2. Schematic representation of the tropical Pacific ocean-atmosphere system during (a) normal conditions, (b) El Niño conditions, and (c) La Niña conditions. On the left-hand side of each panel is Asia/Australia, and on the right is the coast of the Americas. Colors indicate surface temperature (red: warm, blue: cold). Arrows indicate the atmospheric circulation, with longer arrows indicating stronger trade winds. The cross section across the front of the panel shows the equatorial thermocline (the boundary between warm surface waters and colder deep waters), which is normally shallow in the east and deep in the west, while during El Niño conditions, the thermocline deepens in the east.

During an El Niño event, this whole picture changes (Fig. 2-2b). The first step is a reduction of the strength of the trade winds. This causes a reduction in the upwelling, and the eastern Pacific Ocean warms, as noted by the Peruvian fishermen. The atmospheric circulation responds with the rainfall moving into the east with the warm surface temperature. This typically leads to wet conditions in the normally dry coastal Peru and Chile, and dry conditions throughout the western Pacific. There is also another side to the story, which occurs when the trade winds strengthen, which is called La Niña (Fig. 2-2c). In this case, upwelling is enhanced and the eastern Pacific Ocean becomes colder, while in the atmosphere, the rainfall shifts further to the west resulting in anomalously dry conditions over many parts of the central Pacific.

We can keep track of the occurrence of El Niños or La Niñas by simply knowing how the sea surface temperature varies from year to year. Figure 2-3 shows a timeline for El Niño constructed by taking the average surface temperature anomaly in the eastern equatorial Pacific (referred to as the "NINO3 index"). The largest El Niño events of this 20th century occurred in the winters of 1972–1973, 1982–1983, and 1997–1998, as indicated by the large positive values of the NINO3 index, and La Niñas occurred in 1970–1971, 1988–1989, 1998–1999. A moderate strength El Niño event is currently ongoing in the winter of 2006–2007.

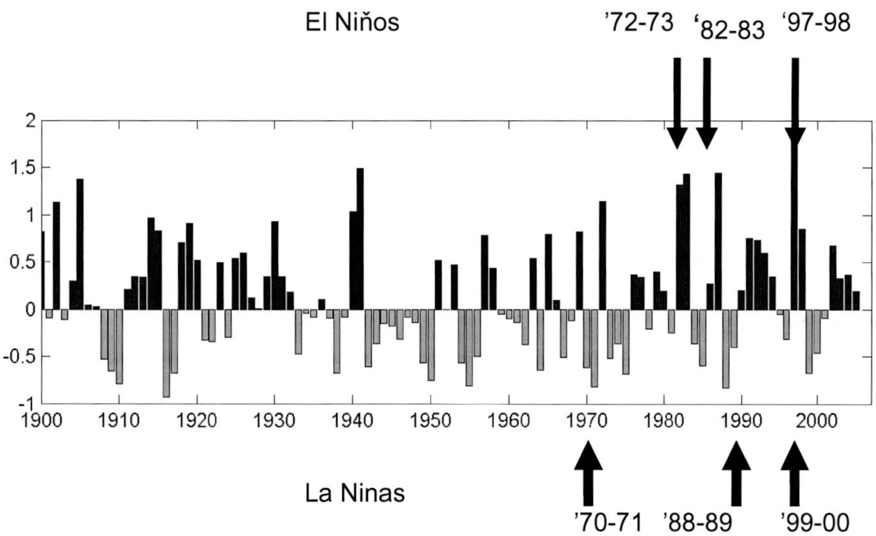

FIGURE 2-3. Timeline for ENSO. Values for the annual mean temperature anomaly for the eastern equatorial Pacific (commonly referred to as the NINO3 region). Positive values (dark gray) indicate warmer than normal temperature, and negative values (light gray) indicate colder than normal temperatures. When the annual mean NINO3 index exceeds 0.5 C, this is typically called an El Niño event, and values lower than –0.5 are La Niña events. Major events of the last several decades are labeled.

What starts an El Niño or La Niña event? Why do the trade winds weaken or strengthen to kick off the event? The answers to these questions are as yet unknown. Some studies suggest that each event carries with it the seeds of the next events. That is, an El Niño event will kick off a subsequent La Niña event a year or two later, and that La Niña will set off an El Niño event a few years down the road (Cane, 2005). Other studies suggest that the events are initiated by weather; that if a collection of storms line up in such a way near the equator in the central Pacific as to reduce the trade winds, an El Niño event will ensue. Both of these ideas have validity, and both may play a role in how events begin, but the theory has not yet been unified.

No matter what the actual cause of the event, if we know that the trade winds are changing, a prediction can be made about the likelihood of an event occurring in the upcoming months. This is essentially what is currently done at modeling centers around the world. An observing network is in place in the tropical Pacific, which includes instruments in the ocean measuring currents and temperature, atmospheric instruments to record winds and sea level pressure, and satellite observations of temperature. These observations are routinely input into climate models, which then are used to make a quantitative prediction about how the Pacific climate will evolve in the 3, 6, 9, 12 months to come. This is done much in the same way as weather forecasting in that real-time observations are put into a computer model, and that model then generates likelihoods of rain, cloud cover, temperature change, and so on.

ENSO events have well-known impacts outside the tropical Pacific. These occur through what are called atmospheric "teleconnections." The shifts of the rainfall east and west in the Pacific have a ripple effect in the global atmospheric circulation. When the rain moves toward the east during an El Niño event, the jet stream over the northern Pacific Ocean shifts to the north and is hence more likely to hit the American coast in Alaska than in the Pacific Northwest as it does in a normal year. Hence, the winters in Alaska during El Niño years are wetter than normal, while in Washington State they are drier. Also, a southern branch of the jet stream develops over the southern United States leading to wetter conditions from New Mexico to Florida. There are other well-known teleconnections throughout the globe, as shown in Figure 2-4. Drier than normal conditions in Northeast Brazil and southern Africa tend to occur during El Niño. La Niña has generally the opposite effects: regions that are dry during El Niño events tend to be wet during La Niña events.

Thus, with some knowledge of what ENSO is doing, a prediction can be made about how regional climates around the world may depart from normal conditions. These so-called seasonal climate forecasts are also routinely made at numerous modeling centers. Predictions are made 3 to 6 months in advance and are generally presented as likelihoods of temperature or precipitation being below normal, normal, or above normal. (For more complete information, see, for example, http://iri.columbia.edu/pred, where the prediction procedure used at the International Research Institute on Climate Prediction is explained in detail.)

It is important to note that these seasonal predictions are probabilistic. The prediction will typically give a percentage likelihood that a season will be drier/wetter or warmer/colder

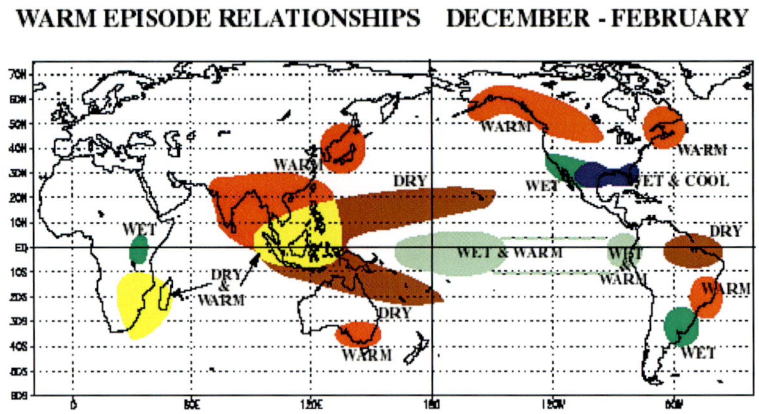

FIGURE 2-4. Global map of the regions affected by El Niño events during winter. From Ropelewski and Halpert (1987).

than normal. Let us return to our example of summer temperatures in Chicago (Fig. 2-1) for a moment to appreciate the probabilistic nature of forecasts. Let us say that there is a 70% chance that the summer will be warmer (as in Fig. 2-1b) during an El Niño year. Day-to-day weather fluctuations will still occur, so that even if the prediction is right (which is only 70%), there will still be relatively cold and warm days. The same holds true for any of these teleconnected regions. Therefore, for example, in New Mexico an El Niño year will tend to be wetter than normal, but that will just shift probabilities rather than resulting in more rain every day.

Longer-Term Changes

Whereas ENSO causes year-to-year variability, there are also longer-term changes in climate that lead to shifts in meteorological conditions around the world. Perhaps the most significant of these changes is global warming. Since the mid-19th century, the average surface temperature over the globe has increased approximately 0.6°C (Fig. 2-5). The five warmest years on record occurred at the turn of the past century, with 2006 being the warmest ever recorded. Over this period, there has also been an exponential increase in the amount of carbon dioxide in the atmosphere (Fig. 2-5), primarily as a result of the burning of fossil fuels. Carbon dioxide is what is called a greenhouse gas. This gas absorbs infrared radiation emitted by the Earth's surface and reradiates it back down to the surface. The net effect of carbon dioxide in the atmosphere, like a greenhouse, is to trap heat, and the expected result is an increase in temperature, consistent with the observations.

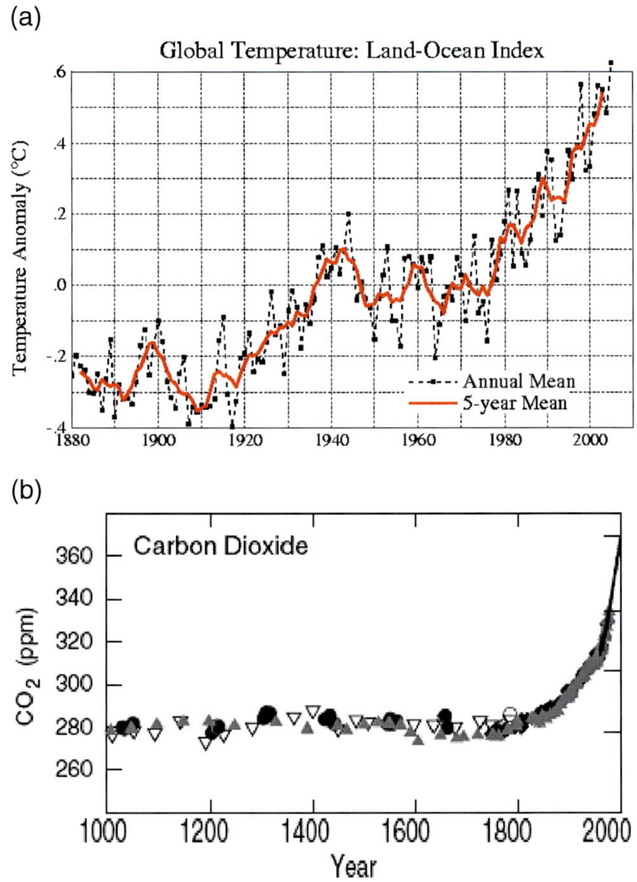

FIGURE 2-5. (a) Annual mean global mean temperature anomaly (from www.giss.nasa.gov). (b) Carbon dioxide concentrations (in part per million) for the past 1000 years both from ice core records and direct measurements (from www.ipcc.ch).

It is now well accepted that the observed warming is anthropogenic (i.e., caused by humans). This has been demonstrated extensively by the work of the Intergovernmental Panel on Climate Change (IPCC). This panel was established in 1988 by the World Meteorological Organization (WMO) and the United Nations Environmental Program (UNEP), and it produces periodic reports synthesizing the current state of scientific knowledge about climate change. The most recent IPPC reports came out in 2001 and 2007 (www.ipcc.ch). In the 2001 report, it was concluded that

"There is new and stronger evidence that most of the warming observed over the last 50 years is attributable to human activity." The most important pieces of evidence to support this statement, in addition to the expected qualitative relationship between carbon dioxide and temperature as illustrated in Figure 2-6, come from models of the climate system.

The state-of-the-art models are referred to as GCMs, which stands for either general circulation models or global climate models (the two are used interchangeably). These models use equations for the laws of physics that are solved on a computer to simulate the climate. There are a number of different GCMs from climate modeling centers around the world. Each model is slightly different in the way that it simulates certain climate phenomena such as clouds and rain. As part of the 2001 report, the IPCC coordinated efforts from these different modeling centers to intercompare the results produced by the GCMs when carbon dioxide levels are increased in the atmosphere. Another set of experiments has also been performed for the 2007 report.

There are two important results from these sets of climate model experiments. First, when climate models are run with observed climate forcings of the 20th century including carbon dioxide as well as "natural" (i.e., nonanthropogenic) forcings such as volcanic eruptions and slight changes in the solar irradiance, the late 20th century warming can only be simulated if greenhouse gases are included (Fig. 2-6). In other words, the recent warming is not part of a natural cycle. The other important result is that the GCMs predict that warming will continue to occur in the future. The IPCC developed scenarios for future increases in carbon dioxide based on different rates of consumption of fossil fuels versus the development of cleaner, or renewable, energy technology. No matter what the scenario, all GCMs predict significant warming in the future, in the range of 1°C to 3°C in the next 50 years (Fig. 2-7).

This may not seem like a large amount of warming, given that we can experience several tens of degrees of temperature change over a day. However, this warming is not distributed uniformly over the globe. The Arctic warmed by several degrees over the 20th century, compared with the less than 1°C warming averaged over the globe, and there are already discernable impacts of this warming on the ecosystems there (Arctic Climate Impact Assessment [ACIA], 2005). Also, as we have seen earlier, a 2°C shift in the PDF of meteorological conditions can lead to significant changes in the likelihood of extreme events, so that heat waves may increase as the world warms.

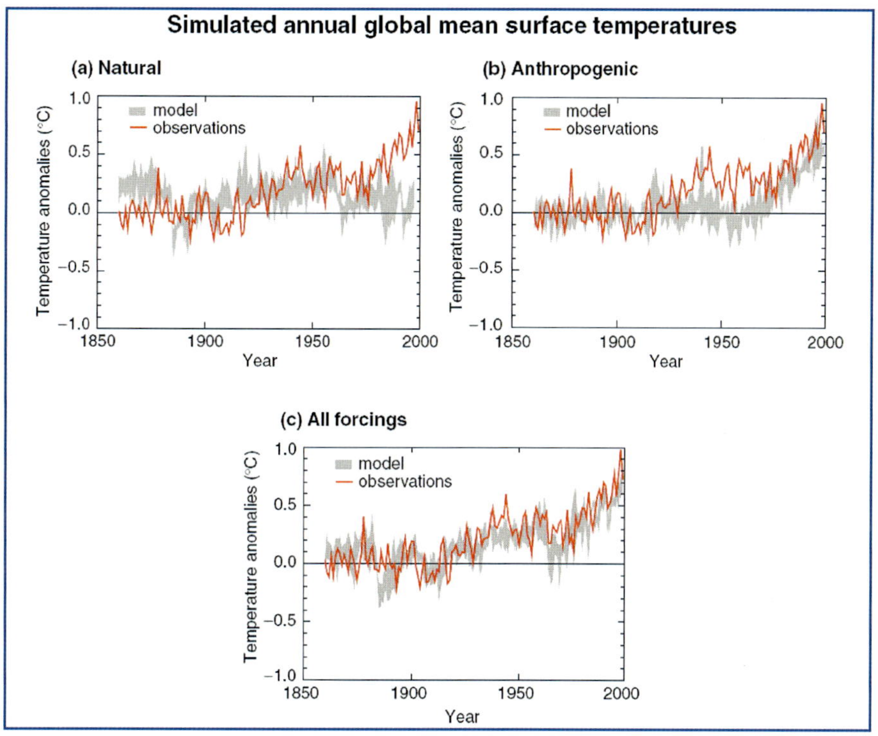

FIGURE 2-6. Simulation of the 20th-century global mean temperature from the IPCC (2001) Summary for Policy Makers. (a) Model simulations (range for all models is indicated by the gray bars), which include only natural climate forcings (variations in the solar irradiance and volcanic aerosols). Observations are shown in red. (b) Model simulations with anthropogenic forcings (primarily carbon dioxide, but others are included such as sulfate aerosols). (c) Combination of natural plus anthropogenic forcings. Together this set of simulations shows that the late 20th-century warming of the globe can only be explained when anthropogenic forcings are included.

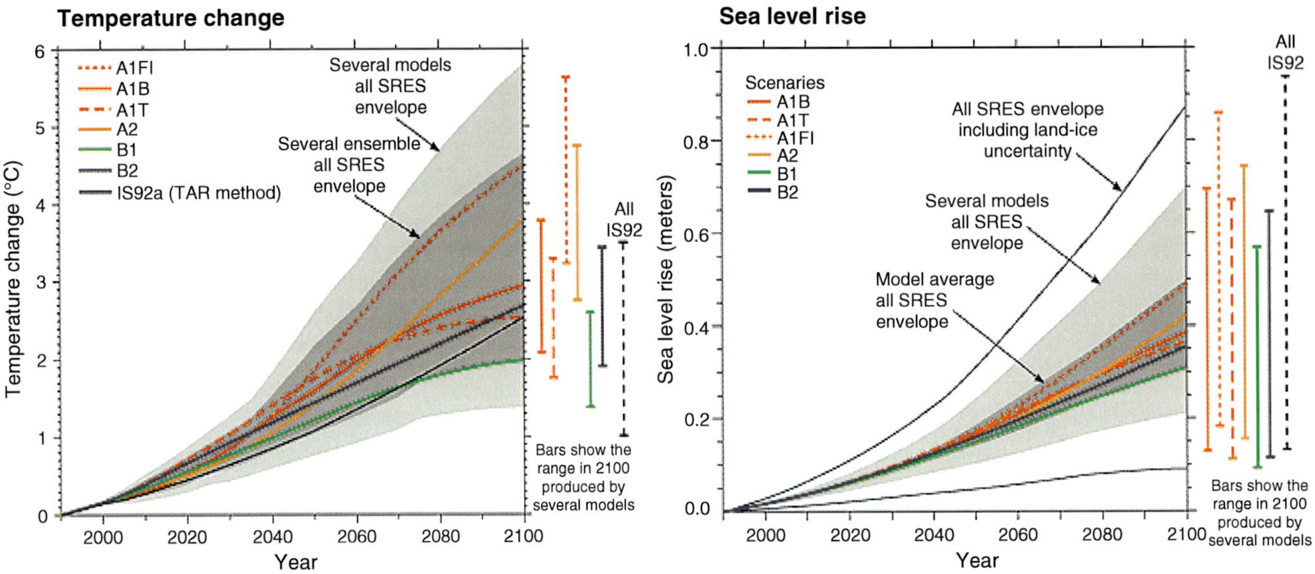

FIGURE 2-7. Future projections of global mean temperature *(left)* and sea level *(right)* from the IPCC (2001) Summary for Policy Makers. The gray region indicates the range simulated by all the climate models. The colored lines indicate the mean of the model simulations for different scenarios of future fossil fuel consumption. A1F1 is the most fossil fuel intensive scenario, whereas the B1 scenario presumes increased use of renewable energy sources.

Another important aspect of global warming is the effect on the global sea level. As the oceans warm, there is a thermal expansion of the water. It is estimated that this has already caused 5 cm of sea level rise over the 20th century (IPCC, 2001). Also, as the high latitudes warm, the land-based ice melts, causing an additional sea level rise. Dramatic examples of such melting have already occurred in Antarctica with the collapse of the Larson B ice shelf in 2002, which had been frozen for thousands of years but collapsed in weeks. Projections of future sea level rise vary considerably from different models (Fig. 2-7), but on the upper end, the increases can be as large as 30 to 40 cm in the next 50 years. Again, although this may seem like a small amount given that, for example, tides can cause variations in the sea level by several meters over one day, the slow but persistent change can have large effects on freshwater systems, which supply water for human consumption or agriculture and also maintain ecosystems; furthermore, this sea level rise can enhance the impact of a storm surge on the coasts.

VULNERABILITY ISSUES

The severity of the physical impact of a climate event is only part of the equation that determines the outcomes for human populations. A complex interaction of social and environmental factors determines how different regions and populations are affected by natural phenomena. This notion has been studied under the heading of "vulnerability" (Kates, 1985; White and Haas, 1975). Vulnerability has been defined as a "system's exposure and sensitivity to stress and its capacity to absorb or cope with the effects of these stressors" (IPCC, 2001). Different research disciplines have molded this definition into their own terms, making it challenging to move from identification of vulnerability in theory to identification of vulnerability in practice (Burton *et al.*, 1978; Eakin and Luers, 2006; Holling, 1973; Timmerman, 1981). Next, we briefly cover some of the different ways vulnerability assessment is approached, particularly in relation to climate and human health.

Natural hazard risk analysts tend to define vulnerability in biophysical terms. They seek to identify which specific biophysical hazards human populations are vulnerable to, what consequences might be expected as a result of these hazards, and where and when those impacts can be expected to occur (Jones, 2001; Timmerman, 1981). In fields of health, hazard and risk researchers may analyze the likelihood of certain diseases occurring in a specific region under changing climate conditions or under conditions caused by climate variability (Epstein *et al.*, 1998; Martens *et al.*, 1995; McMichael *et al.*, 1996). This analysis may include, for example, the examination of increased incidences of malaria that occur during El Niño events in parts of South America, South Asia, and sub-Saharan Africa (Haines *et al.*, 2006; Martens *et al.*, 1997). Increased precipitation in these regions has been associated with increases in malaria vectors, and therefore greater incidence and thus regional vulnerability (in theory) to the disease during these episodes.

A different approach toward examining vulnerability is taken by social science researchers who focus on the capacity of individuals or groups to deal with the impacts of natural hazards (Bowden *et al.,* 1981; Patz *et al.,* 2001). They are interested in understanding sociopolitical, socioeconomic, and cultural factors that make people more or less vulnerable to different risks, as well as differential abilities to cope with impacts. These can include factors such as access to treatment facilities and health care, access to freshwater sources and sanitation, population migration patterns, behaviors and lifestyles, availability of household level economic resources, and extent of social networks (Bohle *et al.,* 1994; Furgal and Seguin, 2006; Liverman, 1994). Institutional design, governing mechanisms, and policies also contribute to vulnerability and are typical subjects of such studies.

Social scientists often collaborate with epidemiologists. An example of results of such collaborations is the identification of specific populations that have been repeatedly exposed to malaria developing acquired immunity to the disease. Although they can be infected again, these individuals are less likely to develop severe symptoms. Thus, these populations may be less vulnerable than populations who have never been exposed to the disease before (Centers for Disease Control and Prevention [CDC], 2004; Haines *et al.,* 2006). Societal factors are also key to determining the vulnerability of certain populations. For example, wealthier regions with the capacity to monitor and control vectors (such as mosquitoes) might be less vulnerable than poorer regions without such capacity (Gallup and Sachs, 2001; Smith, 2004).

Making the shift from vulnerability theory to practice, and attempting to streamline these divergent approaches into a unified assessment methodology, has been the focus of recent research efforts (Eakin and Luers, 2006; Morlot *et al.,* 2005). Comprehensive vulnerability assessments aim to answer all of the questions: who are the vulnerable populations, to what are they vulnerable, and why are some populations more or less vulnerable than others (Patz *et al.,* 2001; Smith, 2004)? Because vulnerability is a biological, cultural, political, social, and physical concept, these assessments must be context driven (Adger, 1999; Brooks *et al.,* 2005)—that is, they should be designed specifically for a particular place and with a particular group of decision makers in mind. This adds to the challenge of accurately describing vulnerability.

In addition to the complexities in identifying current vulnerabilities to climate events, future uncertainties must also be considered. The geographical range of many diseases may change under proposed impacts of global climate change (Hogrefe *et al.,* 2004; Kovats *et al.,* 2001). Diseases like malaria, dengue fever, viral encephalitis, and yellow fever may spread into new regions as conditions make these areas more hospitable for various disease vectors. Additionally, changing patterns of migration and population densities are plausible scenarios of climate change (Haines *et al.,* 2006; Parry *et al.,* 2004; Patz, *et al.,* 2005) and are likely to impact future vulnerabilities.

Although development of scenarios of disease characteristics under changing climate conditions is valuable for policy consideration, it is fraught with uncertainty in our ability to predict (1) the details of climate and weather patterns and their interaction with diseases at the spatial scale on which much human decision making occurs and (2) human behavior in terms of migratory patterns, civil strife, and changes in technology. Unfortunately, some human factors (including poverty, limited access to clean water, and minimal education about risky behaviors) seem to be endemic to many less industrialized countries. We know with certainty that these contribute to vulnerability to current *and* future conditions, and thus improving these conditions should arguably be the first step in preparing for an uncertain future.

References

Adams, D.K., Comrie, A.C., 1997. The North American Monsoon. Bull. Am. Meteorological Soc. 78, 2197–2213.

Adger, W.N., 1999. Social vulnerability to climate change and extremes in coastal Vietnam. World Dev. 27, 249–269.

Agency Technical Work Group for the State of New Mexico, 2005. Potential Effects of Climate Change on New Mexico, December 30, 2005, www.nmenv.state.nm.us/cc/Potential_Effects_Climate_Change_NM.pdf. (accessed February 20, 2006).

Alwang, J., Siegel, P.B., 1999. Labor shortages on small landholdings in Malawi: Implications for policy reforms. World Dev. 27, 1461–1475.

Arctic Climate Impact Assessment (ACIA), 2005. Scientific Report. Cambridge, Cambridge University Press.

Bohle H.G., Downing T.E., Watts M.J., 1994. Climate-change and social vulnerability: Toward a sociology and geography of food insecurity. Glob. Environ. Change 4, 37–48.

Bowden, M.J., Kates, R.W., Kay, P.A., Riebsame, W.E., Warrick, R.A., Johnson, D.L., Gould, H.A., Weiner, D., 1981. Climate and history: The effect of climate fluctuations on human populations: Two hypotheses, pp. 497–513. Cambridge, UK, Cambridge University Press.

Broad, K., Agrawala S., 2000. The Ethiopia food crisis: Uses and limits of climate forecasts. Science 289, 1693–1694.

Brooks, N., Adger, W.N., Kelly, P.M., 2005. The determinants of vulnerability and adaptive capacity at the national level and the implications for adaptation. Glob. Environ. Change 15, 151–163.

Burton, I., Kates, R.W., White, G.F., 1978. The Environment as Hazard. New York, Oxford University Press.

Cane, M. A. 2005. The evolution of El Niño, past and future. Earth Planet. Sci. Lett., 230, 227–240.

Casimiro, E., Calheiros, J., Santos, F. D., Kovats, S., 2006. National assessment of human health effects of climate change in Portugal: Approach and key findings. Environ. Health Perspect. 114, 1950–1956.

Centers for Disease Control (CDC), 1995. Morbidity and Mortality Weekly Report, August 11, 1995, 44, 577–579.

Centers for Disease Control and Prevention (CDC), 2003. Morbidity and Mortality Weekly Report, 52, 610–613.

Centers for Disease Control and Prevention (CDC), 2006a. About Extreme Heat, www.bt.cdc.gov/disasters/extremeheat/about.asp (accessed November 7, 2006).

Center for Disease Control and Prevention (CDC), 2006b. All About Hantaviruses. Home page, October 26, 2006, www.cdc.gov/ncidod/diseases/hanta/hps (accessed November 14, 2006). Centers for Disease Control and Prevention (CDC), National Center for Infectious Diseases, Division of Parasitic Diseases, 2004. Malaria: Human factors and epidemiology for malaria. Available online at www.cdc.gov/malaria/distribution_epi/human_epidemiology.htm.

Childs, J.E., Ksiazek, C.F., Spiropoulou, C.F., Krebs, J.W., Morzunov, S., Maupin, G.O., Gage, K.L., Rollin, P., Sarisky, J., Enscore, R., Peters, C.J., Nichol, S.T., 1994. Serologic and genetic identification of peromyscus maniculatus as the primary rodent reservoir for a new hantavirus in the southwestern United States. J. Infect. Dis. 169, 1271–1280.

Chima, R.I., Goodman C.A., Mills A., 2003. The economic impact of malaria in Africa: A critical review of the evidence. Health Policy 63, 17–36.

Colwell, R.R., 1996. Global climate and infectious disease: The cholera paradigm. Science 274, 2025–2031.

Committee on International Science, Engineering and Technology Policy (CISET). Global Microbial Threats in the 1990s, http://clinton1.nara.gov/White_House/EOP/OSTP/CISET/html/iintro.html (accessed December 12, 2005).

Craig, M.H., Snow, R.W., Le Sueur, D., 1999. A climate-based distribution mode of malaria transmission in sub-Saharan Africa. Parasitology Today 15, 105–111.

DaSilva, J., Garanganga, B., Teveredzi, V., Marx, S.M., Mason, S.J., Connor, S.J., 2004. Improving epidemic malaria planning, preparedness and response in southern Africa, Malaria J. 3, 37.

Davies, S., Buchanan-Smith, M., Lambert, R., 1991. Early warning in the Sahel and Horn of Africa: The state of the art. A review of the literature. Volume One. Brighton, Institute of Development Studies.

Dhainaut, J.F., Claessens, Y.E., Ginsburg, C., Riou, B., 2004. Unprecedented heat-related deaths during the 2003 heat wave in Paris: Consequences on emergency departments. Crit. Care 8, 1–2.

Dilley, M., 2000. Reducing vulnerability to climate variability in southern Africa: The growing role of climate information. Clim. Change 45, 63–73.

Dorward, A., Kydd, J., 2004. The Malawi 2002 food crisis: The rural development challenge, J. Mod. Afr. Stud. 42, 343–361.

Duchin, J.S., Koster, F.T., Peters, C.J., Simpson, G.L., Tempest, B., Zaki, S.R., Ksiazek, T.G., Rollin, P.E., Nichol, S., Umland, E.T., Moolenaar, R.L., Reef, S.E., Nolte, K.B., Gallaher, M.M., Butler, J.C., Breiman, R.F, for the Hantavirus Study Group, 1994. Hantavirus pulmonary syndrome: A clinical description of 17 patients with a newly recognized disease. N. Engl. J. Med. 330, 949–955.

Eakin, H., Luers, A.L., 2006. Assessing the vulnerability of social-environmental systems. Ann. Rev. Environ. Resour. 31, 365–394.

Ebi, K., Lewis, N., Corvalan, C., 2006. Climate variability and change and their potential health effects in small island states: Information for adaptation planning in the health sector. Environ. Health Perspect. 114, 1957–1963.

Engelthaler D.M., Mosley, D.G., Cheek, J.E., Levy, C.E., Komatsu, K.K., Ettestad, P., Davis, T., Tanda, D.T., Miller, L., Frampton, J.W., Porter, R., Bryan, R.T., 1999. Climatic and environmental patterns associated with hantavirus pulmonary syndrome, Four Corners Region, United States. Emerg. Infect. Dis. 5, 87–94.

Epstein, P., 1997. Climate, ecology, and human health. Consequences. 3, 2–19.

Epstein, P.R., Diaz, H.F., Elias, S., Grabherr, G., Graham, N.E., Martens, W.J.M., Mosley-Thompson, E., Susskind, J., 1998. Biological and physical signs of climate change: Focus on mosquito borne diseases. BAMS 79, 409–417.

Franke C.R., Ziller, M., Staubach, C., Latif, M., 2002. Impact of the El Niño/Southern Oscillation on visceral leishmaniasis, Brazil. Emerg. Infect. Dise. 8, 914–917.

Furgal, C., Seguin, J., 2006. Climate Change Health and Vulnerability in Canadian Northern Aboriginal Communities. Environ. Health Perspect. 114, 1964–1970.

Gallup, J.L., Sachs, J.D., 2001. The economic burden of malaria. Am. J. Trop. Med. Hyg. 64, 85–96.

Gill, C.A., 1920. The role of meteorology and malaria. Indian J. Med. Res. 8, 633–693.

Gill C.A., 1921. The role of meteorology in malaria. Indian J. Med. Res. 8, 633–693.

Glantz, M.H., 2001. Currents of Change: Impacts of El Niño and La Niña on Climate and Society, 2nd ed., preface. Cambridge, Cambridge University Press.

Glantz, M.H., Streets, D.G., Stewart, T.R., Bhatti, N., Moore, C.M., Rosa, C.H., 1998. Exploring the concept of climate surprises: A review of the literature on the concept of surprise and how it is related to climate change. Prepared for the U.S. Department of Energy, Office of Energy Research, October 1998.

Glass, G.E., Shields, T.M., Parmenter, R.R., Goade, D., Mills, J.N., Cheek, J., Cook, J., Yates, T.L., 2006. Predicted hantavirus risk in 2006, for the Southwestern U.S. Occasional Papers, Museum of Texas Tech University 255, 1–16.

Gubler, D.J., Reiter, P., Ebi, K.L., Yap, W., Nasci, R., Patz, J.A., 2001. Climate variability and change in the United States: Potential impacts on vector- and rodent-borne diseases. Environ. Health Perspect. 109, 223–233.

Haines, A., Kovats R.S., Campbell-Lendrum, D., Corvalan C., 2006. Climate change and human health: Impacts, vulnerability, and public health. Public Health 120. 585–596.

Hansen, J.W., Dilley, M., Goddard, L., Ebrahimian, E., Ericksen, P. Climate Variability and the Millennium Development Goal Hunger Target. International Research Institute for Climate Prediction, The Earth Institute of Columbia University. IRI Technical Report No. 04–04, September 12, 2004.

Hogrefe, C., Biswas, J., Lynn, B., Civerolo, K., Ku, J.Y., Rosenthal, J., Rosenzweig, C., Goldberg, R., Kinney, P.L., 2004. Simulating regional-scale ozone climatology over the eastern United States: Model evaluation results. Atmos. Environ. 38, 2627–2638.

Holling, C.S., 1973. Resilience and stability in ecological systems. Ann. Rev. Ecol. Systemat. 4, 1–23.

Hoskins, B., Neale, R., Rodwell, M., Yang, G., 1999. Aspects of the large-scale tropical atmospheric circulation, Tellus Series a-Dynamic Meteorol. Oceanogr. 51, 33–44.

Intergovernmental Panel on Climate Change (IPCC), 2001. Working Group II. Impacts, Adaptation, and Vulnerability. Third Assessment Report.

International Research Institute for Climate and Society (IRICS), 2006. A Gap Analysis for the Implementation of the Global Climate Observing System Programme in Africa. IRI Technical Report IRI-TR 06–01. Palisades, NY, International Research Institute for Climate and Society.

Jones, R.N., 2001. An environmental risk assessment/management framework for climate change impact assessments. Nat. Hazards 23, 197–230.

Kates, R.W., 1985. Climate Impact Assessment, chapter 1: The Interaction of Climate and Society. New York, John Wiley & Sons.

Keller D., Ettestad, P., Sewell, C.M., Rains, R., Englender, S., Woods, T., ScD, Pape, J., Tanda, D., Reynolds, J., Hoffman, R., MD, Mertz, G., Scully, G., Mapel, D., Koster, F., DeLury, J., Hansbarger, C., Hjelle, B., Yates, T., Iralu, J., Freeman, C., Hroch, B., Nunes, A., Loretto, D., Peter, D., Fulgham, R., Courtois, L., Mattson, M., Charleston, R., Cheek, J., 1998. Hantavirus pulmonary syndrome: Colorado and New Mexico, 1998. MMWR Weekly 47, 449–452, www.cdc.gov/mmwr/preview/mmwrhtml/00053320.htm (accessed December 2, 2005).

Klinenberg, E., 2002. Heat Wave: A Social Autopsy of Disaster in Chicago. Chicago, University of Chicago Press.

Kolivras, K.N., Comrie, A.C., 2004. Climate and infectious diseases in the southwestern United States. Prog. Phys. Geography 28, 387–398.

Kovats, R.S., Campbell-Lendrum, D., McMichael, T., Woodward, A., Cox, J.S., 2001. Early effects of climate change: Do they include changes in vector-borne disease? Philosop. Trans. Royal Society B 356, 1057–1068.

Kunkel, K.E., Changnon, S.A., Reinke, B.C., Arritt, R.W., 1996. The July 1995 heat wave in the Midwest: A climatic perspective and critical weather factors. Bull. Am. Meteorological Soc. 77, 1507–1518.

Kunkel, K.E., Pielke, P.A., Changnon, S.A., 1999. Temporal fluctuations in weather and climate extremes that cause economic and human health impacts: A review. Bull. American Meteorological Soc. 80, 1077–1098.

Larsen, J., 2006. Setting the Record Straight: More than 52,000 Europeans died from heat in summer 2003. Earth Policy Institute, downloaded from www.earth-policy.org/Updates/2006/Update56.htm#top, on November 27, 2006.

Liverman, D., 1994. Vulnerability to global environmental change. In Cutter S.L. (ed.), Environmental Risks and Hazards, pp. 326–342. Englewood Cliffs, NJ, Prentice Hall.

Liverman, D.M., Merideth, R., 2002. Climate and society in the U.S. southwest: The Context for a Regional Assessment. Climate Res. 21, 199–218.

Lowry, W., 1967. The climate of cities. Sci. Am. 217, 15–23.

Malaney, P., Spielman, A., Sachs, J., 2004. The malaria gap. Am. J.Tropical Med. Hyg. 71(2 suppl), 141–146.

Martens, W.J.M., Jetten, Theo H., Focks, Dana A., 1997. Sensitivity of Malaria, schistosomiasis and dengue to global warming. Climatic Change 35, 145–156.

Martens, W.J.M., Niessen, L.W., Rotmans, J., Jetten, T.H., McMichael, A.J., 1995. Potential risk of global climate change on malaria risk. Environ. Health Perspect. 103, 458–464.

McMichael, A.J., Haines, A., Slooff, R., Eds., 1996. Climate Change and Human Health. World Health Organization, 297 pp.

Meehl, G.A., and Tebaldi, C., 2004. More intense, more frequent, and longer lasting heat waves in the 21st century. Science 305, 994–997.

Mills, J.N., Corneli, A., Young, J.C., Garrison, L.E., Khan, A.S., Ksiazek, T.G., 2002. Hantavirus pulmonary syndrome—United States: Updated recommendations for risk reduction. MMWR, CDC. 51(RR09), 1–12, www.cdc.gov/MMWR/preview/mmwrhtml/rr5109a1.htm (accessed November 14, 2006).

Morlot, J.C., Smith, J., Agrawala, S., Franck, T., 2005. Long-term goals and post-2012 commitments: Where do we go from here with climate policy? Climate Policy 5, 3.

National Oceanic and Atmospheric Administration (NOAA). 2006. Annual Climate Review U.S. Summary. Home page, www.ncdc.noaa.gov/oa/climate/research/2005/ann/us-summary.html (accessed November 14, 2006).

National Research Council (NRC), 2001. Under the Weather: Climate, Ecosystems, and Infectious Disease. Washington DC, National Academy Press.

New Mexico Department of Health, 2006. Hantavirus home page, www.health.state.nm.us/hanta.html (accessed November 15, 2006).

Orr. A., 2000. Green gold? Burley tobacco, smallholder agriculture, and poverty alleviation in Malawi. World Dev. 28, 347–363.

Pan American Health Organization (PAHO), 1999. Hantavirus in the Americas: Guidelines for Diagnosis, Treatment, Prevention, and Control, www.paho.org/English/AD/DPC/CD/hantavirus-americas.htm (accessed December 1, 2005).

Parry, M.L., Rosenzweig, C., Iglesias, A., Livermore, M., Fischer, G., 2004. Effects of climate change on global food production under SRES emissions and socio-economic scenarios. Glob. Environ. Change 14, 1.

Pascual, M., Bouma, M.J., Dobson, A.P., 2002. Cholera and climate: Revisiting the quantitative evidence. Microbes Infect. 4, 237–245.

Patz, J.A., Campbell-Lendrum, D., Holloway, T., Foley, J.A., 2005. Impact of regional climate change on human health. Nature 438, 310–317.

Patz, J.A., McGeehin, M.A., Bernard, S.M., Ebi, K.L., Epstein, P.R., Grambsch, A., Gubler, D.J., Reiter, P., Romieu, I., Rose, J.B., Samet, J.M., Trtanj, J., 2001. The potential health impacts of climate variability and change for the United States, executive summary of the report of the health sector of the US national assessment. J. Environ.l Health 64, 2; Health Module.

Peters, P.E., 2002. The limits of knowledge: securing rural livelihoods in a situation of resource scarcity. In Natural Resources Management in African Agriculture: Understanding and Improving Current Practices, Christopher B. Barrett (ed.). CABI, Wallingford, UK.

Peters, P.E., 2006. Rural income and poverty in a time of radical change in Malawi. J. Dev. Stud. 42, 322–345.

Pini, N., Levis, S., Calderon, G., Ramirez, J., Bravo, D., Lozano, E., Ripio, C., St. Jeor, S., Ksiazek, T.G., Barquez, R.M., Enria, D., 2003. Hantavirus infection in humans and rodents, Northwestern Argentina. Emerg. Infect. Dis. 9, 1070–1076.

Pottinger, R., 2005. Hantavirus in Indian country: The first decade in review. Am. Indian Culture Res. J. 29, 35–56.

Rogers L., 1926. Small-pox and climate in India: Forecasting of epidemics. Med. Res. Council Rep. 101, 2–22.

Ropelewski, C.F., Halpert, M.S., 1987. Global and regional scale precipitation patterns associated with El Niño/Southern Oscillation, Mon. Wea. Rev. 115,1606–1626.

Rosenthal, N.E., Sack, D.A., Gillin J.C., Lewy, A.J., Goodwin, F.K., Davenport, Y., Mueller, P.S., Newsome, D.A., Wehr, T.A., 1984. Seasonal affective disorder: A description of the syndrome and preliminary findings with light therapy. Arch. Gen. Psychiatry 41, 1.

Rosenzweig, C., Parry, M.L., 1994. Potential impact of climate change on world food supply. Nature 367, 133–138.

Sachs, J., Malaney, P., 2002. The economic and social burden of malaria. Nature 415, 680–685.

Seager, R., 2006. The source of Europe's mild climate. Am. Sci. 94, 334–341.

Sen, A., 1981. Poverty and Famines: An Essay on Entitlement and Deprivation. Oxford, Oxford University Press.

Sheppard, P.R., Comrie, A.C., Packin, G.D., Angersbach, K., Hughes, M.K., 2002. The climate of the U.S. southwest. Climate Res. 21, 219–238.

Smith, Joel B., 2004. A synthesis of potential climate change impacts on the U.S. Report Prepared for the Pew Center on Global Climate Change.

Swaroop, S., 1949. Forecasting of epidemic malaria in the Punjab, India. Am. J. Tropical Med. 29, 1–17.

Thomson, M.C., Abayomi, K., Barnston, A.G., Levy, M., Dilley, M., 2003. El Niño and drought in southern Africa. Lancet. 361, 437–438.

Thomson, M.C., Connor, S.J., Milligan, P.J.M., Flasse, S.P., 1996. The ecology of malaria: As seen from Earth-observation satellites. Ann. Trop. Med. Parasitol. 90, 243–264.

Thomson, M.C., Doblas-Reyes, F.J., Mason, S.J., Hagedorn, R., Connor, S.J., Phindela, T., Morse, A.P., Palmer, T.N., 2006. Malaria early warnings based on seasonal climate forecasts from multi-model ensembles. Nature 439, 576–579.

Thomson, M.C., Mason, S.J., Phindela, T., Connor, S.J., 2005. Use of rainfall and sea surface temperature monitoring for malaria early warning in Botswana. Am. J. Tropical Med. Hyg. 73, 214–221.

Timmerman, P., 1981. Vulnerability, Resilience and the Collapse of Society. Environmental Monograph No. 1. Institute for Environmental Studies, University of Toronto, Canada.

Travis, J., 1997. Spying diseases from the sky. Science News 152, 72.

Trenberth, K.E., Hoar, T.J., 1996. The 1990–1995 El Niño-Southern Oscillation event: Longest on record. Geophys. Res. Lett. 23, 57–60.

Vandentorren, F.S., Medina S., Pascal M., Maulpoix, A., Cohen, J-C., Ledrans M., 2004. Mortality in 13 French cities during the August 2003 heat wave. Am. J. Public Health 94, 1518–1520.

University of New Mexico, 2005. Unrevised County Population from the Census Bureau. University of New Mexico home page, www.unm.edu/~bber/demo/cntypop.htm (accessed November 15, 2005).

Wells, R.M., Estani, S.S., Yadon, Z.E., Enria, D., Padula, P., Pini, N., Mills, J.N., Peters, C.J., Segura, E.L., and the Hantavirus Pulmonary Syndrome Study Group for Patagonia 1997. An unusual hantavirus outbreak in Southern Argentina: Person-to-person transmission. Emerg. Infect. Dis. 3, 171–174.

Wenzel, R.P., 1994. A New Hantavirus Infection in North America. N Engl. J. Med. 330, 1004–1005.

Western Regional Climate Center (WRCC). 2005. Climate of New Mexico, www.wrcc.dri.edu/narratives/NEWMEXICO.htm (accessed December 1, 2005).

White, G.F., Haas, J.E., 1975. Assessment of Research on Natural Hazards. Cambridge, MA, MIT Press.

World Bank, 2006. World Development Indicators database, April 2006, http://devdata.worldbank.org/external/CPProfile.asp?CCODE=MWI&PTYPE=CP (accessed October 31, 2006).

World Health Organization, 2002. The World Health Report-Reducing Risks, Promoting Healthy Life. Leysin, Switzerland; World Health Organization.

World Health Organization, 2003. Africa Malaria Report, chapter 1: The Burden of Malaria. Geneva, World Health Organization; and MARA (Mapping Malaria Risk in Africa), www.mara.org.za.

World Health Organization, 2004a. Prevention and Control of Malaria Epidemics: Forecasting, Prevention, Early Detection and Control: From Policy to Practice. Report of an Informal Consultation, Leysin, Switzerland. Geneva; World Health Organization.

World Health Organization, 2004b. Using Climate to Predict Infectious Disease Outbreaks: A Review. Communicable Diseases Surveillance and Response Protection of the Human Environment, Roll Back Malaria, Geneva.

Worrall, E., Rietveld, A., Delacollette, C., 2004. The burden of malaria epidemics and cost-effectiveness of interventions in epidemic situations in Africa. Am. J. Tropical Med. Hyg., 71(2 suppl), 136–140.

Wyman, Richard L. (ed.), 1991. Global climate change and life on earth. New York, Routledge, Chapman & Hall.

Yates, T.L., Mills, J.N., Parmenter, C.A., Ksiazek, T.G., Parmenter, R.R., Vande Castle, J.R., Calisher, C.H., Nichol, S.T., Abbott, K.D., Young, J.C., Morrison, Michael, L., Beaty, B.J., Dunnum, J.L., Baker, R.J., Salazar-Bravo, J., Peters, C.J., 2002. The ecology and evolutionary history of an emergent disease: hantavirus pulmonary syndrome. BioScience 52, 989–998.

Zellicoff, A.P., Bellomo, M., 2005. Microbes: Are We Ready for the Next Plague? New York, Amacom.

STUDY QUESTIONS

1. What role might the El Niño /Southern Oscillation (ENSO) cycle play in affecting Hantavirus outbreaks in New Mexico? What are the physical mechanisms linking climate and ecology that explain this connection?
2. What environmental, socioeconomic, and cultural conditions create vulnerability to Hantavirus in New Mexico? Describe these conditions, and explain what various types of solutions might alleviate this vulnerability.
3. How did the Hantavirus public health campaign address differential vulnerabilities among citizens?
4. How are scientists using seasonal climate forecasts to predict malaria risk? Describe the role of early warning systems in reducing the risk of malaria.
5. Discuss some of the factors that increase or decrease human vulnerability to malaria. How do antimalarial drugs influence vulnerability?
6. How have the combined impacts of climate and malaria contributed to the socioeconomic conditions across Africa?
7. Why are the conditions of heat waves exacerbated in urban areas? What populations are most vulnerable to urban heat waves, and what factors contribute to this increased vulnerability?
8. How might the World Bank's drought insurance program reduce vulnerability to food shortages in Malawi?
9. Discuss how the Malawi case study demonstrates the potential use of climate information for improving health through increased economic opportunities.
10. Choose a case from your community, and describe why it is important to consider climate impacts (in addition to social, economic, and political factors) in planning for public health policy. Include in your discussion thoughts on how future changes in each of these factors might impact risk to human health.

CASE STUDIES

2

In the following section, we present four case studies as examples of the diversity of interaction between humans and different climatic factors. The cases also illustrate the differential ability and approaches to coping with climatic variability.

CASE STUDY 1: HANTAVIRUS AND EL NIÑO IN NEW MEXICO

The Hantavirus pulmonary syndrome (HPS) is an emerging infectious disease that was first recognized in the early 1990s in the Four Corners region in the U.S. Southwest. Since then, HPS has become endemic in several states with a total of 453 confirmed cases. The outbreak began in May of 1993 in New Mexico following heavy rains associated with ENSO (CDC, 2006b). By the end of that year, cases were reported across 14 states. New Mexico accounts for 15% of all HPS cases nationwide, and within that state, Native Americans account for 66% of those cases (CDC, 2006b; New Mexico Department of Health, 2006). Anomalous rains and a change in the ecology of deer mice (the reservoirs of HPS) have been attributed to the outbreak. Others suggest that social and economic factors may play a role as well and place certain groups at higher risk. This case study examines the relationship between climate and the Hantavirus in New Mexico and presents some of the ecological and social risk factors that may have contributed to the HPS cases in that state.

Description

Hantaviruses are rodent-borne diseases that are divided up into two large groups: hemorrhagic fever with renal syndrome virus (HFRS) found in Asia and Europe, and Hantavirus pulmonary syndrome (HPS) found in North and South America. The virus was originally named after the Hantaan River in South Korea where the virus prototype was discovered (Wenzel, 1994).

HPS is the first disease in North America classified as a respiratory distress syndrome that has a high occurrence in adults (Wenzel, 1994). In the Untied States, *Sin Nombre* (meaning without name) is the virus that causes the HPS disease (Duchin *et al.,* 1994). Early symptoms are flulike, including fevers, aches, abdominal pain, chills, vomiting, diarrhea, and shortness of breath. In days, the disease can progress to severe shortness of breath and coughing as the lungs are filled with fluid. It can eventually result in respiratory failure (National Research Council [NRC], 2001; New Mexico Department of Health, 2006). Although considered rare to the general public, it is deadly with a case fatality rate of 35%, and there is no known cure (CDC, 2006b).

HPS is primarily spread in the United States, through the deer mouse species, *Peromyscus maniculatus* (Childs *et al.,* 1994). This small rodent is commonly found in rural parts of the Southwest and throughout much of the United States (Mills *et al.,* 2002). They are very small with large eyes and ears, and seem quite harmless (CDC, 2006b). Transmission to humans occurs through the inhalation of deer mice feces, urine, and saliva. The virus is shed when the deer mice defecate or when they come into contact with each other. Although the rodents can infect each other, they are not harmed by the virus (Mills *et al.,* 2002). Fortunately, no known case of person-to-person transmission has been reported in the United States (CDC, 2006b). However, reports of this route of transmission of a related-strain of HPS were investigated in Argentina with contradicting results (Pini *et al.,* 2003; Wells *et al.,* 1997).

Climate and Ecological Connections: Climate in New Mexico

New Mexico's climate is influenced by subtropical and midlatitude circulations and varies according to topography, elevation, and land cover. During the summer months, on average it receives half of its annual precipitation when the North American monsoon, a surge of southeasterly airflows, brings moisture from the gulfs of California and Mexico (Adams and Comrie, 1997; Sheppard *et al.,* 2002). Thunderstorms and flash floods are prevalent during this time (Adams and Comrie, 1997). In the winter, New Mexico experiences a second maximum of precipitation, which is important for water reservoirs (Sheppard *et al.,* 2002). In northwestern New Mexico (where the most HPS cases have been documented), higher precipitation rates compared to the state average are experienced during the warmest months of the year, typically May through October (Western Regional Climate Center [WRCC], 2005).

Every few years the region's climate is influenced by ENSO, but the level of influence varies from event to event and depends on local climate conditions (WRCC, 2005). During El Niño and La Niña years, precipitation anomalies have a tendency to occur in the wintertime. On average, El Niño increases precipitation and cools temperatures. During La Niña, opposite anomalies occur (Sheppard *et al.,* 2002).

In 1992, after enduring 6 years of drought, the region was impacted by a moderate El Niño, bringing unusually high rainfall amounts to New Mexico (Agency Technical Work Group, State of New Mexico, 2005) that would last until 1995 (Trenberth and Hoar, 1996). Another El Niño would strike in 1997/1998, and again New Mexico experienced abnormally high rainfall (Sheppard *et al.,* 2002); reportedly, this was one of the strongest El Niños in 100 years (Glantz, 2001).

During this time, HPS cases continued to be reported in each of the years between the moderate El Niño (1991/1992) and the major El Niño (1997/1998). There were no case fatalities until 1998 when the number of annual cases would markedly increase from 2 to 6, and then 10 cases annually in 1999 and 2000, respectively. In 2006, New Mexico again encountered above average number of HPS cases (New Mexico Department of Health, 2006). Again, unusually high rainfall in New Mexico, possibly associated with a weak El Niño event in 2004/2005 (National Oceanic and Atmospheric Administration [NOAA], 2006), preceded the spike in annual cases (New Mexico Department of Health, 2006). (See Fig. 2-8 for annual number of HPS cases.)

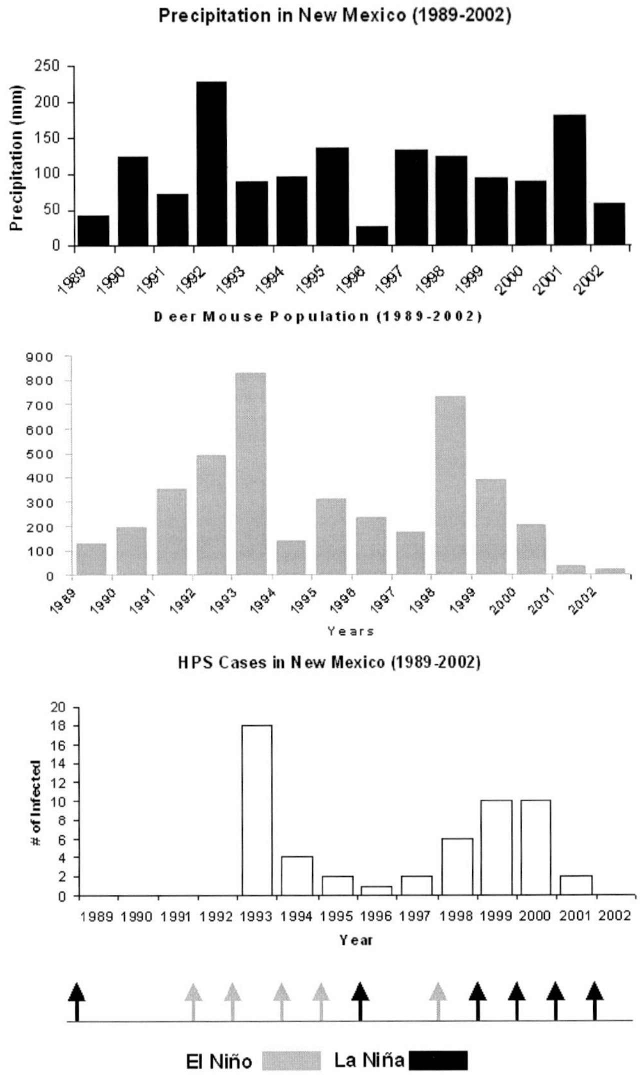

FIGURE 2-8. This schematic illustrates the relationship between (a) precipitation (fall to spring in New Mexico), (b) deer mouse population (springtime density), and (c) HPS cases in New Mexico. There is a 1-year time lag between rainfall and deer mouse population and infections. El Niño (red) and La Niño (blue) years are given for reference. Source: Precipitation and deer mouse population data from Sevilleta Long Term Ecological Research Project: http://sevilleta.unm.edu/; HPS cases (CDC); El Niño, La Niña years. (Trenberth, 1996; L. Goddard, International Research Institute for Climate and Society, personal communication.)

Ecological Connections

El Niño-induced rains have been associated with the outbreaks and the emergence of HPS. Many believe that the rains created a favorable environment for deer mice by increasing their food supply (vegetation) during the fall and winter. In the spring, the rodent population flourished, and likely increased exposure to humans (Engelthaler *et al.,* 1999). As Figure 2-9 shows, HPS cases are more prevalent in the springtime. Engelthaler *et al.* (1999) further posited that competition among the rodents would have ensued as their population grew, and hence, rodent-to-rodent contact would have increased and possibly resulting in more HPS-infected mice.

Indeed, deer mice populations in 1993 had increased dramatically by almost ten-fold (Kolivras and Comrie, 2004).

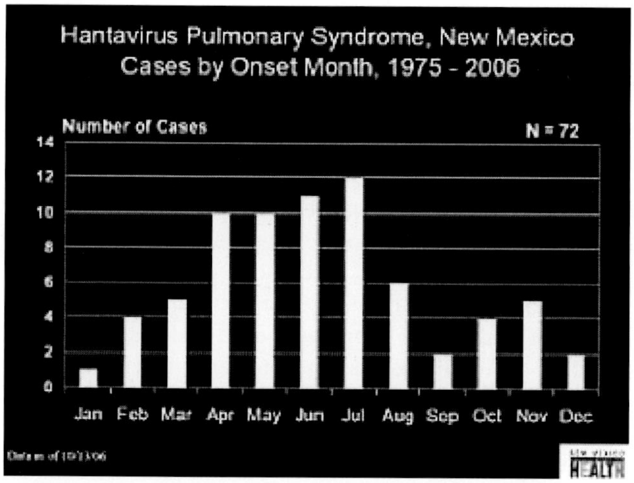

FIGURE 2-9. HPS cases in New Mexico by month. From the Department of Health, New Mexico.

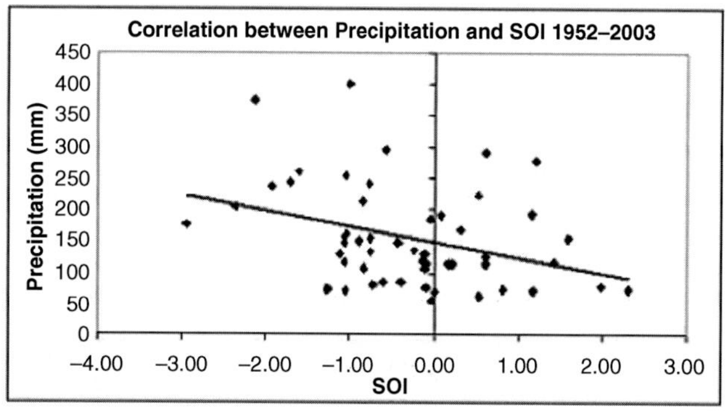

FIGURE 2-10. This figure represents the relationship between El Niño, La Niña years and precipitation (annual) in New Mexico from 1952 to 2003. The correlation between El Niño (Southern Oscillation Index) and precipitation is 35%; though significant (z = 0.0062), this number does not make a strong case for attribution. Figure constructed by Ivan Ramirez and Rachel Harris.

The first major outbreak took place in late spring of 1993 in the northwestern corner of New Mexico, a region that experienced high precipitation rates that previous winter. By spring of 1996, a La Niña year, deer mice density had diminished, and by summer HPS cases had declined (Yates *et al.*, 2002). Unfortunately, the following year, El Niño returned and increased rains again persisted. Vegetation grew markedly reaching a 10-year high, and deer mice doubled from March 1997 to March 1998 (Keller *et al.*, 1998). In northwestern Mexico, the density of mice reached 30 per hectare (Yates *et al.*, 2002). HPS cases soared again, but did not reach the levels of infection of 1993 (CDC, 2006b). However, the fatality rate was higher. Moreover, HPS cases almost doubled from 1998 to 1999, even though there was a lower density of mice. It was later discovered that although the number of rodents had diminished, the number of infected mice had increased in the region (Yates *et al.*, 2002). Cases continued to occur until 2001 and then resurged in 2003 (New Mexico Department of Health, 2006).

Is El Niño to blame? By observing the schematic of relationships between mice, rain, HPS cases, and El Niños (La Niñas), it seems most evident in the years 1993, 1998, and 1999 (see Fig. 2-8). Certainly HPS cases were documented and people died. However, El Niños and La Niñas have occurred in the past, yet, HPS outbreaks did not take place. When examining the relationship between El Niño (represented by the Southern Oscillation Index [SOI]) and precipitation in New Mexico), the correlation is significant, but weak (Fig. 2-10). It suggests a connection, but the climate relationship warrants further study.

Other geoenvironmental factors that seem to be important in HPS risk exposure are location, such as elevation and rural settings, and drought (Engelthaler *et al.*, 1999; Glass *et al.*, 2006; Kolivras and Comrie, 2004; Wenzel, 1994). Engelthaler *et al.* (1999) found that exposure to HPS in 66% of the infected people in the region was at elevations between 1800 m and 2500 m. A study by Kolivras and Comrie (2004) further demonstrated that exposure to HPS was less likely at low desert areas. As mentioned earlier, multiyear drought followed by extreme precipitation had preceded the first outbreak, and therefore may have contributed to the changes in deer mice ecology. Wenzel (1994) suggested that this may have adversely affected the natural predators of the deer mice, further allowing rodents to propagate.

Social and Economic Factors

Native Americans represent the highest numbers of HPS cases in New Mexico (New Mexico Department of Health, 2006). (See Fig. 2-11 for HPS cases in New Mexico by county and Native American populations.) They disproportionately live in poverty in the Southwest region (Liverman and Meredith, 2002). In 1999, 36.2% of all Native Americans lived below the poverty line in New Mexico; in particular, 31.4% lacked telephone service in the household, and 14.8% lacked plumbing services (University of New Mexico, 2005). Those without phone service may not have access to the Internet, an important source for public health information about HPS. Furthermore, lower economic resources could also indicate less access to adequate health care. According to Pottinger (2005), these conditions may place Native Americans at higher risk. Poor housing quality and workplace environment have been strongly associated with increased exposure to HPS (CDC, 2006b). Many of those infected reportedly lived in trailers. It is difficult to mouse-proof these homes because the deer mice are able to pass through tiny holes the size of a nickel (Zellicoff and Bellomo, 2005).

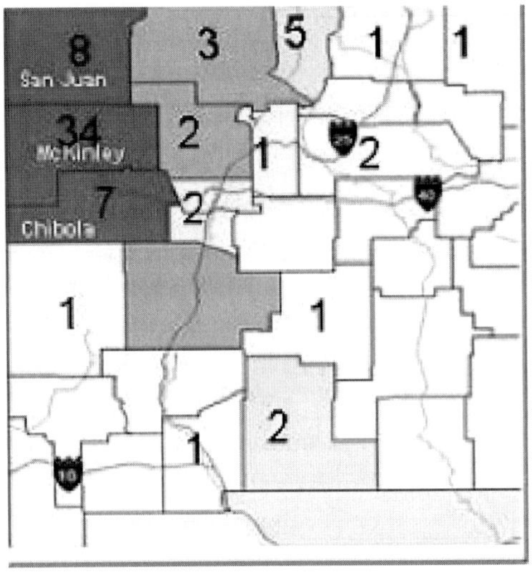

FIGURE 2-11. New Mexico counties and HPS cases from 1975 to 2006 (October). The darker green areas show the higher prevalence of Native Americans; the numbers represent the cases of HPS. This figure, illustrates the high occurrence of HPS cases and high Native American populations. The counties of San Juan, McKinley, and Chibola have the highest HPS cases in New Mexico. From Census 2000 and Departments of Health, New Mexico.

Scientific and Public Policy Responses

Public health authorities realized the urgency of an emerging infectious disease in the region and acted. Medical physicians and staff developed treatment methods that helped stabilize patients and decrease mortality rates (Yates *et al.,* 2002). The state allocated several million dollars for research and development (Committee on International Science, Engineering and Technology Policy [CISET], 2005). Public education campaigns and rodent control plans soon followed.

First, physicians, staff, and state epidemiologists were informed and familiarized with state plans for containment and control. Because Hantavirus patients exhibit symptoms similar to that of the flu, medical staff were trained to identify HPS-like cases and prevent any misdiagnosis. Informational videos on the Hantavirus were distributed and supported by training classes for diagnosis, and audio conferences between medical centers were facilitated. Second, the general public was informed of the symptoms and prevention measures through the Internet, videocassette distribution, brochures, and print, TV, and radio media. A national hotline was also created. Some of the outreach tools used for the general public were bilingual. For example, the videos, Web site, and the phone hotline were produced in English and Spanish. According to Pan American Health Organizations (PAHO), one of the goals of health officials was to target programs for the communities in the most affected regions. However, the report does not indicate whether this was accomplished (PAHO, 1999).

One of the main prevention campaigns launched by the U.S. Centers for Disease Control and Prevention (CDC) was the "Seal Up! Trap Up! Clean Up! Program." It informs the public how to prevent rodents from entering the home, and how to deal with them if their homes are breached. The first step is to seal up your home by closing holes and gaps with cement, hardware cloth, and sheet metal. The second step is to trap all rodents in the home, if any. A quick kill is best so bait traps are recommended because other traps may frighten the mice into urinating and risk transmission. Lastly, cleaning urine and rodent droppings, dead rodents, nests, and clearing food sources are recommended with extreme caution. This seems to be the most complex of all the steps because of the multi-step process and materials required in order to minimize transmission. For example, before sweeping an area of rodent excreta, it is recommended that you wear gloves and an air mask, then proceed to apply bleach and water to the excreta, and wait 5 minutes. Latex gloves, air-purifying mask respirators, bleach, and water are some of the tools necessary for action (CDC, 2006b).

A tool that is currently being developed in order to predict and prevent future HPS outbreaks is risk analysis mapping with the use of Geographic Information Systems (GIS) and satellite imagery. Glass *et al.* (2006) were able to predict likely HPS exposure sites in 2006. They combined landscape level surveillance with satellite data and epidemiological data to project suitable environmental sites for HPS.

Conclusions

The specific contribution of ENSO is still unknown, but the link between climate anomalies typically associated with El Niño events and HPS does suggest a connection. Year-to-year rainfall variability appears to be the major influence on disease transmission through its effect on rodent ecology. If these relationships can be better understood, then climate information along with the efforts of risk analysis mapping may help to prevent outbreaks. However, social and economic conditions should not be ignored and should always be considered in policy decisions and applications.

CASE STUDY 2: MALARIA EPIDEMICS

Introduction

Malaria is caused by a parasite (*Plasmodium* spp.) that is transmitted by mosquitoes. Symptoms include fever, chills, headache, nausea and vomiting, muscle pain, anemia, and jaundice. Transmission occurs from one human to another by the bite of an infected vector (*Anopheles* mosquitoes), from a mother to her unborn baby (i.e., transmitted congenitally), and by blood transfusions. First symptoms usually occur 10 days to 4 weeks after infection, although they can appear as early as 8 days or as late as a year later. People who are regularly exposed to the parasite can build immunity, which is lost after years of nonexposure. Antimalarial medications exist, but resistance to existing drugs has caused crises in many parts of the world. Traditional treatments for malaria are becoming less and less effective. Vaccines have been under development for many years but are currently not available for widespread use.

The Burden of Malaria

Malaria is a major public health problem for many developing countries across the globe, and it is associated with great social and economic hardship. Africa carries the largest burden of morbidity and mortality from malaria; close to 86% of all malaria deaths occur in Africa (Table 2-1).

TABLE 2-1. Malaria-related deaths (in 1000s) in the year 2001.

Location	Malaria deaths	Percentage of worldwide malaria deaths
World	1,124	
Africa	963	(85.7%)
Americas	1	—
Eastern Mediterranean	55	(4.9%)
Europe	0	—
Southeast Asia	95	(8.5%)
Western Pacific	10	(0.9%)

From the World Health Organization World Health Report, 2002.

The economic burden of malaria manifests itself in the short term as much as in the long term. Short-run costs include economic losses associated with infant and child mortality and morbidity, the costs of treatment and prevention, and the indirect cost of productive time, and labor lost. Malaria impedes economic growth and long-term development by hampering the flows of trade, foreign investment, and commerce, hence impacting the population nationwide (Chima *et al.,* 2003; Gallup and Sachs, 2001; Malaney *et al.,* 2004; Sachs and Malaney, 2002).

Who Is at Risk?

Malaria is endemic in many areas (i.e., infection is maintained in a population without need for external input). In these areas, transmission usually follows a seasonal pattern. In this case study, we will focus on the epidemic occurrence of the disease (i.e., appearing as new cases, during a given period, at a rate that substantially exceeds what is considered normal for a given location).

Malaria epidemics can be caused by climate anomalies, such as excessive or prolonged rainfall or unusual increases in temperature, occurring in arid or semiarid zones where the human population has had little exposure to the parasite and hence lacks immunity. Interannual variation are caused by phenomena such as strong ENSO events commonly associated with drought or floods and changes in average temperatures. Malaria epidemics may also develop in situations where normal (endemic) malaria transmission might grow worse because of unexpected human migration, war, and political instability. Similarly, a failure of, or lack of attention to, control activities can give rise to epidemics in areas where malaria was controlled until that time.

Yet even in epidemics arising from complex sociopolitical crises or as a result of a breakdown in control activities (causes that are not immediately related to ocean-atmosphere coupled systems), the impact of malaria epidemics is greatest when they follow periods of drought and famine, which make human populations more vulnerable. In the following, we focus on epidemics occurring as a result of abnormal meteorological conditions.

Estimates of the population at risk of climate dependent malaria epidemics in Africa vary depending on the method and data used (Figure 2-12). The most recent estimate of the population at risk of climate related malaria epidemics in Africa is 124.7 million, based on United Nations population data for 2001 (Worrall *et al.,* 2004).

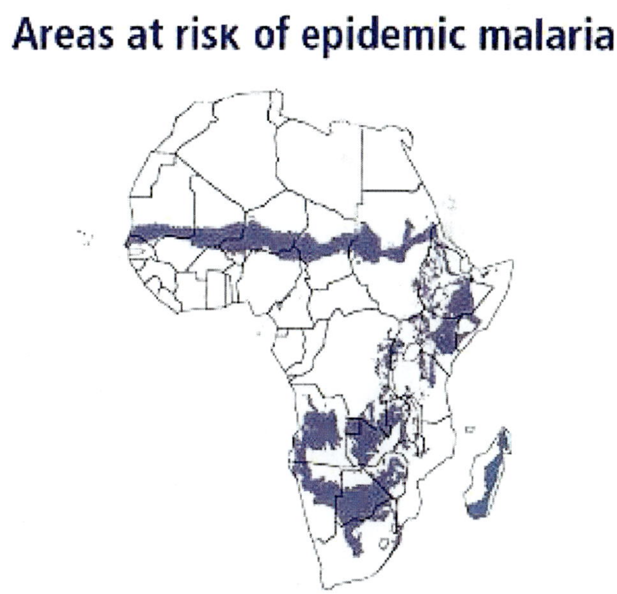

FIGURE 2-12. Spatial distribution of epidemic risk. Adapted from WHO (2003).

Climate and Malaria Epidemics

Meteorological and climatic factors influence the transmission and incidence of malaria. Interannual climate variability is an important determinant of epidemics as climate drives both mosquito vector dynamics and parasite develop-

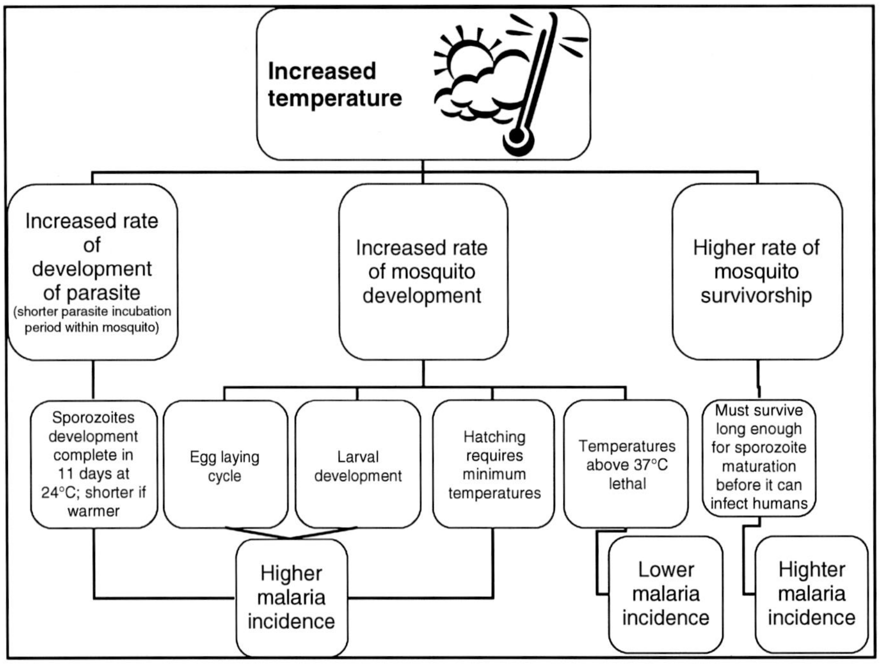

FIGURE 2-13. Influence of temperature on mosquitoes, parasites, and malaria transmission.

ment rates through rainfall, temperature, and humidity (Thomson *et al.,* 1996).

Temperature influences the development rates of the malaria parasite (*plasmodium*). For instance, part of the parasite life cycle taking place within the mosquito consists of the development of sporozoites. This process can be completed in 11 days if the temperature is 24°C, taking longer if cooler and shorter if warmer. Temperatures lower than 18°C compromise the parasite development (Craig *et al.,* 1999). In general, higher temperatures shorten the parasite life cycle within the mosquito host and increase the stability of disease transmission.

Temperature and rainfall influence the development rate of the mosquito (which is both a host to the parasite and a vector). Elevated rainfall levels increase the availability of mosquito breeding sites, leading to larger malaria vector populations, if temperature is favorable. Temperature affects the egg-laying cycle, the rate of larval development, and their emergence. For example, hatching requires minimum temperatures, yet temperatures above 37°C are lethal (Fig. 2-13).

Increased rainfall is associated with increases in humidity, which affects mosquito survivorship. This in turn has an effect on parasite survivorship: the mosquito must survive long enough for the parasite's maturation process, which can last from 1 to 3 weeks and must be completed before it can infect humans. Therefore, higher adult vector survivorship results in greater probability of malaria transmission. Rainfall much above normal can lead to "flushing out" of mosquito breeding sites and can potentially lead to lower malaria incidence (Fig. 2-14).

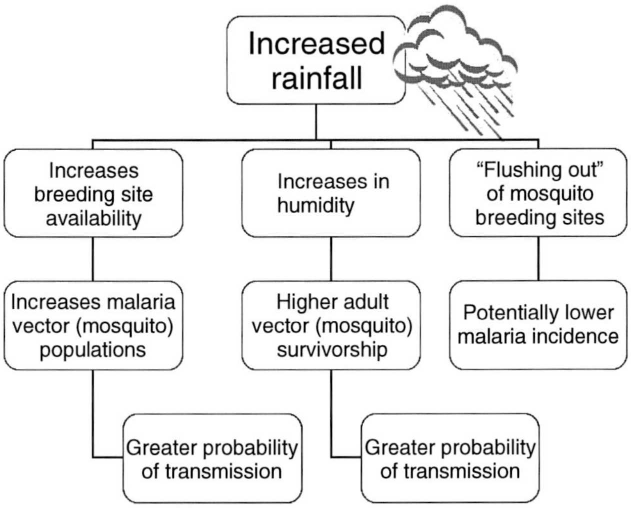

FIGURE 2-14. Influence of precipitation on mosquitoes, parasites, and malaria transmission.

Botswana Case Study

For most of southern Africa, El Niño is associated with lower than normal rainfall, and La Niña is associated with above-normal rainfall during the rainy season. Hence, a rainy season shaped by La Niña conditions increases the risk of malaria epidemics.

Botswana is semiarid and prone to periodic malaria epidemics. Peak malaria transmission season in Botswana is

March–April, following the rainy season (October through April) with a lag time of a few months. Thomson *et al.* (2005, 2006) provided evidence for the relationship between malaria incidence and precipitation anomalies from 1982 to 2002 (Fig. 2-15). Using observed rainfall data, they showed that rainfall accumulations between December and February (Fig. 2-16) were most important in determining malaria incidence in March and April (incidence adjusted for nonclimate related trends such as drug resistance, health policy changes, etc.).

Seasonal Climate Forecasts and Malaria Early Warning Systems

December–January–February national rainfall totals are not available until the beginning of March. Yet successful preparation and response require lead times of several months. Thomson *et al.* (2005, 2006) conducted a retrospective forecast using multiple GCMs and demonstrated that sea surface temperatures during November to February measured in the central equatorial Pacific provided significant predictive skill for the malaria season in advance of its peak season. Warnings of changes in malaria risk 3 to 5 months in advance are possible using current multimodel seasonal rainfall forecasts (available in early November) with only a small loss of precision.

Seasonal climate forecasts using GCMs are now being adopted and implemented as part of the World Health Organization's (WHO) integrated framework for improved early warning and early detection of malaria epidemics. The first such system is in operation in Botswana (DaSilva *et al.*, 2004) and has great potential for extension into other regions of Africa (Fig. 2-17) and the world.

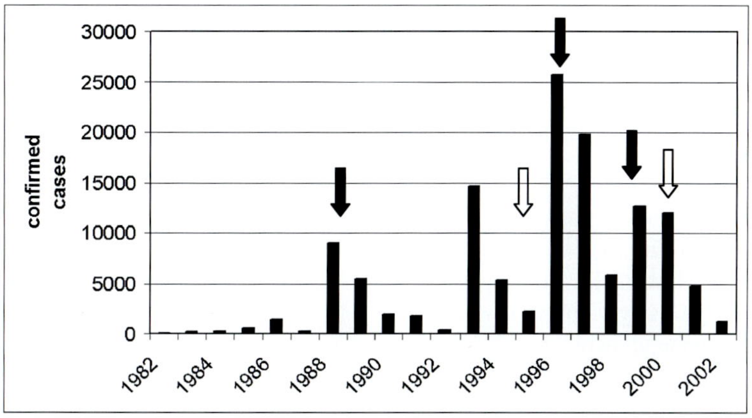

FIGURE 2-15. Time series of malaria incidence in Botswana, 1982–2002 (black arrows represent El Niño events, white arrows represent La Niña events). Adapted from Thomson *et al.* (2005).

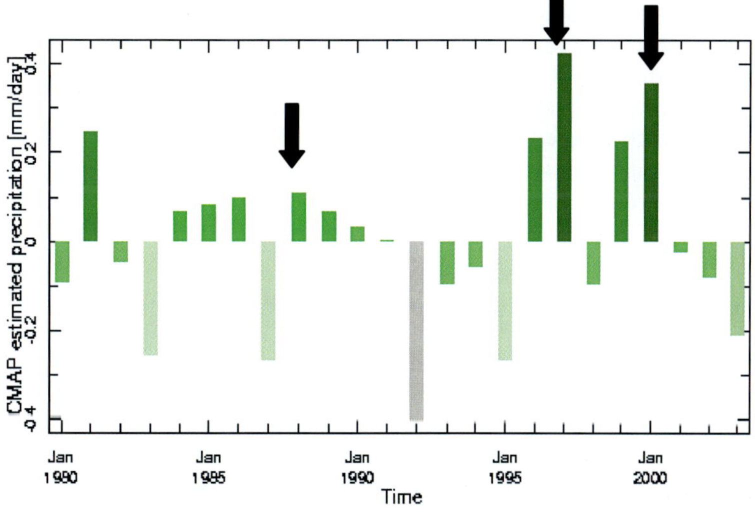

FIGURE 2-16. Estimated rainfall anomalies in mm per day in Botswana, 1980–2003. Output based on rainfall data from stations in Botswana, IRI data library, http://iridl.ldeo.columbia.edu/index.html. Solid arrows indicate years with more than normal rainfall that where followed by higher than normal malaria incidence.

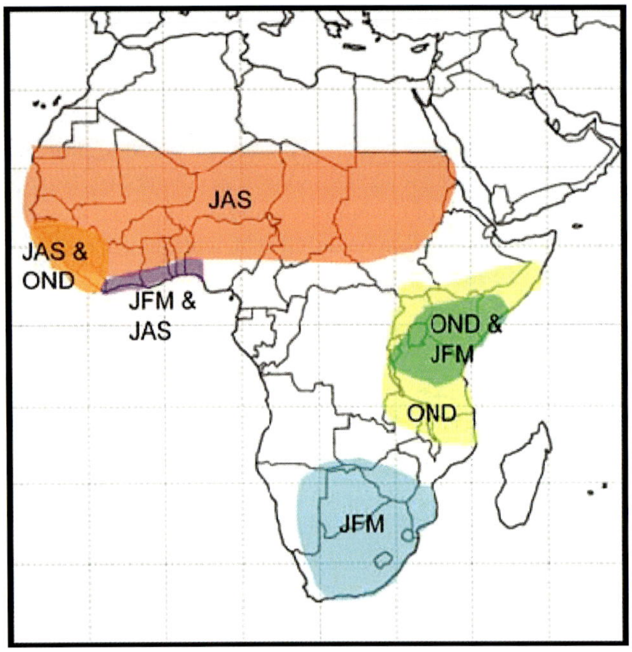

FIGURE 2-17. Regions and seasons with greatest potential predictability in Africa. JAS = July–September, OND = October–December, JFM = January–March. From International Research Institute for Climate and Society (IRICS) (2006).

Seasonal climate forecasts will not eliminate the need for precipitation monitoring, vulnerability monitoring, research into new treatments, and so on. Nevertheless, they can contribute to a more comprehensive early warning system by providing probabilities of anomalously high rainfall averages.

CASE STUDY 3: URBAN HEAT WAVES AND URBAN MORTALITY

The urban heat wave is a frequently ignored climate hazard, yet excessive heat spells can have significant and deadly impacts on the health of urban residents. Extreme heat in urban areas is a quiet disaster, overlooked in part because it is rarely spectacular enough for media coverage and in part because its financial ramifications are more limited than more physically destructive hazards (e.g., floods, earthquakes, or tsunamis). Equally important, its victims, who are generally solitary, poor, chronically ill, and elderly, are frequently "invisible" in the city. Yet although the destructive potential of heat waves is not widely recognized, the Centers for Disease Control (CDC, 2006a) reports that there were more than 8000 deaths related to heat waves in the United States between 1979 and 1999, an average of approximately 400 deaths per year resulting from heat exhaustion, heat stroke, or related illnesses (CDC, 2003). Children and pets left in enclosed vehicles are also at increased risk of death during urban heat waves. In the United States, victims of heat waves outnumber the victims of hurricanes, lightening, tornadoes, floods, and earthquakes combined (CDC, 2006a). The impact of this quiet climate hazard lasts beyond the heat wave itself, for even the elderly who survive heat-related illnesses may experience changed mental states and the loss of capacity to function independently.

Climate conditions likely to cause heat-related mortality involve high daytime temperatures combined with little to no cooling at night, a pattern sustained for multiple days at a time. Scientists have known for some time that cities experience different microclimates than their surrounding areas and call this the "urban heat island effect." Therefore, urban areas are at greater risk of lethal heat buildup than other types of land use. William Lowry describes five causes for urban heat islands, including (1) the capacity of urban buildings (and other structures such as roads and parking lots) to conduct and absorb heat; (2) the capacity of these structures to reflect heat that they do not absorb to other surfaces; (3) the existence of additional sources of heat produced within cities (such as industrial, automotive, and air conditioning exhaust); (4) the use of closed pipes and other containers for sanitary disposal of liquids, minimizing the potential for evaporation and cooling of the air by the liquids; and (5) contaminants and pollutants in urban air that trap and retain heat (Lowry, 1967).

In the 1980s and 1990s, there were a number of major heat waves in American cities, including New York City (1984), St. Louis (1980), Philadelphia (1993), Dallas (1998), and Milwaukee (1995). But the most deadly heat wave during this period was in Chicago in 1995, when 739 heat-related deaths occurred in a single week in July. Eric Klinenberg, compared urban mortality from the heat wave to other natural disasters and found that the 1995 heat wave in Chicago killed twice as many people as the Chicago fire of 1871, 10 times as many people as the Northridge (California) earthquake in 1994, 12 times as many as the San Francisco earthquake of 1989, and 20 times as many as Hurricane Andrew in 1992 (Klinenberg, 2002).

The Chicago heat wave was anticipated, but its length and its fatal consequences for the urban population were not foreseen (Kunkel *et al.,* 1996). Forecasters warned the population that the city would experience unusually high temperatures that week, and many residents left the city for cooler areas in the country or retreated indoors into air-conditioned rooms. In the Chicago heat waves of the 1930s and the 1950s, when there was little or no air conditioning, the city's population slept in the parks or along Lake Michigan to cool off at night; but in 1995, a fear of crime kept the population behind locked doors despite the heat. Going away or turning down the air conditioning was not an option for many of the city's elderly poor. Not only did many of them lack access to air conditioning (or the funds to pay for its operation), but they were also fearful of going out in the street or even leaving their windows open, meaning that they spent their time in air-tight rooms that became even hotter than temperatures outside. The resulting heat-related illnesses and deaths strained the capacity of city officials to deal with them. Health care workers could not meet the demands for their help, and at least 20 hospitals were forced to close their doors to new patients during the heat wave (Klinenberg, 2002). The Chicago morgue was quickly filled to capacity and a local meat-packing plant actually donated refrigerated trucks to the city to hold the overflow bodies of victims. The impact of the heat was also felt in nearby Milwaukee (Wisconsin) a city 90 miles north of Chicago. However, Milwaukee had a smaller total population, and the total number of fatalities in the city was less (Fig. 2-18).

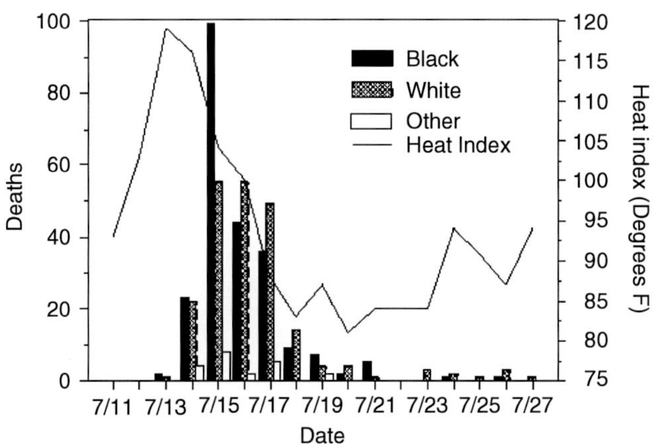

*n = 465.
†The Cook County Medical Examiner's Office categorizes race of decedents as black, white, or other.

FIGURE 2-18. Number of heat related deaths by date of occurrence and race of decedent, and heat index, by date: July 11–27, 1995. From CDC (1995).

A few years later in 2003, an usually hot summer in Europe was initially estimated to have resulted in 35,000 excess heat-related deaths, but subsequent estimates suggest that the actual number of fatalities was in excess of 52,000 (Larsen, 2006). In London alone, officials estimated that 900 people died in August from heat-related conditions. However, in France, where temperatures reached 104°F (40°C) and remained high for 2 weeks, excess heat-related deaths were even higher. Before the heat wave cooled down in September, nearly 15,000 people were reported dead, mostly elderly people, many of them living alone in urban areas. In the Paris region, there was a 142% increase in expected mortality during the heat wave (Dhainaut *et al.*, 2004; Vandentorren *et al.*, 2004). Thousands of farm animals also perished in the heat. The situation was similar to that in Chicago 8 years earlier and resulted in a similar questioning of the political, medical, and administrative response to the heat emergency (Table 2-2).

Conclusions

What can we expect in the future? Climate change is likely to increase the risk of urban heat waves and heat-related mortality and morbidity in 21st century. Scientists predict that future climate change will result in higher average temperatures and more severe and higher probability of periods of excessive heat, and that this will be a particular problem in northern cities (Meehl and Tebaldi, 2004). Meehl and Tebaldi (2004) have attributed the cause to the continued buildup of greenhouse gases and their influence on atmospheric circulation patterns, leading to prolonged periods of extreme heat on Earth's surface. When coupled with the urban heat island effect, this could have a significant impact on the elderly in urban centers in Europe and North America. As urbanization trends in developed countries continue upward (resulting in more "megacities") and the population continues to age, the number of vulnerable individuals will likely increase.

TABLE 2-2. The human toll of the 2003 European heat wave: comparison of 2003 and 2006 fatality estimates.

Country	Number of Fatalities from Heat		
	October 2003 Estimates	July 2006 Estimates	Difference
Italy	4,175	18,257	14,082
France	14,802	14,802	0
Germany	7,000	7,000	0
Spain	4,230	4,130	−100
England and Wales	2,045	2,139	94
Portugal	1,316	2,099	783
Netherlands	1,400	1,800	400
Belgium	150	1,250	1,100
Switzerland	N/A	975	975
Total of above countries	35,118	52,452	17,334

Note: This table compares data on the number of fatalities in the European Summer 2003 heat wave as published by the Earth Policy Institute in October 2003 and in July 2006 based on available information at the time of release. Data are not strictly comparable for 2003 and 2006 because the period for measuring fatalities shifted during the interim. For example, data for Italy in the 2003 tabulation was for July 16–August 15 only, whereas those for the 2006 tabulation include July–September.

Compiled by Janet Larsen, Earth Policy Institute, July 2006.

The unexpected excess mortality in Chicago and Paris during recent heat waves is a reminder that climatc is as much a social and public health issue as a meteorological issue. Studies of the impacts of these heat waves suggest that emergency response to reduce the adverse health impacts on urban populations needs to involve improved forecasts of heat waves and better understanding of their potential adverse impacts on vulnerable populations in the city. The urban population of all ages and economic backgrounds must understand the risks and be trained to respond appropriately to extreme heat, particularly to the impact of extreme heat on the highly vulnerable. The Centers for Disease Control recommends that air-conditioning be recognized as a requirement in hot urban areas, not a luxury, and that public cool spaces should be provided for those who cannot afford their own private cooling systems.

CASE STUDY 4: DROUGHT INSURANCE IN MALAWI

Weather and food security are inextricably linked; drought, flooding, or other weather-related events can decimate food supplies and leave an area prone to famine, especially if the population is particularly vulnerable to such shocks because of weak or absent infrastructure or other forms of support. Following Sen's (1981) proposal that famine results from the lack of access to the means to produce or purchase food, rather than from food unavailability, food security and aid programs have focused on ways to increase access to resources, including agricultural inputs and new technologies. Yet for many people, access to seeds, fertilizers, and other means of food production is restricted by socioeconomic and political constraints; banks will not loan funds without collateral, governments are often unable to provide support, and international banks are increasingly critical of subsidies that might provide some assistance.

In the case of the farmers in Malawi, the ability to provide food security for themselves, their families, and the larger community depends largely on the amounts of rainfall that provide the only source of water for their crops. Rainfall amounts in Malawi, as with the rest of Southern Africa, are strongly tied to ENSO cycles; with El Niño conditions, drought often follows (Thomson *et al.,* 2003). In drought years, lack of rainfall has led to widespread famine and death. In the 1990s, three periods of drought led to widespread hunger; another drought in 2002 again led to food shortages. In combination with low and fragile incomes, largely agricultural in origin (Dorward and Kydd, 2004), low education levels (a literacy of 64.1% [World Bank, 2006]), and a high rate of HIV/AIDS infections (14.1% prevalence of HIV/AIDS for 15- to 49-year-olds [World Bank, 2006]), smallholder farmers (those who farm on only a few acres of land) have become highly vulnerable to food insecurity. In a longitudinal study starting in 1986 and ending in 1997, Peters found that "the common aim of food self-sufficiency is more a dream than reality since the ability to reach it is beyond most families" (Peters, 2006, p. 328). Several solutions to food insecurity in Malawi have been

FIGURE 2-19. Malawi map and country data. Full name: The Republic of Malawi; 2005 Population: 12.9 million (World Bank, 2006); Capital: Lilongwe; Area: 118,500 sq km (World Bank, 2006); Major languages: English, Chichewa (both official); Major religions: Christianity, Islam; 2004 Life expectancy: 40.2 years (World Bank, 2006); 2006 Monetary unit: 137 Malawi kwacha (MK) = $1US; Main exports: Tobacco, tea, sugar, cotton; GNI per capita: $160US (World Bank, 2006). Image from www.planiglobe.com (licensed under a Creative Commons Attribution 2.5 License).

suggested, including access to financing and credit (Alwang and Siegel, 1999), better crops (Orr, 2000), extension programs (Peters, 2002), and institutional changes that respond to local issues (Dorward and Kydd, 2004) (Fig. 2-19).

In 2005, the World Bank initiated a drought insurance and loan program that has attempted to increase the food security of Malawi smallholder farmers. It enables farmers to get access to needed inputs, including fertilizer and more drought-tolerant seeds, by accepting loans that are insured against drought. Farmers are given a loan that covers the costs of improved groundnut seeds, an insurance policy, and other fees and taxes associated with the loan. The insurance policy was developed as index insurance; rather than cover crop losses that would need to be verified and are also subject to fraud, the insurance policies pay out under drought conditions as measured by rainfall amounts at a nearby meteorological station. When rainfall is within a suitable range, farmers are expected to repay the loan in full, using the proceeds of their groundnut sales. In this way, farmers can access improved seeds without initial collateral or capital, increase their yields under most climatic conditions, and repay the loans when money is available at the end of the season.

Farmers apply to this program as members of the National Smallholder Farmers' Association of Malawi (NASFAM), a donor-funded extension agency. Each NASFAM administrative area is divided into several associations, each of which has a number of farming clubs that have an average of 10 farmers each, ranging from 4 to 31 farmers. These clubs take out their loans jointly, in an effort to increase compliance to the loan conditions through social pressure and shared responsibility. In addition, joint liability helps insure that nonrainfall related crop problems can be more easily absorbed than by an individual, and that "lazy farmers," identified as a major problem in Malawi by farmers themselves, can be encouraged to help with a group effort to repay the loans.

Another crucial element of this program is the NASFAM organization itself. Often working as the program representatives of the farmers in meetings with banks and insurers, NASFAM employees use their knowledge of farming practices, risks, and resources to tailor the loan program to the farmers' needs and abilities to repay. For example, in designing the packaged groundnut and maize loan for 2006–2007, NASFAM field agents suggested that reducing the field size from 3 to 1.5 acres would allow smaller farms to participate. NASFAM also handles the loan, seed, and input disbursements, as well as groundnut sales.

This program is in high demand by farmers and is expected to grow from about 700 to close to 3000 farmers. The banking and insurance companies involved, as well as the World Bank, are also very satisfied with the program. In summer 2006, workshops with farmers, bankers, insurers, and others indicated that the program was viewed as transparent and fair. In fact, farmers often asked for an extension of the insurance to cover other kinds of losses, and the 2006–2007 loan package will include an option for farmers to obtain a loan for maize seed and fertilizer; the groundnut-maize package is still indexed to rainfall amounts, and loan repayment will occur through groundnut sales, as maize is largely a subsistence crop. The details of the insurance program are showed in Table 2-3 for one of the five areas, including the sum insured, premium costs, and taxes.

The loan/insurance package was developed from a variety of information sources, including historical rainfall amounts at the meteorological stations, the price farmers could afford to borrow given expected yields and prices for groundnuts, and the preference farmers expressed for small and frequent insurance payouts (as opposed to larger, rare payouts for the most extreme drought events). The final insurance package

TABLE 2-3. Loan and insurance amounts for Lilongwe area (in Kwacha, the local currency, 137 Kwacha = $1US).

Amounts	Groundnuts Alone	Groundnuts and Maize
Interest for the period	27.50%	27.50%
Input cost	3200	4500
Surtax rate	17.50%	17.50%
Insurance rate with tax	9.17%	11.16%
Sum insured	4619.85	4927.21
Input cost	3200.00	4500.00
Premium without tax	360.35	468.08
Surtax	63.06	81.91
Premium with surtax	423.41	550.00
Loan principal	3623.41	5050.00
Interest	996.44	1388.75
Total loan	4619.85	6438.75

From N. Peterson, Columbia University, unpublished data.

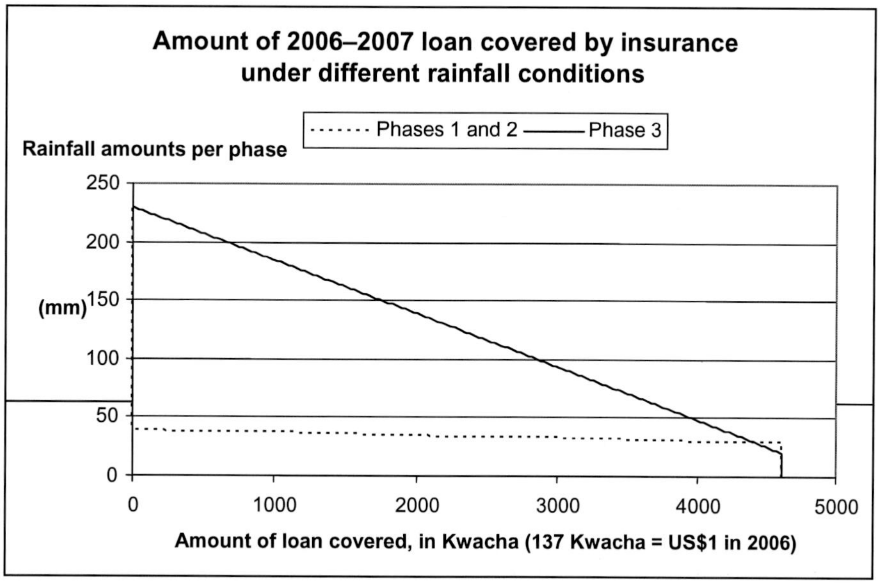

FIGURE 2-20. Amount of 2006–2007 loan covered by insurance under different rainfall conditions.

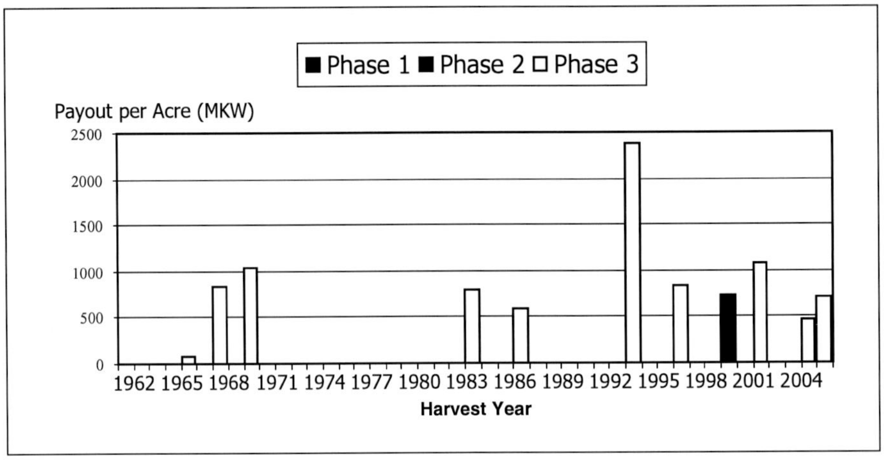

FIGURE 2-21. Performance of 2006–2007 contract using historical rainfall data.

for 2006–2007 groundnuts thus includes three phases of groundnut growth, with varying sensitivities to rainfall and consequently differing payout scales. The third phase provides the bulk of the payouts, insuring the more even distribution of payouts desired by the farmers (Fig. 2-20).

Conducting a retrospective analysis, we find that historically, this loan package would have provided payouts in 11 years since 1961, most in the third phase (Fig. 2-21). This program thus has been able to finance needed inputs and seed improvements for some of the most vulnerable farmers in Malawi, while also improving the farmers' credit histories, incomes, and organizational capabilities. In terms of climate risk, this program also has the potential to design forecast-specific crop packages that minimize potential risks from drought and also capitalize on good rainfalls. Future plans for the program include modifying it for climate predictions (as noted above, droughts often occur in El Niño years in this part of Africa) and expanding it to other parts of Africa, which could further reduce climate vulnerability by lowering the risk of famine and reducing the cost of the loans for farmers.

CONCLUSIONS

The case studies presented clearly indicate that climate events can directly and indirectly impact large groups of people, but they also show how human behavior can make different groups more or less vulnerable to similar events. In this chapter, we have focused on just a few examples. If we extend our thought experiments to other potential human impacts resulting from climate-induced changes in land cover (e.g., increased flooding), loss of biodiversity (e.g., reduction in natural remedy sources), or increased conflict over scarce resource (e.g., "water wars"), the hypothetical scenarios unfortunately are endless. From a risk perception perspective, we humans are cognitively less adept at envisioning impacts that we have not personally experienced. Furthermore, there are tendencies to both overweigh low probability/high impact events and give less attention to more likely but seemingly less dramatic potential events.

Beyond cognitive tendencies, our inability to plan in detail for specific impacts is understandable for additional reasons: (1) medium and longer-term climate forecasting skill is insufficient for many local scale decisions, both from a spatial and temporal perspective; (2) the complex linkages and feedbacks between ecological processes and climate fluctuations are notoriously difficulty to quantify with much precision; and (3) shorter-term basic human needs and political ones often trump longer-term planning initiatives.

Thus, unprecedented climate change has the potential to present humankind with surprises of all sorts. Some may be partially beneficial, for example, reduction in habitat suitable for some parasites. Overall, however, the consensus in the scientific community is that the potential health risks far outweigh the potential gains. Acknowledging our scientific limitations for prediction, political myopia, and pressing current needs should lead us to seriously consider further developing policy approaches that make us better able to cope with a diversity of future scenarios. Such policies should include the design of flexible institutions and regulations that allow course changes as environmental surprises and scientific breakthroughs occur; acknowledgment of the rights of humans to fulfill their basic needs in terms of water, food, and education; and aggressive pursuit of solutions to the conditions that put certain groups at significantly higher risk to current *and* potential environmental perturbations. These recommendations are not novel but need repeating and reinvigoration whenever possible. Perhaps one positive, unintended consequence of the looming threat of global climate change is that it may force citizens of diverse geographic, political, and religious spheres to recognize a common global connection and to work together toward a solution.

CHAPTER

3

The Geological Perspective

Hazards in the Oceanic Environment from a Dynamic Earth

TIM DIXON AND EMILE OKAL

INTRODUCTION

In the theory of plate tectonics, the Earth's rigid outer shell is considered a series of "plates," approximately 100 km thick, moving relative to one another in response to convection in the underlying mantle. Convection in the Earth's mantle is driven by the planet's internal heat, some of it left over from the considerable temperatures reached during its accretion from the solar nebula and some being generated by the natural radioactive decay of uranium, thorium, and an unstable isotope of potassium, all present in small concentrations in mantle rocks. Mantle convection and the corresponding plate motion and volcanism are the main ways that the Earth dissipates both its primordial and its radiogenic heat.

Part of the success of the plate tectonics theory lies in its simplicity: most of the main geological processes at the Earth's surface can be described by a system of slightly more than a dozen such plates (Fig. 3-1). The largest is the Pacific plate, which makes up most of the floor of the Pacific Ocean, west of the East Pacific Rise. The North American plate is a medium-sized plate, comprising the continent of North America east of the San Andreas Fault and the northwestern Atlantic Ocean.

Most active geological processes, including ones that affect human health, occur on or near the boundaries between plates. The San Andreas Fault in California is the boundary between the Pacific and North American plates. Relative motion between the Pacific and North American plates occurs at a rate of 4 to 5 cm/yr, of which about 3 cm/yr occurs on the San Andreas Fault. If the fault were a frictionless boundary, this motion would occur as continuous creep. Unfortunately, along most of its segments, the fault is not frictionless and strain builds up, releasing every few hundred years in a process known as an earthquake, with obvious impacts on human health.

Most plate boundaries, however, are located in the ocean basins, so it is appropriate to discuss the impact of geological processes on human health in this textbook. The eastern boundary of the North American plate is the Mid-Atlantic Ridge, a mountain chain (with a cleft in the middle) that runs more or less up the middle of the Atlantic Ocean, separating the North American plate from the Eurasian plate. The reason for the cleft in the middle of the ridge is that North America and Eurasia plates are pulling apart at a rate of about 2 cm/yr, and volcanism does not quite fill the resulting gap. Although most of this boundary is submarine, a small part of it is exposed in Iceland. As a direct result of their proximity to this plate boundary, Icelanders are exposed to many volcanic eruptions and earthquakes, can make power from geothermal sources, can smelt aluminum from bauxite ore (a hydrated oxide of aluminum) economically, and have an ever-expanding supply of real estate (literally). They also have to be concerned with excess natural fluorine in their environment, which can lead to fluorosis. Thus, the Earth's geological processes play both positive and negative roles in human health and welfare.

The goal of this chapter is to give the reader a deeper understanding of these roles. We provide a basic background in marine geology and highlight a few examples where marine and coastal geological processes have direct and indirect impacts on human health.

Plate Boundaries and Hazards

There are three types of plate boundary, classified by the three types of relative motion that are possible (Fig. 3-2):

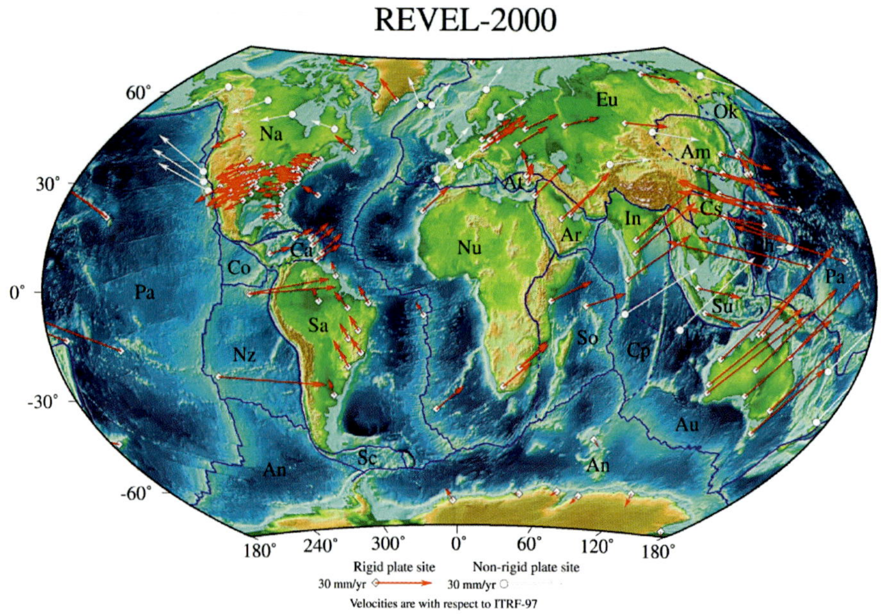

FIGURE 3-1. Map of Earth's major plates. Arrows indicate motion of selected sites, as measured by high precision GPS, relative to an arbitrary global reference frame. Red arrows are plate interior sites, white arrows are sites near plate boundaries. Plate names are abbreviated, (e.g., Pa is Pacific, Na is North America, Sa is South America, Co is Cocos). From Sella *et al.*, 2007.

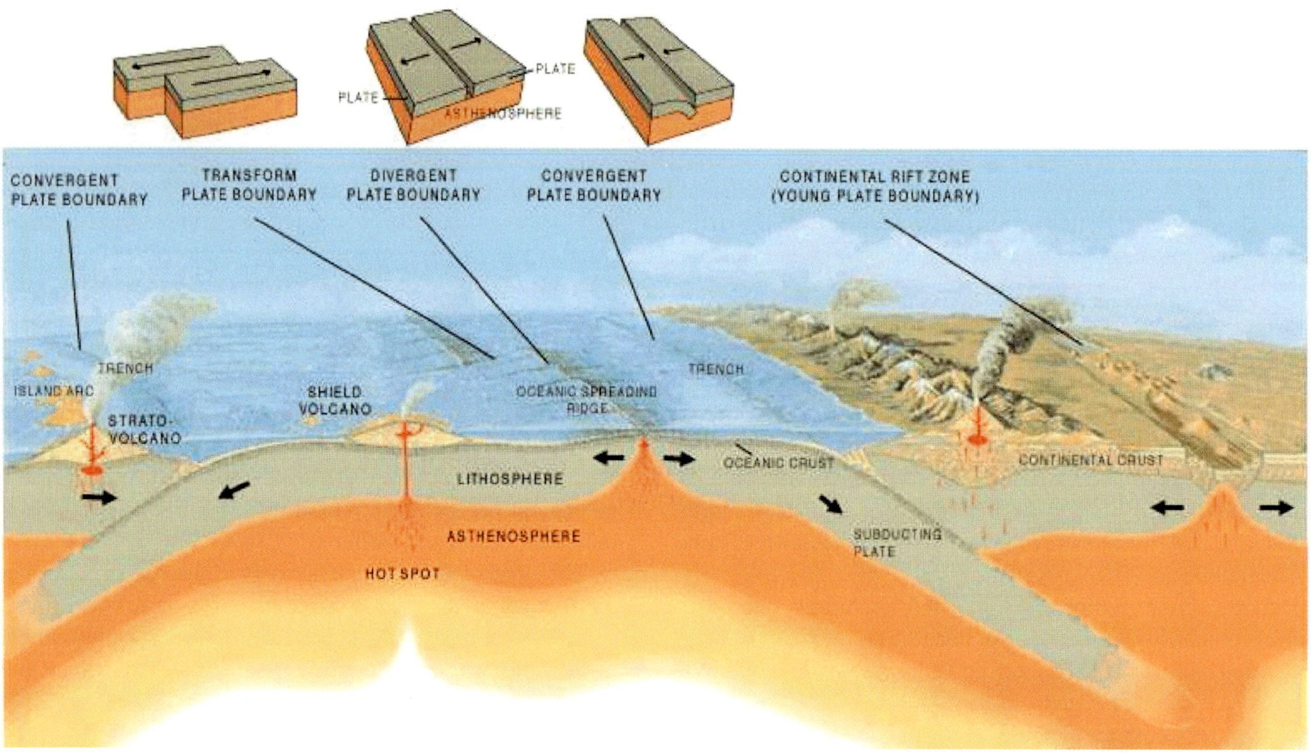

FIGURE 3-2. Cartoon illustrating the three types of plate boundaries. Note that convergent plate boundaries include both ocean-continent boundaries (right-hand side) resulting in volcanoes on the continental edge, as in Central and South America, or ocean-ocean boundaries (left-hand side) resulting in island arc volcanoes, as in the western Pacific. In both cases, a deep trench marks the surface expression of the convergent boundary. Courtesy of U.S. Geological Survey.

1. Plates can move away from one another, at a boundary termed a rift or spreading ridge (e.g., the Mid-Atlantic Ridge). This is the process by which new sea floor is formed: upwelling magma (molten rock) is extruded and cools along the spreading ridge, then it moves away as the plates separate as if the cooled magma were on a conveyor belt.
2. One plate can move toward another plate, termed a convergent or subduction zone boundary (e.g., the Middle America Trench and the Peru-Chile Trench, on the west coasts of North and South America, respectively). This is the process by which seafloor is destroyed.
3. Plates can move past each other, termed a transform boundary (e.g., the San Andreas Fault).

Because of their potential for large-scale destruction of infrastructure, the principal geological hazards that most directly affect human health are earthquakes, tsunamis, and volcanic eruptions. Health effects of these processes may be direct (e.g., death by drowning in a tsunami) or indirect, (e.g., disease and malnutrition associated with loss of crops, livestock, and civil infrastructure as a result of a tsunami).

All three types of plate boundaries can generate earthquakes, whereas type 2 generates most tsunamis. Both types 1 and 2 can generate volcanoes. Volcanoes that are not associated with a plate boundary (midplate or "hot spot" volcanoes such as Hawaii or Yellowstone) may also occur (Fig. 3-2).

Sea level rise and coastal subsidence are not generally related to plate boundary processes (with a few notable exceptions) but are geological processes that can have profound effects on human health through their role in exacerbating flooding during tropical storms and hurricanes. This was demonstrated most recently in August 2005 when Hurricane Katrina struck New Orleans. The geological background to these issues is discussed at the end of this chapter.

EARTHQUAKES

Earthquakes are brittle ruptures between or inside blocks of rock in the Earth's crust or upper mantle. They occur when stress accumulated by the planet's tectonic forces overcomes the local strength of rocks. Most faults (e.g., at plate boundaries) are locked, meaning that rather than being able to accommodate plate motion through continuous creep or sliding, the two walls of the fault remain attached (locked) while deforming, locally accumulating tectonic stresses that build up with time. When this deformation becomes too large, the rock cracks (as would a window pane if pressure were applied from the edges toward the middle), the two sides of the fault slip past one another (this constitutes the earthquake), and then the rock returns to an undeformed state. The process is then repeated to build the next earthquake, resulting over long timescales in a *stick-and-slip* phenomenon analogous in some ways to the screeching of a car's tires around a corner. The recurrence time between major earthquakes at plate boundaries is a complex, irregular, and poorly understood function of the geological environment, ranging typically between 100 and 10,000 years.

In the oceanic environment, the largest earthquakes occur at subduction boundaries (type 2), such as the western coast of South America and the Japanese and Indonesian arcs in the western Pacific. At such locations, an oceanic plate penetrates down ("subducts") into the mantle beneath an overlying plate that can be continental (e.g., in Chile) or oceanic (e.g., at the Tonga-Kermadec arc). Along midoceanic spreading centers (type 1 boundaries), earthquakes occur only on slow-spreading segments, such as the Mid-Atlantic Ridge; they are absent from fast spreading ones, such as the East Pacific Rise. Even so, only moderate earthquakes, not exceeding magnitude 6, take place along slow-spreading centers. In general, they carry little hazard to humans because of their small size and underwater locations. Earthquakes at type 3 boundaries occur along large lateral offsets in the midoceanic ridge system, called transform faults, which can occasionally extend onto land, as in the case of the San Andreas Fault, which connects spreading centers at the opposite ends of California, or the Sumatra Fault, which helps partition (with the nearby subduction zone) the oblique convergence between the Australian and Eurasian plates. Transform fault earthquakes can reach magnitude 8 both on land (e.g., San Andreas, 1857) and at sea (Southwest Indian Ocean Ridge, 1942). Finally, earthquakes can occur inside rather than between plates. Such events have been documented at the magnitude 7 level in the Pacific Basin, and complex earthquakes occurring in the vicinity of plate boundaries (but not exactly along them) regularly reach magnitude 8. The destructive earthquakes at the Sanriku coast of Japan in 1933 and at Sumbawa, Indonesia, in 1977 were of that type.

The hazardous nature of earthquakes rests in their ability to inflict damage or destroy buildings and infrastructure. An earthquake source does not change the physical parameters of the environment to an extent that constitutes a health hazard to humans. The maximum accelerations of the Earth's surface created by earthquakes remain on the order of 1 *g* (or about 10 m/s^2), much less than those experienced by a fighter pilot. In this respect, *earthquakes do not kill people*; their combined effects, primarily building collapses but also fires, landslides, or tsunamis, do. Table 3-1 lists events with the greatest reported death tolls. Although most casualties in the 2004 Sumatra, 1755 Lisbon, and 1896 Sanriku events were victims of the tsunami triggered by the earthquake, building and infrastructure collapse remain the greatest

TABLE 3-1. Death toll in major earthquakes.

Year	Region	Death Toll: Absolute	Death Toll: Scaled to Global Population
1556	Shansi, China	800,000	1/625
1780	Iran	280,000	1/3000
1976	Tangshan, China	250,000	
2004	**Sumatra**	**250,000**	**1/24,000**
1920	Kansu, China	200,000	
1923	Tokyo, Japan	200,000	
1927	Tsinghai, China	200,000	
2005	**Pakistan**	**80,000**	**1/75,000**
1755	Lisbon	70,000	1/10,000
1908	Messina, Sicily	70,000	
1970	Ancash, Peru	66,000	
1999	Izmit, Turkey	40,000	
2003	Bam, Iran	40,000	
1896	Sanriku, Japan	30,000	
1989	Armenia	30,000	
1939	Chillan, Chile	25,000	
1906	Valparaiso, Chile	20,000	
2001	Bhuj, India	20,000	
.....			
1906	San Francisco	3,000	
.....			
1989	Loma Prieta, California	68	
For reference:	Hiroshima:	200,000	
	WWI + Influenza epidemic:	30,000,000	1/60

hazard induced by most earthquakes, especially moderate ones. Damage to buildings and their eventual collapse results from the shaking and deformation they undergo during the earthquake, processes that are controlled by the ground acceleration provoked by the earthquake waves. This acceleration is a function of the earthquake's source (principally its "size," which can be expressed as a magnitude) and of the history of the seismic waves as they travel along the path between the source and the receiver (a building), as well as the site response at the location of the individual building. Other things being equal, an earthquake wave reaching soil with poor mechanical properties (sand, mud, clay) will displace it more efficiently than a stiffer geological structure, such as a hard rock made of granite. This enhanced site response in loose sedimentary environments was illustrated in a spectacular fashion during the 1989 Loma Prieta earthquake: the Golden Gate Bridge in San Francisco, solidly anchored in strong plutonic and metamorphic rock formations, was unaffected by the shaking, whereas a section of the Oakland Bay Bridge just a few kilometers away collapsed, because it was built on mud flats at the bottom of the bay.

In addition, a phenomenon known as liquefaction can result in the loss of rigidity when soils contain a significant amount of water. Under the shaking created by the earthquake, the solid sedimentary matrix loses its cohesion, and the soil then behaves essentially as a liquid. The soil becomes unable to support structures because it loses its shear strength; unlike solids, liquids, by definition, have no shear strength (i.e., no resistance to sideways forces). Figure 3-3 shows a spectacular example of soil liquefaction during the 1964 Niigata, Japan, earthquake. Note that the building at the center has no structural damage from the earthquake; it simply tipped over on its side when the liquefied soil could no longer support its foundation. Such soil conditions are especially common in low-lying coastal areas, such as river deltas (described later in this chapter) and in former lake beds, such as Mexico City.

FIGURE 3-3. Example of damage caused by soil liquefaction, following the 1964 earthquake in Niigata, Japan. Note that the buildings are structurally intact but have tipped over because the liquefied soil failed to support their foundations. From the National Geophysical Data Center.

These kinds of hazards can be mitigated through the use of appropriate zoning and building codes, the latter concept having been introduced in California following the 1925 Santa Barbara and 1933 Long Beach earthquakes [both magnitude (M) = 6.3; with 10 and 120 deaths, respectively] and systematically revised since then following each significant earthquake. Ideally, building on soils with poor mechanical properties or easily subject to liquefaction (e.g., landfills) should be prohibited and construction should use designs and materials minimizing the risk of structural collapse. Solutions can be as simple as triangular bracing or as innovative as decoupling the building from the ground and letting it ride the earthquake on an isolating rubber cushion, as pioneered at the National Museum of New Zealand in Wellington. In practice, many existing buildings and structures can and must be retrofitted to higher standards, which in itself constitutes a separate engineering discipline. As a rule of thumb, the most hazardous building structure is the rural adobe house whose walls fail during shaking, provoking the collapse of the roof on the occupants. Brick and unreinforced concrete buildings in urban areas are also dangerous. In contrast, simple dwellings woven from tree branches and leaves can provide surprisingly good protection, as their flexibility accommodates torsion during the impact of the most destructive shear waves.

Because earthquakes cannot be predicted, the population at risk must rely on individual reaction to shaking for personal mitigation; this includes taking refuge under tables from falling objects, avoiding crossing entryways in or out of buildings, as well as adequate preparedness in areas at risk, such as advance stocking of food, water, and first-aid supplies.

Other effects of earthquakes include large-scale fires, set up in urban areas by the often simultaneous rupture of gas and power lines, with the most tragic examples being San Francisco (1906) and Tokyo (1923). This hazard remains present in modern times, as experienced in Kobe (1995), and mitigation efforts can aim, for example, at shutting off gas lines automatically above a predetermined threshold of shaking, as recorded by strong motion detectors.

In addition, earthquakes can trigger landslides in gravitationally unstable environments. In high relief terrains, earthquake shaking can induce landslides either directly or by the liquefaction process described earlier. Landslides can be subaerial (i.e., taking place above sea level), submarine, or a combination of both (the subaerial material falling into the sea at the shore line). Subaerial landslides constitute a separate hazard, as they not only destroy impacted structures but can simply bury them, causing immediate death to any occupants. Their dynamics, comparable to those of snow avalanches, fall into the realm of fluid dynamics. The most catastrophic example of an earthquake-triggered landslide took place on May 31, 1970, in the Ancash district of northern Peru, leading to the eradication of the city of Huaraz with a death toll of 70,000. Significant landslides were also observed during the 1999 Chi-chi earthquake in Taiwan and the 2001 event in El Salvador. Underwater landslides are, of course, much less well known but can lead to lethal tsunamis, such as in Papua New Guinea in 1998 (2200 deaths; see the following discussion).

TSUNAMIS

General Properties

Tsunamis (a Japanese idiom literally meaning "harbor wave") are gravitational oscillations of the entire body of an oceanic basin following the deformation of the sea floor (or exceptionally of the surface). They take the form of waves propagating at a speed c controlled by the depth h of the water column according to the formula $c \cong \sqrt{gh}$, where g is the Earth's gravity. In a 5-km deep ocean, this amounts to 220 m/s, or the speed of a modern jetliner.

Generation

Tsunamis are most often triggered by major earthquakes and are capable of exporting death and destruction to distant shores, their propagation being limited only by the size of the oceanic basin. For example, the 1960 Chilean earthquake resulted in upward of 150 deaths in Japan, 24 hours after the origin of the earthquake. Tsunamis can also be generated by landslides (e.g., Papua New Guinea, 1998), volcanic eruptions in the sea (e.g., Santorini, 1650 B.C.; Krakatau, 1883), and even bolide impacts (e.g., Chicxulub, Yucatan, 65 million years ago).

Tsunamis generated by major earthquakes result from the displacement of large masses of rock (and consequently of large volumes of water) over relatively small distances. For example, during the 2004 Sumatra event, the fault extended for 1200 km north of the epicenter, but the seismic slip on that fault (the amount of displacement on the fault that occurred during the earthquake) did not exceed 20 m. As a result, a tsunami wave on the high seas, far from its source and from any shoreline, has a small amplitude (70 cm according to a satellite measurement during the 2004 event) spread out over a considerable wavelength (typically on the order of 300 km). As the tsunami approaches a coastline, it *shoals*—that is, its energy is concentrated over a shallower water column, its propagation speed slows down considerably, its wavelength shortens, and the amplitude of the wave is increased many times. In Sumatra, waves coming ashore reached 32 m locally, whereas the corresponding water currents reached 30 m/s or 100 km/h. Even in the far field, many thousands of kilometers from the source, runup amplitudes (flood heights) of 10 m are not uncommon. In the Sumatra example, even the coast of Somalia, more than 5000 kilometers across the Indian Ocean from Indonesia, was affected by the tsunami wave.

Such runup amplitudes and current velocities give tsunamis exceptional powers of devastation. In addition to drowning, tsunami fatalities come from the complete destruction of buildings and infrastructure on the affected shores, with masonry walls obstructing the wave's flow being destroyed systematically. Furthermore, tsunami waves can lift objects as large as buses, locomotives, and small boats, and transform them into projectiles inflicting further damage onshore. During the 2004 Sumatra tsunami, many victims were run over by vehicles moving with the tsunami at speeds comparable to highway traffic; in Banda Aceh, a 10,000-ton barge was moved 3 km inland. Once the tsunami reaches its maximum penetration inland, the water regresses back to the sea, and the draw-down currents can be devastating also because as soil is eroded, roadbeds are further scoured, and debris and people are simply washed out to sea. Finally, during the shoaling process, the wave erodes the sea floor, occasionally transforming the underwater landscape and affecting fisheries for considerable periods of time thereafter.

Underwater landslides can also generate locally devastating tsunamis. These processes differ from earthquakes in that they move volumes of rock with much smaller horizontal dimensions (rarely exceeding 30 km in linear dimension) over considerable distances. For example, a landslide at the head of a canyon may be several tens of kilometers in lateral dimension but flow several hundred kilometers. Large underwater flows of such sediments are sometimes called turbidity currents. The waves of such tsunamis have considerable heights but much shorter wavelengths than those generated by earthquakes, and as a result, they do not propagate efficiently to the far field. Examples include the 1929 Grand Banks tsunami whose waves ran up to 27 m and killed 27 people on the nearby coast of Newfoundland, and the dramatic 1998 event that caused 2200 deaths along the Northern coast of Papua New Guinea. That tsunami reached 15 m and eradicated several villages, but its effects were confined to a 35-km stretch of shoreline, and the tsunami was hardly noticed elsewhere in the Pacific Basin. Tsunamigenic landslides are generally triggered by earthquakes, but their mechanism can involve a delay, expressing the nonlinearity inherent in the release of the precarious material. In the case of the Papua New Guinea tsunami, Synolakis *et al.* (2002) established that the slide took place 13 minutes after the seismic source. In other cases, there may be no delay, and the evidence for the landslide rests in occurrences of dramatic, locally enhanced waves, as for example during the 1946 Aleutian tsunami where a 42-m runup eradicated the lighthouse at Scotch Cap (Okal *et al.*, 2003).

Note also that even earthquakes whose rupture occurs on land can generate tsunamis by triggering submarine landslides through the shaking of underwater sedimentary structures located at some distance from the seismic source. Examples include the 1989 Loma Prieta earthquake, which generated a local tsunami at Moss Landing, California (40 km away), and reportedly the 1910 Rukwa earthquake along the African Rift, which triggered a landslide in the Indian Ocean more than 800 km away.

When its seismic trigger is too small to be recorded, the landslide appears "aseismic," as was the case for example in 1994, when a 3 million-m^3 underwater landslide generated an 11-m wave a few km away in Skagway, Alaska, killing one dock worker. Subaerial landslides falling into the sea have also given rise to damaging tsunamis that can reach gigantic proportions when the landslide falls into a shallow body of water. The record for this type of tsunami is the 525-m runup at Lituya Bay, Alaska, following a major strike-slip earthquake on the Fairweather fault on July 10, 1958. This tsunami, however, failed to penetrate the body of the Pacific Ocean with a meaningful amplitude (Miller, 1960). Tsunamis generated by volcanic explosions or bolide impacts share the characteristics of landslide-generated ones. For example, the 1883 Krakatau tsunami was devastating locally (with waves reaching 15 m and killing 34,000 people), but it did no damage in the far field, even though it was recorded worldwide on tidal gauges.

Mitigation

Direct mitigation of tsunami damage generally involves building seawalls specifically engineered to reflect the waves back to sea; such structures are used systematically along the coast of Japan. However, they are only as good as their maximum height; 6-m walls were ineffective against the 10-m waves hitting Okushiri Island in 1993 (Fig. 3-4).

The only way for humans to avoid the effects of tsunamis is to take refuge away from the wave's reach. In areas where it is difficult to reach more inshore of the inundation line (isthmi, land spits, congested urban areas), the concept of vertical evacuation must be emphasized; it can be as simple as climbing up deeply rooted trees (in general, palm trees are to be avoided), a strategy saving many lives in Papua New Guinea in 1998, or taking shelter atop high-rise buildings (of course, not using elevators, which are at risk for getting trapped at inundated levels). In this respect, pillared structures are particularly valuable, as they provide essentially free passage to the waves through the first floors of construction, offering no cross-section for destructive impact. This concept is being used in the building of evacuation platforms in Japanese harbors.

Tsunami Warning

As tsunami waves propagate relatively slowly (about 200 m/s) and in particular much slower than seismic waves (typically several km/s), it may be possible to provide advance notice of the existence and progress of the tsunami to distant shores. The efficiency of such warnings obviously increases with distance, as more time becomes available for the evacuation of threatened shores.

In practice, a tsunami-warning center analyzes seismic data from the parent earthquake and advises government officials when the event's magnitude exceeds a particular threshold, depending on regional conditions. Regardless of the intrinsic difficulty of measuring in real time the size of the greatest earthquakes, special challenges are posed by certain earthquakes featuring an anomalously slow release of energy, which makes their conventional seismic waves deceptively weak (e.g., Nicaragua, 1992; Java, 1994 and 2006), or by contrast by events such as the large 2005 Nias,

FIGURE 3-4. View of the small town of Aonae, on the island of Okushiri, Japan, in the aftermath of the Japan Sea tsunami of July 12, 1993. Note the devastation wrought by the tsunami wave; all housing in the left part of the photograph has been destroyed and the rubble washed out into the harbor. Note also the fishing boats carried inland and the fires, still burning in this next-day photograph. From Y. Tsuji.

Indonesia earthquake, whose far-field tsunami turned out to be unexpectedly small because of shallow bathymetry and the presence of large islands in its source rupture area.

These remarks illustrate the difficulty of assessing, especially in real time, the tsunamigenic character of a large earthquake. For this reason, efforts aim at incorporating direct observations of the tsunami wave itself into the warning algorithms. In addition to shore-based records on tidal gauges (which suffer from the nonlinear response of these devices and the often nonoptimal ports where they are deployed), a number of modern technologies are starting to provide direct measurements of the waves on the high seas. These technologies include (1) detection by ocean-bottom microbarographs of the overpressure created by the wave and the relaying of this information via offshore buoys, of which a growing number are being deployed in the aftermath of the Sumatra event; (2) direct imaging of the wave at the surface of the sea by satellite altimetry (a proven technique suffering from the aleas of geographic sampling); and (3) over-the-horizon radar exploiting the subtle coupling of the tsunami wave to the overlying atmospheric column (still at the prototype level). Once a warning is issued, evacuation procedures become the responsibility of civil defense authorities and must have been properly designed and tested in advance of the event. The catastrophic death toll of the 2004 Sumatra tsunami in the far field was due principally to the lack of established communication channels between the Pacific Tsunami Warning Center in Hawaii and the authorities of the countries at risk, and of appropriate response procedures in those countries.

Because of the limited amount of time available, centralized warnings are of limited value to humans in the near field, where self-evacuation must be the rule for any individual feeling an earthquake on the shoreline or observing any anomalous behavior of the sea, in particular a large draw down that exposes the sea floor beyond the line of lowest tides. As a gross rule of thumb, and barring exceptionally large tsunamis, evacuation to an altitude of 15 m for at least several hours following the maximum disturbance of the sea is appropriate. Automatic actions, including the closing of sluices where rivers enter the sea, remain valuable mitigation options in the near field.

Finally, we wish to emphasize the value of education of populations at risk as an effective way to minimize the impact of tsunamis on humans. Education can be formal in the classroom, parental through the transmission of ancestral knowledge, or highlighted by media programs and drills. In all cases, education should emphasize a few simple points. Tsunamis are natural phenomena occurring as part of the geological activity of the Earth, and as such they can and will recur; an orderly evacuation following either an official warning or the feeling of shaking along the water line or the observation of any anomalous behavior of the sea will save lives. Success stories as diverse as the evacuation of the village of Baie Martelli in Vanuatu in 1999 (Caminade *et al.*, 2000), the case of the Moken tribe in the Andaman Islands in 2004, or the well-publicized story of Tilly Smith, a 10-year English girl vacationing in Phuket, all confirm, if need be, that education *does* indeed work.

VOLCANOES

Volcanoes are the surface manifestation of a process that caused melting of rock at depth. Molten rock (magma) tends to be less dense than its surroundings and hence rises. Volcanoes represent surface accumulation of erupted magmatic products, which include lava (magma at the surface) and volcanic ash resulting from explosive eruptions, particles of rapidly quenched magma, and particles of the surrounding "country rock." Though midocean ridges are volumetrically the largest form of volcanism on the planet, they have little short-term impact on human health. For this chapter, we will consider volcanoes in two other geological settings, namely hot spots and subduction zones

Hot Spots

As the name implies, a hot spot is a region in the mantle that is anomalously hot. It need not be associated with a plate boundary, and in fact the best-known example of hot spot volcanism is Hawaii, near the middle of the Pacific plate (Fig. 3-5). Although the details of what makes and maintains a hot spot are not well known, they likely represent regions in the mantle that are undergoing focused convective upwelling, which provides a way for the Earth to dissipate radioactively generated heat (broader scale convection also drives plate motion). Because the viscosity (resistance to flow) in the mantle is a strong function of temperature, once established, a hot spot plume will tend to stay focused in one region for long periods of time. The Hawaiian plume is known to have been active for more than 50 million years, resulting in an age-progressive chain of islands (hot spot track) across the Pacific sea floor, aligned in the direction of plate motion. A change in the trend of the island/seamount chain, from west-northwest to northwest, is interpreted to indicate a change in the direction of Pacific plate motion ~40 million to 50 million years ago (e.g., Clague and Dalrymple, 1989). The total age of the chain and hot spot is unknown, since the distal end of the chain may have been subducted near Kamchatka (Fig. 3-5).

In a long-term geological sense, one implication of hot spot volcanism for human health is overwhelmingly positive: it provides land surface within a large ocean for many new species (including humans) to colonize, evolve, and take holidays. On short timescales, of course, lava flows, as well as landslides, tsunamis, and earthquakes associated

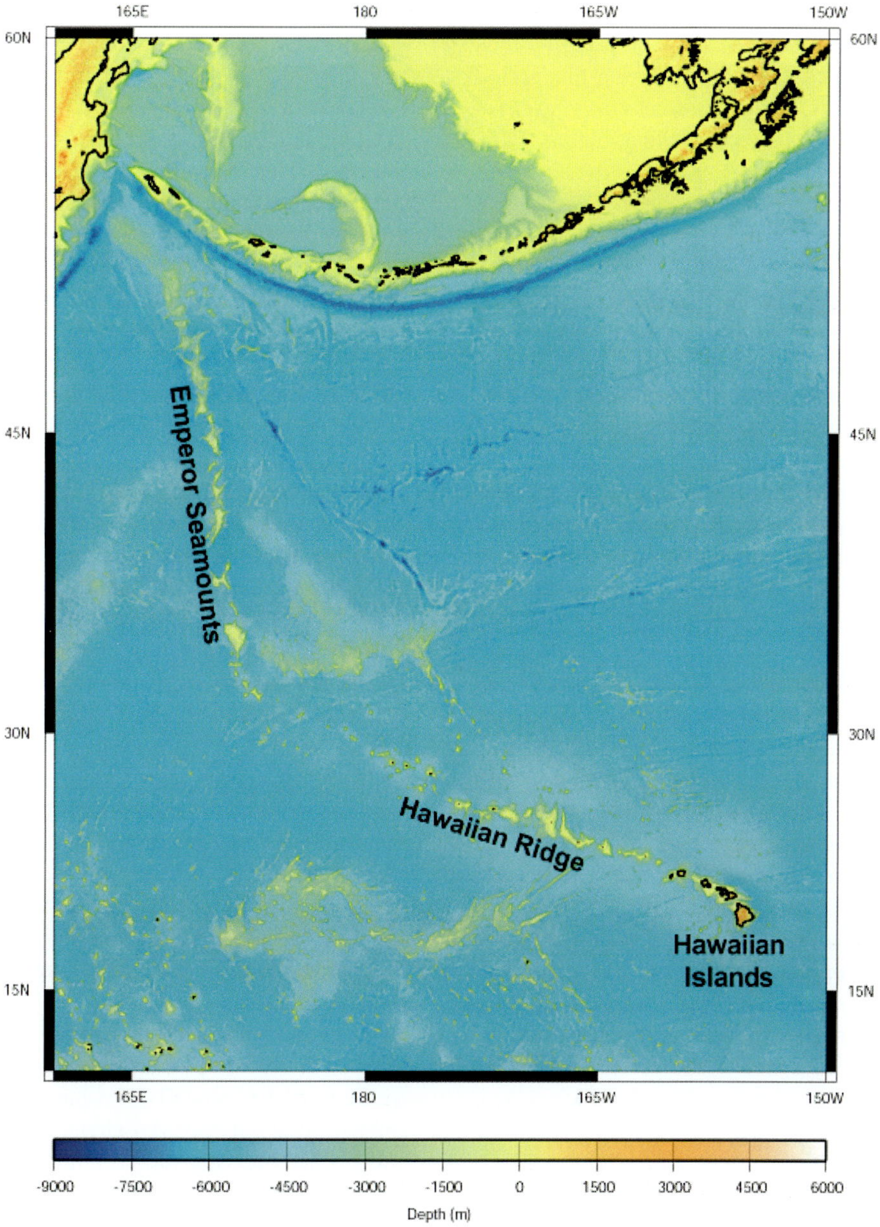

FIGURE 3-5. Bathymetric map of the Hawaii-Emperor seamount chain. The bend between the Hawaiian and Emperor chains occurred at about 43–50 Ma (*Clague and Dalrymple,* 1989; *Clague*, personal communication). Image courtesy of D. Clague and the Monterey Bay Aquarium Research Institute.

with volcanic processes can prove deleterious to human health. Lava flows themselves are not particularly hazardous: they are generally slow moving, and there is usually sufficient warning of their approach. Landslides, tsunamis, and associated earthquakes (e.g., Amelung *et al.*, 2007; Owen and Bürgmann, 2006) can be more disruptive, and on "hot spot" volcanoes such as Hawaii, generally result from the topographic instability associated with a growing volcanic pile and related dike intrusion. A dike is a volcanic formation that results from the intrusion of magma in a vertical wall. In the case of Hawaii, dike growth may indicate seaward motion of a large block of rock on the flank of the volcanic cone because of gravitational sliding; magma fills the resulting vertical crack. Landslides extending up to 80 km offshore suggest that this process can occur catastrophically and lead to locally generated tsunamis (Moore *et al.*, 1995). Problems associated with the large 1783 Lakagigar eruption on Iceland in 1783, another hot spot, caused numerous deaths between 1783 and 1785, primarily the result of starvation. This occurred both from direct effects of the eruption as well as contamination of ground water and pasture land, with subsequent loss of livestock upon which many early settlers depended (Stothers, 1996; Thordarson and Self, 1993; Vasey, 1991). Studies of the livestock remains and graves of the settlers killed during this period suggest that the effect of fluorosis from excess fluorine emissions from the volcano was a contributing factor (Pain, 2005).

Subduction Zones

Volcanoes associated with subduction zones are common along the circum-Pacific "ring of fire." Along the west coast of North and South America, these volcanoes occur exclusively on land, but in the western Pacific and Aleutian Islands, they occur as discrete islands, often in an arc-shaped chain, and for this reason are sometimes termed *island arc volcanoes* (Fig. 3-2). They probably represent the greatest volcanic hazard to human health, for several reasons. First, they are more numerous than other types of volcanoes, at least in terms of subaerial volcanoes (volcanoes located on the sea floor, associated with spreading ridges, are probably more numerous but pose essentially no health risk). Second, for reasons that will be discussed, these volcanoes can erupt catastrophically with large explosions, affecting large areas. Large intracontinental calderas such as Yellowstone may impact even larger areas, but these are thought to erupt rarely, perhaps once in 10^6 years, and are not considered here. In contrast, a typical island arc volcano may erupt every few hundred years. Depending on the criteria used for assessing activity, there are approximately 500 to 1000 such volcanoes that are considered active, meaning that, on average, we can expect one or more such eruptions per year on the Earth. Not all of these will be major eruptions, however. Although we review the human health impacts of such eruptions, it should be noted that from a long-term environmental and geological perspective, this class of volcanism is important for the growth of continental crust, the maintenance of soil fertility in many agricultural regions, and generation of numerous natural resources, especially base metals.

How does subduction generate volcanoes? The subducting plate, which is relatively cool, tends to cool the surrounding mantle, so at first glance it seems counterintuitive that volcanoes, which require melting of mantle material, should form here at all. The cause is melting point depression. Recall that subducted oceanic plates include seafloor formed many million of years before subduction, back at the midocean ridge (Fig. 3-2). During their long journey across the ocean to a subduction zone, and especially shortly after formation at the midocean ridge, the rocks that constitute the upper few kilometers of the oceanic plate become hydrated, as new minerals form that incorporate H_2O into their mineral structure (the original minerals formed at the ridge are anyhydrous). As the plate is subducted, these hydrated minerals are subjected to great pressures and release their water, much as water is squeezed from a sponge. Much of this water is squeezed out within the first 5 to 10 kilometers depth, thereby returning to the ocean, but some is released at greater depths, becoming entrained in the surrounding mantle and moving downward with the plate. This entrained water then lowers the melting temperature of the mantle rocks, much in the same way as salt is used to melt snow on roads, and the ambient mantle conditions can be hot enough for partial melting of hydrated rock even though the temperatures would be insufficient to melt dry rock. The newly generated melt is less dense that its surroundings and rises, eventually erupting to form a volcano. Most subduction zone volcanoes occur directly above the point where the surface of the subducting plate reaches a depth of 100 ± 10 km.

It should come as no surprise that the resulting magmas are water rich. This is one of the main reasons why these volcanoes are so hazardous. Like bubbles in a corked bottle of champagne, the water and other gases in the magma (including CO_2 and SO_2) remain dissolved in the magma at depth, where the confining pressure is high. As the magma approaches the surface, however, the pressure lessens, allowing the gases to come out of solution, forming bubbles that may not be able to escape the magma because of its high viscosity. In the last few kilometers of the magma's rise, as pressure falls rapidly, additional gases are released (their solubility decreases with decreasing pressure), and the volume of existing bubbles expands dramatically, leading to an explosive eruption. The explosion and resulting fragmentation of the lava leads to formation of pyroclastic deposits (from *pyro* meaning fire), including fine ash and a range of larger materials.

Volcanologists classify pyroclastic deposits in a number of ways, including particle size, composition, whether they are deposited as air-fall or as landslides, and whether they are deposited hot or cold. Lahars, for example, are generally fine-grained deposits, representing ash and other pyroclastic debris that initially accumulates on the upper slopes of the volcano from explosive eruptions but then gets saturated with rainfall or melted snow and begins to move downslope, to be redeposited at lower elevations. Lahars can be quite fast moving (they have consistencies that range from thin slurries to that of wet cement) and can quickly entomb entire towns and villages unfortunate enough to be located in the lahar channel, typically an existing river or stream channel. A relatively small eruption of Nevado del Ruiz volcano in Colombia in 1985 killed approximately 23,000 people as a result of erupted ash and lava mixing with snow and ice at the volcano summit (Pierson *et al.*, 1990). The lahars impacted populated areas several hours after the main eruption began. People died both directly from the lahar (by suffocation or internal injuries from the initial flow and devastation) and indirectly by dehydration or infection within several days of the event, as they were trapped in the warm, sticky deposit, and rescuers were unable to reach them.

Lahars are a largely avoidable hazard, in the sense that they usually come with some long-term warming (they typically occur days to months after the initial volcanic eruption), they occur in predictable places (often an existing channel), and while they can be fast-moving (up to ~50 km/hr), there is generally some short-term warning. In the case of Nevado del Ruiz the initial eruption was felt by inhabitants, but civil and religious authorities had assured them that all was well. Previous lahars had occurred in essentially the same location, in 1595 and 1845. Hazard maps warned of the danger of lahars for the town of Armero, destroyed by the 1985 flow (Fig. 3-6). The town was actually built on top of the previous town destroyed by the 1845 lahar.

In contrast, pyroclastic deposits that are a direct result of an eruption may come with much less warning, other than general signs of volcanic unrest, such as local earthquakes. Although volcanologists recognize many different types, all such deposits result from several characteristics of arc volcanoes. First, these volcanoes represent stratified layers of older lava and pyroclastic deposits that can build to high elevation, resulting in the classic "Mount Fuji" conical profile. These may be considered gravitationally unstable piles of relatively weak material, prone to landslides. Second, if such a volcano is active, it can have a large mass of molten, gas-charged rock perched in a magma chamber at relatively high elevation, just beneath the summit of the volcano. A large landslide from such an edifice can flow out many tens of km from the edifice. Moreover, if such a landslide happens to de-pressurize ("uncork") a gas-charged volume of magma, the magma can erupt violently. This happened at Mt. St Helens in 1980. The resulting mass of hot magma, ash, and rock debris initially erupts upward,

FIGURE 3-6. Lahar from Nevado del Ruiz, Colombia. The former town of Armero is located beneath the lahar, approximately in the image center. Approximately three quarters of the town's original 29,000 inhabitants were killed by the lahar and lie entombed within it. Photo by J. Marso, taken in late November of 1985.

then falls to Earth and down the slopes of the volcano, again for many tens of kilometers. The resulting flow of hot, gas-charged ash and debris moves quite rapidly, killing everything in its path. Such an event occurred on the island of Martinique in the Caribbean in 1902 when Mt. Pelée erupted, killing 29,000 people. The phenomenon is termed *nuées ardentes* (literally, "searing cloud"), also known as a pyroclastic flow. A similar phenomenon destroyed the city of Pompeii, in Italy in 79 A.D., when Mount Vesuvius erupted near the modern city of Naples.

A longer-term health effect from such eruptions is the formation and widespread dispersion of fine-grained volcanic ash, which can travel hundreds of even thousands of kilometers from the eruption site. Certain components in the ash, especially the mineral cristobalite, a polymorph of quartz (SiO_2), are known to be highly irritating to lung tissue and may contribute to chronic lung diseases such as silicosis (Baxter *et al.*, 1999; Horwell *et al.*, 2003; Housley *et al.*, 2002; Wilson *et al.*, 2000).

COASTAL SUBSIDENCE AND FLOODING

It is generally agreed that sea level is rising in response to global warming, because of a combination of thermal expansion (ocean volume increase) and melting of land ice (ocean mass increase). The rate of increase averaged over the past hundred years is ~2 mm/yr (e.g., Miller and Douglas, 2004), although this is likely to increase in the future as melting of land ice (Greenland, Antarctica, and mountain glaciers) accelerates. As discussed elsewhere in this volume, global sea level rise will have several deleterious impacts on human populations, especially in coastal areas, including loss of freshwater aquifers because of saltwater intrusion and increased susceptibility of coastal infrastructure to flood damage associated with tropical storms and hurricanes. As tragically demonstrated in New Orleans in the aftermath of Hurricane Katrina in 2005, such flooding, although predictable, can nevertheless have catastrophic consequences, leading to immediate death by drowning and subsequent fatalities from dehydration, disease, untreated chronic conditions, and other causes associated with the breakdown of infrastructure and civil society.

Certain coastal areas are at much greater risk from these processes because of land subsidence; they are experiencing higher rates of *relative* sea rise because of a *combination* of rising ocean levels *and* falling land levels. Such regions are in some ways like the proverbial canary in a coal mine, illustrating what may happen to many areas in a few hundred years when global sea level rise results in widespread inundation of highly populated coastal regions. Before considering these impacts further, we need to consider the geological and anthropogenic causes of subsidence and the technical challenges involved in measuring land subsidence and sea level rise.

Subsidence can be defined as vertical motion of the land surface toward Earth's center of mass. When we consider tide gauge data (a common method of measuring present-day sea level rise; Fig. 3-7) and geological indicators of past sea level location such as the height and age of coastal geomorphic features (which can give longer term indicators of relative sea level rise or fall; Fig. 3-8), it is important to realize that both are relative sea level indicators and do not directly distinguish land subsidence (a local process) from local and global sea level rise (ocean expansion that results from mass or volume increase, for example, because of glacier melting or

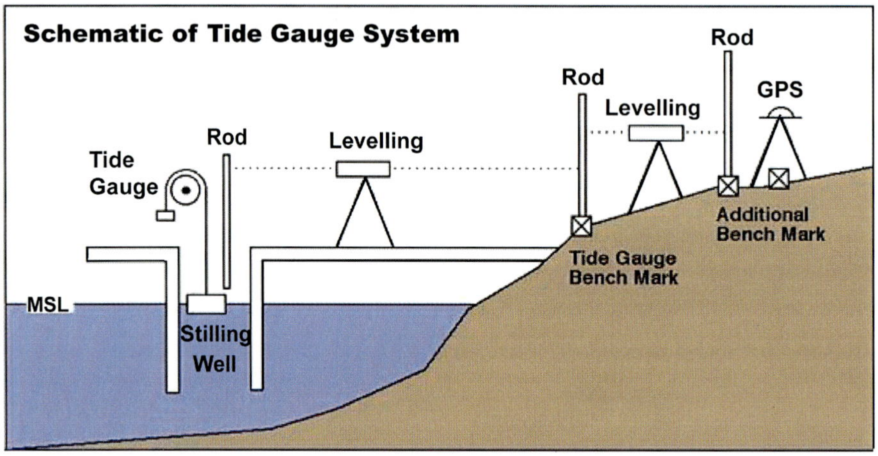

FIGURE 3-7. Tide gauges, precise leveling, and GPS are used to measure changes in sea level and relate these changes to a stable reference, such as Earth's center of mass. The tide gauge can basically be considered a stick that records water level. The system includes a stilling well, to damp out short-term wave motions, and a recording device. Leveling may be performed to connect the tide gauge, often mounted on a dock or pier, to a geodetic benchmark on land. GPS can be used to tie this benchmark to a global reference frame to record long-term changes relative to Earth's center of mass.

FIGURE 3-8. Wave cut coastal terraces at Point Arena, California, marking the positions of former sea level stands. Note the modern coastline at the base of the cliffs. The base of each terrace cliff marks sea level at several times in the past; the "flight" of terraces at this location indicates relative sea level fall over an extended period of time (i.e., land has uplifted at rates greater than the sea level rise because of tectonic forces). Image courtesy of U.S. Geological Survey.

thermal expansion, and changes in coastal currents). All of these processes may contribute to relative sea level rise. It is, of course, the sum of these effects that is important in terms of flood hazard.

Space Geodesy

Distinguishing land subsidence from true sea level rise has been made much easier by the advent of space geodesy. Geodesy is the science of measuring locations on the Earth's surface, and changes in those locations. Space geodesy exploits satellites, both artificial satellites such as the Global Positioning System (GPS) and Earth's natural satellite (the Moon), to establish an external reference frame for such measurements (e.g., Dixon, 1991; Seeber, 2003). For example, we can look at changes of the Earth's surface relative to the center of mass of the Earth, removing the ambiguity inherent in a relative sea level indicator such as a tide gauge. Since the 1990s, GPS measurements of the stability of coastal regions have clarified the relative importance of land subsidence and increasing ocean volume/mass increase in the interpretation of relative sea level rise data from tide gauges (Snay *et al.*, 2007). In effect, the GPS is used to calibrate tide gauges for ground motion (Figs. 3-7 and 3-9).

How It Works

GPS can be thought of as a timing device in the sky, measuring distance between the phase centers of the GPS satellite and ground receiver antennas through measurements of time. In principle, if we know the distance between at least three satellites and a given ground receiver, we can calculate the position of the ground receiver in three dimensions. In practice, at least four satellites are required, because rather than measure distance directly, we measure time (actually time delay, $\Delta\tau$) between the transmission of a satellite signal and its subsequent reception on the ground and use knowledge of the speed of light, *c*, to estimate the distance *d* ($d = \Delta t$). Because of the possibility of timing errors, four or more satellites are required to solve for the three position components and the clock offset. In a typical handheld GPS receiver, a simple measurement of "group delay" is performed, using a signal code that is modulated on one of the sinusoidal carrier waves transmitted by the satellite. To achieve the high precision necessary to investigate subsidence and other geophysical phenomena, special GPS receivers and antennas are required. These receivers use two frequencies (to compensate for ionospheric fluctuations that affect the speed of light) and make highly precise time/distance measurements through the use of phase measurements on the carrier itself, in addition to the group delay from the modulated code. Subsequent analysis of the phase and group delay data must account for a number of geophysical effects, such as tides and atmospheric perturbations, which may also cause *c* to vary, using sophisticated models. Dixon (1991) and Seeber (2003) provided additional details. Figure 3-9 shows an example time series of GPS data spanning several years, showing long-term subsidence at this coastal site near New Orleans, Louisiana, as well as short-term annual and semiannual variations, possibly related to Mississippi River flooding.

Synthetic Aperture Radar (SAR) is an active remote sensing technique that produces its own illumination

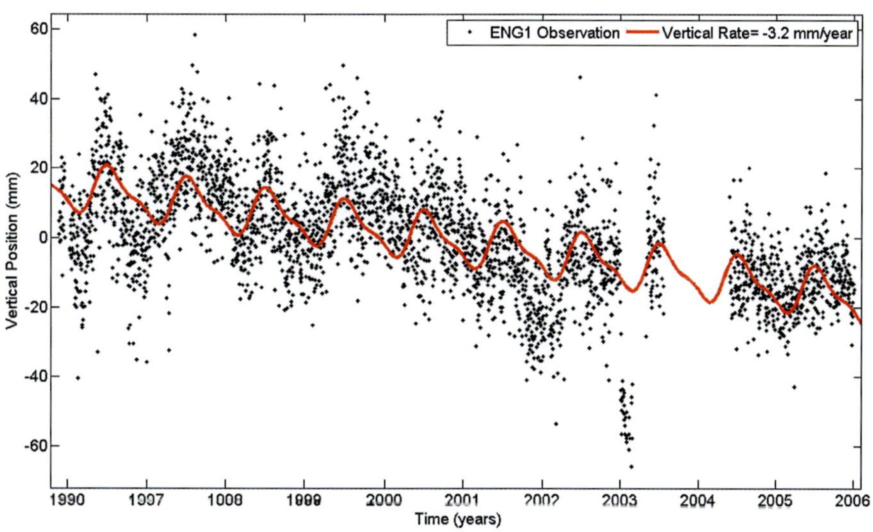

FIGURE 3-9. Example of a high-precision GPS time series, showing daily height estimates (black dots) from a site near English Turn in the Mississippi River, New Orleans. The red line is a model that assumes both annual and semiannual fluctuations, as well as a long-term linear rate. Note the annual fluctuation (probably from loading associated with the annual spring flood of the Mississippi River) and the longer-term trend of subsidence.

FIGURE 3-10. Danube River delta, imaged by the National Aeronautics and Space Administration's Landsat satellite. Image courtesy of NASA.

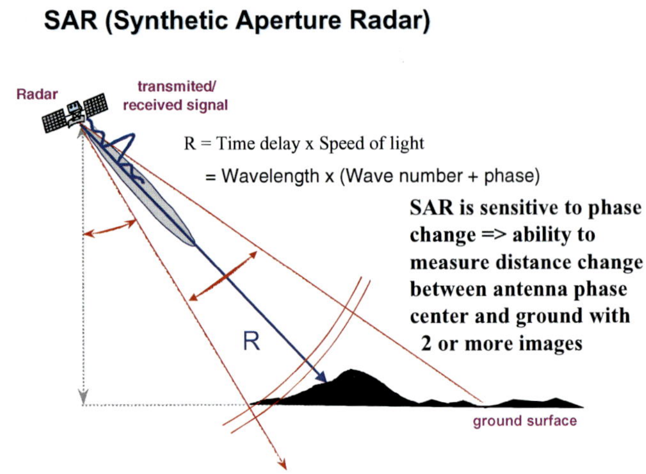

FIGURE 3-11. Principle of Synthetic Aperture Radar (SAR) and Interferometric SAR (InSAR). The space-based radar sensor measures both the amplitude and phase of the reflected signal reflected from the ground. By comparing phase change between successive SAR images (which may be acquired several months apart) and having accurate knowledge of the location of the spacecraft, it is possible to detect changes in the ground surface as small as several millimeters, a fraction of the SAR wavelength.

(microwave energy), and hence works in all weather, day or night (Fig. 3-10). SAR is a coherent imaging system, meaning that both amplitude and phase information in the reflected signal can be used in the data analysis. In effect, the original sinusoidal shape of the transmitted pulse is preserved after reflection from the surface and reception at the antenna. Although the amplitude data produces the image, the phase data in SAR are the key to using this technique for more than just imaging, turning it into a precise space geodetic technique analogous in some respects to GPS. Phase comparison of successive SAR images (interferometric SAR, or InSAR) allows an estimate of the change in the position of the surface to an accuracy of about 5% of the SAR wavelength. A typical SAR wavelength is about 6 cm, so the precision of the surface change measurement is several mm.

Plate-Boundary-Related Subsidence

Many processes can contribute to ground subsidence. For this discussion we will focus on those likely to affect coastal areas. The major plate boundary process that contributes to coastal subsidence is the cyclical pattern of strain accumulation and release associated with the earthquake cycle at subduction zones. As discussed in a previous section, during the interseismic phase, the down-going (usually oceanic) plate is coupled to the overriding plate, forcing it down and landward. This motion is "recovered" during the subsequent earthquake. The cycle is highly variable in length, but periods in the range 50 to 500 years probably describe most subduction zone earthquake cycles. Vertical motions (mainly down during the long interseismic phase, up during the earthquake) may exceed several meters in amplitude (e.g., Barrientos and Ward, 1990; Holdahl and Sauber, 1994). However, because of the cyclical nature of this motion, the long-term impact is usually small. Exceptions would occur in coastal areas that lack zoning regulations (or have poorly enforced regulations), allowing significant development near the beginning of an earthquake cycle (taking advantage of all the extra coastal area). The region would then experience progressive subsidence in subsequent decades as the shoreline newly created by the previous earthquake is progressively drowned and the region becomes increasingly susceptible to storm-related flooding.

Nonplate Boundary Subsidence

One of the most geologically active coastal areas not associated with plate boundary processes is the deltaic environment, where a river meets the ocean. As the velocity of the river slows upon entering the ocean, suspended sediments are deposited, unless strong ocean currents move and disperse the sediments along the coast. In the absence of such currents, the resulting deposits often have an approximately triangular shape in plan (map) view, hence the term *delta,* named after the Greek letter Δ (Fig. 3-11). Many cities are built on or near deltas, representing the historical importance of transportation hubs that connect oceangoing ship transport with barges and other river vessels that can access continental interiors. Deltas also tend to be fertile regions with abundant fresh water for irrigation, promoting agriculture. For both these reasons, many of the world's deltas are densely populated. Examples include New Orleans, near the mouth of the Mississippi River in the United States; Alexandria, near the mouth of the Nile River in Egypt; Venice, near the mouth of the Po River in Italy, and Calcutta and Dhaka in India/Bangladesh, at the combined mega-delta of the Ganges and Brahmaputra rivers.

Nichols and Leatherman (1995) suggested that a 1-m change in relative sea level will displace 6 million people in the Nile Delta. As populations increase, this number will certainly grow, particularly in third world countries with rapid rates of population increase. Bangladesh has an even greater number of people at risk. One meter of sea level rise will take several hundred years if current rates continue, but there are other factors to consider:

1. Will the current rate of sea level rise remain constant?
2. Will global warming impact the number or intensity of future storms and hurricanes?
3. Will the delta subside independently of sea level rise, increasing the rate of *relative* sea level rise?
4. Will land loss and population displacement take place gradually, or catastrophically? How will society react to these changes in our environment? What are the human health impacts?

The combination of dense population, low elevation, land subsidence, rising sea level, and possibly more frequent, more intense hurricanes and resulting storm surges clearly suggests the potential for the world's densely populated deltas to undergo major disasters in the future, with significant implications for human health. Let us look at these questions in more detail and consider a case history.

With regard to question 1, the rate of sea level rise over the next 50 to 100 years is largely determined by the fate of the Greenland and Antarctic ice sheets, which represent a huge reservoir of freshwater perched above current sea level (Alley *et al.*, 2005). Until recently, it was assumed that the thermal inertia of these large ice bodies was sufficiently high that current global warming would not significantly impact their melting rate for hundreds or even thousands of years. This was based on simple thermal conduction theory, which suggests that it takes a long time for temperature changes at the surface of the ice to impact the main ice mass, because water in all its forms is a poor heat conductor. However, observations of outlet glaciers (e.g., Rignot and Thomas, 2002) and the changing mass of Greenland ice (Velicogna and Wahr, 2006) suggest that Greenland and possibly Antarctica are melting at much faster rates than initially predicted. This may reflect advective (as opposed to conductive) thermal processes, whereby melt water at the surface of the ice penetrates cracks, allowing it to flow to the base of the glacier. In effect, this advects heat deep within the glacier and may even lubricate the basal ice, allowing faster outward flow of ice to the ocean, facilitating more rapid break up of the ice sheet.

With regard to question 2, this is an area of much current research. There are some indications that global warming may lead to both increased numbers of storms (Webster *et al.*, 2005) as well as increased storm intensity (Emanuel, 2005), but these are not settled questions, and more research is needed. What seems clear is that we should not assume that storm statistics based on the record of the past 100 years can be used to predict what will happen in the next 100 years.

With regard to questions 3 (delta subsidence), all deltas subside, for a variety of natural and artificial reasons, as outlined later. However, in a natural delta, subsidence is compensated by ongoing sediment deposition. As the delta subsides, the sediment-water interface is maintained near sea level by seasonal flooding and new sediment deposition, at least averaged over several years. In this way, sedimentary deposits in deltas may accumulate over thousands or millions of years and become thousands of meters thick. However, if this process is interrupted, for example, by channelizing the river with artificial levees to reduce or eliminate seasonal overbank flooding, the surface on which the city is built will continue to subside without compensating sediment deposition. Over time, this can result in extremely low elevations (in some cases several meters below mean sea level) and a consequent extreme flood hazard.

Three natural processes and one anthrogenic process are thought to contribute to delta subsidence:

1. Compaction of sediments by the weight of overburden (younger sediments deposited above) as pore water is gradually expelled and sediment density increases.
2. Isostatic adjustment as the crust and upper mantle of the Earth deflect and adjust to the weight of the delta.
3. Gravity sliding as the mass of recently deposited deltaic sediment slides down the continental slope into deeper water.
4. Fluid pumping (extraction of ground water, oil, or natural gas) may also contribute to localized subsidence as reservoir pressure drops, leading to excess compaction of the material comprising the reservoir.

In addition, sediment compaction may be exacerbated by human activities, especially in the case of organic rich soils that are drained for irrigation or urbanization. At this point, they become dessicated, and carbon-rich material is exposed to air, oxidizing to form CO_2, a gas which diffuses into the atmosphere, with significant mass and volume loss in the soil column.

The rates at which these various processes occur may vary significantly from delta to delta, may vary within a given delta, may also vary with time, and in general are poorly known. Of course, the rates at which these processes happen will affect the extent to which they matter on human timescales. Thus, to address question 4 (human impacts), we need to have some quantitative understanding of these various processes.

Let's look at a recent example from New Orleans and the Mississippi Delta. We have seen that the construction of artificial levees that serve to channel the river and reduce or eliminate overbank flooding, as has been done to the Mississippi River in the vicinity of New Orleans, will eventually lead to low elevation of the delta and increased flood hazard. However, in the short term, there are two advantageous consequences. First, the urban region is temporarily protected from flooding. Second, it provides a stable surface upon which geologists and geodesists can make measurements to quantify and better understand the subsidence process. Of course, the engineers who construct levees and the taxpayers who fund them probably did not intend the second consequence, but scientists play the hand they are dealt!

Current subsidence in the Mississippi Delta was first described by Penland and Ramsey (1990), who presented tide gauge data for Florida and other parts of the Gulf Coast including the Mississippi Delta. Along stable coastal areas with negligible ground subsidence such as Florida, relative sea level rise is occurring at rates of about 2 mm/yr. This presumably represents a global sea level signal associated with warming in the past 50 to 100 years. In contrast, along the Mississippi Delta, higher rates (~10 to 12 mm/yr) are observed, presumably representing a combination of ~2 mm/yr of sea level rise and ~8 to 10 mm/yr of land subsidence. The highest rates of subsidence tend to correlate with known regions of high land loss in the past few decades and of thickest Holocene deposits.

Subsidence in the region is best defined in New Orleans. This large urban area has understandably been the focus of a number of investigations because of the flood hazard facing this vulnerable community. In particular, we can use GPS and InSAR data.

GPS Data

The decade-long GPS time series at ENG-1, a site near New Orleans, reliably characterizes the average rate of surface subsidence at that location since the 1990s (3.1 ± 0.9 mm/yr) within its measurement uncertainty (Fig. 3-9). This time series also exhibits annual fluctuations, which may be related to annual fluctuations of the surface, perhaps because of water table fluctuation, differential seasonal loading of the delta, or unmodeled atmospheric processes that affect the signal.

SAR Data

High subsidence rates are observed today in parts of New Orleans using a variant of the InSAR technique, called PSInSAR (Fig. 3-12, from Dixon *et al.*, 2006). The technique exploits strong radar reflectors in the urban environment, such as buildings, or "Permanent Scatterers," and estimates their motion averaged over several years. The average subsidence rate in the city and surrounding suburban areas based on PSInSAR is 6 ± 2 mm/yr. This is equivalent within uncertainty to the average rate of 5 mm/yr reported by Burkett *et al.* (2003) from leveling data and the average for the larger delta reported by Dokka *et al.* (2006) from GPS (5 ± 2 mm/yr). Parts of New Orleans experience even higher subsidence rates, 20 mm/yr or more. These high subsidence areas tend to be located in former wetlands, drained for agriculture and urbanization, and hence are subject to desiccation and oxidation.

Based in part on the new space geodetic results and comparison to older leveling results, surface elevation data, soil maps, and a variety of other geophysical studies, we can now estimate quantitatively the rates of subsidence associated with individual processes.

Compaction of Young (Holocene; Less Than 10,000 Years Old) Sediments

These comprise the upper few meters to several tens of meters for most parts of the Mississippi Delta; maximum Holocene thickness is about 100 meters. In general, subsidence rates will be highest where the thickness of Holocene sediments is greatest, and they may reach several mm/yr. Older sediments also compact, but in general the rates are lower. Organic-rich marshy sediments may achieve the highest subsidence rates because of oxidation. Rates of compaction in this case can exceed 20 mm/yr, and the process may continue for many decades until a highly compacted, low-carbon state is achieved.

Subsidence of the Delta Because of Mass Loading

If sediment flux is steady, the delta will attain a state close to isostatic equilibrium. However, the Mississippi Delta received a large sediment load near the end of the last pulse of Holocene glaciation. Because of the delayed response of the viscous upper mantle, the delta may still be adjusting to this additional load, with subsidence rates as high as several mm/yr (Ivins *et al.*, 2007). A more recent perturbation may have increased erosion rates within the Mississippi drainage basin and consequent increased deposition in and near the delta, as the continental interior was cleared for agriculture in the past 150 years.

Tectonic Subsidence

Large-scale motion of the delta down and to the south occurs as it undergoes gravity sliding into the Gulf of Mexico. Many deltas probably do this, but for the Mississippi Delta, new GPS data have quantified the rate as ~2 ± 1 mm/yr to the south (Dokka *et al.*, 2006). Subsidence asso-

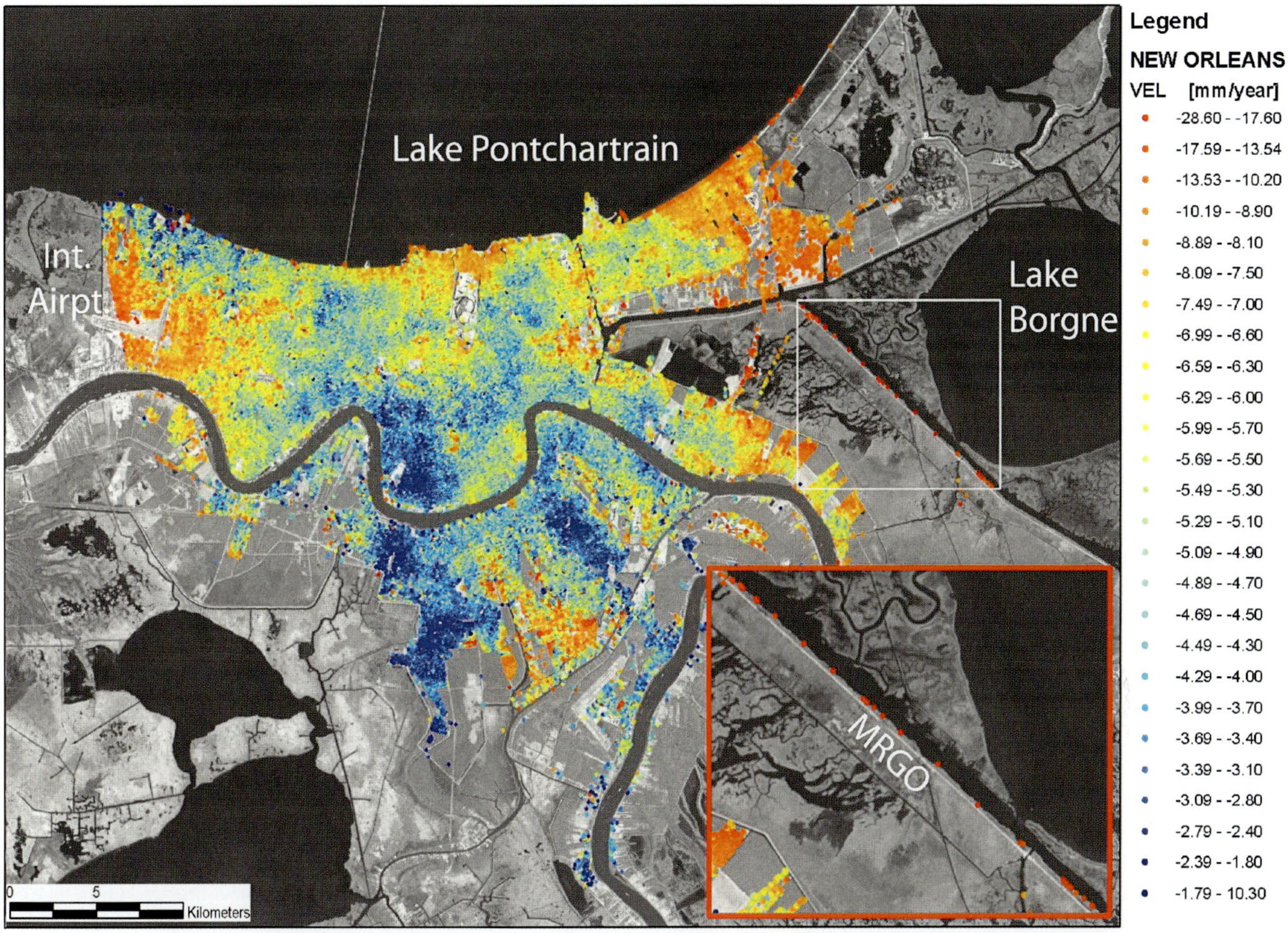

FIGURE 3-12. Subsidence map of New Orleans, Louisiana, based on Radarsat data from 2002–2005. Note high subsidence rates (>15 mm/year) near some coastal areas (recently backfilled), International Airport (town of Kenner, former marshland drained in the 1920s and 1930s for agriculture and urbanization), and adjacent to the Mississippi River-Gulf Outlet (MRGO) canal (inset). From Dixon *et al.* (2006).

ciated with this horizontal motion is less precisely known (and may vary as a function of distance from active normal faults that accommodate the motion) but is probably in the range 2 to 4 mm/yr. Note that the mean rate of GPS-measured delta subsidence reported by Dokka *et al.* (2006) (5 ± 2 mm/yr) represents the sum of several effects, including tectonic subsidence, mass loading, and some sediment compaction.

Fluid Withdrawal

This process typically produces subsidence "cones" within a few kilometers of the point of withdrawal, but it may be more widespread depending on the nature and depth of the reservoir and the rate, magnitude, and timing of production. Onshore hydrocarbon production slowed significantly after the 1970s in Louisiana and is probably not a significant factor in New Orleans or most of the Mississippi Delta.

Elevation and Flooding

Are the high rates of subsidence measured today (e.g., 2002–2005 from the PSInSAR results) typical of subsidence over the past 100 to 150 years? Can they explain the current low elevation of the city? Some parts of New Orleans currently lie 3.0 meters or more below sea level. Major drainage and levee construction in the region began after 1850. Assuming low elevations are somehow related to levee construction and assuming starting elevations close to sea level, average subsidence rates of at least 20 mm/yr over the past 150 years are required to achieve these low elevations. Thus, we conclude that the rapid rates of subsidence measured today in parts of New Orleans could explain the low elevation of parts of the city, especially if the lowest lying areas are characterized by organic rich soils, typical of former marshes. Inspection of soil maps suggests that this is indeed the case.

The main consequence of such high subsidence rates, if sustained over many decades, is, of course, low elevation.

FIGURE 3-13. Flooding in New Orleans Louisiana, approximately 12 hours after passage of Hurricane Katrina, August 29, 2005. Lake Pontchartrain is at the top of the image, the Mississippi River is the dark sinuous band along the bottom third of the image, the white areas are clouds, and the dark areas on land represent flooded areas. The black line marks the approximate maximum flood extent. The yellow numbers show the elevation of the flood line in meters (negative numbers denotes elevations below sea level). Image from SPOT-2 satellite, downloaded and processed at Center for Southeastern Tropical Advanced Remote Sensing (CSTARS), University of Miami. Courtesy of SPOT Image Corp. and D. Whitman, Florida International University.

This was tragically demonstrated in late August and September of 2005 when Hurricane Katrina struck New Orleans, overtopped several levees, and flooded the low-lying parts of the city (Fig. 3-13). Immediate fatalities, largely because of drowning, exceeded 1000 people. Most drowning fatalities were restricted to parts of the city where elevations were more than 2 meters below sea level. By definition, these areas had experienced high subsidence rates for long periods of time (e.g., 20 mm/yr for 100 years). Longer-term health consequences have been profound and are related to loss of infrastructure, loss of livelihood, and consequent loss of access to health care, interruptions to education, increased poverty, and increased susceptibility to disease. As of mid-2007, the population of the city has been reduced by nearly 50% from the prestorm value. On the other hand, immediate health consequences for the survivors were relatively benign. Although floodwaters were highly polluted, with high levels of fecal indicator bacteria and microbial pathogens, concentrations of key indicator bacteria in Lake Pontchartrain, where the floodwaters were eventually pumped, returned to background concentrations within a few months (Sinigalliano *et al.*, 2007).

References

Alley, R.B., Clark, P.U., Huybrechts, P., Joughin, I., 2005. Ice-sheet and sea-level changes. Science 21, 310, 456–460.

Amelung, F., Yun, S.H., Walter, T., Kim, S.W., 2007. Stress control of deep rift intrusion at Mauna Loa volcano, Hawaii. Science 316, 1026–1030, doi: 10.1126/science.1140035.

Barrientos, S.E., Ward, S., 1990. The 1960 Chile earthquake, inversion for slip distribution from surface deformation. Geophys. J. Int. 103, 589–598.

Baxter, P.J., Bonadonna, C., Dupree, R., Hards, V.L., Kohn, S.C., Nichols, M.D., Nicholson, R.A., Norton, G., Searl, A. Sparks, R.S.J., Vickers, B.P., 1999. Cristobalite in volcanic ash of the Soufriere Hills Volcano, Montserrat, British West Indies. Science 283, 1142–1145.

Burkett, V.R., Zilkowski, D.B., Hart, D.A., 2003. Sea level rise and subsidence: Implications for flooding in new Orleans, Louisiana. In Prince, K.R., and Galloway, D.L. (eds.), Subsidence International Group Conference, Galveston, Texas, Nov. 27–29, 2001, 63–70, Rpt 03–308, US Geol. Surv. Water Resour. Div.

Caminade, J.P., Charlie, D., Kánoğlu, U., Koshimura, S., Matsutomi, H., Moore, A., Ruscher, C., Synolakis, C., Takahashi, T., 2000. Vanuatu earthquake and tsunami cause much damage, few casualties, Eos, Transactions. Am. Geophys. Union 81(52), 641, 646–647.

Clague, D.A., Dalrymple, G.B., 1989. Tectonic, geochronology and origin of the Hawaii-Emperor Chain. In Winterer, E.L., Hussong, D.M., and Decker, R.W. (eds.), The Geology of North America, vol. N. Boulder, CO, The Eastern Pacific Ocean and Hawaii, Geological Society of America.

Dixon, T.H., 1991. An introduction to the Global Positioning System and some geological applications. Rev. Geophys. 29, 249–276.

Dixon, T.H., Amelung, F., Ferretti, A., Novoli, F., Rocca, F., Dokka, R., Sella, G., Kim, S.-W., Wdowinski, S., Whitman, D., 2006. New Orleans subsidence: Rates and spatial variation measured by permanent scatterer interferometry. Nature, 441, 587–588.

Dokka, R.K., Sella, G., Dixon, T.H., 2006. Tectonic control of New Orleans subsidence and coupled southward displacement of Southeast

Louisiana. Geophys. Res. Lett. 33, L23308, doi 10.1029/2006 GL027250.

Emanuel, K., 2005. Increasing destructiveness of tropical cyclones over the past 30 years. Nature, 436, 686–688.

Holdahl, S., Sauber, J., 1994, Coseismic slip in the 1964 Prince William Sound earthquake: A new geodetic inversion. Pure Appl. Geophys. 142, 55–82.

Horwell, C.J., Sparks, R.S.J., Brewer, T.S., Llewellin, E.W., Williamson, B.J., 2003. Characterization of respirable volcanic ash from the Soufrière Hills volcano, Montserrat, with implications for human health hazards. Bull. Volcanology 65, doi 10.1007/s00445–002–0266–6.

Housley, D.G., Bérubé, K.A., Jones, T.P., Anderson, S., Pooley, F.D., Richards, R.J., 2002. Pulmonary epithelial response in the rat lung to instilled Montserrat respirable dusts and their major mineral components. Occup. Environ. Med. 59, 466–472.

Ivins, E.R., Dokka, R.K., Blom, R.G., 2007. Post-glacial sediment load and subsidence in coastal Louisiana. Geophys. Res. Lett. 34, L16303, doi:10.1029/2007GL030003.

Miller, D.J., 1960. The Alaska earthquake of July 10, 1958: Giant wave in Lituya Bay. Bull. Seismolog. Soc. Am. 50, 253–266.

Moore, J.G., Bryan, W.B., Beeson, M.H., Normark, W.R., 1995. Giant blocks in the South Kona landslide, Hawaii. Geology 23, 125–128.

Nichols, R.J., Leatherman, S.P., 1995. Potential impacts of accelerated sea level rise on developing countries, J. Coastal Res., Special Issue 14.

Okal, E.A., Plafker, G., Synolakis, C.E., Borrero, J.C., 2003. Near-field survey of the 1946 Aleutian tsunami on Unimak and Sanak Islands. Bull. Seismolog. Soc. Am. 93, 1226–1234.

Owen, S.E., Bürgmann, R., 2006. An increment of volcano collapse: Kinematics of the 1975 Kalapana, Hawaii, earthquake, J. Volcanology Geothermal Res. 150, 163–185.

Pain, S., 2005. Apocalypse then. New Sci. 186, 56–57.

Penland, S., Ramsey, K.E., 1990. Relative sea level rise in Louisiana and the Gulf of Mexico, 1908–1988. J. Coastal Res. 8, 323–342.

Pierson, T.C., Janda, R.J., Thouret, J.-C., Borrero, C.A., 1990. Perturbation and melting of snow and ice by the 13 November 1985 eruption of Nevado del Ruiz, Colombia, and consequent mobilization, flow, and deposition of lahars. J. Volcanology Geothermal Res. 41, 17–66.

Rignot, E., Thomas, R.H., 2002. Mass balance of polar ice sheets. Science 30, 297, 1502–1506.

Seeber, G., 2003. Satellite Geodesy, 2nd ed. Berlin, W. DeGruyter.

Sella, G., Stein, S., Dixon, T.H., Craymer, M., James, T., Mazzotti, S., Dokka, R.K., 2007. Observation of glacial isostatic adjustment in "stable" North America with GPS, Geophys. Res. Lett. 34, L02306, doi:10:1–29/2006GL027081.

Sinigalliano, C.D., Gidley, M. L., Shibata, T., Whitman, D., Dixon, T.H., Laws, E., Hou, A., Bachoon, D., Brand, L., Amaral-Zettler, L., Gast, R.J., Steward, G.F., Nigro, O.D., Fujioka, R., Betancourt, W.Q., Vithanage, G., Mathews, J., Fleming, L.E., Solo-Gabriele, H.M., 2007. Impacts of Hurricanes Katrina and Rita on the microbial landscape of the New Orleans area. Proceedings, National Academy of Sciences, 104, 9029–9034.

Snay, R., Cline, M., Dillinger, W., Foote, R., Hilla, S., Kass, W., Ray, J., Rohde, J., Sella, G., Soler, T., 2007. Using global positioning system-derived crustal velocities to estimate rates of absolute sea level change from North American tide gauge records. Journal of Geophysical Research, 112, B04409, doi:10.1029/2006JB004606.

Stothers, R. B., 1996. The great dry fog of 1783. Clim. Change 32, 79–89.

Synolakis, C.E., Bardet, J.-P., Borrero, J.C., Davies, H.L., Okal, E., Silver, E.A., Sweet, S., Tappin, D.R., 2002. The slump origin of the 1998 Papua New Guinea tsunami. Proceedings of the Royal Society, London, Series A, 458, 763–789.

Thordarson, T., Self, S., 1993. The Laki (Skaftár Fires) and Grímsvötn eruptions in 1783–1785, Bull. Volcanology 55, 233–263.

Vasey, D.E., 1991. Population, agriculture, and famine: Iceland, 1784–1785, Hum. Ecol 19, DOI 10.1007/BF0088898.

Velicogna, I., Wahr, J., 2006. Measurements of time-variable gravity show mass loss in Antarctica. Science 24, 311, 1754–1756.

Webster, P.J., Holland, G.J., Curry, J.A., Chang, H.-R., 2005. Changes in tropical cyclone number, duration, and intensity in a warming environment. Science 309, 1844–1846.

Wilson, M.R., Stone, V., Cullen, RT., Searl, A., Maynard, R.L., Donaldson, K., 2000. *In vitro* toxicology of respirable Montserrat volcanic ash. Occup. Environ. Med. 57, 727–733.

STUDY QUESTIONS

1. Assume you own a piece of real estate in Iceland measuring 1 km by 1 km that spans the main boundary between the North American and Eurasian plate. Assuming you hold on to your investment for 50 years, how much area have you gained?
2. Assuming all the ice in Greenland and Antarctica melts, how much would the global sea level rise? Assume Greenland ice is 1 km thick and Antarctic ice is 2 km thick.
3. You have been asked to travel to an oceanic island to investigate an outbreak of fluorosis and recommend solutions. Once you arrive, local authorities assure you that that the problem has been traced to a plant that manufactures toothpaste and has been remedied. However, the island also has an active volcano. How could you determine if fluorosis is actually endemic to the island because of volcanism, but has only recently been recognized and reported?
4. Calculate the time required for an electromagnetic signal to travel from a GPS satellite to a receiver on the Earth's surface. Assume that the satellite is 20,000 km away from the receiver.
5. (a)Assuming a coastline has a constant slope of 1% (1 m vertical drop for each 100 m horizontal distance), how far inland will a 5 m storm surge travel (ignore complexities associated with vegetation or other barriers and wave dynamics)? (b) How much will this answer change in 100 years if sea level rises at an average rate of 5 mm/yr and coastal subsidence occurs at a rate of 10 mm/yr. For part (b), assume that levee construction temporarily holds back the water until the storm surge.

CHAPTER

4

Overview of Atlantic Basin Hurricanes

BARRY D. KEIM AND ROBERT A. MULLER

INTRODUCTION

Tropical storms and hurricanes have received considerable attention in recent decades because of the extraordinary monetary damage and loss of life these storms have caused. From the heavy rainfall produced by Tropical Storm Allison in 2001, to the catastrophic damage from wind and surge in Hurricanes Andrew, Katrina, and others, the media attention given these events is warranted and has raised awareness worldwide about a major coastal hazard. These events represent the most violent meteorological hazards over the oceans; these storms also inflict considerable damage to both natural environments and cultural landscapes within the coastal zone. This chapter reviews Atlantic Basin hurricanes from the following perspectives:

- Where and why the storms form and their seasonality
- The deadliest storms in the Atlantic Basin
- The spatiotemporal patterns of strikes and hurricane return periods along the Atlantic coast from Maine to Texas
- A review of the most active season on record—2005
- A synthesis of the debate about the impacts of potential global warming on hurricanes

We note that North Atlantic hurricanes constitute 11% of global hurricanes and that U.S. land-falling hurricanes make up only 25% of North Atlantic hurricanes.

HURRICANE FORMATION

Tropical storms and hurricanes form for the purpose of redistributing energy. They form in tropical oceanic regions where heat accumulates during the high-Sun season; this excess heat is then moved poleward by atmospheric and oceanic currents. In essence, hurricanes and tropical storms serve to maintain some semblance of an annual balance of heat and energy within the global ocean-atmosphere environment.

Tropical storm and hurricane formation requires specific oceanographic and meteorological conditions. First, with regard to the ocean, sea-surface temperatures (SSTs) ≥80°F are needed to provide energy for a storm. The warmer the ocean surface, the greater the potential for development. Hurricane seasons that have exceptionally warm SSTs (e.g., 2005) tend to have higher frequencies of storms. Second, evaporation rates off the ocean surface must be high. As water evaporates, energy is consumed, which is stored as latent (stored) energy in the atmosphere of the storms. This energy later is unleashed as sensible heat when condensation occurs (clouds form), thereby making the tropical system even more powerful. Third, upper airflow at approximately 25,000 to 50,000 feet must be favorable in that it allows the converging and rising moist tropical air in the system to vent aloft. Otherwise, if strong winds exist at high levels of the atmosphere, wind shear tears apart the rising cells and inhibits development of the storm. Finally, storms tend to form between 5° to 25° latitude in both the northern and southern hemispheres. They do not form close to the equator because there is weak to nonexistent Coriolis forcing (see Chapter 1), thereby preventing the rotation necessary to form closed circulations that become the initial stages of hurricane development. They tend not to form poleward of 25° latitude because the general circulation of the atmosphere tends to subside in a latitudinal belt between 25° to 30° in both hemispheres. When all of these factors are in place, the potential is high for the formation of a tropical storm or hurricane within the designated regions.

Geographical regions where tropical storms and hurricanes form include the North Atlantic Ocean, including the

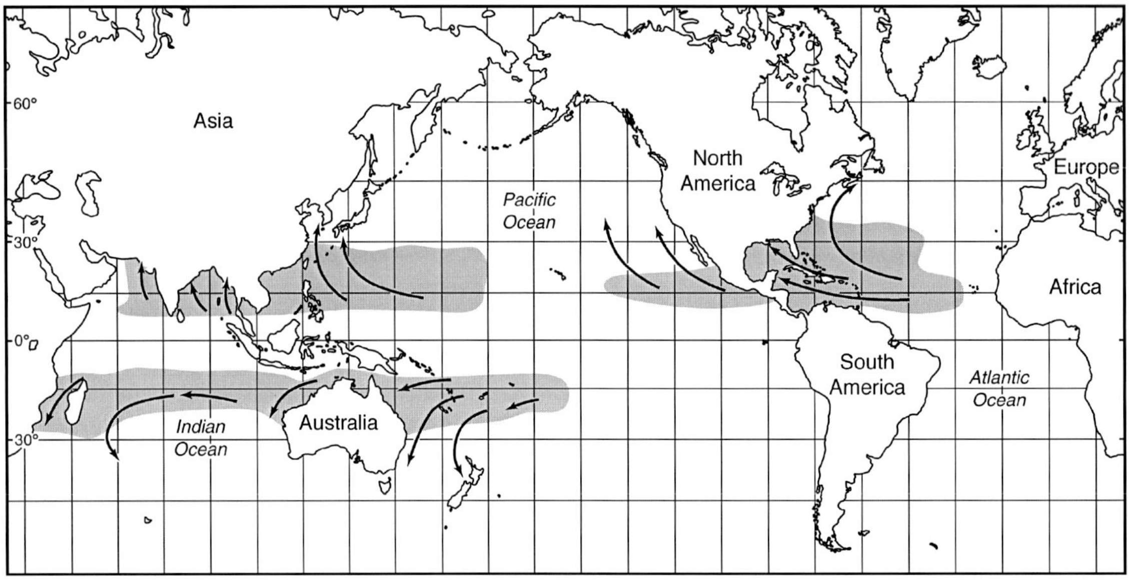

FIGURE 4-1. Breeding grounds (in gray) for tropical storms and hurricanes around the world.

Caribbean Sea, and Gulf of Mexico—all the focus of this chapter. Tropical storms and hurricanes also form in the North and South Pacific and the Indian Ocean (Fig. 4-1), but because they do not tend to form in the South Atlantic Ocean or in the southeastern Pacific Ocean, a storm striking South America is a rare event. In North America, in both the North Atlantic and Pacific Oceans, we call these storms hurricanes—named by the Carib Indians as the God of Evil, called *hurican*. In the western North Pacific Ocean, they are called typhoons, and in the South Pacific and Indian Oceans, they are called tropical cyclones. Regardless of where they form and what they are called, they are all the same meteorological phenomena; they all require the same oceanographic and meteorological elements noted earlier. Also, they are all warm cored in that they consist entirely of warm-humid tropical air and are driven by latent heat exchanges. This is in stark contrast to extratropical systems—like a nor'easter—that have frontal boundaries and are driven by thermal contrasts between opposing air masses.

The typical sequence for the formation of a hurricane begins with a cluster of thunderstorms. Sometimes, these thunderstorms can even originate over land (e.g., in the Sahel in Africa), but most often they form over water. After a cluster of thunderstorms begins to organize, it typically forms an area of weak low pressure causing a perturbation in the pressure pattern, called a tropical wave. In a typical year, the North Atlantic Basin alone may have more than 100 tropical waves, any one of which may develop into a hurricane. After a closed circulation forms, it is called a tropical depression, indicating that the collection of thunderstorms then has a central location anchoring the circulation. Once wind speeds around this center of low pressure reach 38 miles per hour (mph), the storm becomes a tropical storm,

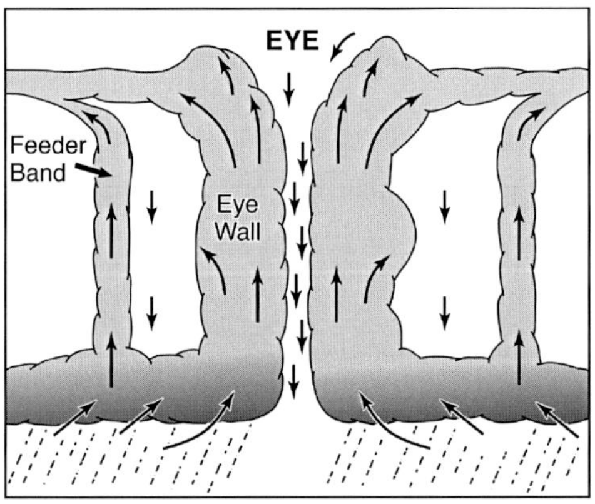

FIGURE 4-2. Cross section of a hurricane.

and when wind speeds reach 74 mph, it forms a calm central eye and becomes a hurricane. These hurricanes consist of concentric stormy convective feeder bands rotating around the eye, with an eye wall around the eye, and with weak downdrafts in between the feeder bands (Fig. 4-2). Once a storm reaches hurricane strength, it is then measured on the Saffir-Simpson Hurricane Scale, ranging from Category 1 to Category 5 (Table 4-1).

HURRICANE SEASONALITY

The seasonality of tropical storms and hurricanes primarily depends on SSTs. Hurricane season in the North Atlantic Basin officially begins on June 1 and extends to November 30; hence half the year is "in season." The first of June is not really a magical date when storms suddenly begin

TABLE 4-1. Saffir/Simpson Hurricane Scale.

Scale Number		Typical characteristics of hurricanes by category			
(Category)	Winds (Mph)	(Millibars)	(Inches)	Surge (Feet)	Damage
1	74–95	>979	>28.91	4 to 5	Minimal
2	96–110	965–979	28.50–28.91	6 to 8	Moderate
3	111–130	945–964	27.91–28.47	9 to 12	Extensive
4	131–155	920–944	27.17–27.88	13 to 18	Extreme
5	>155	<920	<27.17	>18	Catastrophic

appearing. In fact, in the North Atlantic Basin, 15 storms have formed in May from 1886 to 2006 (Fig. 4-3), and there have been others from January through April. The season also does not suddenly come to a halt on the last day in November; nine storms have formed in December over the same period (Fig. 4-3). Frequencies (and intensities) of storms increase from May through September then begin to taper off. The peak in September, near September 10, generally corresponds to the time in the summer season with the warmest SSTs. The heart of the season ranges from mid-August and extends through the first week of October. It is during this period that one can expect the highest frequencies of storms. Most of the famous Atlantic hurricanes occurred during this period, including the Galveston Hurricane of 1900, Camille, Hugo, Andrew, and Katrina.

During development in the North Atlantic Ocean, these storms tend to move westward at tropical latitudes steered by persistent trade winds that blow from east to west. Later, storms often drift northward, tracking around the Atlantic Subtropical High (also called the Bermuda High or the Azores High), affecting areas within the Caribbean Sea, Gulf of Mexico, and western Atlantic Ocean. Occasionally these storms will remain over the Gulf Stream and North Atlantic Drift to affect western Europe (Fig. 4-3) (e.g., Tropical Storm Andrew in 1986 and Hurricane Bob in 1991). Figure 4-3 depicts all the tropical storm and hurricane tracks from 1886 to 2006 for each month from May through December. The classic parabolic curved path is clearly evident here as storms drift from east to west in the trade wind belt, then drift poleward until steered by the westerlies, where they travel from west back to east at higher latitudes. In May and June, most Atlantic Basin storms form either in the Caribbean Sea or Gulf of Mexico, as these shallow waters tend to warm more quickly than the deep tropical Atlantic Ocean. In July, storms begin forming farther out in the Atlantic Ocean, and the classic parabolic curve around the Atlantic Subtropical High becomes more apparent. These characteristics are even more apparent in August and September, when storms can be tracked all the way to western Africa. Storms that can be tracked almost completely across the Atlantic are frequently called Cape Verde Hurricanes because they form near the Cape Verde Islands. As SSTs cool and the Atlantic Subtropical High weakens in October, the tracking pattern becomes more erratic, with more storms remaining in the middle of the Atlantic Ocean. By November and December, the storm tracks become very irregular, with little predictability in steering currents.

DEADLIEST STORMS

Many factors must be considered when trying to understand death statistics from hurricanes. Three of the most obvious questions that arise are (1) when in history did the storm occur, (2) which country did it occur in, and (3) how vulnerable is that specific portion of the coastline? Within the United States, the vast majority of the 50 deadliest storms since 1851 have occurred in the 1880s and early 1900s (Table 4-2). Note that aircraft reconnaissance of hurricanes began in 1944 and geostationary satellite coverage began in 1966. Before these observational capabilities were implemented, forecasting techniques were rather crude, and surveillance of the storms and warnings were minimal.

Internationally, one of the most vulnerable locations is Bangladesh. Here, many millions of people live on a low-lying river delta, which is easily flooded by storm surge. There have been several instances where these surges have killed more than a 100,000 people in a single event. A particular tropical cyclone event in 1970 is estimated to have killed between 300,000 and 500,000 people. Several other events have occurred in Bangladesh and in India, with staggering losses of life, losses that dwarf all casualty data from the North Atlantic Basin.

Perhaps the deadliest hurricane in the Atlantic Basin occurred in 1780. It is believed that the Great Hurricane of 1780 killed more than 22,000 people, mostly across the Lesser Antilles, including Barbados, Martinique, St. Lucia, and St. Eustatius. One reason for the great loss of life is the total lack of hurricane forecasting at the time, in addition to housing structures ill-equipped to handle hurricane-force winds.

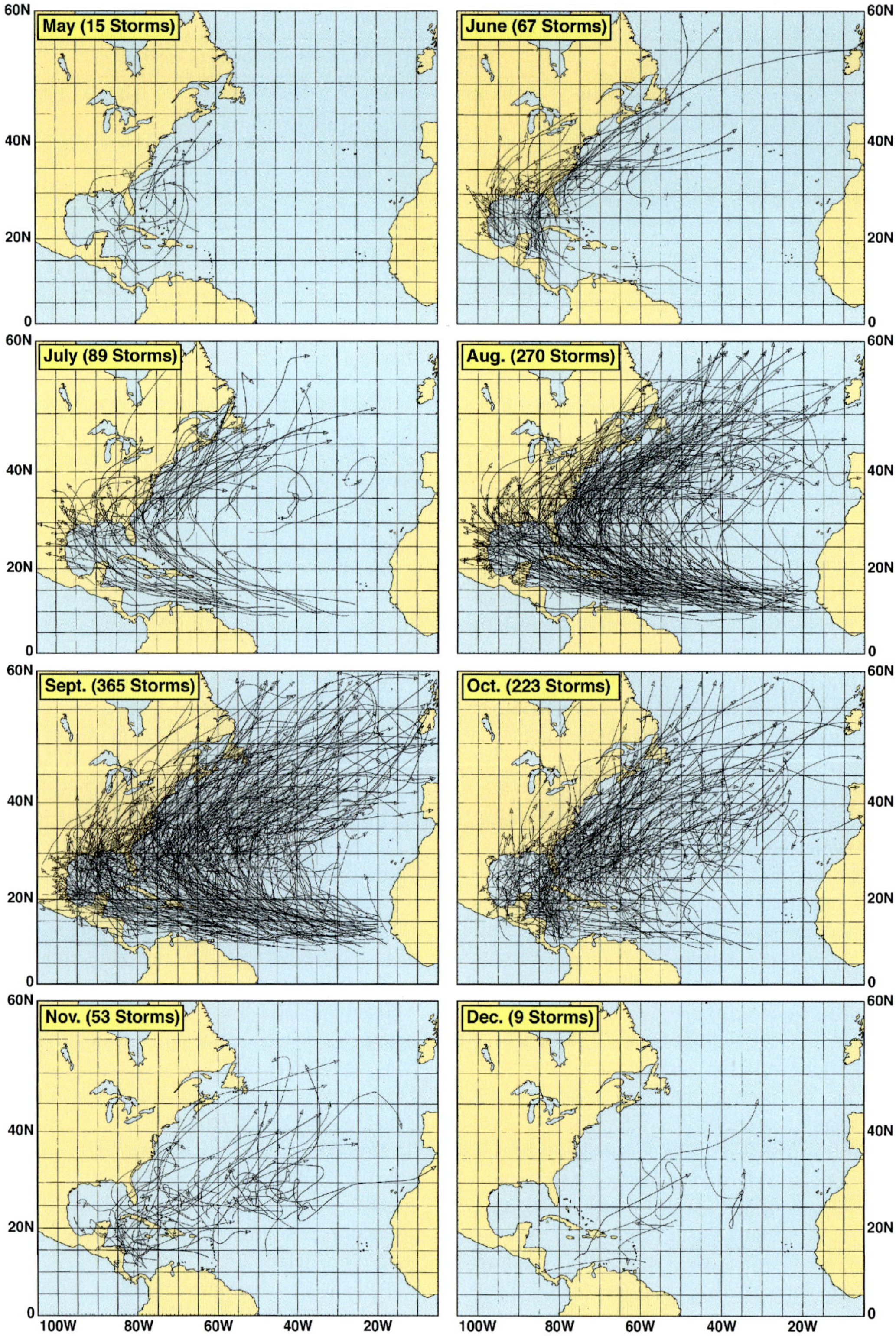

FIGURE 4-3. Tropical storm and hurricane tracks of the North Atlantic Basin from 1886 through 2006 by month, May through December. Adapted from Neumann (1993).

TABLE 4-2. Top 50 deadliest tropical storms and hurricanes to hit the United States.

RANK	HURRICANE	YEAR	CATEGORY	DEATHS
1	TX (Galveston)	1900	4	8000[a]
2	FL (SE/Lake Okeechobee)	1928	4	2500[b]
3	*KATRINA* (SE LA/MS)	2005	3	1500
4	LA (Cheniere Caminanda)	1893	4	1100–1400[c]
5	SC/GA (Sea Islands)	1893	3	1000–2000[d]
6	GA/SC	1881	2	700
7	*AUDREY* (SW LA/N TX)	1957	4	416[h]
8	FL (Keys)	1935	5	408
9	LA (Last Island)	1856	4	400[e]
10	FL (Miami)/MS/AL/ Pensacola	1926	4	372
11	LA (Grand Isle)	1909	3	350
12	FL (Keys)/S TX	1919	4	287[j]
13	LA (New Orleans)	1915	4	275[e]
13	TX (Galveston)	1915	4	275
15	New England	1938	3	256
15	*CAMILLE* (MS/SE LA/VA)	1969	5	256
17	*DIANE* (NE U.S.)	1955	1	184
18	GA, SC, NC	1898	4	179
19	TX	1875	3	176
20	SE FL	1906	3	164
21	TX (Indianola)	1886	4	150
22	MS/AL/Pensacola	1906	2	134
23	FL, GA, SC	1896	3	130
24	*AGNES* (FL/NE U.S.)	1972	1	122[f]
25	*HAZEL* (SC/NC)	1954	4	95
26	*BETSY* (SE FL/SE LA)	1965	3	75
27	Northeast U.S.	1944	3	64[g]
28	*CAROL* (NE U.S.)	1954	3	60
29	*FLOYD* (Mid Atlantic & NE U.S.)	1999	2	56
30	NC	1883	2	53
31	SE FL/SE LA/MS	1947	4	51
32	NC, SC	1899	3	50[h,i]
32	GA/SC/NC	1940	2	50
32	*DONNA* (FL/Eastern U.S.)	1960	4	50
35	LA	1860	2	47[h]
36	NC, VA	1879	3	46[h,j]
36	*CARLA* (N & Central TX)	1961	4	46
38	TX (Velasco)	1909	3	41
38	*ALLISON* (SE TX)	2001	TS[k]	41
40	Mid-Atlantic	1889	none[l]	40[h,j]
40	TX (Freeport)	1932	4	40
40	S TX	1933	3	40
43	*HILDA* (LA)	1964	3	38
44	SW LA	1918	3	34
45	SW FL	1910	3	30
45	*ALBERTO* (NW FL, GA, AL)	1994	TS[k]	30
47	SC, FL	1893	3	28[m]
48	New England	1878	2	27[h,n]
48	Texas	1886	2	27[h]
50	*FRAN* (NC)	1996	3	26

Notes: [a]Could be as high as 12,000
[b]Could be as high as 3000
[c]Total including offshore losses near 2000
[d]August
[e]Total including offshore losses is 600
[f]No more than
[g]Total including offshore losses is 390
[h]At least
[i]Puerto Rico 1899 and NC, SC 1899 are the same storm
[j]Could include some offshore losses
[k]Only of Tropical Storm intensity
[l]Remained offshore
[m]Mid-October
[n]Four deaths at shoreline or just offshore
[o]Possibly a total from two hurricanes
Adapted from Blake *et al.* (2007).

The second most deadly storm in the Atlantic Basin was Hurricane Mitch in October/November 1998. Because of the storm's slow forward movement, Mitch produced heavy rainfall over Honduras and Nicaragua. Flooding, combined with numerous landslides in this region, led to 11,000 deaths with another 8000 missing. Although this storm occurred in modern times, there was some uncertainty in the forecasting before the hurricane made landfall, complicating evacuations across Central America. Furthermore, the rugged terrain, used intensely in some places for agriculture, was highly susceptible to mudslides.

The third deadliest hurricane in the North Atlantic Basin, and the overall deadliest within the United States, was the Galveston Hurricane of 1900 (Table 4-2). This storm made landfall at Galveston on September 8, 1900. The U.S. Weather Bureau was aware that this storm had made landfall over Cuba, but headquarters in Washington, D.C., was convinced the storm would curve and would affect Florida and the Mid-Atlantic Region. They were wrong. As a result, those on Galveston Island were caught by surprise as the 15-foot storm surge overwhelmed the barrier island, which had a high point of only 8.7 feet above mean sea level in 1900. Between 6000 to 12,000 people perished in this event. As a result, residents have since built a 17-foot seawall to protect the island, and fill from the Gulf of Mexico was brought in to elevate most of the city when it was rebuilt.

More recent storms in Table 4-2 do not have poor surveillance as an excuse for the loss of life. For example, there was nearly a 48-hour warning in southeastern Louisiana before Hurricane Katrina made landfall. Remarkably, Louisiana officials were able to evacuate more than a million people from metropolitan New Orleans before landfall. However, tens of thousands of people remained behind; many of them were either elderly or from lower-income families who had no means of transportation. That, combined with the unique setting of metropolitan New Orleans, much of which lies below mean sea level and is protected by a levee system, led to numerous flooding deaths when the levees failed. A disproportionate number of these 1500 deaths were elderly people, who were unable to endure the miserable conditions in the aftermath of the storm. In addition to the great loss of life, this storm was by far the most costly in the United States with more than \$81 billion in damages in both Louisiana and Mississippi.

Another notable storm in the top 50 deadliest is Tropical Storm Allison in 2001. This storm caused 41 deaths, mostly in the Houston area. The storm made landfall near Houston on June 4, with little warning from the National Weather Service. The storm was not all that strong, but it's slow and winding path over southeastern Texas allowed the storm to produce more than 40 inches of rainfall over parts of Houston. The storm also caused flooding in Louisiana and later in Pennsylvania. Homes and businesses were inundated, thereby causing more than \$6 billion in damages. Allison is the only storm to have had its name retired by the World Meteorological Organization (WMO) after having attained only tropical storm status.

SPATIOTEMPORAL PATTERNS OF LAND-FALLING STORMS IN THE UNITED STATES

Figure 4-4 provides an assessment of tropical storm and hurricane strikes at 45 locations along the Gulf and Atlantic Coasts by merging time, geography, and storm intensity. Based on these time series data from 1901 through 2006, three broad geographical areas are highly active: (1) south Florida including both the Gulf of Mexico and Atlantic Coasts; (2) North Carolina, especially the Outer Banks; and (3) the coast of the northern Gulf of Mexico from Galveston, Texas, eastward to the Florida Panhandle. Two coastal sections between these hyperactive coasts have noticeably fewer strikes. The first is from Apalachicola, Florida, southeastward to south of Cedar Key, Florida, on the Gulf Coast, known as the Big Bend Region of Florida. The second is on the Atlantic Coast from northern Florida northeastward to include the Georgia coast. Interestingly, the Georgia coastal region was hit by a storm in 1881 that caused 700 fatalities and was hit by another in 1893 that caused between 1000 and 2000 fatalities, representing the fifth and sixth most deadly storms in the United States since 1851 (Table 4-2). From Virginia through New England, tropical cyclone strikes become increasingly uncommon, except for a modest increase along the more-exposed coasts of eastern Long Island and southern New England. The figure shows no major (Category 3 and above) hurricane strikes north of Cape Hatteras. Although hurricane frequencies decrease northward from Cape Hatteras, there is at the same time an increasing frequency of geographically large destructive winter and spring storms, often called nor'easters.

Figure 4-4 shows that the most dramatic interdecadal variability has occurred in South Florida on both the east and west coasts. There were frequent hurricane strikes for the 25 years between the mid-1920s and 1950, and several hurricanes again during the mid-1960s. However, with the exception of Hurricane Andrew in 1992, there are almost no hurricane strikes in this region and only a few tropical storm strikes since the 1960s—that is, until 2004 and 2005. In North Carolina, two temporal clusters of strike events are evident: in the 1950s and again in the 1990s. The time series also suggests that there are clusters of strikes in one region that may be associated with infrequent strikes in adjacent regions. One example is the clusters of strikes in southern Florida from 1926 to 1950 and 1964 to 1966, while the Outer Banks of North Carolina was inactive. During the 1950s, when the Outer Banks of North Carolina was very

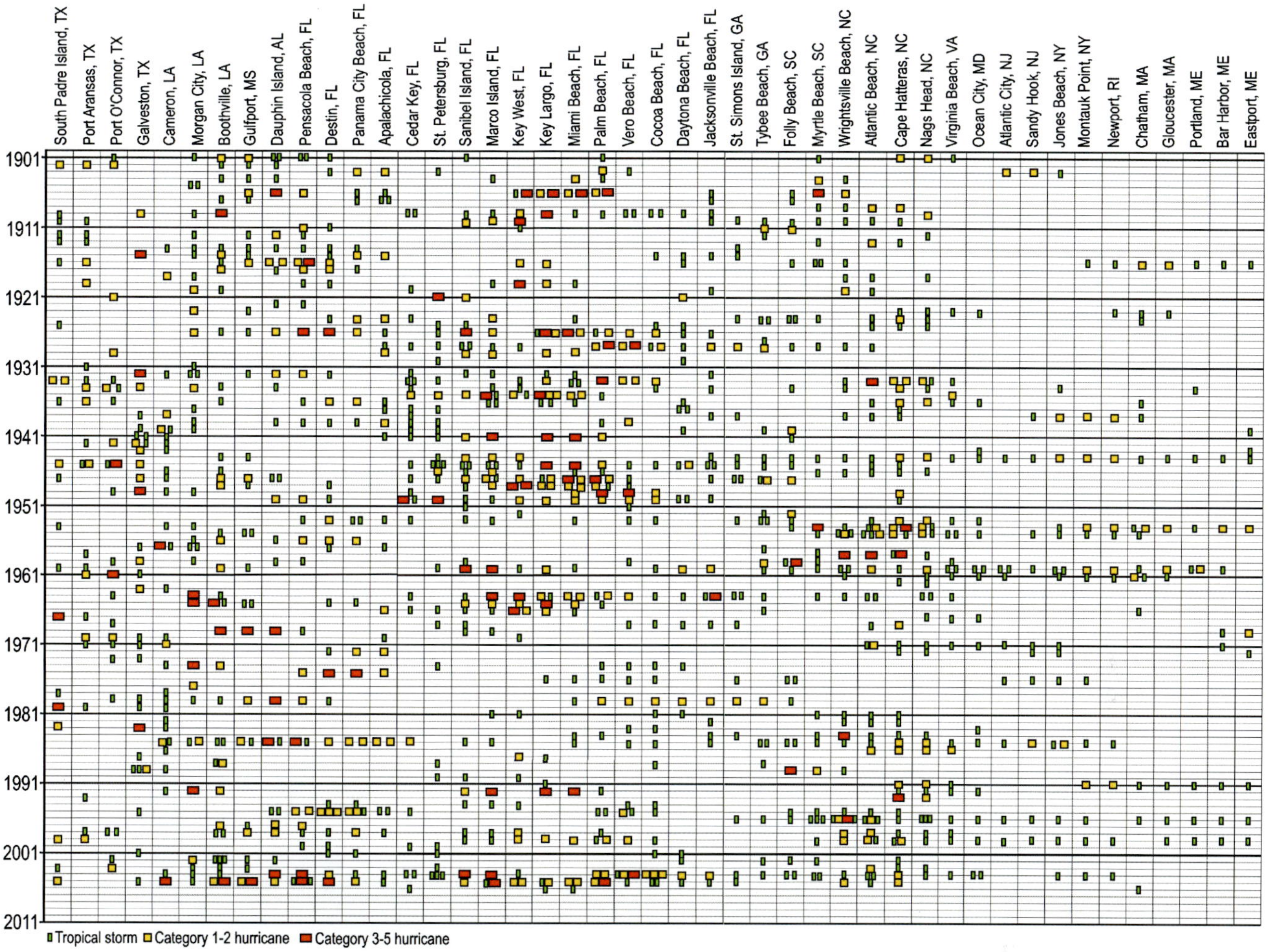

FIGURE 4-4. Spatiotemporal patterns of tropical storm, hurricane, and severe hurricane strikes at 45 locations from Texas to Maine. Adapted from Keim *et al.* (2007).

active, South Florida was relatively quiet. Another example is shown in Figure 4-5, where severe hurricane strikes (Category 3 to 5) in the 1950s were concentrated along the East Coast well north of Florida, whereas in the 1960s, the Central Gulf Coast was the primary target.

The total dataset shows a temporal pattern of active landfalling storms in the United States beginning in the mid-1920s and extending into the mid-1960s. A relatively inactive period for strikes in the United States is evident from the mid-1960s through the mid-1990s, followed again by the extremely active seasons from 1995 to 2006. These patterns are consistent with documented fluctuations in the Atlantic Multidecadal Oscillation (AMO). The AMO serves as an index of sea surface temperatures in the North Atlantic Ocean, where a positive (negative) index indicates warmer (colder) than normal sea-surface temperatures. Although the observed hurricane patterns are consistent with the AMO (i.e., more hurricanes occurred in the positive phase than in the negative phase), this approach to understanding storm strikes does not take into consideration the longevity of the storms or their intensities before landfall. It is these two metrics that are often brought into the debate regarding the impacts of recent climate change on the hurricane climatology.

STORM RETURN PERIODS

Return periods for tropical storms, hurricanes, and severe hurricanes provide a useful baseline for purposes of planning in the coastal zone and adjacent continental shelf areas. For that reason, return periods were calculated for each of the 45 locations along the Gulf of Mexico and East Coast studied in Figure 4-4 (Fig. 4-6). The average return periods depicted here are derived by dividing the 105-year period (1901 through 2005) of record by the total number of strikes at the respective locations. Return periods for all tropical storms and hurricanes are represented in the inner tier, the

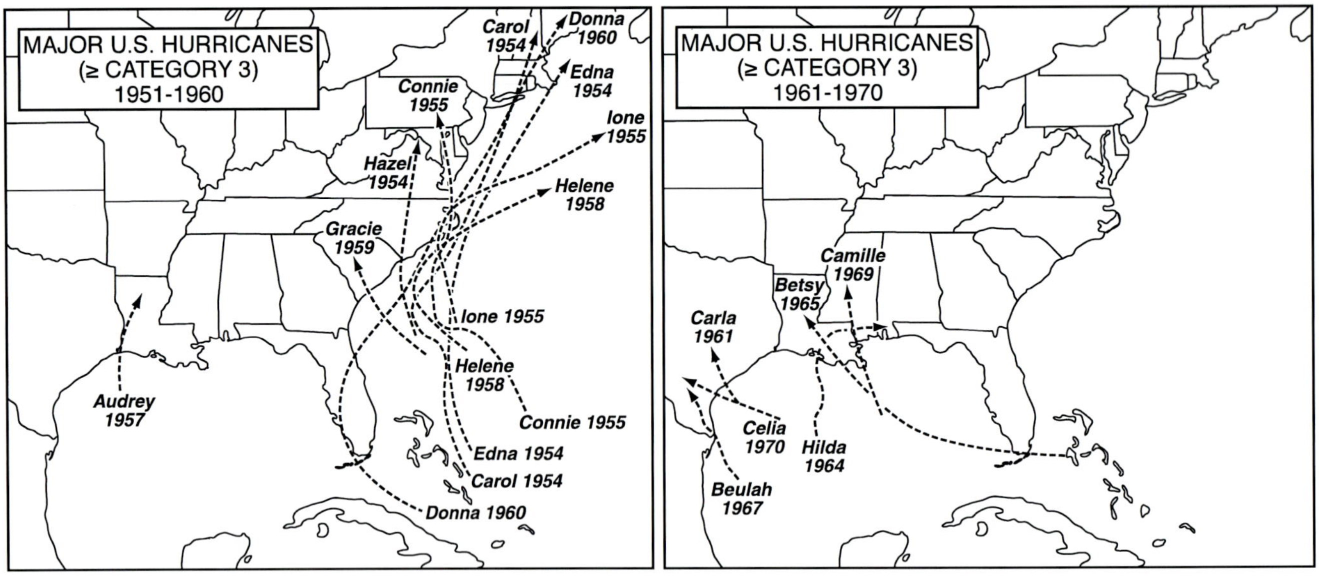

FIGURE 4-5. Comparison of the 1950s with the 1960s in terms of severe hurricane (Categories 3 through 5) strikes in the United States. From Blake *et al.* (2007).

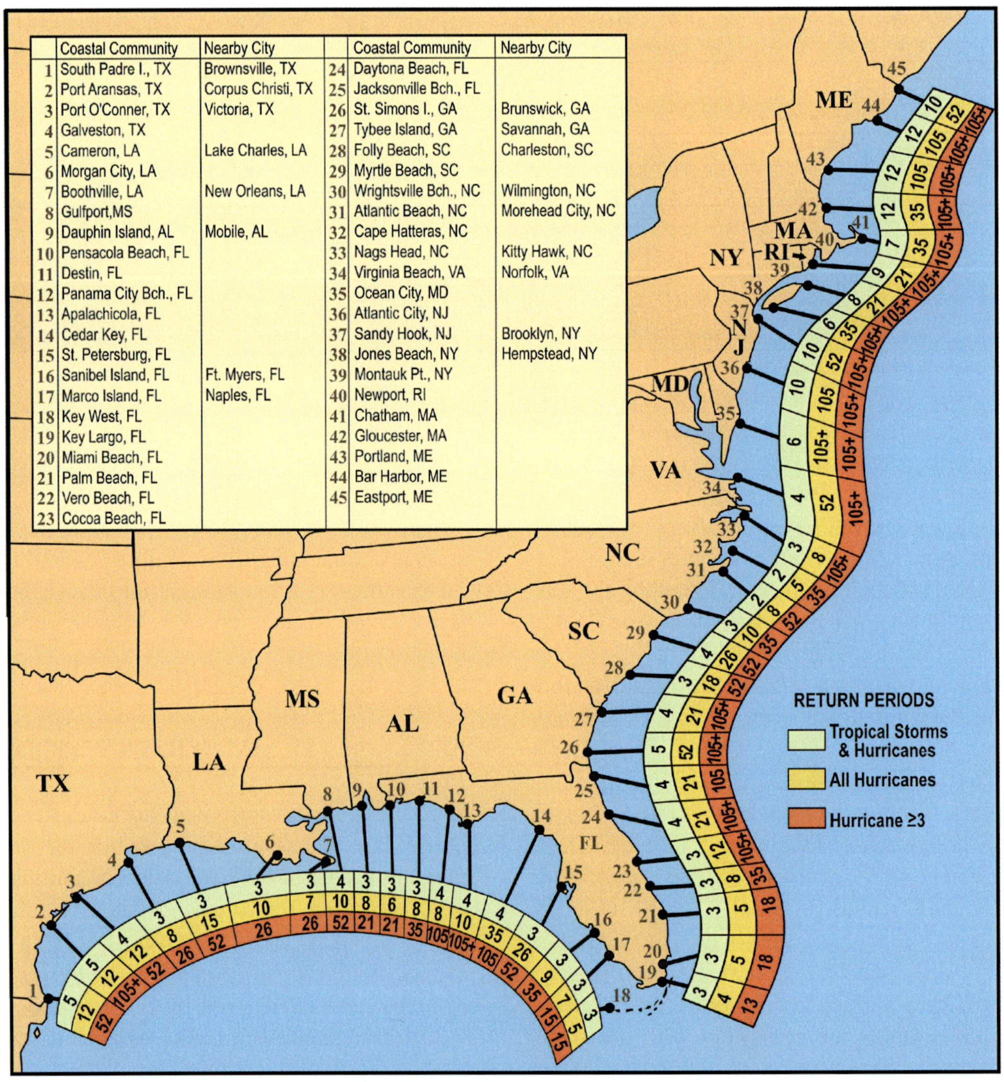

FIGURE 4-6. Return periods of tropical storms, hurricanes, and severe hurricanes (in numbers of years) at 45 locations from Texas to Maine. From Keim *et al.* (2007).

middle tier for all hurricanes; the outer tier only for major hurricanes—Categories 3, 4, and 5. Tropical storm return periods (measured in numbers of years) are short, hence they occur with greater frequency than hurricanes and severe hurricanes. They occur once every 2 years, on average, along the Outer Banks of North Carolina, and once every 3 years along the north-central Gulf Coast as well as in southern Florida. Tropical storm return periods are 10 years or greater in southern Georgia, New Jersey, and northern New England.

For all hurricanes, return periods are shortest, about 5 years, in southern Florida and the Outer Banks of North Carolina, and 10 years or less at most locations along the northern Gulf Coast. Less frequent hurricane strikes, and much longer return periods, are evident in the Big Bend Region of Florida (35 years) and at St. Simons Island on the coast of Georgia (52 years). Both coastal regions are often sheltered from direct storm strikes by the configurations of adjacent coastlines. North of Nags Head, North Carolina, return periods for all hurricanes range between 21 and 35 years at more exposed locations. Return periods extend to 52 or more years at more sheltered portions of this coastline.

For major hurricanes only, return periods are shortest (more frequent hurricanes) in southern Florida at 13 to18 years on average and 21 to 26 years from Galveston eastward to Pensacola Beach; exceptions are at Cameron and Gulfport, locations protected by the extended delta of the Mississippi River. Major hurricane return periods are longer, 35 years, on the Outer Banks in North Carolina.

THE 2005 HURRICANE SEASON

The 2005 Atlantic hurricane season was one for the record books, having set many records including the greatest number of named tropical storms and hurricanes at 27, the most hurricanes at 15, the most Category 5 Hurricanes at 3, and most severe hurricanes (Categories 3 to 5) making landfall in the United States at 4. Each of the Category 5 hurricanes traversed the Gulf of Mexico and passed over the Loop Current. The Loop Current is a warm-water current that bulges northward into the Gulf, coming from the Caribbean Sea between the Yucatan Peninsula and Cuba. It provides enormous energy to fuel rapid growth and intensity in tropical storms and hurricanes.

The 2005 season began early and remained active over the course of the season, breaking other records as well. Figure 4-7 depicts the dates of occurrence of storms in sequence over the past 155 hurricane seasons to place the 2005 season into perspective, relative to early, median, and late arrival dates of storms. Here, dates of the *n*th storm (first, second, third, etc.) of the 2005 season are compared with the *n*th storm of all other seasons in the North Atlantic Basin since 1851. Box and whisker plots of tropical cyclone date distributions by each *n*th storm in sequence are shown, where the box depicts the median, 25th, and 75th percentiles of the distribution and the whiskers of the 5th and 95th percentiles. Five seasons (1887, 1933, 1936, 1995, and 2005) that were particularly active with early dates of occurrence are highlighted. The year 2005 stands out as a season that set many records for earliest occurrence dates for the *n*th named storms. The 4th named storm of 2005, Dennis, formed on July 5 becoming the earliest occurring 4th storm on record, nearly 2 months earlier than the median date for all 4th named storms. This event began a run of earliest date records for the 4th through 11th named storms, with the 11th storm being Hurricane Katrina. (Note the impact of Hurricane Katrina's aftermath on the childhood home of one of the authors [BDK] in Fig. 4-8.) The 12th storm formed only 5 days later, as Katrina made landfall in Louisiana, but was later than the 12th storms of 1995 and 1933. The 13th storm of 2005 formed and again was a record date. The 2005 season set every earliest record thereafter, with the exception of the 18th named storm, which missed by one day to 1933. Since no other season experienced more than 21 named storms, the 2005 occurrence for 22nd through 28th storms set records automatically. However one chooses to analyze the 2005 season in this basin, it was remarkable in almost every way. However, it begs the question as to whether this season serves as a harbinger for the future.

WHAT LIES AHEAD?

The empirical record over the past century or more in the North Atlantic Basin, including the Gulf of Mexico and Caribbean Sea, shows multidecadal scale variability but no discernable trends that can be linked conclusively to natural or anthropogenic factors. This variability in hurricane activity includes a period of relatively low frequencies from the early 1900s to the 1920s, followed by an active period in the late 1920s to the 1960s, which was followed by a period of lower activity in the 1970s through the early 1990s. These multidecadal patterns in hurricane activity appear attributable to shifts in SSTs, where above normal SSTs in the North Atlantic Basin combined with below normal SSTs in the southern Atlantic Ocean (south of the equator) are most conducive to intense hurricane formation. There are signs that hurricanes may now be increasing in frequency again, thereby returning hurricane frequencies to what was experienced during the four decades before 1965. For example, since 1995, every year, with the exception of the El Niño years of 1997 and 2006, had higher than average numbers of named storms in the North Atlantic Ocean, including 19 named storms in 1995, 16 in 2003, and 15 in both 2001 and 2004, not to mention the record breaking season of 2005.

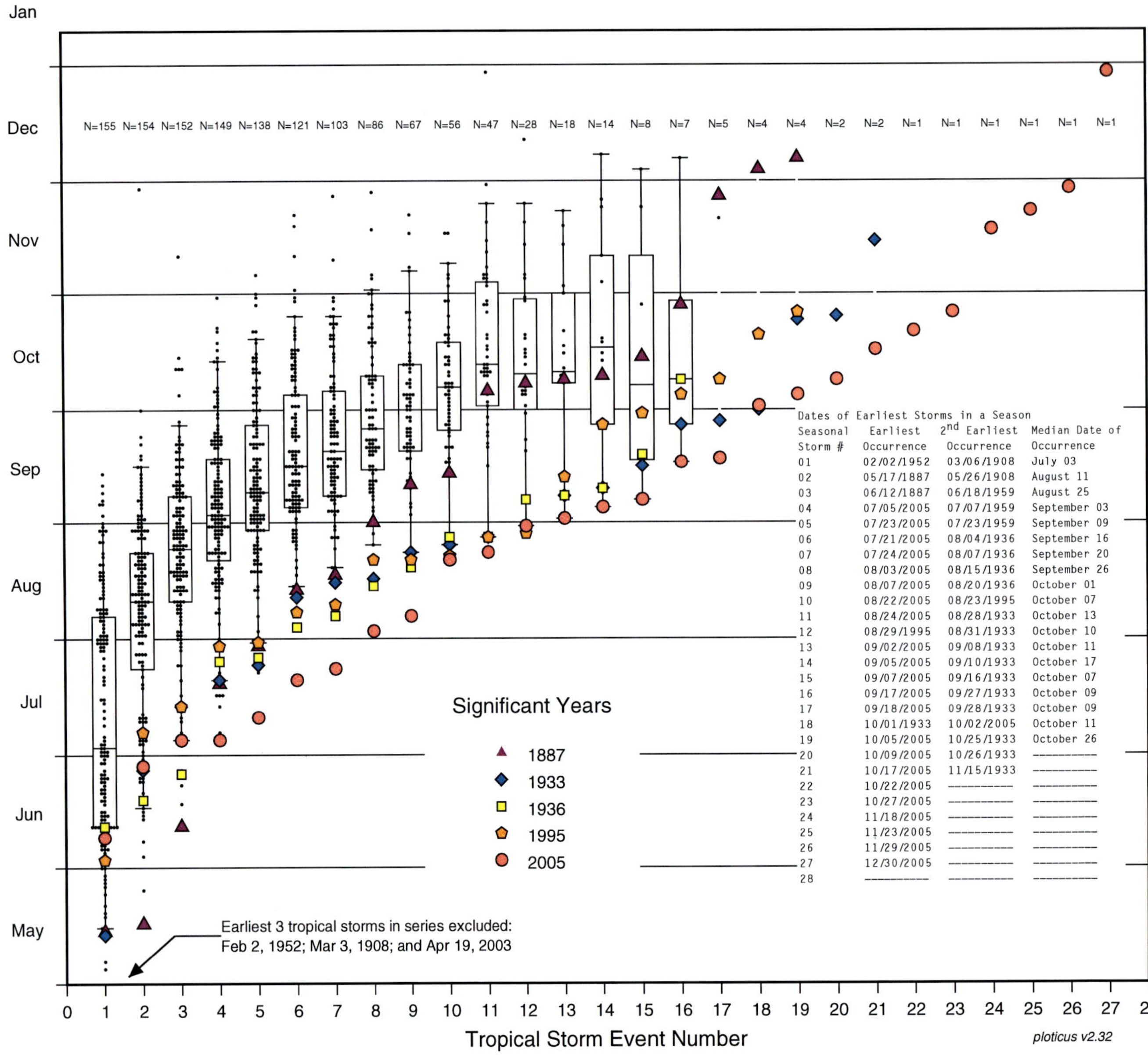

FIGURE 4-7. Dates of occurrence of all tropical storms from 1851 through 2005 ordered in sequence for each hurricane season. From Keim and Robbins (2006).

Roughly 10 named storms would be considered average. The 1997 season had only seven named storms and coincided with one of the most intense El Niño events on record. El Niños produce storm-inhibiting westerly wind shear aloft. The 2006 season was also relatively quiet with nine named storms and was also an El Niño year. Despite these two quiet seasons associated with El Niño, there are indications that the hurricane climatology of the North Atlantic Basin may remain hyperactive for decades to come.

Looking beyond the next few decades, numerous approaches have been tried to predict potential changes in hurricane frequencies and intensities under global warming scenarios using general circulation models (GCMs). Results are mixed, but there is a suggestion that overall frequencies of tropical storms could decrease somewhat, whereas the formation of intense hurricanes could increase. Regardless of whether global warming is occurring and is affecting the hurricane climatology, we still face many issues regarding these powerful storms, including the following:

- Ensuring residential and commercial structures in the coastal zone
- Maintaining shipping lanes to avoid disruptions in commerce

FIGURE 4-8. Chalmette, Louisiana, on August 29, 2005. One of the authors (BDK) grew up in the two-story white brick home. Photo taken by Arnold Crabtree.

- Runoff from swine and cattle feedlots into rivers and estuaries from heavy rainfall
- Saltwater storm surges killing wetland vegetation and damaging ecosystems and agricultural fields
- Removal of recreational beach sand from famous beaches
- New inlets cut into barrier islands
- Overwash sands filling lagoons and sounds
- Outbreaks of disease in people and animals
- Destruction of roads, bridges, water and sewer, electricity, and other infrastructure
- Planning for short- and long-term evacuations of coastal residents
- Limiting development in coastal zones
- Improving zoning regulations in coastal zones

This list is not intended to be comprehensive but serves as a starting point to generate discussion on coping with the ravages of tropical storms and hurricanes.

References

Blake, E.S., Rappaport, E.N., Landsea, C.W. 2007. The deadliest, costliest, and most intense United States tropical cyclones from 1851 to 2006 (and other frequently requested hurricane facts). Miami, National Weather Service, National Hurricane Center.

Bengtsson, L., Botzet, M., Esch, M., 1996. Will greenhouse gas-induced warming over the next 50 years lead to higher frequency and greater intensity of hurricanes? Tellus 48A, 57–73.

Bove, M.C., Elsner, J.B., Landsea, C.W., Niu, X., O'Brien, J.J., 1998. Effect of El Niño on U.S. landfalling hurricanes revisited. Bull. Am. Meteorological Soc. 79, 2477–2482.

Curry, J.A., Webster, P.J., Holland, G.J., 2006. Mixing politics and science in testing the hypothesis that greenhouse warming is causing a global increase in hurricane intensity. Bull. Am. Meteorological Soc. 87, 1025–1037.

Elsner, J.B., Kara, B., 1999. Hurricanes of the North Atlantic: Climate and Society. Oxford, Oxford University Press.

Emanuel, K., 2005. Increasing destructiveness of tropical cyclones over the past 30 years. Nature 436, 686–688.

Gray, W.M., 1999. On the causes of multi-decadal climate change and prospects for increased Atlantic Basin hurricane activity in coming decades. 10th Symposium on Global Change Studies, Preprints, American Meteorology Society, pp. 183–186.

Henderson-Sellers, A., Zhang, H., Berz, G., Emanuel, K., Gray, W., Landsea, C. Holland, G., Lighthill, J., Shieh, S-L., Webster, P., McGuffie, K., 1998. Tropical cyclones and global climate change: A post-IPCC assessment. Bull. Am. Meteorological Soc. 79, 19–38.

Keim, B.D., Muller, R.A. Stone, G.W., 2004. Spatial and temporal variability if coastal storms in the North Atlantic Basin. Mar. Geol. 210, 7–15.

Keim, B.D., Muller, R.A. Stone, G.W., 2007. Spatiotemporal patterns and return periods of tropical storm and hurricane strikes from Texas to Maine. J. Climate 20, 3498–3509.

Keim, B.D., Robbins, K.D., 2006. Occurrence dates of North Atlantic tropical storms and hurricanes: 2005 in perspective. Geophysical Research Letters 33, L21706, doi:10.1029/2006GL027671.

Landsea, C.W., 2005. Hurricanes and global warming. Nature 438, E11–E13.

Landsea, C.W., Pielke, R.A., Jr., Mestas-Nunez, A.M., Knaff, J.A., 1999. Atlantic basin hurricanes: Indices of climate changes. Climatic Change 42, 89–129.

Muller, R.A., Stone, G. W., 2001. A climatology of tropical storm and hurricane strikes to enhance vulnerability prediction for the southeast US coast. J. Coastal Res. 17, 949–956.

Neumann, C.J., Jarvinen, B.R., McAdie, C.J., Elms, J.D., 1993. Tropical Cyclones of the North Atlantic Ocean, 1871–1992. Historical Climatology Series 6–2. Asheville, NC, National Climatic Data Center.

Pielke, R.A., Jr., Landsea, C.N., 1999. La Niña, El Niño, and Atlantic hurricane damages in the United States. Bull. Am. Meteorology Soc. 80, 2027–2033.

C H A P T E R

5

Oceans and Human Health

Human Dimensions

DAVID LETSON

INTRODUCTION

In 2005, the seafood community in Franklin County (Florida) endured dislocations, personal hardships, and lapses of trust in resource management officials. The Franklin County oyster fishery lost an estimated $6 million in the latter half of the year after regulators closed shellfish harvesting in July and August in response to bacterial problems following Hurricane Dennis and from September to November in response to harmful algal blooms following Hurricane Katrina (Ritchie, 2005). Minimizing the human impacts of marine resource management requires an understanding of its often unintended sociocultural consequences. Social science research is needed to inform the decisions of regulating agencies and coastal communities.

Human health and oceanic health are fundamentally linked. Because the oceans are enormous, are increasingly proximate to humans, and are shared by nations, this close relationship matters to us all (Ballard *et al.,* 2004; Curran *et al.,* 2002; Fleming and Laws, 2006; Harvell *et al.,* 1999; 2002; Sandifer *et al.,* 2004). Human activities, such as pollutant discharges and fishing, influence marine ecosystem health. Conversely, human health depends on the functions, products, and services of marine ecosystems.

In 1967, Marshall McLuhan introduced the phrase "the global village" to describe the ever speedier communications that make our world a smaller place. Naturally, where information has flowed, business has followed, and today globalization has come to be associated with growing economic integration and liberalization, trade deregulation, convergence of macroeconomic policies, modification of the role and concept of nation state, proliferation of supranational agreements and regulatory bodies, and globalization of information systems. As economic forces stimulate our demands on oceans, for both resource use and waste disposal, the approaching closure of what once seemed a vast and open frontier has important implications for humanity.

The risks that oceans pose to human health are compounded by increasing coastal populations. The world population is estimated to increase from about 6 billion currently to 8.3 billion by 2025, with 90% of this growth occurring in subtropical and tropical countries. More than 2 billion people worldwide rely on seafood as a major source of protein in their diet, and seafood consumption continues to increase worldwide (Food and Agriculture Organization [FAO], 2004). In the United States, coastal watershed counties cover less than 25% of the land area but contain more than 52% of its population, according to NOAA's "Report on Coastal Population Trends" (Crossett *et al.,* 2004). Population density in coastal counties is about 300 persons per square mile compared to 98 for the entire United States. About 3600 people a day move to coastal counties in the United States, a trend that suggests our coastal population will reach 165 million in 2015. With another 180 million annual visitors, the pressure on our oceans is growing. Population growth and tourism benefit coastal communities in a number of ways, including new jobs and an enhanced quality of life. However, growth also raises public health concerns because it puts more people and property at risk from coastal hazards, reduces and fragments wildlife habitats, alters sediment and water flows, and contributes to coastal water pollution.

One challenge in the study of oceans and human health is that it is inherently a multidisciplinary endeavor, drawing not only from oceanography, marine biology, epidemiology, meteorology, and statistics, but also from the social sciences. The study of oceans and human health is in part a social and economic problem, because it commits scarce resources to improve societal well-being. To do so wisely requires an active research community that recognizes links between human and ocean health and facilitates cooperative studies in physical, life, and social sciences.

In the past, development did not appear to threaten many marine resources. With continued coastal population growth and expanding resource usage, however, the consequences of development are more evident. Balancing the economic uses of the marine resources with our desire to preserve their natural character will require that we think carefully about our priorities. Social science research can help in this regard.

THE WEALTH OF OCEANS

Marine ecosystems are forms of natural capital that provide value in stocks and flows of goods and services, many of which relate to human health. Goods are used directly as seafood, pharmaceuticals, oils, and additives. From recreational opportunities (such as boating and wildlife observation) to the harvesting of fish and other seafood for human consumption, marine habitats provide us with many direct benefits.

Many important medical treatments and biotechnologies are based on chemicals derived from marine organisms, such as sponges, soft corals, mollusks, bacteria, and algae. Marine bioproducts with anti-inflammatory and cancer-fighting properties, for example, are among the medical advances found at sea. Other natural compounds yet to be discovered might serve not only as pharmaceuticals but also as nutritional supplements, medical diagnostics, cosmetics, agricultural chemicals, enzymes and chemical probes for disease research, and many other applications.

Marine ecosystems also provide many services, some of which are critical to human health, such as ecosystem resilience, genetic diversity, and aesthetic appreciation. Indirect benefits (e.g., biological support and water and air purification) and passive-use benefits (e.g., the satisfaction that we get from knowing that the environment has been preserved for future generations) influence our decisions to exploit or conserve marine resources.

One notable service that oceans provide is waste assimilation. From the industrial and municipal sewage that flows into the seas to the carbon dioxide emitted by motor vehicles, the oceans serve as the ultimate repository of many waste products. However, the oceans have only a limited capacity to absorb waste and convert it into biologically benign material. If the oceans' assimilative capacity exceeds the volume of waste received, the environment remains unharmed. Unfortunately, however, the volume of waste often exceeds the assimilative capacity, and damage occurs. Environmental damage reduces the future productive capacity of marine ecosystems and adversely affects indirect use and amenity values, including human health. Waste resulting in the decline of the environment has an adverse impact on individual and societal utility.

MANAGING MARINE RESOURCES REQUIRES ECONOMIC CHOICES

Not all benefits of marine resources are as obvious as those associated with tourism or shipping. Some resource uses and values occur at a distance from the resource itself. For example, sea grass and wetland systems provide critical habitat for marine and estuarine fish, shellfish, and mammals, including many highly valued recreational and commercial species. Also, dune systems buffer inland areas from the effects of strong storms. Efficient marine resource management considers offsite ecosystem and flood control benefits, despite their diffuse nature.

Because many marine resource uses are not traded in markets, their values cannot be measured in traditional ways (Letson and Milon, 2006). For example, a proposed housing development that could damage the ecological integrity of wetlands might indirectly hurt recreational and commercial fishing as well. The values of ecosystem "services" are frequently intangible but may also be important. Such values were often unaccounted in the past because economists could not estimate them; consequently, in comparing costs and benefits of public works or other coastal development (e.g., housing, new industry, recreational facilities), these values were often ignored. Since the 1970s, however, economists have developed a variety of techniques for estimating values of nonmarket goods and services.

Among the social sciences, economics in particular contributes an ability to quantify changes in social welfare resulting from changes in the condition or availability of resources. Oceans and human health research is hardly free, and attaching a monetary value to these publicly available goods helps public officials determine if further investments are worthwhile. Improving our knowledge of economic values informs policy making by identifying, or at least approximating, what may be the best choice between alternative investment options. Economists use market and nonmarket information to assess options and suggest priorities for decision makers.

Why do we need to know economic values? The reason is that the availability of marine resources is scarce relative to the demands we place on them. Because marine resources are scarce, managing them is partly an economic problem. Economics can inform decision makers about the values of alternative, and in some cases, the competing uses of our marine resources.

If marine resources like wetlands and fisheries were available for everyone in any quantity, no economic problem would exist. We could all have what we want, without having to choose. But resources such as fish stocks are not unlimited. While resource management agencies may develop harvesting and creel regulations, other factors, for instance, the loss of wetlands, industrial discharges, and runoff from new development can also affect fishery productivity. In public policy (as in our daily lives), we frequently must make choices. Because we must choose, we should consider which resource uses are most highly valued. The key notion here is what economists call "opportunity cost," the idea that the resource use choices we make can restrict other opportunities.

In economics, the term *value* means the price individuals are willing to pay to obtain goods and services. Estimating the value of goods and services not traded in markets (for example, recreational fishers' willingness to pay for more abundant fish stocks) can be indirect and sometimes controversial. Because some economic values of marine resources have been difficult to measure, policy makers have sometimes ignored these values in the past. However, we *need* to know the values of alternative uses of such resources because managing them is in part an economic problem.

ECONOMIC PROBLEM STATEMENT

"Thanking the captain, I went up to the shelves. Works of science, ethics, literature, in many languages, were in abundance; but I did not see a single book on economics, apparently a subject strictly proscribed on board."

—Professor Pierre Aronnax in *20,000 Leagues under the Sea*

Economics provides a set of tools for evaluating tradeoffs in how we might improve societal well-being. Available economic modeling approaches represent good ways with which to value marginal (i.e., small) improvements in environmental quality. Here we offer a stylized individual decision model to illustrate the central concepts of the economic theory of ocean resource use. The so-called rational choice model is prescriptive in that it describes how individuals or policy makers should act if they wish to maximize expected utility or value. Although the model has considerable analytical power, it may or may not accurately describe how individuals actually make decisions. The framework can be applied either to an individual or to society as a whole. At the individual level, the payoffs would take into account individual preferences for a host of ocean resource uses. At the societal level, the framework could be used to estimate the value to society of an improvement in environmental quality.

Economists assume that an individual's utility is a function of commodities that the individual consumes. The individual is assumed to be rational and a utility *maximizer*. The utility function allows for any type of preferences. For purposes of this description, utility is a function of consumption of commodities, X. The level of utility is also dependent on the level of environmental quality, EQ:

$$U = U(X, EQ)$$

The individual has a certain income, Y; faces a vector of commodity prices, which we assume constant, and we therefore suppress in our notation; and has a given level of environmental quality, EQ_1. The person will choose the level and composition of X that maximizes utility. The optimal level of utility that the individual can achieve given these constraints, U^*, can thus be written as a function of income and environmental quality:

$$U^* = V(Y, EQ_1)$$

The term V indicates that utility is now measured as a function of income and environmental quality using what economists call an *indirect utility function* rather than a *direct utility function*, U, which is a function of the consumption of commodities, X.

Given the option between different levels of environmental quality, EQ_1 and EQ_2, where we assume EQ_2 to be a higher quality level, we can define "willingness to pay (WTP)" as the maximum amount an individual is willing to pay to ensure that a welfare-increasing activity takes place, or the maximum the person is willing to pay to prevent a welfare-decreasing activity from being implemented:

$$U^* = V(Y, EQ_1) = V(Y - WTP, EQ_2)$$

In other words, it is the amount of money that can be taken away from income, Y, given the fixed and exogenous prices, while keeping the individual at the same level of utility, U^*, that she or he had before forecast quality was improved from EQ_1 to EQ_2.

A variety of estimation methods may be used to gauge empirical magnitudes for improved human health or environmental quality. In general, economics offers two classes of methods for the empirical estimation of values for public goods (such as hurricane forecasts). The first, "revealed preference" (RP) methods, generally relies on observations of actual behavior either directly in markets (e.g., buying a hamburger) or indirectly in decisions that reveal preferences (e.g., paying more for a house with a great view than for an identical house without the view implies the revealed value of the scenic view). The second empirical approach, "stated preference" (SP) methods, attempts to directly measure individuals' WTP for an amenity (such as improved health) without necessarily having to rely on a complete model of the individuals' utility function. SP methods are generally survey-based approaches that ask individuals to make choices in hypothetical situations that either directly or indirectly indicate the individual's values for an amenity (where *amenity* is broadly defined to include environmental attributes to preferences in terms of health outcomes).

Each empirical approach has its strengths and weaknesses. SP approaches can also allow the individual to state his or her benefits from the change without the researcher limiting these benefits by preconceived constraints on the individual values. The WTP approach, when combined with an SP method, thus attempts to capture all of the values for all potential impacts on the individual. SP valuation is necessary (as opposed to RP valuation) when there are no observable market values for a commodity or no behavioral trails that permit RP approaches to be used. SP methods can also be used for valuing hypothetical commodities such as improved environmental quality (amenities for which few market or RP data exist).

The reliability and validity of SP methods (like contingent valuation) depend on the extent to which they measure true values. In particular, critics have suggested that SP value estimates will not be comparable to RP value estimates that rely on information from observed market exchanges. In response to this critique, Carson *et al.* (1996) reviewed comparisons between SP and RP value results (primarily travel cost and hedonic prices) for valuation of comparable quasi-public goods. The authors concluded that, on average, the SP results are comparable to or slightly lower than the RP results for similar amenities. A number of books have also reviewed issues in the implementation of SP studies as progress has been made in developing SP value methods (see Bjornstad and Kahn, 1996; Cummings *et al.,* 1986; Kopp *et al.,* 1997; Mitchell and Carson, 1989).

MARKET FAILURES

The market system is a decentralized exchange mechanism that often enables society to allocate resources efficiently. However, markets can fail to allocate natural resources efficiently through the pricing mechanism if they are unable to accurately capture the full social costs of exploiting the marine resource. When markets do not exist or are otherwise unable to allocate marine resources efficiently, a market failure is said to occur.

Why do markets not function well or exist at all for many uses of marine resources? In part, market transactions never materialize because many marine resources are "common pool resources." That is, they are subject to "rivalry" in consumption, and are "nonexcludable" in provision. Both features are crucial to understanding the economic nature of user conflicts. Rivalry occurs when one person's consumption of a good diminishes others' ability to consume that good. Rivalry often leads to user conflicts. In nearshore waters, for example, conflicts might arise between jet skiers and commercial fishermen, or between recreational anglers and aquaculture operations. Nonexcludability refers to a situation where a resource owner cannot prevent anyone else from using the resource, as in cases where public officials have difficulty enforcing fishing regulations. Taken together, the rivalry and nonexcludability features explain why markets do not develop for many marine resources. Users will not pay for what they can use for free; and without a price to ration access, crowding and conflict will result. Because their use is often free and rival, many coastal resources suffer from overuse.

Another instance of market failure is in the provision of "public goods," which are both nonexcludable in provision and nonrival in consumption. Oceans offer an example, because they may not diminish in quality, even if many individuals enjoy their benefits. In addition, one consumer often cannot prevent another from enjoying the recreational and amenity values an ocean provides. Other examples of public goods include clean air and water. One characteristic of public goods is that they may be underprovided, as compared to a competitive market allocation, because individuals have an incentive to "free ride"—that is, to receive benefits without paying for them.

Many economic activities may impose spillover costs to individuals and to society, yet another type of market failure is known as "externalities." A negative externality occurs when the by-product of an economic activity imposes a cost on society not captured in the market. For instance, motor vehicle emissions (e.g., carbon dioxide, nitrogen oxide) contribute to the warming of the troposphere through the greenhouse effect. In turn, atmospheric warming contributes to rising sea levels. Rising seas, in turn, may lead to flooding. These costs are currently not captured through registration fees or the price of gasoline.

The scarcity of resources in relation to human demands implies choices, and thus trade-offs. In markets, we can make informed choices. Products are visible, have well known characteristics, and carry designated prices. In contrast, while they spawn a great deal of economic activity, uses of marine resources (such as commercial fisheries) usually are not themselves transacted in markets. Registration fees for boaters, for example, while not negligible, are intended to cover administrative costs and do not represent users' willingness to pay for boating access. Consequently, much less information exists about these resource uses. Posted prices are lacking that would reflect user values.

The lack of markets for many marine resources implies a lack of information for decision makers. We do not know as much as we would like about which uses of, say, fish stocks are most important to protect. Yet if we are to make informed choices, we must have some measures of the economic values we are trading off.

We usually think of the economy in terms of market economic values such as spending, sales, output, income, employment, and tax revenues generated. However, the economic values we observe in markets may well be conditioned by others not directly transacted in markets. Beach use, for instance, often is not allocated by markets, although beach recreation directly and indirectly may generate a great deal of market-based economic activity.

Many of the most pressing marine resource issues for human health involve market failures. Human mortality from natural hazards stems from systematic underprovision of public goods (such as stricter building codes and land-use zoning), whereas harmful algal blooms, seafood contamination, and global change all stem from externalities. Economists tend to agree that markets are generally most well suited for efficiently allocating society's resources. Markets can fail, however, imposing costs on society not captured in the market. To combat environmental externalities and other forms of market failure, government agencies use regulations and other mechanisms to correct these market imperfections and protect the ocean environment and human health. Finding an appropriate balance between the demands of the marketplace and the need to preserve natural resources enables our society both to expand the economy and to protect ocean and human health.

HURRICANES AND NATURAL HAZARDS

Natural hazards, including tropical cyclones or hurricanes, cause extensive loss of life in underdeveloped countries and economic loss in the United States (see Chapters 3 and 4). A tropical cyclone in 1971 hit Bangladesh with a 30-foot-high storm surge that inundated the low-lying, densely populated country, causing at least 250,000 deaths. In 1998, Hurricane Mitch idled for several days over the mountains of Honduras and Nicaragua, ending more than 10,000 lives, and it nearly destroyed the economies and social infrastructure of these countries. The 2004 Indian Ocean earthquake triggered a series of devastating tsunamis that spread throughout the Indian Ocean, killing large numbers of people and inundating coastal communities across South and Southeast Asia, including parts of Indonesia, Sri Lanka, India, and Thailand. The worldwide death toll from this event included a total of 229,866 people lost, with 186,983 dead and another 42,883 missing.

Natural disasters disproportionately harm poor countries because they have densely populated coasts, poor construction, sparse infrastructure, and inadequate health care. Poor land use leads to widespread environmental degradation (such as deforestation and wetlands destruction), which in turn worsens flooding and landslides. Emergency preparation and response are often inadequate, and hazard insurance is rarely available, further slowing recovery. Although total economic losses from natural disasters are increasingly concentrated in the rich world, losses as a percentage of incomes in poor countries can be many times greater. Damages from Hurricane Mitch, for example, were estimated at between \$5 and \$7 billion (almost the annual combined total economic activity of Honduras and Nicaragua); their economies have yet to recover (Pielke, 2006).

Property damages in the developed world from hurricanes have been rapidly increasing since the 1970s. The best indication of our increasing economic vulnerability is insured losses, as the United States and other nations do not consistently collect and compile property loss data, let alone estimates of the costs associated with social dislocation or the destruction of natural environments. Between 1967 and 1996, insurance payouts (which cover only a fraction of losses) rose steadily from \$1 billion between 1967 and 1971, to \$61 billion between 1992 and 1996, roughly doubling every 5 years. To date, insured losses from Katrina alone have exceeded \$40 billion. In the United States, where stricter building codes, improved forecasts, and early warning systems have helped save lives, deaths from natural hazards are expected to rise along with development and population along the nation's coasts. In addition, climate change may increase storms and sea-level rise, making the coasts throughout the world increasingly vulnerable.

The public goods needed to reduce disaster vulnerability for the most part already exist, such as warning systems, risk assessment techniques, better building codes and code enforcement, land-use standards, and emergency-preparedness plans. Hazards arise from the intersection of the natural and built worlds and human communities (Mileti, 1999). Mileti went on to argue that disaster losses are the predictable results of interactions among three major systems: the physical environment (e.g., hazardous events); the social and demographic characteristics of the communities that experience them; and buildings, roads, bridges, and other components of the constructed environment. Somehow, the crisis of growing disaster vulnerability only becomes news *after* disaster strikes. Yet we know that effective action is possible to reduce disaster losses even in the face of poverty and dense population.

Disparities in disaster vulnerability between rich and poor will continue to grow. The vast majority of population growth is occurring in the developing world. This growth, in turn, drives urbanization and coastal migration in these poorer countries. As a result, in the next two decades, the population of urban areas in the developing world will likely increase by two billion people. This population is being added to cities that are mostly located on coastal or flood plains (or in earthquake zones) and are unable to provide the quality of housing, services, infrastructure, and environmental protection that can help reduce vulnerability. Uncontrolled urban growth exacerbates hazards and urban growth.

CLIMATE CHANGE

In addition to the issue of increasing sea levels and increased impacts of disasters in rapidly growing coastal communities worldwide described earlier, as the climate changes, marine ecosystems will be destabilized, posing a number of additional risks to human health (also see Chapter

2). In particular, global climate change may facilitate the spread of human diseases (such as cholera) via the marine environment. Pathogens now limited to tropical waters could move toward the poles as sea-surface temperatures rise. For example, the bacterium that causes cholera (*Vibrio cholerae*) has been implicated in disease outbreaks fueled by the warming of coastal surface water temperatures (Ballard *et al.,* 2004).

A key behavioral consideration will be our ability to adapt to these risks for our future health and livelihoods. A range of negative health impacts is possible from climate change, but adaptation is likely to help protect much of the developed world population. The capability of humanity to adapt will depend a great deal on social factors, such as geography and economic status. Human populations will continue to differ in vulnerability. Factors such as crowding, food scarcity, poverty, and local environmental decline will make populations in some developing countries especially vulnerable. In addition, air and water pollution, habitat fragmentation, wetland loss, coastal erosion, and reductions in fisheries are likely to be compounded by climate-related stresses. Increasing sea level and temperature will especially threaten the rising populations that live or work close to seas. Likewise, in developed countries, the demographic trend toward an aging population will raise the health risks. Maintaining and improving public health and community infrastructure, from water treatment systems to emergency shelters, will be important for minimizing the impacts of water-borne diseases, heat stress, air pollution, extreme weather events, and diseases transmitted by insects, ticks, and rodents.

HARMFUL ALGAL BLOOMS

Much of the world's coastlines have experienced increases in the number, frequency, and type of harmful algal blooms (HABs) in recent years (see Section I.C.). HABs are destructive concentrations of certain algal species, usually in ocean waters, although freshwaters are also experiencing HAB events (Bauer, 2006). Scientists have reported that excessive human releases of nutrients and a subtle rise in ocean surface temperatures are causing an increase in pathogens, primarily bacteria and viruses (Harvell, 1999, 2002). Excessive nutrient inputs, often from nonpoint sources, can lead to certain harmful algal blooms that are toxic to fish and humans and can result in oxygen depletion, which kills marine organisms and diminishes recreational and commercial fishing.

Annually, HABs are believed to cost the U.S. fishing and tourism industries more than $80 million in public health costs (divided between shellfish poisoning and ciguatera fish poisoning), commercial fishing losses, recreation and tourism losses, and monitoring and management costs (Hoagland and Scatasta, 2006). The authors believe their estimate may be conservative because it does not include considerations such as diminished recreational opportunities, unreported illnesses and medical costs, and reduced property values. Although small in national terms, damages can be catastrophic to low-income fishing communities, as witnessed in Maryland in 1997 during an outbreak of *Pfiesteria piscicida* (a species of dinoflagellate) associated with widespread fish kills. Overall, tourism was hurt by news coverage of seafood poisonings, and reports of red tides had a swift and chilling effect on ocean-side resort visits, beach going, and boating (Hoagland *et al.,* 2002). Aquaculture can also be severely damaged by HABs, which can cause rapid fish kills and result in harvesting bans.

SEAFOOD SAFETY

Globally, more than a billion people rely on seafood as their main source of animal protein (FAO, 2004). Contaminated seafood is one of the most frequent causes of human diseases contracted from the ocean, including both pathogenic contamination and chemical contamination. Human effluents impair marine ecosystem health, which in turn affects human health. At root are externalities. Wastewater and stormwater discharges can contaminate water and marine organisms, leading to viral and bacterial diseases outbreaks. Pollutants also may enter the oceans from rivers and from atmospheric deposition. Nutrient-rich coastal waters provide ideal conditions for the growth of these pathogens including certain HABs. Chemicals (such as mercury and dioxins) that exist as environmental contaminants and are concentrated in food chains continue to be a health concern for humans, especially in terms of reproductive and developmental problems. With increasing coastal watershed populations, waste and pollution have increased to a level that creates negative environmental and human health-related consequences (Ballard *et al.,* 2004).

CONCLUSIONS

We have overused marine resources for hundreds of years, at substantial cost to our own health and that of marine ecosystems as well. Why do we continue? Are we greedy? Weak? Are we blinded by the romance of freedom, adventure, and courage that a livelihood derived from the oceans represents?

The reason overuse continues is not because people are weak, greedy, or overly romantic, but because it is a rational thing to do. Marine resources that you or I do not capture today may well be taken by someone else. Why invest in the long-term sustainability of marine resources if what happens tomorrow or next week or next year is highly uncertain? That is not rational for the individual decision maker. The only rational thing for the individual to do is race for marine resources, early and often, and to build a boat that will outrace competitors. An understanding of the economics of marine resources can be used to protect the oceans and human health by reversing the incentives for both the individual and society for overuse.

Marine ecosystems are valuable assets in the number and variety of services they offer to humanity, many of which contribute to our health. Because of externalities, and the public good and common pool characteristics of many of these services, market forces can be relied on neither to guide them to their most highly valued uses nor to reveal prices that reflect their true social values. Failures of the market system to allocate and price marine ecosystems services correctly create the need for economic analysis to guide decision making.

As individuals and citizens, we have difficult choices to make as to how to use ocean science to improve public health. In the United States, the catastrophic 2004 and 2005 hurricane seasons serve as painful examples that have raised many important questions. Economics can contribute to this important, multidisciplinary discussion, because resources available for ocean-related public health protection are scarce. A systemic economic framework helps organize how we think about an optimal mix of strategies to protect ocean and human health and how new scientific discoveries may affect that mix.

Economics Case History: Ballast Water and Human Health

Ballast water transfers and aquatic invasive species are perhaps the biggest environmental challenge facing the global shipping industry (Raaymakers, 2002). Modern shipping cannot operate without ballast water, which gives balance and stability to unladen vessels. A potentially serious environmental problem arises when discharged ballast water contains aquatic life. Thousands of aquatic species may be carried in ships' ballast water, including bacteria and other microbes, microalgae, and small invertebrates, as well as the eggs, spores, seeds, cysts, and larvae of various aquatic plant and animal species (Carlton, 1999). The use of water as ballast and the development of larger, faster ships completing their voyages in ever shorter times, combined with rapidly increasing world trade, means the natural barriers to the dispersal of species across the oceans are disappearing. Aquatic bioinvasions have important consequences for human health.

Global shipping traffic has long served as a conduit for disease. Casale (2002) stated, "For six hundred years leaders in the health and maritime industries have recognized the international transport of disease as a public health threat." As early as the 14th century, it was understood that plague epidemics moved along maritime trade routes. The concept of quarantine originated in Venice. Ships were required to stay at anchor offshore for 40 days and were not allowed to enter the port until there was reasonable assurance the ship was disease-free.

Although little understanding existed of disease vectors in the 1300s, the effects of disease transmission were well known. In 1347, several ships returned to Venice from Constantinople carrying rats with fleas infected with bubonic plague, or Black Death, to a population that was immunologically vulnerable. By 1348, the disease had spread to Paris and was transmitted to London within a few months. During the course of this epidemic, the mortality rate reached over 65% in many cities. Historians report that it had significant effects on the economy of Europe for the next 200 years (University of Virginia, 1994).

Despite modern quarantine procedures that have addressed the conventional modes of ship-transported human diseases, ballast water remains a significant vector for pathogens and toxic organisms (Ruiz *et al.*, 2000). Public health professionals were astonished to discover that *Vibrio cholera* could invade some species of algae, then enter a dormant state awaiting favorable conditions that facilitate its reemergence as an infectious agent (Monroe and Colwell, 1996). Some cholera epidemics appear to be directly associated with ballast water. One example is an epidemic that began simultaneously at three separate ports in Peru in 1991, sweeping across South America and affecting more than a million people and killing more than 10,000 by 1994. This strain had previously been reported only in Bangladesh.

In addition to bacteria and viruses, ballast water can also transfer a range of species of microalgae, including toxic species that form harmful algal blooms. The public health impacts of such outbreaks are well documented. For example, paralytic shellfish poisoning (PSP) is associated with the consumption of shellfish contaminated with natural toxins (saxitoxin) of the dinoflagellate *Alexandrium* and other species; PSP can cause severe illness and death in humans who consume the contaminated shellfish. There have been reports of the transfer of these toxic dinoflagellates in ballast water with subsequent cases of PSP in new geographic areas.

The rate of aquatic bioinvasions has increased exponentially over the past 200 years (Carlton, 2001). As a result of globalization, free trade arrangements, and economic development, maritime transport is rapidly increasing. The problem of ballast water and aquatic bioinvasions must be addressed on an international basis because coastal states are inextricably linked by currents and shipping. Despite the already significant extent of aquatic bioinvasions, the risk of further invasions (and their economic impact) is increasing.

References

Ballard, R.D., Beattie, T.A., Borrone, L., Coleman, J.M., D'Amato, A., Dickerson, L., Gaffney II, P.G., Hershman, M., Koch, C., Kelly, P.L., Muller-Karger, F., Rasmuson, E.B., Rosenberg, A.A., Ruckelshaus, W.D., Sandifer, P.A., Watkins, J.D., 2004. Connecting the oceans and human health. In U.S. Commission on Ocean Policy Final Report: An Ocean Blueprint for the 21st Century, pp. 338–351. Accessed December 29, 2006, at www.oceancommission.gov/documents/full_color_rpt/welcome.html#full.

Bauer, M. (ed.), 2006. Harmful Algal Research and Response: A Human Dimensions Strategy. Woods Hole, MA, National Office for Marine Biotoxins and Harmful Algal Blooms, Woods Hole Oceanographic Institution.

Bjornstad, D., Kahn, J. (eds.), 1996. The Contingent Valuation of Environmental Resources: Methodological Issues & Research Needs. Brookfield, VT, Edward Elgar.

Carlton, J.T., 1999. The scale and ecological consequences of biological invasions in the world's oceans, pp. 195–212. In Sandlund, O.T., Schei, P.J., and Viken, A. (eds.), Invasive Species and Biodiversity Management. Dordrecht, The Netherlands, Kluwer Academic.

Carlton, J.T., 2001. Introduced species in US coastal waters: Environmental impacts and management priorities. Arlington, VA, Pew Oceans Commission.

Carson, R., Flores, N., Martin, K., Wright J., 1996. Contingent valuation and revealed preferences: Comparing estimates for quasi-public goods. Land Econ. 72, 80–99.

Casale, G.A., 2002. Ballast water: A public health issue? Ballast Water News, Issue 8: Jan–March 2002. GloBallast Programme, IMO London.

Crossett, K.M., Culliton, T.J., Wiley, P.C., Goodspeed, T.R., 2004. Population Trends along the Coastal U.S.: 1980–2008. NOAA/NOS. Accessed December 27, 2006, at www.oceanservice.noaa.gov/programs/mb/pdfs/coastal_pop_trends_complete.pdf.

Cummings, R., Brookshire, D., Schulze, W., 1986. Valuing Environmental Goods: An Assessment of the Contingent Valuation Method. Savage, MD, Rowman & Littlefield.

Curran, S., Kumar, A., Lutz W., Williams, M., 2002. Interactions between coastal and marine ecosystems and human population systems: Perspectives on how consumption mediates this interaction. Ambio 31, 264–268.

Fleming, L.E., Laws, E. (eds.), 2006. Overview: Special issue on the oceans and human health. Oceanography 19, 18–23.

Food and Agriculture Organization (FAO), 2004. State of the World Fisheries and Aquaculture. Rome, FAO Fisheries Dept.

Harvell, C.D., Kim, K., Burkholder, J.M., Colwell, R.R., Epstein, P.R., Grimes, J., Hofmann, E.E., Lipp, E.K., Osterhaus, A.D.M.E., Overstreet, R.M., Porter, J.W., Smith, G.W., Vasta, G.R., 1999. Emerging marine diseases-climate links and anthropogenic factors. Science 285, 1505–1510.

Harvell, C.D., Mitchell, C.E., Ward, J.R., Altizer, S., Dobson, A.P., Ostfeld, R.S., Samuel, M.D., 2002. Climate warming and disease risks for terrestrial and marine biota. Science 296, 2158–2162.

Hoagland, P., Anderson, D.M., Kaoru, Y., White, A.W., 2002. Average annual economic impacts of harmful algal blooms in the united states: Some preliminary estimates. Estuaries 25, 677–695.

Hoagland, P., Scatasta, S., 2006. The economic effects of harmful algal blooms. In Graneli, E., and Turner, J. (eds.), Ecology of Harmful Algae, Ecology Studies Series, chapter 29. Dordrecht, The Netherlands, Springer-Verlag.

Kopp, R., Pommerehne, W., Schwarz, N. (eds.), 1997. Determining the Value of Non-Market Goods. Kluwer Academic Publisher. Boston, MA.

Letson, D., Milon, J.W. (eds.), 2006. Florida Coastal Environmental Resources: A Guide to Economic Valuation and Impact Analysis, Gainesville: Florida Sea Grant, (2002). Accessed December 27, 2006, at http://nsgl.gso.uri.edu/flsgp/flsgph02002.pdf.

Mileti, D., 1999. Disaster by Design: A Reassessment of Natural Hazards in the United States. Washington, DC, Joseph Henry Press.

Mitchell, R., Carson, R., 1989. Using Surveys to Value Public Goods: The Contingent Valuation Method. Washington, DC, Resources for the Future.

Monroe P.M., Colwell, R.R., 1996. Fate of Vibrio cholerae 01 in seawater microcosms. Water Research 30,47–50.

Pielke, R.A., Jr., 2006. Seventh annual Roger Revelle commemorative lecture: Disasters, death, and destruction: Making sense of recent calamities. Oceanography 19, 138–147.

Raaymakers, S., 2002. The Ballast water problem: Global ecological, economic and human health impacts. Paper Presented at the RECSO/IMO Joint Seminar on Tanker Ballast Water Management & Technologies, Dubai, UAE.

Ritchie, B., 2005. Bay's closure hurts families financially: County suspicious of state's oystering ban. Tallahassee Democrat, November 2, 2005(A1), final edition.

Ruiz, G. M., Rawlings, T.K., Dobbs, F.C., Drake, L.A., Mullady, T., Huq, A., Colwell, R.R., 2000. Global spread of microorganisms by ships. Nature 408, 49.

Sandifer, P.A., Holland, A.F., Rowles, T.K., Scott, G.I., Tyson, F.L., Rice, D.L., Dearry, A., 2004. Guest editorial: The oceans and human health. Environmental Health Perspectives 112, A454–A456.

University of Virginia, 1994. Plague and Public Health in Renaissance Europe. Charlottesville, VA, The Institute for Advanced Technologies in the Humanities, University of Virginia.

STUDY QUESTIONS

1. The scarcity of marine resources in relation to human demands implies choices, and thus trade-offs. Discuss the ways in which marine resources contribute to our economic well-being and, in particular, how they contribute to human health. How might some uses of marine resource be in conflict with others? How should we resolve these conflicts?
2. Explain globalization, and discuss what it may mean for oceans and human health.
3. Discuss how coastal migration compounds many risks that oceans pose for human health.
4. Explain what economists mean when they refer to fisheries as a "common pool resource." What about common pool resources makes them difficult to manage?
5. Explain what economists mean when they refer to greenhouse gas emissions as an "externality." Why do externalities occur, and what can be done about them?
6. Antarctica suffers from the most extreme weather hazards on Earth, yet it has suffered few disasters. Besides an environmental event, such as a hurricane, what else is necessary for a disaster to occur?
7. Minimizing the human impacts of marine resource management requires an understanding of its often unintended sociocultural consequences. What human impacts should policy officials try to avoid when issuing public advisories for harmful algal blooms? Hurricanes?

SECTION I

B. Effects of Anthropogenic Substances

CHAPTER

6

Background Toxicology

KEITH B. TIERNEY AND CHRISTOPHER J. KENNEDY

INTRODUCTION

The oceans have long been associated with toxic substances. Approximately 1200 species of marine organisms widely distributed throughout the marine fauna in almost all seas and oceans produce *toxins* (toxic substances produced by living organisms). The oceans are also the source of toxic substances that are not produced by organisms. For example, petroleum seeps from the ocean bottom and the erosion of sedimentary rock releases approximately 235 million liters of oil into the oceans and contributes significantly to marine contamination. Evolution has allowed for biota in the oceans to adapt to *xenobiotics* (foreign chemicals) from both of these sources by using various chemical defense mechanisms. The relatively recent phenomenon of industrial pollution and the release of novel synthetic compounds into the oceans at high concentrations over short time periods has not allowed for evolutionary adaptation of chemical defenses to all of these compounds and presents a significant and serious challenge to inhabitants of the oceans and to those relying on them.

The German-Swiss alchemist and physician Theophrastus Phillippus Aureolus Bombastus von Hohenheim, also known as Paracelsus, wrote one of the fundamental tenets of toxicology in the 16th century: "All things are poison and nothing is without poison, only the dose permits something not to be poisonous." Unfortunately, Paracelsus' quote has been erroneously applied on ecological and geographical scales for the management of waste. The idea that "dilution is the solution to pollution" (i.e., without an appreciable dose there are no effects) has allowed the vast oceans to serve as a repository or sink of environmental contaminants, a practice that continues today.

As a subdiscipline of science, toxicology (the study of the adverse effects of chemicals or physical agents on biological systems) borrows freely from several scientific disciplines, some of which are listed in Figure 6-1. In toxicology, knowledge gained on both environmental and biological fate combined with that on toxic effects allows for predictions to be made and for environmental risk to be assessed. To be effective, the output of the science needs to feed into regulatory bodies and become practice.

This chapter is an introduction to background toxicology and has been categorized according to the toxic action paradigm (Fig. 6-2), which consists of three phases: the *exposure*, the *toxicokinetic,* and the *toxicodynamic* phases. The concepts and key facts presented for each of these phases form a knowledge base that answers these essential and fundamental questions: (1) why do chemicals end up where they do in the environment, (2) how are they taken up and processed by biota, and (3) what are the potential effects that can result from the interactions between chemicals and organisms? Armed with this information, we describe the various methods and techniques that underlie toxicity testing and other toxicological assessments. We then take a practical approach, illustrating how this knowledge is used in a modern application known as environmental risk assessment, which endeavors to understand, assess, and predict the risks to humans and the ecosystem when they are exposed to xenobiotics.

ENVIRONMENTAL FATE

Kitimat Arm, on the north coast of British Columbia, is alleged to be polluted with polycyclic aromatic hydrocarbons (PAHs) from an aluminum smelter (Eickhoff *et al.*, 2003). These PAHs are found associated with bottom sediments in this marine environment and, therefore, the majority of toxicological research focuses on the Dungeness crab,

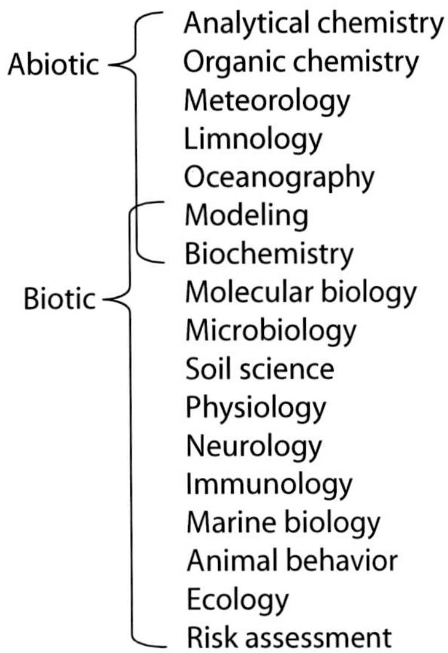

FIGURE 6-1. Some of the biotic and abiotic disciplines involved in ecotoxicology. Adapted from Landis and Yu (1999).

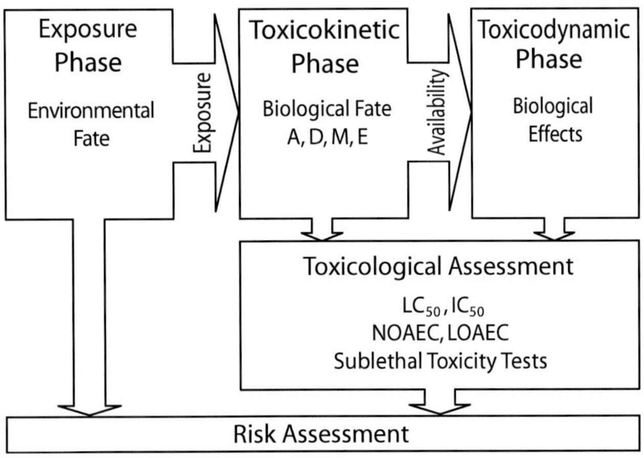

FIGURE 6-2. The classic three-part toxicology paradigm and how it relates to the applied disciplines of toxicity testing and risk assessment. The toxicokinetics abbreviations are A—absorption, D—distribution, M—metabolism, and E—elimination.

Cancer magister, with less attention given to other organisms such as pelagic fish species. This focus is not misplaced; there is less likelihood of direct toxic effects elicited by an environmental *contaminant* (a chemical not normally found in an environment) on a biological system if there is no contact or contaminant pathway between them. Thus, predicting or assessing the toxic outcomes from environmental chemicals begins with an understanding of environmental fate and a determination of potential exposure.

Environmental Compartments and Phases

Environmental chemistry and toxicology are branches of environmental science that borrow freely from several other scientific disciplines for the purpose of describing the behavior of chemicals in the environment both qualitatively and quantitatively. The fate and distribution of chemicals in the environment are determined by a complex interplay between the physicochemical properties of the compound and a multifarious environment. The complexities and subtleties of the environment and a chemical's behavior in it must be simplified to be understood, modeled, and eventually predicted.

The first step in this simplification is to compartmentalize the environment into four phases: air, water, biota, and sediments/soil, which may or may not be in contact with each other and which may be continuous (e.g., water in a lake) or discontinuous (e.g., sediment particles suspended in water). These phases are sometimes heterogeneous, but for the purposes of modeling chemical fate, these phases are usually assumed to be homogeneous. Traditionally, the movement of a chemical from one compartment to another has been illustrated for a single phase (e.g., water to water separated by a semipermeable membrane), with the driving force for movement being a concentration gradient. At equilibrium, the net movement between two compartments of similar phase is zero, and the concentration in one compartment is equal to that in the other.

The "preference" of a chemical for a particular phase may be evident when two dissimilar phases are in contact, such as water and an organic solvent (e.g., 1-octanol). Many organic molecules such as DDT, PCBs, many pesticides, dioxins, and so on are *hydrophobic* and have limited water solubility. Therefore, these types of compounds tend to move from a water phase into an organic phase, such as an organic solvent or lipid. In the real environment, hydrophobic chemicals tend to *partition* (move) into sediments and biota, which act much like an organic phase. The term *fugacity* refers to the property of a chemical in a phase and is defined as the "tendency to flee." In water, hydrophobic organic molecules have a high fugacity. Sediments and biota have a high *fugacity capacity* for these chemicals. For example, if DDT is sprayed onto a lake, it will not dissolve well into the water and so will partition into sediments and the lake's organisms. In this example, at equilibrium, the net movement of DDT between water, sediments, and biota will be zero. However, the measured concentrations of DDT in these environmental compartments will likely be $[\text{sediment}]_c >> [\text{biota}]_c >>>>>> [\text{water}]_c$, i.e., the concentrations will not be equal. At equilibrium in this situation, the fugacities of the chemical in each compartment are equal: $\text{sediment}_f = \text{biota}_f = \text{water}_f$.

The other environmental compartment of note in this description is the atmosphere. Here, one should think of a chemical's "atmospheric solubility" in determining if it will partition into the air. This is determined by a chemical's vapor pressure, where compounds with high vapor pressures will partition more into the air than other compartments. For modeling purposes, vapor pressures are readily available from the literature. Determining how a chemical may partition between aqueous (water) and organic (sediment/lipid/biota) phases is not as simple. Partition coefficients are ratios of solubilities in one phase (e.g., 1-octanol, hexane, fat, lipids, etc.) to that in another (e.g., water). These are determined empirically by shaking volumes of the two phases and adding the chemical to the mixture. When equilibrium is reached, the concentrations of the chemical in the organic phase and in the aqueous phase are measured. The most important and frequently used descriptor of partitioning is the *octanol-water partition coefficient* or K_{ow}. K_{ow} is a measure of hydrophobicity (i.e., the tendency to "hate" or partition out of water). Because K_{ow} varies from approximately 10^{-1} to 10^{7}, it is commonly expressed as log K_{ow}. Log K_{ow} values of 4 to 7 typically represent hydrophobic chemicals that partition into sediments and biota preferentially over water.

Special terminology is used to describe the uptake of chemicals into biota. *Bioaccumulation* is defined as the uptake of chemical from the *abiotic* (e.g., water) or *biotic* (e.g., food) environment. *Bioconcentration* refers to the accumulation of chemical from the abiotic environment into an organism, resulting in concentrations higher than in the environment. *Biomagnification* is the term used to describe the accumulation of chemical from its biotic environment (i.e., its food) to higher concentrations than are found in its prey. Biomagnification can result in increases in the concentrations of some chemicals through successive trophic levels. Two factors are primarily responsible for bioconcentration and biomagnification: a high partition coefficient and recalcitrance toward all types of degradation.

In the oceans, biomagnification is evident in top-trophic level consumers. For example, beluga and killer whales are among the most contaminated animals in the world. In fact, their load of persistent organic pollutants (POPs) is so severe that they could be considered toxic waste according to contaminant regulations of some countries. The transient, seal-consuming killer whales in British Columbia have 200 mg/kg of *legacy* PCB POPs. Even at this concentration *acute effects* (i.e., those occurring in less than 5 days) are absent. Rather, POPs tend to act as endocrine disrupting chemicals (EDCs), causing negative effects on various systems, such as the immune system, over much longer time frames. For example, 17 mg/kg of POPs is the threshold for causing *immunocompromization* in the harbor seal, another marine mammal (Ross, 2006). Populations with high POP concentrations may suffer higher rates of infection and so may experience high mortality rates.

For human communities that depend on the marine environment for much of their diet, POP concentrations may be similarly elevated and harmful. The indigenous Inuit populations of northern Quebec confirm this prediction, as in the early 1990s, fat-rich breast milk was found to contain 3 mg/kg lipid of PCBs, or approximately five times more PCBs than populations in southern Quebec (Dewailly *et al.*, 1992). These levels may have contributed to an increased susceptibility to *otitis media,* an infection of the inner ear (Dewailly *et al.*, 2000) (see Chapter 10). For these distinct populations, there is no *sentinel* or *indicator species* to warn them that their food source may be harmful.

BIOLOGICAL FATE

Toxicity is dependent on the actual chemical concentration in the *target organ* (organ of damage) or more specifically, at a *target site* or *receptor* (biological entity affected) where the toxic effects occur, although this is often difficult or impossible to measure. The concentration at the site of action is dependent on a chemical's disposition (i.e., its *absorption, distribution, metabolism,* and *excretion*) within an organism. Collectively, these processes are known as *toxicokinetics* (Fig. 6-2).

Absorption

Absorption is the process by which a chemical enters the body or cell from the environment. Four characteristics of epithelial tissue (tissue in contact with the environment) determine the rate of xenobiotic uptake: (1) the biological makeup of the epithelia, which will determine the *permeability* of the compound; (2) the surface area; (3) the diffusion distance; and (4) the blood flow to the tissue. A high permeability factor, small diffusion distance, and large blood flow all lead to enhanced uptake of chemical. The major sites of uptake or absorption of xenobiotics are integuments, respiratory systems, and digestive systems. *Integuments* or skin can act, in part, as barriers to the outside environment and limit the absorption of xenobiotics. Conversely, respiratory systems and digestive systems are designed to enhance the transfer of gases and nutrients, respectively, with the environment and have all of the prerequisites for rapid xenobiotic uptake.

Several mechanisms exist by which chemicals can move into cells (Fig. 6-3). *Simple passive diffusion* is the primary mechanism by which most lipophilic toxicants move into cells. The lipid bilayer of the cell membrane can act as a barrier to most nonionized hydrophilic molecules and ions. Small hydrophilic molecules (including small ionized/dissociated molecules) of approximately MW < 100 (Schanker, 1961) can traverse this barrier by diffusing through aqueous channels (approx. 4 nm) that exist in many membranes. Filtration occurs at the tissue level (e.g., in the glomeruli of the human kidney) in which solutes move with the bulk flow of water through pores (approx. 70 nm). This type of tissue-specific movement is more important in the elimination of toxicants than in their absorption.

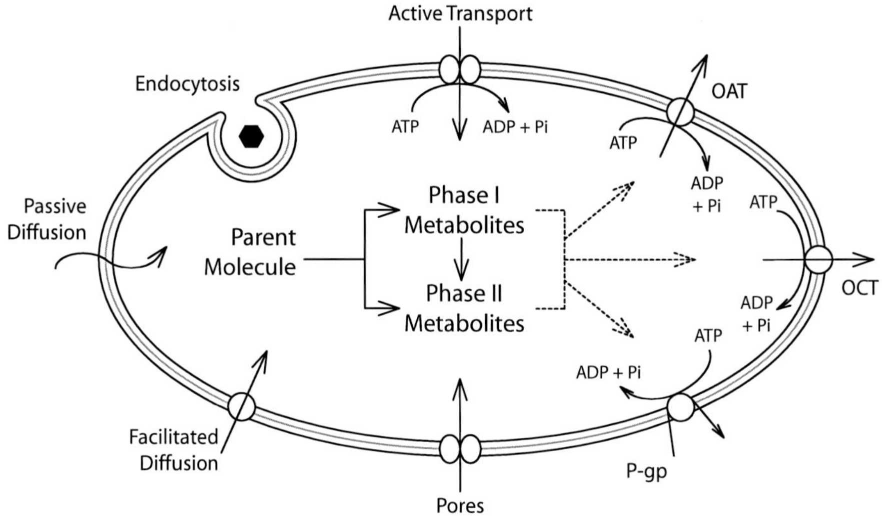

FIGURE 6-3. Summary diagram of uptake, biotransformation, and cellular elimination processes for toxicants. OAT—organic anion transporter, OCT—organic cation transporter, P-gp—P-glycoprotein.

Most organic toxicants or other hydrophobic/lipophilic molecules can directly diffuse through the lipid phase of cell membranes. Their rate of transfer across cell membranes is dependent on their hydrophobicity. Because most organic molecules fall within a similar range of lipophilicity, it is their hydrophobic nature that determines how fast they will partition into a membrane. The partition coefficient correlates fairly well with the extent of uptake through membranes where compounds with higher log K_{ow} values tend to accumulate more rapidly in cells.

Toxicants that are too water soluble to diffuse through cell membranes and too large to flow through aqueous channels still gain entry into cells through active transport and *facilitated diffusion* (Fig. 6-3), where a membrane-associated carrier protein provides the vehicle to transport the compound through the lipid barrier of the membrane. *Active transport* systems are linked either to energy producing enzymes (e.g., ATPase) or to the *co-transport* of other molecules (e.g., sodium ions). In facilitated diffusion, the energy for transport is the concentration gradient that exists for the compound. Because a carrier protein is used in both active transport and facilitated diffusion of toxicants, uptake by these systems can become saturated.

Endocytosis includes both *pinocytosis* (liquids) and *phagocytosis* (solids), which are processes where the cell membrane flows around and engulfs droplets or solid particles. The overall importance of this specialized transport is unclear, but for some toxicants it is the primary route of uptake. For example, botulinum toxin exerts its effects after entering nerve cells by receptor-mediated endocytosis.

Distribution

The distribution of a toxicant can determine its toxicity as the toxicant must reach the site of action at a high enough concentration and for a sufficient period of time, to elicit a response. Once a chemical enters the circulatory system, it may be accumulated at the site of toxic action, be transferred to a site of storage, or be transported to organs of biotransformation or elimination.

Water-soluble xenobiotics will have sufficient solubility in the aqueous component of circulatory fluids (plasma) to be transported as dissolved chemical; however, hydrophobic compounds are often transported in association with plasma proteins. Release of the protein-bound chemical occurs when the affinity of another biomolecule or tissue component is greater than that of the plasma protein. Overall body distribution is dependent on several factors including the physicochemical properties of the chemical and the affinity of the chemical for tissue constituents. Toxicants that do not readily pass through cell membranes or make use of specialized transport mechanisms have a restricted distribution, whereas other toxicants that readily pass through cell membranes can become distributed widely throughout the body. Most chemicals will distribute to all tissues to some degree. Tissues in which compounds distribute but do not elicit a toxic response are often called *depots* or *sinks.* In some cases, compounds may be stored in such tissues and slowly released back into the systemic circulation for elimination, thus protecting the organism from acute adverse effects. However, increases in the overall residence time because of storage can lead to chronic toxicity.

Biotransformation

Continuous exposure of humans and wildlife to xenobiotics of natural or anthropogenic sources, even at very low concentrations, could result in the accumulation of these chemicals to toxic levels. This is particularly true of those compounds that are lipophilic and readily absorbed and sequestered by the body. Many excretion routes exist, however, and the ease with which these compounds are eliminated depends on their water solubility. Lipophilic compounds that make their way into excretory fluids will be reabsorbed. This is why lipophilic compounds tend to accumulate: they are easily absorbed and poorly excreted.

Biotransformation is the conversion of xenobiotic chemicals into different chemical structures with the aid of endogenous enzymes. The term *metabolism* is often used interchangeably for this process, as is *detoxification*. However, as will be illustrated, biotransformation does not always yield a less toxic product, rather it may convert a xenobiotic into a more toxic metabolite by a process called *bioactivation* (see, for example, Chapter 32). The capacity of an organism to metabolize xenobiotics has substantial effects on the tissue levels of toxicants, metabolite patterns, and xenobiotic half-life, all of which may affect the severity and duration of a toxic response (Buhler and Williams, 1989). For example, the inability of the sea lamprey to metabolize the 3-trifluoromethyl-4-nitrophenol makes this compound uniquely toxic to this invasive species, and so it has been used as a lampricide in several jurisdictions.

The general purpose of biotransformation is to convert lipophilic parent xenobiotics into more water-soluble metabolites in order to limit the distribution of these chemicals in the body (water-soluble metabolites will less likely be taken up by tissues) and to ultimately enhance their excretion. The excretion of water-soluble compounds is an efficient process, because once in excretory fluids, little reabsorption occurs because of the membrane barriers lining excretory routes.

Biotransformation reactions are enzymatic in nature, and a single chemical may undergo several transformation reactions. The parent molecule may undergo chemical modification at a number of sites, and the products, or metabolites, may themselves undergo further biotransformation reactions, producing distinct end products. The same enzyme systems that perform these reactions are also used inmodifying many endogenous biomolecules such as steroid hormones.

All organisms have some ability to metabolize foreign compounds that are taken up; however, their evolved abilities in this regard vary widely. The highest biotransformation ability is usually found in mammals and birds, followed by fish and reptiles, and then invertebrates. The major sites of biotransformation in mammals are the liver, lung, nasal mucosa, skin, and gastrointestinal (GI) tract (all sites of potential entry of xenobiotics). One of the primary functions of the liver is to metabolize foreign chemicals before they enter the general circulation; therefore, it is not surprising to find that the liver has a high lipid content (that enhances partitioning of chemicals to this tissue), a high blood flow from the GI tract (for delivery of chemicals), and very high concentrations of biotransformation enzymes.

Xenobiotic metabolism generally consists of two phases. In phase 1 reactions, a polar reactive functional group (e.g., -OH, -SH, $-NH_2$, -COOH) is introduced to or exposed in the parent molecule, rendering it more water-soluble and potentially ready for excretion, as well as more suitable for phase 2 enzymatic reactions. In phase 2 reactions, the phase 1 metabolite is conjugated (joined, or covalently bonded) to various endogenous molecules such as sugars, amino acids, and sulfate, forming exceptionally water-soluble products (usually undergoing significant ionization at physiological pH). These endogenous conjugating moieties (groups) are normally added to the phase 1 metabolite to promote secretion at various epithelia whose transport systems specifically recognize the conjugating moiety. Thus, the excretion of phase 2 conjugates is enhanced through two bestowed properties: increased water solubility and participation in active secretion. Phase 2 reactions almost always result in a less toxic metabolite; however, the products of phase 1 reactions can be more reactive than the parent molecule. The phase 1–phase 2 relationship is shown in Figure 6-3.

Phase 1 Reactions and Microsomal Monooxygenations

Phase 1 reactions are the predominant biotransformation pathway for most xenobiotics and include microsomal monooxygenations, cytosolic and mitochondrial oxidations, co-oxidations in the prostaglandin synthetase reaction, reduction reactions, hydrolysis reactions, and epoxide hydration. All of these reactions have a common theme: they either unmask or introduce a polar functional group in or onto the xenobiotic.

For the purposes of this chapter, the most dominant oxidation reaction catalyzed by the *cytochrome P-450-dependent mixedfunction oxidase* (*MFO*) system will be discussed. The MFO system is responsible for the oxidation of many endogenous compounds (e.g., steroids, vitamins, and fatty acids), but it is also responsible for catalyzing the initial oxidation of exogenous compounds (e.g., pesticides and PAHs). Monooxygenation reactions are those in which one atom of a molecule of oxygen is incorporated into the substrate while the other is reduced to water. The electrons involved in the reaction are derived from NADPH, and therefore the overall reaction is written as follows:

$$RH + O_2 + (NADPH + H^+) \rightarrow ROH + H_2O\ (NADP^+)$$

The MFO "system" consists of lipids of the smooth endoplasmic reticulum (ER) of the cell, and two enzymes: cytochrome P450 (CYP P450) and NADPH-cytochrome P450 reductase. These enzymes are embedded in the phospholipid matrix of the ER, which plays two crucial roles: it facilitates an interaction between the two enzymes and also holds lipophilic xenobiotic substrates in place to be acted upon. NADPH-cytochrome P450 reductase uses NADPH as a cofactor, which, after binding of the substrate to oxidized

CYP P450, transfers electrons to CYP P450, thereby reducing it, allowing for several steps whereby one atom of oxygen from O_2 is transferred onto the xenobiotic and the other oxygen is used to form water. One unique and important feature of the CYP P450 oxidation system is in the broad spectrum of xenobiotics that can serve as substrates for this enzyme; being broadly specific allows an organism to successfully metabolize a large class of unknown substrates that it may absorb from its environment.

Phase 2 Reactions

The prerequisite for compounds to undergo phase 2 or *conjugation reactions* is that a parent xenobiotic or its phase 1 metabolite contain a polar functional group. In phase 2 reactions, a substrate is conjugated, or joined, to an endogenous molecule such as a sugar, amino acid, sulfate group, glutathione, and so on. The products of conjugation reactions are almost always less toxic and more water soluble than the original substrate. Conjugates are also substrates for specific transport proteins in epithelial tissues and are therefore more easily excreted. There are two general types of phase 2 reactions that are distinguished by the participation of ATP and the "activation" of either the substrate or the coenzyme. In the first type, ATP is used to form an activated intermediate that is a reactive endogenous molecule. This activated coenzyme will be conjugated to the substrate at the site of one of the functional groups listed earlier on the molecule. The enzymes that catalyze the reaction between the substrate and activated coenzyme are collectively called *transferases*. In the second type of conjugation reaction, the substrate itself is activated which then combines with an unactivated coenzyme (e.g., an amino acid) forming the phase 2 metabolite.

The important role of biotransformation in the toxicity of a chemical of a given dose cannot be understated. Biotransformation reduces toxicant half-life and generally results in reduced toxicity through the generation of less toxic metabolites, although metabolism to more toxic products (bioactivation) is a principal modifier of toxic potential. For example, the PAH benzo[a]pyrene is a *pro-carcinogen* (i.e., it does not produce cancer at the site of application, but rather is carcinogenic to distant tissues where metabolic activation occurs). Through a series of oxidation and hydrolysis reactions, a highly carcinogenic metabolite (+)-benzo[a]pyrene-7,8-diol-9,10-epoxide is formed, which can react with cellular DNA and initiate the carcinogenic process. Thus, any *intrinsic* or *extrinsic factors,* which can modify an organism's ability to biotransform xenobiotics, may have great bearing on the risk posed by contaminant exposure. Some of the more important intrinsic factors include species, strain and other genetic variations, development, sex, age, hormones, pregnancy, disease, circadian rhythms, nutritional effects, and tissue injury. Extrinsic factors that can modify biotransformation include preexposure to xenobiotics resulting in either inhibition of biotransformation enzyme systems or induction (increases) in enzyme concentrations following exposure. In ectothermic organisms, environmental temperature and the rate of temperature change can play a large role in altering xenobiotic metabolism.

Excretion

The ability of almost all organisms to eliminate an astonishing array of natural and synthetic compounds is indicative of the unknown chemical challenges excretory systems have faced through evolutionary time and a tribute to their adaptability. The underlying strategy of these systems is elegant; regardless of chemical structure and a multitude of possible physicochemical characteristics, excretory systems rely on the conversion of xenobiotics to a form similar to endogenous molecules marked for excretion (i.e., to make them water soluble). Most excretory systems are based on the water solubility of chemicals for efficient elimination, which mitigates the dilemma of reabsorption. Two systems are used to overcome the problem that water-soluble chemicals do not pass through membranes: filtration and secretion, which will be discussed later in more detail.

There are several major and minor sites of xenobiotic elimination in organisms that are species specific (i.e., some of the organs of excretion in one species do not exist) or perform other functions in another species. In mammals, the major sites of toxicant elimination are the lungs (for volatile compounds that exist as gases at physiological temperatures, e.g., ethanol), the kidneys, and the hepatobiliary system. Water solubility plays no appreciable role in lung excretion, but it is the predominant chemical characteristic for successful elimination at both the kidney and liver. Other minor excretory routes include milk (e.g., halogenated hydrocarbons), eggs (e.g., mirex), fetus (e.g., teratogens thalidomide and diethylstilbestrol), alimentary elimination (via secretions such as saliva, e.g., opiate narcotics, DDT), sweat (e.g., metals), sebaceous glands (e.g., halogenated insecticides, PCBs), hair (e.g., Se, Hg, As), feathers (e.g., Hg), scales (e.g., tetracyclines), and nails.

Renal Excretion

Three processes are involved in renal excretion in kidneys: filtration, passive tubular diffusion, and active tubular secretion. Glomerular filtration allows for the passive filtering of both water-soluble and hydrophobic xenobiotics, which pass from the plasma into the ultrafiltrate in the kidney tubule lumen through glomerular pores. Passive tubular reabsorption of hydrophobic xenobiotics can occur at any site along the tubules. In addition, special mechanisms for the reabsorption and conservation of important biomolecules such as proteins exist, and these mechanisms may reuptake certain xenobiotics and lead to their accumulation in the kidney and result in renal toxicity (e.g., Cd^{2+} bound to the protein metallothionein).

Hydrophobic chemicals can also enter the lumen of the tubules by simple passive diffusion through the cells into the lumen; however, because the ultrafiltrate is an aqueous environment, this diffusion is limited, and any hydrophobic molecule that does move into the lumen will likely be reabsorbed by the cells lining the tubule. Under some conditions, weak acids and bases are unionized in the pH environment of the interior of cells and can diffuse freely through to the exterior of the membrane where, upon encountering a different pH of the urine, they become ionized and move out into that aqueous environment. This is often called "ion trapping" and is a mode of excretion of some weak acids such as the broad spectrum biocide pentachlorophenol at the surface of teleost gills (Kennedy and Law, 1986)

The third means by which xenobiotics can be excreted at the kidneys is by active tubular secretion via ATP-binding cassette (ABC)-transporters which mediate the ATP-dependent transport of conjugates of lipophilic compounds (phase 2 products). It is the added endogenous moiety of the conjugated xenobiotic that is recognized by these transport proteins located in both the basolateral and apical membranes (both directed toward the lumen of the tubule) of kidney tubule cells. Active transport of these conjugates and other like molecules is a significant route of excretion, but the process is a saturable one, unlike diffusion. There exist two tubular secretory processes, one for organic anions (acids) and the other for organic cations (bases) (Fig. 6-3), which are members of the multidrug resistance-associated protein (MRP) family.

Hepatobiliary Excretion

Hepatobiliary excretion is a major route of toxicant excretion. The liver is in an advantageous position to excrete xenobiotics, particularly those that have been absorbed from the diet in the gastrointestinal tract. The liver receives most of the blood flow from the GI tract before it enters the general systemic circulation, has a high lipid content enhancing the partitioning of lipophilic toxicants, and has an effective biotransformation system that effectively prevents intracellular accumulation (which would reduce blood:cell gradients). No filtration occurs in the liver; passive diffusion and special transporters are the sole means for the elimination of chemicals. Hepatocytes are the major sites of biotransformation in the body and have the highest concentrations of MRPs, the family of conjugate-transporting ATPases. The direction of transport is to either the blood (for excretion at the kidney) or to the bile for excretion in the feces. Compounds are secreted into small tubules called bile canaliculi that flow into the finest branches of the bile duct, the cholangioles. These in turn empty into the hepatic duct, which

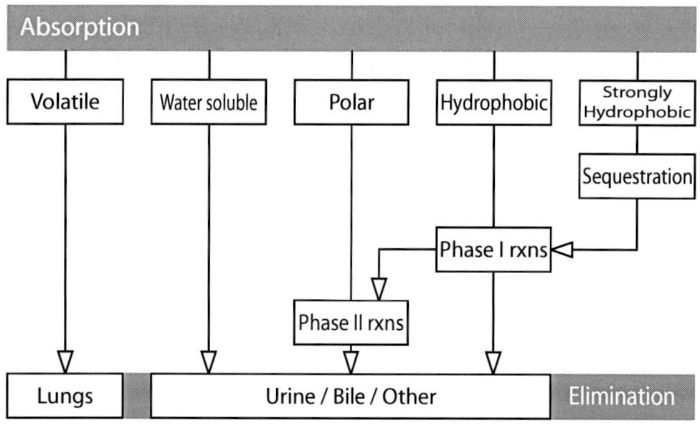

FIGURE 6-4. Pathways of biotransformation and elimination routes taken by toxicants with differing physicochemical characteristics.

carries the bile to the gallbladder. Xenobiotics or their metabolites are held in the gallbladder as a reservoir until release when the organism eats a meal. The chemicals then exit the organism in the fecal material. In some species (e.g., rat, whales, deer), there is no gallbladder and bile is released directly from the bile duct into the duodenum.

Figure 6-4 summarizes the various pathways from absorption to elimination of xenobiotics of varying physicochemical properties. The preceding examples of excretory routes presuppose that xenobiotics have entered the body or a particular tissue and must be dealt with because they are lipophilic and difficult to excrete in that form. One defense mechanism is known that attempts to remove such compounds *before* they enter a cell or tissue. This line of xenobiotic defense is a multixenobiotic resistance (MXR) mechanism. MXR activity is mediated by the expression of a variety of transmembrane transport proteins, the most common among them being P-glycoprotein (P-gp) (Fig. 6-3) (Bard, 2000). P-gp acts as an energy-dependent pump mediating the cellular efflux of a large number of moderately hydrophobic compounds (Kurelec *et al.*, 1998) including endogenous compounds (Naito *et al.*, 1989), drugs, natural products (Gottesman and Pastan, 1988), anthropogenic chemicals (Bain and LeBlanc, 1996; Cornwall *et al.*, 1995), and the products of phase 1 reactions (Bard, 2000). P-gps also confer multidrug resistance (MDR) to tumor cell lines (Juliano and Ling, 1976) and tumors of human patients (Gerlach *et al.*, 1986) by preventing the cellular accumulation of chemotherapy drugs.

BIOLOGICAL EFFECTS

The toxicodynamic phase of toxic action includes the interactions between a toxic agent and a biomolecule (its receptor) and the resulting biological effects that ensue (sequelae) (Fig. 6-2). This chapter uses a classification scheme for toxic effects that includes categories based on

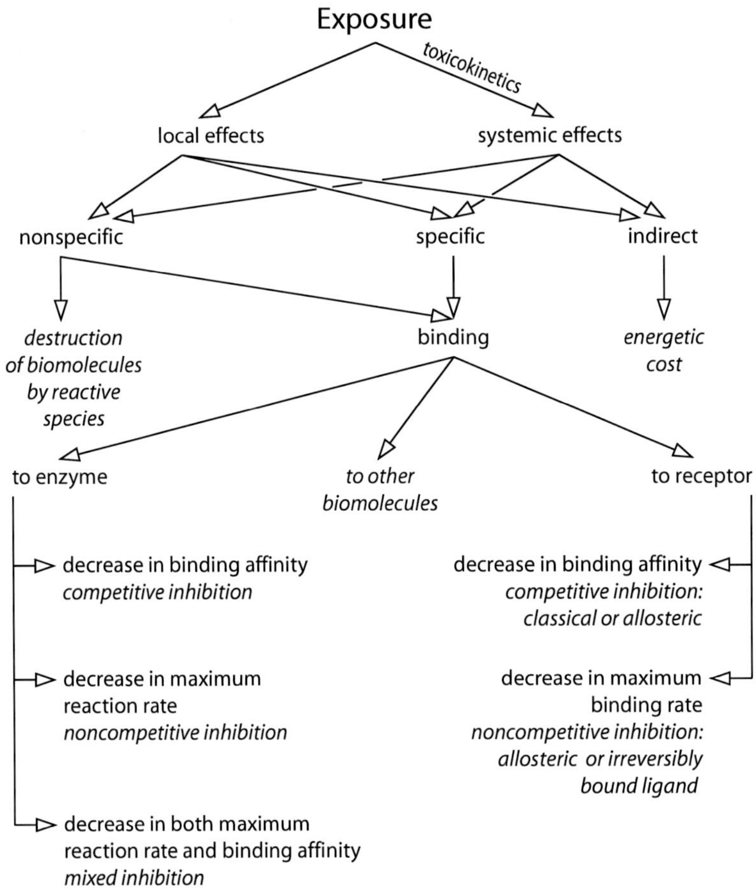

FIGURE 6-5. A flowchart from the exposure of a toxic agent through to its potential toxicodynamic interactions with various biomolecules. Italics are used to indicate the resulting biological effects.

the location of action, the particular organ/organelle/biomolecule affected, and the molecular events that occur (Fig. 6-5). Simplified, biological effects can be classified as being local or systemic. Effects are *local* when toxicity occurs at the initial site of absorption (e.g., skin or mucosa) or site of administration (e.g., intramuscular). If a toxicant is transported or diffuses to a more distant site and interacts with an internal target, it is called a *systemic effect.* Both local and systemic effects can be classified into three general categories: *indirect*, *nonspecific*, or *specific*.

Indirect Effects

Some environmental contaminants have low inherent toxicity—that is, they do not interact negatively with biomolecules until very high doses are reached. Nevertheless, these compounds may need to be biotransformed and excreted; these processes require energy. In addition, chemical exposure can cause stress in organisms and result in energy use increases. When nutrients and energy are limited (e.g., in winter), expenditures used in dealing with foreign chemicals may not be available for other processes such as growth or reproduction. The magnitude and significance of this extra energy use, however, and its effects on processes such as growth, reproduction, and survival are currently unknown. Cresswell *et al.* (1992) were the first to demonstrate unequivocally that detoxification of a toxin resulted in an increase in energy metabolism in an insect. Work in birds (Guglielmo *et al.*, 1996), marsupials (Dash, 1988; Foley, 1992), and mammals (Lindroth and Batzli, 1984) have directly measured metabolite excretion and *estimated* the associated energy losses to be 10% to 14% of metabolizable energy intake.

Nonspecific Effects

Nonspecific effects are those caused by toxicants that affect multiple targets. Such effects can be divided into the two subcategories: the *alteration of biomolecules by reactive species* and *nonspecific binding to biomolecules.*

Alteration of Biomolecules by Reactive Species

Reactive species include *free radicals*, which are typically small molecules with unpaired electrons, and *reactive oxygen species* (*ROS*). ROS are biologically produced and include superoxide (O_2^-), hydrogen peroxide (H_2O_2), organic peroxide (ROOH) and its radical (ROO^-), alkoxy radicals

(RO^-), hydroxyl (OH^-), nitric oxide (NO), hypochlorous acid (HOCl), and peroxynitrite (NO_3^-). Some of these are metabolic by-products and others, such as superoxide and hypochlorous acid, are specifically produced by neutrophils for their biocidal properties (bacteria killing), whereas NO is an important signaling molecule. Environmental stress and some toxic agents can cause an excess generation of reactive species and cause oxidative stress. Here, ROS may damage cells through a variety of mechanisms, including protein damage, lipid peroxidation (oxidation of membrane lipids), DNA base modification, and even strand breakage.

The metal cadmium (Cd^{2+}), for instance, a common contaminant of surface waters (Lyndersen *et al.*, 2002; May *et al.*, 2001), can bring about oxidative stress indirectly. Cd^{2+} can inhibit antioxidant *glutathione*-dependent enzymes that essentially scavenge ROS. When depletion of these enzymes occurs, concentrations of ROS such as H_2O_2 increase and can cause *DNA fragmentation*, which may induce apoptosis, also called programmed cell death (Watjen and Beyersmann, 2004).

Nonspecific Binding

Some agents can form ionic, covalent, hydrogen, or other bonds with biomolecules and interfere with their structure and function. Some compounds will bind to a broad range of biomolecules and therefore do not have a specific target. Such binding can result in a diverse array of toxic effects for an organism. For example, nonspecific protein binding is a hallmark of heavy metal toxicity. Metals such as mercury (Hg^{2+}) and cadmium (Cd^{2+}) bind nonspecifically to many proteins, particularly at sulfhydryl (–SH) groups, which are common functional groups on many proteins. This binding degrades the functionality of multiple proteins, which is why heavy metal toxicity can result in multiple symptoms in humans (see Chapters 8 and 10).

Specific Binding to Receptors, Enzymes, or Other Biomolecules

Many toxic agents exert their effects through binding to a limited number of target sites or biomolecules (i.e., specific binding). Such binding can be divided into three categories based on the type of binding that occurs: *covalent binding* (*adducts*) or *noncovalent binding* (*intercalation*), and general binding to other biomolecules. In this chapter, binding is classified by the site of binding: either to enzymes, receptors, or other biomolecules.

Binding to Other Biomolecules

Adducts

When two distinct chemical entities covalently join, an adduct (**add**ition pro**duct**) is formed. In toxicology, the term *adduct* is usually used to describe the product of a toxic agent covalently bound to DNA, although adducts may also arise from the binding to other biomolecules such as proteins. Typically, adducts are generated between reactive metabolites (formed during phase 1 biotransformation reactions) and biomolecules. These reactive metabolites are electrophilic (electron seeking) compounds that bind to nucleophiles that are electron rich (i.e., molecules that donate/share electrons) such as DNA.

Benzo(a)pyrene (*BaP*), a *polycyclic aromatic hydrocarbon* (*PAH*), causes cancer by binding to DNA. However, BaP needs to be metabolically activated before binding will occur. Phase 1 metabolism of BaP produces reactive metabolites that are electrophilic and readily form DNA adducts. These adducts can cause DNA copy or transcriptional errors, which can potentially affect all downstream products. Nowhere is such modification of more profound impact than if adducts occur in sections of the genome that code for growth. For example, aberrant tissue growth (tumors) can result from adducts with *proto-oncogenes*, which code for proteins involved with promoting cell growth or differentiation, or through alteration of *tumor suppressor genes*, which inhibit cell growth. Bottom fish such as the English sole living in areas of high urban and industrial activity show increased frequency of neoplasms likely attributable to PAH pollution (Varanasi *et al.*, 1989).

Intercalation

Small molecules may become inserted in between the planar adjacent base pairs of DNA. These noncovalent interactions, known as intercalation or groove-binding, may cause a three-dimensional change in the DNA that can increase the length of the strand, cause it to unwind, or break a chromosome. Subsequent DNA transcription or replication may be impaired, and frameshift mutagenesis (i.e., insertion of an extra DNA base) may occur (Snyder *et al.*, 2005). A well-known example of an intercalating compound is acridine, a planar, three-ringed carbon molecule containing a single central nitrogen atom, often found in coal tar. Acridine intercalates DNA and causes frameshifts, which may lead to carcinogenesis. Intercalation or nonbinding associations may occur with biomolecules other than DNA; however, the contribution of such interactions to toxicology are presently unknown.

Enzyme and Receptor Binding

Many known toxic effects are mediated through altering the performance of enzymes and receptors. The focus of this section is on these proteins in particular because toxic effects mediated through these molecules are a well-researched and important area of toxicology. Proteins that *catalyze* chemical reactions and are not reversibly changed by the reaction are called enzymes. Proteins where the binding of a *ligand* (typically a small effector molecule) gives rise to *transduction* (conversion of one signal to another) are called receptors.

The actions of exogenous chemical agents on enzymes and receptors are typically divided into two categories. For enzymes, agents that enhance or activate a response are *activators*, whereas those that reduce or inhibit a response are *inhibitors*. For receptors, agents that enhance or activate are referred to as *agonists*, and those that decrease or abolish the ability of a ligand to evoke a response are *antagonists/inhibitors*. Although we have focused our attention on inhibitors, it should be noted that activators and agonists can cause important toxic effects in organisms. For example, the synthetic estrogen ethynylestradiol (EE2) is a potent estrogen receptor agonist. EE2 is such because it binds to the estrogen receptor more strongly than estrogen (Denny *et al.*, 2005). As a result, sewage effluents containing EE2 discharged into the environment can cause the feminization of male fishes (Palace *et al.*, 2006) (see also Chapter 9). Overall, agents can act at the same site as the natural receptor ligand(s) or enzyme substrate(s), whereas others will act away from this site, at an *allosteric site*.

Enzymes and Enzyme Inhibition

Michaelis-Menten kinetics describes enzyme reaction rate characteristics and are used to determine how toxicants are affecting enzymes. Normally, when a fixed amount of enzyme is present, the velocity of the conversion of the substrate to product increases as the concentration of the substrate is increased until the enzyme is working at its maximum velocity (V_{max} [Fig. 6-6b]). The ability of the enzyme to bind to its substrate is described by K_m, which is the substrate concentration that yields half of the V_{max}. Higher K_m values indicate a lower binding affinity (i.e., you have to add more substrate to get half V_{max}). Analysis of enzyme-substrate interactions are examined by using double-reciprocal plots of the previous figure (Fig. 6-6b) and are known as a *Lineweaver-Burk plot*s. K_m is determined by the negative inverse of the x-axis intercept, and V_{max} is determined by the inverse of the y-axis intercept.

Three major categories exist to describe enzyme inhibition by toxicants: *competitive*, *noncompetitive*, and *mixed inhibition*. The changes that each type of inhibition elicits in enzymes can be visualized as changes in the x- and y-intercepts of the Lineweaver-Burk plots (Figs. 6-6c–e). In short, competitive inhibitors cause an increase in K_m (i.e., a decrease in binding affinity), without appreciably altering V_{max} (Fig. 6-6c), by competing with the natural substrate for the active site of the enzyme. The effects of commonly used *organophosphorus* (*OP*) and *carbamate* pesticides on *acetylcholinesterase* (AChE) are pertinent examples of competitive enzyme inhibition. AChE breaks down the neurotransmitter acetylcholine (ACh) following its release from a presynaptic nerve ending, stopping the signal from continuing once the nerve signal has been received. Both OPs and carbamates compete with ACh for the active site on the AChE enzyme and slow down the breakdown of ACh in nerves, resulting in a continued nerve firing and possible paralysis.

Conversely, *noncompetitive inhibitors* cause a decrease in the ability of enzymes to catalyze the reaction (lower V_{max}), without altering the binding affinity of the substrate (Fig. 6-6d). Mixed inhibition occurs with agents that cause a combination of effects: they cause a decrease in V_{max} and an increase in K_m (Fig. 6-6e). In some rare cases, agents will bind to the enzyme-substrate complex, and this causes a decrease in K_m (i.e., an increase in binding affinity) and a decrease in V_{max}. These agents are referred to as *uncompetitive inhibitors*. An example of uncompetitive inhibition is the action of epristeride (a pharmacologic agent) on 5 α-reductase, which catalyzes testosterone's conversion to dihydrotestosterone (Levy *et al.*, 1994). Because dihydrotestosterone triggers growth in prostate cells, people suffering from prostatic cancer have been treated with this enzyme inhibitor.

Receptors and Receptor Inhibition

The binding characteristics of receptors are known as *receptor-ligand kinetics*. Increasing the concentration of the receptor ligand (free ligand) with a fixed number of receptors will yield a maximum amount of receptor binding (bound ligand) (B_{max}) (Fig. 6-7a). The affinity of the receptor for its ligand is described by K_d, which is the ligand concentration that results in half of the B_{max}. Analysis of receptor-ligand interactions is typically carried out with its own procedure for a linear transformation of the data, known as *Scatchard analysis* (Fig. 6-7b). Here, the y-axis is the ratio of the bound/free ligand, and the x-axis is simply the amount of bound ligand. Using Scatchard plots, the slope of the line is equal to $-1/K_d$ and the x-intercept is B_{max}. For this reason, Scatchard plots provide an excellent method to determine how toxicants can change binding affinity (K_d) or maximum binding (B_{max}).

As with enzymes, changes in the ability of the ligand to bind its intended site help classify receptor inhibition. As mentioned previously, Scatchard analysis can show changes in K_d and B_{max}, allowing the mechanism of inhibition to be determined. As with enzymes, both competitive and noncompetitive inhibition exists for receptors, although their meaning is not synonymous with those for enzymes.

Receptor competitive inhibition results in an increase in K_d (i.e., decreased binding affinity), which may be brought about in a *classical* way (i.e., through a toxicant binding reversibly to the ligand site) or through *nonclassical, allosteric* inhibition. In *nonclassical, allosteric* inhibition, the toxicant binds to an allosteric site causing an alteration in the ligand binding (orthosteric) site and increasing K_d (Fig 6-7c). The antagonist action of atropine, an alkaloid isolated initially from the nightshade plant *Atropa belladonna*, on

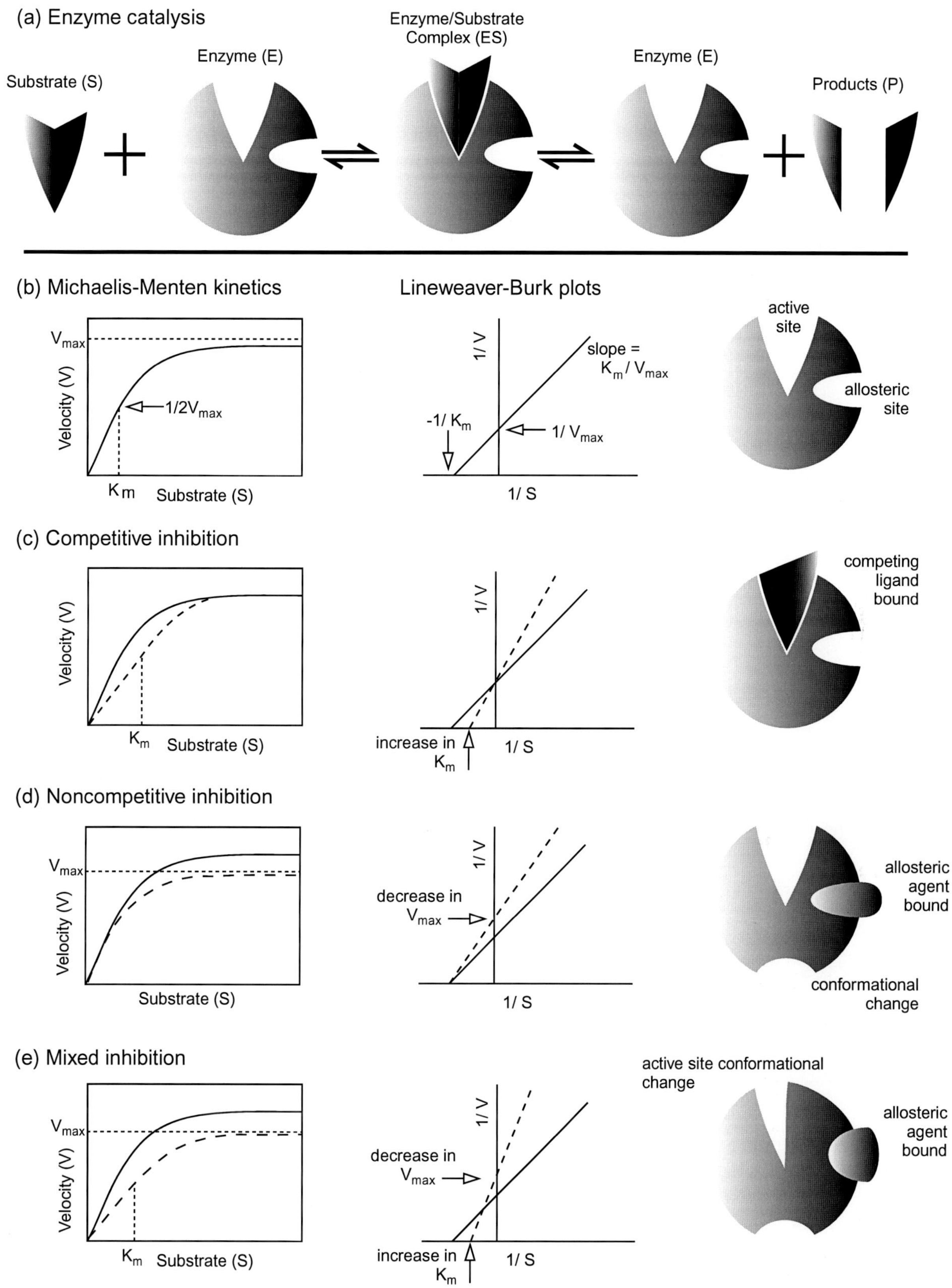

FIGURE 6-6. (a) A typical enzyme-substrate model depicting reaction catalysis. (b) The accompanying Michaelis-Menten enzyme kinetics, where the dissociation constant (K_m) is equal to half of the maximum reaction rate (V_{max}) and the common Lineweaver-Burk linear transformation of this data. (c) The increase in K_m typical of competitive inhibition. (d) An agent bound at an allosteric site causing a decrease in the functionality of this enzyme (illustrated as a missing notch from the enzyme) such that the reaction rate is slowed. (e) Also indicates the binding of an agent at an allosteric site; however, here the enzyme's active site has also been altered in shape, which will cause a decrease in binding affinity.

acetylcholine receptors is a good example of classic competitive inhibition in receptors. Atropine increases the K_d of acetylcholine receptors for ACh by binding more strongly than this natural ligand.

Similarly, there are two mechanisms by which noncompetitive inhibitors affect receptors, both of which will be evident as a change in B_{max} (Fig. 6-7d). An agent that binds at an allosteric site may cause a decrease in B_{max}. However, an agent that binds irreversibly at the ligand's site may cause a similar change in B_{max} by effectively reducing the receptor number. Cyclothiazide, a commonly used diuretic, is an example noncompetitive allosteric inhibitor of certain glutamate receptors. Cyclothiazide binds to a different site than glutamate and in so doing inhibits the receptors functionality (Surin *et al.*, 2007).

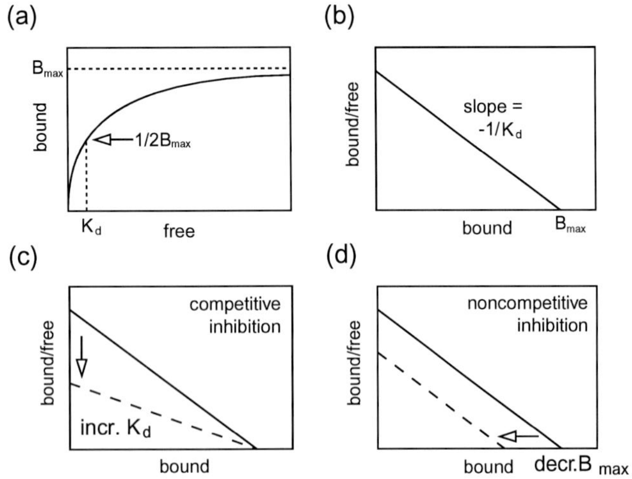

FIGURE 6-7. Receptor-ligand kinetics, depicting (a) the relationship between bound and free ligand at equilibrium and how this relates to the maximum receptor binding (B_{max}) and the receptor affinity (K_d) (equal to the ligand concentration at half of the B_{max}). (b) The Scatchard linear transformation of this data. (c) The increase in K_d observed with competitive inhibition and (d) decrease in B_{max} observed with noncompetitive inhibition.

Downstream Effects in the Whole Organism

Biological systems tend to operate close to a performance optimum that remains fairly constant through time (excepting that there will be changes with growth, development, seasons, etc.) (Fig. 6-8). This stability can be altered by xenobiotic exposure. In terms of performance alterations (of electron transport, biochemical reactions, nerve transmission, muscle function, cell growth, reproduction, etc.), there are two types: those that inhibit performance (i.e., are *inhibitory*) and those that stimulate it (i.e., are *stimulatory*). Both types move the animal away from its preferred state.

Increases in chemical doses will generally move organisms away from the optimum, causing performance alterations in a dose-dependant manner. Over time, *adaptation* (Fig. 6-8) could return performance to baseline levels. These processes can restore the performance optima through genetic, physiological or behavioral means. If, however, the dose surpasses a *threshold* where adaptation is not possible (i.e., negative effects exceed the animals ability to repair or compensate for toxic effects), survival may still be possible but the organism will operate in a suboptimal condition (Fig. 6-8). Adaptation and living in a suboptimal condition can incur significant energy costs to the animal. With sustained stimulation without adaptation, *debt repayment* may ensue (Fig. 6-8). This debt repayment may take an organism to an exhaustive (i.e., energy depleted) state. This may result in organisms of poor quality and eventually result in lower reproductive output and hence overall fitness. As an example of such costs, salmon exposed to moderate level of the AChE inhibitor chlorpyrifos did not appear to lose any swimming ability, which was unexpected because AChE is a major factor in maintaining muscle performance (Tierney *et al.*, 2007a). The exposed salmon retained their swimming ability by using anaerobically poised white muscle to make up for the inhibited aerobically poised red muscle. However, this resulted in an oxygen debt, as evidenced by elevated

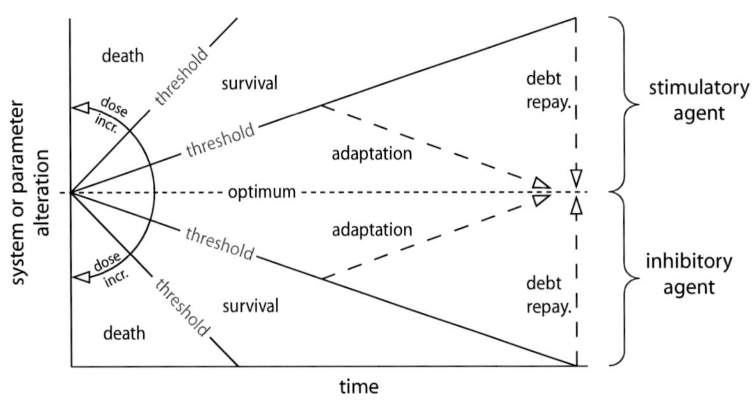

FIGURE 6-8. The concept of how increasing concentrations of a contaminant may affect the optimum performance of a system or a parameter.

concentrations of the anaerobic metabolite lactate. This debt may have impaired any subsequent exercise. If chlorpyrifos exposure had been continued for these salmon, a second, lethal threshold may have been surpassed (Fig. 6-8). Here coping and adapting to the toxic insult would not have been possible and mortality would have ensued.

TOXICOLOGICAL ASSESSMENT

The extent of exposure and the magnitude of effects in an organism form a correlative relationship that is at the foundation of toxicology and is called the *dose-response (DR)* or *concentration-response (CR) relationship.* Typically, the term *dose* is used when a known amount of a chemical is administered (e.g., intravenous injection) and *concentration* is used when the amount entering an organism is unknown (e.g., a fish exposed to a specific concentration of a chemical in an aquarium). The DR-CR relationship assumes that exposure to the chemical is causing the responses seen. It also assumes that the dose administered or exposure concentration is related to the concentration at the target site, and as the toxicant exposure is increased, effects will also increase. Often a *threshold concentration* exists, which is a concentration below which no effects are seen (Fig. 6-9b). However, the existence of a threshold may reflect the limitation of the measurement technique more than the absence of effects at lower concentrations. For example, if lethality is being measured as the response and a threshold concentration exists below which no death is observed, sublethal effects may be occurring that are not observed (e.g., a biochemical alteration may be occurring).

DR-CR curves are typically sigmoidal in shape (Fig. 6-9b). This shape can be attributed to the underlying normal distribution of traits that determine the susceptibility or tolerance of individuals (Fig. 6-9a). In any group, genotypic and phenotypic differences will give rise to a range of toxicant susceptibilities, such that a few organisms in the group are very susceptible, while the majority in the group have an "average" susceptibility, and a few are tolerant. This principle of variation in susceptibility is true across levels of biological organization, as molecules, cells, and tissues will also have a range of susceptibilities. For example, in a toxicity test, responses will rise slowly at first (from 0% response) as the concentration of a toxicant is increased because there are only a few sensitive individuals. When the concentration reaches a level to which the "average" is susceptible, the response rises steeply as there are many of the group that have this level of vulnerability. Then the response rises slowly again as the concentration continues to increase, as there are few tolerant individuals remaining, until a concentration is reached at which 100% of the test subjects are affected. Typically, toxicities are reported as the concentration or dose that yields 50% of a response. At this point, the error associated with this toxicity estimate is lowest, whereas the confidence highest. Additionally, consistent reporting of the 50% value facilitates the comparison of toxicity values across studies and between chemicals.

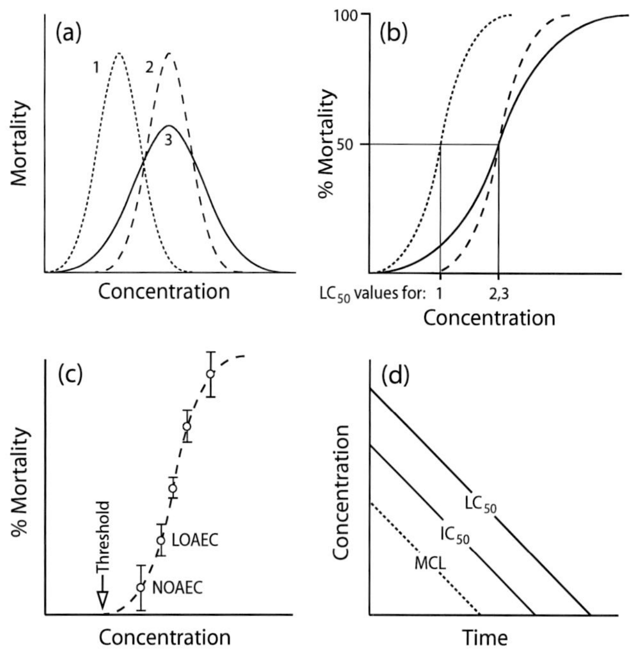

FIGURE 6-9. Toxicity tests, their derivation and their implications. (a) The normal (Gaussian) distributions of the mortality observed in three different species exposed to a toxic chemical. (b) A cumulative transformation of this data yields the typical sigmoidal distribution of most lethality curves. Example LC_{50} values are given to show how differences in susceptibility relate to curve shape and location. (c) Example data for curve 2, depicting the toxicity threshold, the no observed adverse effect concentration (NOAEC), the highest concentration where the response is not significantly different from zero, and also the lowest observed adverse effect concentration (LOAEC), the lowest concentration where the response is significantly different from zero. (d) The relationship between exposure time and the concentration needed to cause 50% of the effect. For example, with extended time periods, LC_{50} concentrations are lower, indicating that chronic exposure at low concentrations can yield similar toxic effects as short exposures to high concentrations. This is true for both lethal (e.g., LC_{50}) and sublethal (IC_{50} and other) effects. A dashed line is used to suggest a position for an environmental maximum contaminant level (MCL) where organisms would be insulated from any known toxic effects.

There are a variety of *biological endpoints* that have been used in toxicity tests to capture responses at all levels of biological organization (Table 6-1). The best toxicity endpoints are those that are closely associated with the mode of action, unequivocal and clearly relevant to the compounds toxic effects, or relevant to the environmental concern. The duration of many toxicity tests are categorized as *acute, subchronic,* or *chronic*. The first two of the three categories are irrespective of the animal's life span. They are simply defined as up to or beyond 5 days in duration (i.e., *96-hr* acute versus *5-d* subchronic). This terminology is believed to be at least partially due to the length of a typical work-

TABLE 6-1. Toxicological endpoints as they relate to biological organization and endpoint class. The relevance of these endpoints to a mechanistic understanding of toxic action and to significance at the ecosystem level are also indicated (i.e., mechanistic understanding is highest at the molecular level).

	Biological Endpoint	Biological Level of Organization	Endpoint Class	Endpoint Relevance
1	Molecular	Subcellular		Mechanistic
2	Biochemical	Subcellular and cellular	Biomarkers (suborganismal)	
3	Physiological	Tissue and organ		
4	Behavioral	Organism	Bioindicators (organismal)	
5	Ecological	Population Community Ecosystem		Ecorelevance

week—acute testing can be completed within a week. In contrast, chronic testing is longer term and often proportional to the organism's life span, potentially representing 10% or greater of the life span. These exposure lengths conceivably have environmental relevance, as acute exposure may simulate a contaminant discharge that occurs from a *point source*, meaning that the discharge is easily quantified and possibly short lived. Conversely, chronic testing may better simulate *nonpoint source* discharges (i.e., those inputs that tend to be difficult to quantify in space and time) and typically result in low contaminant concentrations over long time periods.

Lethality Endpoints

The most commonly measured endpoint of toxicity is mortality as measured in *lethality* tests because it is *binary* or *quantal* (i.e., there is no gray area in determining the response) and because survival is an ecologically relevant endpoint. Data for these tests are collected through subjecting several groups of organisms to a series of discrete chemical exposure concentrations, such that at least three and preferably more concentrations result in some deaths and some survivals (i.e., kill less than 100% and more than 0% of a test group). The resulting concentration and mortality response data (usually plotted as percentage mortality and not the number of dead animals) is plotted and statistics are used to calculate the concentration or dose that relates to 50% mortality in the test animals (Fig. 6-9b). The standardized nomenclature of lethality tests makes them readily identifiable; for example, in mammalian studies or those where the chemical is administered, the dose causing mortality in 50% of the test population is abbreviated LD_{50} (for *lethal dose*). Units for this measure are mass of agent per mass of organism (e.g., mg/kg). For an organism exposed externally to an agent (such as daphnid in a beaker), this toxicity point is described in terms of a concentration, the LC_{50}, which has units of mass per unit volume (mg/L). For example, the oral LD_{50} for brevetoxin, the toxin associated with neurotoxic shellfish poisoning from red tides, was 520 μg/kg in the mouse (Baden and Mende, 1982). By comparison, a 24-hr LC_{50} (mortality measured after 24-hr exposure) was 21 μg/L for the striped mullet (Pierce, 1993). Given these data, it is difficult to say whether this neurotoxin is more toxic to mammals or fish because the LD_{50} and LC_{50} values, respectively, have different units; the actual dose that was absorbed by the fish is unknown. Regardless, both of these values are low compared to many compounds; a low LC_{50} or LD_{50} value means that even small amounts of chemical can result in toxic effects. Two other values of note on lethality curves are the no observed adverse effect concentration (NOAEC), which is the highest concentration not significantly different from zero, and the lowest observed adverse effect concentration (LOAEC), which is the lowest concentration that is significantly different from zero.

Sublethal Endpoints

Toxicity testing using lethality as an endpoint has achieved wide acceptance because it produces unambiguous and easily obtained toxicity data. However, the relevance of lethality tests may be limited to short-term, high concentration exposures. Exposure to toxic agents may more typically bring about alterations in the "health" or condition of an organism. In situations where chemical concentrations are lower than those that can cause lethality, particularly under chronic exposure conditions, other effects may occur that do not kill the organism. Any *sublethal effects* that alter growth, immune system function, swimming ability, behaviors, reproduction and so on reduce the fitness of organisms. For this reason, environmental relevance may be improved by using tests that measure changes in *biomarkers* (*suborgan-*

ismal level endpoints) or *bioindicators* (*organismal* level or above endpoints). There are a variety of such endpoints, and these can be categorized according to increasing levels of biological organization (Table 6-1). Each successive biological order integrates multiple lower levels and typically has a higher buffering capacity. Examples of each are provided next, but note that a suite of such measures should be employed to estimate the impact of a toxic agent.

Molecular Endpoints

These endpoints are based on *molecular biology*, which includes the study of genes, genomes, RNA, proteins, and the interactions between them. Subdisciplines that have respective foci in these areas are *genomics*, *transcriptomics* (mRNA expression), and *proteomics* (protein expression). All three subdisciplines, together in concert with the action of toxic agents on them, comprise one of the more topical areas in toxicology: *toxicogenomics*. In general, molecular endpoints have the advantage of rapidly providing valuable insight into the basic mechanisms by which compounds cause toxicity. However, because these endpoints represent the first level of toxicity endpoints (Table 6-1), they also potentially provide the least ecological relevance/information.

As mentioned earlier, reactive species may bind to genetic material (DNA) and form adducts. Although quantifying DNA adducts is not a direct measure of a toxic effect, it can be an indicator of potential effects such as altered transcriptional products and replication. New developments in molecular biology have allowed toxicologists to move forward from adduct measurement to measuring the effects of toxicants on the expression of multiple genes; the tool is the *DNA microarray* or *genechip*. Microarrays are commercially available in some cases for entire genomes, such as yeast and humans. Specific arrays can also be manufactured to contain genes from particular biochemical pathways (e.g., estrogenic responses; Larkin *et al.*, 2002). Gene chips enable quantification of up- or down-regulated genes in treated versus control organisms. For example, 95 genes on an Atlantic salmon/trout microarray consisting of 16,000 gene probes were found to be altered in rainbow trout exposed to 1 μg/L BaP for 7 days (Hook *et al.*, 2006). Not surprisingly, many of the expression changes were associated with protective and stress response proteins such as detoxification enzymes and heat-shock proteins.

Biochemical Endpoints

Biochemical endpoints include various measurements of proteins (i.e., the final level of biological organization of the preceding section) and other biomolecules, such as carbohydrates and lipids. Unlike the previous section, which dealt largely with interactions between DNA or RNA and protein, biochemical endpoints tend to explore processes, such as enzymatic reactions. For example, the inhibition of enzymes is often used in biochemical tests of pesticide potency. Here inhibition may be quantified in terms of IC_{50}, or the concentration of the toxic agent causing 50% enzyme inhibition. For example, the IC_{50} for rat brain acetylcholinesterase by the insecticide chlorpyrifos was 10 nM (Mortensen *et al.*, 1998).

Biochemical changes may also be manifest at the cellular level, and so cellular responses in a system can be used as endpoints. For example, exposure to some environmental contaminants can evoke *apoptosis*, or programmed cell death. Apoptotic cell death is characterized by a defined sequence of events: chromatin condensation, DNA fragmentation, membrane blebbing, compartmentalization of cellular contents, and finally phagocytosis of dying cells. Several techniques can be used to detect apoptosis in exposed animal tissues by measuring DNA fragmentation and proteolytic enzymes (enzymes that break apart protein) known as caspases.

Physiological Endpoints

These *suborganismal* and *organismal* endpoints are measures of alterations in the mechanics and functioning of tissues, organs, or whole organisms. Proper functioning of complex systems relies on intact and integrated components. The utility of these tests are that toxic insult on any subcomponent will reduce overall performance producing a measurable effect. Function can be measured either *in vitro* or *in vivo*. *In vitro* assessment may consist of isolated cells; for instance, testicular performance after toxicant exposure can be assessed using sperm motility rate (Moline *et al.*, 2000). Organ performance can be assessed using isolated tissues or *in vivo* (i.e., directly in the animal). For example, to determine if pesticides can affect olfactory performance in fish, the electrical response of olfactory neurons to different odorants can be assessed before and after exposure. Results from one study showed that exposure of juvenile salmon to the phenylurea herbicide linuron at 10 μg/L for only 2 minutes will reduce the ability of their olfactory neurons to detect odorant cues associated with predators (Tierney *et al.*, 2007b).

Physiologic performance can also be assessed using whole organism responses. For example, chlorpyrifos exposure impairs acetylcholinesterase, an enzyme critical to the functioning of the cholinergic nervous system. Because this part of the nervous system controls musculoskeletal function, swimming performance would be expected to decline in chlorpyrifos-exposed salmon (Tierney *et al.*, 2007a). Because swimming performance is vital to salmon survival, this endpoint has the advantage of taking a known biochemical inhibition and applying it to a more ecologically relevant endpoint. One drawback to physiological endpoints is that they are typically more costly or difficult to attain. For example, to test swimming performance, several whole fish need to be tested in an expensive and sizable swim tunnel.

Behavioral Endpoints

Behavior consists of the actions that an organism takes within or to its biotic and abiotic environment. Behavior results from an integration of sensory input, the cognitive response(s) the input may evoke, which perhaps includes decision making, and any resulting motor coordination. Toxicants can alter behavior, and *behavioral toxicology* or *ethotoxicology* has emerged to study how normal behavior may be changed. This developing field is of boundless potential, because the number of potential behaviors is unlimited—especially considering that many responses are learned. Nevertheless, all actions are similar in that they are designed to place an organism in favorable conditions. This principle may be exploited in a variety of ways to determine the effects of toxic agents.

Like any other toxicity endpoint, a behavioral endpoint should be measurable, repeatable, sensitive, and ecologically relevant (i.e., have explicit relevance to survival). Repeatability can arise from a strong innate response, such as salinity preference or avoidance in aquatic organisms, or through training, such as by reward-based feeding. In fact, trainability itself may be a good metric. Other potential endpoints could be based on food location, predator avoidance, social interactions, or reproductive ability. Locomotor activity is sometimes included; however, it only applies if more than basic performance is assessed (as in the example of swimming performance from the previous section).

Although the majority of behavioral tests are acute, important alterations in organism behavior may also be apparent after chronic exposures or even *multigenerational exposures*. For example, female mice that had been fed 10 μg/kg of the endocrine disrupting chemical (EDC) bisphenol A during days 10 to 14 of their gestation, spent less time in the nest and nursing new pups when they were themselves adults (Palanza *et al.*, 2002). Learning and memory may also be affected by early exposure. Adult female rats that had been exposed to three PCB congeners (28, 118, and 153) *in utero* and during lactation, were slower to acquire memory (Schantz *et al.*, 1995). Such ethological differences may conceivably translate to population level changes.

Population to Ecosystem Endpoints

Extrapolation of the previous toxicity endpoints to population-level effects can be difficult because of a number of factors. One of the primary ones is the lack of adequate inclusion of individual variability, which can be accomplished through the use of a representative sample of the population. Lethality tests discussed earlier may achieve this, but the endpoint fails to capture sublethal effects. Other limitations salient to lab-based work—such as environmental modifying factors of toxicity—can likewise be reduced by conducting fieldwork. Limitations of such work are obvious and include time, money, and ability to or difficulty in manipulating natural environments.

It can be argued that the effects of toxicants on individual organisms are not meaningful unless they impact population dynamics. Population-level effects may be evident as alterations in a population's number of individuals, density, age structure, age structure cycling, or genetic structure (e.g., changes in gene frequencies of certain alleles). In some cases, measures of organismal fitness (i.e., how well individuals contribute to future generations) can be assessed and extrapolated to the population level using multigeneration toxicity tests and modeling. The direct effects of pollutants that impact growth, reproduction, and survivorship are the types of changes that can be most easily translated into population-level responses through standard population models. For example, the effects of Aroclor 1254, a PCB mixture used extensively in North America into the 1970s and the source of many legacy PCBs, were examined by feeding three generations of mice 5 mg/kg of PCBs (McCoy *et al.*, 1995). Results were that not only were the first and second-generation offspring significantly smaller, but by the second generation, litter sizes were reduced and fewer offspring survived weaning. This finding demonstrates how sublethal effects that directly affect population sizes may be manifest after a few generations. These assessment techniques help bridge the gap between exposure, toxicological effects at the sub- or organismal level, and risk assessment at the population level (Rose *et al.*, 1999).

Attrill (2002) defined a community as "a group of organisms occurring in a particular environment, presumably interacting with each other and with the environment, and separable by means of ecological survey from other groups." The advantages to community assessments include ecological relevance, a multispecies response, integrating conditions over long time periods and sometimes complex mixtures of toxicants (Attrill, 2002). Micheli *et al.* (1999) categorized community-level toxic assessment measurements into two types: compositional variability, which are changes in the relative abundance or biomass of component species, or aggregate variability, which are changes in summary properties of the community such as total abundance, biomass, and richness. A number of experimental designs are employed to carry out these measurements in multispecies toxicity tests, and these range from laboratory-based (microcosms, mesocosms) to field-based community investigations. There has been much scientific argument regarding the use of both lab-based and field-based investigations (Attrill, 2002; Landis and Yu, 1999). Microcosms and mesocosms are multispecies test systems that include simplified naturally assembled ecological structures that can provide replicated, controlled, and repeatable conditions (which has as a drawback in that natural systems include both spatial and temporal heterogeneity). Field-based community-level investigations have also been criticized (e.g., for lacking appropriate control or reference sites). It is essential that all community-level investigations include appropriate spatial and temporal replication and, in particular, replicated control locations (Atrill, 2002).

At the ecosystem level, both its structure (community, habitat, etc.) and function can be examined. For example, the effects of contaminant stress on ecosystem productivity and respiration have been measured. Perhaps partially prompted by concerns over global warming (see Chapter 1), one study measured the flux of CO_2 from 18 European forests (Janssens *et al.*, 2001). An interesting finding was that soil disturbance from anthropogenic activity can cause a forest to be a net producer of carbon (i.e., further global warming).

Obtaining a firm understanding of the effects of xenobiotics on organisms can begin at any level of the biological hierarchy; however, most often studies have been initiated following the observation of toxic effects in the wild, in response to declines in the abundance of species or habitat. For example, studies on the effects of bleached Kraft pulp mill effluent began following observations of altered reproductive status and states in the white sucker *Catostomus commersoni* in Lake Superior (Munkittrick *et al.*, 1992; Servos *et al.*, 1996). As mentioned earlier, studies at lower levels in the hierarchy yield mechanistic understanding, whereas those at higher levels have more ecological relevance. Therefore, a suite of studies spanning several levels is desired, and the program of study should be designed with the following goals in mind (modified from Johnson and Collier, 2002): (1) to establish causality between the toxicant and effects; (2) to link effects with underlying biochemical/physiological events, thereby providing insight into mechanism of action and enhancing prediction of more severe whole-organism impacts (for early warning systems); (3) to predict how effects on individuals might affect populations characteristics; and (4) to assess community and ecosystem-level effects of population change by incorporating any modifying factors such as the interactions between species.

Environmental Risk Assessment

Much of environmental toxicology, although not exclusively, is conducted with application in mind. One such application is *environmental risk assessment* (*ERA*). Environmental risk assessment is a tool used to estimate or *measure the probability* that a *hazard* (i.e., a chemical present in the environment) will cause harm to human or ecological receptors (targets). In risk assessment, to estimate the potential for a chemical to present an unacceptable risk to a receptor, three factors are considered: (1) the characteristics of the chemical (i.e., concentration and chemical properties), (2) the potential for a receptor to be exposed to the chemical, and (3) the potential for toxic effects (Fig. 6-10). These factors must be considered collectively as they are all important in estimating risk. For example, the mere presence of a chemical in the environment does not necessarily imply there is a risk to a receptor; the receptor must first be exposed to the chemical. Likewise, the receptor's exposure to the chemical does not necessarily equate to unacceptable risk, the toxicity of the chemical must be evaluated. In risk assessment, all available information is compiled to characterize potential exposures and effects and to then integrate them into an understanding of risk (probability of harm).

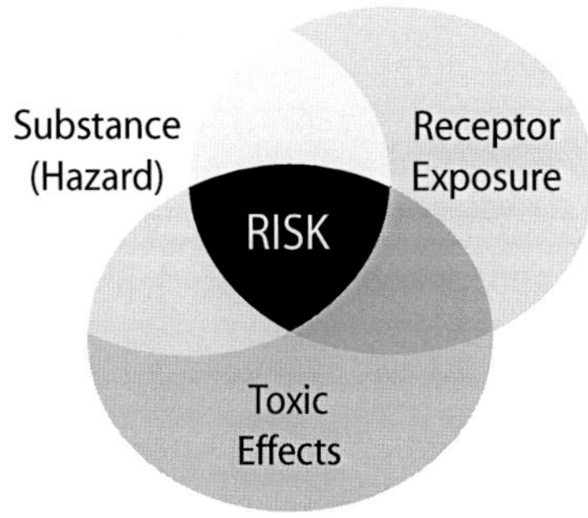

FIGURE 6-10. The environmental risk assessment paradigm.

The adoption of risk assessment as a fundamental component of environmental assessment and decision making has been stimulated by the recognition that eliminating the environmental effects of human activities is an impossibility and that based on the widespread nature of chemical impacts to the environment, there is a need for a framework that allows for the prioritization of the impacts for decision and policy-making purposes (e.g., to remediate a contaminated site or not?). Risk assessment may be performed at many scales depending on the management goals. It may be conducted to evaluate risks to an individual or a population, and it may evaluate local (e.g., a hazardous waste site), regional (e.g., Chesapeake Bay) or global (e.g., global warming) contamination.

Both human health and ecological risk assessments are performed within clear guidelines or frameworks that continue to develop as information evolves. For brevity, the steps involved in a typical ecological risk assessment (ERA) are described (U.S. Environmental Protection Agency [USEPA], 1998). Step 1, *problem formulation*, involves identification of chemicals of potential concern (COPC), as well as the development of an analysis plan, which defines goals, endpoints, and measures of effect to be evaluated at the site. Step 2, the *analysis phase*, involves the characterization of exposure and the characterization of effects. This phase involves the identification of populations of organisms (including humans) with the potential to be exposed to the COPCs, as well as the identification of exposure pathways by which the receptors could be exposed. Characterization of effects may include the use of models to estimate the dose the receptor is exposed to, or it may include toxicity tests using contaminated media (i.e., sediment or surface water). The toxicity assessment component of this step involves the compilation and evaluation of toxicological information for the COPCs in order to evaluate the potential for the chemicals to cause adverse effects in exposed individuals. The compiled information is utilized in step 3 of the risk assessment, *risk characterization.* In risk characterization, the results of the exposure and toxicity assessments are summarized and integrated into quantitative or qualitative expressions of risk. Risks are estimated by comparing the estimated chemical intakes (dose) to an "acceptable" level of exposure provided by a reference concentration for the COPC (i.e., a concentration deemed safe by toxicity testing). This comparison is called a hazard quotient; hazard quotients for COPCs are compared to the risk-based standard (usually a value of 1) to determine if the COPC poses an unacceptable level of risk to the receptor.

Risk assessment is one input to environmental management decisions. Other inputs include stakeholder concerns, availability of technical solutions, benefits, equity, costs, legal mandates, considerations of ecological values as well as ecosystem-based science, and political issues.

References

Attrill, M.J., 2002. Community-level indicators of stress in aquatic ecosystems. In Adams, S.M. (ed.), Biological Indicators of Aquatic Ecosystem Stress. Bethesda, MD, American Fisheries Society.

Baden, D.G., Mende, T.J., 1982. Toxicity of two toxins from the Florida red tide marine dinoflagellate, *Gymnodinium breve*. Toxicon 20, 457–461.

Bain, L.J., LeBlanc, G.A., 1996. Interaction of structurally diverse pesticides with the human MDR1 gene product P-glycoprotein. Toxicol. Appl. Pharmacol. 141, 288–298.

Bard, S.M., 2000. Multixenobiotic resistance as a cellular defense mechanism in aquatic organisms. Aquat. Toxicol. 48, 357–389.

Buhler, D.R., Williams, D.E., 1989. Enzymes involved in metabolism of PAH by fishes and other aquatic animals: Oxidative enzymes (or phase I enzymes). In Varanasi, U. (ed.), Metabolism of Polycyclic Aromatic Hydrocarbons in the Aquatic Environment, pp. 151–184. Boca Raton, FL, CRC Press.

Cornwall, R., Toomey, B.H., Bard, S., Bacon, C., Jarman, W.M., Epel, D., 1995. Characterization of multixenobiotic/multidrug transport in the gills of the mussel *Mytilus californianus* and identification of environmental substrates. Aquat. Toxicol. 31, 277–296.

Cresswell, J.E., Merritt, S.Z., Martin, M.M., 1992. The effect of dietary nicotine on the allocation of assimilated food to energy metabolism and growth in fourth-instar larvae of the southern army worm, *Spodoptera-eridania* Lepidoptera Noctuidae. Oecologia 89, 449–453.

Dash, J.A., 1988. Effects of dietary terpenes on glucuronic acid excretion and ascorbic acid turnover in the brushtail possum, *Trichosurus vulpecula*. Comp. Biochem. Physiol. 89B, 221–226.

Denny, J.S., Tapper, M.A., Schmieder, P.K., Hornung, M.W., Jensen, K.M., Ankley, G.T., Henry, T.R., 2005. Comparison of relative binding affinities of endocrine active compounds to fathead minnow and rainbow trout estrogen receptors. Environ. Toxicol. Chem. 24, 2948–2953.

Dewailly, E., Ayotte, P., Bruneau, S., Gingras, S., Belles-Iles, M., Roy, R., 2000. Susceptibility to infections and immune status in Inuit infants exposed to organochlorines. Environ. Health Perspect. 108, 205–211.

Dewailly, E., Nantel, A., Bruneau S, Laliberte C, Ferron L, Gingras S., 1992. Breast milk contamination by PCDDs, PCDFs and PCBs in Arctic Quebec: A preliminary assessment. Chemosphere 25, 1245–1249.

Eickhoff, C.V, Gobas, F.A.P.C, Law, F.C.P., 2003. Screening pyrene metabolites in the hemolymph of Dungeness crabs (*Cancer magister*) with synchronous fluorescence spectrometry: Method development and application. Environ. Toxicol. Chem. 22, 59–66.

Foley, W.J., 1992. Nitrogen and energy retention and acid-base status in the common ringtail possum, *Pseudocheirus peregrinus*, evidence of the effects of absorbed allelochemicals. Physiol. Zool. 65, 403–421.

Gerlach, J.H., Kartner, N., Bell, D.R., Ling, V., 1986. Multidrug resistance. Cancer Surv. 5, 25–46.

Gottesman, M.M., Pastan, I., 1988. Resistance to multiple chemotherapeutic agents in human cancer cells. TIPS 9, 54–58.

Guglielmo, C.G., Karasov, W.H., Jakubas, W.J., 1996. Nutritional costs of a plant secondary metabolite explain selective foraging by ruffed grouse. Ecology 77, 1103–1115.

Hook, S.E., Skillman, A.D., Small, J.A., Schultz, I.R., 2006. Gene expression patterns in rainbow trout, *Oncorhynchus mykiss*, exposed to a suite of model toxicants. Aquat. Toxicol. 77, 372–385.

Janssens, I.A., Lankreijer, H. Matteucci, G., Kowalski, A.S., Buchmann, N., Epron, D., Pilegaard, K., Kutsch, W., Longdoz, B., Grünwald, T., Montagnani, L., Dore, S., Rebmann, C., Moors, E. J., Grelle, A., Rannik, Ü., Morgenstern, K., Oltchev, S., Clement, R., Guðmundsson, J., Minerbi, S., Berbigier, P., Ibrom, A., Moncrieff, J., Aubinet, M., Bernhofer, C., Jensen, N. O., Vesala, T., Granier, A., Schulze, E.-D., Lindroth, A., Dolman, A. J., Jarvis, P. G., Ceulemans, R., Valentini, R. 2001. Productivity overshadows temperature in determining soil and ecosystem respiration across European forests. Glob. Change Biology 7, 269–278.

Johnson, L.L., Collier, T.K., 2002. Assessing contaminant-induced stress across levels of biological organization. Adams, S.M. (ed.), Bethesda, MD, American Fisheries Society.

Juliano, R.L., Ling, V., 1976. A surface glycoprotein drug permeability in Chinese hamster ovary cell mutants. Biochim. Biophys. Acta 455, 152–154.

Kennedy, C.J., Law, F.C.P., 1986. Toxicokinetics of chlorinated phenols in rainbow trout following different routes of chemical administration. Can. Tech. Report Fish. Aquat. Sci. 1480, 124–125.

Kurelec, B., Britvic, S., Pivcevic, B., Smital, T., 1998. Fragility of multixenobiotic resistance in aquatic organisms enhances the complexity of risk assessment. Mar. Environ. Res. 46, 415–419.

Landis, W.G., Yu, M.-H. (eds.), 1999. Introduction to Environmental Toxicology: Impacts of Chemicals upon Ecological Systems, Boca Raton, FL, Lewis Publishers, CRC Press.

Larkin, P., Sabo-Attwood, T., Kelso, J., Denslow, N.D., 2002. Gene expression analysis of largemouth bass exposed to estradiol, nonylphenol, and p,p9-DDE. Comp. Biochem. Physiol. 133B, 543–557.

Levy, M.A., Brandt, M., Sheedy, K.M., Dinh, J.T., Holt, D.A., Garrison, L.M., Bergsma, D.J., Metcalf, B.W., 1994. Epristeride is a selective and specific uncompetitive inhibitor of human steroid 5α-reductase isoform 2. J. Steroid Biochem. Mol. Biol. 48, 197–206.

Lindroth, R.L., Batzli, G.O., 1984. Plant phenolics as chemical defenses effects of natural phenolics on survival and growth of prairie voles, *Microtus orchrogaster*. J. Chem. Ecol. 10, 229–244.

Lyndersen, E., Lçfgren, S., Arnese, R.T., 2002. Metals in Scandinavian surface waters: Effects of acidification, liming, and potential reacidification. Crit. Rev. Env. Sci. Technol. 32, 165–180.

May, T.W., Wiedmeyer, R.R., Gober, J., Larson, S., 2001. Influence of mining-related activities on concentration of metals in water and sediment from streams of the Black Hills, South Dakota. Arch. Environ. Contam. Toxicol. 40, 1–9.

McCoy, G., Finlay, M.F., Rhone, A., James, K., Cobb, G.P., 1995. Chronic polychlorinated biphenyls exposure on three generations of oldfield mice (*Peromyscus polionotus*): Effects on reproduction, growth, and body residues. Arch. Environ. Contam. Toxicol. 28, 431–435.

Micheli, F., Cottingham, K.L. Bascompte, J., Bjørnstad, O.N., Eckert, G. L., Fischer, J.M., Keitt, H., Kendall, B.E., Klug, J.L., Rusak, J.A., 1999. The dual nature of community variability. Oikos 85, 161–169.

Moline, J.M., Golden, A.L., Bar-Chama, N., Smith, E., Rauch, M.E., Chapin, R.E., Perreault, S.D., Schrader, S.M., Suk, W.A., Landrigan, P.J., 2000. Exposure to hazardous substances and male reproductive health: A research framework. Environ. Health Perspect. 108, 803–813.

Mortensen, S.R., Brimijoin, S., Hooper, M.J., Padilla, S., 1998. Comparison of the in vitro sensitivity of rat acetylcholinesterase to chlorpyrifos-oxon: What do tissue IC50 values represent? Toxicol. App. Pharmacol. 148, 46–49.

Munkittrick, K.R., McMaster, M.E., Portt, C.B., Van der Kraak, G.J., Smith, I.R., Dixon, D.G., 1992. Changes in maturity, plasma sex steroid levels, hepatic mixed-function oxygenase activity, and the presence of external lesions in lake whitefish (*Coregonus clupeaformis*) exposed to bleached kraft mill effluent. Can. J. Fish. Aquat. Sci. 49, 1560–1569.

Naito, M., Yusa, K., Tsuruo, T., 1989. Steroid hormones inhibit binding of Vinca alkaloid to multidrug resistance related P-glycoprotein. Biochem. Biophys. Res. Commun. 158, 1066–1071.

Palace, V.P., Wautier, K.G., Evans, R.E., Blanchfield, P.J., Mills, K.H., Chalanchuk, S.M., Godard, D., McMaster, M.E., Tetreault, G.R., Peters, L.E., Vandenbyllaardt, L., Kidd, K.A., 2006. Biochemical and histopathological effects in pearl dace (*Margariscus margarita*) chronically exposed to a synthetic estrogen in a whole lake experiment. Environ. Toxicol. Chem. 25, 1114–1125.

Palanza, P., Howdeshell, K.L., Parmigiani, S., vom Saal, F.S., 2002. Exposure to a low dose of bisphenol A during fetal life or in adulthood alters maternal behavior in mice. Environ. Health Perspect. 10s3, 415–422.

Pierce, R.H., 1993. Mote Marine Laboratory Red Tide Research. Technical Report no. 284. St. Petersburg, Florida Department of Natural Resources, Mote Marine Laboratory.

Rose, K.A., Brewer, L.W., Barnthouse, L.W., Fox, G.A., Gard, N.W., Mendonca, M., Munkittrick, K.R., Vitt, L.J., 1999. Ecological responses of oviparous vertebrates to contaminant effects on reproduction and development. In DiGiulio, R.T., Tillit, D.E. (eds.), Reproductive Developmental Effects of Contaminants in Oviparous Vertebrates, pp. 225–281. Pensacola, FL, SETAC Press.

Ross, P.S., 2006. Fireproof killer whales (*Orcinus orca*): Flame-retardant chemicals and the conservation imperative in the charismatic icon of British Columbia, Canada. Can. J. Fish. Aquat. Sci. 63, 224–234.

Schanker, L.S., 1961. Mechanisms of drug absorption and distribution. Ann. Rev. Pharmacol. 1, 29–44.

Schantz, S.L, Moshtaghian J., Ness, D.K., 1995. Spatial learning deficits in adult rats exposed to ortho-substituted PCB congeners during gestation and lactation. Toxicol. Sci. 26, 117–126.

Servos, M.R., Munkittrick, K.R., Carey, J., Van der Kraak, G. (eds.), 1996. Environmental effects of pulp mill effluents. Boca Raton, FL, St. Lucie Press.

Snyder, R.D., McNulty, J., Zairov, G., Ewing, D.E., Hendry, L.B., 2005. The influence of N-dialkyl and other cationic substituents on DNA intercalation and genotoxicity. Mut. Res. 578, 88–99.

Surin, A., Pshenichkin, S., Grajkowska, E., Surina, E., Wroblewski, J.T., 2007. Cyclothiazide selectively inhibits mGluR1 receptors interacting with a common allosteric site for non-competitive antagonists. Neuropharmacology 52, 744–754.

Tierney, K.B., Casselman, M., Takeda, S., Farrell, A.P., Kennedy, C.J., 2007a. The relationship between cholinesterase inhibition and two types of swimming performance in chlorpyrifos-exposed coho salmon. In press, December 15, 2006, Environ. Toxicol. Chem. 26(5) [May 2007].

Tierney, K.B., Ross, P.S., Kennedy, C.J., 2007b. Linuron and carbaryl differentially impair baseline amino acid and bile salt olfactory responses in three salmonids. In press, December 3, 2006, Toxicology, doi:10.1016/j.tox.2006.12.001.

U. S. Environmental Protection Agency (USEPA), 1998. Guidelines for Ecological Risk Assessment. U.S. Environmental Protection Agency, Risk Assessment Forum, Washington, DC, EPA/630/R095/002F.

Varanasi, U., Reichert, W.L., Le Eberhart, B.T., Stein, J.E., 1989. Formation and persistence of benzo[a]pyrene-diolepoxide-DNA adducts in liver of English sole (*Parophrys vetulus*). Chem. Biol. Interact. 69, 203–216.

Watjen, W., Beyersmann, D., 2004. Cadmium-induced apoptosis in C6 glioma cells: influence of oxidative stress. Biometals. 17, 65–78.

STUDY QUESTIONS

A forestry company wishes to apply a pesticide to a forest for the control of a beetle infestation. The pesticide contains a carbamate insecticide and a surfactant, a compound that is believed to be a strong estrogen mimic. Although the forest canopy is the target of the spraying, many salmon-bearing streams will likely receive overspray. The company has contacted you to identify some risks to the salmon and the mechanisms by which these fish may be affected. In your response, consider these questions:

1. For the carbamate:
   a. What enzyme might be negatively affected?
   b. What type of inhibition is this considered?
   c. How might the K_m of the enzyme be altered?
   d. What symptoms would you expect to see in whole salmon accidentally poisoned by the carbamate, and how long after exposure might toxic effects be expected?
2. For the surfactant:
   a. What receptor(s) might the surfactant affect?
   b. What type of effect would occur?
   c. How would the K_d of the receptor be altered?
   d. What symptoms would you expect to see in whole salmon accidentally poisoned by the surfactant, and how long after exposure might toxic effects be expected?

After you supply your response, the company asks you to build a small research program to better study the potential problem of fish poisoning. Use questions 3 to 5 to guide a small proposal.

3. Using your knowledge of the dose-response curve, what types of experiments could you propose to test how this pesticide may affect the salmon? Include at least five endpoints that would be relevant to determining toxic effects.
4. Would you like to consider the carbamate and the surfactant separately or together?

5. Lastly, what recommendation could you offer about a maximum contaminant level (MCL) in the water of areas where spraying is to occur?
6. The polycyclic aromatic hydrocarbon benzo[a]pyrene has a low vapor pressure, a log K_{ow} value of approximately 6, and is readily metabolized by most organisms. Based solely on this information, design a sampling regime for a university research team studying the PAH contamination of a major coastal waterway. In what organisms would you expect to find the toxic effects of this chemical?
7. The immune systems of animals protect them from pathogens with a specific defense mechanism using antibodies. Explain the different defense strategy used by organisms in protecting themselves from xenobiotics. What reasons might be given for the differences in the strategies used for protection against biological versus chemical invaders?
8. It is proposed that a site that housed a gas station be used to build residential condominiums. An analysis of the soil reveals that it is contaminated from this use. You have been hired to conduct a human health risk assessment for the site, and it is necessary to determine if the soil needs to be remediated or removed before construction. Which chemicals are of the utmost toxicological concern? Outline the steps you would take in the assessment process.

CHAPTER

7

Organic Pollutants

Presence and Effects in Humans and Marine Animals

CHRISTOPHER M. REDDY, JOHN J. STEGEMAN, AND MARK E. HAHN

INTRODUCTION

During the course of the 20th century, the planet became and is now chemically different from any previous time. The difference we speak of is that resulting from the introduction of synthetic chemicals that had not existed before, or from chemicals formed by natural processes that previously had existed only in trace amounts but now occur in greatly increased abundance because of human activity. The abundance, persistence, and distribution pathways mean that many such chemicals often end up in the oceans. Depending on chemical use and disposal practices, this often means the coastal ocean. This chapter focuses on those organic chemicals, which because of abundance, persistence, and biological activity (intended or unintended) may pose risks to the health and well-being both of organisms in the sea as well as humans potentially exposed to these chemicals via contaminated marine resources. Concerns about possible effects of these chemicals thus can be viewed under the rubric of "oceans and human health" or, conversely, of "humans and ocean health."

Many thousands of different chemicals have been made since the chemical industries became established (Muir and Howard, 2006). One may distinguish these chronologically, according to when they were made. The period of the 1930s through the 1960s saw new structures, including many pesticides (insecticides, fungicides, and herbicides). Common insecticides include 1,1,1-trichloro-2,2-bis(*p*-chlorophenyl)ethane (dichlorodiphenyltrichloroethane or DDT) and toxaphene, and herbicides include 2,4-dichlorophenoxyacetic acid (2,4-D) and 2,4,5-trichlorophenoxyacetic acid (2,4,5-T) (Table 7-1; Fig. 7-1). This period also saw growth in the use of chemicals designed for industrial applications, with no intended biological activity. Most well known among these are the polychlorinated biphenyls (PCBs), a group of 209 different compounds (congeners) (Table 7-1; Fig. 7-1). Also important are compounds such as chlorinated dibenzo-*p*-dioxins (PCDDs) and chlorinated dibenzofurans (PCDFs) that occur as unintended byproducts during the synthesis of industrial compounds. As unintended biological effects of pesticides, PCBs, and other compounds were identified in wildlife, concerns for effects in humans grew as well. Around the 1970s, some chemicals began to be banned from production or sale in some parts of the world. Replacements often were identified, and while now most are tested for possible health effects to animals, these chemicals are not free of effects.

The chemicals of concern include natural as well as anthropogenic chemicals. Natural chemicals include some such as polycyclic aromatic hydrocarbons (PAHs). These are formed by transformation of organic matter, in diagenetic processes resulting in petroleum, and during incomplete combustion of organic matter. Whereas natural fires are such a source, anthropogenic sources such as combustion engines and power plants also produce large quantities of PAH. Analysis of sediments from before 1900 indicates a natural origin of dioxins and dibenzofurans as well (Green *et al.*, 2004). There also are biosynthesized halogenated natural products, for example halogenated dimethyl bipyrroles, which can have biological activities like some of the synthetic chemicals.

Technology for detection of chemicals, and determining anthropogenic or natural origin, has progressed immensely since the discovery of PCBs in the environment in the mid-1960s. This technology has resulted in the ability to detect ever smaller amounts of chemicals in the environment and

TABLE 7-1. Listing of some organic pollutants found in the marine environment and humans. See Figure 7-1 for representative structures.

Compound Name[a,b]	MW (g $mole^{-1}$)	Elemental Composition	Kow[c] [(mol L^{-1} octanol)/ mol L^{-1} water)]	Date of First Usage
Aldrin*	364.9	$C_{12}H_8Cl_6$	5.6×10^6	Early 1950s
Chlordane*	409.8	$C_{10}H_6Cl_8$	2.0×10^6	Late 1940s
Dieldrin*	380.9	$C_{12}H_8Cl_6O$	2.8×10^5	Late 1940s
Dioxins* (2,3,7,8-tetrachlorodibenzo-*p*-dioxin)	321.9	$C_{12}H_4Cl_4O_2$	7.9×10^6	NA[d]
DDT*	354.9	$C_{14}H_9Cl_5$	6.2×10^6	1942
Endrin*	380.9	$C_{12}H_8Cl_6O$	2.8×10^5	Late 1940
Furans* (2,3,7,8-tetrachlorodibenzofuran)	305.9	$C_{12}H_4Cl_4O$	2.0×10^6	NA[d]
Heptachlor*	373.3	$C_{10}H_5Cl_7$	7.2×10^5	Late 1940s
Hexachlorobenzene*	284.8	C_6H_6	7.2×10^5	~1940
Mirex*	545.6	$C_{10}Cl_{12}$	1.0×10^7	1950s
PCBs* (3,3′,4,4′,5,5′-hexachlorobiphenyl)	360.9	$C_{12}H_4Cl_6$	4.2×10^7	1929
Toxaphene*	413.8	$C_{10}H_{10}Cl_8$	6.2×10^6	Late 1940s
PAHS (Benzo[*a*]pyrene)	252.3	$C_{20}H_{12}$	1.1×10^6	NA[d]
PBDEs (decabromodiphenyl ether)	943.2	$C_{12}OBr_{10}$	$>1.0 \times 10^{7}$[c]	1960

[a]Many of these compounds were not produced as pure compounds but rather as technical mixtures. We have tried to choose ideal compounds that represent the many compounds that could be found in these technical mixtures. For example, for the dioxins, we chose 2,3,7,8-tetrachlorodibenzo-*p*-dioxin because it is one of the most toxic compounds known.

[b]Compounds marked with an * are the "The Dirty Dozen" or persistent organic pollutants (POPs) recognized by the United Nation Environmental Program.

[c]There have been many studies on such measurements. For consistency, we used an online program that calculated the Kow based on chemical structure. It is available at www.syrres.com/esc/kowdemo.htm. The only exception is decabromodiphenyl ether, which is difficult to predict or measure because its Kow is so large.

[d]These compounds have been produced by combustion and predate human synthesis ether advertently or inadvertently.

in human tissues, as well as the ability to detect new chemicals not previously recognized as environmental contaminants. Unfortunately, our ability to assess the risk posed by exposure to these chemicals at environmentally relevant levels has not kept pace with the progress in analytical chemistry. Thus, the impact of these exposures on human and environmental health remains a topic of great uncertainty.

The concern that chemicals in the oceans may have health effects is based on now abundant evidence from animal studies, and compelling but less definitive evidence from human epidemiology. Thus, many of the organic chemicals introduced in the oceans are capable of adversely affecting the health of humans and marine organisms exposed to them. We summarize here the processes and trends in the distribution of organic chemicals in the oceans and some of the mechanisms by which these contaminants cause toxicity, and we discuss the extent to which such knowledge can be extrapolated to predict the impact of these compounds on humans and marine animals.

ORGANIC CHEMICALS IN THE OCEANS

Analysis

A brief note on the analytical methods for chemical detection is important. Concern about possible health consequences from organic chemicals in the oceans depends in part on knowing their presence and abundance. The technology for detection has progressed over the past few decades, with sensitivity having greatly increased. One result is that concentrations can now be determined when once the amounts would have been below the limit of detection. However, the general strategy for measuring amounts of chemical residues in humans, animal, and environmental

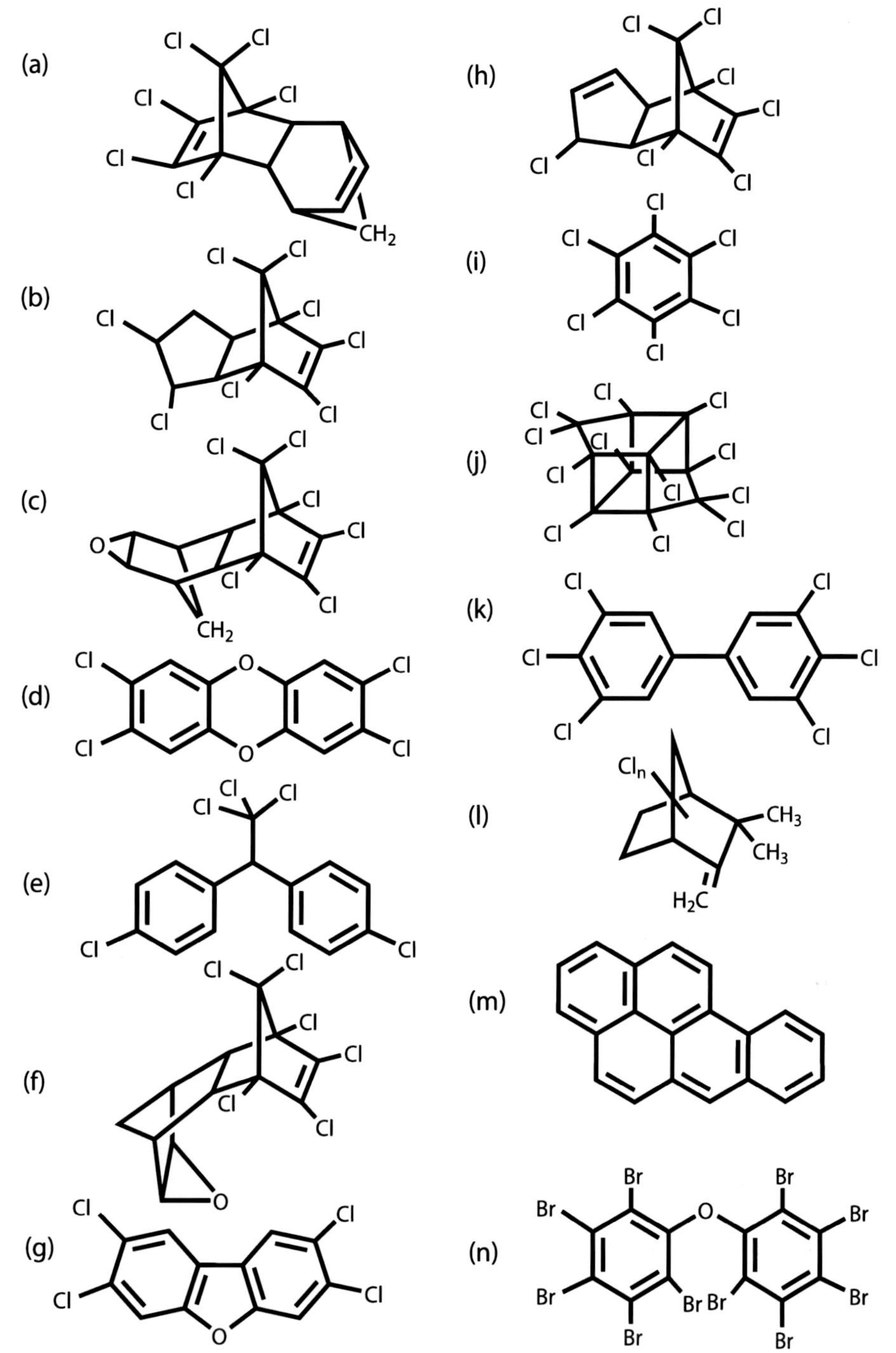

FIGURE 7-1. Structures of select organic pollutants. (a) Aldrin, (b) Chlordane, (c) Dieldrin, (d) 2,3,7,8-tetrachlorodibenzo-*p*-dioxin, (e) DDT, (f) Endrin, (g) 2,3,7,8-tetrachlorodibenzofuran, (h) Heptachlor, (i) Hexachlorobenzene, (j) Mirex, (k) 3,3′,4,4′,5,5′-hexachlorobiphenyl, (l) generic Toxaphene congener, (m) Benzo[*a*]pyrene, and (n) decabromodiphenyl ether. Please refer to Table 7-1 for additional information. Because Toxaphene has so many congeners, we only show a generic form.

matrices has changed very little (Erickson, 1997). Samples are extracted with an organic solvent and then additional chromatographic or "cleanup" steps are used to remove fats and lipids. The latter steps include silica gel or alumina chromatography, both of which take advantage of the polarity (or how the electrons are distributed in the molecules). Other more sophisticated methods involve gel permeation chromatography, which relies on the different sizes of molecules in the extracts. Fats and other lipids elute earlier from these columns. Appropriate fractions eluting from

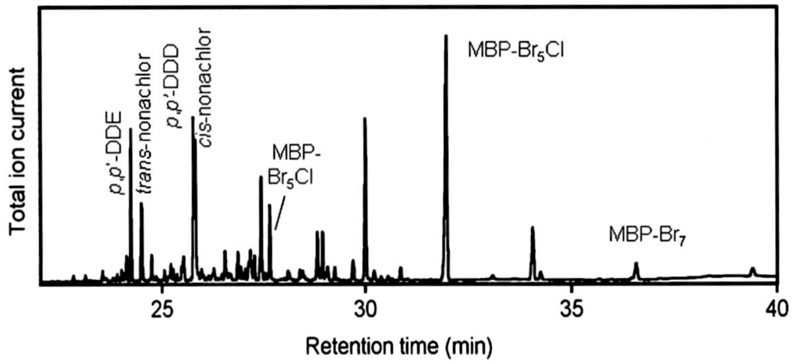

FIGURE 7-2. Partial GC-MS trace of an extract from common dolphin blubber. Some compounds are annotated, and the others are PCBs and other pesticides. A glass capillary column separated these compounds, and they were detected with a mass spectrometer operating in negative ion chemical ionization mode.

these columns are then analyzed by gas chromatography with a variety of different detectors. Initially in the 1960s to the early 1980s, samples were separated with packed columns (0.5 cm wide) and detected with flame ionization or elector capture detectors. These approaches were useful but did not provide the necessary resolution to separate the vast number of compounds that occur in extracts. This led to employing glass capillary (0.25 mm wide) columns that have better separation power and mass spectrometers as detectors (GC-MS). For example, Figure 7-2 shows a gas chromatogram with mass spectral detection of chemicals in an extract of common dolphin blubber. Several persistent organic pollutants (POPs) are present as well as some natural halogenated compounds (see later sections). Many of the compounds not identified are PCBs and other POPs.

More recently, liquid chromatography mass spectrometry (LC-MS) has been employed for contaminants that are not as easily analyzed by gas chromatography (Hirsch *et al.*, 1999). Some of these compounds would include personal care products or pharmaceutical products, such as erythromycin (Glassmeyer *et al.*, 2005; Hirsch *et al.*, 1999) (see also Chapter 9).

"Traditional" Contaminants: POPs, PAHs, and Organometals

The chemicals most often associated with marine pollution, and with environmental pollution in general, are those first described as anthropogenic environmental contaminants in the 1960s and 1970s: PCBs, chlorinated insecticides such as DDT and its breakdown products (together, ΣDDT),[1] PAHs derived from combustion or from oil spills and organometals such as methylmercury and organotins. The organometals are described in detail in Chapter 8 and will not be considered further here. Many chemicals of concern have been reviewed extensively in the literature and are well known. Here, we focus on some of the key aspects of their behavior and on recent geographical and temporal trends.

Persistent Organic Pollutants (POPs)

The two most important factors contributing to concern about environmental contaminants are their innate toxicity and their persistence in the environment. The term *persistence* is not absolute, but rather relative. Persistence does not indicate infinite lifetimes for these chemicals; even highly persistent chemicals are degraded eventually, through biotic and abiotic mechanisms. Biological processes include biotransformation within animals (Stegeman and Hahn, 1994) or bacteria (Alexander, 1999), the latter being of much greater relevance to the overall environmental fate of chemicals. Abiotic processes can degrade many contaminants or remove them from the bioactive pool (sequestration into the deep ocean, sediments, and soils). For example, Jonsson *et al.* (2003) estimated that some PCBs will be deposited into deep continental sediments, with half-lives of 100 years for removal of PCBs from the biologically available pool. Hence, deposition and burial into coastal sediment occurs but is slow.

Some PAHs are susceptible to photodegradation (Gschwend and Hites, 1981). For PCBs and PCDD/Fs, atmospheric reactions of these contaminants with hydroxyl radicals are important but are dependent on the specific type of pollutant. For example, for PCDD/Fs having four to five chlorine atoms, removal rates from the atmosphere via reaction with hydroxyl radicals are similar to those for deposition into soils and the ocean. However, more highly chlorinated PCDD/Fs are much more likely to be deposited into the ocean, and eventually into sediments, than to be degraded in the atmosphere (Lohmann *et al.*, 2006).

[1]DDT can undergo biotic and abiotic transformation to several products, including DDE (dichlorodiphenyldichloroethylene) and DDD (dichlorodiphenyldichloroethane) (Quensen *et al.*, 1998). DDE is much more persistent in the environment, as well as more toxic to biological systems (Alexander, 1999).

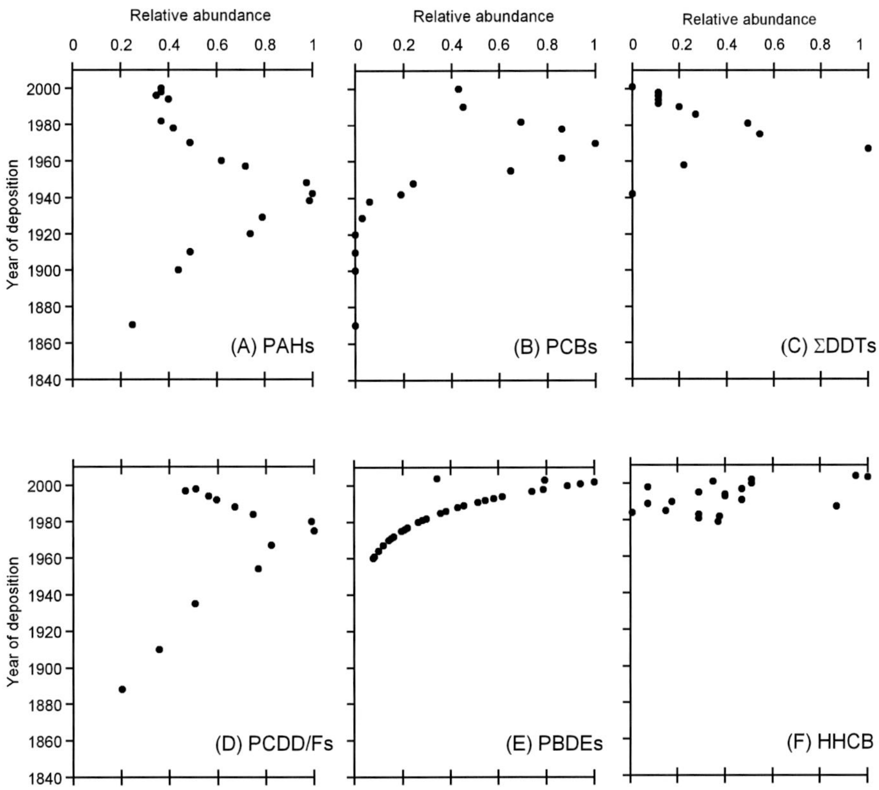

FIGURE 7-3. Recent downcore sediment profiles from the Great Lakes region of the United States. (A) PAHs (Schneider *et al.*, 2001); (B) PCBs (Van Metre and Mahler, 2005); (C), ΣDDTs (Van Metre and Mahler, 2005); (D) PCDD/Fs (Baker and Hites, 2000); (E) PBDEs (Zhu and Hites, 2005), and (F) HHCB (Peck *et al.*, 2006). The data used for preparing this figure were originally expressed either as content or fluxes. For each core, we normalized all values to that of the maximum value for that core.

Temporal Trends

Most organic pollutants found to persist in the ocean have large K_{ow} values[2] (>100,000; Table 7-1). The ability of these compounds to partition into tissues and into organic matter of sediments is directly correlated to the compounds' K_{ow} values (Schwarzenbach *et al.*, 2003). As sediments are buried they can act like tree rings or ice cores, containing a record of chemical conditions when they were deposited (Kawamura and Suzuki, 1994). Hence, sediment cores provide valuable insights into the history of these compounds (Charles and Hites, 1987; Goldberg *et al.*, 1977; Latimer and Quinn, 1996).

To highlight the usefulness of sediment cores, in Figure 7-3 we show sediment profiles or archives of six difference compounds or classes of compounds. In constructing this figure we relied on data from the Great Lakes region. It would have been more appropriate to show data from oceanic records, but such records are patchy and thus provide less informative comparisons. The data used have come from cited references (listed in the figure caption) and were published as either the chemical content or sediment flux. The chronology of these cores, or the age of the layers, was generally determined from abundance of isotopes of radioactive elements such as ^{210}Pb (Appleby and Oldfield, 1978).

What is readily apparent is that the pollution history varies significantly in these records. First, both the PAHs and PCDD/Fs have nonzero values even in 1840, which predate the beginning of the Industrial Revolution in the United Sates. This background signal results from the transformation of organic matter by either microbes or combustion of biomass not related to human activity (Venkatesan, 1988). Note that PAHs also occur in petroleum and refined products, but this source represents a small contribution compared to combustion sources (Lima *et al.*, 2005). Thus, the switch from coal to petroleum as the major fuel source in the late 1950s in the United States matches nicely with the PAH maximum (Lima *et al.*, 2003). The time of the maximum also reflects when source control measures began at power plants and in industry. Some new data on PAH

[2]The octanol-water partition coefficient (K_{ow}) is a measure of the tendency of a compound to bioaccumulate into fatty tissues and other organic matter. The larger the value, the greater the tendency. All of the compounds shown in Table 7-1 have values greater than 10,000. Such values are strong indicators of potential for bioaccumulation.

levels show that, unlike the PCDD/Fs, which are declining in concentration, levels of PAHs no longer are decreasing. Some have argued that PAH levels are no longer decreasing because of the growth of urban areas (Lima *et al.*, 2003; Van Metre *et al.*, 2000).

Levels of both PCBs and ΣDDTs have nearly the same profiles, reflecting the time of original production: 1929 and 1942, respectively (Table 7-1). Maxima for both were in the 1960s at the time that research in Europe and the United States began to warn of the current or potential dangers of these compounds.

The last two compounds are the polybrominated diphenylethers (PBDEs) and the synthetic musk used in fragrances, 1,3,4,6,7,8-hexahydro-4,6,6,7,8,8-hexamethylcyclopenta-γ-2-benzopyran (HHCB); these have much later input times. The PDBEs began to be used in 1960s and rapidly increased for the next four decades (see Case Studies, presented later in the chapter). The levels of some PBDEs in sediment cores are now decreasing because production of penta and octa PBDEs has stopped. The levels of the musk HHCB continue to increase, but it is difficult to predict the future trend as there has not been enough time for this picture to develop (see also Chapter 9).

In summary, sediment cores reflect the history of pollution. In fact, some scientists have used industrial production records, such as the onset of PCBs or DDTs, to date sediment layers (Latimer and Quinn, 1996). As indicated in sections that follow, temporal trends like those revealed in sediment cores also appear in the burden of contaminants in humans.

Geographic Trends

Temporal trends in the abundance of chemicals are superimposed on geographic trends. Chemicals of concern emanate from sources on all continents. However, the magnitude of input varies dramatically, depending on differences in industrial, agricultural, and waste disposal practices. Coastal zones adjacent to population centers tend to have the highest concentrations of chemicals, whose identity reflects the chemicals used in those regions. Regardless of the original sources, however, global atmospheric and hydrographic circulation pathways have distributed all such chemicals around the globe. As suggested previously, quantification of residues in some areas still may be limited by detection methods, and not every location has been examined. Nevertheless, amounts of chemicals have been measured in places around the globe so that we can conclude that they are ubiquitous, that there are no places that are free of such contamination. However, there are some general features that appear.

Chemicals measured in sediments, water, atmosphere, or biota can indicate geographic trends. Contaminants in biota, however, may be the most relevant to assessing health risks from marine contamination, and in some settings they may be easier to obtain and analyze.

Penetration to the deep ocean in many locations remote from direct continental sources is expected to reflect the deposition of residues from the atmosphere. However, there are also geographic differences that can be related to specific water movement, including inputs from continental sources. Outflow from the Mediterranean into the Eastern North Atlantic has been reported (Marti *et al.*, 2001). An example in the Western N. Atlantic involved measurement of PCB and toxaphene residues in a deep sea fish, the rattail *Coryphaenoides armatus*, collected at about 1500 meters from two widely separated locations (Stegeman *et al.*, 1986). There were about ten-fold greater levels of PCBs in fish collected in slope waters downslope from the Hudson Canyon Trough, than in fish downslope the Carson Canyon Trough, off Newfoundland.

Another, somewhat surprising avenue by which contaminants may reach the deep ocean is in whale carcasses ("whale falls" [Haag, 2005]). As indicated in many places in this chapter, marine mammals accumulate and often store large quantities of lipophilic contaminants because of their size, their high blubber content, and their typical position as apex predators. Carcasses reaching the sea floor certainly will carry contaminants that could represent large local inputs around the world. Such events would be discrete and occur more randomly than transport through oceanic currents.

There are intriguing scientific questions concerning possible effects of chemicals in remote regions where concentrations may be low, such as in the deep ocean. In the study of PCBs in deep sea fish cited earlier, there were greater levels of enzymes that are biomarkers of exposure to PCBs (and PCDDs) in Hudson Canyon fish, indicating that biochemical and possibly biological effects may be occurring even in the deep sea. Similar results have been obtained in mid-water fish (Stegeman *et al.,* 2001). Whether further effects might occur is not known. However, populations of animals exposed only to low concentrations could be more susceptible to effects than more highly exposed coastal fishes (See Impact on Marine Organisms/Environmental Health, presented later in the chapter).

Transport of Contaminants to High Latitude Regions and Indigenous Peoples

The transport of organic pollutants from industrial areas to remote oceans and high latitude regions is influenced both by physical forces (air movement) and properties of the chemicals (Wania and Mackay, 1993, 1996). Although most organic pollutants have small vapor pressures and, hence, do not evaporate easily, they nevertheless can partition into the air and be transported. This process is often called the "global distillation effect," as compounds that have the highest vapor pressures are found in the highest latitudes (Wania and Mackay, 1996).

Transport may differ for organic contaminants such as combustion-derived PAHs. They are carried on aerosols in areas downstream of major industrial areas. For example, Windsor and Hites (1979) found that the concentration of PAHs in the Gulf of Maine decreased with distance from Boston, Massachusetts. These PAHs often are so strongly associated with soot particles that global distillation is less important for these compounds; the delivery is across latitudes and there is not a major input to higher latitudes. The strong sorption to particles and eventual deposition in sediments also may limit the bioavailability of PAHs (Gustafsson *et al.*, 1997).

New studies have found that biological transport can also deliver organic pollutants to remote latitudes (Blais *et al.*, 2005, 2007; Evenset *et al.*, 2007; Krummel *et al.*, 2003). Evenset *et al.* (2007) calculated that 80% of the chemical load to Lake Ellasjoen (74°N) in the Bering Sea was delivered by bird droppings. Moreover, the authors calculated that this input term was 30 times greater than atmospheric transport. Another study in Arctic ponds at 76°N determined that levels of organic pollutants and methyl mercury were directly proportional to seabird populations (Blais *et al.*, 2005). In addition to bird droppings, other research reveals that some fish, such as wild salmon, can accumulate contaminants from the ocean and transport them to the inland waters where they spawn and die (Blais *et al.*, 2007; Krummel *et al.*, 2003). Two important points must be made here. The physical transport of pollutants via global distillation is mainly indiscriminate with respect to areas of geographic deposition but rather depends primarily on the chemical properties of the contaminant. Biological transport, however, can be much more localized and often closest to coastal areas where indigenous people live.

Developed Versus Developing Nations

Since the 1970s, some countries have been rapidly increasing the industrial segment of their economies, with increased generation of chemicals of concern. This is apparent in sediment cores collected from coastal China, for example, which show trends different from those in coastal U.S. sediments (Fig. 7-4). These sediment profiles are nearly mirror images. For both PAHs and PCBs, decreasing trends in the United States have occurred after maximum levels several decades ago. China's emissions continue to increase and have not reached an apex. Such results are consistent with changes in the total primary energy production from 1993 to 2003; China's increased by 48% even although it was only half of U.S. production (Fig. 7-5). However, the United States only increased by 12%. Similar differences in trends relative those in U.S sediments can be observed for India and Russia.

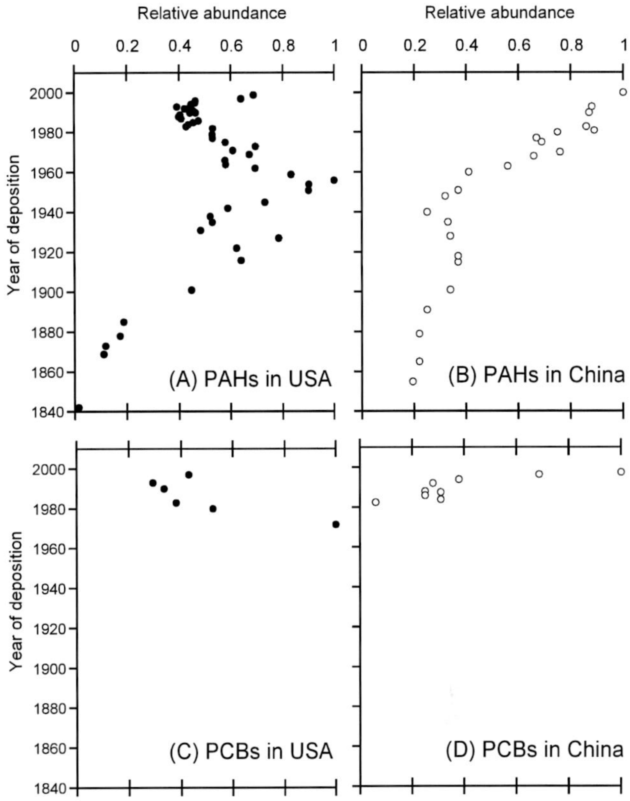

FIGURE 7-4. Downcore sediment profiles from coastal areas in the United States and China for PAHs (A and B) and PCBs (C and D). The U.S. cores were collected from Narragansett Bay (Hartmann *et al.*, 2005; Lima *et al.*, 2003). The cores from China for PAHs and PCBs were collected for PAHs and PCBs near the Yellow Sea Western Current (Guo *et al.*, 2006) for PAHs and PCBs from the Pearl River Estuary (Mai *et al.*, 2005).

There are numerous reasons why these changes in energy usage can affect the oceans and human health. First, combustion of organic matter can lead to emissions of a wide range of PCDDs and PCDFs as well PAHs and soot (Lima *et al.*, 2005). Human health impacts from inhaling combustion-derived materials are well documented (Dockery *et al.*, 1993; Kunzli *et al.*, 2000). These combustion-derived emissions also lead to increased loads of nitrogen to the atmosphere, which in turn, can be deposited in coastal waters. It has been suggested that this input of nitrogen can cause eutrophication and could contribute to harmful algal blooms (Paerl *et al.*, 2002; Spokes and Jickells, 2005). The increased demand for fuel may also increase the amounts of petroleum that must be delivered *via* barges or pipelines near coastal areas, increasing the risk of spills or leaks. Urban runoff (leaking engine oil) from vehicles eventually is deposited in coastal areas. Increased burning of fossil fuels will lead to concomitant increases in carbon dioxide emissions, which will cause the ocean to acidify, endangering coral reefs and other calcareous organisms that play integral roles in the ocean's food web (Doney, 2006).

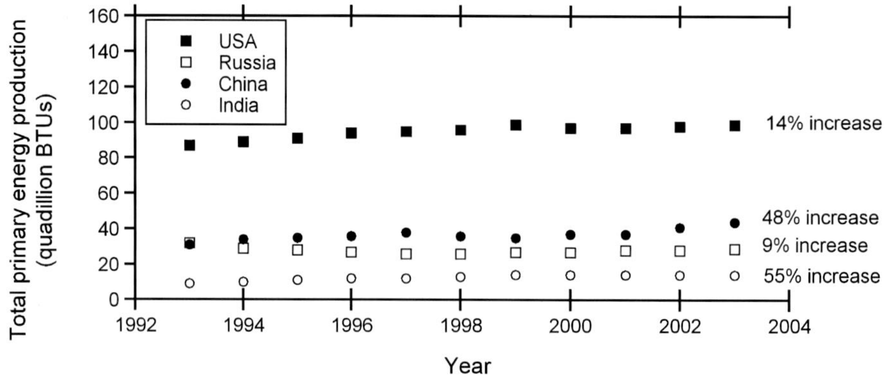

FIGURE 7-5. Total primary energy production (quadrillion BTUs) for the United States and Russia versus China and India. Data obtained from the U.S. Department of Energy Web site (www.eia.doe.gov/emeu/international/energy-production.html).

Emerging Contaminants: Brominated, Fluorinated Compounds, and Pharmaceuticals

As concentrations of "traditional" contaminants such as DDTs and PCBs have declined, the introduction of alternative chemicals, combined with advances in analytical methods, has led to the identification of new, so-called emerging contaminants. Here we highlight a few of particular relevance to marine systems.

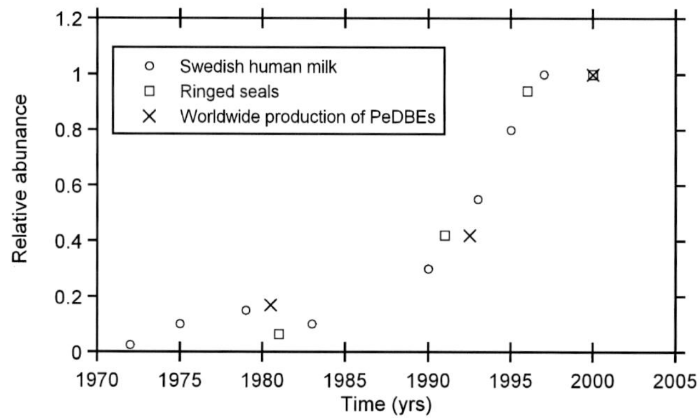

FIGURE 7-6. Relative abundances of PeDBEs in Swedish human milk, mainly BDE-47 (Noren and Meironyte, 2000), male ringed seals from Holman Island from the Northwest Territories, BDE-47 (Ikonomou *et al.*, 2002), and worldwide production of pentabromodiphenyl ethers (PeBDEs; [Ikonomou *et al.*, 2002]).

Brominated Flame Retardants (BFRs)

Since the mid-1990s, the brominated flame-retardants, including polybrominated diphenyl ethers (PBDEs), have been found in human milk and food, terrestrial and marine animals, household dust, lint from clothes dryers, and sediments (DeWit, 2000; Hites, 2004; Huwe and Larsen, 2005; Ikonomou *et al.*, 2002; Rahman *et al.*, 2001; Stapleton *et al.*, 2005). PBDEs have been used in appliances, textiles, plastics, foams, and other products where human exposure was inevitable. Like PCBs, PBDEs are composed of a group of 209 congeners that are hydrophobic and have high K_{ow} values. They are a bit less stable in the environment than PCBs because the carbon to bromine bond is weaker in PBDEs than the carbon to chlorine bond in PCBs. This subtle difference is why PBDEs are used as flame-retardants—at elevated temperatures, the bromine atoms dissociate and quench fires.

Production of PBDEs started in the 1960s, and global sales were ~70,000 metric tons by 2001 (Fig. 7-6). They were sold initially as several different types of mixtures that contained pentabrominated or octabrominated congeners; these two have since been phased out. The biggest seller has been a mixture dominated by the fully brominated congener, decabromodiphenylether (DecaBDE), which continues to be marketed and sold. Toxicological data on PBDEs are not abundant, but some experiments performed on laboratory animals indicate that they act as neurotoxicants and affect endocrine processes (Birnbaum and Cohen Hubal, 2006; Eriksson *et al.*, 2001; Lilienthal *et al.*, 2006).

The awareness of PBDEs was raised in a study in Sweden by Norén and Mieronyté (2000), who showed that the concentration of PBDEs in human milk increased exponentially over a 25-year period beginning in 1970, with a doubling rate of 5 years (Fig. 7-6). Schecter *et al.* (2003) analyzed human milk collected in 2002 in the United States and determined that concentrations of PBDEs were 10 to 100 times greater than those found in Europe. More recent studies also indicate that the intake of PBDEs occurs through sources in addition to the dietary route that characterizes PCBs; this additional source is household dust (Schecter *et al.*, 2006). The latter study also found that among foodstuffs, fish was a larger source of PBDEs than meat or dairy products.

Together, these data indicate that PBDEs must be considered seriously as a potential risk to human health.

Pharmaceuticals and Personal Care Products (PPCPs)

In contrast to compounds such as PCBs and PBDEs, which were synthesized for industrial purposes and were not intended to elicit biological effects, pharmaceuticals (especially) and some personal care products were designed and marketed precisely because of their ability to interact with biological systems. For example, estrogens used in contraceptive pills and antidepressants targeting pharmacological receptors or neurotransmitter transporters are specific and highly potent biomimetics. In a groundbreaking study, Kolpin *et al.* (2002) demonstrated that many of these compounds find their way to rivers and streams. The degree to which these compounds contaminate coastal marine systems is less well understood. The presence and potential effects of PPCPs in aquatic and marine systems is summarized in a recent review (Fent *et al.*, 2006) and elsewhere in this book (Chapter 9).

Fluorine-Containing Compounds

The presence of polyfluoroalkyl compounds in human serum has been known for decades (Taves, 1968). Recently, as these compounds have been measured increasingly in humans and wildlife (Giesy and Kannan, 2001; Inoue *et al.*, 2004; Kannan *et al.*, 2001, 2004; Taniyasu *et al.*, 2003), there has been renewed interest in understanding their source, fate, and impacts. Much of the attention has been focused on perfluorooctanoic acid (PFOA) and perfluorooctane sulfonate (PFOS). The environmental distribution and toxicity of these compounds is described in several recent reviews (Hekster *et al.*, 2003; Houde *et al.*, 2006; Lau *et al.*, 2004).

Natural Halogenated Compounds

In addition to the numerous anthropogenic halogenated organic compounds (HOCs) found in the environment (Hites, 2004; Simonich and Hites, 1995) (Table 7-1; Fig. 7-1), evidence is mounting that some HOCs found in animal tissues, air, humans, worms, and food are not anthropogenic in origin but rather are natural products. For example, two methoxylated polybrominated diphenyl ethers (MeO-PBDEs; Figs. 7-7a,b) isolated from a True's beaked whale were shown to be natural by virtue of their radiocarbon content (discussed later) (Teuten *et al.*, 2005). A similar analysis of an isolated mixed halogenated 2,2′-dimethyl bipyrrole (DMBP-Br_4Cl_2; Fig. 7-7d) revealed that it too was natural (Reddy *et al.*, 2004).

Compound-specific radiocarbon (^{14}C) analysis is a powerful tool for determining whether a compound is naturally derived or a product of industrial activity (Teuten *et al.*, 2005). With the exception of toxaphene (which is produced from plant-derived terpenes), all industrial chemicals are produced with carbon derived from petrochemicals. Because the half-life of ^{14}C is 5730 yrs and petroleum is at least 1 million years old, there is no detectable ^{14}C in industrial (petroleum-derived) compounds. In contrast, natural com-

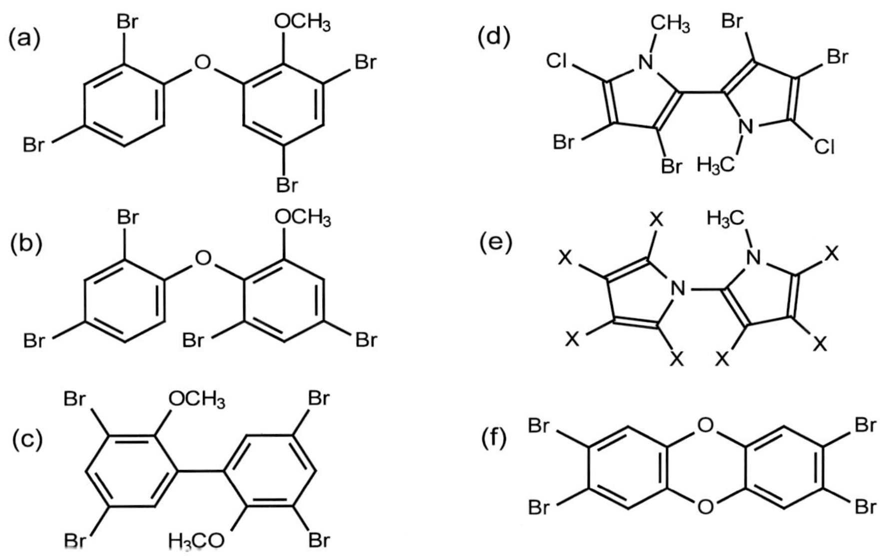

FIGURE 7-7. Structures of several proposed or proven naturally occurring halogenated compounds. (a) 2-(2′,4′-dibromophenoxy)-4,6,dibromoanisole (2′-MeO-BDE68), (b) 2-(2′,4′-dibromophenoxy)-3,5,dibromoanisole (6-MeO-BDE47) (c) 3,3′,5,5′-tetrabromo-2,2′-dimethoxy-1,1′-biphenyl, (d) 3,3′,4,4′-tetrabromo-5,5′-dichloro-1,1′-dimethyl-1*H*,1′*H*-2,2′-bipyrrole (DMBP-Br_4Cl_2), (e) perhalogenated methyl bipyrrole (when X = Cl_7, then the compound is Q1) and, (f) 2,3,7,8-tetrabromodibenzo-*p*-dioxin. See text for additional details.

pounds gain their carbon *via* recently photosynthesized material, which has a normal amount of ^{14}C (from production in the atmosphere and emissions from above ground nuclear weapons testing). Therefore, evidence of a naturally produced compound can be determined by virtue of its ^{14}C content. If there is no ^{14}C, the compound is industrially produced. If the compound has ^{14}C, it is more likely to be natural.

Other possibly natural HOCs include a dimethoxylated polybrominated biphenyl (diMeO-PBBs) (Marsh *et al.*, 2005) (Fig. 7-7c), polybrominated dibenzodioxins (Malmvarn *et al.*, 2005) (Fig. 7-7f), several halogenated dimethyl bipyrroles (Tittlemier *et al.*, 1999) (Fig. 7-7d), and a suite of halogenated 1,2′-methyl bipyrroles (MBPs; Fig. 7-7e) (Teuten *et al.*, 2006b, 2006c; Vetter *et al.*, 2000). The heptachlorinated MBP (referred to as Q1) has been found in breast milk of Faroe islanders who eat a diet rich in whale blubber (Vetter *et al.*, 2000). Other analogs of Q1 can be perbrominated or composed of a mixture of bromine and chlorine atoms (Teuten *et al.*, 2006b, 2006c; Teuten and Reddy, 2007). Although the MeO-PBDEs are biosynthesized by marine sponges (Anjaneyulu *et al.*, 1996; Utkina *et al.*, 2002), the sources of many other apparently natural compounds is not clear. Evidence supporting the idea that these are in fact natural products includes (1) no record of industrial sources, (2) efforts to synthesize them have often been difficult, with low yields (Gribble *et al.*, 1999), (3) the presence of radiocarbon in some of these compounds (Reddy *et al.*, 2004; Teuten *et al.*, 2005) as well as in 2, 4-dibromophenol in an acorn worm *Saccoglossus bromophenolosus* (Teuten *et al.*, 2006a), (4) the geographic distribution of these compounds in marine mammals is distinct from that of anthropogenic compounds (Stapleton *et al.*, 2006; Tittlemier *et al.*, 2002), and (5) many of these compounds were found in an archived whale oil sample collected on the last voyage of the whaling ship the *Charles W. Morgan* (Teuten and Reddy, 2007), which ended in 1921 and predates large-scale industrial manufacture of HOCs that began in the late 1920s (Lipnick and Muir, 2000).

Constraining the sources and cycling of these and other HOCs of unknown origin is important because their occurrence in an assortment of biota, including humans, indicates a widespread distribution in the environment. If these compounds are truly natural, they have likely been present in the environment for a much longer time period than industrially synthesized HOCs. Hence, they could be useful for studying the evolutionary responses of biota to HOCs. They also have chemical and physical properties that are similar to those of industrial HOCs, so the natural compounds could be excellent subjects to study the long-term fate of such compounds. If some of these compounds actually derive from industrial activity, careful consideration would need to be given as to their source, mode of production, fate, impact, and whether their emissions can or need to be controlled.

As noted previously, many of the chemical properties of these natural compounds are similar to those of industrial HOCs (Teuten and Reddy, 2007; Tittlemier *et al.*, 2004). Like industrial HOCs, they may degrade in the environment, although very slowly (Sinkkonen and Paasivirta, 2000). For example, there is some evidence that in some marine mammals the bromochloro versions of Q1 can be dehalogenated (Teuten and Reddy, 2007). With adequate regulations regarding the manufacture and release of the industrial versions, it is likely that in the future natural HOCs, rather than industrial ones, will again be the predominant HOCs found in animal and human tissue (Teuten and Reddy, 2007).

Some of these natural compounds have been detected in recent samples from marine mammals, human milk, and commercially available fish in Canada (Stapleton *et al.*, 2006; Teuten *et al.*, 2006b, 2006c; Tittlemier, 2004; Tittlemier *et al.*, 2002; Vetter *et al.*, 2000). Numerous efforts are underway to identify the sources, ecological functions, and biological activities of these compounds (Tittlemier *et al.*, 2003; Vetter *et al.*, 2005). In addition, efforts should be focused on understanding how exposure to these compounds may have prepared bacteria, plants, animals, and humans for industrial HOCs introduced during the 20th century. It is well known that organisms have evolved defensive mechanisms against chemicals in their environment, and until recently the sources of these chemicals were primarily natural. The importance of natural HOCs in the evolution of these defenses is not yet understood (Stegeman and Hahn, 1994).

HUMAN EXPOSURES AND EFFECTS

Human Exposure *via* Marine Sources

There are numerous routes by which humans may be exposed to a chemical of concern. Some pathways dominate over others, depending on the source and properties of a chemical. The chemicals of concern here are widespread in the environment and thus populations in most parts of the world may be at risk for exposure. While there could be exposure by dermal and respiratory pathways, for most such chemicals the dominant pathway is dietary. Thus, the degree of contamination of food resources will be of major importance in determining the degree of exposure in various human populations.

The issue of food contamination is one that has been addressed in hundreds of scientific publications. Such publications have addressed the levels of chemicals including PAHs, PCBs, PCDDs, heavy metals, petroleum hydrocarbons, pesticides, and, more recently, emerging HOCs including PBDEs. Analyses of different types of food typically show that milk and fish are major contributors of PCDDs to the diet. For example, in a recent review, Papke (1998) examined sources, trends, and relationships among variables in PCDDs and PCDFs over 10 years, principally focusing on studies in Germany. PCDD/Fs in the diet were about one-third each from meat and meat products, milk and milk products, and fish. A similar study in Italy found that milk and fish were the major contributors (Taioli *et al.*, 2005). It is not the intention here to catalog the levels in seafood resources around the globe. A large compilation of the levels of PCDD/Fs and PCBs in various edible marine species from locations around the world, including some estimates of human exposure, was published by Domingo and Bocio (2007). It is sufficient to say that there are concerns about contamination in all parts of the world. Here we highlight some of the general features that arise from the wealth of studies that have been done, providing examples from the literature.

Different types of organisms that are used as food resources are known to accumulate or retain different types of compounds to different levels. This bioaccumulation depends in part on the nature of the chemical and whether it is readily absorbed and whether it can be readily metabolized (i.e., whether the chemical structure is susceptible to enzymatic alteration favoring elimination by the organism). Generally chemicals that are more hydrophobic, with a high Kow (Table 7-1), will tend to accumulate to a greater degree than those that are more water soluble, regardless of whether the animal acquires the chemical in the diet or by uptake across gills (Meador *et al.*, 1995; Stegeman, 1981). However, chemicals that are similarly hydrophobic can differ substantially in their susceptibility to enzymatic attack. Thus, many of the higher molecular weight PAHs and many PCB congeners are similar in molecular weight and are similarly hydrophobic, but the PAHs are much more readily metabolized than are the PCB congeners of greatest concern (those congeners with four or more chlorine substituents) (Matthews and Dedrick, 1984; Stegeman, 1981; White *et al.*, 2000).

Coupled with the susceptibility of the chemical to metabolism, the ability of resource organisms to carry out metabolism and eliminate the chemical will influence the level of contaminant that persists in those organisms. Some moderately hydrophobic chemicals can be eliminated from invertebrates as well as vertebrates by partitioning and by the action of membrane transporter proteins (Kurelec, 1992). However, other compounds are more readily eliminated after metabolism by enzymes collectively referred to as xenobiotic (foreign chemical) metabolizing enzymes, such as the cytochrome P450 enzymes and various conjugating enzymes (Stegeman and Hahn, 1994).

Although such enzymes occur in all types of animals, there are inherent differences in the ability of the most commonly consumed types of organisms to carry out foreign chemical metabolism. Thus, mollusks tend to have a lower capacity for PAH oxidation than do fish, and crustaceans fall somewhere in between (James and Boyle, 1998; Livingstone, 1991). Mammals tend to have even greater capacity for such metabolism. Unlike the PAHs, many of the PCBs are metabolized only slowly, which is reflected in a slow clearance from the body by all types of organisms (Colborn and Smolen, 1996; Matthews and Dedrick, 1984). Studies in some regions have indicated that among persistent organic pollutants, there may be greater concern about PAHs and PCBs than, for example, ΣDDTs (e.g., Binelli and Provini, 2003). However, while most organisms are able to eliminate such compounds, there can be variation in the amounts of particular contaminants in particular groups, and sometimes crustaceans may contain greater amounts than fishes (e.g., Porte and Albaiges, 1994). There may be a generally greater concern about possible consumption of PAH contaminants in shellfish than in fish, while greater amounts of PCBs and related HOCs in the diet may reflect the amounts of fish consumed (e.g., Moon and Ok, 2006). This is reflected in health advisories in some locations, recommending limitations on the consumption of fish over specified times, especially by populations more susceptible to some effects of the chemicals (Nesheim and Yaktine, 2007).

A further factor influencing the amount accumulated from marine foodstuffs is the amount of lipid in the organism or organ being consumed. Thus, fatty fish tend to accumulate greater amounts of lipophilic chemicals than do lean fish, and fat-rich organs tend to accumulate more than lean organs. As highlighted earlier and in Chapter 10, this is may be of substantial concern for groups who consume blubber of marine mammals, for cultural or other reasons. Thus, while marine mammals may tend to have a greater capacity for metabolism and elimination of chemicals, there can be accumulation in some tissues that are preferentially eaten by some groups.

Trends in Human Blood/Milk Levels

Accumulation of contaminants from food resources can be detected by chemical analysis of human tissues and fluids. Adipose tissue and especially blood and milk are frequently assayed for chemicals of interest. Archived samples have been especially important in permitting historical analyses using the most recent analytical methods. Measurements indicate that those groups who have greater amounts of more heavily contaminated food in their diets tend to have greater amounts of those contaminants in their bodies. This has been documented repeatedly in studies of northern peoples who have marine mammal meat and blubber in their diets.

An important and ongoing series of studies has addressed contamination of residents of the Faroe Islands, showing the relationship between contamination levels in components of diet and contaminant levels in local populations (Grandjean *et al.*, 2003; Schantz *et al.*, 2003; Weihe *et al.*, 1996). Such studies in other locations show similar relationships between the composition of the diet, the contaminant levels in those dietary items, and contaminant levels in consumers (e.g., Johansen *et al.*, 2004; Moon and Ok, 2006; Sandanger *et al.*, 2006). The studies of Sandanger *et al.* (2003) examined the content of persistent organic pollutants in residents of the Chukota Peninsula in the Russian arctic. Chemicals were measured in plasma of 50 individuals and related to frequency of types of food in the diet. The combined intake of blubber from walrus, seal, and whale was a significant predictor of the plasma concentrations of total PCBs and borderline for ΣDDTs (Sandanger *et al.*, 2003).

As another notable example, Fangstrom *et al.* (2002) showed that PCBs were present at high levels in serum of Faroese women who consumed marine mammal tissues. They also reported that hydroxylated PCBs were present in serum of these women. Hydroxylated PCBs are products of metabolism that also have biological activity and could come from the whales or be formed by enzymatic action in the women themselves. Many older studies did not examine such metabolites.

Analysis of chemical residues in human milk and blood can reveal important geographic and temporal trends in the levels of contaminants in human populations. Publications concerning this issue sometimes address principally marine food sources, although most studies necessarily reflect the sum of different sources. As indicated in the section on temporal trends, in some regions the levels of PCBs and some other contaminants in the environment have shown a decline. This is reflected also in chemical residues in blood. A study in Sweden (Hagmar *et al.*, 2006) addressed the interindividual as well as temporal variation in blood levels of various contaminants over a 10-year period. Blood samples were drawn from the same 39 individuals in 1991 and 2001. Lipid adjusted serum concentrations of PCB congener 153 (2,2′,4,4′,5,5′-hexachlorobiphenyl, the most abundant PCB congener), DDE and hexachlorobenzene all declined, by 34%, 55%, and 53%, respectively, between 1991 and 2001. Increasing body-mass index was associated with the decrease but explained only 5% to 13% of the variation.

The amounts of chemical residues in human milk have been measured in samples from locations around the world. These studies also reveal trends consistent with the temporal and geographic trends indicated previously. A 30-year perspective in Germany (Furst, 2006) shows that in contrast to PBDEs, which are increasing in human milk (Furst, 2006; see also Organic Chemicals in the Oceans), the levels of most persistent organic chemicals show declines, presumably associated with decreasing use and environmental levels of such chemicals.

An important question regarding contaminants in milk concerns the persistence of chemicals in children who are exposed by nursing, as well as *in utero.* A prospective birth cohort of 1022 participants was established over a 21-month period during 1986–1987 in the Faroe Islands, to examine this issue (Barr *et al.*, 2006). Mothers' intake of blubber was assessed. The children's exposure was assessed by measuring serum content of PCBs (37 PCB congeners), DDT and DDE, at birth, and at about 7 years and 14 years of age. PCB concentrations at 7 years were generally two to three times higher than at 14 years. Umbilical cord PCB concentrations were correlated with PCB concentrations in both 7- and 14-year serum samples. Analyses showed that breast-feeding duration was the primary contributor to serum total PCB concentrations at 7 years, and blubber consumption was the primary contributor at 14 years. The study suggested that exposures from breast-feeding were sufficiently great so that exposures through the diet over succeeding years did not fully dilute the contribution of these early exposures to body burdens in the children.

Mechanisms of Action: Insight from Experimental Models

As described earlier, human exposure to marine-derived contaminants is fairly well documented. A more difficult question is whether humans thus exposed are at risk for toxic effects. The answers to this question must necessarily come from epidemiological studies of exposed populations, as well as through extrapolation of findings obtained in experimental systems. Chapter 10 describes some of the epidemiological data from human populations highly exposed to PCBs and related compounds. Here we address the issue of extrapolation from experimental data.

The process of extrapolation is most accurate when built on a foundation of mechanistic understanding. Thus, substantial efforts have been made to understand the mechanisms by which the contaminants described in this chapter might act to cause toxicity in humans. By necessity, most of these studies have been performed using model systems, most often rodents but also nonhuman primates and a variety of nonmammalian species including birds and fish. In addition, studies have been carried out in cell lines derived from humans and in human tissues in order to determine to what extent mechanisms determined in nonhuman systems can be extra-polated to humans. The general approach and philosophy concerning the combined use of mechanistic research in human and nonhuman experimental systems to predict the human health effects of contaminants have been reviewed (Brent, 2004; Haber *et al.*, 2001; Lehman-McKeeman, 2002).

Despite extensive research over many years, for most of the chemicals mentioned in this chapter, our understanding of mechanisms of action is incomplete—in many cases rudimentary. However, there is appreciable understanding of PAH toxicity, which can include carcinogenesis as well as other toxicities that are dependant on metabolic transformation to more reactive products that can bind to DNA or other biomolecules (Fig. 7-8; see also Chapter 32). These processes occur in mammals and have been implicated in some fish populations where high prevalence of liver tumors had been found (McMahon *et al.*, 1990; Myers *et al.*, 1994).

One important group of marine contaminants for which mechanisms are fairly well understood is the dioxin-like compounds (DLCs). These compounds include the extremely potent 2,3,7,8-tetrachlorodibenzo-*p*-dioxin (TCDD), other 2,3,7,8-substituted PCDDs and PCDFs, and a small number of "planar" (non- and mono-*ortho*-substituted) PCBs (Table 7-1; Fig. 7-1). The dioxin-like PCBs include only about a dozen or so of the 209 possible PCB congeners and make up only a small percentage of the total mass of PCBs in most environmental samples, but they are thought to contribute disproportionately to the overall toxicity of PCB mixtures because of the mechanism by which they act. A brief discussion of what is known about how these compounds cause toxicity and what has been learned about this mechanism in humans will serve as an example both of the power of molecular toxicology and of the difficulties in transferring that knowledge from experimental systems to human risk assessment.

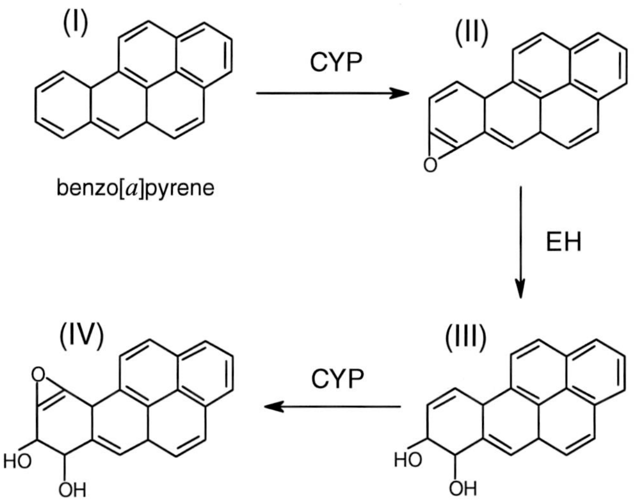

FIGURE 7-8. Example of metabolism of a polynuclear aromatic hydrocarbon. This depicts a well-known pathway for metabolism of the carcinogen benzo[a]pyrene (I), involving the addition of oxygen and water. Initial formation of an epoxide (II) by cytochrome P450 (CYP) is followed by addition of H_2O by epoxide hydrolase (EH) to form a diol (III), and then a second addition of oxygen by CYP to form a diol-epoxide (IV). The path shown is one leading to formation of a carcinogenic derivative and is accomplished similarly in all types of vertebrates, from fish to mammals. Typically, it is CYP1 enzymes, induced via the aryl hydrocarbon receptor, that catalyze the oxygen addition.

Dioxin-like compounds have a variety of effects in experimental animals. From a human health perspective, perhaps the most important are carcinogenicity, immunotoxicity, and reproductive/developmental toxicity (Committee on EPA's Exposure and Human Health Reassessment of TCDD and Related Compounds, 2006). The extreme toxic potency of DLCs in experimental animals is in large part a result of the fact that they act *via* an intracellular receptor, the aryl hydrocarbon receptor (Ah receptor or AHR), which is expressed in most tissues and regulates the expression of a large number of genes. The AHR was first discovered as the protein responsible for a mouse strain difference in sensitivity to dioxins and PAHs (which also act in part through this receptor) (Nebert *et al.*, 2004; Schmidt and Bradfield, 1996). Subsequently, it has been extensively characterized in a variety of species, especially vertebrate species from fish to humans (Hahn, 2002).

The AHR is a latent transcription factor that becomes activated by binding of DLCs. The cascade of protein-protein and protein-DNA interactions that follows, culminating in the induction of gene expression, has been elucidated in some detail (Fig. 7-9). The AHR also represses some genes, although the mechanisms involved are not as well understood. Nevertheless, it is clear that the AHR is necessary for the toxicity of DLCs and that in some cases inter- and intraspecies differences in sensitivity can be explained by differences in the AHR protein (Fernandez-Salguero *et al.*, 1996; Karchner *et al.*, 2006; Poland *et al.*, 1994).

Although the AHR is known to be required for DLC toxicity, the exact mechanism by which activation of the AHR leads to toxicity is not clear. In fact, there may be several mechanisms (i.e., target genes) that operate in different tissues, species, and life stages. A major challenge in understanding mechanisms of AHR-dependent toxicity has been that exposure and toxicity are separated in time; for humans especially, the potential effects of concern are those arising from low-level, chronic exposure to these compounds. Extrapolating results from acute, high-dose experiments in animals to the most relevant human exposure situations is fraught with difficulties.

One area in which an understanding of DLC mechanisms has been valuable is in assessing the potential impact of mixtures of DLCs. Although many PCDDs, PCDFs, and PCBs act through the AHR, they do so with widely disparate potencies, ranging over orders of magnitude. An understanding of the common AHR-dependent mechanism by which they all act has facilitated an approach to summing the concentrations of each compound, after correcting for its potency relative to that of TCDD—the so-called "TCDD-Equivalency" (TEQ) approach (Safe 1990; Van den Berg *et al.*, 2006). This approach is not perfect; for example, relative potencies are somewhat endpoint-dependent and some compounds are partial agonists or antagonists, acting against an

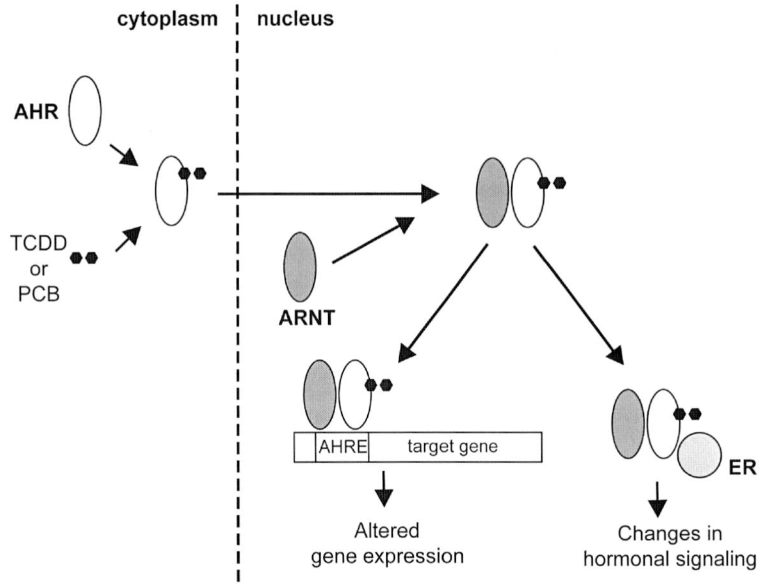

FIGURE 7-9. The aryl hydrocarbon receptor (AHR) pathway involved in gene regulation and toxicity. In mammals and other vertebrate animals, the AHR protein typically is found in the cytoplasm in a complex with accessory proteins. Binding of a compound such as TCDD activates the AHR, causing it to move into the nucleus, where it forms a complex with a related protein, ARNT. The TCDD-AHR-ARNT complex binds to regulatory DNA sequences near target genes such as CYP1A1 and regulates gene expression by recruiting other proteins that help initiate transcription of the target gene. The AHR also is thought to interact with other transcription factors and thereby modulate other signaling pathways, such as that for estrogens, mediated by the estrogen receptor (ER). The AHR is necessary for the toxicity of chlorinated dioxins and certain PCBs, but the specific mechanisms by which AHR-mediated changes in gene expression or interactions with other proteins cause toxicity are not yet known.

additive effect. Nevertheless, the TEQ approach provides a useful first-order approximation for the potential dioxin-like toxicity of a mixture.

Human Risk

As described in the previous section, we have a fairly good idea of at least the initial mechanisms by which PAHs and DLCs cause toxicity in animals. How, then, can we use this knowledge to assess the risk of these compounds to humans that are exposed, for example, through consumption of contaminated fish? Previously, human risk assessment involved extrapolating from effects observed in experimental animals to humans, after accounting both for the lower exposure of humans and for possible species differences in sensitivity. This was a conservative approach and did not incorporate mechanistic information. For example, the risk to humans from consuming PCB-contaminated fish has been calculated using results of carcinogenicity studies in which rodents were exposed to PCB mixtures (Barron *et al.*, 1994; Boyer *et al.*, 1991). This approach does not consider the relative contributions of dioxin-like PCBs versus other PCBs in the mixture and does not incorporate any of the knowledge about how mechanisms determined in rodents might apply to humans.

Susceptible Populations and Molecular Determinants of Susceptibility

Estimates of risk must employ several levels of extrapolation: extrapolation from the high doses typical of animal studies (for statistical reasons) to the lower doses typical of human exposures, extrapolation across species (usually rodent to human), and extrapolation among life-history stages.

High-to-low dose extrapolation is a complex and contentious issue characterized by extreme uncertainty concerning the shape of the dose-response curves for different endpoints and at environmentally relevant exposure levels. A detailed discussion of dose extrapolation for dioxin-like compounds can be found elsewhere (Committee on EPA's Exposure and Human Health Reassessment of TCDD and Related Compounds, 2006).

Extrapolation from one species to another (humans) is facilitated by a fundamental understanding of the comparative biology of the biochemical systems involved in the response to a toxicant. For example, for DLCs, numerous studies have compared properties of AHRs from experimental animals and humans. Studies using human cell lines and human tissues suggest that humans have an AHR that is approximately ten-fold less sensitive to TCDD (Connor and

Aylward, 2006; Okey, 2007). However, there is substantial interindividual variability among humans with respect to the binding affinity of AHR for TCDD, with some individuals appearing to possess high-affinity AHRs similar to those of dioxin-sensitive rodents (Okey *et al.*, 1997). In contrast to this biochemical variability, there is little variation in AHR sequence among individuals (Harper *et al.*, 2002; Okey *et al.*, 2005). Thus, it is not yet possible to predict which humans are likely to be most sensitive to DLCs.

Another important factor in assessing differential susceptibility to marine derived organic contaminants is the life stage at which exposure is occurring. Developing animals are often more sensitive to chemicals, especially those acting through receptor-dependent mechanisms. In particular, the developing nervous system is thought to be especially sensitive to lipophilic toxicants such as PCBs and methylmercury (Roegge and Schantz, 2006; Schantz and Widholm, 2001; Schantz *et al.*, 2003). Neonates may receive substantial exposure to lipophilic compounds through their mother's milk. In addition, developing animals may have protective mechanisms that are not yet fully functional. Thus, consumption advisories often target children and women of child-bearing age, because these are populations that are considered especially susceptible to chemical effects.

IMPACT ON MARINE ORGANISMS/ ENVIRONMENTAL HEALTH

Some of the earliest indications that halogenated organic chemicals could be harmful to nontarget species were effects noticed in wildlife, for example, the well-known phenomenon of eggshell thinning in some birds exposed to DDT and other insecticides during the 1950s and 1960s (Carson, 1962). In more recent times, effects in wildlife—including marine wildlife—have played a similarly prominent role in raising awareness about the potential hazards of chemicals, especially with respect to chemicals capable of interfering with the function of endocrine systems (Colborn *et al.*, 1993; National Research Council, 1999). A few examples serve to illustrate the kinds of effects observed in marine animals.

Marine Birds and Mammals

Animals at the apex of food webs are at greatest risk for accumulation of organic pollutants. Thus, fish-eating birds and seals and whales that consume fish or other mammals are most likely to be impacted by chemical exposure (Colborn and Smolen, 1996; O'Shea, 1999). Yet these top-level marine consumers are extraordinarily difficult to study because of ethical, legal, and logistical challenges, so there remains a great deal of uncertainty about the nature and magnitude of the effects and the specific chemicals involved (Marine Mammal Commission, 1999).

Among the birds thought to exhibit reproductive impairment resulting from organic chemical exposure are terns, cormorants, and other fish-eating birds in the Great Lakes (Giesy *et al.*, 1994) as well as in Europe (Bosveld and Van den Berg, 1994). However, caution must be exercised in ascribing apparent reproductive abnormalities to the effects of chemicals, as natural explanations are possible (Hart *et al.*, 2003).

Among aquatic mammals, odontocete cetaceans (toothed whales) and pinnipeds (seals) accumulate the greatest concentrations of anthropogenic organic contaminants (as well as natural HOCs; see the section titled Natural Halogenated Compounds). The St. Lawrence Estuary population of beluga whale (*Delphinapterus leucas*) is the poster child for chemical impacts on marine mammals. Protected status since the 1980s has failed to reverse the earlier declines in this population, and the high concentrations of organic contaminants accumulated by these whales have been blamed (Deguise *et al.*, 1995). Yet even in this well-studied isolated population, conclusive evidence implicating anthropogenic chemicals in this effect has been elusive. As with most such cases involving wildlife, we must rely on a "weight of evidence" approach to assigning causality (Marine Mammal Commission, 1999; Ross, 2000). Such an approach is enhanced by the incorporation of mechanistic information obtained from comparative studies of proteins involved in toxicity (Haber *et al.*, 2001; Jensen and Hahn, 2001; Lehman-McKeeman, 2002; White *et al.*, 1994).

There is one example in which direct experimental evidence has contributed to an understanding of organic chemical effects in free-ranging marine mammals. In a series of studies performed in the Netherlands, captive harbor seals (*Phoca vitulina*) were fed fish containing high or low concentrations of PCBs. Alterations in both reproductive and immunological functions were observed (Reijnders, 1986; Ross *et al.*, 1996), suggesting that PCBs at concentrations found in Baltic Sea fish were adversely impacting the health of these marine mammals.

In the absence of direct experimental data from the species of interest, risk assessments can be performed using data from surrogate species. For example, data from studies in captive mink have been used in individual- and population-based risk assessment models to predict the risk of reproductive toxicity from PCB exposure in wild bottlenose dolphins (Hall *et al.*, 2006; Schwacke *et al.*, 2002).

Population-Level Effects in Fish

Fish are known to be extremely sensitive to dioxin-like compounds. In perhaps the best example of environmental epidemiology, a long-term study involving toxicologists and environmental chemists provided convincing evidence that contamination of the Great Lakes with DLCs from the 1940s through 1970s was responsible for reproductive failure of lake trout (*Salvelinus namaycush*) populations (Cook *et al.*, 2003).

In contrast to the loss of reproductive capability in lake trout exposed to DLCs, at least one marine species has demonstrated an extraordinary capacity to adapt to these compounds, at the same time demonstrating a different kind of population-level effect.

The Atlantic killifish *Fundulus heteroclitus* inhabits estuaries from the Canadian maritime provinces to Florida and for years has served as a valuable experimental model for environmental adaptation (Burnett *et al.*, 2007). The ability of *F. heteroclitus* populations to adapt to chemicals was first demonstrated with respect to methylmercury (Weis and Weis, 1989). Subsequently, several populations of this species were shown to have developed resistance to DLCs (Nacci *et al.*, 1999, 2002; Prince and Cooper, 1995) or PAHs (Van Veld and Westbrook, 1995). The mechanisms by which the resistance occurs are not well understood but have been hypothesized to involve alterations in AHR-dependent signaling pathways. Allelic variants of AHRs exist and differ among sites (Hahn *et al.*, 2004), but whether these or other changes are responsible for the resistant phenotypes is not yet known. Some evidence suggests that the mechanisms of resistance at different sites may be distinct; for example, at some sites the resistance in heritable through at least two generations, whereas at other sites the resistance is lost in offspring of field-caught fish (van Veld and Nacci, 2007).

Although seemingly beneficial, resistance to DLCs may be associated with costs such as increased sensitivity to other environmental stressors (e.g., hypoxia) (Meyer and Di Giulio, 2002). Interestingly, the evolution of resistance in *F. heteroclitus* at these sites has not been accompanied by an overall loss of genetic diversity (Cohen, 2002; Roark *et al.*, 2005), suggesting that resistant genotypes are maintained in the face of continued gene flow. This would predict that as contaminant loads are reduced through natural and engineered processes, the populations will revert to sensitive phenotypes, as demonstrated in other situations (Levinton *et al.*, 2003).

CONCLUSIONS

The generation and use of organic chemicals, their environmental fate and presence in marine systems, and their possible effects on experimental animals and humans have been the subject of thousands of published papers. In a brief chapter such as this, we can only highlight some of the information, concerns, and uncertainties that exist for these compounds. Some of the main points that we hope are evident from the previous sections include the following:

- Human activities have altered the chemical environment of the oceans, with potential impacts on both humans and marine organisms. The oceans and marine organisms also contain natural products that are structurally similar to some toxic anthropogenic contaminants but whose health effects are essentially unknown.
- Concentrations of many "traditional" contaminants (PCBs, PCDD/Fs, PAHs) show declining trends whereas those of "emerging" contaminants such as brominated flame-retardants and polyfluoroalkyl compounds are increasing.
- Organic chemicals are distributed by physical and biological processes, and concentrations and temporal trends vary geographically.
- Our ability to measure organic chemicals in humans and the marine environment surpasses our ability to understand the impacts of the concentrations measured, especially the low-to-moderate concentrations that characterize remote regions and most humans.
- Possible human health impacts can be inferred from epidemiological studies as well as by extrapolating from studies in experimental animals. Extrapolation is more accurate if informed by mechanistic information obtained from studies in the target species.
- Marine animals at higher trophic levels and living near coastal regions are the most highly exposed and thus at greatest risk for effects of contaminants. Like humans, many of these top consumers, such as some marine mammals, cannot be studied directly. Thus, approaches to infer risk parallel those used in humans: eco-epidemiology and extrapolation from studies in surrogate species.
- Some marine species have evolved adaptive mechanisms that help them survive exposure to contaminants. However, such adaptations may be accompanied by ecological costs and could result in greater accumulations of chemicals in such animals, increasing risk to consumers.
- Case studies demonstrate how difficult it can be to balance the risks and benefits associated with consuming contaminated seafood.

Acknowledgments

Preparation of this chapter was supported by grants establishing the Woods Hole Center for Oceans and Human Health (NIEHS P50ES012742 and NSF-OCE-0430724) as well as by the Superfund Basic Research Program grant P42ES007381 (MEH and JJS), National Science Foundation grant OCE-0550486 (CMR) and a grant from the WHOI Ocean Life Institute (CMR and MEH). We also acknowledge with gratitude a large number of talented students, postdocs, colleagues, and collaborators who have been our partners over the years in exploring questions and issues concerning the presence and effects of organic chemicals in the oceans.

References

Alexander, M., 1999. Biodegradation and Bioremediation, 2nd ed. San Diego CA Academic Press.

Andersson, Ö., Blomkvist, G., 1981. Polybrominated aromatic pollutants found in fish in Sweden. Chemosphere 10, 1051.

Anjaneyulu, V., Nageswara Rao, K., Radhika, P., Muralikrishna, M., Connolly, J.D., 1996. A new tetrabromodiphenyl ether from the sponge *Dysidea herbacea* of the Indian Ocean. Indian J. Chem. 35B, 89–90.

Appleby, P., Oldfield, F., 1978. The calculation of lead-210 dates assuming a constant rate of supply of unsupported 210Pb to the sediment. Catena 5, 1–8.

Baker, J.I., Hites, R.A., 2000. Siskiwit Lake revisited: Time trends of polychlorinated dibenzo-p-dioxin and dibenzofuran deposition at Isle Royale, Michigan. Environ. Sci. Technol. 34, 2887–2891.

Barr, D.B., Weihe, P., Davis, M.D., Needham, L.L., Grandjean, P., 2006. Serum polychlorinated biphenyl and organochlorine insecticide concentrations in a Faroese birth cohort. Chemosphere 62, 1167–1182.

Barron, M.G., Yurk, J.J., Crothers, D.B., 1994. Assessment of potential cancer risk from consumption of PCBs bioaccumulated in fish and shellfish. Environ. Health Perspect. 102, 562–567.

Bell, J.G., McGhee, F., Dick, J.R., Tocher, D.R., 2005. Dioxin and dioxin-like polychlorinated biphenyls (PCBs) in Scottish farmed salmon (Salmo salar): Effects of replacement of dietary marine fish oil with vegetable oils. Aquaculture 243, 305–314.

Berntssen, M.H., Lundebye, A.K., Torstensen, B.E., 2005. Reducing the levels of dioxins and dioxin-like PCBs in farmed Atlantic salmon by substitution of fish oil with vegetable oil in the feed. Aquac. Nutr. 11, 219–231.

Binelli, A., Provini, A., 2003. POPs in edible clams from different Italian and European markets and possible human health risk. Mar. Pollut. Bull. 46, 879–886.

Birnbaum, L.S., Cohen Hubal, E.A., 2006. Polybrominated diphenyl ethers: A case study for using biomonitoring data to address risk assessment questions. Environ. Health Perspect. 114, 1770–1775.

Blais, J.M., Kimpe, L.E., McMahon, D., Keatley, B.E., Mallory, M.L., Douglas, M.S., Smol, J.P., 2005. Arctic seabirds transport marine-derived contaminants. Science 309, 445.

Blais, J.M., Macdonald, R.W., Mackey, D., Webster, E., Harvey, C., Smol, J.P., 2007. Biologically mediated transport of contaminants to aquatic systems. Environ. Sci. Technol. 41, 1075–1084.

Bosveld, A.T.C., Van den Berg, M., 1994. Effects of polychlorinated biphenyls, dibenzo-p-dioxins, and dibenzofurans on fish-eating birds. Environ. Rev. 2, 147–166.

Boyer, I.J., Kokoski, C.J., Bolger, P.M., 1991. Role of FDA in establishing tolerable levels for dioxin and PCBs in aquatic organisms. J. Toxicol. Environ. Health 33, 93–101.

Brent, R.L., 2004. Utilization of animal studies to determine the effects and human risks of environmental toxicants (drugs, chemicals, and physical agents). Pediatrics 113, 984–995.

Burnett, K.G., Bain, L.J., Baldwin, W.S., Callard, G.V., Cohen, S., Giulio, R.T.D., Evans, D.H., Gómez-Chiarri, M., Hahn, M.E., Hoover, C.A., Karchner, S.I., Katoh, F., MacLatchy, D.L., Marshall, W.S., Meyer, J.N., Nacci, D.E., Oleksiak, M.F., Rees, B.B., Singer, T.P., Stegeman, J.J., Towle, D.W., Veld, P.A.V., Vogelbein, W.K., Whitehead, A., Winn, R.N., Crawford, D.L., 2007. Fundulus as the premier teleost model in environmental biology: Opportunities for new insights using genomics. Comp. Biochem. Physiol., doi:10.1016/j.cbd.2007.09.001

Carlson, D.L., Hites, R.A., 2005. Polychlorinated biphenyls in salmon and salmon feed: Global differences and bioaccumulation. Environ. Sci. Technol. 39, 7389–7395.

Carson, R., 1962. Silent Spring. Boston, Houghton Mifflin.

Charles, M.J., Hites, R.A., 1987. Sediments as archives of sediment pollution trends. In Hites, R.A., and Eisenreich, S.J. (eds.), Sources and Fates of Aquatic Pollutants: Advances in Chemistry Series 216, Washington, DC, American Chemical Society.

Cohen, S., 2002. Strong positive selection and habitat-specific amino acid substitution patterns in MHC from an estuarine fish under intense pollution stress. Mol. Biol. Evol. 19, 1870–1880.

Colborn, T., Saal, F.S.V., Soto, A.M., 1993. Developmental Effects of Endocrine-Disrupting Chemicals in Wildlife and Humans. Environ. Health Perspect. 101, 378–384.

Colborn, T., Smolen, M.J., 1996. Epidemiological analysis of persistent organochlorine contaminants in cetaceans. Rev. Environ. Contam. Toxicol. 146, 91–172.

Committee on EPA's Exposure and Human Health Reassessment of TCDD and Related Compounds, N.R.C., 2006. Health risks from dioxin and related compounds: Evaluation of the EPA reassessment. National Academy Press, Washington DC.

Connor, K.T., Aylward, L.L., 2006. Human response to dioxin: Aryl hydrocarbon receptor (AhR) molecular structure, function, and dose-response data for enzyme induction indicate an impaired human AhR. J. Toxicol. Environ. Health B. Crit. Rev. 9, 147–171.

Cook, P.M., Robbins, J.A., Endicott, D.D., Lodge, K.B., Guiney, P.D., Walker, M.K., Zabel, E.W., Peterson, R.E., 2003. Effects of aryl hydrocarbon receptor-mediated early life stage toxicity on lake trout populations in Lake Ontario during the 20th century. Environ. Sci. Technol. 37, 3864–3877.

Deguise, S., Martineau, D., Beland, P., Fournier, M., 1995. Possible mechanisms of action of environmental contaminants on St. Lawrence beluga whales (*Delphinapterus leucas*). Environ. Health Perspect. 103, 73–77.

DeWit, C.A., 2000. Brominated Flame Retardants. Report 5605: Swedish Environmental Protection Agency, 1–96.

Dockery, D.W., Pope 3rd, C.A., Xu, X., Spengler, J.D., Ware, J.H., Fay, M.E., Ferris, B.G., Jr., Speizer, F.E., 1993. An association between air pollution and mortality in six U.S. cities. N. Engl. J. Med. 329, 1753–1759.

Domingo, J.L., Bocio, A., 2007. Levels of PCDD/PCDFs and PCBs in edible marine species and human intake: A literature review. Environ. Int. 33, 397–405.

Doney, S.C., 2006. The dangers of ocean acidification. Sci. Am. 294, 58–65.

Erickson, M., 1997. The Analytical Chemistry of PCBs, 2nd ed. Boca Raton, FL, CRC Press.

Eriksson, P., Jakobsson, E., Fredriksson, A., 2001. Brominated flame retardants: A novel class of developmental neurotoxicants in our environment? Environ. Health Perspect. 109, 903–908.

Evenset, A., Carroll, J., Christensen, G.N., Kallenborn, R., Gregor, D., Gabrielsen, G.W., 2007. Seabird guano is an efficient conveyer of persistent organic pollutants (POPs) to Arctic lake ecosystems. Environ. Sci. Technol. 41, 1173–1179.

Fangstrom, B., Athanasiadou, M., Grandjean, P., Weihe, P., Bergman, A., 2002. Hydroxylated PCB metabolites and PCBs in serum from pregnant Faroese women. Environ. Health Perspect. 110, 895–899.

Fent, K., Weston, A.A., Caminada, D., 2006. Ecotoxicology of human pharmaceuticals. Aquat. Toxicol. 76, 122–159.

Fernandez-Salguero, P., Hilbert, D.M., Rudikoff, S., Ward, J.M., Gonzalez, F.J., 1996. Aryl-hydrocarbon receptor-deficient mice are resistant to 2,3,7,8-tetrachlorodibenzo-p-dioxin-induced toxicity. Toxicol. Appl. Pharmacol. 140, 173–179.

Furst, P., 2006. Dioxins, polychlorinated biphenyls and other organohalogen compounds in human milk. Levels, correlations, trends and exposure through breastfeeding. Mol. Nutr. Food Res. 50, 922–933.

Giesy, J.P., Kannan, K., 2001. Global distribution of perfluorooctane sulfonate in wildlife. Environ. Sci. Technol. 35, 1339–1342.

Giesy, J.P., Ludwig, J.P., Tillitt, D.E., 1994. Deformities in birds of the Great Lakes Region: Assigning causality. Environ. Sci. Technol. 28, 128A-135A.

Glassmeyer, S.T., Furlong, E.T., Kolpin, D.W., Cahill, J.D., Zaugg, S.D., Werner, S.L., Meyer, M.T., Kryak, D.D., 2005. Transport of chemical and microbial compounds from known wastewater discharges: Potential for use as indicators of human fecal contamination. Environ. Sci. Technol. 39, 5157–5169.

Goldberg, E., Gamble, E., Griffin, J., Koide, M., 1977. Pollution history of Narragansett Bay as recorded in its sediments. Environ. Sci. Technol. 5, 549–561.

Grandjean, P., White, R.F., Weihe, P., Jorgensen, P.J., 2003. Neurotoxic risk caused by stable and variable exposure to methylmercury from seafood. Ambul. Pediatr. 3, 18–23.

Green, N.J., Hassanin, A., Johnston, A.E., Jones, K.C., 2004. Observations on historical, contemporary, and natural PCDD/Fs. Environ. Sci. Technol. 38, 715–723.

Gribble, G.W., Blank, D.H., Jasinski, J.P., 1999. Synthesis and identification of two halogenated bipyrroles present in seabird eggs. Chem. Commun. 21, 2195–2196.

Gschwend, P.M., Hites, R.A., 1981. Fluxes of polycyclic aromatic hydrocarbons to marine and lacustrine sediments in the northeastern United States. Geochimica et Cosmochimica Acta 45, 2359–2367.

Guo, Z., Lin, T., Zhang, G., Yang, Z., Fang, M., 2006. High-resolution depositional records of polycyclic aromatic hydrocarbons in the central continental shelf mud of the East China Sea. Environ. Sci. Technol. 40, 5304–5311.

Gustafsson, O., Haghseta, F., Chan, C., Macfarlane, J., Gschwend, P.M., 1997. Quantification of the dilute sedimentary soot-phase: Implications for PAH speciation and bioavailability. Environ. Sci. Technol. 31, 203–209.

Haag, A., 2005. Marine biology: Whale fall. Nature 433, 566–567.

Haber, L.T., Maier, A., Zhao, Q., Dollarhide, J.S., Savage, R.E., Dourson, M.L., 2001. Applications of mechanistic data in risk assessment: The past, present, and future. Toxicol. Sci. 61, 32–39.

Hagmar, L., Wallin, E., Vessby, B., Jonsson, B.A., Bergman, A., Rylander, L., 2006. Intra-individual variations and time trends 1991–2001 in human serum levels of PCB, DDE and hexachlorobenzene. Chemosphere 64, 1507–1513.

Hahn, M.E., 2002. Aryl hydrocarbon receptors: Diversity and evolution. Chem. Biol. Interact. 141, 131–160.

Hahn, M.E., Karchner, S.I., Franks, D.G., Merson, R.R., 2004. Aryl hydrocarbon receptor polymorphisms and dioxin resistance in Atlantic killifish (*Fundulus heteroclitus*). Pharmacogenetics 14, 131–143.

Hall, A.J., McConnell, B.J., Rowles, T.K., Aguilar, A., Borrell, A., Schwacke, L., Reijnders, P.J., Wells, R.S., 2006. Individual-based model framework to assess population consequences of polychlorinated biphenyl exposure in bottlenose dolphins. Environ. Health Perspect. 114 Suppl 1, 60–64.

Harper, C.R., Jacobson, T.A., 2001. The fats of life: The role of omega-3 fatty acids in the prevention of coronary heart disease. Arch. Intern. Med. 161, 2185–2192.

Harper, P.A., Wong, J.M.Y., Lam, M.S.M., Okey, A.B., 2002. Polymorphisms in the human AH receptor. Chemico-Biological Interactions 141, 161–187.

Hart, C.A., Nisbet, I.C.T., Kennedy, S.W., Hahn, M.E., 2003. Gonadal feminization and halogenated environmental contaminants in common terns (*Sterna hirundo*): Evidence that ovotestes in male embryos do not persist to the prefledgling stage. Ecotoxicology 12, 125–140.

Hartmann, P.C., Quinn, J.G., Cairns, R.W., King, J.W., 2005. Depositional history of organic contaminants in Narragansett Bay, Rhode Island, USA. Mar. Pollut. Bull. 50, 388–395.

Hekster, F.M., Laane, R.W., de Voogt, P., 2003. Environmental and toxicity effects of perfluoroalkylated substances. Rev. Environ. Contam. Toxicol. 179, 99–121.

Hirsch, R., Ternes, T., Haberer, K., Kratz, K.L., 1999. Occurrence of antibiotics in the aquatic environment. Sci. Total. Environ. 225, 109–118.

Hites, R.A., 2004. Polybrominated diphenyl ethers in the environment and in people: A meta-analysis of concentrations. Environ. Sci. Technol. 38, 945–956.

Hites, R.A., Foran, J.A., Carpenter, D.O., Hamilton, M.C., Knuth, B.A., Schwager, S.J., 2004a. Global assessment of organic contaminants in farmed salmon. Science 303, 226–229.

Hites, R.A., Foran, J.A., Carpenter, D.O., Hamilton, M.C., Knuth, B.A., Schwager, S.J., 2004b. Response to letters. Science 305, 475–478.

Houde, M., Martin, J.W., Letcher, R.J., Solomon, K.R., Muir, D.C., 2006. Biological monitoring of polyfluoroalkyl substances: A review. Environ. Sci. Technol. 40, 3463–3473.

Huwe, J.K., Larsen, G.L., 2005. Polychlorinated dioxins, furans, and biphenyls, and polybrominated diphenyl ethers in a U.S. meat market basket and estimates of dietary intake. Environ. Sci. Technol. 39, 5606–5611.

Ikonomou, M.G., Rayne, S., Addison, R.F., 2002. Exponential increases of the brominated flame retardants, polybrominated diphenyl ethers, in the Canadian Arctic from 1981 to 2000. Environ. Sci. Technol. 36, 1886–1892.

Inoue, K., Okada, F., Ito, R., Kato, S., Sasaki, S., Nakajima, S., Uno, A., Saijo, Y., Sata, F., Yoshimura, Y., Kishi, R., Nakazawa, H., 2004. Perfluorooctane sulfonate (PFOS) and related perfluorinated compounds in human maternal and cord blood samples: Assessment of PFOS exposure in a susceptible population during pregnancy. Environ. Health Perspect. 112, 1204–1207.

James, M.O., Boyle, S.M., 1998. Cytochromes P450 in crustacea. Comp. Biochem. Physiol. 121C, 157–172.

Janssen, B., Asplund, L., Olsson, M., 1987. Brominated flame retardants: Ubiquitous environmental pollutants? Chemosphere 16, 2343.

Jensen, B.A., Hahn, M.E., 2001. cDNA cloning and characterization of a high affinity aryl hydrocarbon receptor in a cetacean, the beluga, *Delphinapterus leucas*. Toxicol. Sci. 64, 41–56.

Jensen, S., 1966. Report of a new chemical hazard. New Scientist 32, 612.

Johansen, P., Muir, D., Asmund, G., Riget, F., 2004. Human exposure to contaminants in the traditional Greenland diet. Sci. Total Environ. 331, 189–206.

Jonsson, A., Gustafsson, O., Axelman, J., Sundberg, H., 2003. Global accounting of PCBs in the continental shelf sediments. Environ. Sci. Technol. 37, 245–255.

Kannan, K., Corsolini, S., Falandysz, J., Fillmann, G., Kumar, K.S., Loganathan, B.G., Mohd, M.A., Olivero, J., Van Wouwe, N., Yang, J.H., Aldoust, K.M., 2004. Perfluorooctanesulfonate and related fluorochemicals in human blood from several countries. Environ. Sci. Technol. 38, 4489–4495.

Kannan, K., Koistinen, J., Beckmen, K., Evans, T., Gorzelany, J.F., Hansen, K.J., Jones, P.D., Helle, E., Nyman, M., Giesy, J.P., 2001. Accumulation of perfluorooctane sulfonate in marine mammals. Environ. Sci. Technol. 35, 1593–1598.

Karchner, S.I., Franks, D.G., Kennedy, S.W., Hahn, M.E., 2006. The molecular basis for differential dioxin sensitivity in birds: Role of the aryl hydrocarbon receptor. Proc. Natl. Acad. Sci. USA 103, 6252–6257.

Kawamura, K., Suzuki, I., 1994. Ice core record of polycyclic aromatic hydrocarbons over the past 400 years. Naturwissenschaften 81, 502–505.

Kolpin, D.W., Furlong, E.T., Meyer, M.T., Thurman, E.M., Zaugg, S.D., Barber, L.B., Buxton, H.T., 2002. Pharmaceuticals, hormones, and other organic wastewater contaminants in U.S. streams, 1999–2000: A national reconnaissance. Environ. Sci. Technol. 36, 1202–1211.

Krummel, E.M., Macdonald, R.W., Kimpe, L.E., Gregory-Eaves, I., Demers, M.J., Smol, J.P., Finney, B., Blais, J.M., 2003. Aquatic ecology: Delivery of pollutants by spawning salmon. Nature 425, 255–256.

Kunzli, N., Kaiser, R., Medina, S., Studnicka, M., Chanel, O., Filliger, P., Herry, M., Horak, F., Jr., Puybonnieux-Texier, V., Quenel, P., Schneider, J., Seethaler, R., Vergnaud, J.C., Sommer, H., 2000. Public-health impact of outdoor and traffic-related air pollution: A European assessment. Lancet 356, 795–801.

Kurelec, B., 1992. The multixenobiotic resistance mechanism in aquatic organisms. Crit. Rev. Toxicol. 22, 23–43.

Latimer, J.S., Quinn, J.G., 1996. Historical trends and current inputs of hydrophobic organic compounds in an urban estuary: The sedimentary record. Environ. Sci. Technol. 30, 623–633.

Lau, C., Butenhoff, J.L., Rogers, J.M., 2004. The developmental toxicity of perfluoroalkyl acids and their derivatives. Toxicol. Appl. Pharmacol. 198, 231–241.

Lehman-McKeeman, L.D., 2002. Incorporating mechanistic data into risk assessment. Toxicology 181–182, 271–274.

Levinton, J.S., Suatoni, E., Wallace, W., Junkins, R., Kelaher, B., Allen, B.J., 2003. Rapid loss of genetically based resistance to metals after the cleanup of a Superfund site. Proc. Natl. Acad. Sci. USA 100, 9889–9891.

Lilienthal, H., Hack, A., Roth-Harer, A., Grande, S.W., Talsness, C.E., 2006. Effects of developmental exposure to 2,2 ,4,4 ,5-pentabromodiphenyl ether (PBDE-99) on sex steroids, sexual development, and sexually dimorphic behavior in rats. Environ. Health Perspect. 114, 194–201.

Lima, A.L., Eglinton, T.I., Reddy, C.M., 2003. High-resolution record of pyrogenic polycyclic aromatic hydrocarbon deposition during the 20th century. Environ. Sci. Technol. 37, 53–61.

Lima, A.L., Farrington, J.W., Reddy, C.M., 2005. Combustion-derived polycyclic aromatic hydrocarbons in the environment: a review. Environ. Forensics 6, 109–131.

Lipnick, R.L., Muir, D.C.G., 2000. History of persistent, bioaccumulative and toxic chemicals. In Lipnick, R.L., Hermens, J.L.M., Jones, K.C., Muir, and D.G. (eds.), ACS Symposium Series 773. Chemicals in the Environment: Fate, Impact and Remediation. Washington, DC, American Chemical Society.

Livingstone, D.R., 1991. Organic xenobiotic metabolism in marine invertebrates. In Gilles, R. (ed.), Advances in Comparative and Environmental Physiology, vol. 7. Berlin, Heidelberg, Springer-Verlag.

Lohmann, R., Jurado, E., Dachs, J., Lohmann, U., Jones, K.C., 2006. Quantifying the importance of the atmospheric sink for polychlorinated dioxins and furans relative to other global loss processes. J. Geophys. Res. Atmos. 111, D21303.

Lund, E., Engeset, D., Alsaker, E., Skeie, G., Hjartaker, A., Lundebye, A.K., Niebor, E., 2004. Cancer risk and salmon intake. Science 305, 477–478; author reply 477–478.

Mai, B., Zeng, E.Y., Luo, X., Yang, Q., Zhang, G., Li, X., Sheng, G., Fu, J., 2005. Abundances, depositional fluxes, and homologue patterns of polychlorinated biphenyls in dated sediment cores from the Pearl River Delta, China. Environ. Sci. Technol. 39, 49–56.

Malmvarn, A., Zebuhr, Y., Jensen, S., Kautsky, L., Greyerz, E., Nakano, T., Asplund, L., 2005. Identification of polybrominated dibenzo-p-dioxins in blue mussels (*Mytilus edulis*) from the Baltic Sea. Environ. Sci. Technol. 39, 8235–8242.

Marine Mammal Commission, 1999. Marine Mammals and Persistent Ocean Contaminants: Proceedings of the Marine Mammal Commission Workshop, Keystone, Colorado, Marine Mammal Commission, Bethesda, MD. October 12–15, 1998. In O'Shea, T.J., Reeves, R.R., and Long, A.K. (eds.).

Marsh, G., Athanasiadou, M., Athanassiadis, I., Bergman, A., Endo, T., Haraguchi, K., 2005. Identification, quantification, and synthesis of a novel dimethoxylated polybrominated biphenyl in marine mammals caught off the coast of Japan. Environ. Sci. Technol. 39, 8684–8690.

Marti, S., Bayona, J.M., Albaiges, J., 2001. A potential source of organic pollutants into the northeastern Atlantic: The outflow of the Mediterranean deep-lying waters through the Gibraltar Strait. Environ. Sci. Technol. 35, 2682–2689.

Matthews, H.B., Dedrick, R.L., 1984. Pharmacokinetics of PCBs. Annu. Rev. Pharmacol. Toxicol. 24, 85–103.

McMahon, G., Huber, L.J., Moore, M.J., Stegeman, J.J., Wogan, G.N., 1990. Mutations in c-Ki-ras oncogenes in diseased livers of winter flounder from Boston Harbor. Proc. Natl. Acad. Sci. USA 87, 841–845.

Meador, J.P., Stein, J.E., Reichert, W.L., Varanasi, U., 1995. Bioaccumulation of polycyclic aromatic hydrocarbons by marine organisms. Rev. Environ. Contam. Toxicol. 143, 79–165.

Meyer, J.N., Di Giulio, R.T., 2002. Heritable adaptation and fitness costs in killifish (*Fundulus heteroclitus*) inhabiting a polluted estuary. Ecol. Appl. 13, 490–503.

Moon, H.B., Ok, G., 2006. Dietary intake of PCDDs, PCDFs and dioxin-like PCBs, due to the consumption of various marine organisms from Korea. Chemosphere 62, 1142–1152.

Mozaffarian, D., Rimm, E.B., 2006. Fish intake, contaminants, and human health: Evaluating the risks and the benefits. JAMA 296, 1885–1899.

Muir, D.C., Howard, P.H., 2006. Are there other persistent organic pollutants? A challenge for environmental chemists. Environ. Sci. Technol. 40, 7157–7166.

Myers, M.S., Stehr, C.M., Olson, O.P., Johnson, L.L., McCain, B.B., Chan, S.-L., Varanasi, U., 1994. Relationships between toxicopathic hepatic lesions and exposure to chemical contaminants in english sole (*Pleuronectes vetulus*), starry flounder (*Platichthys stellatus*) and white croaker (*Genyonemus leneatus*) from selected marine sites on the Pacific coast, USA. Environ. Health Perspect. 102, 200–214.

Nacci, D., Coiro, L., Champlin, D., Jayaraman, S., McKinney, R., Gleason, T., Munns, W.R., Jr., Specker, J.L., Cooper, K., 1999. Adaptation of wild populatons of the estuarine fish *Fundulus heteroclitus* to persistent environmental contaminants. Marine Biol. 134, 9–17.

Nacci, D.E., Champlin, D., Coiro, L., McKinney, R., Jayaraman, S., 2002. Predicting the occurrence of genetic adaptation to dioxinlike compounds in populations of the estuarine fish *Fundulus heteroclitus*. Environ. Toxicol. Chem. 21, 1525–1532.

National Research Council, 1999. Hormonally Active Agents in the Environment. National Academy Press, Washington DC.

Nebert, D.W., Dalton, T.P., Okey, A.B., Gonzalez, F.J., 2004. Role of aryl hydrocarbon receptor-mediated induction of the CYP1 enzymes in environmental toxicity and cancer. J. Biol. Chem. 279, 23847–23850.

Nesheim, M.C., Yaktine, A.L., 2007. Seafood Choices. Balancing Benefits and Risks. In The National Academies Press, Washington, DC.

Noren, K., Meironyte, D., 2000. Certain organochlorine and organobromine contaminants in Swedish human milk in perspective of past 20–30 years. Chemosphere 40, 1111–1123.

O'Shea, T.J., 1999. Environmental contaminants and marine mammals. In Reynolds, J.E., and Rommel, S.A. (eds.), Biology of Marine Mammals. Washington, DC, Smithsonian Institution Press.

Okey, A.B., 2007. An AH Receptor Odyssey to the Shores of Toxicology: The Deichmann Lecture. Toxicol. Sci. 98, 5–38.

Okey, A.B., Boutros, P.C., Harper, P.A., 2005. Polymorphisms of human nuclear receptors that control expression of drug-metabolizing enzymes. Pharmacogenet Genomics 15, 371–379.

Okey, A.B., Giannone, J.V., Smart, W., Wong, J.M., Manchester, D.K., Parker, N.B., Feeley, M.M., Grant, D.L., Gilman, A., 1997. Binding of 2,3,7,8-tetrachlorodibenzo-p-dioxin to AH receptor in placentas from normal versus abnormal pregnancy outcomes. Chemosphere 34, 1535–1547.

Paerl, H.W., R.L., D., Whitall, D.R., 2002. Atmospheric deposition of nitrogen: Implications for nutrient over-enrichment of coastal waters. Estuaries 25, 677–693.

Papke, O., 1998. PCDD/PCDF: Human background data for Germany, a 10-year experience. Environ. Health Perspect. 106 Suppl 2, 723–731.

Peck, A.M., Linebaugh, E.K., Hornbuckle, K.C., 2006. Synthetic musk fragrances in Lake Erie and Lake Ontario sediment cores. Environ. Sci. Technol. 40, 5629–5635.

Poland, A., Palen, D., Glover, E., 1994. Analysis of the four alleles of the murine aryl hydrocarbon receptor. Mol. Pharmacol. 46, 915–921.

Porte, C., Albaiges, J., 1994. Bioaccumulation patterns of hydrocarbons and polychlorinated biphenyls in bivalves, crustaceans, and fishes. Arch. Environ. Contam. Toxicol. 26, 273–281.

Prince, R., Cooper, K.R., 1995. Comparisons of the effects of 2,3,7,8-tetrachlorodibenzo-p-dioxin on chemically impacted and nonimpacted subpopulations of *Fundulus heteroclitus*: I. TCDD toxicity. Environ. Toxicol. Chem. 14, 579–587.

Quensen 3rd, J.F., Mueller, S.A., Jain, M.K., Tiedje, J.M., 1998. Reductive dechlorination of DDE to DDMU in marine sediment microcosms. Science 280, 722–724.

Rahman, F., Langford, K.H., Scrimshaw, M.D., Lester, J.N., 2001. Polybrominated diphenyl ether (PBDE) flame retardants. Sci. Total Environ. 275, 1–17.

Reddy, C.M., Xu, L., O'Neil, G.W., Nelson, R.K., Eglinton, T.I., Faulkner, D.J., Norstrom, R., Ross, P.S., Tittlemier, S.A., 2004. Radiocarbon evidence for a naturally produced, bioaccumulating halogenated organic compound. Environ. Sci. Technol. 38, 1992–1997.

Reijnders, P.J.H., 1986. Reproductive failure in common seals feeding on fish from polluted coastal waters. Nature 324, 456–457.

Risebrough, R.W., Reiche, P., Peakall, D.B., Herman, S.G., Kirven, M.N., 1968. Polychlorinted biphenyls in the global ecosystem. Nature 220, 1098–1102.

Roark, S.A., Nacci, D., Coiro, L., Champlin, D., Guttman, S.I., 2005. Population genetic structure of a nonmigratory estuarine fish (*Fundulus heteroclitus*) across a strong gradient of polychlorinated biphenyl contamination. Environ. Toxicol. Chem. 24, 717–725.

Roegge, C.S., Schantz, S.L., 2006. Motor function following developmental exposure to PCBS and/or MEHG. Neurotoxicol. Teratol. 28, 260–277.

Ross, P., de Swart, R., Addison, R., Van Loveren, H., Vos, J., Osterhaus, A., 1996. Contaminant-induced immunotoxicity in harbour seals: Wildlife at risk? Toxicology 112, 157–169.

Ross, P.S., 2000. Marine mammals as sentinels in ecological risk assessment. Human Ecol. Risk Assess. 6, 29–46.

Safe, S., 1990. Polychlorinated biphenyls (PCBs), dibenzo-p-dioxins (PCDDs), dibenzofurans (PCDFs), and related compounds: Environmental and mechanistic considerations which support the development of toxic equivalency factors (TEFs). CRC Crit. Rev. Toxicol. 21, 51–88.

Sandanger, T.M., Brustad, M., Odland, J.O., Doudarev, A.A., Miretsky, G.I., Chaschin, V., Burkow, I.C., Lund, E., 2003. Human plasma levels of POPs, and diet among native people from Uelen, Chukotka. J. Environ. Monit. 5, 689–696.

Sandanger, T.M., Brustad, M., Sandau, C.D., Lund, E., 2006. Levels of persistent organic pollutants (POPs) in a coastal northern Norwegian population with high fish-liver intake. J. Environ. Monit. 8, 552–557.

Schantz, S.L., Widholm, J.J., 2001. Cognitive effects of endocrine-disrupting chemicals in animals. Environ. Health Perspect. 109, 1197–1206.

Schantz, S.L., Widholm, J.J., Rice, D.C., 2003. Effects of PCB exposure on neuropsychological function in children. Environ. Health Perspect. 111, 357–576.

Schecter, A., Papke, O., Harris, T.R., Tung, K.C., Musumba, A., Olson, J., Birnbaum, L., 2006. Polybrominated diphenyl ether (PBDE) levels in an expanded market basket survey of U.S. food and estimated PBDE dietary intake by age and sex. Environ. Health Perspect. 114, 1515–1520.

Schecter, A., Pavuk, M., Papke, O., Ryan, J.J., Birnbaum, L., Rosen, R., 2003. Polybrominated diphenyl ethers (PBDEs) in U.S. mothers' milk. Environ. Health Perspect. 111, 1723–1729.

Schmidt, J.V., Bradfield, C.A., 1996. Ah receptor signaling pathways. Ann. Rev. Cell Dev. Biol. 12, 55–89.

Schneider, A.R., Stapleton, H.M., Cornwell, J., Baker, J.E., 2001. Recent declines in PAH, PCB, and toxaphene levels in the northern Great Lakes as determined from high resolution sediment cores. Environ. Sci. Technol. 35, 3809–3815.

Schwacke, L.H., Voit, E.O., Hansen, L.J., Wells, R.S., Mitchum, G.B., Hohn, A.A., Fair, P.A., 2002. Probabilistic risk assessment of reproductive effects of polychlorinated biphenyls on bottlenose dolphins (*Tursiops truncatus*) from the Southeast United States Coast. Environ. Toxicol. Chem. 21, 2752–2764.

Schwarzenbach, R.P., Gschwend, P.M., Imboden, D.M., 2003. Environmental Organic Chemistry, 2nd ed. New York, Wiley & Sons.

Shaw, S.D., Brenner, D., Berger, M.L., Carpenter, D.O., Hong, C.S., Kannan, K., 2006. PCBs, PCDD/Fs, and organochlorine pesticides in farmed Atlantic salmon from Maine, eastern Canada, and Norway, and wild salmon from Alaska. Environ. Sci. Technol. 40, 5347–5354.

Simonich, S.L., Hites, R.A., 1995. Global distribution of persistent organochlorine compounds. Science 269, 1851–1854.

Sinkkonen, S., Paasivirta, J., 2000. Degradation half-life times of PCDDs, PCDFs and PCBs for environmental fate modeling. Chemosphere 40, 943–949.

Spokes, L.J., Jickells, T.D., 2005. Is the atmosphere really an important source of reactive nitrogen to coastal waters? Continental Shelf Res. 25, 2022–2035.

Stapleton, H.M., Alaee, M., Letcher, R.J., Baker, J.E., 2004. Debromination of the flame retardant decabromodiphenyl ether by juvenile carp (*Cyprinus carpio*) following dietary exposure. Environ. Sci. Technol. 38, 112–119.

Stapleton, H.M., Dodder, N.G., Kucklick, J.R., Reddy, C.M., Schantz, M.M., Becker, P.R., Gulland, F., Porter, B.J., Wise, S.A., 2006. Determination of HBCD, PBDEs and MeO-BDEs in California sea lions (*Zalophus californianus*) stranded between 1993 and 2003. Mar. Pollut. Bull. 52, 522–531.

Stapleton, H.M., Dodder, N.G., Offenberg, J.H., Schantz, M.M., Wise, S.A., 2005. Polybrominated diphenyl ethers in house dust and clothes dryer lint. Environ. Sci. Technol. 39, 925–931.

Stegeman, J.J., 1981. Polynuclear aromatic hydrocarbons and their metabolism in the marine environment. In Gelboin, H.V., and Ts'o, P.O.P. (eds.), Polycyclic Hydrocarbons and Cancer, vol. 3. New York, Academic Press.

Stegeman, J.J., Hahn, M.E., 1994. Biochemistry and molecular biology of monooxygenases: Current perspectives on forms, functions, and regulation of cytochrome P450 in aquatic species. In Malins, D.C., and Ostrander, G.K. (eds.), Aquatic Toxicology: Molecular, Biochemical and Cellular Perspectives, Boca Raton, FL, CRC/Lewis.

Stegeman, J.J., Kloepper-Sams, P.J., Farrington, J.W., 1986. Monooxygenase induction and chlorobiphenyls in the deep-sea fish *Coryphaenoides armatus*. Science 231, 1287–1289.

Stegeman, J.J., Schlezinger, J.J., Craddock, J.E., Tillitt, D.E., 2001. Cytochrome P450 1A expression in mid-water fishes: potential effects of chemical contaminants in remote oceanic zones. Environ. Sci. Technol. 35, 54–62.

Taioli, E., Marabelli, R., Scortichini, G., Migliorati, G., Pedotti, P., Cigliano, A., Caporale, V., 2005. Human exposure to dioxins through diet in Italy. Chemosphere 61, 1672–1676.

Taniyasu, S., Kannan, K., Horii, Y., Hanari, N., Yamashita, N., 2003. A survey of perfluorooctane sulfonate and related perfluorinated organic compounds in water, fish, birds, and humans from Japan. Environ. Sci. Technol. 37, 2634–2639.

Taves, D.R., 1968. Evidence that there are two forms of fluoride in human serum. Nature 217, 1050–1051.

Teuten, E.L., King, G.M., Reddy, C.M., 2006a. Natural 14C in *Saccoglossus bromophenolosus* compared to its surrounding sediments. Marine Ecology Progress Series 324, 167–172.

Teuten, E.L., Pedler, B.E., Hangsterfer, A.N., Reddy, C.M., 2006b. Identification of highly brominated analogues of Q1 in marine mammals. Environ. Pollut. 144, 336–344.

Teuten, E.L., Reddy, C.M., 2007. Halogenated organic compounds in archived whale oil: A pre-industrial record. Environ. Pollut. 145, 668–671.

Teuten, E.L., Saint-Louis, R., Pedler, B.E., Xu, L., Pelletier, E., Reddy, C.M., 2006c. Expanding the range of brominated Q1 analogues. Mar. Pollut. Bull. 52, 578–582.

Teuten, E.L., Xu, L., Reddy, C.M., 2005. Two abundant bioaccumulated halogenated compounds are natural products. Science 307, 917–920.

Tittlemier, S., Borrell, A., Duffe, J., Duignan, P.J., Fair, P., Hall, A., Hoekstra, P., Kovacs, K.M., Krahn, M.M., Lebeuf, M., Lydersen, C., Muir, D., O'Hara, T., Olsson, M., Pranschke, J., Ross, P., Siebert, U., Stern, G., Tanabe, S., Norstrom, R., 2002. Global distribution of halogenated dimethyl bipyrroles in marine mammal blubber. Arch. Environ. Contam. Toxicol. 43, 244–255.

Tittlemier, S.A., 2004. Dietary exposure to a group of naturally produced organohalogens (halogenated dimethyl bipyrroles) via consumption of fish and seafood. J. Agric. Food Chem. 52, 2010–2015.

Tittlemier, S.A., Braekevelt, E., Halldorson, T., Reddy, C.M., Norstrom, R.J., 2004. Vapour pressures, aqueous solubilities, Henry's Law constants, and octanol/water partition coefficients of a series of mixed halogenated dimethyl bipyrroles. Chemosphere 57, 1373–1381.

Tittlemier, S.A., Kennedy, S.W., Hahn, M.E., Reddy, C.M., Norstrom, R.J., 2003. Naturally-produced halogenated dimethyl bipyrroles bind to the Ah receptor and induce cytochrome P4501A and porphyrin accumulation in chicken embryo hepatocytes. Environ. Toxicol. Chem. 22, 1497–1506.

Tittlemier, S.A., Simon, M., Jarman, W.M., Elliott, J.E., Norstrom, R.J., 1999. Identification of a novel C10 H6 N2 Br4 Cl2 heterocyclic compound in seabird eggs: A bioaccumulating marine natural product? Environ. Sci. Technol. 33, 26–33.

Trushenski, J.T., Kasper, C.S., Kohler, C.C., 2006. Challenges and Opportunities in Finfish Nutrition. North Am. J. Aquaculture 68, 122–140.

Tuomisto, J.T., Tuomisto, J., Tainio, M., Niittynen, M., Verkasalo, P., Vartiainen, T., Kiviranta, H., Pekkanen, J., 2004. Risk-benefit analysis of eating farmed salmon. Science 305, 476–477; author reply 476–477.

Utkina, N.K., Denisenko, V.A., Virovaya, M.V., Scholokova, O.V., Prokof'eva, N.G., 2002. Two new minor polybrominated dibenzo-p-dioxins from the marine sponge *Dysidea dendyi*. J. Nat. Prod. 65, 1213–1215.

Van den Berg, M., Birnbaum, L.S., Denison, M., De Vito, M., Farland, W., Feeley, M., Fiedler, H., Hakansson, H., Hanberg, A., Haws, L., Rose, M., Safe, S., Schrenk, D., Tohyama, C., Tritscher, A., Tuomisto, J., Tysklind, M., Walker, N., Peterson, R.E., 2006. The 2005 World Health Organization reevaluation of human and Mammalian toxic equivalency factors for dioxins and dioxin-like compounds. Toxicol. Sci. 93, 223–241.

Van Metre, P.C., Mahler, B.J., 2005. Trends in hydrophobic organic contaminants in urban and reference lake sediments across the United States, 1970–2001. Environ. Sci. Technol. 39, 5567–5574.

Van Metre, P.C., Mahler, B.J., Furlong, E., 2000. Urban sprawl leaves its PAH signature. Environ. Sci. Technol. 34, 2495–2503.

van Veld, P.A., Nacci, D.E., 2007. Chapter 13. Toxicity Resistance. In Di Giulio, R.T., Hinton, D.E. (eds.), The Toxicology of Fishes, Taylor & Francis, Boca Raton, FL.

Van Veld, P.A., Westbrook, D.J., 1995. Evidence for depression of cytochrome P4501A in a population of chemically resistant mummichog (*Fundulus heteroclitus*). Environ. Sci. 3, 221–234.

Venkatesan, M.I., 1988. Occurrence and possible sources of perylene in marine sediments—a review. Mar. Chem. 25, 1–27

Vetter, W., Alder, L., Kallenborn, R., Schlabach, M., 2000. Determination of Q1, an unknown organochlorine contaminant, in human milk, Antarctic air, and further environmental samples. Environ. Pollut. 110, 401–409.

Vetter, W., Hahn, M.E., Tomy, G., Ruppe, S., Vatter, S., Chahbane, N., Lenoir, D., Schramm, K.-W., Scherer, G., 2005. Biological activity and physico-chemical parameters of the marine halogenated natural products 2,3,3′,4,4′,5,5′-heptachloro-2′-methyl-1,2′-bipyrrole (Q1) and 2,4,6-tribromoanisole (TBA). Arch. Environ. Contam. Toxicol. 48, 1–9.

Wania, F., Mackay, D., 1993. Global fractionation and cold condensation of low volatility organochlorine compounds in polar regions. Ambio 22, 10–18.

Wania, F., Mackay, D., 1996. Tracking the distribution of persistent organic pollutants. Environ. Sci. Technol. 30, 390A-396A.

Weihe, P., Grandjean, P., Debes, F., White, R., 1996. Health implications for Faroe islanders of heavy metals and PCBs from pilot whales. Sci. Total Environ. 186, 141–148.

Weis, J.S., Weis, P., 1989. Tolerance and stress in a polluted environment. Bioscience 39, 89–95.

White, R.D., Hahn, M.E., Lockhart, W.L., Stegeman, J.J., 1994. Catalytic and immunochemical characterization of hepatic microsomal cytochromes P450 in beluga whales (*Delphinapterus leucas*). Toxicol. Appl. Pharmacol. 126, 45–57.

White, R.D., Shea, D., Schlezinger, J.J., Hahn, M.E., Stegeman, J.J., 2000. In vitro metabolism of polychlorinated biphenyl congeners by beluga whale (*Delphinapterus leucas*) and pilot whale (*Globicephala melas*) and relationship to cytochrome P450 expression. Comp. Biochem. Physiol. 126, 267–284.

Willett, W.C., 2005. Fish: Balancing health risks and benefits. Am. J. Prev. Med. 29, 320–321.

Windsor, J.G., Hites, R.A., 1979. Polycyclic aromatic hydrocarbons in Gulf of Maine sediments and Nova Scotia soils. Geochimica et Cosmochimica Acta 43, 27–33.

Zhu, L.Y., Hites, R.A., 2005. Brominated flame retardants in sediment cores from Lakes Michigan and Erie. Environ. Sci. Technol. 39, 3488–3494.

STUDY QUESTIONS

1. If you were asked to testify for Congress on the effects that humans have had on the environment during the past 150 years, what type of data would you use?
2. What would you advise a chemical manufacturer about the chemical and biochemical properties that should be avoided in their products? Consider this advice with the understanding that eventually some fraction of these chemicals will be released into the environment.
3. List two ways that chemicals can be transported to remote locations.
4. Humans can be exposed to chemicals by several means. List a few, and highlight the most dominant one.
5. Describe how a chemical can be bioaccumulated and persist in an animal. Why are some chemicals less persistent in animals than others?
6. Evaluating the risk of toxic effects for humans often relies on extrapolating experimental data. Why must the results of this approach be viewed with caution? What are the uncertainties?
7. One piece of direct evidence about chemical effects in free-ranging marine mammals was from studies that knowingly fed captive harbor seals fish containing high or low concentrations of PCBs. This study showed that PCBs negatively affected reproductive and immunological functions of the seals. Clearly, this was an important study, but would you conduct a similar experiment?

CASE STUDY

7

CASE STUDY 1: CONTAMINANTS IN FARMED VERSUS WILD FISH

Human health, marine environmental health, and organic chemicals are inextricably linked in a variety of ways. One case study serves to illustrate the complexity of the interconnectedness and how a variety of scientific, social, economic, and political concerns are involved on many levels.

Numerous studies have shown that eating salmon has many health benefits, especially lowering risks of cardiovascular disease from the uptake of omega-3 fatty acids (Harper and Jacobson, 2001). These positive attributes and difficulties in catching and sustaining wild salmon as a sole source has led to salmon farming, which has increased salmon yields by over a factor of 40 during since the 1980s (based on data from the United Nations). Current estimates are that the major markets of salmon rely on more than half to be farmed.

Hites *et al.* (2004a) published extensive data on the levels and sources of organic contaminants in farmed and wild salmon and suggested dietary guidelines for these fish. The authors presented data from approximately 700 fish from Europe, North America, and South America that were analyzed for several pesticides, PCBs, and PCDDs. They showed that farmed salmon were significantly more burdened with most of these contaminants. Moreover, there was a clear trend that farmed salmon from Europe had higher levels than those from North and South America. To investigate these differences, Hites *et al.* (2004a) also analyzed the feed used in salmon farming, mainly smaller fish, and found levels similar to that of the wild salmon, indicating the most likely source. They dismissed the idea that bioaccumulation of the contaminants from the water column into the fish was a substantial route of exposure relative to feeds. Hites *et al.* (2004a) stressed the need for efforts to accurately inform consumers about the sources of wild and farmed salmon. These results and arguments have been well received and catalyzed the analysis of fish-based feeds (Carlson and Hites, 2005) and consideration of vegetable oils as feeds (Bell *et al.*, 2005; Berntssen *et al.*, 2005; Trushenski *et al.*, 2006). One study has found that farmed salmon collected and analyzed over a 2-year period had decreased contaminant levels with time and suggested that it was the result of successes in monitoring feeds (Shaw *et al.*, 2006).

Using a model developed by U.S. Environmental Protection Agency, Hites *et al.* (2004a) performed a risk assessment for human health impacts of contaminants in farmed salmon. Their intention apparently was to highlight the risks of salmon consumption without considering the cardiovascular benefits. In performing the risk assessment, they considered only cancer as an endpoint. They noted that individually, the concentrations of PCBs and dieldrin did not exceed U.S. Food and Drug Administration tolerances for these contaminants in seafood. However, they found that using EPA guidelines, which include calculations for combining risks for multiple chemicals, the combined concentrations of three of the contaminants (PCBs, toxaphene, dieldrin) would trigger fish consumption advisories, and those for the farmed salmon would be more restrictive. According to this assessment, consumption of farmed salmon should be limited to no more than <0.5 to 2 meals per month (depending on the sample) to minimize potential negative health effects associated with exposure to organic contaminants in these fish.

Many criticized the approach and calculations of Hites *et al.* (2004a). Their conclusions were well publicized in the press, which may have sensationalized them to some extent. One public health scientist even suggested that the Hites *et al.* paper "likely caused substantial numbers of premature deaths" by contributing to a reduction in fish consumption (Willett, 2005). The EPA calculations used by Hites *et al.* involve a variety of conservative assumptions (such as using the upper confidence intervals for risk estimates) and uncertainties (Tuomisto *et al.*, 2004). In reality, little is known about the interactive effects of contaminants such as PCBs, toxaphene, and dieldrin, for example, whether cancer risks are indeed additive. Other arguments against the conclusions of Hites *et al.* (2004a) included that the benefits of dietary omega-3 fatty acids in reducing the risk of coronary death would outweigh the risk from exposure to organic contaminants (Mozaffarian and Rimm, 2006; Tuomisto *et al.*, 2004) and that the advisory was inconsistent with Norwegian cancer registries that showed no differences in cancer for women who ate higher doses of salmon, mainly farmed, than women who ate less (Lund *et al.*, 2004). Hites and colleagues responded with vigor and stressed that other effects besides cancer must be considered, that the population was too small when Lund *et al.* (2004) assessed incidence of cancer for the Norwegian woman, and that their initial study did educate the public about such risks (Hites *et al.*, 2004b).

These debates continue and reveal that it is easier to measure contaminants in salmon than to assess the risks as compared to the benefits of consumption. Whatever the case, it is important to note that the controversy regarding farmed versus wild salmon did stimulate efforts to reduce contaminant loading in farmed salmon, which ultimately may lead to reduced exposure of humans to organic contaminants through this important food source. In addition, the Hites *et al.* paper and similar reports helped trigger an extensive discussion of how best to balance the benefits and risks of consuming seafood (Nesheim and Yaktine, 2007).

CASE STUDY 2: WHY WERE PBDES USED IN THE 1980s AND 1990s?

In late 2004, manufacturers of PBDEs voluntarily stopped selling two types of flame-retardants in the United States. This phaseout was the result of government pressure and threatening data about the environmental fate and health effects of these compounds. Only sales of decabromodiphenyl ether (DecaBDE; Table 7-1) continue.

It is reasonable to ask, in light of experiences with organochlorines in the 1960s and 1970s, why PBDEs continued to be used into the 1980s and 1990s. It would appear that, by ignoring the history of PCBs and DDTs, manufacturers of PBDEs failed to recognize the potential problems of using these compounds, which were predictable from knowledge of their chemical properties (Table 7-1).

Production of PCBs began in the late 1920s (Table 7-1). In the United States, sales paralleled the industrial growth following World War II. In the mid-1960s, while attempting to analyze for DDT in samples, Swedish chemist Soren Jensen found PCBs in fish and bird samples (Jensen, 1966). Jensen's work and studies by others revealed by the late 1960s that PCBs were clearly organic chemicals of environmental concern (Risebrough *et al.*, 1968). Even with these findings, sales of PCBs by Monsanto in the United States continued, peaking in 1970. Monsanto voluntarily ceased production and sales in 1977. Thus, about a 10-year lag occurred between the discovery of PCBs in the environment and a major restriction of their use.

PBDEs were first manufactured for use as flame-retardants in 1960. In the late 1970s and early 1980s, PBDEs began to be detected in environmental samples. Swedish scientists Andersson and Blomkvist (1981) were the first to find these compounds in lake fish far from any known chemical plants. Additional studies continued around the world in the 1980s. By 1987, manuscripts such as that titled "Brominated Flame Retardants: Ubiquitous Environmental Pollutants?" (Janssen *et al.*, 1987) documented the global occurrence of PBDEs and stressed that this was an important topic for study. Regardless of these findings and the histories of PCBs and DDT (not discussed but similar to PCBs), sales of PBDEs continued through the 1980s and 1990s.

Beginning in the late 1990s, work by Norén and Mieronyté (2000), including analysis of milk archived since the 1970s, raised awareness of human exposure to PBDEs (see the earlier section titled Brominated Flame-Retardants). Because congeners from mixtures of pentabrominated and octabrominated diphenyl ethers were those most often measured in human samples, a voluntary phaseout of PBDE mixtures containing these congeners began in 2004—nearly two decades after the publication of the report by Jansson *et al.* (1987). In 2005, DecaBDE was the only PBDE on the market. Manufacturers argue that DecaBDE is not usually detected in environmental matrices. However when Stapleton *et al.* (2004) fed DecaBDE-contaminated food to carp, the fish did not accumulate it but did generate lower brominated congeners, which are detected in humans and fish. Thus, use of decaBDE may result in continued input of lower chlorinated PBDEs to the environment and especially biota.

As flame-retardants, PBDEs are life-saving organic compounds, but their use carries risks. Numerous research groups and agencies are closely examining the environmental fate of DecaBDE and a wide range of possible biological effects that it may exert on humans and animals.

More generally, only a fraction of the most highly produced industrial chemicals have been evaluated for bioaccumulation, environmental fate, and toxicity. To avoid other foreseeable problems like those encountered with PBDEs, efforts are under way to increase testing in the United States. As mentioned earlier, one of the key parameters used for predicting whether chemicals will persist in the environment is the K_{ow}. PCBs, DDT, and PBDEs have large K_{ow} values. The scientific concept of using K_{ow} values to gauge bioaccumulation potential was widely discussed in the scientific literature in the late 1970s to the mid-1980s. By the mid-1990s, textbooks on environmental chemistry had whole chapters devoted to this topic (Schwarzenbach *et al.*, 2003). Considering that the importance of K_{ow} has been known for decades and the lessons learned from the use of highly persistent organochlorines, the continued use of PBDEs is puzzling.

C H A P T E R

8

Metals

Ocean Ecosystems and Human Health

JOANNA BURGER AND MICHAEL GOCHFELD

INTRODUCTION

The public, health professionals, scientists, regulators, and public policy makers are interested in metal concentrations in the ocean because metals can bioaccumulate in marine organisms, eventually harming other organisms that eat them, including people. Often, the adverse effects of metals are most severe on developing young and on young children, although exposure to high levels can harm healthy adults as well. It is the potential for causing harm, and even death, in marine organisms and in people that makes it important to understand where metals come from and how they move through the marine ecosystem.

Understanding the levels of metals in marine organisms, how metals are transferred from organisms to organism, how they accumulate in people, and the effects of metals on people is an extremely important task for scientists. Armed with such knowledge managers, regulators, and legislatures can develop guidelines, regulations, and laws to reduce levels in marine organisms. Health professionals can reduce human exposure by providing advice on consumption patterns and can develop methods of reducing the effects on people who have been exposed to harmful levels.

Humans rely heavily on food from the sea, including algae; shellfish; other invertebrates such as crustaceans, octopus and squid; fish, birds, and their eggs; and marine mammals. Even though most people in developed countries think of seafood as fish, lobster, and shellfish, seafood includes algae (used for food and medicine), birds and their eggs, and marine mammals (subsistence peoples). Although people who live near the sea eat more seafood than those who live in the interior of countries, seafood, fresh and frozen, has become increasingly available and is gaining in popularity throughout the world. Food from the sea is a staple in the diets of people in many countries, particularly those living on islands or where a large percentage of people live in coastal communities. In many parts of the world, more than half the people live in coastal communities where fish is prominent in their diets. Even in the United States, 53% of the population lives in coastal counties, which make up only 17% of counties (National Oceanic & Atmospheric Administration [NOAA], 2006). In the United States, for example, the average person consumes nearly 20 pounds of fish a year, compared to beef at nearly 70 pounds and chicken at 80 pounds (NOAA, 2004b). Consumption is rising yearly and is much higher for Native Americans and some minorities; the upper 95% of Americans consume an average of nearly 80 pounds (Jacobs *et al.,* 1998). The top seafood consumers per capita, in descending order, are Japan, China, the United States, Indonesia, and Russia; Japanese consume three times as much seafood as those residing in the United States (Department of Agriculture, 2006).

In this chapter, we discuss what metals are of concern, where they come from, how they move through the food chain, other factors that affect metals levels in biota, how metals can harm people, and what can be done to reduce metal levels in marine organisms and reduce harm to people eating them. Risk management is a critical part of the process of understanding metals and human health. Like most other choices that people make, eating seafood involves balancing the risks and benefits. Seafood provides a healthy source of protein and other nutrients, but these benefits must be balanced against any potential risks because of contaminants. The choices people face involve what kinds and how much seafood to eat, as well as what kinds of other protein to consume. This aspect will be discussed in depth. Mercury is the key metal of concern in seafood, but others will be discussed, including arsenic, cadmium, and lead. Although other contaminants, such as persistent organic compounds and radionuclides, can be a concern in fish, we concentrate on metals in this chapter.

THE SOURCE OF METALS IN THE SEA

Metals in the marine environment come from geological processes (volcanos and erosion) and from anthropogenic sources such as point sources, runoff, discharges from rivers, and atmospheric deposition. Mercury is transported in the atmosphere mainly as gaseous elemental mercury and is deposited through wet (rainfall) and dry (particulate) deposition processes as inorganic mercury (Mason *et al.,* 1994). Seawater contains trace quantities of most of the elements. Among the main toxic heavy metals in seawater, arsenic (2.6 ppb in seawater), selenium (0.9 ppb), manganese (0.4 ppb), chromium (0.2 ppb), mercury (0.15 ppb), and cadmium (0.11 ppb) are of concern (Turekian, 1968). Currents, stratification, and upwellings change the distribution and availability of metals and other nutrients. Sediments are considered the ultimate sink for metals in aquatic environments, including bays, estuaries, and the ocean (Clements, 1991). Disturbances to the sediments in bays and estuaries, by dredging or severe storms, can result in the resuspension of metals in the water column, presenting a problem for organisms at many levels on the food chain.

FOOD CHAIN TRANSFER

Inorganic mercury in aquatic systems can be converted to methylmercury by microorganisms in sediment (Brezonik *et al.,* 1991; Jensen and Jernelov, 1969; Spry and Wiener, 1991; Zillious *et al.,* 1993). These microorganisms are then eaten by larger species, and the methylmercury is transferred up the aquatic food web (Watras *et al.,* 1998). At each trophic level, organisms sequester some metal, excreting less than they consume, resulting in increasing tissue concentration at higher trophic levels, a process called biomagnification.

The bioavailability and toxicity of metals to aquatic organisms depends on the chemical form of the metal, particularly whether or not it is part of an organic compound. Transformation among species depends on environmental conditions, such as pH, redox conditions, ligands, temperature, and the concentrations and nature of particulate matter (Brezonik *et al.,* 1991). Inorganic and organic metals have different effects. Methylmercury is the form of mercury that has the greatest adverse effects at low doses on marine organisms and their predators, including people. By contrast, organic arsenic compounds are less toxic than inorganic compounds. Many studies have shown that almost all of the mercury in fish muscle is methylmercury, and 90% is a reasonable approximation of this proportion, which does vary somewhat among fish types and laboratories, but not by age of the fish (Lansens *et al.,* 1991). Bloom (1991) even suggested that over 95% of mercury present in fish is methylmercury, and that lower levels have been biased by analytical and homogeneity variability.

Metals exist in solution, bound to particulate matter or incorporated in biota. Once in the water column or sediment, metals can accumulate in organisms. Whereas rooted aquatic plants take up metals through their roots, algae take up contaminants through the water directly by adsorption and absorption, as do zooplankton and phytoplankton. Macroalgae take up metals against a concentration gradient (Phillips, 1990), which can result in levels many times those found in surrounding waters (Ragan *et al.,* 1979). For example, some shellfish can concentrate cadmium up to 4000 times above the level in seawater yielding a bioconcentration factor (BF = 400) (Pesch and Steward, 1980). Benthic invertebrates readily take up and accumulate metals from their surrounding environment, including water, sediment, interstitial water, and food. Depending on where benthic organisms feed, levels in their tissues reflect primarily the levels in the sediment from absorption or adsorption, the levels in the water from absorption, or levels in their foods from ingestion. For many benthic invertebrates, the primary route of exposure is directly from water, although food may be significant for some species (Burrows and Whitton, 1983).

Accumulation refers to the tendency for organisms to build up levels of metals in their tissues (see Table 8-1). With time, the levels in these benthic organisms can reach levels that are several orders of magnitude higher than levels found in the surrounding water or sediment (Clements, 1991). Benthic invertebrates are often the major component in the diet of many fish species, making them an important link in the food chain to higher trophic levels.

Fish similarly accumulate metals from their food, as well as water that passes over their gills and through their gut, and some laboratory studies have shown that water is the primary route of exposure (Clements, 1991; Prosi, 1979). This finding is controversial. Other studies indicate that food is the primary route of exposure (Douben, 1989; Harrison and Klaverkamp, 1989), and dietary uptake probably accounts for more that 90% of the total uptake in fish (Wiener *et al.,* 2003). The assimilation efficiency for the uptake of dietary methylmercury in fish is 65% to 80% (Wiener and Spry, 1996). For marine birds and mammals, the main route of exposure is through food, which, in many cases, has already accumulated high levels of heavy metals. Ultimately, the correlation between mercury levels in high trophic level predators, such as large fish, is correlated more highly with sediment than with water column concentrations (Kannan *et al.,* 1998; South Florida Water Management District [SFWMD], 2001, 2002), indicating that the food chain is a greater contributor to body burden than direct water contact.

Biomagnification refers to the accumulation of higher levels of contaminants with increasing trophic level and position on the food web (Table 8-1). It results when organisms take up metal and sequester it, making their output less than their intake. Food webs, which can be plotted for specific ecosystems, refer to the complexity of producers and consumers, including herbivores and carnivores. Food chain usually refers to examining the specific links to a given top-level predator. Biomagnification has been clearly demonstrated for mercury (Bryan and Langston, 1992; Burger *et al.*, 2001b) and has been suggested for cadmium (Burger *et al.*, 1984; Gochfeld and Burger, 1982), but considerable research is needed to examine this aspect for other metals.

Biomagnification, also called bioamplification, has important consequences for human exposure because people can eat seafood at several different trophic levels. The potential for selection among different species of seafood thus allows consumers to reduce or minimize their exposure to contaminants, as well as maximizing their energy and nutrient intake. For example, Aleuts in the Bering Sea ecosystem, the resident Native Americans that rely heavily on food they can gather, hunt or fish, have a choice of foods at different trophic levels (Fig. 8-1); mercury exposure will vary accordingly.

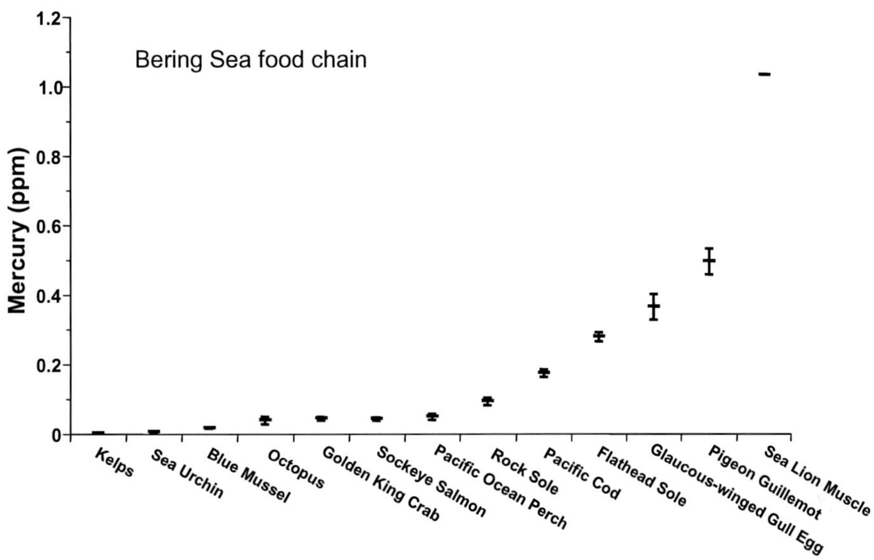

FIGURE 8-1. Mercury levels (mean ppm with standard error) vary by position on the food chain in organisms from the Bering Sea (Burger and Gochfeld, unpublished data).

TABLE 8-1. Definition of terms that are essential for understanding why metals provide a threat to marine organisms, including people who consume seafood.

Term	Definition	Example
Accumulation	Metals build up in media, such as soil or water	Over time, metals in runoff from municipal waste can build up higher and higher levels in the sediment of bays and estuaries
Bioavailable	The metal is in a form that can be taken up by organisms	Methylmercury can be taken up, or absorbed by fish from invertebrates
Food chain or web	Species that eat one another	Algae is eaten by small fish, who are eaten by larger fish, who are eaten by still larger fish; people eat fish of different sizes
Bioaccumulation	Metals build up in organisms	Over time, organisms such as mussels that eat by filter-feeding can accumulate metals from the sediment or microorganisms living in the sediment
Biomagnify	With each step in the food chain, organisms can bioaccumulate higher levels because each organism has concentrated the metal	Metals are stored in tissues and accumulate over time, building up higher levels; a predator eats prey having a certain level and has the potential to accumulate many times that level in its own tissues; A predator may accumulate levels many times higher than those of its prey

OTHER FACTORS AFFECTING METAL LEVELS IN SEAFOOD

In addition to food chain effects, a range of other factors affect metal levels in seafood, including age, gender, size, foraging location, movement patterns, and the migratory behavior of the seafood species in question. Some factors will not directly affect human exposure, as consumers often cannot identify seafood by gender or its habitat. However, consumers can take some features into account, especially fish size.

Age, Size, and Sex

Mercury bioaccumulates and biomagnifies with size and age of fish (Bidone *et al.,* 1997; Braune, 1987; Burger *et al.,* 2001b; Lacerda *et al.,* 1994; Lange *et al.,* 1994; Phillips *et al.,* 1980), especially for large marine predatory fish. Storelli *et al.* (2002) reported that size and mercury levels were highly correlated for swordfish (*Xiphias gladius*) and bluefin tuna (*Thunnus thynnus*) from the Mediterranean Sea. However, whereas yellowfin tuna (*Thunnus albacares*) showed increasing mercury levels with increasing size (length and weight), albacore tuna (*Thunnus alalunga*) did not (Fiji, 2006). Luten *et al.* (1987) also showed increasing mercury levels with increasing size in cod. For most fish, the age is unknown and size is used as a surrogate for age (Boening, 2000). However, Braune (1987) found that in known-aged herring (*Clupea harengus*), mercury level was more strongly correlated with age than with weight or length. Similar relationships have been found for other metals in some fish, such as selenium (Burger *et al.,* 2001a), and for arsenic, cadmium, and chromium (Burger *et al.,* 2002). The age of fish is determined by examining the otoliths (inner ear bone), a process that takes time and expertise. In addition to eating larger prey items, and living longer allowing a longer accumulation period, Trudel and Rasmussen (1997) found that the elimination rate is negatively correlated with size, suggesting another reason for larger fish to have significantly higher mercury levels.

Few scientists report metal levels by sex, and consumers usually have no idea of the sex of seafood they consume. However, some sex differences have been identified. For example, female lobsters (*Panulitrus inflatus*) had higher muscle levels of copper, manganese, and nickel than males (Paez-Osuna *et al.,* 1995). In the echinoderm asteroid (*Asterias rubens*), females had significantly higher levels of zinc and cadmium, and males had higher levels of chromium and iron (Temara *et al.,* 2003). Some studies have failed to find a sex difference in metal levels, including studies with king crab (*Pseudocarcinus gigas*; Turoczy *et al.,* 2001), a gastropod (*Bembicium nanum*; Gay and Maher, 2003), and horned octopus (*Eledone cirrhosa*; Rossi *et al.,* 1993). Overall, these studies do not present a clear picture of how gender is affecting contaminant loads, although it seems as if males have higher levels of more contaminants in invertebrates than females. In some species of fish, males had higher mercury levels than females of equal age (World Health Organization [WHO], 1989), although most species do not show gender differences. Sex differences have also been found in some marine birds and mammals that may form part of subsistence diets. Male Gentoo penguins (*Pygoscelis papua*) in South Georgia had higher levels of mercury in feathers than females, although female northern giant petrel (*Macronectes halli*) had higher levels; there were no differences for southern giant petrel (*M. giganteus*), gray-headed albatross (*Diomedea chrysostoma*), and black-browed albatross (*Diomedea malanophris*; Becker *et al.,* 2002). For a wide range of metals, there were either no significant sex-related differences for any tissues or females had higher levels for most metals in birds (Burger, 1993). In spinner dolphins (*Stenella longirostris*) stranded in California, females had higher mean levels of mercury in liver, kidney, and muscle than did males (Ruelas *et al.,* 2000). In contrast, male hooded seals (*Cystophora cristata*) from Greenland had higher levels of mercury and selenium than females in all tissues (muscle, liver, kidney), but there were generally no differences for harp seals (*Pagophilus groenlandicus*; Julshamn and Grahl-Nielsen, 2000).

Habitat and Mobility

Metal levels in water and sediment differ as a function of habitat. As might be expected, metal levels are higher in bays and estuaries than in the open ocean, largely because of their proximity to runoff from the land and river discharges. Marine organisms that spend more time along coasts should have higher levels than those living in the open ocean, given the same or similar species of the same sizes. For many marine fish, juveniles live in inshore habitats in kelp or eelgrass beds, eating small invertebrates, while the adults move offshore (Able and Fahay, 1998). Thus, exposure to contaminants differs between young and adults.

Kelp and other plants are sedentary, as are some invertebrates; however, for most marine organisms, mobility patterns affect metal levels. Mobility can refer to the vertical mobility of some organisms by time of day, seasonal inshore-offshore movement, seasonal movement along the shore, and migrations that may be hundreds of km. Such movements take marine organisms into different habitats and locales where they are exposed to both different metals and different concentrations of metals in the water and their prey. This variability contributes to the uncertainly and unpredictability in metal levels in seafood, especially where consumers are unable to ascertain where fish or shellfish were captured.

RISK TO HUMANS FROM METALS IN SEAFOOD

Risk is the degree to which a person is vulnerable to a hazard, such as heavy metals. It is a function of exposure and the levels of contaminants in seafood. Levels can be very high in algae, shellfish, fish, birds, and marine mammals, but if no one eats them, then there is no risk. Exposure can also be blocked if the metals are not available to the consumers; the metals have to be bioavailable to be absorbed into the digestive track. Determining human risk is thus a function of understanding the levels of metals in seafood, bioavailability for humans, exposure, and individual susceptibility or vulnerability.

Receptor Groups

There are three main human receptor groups for seafood consumption: subsistence hunters and fishermen, recreational fishermen, and commercial seafood consumers, although some health-conscious people use kelp as a nutritional supplement. Subsistence people are those who depend on self-caught or hunted species as their main or only source of protein, whereas recreationists are those people who hunt or fish for pleasure and do not require this food for sustenance. This distinction, however, is arbitrary and has been blurred in our modern world because some subsistence people, defined culturally, do not eat as much wild-caught foods as some recreationists who happen to love self-caught fish and game. The main issue should be how much wild-caught fish and game people eat and what is their risk, not what label is applied to them (Burger, 2002a, 2002b).

Most attention to metals and other contaminants in seafood has been devoted to self-caught fish and game for "subsistence" peoples because they were thought to be the most highly exposed. Fishing is a popular pastime in both urban and rural areas of the United States, as well as throughout the world (Burger *et al.,* 1992, 1993, 1999a, 2001a, 2001b; Ramos and Crain, 2001; Toth and Brown, 1997). In many places, people fish throughout the year. Even in northern regions, people ice fish in the winter. Fishing not only provides fish and shellfish to eat but confers a range of other social benefits that include interactions with family or friends, provision of seafood for friends and fish fries, a respite from the stresses of urban or rural life, and communion with nature (Burger, 2002a; Fleming *et al.,* 1995; Toth and Brown, 1997). Increasing attention to health and nutrition has increased the public's consumption of fish, even among those who never fish themselves (NOAA, 2004a).

Commercial consumers, the vast majority in the world, are people who obtain their seafood from commercial sources. Relatively little attention has been devoted to metals in commercially available seafood, largely because state governments have concentrated on freshwater lakes directly amenable to their control and regulations. States can issue consumption advisories for their freshwaters, and they can impose standards to reduce metal contamination of their waters.

Sensitive and Vulnerable Populations

Not all people are equally susceptible to the effects of heavy metals such as mercury, arsenic, cadmium, and lead that can occur in seafood. Fetuses and young children are the most sensitive to the effects of metals, but adults can be affected as well (Gochfeld, 2003; Hightower and Moore, 2003). Even within life stages (fetus, child, adult), there are differences in sensitivity or susceptibility to contaminants.

People vary in their exposure because of a variety of factors, including economic status, cultural values, availability, and residence location. People can rely on self-caught seafood, for example, because it is part of their culture or tradition (Native Americans; Harris and Harper, 1997), ethnic values (black traditions in the south, Burger *et al.,* 1999b; Toth and Brown, 1997), or economic status (Duhaime *et al.,* 2004). People living close to the sea are particularly vulnerable when a high proportion of their food comes from the sea. Examples include the Aleuts living on remote islands in the Bering Sea, and people living on the Seychelles and Faroes Islands, among others (discussed later).

VULNERABLE MARINE ECOSYSTEMS

Metals can adversely impact marine organisms themselves, and human exploitation of marine biota can seriously impact populations. Both of these in turn affect people because they reduce the overall food available for human consumption. All organisms in the marine environment are potentially impacted by metal contamination, particularly in estuarine and coastal environments where runoff and riverine discharges can markedly increase levels. Whereas most marine organisms have evolved with the natural levels of heavy metals in the ocean that result from geological process, they have not evolved with the higher levels present in some bays and estuaries.

There is serious concern for populations of marine organisms, especially fish and some shellfish (Safina *et al.,* 2005). Although overfishing is not the focus of this chapter, overfishing will affect human use of the ocean's resources and, given that fish are a healthy source of protein, could adversely affect human health.

HUMAN HEALTH CONSEQUENCES

In the United States, there has been a general upward trend in seafood consumption since the 1960s, despite an increase in price relative to meat. The trend has waxed with nutritional advice and waned with hazard advisories (Food and Agricultural Organization [FAO], 1998), but it continues to increase, now exceeding 7.4 Kg/capita per year (NOAA, 2004a). The top 10 species of fish and shellfish consumed in the United States are, in descending order (in pounds/person), shrimp (4.2), canned tuna (3.3), salmon (2.1), pollock (1.3), catfish (1.1), tilapia (.70), crab (0.63), cod (0.60), clams (0.47), and flatfish (0.33) (NOAA, 2004b). Of these, salmon has the highest concentration of omega-3 (n-3) fatty acids, followed by pollack and then shrimp and clams. Salmon contain over five times as much omega-3 as the other species in the top 10 (Department of Agriculture, 2006).

Human use of shellfish and fish requires little description. Kelp and other algae, however, are usually not included in seafood analyses, yet they are consumed in Alaska and Canada (Chan *et al.,* 1995; Garza, 2005; Sharp *et al.,* 1988) and are eaten extensively in Asia (Phaneuf *et al.,* 1999). Kelp is also used in folk medicine and has appeared as tablets in U.S. health food stores (vanNetten *et al.,* 2000).

Methods of Determining Health Effects

Determining the health consequences of metals is difficult because there are only three main methods of assessing effects: animal models of exposure and effects, cases of acute metal poisoning, and epidemiological studies. Animal models are experiments conducted in laboratories or the field where animals are exposed to known levels of metals and effects can be measured. These studies are useful for defining and understanding the range of effects and for examining the effects of different doses (levels given). The disadvantages of laboratory studies are that we cannot be sure if the effects will be similar in people at the same dosages. Nonetheless, such experiments are essential for demonstrating the effects of metals in mammalian systems.

Cases of acute metal poisoning are identified because individuals have immediate and often dramatic effects, which are easily recognized and require identification before treatment. Such poisonings are generally confined to occupational or deliberate exposures. Although seafood with microorganism contamination can produce acute disease, metal poisoning from ingestion is usually insidious with a long latency before some cumulative threshold is exceeded. The classic case for a metal is Minamata disease, discussed later. Less severe cases for mercury poisoning have been identified in people who have regularly consumed high quantities of predatory fish (Gochfeld, 2003; Hightower and Moore, 2003).

Epidemiological studies examine groups of people and relate them to different exposures. They are useful in ascertaining relationships where exposures are variable and significant, but they suffer from the problem of being unable to separate confounding variables. That is, people eating fish are exposed to a range of contaminants, such as PCBs, mercury, and other contaminants.

There are remarkably few studies linking particular mercury levels or consumption amounts with symptoms of mercury poisoning. Predatory fish (swordfish, shark) can exceed 1 ppm (ug/g MeHg on wet weight basis) in their flesh. At 1 ppm, a person who eats one 228 g (8 oz) meal a day would therefore take in 228 ug/day, resulting in a graduate buildup of tissue concentrations. The metric on which fish advisories are based is intake in micrograms per kilogram of body weight per day, and allowable levels range from 0.1 to 0.4 ug/kg/day. At 228 ug/day, a 70–kg person is consuming more than 3 ug/kg/day, excessive by any standard.

There are biomarkers of both exposure and effects, and both are useful in ascertaining health effects. The National Research Council (NRC, 1987) recognizes three categories of biomarkers: exposure, effect, and susceptibility. Biomonitoring refers to the routine and periodic measurement of some marker in persons who are at risk or exposed to a contaminant. For metals, the main biomarkers are actual measurement of metal concentrations in blood, urine, or tissues. Organic metal compounds are excreted mainly in feces, whereas inorganic compounds conjugated to carrier molecules are excreted mainly in urine. Metals have an affinity for sulfhydryl groups, such as those found in the amino acids cysteine, and cross links between the sulfurs atoms on two cysteines form the disulfide bridges that impart rigidity and functional features to proteins. Metals with high affinity for sulfur can break these bridges and bind to the sulfur, compromising the function of proteins, particularly enzymes.

Minamata Disease

The classic case for the effects of a metal derived from the consumption of seafood is Minamata disease. In the 1950s, many residents on the shores of Minamata Bay, Kyushu Island, Japan, suffered a strange, progressive, and debilitating neurological disease. Babies were born with profound mental and physical retardation. By 1960, the cause was evident. Over a period of years, the victims had consumed large quantities of fish contaminated with methylmercury emitted by the Chissos Chemical plant into the bay. Inorganic mercury emitted by the plant was converted to methylmercury, although this phenomenon was not elucidated until Jensen and Jernelov (1969) demonstrated methylation by anaerobic bacteria in aquatic sediments. Fish accumulated mercury to very high levels, >50 ppm. The disease was not reversible, although halting the consumption of fish from the bay slowed progression and prevented new cases. In 1965, a second similar epidemic occurred at

Niigata, Honshu, Japan (Eto, 1997). Other outbreaks of methylmercury poisoning have been attributed to the treatment of grain with organomercurial fungicides, with large outbreaks in Guatemala and Iraq (Elhassani, 1982).

EFFECTS OF MERCURY AND OTHER HEAVY METALS

Although many metals can have adverse effects on humans and other receptors, it is mercury, cadmium, lead, and arsenic in marine organisms that pose the main risk to people. Cases of acute metal poisoning, epidemiological studies, and animal models have all indicated that there can be severe effects from exposure to high levels of these chemicals, except that the effects of arsenic are muted when it is present in organic form. The toxic effects of these elements are described in many sources, are well known, and will be summarized here only briefly.

Mercury

Methylmercury is among the most toxic of the mercury species and predominates in seafood. Fish consumption is the only significant source of methylmercury exposure for the public (Rice *et al.,* 2000). Methylmercury is reported to counteract the cardioprotective effects of fish consumption (Rissanen *et al.,* 2000; Salonen *et al.,* 1995) and to damage developing fetuses and young children (NRC, 2000). Maternal exposures can threaten the fetus because chemicals can be transferred to the developing fetus (Gulson *et al.,* 1998).

The public health messages to increase fish consumption for its health benefits encouraged people to eat fish frequently. Hightower and Moore (2003) reported on a group of patients with frequent fish consumption who manifested signs of organomercury poisoning coupled with elevated concentrations of mercury. Similarly, Gochfeld (2003) found impaired neurobehavioral performance associated with high levels of mercury in hair; patients recovered when they ceased eating fish with high levels of mercury.

The growing concern over widespread methylmercury contamination of fish prompted several investigations including two well-funded, well-designed, and well-executed prospective studies of child development, when prenatal and postnatal exposure to contaminants were well known. The Seychelles in the tropical Indian Ocean has a population of diverse ethnicity. Reef fish play a major role in the diet, which is supplemented by a variety of fruits and vegetables. The Faroes, in the North Atlantic, has mainly a Scandinavian population. Marine fish, supplemented periodically by marine mammals, play an important role in the diet, whereas fresh fruits and vegetables are at a premium.

Neurodevelopmental studies in the Faroes identified performance decrements associated with increasing mercury levels in the mother or in cord blood. Studies with 14–year-olds, followed since birth, have yielded important results. Methylmercury exposure, even at moderate levels prenatally, affected several neuropsychological domains, including finger-tapping speed, reaction time on a continued performance task, and cued naming (Debes *et al.,* 2006). The Seychelles study did not provide clear evidence of performance or developmental deficits (Davidson *et al.,* 2006). The lack of an effect may be because the levels of mercury in fish were not as high as those in the Faroes or in Japan and the Seychelles diet is rich in vegetables and fruits. Cultural differences could have also impacted performance on tests.

Other Metals

Cadmium, in humans, accumulates throughout life, mainly in the kidney, with a slight decline in concentration in old age. Cadmium is carcinogenic, but the main documented effects have been on the kidney tubule, and through the loss of calcium, it impacts the bone. Cadmium toxicity in Japan resulted in itai-itai, a painful disease characterized by weakened bones and pathologic fractures. Cadmium is readily absorbed from food, particularly in women who may have low iron saturation (Agency for Toxic Substances and Disease Registry [ATSDR], 1999), which increases the expression of a divalent cation transporter, inadvertently increasing cadmium uptake.

Lead in humans causes neurobehavioral and cognitive dysfunction (Bellinger *et al.,* 1987; Mitchell, 1987; Needleman *et al.,* 1990; Rice, 1984) and retarded psychomotor development (Schwartz and Otto, 1987). The effects of lead on cognition are even evident in middle-aged and elderly people (Payton *et al.,* 1998). Lead also causes hypertension (Schwartz, 1991).

Most arsenic in seafood is organic arsenic, which is less toxic than inorganic arsenic species (ATSDR, 2000; Eisler, 1994). The main toxic species are the inorganic arsenites and arsenates. However, some interconversion does occur. Inorganic arsenic is toxic to most organs and most species, and arsenic is a known human carcinogen (ATSDR, 2005).

Selenium is known to have a protective effect on mercury exposure (Satoh *et al.,* 1985) and it plays an antioxidative role (Hansen, 1988). The body produces seleno-proteins, which may bind cations. Selenium toxicity is generally not a concern for consumption of fish and shellfish.

Contaminant Mixtures in Seafood

Most people are exposed to mixtures of contaminants, which can include mercury, PCBs, other persistent organic compounds, and radionuclides, making it difficult to assign the effects of only one contaminant. Although it is essential to understand the effects of individual contaminants for treatment purposes, and for reducing the levels in seafood and other organisms in marine ecosystems, it may only be possible to identify the main contaminant of concern for most seafood. In some cases, there may be sufficiently high levels of two or more contaminants to make such assignment impossible.

The recognition that contaminant levels in some fish are sufficiently high to cause adverse human health effects is troubling. Adverse health effects include counteracting the cardioprotective effects of omega-3 oils (Guallar *et al.*, 2002), damaging unborn babies and young children (ATSDR, 1996; Consumer Reports, 2003; Institute of Medicine [IOM], 1991, 2006; Iso and Rexrode, 2001; Lonky *et al.*, 1996; Moya, 2004; Nestel, 2001; Neuringer *et al.*, 1994; Olsen and Secher, 2002), and adversely affecting adult behavior and physiology (Hightower and Moore, 2003; Hites *et al.*, 2004). There is a positive relationship between mercury and polychlorinated biphenyl (PCB) levels in fish, fish consumption by pregnant women, and deficits in neurobehavioral development in children (IOM, 1991; Jacobson and Jacobson, 1996; NRC, 2000; Schantz, 1996; Schantz *et al.*, 2003; Sparks and Shepherd, 1994). There is a decline in fecundity in women who consume large quantities of contaminated fish from Lake Ontario (Buck *et al.*, 2000). There is also a suggestion that mercury affects blood pressure (Vupputuri *et al.*, 2005). Generally, there is a positive relationship between mercury levels in people and fish consumption (Johnsson *et al.*, 2005; Knobeloch *et al.*, 2005). The extensive discussion about what the "safe" level of exposure is may be partly political and is surely controversial (NRC, 2000; Stern, 1993; Stern *et al.*, 2004). Most of the studies noted here, however, dealt with fish from freshwater lakes and rivers and not with marine seafood.

Time Course of Exposure

The role of occasional peak exposures versus chronic lower level exposures to methylmercury, for example, requires closer attention, especially for pregnant women. It may be essential to develop single-meal fish consumption advisories, especially for fish species high in methylmercury, PCBs, or other contaminants (Ginsberg and Toal, 2000).

HUMAN HEALTH GUIDELINES FOR SEAFOOD SAFETY

Despite the importance of seafood in the diet of people worldwide, there is no uniform source of guidance or standards for most metals in fish or shellfish tissue. There is not even a single reference for acceptable levels of most metals in fish. Human health standards, guidelines, or action levels exist mainly for mercury. The USFDA has an action level of 1.0 mg/kg (ppm), wet weight for methylmercury in fish (Food and Drug Administration [FDA], 2001), but not for any other metals; the level of 1.0 is a regulatory action level, rather than a risk level. Originally the FDA had set 0.5 ppm total mercury as the action level, comparable to many other nations, but this was relaxed to 1.0 ppm. The United Kingdom and the European Union have established criteria for certain metals in fish (e.g., the level for mercury is 0.5 ppm in edible fish, with up to 1 ppm allowed for certain exempt predatory fish species). China has set standards for methylmercury in canned fish (ppm wet weight) of 0.5 ppm (except 1 ppm is allowed in shark, sailfish, tuna, pike, and other high-mercury fish (Burger and Gochfeld, 2004). In 1982, the European Commission set an environmental quality standard for mercury, stating that the mean concentration of mercury in a representative sample of fish should not exceed 0.3 mg/kg (wet weight). The U.S. Environmental Protection Agency (EPA) promulgated this value as an ambient water quality standard in 2001 (EPA, 2001).

It is possible to assemble some guidance for other metals using the U.S. Environmental Protection Agency's oral reference dose for chronic exposure. The chronic oral RfD (in mg/kg/day) is as follows: arsenic (0.0003), cadmium (0.01), mercury (0.0001), and selenium (0.005) (Burger and Gochfeld, 2005). By comparing the daily intake (concentration in fish times the amount consumed) with the chronic oral RfD, it is possible to determine whether a person is exceeding acceptable health guidance levels. Unfortunately, many of the standards were last revised in the early 1980s, suggesting an urgent need for regulators, scientists, and managers to address these issues.

RISK MANAGEMENT OF METALS IN MARINE FOODS

Risk is a function of exposure and hazard levels. There is no risk if there is no exposure, and there is no risk if the hazard levels are well below any human health effects levels. Risk management of metals in marine foods, then, can involve reducing the concentrations by attacking the sources, reducing exposures, or both. The agencies and governments responsible for reducing hazard levels and for reducing exposures are different.

Hazard Reduction

The risk to humans from consuming seafood with high metal levels can be addressed by reducing contaminant levels in marine ecosystems, thereby reducing the levels in kelp, shellfish, fish, and marine birds and mammals. Metals in the oceans come from natural geological processes and from anthropogenic sources. Although we cannot change the geological processes, we can decrease the anthropogenic sources, which include mainly runoff, riverine effluent, and atmospheric deposition (Campbell, 1994). Determining the relative contribution of each type of anthropogenic source is a first step in reducing these sources.

Reducing the input of contaminants into aquatic systems ultimately reduces the levels in fish and shellfish, but there is a lag time. For example, in the Everglades of Florida, reductions in mercury inputs were evident in declines in mercury in fish tissue within 8 years (SFWMD, 2004), but in other places, decreases have been much slower. The rapid response in Florida occurred because the prevailing winds are from east to west, carrying metals from the power plants of the Gold Coast of Florida to the Everglades. Such a wind pattern does not occur throughout the rest of the Unites States, making the control of atmospheric contaminants more difficult.

Reducing contaminants in aquatic ecosystems requires a concerted regulatory effort on a regional scale, as well as on an international scale. New Jersey regulations reducing emissions from garbage incinerators were very effective in controlling a major source (New Jersey Mercury Task Force, 2001). On the other hand, in the northeastern part of North America, atmospheric transport of mercury, mainly from coal-fired power plants in the Midwestern states, is the major controllable input (Northeast States for Coordinated Air Use Management and Canadian Ecological Monitoring and Assessment Network [NESCAUM], 1998). The Environmental Protection Agency's Clear Skies initiative relies heavily on emission trading but does not take full effect until 2018 (EPA, 2003) and will have little effect on the atmospheric transport of mercury.

The atmospheric deposition problem, however, is global and must be solved globally. Industrialized countries worldwide are emitting mercury and other contaminants into the air from fossil fuel, metal smelters, chloralkalai plants, instruments, batteries, and switches (New Jersey Mercury Task Force, 2001). Once airborne, the mercury circulates globally, allowing atmospheric fallout even in remote polar regions. Efforts to eliminate mercury from many branches of commerce will reduce this source, but recycling of mercury provides only temporary relief. Ultimately, however, it is the responsibility of governments to reduce contaminants in the environment so that seafood is safe for human consumption and so that marine ecosystems are healthy generally.

Global regulations and agreements to reduce atmospheric levels of heavy metals such as mercury, however, will not result in immediate effects because of the response lag time. Heavy metals already accumulated in the system will not disappear and will be relegated to sediment sinks at the bottom of the ocean only after years and decades.

Exposure Reduction

Another approach to risk reduction is to shift the burden from environmental protection to personal behavior (Halkier, 1999; Jakus *et al.,* 1997), to issue consumption advisories, and to assume that personal behavior will change accordingly. This approach can include providing individuals with sufficient information about contaminant levels in different species of fish or other organisms (Fig. 8-2). Then consumers can select different foods, depending on their risk vulnerability (such as age or pregnancy status) or on their personal risk aversiveness.

State, federal, and tribal agencies have responded to potential health risks from contaminants in fish by issuing consumption advisories. In general, it is state agencies that are responsible for the health and safety of their citizens, and thus, they issue the advisories. The number of states issuing consumption advisories has increased dramatically since the mid-1990s, due partly to increased sensitivity of measurements but also to increased contaminant levels in fish. Forty-eight states have issued consumption advisories, primarily because of mercury and PCBs (EPA, 2005). Wyoming and Alaska are the only states in the United States without fish consumption advisories (EPA, 2004). Alaska has taken a strong antiadvisory stance that nutritional benefits of fish outweigh the risks, especially for subsistence fishers (Egeland and Middaugh, 1997), although Burger *et al.* (2007) examine this in more detail. The fishing industry counters with advisories aimed at increasing fish consumption during pregnancy. However, most of these advisories are for freshwater fish, and little attention has been devoted to marine seafood. Further, most states distribute fish consumption guidance with fishing licenses, which are usually not required for marine fish or for Native Americans.

The U.S. Food and Drug Administration (FDA, 2001, 2003) issued consumption advisories based on methylmercury, which advised pregnant women and women of childbearing age who may become pregnant to entirely avoid eating four types of marine fish (shark, swordfish, king mackerel, and tilefish) and limit their consumption of all other fish to just 12 ounces per week (FDA, 2001). This advisory was recently amended to add that people should "mix up the types of fish and shellfish" they eat and avoid eating the same type of fish or shellfish more than once a week (FDA, 2004). Advisories for tuna, particularly canned tuna, remain controversial (Burger and Gochfeld, 2004), and there is little advice and advisories for fish available commercially (Burger *et al.,* 2004, 2005), although the FDA advisory does indicate that white tuna has higher mercury than light tuna (FDA, 2004). The FDA advice is for both commercial and self-caught fish, whereas most state advisories relate only to self-caught fish, although this is gradually changing.

Compliance with consumption advisories is sometimes low, leading to questions about the efficacy of advisories as

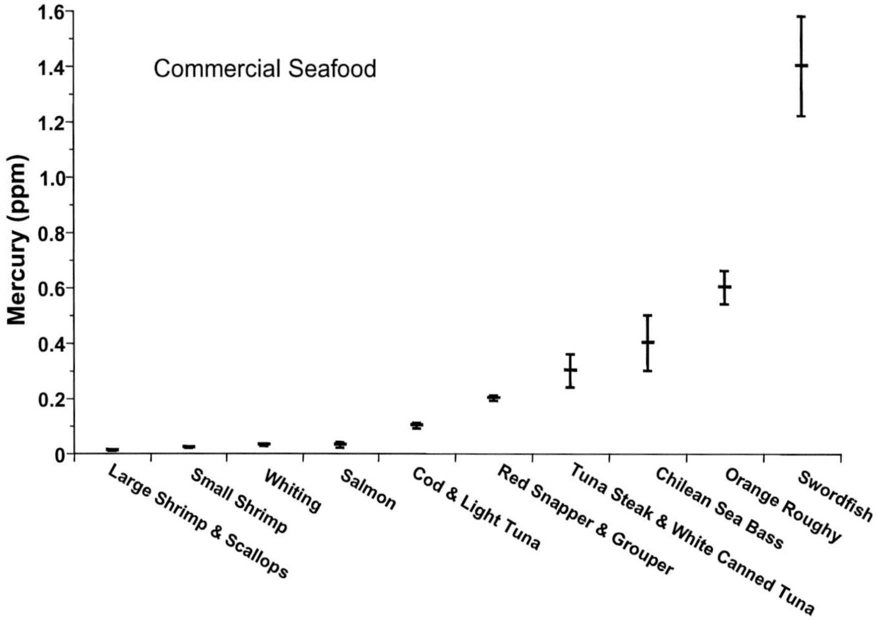

FIGURE 8-2. Mercury levels (mean ppm with standard error) for commercial fish obtained from supermarkets in New Jersey and Illinois. From Burger and Gochfeld (2005, 2006b).

a public health policy (Burger, 2000; Connelly and Knuth, 1998; Jardine, 2003; Reinert *et al.,* 1991, 1996). However, one study from Newark Bay, New Jersey, reported that Latino fishermen showed a willingness to change their consumption behavior when presented with clear risk information (Burger *et al.,* 1999a; Pflugh *et al.,* 1999), and another study reported a decline in fish consumption among pregnant women following a federal mercury advisory issued in January 2001 (Oken *et al.,* 2003). However, the authors found that although many people in the study population in New Jersey had heard about advisories concerning tuna, less than a third knew about advisories concerning shark and swordfish, and most did not have specific information about the basis for such warnings (Burger, 2005). Compliance may be low because advisories do not take into account the balancing of risks and benefits that people engage in every day when making decisions.

Risk Balancing

When people select foods for consumption, they do so with many factors in mind. Whether consciously or unconsciously, they balance the risks and benefits of particular foods against others. Decisions include whether to eat fish, poultry, beef or some other source of protein, what kinds of each type to eat, and how much to eat. They balance the good and bad aspects of fish consumption (Gochfeld and Burger 2005; Knuth *et al.,* 2003; Sidhu, 2003). However, a wide range of other factors enter such decisions, including availability, cost, personal likes, cultural values, nutritional information, contaminant information, ease of preparation (Fig. 8-3), and can be managed both by individuals, health professionals, and governmental agencies.

Scientists and health professionals tend to concentrate on the risks and benefits of fish consumption, without necessarily thinking about other trade-offs. Stakeholders clearly should be involved in the entire process, from risk determination from metals to risk management (Ebert, 1996). Since the mid-1990s, scientists and health professionals have devoted considerable attention to understanding the benefits and risks from consuming fish, particularly for self-caught fish (Anderson and Wiener, 1995; Burger *et al.,* 2001b; Egeland and Middaugh, 1997, 1998; Gochfeld and Burger, 2005; Lange *et al.,* 1994). Fish are a healthy source of protein, provide omega-3 (n-3) fatty acids that are generally accepted to reduce cholesterol levels, and reduce the incidence of heart disease, stroke, and preterm delivery (Albert *et al.,* 2002; Anderson and Wiener, 1995; Daviglus *et al.,* 2002; Hu *et al.,* 2002; Patterson, 2002), although Garcia-Closas *et al.* (1993) did not find a negative association between fish consumption and ischemic heart disease mortality. Further, Iribarren *et al.* (2004) showed a positive relationship between consumption of fish with high n-3 fatty acids and a lower likelihood of high hostility in young adults. The public seems to better understand the health benefits from eating seafood than the associated health risks (Burger, 2005).

Most studies examining the risks or benefits from fish or shellfish consumption examine only one benefit or one risk, and usually from only one health endpoint or contaminant. Willet attempted to deal with the risk/benefit questions by

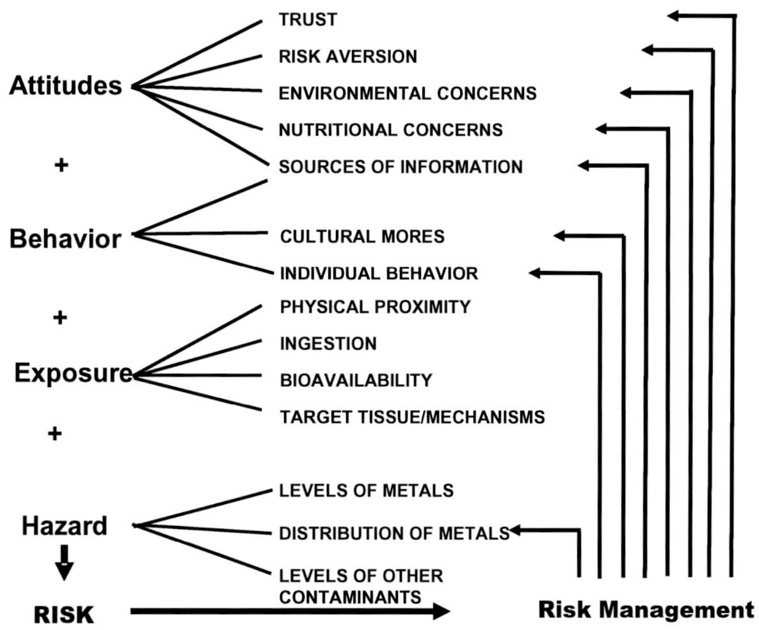

FIGURE 8-3. A framework for risk management of fish, both self-caught and commercially available. Partly adapted from Burger and Gochfeld (2006a).

examining together a series of studies that addressed the benefits of fish consumption on a wide range of public health endpoints (Willett, 2005) and concluded that where there are potential risks and benefits, both risk and benefit information should be provided. A 2006 study by the Institute of Medicine (2006) concluded that for most people, the health benefits of eating fish and shellfish outweigh any risks from contamination by toxic chemicals. This begs the question, however, because it does not address the concerns of sensitive populations, sensitive human life stages, or vulnerable people, those who consume large quantities of fish—outliers in the consumption distribution—whether they are subsistence people, recreationists, or simply consume unusually high levels (Burger *et al.,* 1999b). Recommendations for fish consumption during pregnancy remain controversial in 2007.

References

Able, K.W., Fahay, M.P., 1998. The first year in the life of estuarine fishes in the Middle Atlantic Bight. New Brunswick, NJ, Rutgers University Press.

Agency for Toxic Substances and Disease Registry (ATSDR), 1996. Mercury awareness for the public. Hazard Subst. Public Health 6, 4–5.

Agency for Toxic Substances and Disease Registry (ATSDR), 1999. Toxicologic profile: Cadmium. Atlanta, GA, Centers for Disease Control: Agency for Toxic Substances and Disease Registry.

Agency for Toxic Substances and Disease Registry (ATSDR), 2000. Toxicologic Profile for Arsenic. Atlanta, GA, Centers for Disease Control: Agency for Toxic Substances and Disease Registry.

Agency for Toxic Substances and Disease Registry (ATSDR), 2005. Toxicologic profile: Arsenic. Atlanta, GA, Centers for Disease Control: Agency for Toxic Substances and Disease Registry.

Albert, C.M., Canoism, G.M., Stampfer, M.J., Ridker, P.M., Manson, J.E., Willett, W.C., Ma, J., 2002. Blood levels of long-chain n-3 fatty acids and the risk of sudden death. N. Engl. J. Med. 235, 1113–1118.

Anderson, P.D., Wiener, J.B., 1995. Eating fish. In Graham J.D., and Wiener, J.B. (eds.), Risk versus Risk: Tradeoffs in Protecting Health and the Environment. Cambridge, MA, Harvard University Press.

Becker, P.H., Gonzalez-Solis, J., Behrends, B., Croxall, J., 2002. Feather mercury levels in seabirds at South Georgia: Influence of trophic position, sex, and age. Mar. Ecol. Progr. Series 243, 261–269.

Bellinger, D., Leviton, A., Waternaux, C., Needleman, H., Rabinowitz, M., 1987. Longitudinal analyses of prenatal and postnatal lead exposure and early cognitive development. N. Engl. J. Med. 316, 1037–1043.

Bidone, E.D., Castilhos, Z.C., Santos, T.J.S., Souza, T.M.C., Lacerda, L.D., 1997. Fish contamination and human exposure to mercury in Tartarugalzinho River, Northern Amazon, Brazil. A screening approach. Water, Air, Soil Poll. 97, 9–15.

Bloom, N.S., 1991. On the chemical form of mercury in edible fish and marine invertebrate tissue. Can. J. Fish. Aquat. Sci. 49, 1010–1017.

Boening, D.W., 2000. Ecological effects, transport, and fate of mercury: A general review. Chemosphere 40, 1335–1351.

Braune, B.M., 1987. Mercury accumulation in relation to size and age of Atlantic herring (*Clupea harengus harengus*) from the southwestern Bay of Fundy, Canada. Arch. Environ. Contam. Toxicol. 16, 311–320.

Brezonik, P.K., King, S.O., Mach, C.E., 1991. The influence of water chemistry on trace metal bioavailability and toxicity in aquatic organisms. In Newman, M.D., and McIntosh, A.W. (eds.), Metal ecotoxicology. Boca Raton, FL, Lewis.

Bryan, G.W., Langston, W.J., 1992. Bioavailability, accumulation and effects of heavy metals in sediments with special reference to the United Kingdom estuaries: A review. Environ. Pollut. 76, 89–95.

Buck, G.M., Vena, J.E., Schisterman, E.F., Dmochowski, J., Mendola, P., Sever, L.E., Fitzgerald, E., Kostyniak, P., Greizerstein, H., Olson, J., 2000. Parental consumption of contaminated sport fish from Lake Ontario and predicted fecundability. Epidem. 11, 388–393.

Burger, J., 1993. Metals in avian feathers: Bioindicators of environmental pollution. Rev. Environ. Toxicol. 5, 197–306.

Burger, J., 2000. Consumption advisories and compliance: The fishing public and the deamplification of risk. J. Environ. Plan. Manage. 43, 471–488.

Burger, J., 2002a. Consumption patterns and why people fish. Environ. Res. 90, 125–135.

Burger, J., 2002b. Daily consumption of wild fish and game: Exposures of high end recreationists. Intern. J. Environ. Health Res. 12, 343–354.

Burger, J., 2005. Fishing, fish consumption and knowledge about advisories in college students and others in central New Jersey. Environ. Res. 98, 268–275.

Burger, J., Cooper, K., Gochfeld, M., 1992. Exposure assessment for heavy metal ingestion from a sport fish in Puerto Rico: Estimating risk for local fishermen. J. Toxicol. Environ. Health 36, 355–365.

Burger, J., Gaines, K.F., Boring, C.S., Stephens, Jr. W.L., Snodgrass, J., Gochfeld, M., 2001a. Mercury and selenium in fish from the Savannah River: Species, trophic level, and locational differences. Environ Res. 87, 108–118.

Burger, J., Gaines, K.F., Gochfeld, M., 2001b. Ethnic differences in risk from mercury among Savannah River fishermen. Risk Anal. 21, 533–544.

Burger, J., Gaines, K.F., Boring, C.S., Stephens, W.L., Snodgrass, J., Dixon, C., McMahon, M., Shukla, S., Shukla, T., Gochfeld, M., 2002. Metal levels in fish from the Savannah River: Potential hazards to fish and other receptors. Environ. Res. 89, 85–97.

Burger, J., Gochfeld, M., 2004. Mercury in canned tuna: White versus light and temporal variation. Environ. Res. 96, 239–249.

Burger, J., Gochfeld, M., 2005. Heavy metals in commercial fish in New Jersey. Environ. Res. 99, 403–412.

Burger, J., Gochfeld, M., 2006a. A framework and information needs for the management of the risks from consumption of self-caught fish. Environ. Res. 101, 275–285.

Burger, J., Gochfeld, M., 2006b. Mercury in fish available in supermarkets in Illinois: Are there regional differences. Sci. Total Environ. 367, 110–116.

Burger J., Gochfeld M., Leck C., 1984. Waterbirds on Raritan Bay: A preliminary analysis of their distribution and heavy metal levels. Bull. NJ. Acad. Sci. 29, 133–142.

Burger, J., Gochfeld, M., Jeitner, C., Burke, S., Stamm, T., Snigaroff, R., Snigaroff, D., Patrick, R., Weston, J., 2007. Mercury levels and potential risk from subsistence foods from the Aleutians. Science Total Environ. 384, 93–105.

Burger, J., Pflugh, K.K., Lurig, L., von Hagen, L.A., von Hagen, S.A., 1999a. Fishing in urban New Jersey: Ethnicity affects information sources, perception, and compliance. Risk Anal. 19, 217–229.

Burger, J., Staine, K., Gochfeld, M., 1993, Fishing in contaminated waters: Knowledge and risk perception of hazards by fishermen in New York City. J. Toxicol. Environ. Health 3, 95–105.

Burger, J.W., Stephens, C.S., Boring, C.S., Kuklinski, M., Gibbons, J.W., Gochfeld, M., 1999b. Factors in exposure assessment: Ethnic and socioeconomic differences in Fishing and Consumption of fish caught along the Savannah River. Risk Anal.19, 427–438.

Burger, J., Stern, A.H., Dixon, C., Jeitner, C., Shukla, S., Burke, S., Gochfeld, M., 2004. Fish availability in supermarkets and fish markets in New Jersey. Sci. Total Environ. 333, 89–97.

Burger, J., Stern, A.H., Gochfeld, M., 2005. Mercury in commercial fish: Optimizing individual choices to reduce risk. Environ. Health Perspect. 113, 266–270.

Burrows, I.G., Whitton, B.A., 1983. Heavy metals in water, sediment and macroinvertebrates from a metal-contaminated river free of organic pollution. Hydrobiologia 106, 263–273.

Campbell, K.R., 1994. Concentrations of heavy metals associated with urban runoff in fish living in stormwater treatment ponds. Arch. Environ. Contam. Toxicol. 27, 352–356.

Chan, H.M., Kim, C., Khoday, K., Receveur, O., Kuhnlein, H.V., 1995. Assessment of dietary exposure to trace metals in Baffin Inuit food. Environ. Health Perspect. 103, 740–746.

Clements, W.H., 1991. Community responses of stream organisms to heavy metals: A review of observational and experimental approaches. In Newman, M.C., and McIntosh, A.W. (eds.), Metal Ecotoxicity: Concepts and Applications, pp. 363–391. Chelsea, MI, Lewis.

Connelly, N.A., Knuth, B.A., 1998. Evaluating risk communication: Examining target audience perceptions about four presentation formats for fish consumption health advisory information. Risk Anal. 18, 649–659.

Consumer Reports., 2003. American's fish: Fair or foul? Available at www.consumerreports.org/special/consumerInteret/Reports/0102fis0.html (accessed April 1, 2004). New York, Consumers Union.

Davidson, P.W., Myers, G.J., Weiss, B., Shamlaye, C.F., Cox, C., 2006. Prenatal methylmercury exposure from fish consumption and child development: A review of evidence and perspectives from the Seychelles child development study. Neurotoxicol. 27, 1106–1109.

Daviglus, M., Sheeshka, J., Murkin, E., 2002. Health benefits from eating fish. Comments Toxicol. 8, 345–374.

Debes F., Budtz-Jorgensen, E., Weihe, P., White, R.F., Grandjean, P., 2006. Impact of prenatal methylmercury exposure on neurobehavioral function at age 14 years. Nerutoxicol. Teratol. 28, 536–547.

Douben, P.E.T., 1989. Metabolic rate and uptake and loss of cadmium from food by the fish *Noemacheilus barbatulus L.* (Stone Loach). Environ. Pollut. 59, 177–202.

Duhaime, G., Chabot, M., Frechette, P., Robichaud, V., Proulx, S., 2004. The impact of dietary changes among the Inuit of Nunavik (Canada): A socioeconomic assessment of possible public health recommendations dealing with food contamination. Risk Anal. 24, 1007–1018.

Ebert, E.S., 1996. Fish consumption and human health: Developing partnerships between risk assessors and resource managers. Am. Fish. Soc. Symp. 16, 261–270.

Egeland, G.M., Middaugh, J.P., 1997. Balancing fish consumption benefits with mercury exposure. Science 278, 1904–1906.

Egeland, G.M., Middaugh, J.P., 1998. The use of traditional foods in a healthy diet in Alaska. Alaska Epidemiol. Bull. 2, 1–140.

Eisler, R., 1986. Chromium hazards to fish, wildlife, and invertebrates: A synoptic review. US Fish and Wildlife Service Rep, 85(1.6), Washington, DC.

Elhassani, S.B., 1982. The many faces of methylmercury poisoning. Toxicol. Clin. Toxicol. 19, 875–906.

Environmental Protection Agency (EPA), 2001. Environmental water quality standard: Freshwater, www.epa.gov/fedrgstr/EPA-WATER/2001/January/Day-08/w217.htm (accessed Oct. 23, 2007).

Environmental Protection Agency (EPA), 2003. Clear Skies Act of 2003. Washington, DC. Environmental Protection Agency, www.epa.gov/air/clearskies/fact2003.html (accessed Oct. 23, 2007).

Environmental Protection Agency (EPA), 2004. National listing of fish advisories. Cincinnati, OH, U.S. Environmental Protection Agency. Available at http//map1.epa.gov (accessed Oct. 30, 2004).

Environmental Protection Agency (EPA), 2005. National listing of fish advisories. Cincinnati, OH: U.S. Environmental Protection Agency. Available at http//map1.epa.gov (accessed Oct. 23, 2007).

Eto, K., 1997. Pathology of Minamata disease. Toxicol. Pathol. 25, 632–634.

Fiji, 2006. Mercury levels in Fijian seafood and potential public health implications. Report for World Health Organization. http://www.wpro.who.int/NR/rdonlyres/12A08026-97EF-4E01-9C7E-91CCD4A96

CA8/0/mercury_level_fijian_seafoods.pdf (accessed Oct. 23, 2007). www.wpro.who.int/NR/rdonlyres/437775AO-671F-41AE-A068 (accessed Feb. 10, 2006).

Fleming, L.E., Watkins, S., Kaderman, R., Levin, B., Ayar, D.R., Bizzio, M., Stephens, D., Bean, J.A., 1995. Mercury exposure in humans through food consumption from the Everglades of Florida. Water, Air, Soil Pollut., 80, 41–48.

Food and Agricultural Organization [FOA], 1998. Seafood safety: Economics of hazard analysis and critical control point (HACCP) programmes. Rome, Italy. United Nations Food and Agricultural Organization, Fisheries Technical Paper, Seafood Safety-Economics of Hazard Analysis and Critical Control Point (HA http://www.fao.org/docrep/003/x0465e/x0465e00.htm (accessed Oct. 23, 2007).

Food and Drug Administration (FDA), 2001. FDA consumer advisory. Available at www.fda.gov/bbs/topics/ANSWERS/2001/advisory.html. Washington, DC: U.S. Food and Drug Administration (accessed Oct. 23, 2007).

Food and Drug Administration (FDA), 2003. FDA consumer advisory. FDA's Advisory on Methylmercuryin Fish. www.cfsan.fda.gov/~lrd/tpmehg2.html (accessed Oct. 23, 2007).

Food and Drug Administration (FDA), 2004. Mercury levels in commercial fish and shellfish. Available at http://vm.cfsan.fda.gov/-frf/sea-mehg.html. Washington, DC: U.S. Food and Drug Administration (accessed Oct. 23, 2007).

Garcia-Closas, R., Serr-Majem, L., Segura, R., 1993. Fish consumption,-3 fatty acids and the Mediterranean diet. Eur. J. Clin. Nutr. 47, s85–s95.

Garza, D., 2005, Common edible seaweeds in the Gulf of Alaska. Fairbanks, AK, Alaska Sea Grant Program.

Gay, D., Maher, W., 2003. Natural variation of copper, zinc, cadmium and selenium concentrations in *Bembicium nanum* and their potential use as a biomonitor of trace metals. Water Res. 37, 2173–2185.

Ginsberg, G.L., Toal, B.F., 2000. Development of a single-meal fish consumption advisory for methylmercury. Risk Anal. 20, 41–47.

Gochfeld, M., 2003. Cases of mercury exposure, bioavailability, and absorption. Ecotoxicol. Environ. Safety 56, 174–179.

Gochfeld, M., Burger, J., 1982. Biological concentration of cadmium in estuarine birds of the New York Bight. Colon. Waterbirds 5, 116–123.

Gochfeld, M., Burger, J., 2005. Good fish/bad fish: A composite dose and benefit-risk curve. Neuro Toxicol. 26, 511–520.

Guallar, E., Sanz-Gallardo, M.I., van't Veer, P., Bode, P., Aro, A., Gomez-Aracena, J., 2002. Heavy metals and Myocardial Infarction Study Group: Mercury, fish oils, and the risk of myocardial infarction. N. Engl. J. Med. 347, 1747–1754.

Gulson, B.L., Mahaffey, K.R., Jameson, C.W., Mizon, K.J., Korsch, M.J., Cameron, M.A., 1998. Mobilization of lead from the skeleton during the postnatal period is larger than during pregnancy. J. Lab. Clin. Med. 131, 324–329.

Halkier, B., 1999. Consequences of the politicization of consumption: The example of environmentally friendly consumption practices. J. Environ, Policy Plan. 1, 25–41.

Hansen, J.C., 1988. Has selenium a beneficial role in human exposure to organic mercury? Med. Hypotheses 25, 45–53.

Harris, S.G., Harper, B.L., 1997. A Native American exposure scenario and a tribal risk model. Risk Anal. 17, 789–795.

Harrison S.E., Klaverkamp, J.F., 1989. Uptake, elimination and tissue distribution of aqueous cadmium by rainbow trout (*Salmo gairdneri* Richardson) and lake whitefish (*Coregonus clupeaormis* Mitchell). Environ. Toxicol. Chem. 8, 87–97.

Hightower, J.M., Moore, D., 2003. Mercury levels in high-end consumers of fish. Environ. Health Perspect. 111, 604–608.

Hites, R.A., Foran, J.A., Carpenter, D.O., Hamilton, M.C., Knuth, B.A., Schwager, S.J., 2004. Global assessment of organic contaminants in farmed salmon. Sci. 303, 226–229.

Hu, F.B., Bronner, L., Willet, W.C., Stampfer, M.J., Rexrode, K.I.M., Albert, C.M., Hunter, D., Manson, J.E., 2002. Fish and omega-3 fatty acid intake and risk of coronary heart disease in women. JAMA 287, 1815–1821.

Institute of Medicine (IOM), 1991. Seafood Safety. Washington, DC, National Academy Press.

Institute of Medicine (IOM), 2006. Seafood Choices: Balancing benefits and risks. Washington, DC, National Academy Press.

Iribarren, C., Markovitz, J.H., Jacobs, D.R., Jr., Schreiner, P.J., Daviglus, M., Hibbeln, J.R., 2004. Dietary intake of n-3, n-6 fatty acids and fish: Relationship with hostility in young adults: The CARDIA study. Eur. J. Clin. Nutr. 58, 24–31.

Iso, H., Rexrode, K.M., 2001. Intake of fish and omega-3 fatty acids and risk of stroke in women. JAMA 285, 304–312.

Jacobs, H.L., Kahn, H.D., Stralka, K.A., Phan, D.R., 1998. Estimates of per capita fish consumption in the US based on the Continuing Survey of Food Intake by Individuals (CSFII). Risk Anal. 18, 283–291.

Jacobson, J.L., Jacobson, S.W., 1996. Intellectual impairment in children exposed to polychlorinated biphenyls in utero. N. Engl. J. Med. 335, 783–789.

Jakus, P.M., Downing, M., Bevelhimer, M.S., Fly, J.M., 1997. Do sportfish consumption advisories affect reservoir anglers' site choice? Agricult. Res. Econ. Rev. 26, 196–204.

Jardine, C.G., 2003. Development of a public participation and communication protocol for establishing fish consumption advisories. Risk Anal. 23, 461–471.

Jardine, C.G., Hrudey, S.E., Shortreed, J.H., Craig, L., Krewski, D., Furgal, C., McColl, S., 2003. Risk management frameworks for human health and environmental risks. J. Toxicol. Environ. Health, Part B. 6, 569–641.

Jensen, S., Jernelov, A., 1969. Biological methylation of mercury in aquatic organisms. Nature 223, 753–754.

Johnsson, C., Schutz, A., Sallsten, G., 2005. Impact of consumption of freshwater fish on mercury levels in hair, blood, urine, and alveolar air. J. Toxicol. Environ. Health 68, 129–141.

Julshamn, K., Grahl-Nielsen, O., 2000. Trace elements in harp seal (*Pagophilus groenlandicus*) and hooded seal (*Cystophora cristata*) from the Greenland Sea: A multivariate approach. Sci. Total Environ. 250, 123–133.

Kannan, K., Smith, R.G., Jr., Lee, R.F., Windom, H.L., Heitmuler, P.T., Macauley, J.M., Summers, J.K., 1998. Distribution of total mercury and methylmercury in water, sediment and fish from South Florida estuaries. Arch. Environ. Contam. Toxicol. 34, 109–118.

Knobeloch, L., Anderson, H.A., Imm, P., Peters, D., Smith, A., 2005. Fish consumption, advisory awareness, and hair mercury levels among women of childbearing age. Environ. Res. 97, 220–227.

Knuth, B., Connelly, N.A., Sheeshka, J., Patterson, J., 2003. Weighing health benefits and health risk information when consuming sport-caught fish. Risk Anal. 23, 1185–1197.

Lacerda, L.D., Bidone, E.D., Giumaraes, A.F., Pfeiffer, W.C., 1994. Mercury concentrations in fish from the Itacaiunas-Parauapebas River System, Carajas region, Amazon. Ann. Acad. Bras. Sci. 66, 373–379.

Lange, T.R., Royals, H.E., Connor, L.L., 1994. Mercury accumulation in largemouth bass (*Micropterus salmoides*) in a Florida Lake. Arch. Environ. Contam. Toxicol. 27, 466–471.

Lansens, P., Leermakers, M., Vaeyens, W., 1991. Determination of methylmercury in fish by headspace-gas chromatography with microwave-induced-plasma detection. Water, Air, Soil Pollut. 56, 103–115.

Lonky, E., Reihman, J., Darvill, T., Mather, Sr. J., Daly, H., 1996. Neonatal behavioral assessment scale performance in humans influenced by maternal consumption of environmentally contaminated fish. J. Great Lakes Res. 22, 198–212.

Luten, J.B., Bouquet, W., Riekwel-Booy, G., Rauchbaar, A.B., Scholte, M.W.M., 1987. Mercury in flounder, *Platichtys flesus*, Cod, *Gadus morhua*, and Perch, *Perca fluviatilis*, in relation to their length and environment. Bull. Environ. Contam. Toxicol. 38, 318–323.

Mason, R.P., Fitzgerald, W.F., Morel, F.M.M., 1994. Aquatic biogeochemical cycling of elemental mercury: an anthropogenic influence. Geochem. Cosmochim. 58, 3191–3198.

Mitchell, J.W. (ed.), 1987. Occupational medicine focus: Lead toxicity and reproduction. J. Occup. Med. 29, 397–399.

Moya, J., 2004. Overview of fish consumption rates in the United States. Human Ecol. Risk Assess. 10, 1195–1211.

National Oceanic & Atmospheric Administration (NOAA), 2004a. Seafood consumption rose again in 2003. NOAA Magazine, U.S. Commerce Department, National Oceanographic and Atmospheric Administration, www.noaanews.noaa.gov/stories2004/s2322.htm (accessed Oct. 23, 2007).

National Oceanic & Atmospheric Administration (NOAA), 2004b. US 2004 seafood consumption reaches record 16.6 pounds per person and the most popular seafood species consumed in the United States, www.noaaanews.noaa.gove/stories2005/pdf/seafood-consumption11-09-2005.pdf (accessed December 2006).

National Oceanic & Atmospheric Administration (NOAA), 2006. Coastal trends report series: Population trends along the coastal United States 1980–2008, www.oceanservice.noaa.gov/programs/mb/supp_cstl_population.html (accessed Oct. 23, 2007).

National Research Council (NRC), 1987. Biological markers in environmental health research. Environ. Health Persp. 74, 3–9.

National Research Council (NRC), 2000. Toxicological effects of methylmercury. Washington, DC, National Academy Press.

Needleman, H.L., Schell, A., Bellinger, D., Leviton, A., Allred, E.N., 1990. The long-term effects of exposure to lead in childhood: An 11 year follow-up report. N. Engl. J. Med 322, 83–88.

NESCAUM, 1998. Northeast States and Eastern Canadian Provinces Mercury Study: A Framework for Action. Boston: Northeast States for Coordinated Air Use Management and Canadian Ecological Monitoring and Assessment Network.

Nestel, P., 2001. Effect of fish oil and fish on cardiovascular diseases. Current Artherosceral Res. 3, 68–73.

Neuringer, M., Reisbick, S., Janowsky, J., 1994. The role of n-3 fatty acids in visual and cognitive development: Current evidence and methods of assessment. J. Pediatr. 125, 39–47.

New Jersey Mercury Task Force, 2001. Report of the New Jersey Mercury Task Force. Trenton, NJ, New Jersey Department of Environmental Protection (NJ DEP). Available at http://www.state.nj.us/dep/dsr/nj-mercury-volume1.pdf (accessed Oct. 23, 2007).

Oken, E., Kleinman, K.P., Berland, W.E., Simon, S.R., Rich-Edwards, J. W., Gillman, M.W., 2003. Decline in fish consumption among pregnant women after a national mercury advisory. Obstet. Gynecol. 102, 346–351.

Olsen, S.F., Secher, N.J., 2002. Low consumption of seafood in early pregnancy as a risk factor for preterm delivery: Prospective cohort study. Brit. Med. J. 324, 447–450.

Paez-Osuna, F., Perez-Gonsalez, R., Izaguirre-Fierro, G., Zazueta-Padilla, H.M., Flores-Campana, L.M., 1995. Trace metal concentrations and their distribution in the lobster *Panulirus inflatus* (Bouvier, 1895) from the Mexican Pacific coast. Environ. Poll. 90, 163–170.

Patterson, J., 2002. Introduction—comparative dietary risk: Balance the risks and benefits of fish consumption. Comments Toxicol. 8, 337–344.

Payton, M., Riggs, K.M., Piro III, A., Weiss, S.T., Hu, H., 1998. Relations of bone and blood lead to cognitive function: The VA normative aging study. Neurotoxicol. Teratol. 20, 19–27.

Pesch, G.C., Steward, N.E., 1980. Cadmium toxicity to three species of estuarine invertebrates. Marine Environ. Res. 3, 145–155.

Pflugh, K.K., Lurig, L., vonHagen, L.A., vonHagen, S., Burger, J., 1999. Urban angler's perception of risk from contaminated fish. Sci. Total Environ. 228, 203–218.

Phaneuf, D., Cote, I., Duman, P., Ferron, L.A., LeBlanc, A., 1999. Evaluation of the contamination of marine algae (seaweed) from the St. Lawrence River and likely to be consumed by humans. Environ. Res. 80, 5175–5182.

Phillips, G.R., Lenhart, T.E., Gregory, R.W., 1980. Relations between trophic position and mercury accumulation among fishes from the Tongue River Reservoir, Montana. Environ. Res. 22, 73–80.

Prosi, F., 1979. Heavy metals in aquatic organisms. In Forstner, U., and Whittman, G.T.W. (eds.), Metal Pollution in the Aquatic Environment, pp. 271–323. Berlin, Springer Verlag.

Ragan, M.A., Smidsrod, O., Larson, B., 1979. Chelation of divalent metal ions by brown algal polyphenols. Mar. Chem. 7, 265–271.

Ramos, A.M., Crain, E.F., 2001. Potential health risks of recreational fishing in New York City. Ambulat. Pediatr. 1, 252–255.

Reinert, R.E., Knuth, B.A., Kamrin, M.A., Stober, Q.J., 1991. Risk assessment, risk management, and fish consumption advisories in the United States. Fish 16, 5–12.

Reinert, R.E., Knuth, B.A., Kamrin, M.A., Stober, Q.J., 1996. A review of the basic principles and assumptions used to issue fish consumption advisories. Am. Fish. Soc. Symp. 16, 98–106.

Rice, D.C., 1984. Behavioral deficit (delayed matching to sample) in monkeys exposed from birth to low levels of lead. Toxicol. Appl. Pharmacol. 75, 337–345.

Rice, G., Swartout, J., Mahaffey, K., Schoeny, R., 2000. Derivation of U. S. EPA's oral Reference Dose (RfD) for methylmercury. Drug Chem. Toxicol. 23, 41–54.

Rissanen, T., Voutilainen, S., Nyyssonen, K., Lakka, T.A., Salonen, J.T., 2000. Fish oil-derived fatty acids, docosahexaenoic acid and docosaphentaenoic acid, and the risk of acute coronary events: The Kuopio ischaemic heart disease risk factor study. Circulat. 102, 2677–2679.

Rossi, A., Pellegrini, D., Belcari, P., Barghighiani, C., 1993. Mercury in Eledone cirrhosa from the Northern Tyrrhenian Sea: Contents and relations with life cycle. Mar. Pollut. Bull. 26, 683–686.

Ruelas, J.R., Paez-Osuna, F., Perez-Cortes, H., 2000. Distribution of mercury in muscle, liver and kidney of the spinner dolphin (*Stenella longirostris*) stranded in the southern Gulf of California. Marine Poll. Bull. 40, 1063–1066.

Safina, C., Rosenberg, A.A., Myers, R.A., Quinn, T.O, Collie, J.S., 2005. Oceans, U.S. ocean fish recovery: Staying the course. Science 309, 707–708.

Salonen, J.T., Seppanen, K., Nyyssonen, K., Korpela, H., Kauhanen, J., Kantola, M., Tuomilehto, J., Esterbauer, H., Tatzber, F., Salonen, R., 1995. Intake of mercury from fish, lipid peroxidation, and the risk of myocardial infarction and coronary, cardiovascular, and any death in eastern Finnish men. Circulation 91, 645–655.

Satoh, H., Yasuda, N., Shimai, S., 1985. Development of reflexes in neonatal mice prenatally exposed to methylmercury and selenite. Toxicol. Letters 25, 199–203.

Schantz, S.L., 1996. Developmental neurotoxicity of PCBs in humans: What do we know and where do we go from here? Neurotox. Teratol. 18, 217–227.

Schantz, S.L., Widholm, J.J., Rice, D.C., 2003. Effects of PCB exposure on neuropsychological function in children. Environ. Health Perspect. 111, 357–376.

Schwartz, J., Otto, D., 1987. Blood lead, hearing thresh, and neurobehavioral development in children and youth. Arch. Environ. Health. 42, 153–160.

Schwartz, J., 1991. Lead, blood pressure, and cardiovascular disease in men and women. Environ. Health Perspect. 91, 71–76.

Sharp, G.J., Samant, H.S., Vaidya, O.C., 1988. Selected metal levels of commercially valuable seaweeds adjacent to and distant from point sources of contamination in Nova Scotia and New Brunswick. Bull. Environ. Contam. Toxicol. 40, 724–730.

Sidhu, K.S., 2003. Health benefits and potential risks related to consumption of fish or fish oil. Reg. Toxicol. Pharmacol. 38, 336–344.

South Florida Water Management District (SFWMD)., 2001, 2002, and 2004. Everglades consolidated report. West Palm Beach, FL, SFWMD.

Sparks, P., Shepherd, R., 1994. Public perceptions of the potential hazards associated with food production: An empirical study. Risk Anal. 14, 799–808.

Spry, D.J., Wiener, J.G., 1991. Metal bioavailability and toxicity to fish in low-alkalinity lakes: A critical review. Environ. Pollut. 71, 243–304.

Stern, A.H., 1993. Re-evaluation of the reference dose for methylmercury and assessment of current exposure levels. Risk Anal. 13, 355–364.

Stern, A.H., Jacobson, J.L., Ryan, L., Burke, T.A., 2004. Do recent data from the Seychelles Islands alter the conclusions of the NRC Report on the toxicological effects of methylmercury? Environ. Health 3, 2.

Storelli, M.M., Stuffler, R.G., Marcotrigiano, G.O., 2002. Total and methylmercury residues in tuna-fish from the Mediterranean Sea. Food Additiv. Contam. 19, 715–720.

Temara, A., Warnau, M., Jangoux, M., Dubois, P., 2003. Factors influencing the concentrations of heavy metals in the asteroid *Asterias rubens* L. (Echinodermata). Science Total Environ. 203, 51–63.

Toth, J.F., Jr., Brown, R.B., 1997. Racial and gender meanings of why people participate in recreational fishing. Leisure Sci. 19, 129–146.

Trudel, M., Rasmussen, J.B., 1997. Modeling the elimination of mercury by fish. Environ. Sci. Technol. 31, 1716–1722.

Turekian, K.K., 1968. Oceans. Prentice-Hall, Englewood Cliffs, NJ, www.cfriends.org.nz/oceano/seawater.htm (accessed December 2006).

Turoczy, N.J., Mitchell, B.D., Levings, A.H., Rajendram, V.S., 2001. Cadmium, copper, mercury, and zinc concentrations in tissues of king crab (*Pseudocaricinus gigas*) from southeast Australian waters. Environ. Int. 27, 327–334.

VanNetten, C., Cann, S.A.H., Morley, D.R., VanNetten, J.P., 2000. Elemental and radioactive analysis of commercially available seaweed. Sci. Total Environ. 255, 169–175.

Vupputuri, S., Longnecker, M.P., Daniels, J.L., Xuguang, G., Sandler, D. P., 2005. Blood mercury level and blood pressure among US women: Results from the national Health and Nutrition Examination Survey 1999–2000. Environ Res, 97, 195–200.

Watras, C.J., Back, R.C., Halvorsen, S., Hudson, R.J.M., Morrison, K.A., Wente, S.P., 1998. Bioaccumulation of mercury in pelagic freshwater food webs. Sci. Total Environ. 219, 183–208.

Wiener, J.G., Krabbenhoft, D.P., Heinz, G.H., Scheuhammer, M., 2003. Ecotoxicology of mercury. In Hoffman, D.J., Rattner, B.A., Burton, G.A., Jr., and Cairns, J., Jr., Handbook of Ecotoxicology, pp. 409–463. Boca Raton, FL, Lewis.

Wiener, J.G., Spry, D.J., 1996. Toxicological significance of mercury in freshwater fish. In Beyer, W.N, Heins, G.H., and Redmon-Norwood, A.W. (eds.), Environmental Contaminants in Wildlife, pp. 287–339. Boca Raton, FL: Lewis.

Willett, W.C., 2005. Balancing health risks and benefits. Am. J. Prev. Med. 29, 320–321.

World Health Organization (WHO), 1989. Mercury-environmental aspects. WHO, Geneva, Switzerland.

Zillioux, E.J., Porcella, D.B., Benoit, J.M., 1993. Mercury cycling and effects in freshwater ecosystems. Environ. Toxicol. Chem. 12, 2245–2264.

STUDY QUESTIONS

1. Why does it matter how much mercury or other metals are in shellfish and fish? Is there a cost to society or only to individuals of contaminants in fish?
2. How do metals move up the food chain? Why do some organisms accumulate mercury or other metals, whereas others do not?
3. Define *bioaccumulate* and *biomagnify*, and explain why these mechanisms are important and what factors might affect each?
4. Why does speciation of mercury matter? What form is problematic for marine organisms and for people, and why?
5. What is the relationship between the age and size of a fish and mercury levels? What does this mean for biomagnification and for human health?
6. Scientists often concentrate on either self-caught fish or commercial fish when computing risk to humans from consumption. However, what key factors should be considered when determining risk? Do these risks differ among people of different ages, genders, or other factors?
7. What are the different methods of determining the adverse health effects of exposure to mercury or other metals? What are the advantages of each method?
8. What is the difference between human health guidelines for fish consumption and a formal risk assessment? Can you as a consumer easily use either of these guidelines?
9. What methods are available to consumers of fish (including humans) to reduce the risk from mercury and other metals?
10. If you were asked to conduct a risk balancing discussion with your classmates about seafood consumption, what aspects would you begin with? What would risk balancing entail in the modern world?

CHAPTER

9

The Fate of Pharmaceuticals and Personal Care Products in the Environment

M. DANIELLE MCDONALD AND DANIEL D. RIEMER

INTRODUCTION

Pharmaceuticals and the active ingredients in personal care products make up a large group of toxicants that have, until relatively recently, gone unrecognized. However, with advances in environmental residue analysis and higher detection efficiencies, scientists have now become aware that these contaminants are present in the world's rivers, lakes, and even oceans; and many are found in high enough concentrations not only to cause harm to aquatic organisms but to pose a potential risk to humans. At the same time, humans are at risk voluntarily; many of us use these products on a daily basis, which results in direct exposure. Several compounds fall into this category of emerging toxicants. With respect to pharmaceuticals, groups have been defined as (1) nonsteroidal anti-inflammatory drugs (pain relievers), (2) beta-blockers (blood pressure modulators), (3) blood lipid lowering agents (cholesterol reducers), (4) neuroactive compounds (e.g., antidepressants), (5) steroidal hormones (e.g., contraceptives), and (6) antibiotics. The main concerns with respect to personal care products are (1) surfactants (detergents), (2) musks (perfumes), and (3) UV filters (sunscreens). Their entry into the environment is mainly through human use; however, veterinary medicines and their metabolites are also released into the environment. Human health is potentially at risk through exposure to polluted surface and groundwater and consumption of contaminated drinking water and aquatic organisms. On a more global scale is the danger of ecosystem destruction and collapse. The goal of this chapter is to summarize the ongoing progress in this relatively new area of toxicology, with respect to the chemistry and ecological impact of these emerging toxicants with a direct emphasis on the potential risks on human health.

ENTRY OF PHARMACEUTICALS AND PERSONAL CARE PRODUCTS (PPCPs) INTO THE ENVIRONMENT

Pharmaceuticals and personal care products (PPCPs) are introduced to the ecosystem through a number of routes (Fig. 9-1). PPCPs mainly enter the environment through human consumption, excretion, and the subsequent treatment, or lack thereof, and release by sewage treatment plants (STPs). PPCPs also enter STPs through bathing and by the improper disposal of unused and expired pharmaceuticals. In a survey done in 2006, more than 53% of patients had flushed unused medications down the toilet, and 35% had rinsed them down the sink (Seehusen and Edwards, 2006). Many STPs are not designed to remove pharmaceuticals and do so ineffectively; studies indicate that elimination efficiencies of pharmaceuticals span a large range (0 to 99%) (Fent *et al.*, 2006; Ternes, 1998). In wastewater treatment, two elimination processes are important: adsorption to suspended solids (sewage sludge) and biodegradation (Fent *et al.*, 2006). Adsorption is dependent on the interactions of the pharmaceutical with particulates and microorganisms in the sludge. In general, acidic pharmaceuticals tend not to be eliminated from wastewater in this way; however, basic pharmaceuticals, such as antibiotics, can adsorb to sludge to a significant extent. Although adsorption to sludge can remove a compound from wastewater, degradation in sludge does not always occur, and many compounds, such as steroid hormones, will be present in sludge in measurable amounts. This becomes an environmental problem when the contaminated sludge is then used as land fertilizer (Dizer *et al.*, 2002; Golet *et al.*, 2003).

Biodegradation is the most important elimination process in wastewater treatment for many pharmaceuticals. In

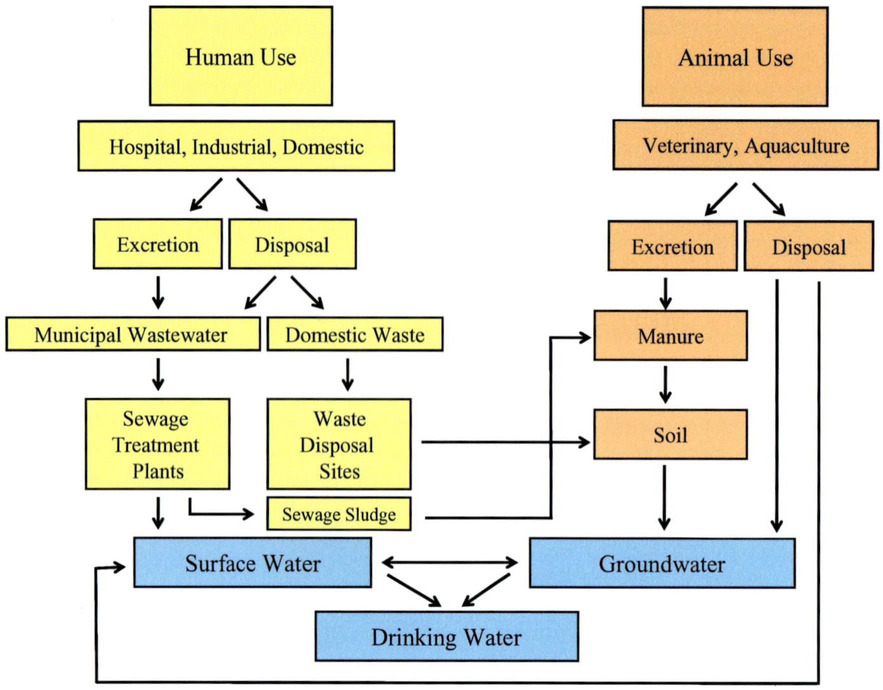

FIGURE 9-1. Schematic showing possible sources and pathways for the occurrence of pharmaceutical residues in the environment. Adapted from Heberer (2002).

general, the biological decomposition of pharmaceuticals increases with the increase in hydraulic retention time and with the age of the sludge in the activated sludge treatment. Studies on elimination rates during the STP process are mainly based on measurements of influent and effluent concentrations in STPs, and they vary according to the construction and treatment technology, hydraulic retention time, season, and performance of the STP. Once in surface waters, biotransformation through biodegradation occurs, but abiotic transformation reactions, such as photodegradation, are probably more important.

Another major way that PPCPs find their way into the aquatic environment is through the agricultural runoff from veterinary-treated livestock (Boxall *et al.*, 2002, 2003). PPCPs can also be released during manufacturing, through irrigation with reclaimed wastewater and by landfill leachates (Balcioglu and Ötker, 2003; Cordy *et al.*, 2004; Daughton and Ternes, 1999; Kinney *et al.*, 2006a, 2006b).

MEASURING PHARMACEUTICALS AND PERSONAL CARE PRODUCTS IN THE ENVIRONMENT

Advances in analytical technologies since the 1990s have provided the capabilities required to probe the natural environment for traces of pharmaceuticals, personal care products, and their metabolites and physiochemical breakdown products (Table 9-1). With the application of these advanced techniques, a broad range of compounds used during normal human behavior are now being observed in the environment. These compounds enter wastewater and receiving water bodies without specific regard for their removal, treatment, or potential ecological effects. Most of these compounds are polar and thus are not amenable to analysis by conventional gas chromatography with mass spectrometric detection (GC/MS). The primary driver allowing the observation of these compounds was the optimization of liquid chromatography with mass spectrometric detection (LC/MS). Current LC/MS instrumentation allows the detection of a wide range of PPCPs to a level of low ng L^{-1}. GC/MS is still a popular technique used for analysis of compounds amenable to derivatization, but the derivatization process can be problematic. Other biochemical techniques, including biosensors and immunoassays, have been developed for specific PPCPs, but broad applicability has not been realized. Extraction, which substantially enhances the sensitivity of the measurement, is generally required for all these analytical techniques. The most common extraction technique is solid phase extraction (SPE). Liquid-liquid extraction is sometimes used, and solid-phase microextraction (SPME), along with its variants, is becoming more popular. Even with the dramatic advancements in analytical technology, our understanding of the presence and behavior of this broad range of

TABLE 9-1. Analytical techniques used for analyzing pharmaceuticals and personal care products in natural waters.

Instrumentation	Compounds	Extraction Method	Detection Limits (ngL^{-1})	Reference
LC/MS				
Tandem quadrupole (ESI/APCI)	Neutral and acidic pharmaceuticals and personal care products	SPE	0.1–1.6 <1	Trenholm *et al.*, 2006; Vanderford *et al.*, 2003
Tandem quadrupole (ESI)	Anti-inflammatory pharmaceuticals	SPE	<1	Marchese *et al.*, 2003
Tandem quadrupole (ESI)	Acidic pharmaceuticals	SPE	6–200	Quintana and Reemtsma, 2004
Tandem quadrupole (ESI)	Neutral and acidic pharmaceuticals and antibiotics	SPE	5–25	Metcalfe *et al.*, 2003; Stolker *et al.*, 2004
Tandem quadrupole (ESI)	Lipid regulators	SPE	0.1–1	Metcalfe *et al.*, 2003
Ion Trap (ESI)	Antibiotics	SPE	30–70	Yang and Carlson, 2004
Ion Trap (ESI)	Hormones and endocrine disruptors	SPE	0.1–20	Benijts *et al.*, 2004
Ion Trap (ESI)	Neutral and acidic pharmaceuticals	SPE	10–50	Hilton and Thomas, 2003
Time of flight (ESI)	Neutral and acidic pharmaceuticals and antibiotics	SPE	5–25	Stolker *et al.*, 2004
GC/MS				
Quadrupole (EI)	Synthetic and Natural Estrogens	SPE	1–3	Quintana *et al.*, 2004
Quadrupole (EI)	Bisphenol-A	SBSE	0.2–1	Kawaguchi *et al.*, 2004
Quadrupole (EI)	Estrogens	SPME	2–378	Yang *et al.*, 2006
Quadrupole (CI)	Estrogens and xenoestrogens	SPME	0.05–1	Lerch and Zinn, 2003
Ion Trap (EI)	Estrogens	SPE	<1	Noppe *et al.*, 2005
Ion Trap (EI)	Phenolic xenoestrogens	SBSE	0.5–5	Kawaguchi *et al.*, 2004
Ion Trap (EI)	Endocrine disruptors and personal care products	Liquid-Liquid	1.1–7.8	Trenholm *et al.*, 2006

potential ecologically sensitive compounds is limited by the sensitivity of the instrumentation employed.

Most observations of PPCPs in the environment have occurred in the following situations: influents to wastewater treatment, at select stages of wastewater treatment, wastewater treatment effluents, and in waters directly adjacent to wastewater treatment or other point sources. The majority of measurements have occurred in freshwater rivers and lakes with little emphasis on marine systems. It is only recently that certain marine ecosystems are being investigated (Aguera *et al.*, 2003; Atkinson *et al.*, 2003; Belfroid *et al.*, 1999; Hektoen *et al.*, 1995; Jelicic and Ahel, 2003; Oros *et al.*, 2003; Thomas and Hilton, 2004). Several reviews comprehensively discuss different aspects of the analysis of pharmaceuticals and personal care products in the environment (Debska *et al.*, 2004; Ingerslev and Halling-Sorensen, 2004; Petrovic *et al.*, 2005; Richardson, 2006; Richardson and Ternes, 2005).

Herein, we will describe the most recent advances in the measurement of pharmaceuticals, personal care products, and their associated metabolites, focusing on LC/MS, GC/MS, and the associated extraction techniques.

Liquid Chromatography/Mass Spectrometry (LC/MS)

Liquid chromatography/mass spectrometry (LC/MS) is the most popular technique for determining pharmaceuticals, personal care products, and their associated metabolites in the environment because of (1) low detection limits, especially with latest generation instrumentation; (2) no need for sample derivatization resulting in less statistical variation in the method; and (3) added selectivity and sensitivity provided by tandem-MS. Most techniques use high performance liquid chromatography (HPLC) to separate components before MS detection. This offers substantial gains in sensitivity and selectivity by further resolving individual components from potential interferents. After ionization, the mass spectrometer detects the compounds of interest by producing ions (charged particles) and using electric and/or magnetic fields to separate the ions based on their mass to charge ratio (m/z).

Several ionization techniques are used in LC/MS, including electropray ionization (ESI), atmospheric pressure chemical ionization (APCI), atmospheric pressure photoionization (APPI), and sonic spray ionization (SSI). ESI and APCI are the most prevalent. All are considered atmospheric pressure ionization techniques (API) in that effluent from the HPLC (mobile phase and the sample) is ionized at atmospheric pressure and the ions are directed into the vacuum chamber of the mass spectrometer through a small orifice.

ESI produces ions in the liquid phase (liquid phase ionization) by passing the sample through a small nebulizer under high voltage. The system can produce both positive and negative ions. APCI evaporates the sample through a heated nebulizer before the gas is ionized by a plasma (corona) discharge (gas phase ionization). Again, both positive and negative ions can be produced. APPI is similar to APCI except photons (high-energy UV) are used to ionize the analyte molecules rather than the corona discharge. SSI produces ions with no voltage or heating. HPLC effluent is passed though a supersonic spray, and charged droplets and gaseous ions are produced at atmospheric pressure because of the development of an unbalanced charge distribution during droplet formation.

Several different mass analyzers may be used for LC/MS instrumentation, specifically quadrupoles, including tandem quadrupoles, ion traps, and time of flight (TOF) analyzers. Quadrupoles, particularly tandem quadrupoles, are most commonly used for environmental analysis of pharmaceuticals and personal care products. Ion traps have some utility, as do TOF analyzers.

Quadrupole mass analyzers use four parallel electrode rods, which generate electric fields that are used to filter ions based on their mass to charge ratio. By altering the electric fields, only ions of a selected mass reach the detector. A full "mass spectrum" can be determined by scanning a low mass to high mass range. As an extension of the single quadrupole technique, tandem quadrupoles allow further enhancement of signal to noise ratios and selectivity in an analysis. Tandem quadrupoles allow a specific "parent" ion to be isolated and fragmented (by the second quadrupole) for scanning by the third quadrupole for "daughter" ions. Ion trap mass analyzers use a spherical electrode to produce an electrical field that captures ions as they enter the analyzer chamber. Trapped ions are processed by an oscillating field and are selectively released to the detector providing a "mass spectrum." Similar to a tandem quadrupole MS, the ion trap can be used to isolate a specific parent ion, fragment, and scan for the daughter ions. Linear ion traps, where a quadrupole is used to trap ions, are also becoming available. These offer the advantages of trapping specific ions like a conventional spherical ion trap with the added advantage of increased trapping capacity, which results in increased sensitivity. Time of flight (TOF) analyzers apply a constant electrical field to all ions and measure the time for the ions to move to the detector. The relation between time and charge to mass ratio gives the mass of the ion, with smaller ions traveling more quickly.

Gas Chromatography/Mass Spectrometry (GC/MS)

Even with the requirements for sample derivatization, GC/MS is still a popular method for analyzing specific classes of pharmaceuticals and personal care products, primarily hormones. Essentially analogous, to LC/MS, GC/MS requires a separation step before mass spectrometric analysis. GC/MS uses gas chromatography to separate individual compounds before detection via mass spectrometry. This requires vaporization of the sample before separation. Two of the most common ionization techniques are electron impact (EI) and chemical ionization (CI). These ionization processes occur within the mass spectrometer after the effluent from the chromatography column is passed into the ion source. Here the individual compounds are ionized and directed toward the mass filter. Chemical ionization technique uses virtually the same ion source device as in EI except CI uses an added reagent gas, which is first subjected to electron impact. Sample ions are formed by the ion-molecule reactions of the reagent gas ions and sample molecules. Positive ions and negative sample ions are formed in the CI process and are then directed toward the mass filter.

As with LC/MS several different mass analyzers may be used, but single quadrupole and ion traps are the most common with some tandem quadrupoles being used. The same mechanism of operation described earlier applies.

Several derivatization procedures have been developed that allow for the analysis of pharmaceuticals and personal care products by GC/MS, although the technique is most commonly used for synthetic and natural estrogens (Quintana *et al.*, 2004). Generally, the derivatization procedure converts the original compound(s) of interest into a less polar, more volatile, and more thermally stable species that is capable of being chromatographed via gas chromatography. The most common derivatization procedure is a silylation, which involves the replacement of an acidic hydrogen on the compound with an alkylsilyl group. The derivatization can also enhance mass spectrometric properties of derivatives by introducing characteristic ions of use with specific ionization techniques. The most prevalent silylation reagents that are used include N-methyl-N-(trimethylsilyl) trifluoroacetamide (MSTFA) and N-(tert-butyldimethylsilyl)-N-methyltrifluoroacetamide (MTBSTFA) (compare with Shareef *et al.*, 2006).

Both LC/MS and GC/MS techniques require an extraction step to isolate the analytes of interest from the aqueous sample. This step also serves to reduce other interfering matrix components and effectively concentrates the analytes by a factor of ~1000 depending on the sample size and final extracted volume. Sample sizes are usually between 500 mL and 2 L. Solid phase extraction (SPE) is the most popular extraction technique with variants of SPE, notably solid phase microextraction (SPME) and stir bar sorptive extraction (SBSE) becoming more prevalent. All are based on copolymer or reversed phase materials contained in cartridges, on surface of filter type disks, on the surface small fibers (SPME), or on the surface of small stirbars. Extraction is optimized to selectively retain compounds of interest from the sample matrix. Examples of SPE materials include alkyl modified bonded silica (LC-18), divinylbenzene-N-vinylpyrrolidone copolymers (Oasis HLB), styrene-divinylbenzene copolymers (LiChrolut EN), and OH-styrene-divinylbenzene copolymers (Isolute ENV+). The general extraction process involves adjusting the pH of the sample, passage of the sample through the cartridge or disk at an appropriate flow rate or exposing the extraction fiber or stir bar to the sample for a specific time period, washing the extraction surface with appropriate solvents to remove interferents, and final elution of the analytes of interest with an appropriate solvent that quantitatively releases the components of interest and is amenable to analysis via LC/MS or GC/MS. The extract is usually further concentrated by evaporating excess solvent to a final volume of 0.5 to 1.0 mL. Aliquots of this final extract are injected into the instrument for analysis. The most comprehensive methods to date allow the analysis of a broad range of pharmaceuticals and personal care products and rely exclusively on SPE followed by LC/MS analysis (Benijts et al., 2004; Trenholm et al., 2006; Vanderford et al., 2003; Vanderford and Snyder, 2006). These techniques all have excellent sensitivity (low ng/L) and allow for the analysis of between 27 and 35 individual compounds from a single water sample.

PPCP ACTIONS AND ECOTOXICOLOGICAL EFFECTS

The effects that result from acute or short-term exposure to either pharmaceuticals or personal care products are generally irrelevant and unrealistic with respect to environmental concentrations. In other words, concentrations of PPCPs needed to elicit an effect within a short period of time are 100 to 1000 times higher than present-day environmental concentrations, suggesting that acute toxicity is only relevant in the case of a toxic spill. What appears to be more critical to understand is the effect of chronic or long-term exposure of low, environmentally realistic concentrations on drug-specific physiological endpoints, as many aquatic organisms are continuously exposed over long periods of time or even over their entire life cycle. Chronic toxicity is usually not lethal, but it is this long-term exposure that can result in changes in the reproductive success, behavior, and general fitness of aquatic biota, all of which ultimately affect the ecosystem. Moreover, chronic, sublethal exposure of aquatic organisms to low concentrations of PPCPs, or to any toxicant for that matter, can have implications on human health, because it allows for the organism to accumulate the toxicant in its tissues, sometimes to very high levels, which could then be transferred to other aquatic organisms and finally to humans through the food chain. Whether these compounds have an effect on humans after consumption or whether PPCPs have reached high enough concentrations within aquatic biota to be a risk to human health are known in some cases but not in all; regardless, the potential risks are there.

Bioaccumulation is the uptake and retention of a compound from the water, food, or substrate and for bioaccumulation to occur the rate of uptake from all sources must be greater than the loss of the chemical from the tissues of the organism. Because of their lipophilicity, many PPCPs are readily taken up by aquatic organisms and have high bioaccumulation factors (BAFs), which is the ratio of the concentration of a chemical in the tissues of an organism to its concentration in all ambient environmental compartments in equilibrium with the organism (Neff, 2002). PPCPs as a group generally have high bioconcentration factors (BCFs), which are based on a special case of bioaccumulation in which a chemical is taken up and retained from the water alone. Bioconcentration is easier to measure and model mathematically than bioaccumulation, and most of our understanding of uptake and release of chemicals by aquatic organisms comes from bioconcentration studies. When an animal is exposed to a lipophilic compound, as is the case for many PPCPs, the compound moves into tissue lipids until equilibrium, approximated by the octanol/water partition coefficient (K_{ow}) for the chemical is reached. The log K_{ow} is a useful parameter commonly reported in toxicological studies to estimate the lipophilicity of a chemical and to predict its BCF in aquatic organisms (Tables 9-2 and 9-3). Essentially, these parameters also help to predict whether a compound has the potential to become a risk to human health. In general, a BCF > 500 or a log $K_{ow} > 4$ are considered high, and compounds with these values show an elevated potential for bioaccumulation and eventual toxicity to the organisms or to humans that consume them.

Pharmaceuticals: Mode of Actions and Effects

Unlike most other environmental toxicants, pharmaceuticals are designed to target specific metabolic and molecular pathways in humans or domesticated animals. When introduced into the environment, these compounds will likely affect the same pathways and have similar side effects in aquatic organisms (i.e., aquatic plants, invertebrates and vertebrates) as they do in mammals because the target organs or receptors are similar. However, this is not always the case. The mechanism of action of the active ingredient may be different or lacking in the aquatic organism. Unfortunately, the lack of a similar mechanism seldom results in a "noneffect" of the compound; instead, a response different

TABLE 9-2. Reported log K_{ow} values for selected pharmaceuticals (value based on chemical structure).†

Pharmaceutical	Log K_{ow}	Reference
1. NSAIDs		
-Diclofenac	4.5	Avdeef *et al.*, 1998
-Ibuprofen	4.0	Avdeef *et al.*, 1998
-Naproxen	3.3	Cleuvers, 2004
-ASA	1.2	Cleuvers, 2004
-Salicylic Acid	2.3	Hansch and Anderson, 1967
-Acetaminophen	0.5	Henschel *et al.*, 1997
2. Blockers		
-Propranolol	3.6	Hardman *et al.*, 1996
-Betaxolol	3.0	Sanderson *et al.*, 2003
-Metoprolol	2.2	Hardman *et al.*, 1996
3. Blood-lipid lowering agents		
-Clofibric acid	2.6†	http://pubchem.ncbi.nlm.nih.gov
-Gemfibrizol	4.8	Sanderson *et al.*, 2003
4. Neuroactive agents		
-Fluoxetine	4.1	Nentwig, 2007
-Sertraline	4.3	Deak *et al.*, 2006
5. Steroidal hormones		
-17α-ethinylestradiol	3.7	Hansch *et al.*, 1995
-17β-estradiol	4.0	Lundberg, 1979
-Estriol	2.5	Leszczynski and Schafer, 1990
-Estrone	3.1	Leszczynski and Schafer, 1990
6. Antibiotics		
-Oxytetracycline	−1.1	Loke *et al.*, 2002
-Erythromycin	2.8	Donovan and Pescatore, 2002
-Sulfonamide	−0.7	Hansch and Anderson, 1967
-Ciprofloxacin	−1.1	Hansch *et al.*, 1995

TABLE 9-3. Reported log K_{ow} values for selected personal care products.

Personal Care Product	logK_{ow}	Reference
1. Surfactants		
-LAS	2.5	Ying, 2006
-nonylphenol	4.5	Ahel and Giger, 1993a,b
-octylphenol	4.1	Ahel and Giger, 1993a,b
2. Musks		
-HHCB	5.9	Eschke, 1994
-AHTN	5.8	Eschke, 1994
-musk xylene	4.9	Balk *et al.*, 2001a
-musk ketone	4.3	Balk *et al.*, 2001a
3. UV filters		
-4-MBC	5.1	Buser *et al.*, 2006
-OC	6.9	Buser *et al.*, 2006
-BP-3	3.8	Balk *et al.*, 2001b
-EHMC	6.0	Balk *et al.*, 2001b

than anticipated is usually measured. Because the specific modes of action in aquatic organisms are not well known for many drugs compounds, toxicity analysis of pharmaceuticals in aquatic organisms has been a challenge. We will now go into some detail of the mode of action and toxicity of the different pharmaceutical groups. However, one should keep in mind that environmental contamination of only one pharmaceutical or personal care product seldom occurs; instead, a contaminated environment usually consists of mixtures of toxicants.

Nonsteroidal Anti-inflammatory Drugs

Nonsteroidal anti-inflammatory drugs (NSAIDs) such as acetylsalicylic acid and salicylate (Aspirin), ibuprofen (Advil), naproxen (Aleve), and diclofenac (Volteran) are commonly used to treat inflammation and pain, to relieve fever, and for long-term treatment of rheumatic diseases. They act by inhibiting one or both of the two isoforms of the cyclooxygenase enzyme (COX-1 and COX-2) that catalyze the synthesis of different prostagladins from arachidonic acid (Vane and Botting, 1998). Thus, NSAIDs inhibit the synthesis of prostaglandins, which are directly involved in the inflammatory and pain response, regulation of blood flow, vascular permeability, coagulation and synthesis of protective gastric mucosa, and kidney function (Smith, 1971; Vane, 1971). Classic NSAIDs inhibit both COX-1 and COX-2 to differing degrees, whereas new NSAIDs act more selectively on COX-2, the form responsible for inflammation. Most side effects of these compounds, such as kidney, stomach, and liver damage, are related to the physiological function of prostaglandins (Sanchez *et al.*, 2002). In the kidney, prostaglandins are involved in maintenance of the equilibrium between vasoconstriction and vasodilation of the blood vessels that supply glomerular filtration. Renal damage and failure after chronic NSAID treatment seems to be triggered by the lack of prostaglandins in vasodilation-induction. Gastric damage is thought to be caused by inhibition of both COX isoforms (Wallace, 1997; Wallace *et al.*, 2000). In contrast, liver damage is apparently caused by buildup of reactive metabolites rather than inhibition of prostaglandin synthesis (Bjorkman, 1998). Prostaglandins are formed in a diverse range of vertebrates and invertebrates. In lower aquatic organisms such as corals, prostaglandin synthesis is independent of COX, involving other enzymes (Song and Brash, 1991). However, COX-1 and COX-2 have been molecularly characterized in both teleost and elasmobranch fishes (Roberts *et al.*, 2000; Zou *et al.*, 1999) and show high homology to their human counterparts.

NSAIDs are widely used and consequently among the most concentrated pharmaceuticals in the environment. For this reason, much research has been done on the acute and chronic effects of NSAIDs in aquatic organisms (Buser *et al.*, 1998b; Fent *et al.*, 2006; Gross *et al.*, 2004; Kolpin *et al.*, 2002; Ternes, 1998). In general, toxicity data vary for each NSAID, but diclofenac appears to have the highest toxicity among the other pharmaceuticals of this class. Acute and more long-term diclofenac exposure has been shown to affect aquatic plants and invertebrates (Cleuvers, 2004; Ferrari *et al.*, 2003, 2004). In fish, side effects similar to those outlined for humans are evident—that is, liver pathology, kidney lesions, and gastric damage have been measured after chronic diclofenac exposure at concentrations that were only slightly higher than those found in the environment (Hoeger *et al.*, 2005; Schwaiger *et al.*, 2004; Triebskorn *et al.*, 2004). Significant alterations have also been measured in trout gill after chronic diclofenac exposure (Hoeger *et al.*, 2005; Schwaiger *et al.*, 2004), demonstrating that NSAIDs can have effects on nontarget organisms that are considerably nonmammalian and are therefore unpredictable. The chronic impact of NSAIDs on fish gill, a nonmammalian organ, could potentially affect oxygen supply, mobility, and ultimately behavior, reproduction, and survival.

Although most of the toxicological work has been done on the effects of diclofenac, other NSAIDs have been shown to substantially affect aquatic biota as well. Acetylsalicylic acid and salicylate have been shown to affect reproduction in aquatic invertebrates and the stress response in trout and tilapia (Gravel and Vijayan, 2006; Marques *et al.*, 2004; Van Anholt *et al.*, 2003). Naproxen appears to be acutely toxic at very high, albeit environmentally unrealistic, concentrations to algae and aquatic invertebrates; however, chronic exposure to low levels of naproxen caused inhibition of population growth in aquatic invertebrates (Isidori *et al.*, 2005a). In fish, acute exposure to relatively low levels of ibuprofen induces heat shock protein 70, which works in defense against stressor-mediated proteotoxicity (Gravel and Vijayan, 2006).

Very low amounts of diclofenac and ibuprofen have been detected in drinking water (Ternes, 1998). However, a more proximate risk to human health both directly and indirectly may be through trophic transfer. NSAIDs, especially diclofenac and ibuprofen, have relatively high log K_{ow} values suggesting a high probability of bioaccumulation (Table 9-2; Avdeef *et al.*, 1998). Along these lines, diclofenac has been revealed as the cause for the unusually high death rate of three species of vultures in India and Pakistan because of renal failure and visceral gout, a consequence of the accumulation of uric acid through the body cavity following kidney malfunction (Oaks *et al.*, 2004; Prakash *et al.*, 2003; Risebrough, 2004). Exposure to diclofenac was through feeding on the carcasses of domestic cattle treated with a normal veterinary dose of drug, which then accumulated in their tissues. Because the lowest observed effect concentration (LOEC) for diclofenac was low in vultures, concern over the high toxicity of this NSAID led India to ban its use in 2005. However, the decline of the vulture populations has had major ecological repercussions that now threaten human health. Specifically, feral dog and rat populations have increased as a consequence of reduced competition over their common food source. As a result, the risk of disease such as rabies has also increased (Dorne *et al.*, 2007). Furthermore, there are social concerns; in smaller Indian communities vultures are used to dispose of human corpses in burying rituals.

Beta-Blockers

The adrenergic system is involved in many physiological functions such as regulation of the heart function, vasodilation, and bronchodilation, as well as carbohydrate and lipid metabolism in response to stressors such as starvation (Jacob *et al.*, 1998). Beta blockers (β-blockers) act by competitively inhibiting β-adrenergic receptors, which are coupled with different G-proteins that ultimately enhance the synthesis of the second messenger signaling molecule cAMP. Thus, β-blockers such as propranolol, bisoprolol and metoprolol, are used in the treatment of high blood pressure (hypertension) and in post–heart attack patients to prevent further attacks. According to medical needs, β-blockers may selectively inhibit one or more β-receptor types and side effects are mainly bronchconstriction and disturbed peripheral circulation. β-adrenergic receptors have been found in aquatic organisms with a high degree of sequence conservation with other vertebrate homologues (Devic *et al.*, 1997; Haider and

Baqri, 2000; Nickerson *et al.*, 2001; Ruuskanen *et al.*, 2005). In fish, these receptors have been found in the heart, liver, red blood cells, and reproductive tissues (Gamperl *et al.*, 1994; Perry and Reid, 1992; Reid and Perry, 1991, 1995; Reid *et al.*, 1992). They are believed to play a similar role in these organisms as they do in humans (Nickerson *et al.*, 2001; Ruuskanen *et al.*, 2005); however, some functional differences have been measured among species; for example, some compounds that are strong β-receptor agonists in mammals will have only a weak agonistic effect in fish, suggesting that the fish receptor has a lower affinity for some of these compounds (Dugan *et al.*, 2003).

In general, β-blockers are found in slightly lower concentrations in wastewater and surface waters than NSAIDs (Fent *et al.*, 2006; Ternes, 1998). The toxicity of β-blockers has not been extensively studied, with the exception of propranolol, which shows the highest acute toxicology and log K_{ow} as compared to other β-blockers suggesting the potential for bioaccumulation (Table 9-2; Doggrell, 1990; Hardman *et al.*, 1996; Huggett *et al.*, 2002; Sanderson *et al.*, 2003). At extreme concentrations, propranolol has been shown to affect aquatic invertebrates on the level of reproduction (Huggett *et al.*, 2002). However, the concentration needed to elicit a response was dramatically above those found in the environment, suggesting that β-blockers are of little risk to invertebrate populations. In contrast, a significant decline in plasma testosterone, the total number of eggs and the number of viable eggs has been measured in the Japanese medaka, *Oryzias latispes*, after 4 weeks of propranolol exposure at relatively low concentrations, suggesting that β-blockers may pose a real risk to the health of future fish populations (Huggett *et al.*, 2002). Whether this disruption in the endocrine system in fish translates to a real risk to human health is still unclear.

Blood Lipid Lowering Agents

There are two types of blood lipid lowering agents, or antilipidemic drugs: statins and fibrates. Both are used to decrease the concentration of circulating cholesterol, but only fibrates also decrease circulating triglycerides. Fibrates have been targeted analytically more often in the aquatic environment than statins, and for that reason, we will focus on the impact of fibrates on aquatic organisms and human health. Fibrates activate the lipoprotein lipase enzyme, which is mainly responsible for the conversion of very low density lipoprotein (VLDL) to high-density lipoprotein (HDL), consequently decreasing plasma triglyceride concentrations by binding to peroxisome proliferatory-activated receptors (PPARs) (Staels *et al.*, 1998). All three subtypes of PPARs are found in fish and amphibians; they show high sequence homology to their mammalian counterparts (Andersen *et al.*, 2000; Ibabe *et al.*, 2002, 2004; Leaver *et al.*, 1998; Ruyter *et al.*, 1997) and appear to respond similarly to fibrates (Ibabe *et al.*, 2005a, 2005b; Ruyter *et al.*, 1997).

Clofibric acid, which is the active metabolite from the lipid regulators clofibrate, etofibrate, and theofibrate, is believed to be the first pharmaceutical ever measured in sewage treatment effluent in the 1970s (Garrison *et al.*, 1976). Since this initial finding, clofibric acid has been found throughout the world's waters (Buser *et al.*, 1998a; Heberer and Stan, 1997; Stumpf *et al.* 1996; Ternes, 1998) and other blood lipid lowering compounds such as bezafibrate, gemfibrozil, fenobibrate, and their metabolites, have also been detected. Despite the early detection of these compounds in the environment, the toxicity of this group of compounds is not extensively reported; however, the studies that have been done indicate that clofibric acid may not be toxic to aquatic invertebrates and vertebrates at environmentally realistic concentrations (Emblidge and DeLorenzo, 2006; Nunes *et al.*, 2005). On the other hand, chronic exposure of goldfish to waterborne gemfibrizol at an environmentally realistic concentration led to a bioconcentration factor of 500 in plasma, corresponding with its high log K_{ow} (Table 9-2; Mimeault *et al.*, 2005; Sanderson *et al.*, 2003). What was even more interesting was that goldfish kept in water with undetectable levels of gemfibrizol had measurable quantities of the compound in the blood, emphasizing the high capacity for this drug to bioaccumulate. Plasma testosterone was reduced by 50% after exposure to low gemfibrizol concentrations, suggesting that this compound may be acting as an endocrine disruptor in fish, similar to the effects of propranolol on Japanese medaka (Huggett *et al.*, 2002), and could have an impact of the reproductive fitness of at least goldfish (Mimeault *et al.*, 2005). Whether or not these endocrinological effects apply to other aquatic organisms is not known. However, the high capacity for gemfibrizol to bioaccumulate in fish at environmentally relevant concentrations is most definitely a cause for concern for human health when considering consumption of contaminated fish and bioaccumulation in our own tissues.

Neuroactive Compounds

Although many neuroactive compounds are being released into the environment (antiepileptics, antipsychotics), much of the environmental focus has been on antidepressants. Selective serotonin reuptake inhibitors (SSRIs) are a major class of widely prescribed antidepressants that include fluoxetine (Prozac), sertraline (Zoloft), fluvoxamine (Luvox), and paroxetine (Paxil), and act by inhibiting the reuptake of the neurochemical, serotonin (5-HT; 5-hydroxytryptamine) by 5-HT-specific transporters (SERT). These transporters are not only found in neurons but also in blood platelets, lymphocytes, and other cells throughout the body. The action of SSRIs essentially mimics that of 5-HT by increasing the 5-HT level in the synaptic space or extracellular fluid, depending on the location of SERT. In humans, 5-HT directly acts on the immune and vascular systems, stimulates the stress response, alters appetite, influences behavior, and modulates sexual function.

Serotonin is a biogenic amine common in both vertebrate and invertebrate systems. SERT, the serotonin transporter that is targeted by SSRIs, as well as 12 of the 14 5-HT receptors found in mammals, have been molecularly and pharmacologically described in fish. The serotonergic system among vertebrates is believed to be well conserved and 5-HT has been documented to mediate countless physiological and behavioral processes in fish (Cerda *et al.*, 1998; Foran *et al.*, 2004; Fritsche *et al.*, 1992, 1993; Hoglund *et al.*, 2002; McDonald and Walsh, 2004; Sundin, 1995, Sundin *et al.*, 1998; Sundin and Nilsson, 2000; Winberg and Lepage, 1998; Winberg and Nilsson, 1993; Winberg *et al.*, 1991; Wood *et al.*, 2003). However, 5-HT also plays a major role in many physiological regulatory processes in aquatic invertebrates. In mollusks, reproduction functions including spawning, oocyte maturation, and parturition are regulated by 5-HT in addition to heartbeat rhythm, feeding/biting, swimming and motor patterns, beating of cilia, and induction of larval metamorphosis (Couper and Leise, 1996; Fong, 1998; Fong *et al.*, 1998).

Of all the pharmaceuticals released into the environment, fluoxetine has been shown to be one of the most potentially toxic human drugs to aquatic species (Brooks *et al.*, 2003a, 2003b). Fluoxetine has been found in streams at relatively low concentrations (Kolpin *et al.*, 2002), and fluoxetine and its metabolites have been detected in fish sampled from the wild (Brooks *et al.*, 2005), demonstrating a capacity for bioaccumulation as reflected in its high log K_{ow} (Table 9-2). In addition to accumulation in the liver, substantial accumulation of fluoxetine has been measured in the brain, which could have neurological and behavioral effects, and muscle, indicating the potential for human health effects on the consumption of fish. In general and common to most other pharmaceuticals studied, SSRIs appear to disrupt the endocrine system in aquatic organisms and changes in reproduction, fecundity and/or hormone levels have been measured in aquatic crustaceans (Brooks *et al.*, 2003b; Henry *et al.*, 2004), mollusks (Fong, 1998; Fong *et al.*, 1998), and fish (Brooks *et al.*, 2003a, b; Foran *et al.*, 2004) after chronic exposure. However, because 5-HT is potentially the most ubiquitous neurochemical in the body, it is predicted that the effects of sublethal SSRI exposure are extensive and the goal of future studies should be to pursue this hypothesis.

Steroidal Hormones

Although many types of synthetic hormone modulators are used for therapeutic purposes such as androgen hormone inhibitors and thyroxine analogs that mimic thyroid hormones, the steroid metabolites that have proven to be most polluting to the aquatic environment are the main active ingredients of contraceptive pills; 17α-ethinylestradiol (EE2), 17 β-estradiol (E2), estriol, and estrone. These synthetic xenoestrogens are also used in estrogen-replacement therapy, in veterinary medicine and in aquaculture. Steroid hormones are fat-soluble and have fairly high log K_{ow} values (Table 9-2); they readily cross the cell membrane to interact with receptor proteins; in the case of estrogens, these are ER-α and ER-β. Evidence suggests a third class, putative ER or ER-γ, is involved in fish and possibly mammals (Dodge *et al.*, 1996; Menuet *et al.*, 2002). The steroid binds to the receptor and the steroid-receptor complex binds to target regions of DNA termed "response elements" activate a cascade of reactions.

Many waterborne pharmaceuticals exert negative effects on the endocrine system of aquatic organisms and are deemed "endocrine disruptors," even though this is not their primary mode of action in humans nor is it their predicted effect in the aquatic environment. What makes steroid hormones unique in this sense is that their target in humans is in fact the endocrine system. For this reason, this class of compounds has had a substantial impact on the aquatic environment and poses an undeniable risk to human health (McLachlan, 2001; McLachlan *et al.*, 2001). Synthetic steroids are frequently prescribed as oral contraceptives, but because of their high pharmacological potency and our high sensitivity, the total amounts annually sold are low compared to other pharmaceuticals. Subsequently, the amount that is measured in aquatic systems is generally an order of magnitude lower than all other pharmaceuticals (Heberer, 2002). Unfortunately, like us, fish and other aquatic organisms are an order of magnitude more sensitive to these compounds; exposure to only 0.1 ng/l of EE2 can provoke the formation of testis-ova or the complete transformation of testis to ovary in some species of male fishes (Purdom *et al.*, 1994), a phenomenon that was first observed for fish in sewage treatment lagoons in the mid-1980s (Routledge *et al.*, 1998). Thus, despite their low environmental concentrations, these compounds are still a cause for concern because of their LOECs.

Over the past several decades, there has been a strong focus on the negative impact of steroid hormones in the environment and the findings of this research effort cannot be summarized completely in this chapter (see Falconer *et al.*, 2006; Hutchinson *et al.*, 2000; Porte *et al.*, 2006). In general, most of the estrogenic metabolites tend to result in a similar ecological effect, which is a significant reduction in reproduction, whether it be due to a decrease in reproductive behavior and fecundity (Nash *et al.*, 2004), an increase in feminization (Hirai *et al.*, 2006; Metcalfe *et al.*, 2001) or a decrease in fertilization rate (Van den Belt *et al.*, 2001). While the changes in reproductive abilities of aquatic biota are devastating ecologically and will ultimately have a global impact if left unchecked, it is unclear whether these changes transfer directly to human health. Environmental estrogens have been linked to the recent decline in the human sex ratio (World Health Organization/International Program on Chemical Safety [WHO/IPCS], 2002), but the evidence for this association is not conclusive (Jongbloet *et al.*, 2001). However, research and discussion support the implication that environmental pollution is leading to a decreased number of male births (Hood, 2005; Mackenzie *et al.*, 2005).

Antibiotics

The widespread and indiscriminant use and misuse of antibiotics is the leading proposed cause of the increased and spreading resistance among bacteria. In general, drug resistance in bacteria is not believed to develop in surface waters, because environmental concentrations of antibiotics are not high enough to be toxic and promote the selection of resistant bacteria. However, once the introduction of antibiotics into the environment reaches levels that are toxic, those bacteria that survive the encounter will multiply, and quickly a resistant strain is developed. So while the current level of disposal and excretion of antibiotic drugs into wastewater does not cause bacterial resistance, whether it serves to maintain bacterial resistance remains unknown. Veterinary and animal husbandry as well as aquaculture play a major role in the introduction of antibiotics into the environment. In fish farms, 70% to 80% of the antibiotics administered as medicated food pellets are released into the aquatic environment via urinary and fecal excretion and in uneaten food. The human health implications of these pharmaceuticals in the environment are many-fold; the detrimental impact of widespread antibiotic resistance could lead to epidemics or the return of infection by bacteria that were believed to be eradicated. More directly, the use of antibiotics in animal husbandry and aquaculture results in eventual human consumption.

Antibiotics work either by chemically interfering with the bacteria's ability to reproduce (bacteriostatic) or by actually killing the bacteria (bactericidial). The macrolides, such as erythrocmycin, are active against gram-positive bacteria whereas the tetracyclines have a broad spectrum of antibacterial activity (i.e., target both gram-negative and gram-positive bacteria); both types of antibiotic inhibit protein synthesis and bacterial growth. The sulphonamides (also known as sulfa drugs), extensively used in the past but less so now because of their harmful side effects, limit the survival of bacteria by interfering with the production and usage of folic acid. The quinolones or fluoroquinolones (i.e., ciprofloxacin) are one of the most important groups of synthetic antibiotics used in aquaculture inhibiting key bacterial enzymes (DNA gyrase and topoisomerase IV) involved in unwinding the DNA helix for replication and transcription, thus preventing DNA replication, repair, recombination and transposition. Early generations of these antibiotics targeted only gram-negative bacteria, but later generations have enhanced antimicrobial activity against gram-negative and have expanded their activity to include gram-positive and anaerobic bacteria.

Although it is well documented that antibiotics affect the gene pool of microorganisms resulting in resistance, the effects of waterborne antibiotics on nontarget aquatic organisms are not well understood. As observed for most other pharmaceutical compounds, studies indicate that antibiotics can be acutely toxic to aquatic organisms, but only at concentrations that far exceed those found in the environment (Isidori *et al.*, 2005b; Robinson *et al.*, 2005). On a more chronic level, both erythromycin and tetracycline have been shown to affect the growth of algae and aquatic invertebrates when exposed to relatively low concentrations (Isidori *et al.*, 2005b; Pomati *et al.*, 2004), suggesting some potential for harmful effects with long-term exposure at environmental concentrations and implications for the ecosystem as a whole. However, the log K_{ow} values of these compounds are very low (Table 9-2), indicating a low capacity for bioaccumulation within biota and suggesting that the greatest risk to human health with respect to antibiotics in the environment is antibiotic resistance.

Personal Care Products: Concentrations and Ecotoxicity

The active ingredients in personal care products differ from pharmaceuticals in their biological activity as they do not have a specific physiological mode of action. Thus, the environmental impact of these compounds is difficult to predict. Nonetheless, many of the active ingredients in personal care products are toxic when organisms are exposed to high enough doses, but perhaps even more surprising are the potential risks we might be exposing ourselves to when using these seemingly harmless compounds.

Surfactants

Surfactants are a diverse group of chemicals that are designed to have cleaning or solubilization properties. They generally consist of a polar head group (either charged or uncharged), which easily dissolves in water, and a nonpolar hydrocarbon tail, which does not dissolve easily in water. Hence, surfactants combine hydrophobic and hydrophilic properties in one molecule. Because of their hydrophobicity, this group of chemicals as a whole has high log K_{ow} values (Table 9-3). Synthetic surfactants are widely used in household cleaning detergents, personal care products, textiles, paints, polymers, pesticide formulations, pharmaceuticals, mining, oil recovery, and the pulp and paper industry (reviewed by Ying, 2006). They are divided into anionic, nonionic, and cationic surfactants. Linear alkylbenzene sulphonates (LASs) are the most popular anionic surfactants and have been used since the 1970s with an estimated global consumption of 2.8 million tons in 1998 (Cserhati *et al.*, 2002; Verge and Moreno, 2000; Ying, 2006). Alkylphenol ethoxylates are common nonionic surfactants that have relatively stable biodegradation products nonylphenol and octylphenol, which have been shown to be toxic to both marine and freshwater aquatic organisms (Comber *et al.*, 1993; Mcleese *et al.*, 1981). Quaternary ammonium based surfactants are cationic surfactants and are used mainly as fabric softeners and antiseptic agents in detergents.

After use, surfactants and their degradation products are discharged to sewage treatment plants or directly to surface waters. Because of their widespread use and high consumption, surfactants and their degradation products have been detected at various concentrations in surface waters, sediments, and sludge-amended soils. Aquatic toxicity data are widely available for surfactants, and, in general, acute and chronic toxicity is achieved by exceedingly high concentrations of surfactant. Surfactants have shown biological activity either by binding to various bioactive macromolecules such as starch (Merta and Stenius, 1999), proteins (Nielsen *et al.*, 2000), peptides and DNA (Lindman *et al.*, 2000) or by inserting into various cell fragments (i.e., phospholipid membranes) causing malfunction (Cserhati *et al.*, 2002). Symptoms of toxicity include nonspecific narcosis in tadpoles (Mann and Bidwell, 2001), hypertrophied lamellar gill epithelia, reduced swimming capacity (Hofer *et al.*, 1995), and estrogen disruption in rainbow trout (Jobling and Sumpter, 1993; Jobling *et al.*, 1996; Pedersen *et al.*, 1999), and reduced egg production and larval survival in fathead minnows (Dorn *et al.*, 1997; Wong *et al.*, 1997). Because sewage sludges, which are found to contain high concentrations of contaminants including surfactants, are being applied on agricultural lands as fertilizer, the terrestrial environment has become a significant sink for these compounds.

Musks

Musk is one of the most important and often used fragrances in perfumery; however, the high cost and limited supply of natural musk has forced the production of synthetic compounds with similar properties to the natural substances. These compounds are not only used in perfume but are employed in fragrance body lotions, laundry detergents, air fresheners, antiperspirants, soaps, hair products, household sanitation products, and other scented personal products at concentrations of up to several milligrams per gram of product (Wong, 2006). Presently, the main synthetic musks are separated into two groups: nitro musks and polycyclic musks. The worldwide production of musks increased from about 7000 to 8000 metric tons/year between 1987 and 1996 with a concurrent production shift from nitro musks to polycyclic musks (Rimkus, 1999). The most widely used polycyclic musks are HHCB (1,3,4,6,7,8-hexahydro-4,5,5,7,8,8-hexamethylcyclopenta[g]-2-benzopyran) and AHTN (7-acetyl-1,1,34,4,6-hexamethyl-1,2,3,4-tetrahydronapthalene), whereas musk xylene (1-tert-butyl-3,5-dimethyl-2,4,6-trinitrobenzene) and musk ketone (4-terrt-butyl-2,6-dimethyl-3,5-dinitrophenylethanon) are the most common nitro musks; however, musk xylene was discontinued in Japan and a voluntary ban is in force in Germany (Käfferlein *et al.*, 1998). Although the use of musk xylene is still high in the United States, it is banned in products with a risk of oral uptake (Luckenbach and Epel, 2005). The actual concentrations of synthetic musks in consumer products will differ as a function of the formulations and may vary by up to five orders of magnitude (Reiner and Kannan, 2006). These compounds are persistent, relatively inexpensive to synthesize, and have a high affinity to bind to fabrics (Reiner and Kannan, 2006; Rimkus, 1999; Sommer and Juhl, 2004).

Daily use and elimination of synthetic musk-containing household products down the drain results in the discharge of these compounds into the sewage system (Bester, 2005; Reiner and Kannan, 2006), and it is this route that is considered to be the primary environmental pathway (Reiner and Kannan, 2006; Salvito *et al.*, 2004). The presence of musks in the aquatic environment was first measured in the 1980s (Yamagishi *et al.*, 1981, 1983); since then, investigations of the distributions of these chemicals in the environment have shown them to be present in low concentrations in surface waters and seawater (Bester *et al.*, 1998; Verbruggen *et al.*, 1999) but two orders of magnitude higher in the sediment (Winkler *et al.*, 1998). These compounds have been detected in aquatic biota such as fish, mussels, shrimp, and benthic invertebrates (Artola-Garicano *et al.*, 2003; Rimkus, 1999), but because acute and chronic toxicity thresholds for musks in invertebrate and fish species are much higher than the environmentally measured levels (Balk and Ford, 1999; Behechti *et al.*, 1998; Breitholtz *et al.*, 2003; Chou and Dietrich, 1999; Wollenberger *et al.*, 2003), the environmental risks posed by musks are assumed to be low. However, since musks are more concentrated in sediment (Winkler *et al.*, 1998), there is likely a greater risk in bottom-dwelling animals (Luckenbach and Epel, 2005). The mode of action of these compounds ranges from necrosis in benthic invertebrates (Artola-Garicano *et al.*, 2003), multidrug transporter inhibition in mussel (Luckenbach and Epel, 2005), disruptions in early-life stage development in zebrafish (Carlsson and Norrgren, 2004), and possible endocrine disruption in trout and frog (Chou and Dietrich, 1999). Furthermore, the hydrophobicity of musks results in bioaccumulation and levels in tissues of aquatic organisms that are 10^1 to 10^4-fold above environmental levels (Table 9-3; Gatermann *et al.*, 2002; Rimkus, 1999).

Most humans are voluntarily exposed to polycyclic musks on a daily basis through application to the skin, and consequently these compounds have been measured in human adipose tissue, mother's milk, and blood plasma at concentrations that are one to two orders of magnitude below the effective concentrations seen in the previously mentioned studies on aquatic organisms (Käfferlein *et al.*, 1998; Rimkus and Wolf, 1996). Based on the daily dermal exposure of musks in most humans, it is unlikely that these compounds are significantly affecting humans through consumption of contaminated seafood or fish; however, inhalation exposure cannot be ruled out as polycyclic musks have also been measured in air (Kallenborn *et al.*, 1999; Reiner and Kannan, 2006). These compounds have been shown to act as potential endocrine disruptors in humans; however, the risk appears to be relatively low (Bitsch *et al.*, 2002).

UV filters

UV filters in sunscreen products and cosmetics protect the skin from damage by blocking out harmful UV radiation; some filters protect against UVB irradiation at 280 to 315 nm and others offer protection against wavelengths in the range 315 to 400 nm. There are two basic types of UV filter: inorganic filters (such as zinc oxide) work primarily by reflecting, scattering, and absorbing UV light, whereas organic filters (4-MBC, 3-(4′-methyl-benzylidene)bornan-2-one; 4-methylbenzylidene camphor; OC, octocrylene; BP-3, benzophenone-3; and EHMC, ethylhexyl methoxy cinnamate) absorb UV light. Organic UV filters are added in concentrations of up to 10% in sunscreen products for skin protection. They are also included in other cosmetics such as beauty creams, lipsticks, skin lotions, hair sprays, hair dyes, shampoos, and bubble baths for product stability and durability. The trend since the 1990s is a greater usage of sunscreen products with higher sun protection factors (SPF > 20) as a result of the growing concern about UV radiation and skin cancer. A higher SPF means a higher concentration of UV filters within the product; typically two or more compounds are used to protect against UVA and UVB, resulting in an increase in the release of these compounds into the environment.

UV filters can enter the environment directly from skin during swimming and indirectly via wastewater by removal of sunscreen residues after bathing or renal excretion after oral uptake (in the case of lipsticks) or absorption through the skin. Like other personal care products, UV filters are highly lipophilic, have fairly high log K_{ow} values, and bioconcentration of these compounds in several fish species has been measured (Table 9-3; Balmer *et al.*, 2005; Buser *et al.*, 2006; Nagtegaal *et al.*, 1997; Schlumpf *et al.*, 2001). These compounds appear to have estrogenic effects; 10 of 23 UV filters tested *in vitro* on rainbow trout estrogen receptors were found to possess estrogenic activity, and three of eight showed estrogenic affects *in vivo* (Kunz *et al.*, 2006). At the concentrations used in this study compared to environmental concentrations of UV filters, exposure to a single UV filter would probably not pose a hazard to fish. However, different UV filters may act additively (Heneweer *et al.*, 2005) as indicated for other endocrine disruptors (Routledge *et al.*, 1998). In addition, long-term exposure to UV filters may affect fish reproduction at much lower concentrations. Evidence suggests that UV filters may significantly contribute to the total body burden of endocrine active compounds (which include pharmaceuticals addressed but also include insecticides, herbicides, PCBs, etc.) in wildlife and may play an ecotoxicologic role in humans through trophic transfer (Schlumpf *et al.*, 2001).

In addition to potential indirect exposure through the consumption of contaminated fish, humans are directly exposed to UV filters by dermal absorption (Aghazarian *et al.*, 1999; Hagedorn-Leweke and Lippold, 1995; Hayden *et al.*, 1997; Jiang *et al.*, 1999). The UV filter, BP-3 and its metabolite have been detected in human urine within 4 hours of application of commercially available sunscreen products to the skin (Felix *et al.*, 1998; Hayden *et al.*, 1997). BP-3 has also been found to be readily absorbed from the gastrointestinal tract (Kadry *et al.*, 1995). Evidence suggests that several UV filters, including BP-3 and 4-MBC, have estrogenic effects in human and rat suggesting an endocrine disruptor function (Schlumpf *et al.*, 2001), the same UV filters that have been found to accumulate in fish and in human milk (Balmer *et al.*, 2005; Hany and Nagel, 1995).

Other Compounds

Many other compounds that fall under the category of pharmaceutical and personal care products have been measured in significant quantities in the environment. Since the 1990s, there has been a trend to add antimicrobials, such as triclosan, to soaps, deodorants, skin creams, toothpastes, and detergents; air fresheners now fight "odor causing bacteria," and antimicrobials are found in shoes, kitchen utensils, and children's toys. Caffeine is so ubiquitous in the environment it can serve as an anthropogenic marker in aquatic systems. Nicotine, antacids, diuretics, X-ray contrast media, antidiabetic compounds, antiepileptic compounds, impotence drugs, and antitumor agents are also found in the environment and could potentially pose a risk to human health (Daughton and Ternes, 1999; Fent *et al.*, 2006; Kolpin *et al.*, 2002; Ternes, 1998; Weigel *et al.*, 2002, 2004).

One should also keep in mind that there is no such environment where only one PPCP is found and aquatic organisms in contaminated environments are chronically exposed to low levels of many of these compounds over the course of their lifetime. Although several studies have looked at the effects of PPCP mixtures on aquatic biota, this concept is still poorly understood and studied. However, one might anticipate synergism by some of these compounds, where an effect can be elicited at orders of magnitude lower concentration that can be predicated by additive action, especially with PPCPs that have similar mode of action or toxic effect, such as chemicals from the same class of pharmaceutical or compounds that show the same ability to disrupt the endocrine system (Arnold *et al.*, 1996, 1997; Falconer *et al.*, 2006; Gaido *et al.*, 1997; McLachlan, 1997; Routledge *et al.*, 1998). At the same time, to evaluate the real ecological impact of any therapeutic pharmaceutical on the aquatic environment and the potential risk to human health, one needs also to consider other factors that will increase or reduce potential toxicity such as their concentrations at the time of maximal discharge during the day, their presence in the sediments, seasonal changes in water temperature, and their stability in water. Abiotic degradation such as hydrolysis and photolysis, the formation of metabolites via urine excretion, and microbial transformations in sewage treatment plants sometimes result in the formation of by-products that are more harmful than the parent compounds. On the flip side, these compounds may bind or form complexes with suspended matter or ions, which will reduce their bioavailability or toxicity.

PREVENTING ENVIRONMENTAL PPCP CONTAMINATION

Essentially, the most straightforward way to reduce the quantity of pharmaceuticals in the environment is to limit their consumption. Educating health care practitioners to ensure they fully understand the importance and environmental implications of selecting the right medication and therapy for each patient is one way to accomplish such reductions. Identifying the lowest effective dosage on an individual basis could also minimize the volume of pharmaceutical waste (Daughton, 2003). The packaging of PPCPs should be considered and reduced, especially of those more prone to being thrown out because they are purchased in quantities too great to be used before they expire. Educating patients of the importance of completing treatments and following their physician's directions precisely will ensure the proper dosing of pharmaceuticals.

In general, unused medications are either disposed of in the trash, are flushed down the toilet or sink, or are shared with other individuals (Kuspis and Krenzelok, 1996; Seehusen and Edwards, 2006). These methods do not only lead to detrimental effects on environmental health, but they have can have a harmful effect on human health directly. Australia, Canada, and many nations within the European Union are the front-runners in taking proactive measures such as unwanted medication collection events and pharmaceutical take-back legislation to reduce the harmful effects of improper PPCP disposal on human and environmental health. Essentially, these programs provide the legal framework and resources required to allow health care facilities, patients, and the public to return unwanted and expired pharmaceuticals where they can be reused or disposed of by incineration. These programs not only reduce the amount of pharmaceutical waste introduced to the environment by keeping these compounds out of landfills and the water supply, but these efforts help to avoid dosing errors, drug abuse, and accidental poisonings that result in the accumulation of unused pharmaceuticals in the home, and they foster patient privacy and prevent identity left by keeping medication vials out of landfills where personal information could be discovered.

References

Aghazarian, V., Tchiakpe, L., Reynier, J.P., Gayte-Sorbier, A., 1999. Release of benzimidazole and bezylidene camphor from topical sunscreen formulations. Drug Dev. Ind. Pharm. 25, 1277–1282.

Aguera, A., Fernandez-Alba, A., Piedra, L., Mezcua, M., Gomez, M., 2003. Evaluation of triclosan and biphenylol in marine sediments and urban wastewaters by pressurized liquid extraction and solid phase extraction followed by gas chromatography mass spectrometry and liquid chromatography mass spectrometry. Anal. Chim. Acta 480, 193–205.

Ahel, M., Giger, W., 1993a. Aqueous solubility of alkylphenols and alkylphenol polyethoxylates. Chemosphere 26, 1461–1470.

Ahel, M., Giger, W., 1993b. Partitioning of alkylphenols and alkylphenol polyethoxylate between water and organic solvents. Chemosphere 26, 1471–1478.

Andersen, O., Eijsink, V.G.H., Thomassen, M., 2000. Multiple variants of the peroxisome proliferator-activated receptor (PPAR) gamma are expressed in the liver of Atlantic salmon (*Salmo salar*). Gene 255, 411–418.

Arnold, S.F., Bergeron, J.M., Tran, D.Q., Collins, B.M., Vonier, P.M., Crews, D., Toscano, W.A.Jr., Mclachlan, J.A., 1997. Synergistic responses of steroidal estrogens *in vitro* (yeast) and *in vivo* (turtles). Biochem. Biophys. Res. Commun. 235, 336–342.

Arnold, S.F., Klotz, D.M., Collins, B.M., Vonier, P.M., Guilette, L.J., Jr., Mclachlan, J.A., 1996. Synergistic activation of estrogen receptor with combinations of environmental chemicals. Science 272, 1489–1492.

Artola-Garicano, E., Sinnige, T.L., van Holsteijn, I., Vaes, W.H.J., Hermens, J.L.M., 2003. Bioconcentration and acute toxicity of polycyclic musks in two benthic organisms (*Chironomus riparius* and *Lumbriculus variegatus*). Environ. Toxicol. Chem. 22, 1086–1092.

Atkinson, S., Atkinson, M., Tarrant, A., 2003. Estrogens from sewage in coastal marine environments. Environ. Health Persp. 111, 531–535.

Avdeef, A., Box, K.J., Comer, J.E., Hibber, C., Tam, K.Y., 1998. pH-metic log P 10. Determination of liposomal membrane-water partition coefficients of ionizable drugs. Pharm. Res. 15, 209–215.

Balcioglu, I.A., Ötker, M., 2003. Treatment of pharmaceutical wastewater containing antibiotics by O_3 and O_3/H_2O_2 processes. Chemosphere 50, 85–95.

Balk, F., Blok, H., Salvito, D.T., 2001a. Environmental risks of musk fragrance ingredients. In Daughton, C.G., and Jones-Lepp, T.L. (eds.), Pharmaceuticals and Personal Care Products in the Environment–Scientific and Regulatory Issues. Washington, DC, American Chemical Society.

Balk, F., Ford, R.A., 1999. Environmental risk assessment for the polycyclic musks, AHTN and HHCB in the EU. II. Effect assessment and risk characterisation. Toxicol. Lett. 111, 81–94.

Balk, F., Schwegler, A., Rijs, G., Gebruik, emissie en aanwezigheid van UV-filters in zonnebrandcrèmes in watersystemen. Een verkennende studie. RIZA -werkdocument No. 2001.206x. 2001b. RIZA.

Balmer, M., Buser, H.R., Müller, M.D., Poiger, T., 2005. Occurence of some organic UV filters in wastewater, in surface waters and in fish from Swiss lakes. Environ. Sci. Technol. 39, 953–962.

Behechti, A., Schramm, K.W., Attar, A., Niederfellner, J., Kettrup, A., 1998. Acute aquatic toxicities of four musk xylene derivatives on *Daphnia magna*. Water Res. 32, 1704–1707.

Belfroid, A., Van der Horst, A., Vethaak, A., Schafer, A., Rijs, G., Wegener, J., Cofino, W., 1999. Analysis and occurrence of estrogenic hormones and their glucuronides in surface water and waste water in The Netherlands. Sci. Total. Environ. 225, 101–108.

Benijts, T., Lambert, W., De Leenheer, A., 2004. Analysis of multiple endocrine disruptors in environmental waters via wide-spectrum solid-phase extraction and dual-polarity ionization LC-ion trap-MS/MS. Analytical Chem. 76, 704–711.

Bester, K., 2005. Polycyclic musks in the Ruhr catchment area: Transport, discharges of waste water, and transformations of HHCB, AHTN and HHCB-lactone. J. Environ. Monitor. 7, 43–51

Bester, K., Huhnerfuss, H., Lange, W., Rimkus, G.G., Theobald, N., 1998. Results of non target screening of lipophilic organic pollutants in the German Bight II: Polycyclic musk fragrances. Water Res. 32, 1857–1863.

Bitsch, N., Dudas, C., Korner, W., Failing, K., Biselli, S., Rimkus, G., Brunn, H., 2002. Estrogenic activity of musk fragrances detected by the E-screen assay using human MCF-7 cells. Arch. Environ. Contam. Toxicol. 43, 257–264.

Bjorkman, D., 1998. Nonsteroidal anti-inflammatory drug-associated toxicity of the liver, lower gastrointeinal tract and esophagus. Am. J. Med. 105, 17S-21S.

Boxall, A.B., Kolpin, D.W., Halling-Sorensen, B., Tolls, J., 2003. Are veterinary medicines causing environmental risks? Environ. Sci. Technol. 37, 286A-294A.

Boxall, A.B.A., Blackwell, P., Cavallo, R., Kay, P., Tolls, J., 2002. The sorption and transport of a sulphonamide antibiotic in soil systems. Toxicol. Lett. 131, 19–28.

Breitholtz, M., Wollenberger, L., Dinan, L., 2003. Effects of four synthetic musks on the life cycle of the harpacticoid copepod *Nitocra spinipes*. Aquat. Toxicol.63, 103–118.

Brooks, B.W., Chambliss, C.K., Stanley, J.K., Ramirez, A., Banks, K.E., Johnson, R.D., Lewis, R.J., 2005. Determination of select antidepressants in fish from an effluent-dominated stream. Environ. Toxicol. Chem. 24, 464–469.

Brooks, B.W., Foran, C.M., Richards, S.M., Weston, J., Turner, P.K., Stanley, J.K., Solomon, K.R., Slattery, M., La Point, T.W., 2003a. Aquatic ecotoxicology of fluoxetine. Toxicol. Lett. 142, 169–183.

Brooks, B.W., Turner, P.K., Stanley, J.K., Weston, J.J., Glidewell, E.A., Foran, C.M., Slattery, M., La Point, T.W., Huggett, D.B., 2003b. Waterborne and sediment toxicity of fluoxetine to select organisms. Chemosphere 52, 135–142.

Buser, H.R., Balmer, M.E., Schmid, P., Kohler, M., 2006. Occurence of UV filters 4–Methylbenzylidene camphor and octocrylene in fish from various Swiss rivers with inputs from wastewater treatment plants. Environ. Sci. Technol. 40, 1427–1431.

Buser, H.R., Muller, M.D., Theobald, N., 1998a. Occurrence of the pharmaceutical drug clofibric acid and the herbicide mecoprop in various Swiss lakes and in the North Sea. Environ. Sci. Technol. 32, 188–192.

Buser, H.R., Poiger, T., Muller, M.D., 1998b. Occurrence and fate of the pharmaceutical drug diclofenac in surface waters: Rapid photodegradation in a lake. Environ. Sci. Technol. 32, 3449–3456.

Carlsson, G., Norrgren, L., 2004. Synthetic musk toxicity to early life stages of zebrafish (*Danio rerio*). Arch. Environ. Contamin. Toxicol. 46, 102–105.

Cerda, J., Reich, G., Wallace, R.A., Selman, K., 1998. Serotonin inhibition of steroid-induced meiotic maturation in the teleost *Fundulus heteroclitus*: Role of cyclic AMP and protein kinases. Mol. Reprod. Dev. 49, 333–341.

Chou, Y.-J., Dietrich, D.R., 1999. Toxicity of nitromusks in early lifestages of South African clawed frog (*Xenopus laevis*) and zebrafish (*Danio rerio*). Toxicol. Lett. 111, 17–25.

Cleuvers, M., 2004. Mixture toxicity of the anti-inflammatory drugs diclofenac, ibuprofen, naproxen, and acetylsalicylic acid. Ecotox. Environ. Safety 59, 309–315.

Comber, M.H.I., Williams, T.D., Stewart, K.M., 1993. The effects of nonylphenol on daphnia-magna. Water Res. 27, 273–276.

Cordy, G.E., Duran, N.L., Bouwer, H., Rice, R.C., Furlong, E.T., Zaugg, S.D., Meyer, M.T., Barber, L.B., Kolpin, D.W., 2004. Do pharmaceuticals, pathogens, and other organic waste water compounds persist when waste water is used for recharge? Ground Water Monit. R. 24, 58–69.

Couper, J.M., Leise, E.M., 1996. Serotonin injections induce metamorphosis in larvae of the gastropod mollusc *Ilyanassa obsoleta*. Biol. Bull. 191, 178–186.

Cserhati, T., Forgacs, E., Oros, G., 2002. Biological activity and environmental impact of anionic surfactants. Environ. Int. 28, 337–348.

Daughton, L.G., 2003. Cradle-to-cradle stewardship of drugs for minimizing their environmental disposition while promoting human health II. Drug disposal, waste reduction, and future directions. Environ. Health Perspect. III, 775–785.

Daughton, C.G., Ternes, T.A., 1999. Pharmaceuticals and personal care products in the environment: Agents of subtle change? Environ. Health Persp. 107, 907–938.

Deak, K., Takacs-Novak, K., Tihanyi, K., Noszal, B., 2006. Physicochemical profiling of anti-depressive sertraline: Solubitility, ionization, lipophilicity. Med. Chem. 2, 385–389.

Debska, J., Kot-Wasik, A., Namiesnik, J., 2004. Fate and analysis of pharmaceutical residues in the aquatic environment. Crit. Rev. Anal. Chem. 34, 51–67.

Devic, E., Paquereau, L., Steinberg, R., Caput, D., Audigier, Y., 1997. Early expression of a beta(1)-adrenergic receptor and catecholamines in *Xenopus* oocytes and embryos. Febs. Lett. 417, 184–190.

Dizer, H., Fischer, B., Sepulveda, I., Loffredo, E., Senesi, N., Santana, F., Hansen, P.D., 2002. Estrogenic effect of leachates and soil extracts from lysimeters spiked with sewage sludge and reference endocrine disrupters. Environ. Toxicol. 17, 105–112.

Dodge, J.A., Lugar, C.W., Fahey, K.J., Cullinan, G.J., Phillips, D.L., Short, L.L., Glasebrook, A.L., Sato, M., Kim, J.K., Yang, N.N., Magee, D.E., Cole, H.W., Byrant, H.U., 1996. LY326315 center dot HCl: A novel estrogen receptor modulator with selectivity for bone over uterus in rats. J. Bone Min. Res. 11, T589.

Doggrell, S.A., 1990. The Membrane Stabilizing and Beta-1–Adrenoceptor Blocking Activity of (+)-Propranolol and (–)-Propranolol on the Rat Left Atria. Gen. Pharm. 21, 677–680.

Donovan, S.F., Pescatore, M.C., 2002. Method for measuring the logarithm of the octanol-water partition coefficient by using short octadecyl-poly(vinyl alcohol) high performance liquid chromatography columns. J. Chromatogr. 952A, 47–61.

Dorn, P.B., Rodgers, J.H., Dubey, S.T., Gillespie, W.B., Lizotte, R.E., 1997. An assessment of the ecological effects of a C9–11 linear alcohol ethoxylate surfactant in stream mesocosm experiments. Ecotox. 6, 275–292.

Dorne, J.L.C.M., Skinner, L., Frampton, G.K., Spurgeon, D.J., Ragas, A. M.J., 2007. Human and environmental risk assessment of pharmaceuticals: Differences, similarities, lessons from toxicology. Anal. Bioanal. Chem. 387, 1259–1268.

Dugan, S.G., Lortie, M.B., Nickerson, J.G., Moon, T.W., 2003. Regulation of the rainbow trout (*Oncorhynchus mykiss*) hepatic beta(2)-adrenoceptor by adrenergic agonists. Comp. Biochem. Physiol. 136B, 331–342.

Emblidge, J.P., DeLorenzo, M.E., 2006. Preliminary risk assessment of the lipid-regulating pharmaceutical clofibric acid, for three estuarine species. Environ. Res. 100, 216–226.

Eschke, H.-D., 1994. Organische Zenobiotika und Schwermetalle in Fischen aud dem Einzugsgebiet der Ruhr. Ruhrwassergüte 1994, Ruhrverband, Essen, Germany, pp. 62–68.

Falconer, I.R., Chapman, H.F., Moore, M.R., Ranmuthugala, G., 2006. Endocrine-disrupting compounds: A review of their challenge to sustainable and safe water supply and water reuse. Environ. Toxicol. 21, 181–191.

Felix, T., Hall, B.J., Brodbelt, J.S., 1998. Determination of benzophenone-3 and metabolistes in water and human urine by solid-phase microextraction and quadrupole ion trap GC-MS. Anal. Chim. Acta 371, 195–203.

Fent, K., Weston, A.A., Caminada, D., 2006. Ecotoxicology of human pharmaceuticals. Aquat. Toxicol. 76, 122–159.

Ferrari, B., Mons, R., Vollat, B., Fraysse, B., Paxeus, N., Lo Giudice, R., Pollio, A., Garric, J., 2004. Environmental risk assessment of six human pharmaceuticals: Are the current environmental risk assessment procedures sufficient for the protection of the aquatic environment? Environ. Toxicol. Chem. 23, 1344–1354.

Ferrari, B., Paxeus, N., Lo Giudice, R., Pollio, A., Garric, J., 2003. Ecotoxicological impact of pharmaceuticals found in treated wastewaters: Study of carbamazepine, clofibric acid, and diclofenac. Ecotox. Environ. Safe. 56, 450.

Fong, P.P., 1998. Zebra mussel spawning is induced in low concentrations of putative serotonin reuptake inhibitors. Biol. Bull. 194, 143–149.

Fong, P.P., Huminski, P.T., D'Urso, L.M., 1998. Induction and potentiation of parturition in fingernail clams (*Sphaerium striatinum*) by selective serotonin re-uptake inhibitors (SSRIs). J. Exp. Zool. 280, 260–264.

Foran, C.M., Weston, J., Slattery, M., Brooks, B.W., Huggett, D.B., 2004. Reproductive assessment of Japanese medaka (*Oryzias latipes*) following a four-week fluoxetine (SSRI) exposure. Arch. Environ. Contam. Toxicol. 46, 511–517.

Fritsche, R., Reid, S.G., Thomas, S., Perry, S.F., 1993. Serotonin mediated release of catecholamines in the rainbow trout *Oncorhynchus mykiss*. J. Exp. Biol. 178, 191–204.

Fritsche, R., Thomas, S., Perry, S.F., 1992. Effects of serotonin on circulation and respiration in the rainbow trout *Oncorhynchus mykiss*. J. Exp. Biol. 173, 59–73.

Gaido, K.W., McDonnell, D.P., Korach, K.S., Safe, S.H., 1997. Estrogenic activity of chemical mixtures: Is there synergism? CIIT Activities 17, 1–7.

Gamperl, A.K., Wilkinson, M., Boutilier, R.G., 1994. Beta adrenoceptors in the trout (*Oncorhynchus mykiss*) heart-characterization, quantification, and effects of repeated catecholamine exposure. Gen. Comp. Endocrinol. 95, 259–272.

Garrison, A.W., Pope, J.D., Allen, F.R., 1976. Analysis of organic compounds in domestic wastewater. In Keith, C.H. (ed.), Identification and Analysis of Organic Pollutants in Water, Ann Arbor, MI, Ann Arbor Science.

Gatermann, R., Biselli, S., Huhnerfuss, H., Rimkus, G.G., Hecker, M., Karbe, L., 2002. Synthetic musks in the environment. Part 1: Peciesdependent bioaccumulation of polycyclic and nitro musk fragrances in freshwater fish and mussels. Arch. Environ. Contam. Toxicol. 42, 437–446.

Golet, E.M., Xifra, I., Siegrist, H., Alder, A.C., Giger, W., 2003. Environmental exposure assessment of fluoroquinolone antibacterial agents from sewage to soil. Environ. Sci. & Technol. 37, 3243–3249.

Gravel, A., Vijayan, M.M., 2006. Salicylate disrupts interrenal steroidogenesis and brain glucocorticoid receptor expression in rainbow trout. Toxicol. Sci. 93, 41–49.

Gross, B., Montgomery-Brown, J., Naumann, A., Reinhard, M., 2004. Occurrence and fate of pharmaceuticals and alkylphenol ethoxylate metabolites in an effluent-dominated river and wetland. Environ. Toxicol. Chem. 23, 2074–2083.

Hagedorn-Leweke, U., Lippold, B.C., 1995. Absorption of sunscreens and other compounds through human skin *in vivo*; Derivation of a method to predict maximum fluxes. Pharm. Res. 12, 1354–1360.

Haider, S., Baqri, S.S.R., 2000. beta-Adrenoceptor antagonists reinitiate meiotic maturation in *Clarias batrachus* oocytes. Comp. Biochem. Physiol. 126A, 517–525.

Hansch, C., Anderson, S.M., 1967. The effect of intramolcular hydrophobic bonding on partition coefficients. J. Org. Chem 32, 2583–2586.

Hansch, C., Leo, A., Hoekman, D., 1995. Exploring QSAR: hydrophobic, electronic and steric constants. Washington, DC, American Chemical Society.

Hany, J., Nagel, R., 1995. Nachweis von UV-filtersubstanzen in muttermilch. aus Rhenland-Pfalz. Deut. Lebensm. -Rundsch. 91, 341–345.

Hardman, J., Limbird, L., Molinoff, P., Ruddon, R., Gilman, A., 1996. Goodman and Gilman's Pharmacological Basis of Therapeutics. New York, McGraw-Hill.

Hayden, C.G.J., Roberts, M.S., Benson, H.A.E., 1997. Systemic adsorption of sunscreen after topical application. Lancet 350, 863–864.

Heberer, T., Stan, H.J., 1997. Determination of clofibric acid and N-(phenylsulfonyl)-sarco-sine in sewage, river, and drinking water. Int. J. Environ. Analytic. Chem. 67, 113–123.

Heberer, T., 2002. Occurrence, fate, and removal of pharmaceutical residues in the aquatic environment: A review of recent research data. Toxicol. Lett. 131, 5–17.

Hektoen, H., Berge, J., Hormazabal, V., Yndestad, M., 1995. Persistence of antibacterial agents in marine sediments. Aquaculture. 133, 175–184.

Heneweer, M., Musse, M., Van den Berg, J., Sanderson, T., 2005. Additive estrogenic effects of mixtures of frequently used UV filters on pS2–gene transcription in MCF-7 cells. Toxicol. Appl. Pharmacol. 208, 170–177.

Henry, T.B., Kwon, J.W., Armbrust, K.L., Black, M.C., 2004. Acute and chronic toxicity of five selective serotonin reuptake inhibitors in Ceriodaphnia dubia. Environ.l Toxicol. Chem. 23, 2229–2233.

Henschel, K.P., Wenzel, A., Diedrich, M., Fliedner, A., 1997. Environmental hazard assessment of pharmaceuticals. Regul. Toxicol. Pharmacol. 25, 220–225.

Hilton, M., Thomas, K., 2003. Determination of selected human pharmaceutical compounds in effluent and surface water samples by high-performance liquid chromatography-electrospray tandem mass spectrometry. *J. Chromatogr.* 1015A, 129–141.

Hirai, N., Nanba, A., Koshio, M., Kondo, T., Morita, M., Tatarazako, N., 2006. Feminization of Japanese medaka (*Oryzias latipes*) exposed to 17b-estradiol: Formation of testis-ova and sex-transformation during early ontogeny. Aquat. Toxicol. 77, 78–86.

Hoeger, B., Kollner, B., Dietrich, D.R., Hitzfeld, B., 2005. Water-borne diclofenac affects kidney and gill integrity and selected immune parameters in brown trout (*Salmo trutta f. fario*). Aquat. Toxicol. 75, 53–64.

Hofer, R., Jeney, Z., Bucher, F., 1995. Chronic effects of linear alkylbenzene sulfonate (LAS) and ammonia on rainbow trout (*Oncorhynchus mykiss*) fry at water criteria limits. Water Res. 29, 2725–2729.

Hoglund, E., Balm, P.H.M., Winberg, S., 2002. Stimulatory and inhibitory effects of 5–HT1A receptors on adrenocorticotropic hormone and cortisol secretion in a teleost fish, the Arctic charr (*Salvelinus alpinus*). Neurosci. Lett. 324, 193–196.

Hood, E., 2005. Are EDCs blurring issues of gender? Environ. Health. Persp. 113, A670–A677.

Huggett, D.B., Brooks, B.W., Peterson, B., Foran, C.M., Schlenk, D., 2002. Toxicity of select beta adrenergic receptor-blocking pharmaceuticals (β-blockers) on aquatic organisms. Arch. Environ. Contam. Toxicol. 43, 229–235.

Hutchinson, T.H., Brown, R., Brugger, K.E., Campbell, P.M., Holt, M., Lange, R., McCahon, P., Tattersfield, L.J., Van Egmond, R., 2000. Ecological risk assessment of endocrine disruptors. Environ. Health. Persp. 108, 1007–1014.

Ibabe, A., Bilbao, E., Cajaraville, M.P., 2005a. Expression of peroxisome proliferator-activated receptors in zebrafish (*Danio rerio*) depending on gender and developmental stage. Histochem. Cell Biol. 123, 75–87.

Ibabe, A., Grabenbauer, M., Baumgart, E., Fahimi, H.D., Cajaraville, M. P., 2002. Expression of peroxisome proliferator-activated receptors in zebrafish (*Danio rerio*). Histochem. Cell Biol. 118, 231–239.

Ibabe, A., Grabenbauer, M., Baumgart, E., Volkl, A., Fahimi, H.D., Cajaraville, M.P., 2004. Expression of peroxisome proliferator-activated receptors in the liver of gray mullet (*Mugil cephalus*). Acta Histochem. 106, 11–19.

Ibabe, A., Herrero, A., Cajaraville, M.P., 2005b. Modulation of peroxisome proliferator-activated receptors (PPARs) by PPAR alpha- and PPAR gamma-specific ligands and by 17 beta-estradiol in isolated zebrafish hepatocytes. Toxicol. in Vitro 19, 725–735.

Ingerslev, F., Halling-Sorensen, B., 2004. Evaluation of analytical chemical methods for detection of estrogens in the environment. Working Report No. 44.

Isidori, M., Lavorgna, M., Nardelli, A., Parrella, A., Previtera, L., Rubino, M., 2005a. Ecotoxicity of naproxen and its phototransformation products. Sci. Total Environ. 348, 93–101.

Isidori, M., Lavorgna, M., Nardelli, A., Pascarella, L., Parrella, A., 2005b. Toxic and genotoxic evaluation of six antibiotics on non-target organisms. Sci. Total Environ. 346, 87–98.

Jacob, S., Rett, K., Henriksen, E.J., 1998. Antihypertensive therapy and insulin sensitivity: Do we have to redefine the role of beta-blocking agents? Am. J. Hyperten. 11, 1258–1265.

Jelicic, I., Ahel, M., 2003. Occurrence of phenazone analgesics and caffeine in Croatian municipal wastewaters. Fres. Environ. Bull. 12, 46–50.

Jiang, R., Roberts, M.S., Collins, D.M., Benson, H.A.E., 1999. Absorption of sunscreens across human skin: An evaluation of commercial products for children and adults. Br. J. Clin. Pharmacol. 48, 635–637.

Jobling, S., Sheahan, D., Osborne, J.A., Matthiessen, P., Sumpter, J.P., 1996. Inhibition of testicular growth in rainbow trout (*Oncorhynchus mykiss*) exposed to estrogenic alkylphenolic chemicals. Environ. Toxicol. Chem. 15, 194–202.

Jobling, S.J., Sumpter, J.P., 1993. Detergent components in sewage effluent are weakly oestrogenic to fish: An in vitro study using rainbow trout hepatocytes. Aquat. Toxciol. 27, 361–372.

Jongbloet, P.H., Zielhuis, G.A., Groenewoud, H.M.M., Pasker-de Jong, P. C.M., 2001. The secular trends in male: Female ratio at birth in postwar industrialised countries. Environ. Health Perspect. 109, 749–752.

Kadry, A.M., Okereke, C.S., Abdel-Rahman, M.S., Friedman, M.A., Davis, R.A., 1995. Pharmacokinetics of benzophenone-3 after oral exposure in male rats. J. Appl. Toxicol. 15, 97–102.

Käfferlein, H.U., Goen, T., Angerer, J., 1998. Musk xylene: Analysis, occurrence, kinetics, and toxicology. Crit. Rev. Toxicol. 28, 431–476.

Kallenborn, R., Gatermann, R., Planting, S., Rimkus, G.G., Lund, M., Schlabach, M., Burkow, I.C., 1999. Gas chromatographic determination of synthetic musk compounds in Norwegian air samples. J. Chromatogr. 846A, 295–306.

Kawaguchi, M., Inoue, K., Yoshimura, M., Sakui, N., Okanouchi, N., Ito, R., Yoshimura, Y., Nakazawa, H., 2004. Trace analysis of phenolic xenoestrogens in water samples by stir bar sorptive extraction with *in situ* derivatization and thermal desorption-gas chromatography-mass spectrometry. J. Chromatogr. 1041A, 19–26.

Kinney, C.A., Furlong, E.T., Werner, S.L., Cahill, J.D., 2006a. Presence and distribution of wastewater-derived pharmaceuticals in soil irrigated with reclaimed water. Environ Toxicol. Chem. 25, 317–326.

Kinney, C.A., Furlong, E.T., Zaugg, S.D., Burkhardt, M.R., Werner, S.L., Cahill, J.D., Jorgensen, G.R., 2006b. Survey of organic wastewater contaminants in biosolids destined for land application. Environ. Sci. Technol. 40, 7207–7215.

Kolpin, D.W., Furlong, E.T., Meyer, M.T., Thurman, E.M., Zaugg, S.D., Barber, L.B., Buxton, H.T., 2002. Pharmaceuticals, hormones, and other organic wastewater contaminants in US streams, 1999–2000: A national reconnaissance. Environ. Sci. Technol. 36, 1202–1211.

Kunz, P.Y., Galicia, H.F., Fent, K., 2006. Comparison of *in vitro* and *in vivo* estrogenic activity of UV filters in fish. Toxicol. Sci. 90, 349–361.

Kuspis, D.A., Krenzelok, E.P., 1996. What happens to expired medications? Vet. Human Toxicol. 38, 48–49.

Leaver, M.J., Wright, J., George, S.G., 1998. A peroxisomal proliferator-activated receptor gene from the marine flatfish, the plaice (*Pleuronectes platessa*). Mar. Environ. Res. 46, 75–79.

Lerch, O., Zinn, P., 2003. Derivatisation and gas chromatography-chemical ionisation mass spectrometry of selected synthetic and natural endocrine disruptive chemicals. J. Chromatogr. 991A, 77–97.

Leszczynski, D.E., Schafer, R.M., 1990. Nonspecific and metabolic interactions between steroid hormones and human plasma lipoproteins. Lipids 25, 711–718.

Lindman, B., Marques, E., Miguel, M., Dias, R., Melnikov, S., Khan, A., 2000. Gel formation and association in systems of cataionic surfactant vesicles and oppositely charged polymers. Abstr. Pap. Am. Chem. Soc. 219, U449.

Loke, M.L., Tjomelund, J., Halling-Sorensen, B., 2002. Determination of the distribution coefficient (Log Kd) of oxytetracycline, tylosin A, loaquindox and metronidazole in manure. Chemosphere 48, 351–361.

Luckenbach, T., Epel, D., 2005. Nitromusk and polycyclic musk compounds as long-term inhibitors of cellular xenobiotic defense systems mediated by multidrug transporters. Environ. Health Perspect. 113, 17–24.

Lundberg, B., 1979. Temperature effect on the water solubility and water-octanol partition of some steroids. Acta Pharm. Suec. 16, 151–158.

Mackenzie, C.A., Lockridge, A., Keith, M., 2005. Declining sex ration in a first nation community. Environ. Health Perspect. 113, 1295–1298.

Mann, R.M., Bidwell, J.R., 2001. The acute toxicity of agricultural surfactants to the tadpoles of four Australian and, two exotic frogs. Environ. Poll. 114, 195–205.

Marchese, S., Perret, D., Gentili, A., Curini, R., Pastori, F., 2003. Determination of non-steroidal anti-inflammatory drugs in surface water and wastewater by liquid chromatography-tandem mass spectrometry. Chromatogr. 58, 263–269.

Marques, C.R., Abrantes, N., Goncalves, F., 2004. Life-history traits of standard and autochthonous cladocerans: I. Acute and chronic effects of acetylsalicylic acid. Environ. Toxicol. 19, 518–526.

McDonald, M.D., Walsh, P.J., 2004. Dogmas and controversies in the handling of nitrogenous wastes: 5–HT_2–like receptors are involved in triggering pulsatile urea excretion in the gulf toadfish, *Opsanus beta*. J. Exp. Biol. 207, 2003–2010.

McLachlan, J.A., 1997. Synergistic effect of environmental estrogens: Report withdrawn. Science 277, 459d-463d.

McLachlan, J.A., 2001. Environmental signaling: What embryos and evolution teach us about endocrine disrupting chemicals. Endocrin. Rev. 22, 319–341.

McLachlan, J.A., Newbold, R.R., Burow, M.E., Li, S.F., 2001. From malformations to molecular mechanisms in the male: Three decades of research on endocrine disrupters. Apmis 109, S1–S10.

Mcleese, D.W., Zitko, V., Sergeant, D.B., Burridge, L., Metcalfe, C.D., 1981. Lethality and accumulation of alkylphenols in aquatic fauna. Chemosphere 10, 723–730.

Menuet, A., Pellegrini, E., Anglade, I., Blaise, O., Laudet, V., Kah, O., Pakdel, F., 2002. Molecular characterization of three estrogen receptor forms in zebrafish: Binding characteristics, transactivation properties, and tissue distributions. Biol. Reprod. 66, 1881–1892.

Merta, J., Stenius, P., 1999. Interactions between cationic starch and mixed anionic surfactants. Colloid. Surface. 149A, 367–377.

Metcalfe, C.D., Metcalfe, T.L., Kiparissis, Y., Koenig, B.G., Khan, C., Hughes, R.J., Croley, T.R., March, R.E., Potter, T., 2001. Estrogenic potency of chemicals detected in sewage treatment plant effluents as determined by in vivo assays with Japanese medaka (*Oryzias latipes*). Environ. Toxicol. Chem. 20, 297–308.

Metcalfe, C.D., Miao, C.X., Koenig, B., Struger, J., 2003. Distribution of acidic and neutral drugs in surface waters near sewage treatment plants in the lower Great Lakes, Canada. Environ. Toxicol. Chem. 22, 2881–2889.

Mimeault, C., Woodhouse, A., Miao, X.S., Metcalfe, C.D., Moon, T.W., Trudeau, V.L., 2005. The human lipid regulator, gemfibrozil bioconcentrates and reduces testosterone in the goldfish, *Carassius auratus*. Aquat. Toxicol. 73, 44–54.

Nagtegaal, M., Ternes, T.A., Baumann, W., Nagel, R., 1997. UV-filtersuvstanzen in wasser und fischen. UWSF-Z. Umweltchem ökotox 9, 79–86.

Nash, J.P., Kime, D.E., Van der Ven, L.T., Wester, P.W., Brion, F., Maack, G., Stahlschmidt-Allner, P., Tyler, C.R., 2004. Long-term exposure to environmental concentrations of the pharmaceutical ethinylestradiol causes reproductive failure in fish. Environ. Health Perspect. 112, 1725–1733.

Neff, J.M., 2002. Bioaccumulation in Marine Organisms. Kidlington, Oxford, UK, Elsevier.

Nentwig, G., 2007. Effects of pharmaceuticals on aquatic invertebrates. Part II: The antidepressant drug fluoxetine. Arch. Environ. Contam. Toxicol. 52, 163–170.

Nickerson, J.F., Dugan, S.G., Drouin, G., Moon, T.W., 2001. A putative beta-adrenoceptor from the rainbow trout (*Oncorhynchus mykiss*): Molecular characterization and pharmacology. Eur. J. Biochem. 268, 6465–6472.

Nielsen, A.D., Borch, K., Westh, P., 2000. Thermochemistry of the specific binding of C12 surfactants to bovine serum albumin. Biochim. Biophys. Acta 1479, 321–331.

Noppe, H., De Wasch, K., Poelmans, S., Van Hoof, N., Verslycke, T., Janssen, C.R., De Brabander, H., 2005. Development and validation of an analytical method for detection of estrogens in water. Anal. Bioanal. Chem. 382, 91–98.

Nunes, B., Carvalho, F., Guilhermino, L., 2005. Acute toxicity of widely used pharmaceuticals in aquatic species: *Gambusia holbrooki, Artemia parthenogenetica* and *Tetraselmis chuii*. Ecotox. Environ. Safe. 61, 413–419.

Oaks, J.L., Gilbert, M., Virani, M.Z., Watson, R.T., Meteyer, C.U., Rideout, B.A., Shivaprasad, H.L., Ahmed, S., Chaudhry, M.J.I., Arshad, M., Mahmood, S., Ali, A., Khan, A.A., 2004. Diclofenac residues as the cause of vulture population decline in Pakistan. Nature 427, 630–633.

Oros, D.R., Jarman, W.M., Lowe, T., David, N., Lowe, S., Davis, J.A., 2003. Surveillance for previously unmonitored organic contaminants in the San Francisco Estuary. Mar. Pollut. Bull. 46, 1102–1110.

Pedersen, S.N., Christiansen, L.B., Pedersen, K.L., Korsgaard, B., Bjerregaard, P., 1999. *In vivo* estrogenic activity of branched and linear alkylphenols in rainbow trout (*Oncorhynchus mykiss*). Sci. Total. Environ. 233, 89–96.

Perry, S.F., Reid, S.D., 1992. The relationship between beta-adrenoceptors and adrenergic responsiveness in trout (*Oncorhynchus mykiss*) and eel (*Anguilla rostrata*) erythrocytes. J. Exp. Biol. 167, 235–250.

Petrovic, M., Hernando, M.D., Diaz-Cruz, M.S., Barcelo, D., 2005. Liquid chromatography-tandem mass spectrometry for the analysis of pharmaceutical residues in environmental samples: A review. J. Chromatogr. 1067A, 1–14.

Pomati, F., Netting, A.G., Calamari, D., Neilan, B.A., 2004. Effects of erythromycin, tetracycline and ibuprofen on the growth of *Synechocystis* sp and *Lemna minor*. Aquat. Toxicol. 67, 387–396.

Porte, C., Janer, G., Lorusso, L.C., Ortiz-Zarragoitia, M., Cajaraville, M.P., Fossi, M.C., Canesi, L., 2006. Endocrine disruptors in marine organisms: Approaches and perspectives. Comp Biochem. Physiol. 143C, 303–315.

Prakash, V., Pain, D.J., Cunningham, A.A., Donald, P.F., Prakash, N., Verma, A., Gargi, R., Sivakumar, S., Rahmani, A.R., 2003. Catastrophic collapse of Indian white-backed *Gyps bengalensis* and long-billed *Gyps indicus* vulture populations. Biol. Conserv. 109, 381–390.

Purdom, C.E., Hardiman, P.A., Bye, V.F., Eno, N.C., Tyler, C.R., Sumpter, J.P., 1994. Estrogenic effects of effluents from sewage treatment works. Chem. Ecol. 8, 275–285.

Quintana, J.B., Carpinteiro, J., Rodriguez, I., Lorenzo, R.A., Carro, A.M., Cela, R., 2004. Determination of natural and synthetic estrogens in water by gas chromatography with mass spectrometric detection. J. Chromatogr. 1024A, 177–185

Quintana, J.B., Reemtsma, T., 2004. Sensitive determination of acidic drugs and triclosan in surface and wastewater by ion-pair reverse-phase liquid chromatography/tandem mass spectrometry. Rapid Comm. Mass Sp. 18, 765–774.

Reid, S.D., Moon, T.W., Perry, S.F., 1992. Rainbow trout hepatocyte beta adrenoceptors, catecholamine responsiveness, and effects of cortisol. Am. J. Physiol. 262, R794–R799.

Reid, S.D., Perry, S.F., 1991. The effects and physiological consequences of raised levels of cortisol on rainbow trout (*Oncorhynchus mykiss*) erythrocyte beta adrenoceptors. J. Exp. Biol. 158, 217–240.

Reid, S.G., Perry, S.F., 1995. The effects of hypoxia, *in vivo*, on red blood cell beta-adrenoceptors in the rainbow trout, *Oncorhynchus mykiss*. Fish Physiol.Biochem. 14, 519–524.

Reiner, J.L., Kannan, K., 2006. A survey of polycyclic musks in selected household commodities from the United States. Chemosphere 62, 867–873.

Richardson, S., 2006. Environmental mass spectrometry: Emerging contaminants and current issues. Anal. Chem. 78, 4021–4045.

Richardson, S., Ternes, T.A., 2005. Water analysis: Emerging contaminants and current issues. Anal. Chem.78, 3807–3838.

Rimkus, G.G., 1999. Synthetic musk fragrances as a new class of environmental pollutants. Abstr. Pap. Am. Chem. Soc. 217, U787–U788.

Rimkus, G.G., Wolf, M., 1996. Polycyclic musk fragrances in human adipose tissue and human milk. Chemosphere 33, 2033–2043.

Risebrough, R., 2004. Conservation biology-fatal medicine for vultures. Nature 427, 596–598.

Roberts, S.B., Langenau, D.M., Goetz, F.W., 2000. Cloning and characterization of prostaglandin endoperoxide synthase-1 and -2 from the brook trout ovary. Mol. Cell. Endocrin. 160, 89–97.

Robinson, A.A., Belden, J.B., Lydy, M.J., 2005. Toxicity of fluoroquinolone antibiotics to aquatic organisms. Environ. Toxicol. Chem. 24, 423–430.

Routledge, E.J., Sheahan, D., Desbrow, C., Brighty, G.C., Waldock, M., Sumpter, J.P., 1998. Identification of estrogenic chemicals in STW effluent. 2. *in vivo* responses in trout and roach. Environ. Sci. Technol. 32, 1559–1565.

Ruuskanen, J.O., Laurila, J., Xhaard, H., Rantanen, V.V., Vuoriluoto, K., Wurster, S., Marjamaki, A., Vainio, M., Johnson, M.S., Scheinin, M., 2005. Conserved structural, pharmacological and functional properties among the three human and five zebrafish alpha(2)-adrenoceptors. Br. J. Pharmacol. 144, 165–177.

Ruyter, B., Andersen, O., Dehli, A., Farrants, A.K.O., Gjoen, T., Thomassen, M.S., 1997. Peroxisome proliferator activated receptors in Atlantic salmon (*Salmo salar*): Effects on PPAR transcription and acyl-CoA oxidase activity in hepatocytes by peroxisome proliferators and fatty acids. Biochim. Biophys. Acta 1348, 331–338.

Salvito, D.T., Vey, M.G.H., Senna, R.J., 2004. Fragrance materials and their environmental impact. Flav. Frag. J. 19, 105–108.

Sanchez, S., Martin, M.J., Ortiz, P., Motilva, V., Herrerias, J.M., de la Lastra, C.A., 2002. Role of prostaglandins and nitric oxide in gastric damage induced by metamizol in rats. Inflam. Res. 51, 385–392.

Sanderson, H., Johnson, D.J., Wilson, C.J., Brain, R.A., Solomon, K.R., 2003. Probabilistic hazard assessment of environmentally occurring pharmaceuticals toxicity to fish, daphnids and algae by ECOSAR screening. Toxicol. Lett. 144, 383–395.

Schlumpf, M., Cotton, B., Conscience, M., Haller, V., Steinmann, B., Lichtensteiger, W., 2001. *In vitro* and *in vivo* estrogenicity of UV screens. Environ. Health Perspect. 109, 239–244.

Schwaiger, J., Ferling, H., Mallow, U., Wintermayr, H., Negele, R.D., 2004. Toxic effects of the non-steroidal anti-inflammatory drug diclofenac Part 1: Histopathological alterations and bioaccumulation in rainbow trout. Aquat. Toxicol. 68, 141–150.

Seehusen, D.A., Edwards, J., 2006. Patient practises and beliefs concerning disposal of medications. J. Am Board Fam. Med. 19, 542–547.

Shareef, A., Angove, M.J., Wells, J.D., 2006. Optimization of silylation using N-methyl-N-(trimethylsilyl)-trifluoroacetamide, N,O-bis-(trimethylsilyl)-trifluoroacetamide and N-(tert-butyldimethylsilyl)-N-methyltrifluoroacetamide for the determination of the estrogens estrone and 17a-ethinylestradiol by gas chromatography-mass spectrometry. J. Chromatogr. 1108A, 121–128.

Smith, J.B., 1971. Aspirin selectively inhibits prostaglandin production in human platelets. Nature 231, 235.

Sommer, C., Juhl, H., 2004. Analysis of macrocyclic musks in alcohol-containing cosmetics. Deutsche Lebensmittel-Rundschau 100, 224.

Song, W.C., Brash, A.R., 1991. Purification of an allene oxide synthase and identification of the enzyme as a cytochrome P450. Science 253, 781–784.

Staels, B., Dallongeville, J., Auwerx, J., Schoonjans, K., Leitersdorf, E., Fruchart, J.G., 1998. Mechanism of action of fibrates on lipid and lipoprotein metabolism. Circulation 98, 2088–2093.

Stolker, A.A.M., Niesing, W., Hogendoorn, E.A., Versteegh, J. F.M., Fuchs, R., Brinkman, U.A.T., 2004. Liquid chromatography with triple-quadrupole or quadrupole-time of flight mass spectrometry for screening and confirmation of residues of pharmaceuticals in water. Anal. Bioanal. Chem. 378, 955–963.

Stumpf, M., Ternes, T.A., Wilkin, R.D., Rodrigues, S.V., Baumann, W., 1999. Polar drug residues in sewage and natural waters in the state of Rio de Janeiro. Brazil. Sci. Total Environ. 225, 135–141.

Sundin, L., 1995. Serotonergic vasomotor control in fish gills. Braz. J. Med. Biol. Res. 28, 1217–1221.

Sundin, L., Davison, W., Forster, M., Axelsson, M., 1998. A role of 5–HT_2 receptors in the gill vasculature of the Antarctic fish *Pagothenia borchgrevinki*. J. Exp. Biol. 201, 2129–2138.

Sundin, L., Nilsson, G.E., 2000. Branchial and circulatory responses to serotonin and rapid ambient water acidification in rainbow trout. J. Exp. Zool. 287, 113–119.

Ternes, T.A., 1998. Occurrence of drugs in German sewage treatment plants and rivers. Water Res. 32, 3245–3260.

Thomas, K., Hilton, M., 2004. The occurrence of selected human pharmaceutical compounds in UK estuaries. *Mar. Poll. Bull.* 49, 436–444.

Trenholm, R.A., Vanderford, B.J., Holady, J.C., Rexing, D.J., Snyder, S.A., 2006. Broad range analysis of endocrine disruptors and pharmaceuticals using gas chromatography and liquid chromatography tandem mass spectrometry. Chemosphere. 65, 1990–1998.

Triebskorn, R., Casper, H., Heyd, A., Eikemper, R., Kohler, H.R., Schwaiger, J., 2004. Toxic effects of the non-steroidal anti-inflammatory drug diclofenac Part II: Cytological effects in liver, kidney, gills and intestine of rainbow trout (*Oncorhynchus mykiss*). Aquat. Toxicol. 68, 151–166.

Van Anholt, R.D., Spanings, T., Koven, W., Bonga, S.E.W., 2003. Effects of acetylsalicylic acid treatment on thyroid hormones, prolactins, and the stress response of tilapia (*Oreochromis mossambicus*). Am. J. Physiol. 285, R1098–R1106.

Van den Belt, K., Verheyen, R., Witters, H., 2001. Reproductive effects of ethynylestradiol and 4t-octylphenol on the zebrafish (*Danio rerio*). Arch. Environ. Contam. Toxicol. 41, 458–467.

Vanderford, B.J., Pearson, R.A., Rexing, D.J., Snyder, S.A., 2003. Analysis of endocrine disruptors, pharmaceuticals, and personal care products in water using liquid chromatography/tandem mass spectrometry. Anal. Chem. 75, 6265–6274.

Vanderford, B.J., Snyder, S.A., 2006. Analysis of pharmaceuticals in water by isotope dilution liquid chromatography/tandem mass spectrometry. Environ. Sci. Technol. 40, 7312–7320.

Vane, J.R., 1971. Inhibition of prostaglandin synthesis as a mechanism of action for aspirin-like drugs. Nature 231, 232.

Vane, J.R., Botting, R.M., 1998. Mechanism of action of antiinflammatory drugs. Int. J. Tiss. React. 20, 3–15.

Verbruggen, E.M.J., Van Loon, W.M.G.M., Tonkes, M., Van Duijn, P., Seinen, W., Hermens, J.L.M., 1999. Biomimetic extraction as a tool to identify chemicals with high bioconcentration potential: An illustration by two fragrances in sewage treatment plant effluents and surface waters. Environ. Sci. Technol. 33, 801–806.

Verge, C., Moreno, A., 2000. Effects of anionic surfactants on *Daphnia magna*. Tenside Surfact. Det. 37, 172–175.

Wallace, J.L., 1997. Nonsteroidal anti-inflammatory drugs and gastroenteropathy: The second hundred years. Gastroenterology 112, 1000–1016.

Wallace, J.L., McKnight, W., Reuter, B.K., Vergnolle, N., 2000. NSAID-induced gastric damage in rats: Requirement for inhibition of both cyclooxygenase 1 and 2. Gastroenterology 119, 706–714.

Weigel, S., Berger, U., Jensen, E., Kallenborn, R., Thoresen, H., Hühnerfuss, H., 2004. Determination of selected pharmaceuticals and caffeine in sewage and seawater from Tromso/Norway with emphasis on ibuprofen and its metabolites. Chemosphere 56, 583–592.

Weigel, S., Kuhlmann, J., Hühnerfuss, H., 2002. Drugs and personal care products as ubiquitous pollutants: Occurence and distribution of clofibric acid, caffeine and DEET in the North Sea. Sci. Total. Environ. 295, 131–141.

WHO/IPCS, 2002. Global assessment of the state-of-the-science of endocrine disruptors. World Health Organization/International Program on Chemical Safety. WHO/PCS/EDE/0. 02. Available at www.who.int/ppcs/emerg_site/edc/global_edc_ch5. pdf.

Winberg, S., Lepage, O., 1998. Elevation of brain 5–HT activity, POMC expression, and plasma cortisol in socially subordinate rainbow trout. Am. J. Physiol. 43, R645–R654.

Winberg, S., Nilsson, G.E., 1993. Roles of brain monoamine neurotransmitters in agonistic behavior and stress reactions, with particular reference to fish. Comp. Biochem. Physiol. 106C, 597–614.

Winberg, S., Nilsson, G.E., Olsen, K.H., 1991. Social rank and brain levels of monoamines and monoamine metabolites in Arctic charr, *Salvelinus alpinus* (L). J. Comp. Physiol. 168A, 241–246.

Winkler, M., Kopf, G., Hauptvogel, C., Neu, T., 1998. Fate of artificial musk fragrances associated with suspended particulate matter (SPM) from the River Elbe (Germany) in comparison to other organic contaminants. Chemosphere 37, 1139–1156.

Wollenberger, L., Breitholtz, M., Kusk, K.O., Bengtsson, B.E., 2003. Inhibition of larval development of the marine copepod Acartia tonsa by four synthetic musk substances. Sci. Total Environ. 305, 53–64.

Wong, C.S., 2006. Environmental fate processes and biochemical transformations of chiral emerging organic pollutants. Anal. Bioanal. Chem. 386, 544–558.

Wong, D.C.L., Dorn, P.B., Chai, E.Y., 1997. Acute toxicity and structure-activity relationships of nine alcohol ethoxylate surfactants to fathead minnow and *Daphnia magna*. Environ. Toxicol. Chem. 16, 1970–1976.

Wood, C.M., McDonald, M.D., Sundin, L., Laurent, P., Walsh, P.J., 2003. Pulsatile urea excretion in the gulf toadfish: Mechanisms and controls. Comp. Biochemi. Physiol. 136B, 667–684.

Yamagishi, T., Miyazaki, T., Horii, S., Akiyama, K., 1983. Synthetic musk residues in biota and water from Tama River and Tokyo Bay (Japan). Arch. Environ. Contam. Toxicol. 12, 83–89.

Yamagishi, T., Miyazaki, T., Horii, S., Kaneko, S., 1981. Identification of musk xylene and musk ketone in freshwater fish collected from the Tama River, Tokyo. Bull. Environ. Contam. Toxicol. 26, 656–662.

Yang, L.H., Luan, T.G., Lan, C.Y., 2006. Solid-phase microextraction with on-fiber silylation for simultaneous determinations of endocrine disrupting chemicals and steroid hormones by gas chromatography-mass spectrometry. J. Chromatogr. 1104A, 23–32.

Yang, S., Carlson, K.H., 2004. Solid-phase extraction-high-performance liquid chromatography-ion trap mass spectrometry for analysis of trace concentrations of macrolide antibiotics in natural and waste water matrices. J. Chromatogr. 1038A, 141–155.

Ying, G.G., 2006. Fate, behavior and effects of surfactants and their degradation products in the environment. Environ. Int. 32, 417–431.

Zou, J., Neumann, N.F., Holland, J.W., Belosevic, M., Cunningham, C., Secombes, C.J., Rowley, A.F., 1999. Fish macrophages express a cyclooxygenase-2 homologue after activation. Biochem. J. 340, 153–159.

STUDY QUESTIONS

1. Given the potential routes for introducing pharmaceuticals and personal care products into the environment, in which ecosystems would you predict there to be the greatest potential for "ecosystem decline?"
2. Understanding the concept of K_{ow} is critical for predicting the behavior of organic compounds in biological systems. It also plays a role in the understanding the behavior of the same compounds in the natural environment. Describe why this might be so.
3. There is substantial controversy surrounding the issue of pharmaceuticals and personal care products in the environment, particularly with respect to endocrine disruption. Discuss the controversy and your thoughts as to the validity of the endocrine disruption issue.
4. Investigate and trace the path resulting from your own pharmaceutical and personal care product usage. Where does your personal effluent go? Moreover, what potential does it have to interact with the environment?
5. How would demographics assist in predicting the potential areas of influence for pharmaceuticals and personal care products in the environment? In what regions of the United States would certain types of compounds be more prevalent? Less prevalent? Expand the discussion to different regions of the world.
6. Of the many existing pharmaceuticals and personal care products, including those not described in this chapter, discuss which you predict may have the most detrimental effect on the future health of our environment.
7. Determine the programs that are in place within your own community for the disposal of unwanted medications, and discuss the merits of these protocols as well as how they can be improved.
8. Outline the different trophic levels within a marine environment. Now imagine, because of PPCP contamination, that all the producers within this marine environment were eliminated. Discuss the repercussions on the other trophic levels. What about the nearby community that survives solely on the fish caught from that ecosystem?

CHAPTER

10

Exposure and Effects of Seafood-Borne Contaminants in Maritime Populations

ÉRIC DEWAILLY, DARIA PEREG, ANTHONY KNAP, PHILIPPE ROUJA, JENNIFER GALVIN, AND RICHARD OWEN

INTRODUCTION

The ocean provides a unique source of support for many aspects of humanity. Degradation of the marine ecosystem poses a threat to the health and survival of humankind. For centuries, coastal communities have faced health risks associated with the presence of highly dangerous, natural marine toxins in their seafood (both fish and shellfish). Today, there is evidence of new anthropogenic (i.e., human-made) toxicants in the marine food chain. Ironically, those people living in places traditionally thought to be pristine, remote maritime ecosystems are some of the most likely to be negatively impacted by global environmental change, particularly by the contamination of the marine food chain.

Contaminants in the aquatic food web threaten fishing communities globally, especially those populations who rely on seafood for their primary source of subsistence. For example, the highest body burdens of the heavy metal methyl mercury and the organochlorines (OCs) (such as DDT and PCBs) have been found in remote maritime populations in the northern and southern hemispheres. Studies report that the highest human exposure concentrations and related health effects were found in children living in the Canadian Arctic (Ayotte *et al.*, 1997; Dewailly *et al.*, 1989; 1992a; 1993), remote Canadian fishing populations (Dewailly *et al.*, 1992b; 1994a, 1994b; Ryan *et al.*, 1997), Greenland (Dewailly *et al.*, 1999; Mulvad *et al.*, 1996), the Faroe Islands (Grandjean *et al.*, 1997), the Seychelles Island (Cernichiari *et al.*, 1995), and in Coastal Peru (Marsh *et al.*, 1995). Various biological (Ayotte *et al.*, 1994; 2005; Lagueux *et al.*, 1999) and clinical effects of these contaminants have been reported on the immune system (Dewailly, 2000) and brain development of these children (Grandjean *et al.*, 1997).

A number of indirect, human health impacts have also been reported and deserve attention. An increase in the documentation of health risks related to seafood contamination can have a profound impact on coastal communities that rely on fish and shellfish for subsistence purposes. For example, reports of mercury intoxication have resulted in a strong decrease in fish consumption among James Bay Cree Indians. This trend was followed by an epidemic of cardiovascular disease and diabetes, which were unknown in this population before the 1970s. Canadian Arctic Inuit are still protected from many "Western diseases" because their traditional diet of fish and sea mammals contains high levels of healthy unsaturated fatty acids and micronutrients, such as selenium. However, recent dietary changes have had observable, negative health consequences. Thus, the contamination of the aquatic food chain is a public health concern, not only directly because of the toxic risks posed by these contaminants but also indirectly because of the significant nutritional benefits lost from a subsequent shift toward a more Western diet.

Communities living on isolated islands and in remote coastal regions are the most sensitive and the most affected by environmental global changes. The United Nations has recognized that because of their greater direct dependence on the health of the ocean and coastal environment, the sustainable development and ultimately the survival of remote maritime communities may hinge on minimizing the adverse impacts of global environmental change.

CONTAMINANT EXPOSURES OF MARITIME POPULATIONS

Mercury

Mercury (Hg) is a metal that enters the environment from both natural and anthropogenic sources (see also Chapter 8). Hg is converted by bacteria to methylmercury (MeHg) in lakes and oceans, and bioaccumulates in the marine food web. Hg is also released into the environment by human activities, mainly the burning of fossil fuels and waste incineration. MeHg is highly fetotoxic (i.e., toxic to the fetus). The developmental neurotoxicity of MeHg first became evident during the 1950s at Minamata Bay (Japan), which was heavily contaminated with Hg from industrial effluents. Infants born to women who had eaten fish from Minamata Bay exhibited a range of central nervous system impairments, including mental retardation, primitive reflexes, cerebral ataxia, and seizures (Harada, 1995). Three prospective, longitudinal studies have since examined the effects of prenatal exposure to low doses of MeHg in maritime populations, including those in New Zealand, the Faroe Islands, and the Seychelles Islands (Davidson *et al.*, 1998; Grandjean *et al.*, 1997; Kjellstrom *et al.*, 1986; Myers *et al.*, 1995a). In the Faroe population, high dietary MeHg exposure came from fish and pilot whale consumption (Grandjean *et al.*, 1992). In the Seychelles (Myers *et al.*, 1995b) and the New Zealand populations (Kjellstrom *et al.*, 1986), pelagic and reef fish consumption were the sources of exposure.

High mercury exposure has been found in many different remote maritime populations worldwide, particularly in South America, Asia, and the Arctic. High blood and hair concentrations of Hg have been found in fishermen from Tyrrhenian and in coastal villages of Madeira (Renzoni *et al.*, 1998).

In Asia, Hg concentrations in human hair from Cambodia ranged from 0.54 to 190 μg/g. About 3% of the samples contained Hg levels exceeding the World Health Organization (WHO) recommendation (50 μg/g), and the levels in some hair samples of women also exceeded the no observable adverse effect level (NOAEL) of 10 μg/g associated with fetus neurotoxicity or harm to the nervous system (Agusa *et al.*, 2005). Fish is the major source of methyl mercury in the diet of Hong Kong residents (Dickman and Leung, 1998). The average person in Hong Kong consumes fish or shellfish four or more times a week, averaging about 60 kg of fish per year. The mean hair Hg concentration for over 200 Hong Kong residents was 3.3 mg/kg, which is more than double the U.S. mean Hg concentration.

In Peru, a prospective study of 131 infant-mother pairs in Mancora showed peak maternal hair MeHg levels during pregnancy from 1.2 μg/g to 30.0 μg/g (geometric mean 8.3). The MeHg was believed to be derived from marine fish in the diet, and there was no increase in the frequency of neurodevelopmental abnormalities measured in early childhood. This may be because of the possible role of selenium or other protective mechanisms in marine fish (Marsh *et al.*, 1995).

In the Arctic, high mercury exposure has been found in various Inuit populations in Greenland and in Arctic Canada. Major sources of exposure are sea mammal meat and predatory fish consumption (Artic Monitoring Assessment Programme [AMAP], 2003). In Qaanaaq in the Thule district of northern Greenland, 43 children were examined in 1995. The subjects varied from 6.2 to 12.0 years of age, with a median age of 8.4 years. The children's hair Hg concentrations varied up to 18.4 μg/g (geometric mean 5.5). Maternal hair samples showed a maximum Hg concentration of 32.9 μg/g (geometric mean 15.5) (Weihe *et al.*, 2002). Results from Canadian studies show that among Inuit women from NWT/Nunavut, 3% exceeded the Canadian Guideline Level of Concern of 20 μg/L for Hg, and 34% exceeded the lower 5.8 μg/L US-based Guideline; Nunavik and Baffin Inuit women had the highest percentage exceedances (16 and 9.7%, respectively). The percentage exceedance of the 5.8 μg/L blood guideline among Canadian Inuit women overall was 34%, and ranged from a low of 16% in Inuvik (Western Arctic) to a high of 68% in Baffin (Eastern Arctic). Among non-Inuit women, none exceeded the 20 μg/L guideline, and only 1% exceeded the 5.8 μg/L guideline (Van Oostdam *et al.*, 1999).

In Alaska, 48% of mothers from Bethel had blood Hg levels greater than or equal to the 5.8 μg/L EPA guideline value, whereas none of the Barrow maternal blood samples exceeded this guideline. Neither group exceeded the 20 μg/L guideline value.

In the Faroes, the geometric average for hair Hg at parturition in 1986/1987 was 4.3 μg/g, 4.0 μg/g in 1994, and 2.1 μg/g in 1998/1999. Public health warnings about pregnant women's consumption of pilot whale meat have reduced the hair mercury concentration significantly, but still the majority of the Faroes population has Hg concentrations above the 1.2 μg/g limit. In 1986/1987, 13% exceeded 10 μg/g, and in 1998/1999, only 3% did. In other words, less than 3% of the Faroes population now exceeds the benchmark dose of 12 μg/g of Hg in hair.

Generally speaking, high exposure to Hg is related to the consumption of predatory fish. In the tropics, pelagic fish (such as sharks, swordfish, and large tunas) are often highly contaminated. A specific situation is observed with the blue marlin, which mostly contains the less toxic inorganic mercury (Schultz *et al.*, 1976). However, consumption of predatory reef fish (such as snappers, barracudas and groupers) can also raise Hg concentrations above the 0.5 μg/g limit for fish concentration. One of the most recent advisories issued by the U.S. Food and Drug Administration (USFDA) informed women of reproductive age to avoid the consumption of four species known to be highly contaminated by MeHg: king mackerel, shark, swordfish, and tilefish (U.S. Department of Health and Human Services and U.S. Environmental Protection Agency, 2004).

Persistent Organic Pollutants

Persistent organic pollutants (POPs) are widespread environmental pollutants and now present global contamination problems (see also Chapter 7). They make up several types of compounds for industrial and domestic use, including organochlorines, organobromines, and perfluorinated compounds. POPs are hazardous because of their persistence in the environment, their bioaccumulation potential up the food chain in the tissues of animals and human, and their toxic properties for humans and wildlife (Fu *et al.*, 2003). POPs are transported for long distances by air, rivers, and currents, and therefore contaminate regions far from their source. Long-range transport of contaminants leads to transboundary problems that require the special attention and coordination of international environmental efforts for their monitoring and control (McKone & McLeod, 2003; Tanabe, 1991; Wania and Mackey, 1993, 1996). In particular, organochlorine (OC) residues (pesticides, polychlorinated biphenyls, etc.) have been detected in air, water, soil, sediment, fish, and seabirds around the world, even years after the ban of their use (Breivick *et al.*, 2004).

The United Nations Environment Program (UNEP) has listed 12 organochlorine POPs, also known as "the dirty dozen" by the Stockholm Convention. They constitute dioxins and furans (polychlorinated dibenzo-p-dioxins, polychlorinated dibenzofurans (PCCD/Fs); polychlorinated biphenyls (PCBs); hexachlorobenzene (HCB); and several organochlorine pesticides, including dichloro-diphenyl-trichloroethane (DDT), chlordane, toxaphene, dieldrin, aldrin, endrin, heptachlor and mirex. The 12 POPs targeted by the Stockholm Convention are of public health concern because they contaminate the environment including food chain bioaccumulation and are highly toxic. Studies on the levels of POPs in the global environment show that since the 1980s, emission sources of a number of POPs, including DDTs and HCBs, have shifted from the industrialized countries of the northern hemisphere to the less developing countries in tropical and subtropical regions, such as India, China, and Caribbean countries. This is probably because of both the relatively recent ban on the production and use of these agents and the fact that they are still being used (both legally and illegally) in agricultural activities and for the control of infectious diseases, such as malaria, typhus, and cholera, as part of vector control, particularly in less developed nations and in subtropical areas (Iwata *et al.*, 1994; Loganathan and Kannan, 1994; Minh *et al.*, 2006; Tanabe, 1991).

POPs have been extensively used in developing countries for decades for important purposes such as vector control. Concerns regarding global contamination by POPs have led to their replacement by other pesticides that are less persistent, such as the pyrethroids. However, because POPs are persistent in the environment, in the biota and in humans, concentrations of these compounds will decrease only slowly in populations that have been exposed over many years. Because significant concentrations may persist in the environment, local seafood contamination may be of concern over a longer period of time than suspected.

Maritime populations exposed to POPs are mostly located in the northern hemisphere. High POP exposure in Arctic residents was first found in the 1980s in Nunavik. Breast milk of Inuit women was found to contain 5 to 10 times higher PCB and pesticides concentrations than in the breast milk from Southern Québec mothers (Dewailly *et al.*, 1989, 1993). Inuit from the east coast of Greenland, who consume large amounts of marine mammals, have the highest population proportion exceeding the guidelines for PCB in blood, followed by west coast Greenland Inuit populations, and then Inuit from the Baffin and Nunavik regions of eastern Canada (AMAP, 1998; Dewailly *et al.*, 1999; Muckle *et al.*, 2001; Mulvad *et al.*, 1996). Another well-known example is the high PCB exposure of Faroe Islands residents who also regularly consume marine mammals, at least in the past (Grandjean *et al.*, 1992).

The influence of the consumption of fatty fish from the Baltic Sea on plasma concentrations of PCBs (and metabolites), DDT (and metabolites), HCB, and PBDEs was assessed in Latvian and Swedish men. Both age and fish consumption were significantly correlated with the concentrations of POPs in blood (Sjodin *et al.*, 2000).

HEALTH EFFECTS OF SEAFOOD CONTAMINANTS IN MARITIME POPULATIONS

Our knowledge of the extent of maritime population exposure to lipophilic (i.e., fat stored) pollutants is relatively recent (mid-1980s). Mercury-related exposure and effects have been known for many years and therefore have been one of the most studied contamination problems. However, few major environmental epidemiological studies on the effects of exposure to seafood-borne contaminants have been conducted in remote maritime communities. Conducting studies in remote communities is extremely difficult, primarily because of their location and extreme climatic conditions but also because of their small population size and varying social and behavioral factors. Because these communities are small and scattered across vast oceanic territories, travel and fieldwork become prohibitively expensive, and ultimately the power of epidemiological studies is weakened as a result of the small sample sizes.

Nevertheless, these studies are necessary because the social and behavioral specificity of these remote coastal populations makes it difficult to apply the conclusions and recommendations drawn from major epidemiological studies conducted in populated, often industrialized, regions. Moreover, because humans are exposed to a mixture of many different substances simultaneously, including both toxicants and nutrients, it is unreasonable and presumably not even possible to deal with the risk of single substances in epidemiological studies. Nevertheless, we should consider seriously the results from available cohort studies in these small unique populations on neurological disorders associated with prenatal MeHg (Faroes, Seychelles, New Zealand) and immune dysfunctions in Inuit children exposed prenatally to POPs (Nunavik).

The following discussion of epidemiology is limited to studies related to MeHg and POPs (see also Chapter 11 for additional background). These contaminants are more often implicated in seafood contamination related issues, and most health advisories have been related to MeHg and POPs in the seafood chain.

Epidemiological studies of health effects related to PCBs and mercury have been oriented over the past decades toward prenatal exposure and children's health. The primary health outcome focus has been on neurological systems with respect to POPs and Hg exposure. However, hormonal effects (for POPs, related to reproduction and reproductive cancers), immune deficiency (for POPs), and cardiovascular effects (for Hg) have recently gained considerable attention.

Neurobehavioral Effects

Mercury

Prospective MeHg studies conducted in New Zealand, the Faroe Islands, and the Seychelles Islands involved children without overt clinical symptoms of MeHg poisoning who were assessed for neurobehavioral developmental effects. Umbilical cord blood Hg was the main indicator of prenatal exposure in the Faroe study, although hair Hg concentration during pregnancy was also documented. Maternal hair Hg concentration during pregnancy was the indicator of prenatal MeHg exposure in the Seychelles and the New Zealand studies. Hair Hg is approximately 90% MeHg and has the advantage of providing an historical record of MeHg exposure, whereas the MeHg half-life in human blood is approximately 50 days (Cox *et al.*, 1989; Sherlock *et al.*, 1984). The average maternal hair mercury concentrations varied from 4.3 to 8.8 μg/g between these studies, and a significant number of infants studied were exposed beyond 10 μg/g.

It should be noted that two other studies looked at the effects of prenatal Hg exposure resulting from fish consumption, the first one in Canada (more specifically, in the James Bay Cree population) and the second one in Peru. However, neurobehavioral outcomes were not assessed in depth in these studies (Marsh *et al.*, 1995; McKeown-Eyssen *et al.*, 1983).

The Faroe Islands study reported associations between maternal hair Hg concentrations corresponding to the pregnancy period, and children's performance on neurobehavioral tests, particularly in the domains of fine motor function, attention, language, visual-spatial abilities, and verbal memory (Grandjean *et al.*, 1997). Those effects were subsequently also found to be associated with cord blood Hg concentration (Grandjean *et al.*, 1999). The New Zealand study, in which the exposure and research design were similar to the Seychelles study, also found adverse effects of prenatal MeHg exposure (Kjellstrom *et al.*, 1986). More specifically, higher hair Hg levels were associated with poorer neurodevelopmental test scores in similar domains to those observed in the Faroe study. However, prenatal MeHg exposure was not related to neurobehavioral effects in the Seychelles Islands study (Davidson *et al.*, 1995, 1998; Myers *et al.*, 1995b).

Several differences in findings between the Faroes and the Seychelles studies have been attributed to study design variations, including differences in the marker of Hg exposure, the particular neurobehavioral test battery administered, age at testing, and even varied sources of exposure. When the New Zealand data are considered those differences no longer seem determinative because the New Zealand study, in which the exposure and research design were similar to the Seychelles Islands study, also found neurobehavioral effects, as did the pilot study conducted in the Seychelles Islands population (National Research Council, 2000).

One limitation of the Faroe Islands study was that, because of the confounding of coexistent prenatal Hg and PCB exposure ($r = 0.41$ to 0.49) (Grandjean *et al.*, 1997, 1999), it was difficult to determine whether several of the neurodevelopmental deficits observed at 7 years, especially those on language and memory functions, were the result of prenatal Hg exposure, to PCB exposure, or to both. However, patterns of neurobehavioral damage produced by developmental Hg exposure in animals resemble those found in humans and include sensory system effects, motor or sensorimotor system effects, and cognitive effects.

Cord blood samples (n = 42) from Qaanaaq (Northwest Greenland) were collected and analyzed in 1982, and the children examined were between ages 7 and 12 years. Clinical neurological examination did not reveal any obvious deficits. However, neurophysiological tests showed possible mercury exposure-associated deficits (i.e., auditory evoked potentials), although only in a few cases reaching statistical significance (Weihe *et al.*, 2002).

A prospective study involving 131 infant-mother pairs was conducted in Mancora (Peru) with peak maternal hair MeHg levels during pregnancy ranging from 1.2 ppm to

30.0 ppm (geometric mean 8.3). The MeHg was believed to be derived from marine fish in the diet. There was no increase in the frequency of neurodevelopmental abnormalities in early childhood. The possible role of selenium or other protective mechanisms in marine fish was a possible explanation (Marsh *et al.*, 1995).

PCBs

The developmental toxicity of heat-degraded polychlorinated biphenyls (PCBs) was first recognized in Japan in the late 1960s and in Taiwan in the late 1970s. In similar industrial accidents in both countries, infants born to women who had consumed rice oil contaminated with mixtures of PCBs and polychlorinated dibenzofurans (PCDFs) had skin rashes and exhibited poorer intellectual functioning during infancy and childhood (Chen *et al.*, 1992; Yu *et al.*, 1991).

Effects of prenatal exposure to background levels of PCBs and other POPs from environmental sources have been studied since the 1980s in prospective longitudinal studies conducted in the Netherlands and in the United States (Michigan, North Carolina, New York). The source of PCB exposure was fish consumption from the Great Lakes in both the Michigan (Schwartz *et al.*, 1983) and the New York (Stewart *et al.*, 1999) studies, and consumption of dairy products in the Netherlands (Koopman-Esseboom *et al.*, 1994). PCB exposure was associated with less optimal newborn behavioral function (reflexes, tonicity, and activity levels) in three of the four studies (Huisman *et al.*, 1995a; Rogan *et al.*, 1986; Stewart *et al.*, 2000). Adverse neurological effects of exposure to PCBs have been found up to 18 months of age in the Netherlands study (Huisman *et al.*, 1995b). In Michigan and the Netherlands, higher cord serum PCBs concentrations were associated with lower birth weight and slower growth rates (Fein *et al.*, 1984; Jacobson *et al.*, 1990a, 1996; Patandin *et al.*, 1998). In Michigan, prenatal PCB exposure was associated with poorer visual recognition memory in infancy (Jacobson *et al.*, 1985, 1990b, 1992), an effect that was confirmed in the Oswego study (Darvill *et al.*, 2000). In North Carolina, deficits in psychomotor development up to 24 months were seen in the most highly exposed children (Gladen *et al.*, 1988; Gladen and Rogan, 1991). In Michigan, prenatal PCB exposure was linked to poorer intellectual function at four and at 11 years (Jacobson *et al.*, 1990b; Jacobson and Jacobson, 1996), a finding confirmed in the Netherlands at 42 months (Patandin *et al.*, 1999). Although much larger quantities of PCBs are transferred to nursing infants by breast-feeding than prenatally across the placenta, virtually all the adverse neurobehavioral effects reported to date have been linked specifically to prenatal exposure, indicating that the embryo and fetus are particularly vulnerable to these substances.

One prospective, longitudinal study that examined the effects of prenatal exposure to low doses of methylmercury (MeHg) resulting from fish and pilot whale consumption was performed in Faroe Islands (Grandjean *et al.*, 1992, 1997). Because pilot whale tissues contain several neurotoxicants, this cohort was also exposed to PCBs. In this study, it was difficult to determine whether several of the neurobehavioral deficits observed at 7 years, especially language and memory functions (Budtz-Jorgensen *et al.*, 1999), were caused by prenatal MeHg exposure, PCB exposure, or both.

Reproductive Effects

Typical organochlorine mixtures found in highly exposed human populations contain a large variety of organochlorine compounds, including substances with estrogenic, antiestrogenic, or antiandrogenic capacities. Blood samples were collected from pregnant women and their partners from Greenland, Warsaw, Kharkiv, and from a cohort of Swedish fishermen's wives. Blood samples were analyzed for PCB (congener 153) and DDE (the main DDT metabolite). Information on the participants' fertility, measured as time to pregnancy (TTP), was collected. In total, 778 men and 1505 women were included in the analyses. The data from Warsaw, Kharkiv, and the Swedish fishermen's wives indicated no effect of either male or female exposure to POPs on TTP. However, among men and women from Greenland, there seemed to be an association between serum concentrations of PCB and DDE and prolonged TTP (Axmon *et al.*, 2006).

Immune System Effects

Several POPs display immunotoxic properties in both laboratory animals and humans. In children and young adults accidentally exposed to PCBs and PCDFs in Taiwan ("Yu-Cheng disease"), serum IgA and IgM concentrations as well as percentages of various immune system blood cells (i.e., total T cells, active T cells, and suppressor T cells) were decreased compared to values of age- and sex-matched controls (Chang *et al.*, 1981). The investigation of delayed type hypersensitivity responses further indicated that cell-mediated immune system dysfunction was more frequent among patients than controls. Infants born to Yu-Cheng mothers had more episodes of bronchitis or pneumonia during their first 6 months of life than unexposed infants from the same neighborhoods (Rogan *et al.*, 1988). The authors speculated that the increased frequency of pulmonary diseases could result from a generalized immune disorder induced by transplacental or breast milk exposure to dioxin-like compounds, more likely PCDFs (Rogan *et al.*, 1988). Eight- to 14-year-old children born to Yu-Cheng mothers were shown to be more prone to middle-ear diseases than matched controls (Chao *et al.*, 1997).

In Nunavik, an epidemiological study investigated whether POP exposure in Inuit infants was associated with the incidence of infectious diseases and with immune dysfunction. The number of infectious disease episodes in 98 breast-fed and 73 bottle-fed infants was compiled during the first year of life. Concentrations of organochlorines were measured in early breast milk samples and used as surrogates to prenatal exposure levels. Otitis media was the most frequent disease with 80.0% of breast-fed and 81.3% of bottle-fed infants experiencing at least one episode during the first year of life. During the second follow-up period, the risk of otitis media increased with prenatal exposure to DDE, hexachlorobenzene (HCB), and dieldrin. The relative risk (RR) for 4- to 7-month old infants in the highest tertile of DDE exposure as compared with infants in the lowest was 1.87 (95% confidence interval [CI], 1.07 to 3.26). The relative risk of otitis media over the entire first year of life also statistically increased with prenatal exposure to DDE (RR = 1.52) and HCB (RR = 1.49). Furthermore, the relative risk of recurrent otitis media (≥3 episodes) was augmented by prenatal exposure to these compounds. The analysis showed the dose response relationship with the milk fat level (i.e., prenatal exposure) and not with the calculation of post natal exposure from breast feeding (i.e., weeks of lactation × milk concentration). It was concluded that prenatal organochlorine exposure could be a risk factor for acute otitis media in Inuit infants (Dewailly *et al.*, 2000).

In another cohort conducted in Nunavik between 1997 and 2000, the risk of experiencing frequent infectious diseases episodes was assessed in 89 children exposed to PCBs and DDT during their first year of life. The risks were put in relation with maternal PCB and DDT blood level during pregnancy. Ratios were estimated using logistic regression and the results were adjusted for maternal smoking during pregnancy, the number of smokers in the house, crowding, breast-feeding duration, and sex. This study supported the hypothesis that the high incidence of infections observed in Inuit children (mostly respiratory infections) is associated in part with high prenatal exposure to POPs (Dallaire *et al.*, 2004, 2006a, 2006b).

In the Faroe Islands, a study was done to assess whether prenatal and postnatal exposure to PCBs had impacts on the antibody response to childhood immunizations. Following routine childhood vaccinations against tetanus and diphtheria, 119 children were examined at 18 months and 129 children at age 7 years, and their serum samples were analyzed for tetanus and diphtheria toxoid antibodies and for PCBs. The antibody response to diphtheria toxoid decreased at age 18 months by 24.4% for each doubling of the cumulative PCB exposure at the time of examination. The diphtheria response was lower at age 7 years and was not associated with the exposure. However, the tetanus toxoid antibody response was affected mainly at age 7 years, decreasing by 16.5% for each doubling of the prenatal exposure. These results suggested that perinatal exposure to PCBs may adversely impact on immune responses to childhood vaccinations (Heilman *et al.*, 2006).

Cardiovascular Effects

Although there are no studies that report an association between cardiovascular disease and POPs, Salonen *et al.* (1995) suggested that the high mortality from cardiovascular disease observed among fish eaters from Finland could be explained by high mercury content in fish (mainly nonfatty, freshwater species). This group noted a significant association between mercury concentration in the hair of Eastern Finnish men and the risk of coronary heart diseases (CHD). Mercury can promote the peroxidation of lipids, resulting in more oxidized low-density lipoprotein (LDL), which has been implicated as an initiator of arteriosclerosis. Salonen previously observed in the same population an enhanced risk of CHD death in subjects with low serum selenium concentrations, an antioxidant that can block the mercury-induced lipid peroxidation (Salonen *et al.*, 1982).

The ability of both mercury and selenium to modulate CHD risk is also suggested by observations in fish-eating coastal populations such as the Inuit living in Arctic regions. The Inuit consume large amounts of fish and marine mammals and consequently receive large doses of mercury. However, contrary to the situation in Eastern Finland, the mortality rate from CHD in Inuit is low (Dewailly *et al.*, 2001). This could be the result of this population's consumption of traditional food items and the subsequent intake of fat from marine mammals and fish rich in selenium, such as *muktuk* (beluga and narwhal skin) and marine mammal liver, or from polyunsaturated fatty acides (PUFA).

Blood pressure in childhood is an important determinant of hypertension risk later in life, and methylmercury exposure is a potential environmental risk factor. A birth cohort of 1000 children from the Faroe Islands was examined for prenatal exposure to methylmercury, and at age 7 years, their blood pressure, heart rate, and heart rate variability were determined (Sorensen *et al.*, 1999a). After adjustment for body weight, diastolic and systolic blood pressure increased by 13.9 mmHg [95% confidence limits (CL) = 7.4, 20.4] and 14.6 mmHg (95% CL = 8.3, 20.8), respectively, when cord blood mercury concentrations increased from 1 to 10 μg/liter. Above this level, which corresponds to a current exposure limit, no further increase was seen. Birth weight acted as a modifier, with the mercury effect being stronger in children with lower birth weights. In boys, heart rate variability decreased with increasing mercury exposures, particularly from 1 to 10 μg/liter cord blood, at which the variability was reduced by 47% (95% CL = 14%, 68%). These findings suggested that prenatal exposure to methylmercury might affect the development of cardiovascular homeostasis.

It should be pointed out that more sensitive health endpoints other than neurotoxicity could be of greater relevance in understanding some areas of cardiovascular effects. For example, the low incidence of CHD in Greenland Inuit proposed to be because of the fatty acid composition of their diet could be attenuated by high mercury exposure since studies have indicated that Hg can have a negative effect on the cardiovascular system (Rissanen *et al.*, 2000). The reason is still unknown, but Hg may inhibit important anti-oxidative mechanisms in humans and could promote the peroxidation of unsaturated fatty acids such as DHA and DPA.

Regarding cardiovascular toxicity at low-level methylmercury exposures, the first Faroes cohort showed that blood pressure tended to increase and the heart rate variability tended to decrease when prenatal mercury exposures increased in the low-dose range (Sorensen *et al.*, 1999b). Alkyl-mercury poisoning is associated with increased blood pressure (Höök *et al.*, 1954), and children with mercury poisoning often have increased heart rate and blood pressure (Warkany and Hubbard, 1953). Experimental evidence shows that methylmercury toxicity results in irreversible hypertension that remains many months after cessation of exposure (Wakita, 1987). Although insufficient for risk assessment purposes, this evidence suggests that the cardiovascular system should be considered a potential target for methylmercury. Even a slight negative impact on the cardiovascular system could be of greater public health relevance than a slight impact on the central nervous system because of the high incidence of cardiovascular disease compared to nervous system disease in human populations.

BALANCING THE RISKS AND BENEFITS OF SEAFOOD CONSUMPTION

Because selenium and omega-3 fatty acids are important nutrients found in fish, health benefits (notably cardiovascular benefits and cancer protection) provided by these nutrients may counterbalance toxic risk associated with contaminants. Omega-3 fatty acids are present in fatty fish (mackerel, salmon) and selenium is concentrated in fish skin and liver. Whole fish consumption (versus fillet only) is important especially in the tropics where most of the fat is not located in the flesh but in the abdominal cavity (Rouja *et al.*, 2003). Fish fatness could also change seasonally with water temperature (Rouja *et al.*, 2003).

Several health organizations recommend eating fish twice a week for the general population (Harris, 2004; Kris-Etherton *et al.*, 2002). Fish consumption is largely recognized as beneficial for brain development (Cunnane *et al.*, 2000; Uauy *et al.*, 2001) and protective against cardiovascular diseases (Bucher *et al.*, 2002; He *et al.*, 2004), mental disorders (Casper, 2004; Emsley *et al.*, 2003), and various inflammatory conditions, such as bowel diseases, asthma, and arthritis (Ruxton *et al.*, 2004). Long-chain omega-3 polyunsaturated fatty acids, more specifically eicosapentaenoic acid (EPA) and docosahexaenoic acid (DHA), are arguably the most important nutrients in fish (Harper and Jacobson, 2001).

Neurotoxic effects of MeHg might be attenuated by the protective effects of nutrients such as selenium (Se) and n-3 polyunsaturated fatty acid (n-3 PUFA). Increased intake of these nutrients would be expected in a population (such as the Inuit) who consume relatively large quantities of fish and marine mammals. Although the protective effects of Se on MeHg toxicity have not been adequately documented in humans (National Research Council, 2000), there is strong evidence from animal studies that selenium can influence the deposition of MeHg in the body and some evidence that Se can protect against Hg toxicity (Ganther *et al.*, 1972; Whanger, 1992). n-3 PUFA, especially docosahexaenoic acid (DHA), are essential for brain development (Crawford *et al.*, 1976). DHA deficiency impairs learning and memory in rats (Greiner *et al.*, 1999). Studies have shown that supplementation of n-3 PUFA can enhance visual acuity and brain development in preterm infants (Uauy *et al.*, 2001), but it is not clear whether increased levels of these nutrients during the fetal period can protect full-term infants against neurotoxicity associated with prenatal exposure to environmental contaminants.

CONCLUSIONS

Except for Hg- and OC-induced neurodevelopmental effects studied in the Faroe Islands and in the Seychelles Islands, as well as the studies of the association between POPs and the immune system effects in Nunavik (Canada), few major environmental epidemiological studies have been conducted in remote maritime communities. As previously mentioned, there are several reasons for the lack of research attention on these populations, ranging from physical challenges to unique cultural considerations.

Patterns of exposure could be influenced by fishing seasons, particularly in Arctic regions, and constant exposure versus occasional high exposure may have different toxic consequences. Coastal people consume wild animals. Traditional foods contain specific nutrients, which could influence or counteract the toxicity of contaminants. For example, Inuit are exposed to similar amounts of mercury as the Faroe's people, but the Inuit selenium intake is much higher; the Inuit and Faroe Islanders are both exposed to POPs and Hg, but Seychellois are not exposed to POPs (com. Pers. T Clarkson).

Any recommendations or actions by public health authorities intended to reduce exposure of coastal subsistence populations to environmental contaminants, through, for example, the adoption of new dietary guidelines, should carefully weigh cultural considerations and the possible additional negative implications of lifestyle changes for their target population. Saltwater people are intimately connected to the ocean. Ocean contamination has a direct impact on the health and subsistence of remote maritime communities. This reinforces the need to protect our oceans and increase awareness that human health depends on a healthy ocean environment.

References

Agency for Toxic Substances and Disease Registry (ATSDR), 2000. Toxicological profile for polychlorinated biphenyls (PCBs), U.S. Department for Health and Human Services, Public Health Service.

Agusa, T., Kunito, T., Iwata, H., Monorith, I., Tana, T.S., Subramanian, A., Tanabe, S., 2005. Mercury contamination in human hair and fish from Cambodia: Levels, specific accumulation and risk assessment. Environ. Pollut. 134, 79–86.

Artic Monitoring Assessment Programme (AMAP), 2003. AMAP Assessment 2002: Human Health in the Arctic. Oslo, Norway, Arctic Monitoring and Assessment Programme (AMAP).

Axmon, A., Thulstrup, A.M., Rignell-Hydbom, Pedersen, H.S., Zvyezday, V., Ludwicki, J.K., Jonsson, B.A., Toft, G., Bonde, J.P., Hagmar, L., 2006. Time to pregnancy as a function of male and female serum concentrations of 2,2′4,4′5,5′-hexachlorobiphenyl (CB-153) and 1,1–dichloro-2,2–bis (p-chlorophenyl)-ethylene (p,p′-DDE). Hum. Reprod. 21, 657–659.

Ayotte, P., Dewailly, É, Bruneau, S., Poirier, G., Weber, J.P., Tanguay, R.M., Gagné, R., Roy, R., Feeley, M., Iverson, F., 1994. Biomarkers of early biological responses induced by organochlorines Environ. Stud. 72, 392–396.

Ayotte, P., Dewailly, É., Lambert, G.H., Perkins, S.L., Poon, R., Feeley, M., Larochelle, C., Pereg, D., 2005. Biomarker measurements in a coastal fish-eating population environmentally exposed to organochlorines. Environ. Health Perspect. 113, 1318–1324.

Ayotte, P., Dewailly, É., Ryan, J.J., Bruneau, S., Lebel, G., 1997. PCBs and dioxin-like compounds in plasma of adult Inuit living in Nunavik (Arctic Quebec). Chemosphere 34, 1459–1468.

Barr, D.B., Weihe, P., Davis, M.D., Needham, L.L., Grandjean, P., 2006. Serum polychlorinated biphenyl and organochlorine insecticide concentrations in a Faroese birth cohort. Chemosphere 62, 1167–1182.

Breivik, K., Alcock, R., Li, Y-F., Bailey, R.E., Heidelore, F., Pacyna, J.M., 2004. Primary sources of selected POPs: Regional and global scale emission inventories. Environ. Pollut. 128, 3–16.

Bucher, H.C., Hengstler, P., Schindler, C., Meier, G., 2002. N-3 polyunsaturated fatty acids in coronary heart disease: A meta-analysis of randomized controlled trials. Am. J. Med. 112, 298–304.

Budtz-Jorgensen, E., Keiding, N., Grandjean, P., White, R.F., Weihe, P., 1999. Methylmercury neurotoxicity independent of PCB exposure. Environ. Health Perspect. 107, 236–237.

Butler Walker, J., Houseman, J., Seddon, L., Mcmullen, E., Tofflemire, K., Mills, C., Corriveau, A., Weber, J.P., LeBlanc, A., Walker, M. Donaldson, S.G., Van Oostdam, J., 2006. Maternal and umbilical cord blood levels of mercury, lead, cadmium, and essential trace elements in Arctic Canada. Environ. Res. 100, 295–318.

Butler Walker, J., Seddon, L., McMullen, E., Houseman, J., Tofflmire, K., Corriveau, A., Weber, J.P., Mills, C., Smith, S., Van Oostdam, J., 2003. Organochlorine levels in maternal and umbilical cord blood plasma in Arctic Canada. Sci. Total Environ. 302, 27–52.

Casper, R.C., 2004. Nutrients, neurodevelopment, and mood. Curr/Psychiatry Rep. 6, 425–429.

Cernichiari, E., Toribara, T.Y., Liang, L., Marsh, D.O., Berlin, M.W., Myers, G.J., Cox, C., Shamlaye, C.F., Choisy, O., Davidson, P., 1995. The biological monitoring of mercury in the Seychelles study. Neurotoxicology 16, 613–628.

Chang, K.J., Hsieh, K.H., Lee, T.P., Tang, S.Y., Tung, T.C., 1981. Immunologic evaluation of patients with polychlorinated biphenyl poisoning: Determination of lymphocyte subpopulations. Toxicol. App. Pharmacol. 61, 58–63.

Chao, W.Y., Hsu, C.C., Guo, Y.L.L., 1997. Middle-ear disease in children exposed prenatally to polychlorinated biphenyls and polychlorinated dibenzofurans. Arch. Environ. Health 52, 257–262.

Chen, Y.C.J., Guo, Y.L., Hsu, C.C., Rogan, W.J., 1992. Cognitive development of Yu-Cheng ("Oil Disease") children prenatally exposed to heat-degraded PCBs. JAMA. 268, 3213–3218.

Cox, C., Clarkson, T.W., Marsh, D.O., Zaki, L., Tikriti, S., Myers, G.G., 1989. Dose-response analysis of infants prenatally exposed to methylmercury: An application of a single compartment model to single-strand hair analysis. Environ. Res. 49, 318–332.

Crawford, M.A., Hassam, A.G., Williams, G., Whitehouse, W.E., 1976. Essential fatty acids and fetal brain. Lancet i, 452–453.

Cunnane, S.C., Francescutti, V., Brenna, J.T., Crawford, M.A., 2000. Breast-fed infants achieve a higher rate of brain and whole body docosahexaenoate accumulation than formula-fed infants not consuming dietary docosahexaenoate. Lipids 35, 105–111.

Dallaire, F., Dewailly, É., Laliberté, C., Muckle, G., Ayotte, P., 2002. Temporal trends of organochlorine concentrations in umbilical cord blood of newborns from the Lower North Shore of the St.Lawrence River (Québec, Canada). Environ. Health Perspect. 110, 835–838.

Dallaire, F., Dewailly, É., Muckle, G., Ayotte, P., 2003. Time trends of persistent organic pollutants and heavy metals in umbilical cord blood of Inuit infants born in Nunavik (Québec, Canada) between 1994 and 2001. Environ. Health Perspect. 111, 1660–1664.

Dallaire, F., Dewailly, É., Muckle, G., Vézina, C., Jacobson, S.W., Jacobson, J.L., Ayotte, P., 2004. Acute infections and environmental exposure to organochlorines in Inuit infants from Nunavik. Environ. Health Perspect. 112, 1359–1365.

Dallaire, F., Dewailly, É., Vézina, C., Bruneau, S., Ayotte, P., 2006a. Portrait of outpatient visits and hospitalizations for acute infections in Nunavik preschool children. Can. J. Public Health 97, 362–368.

Dallaire, F., Dewailly, É., Vézina, C., Muckle, G., Weber, J.P., Bruneau, S., Ayotte, P., 2006b. Effect of prenatal exposure to polychlorinated biphenyls on incidence of acute respiratory infections in preschool Inuit children. Environ. Health Perspect. 114, 1301–1305.

Darvill, T., Lonky, E., Reihman, J., Stewart, P., Pagano, J., 2000. Prenatal exposure to PCBs and infant performance on the Fagan test of infant intelligence. Neurotoxicology 25, 11–22.

Davidson, P.W., Myers, G.J., Cox, C.C, Axtell, C., Shamlaye. C., Sloane-reeves. J., Cemichiari, E., Needham. L., Choi, A., Wang, Y.N., 1998. Effects of prenatal and postnatal methylmercury exposure from fish consumption on neurodevelopment. JAMA 280, 701–707.

Davidson, P.W., Myers, G.J., Cox, C.C., Shamlaye, C., Marsh, D.O., Tanner, M.A., Berlin, M., Sloane-Reeves, J., Cemichiari, E., Choisy. O., Choi, A., Clarkson, T.W., 1995. Longitudinal neurodevelopmental study of Seychellois children following in utero exposure to methylmercury from maternal fish ingestion: Outcomes at 19 and 29 months. Neurotoxicology 16, 677–688.

Dewailly, É., Ayotte, P., 1994a. Biomarkers of exposure and effects among a fishing populations of the Lower North Shore of the St. Lawrence River exposed to dioxin-like compounds. Pharmacol. Toxicol. 74 (supp. II), 18.

Dewailly, É., Ayotte, P., Bruneau, S., Gingras, S., Belles-Isles, M., Roy, R., 2000. Susceptibility to infections and immune status in Inuit infants exposed to organochlorines. Environ. Health Perspect. 108, 205–211.

Dewailly, É., Ayotte, P., Bruneau, S., Laliberté, C., Muir, D.C.G., Norstrom, R., 1993. Inuit exposure to organochlorines through the aquatic food chain in Arctic Québec. Environ. Health Perspect. 101, 618–620.

Dewailly, É., Blanchet, C., Lemieux, S., Sauvé, L., Gingras, S., Ayotte, P., Holub, B.J., 2001. N-3 Fatty acids and cardiovascular disease risk factors among the Inuit of Nunavik. Am. J. Clin. Nutr. 74, 464–473.

Dewailly, É., Laliberté, C., Sauvé, L., Ferron, L., Ryan, J.J., Gingras, S., Ayotte, P., 1992b. Sea-bird egg consumption as a major source of PCB exposure for communities living along the gulf of St-Lawrence. Chemosphere 25, 1251–1255.

Dewailly, É., Mulvad, G., Pedersen, H.S., Ayotte, P, Demers, A, Weber, J.P., Hansen, J.C., 1999. Concentration of organochlorines in human brain, liver, and adipose tissue autopsy samples from Greenland. Environ. Health Perspect. 107, 823–828.

Dewailly, É., Nantel, A., Bruneau, S., Laliberté, C., Ferron, L., Gingras, S., 1992a. Breast milk contamination by PCDDs, PCDFs and PCBs in Arctic Quebec: A preliminary assessment. Chemosphere 25, 1245–1249.

Dewailly, É., Nantel, A., Weber, J.-P., Meyer, F., 1989. High levels of PCBs in the breast milk of Inuit women from Arctic Quebec. Bull. Environ. Contam. Toxicol. 43, 641–646.

Dewailly, É., Ryan, J.J., Laliberté, C., Bruneau, S., Weber, J.P., Gingras, S., Carrier, G., 1994b. Exposure of remote maritime populations to coplanar PCBs. Environmental Health Perspective. 102 (suppl. 1), 205–209.

Dickman, M.D., Leung, K.M., 1998. Mercury and organochlorine exposure from fish consumption in Hong Kong. Chemosphere 37, 991–1015.

Emsley, R., Oosthuizen, P., van, Rensburg, S.J., 2003. Clinical potential of omega-3 fatty acids in the treatment of schizophrenia. CNS Drugs 17, 1081–1091.

Fein, G.G., Jacobson, J.L., Jacobson, S.W., Schwartz, P.M., Dowler, J.K., 1984. Prenatal exposure to polychlorinated biphenyls: Effects on birth size and gestational age. J. Pediatr. 105, 315–320.

Fu, J., Mai, B., Sheng, G., Zhang, G., Wan, X., Peng, P., Xiao, X., Ran, R., Cheng, F., Peng., X., Wang, Z., Tang, U.W., 2003. Persistent organic pollutants in environment of the Pearl River Delta, China: An overview. Chemosphere. 52, 1411–1422.

Ganther, H.E., Goudie, C., Sunde, M.L., Kopecky, M.J., Wagner, P., 1972. Selenium: Relation to decreased toxicity of methylmercury added to diets containing tuna. Science 175, 1122–1124.

Gladen, B.C., Rogan, W.J., 1991. Effects of perinatal polychlorinated biphenyls and dichlorodiohenyl dichloroethene on later development. J. Pediatr. 119, 58–63.

Gladen, B.C., Rogan, W.J., Hardy, P., Thullen, J., Tingelstad, J., Tully, M., 1988. Development after exposure to polychlorinated biphenyls and dichlorodiphenyl dichloroethene transplacentally and through human milk. J. Pediatr.113, 991–995.

Grandjean, P., Budtz-Jorgensen E., Jorgensen, P.J., Weihe, P., 2005. Umbilical cord mercury concentration as biomarker of prenatal exposure to methylmercury. Environ. Health Perspect. 113, 905–908.

Grandjean, P., Budtz-Jorgensen, E., White, R.F., Jorgensen, P.J., Weihe, P., Debes, F., Keiding, N., 1999. Methylmercury exposure biomarkers as indicators of neurotoxicity in children aged 7 years. Am. J. Epidemiol. 150, 301–305.

Grandjean P., Weihe, P., Jørgensen, P.J., Clarkson, T., Cernichiari, E., Viderø, T., 1992. Impact of maternal seafood diet on fetal exposure to mercury, selenium, and lead, Arch. Environ. Health 47, 185–195.

Grandjean, P, Weihe, P., White, R.F., Debes, F., Araki, S., Yokoyama, K., Murata, K., Sørensen, R, Dahl, R., Jørgensen, R.J., 1997. Cognitive deficit in 7-year-old children with prenatal exposure to methylmercury. Neurotoxicol. Teratol. 19, 417–428.

Greiner, R.S., Moriguchi, T., Hutton, A., Slotnick, B.M., Salem, N., Jr., 1999. Rats with low levels of brain doxosahexaenoic acid show impaired performance in olfactory-based and spatial learning tasks. Lipids. 34 (suppl.), S239–S243.

Harada, M., 1995. Minamata disease: Methylmercury poisoning in Japan caused by environmental pollution. Crit. Rev. Toxicol. 25, 1–25.

Harper, C.R., Jacobson, T.A., 2001. The role of omega-3 fatty acids in the prevention of coronary heart disease. Arch. Intern Med. 161, 2185–2192.

Harris, W.S., 2004. Are omega-3 fatty acids the most important nutritional modulators of coronary heart disease risk? Curr. Atheroscler. Rep. 6, 447–452.

He, K., Song, Y., Daviglus, M.L., Liu, K., Van Horn, L., Dyer, A.R., Goldbourt, U., Greenland, P., 2004. Accumulated evidence on fish consumption and coronary heart disease mortality: A meta-analysis of cohort studies. Circulation 109, 2705–2711.

Heilmann, C., Grandjean, P., Weihe, P., Nielson, F., Budtz-Jorgensen, D., 2006. Reduced antibody responses to vaccinations in children exposed to polychlorinated biphenyls. PLoS Med. 3, e311.

Höök, O., Lundgren, K.-D., Swensson, A., 1954. On alkyl mercury poisoning, with description of two cases. Acta Med. Scand. 150, 131–137.

Huisman, M., Koopman-Esseboom, C., Fidler, V., Hadders-Algra, M., Van der Paauw, C.G., Tuinstra, LGMTh, Weisglas-Kuperus, N., Sauer, P. J.J., Touwen, B.C.L., Boersma, E.R., 1995a. Perinatal exposure to polychlorinated biphenyls and dioxins and its effect on neonatal neurological development. Early Hum. Dev. 41, 111–127.

Huisman, M., Koopman-Esseboom, C., Lanting, C.I., Van der Paauw, C.G., Tuinstra, LGMTh, Fidler, V., Weisglas-Kuperus, N., Sauer, P.J.J., Boersma, E.R., Touwen, B.C.L., 1995b. Neurological condition in 18–month-old children perinatally exposed to exposure to polychlorinated biphenyls and dioxins. Early Hum. Dev. 43, 165–176.

Iwata, H., Tanabe, S., Sakai, N., Nichimura, A., Tatsukawa, R., 1994. Geographical distribution of persistent organochlorines in air, water and sediments from Asia and Oceania and their implications from global redistribution from lower latitudes. Environ. Pollut. 85, 15–33.

Jacobson, J.L., Jacobson, S.W., 1996. Intellectual impairment in children exposed to polychlorinated biphenyls in utero. N. Engl. J. Med. 335, 783–789.

Jacobson, J.L., Jacobson, S.W., Humphrey, H.E.B., 1990a. Effects of in utero exposure to polychlorinated biphenyls and related contaminants on cognitive functioning in young children. J. Pediatr. 116, 38–45.

Jacobson, J.L., Jacobson, S.W., Humphrey, H.E.B., 1990b. Effects of exposure to PCBs and related compounds on growth and activity in children. Neurotoxicol. Teratol. 12, 319–326.

Jacobson, J.L., Jacobson, S.W., Padgett, R.J., Brumitt, G.A., Billings, R. L., 1992. Effects of prenatal PCB exposure on cognitive processing efficiency and sustained attention. Dev. Psychol. 28, 297–306.

Jacobson, S.W., Fein, G.G., Jacobson, J.L., Schwartz, P.M., Dowler, J.K., 1985. The effects of intrauterine PCB exposure in visual recognition memory. Child Dev. 56, 853–860.

Johansen, P., Muir, D., Asmund, G., Riget, F., 2004. Human exposure to contaminants in the traditional Greenland diet. Sci. Total Environ. 331, 189–206.

Kinloch, D., Kuhnlein, H., Muir, D.C., 1992. Inuit foods and diet: A preliminary assessment of benefits and risks. Sci. Tot. Environ. 122, 247–278.

Kjellstrom, T., Kennedy, P., Wallis, S., Mantell, C., 1986. Physical and mental development of children with prenatal exposure to mercury from fish. Stage 1: Preliminary test at age 4. Sweden, National Swedish Environmental Protection Board.

Koopman-Esseboom, C., Morse, D.C., Weisglas-Kuperus, N., Lutkeschipholt, I.J., Van der Paauw, C.G., Tuinstra, L.G., Brouwer, A., 1994. Effects of dioxins and polychlorinated biphenyls on thyroid hormone status of pregnant women and their infants. Pediatr. Res. 36, 468–473.

Kris-Etherton, P.M., Harris, W.S., Appel, L.J., 2002. AHA Scientific Statement. Fish consumption, fish oil, omega-3 fatty acids, and cardiovascular disease. Circulation 106, 2747–2757.

Lagueux, J., Pereg, D., Ayotte, P., Dewailly, É., Poirier, G.G., 1999. Cytochrome P450 CYP1A1 enzyme activity and DNA adducts in placenta of women environmentally exposed to organochlorines. Environ Res. 80, 369–382.

Loganathan, B.G., Kannan, K., 1994. Global organochlorine contamination trends: An overview. Ambio 23, 187–191.

Marsh, D.O., Turner, M.D., Smith, J.C., Allen, P., Richdale, N., 1995. Fetal methylmercury in a Peruvian fish-eating population. Neurotoxicology. 16, 717–726.

McKeown-Eyssen, G.E., Ruedy J., Neims, A., 1983. Methyl mercury exposure in Northern Quebec. II. Neurologic findings in children. Am. J. Epidemiol. 118, 470–479.

McKone, T.E., McLeod, M., 2003. Tracking multiple pathways of human exposure to persistent multimedia pollutants: Regional, continental and global scale models. Annu. Rev. Environ. Resour. 28, 463–492.

Minh, N.M., Minh, T.B., Kajiwara, N., Kunisue, T., Subramanian, A., Iwata, H., Tana, T.S., Baburajendran, R., Karuppiah, S., Viet, P.H., Tuyen, B.C., Tanabe, S., 2006. Contamination by persistent organic pollutants in dumping sites of Asian developing countries: Implication of emerging pollution sources. Arch. Environ. Contamin. Toxicol. 50, 474–481.

Muckle, G., Ayotte, P., Dewailly, É., Jacobson, S.W., Jacobson, J.L., 2001. Determinants of polychlorinated biphenyls and methylmercury exposure in Inuit women of childbearing age. Environ. Health Perspect. 9, 957–963.

Mulvad, G., Pedersen, H.S., Hansen, J.C., Dewailly, É, Jul, E., Pedersen, M.B., Bjerregaard, G.T., Malcom, Y., Degushi, Y., Middaugh, J.P., 1996. Exposure of Greenlandic Inuit to organochlorines and heavy metals through the marine food-chain: An international study. Sci. Total Environ. 186, 137–139.

Myers, G.J., Marsh, D.O., Cox, C., Davidson, P.W., Shamlaye C.F., Tanner, M.A., Choi, A., Cernichiari, E., Choisy, E., Clarkson, T.W., 1995a. A pilot neurodevelopmental study of Seychellois children following in utero exposure to methylmercury from a maternal fish diet. Neurotoxicology 16, 629–638.

Myers, G.J., Marsh, D.O., Davidson, P.W., Cox, C., Shamlaye, C.F., Tanner, M.A., Choi, A., Cernichiari, E., Choisy, E., Clarkson, T.W., 1995b. Main neurodevelopmental study of Seychellois children following in utero exposure to methylmercury from a maternal fish diet: Outcome at six months. Neurotoxicology 16, 653–664.

National Research Council (NRC), 2000. Toxicological Effects of Methylmercury. Washington, DC, National Academy Press.

Patandin, S., Koopman-Esseboom, C., De Ridder, M.A.J., Weisglas-Kuperus, N., Sauer, P.J.J., 1998. Effects of environmental exposure to polychlorinated biphenyls and dioxins on birth size and growth in Dutch children. Pediatr. Res. 44, 538–545.

Patandin, S., Lanting, C.I., Mulder, P.G., Boersma, E.R., Sauer, P.J., Weisglas-Kuperus, N., 1999. Effects of environmental exposure to polychlorinated biphenyls and dioxins on cognitive abilities in Dutch children at 42 months of age. J. Pediatr. 134, 33–41.

Quinn, B., Gagne, F., Weber, J.P., Blaise C., 2005. Ecotoxicological effects of a semi-submerged municipal dump (Castle Harbour, Bermuda) on the Calico scallop Argopecten gibbus. Mar Pollut Bull. 51, 534–544.

Renzoni, A., Zino, F., Franchi, E., 1998. Mercury levels along the food chain and risk for exposed populations. Environ. Res. 77, 68–72.

Rhainds, M., Levallois, P., Dewailly, É., Ayotte, P., 1999. Lead, mercury, and organochlorine compound levels in cord blood in Quebec, Canada. Arch. Environ. Health 54, 40–47.

Rissanen, T., Voutilainen, S., Nyyssonen, K., Lakka, T.A., Salonen, J.T., 2000. Fish oil-derived fatty acids, docosahexaenoic acid and docosapentaenoic acid, and the risk of acute coronary events. Circulation. 102, 2677–2679.

Rogan, W., Gladen, B.C., Hung, K., Koong, S., Shih, L., 1988. Congenital poisoning by polychlorinated biphenyls and their contaminants in Taiwan. Science 241, 334–336.

Rogan, W., Gladen, B.C., McKinney, J.D., Carreras, N., Hardy, P., Thullen, J., Tinglestad, J., Tully, M., 1986. Neonatal effects of transplacental exposure to PCBs and DDE. J. Pediatr. 109, 335–341.

Rouja, M.P., Dewailly, É., Blanchet, C., the Bardi Community, 2003. Fat, fishing patterns and health among the Bardi People of North Western Australia. Lipids. 38, 399–405.

Ruxton, C.H., Reed, S.C., Simpson, M.J., Millington, K.J., 2004. The health benefits of omega-3 polyunsaturated fatty acids: A review of the evidence. J. Hum. Nutr. Diet 17, 449–459.

Ryan, J.J., Dewailly, É., Gilman, A., Laliberté, C., Ayotte, P., Rodrigue J., 1997. Dioxin-like compounds in fishing people from the Lower North Shore of the St. Lawrence River, Quebec, Canada. Arch. Environ. Health. 52, 309–316.

Salonen, J.T., Alfthan, G., Huttunen, J.K., Pikkarainen, J., Puska, P., 1982. Association between cardiovascular death and myocardial infarction and serum selenium in a matched-pair longitudinal study. Lancet 175–179.

Salonen, J.T., Seppanen, K., Nyyssonen K., Korpela, H., Kauhanen, J., Kantola, M., Tuomilehto, J., Esterbauer, H., Tatzber, F., Salonen, R., 1995. Intake of mercury from fish, lipid peroxidation, and the risk of myocardial infarction and coronary, cardiovascular and any death in Eastern Finnish men. Circulation 91, 645–655.

Schultz, C.D., Crear, D., Pearson, J.E., Rivers, J.E., Hylin, J.W., 1976. Total and organic mercury in the Pacific blue marlin. Environmental Contamination Toxicology 15, 230–234.

Schwartz, P.M, Jacobson, S.W., Fein, G., Jacobson, J.L., Price, H.A., 1983. Lake Michigan fish consumption as a source of polychlorinated biphenyls in human cord serum, maternal serum, and milk. Am. J. Public Health 73, 293–296.

Sherlock, J., Hislop, J., Newton, D., Topping, G., Whittle, K., 1984. Elevation of mercury in human blood from controlled chronic ingestion of methylmercury in fish. Hum. Toxicol. 3, 117–131.

Sjodin, A., Hagmar, L., Wehler, E., Bjork, J., Bergman, A., 2000. Influence of the consumption of fatty Baltic Sea fish on plasma levels of halogenated environmental contaminants in Latvian and Swedish men. Environ. Health Perspect. 108, 1035–1041.

Sorensen, N., Murata, K., Budtz-Jørgensen, E., Weihe, P., Grandjean, P., 1999a. Blood pressure and heart rate in 7–year-old children with prenatal exposure to methylmercury. Epidemiology 10, 370–375.

Sorensen, N., Murata, K., Budtz-Jorgensen, E., Weihe, P., Grandjean, P., 1999b. Prenatal methylmercury exposure as a cardiovascular risk factor at seven years of age. Epidemiology 10, 370–375.

Stewart, P., Darvill, T., Lonky, E., Reihman, J., Pagano, J., Bush, B., 1999. Assessment of prenatal exposure to PCBs from maternal consumption of Great Lakes fish: An analysis of PCB pattern and concentration. Environ. Res., Section A 80, S87–S96.

Stewart, P., Reihman, J., Lonky, E., Darvill, T., Pagano, J., 2000. Prenatal PCB exposure and neonatal behavioral assessment scale (NBAS) performance. Neurotoxicol. Teratol. 22, 21–29.

Tanabe, S., 1991. Fate of Toxic Chemicals in the Tropics, Mar. Pollut. Bull. 22, 259–260.

Uauy, R., Hoffman, D.R., Peirano, P., Birch, D.G., Birch, E.E., 2001. Essential fatty acids in visual and brain development. Lipids 36, 885–895.

U.S. Department of Health and Human Services and U.S. Environmental Protection Agency. What You Need to Know About Mercury in Fish and Shellfish, 2004. EPA and FDA Advice For: Women Who Might

Become Pregnant Women—Who Are Pregnant Nursing Mothers—Young Children. EPA-823–R-04–005.

Van Oostdam, J.V., Gilman, A., Dewailly É., Usher, P., Wheatley, B., Kuhnlein, H., Neve, S., Wlaker, J., Tracy, B., Feeley, M., Jerome, V., Kwavnick, B., 1999. Human health implications of environmental contaminants in Arctic Canada: A review. Sci. Total Environ. 230, 1–82.

Wakita, Y., 1987. Hypertension induced by methyl mercury in rats. Toxicol. Appl. Pharmacol. 89, 144–147.

Wania, F., Mackay. D., 1993. Global fractionation and cold condensation of low volatility organochlorine compounds in polar regions. Ambio 22, 10–18.

Wania, F., Mackay, D. 1996. Tracking the distribution of persistent organic pollutants. Environ Sci. Technocol. 30, 390A-396A.

Warkany, J., Hubbard, D.M., 1953. Acrodynia and mercury. J. Pediatr. 19, 365–386.

Weihe, P., Hansen, J.C., Murata, K., Debes, F., Jorgensen, P., Steuerwald, U., White, R.F., Grandjean, P., 2002. Neurobehavioral performance of Inuit children with increased prenatal exposure to methylmercury. Int. J. Circumpolar Health 61, 41–49

Whanger, P.D., 1992. Selenium in the treatment of heavy metal poisoning and chemical carcinogenesis. J. Trace Elem. Electrolytes Health Dis. 6, 209–221.

World Health Organization (WHO), 2003. WHO Technical series 922. Evaluation of certain food additives and contaminants food additives and contaminants. Sixty-first report of the Joint FAO/WHO Expert Committee on Food Additives. Geneva, World Health Organization.

Yu, M.L., Hsu, C.C., Gladen, B.C., Rogan, W.J., 1991. In utero PCB-PCDF exposure: Relation of developmental delay to dysmorphology and dose. Neurotoxicol. Teratol. 13, 195–202.

STUDY QUESTIONS

1. Explain why and how mercury and POPs accumulate in the aquatic food chain.
2. With regard to the effects of prenatal mercury exposure, try to explain why the Seychelles Island and the Faroe Island studies have contradictory results.
3. What are the positive benefits of the consumption of seafood for humans?
4. What are the main health effects related to POP exposure during pregnancy?
5. What advice could you give a pregnant woman to maximize the benefits and minimize the risks associated with fish consumption both during and after her pregnancy?
6. How can POP and mercury contamination of the aquatic food chain be prevented?
7. If you could design your own mobile response laboratory similar to the Atlantis, what sort of capabilities would you include and why?

C A S E S T U D Y

10

A Case Study in Bermuda: POPs and Heavy Metals in Newborns and Fish

In remote islands, the lack of appropriate environmental monitoring related to the unavailability of appropriate expertise and facilities onsite increases the risk of undetected environmental contamination and associated health hazards. The Atlantis Mobile Laboratory was created as a novel, unique infrastructure to meet the challenges of studying the interplay between coastal ecosystems and human health in remote settings.

THE ATLANTIS MOBILE LABORATORY

Atlantis belongs to Laval University, Quebec, Canada, and was co-funded by the Canadian Foundation for Innovation, the Quebec government and various other partners. It is a self-sufficient laboratory designed to travel by boat or ground transportation to virtually any destination that needs complete laboratory facilities for field-based environmental monitoring, environmental health studies or research activities in related disciplines. The facility comprises six modules built in standard size containers. Three modules are laboratories (microbiology, analytical chemistry/toxicology, and biochemistry) and three are dedicated to support the facility and the transport of material.

The microbiology module includes rapid assessment technologies and conventional microbiological tools to help identify and quantify fecal indicators and pathogens in seawater, drinking water, and seafood. It contains all equipment necessary for classical microbiology (membrane filtration, bacterial culture, etc.), and molecular microbiology (PCR cabinet, thermal cycler, etc.). A geomatic unit is also located in this module, allowing the integration of all data generated by Atlantis' scientists.

The analytical chemistry/toxicology module is dedicated to the measurements of environmental contaminants and other chemicals in diverse matrices (human and animal biological samples, food, water, sediments, etc.). It contains equipment found in other cutting-edge analytical toxicology facilities, including a gas chromatograph mass spectrometer (GC-MS) to measure organic compounds (PCBs, pesticides, etc.), a graphite furnace atomic absorption spectrometer (AAS) to analyze metals and organometals (cadmium, lead, arsenic, butyltins, etc.), and a specialized cold vapor atomic absorption spectrometer for mercury analyses.

The biochemistry/toxicology module is dedicated to the study of the biological activity/toxicity of environmental contaminants, either alone or in mixtures. To allow the widest range of bioassays and biochemical techniques to be performed, the biochemistry/toxicology laboratory is equipped with a complete cell culture room, a separate incubator for bacterial culture, and a wide range of laboratory instruments.

THE BERMUDA EXPEDITION

The Atlantis Mobile Laboratory was first field-tested in Bermuda, a remote island location that could offer ancillary support to the laboratory through the Bermuda Institute of Ocean Sciences (BIOS), formerly known as the Bermuda Biological Station for Research (BBSR). Bermuda is an archipelago located in the Atlantic Ocean approximately 586 miles from Cape Hatteras, North Carolina, and roughly 770 miles from New York. It includes several islands that together cover approximately 21 square miles in area and support a population of about 62,000 (52% female, 48% male), amounting to a population density of nearly 3000 people per square mile. Overall, the quality of life is quite high for Bermudians, with a low infant mortality rate of 3.6 deaths per 1000 births (Census 2000) and life expectancies of 75 years for men and 79 years for women. The country's economy has come to rely on the revenue of the tourist industry, as well as on international business.

The Atlantis research program included a suite of projects tailored to assess the environmental health issues of specific interest in Bermuda. It covered three areas of research: (1) marine ecotoxicology, (2) microbiology, and (3) human toxicology. Within all research projects, the geomatic unit implemented advanced geo-referencing and database management systems. In addition to the research component, educational outreach was a principal goal of Atlantis (Fig. 10-1). All these projects were organized in collaboration with Bermuda public health and environmental protection authorities.

Ecotoxicology

This study aimed to evaluate the toxicity exerted by a dumpsite that is partially submerged in a harbor. This dumpsite illustrates the inherent difficulties encountered in managing waste in remote settings. Scallops were caged and exposed at different distances from the dumpsite and harvested after exposure for nearly 2 months. Scallops were dissected (Fig. 10-1), and several biomarkers were assessed. Results showed that genotoxic effects (i.e., formation of micronuclei) could be observed, and that this effect was directly related to the distance from the dump. Other effects were also observed, including possible endocrine-disrupting effects, metallothioneins induction (induced by heavy metals), and lipid peroxidation (Quinn *et al.*, 2005).

Microbiology

As in many other remote maritime settings, freshwater is a scarce resource in Bermuda. The most frequent source of drinking water in Bermudian households is rainwater collected from rooftops and stored in individual household water tanks. Given the high potential for microbial contamination in such water collection systems, a study was set up

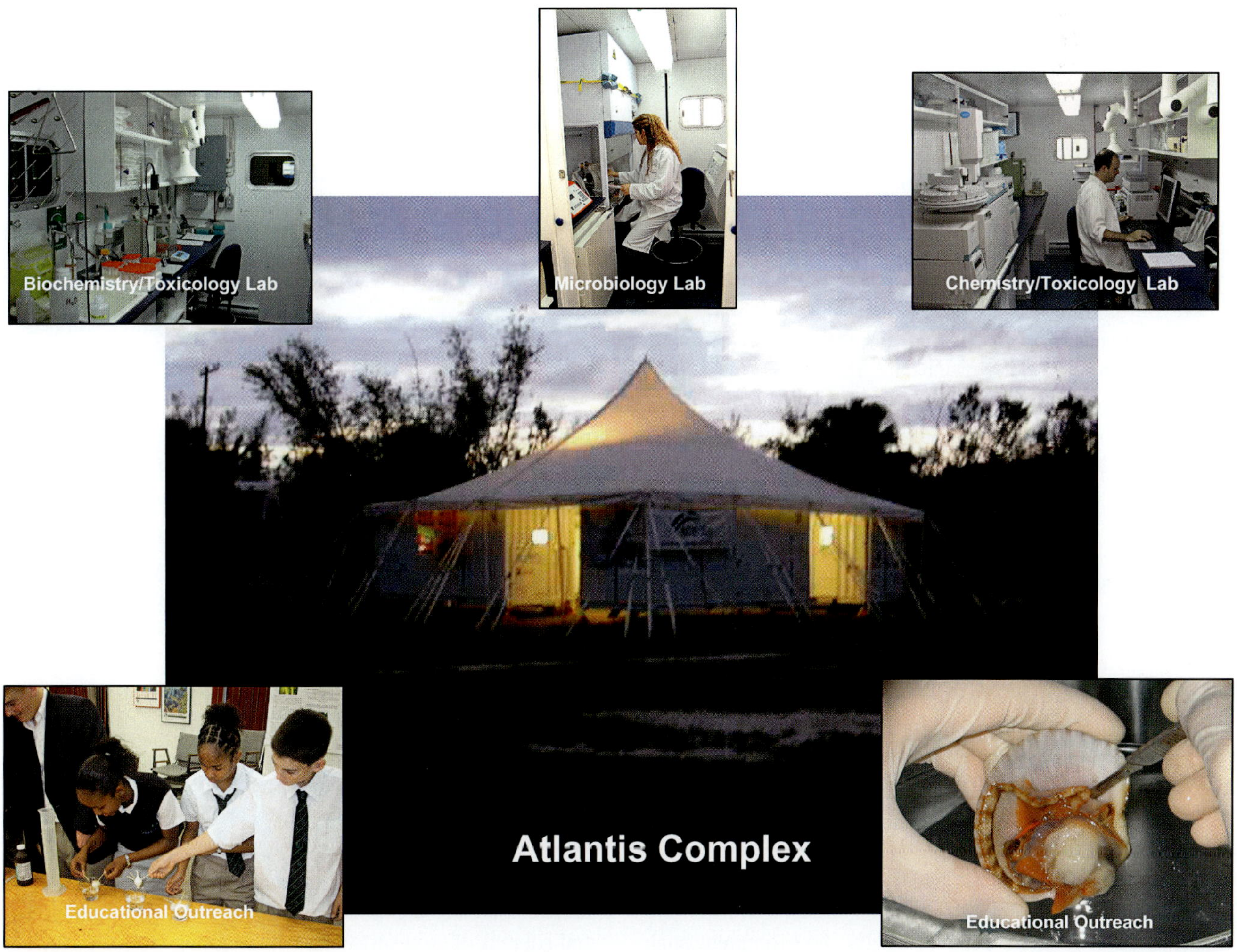

FIGURE 10-1. The Atlantis complex.

to evaluate the microbial quality of household tank water and the efficiency of protective procedures to avoid contamination. The results revealed that tank water was frequently contaminated with coliform bacteria (as measured by Colilert). The only preventive procedure that was efficient was the frequent emptying and cleaning of the tanks. The other preventive measures that were mentioned by study participants (chlorination, roof cleaning, filtration) did not appear to prevent contamination.

Human Toxicology

Exposure to environmental contaminants may occur through dietary sources, especially with the consumption of wild fish and game. In remote areas, the prohibitive importation costs favor the exploitation of local food sources rather than imported goods, hence increasing the risk of exposure to environmental contaminants found in wild foodstuffs. Because prenatal life is a critical period with regards to adverse effects on physical and neurological development, in relation to environmental exposures to toxicants this study aimed at (1) evaluating local fishes as potential sources of mercury and (2) monitoring prenatal exposure to mercury and organochlorines. The results of this study are presented next.

MONITORING MERCURY CONTENT IN FISH CONSUMED BY BERMUDIANS

For Bermudian residents, sport fishing is an integral part of the local culture, even when fishing does not represent an essential food source for sustenance. Because eating fish is full of health benefits, Atlantis researchers and the Bermudian Ministry of the Environment undertook the task of substantiating the mercury profile of fish in Bermuda.

Methods

Fish samples were collected from species generally caught at the top of the food chain (e.g., predatory reef fish, pelagic fish) and those fish that local people might commonly consume. Fishermen caught fish and gave them to the researchers, or the researchers bought pieces of local fresh fish at the store. A total of 88 samples were collected from 25 fish species (i.e., 81 flesh samples, 5 liver samples, 1 roe sample, and 1 fat sample).

Samples were digested in nitric acid in pressured vials and analyzed by cold vapor atomic absorption mass spectrometry in a Pharmacia mercury monitor, Model 100. Results were reported on a wet weight basis, and the limit of detection was 0.05 μg/g wet weight. For computational purposes, samples with undetected levels of mercury were given a value equal to half the detection limit (0.025 μg/g). Means were computed for species with more than one sample available.

Results

Table 10-1 shows the mercury concentrations found in all flesh samples analyzed (μg/g wet weight). Results are expressed as arithmetic means ± standard deviation, and the range of value is also presented. Results for liver, fat and roe are shown in Table 10-2. Mean mercury concentrations can be compared to action limits determined by various international agencies (e.g., 0.5 to 1 μg/g). All samples showing a concentration above the 0.5 μg/g limit are indicated in bold. For some species, the mean mercury concentration was below that threshold, but the range showed that some samples had higher values. The range for these species is indicated in bold. As expected, predatory species showed higher content of mercury than other species. Moreover, fish liver accumulated higher mercury content, especially the shark's liver. On the other hand, these data also indicated that a wide variety of fish species showed low mercury content and might therefore be safer to include in the local diet.

MONITORING PRENATAL EXPOSURE OF BERMUDIANS TO ENVIRONMENTAL CONTAMINANTS

Most epidemiological and experimental studies on health effects related to toxic metals (Pb, Hg) and POPs exposure (mainly PCBs) suggest that prenatal life is the most susceptible period for induction of adverse effects on physical and neurological development. As described earlier, the consumption of certain types of fish may be associated with exposure to MeHg, as well as to Pb and some POPs that accumulate in fish fatty tissues. For this reason, a study was conducted at the King Edward VII Memorial Hospital (KEMH) in Bermuda in order to provide baseline data on prenatal exposure to MeHg and POPs (pesticides and PCBs).

Methods

Women (n = 42) were recruited at KEMH and informed consent was obtained from all who agreed to participate in the study. At birth, cord blood was collected in EDTA-containing Vacutainers; after delivery, questionnaires were administrated to the mothers to gather information on potential sources of environmental contaminant exposure, such as diet and other daily lifestyle habits.

TABLE 10-1. Mercury concentrations (ug/g ww) in the flesh of local fish species in Bermuda.

Common Name	Species	*n*	Range	Mean ± Standard Deviation
Swordfish	*Xiphias gladius*	1		**3.31**
Bermuda mackerel	*Euthynnus alletteratus*	2	**2.10–2.30**	**2.20 ± 0.14**
Amberjack	*Seriola sp.*	1		**1.55**
Black grouper	*Mycteroperca bonaci*	3	**0.64–1.55**	**1.21 ± 0.50**
Sixgill shark	*Hexanchus griseus*	1		**0.92**
Deep water red snapper	*Etelis oculatus*	1		**0.52**
Barracuda	*Sphyraena barracuda*	3	**0.18–0.85**	**0.52 ± 0.33**
Rockfish		2	**0.35–0.65**	**0.50 ± 0.21**
Wahoo	*Acanthocybium solandri*	7	**0.15–1.00**	0.48 ± 0.31
Yellowfin tuna	*Thunnus albacares*	15	**0.20–1.10**	0.34 ± 0.28
Yellowtail snapper	*Ocyurus chrysurus*	5	**0.05–0.65**	0.34 ± 0.26
Ocean robin	*Decapterus macarellus*	4	0.30–0.35	0.31 ± 0.02
Tuna (unidentified)	*Thunnus sp.*	4	0.21–0.36	0.26 ± 0.07
Bermuda spiny lobster	*Panulirus argus*	4	0.15–0.30	0.21 ± 0.08
Bermuda bonita	*Seriola rivoliana*	1		0.20
Blackfin tuna	*Thunnus atlanticus*	2	0.20–0.20	0.20 ± 0.00
Coney	*Cephalopholis fulva*	12	0.10–0.35	0.19 ± 0.08
Red hind	*Epinephelus guttatus*	2	0.15–0.18	0.16 ± 0.02
Bonito		2	<0.05–0.30	0.16 ± 0.19
Tapioca fish		1		0.15
Grey triggerfish	*Balistes capriscus*	1		0.10
Hogfish	*Lachnolaimus maximus*	3	<0.05–0.15	0.09 ± 0.06
Bermuda dolphinfish	*Coryphaena sp.*	1		0.05
Bermuda chub	*Kyphosus sectator*	1		0.025
Black grouper	*Mycteroperca bonaci*	1		0.025
Barber	*Paranthias furcifer*	1		0.025

TABLE 10-2. Mercury concentrations (ug/g ww) in the liver, fat, and roe of some local fish species in Bermuda.

Common Name	Species	Tissue	*n*	Range	Mean ± Standard Deviation
Sixgill shark	*Hexanchus griseus*	Liver	1		**19.10**
Red hind	*Epinephelus guttatus*	Liver	1		**0.63**
Hogfish	*Lachnolaimus maximus*	Liver	2	0.15–0.30	0.22 ± 0.11
Bonito		Liver	1		<0.05
Hogfish	*Lachnolaimus maximus*	Roe	1		<0.05
Coney	*Cephalopholis fulva*	Fat	1		0.25

Mercury was assessed by cold vapor atomic absorption spectrometry. The limit of detection for mercury dosage in whole blood by this method was 2 nmol/L.

Fourteen PCB congeners (IUPAC # 28, 52, 99, 101, 105, 118, 128, 138, 153, 156, 170, 180, 183, and 187) and 11 pesticides (aldrin, dieldrin, mirex, α-chlordane, γ-chlordane, oxychlordane, heptachlor epoxide, cis-nonachlor, trans-nonachlor, *p,p'*-DDE, *p,p'*-DDT, hexachlorobenzene, and β-BHC) were measured in the plasma extracts by high-resolution gas chromatography (HRGC) with negative chemical ionization detection. Limits of quantification were 0.6 μg/L for PCB-52 and total PCBs expressed as Aroclor 1260;

0.2 μg/L for heptachlor epoxide and dieldrin, 0.1 μg/L for PCB-28, β-BHC and *p,p′*-DDT, and 0.06 μg/L for all other congeners and pesticides.

Results

Of the 42 women recruited for this study, only 40 returned completed questionnaires. The age range for the whole study group was 18 to 41 years; mean maternal age was 29.3 ± 0.9 years. Among the subjects, 37% of the women were primigravida (first time pregnant), 32% were pregnant for the second time, and 31% had delivered between two and six times previously. Women came from all parishes of Bermuda. There were no twin pregnancies, no intrauterine growth retardation, and no cardiac problems or other medical problems with the newborns; 8 of 40 deliveries (20%) were by Caesarian section. Of all newborns, 18 were boys and 22 were girls. APGAR scores (a measure of health at birth) at 5 minutes ranged between 8 and 10, with only 1 occurrence of a lower score. Birth weights ranged between 2490 and 4281 grams (mean ± standard deviation = 3283 ± 64 grams). Birth heights ranged between 45 and 52 cm; 48 ± 0.3 cm. All of these parameters suggested normal, healthy pregnancies.

Data on PCB and other organochlorine levels in samples of umbilical cord plasma are presented in Table 10-3. Values are not presented for analytes undetected in all samples (i.e., PCB-28, 52, 99, 105, 118, 128, 156, 170, 183, 187; as well as aldrin, isomers and metabolites of chlordane, cis- and trans-nonachlor, HCB, and mirex). Only percentage detection and the range of values are shown for analytes that were detected. Mean value is shown only for *P,P′*-DDE, which was detected in 100% of samples. Results are expressed as μg/L.

PCB congeners showing detectable levels included C-101, C-138, C-153, and C-180. Aroclor 1260 is an estimate measure of the most persistent mixture of PCB congeners based on C-153 and C-138. It was detected in only 23% of the samples, with concentrations ranging from undetectable levels to 0.9 μg/L. Ten of 14 individual PCB congeners analyzed showed undetectable levels in all samples, and 9 of the 11 other organochlorine compounds measured were not detected. *p,p′*-DDE was detected in 100% of the samples and showed a mean level of 0.43 ± 0.04 μg/L, whereas β-BHC was detected in only 2.6% of the samples.

All cord blood samples but one showed detectable levels of mercury (n = 42). The mean mercury content of cord blood was 41.3 ± 4.7 nmol/L, with levels ranging from undetected values to 160 nmol/L. These levels were higher than expected. Based on two samples for which inorganic mercury was measured, it was estimated that 85% of total mercury measured in cord blood could be attributed to methylmercury (MeHg), suggesting that contaminated seafood was the main source of MeHg exposure.

The association of mercury levels with fish consumption was tested using Pearson's correlation coefficient. Fish consumption was estimated by the number of meals per month that each patient reported eating on a regular basis. Fish species in the questionnaire included wahoo, snapper, tuna, mahi-mahi, salmon, unknown filet, and "other (unspecified) fish." The percentages of women reporting consuming fish on a regular basis is presented in Table 10-4, with distinctions made between local or imported fish. The species most often consumed included wahoo, snapper, tuna, and salmon. Whereas wahoo and snapper were clearly of local origin, tuna was most often imported (no distinction was made regarding canned versus fresh tuna), and most people who reported consuming salmon seemed unsure of its origin. For this reason, the consumption of local fish was analyzed separately from that of total fish consumption. Local fish consumption was estimated by pooling the number of meals per month of wahoo and snapper that each woman reported eating. Separate analyses were also conducted on each fish species (salmon, wahoo, snapper, tuna), despite the low statistical power of some of these analyses.

A significant association was seen between Hg levels and the consumption of total local fish (Spearman's r = 0.584, p

TABLE 10-3. Organochlorines in umbilical cord plasma of Bermudians (μg/L; n = 42).

Mean ± SE	*n*	% Detected	Range	Mean ± Standard Deviation
Arcolor 1260	39	23.1%	<0.6–0.90	NA
PCB-101	39	2.6%	<0.06–0.06	NA
PCB-138	39	7.7%	<0.06–0.07	NA
PCB-153	39	30.8%	<0.06–0.11	NA
PCB-180	39	5.1%	<0.06–0.06	NA
β-BHC	39	2.6%	<0.1–0.14	NA
p,p′-DDE	39	100%	<0.06–1.36	0.43 ± 0.04

NA = nonapplicable.

TABLE 10-4. Proportion of the study population who declared consuming fish on a regular basis (n = 42).

Fish Species	Local Fish	Imported Fish	Unknown Origin	Total
Wahoo	19.0%	2.4%	—	21.4%
Snapper	16.7%	—	—	16.7%
Tuna	11.9%	19%	4.8%	35.7%
Mahi-mahi/dolphin	2.4%	—	—	2.4%
Salmon	4.8%	4.8%	7.1%	16.7%
Filet	4.8%	—	—	4.8%
Other fish	4.8%	2.4%	2.4%	9.5%

< 0.001; $n = 40$), mainly wahoo (r = 0.502, p = 0.001; $n = 40$) and snapper (r = 0.499, p = 0.001; $n = 40$). The consumption of other fish species was not related to Hg content of cord blood. Total fish consumption was computed by pooling all fish meals per month that women reported eating any fish; this variable was not significantly correlated with Hg content of cord blood.

Organochlorine levels were very low, and most samples yielded no detectable levels, therefore, no further analyses were carried out to determine sources of exposure. The limit of detection for most OC analytes was 0.6 µg/L. Mean concentrations of PCBs in cord plasma have been studied in many different populations, including citizens of Faroe Island (Barr *et al.*, 2006), Nunavik Inuit (Dallaire, 2003), Quebec North Shore population (Dallaire *et al.*, 2002), and Michigan anglers (United States) (Schwartz, 1983). Levels found in these populations are reflective of dietary sources of exposure, with one of the most significant being sea mammal fat consumption in the Inuit (Dewailly *et al.*, 1992a, 1993; Johansen *et al.*, 2004; Kinloch *et al.*, 1992; Muckle *et al.*, 2001). Indeed, levels measured in Canadian Inuit (mean [standard deviation] = 1.88 [2.65] µg/L Aroclor 1260) (Butler Walker *et al.*, 2003) were higher than those measured in Bermuda (ranging from 0.6 to 0.9 µg/L), and higher than those reported for the general American population (reported range of 0.9 to 1.5 µg/L) (Agency for Toxic Substances and Disease Registry [ATSDR], 2000). Similarly, other organochlorines measured in cord plasma showed very low levels, and none displayed levels that raise health concerns.

However, mercury results were higher than expected. For comparison, in southern Quebec ($n = 1109$), the geometric mean cord blood Hg concentration was 4.82 nmol/L (concentrations ranging from 1.00–67.00 nmol/L), representing baseline exposure levels for North American populations (Rhainds *et al.*, 1999). In the Bermuda population, the mean level observed for mercury was 41.3 nmol/L, a little more than eightfold above the baseline exposure level. Fish-eating populations such as the Canadian Inuit ($n = 169$) have exhibited a mean cord blood level of 11 µg/L, corresponding to 54.9 nmol/L (Butler Walker *et al.*, 2006), and residents of the Faroe Islands have shown a geometric mean cord blood level of 22.4 µg/L, corresponding to 111.4 nmol/L (Grandjean *et al.*, 2005). In the Bermuda study, mercury could be detected in all samples analyzed except for one. Twenty-five of 42 samples (59%) were above the 30 nmol/L EPA guideline for mercury in cord blood, and 3 of 42 samples had cord blood mercury above the World Health Organization (WHO) guideline of 90 nmol/L. Fortunately, no newborns exhibited cord blood mercury levels that exceeded 200 nmol/L, the lowest concentration associated with 5% risk of developing psychomotor retardation established by the WHO (WHO, 2003).

In this study, mercury exposure can confidently be associated with fish consumption based on the ratio of inorganic mercury over total mercury. This is corroborated by the positive correlation found between the consumption of local fish and cord blood mercury levels. Data suggested that the consumption of wahoo and snapper, both predatory fish, are the main source of exposure to MeHg. This finding is important as it shows that local fish can be contaminated with significant levels of mercury (as shown in the previous section), even in an apparent pristine environment. Furthermore, the results suggest that such contamination is transferred to humans, including the fetus in the case of pregnant women, through consumption of certain fish species. Another key finding was that cord blood levels of mercury were not associated with the consumption of other fish species, suggesting that appropriate Bermuda fish consumption guidelines could be issued to limit exposure to mercury while still allowing consumers to benefit from the important nutrients found in fish.

In summary, the results of this pilot study on prenatal exposure of the Bermudian population to environmental contaminants showed that exposure to organochlorines do not seem to be a matter of concern in the Bermudian population at this time, whereas mercury exposure during pregnancy raises concerns and is mostly associated with the consumption of predatory fishes, an avoidable source of exposure if appropriately documented.

In view of these preliminary results, the Bermudian government has decided to expand the monitoring of mercury in fish in order to better inform the general Bermuda population, particularly pregnant women.

This pilot mission in Bermuda showed that the Atlantis Mobile Marine Laboratory can help remote maritime communities by monitoring their coastal environment and health. The mission of Atlantis is to help manage local risks, increase the awareness of the globalization and impact of contaminants, and support environmental health public policies that might benefit coastal people and ultimately all people worldwide.

SECTION I

C. Effects of Harmful Algal Blooms and Toxins

CHAPTER

11

Epidemiological Tools for Investigating the Effects of Oceans on Public Health

LORRAINE C. BACKER AND LORA E. FLEMING

INTRODUCTION

In the United States, 53% of the population lives in the 17% of the land that is coastline. This coastal population is expected to increase by 25 million people by 2015 (Pew Commission, 2003, and see review by Fleming *et al.,* 2006). Worldwide, 60% of the population lives in coastal areas, and 4 billion people live within 4 km of a coastline (U.S. Ocean Commission, 2004). In the near future, the world's population growth is predicted to center along subtropical and tropical coastlines, making increasing numbers of people vulnerable to ocean-based health risks, such as severe storms and waterborne diseases.

Coastal living does not entail only risks, however. Benefits from ocean- or coast-based activities include ports for economic exchange, food resources, natural products with pharmaceutical applications, and venues for recreation. Understanding the complex interactions among the oceans, coasts, and human populations is a daunting but necessary undertaking if we are to fully benefit from the ocean's bounty while, at the same time, protecting ourselves from ocean-borne risks.

In this chapter, we will present current epidemiological tools for use in discovering, assessing, and interpreting human health risks and benefits from the oceans (including the coasts and the Great Lakes). The chapter is organized into three parts. Part 1 is a brief epidemiology primer, and Part 2 is a series of case studies that will demonstrate how to use this valuable tool to assess human health effects from interactions with our oceans. Part 3 is a brief discussion of emerging issues in oceans and human health and their implications for the future practice of epidemiology and public health.

PART 1: EPIDEMIOLOGY PRIMER

Epidemiology is the application of the scientific method to studying the occurrence or risk of adverse health outcomes in populations—that is, in groups of living things. The tools of epidemiological science have evolved with the natural history of humans as astute health care providers and researchers have uncovered the causative links between exposures and subsequent diseases. Some examples of this scientific progression include the insight of John Snow in identifying the relationship between drinking contaminated water and contracting cholera in London in 1854 (Box 11-1; Fig. 11-1), the observations of Goldberg that led to understanding the importance of niacin in preventing pellagra (Terris, 1964), and the studies of Doll and Hill that defined the role of cigarette smoking as a risk factor for lung cancer (Doll and Hill, 1952). Although the applications of epidemiology vary, the common thread is the study of disease occurrence in relation to characteristics of individuals and their environment (Ahlborn and Norell, 1990).

Defining Disease

Before we can assess the occurrence of disease, we must be able to determine which individuals have the disease or a health outcome of interest. We examine individuals with regard to symptoms and signs, use tests to verify biological and chemical changes, and compare our observations to established diagnostic criteria (Ahlbom and Norell, 1990). Symptoms are subjective manifestations, such as pain, that the person under examination observes and reports. Signs are objective manifestations, such as rash or swelling, which

BOX 11-1. John Snow, "Father of Modern Epidemiology"

When the first cholera epidemic hit England in 1831, people thought it was spread by miasma in the air (Summers, 1989). Medical knowledge had not significantly progressed by the time of the outbreak in Soho 34 years later. However, during that time, an anesthesiologist named Dr. John Snow published a paper speculating that cholera was spread through water. He was convinced that the disease spread from person to person via sewage-tainted water. During the Soho outbreak, he observed that many victims' families lived near the Broad Street drinking water pump (see Fig. 11-1). His own observations about contaminants in water from the Broad Street pump led him to recommend to the Parish that officials remove the handle from the pump to prevent more cases of illness. When the pump handle was removed, the spread of cholera stopped. After the outbreak subsided, Snow's theories were still to be proved, and he was met with considerable resistance. However, his disease sleuthing and medical cartography helped set the stage for the development of a new science—epidemiology.

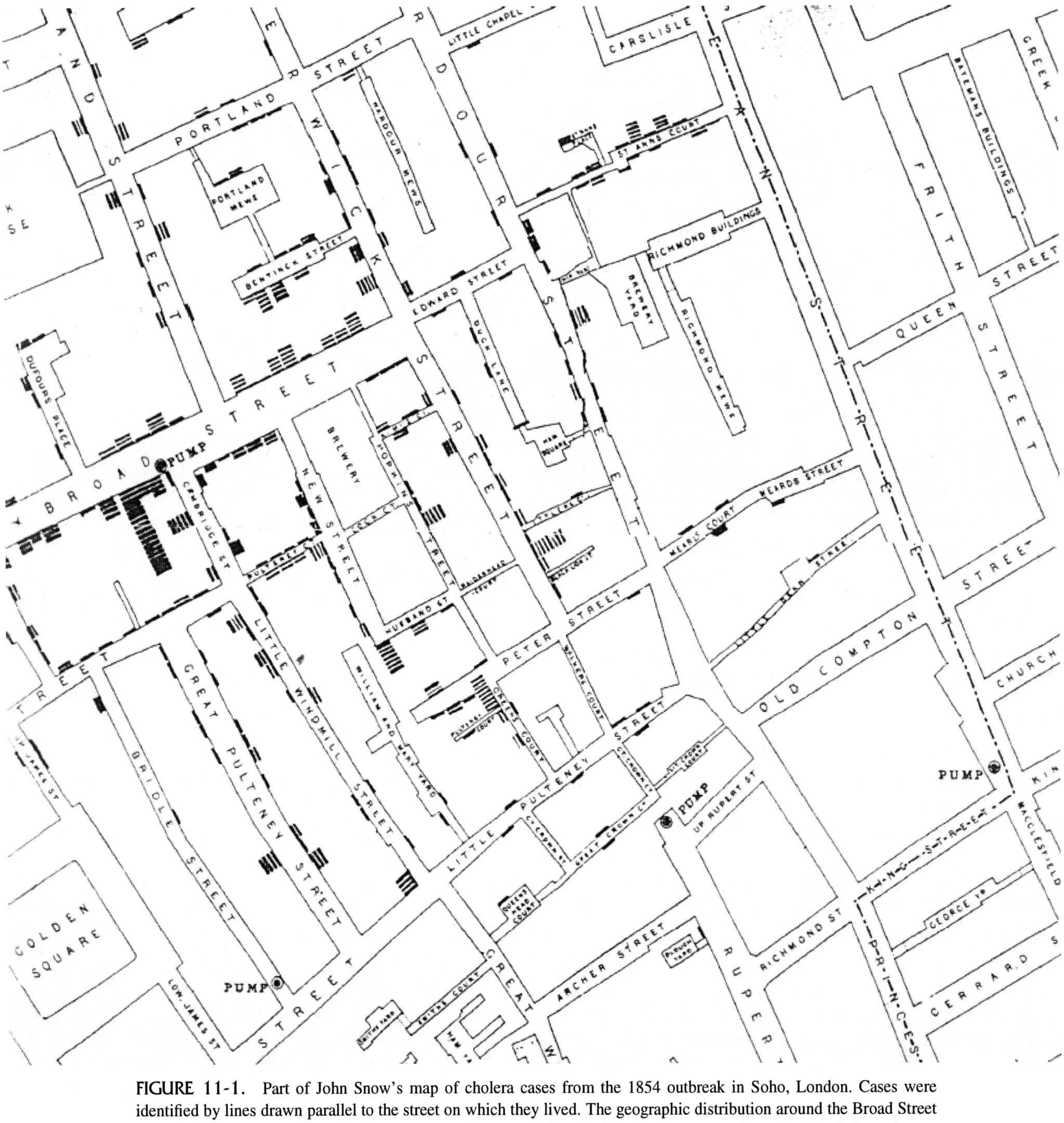

FIGURE 11-1. Part of John Snow's map of cholera cases from the 1854 outbreak in Soho, London. Cases were identified by lines drawn parallel to the street on which they lived. The geographic distribution around the Broad Street pump led Snow to speculate that the pump was the source of contagion. From www.ncgia.ucsb.edu/pubs/snow/snow.html.

can be observed by an examiner such as a health care provider. Tests are objective manifestations that can be measured and read from an instrument. Signs, symptoms, and tests can be used to diagnose an ocean-related illness, such as in ciguatera fish poisoning. In this case, symptoms include numbness around the mouth, fingers, and toes and reversal of hot and cold temperature sensations (Bagnis, 1968). Signs include vomiting and diarrhea, and tests include an analytical assay for the presence of ciguatoxin (the toxin associated with this disease in people) in fish left over from a meal. Because there is rarely leftover fish to test, the final diagnosis for ciguatera fish poisoning is one of exclusion—that is, if the person has the signs and symptoms and reports having eaten a large reef fish, then the person is diagnosed with ciguatera fish poisoning. Other characterized diseases associated with exposures to harmful algae and their toxins are described in Table 11-1.

Descriptive Epidemiology

We use descriptive epidemiology to identify how many cases of disease are occurring, as well as when, where, and to whom (Goodman and Peavy, 1996). The easiest way to examine the extent of an illness is to count the number of cases and compare these counts with historical records to search for unusual patterns or outbreaks. The number of cases also needs to be compared with the number of people in the population in which the cases occurred. Rates are measures for relating the number of cases to the population over a specified period of time. For example, morbidity rates describe the frequency of illness in a population over time, and mortality rates describe the frequency of death over time.

To assess the occurrence of diseases in populations, we must specify how to measure the disease or health outcome

TABLE 11-1. Characteristics of known diseases associated with human exposures to algae and their toxins.

Disease	Cause	Associated Vectors or Exposures	Signs and Symptoms
Amnesiac shellfish poisoning (ASP)	Domoic acid	Bivalve shellfish, primarily scallops, mussels, clams, oysters and (possibly) fish	Vomiting, diarrhea, abdominal pain and neurological effects (confusion, memory loss, disorientation, seizure, coma)
Diarrheic shellfish poisoning (DSP)	Okadaic acid and dinophysistoxins (6+)	Bivalve shellfish, primarily scallops, mussels, clams, oysters	Diarrhea, abdominal pain, chills, headache, fever
Neurotoxic shellfish poisoning (NSP)	Brevetoxins (10+)	Bivalve shellfish, primarily mussels, oysters, scallops, and (possibly) fish	Tingling and numbness of lips, tongue, and throat; muscular aches; dizziness; reversal of the sensations of hot and cold; diarrhea, vomiting
Paralytic shellfish poisoning (PSP)	Saxitoxins (20+)	Scallops, mussels, clams, oysters, cockles, and certain herbivorous fish (e.g., pufferfish), and crabs	Diarrhea, nausea, vomiting leading to paresthesias of mouth and lips, weakness, dysphasia, dysphonia, respiratory paralysis
Ciguatera fish poisoning	Ciguatoxins (10+), Maitotoxin, Scaritoxin	Large reef fish, such as grouper, red snapper, amberjack, barracuda (most common)	*2–6 hrs*: abdominal pain, nausea, vomiting, diarrhea. *3 hrs*: paresthesias, reversal of hot/cold, pain, weakness. *2–5 days*: bradycardia, hypotension, increase in T wave abnormalities *Weeks–months (untreated)*: paresthesias
Azaspiracid shellfish poisoning (AZP)	Azaspiracid	Bivalve shellfish	Nausea, vomiting, liver damage
Florida red tide respiratory irritation	Brevetoxins (10+)	Aerosolized seawater	Acute eye irritation, acute respiratory distress (nonproductive cough, rhinorrhea), asthma acerbation
Blue Green Algae (Cyanobacteria)	Anatoxins (3+), Saxitoxins (2+), Microcystins, Nodularins, Cylindrospermopsins	Direct contact with water, inhaling aerosols, drinking water, and possibly seafood	Skin and eye irritation, nausea, vomiting, respiratory distress, liver damage

Adapted from Backer *et al.*, 2003a; Fleming *et al.*, 2002. The numbers after the toxins indicate the number of congeners that have been identified.

(Ahlbom and Norell, 1990). The measures used in epidemiology can describe either the pool of existing cases of disease (i.e., prevalence) or the occurrence of new cases (i.e., incidence). Disease prevalence describes what proportion of the population has the disease of interest at any given time. The numerator is the number of people having the disease at a specific time, and the denominator is the number of people in the population at that point in time.

Two measures of disease incidence are incidence rate and cumulative incidence. To calculate incidence rate, the numerator is the number of cases of the disease that occurred in a population during a period of time, and the denominator is the sum of the length of time each individual in the population is at risk for getting the disease. The time periods in the denominator are person-time, such as person-years. For calculation of cumulative incidence, the numerator is the number of individuals who get the disease during a certain time period and the denominator is the number of individuals in the population at the beginning of that time period. Like prevalence, cumulative incidence is a proportion, whereas incidence is a rate. An example of how to calculate these measures, again using ciguatera fish poisoning, is presented in Box 11-2.

BOX 11-2. Examples of Prevalence, Cumulative Incidence, and Incidence Rate Calculations

In a mass screening of 800 island residents, 18 were found to have ciguatera fish poisoning. Over the next 2 years, 10 more of these island residents developed the disease. What measures of disease occurrence can be calculated?

In this example, 18 people out of 800 had ciguatera fish poisoning.

$$\text{The prevalence} = 18/800 = 0.023$$

Prevalence is a proportion, and it is thus dimensionless. Prevalence cannot take on a value of less than 0 or greater than 1.

In this example, 10 people out of the 788 who did not have ciguatera fish poisoning at the beginning of the time period became ill with ciguatera in 2 years.

$$\text{The cumulative incidence} = 10/\,(800 - 18) = 0.013 \text{ during the 2-year period}$$

Cumulative incidence is also a proportion and is thus dimensionless. Cumulative incidence cannot take on a value of less than 0 or greater than 1.

In this example, 800 × 2 (or 1600) person-years constitute the period of interest. During the 2-year period, 10 people become ill with ciguatera fish poisoning.

$$\text{The incidence rate} = 10/\,(800 \times 2) = 0.0063 \text{ per year}$$

Incidence is not a proportion; it has the dimension of time. Incidence cannot take on a value less than 0.

Measures of Association

After defining the disease of interest and assessing its occurrence in the population, we can compare the frequency of disease occurrence among a group of individuals who have a certain characteristic (i.e., the exposed population) to the frequency among individuals without that characteristic (i.e., the unexposed, or control, population). We use measures of association to quantify the strength of the statistical association between the exposure and the health outcome of interest. The calculated measure of association depends on the study design used. These measures of association are described in Dicker (1996) and are briefly discussed here and in Box 11-3 in the context of a hypothetical outbreak of ciguatera fish poisoning.

The relative risk is the risk (for disease or other adverse health outcome) in the exposed group compared to the risk in the unexposed group (i.e., background risk). The excess risk is expressed as a ratio. To continue with our example describing ciguatera fish poisoning among island-dwellers, investigators interviewed island residents during a survey of the population designed to assess the occurrence of ciguatera, and they used the information to calculate relative risk (Box 11-3).

Box 11-3. Example of Relative Risk Calculation

Among the population of 800 islanders, 450 people reported eating grouper within the previous year. Among those 450 people, 16 became ill with ciguatera during that same time period. Among the 350 people who did not report eating grouper, 5 became ill with ciguatera. The investigators summarized their data in a two-by-two table, which is presented here. The relative risk of getting ciguatera after eating grouper was (16/450)/(5/350), or 2.54. That is, people who ate grouper (were exposed) were 2.54 times more likely to get sick than those who did not (were unexposed).

		Exposure (Ate Grouper)		
		Yes	***No***	***Total***
Disease (Ciguatera Fish Poisoning)	Yes	16	5	21
	No	434	345	779
	Total	450	350	800

In most studies comparing the characteristics of cases with those of noncases (case-control studies), the investigator does not know the true size of the exposed and unexposed populations, and no denominator is available for calculating an attack rate or a relative risk. However, if the disease is rare, the relative risk can be approximated by the

BOX 11-4. Ciguatera Fish Poisoning Outbreak

The local department of health reported 15 cases of ciguatera fish poisoning following a local banquet. A total of 50 people attended the banquet. Among the cases of ciguatera, 13 reported having eaten grouper at the banquet. Among the 35 people who did not become ill, 8 reported eating the grouper. Below is a two-by-two table showing the numbers of cases and controls exposed or not exposed to grouper.

		Exposure (Ate Grouper)		
		Yes	*No*	*Total*
Disease (Ciguatera Fish Poisoning)	Yes	13	2	15
	No	8	27	35
	Total	21	29	50

BOX 11-5. Neurological Symptoms Associated with Eating Razor Clams

The following data were collected as part of an epidemiological investigation: 20 of 100 people who reported an exposure (e.g., ate razor clams harvested from a specific beach during a specific time) became ill with neurological symptoms, and 15 of 280 people who reported no exposure (i.e., did not eat razor clams from that beach during that time) became ill. The data are arranged in the following two-by-two table.

		Exposure (Ate Razor Clams)		
		Yes	*No*	*Total*
Disease (Neurological Symptoms)	Yes	20	15	35
	No	80	265	345
	Total	100	280	380

odds ratio. The odds ratio describes the chance of being exposed to a specific risk factor given disease status (ill or well), and, like the relative risk, it is calculated from data arranged in a two-by-two table (Box 11-4). In the hypothetical example of ciguatera fish poisoning described previously, the odds ratio would be $(13 \times 27)/(8 \times 2)$ or 21.9. Thus, the odds of having been exposed (i.e., having eaten grouper) are 21.9 times higher for people who were ill (i.e., cases) than for people who were not ill (i.e., controls).

Cross-sectional studies, or surveys, measure the prevalence (the number of existing cases) of a particular outcome or condition rather than the incidence (the number of new cases). The prevalence measures of association analogous to the risk ratio and odds ratio are the prevalence ratio and the prevalence odds ratio. For example, suppose that 20 of 100 people who reported an exposure (e.g., ate razor clams harvested from a specific beach during a specific time) became ill, and 15 of 280 people who reported no exposure (i.e., did not eat razor clams from that beach during that time) became ill. The data are arranged in a two-by-two table in Box 11-5. The prevalence ratio would be $(20/100)/(15/280) = 3.8$; that is, exposed people are 3.8 times more likely than unexposed people to have the illness. The prevalence odds ratio is $(20 \times 265)/(80 \times 15) = 2.75$. The odds of having disease are 2.75 times higher for the exposed group than for the unexposed group.

Epidemiological Research

To understand an epidemiological problem, we must first understand relevant characteristics of the host (e.g., age, sex, race, occupation, susceptibility), the exposure (e.g., biological, physical, chemical properties), and the environment (e.g., physical, biological, social, and economic factors) (Last, 1988). When we understand these characteristics, epidemiological research (including surveillance, case-control studies, cross-sectional studies, cohort studies, clinical trials, and intervention studies) helps us determine the interactions among host, exposure, and environment. Surveillance is described next, and the characteristics of planned studies (see Kleinbaum *et al.,* 1982) are presented in Table 11-2.

Surveillance

Surveillance includes the processes of collecting, analyzing, and disseminating data on specific health effects (e.g., disease, disability) (Thacker and Berkelman, 1998). To be most effective, a surveillance system must have the capacity to collect and analyze data as well as the capability to disseminate data in support of public health decision making. Surveillance is an important tool in public health practice, and surveillance data can be applied to many public health activities (Box 11-6).

In the United States, requirements for reporting diseases and conditions are mandated by state and territorial laws or regulations (Centers for Disease Control and Prevention [CDC], 2001). The Council of State and Territorial Epidemiologists (CSTE) and the Centers for Disease Control and Prevention (CDC) collaborate to determine which diseases must be reported to CDC, and these are called "notifiable diseases." For example, physicians are required by law to report foodborne disease outbreaks and cases of conditions of particular public health importance (including birth defects, anthrax, and tuberculosis). With the exception of paralytic shellfish poisoning (PSP), physicians are not required by CDC to report illnesses associated with

TABLE 11-2. Characteristics of descriptive epidemiological studies.

Characteristic	Study Design: Case-Control Study	Cross-Sectional Study	Cohort Study
Alternative Name	Case-referent Case history Retrospective study	Survey Prevalence study	Follow-up study Incidence study Prospective study
Objective	Define disease etiology	Describe disease or define disease etiology	Describe disease or define disease etiology
Unit of Study	Individual	Individual	Individual
Study Population	Selected from separate populations of available cases of disease and noncases (controls)	Representative sample from a single target population	Population at risk for developing the outcome
Exposure	Unknown at start of study	Measured before the follow-up period	Known for all subjects at start of study
Health Outcome	Rare events (e.g., cancer)	Conditions that are quantitatively measured and can vary over time (e.g., blood pressure) or are common diseases with long durations (e.g., chronic bronchitis)	Common events (e.g., coronary heart disease)
Follow-up	Retrospective	Nondirectional or retrospective, no follow-up period	Prospective, retrospective, or both
Measure of Association between Exposure and Outcome	Odds ratio		
Advantages	Applicable to studies of rare events Efficient use of resources	Used to generate new etiological hypotheses	Subjects enrolled before disease onset, investigator knows hypothesized cause preceded the health outcome

From Kleinbaum *et al.*, 1982.

BOX 11-6. Applications of Surveillance Data in Public Health Activities (Teutsch, 1994)

1. Estimating the magnitude of a health problem in the population at risk
2. Informing the population at risk of exposure or illness
3. Informing managers and organizations responsible for immediate control measures and other interventions
4. Understanding the natural history of an illness or injury
5. Detecting outbreaks or epidemics
6. Documenting the distribution and spread of a health event
7. Testing hypotheses about etiology
8. Evaluating control strategies
9. Monitoring changes in exposure
10. Identifying research needs
11. Facilitating epidemiological and laboratory research
12. Facilitating planning, decision making, and policy
13. Assessing public health status
14. Establishing public health priorities
15. Evaluating public health programs
16. Conducting research

exposure to marine microalgae or their toxins unless they are part of a foodborne outbreak. However, some state health agencies require that physicians or diagnostic laboratories report cases of these illnesses. For example, Florida requires physicians to report neurotoxic shellfish poisoning (NSP), which is associated with the harmful algal bloom (HAB) called Florida red tide, to the state health department (Florida Department of Health, 2001).

An example of surveillance for newly emerging ocean-related public health issues is the Harmful Algal Bloom-related Illness Surveillance System (HABISS). HABISS evolved from the Possible Estuary-Associated Syndrome

(PEAS) surveillance system (CDC, 1999), which was designed to collect data on human health effects from exposure to *Pfiesteria piscicida* and *Pfiesteria*-like organisms. HABISS is being expanded to collect data not only on human health effects, but also on animal health effects, algal bloom characteristics, and environmental parameters.

Field Investigations

Although planned epidemiological studies and surveillance are important in evaluating human illnesses associated with exposure to ocean-borne health threats, the typical response to a report of a foodborne illness or environmental exposure is a field (or outbreak) investigation. Outbreak investigations are different from planned epidemiological studies in a number of ways. For example, the extent of the investigation may be limited because of the need for timely intervention to prevent additional cases of illness. Also, outbreak investigations typically involve a small number of cases, thus limiting both the study design and statistical power. Finally, assessing exposure in the field is difficult because field investigations often begin after the onset of illnesses, so that suspected food items may have been consumed, or the source of an environmental exposure (such as a HAB) may have disappeared (Goodman and Buehler, 1996).

Gregg (1996) summarized how to conduct a field investigation in 10 basic steps (Box 11-7). Several of these steps

BOX 11-7. The 10 steps of an Epidemiological Field Investigation (Gregg, 1996)

1. *Determine the existence of an outbreak.* Local health officials know whether more cases occur of a particular disease than they would normally see. Surveys of local physicians and other health care providers, emergency departments, or poison information centers may provide evidence of an outbreak or epidemic.
2. *Confirm the diagnosis.* In a typical field investigation, the epidemiologist will confirm a clinical diagnosis with standard laboratory tests or a patient exposure history.
3. *Define a case and count cases.* The investigating team must now create a working case definition, ascertain cases, and count them. Existing criteria may be available for developing a case definition, or a new group of symptoms may have to be identified. The case definition must be applied in exactly the same manner to all people (cases and controls) under investigation.
4. *Orient the data in terms of time, place, and person.* Once multiple cases have been identified, the disease can be characterized in terms of person, place, and time. An epidemic is characterized in time using an epidemic curve, which is a histogram plot of the time of illness onset (in an appropriate time interval) (x-axis) and the number of cases (y-axis). The epidemic curve can provide information about the magnitude of the outbreak, the mode of spread, timing, and possible duration of the epidemic.
5. *Determine who is at risk of becoming ill.* The human characteristics that can affect an individual's response to a given exposure may include age, sex, race, nutritional status, or even specific polymorphisms in toxin-metabolizing enzymes. When the field investigation team understands how many people are ill, when and where they were when they became ill, and what their general characteristics are, the team will have at least a good working diagnosis of the disease and can identify people at risk.
6. *Develop a hypothesis explaining the specific exposure that causes disease, and test this hypothesis by appropriate statistical methods.* Once the epidemic has been described in terms of person, place, and time, the exposure that caused the disease needs to be identified. This step requires obtaining information about exposure histories from all persons who became ill and comparing them to those of people who did not become ill. If the exposures of people who became ill are not significantly different from those who remained well, a new hypothesis must be developed.
7. *Compare the hypothesis with the established facts.* Once the investigating team identifies the most probable exposure, the epidemic characteristics must be compared with the known facts of the disease.
8. *Plan a more systematic study.* There may be future opportunities and resources to develop more systematic studies to improve scientific understanding of the disease or exposures.
9. *Prepare a written report.* Usually, a field team's final responsibility is to prepare a written report documenting all aspects of the investigation. The report should be written and disseminated as soon as possible because it may provide the impetus or justification for control and prevention efforts.
10. *Execute control and prevention measures.* This step is based on the knowledge gained in the outbreak investigation; it involves mobilizing local public health capacity to mitigate exposures and prevent the occurrence of more cases.

can be done simultaneously or they can be done in a different order, depending on the needs of the outbreak investigation and response efforts. For example, step 10 is executing control and prevention measures. But if control and prevention measures are readily available, they could be put into place as step 3 as a precautionary measure as soon as a case is diagnosed.

A Note of Caution

Assessing how a specific exposure causes a specific health outcome is the ultimate goal of epidemiological studies. A single epidemiological study can evaluate relationships between exposure to risk factors and subsequent development of disease; however, it cannot prove causality. Rather, causality can only be attributed on the basis of the preponderance of evidence from a body of epidemiological research and after considering a number of criteria, such as biological plausibility. The criteria for causality (see Schlesselman, 1982) are summarized in Box 11-8.

Prevention

From the public health perspective, disease prevention comprises primary, secondary, and tertiary activities.

> **Box 11-8.** Observational Criteria for Causation in Epidemiological Studies (Schlesselman, 1982)
>
> TEMPORAL SEQUENCE
> Ideally, the factors thought to be causal must precede development of the disease in question. This temporal sequence may be difficult to establish for some diseases. For example, it is unclear whether hypertension is a cause of or effect from renal disease.
> CONSISTENCY
> The results from diverse studies should all support the hypothesis of causation.
> STRENGTH OF ASSOCIATION
> The larger the calculated measures of association are, the more likely the association is real.
> DOSE RESPONSE
> The existence of a dose response (i.e., a greater dose leads to greater effects) makes the causal relationship more plausible.
> SPECIFICITY OF EFFECT
> If, in the presence of a specific factor, the disease occurs, whereas in the absence of the factor, there is no disease, then there is support for a causal relationship.
> BIOLOGICAL PLAUSIBILITY
> The proposed association must make biological and physiological sense.

Primary prevention, or the actual decrease in disease incidence, would be ideal (World Health Organization [WHO], 1983). However, incomplete scientific information, inadequate tools with which to assess exposure and disease, and limited resources can make primary prevention unachievable (Fleming *et al.,* 2002).

As described earlier, the prevention of the ocean-related diseases in human populations is based on monitoring and preventing exposures, as well as surveillance of the diseases in human populations. For example, for HAB-related diseases such as NSP, the primary prevention activity is shellfish bed monitoring (Backer *et al.,* 2003a; Baden *et al.,* 1995). When the cell concentration of *Karenia brevis*, the dinoflagellate responsible for Florida red tides, in water samples becomes high or the concentration of brevetoxins in shellfish becomes dangerously high, the beds are closed to harvesting. Shellfish are sampled and the brevetoxin concentrations are monitored until the concentration drops to levels that are safe for human consumption. The shellfish beds are then reopened for harvesting.

Secondary prevention activities decrease the prevalence of disease by instituting early detection. For example, secondary prevention of heat-related illnesses could include monitoring environmental conditions (i.e., temperature and humidity), surveillance of heat-related morbidity and mortality in vulnerable target populations (such as the elderly and outdoor workers), and intervening early in heat waves (Huynen, 2001). Secondary prevention for ocean-related illnesses could include surveillance of gastrointestinal illnesses in people who eat seafood or who frequent at-risk recreational waters. Additional secondary prevention activities could include developing educational materials and monitoring systems for future primary prevention.

The goal of tertiary prevention is to reduce the complications resulting from actual human illness and disease. This would involve early treatment of the acute clinical disease, such as using intravenous mannitol to both treat acute ciguatera fish poisoning and prevent subsequent chronic illness (Baden *et al.,* 1995; Blythe *et al.,* 1994).

All forms of public health prevention require the existence of an infrastructure that includes knowledgeable public health professionals (both in environmental health and epidemiology) who collect surveillance information and conduct environmental monitoring (Fleming *et al.,* 2002). It is also important to educate health care and public health personnel about the diagnosis, treatment, and reporting of ocean-related human illnesses. Finally, at-risk populations and resource managers should be educated to recognize ocean-related illnesses and to assist in preventing exposures (e.g., warning people not to eat contaminated shellfish or swim when there is microbial or harmful algal bloom contamination in coastal waters).

In summary, in Part 1, we discussed the basic tools of epidemiological investigation and research. The application of these tools is demonstrated in Part 2, which includes examples of disease surveillance, outbreak investigations, and planned studies to investigate the effects of oceans on human health.

PART 2: EPIDEMIOLOGICAL TOOLS APPLIED TO OCEANS AND HUMAN HEALTH

Theoretical Surveillance and Outbreak Investigation Example: Respiratory Symptom Reports during Florida Red Tide

During May 2004, 65 of 128 calls to the Aquatic Toxins Hotline of the Florida Poison Information Center (FPIC) in Miami were from South Beach visitors reporting respiratory irritation. Although respiratory irritation is a complaint commonly reported to the FPIC, the staff suspected this number of calls was unusual. They contacted the Florida Department of Health, which conducted a field investigation using the 10 steps outlined by Gregg (1996):

Step 1: Determine the existence of an outbreak. Cough, runny nose, eye irritation, and shortness of breath were the most commonly reported symptoms. These symptoms could be caused by many different exposures and illnesses, and therefore the investigating team needed to know whether it was unusual, during one month, for 65 people to call the hotline about respiratory irritation. The investigating team interviewed hotline staff and was told that, on average, people complaining about similar symptoms make 12 calls per month. The increase from 12 calls per month to 65 suggested this might be an unusual occurrence, perhaps an outbreak.

Step 2: Confirm the diagnosis. The reported symptoms (cough, runny nose, eye irritation, and shortness of breath) were consistent with those reported at times people are on the beach when there is an onshore wind and an offshore Florida red tide bloom (Backer *et al.*, 2003b, 2005; Kirkpatrick *et al.*, 2004). In addition, the investigating team called a random sample of the original callers to confirm their symptoms.

Step 3: Define a case and count cases. The investigating team wanted to know the number of calls about respiratory irritation reported during a time when there was no Florida red tide and during a time when there was a Florida red tide. Data on Florida red tides were available from the local marine research laboratory. The laboratory provided the following objective data for South Beach during May 2004 and May 2003: (1) concentrations of Karenia brevis, the microorganism that forms Florida red tides, in seawater samples (i.e., cell counts) and (2) wind speed and direction. The data demonstrated that an offshore red tide and onshore winds were present at South Beach during May 2004, but not during May 2003. For these 2 months, the team compared the number of calls received by the Marine Toxin Hotline from people who had visited South Beach and called to report respiratory irritation symptoms.

Step 4: Orient the data in terms of time, place, and person. The investigating team defined the time (May 2003 or May 2004), place (South Beach), and person (calls to the Marine Toxin Hotline) for this investigation.

Step 5: Determine who is at risk of becoming ill. The investigating team defined people who visited South Beach during May 2004 as those at risk for respiratory irritation.

Step 6: Develop a hypothesis explaining the specific exposure that causes disease. The investigating team hypothesized that more people called from South Beach to complain about respiratory irritation during the documented Florida red tide (May 2004) than during the comparison time when there was no Florida red tide (May 2003). During the documented Florida red tide, people on South Beach could have been exposed to Karenia brevis cells or toxins through contact with seawater or through inhaling aerosols blown onto the beach. The team constructed the two-by-two table presented in Box 11-9. Using the data in the table, the team used a specific statistical test called a chi-square analysis to assess whether the proportion of calls about respiratory complaints from South Beach visitors in May 2004 was greater than the proportion in May 2003. The chi-square is a statistical analysis that allows us to examine a particular characteristic in each of two samples to determine if the two samples were drawn from different populations (see Sokal and Rohlf, 1981). Their answer was as follows: chi-square = 48.98, $p < 0.001$—that is, the probability that the number of complaints reported by South Beach visitors in May

BOX 11-9.

Table of the Data Collected for the Investigation Respiratory Complaints Recorded by FPIC.

		Exposure (Florida Red Tide)		
		Yes (May 2004)	*No (May 2003)*	*Total*
Disease (Respiratory Illness)	Yes	65	63	128
	No	3	67	70
	Total	68	130	198

2004 was the same as the number reported in May 2003 is < 0.001 (i.e., very small). Therefore, we can conclude that there were significantly more calls in May 2004 than in May 2003.

Step 7: Compare the hypothesis with the established facts. The investigating team had already confirmed the potential exposure (i.e., the presence of both a Florida red tide bloom and onshore winds at South Beach during May 2004), and it hypothesized that aerosolized toxins associated with the bloom caused the reported symptoms. Thus, the physical presence of the tide, the specific reported symptoms, and the temporal association supported the hypothesis that the presence of a Florida red tide was associated with respiratory symptom reporting.

Step 8: Plan a more systematic study. The investigation team was intrigued by its findings and decided to pursue funding to conduct a more systematic study of the human health effects associated with exposure to Florida red tide. Such a study would include the following: (1) objective measurements of brevetoxins produced by Karenia brevis in seawater and aerosol and (2) objective and subjective measurements of respiratory illness and other biological effects in exposed people and in an unexposed comparison group (Backer *et al.,* 2005; Fleming *et al.,* 2005b).

Step 9: Prepare a written report. The team met with the FPIC staff and presented its results. The study results were also used to prepare an education seminar for community pulmonary physicians and emergency room personnel.

Step 10: Execute control and prevention measures. On the basis of the results of the investigation, the Florida Department of Health initiated an education and outreach campaign to inform the local community and visitors about potential health effects from exposure to Florida red tide (Reich *et al.,* 2006).

Theoretical Surveillance and Outbreak Investigation Example: A Day at the Beach—What's Bugging Our Swimmers?

Now let us consider an example of conducting a surveillance and outbreak investigation. We will move away from the red tide incident to an unrelated example of a disease outbreak.

On Tuesday, May 26, a local county department of health began getting calls about diarrheal illnesses from people who had been swimming at South Beach on Sunday, May 24, and Monday, May 25. Investigators from the County Department of Health rapidly initiated an investigation beginning on May 26, following the steps outlined by Gregg (1996):

Step 1: Determine the existence of an outbreak. Diarrhea is typically a self-limiting illness that can be treated at home, and people who did become ill would probably not report their illness to the health department, their health care provider, or an emergency room. In addition, although people exposed to pathogen-contaminated beach water can get diarrhea (Dufour and Weimer, 2006; Fleisher *et al.,* 1998; Pruss, 1998), typically, a day or so will go by before people become ill. Thus, the team would need to conduct active retrospective case finding for diarrhea among people who visited the beach between May 24 and May 25.

Step 2: Confirm the diagnosis. In this investigation, it would not be likely that the team could confirm cases of diarrhea with a specific clinical test. The information about diarrhea would be self-reported by the cases and controls recruited for the study.

Step 3: Define a case and count cases. The team defined a "case" as anyone who visited South Beach between May 24 and May 25 who had diarrhea within 72 hours of visiting the beach and going into the water. The team posted flyers and used radio and newspaper ads to recruit people who had been at South Beach on those days. Individuals in the study population were interviewed within 72 hours of their visit to South Beach, with a second interview 10 days later. The follow-up interview helped the team identify who had diarrhea after visiting South Beach. The assembled team conducted active surveillance for an unexposed, or control, population on South Beach during June 1 and 2. Individuals were interviewed during their beach visit and a second time 10 days later. The team planned to compare the frequency of cases of diarrhea in the presumed exposed (South Beach May 24 and 25) and presumed unexposed (South Beach June 1 and 2) populations.

Step 4: Orient the data in terms of time, place, and person. The team identified the components of their investigation as follows: (1) time was the period from May 24 through May 25, (2) place was South Beach, and (3) person was the beach visitor who went into the water.

Step 5: Determine who is at risk of becoming ill. The team defined who was at risk of becoming ill as those swimming at South Beach on May 24 or 25.

Step 6: Develop a hypothesis explaining the specific exposure that causes disease, and test this hypothesis by appropriate statistical methods. The team developed the hypothesis that visitors to South Beach on May 24 and 25 were at risk for exposure to a waterborne pathogen and that swimming at South Beach would increase an individual's risk of having diarrhea within 72 hours of the visit. The data from the team's

BOX 11-10.

Table of Data Collected for the Outbreak Investigation at South Beach

		Exposure (South Beach)		
		Yes (May 24–25)	*No (Jun 1–2)*	*Total*
Disease (Diarrhea)	Yes	20	2	22
	No	630	298	928
	Total	650	300	930

investigation are presented in Box 11-10. The team calculated a relative risk as follows:

$$(20/650)/(2/300) = 0.03/0.006 = 5$$
$$\text{chi-square} = 7.08,\ p = 0.008$$

The team concluded that there was a significant risk of diarrhea from swimming at South Beach on May 24 and 25. Additional investigation found that an unreported failure at a local sewage plant caused a raw sewage overflow upstream of South Beach on May 23. Thus, sewage contamination was the likely source of the outbreak of gastrointestinal illness at South Beach. Typically, local monitoring would have detected sewage contamination in the water, but a local power failure prevented the Department of Environmental Quality from analyzing the South Beach samples taken on May 23.

Step 7: Compare the hypothesis with the established facts. The bacteria and viruses present in human sewage are associated with many diseases, including diarrhea, in those exposed.

Step 8: Plan a more systematic study. The investigators did not need an additional study to make a public health recommendation.

Step 9: Prepare a written report. A report was provided to the local health officials. The report included recommendations for better communication between the sewage treatment plant operators and local public health officials responsible for the beach.

Step 10: Execute prevention and control measures. The team chose to use the results from its investigation to improve the extent and timing of warnings about possible health effects from exposure to sewage contamination at all local beaches. The team also recommended that the local Department of Environmental Quality develop and implement backup procedures to ensure that all water samples are tested in a timely manner.

PART 3. EMERGING ISSUES IN OCEANS AND HUMAN HEALTH: APPLICATIONS OF EPIDEMIOLOGY

In Parts 1 and 2 of this chapter, we discussed epidemiological tools and their application in public health issues related to human interactions with the oceans. In Part 3, we discuss current and emerging challenges in oceans and human health and how epidemiological tools can be applied to meet these challenges.

One challenge is the magnitude and scope of ocean-related processes and changes that are likely to affect human health. These processes and changes include global climate change, extreme weather events, coastal development, movement of pathogenic microorganisms and chemical contaminants through the oceans, contamination and depletion of the ocean-based resources, and expansion of exposure to harmful algal bloom (HAB) toxins. Another challenge in assessing the association between oceans and human health is how to integrate the enormous volumes of environmental data collected by the Global Ocean Observing System (GOOS) into public health decision making (Rice *et al.*, 2004; Sandifer *et al.*, 2004; Tyson *et al.*, 2004). Not only should we expect to see the application of traditional epidemiological tools to these new and emerging issues of oceans and human health, but we should also engage in a discussion of how emerging challenges will affect both the science of epidemiology and the substance of public health practice.

Global Climate Change

Several lines of evidence suggest that changes in the worldwide climate system are already affecting human health on a global scale (Haines *et al.,* 2000; Patz *et al.,* 2005) (see Chapter 2). For example, the World Health Organization (WHO) has estimated that since the 1970s, global climate change was associated with over 150,000 deaths annually (WHO, 2003). The excess deaths were because of increasing mortality and morbidity from extreme heat, cold, drought, and storms; significant decrements in air and water quality; and changes in the ecology of a wide range of microbial diseases (Patz, *et al.,* 2006). Many of these deaths occurred in low-lying coastal areas and small island nations that are particularly at risk from rising sea levels, intense storms, and emerging ocean-borne microbiological threats. In addition to geographic position, vulnerability to adverse effects from climate change is a function of societal characteristics and other phenomena. For example, communities in coastal areas of the developing world and any communities with limited emergency response resources are particularly vulnerable. However, even developed nations are not immune to local effects initiated by global-scale changes, as demonstrated by the devastation wrought by Hurricane

Katrina on the U.S. Gulf coast in 2005 (Averhoff *et al.*, 2006; Daley *et al.*, 2006; Nelson *et al.*, 2006).

Epidemiological studies can be used to assess the human health effects from changes in climate and weather. For example, Box 11-11 discusses how epidemiology has been used to investigate the public health consequences of heat waves.

Extreme Events

Tropical cyclones (also known as typhoons and hurricanes) and tsunamis are the most destructive short-term marine hazards (see also Chapters 3 and 4). Tropical cyclones have caused an estimated 1.9 million deaths worldwide during the past two centuries. During 1980–2000, an average of 11,800 deaths per year were attributed to cyclones (Keim, 2006). The three deadliest recorded cyclones resulted in catastrophic loss of life. In Bangladesh, 300,000 people died in the cyclone of 1970, and 138,000 died in the cyclone of 1991. In China, 100,000 died in the typhoon of 1922. Sixteen of the 18 deadliest tropical cyclones occurred in the Asia-Pacific region. Hurricane Katrina (2005) will probably be the most costly hurricane to date; it was responsible for billions of dollars of damage to the Louisiana and Mississippi coasts.

Tsunamis have the potential to severely impact coastal community public health. Since 1945, more people have been killed by tsunamis than by earthquakes. In a 100-year period from 1895 to 1995, 454 tsunamis were recorded in the Pacific, putting millions of coastal and island communities at risk. The 2004 Indian Ocean tsunami killed nearly 300,000 people and affected another 2 million people spread among 12 nations (Keim, 2006; Miller *et al.*, 2006; Walker *et al.*, 2006).

Box 11-12 discusses how disaster epidemiology can be used to study the human health consequences of these large-scale natural disasters.

Coastal Development

As noted at the beginning of this chapter, a significant proportion of the world's population lives, and an even greater proportion of the world's future population will live, within a short distance of a coastline (Bowen *et al.*, 2006; Crossett *et al.*, 2004). This intense human coastal migration and settlement has significantly degraded coastal resources,

BOX 11-11. Global Climate Change and Increased Heat-Related Morbidity and Mortality

Over the past few decades, global-scale climate change has resulted in unusually high temperatures on many continents (Kovats and Haines, 2005). In the United States and Europe, the numbers of deaths recorded after typical hot summer weather and in association with unusual summer heat waves have increased (Huynen, 2001). We present two examples of how epidemiological methods were used to investigate two recent heat waves and how the investigations have enhanced our knowledge of the human health consequences of extreme heat.

The first example occurred in Phoenix, Arizona, in the United States. Maricopa County (Arizona) routinely experiences high temperatures, and most indoor environments are artificially cooled. In 2005, the number of deaths from all causes was 182% higher than that reported for any of the previous 4 years. Historically, people most at risk for illness or death from exposure to excess heat are those whose health is already compromised and those who cannot afford to (or who are unwilling to) air-condition their homes (Basu and Samet, 2002). Consistent with these findings, in Maricopa County, Yip *et al.* (in preparation) reported that 82% of the decedents who were found indoors were elderly (≥ 65 years old). However, 62 (66%) of the reported decedents were found outdoors, and 48 (77%) of these were younger than 65 years of age. Thus, many victims of this heat wave were much younger than expected and were probably working outdoors without adequate water or periodic access to someplace cool. This investigation identified previously unreported risk factors for heat-related mortality in a community that is acclimated to seasonal high temperatures.

The second example occurred in Europe. More than 20,000 excess deaths (i.e., 20,000 more deaths than expected based on historical mortality data) were reported during the August 2003 heat wave (Kovats and Haines, 2005). In Paris, the number of deaths reported during the heat wave was 140% of what was typical for past years (Kovats and Haines, 2005). Le Tertre *et al.* (2006) examined how the heat wave impacted all-cause mortality in nine French cities (including Paris and Lyon) that participate in a program to monitor the health effects of air pollution. For the nine cities, investigators used a regression model to analyze the correlation between the heat wave variable and all-cause mortality. They chose to look at all mortality because high temperatures are a contributing factor in deaths from many causes. They created a model to predict the number of excess deaths caused by the heat wave that included long-term and seasonal time trends, the usual effects of temperature and air pollution, and air quality and temperature data. They estimated that 3096 extra deaths were caused by the unprecedented heat wave during the summer of 2003.

BOX 11-12. Disaster Needs Assessment

Epidemiology can be used to measure and describe the adverse health effects resulting from natural or human-caused disasters and the factors that contribute to those effects (Noji, 1997). The overall objectives of this type of epidemiological investigation are to assess the current needs of affected communities and use available resources to meet the most immediate needs and to prevent further adverse health effects. The types of epidemiological studies that can be applied to natural disasters include the following: rapid needs assessment, surveillance, evaluation of how disasters impact public health, evaluation of the natural history of acute health effects, and analytical studies of risk factors for adverse health effects (Noji, 1997). For example, after Hurricane Katrina struck the U.S. Gulf coast in 2005, neighboring states were requested to shelter evacuees from the affected states (Dippold *et al.,* 2006). The Tri County Health Department in the Denver area of Colorado conducted a rapid needs assessment of a sample of newly arriving evacuees from the area hit hardest by Katrina. The assessment identified the most common acute and chronic medical conditions and the number of people needing prescription medications, medical services, housing assistance, and clothing. Based on the assessment, recommendations were made to better meet the needs of evacuees arriving in subsequent weeks, including providing education about altitude sickness and how to locate and apply for existing assistance services.

putting at risk populations dependent on the oceans as a source of food, livelihood, or recreation. Epidemiological studies are essential for assessing how the changing interactions between oceans and coastal communities associated with increasing coastal development impact public health. Specifically, we can use epidemiology to assess the following important public health risks: (1) effects of microbial and chemical contamination on the ocean food web and people as top-level consumers, (2) destruction of important coastal habitats, and (3) effects of deteriorating fisheries on the quantity and quality of available seafood.

Expanding Microbial and Chemical Contamination of Seafood

Historically, most reported seafood-poisoning outbreaks have been associated with bacterial contamination of fish or shellfish and have been restricted to specific geographic areas or a limited number of human communities. More recently, international trade in both wild-harvested and aquaculture seafood products has facilitated introducing pathogens into new geographic areas, new seafood ecosystems, and previously unaffected communities. Further, sophisticated laboratory testing has identified viruses, bacteria, and parasitic organisms in seafood that has historically been safe to eat. For example, the bacteria *Salmonella agona* was introduced to Europe through importing Peruvian fishmeal. *S. agona* spread rapidly into other food production streams, resulting in increasing numbers of human disease outbreaks (D'Aoust, 1994). In addition to exposure to microbial contaminants, people consuming aquaculture products may also be exposed to accumulated antibiotics or to bacteria resistant to multiple antibiotics (Fleming *et al.,* 2006; Park *et al.,* 1994). Epidemiological studies of communities located across wide geographic areas and performed in conjunction with comprehensive seafood monitoring will allow us to assess how expanding seafood contamination impacts global public health.

Whereas microbial contamination is an important emerging issue in tropical and temperate climates, chronic exposure to anthropogenic contaminants is a major health issue for native populations of the Arctic Circle, who are dependent on subsistence hunting and fishing (Dewailly and Knap, 2006). The Inuit diet comprises Arctic marine mammals, fish, and terrestrial wild game. However, all of these animals accumulate high loads of heavy metals (such as mercury); persistent organic pollutants (POPs), such as polychlorinated biphenyls (PCBs); chloro-diphenyl-trichloroethane (DDT); and brominated flame-retardants (polybrominated diphenyl ethers, PBDEs). These populations are faced with the unappealing choice of either accepting an alternate (possibly less healthy) diet or continuing their traditional, but now possibly health-threatening, practice of eating locally caught foods. Epidemiological studies will be needed to assess the nutritional status and overall health of populations whose subsistence food sources are threatened.

Depletion of Ocean-Based Food and Other Marine Resources

Seafood contaminated with anthropogenic pollutants and microbes directly affects human health and the overall burden of disease. Another seafood issue that directly impacts human health is worldwide decimation of seafood stocks as a result of overfishing and habitat destruction. Loss of seafood stocks means loss of community income and a critical source of protein and other nutrients to a significant portion of the world's population. In addition to directly affecting the world's seafood supply, destruction and contamination of oceans and related habitats leads to loss of other potentially useful products, including antibiotics and chemotherapy drugs, and marine models of human disease (Fenical, 2006; Grosell and Walsh, 2006). Although epidemiological methods may be limited in assessing economic effects from loss of seafood stocks, studies of changes in nutritional and health status of populations that rely on seafood will help identify the extent of this emerging public health problem.

Unique Exposures: Marine Aerosols Containing Harmful Algal Bloom Toxins

There is evidence that increases in frequency, intensity, and geographic distribution of harmful algal blooms (HABs) is supported by increased levels of nutrients in the oceans and other water bodies, the artificial transport of microorganisms in ship ballast, and favorable global climate changes (Peperzak, 2005). Historically, marine HABs have affected public health by producing toxins that accumulate in seafood. More recently, epidemiological research has documented that toxins produced by *Karenia brevis*, the organism responsible for Florida red tides, can become aerosolized and inhaled. This exposure induces respiratory irritation in healthy people and acute attacks in people with asthma (Backer *et al.,* 2003a, 2003b, 2005; Fleming *et al.,* 2005a, 2005b). These studies have influenced public health practice by not only identifying an emerging public health issue but also by providing data to support development of public health messages aimed at protecting the general public and more sensitive asthmatics (Abraham and Baden, 2006; Backer and McGillicuddy, 2006; Stumpf *et al.,* 2003). Box 11-13 describes an evaluation of public calls for information about Florida red tide from the Florida Poison Information Center (Quirino *et al.,* 2004).

BOX 11-13. Poison Information Center Red Tide Surveillance (Quirino *et al.,* 2004)

From August 2001 to May 2002, 192 people called the Florida Poison Information Center about exposure to a Florida red tide (Quirino *et al.,* 2004). Investigators conducted a follow-up study to assess the complaints of these callers. They were able to contact 36 people who called about the Florida red tide (exposed) and a control group of 36 others who called on the same days but from geographic areas that did not experience a Florida red tide (unexposed). The two groups were similar in age and sex. The exposed group reported an average of 3.25 + 1.78 respiratory symptoms (i.e., cough, runny nose, itching eyes), whereas the control group reported an average of less than 1 (0.36 + 0.89) respiratory symptom. A McNemar's chi-square (Kleinbaum *et al.,* 1982) comparison found that the number of respiratory symptoms reported by these two groups was significantly different ($p < 0.001$). More specifically, exposed people had a significantly increased relative risk of reporting the following symptoms when compared to controls as follows (relative risk, RR; p-value): cough (7.75, $p < 0.001$), runny nose (7.00; $p < 0.005$), eye irritation (5.25; $p < 0.025$), sore throat (10.00; $p < 0.001$), shortness of breath (16.00; $p < 0.001$), and wheezing (15.00; $p < 0.001$). Finally, symptoms reported by the exposed group lasted for a significantly longer time than those reported by the controls. For example, cough lasted 12.84 ± 25.4 days for those calling about Florida red tide and only 2 ± 1.4 days for those calling from other areas ($p < 0.03$). Finally, the exposed group was significantly more likely than the control group to report seeking medical care for their symptoms (RR = 3.00, $p < 0.025$).

Integration of Public Health and the Worldwide Ocean Observing System

Since the 1990s, the national and international oceanic communities have dedicated extensive resources to develop integrated ocean observing systems (IOOSs) (see Chapter 1, Case Study 1). Currently, IOOSs incorporate and integrate data collected about the Earth for a number of purposes and from a variety of sources, including the following: (1) satellite and remote sensors, (2) floating and moored monitoring buoys, (3) autonomous data-collection instruments and robots, (4) routine monitoring data collection programs, and (5) oceanographic research. Data recorded by IOOS-sponsored systems have been instrumental in establishing the validity of global climate change and in demonstrating its effects on the Earth's environment (Haines *et al.,* 2000).

Recently, scientists and resource managers have begun to tackle the issue of utilizing IOOS data in public health decision making (Ocean.US, 2006). One example of how IOOS data can be applied is the National Hurricane Warning System (Walker *et al.,* 2006). Using satellite and remote sensing data, direct hurricane monitoring in the air, and oceanographic modeling, the National Hurricane Warning System predicts the probable path and impact of hurricanes to shipping and coastal communities, allowing those communities to react in advance to save lives and property. In addition to forecasting climate changes and extreme weather events, scientists expect IOOSs to have the capacity to predict and track harmful algal blooms, sewage contamination, oil spills, and other events in a way that will enable public health officials to mitigate exposure and adverse health effects in affected communities. Box 11-14 describes the development of a local Florida Red Tide information system that has been implemented on Florida's Gulf coast (National Oceanic and Atmospheric Administration [NOAA], 2005; Reich *et al.,* 2006; Stumpf *et al.,* 2003).

How Emerging Oceans and Human Health Issues Impact Epidemiology and Public Health Practice

The emerging issues in oceans and human health present new challenges to the substance and practice of epidemiol-

BOX 11-14. Florida Red Tide and Public Health

From Remote Sensing to Research to Public Health Protection

In a unique collaboration, the Florida Department of Health, the Centers for Disease Control and Prevention (CDC), and public and private partners have established a linked network of oceanographic data, exposure, and disease surveillance on Florida red tide and public health education (National Oceanic and Atmospheric Administration [NOAA], 2005; Reich *et al.,* 2006; Stumpf *et al.,* 2003). As part of the network, the NOAA and the Florida Wildlife Research Institute produce weekly reports of Florida red tide that are available by phone or online and include geographic location of any current blooms and concentrations of *Karenia brevis* cells in seawater (coastwatch.noaa.gov/hab/bulletins_ns.htm). Other network partners, Mote Marine Laboratory and the Sarasota County Lifeguards, now conduct real-time beach condition reporting (http://coolgate.mote.org/redtide). This information is linked to the South Florida Poison Information Center Hotline (telephone number 888-232-8635), which provides 24-hour/day toll-free health information in many languages. The hotline reports cases of HAB-related health events to the Florida Department of Health and to CDC, and the data are incorporated into CDC's harmful algal bloom-related disease surveillance system (HABISS). Members of the Florida red tide network also collaborate with the University of Miami's National Science Foundation (NSF) National Institute for Environmental Health Sciences (NIEHS) Oceans and Human Health Center and the Marine and Freshwater Biomedical Science Center (www.rsmas.miami.edu/groups/ohh and www.rsmas.miami.edu/groups/niehs) to conduct research studies of the causes of Florida red tide and its impacts on public health. Finally, the network has collaborated with the local community organization, Solutions to Avoid Red Tide (START) (www.start1.com) and the Florida Department of Health (www.doh.state.fl.us/ENVIRONMENT/community/aquatic) to create public education materials including beach signs, museum displays, information cards, health care provider information, public health recommendations and response plans, and a traveling exhibit to provide this information to Florida's tourists and residents.

ogy and public health in part because of their immense time and spatial scales (McMichael and Powles, 1999). Traditional epidemiology and public health activities have tended to focus on monitoring single exposures and disease entities in relatively circumscribed populations. By contrast, the emerging issues of oceans and human health cross national and international boundaries and will likely take decades of collaborative effort to identify and investigate. One unique challenge for environmental epidemiology and public health is the need not only to study global health issues to assess causes, risk factors, and prevention, but also to act simultaneously on the basis of available limited data to protect future public health.

Another challenge for public health practice is applying new data resources to human circumstances. For example, careful monitoring of changes in the distribution of disease vectors, such as seafood that harbors pathogenic microorganisms, could provide a projection of future human health consequences (Haines and McMichael, 1997). In addition to using nonhuman animal data, epidemiologists and public health practitioners must become comfortable with the science of forecasting, which will use large complex models integrating data describing human health endpoints with data from ecosystem assessments, environmental and oceanographic monitoring (including IOOS), and health assessments of more sensitive or "sentinel" species. These models can predict how extreme events might initiate or support changes in geographic distribution and seasonality of vector-borne diseases, compromise freshwater and food supplies, and induce major movements and disruption of human populations (Haines and McMichael, 1997; McMichael, 1999; McMichael and Beaglehole, 2000).

The greatest challenges for the science of epidemiology and public health practice as they apply to risks and benefits from the sea include the inherent demands for interdisciplinary research and training and the appropriate application of results to protect public health. Skills and resources as diverse as oceanography and epidemiology, remote sensing and genomics, phytoplankton ecology and microbiology, and engineering and medicine will be needed; and researchers, managers, and public health practitioners must work collaboratively to deliver this new information and technology in an appropriate and timely fashion.

CONCLUSIONS

Many of our most critical environmental health issues are related to human interactions with oceans. Epidemiology will continue to be an important tool for studying environmental health issues, including how to mitigate and prevent adverse human health consequences. Those epidemiologists, public health practitioners, and researchers who are able to work with new technologies and in an interdisciplinary fashion will be the scientists and managers who ultimately solve the challenges of oceans and human health. It is our hope that those who read this book will become the new researchers and managers prepared to meet the challenges of oceans and human health.

Acknowledgments

This work was funded in part from the following sources: the National Center for Environmental Health, the Centers for Disease Control and Prevention, the National Science Foundation, the National Institute of Environmental Health Sciences Oceans and Human Health Center at the University of Miami Rosenstiel School (NSF 0CE0432368; NIEHS 1 P50 ES12736), and the National Institute of Environmental Health Sciences Marine and Freshwater Biomedical Sciences Center at the University of Miami Rosenstiel School (NIEHS P30ES05705).

Disclaimer

The findings and conclusions in this report are those of the author(s) and do not necessarily represent the views of the Centers for Disease Control and Prevention or the Agency for Toxic Substances and Disease Registry.

References

Abraham, W.M., Baden, D.G., 2006. Case study: Aerosolized Florida red tide toxins and human health effects. Oceanography 19, 107–109.

Ahlborn, A., Norell, S., 1990. Introduction to Modern Epidemiology. Chestnut Hill, MA, Epidemiology Resources.

Averhoff, F., Young, S., Mott, J., Fleischaure, A., Brady, J., Straif-Bourgeois, S., Valadez, A., Lurie, D., Palacio H., Buhner, D., Persee, D., Guerra F., Jorgan, J., Moolenaar, R., McNeill, K.M., Byers, P., Kittle, T.S., Chavez, G., Phillips, M., Koo, D., Groseclose, S., Hicks, P., Jones, N.F., Kenneson, A., Vranken, P., Sergienko, E., Bitsko, R.H., Lorick, S.A., 2006. Morbidity surveillance after Hurricane Katrina—Arkansas, Louisiana, Mississippi, and Texas, September 2005. MMWR 55, 727–731.

Backer, L.C., Fleming, L.E., Rowan, A.D., Baden, D.G., 2003a. Epidemiology and public health of human illnesses associated with harmful marine algae. In Hallegraeff, G.M., Anderson, D.M., and Cembella, A.D. (eds.), IOC Manual on Harmful Marine Microalgae, pp. 723–749. Landais, France, UNSECO.

Backer, L.C., Fleming, L.E., Rowan, A., Cheng, Y.-S., Benson, J., Pierce, R., Zaias, J., Bean, J., Bossart, G.D., Johnson, D., Quimbo, R., Baden, D.G., 2003b. Recreational exposure to aerosolized brevetoxins during Florida red tide events. Harmful Algae 2, 19–28.

Backer, L.C., Kirkpatrick, B., Fleming, L.E., Cheng, Y.-S., Pierce, R., Bean, J.A., Clark, R., Johnson, D., Wanner, A., Tamer, R., Baden, D., 2005. Occupational exposure to aerosolized brevetoxins during Florida red tide events: Impacts on a healthy worker population. Environ. Health Perspect. 113, 644–649.

Backer, L.C., McGillicuddy, D.J., 2006. Harmful algal blooms: At the interface between coastal oceanography and human health. Oceanography 19, 94–106.

Baden, D., Fleming, L.E., Bean, J.A., 1995. Marine toxins. In deWolff. F.A. (ed.), Handbook of Clinical Neurology: Intoxications of the Nervous System Part II. Natural Toxins and Drugs. Amsterdam, Elsevier Press, 21(65), 141–175.

Bagnis, R., 1968. Clinical aspects of ciguatera (fish poisoning) in French Polynesia. Hawaii Med. J. 28, 25–28.

Basu, R., Samet, J.M., 2002. Relation between elevated ambient temperature and mortality: A review of the epidemiologic evidence. Epidemiolog. Rev. 24, 190–202.

Blythe, D.G., Fleming, L.E., Ayyar, D.R., deSylva, D., Baden, D., Schrank K., 1994. Mannitol therapy for acute and chronic ciguatera fish poisoning. Memoirs of the Queensland Museum 34, 465–470.

Bowen, R.E., Frankic, A., Davis, M.E., 2006. Human development and resource use in the coastal zone: Influences on human health. Oceanography 19, 62–71.

Centers for Disease Control and Prevention, 1999. Possible estuary-associated syndrome. MMWR 48, 381.

Centers for Disease Control and Prevention, 2001. Diagnosis and management of foodborne illnesses: A primer for physicians. MMWR 50, 25–27.

Crossett, K.M., Culliton, T.J., Wiley, P.C., Goodspeed, T.R., 2004. Population trends along the coastal United States: 1980–2008, pp. 1–3. Washington, DC, U.S. Department of Commerce, National Oceanic and Atmospheric Administration, National Ocean Service.

Daley, W.R., 2006. Public health response to Hurricanes Katrina and Rita—Louisiana, 2005. MMWR 55, 29–30.

D'Aoust, J.-Y, 1994. Salmonella and the international food trade. Int. J. Food Micro., 24, 11–31.

Dewailly, E., Knap, A., 2006. Food from the oceans and human health: Balancing risks and benefits. Oceanography 19, 84–93

Dicker, R.C., 1996. Analyzing and interpreting data. In Gregg, M.B. (ed.), Field Epidemiology, pp. 92–131. New York, Oxford University Press.

Dippold, L., Patnaik, J.L., Vogt., R.L., Gershman, K., Burnite, S., Ghosh, T., Bamberg W., 2005. Illness surveillance and rapid needs assessment among Hurricane Katrina evacuees—Colorado, September 1–23, 2005. MMWR 55, 244–247.

Doll, R., Hill, A.B., 1952. A study of the aetiology of carcinoma of the lung. Br. Med. J. 2, 1271–1286.

Dufour, A.P., Wymer, L.J., 2006. Microbes, monitoring and human health. Oceanography 19, 72–80.

Fenical, W., 2006. Marine pharmaceuticals: Past, present and future. Oceanography 19, 110–119.

Fleisher, J.M., Kay, D., Wyer, M.D., Godfree, A.F., 1998. Estimates of the severity of illnesses associated with bathing in marine recreational waters contaminated with domestic sewage. Int. J. Epidemiol. 27, 722–726.

Fleming, L.E., Backer, L., Rowan, A., 2002. The epidemiology of human illnesses associated with harmful algal blooms. In Baden, D., and Adams, D., (eds.), Neurotoxicology Handbook, vol. 1, pp. 363–381. Totowa, NJ, Humana Press.

Fleming, L.E., Backer, L.C., Baden, D.G., 2005a. Overview of aerosolized Florida red tide toxins: Exposures and effects. Environ. Health Perspect. 113, 618–620.

Fleming, L.E., Broad, K., Clement, A., Dewailly, E., Elmir, S., Knap, A., Pomponi, S.A., Smith, S., Solo Gabriele, H., Walsh, P., 2006. Oceans and human health: Emerging public health risks in the marine environment. Mar. Pollut. Bull. 53, 545–560.

Fleming, L.E., Kirkpatrick, B., Backer, L.C., Bean, J.A., Wanner, A., Dalpra, D., Tamer, R., Zaias, J., Cheng, Y.-S., Pierce, R., Naar, J., Abraham, W., Clark, R., Zhou, Y., Henry, M.S., Johnson, D., Van de Bogart, G., Bossart, G.D., Harrington, M., Baden, D.G., 2005b. Initial evaluation of the effects of aerosolized Florida red tide toxins (brevetoxins) in persons with asthma. Environ. Health Perspect. 113, 650–657.

Florida Department of Health, 2001, www.doh.state.fl.us/disease_ctrl/epi/topics/surv.htm.

Goodman, R.A., Buehler, J.W., 1996. Field epidemiology defined. In Gregg, M.B. (ed.), Field Epidemiology, pp. 3–8. New York, Oxford University Press.

Goodman, R.A., Peavy, J.V., 1996. Describing epidemiologic data. In Gregg, M.B. (ed.), Field Epidemiology, pp. 60–80. New York, Oxford University Press.

Gregg, M.B., 1996. Conducting a field investigation. In Gregg, M.B. (ed.), Field Epidemiology, pp. 44–59. New York, Oxford University Press.

Grosell, M., Walsh, P.J., 2006 Benefits from the sea: Sentinel species and animal models of human health. Oceanography 19, 126–133.

Haines, A, McMicahel, A.J., 1997. Climate change and health: Implications for research, monitoring and policy. BMJ 315, 870–874.

Haines, A., McMichael, A.J., Epstein, P.R., 2000. Environment and health: 2. Global climate change and health. CMAJ 2000. 163, 729–734.

Huynen, M.M.T.E., 2001. The impact of heat waves and cold spells on mortality rates in the Dutch population. Environ. Health Perspect. 109, 463–471.

Keim, M.E., 2006. Cyclones, tsunamis and human health: The key role of preparedness. Oceanography 19, 40–49.

Kirkpatrick, B., Fleming, L., Squicciarini, D., Backer, L.C., Clark, R., Abraham, W., Benson, J., Cheng, Y.-S., Johnson, D., Pierce, R., Zaias, J., Bossart, G., Baden, D.G., 2004. Literature review of Florida red tide: Implications for human health. Harmful Algae 3, 99–115.

Kleinbaum, D.G., Kupper, L.L., Morgenstern, H., 1982. Epidemiologic Research Principles and Quantitative Methods, pp. 63–70. New York, Van Nostrand Reinhold.

Kovats, R.S., Haines, A., 2005. Global climate change and health: Recent findings and future steps CMAJ 172, 501–502.

Last, J.M., (ed.), 1988. Dictionary of epidemiology, 2nd ed. New York, Oxford University Press.

Le Tertre, A., Lefranc, A., Eilstein, D., Declercq, C., Medina, S., Blanchard, M., Chardon, B., Fabre, P., Filleul, L. Justo, J.-F., Pascal, L., Prouvost, H., Cassadou, S., Ledrans, M., 2006. Impact of the 2003 heatwave on all-cause mortality in 9 French cities. Epidemiology 17, 75–79.

McMichael, A.J., 1999. From hazard to habitat: Rethinking environment and health. Epidemiology 10, 460–464.

McMichael, A.J., Beaglehole, R., 2000. The changing global context of public health. Lancet 356, 495–499.

McMichael, A.J., Powles, J.W., 1999. Human numbers, environment, sustainability and health. BMJ 319, 977–980.

Miller, F., Thomalla, F., Downing, T., Chadwick, M., 2006. Case study: Resilient ecosystems, healthy communities: Human health and sustainable ecosystems after the tsunami. Oceanography 19, 50–51.

National Oceanic and Atmospheric Administration (NOAA), 2005. Annual Report of the Gulf of Mexico Harmful Algal Bloom Operational Forecast System (GOM HAB OFS): October 1, 2004, to September 30, 2005 (Operational year 1). Washington, DC, NOAA Technical Report NOS CO-OPS 047. NOAA.

Nelson, S., Luten, J., Jones, J., Ragan, P., Riddick, L., Hart, E., Schulte, J., Corrales, L., Combs, D., Wolkin, A., Strosnider, H., Batts, D., Bayleyegn, T., 2006. Mortality associated with Hurricane Katrina—Florida and Alabama, August–October 2005. MMWR 55, 239–242.

Noji, E.K., 1997. The use of epidemiologic methods in disasters. In Noji, E.K. (ed.), The Public Health Consequences of Disasters, pp. 21–36. New York, Oxford University Press.

Ocean.US., 2006. Public Health Risks: Coastal Observations for Decision-Making, Publication 15. Arlington, VA, Ocean.US.

Park, E.D., Lightner, D.V., Park, D.L., 1994. Antimicrobials in shrimp aquaculture in the United States: Regulatory status and safety concerns. Rev. Environ. Contam. Tox., 138, 1–20.

Patz, J.A., Campbell-Lendrum, D., Holloway, T., Foley, J.A., 2005. Impact of regional climate change on human health. Nature. 438, 310–317.

Patz, J.A., Olson, S.H., Gray, A.L., 2006. Climate change, oceans, and human health. Oceanography 19, 52–59.

Peperzak, L., 2005. Future increase in harmful algal blooms in the North Sea due to climate change. Water Sci. Technol. 51, 1–36.

Pew Oceans Commission, June 2003. America's Living Ocean: Charting a Course for Sea Change, www.pewtrusts.org/pdf/env_pew_oceans_final_report.pdf.

Pruss, A., 1998. Review of epidemiological studies on health effects from exposure to recreational water. Int. J. Epidemiol. 27, 1–9.

Quirino, W., Fleming, L.E., Weisman, R., Backer, L., Kirkpatrick, B., Clark, R., Dalpra, D., Van de Bogart, G., Gaines, M., 2004. Follow up study of red tide associated respiratory illness. Fl. J. Env. Health 186, 18–22.

Reich, A., Backer, L.C., Kirkpatrick, B., Stumpf, R., Fleming, L.E., Stephan, W.B., Weisman, R., Jerez, E., Heil,, C., Steidinger, K., Landsberg, J., Connor, J., DeThomasis, J., Baden, D.G., 2006. Public health and Florida red tide: From remote sensing to poison information (Published Abstract). American Public Health Association Annual Meeting, Boston, MA.

Rice, D., Dearry, A., Garrison, D., 2004. Pioneering research initiatives for oceans and human health. Ecohealth 1, 220–225.

Sandifer, P.A., Holland, A.F., Rowles, T.K., Scott, G.I., 2004. The oceans and human health. Environ. Health Perspect. 112, A454–A455.

Schlesselman, J.J., 1982. Case Control Studies Design, Conduct, Analysis, pp. 22–25. New York, Oxford University Press.

Sokal, R.R., Rohlf, F.J. 1981. Biometry, 2nd ed. San Francisco, W.H. Freeman and Company.

Stumpf, R.P., Culver, M.E., Tester, P.A., Tomlinson, M., Kirkpatrick, G. J., Pederson, B.A., Truby, E., Ransibrahmanakul, V., Soracco, M., 2003. Monitoring *Karenia brevis* blooms in the Gulf of Mexico using satellite ocean color imagery and other data. Harmful Algae 2, 147–160.

Summers, J., 1989. Soho: A History of London's Most Colourful Neighborhood, pp. 113–117. London, Bloomsbury.

Terris, M., 1964. Goldberger on Pellagra. Baton Rouge, LA, Louisiana State University Press.

Teutsch, S.M., 1994. Considerations in planning a surveillance system. In Teutsch, S.M., and Churchill, R.E. (eds.), Principles and Practice of Public Health Surveillance, pp. 18–28. Oxford, Oxford University Press.

Thacker, S.B., Berkelman, R.L. 1998. Public health surveillance in the United States. Epidemiol. Rev. 10, 164–190.

Tyson, F.L., Rice, D.L., Dearry, A., 2004. Connecting the oceans and human health. Environ. Health Perspect. 112, A455–A456.

U.S. Oceans Commission, Sept 2004. An Ocean Blue Print for the 21st Century. U.S. Oceans Commission, www.oceancommission.gov/documents/full_color_rpt/000_ocean_full_report.pdf.

Walker, N., Haag, A., Balasubramanian, S., Leben, R., van Heerden, I., Kemp, P., Mashriqui, H., 2006. Hurricane prediction: A century of advances. Oceanography 19, 24–36.

World Health Organization (WHO), 2003. Climate change and human health: Risks and responses. Summary. Geneva, WHO.

Yip, F., Flanders, W.D., Wolkin, W., Engelthaler, D., Humble, W., Neri, A., Lewis, L., Backer, L., Rubin, C., (in preparation). Impact of excess heat events in Maricopa County, Arizona: 2000–2005.

STUDY QUESTIONS

1. Describe the differences among signs, symptoms, and tests in creating case definitions for epidemiology studies.
2. Explain why it is important to assess baseline measurements of disease in study populations.
3. Describe how public health surveillance can be used as a tool to investigate human health risks associated with our oceans.
4. Describe the steps in conducting a field epidemiology investigation.
5. Describe the three levels of disease prevention and how each might apply to a human health risk associated with our oceans.

6. Discuss three current issues associated with oceans and human health, including which issue you believe has the most impact on human health and why.
7. Describe the challenges of oceans and human health issues with regard to traditional epidemiology.
8. Worldwide, the increasing human population movement to the coasts is putting pressure on the coastal marine environment. What oceans and human health issues derive directly and indirectly from this human coastal migration?
9. You are an epidemiologist of a small tropical island, and you receive a report of an outbreak of illnesses associated with fish consumption at a banquet. Briefly describe the steps you would take to ascertain the cause of the illnesses.
10. What do you think will the greatest oceans and human health issue in the future and why?

CHAPTER

12

Toxic Diatoms

VERA L. TRAINER, BARBARA M. HICKEY, AND STEPHEN S. BATES

INTRODUCTION

A tragic event in 1987 changed forever the way diatoms are viewed. Most diatoms are beneficial to the oceans' ecosystems and ultimately to human health. However, a unique food poisoning in eastern Canada led to the illness of more than 100 people and the death of four elderly individuals. The potent neurotoxin domoic acid (DA) was found to be the culprit, and its source was the marine diatom *Pseudo-nitzschia multiseries*. This event resulted in intense scientific interest in this genus, because it was the first time that a diatom had been shown to produce a toxin. The clinical syndrome was subsequently called amnesic shellfish poisoning (ASP) because of one of its reported symptoms, memory loss. Here, we introduce the general biology of diatoms and focus specifically on toxic diatoms of the genus *Pseudo-nitzschia*. We also discuss DA and its mode of action, describe oceanographic factors that lead to the growth of this toxic diatom, and include specific examples of *Pseudo-nitzschia* blooms and related physiology.

GENERAL BIOLOGY OF DIATOMS

Diatoms are microscopic single-celled plants that live in freshwater and marine ecosystems, from the poles to the tropics. They are found in sea ice and snow; some are even blown into the air. Benthic diatoms live in or on the sediments and on the surface of other plants (epiphytic), animals (epizootic), or rocks (epilithic). Those that live in the water column, together with the dinoflagellates (Chapter 13), other flagellates, and blue-green algae (cyanobacteria) (Chapter 15), among others, make up the phytoplankton (Greek *phyton* = "plant," *planktos* = "wanderer"), also called *micro-algae*. Diatoms are the most abundant of the phytoplankton based on biomass and numbers of species; there are estimated to be greater than 50,000 species in the aquatic ecosystem, and up to 18,000 in the marine environment (Fryxell and Hasle, 2003).

Some diatoms are found as single cells; others form long "chains" by linking to the adjacent cell, either by abutting end to end or by joining their protruding spines or setae. Like all plants, diatoms contain chlorophyll and other pigments that capture the energy of sunlight to convert carbon dioxide and water molecules into carbohydrates via photosynthesis. Their survival also requires nutrients such as nitrogen, phosphorus, silicon, and trace metals. Diatoms are a major source of the world's oxygen, producing amounts equivalent to all the tropical rain forests (Field *et al.*, 1998). They are also responsible for 40% to 45% of the total production of organic carbon compounds in the oceans (Mann, 1999), or 20% to 25% for the entire Earth. As such, they are a significant source of food for other organisms within the marine food web and play an important role in the biogeochemical cycling of elements (Sarthou *et al.*, 2005). Because of their large cell size and rapid sinking rate, they also play an important role as part of the "biological pump" that removes carbon dioxide from the atmosphere and places it at depth, thus helping to moderate the increasing levels of carbon dioxide associated with climate change.

The name *diatom* comes from the Greek word *diatomos* (*dia* = "through" + *temnein* = "to cut"), meaning "cut in two." This is because a diatom cell is composed of two overlapping halves (thecae) that fit together like a petri dish: the upper half is called the "epitheca" and the lower half the "hypotheca" (Fig. 12-1). The theca is composed of an upper (epi-) or lower (hypo-) valve, plus girdle bands that encircle the middle of the cell to hold the two halves together. Each valve possesses a slit, called a "raphe," that runs along the whole length of the valve and is reinforced on the interior

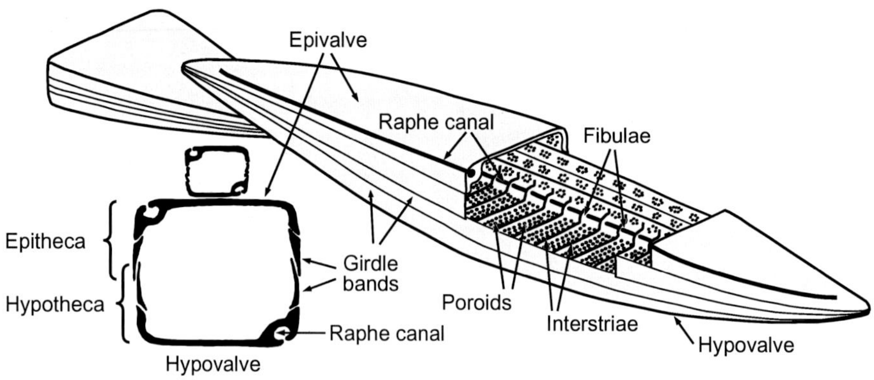

FIGURE 12-1. Silica frustules of *Pseudo-nitzschia multiseries* shown diagrammatically as one cell overlapping part of another, with cell length much reduced for illustration purposes. Raphe is shown on edge of upper valve (epivalve) and opposite corner of lower valve (hypovalve). Section of epitheca removed to show raphe canal, interstriae, three to four rows of poroids in the stria areas, and several poroids on the inner side of girdle bands that encircle the frustule. Diagram at left shows cross-section of two overlapping cells, with raphe canal in diagonally opposite position. Redrawn from MacPhee *et al.* (1992).

side with bridges, called "fibulae." The valves are ornamented with striae (strips containing rows of small poroids), alternating with narrower riblike strips called "interstriae." The entire diatom shell is called the "frustule" and is composed of silica (SiO_2 = glass) that is covered by an organic membrane through which nutrients pass for cell growth. Based on how the silica ribs on the valve radiate, diatoms (division Heterokontophyta, class Bacillariophyceae) are placed into two orders: radially symmetric (round) cells are called centric diatoms and longitudinally symmetric (long, narrow) cells are pennate (from the Latin *penna*, meaning "feather") diatoms. *Pseudo-nitzschia* species are pennate diatoms.

Centric diatoms have flagellated gametes but are otherwise swept passively by currents. Pennate diatoms are capable of limited motility. With *Pseudo-nitzschia*, single cells, or an entire chain of cells, can be seen gliding across a solid surface (such as a microscope slide), sometimes stopping and then reversing direction. The role of this motility in *Pseudo-nitzschia* is uncertain because these cells are found mostly suspended in the water column. Motility is facilitated by a raphe, the longitudinal slit that runs along an edge of the valve face (Fig. 12-1). Mucus is secreted through the raphe and attaches to particles or substrata; as the diatoms move, they leave a trail of this sticky substance. The details of what causes the motive force are still somewhat mysterious, but filaments of the protein actin are thought to change shape, thus moving the cell along.

Diatoms encounter a serious problem because their frustule is composed of solid silica and the hypotheca is slightly smaller than the epitheca. During vegetative cell division, the daughter cell formed on the side of the hypotheca is thus slightly smaller than the parent cell. As cell division continues, the average cell size of the population therefore decreases. Eventually, the cells can become so small that they are no longer able to divide, and they die. This occurs in clonal cultures (i.e., started with one cell) of *Pseudo-nitzschia* and other pennate diatoms. However, cells of many sizes, including large cells, are found in the ocean; most diatoms are able to restore their large size via sexual reproduction. In centric diatoms, the same cell can produce both small flagellated male gametes plus large nonmotile female gametes (i.e., they are homothallic and oogamous).

In pennate diatoms, clones of opposite mating type are needed (i.e., they are heterothalic), and they produce nonflagellated gametes of the same size (i.e., they are isogamous). Thus, mating will not occur in a clonal culture; a clone of opposite mating type must be added. In both types of diatoms, the gametes fuse, and an oval (in centrics) or elongated (in pennates) structure, called an auxospore, is produced. A large initial cell is formed within the fully expanded auxospore and is eventually released, thus restoring the large cell size of the diatom cell. The sexual reproduction of *Pseudo-nitzschia* species can affect bloom dynamics (population growth) and perhaps even toxicity (e.g., Davidovich and Bates, 1998). For a more detailed description of the evolution, morphology, sexuality, systematics, taxonomy, and biology of diatoms, see Round *et al.* (1990) and Hasle and Syvertsen (1997).

Because they are abundant, geographically widespread, live in a wide variety of environments, and are generally well preserved, modern and fossilized diatoms have many uses by human society (Stoermer and Smol, 1999). Fossilized diatom frustules form diatomaceous earth, also called diatomite, which is commonly used as a filter medium in swimming pools and aquaria, a mild abrasive in metal polishes and toothpaste, cat litter, insulating material, chromatographic separation material such as silica gel, a mechanical insecticide (it works by cutting, and thus dehydrating the insects), an absorbent for toxic spills, and in the manufacture of dynamite, among other uses. Fossilized diatoms are also used in paleolimnology and paleoceanography to reconstruct past climate changes. Modern diatoms are used as an indicator of eutrophication (i.e., the enrichment of water by nutrients, especially nitrogen and phosphorus, causing an accelerated growth of algae and leading to undesirable disturbances in the balance of organisms and in water quality) and acidification (i.e., the negative consequences of acid rain). Forensic investigators look for diatom frustules in the lungs to determine if a person died from drowning; they can also tell if the drowning occurred in a marine or freshwater environment (Pollanen, 1997). Nanotechnologists are studying how diatoms lay down their intricate frustules in order to apply this to building smaller silicon computer chips (Bradbury, 2004).

Diatoms can be grown in mass culture in order to extract natural products for biotechnological applications: total lipids for biodiesel fuel, amino acids for cosmetics, antibiotics and antiproliferative agents for the medical field, and silicon derived from frustules for use in nanotechnology (Lebeau and Robert, 2003). The diatom, *Nitzschia laevis*, is grown on a semi-industrial scale for the production of eicosapentaenoic acid (EPA), a polyunsaturated omega-3 fatty acid widely recognized for its beneficial effects on human health (see Chapter 10). Mass cultures containing up to 200 species of marine diatoms are grown for the production of natural supplements, called phytonutrients, reported to boost human health. Thus, diatoms are an essential source of food for aquatic organisms, of products for human needs, and of oxygen for all living creatures.

Because diatom cells are made of silica, they are generally denser than seawater. Therefore, they require special oceanographic conditions, such as mixing and upwelling (i.e., the wind-driven movement of dense, cool, and usually nutrient-rich water toward the ocean surface to replace the warmer, usually nutrient-depleted surface water) that allow them to grow within the sunlit upper layer of the ocean. Currents may also benefit the cells by transporting them to areas that may be more conducive to cell growth. The cells may then proliferate rapidly, causing a "bloom." Most diatom blooms are beneficial. However, when the concentration of some diatom species reaches bloom proportions, their pigments may color the water. Such blooms are popularly known as "red tides," which can also be composed of dinoflagellates or other types of phytoplankton. Scientists now prefer the term "harmful algal bloom" (HAB), because these blooms do not always color the water red (they may be other colors or even colorless), and they are certainly not tides (Glibert *et al.*, 2005a).

Some HABs contain diatom species that can harm other marine organisms by physical means (e.g., they have spines or setae that can damage the gills of finfish) or by removing oxygen from the water because of decomposition when they die. Such diatoms have no direct impact on human health and are discussed elsewhere (Fryxell and Hasle, 2003). However, a few (<20) diatom species cause harm by producing toxins that affect molluscan shellfish, marine mammals, birds, and even humans. *Pseudo-nitzschia* was the first diatom genus known to produce a compound that is toxic to humans and other animals. This naturally occurring toxin is called domoic acid (DA), and the syndrome of DA poisoning in humans is called ASP. The remainder of this chapter will focus on toxic diatoms of the genus *Pseudo-nitzschia*: their production of DA, the effects of this toxin on human health, and the importance of oceanography in triggering, maintaining, and dissipating *Pseudo-nitzschia* blooms.

THE DISCOVERY OF AMNESIC SHELLFISH POISONING

Molluscan shellfish, such as mussels, clams, oysters, and scallops, feed directly on phytoplankton cells suspended in the water column; some individuals filter up to 200 liters of seawater per day. Most food items are beneficial. However, a small portion of marine phytoplankton, only about 100 out of 5000 species (or about 2% of all species), produces algal toxins (phycotoxins) that can be deadly. These phycotoxins become concentrated in the digestive tract of molluscan shellfish as a result of their filter feeding and are thus easily passed on to humans, or other animals, who consume them. Although most shellfish are not affected adversely by the phycotoxins, humans can become seriously sick or even die as a result of consuming these contaminated shellfish. Fortunately, most countries around the world have public health agencies that monitor for the presence of phycotoxins in molluscan shellfish, or the numbers of toxic algal cells in the seawater (see below). When levels of a phycotoxin within the shellfish reach the regulatory limit, harvesting is prohibited until the toxin decreases to acceptable levels after the toxic bloom has dissipated. This has prevented human poisonings, except in cases where individuals illegally harvest shellfish from an area closed to harvesting.

Before 1987, the group of toxin-producing phytoplankton of concern to human health included mostly dinoflagellates; different species of these flagellated cells produce toxins causing paralytic shellfish poisoning (PSP; saxitoxins), diarrhetic shellfish poisoning (DSP; okadaic acid and dinophysis toxins), neurotoxic shellfish poisoning (NSP; brevetoxins) (Chapter 13), and ciguatera fish poisoning (CFP; ciguatoxin) (Chapter 14) (Landsberg *et al.*, 2005). Then, in late autumn, 1987, an outbreak of an unusual human poisoning in Canada was traced to blue mussels (*Mytilus edulis*) that originated at aquaculture sites in bays of eastern Prince Edward Island (Bates *et al.*, 1989). The known shellfish poisoning toxins were quickly ruled out; it was suspected that something new was occurring.

Within a span of only a few days in late November and early December, more than 250 individuals who had consumed these mussels were admitted to hospitals in Quebec and New Brunswick. They had varying degrees of an acute illness characterized by gastrointestinal problems and unusual nervous system abnormalities (Perl *et al.*, 1990). A case was defined as the occurrence of gastrointestinal symptoms within 24 hours, or of neurological symptoms within 48 hours, of consuming the mussels. Of those 107 patients who met this strict case definition, most showed symptoms of vomiting (76%), followed by abdominal cramps (50%), diarrhea (42%), headache—often incapacitating (43%)—and loss of short-term memory (25%). Nineteen patients remained in hospital; 12 required intensive care because of seizures, profuse respiratory secretions, or unstable blood pressure. Approximately equal numbers of males and females were affected (47 and 60 patients, respectively); 46% were 40 to 59 years of age, and 36% were 60 or older. Three elderly male patients (ages 71, 82, and 84) died in hospital, and a fourth (84 years of age) died 3 months after eating the mussels. A closer evaluation of 14 of the more severely affected patients detailed the degree of confusion and disorientation that occurred within 1.5 to 48 hours after consuming the contaminated mussels (Teitelbaum *et al.*, 1990). Most of these patients had difficulty remembering ongoing events occurring since consuming the toxic mussels (i.e., they had a predominantly anterograde amnesia or short-term memory loss). However, some seriously affected patients also had difficulty remembering events that happened several years before the mussel-induced intoxication (i.e., they had retrograde amnesia). Because of the memory problems, the term *ASP* was later given to this clinical syndrome. However, *DA poisoning (DAP)* is sometimes used because shellfish are not always the vector. For a critical review of studies linking HAB illnesses with neuropsychological impairments, see Friedman and Levin (2005).

Identification of Domoic Acid, Its Structure and Mechanism of Action

During these hospitalization episodes, an intensive search began in the mussels for the causative toxin. A bioassay-directed strategy was taken, whereby extracts of both toxic and control mussels were injected into mice to determine which fraction elicited a toxic response (Quilliam and Wright, 1989). Various chromatographic techniques were used to characterize and eventually identify the toxic compound. After an unprecedented 4-day round-the-clock investigation, the molecule in the toxic fraction was pinpointed as a known compound, DA. This was at first met with disbelief, as the literature showed that DA had been used as a treatment in Japan to remove intestinal worms in children. Tests performed in the mid-1950s on the antihelmintic (vermifuge) properties of DA found that a single dose of 20 mg could be given to children and adults without harmful effects. However, in the 1987 event, the two persons with the most severe neurological symptoms ingested 290 mg (i.e., an order of magnitude greater than that given to the Japanese children); those who were unaffected ingested from 15 to 20 mg per person (Perl *et al.*, 1990). As well, those affected in 1987 were elderly and had preconditions that made them more vulnerable (see below). Assuming an average body weight of 50 to 70 kg, DA would have no effect at 0.2 to 0.3 mg/kg, mild effects at 0.9 to 2.0 mg/kg and serious effects from 1.9 to 4.2 mg/kg (Tasker, 2002).

DA was originally isolated from a species of red macroalga (*Chondria armata*); its name comes from "domoi" (or "doumoi"), a local name for that seaweed in Japan (see Bates *et al.*, 1998). It has since been isolated from five other species of red macroalgae. Japanese scientists observed that flies that were attracted to and contacted *C. armata* drying on the seashore died shortly afterward. DA was later shown to have insecticidal properties and was 14 times more potent than DDT when administered into the abdomen of the American cockroach. Another red macroalga, *Digenea simplex*, isolated in southern Japan, was found to contain kainic acid (Fig. 12-2). Both of these "toxins" had been used as a vermifuge for Japanese children.

DA (also called "domoate" in the neurophysiological literature) is a low-molecular-weight (311 daltons), water-soluble, heat-stable, secondary amino acid (Fig. 12-2). It is a member of the kainoid class of organic compounds that includes kainic acid, the marine toxin described earlier. Both domoic and kainic acids contain a domain that is structurally identical to the amino acid, glutamate (glutamic acid), a compound important for proper functioning of the nervous system (Fig. 12-2). Glutamate is the major excitatory neurotransmitter in the mammalian central nervous system; it is responsible for many of the functions within the brain, including cell-to-cell communication and hippocampal long-term potentiation, a process important in learning and memory. As a neurotransmitter, it takes part in the transmission of a nerve impulse from one neuron to another. Simply put, glutamate binds onto specific receptor sites (i.e., NMDA, AMPA, and kainate sites, named after the molecules that stimulate them) on the membrane of the nerve fiber (dendrite). This causes the receptor molecule to undergo a change in its conformation, or shape. In turn, this opens up a microscopic channel (like a gate) in the membrane, which allows the influx of sodium or calcium into the axon (Fig. 12-3). As a result, the neuron is triggered, sending an impulse down its axon fiber via Na^+ and K^+ mediated action potentials (see Chapter 28).

Glutamate is released by the presynaptic neuron into the synaptic cleft (the narrow space between the two abutting nerve cells) and subsequently binds to the receptor sites on the dendrite of the postsynaptic neuron. This causes the next

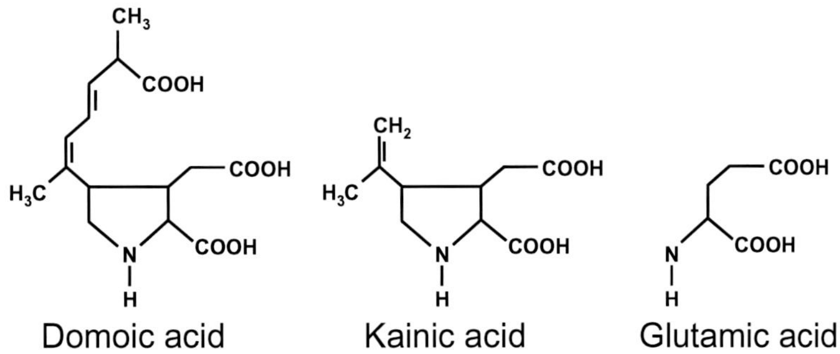

FIGURE 12-2. Structure of domoic acid and its analogues. From Bates *et al.* (1998).

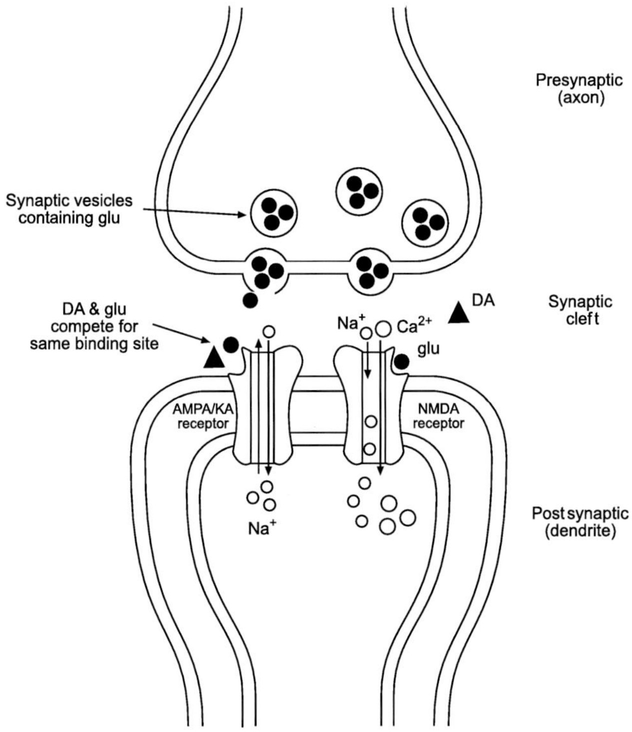

FIGURE 12-3. Diagram of a nerve showing its axon and dendrite. Glutamate (glu) contained in synaptic vesicles is released from the nerve terminal into the synaptic cleft. In this example, DA and glu are competing for the same binding site on the AMPA or kainate (KA) receptor. The activation of AMPA/kainate receptors upon DA binding leads to nerve depolarization and coactivation of NMDA receptors by glutamate binding. Calcium enters through activated NMDA receptors. The sum of this prolonged activation results in ion disturbances and swelling, eventually resulting in DA-induced nerve cell death.

neuron to fire, sending its message down the line. The neuronal firing is controlled because the glutamate is rapidly reabsorbed back into the neuron bulb or is inactivated by special enzymes; this removal closes the sodium and calcium channels. However, cell damage and cell death can occur when excessive amounts of glutamate are released from neuronal cells and cannot be removed.

Blaylock (1994) provided a layperson's explanation of this phenomenon, as well as a description of the toxicity of monosodium glutamate (MSG; the flavor-enhancing food additive, which is simply a glutamate molecule with a sodium ionically bound), and how this relates to the neurophysiology of DA toxicity. Blaylock uses a "key" and "lock" analogy: because the glutamate and DA molecules ("keys") are structurally similar, they both fit into the same receptor (the "lock") on the neuron surface. The five-sided ring structure of DA makes it less flexible than glutamate, which causes it to bind more tightly, resulting in a 30 to 100 times more powerful effect per molecule than seen with glutamate. Whereas glutamate, at low concentration, is rapidly removed, DA is not. Thus, DA affects the brain in a way similar to glutamate neurotoxicity; the neuron becomes over stimulated, meaning that calcium continues to flood into the cell, ATP energy reserves become depleted in an attempt to pump out the excess calcium, and then the neuron begins to swell with water, causing it to burst (as in hepatic encephalopathy, see Chapter 29). DA binds with high affinity (i.e., at low nanomolar concentrations) to kainate receptors that contain either of the protein subunits designated as "GluR5" and "GluR6." These subunits are both highly expressed in the hippocampus, a part of the brain associated with processing and laying down new memories; this explains the memory deficits shown by the patients poisoned with DA (hence the name amnesiac shellfish poisoning or "ASP"). Doble (2000) provided a review of the pharmacology of DA, and Ramsdell (2007) reviewed the molecular and integrative basis of DA toxicity.

Organismal Susceptibility to Domoic Acid

It is remarkable that the major brain regions affected by DA and the resulting behavioral effects of ASP are consistent among different mammalian species (Jeffery *et al.*, 2004). For this reason, experimental organisms (such as mice, rats, rabbits, or monkeys) can be used experimentally to understand the mechanisms of toxicity in humans. There are, however, some species differences in the potency of DA. For example, mice are less sensitive than rats, which in turn are less sensitive than monkeys and humans. It is notable that molluscan shellfish (including oysters, clams and most mussels) appear to be resistant to even huge doses of DA (1000 μg/g or mg/kg = ppm), in contrast to humans who appear to be sensitive to DA at much lower doses (as discussed previously). It has been speculated that some shellfish may contain binding proteins that sequester toxins away from their nerves or that their receptors are altered by mutation at the site of toxin binding (Trainer and Bill, 2004).

The effects of long-term exposure of humans to low concentrations of DA in shellfish are poorly understood. However, there is concern that populations that subsist heavily on shellfish may be at increased risk of chronic, long-term exposure. An epidemiological study of DA exposure in a Native American tribe in Washington State has suggested that infants born in years when DA levels in coastal razor clams were above the regulatory limit of 20 μg/g had lower mental development indices than infants born in other years (Grattan *et al.*, 2003). Further studies must be carried out to assess effects of long-term exposure to DA in humans.

Exposure of DA in mice and rats has indicated an age dependence of neurotoxicity (Ramsdell, 2007). Neonatal rats appear to be up to 40 times more susceptible to DA exposure than adults, presumably because of insufficient clearance of toxin in the poorly developed renal system and to incomplete development of the blood-brain barrier. This is significant because at ecologically relevant levels, DA can be transferred to the neonate via breast milk. However, the amount of toxin transferred appears to be well below symptomatic levels.

Older humans, especially males, may be particularly sensitive to the effects of DA, as evidenced by the four males over the age of 70 who died in the 1987 intoxication event in Canada (Perl *et al.*, 1990; Teitelbaum *et al.*, 1990). Dose-responses for seizures upon DA exposure are about three times greater in aged rats compared to young rats. The mechanism for increased sensitivity to domoic and kainic acids in elderly humans has not been precisely determined. However, decreased renal clearance and changes in excitatory and inhibitory pathways are known to accompany aging. In addition, the loss of a tolerance to DA that diminishes with age appears to be related to a reduction in constitutive GTPase activity in the rat hippocampus.

Identification of *Pseudo-nitzschia* Species

The challenge in the 1987 DA episode was to identify the source of the DA that contaminated the mussels in Cardigan Bay (eastern Prince Edward Island). Rapidly, the known sources of DA (i.e., several species of red macroalgae) were ruled out, as there were not enough of these seaweeds to account for the estimated 1000 kg of DA produced in Cardigan Bay (Bates *et al.*, 1989). Examination of the seawater revealed an almost monospecific bloom of a chain-forming diatom, *Pseudo-nitzschia multiseries* (then called *Nitzschia pungens* forma *multiseries*), reaching concentrations as high as 15×10^6 cells per liter. Furthermore, the digestive tract of the mussels was engorged with identifiable fragments of this same diatom. Finally, cells from Cardigan Bay were isolated into culture and were confirmed to produce DA.

The initial research focused on the unambiguous identification *Pseudo-nitzschia.* As pennate diatoms, the cells have a "lanceolate" shape (i.e., they are long and narrow, gradually tapering toward the ends). What distinguishes them from all other pennate diatoms is that the tips overlap slightly so that the cells form chains ("stepped" colonies, Figs. 12-4 and 12-5a). If left undisturbed (e.g., in a petri dish or in calm ocean waters), chains composed of dozens of cells can be formed. Because the cells narrow toward their tips, each cell that attaches to the next one near the tip is at a slight angle to it; the result is that the chains can form long spirals. When some *Pseudo-nitzschia* species cease growing, however, the chains may fall apart into single cells. Hasle and Syvertsen (1997) and Fryxell and Hasle (2003) provided a guide to the identification and distribution of marine diatoms, including *Pseudo-nitzschia* species.

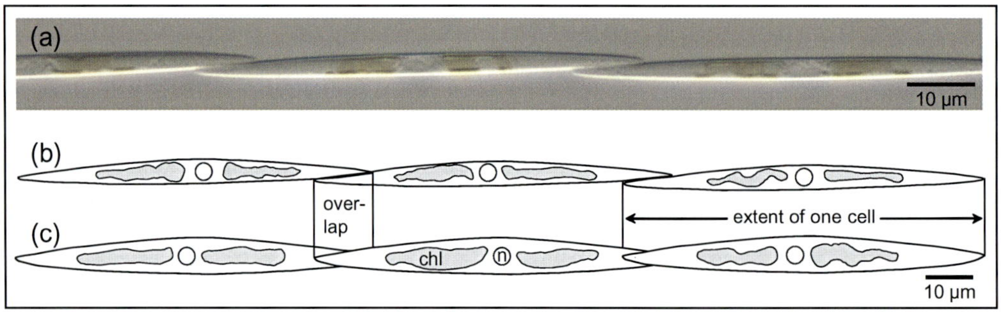

FIGURE 12-4. Chain of *Pseudo-nitzschia* cells. (a) light microscope image of *P. multiseries*, showing girdle (side) view (photo courtesy of Karie Holtermann, University of Washington). (b) Drawing of girdle view. (c) Drawing of valve (top) view. The general shape of the cells and the degree of cell overlap are characteristics used for identifying the species. chl = chloroplast; n = nucleus. Redrawn from Fehling (2004).

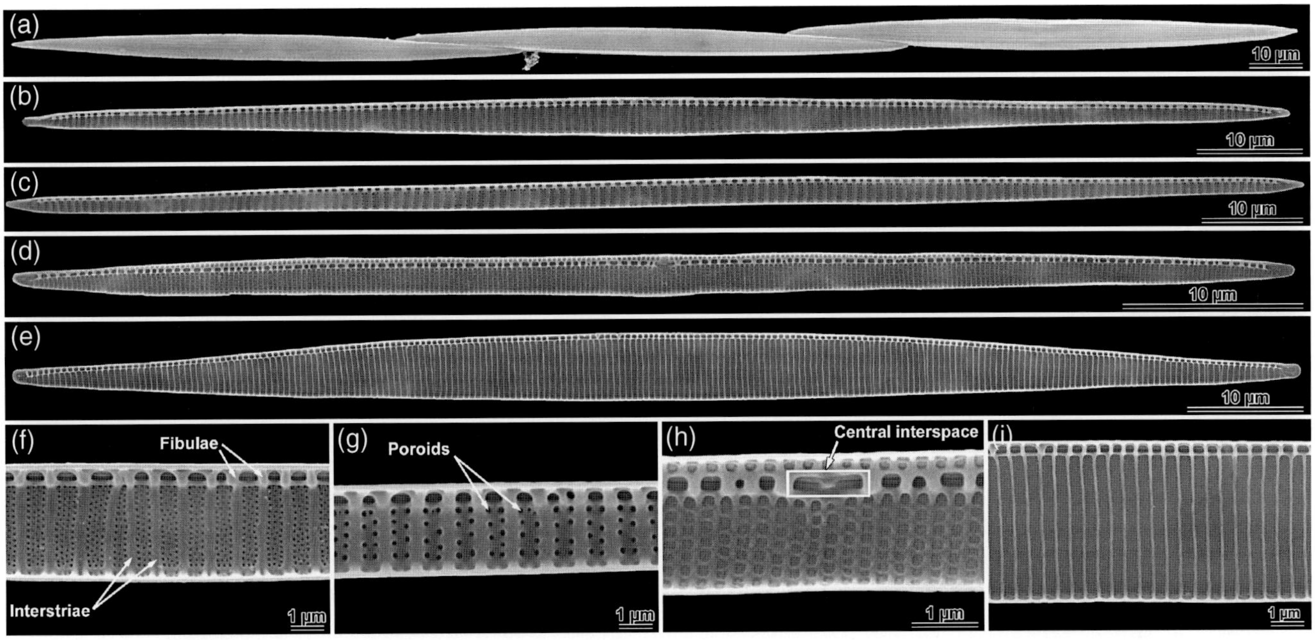

FIGURE 12-5. (a) Scanning electron micrograph (SEM) of ethanol-dehydrated/freeze-dried *Pseudo-nitzschia pungens* in girdle view, showing a chain of cells. (b–e) SEMs of acid-cleaned *Pseudo-nitzschia* species at low magnification, showing the entire cell; and (f–i) at high magnification, showing the central part of the cell. (b, f) *P. multiseries*; (c, g) *P. pungens*; (d, h) *P. calliantha*; and (e, i) *P. seriata*. Notice differences in width of the cells, size, and shape of poroids (but too small to be seen in *P. seriata* image i), number of interstriae and fibulae per unit distance, and the presence (h) or absence (f, g, i) of a central interspace (the large oval structure); all of these characteristics are used to distinguish among species. SEMs from James Ehrman, Digital Microscopy Facility, Mount Allison University, Canada.

From the point of view of human health, it is particularly important to be able to distinguish among the approximately 30 species of *Pseudo-nitzschia*. Of those species, at least 12, thus far, have been shown to produce DA in laboratory culture (Table 12-1). However, not all strains (i.e., organisms of the same species having minor genetic or morphological differences but not considered separate species) produce toxin at detectable levels, while others produce only small amounts of DA. To be able to demonstrate DA production by a species, a single cell (or a chain of cells) must be isolated using a capillary pipette and placed into a seawater medium that contains added nutrients, trace metals, and vitamins to enable their growth. Cells are then harvested at different times from the culture flask: first when they are dividing exponentially and then later when they run out of nutrients and stop dividing during the stationary phase; most species of *Pseudo-nitzschia* produce DA primarily when their growth is slowed (Bates, 1998). They are then analyzed for DA content using any of several available analytical techniques (Quilliam, 2003) or *in vivo* assays (Cembella *et al.*, 2003). This approach assures that the species in question is the source of the toxin; information is also gained about the growth conditions that promoted toxin production.

In general, *Pseudo-nitzschia* species cannot be identified definitively using only light microscopy because of the need to see the fine structure of the frustule. *Pseudo-nitzschia* cells in field samples can initially be divided into those that are wider than 3 μm (the "*seriata* group") and narrower than 3 μm (the "*delicatissima* group") using light microscopy (Hasle and Syvertsen, 1997). The degree of cell overlap can also be characteristic (Fig. 12-4). In some waters, it is possible to divide cells into three groups: (1) *multiseries/pungens*, (2) *australis/fraudulenta/heimii*, and (3) *pseudodelicatissima/delicatissima*. This is because the symmetrically wide, long shape of *P. multiseries/pungens* group can be distinguished by light microscopy from that of the asymmetrically wide, shorter *P. australis/fraudulenta/heimii* group and the much smaller *pseudodelicatissima/delicatissima* group (Trainer and Suddleson, 2005).

However, a definitive identification can only be made by examining the cell with a transmission electron microscope (TEM), or a scanning electron microscope (SEM). The cells are first cleaned with concentrated acids in order to remove the organic cell content as well as the outer organic layer. This exposes the silicon valves, which have intricate structural elements, including poroids, fibulae, interstriae, and, if present, a central interspace (Fig. 12-5). The number and spacing of these genetically fixed ornaments, as well as the shape, length and width of the cell, are used to identify each species (Hasle and Syvertsen, 1997).

Although morphological characteristics are effective for identifying and enumerating *Pseudo-nitzschia* species, the

TABLE 12-1. Species of *Pseudo-nitzschia* proven to produce domoic acid (DA).

Species	Year	DA (pg DA/cell)	Cell Size (μm) Length	Cell Size (μm) Width	Comments
*P. australis**	1992	0.021–37	100	7.1	Some New Zealand and California isolates below DA detection level
*P. calliantha**	1990	0.007–0.221	71	1.6	Previous records may have been reported as *P. pseudodelicatissima*
*P. cuspidata**	2005	1.21×10^{-4}–0.029	48	1.8	Previous records may have been reported as *P. pseudodelicatissima*; the name of the species is in a state of flux; only two reports of toxicity, from Washington State; limited number of isolates tested for toxigenicity
*P. delicatissima**	1990	0.0002–0.12	56	1.8	Toxicity reported only for some isolates from Prince Edward Island, Washington State, and New Zealand; most isolates below DA detection level
*P. fraudulenta**	1998	0.03	90	5.9	Toxicity reported only for two isolates from New Zealand
*P. galaxiae**	2005	7.8×10^{-7}–3.6×10^{-4}	35	1.5	Only one report of toxicity, from the Gulf of Naples
P. multiseries	1988	0.1–67	99	4.2	All isolates shown to be toxic
*P. multistriata**	2002	0.001–0.697	52	3.3	Only one report of toxicity, from the Gulf of Naples
*P. pseudodelicatissima**	1990	0.007–0.221	74	1.4	May be confused with *P. calliantha* and *P. cuspidata*
*P. pungens**	1996	0.0018–0.47	110	3.6	Only reports of toxicity are for some isolates from Washington State and New Zealand; most isolates below DA detection level
*P. seriata**	1994	0.16–33.6	120	6.4	Toxicity reported only for isolates from the north Atlantic
*P. turgidula**	1996	0.033	5468	3.0	Toxicity only reported for some isolates from New Zealand; identification uncertain

Shown are the year first reported, the range in DA concentration found in culture, and mean cell size. An asterisk (*) indicates that DA is below the limit of detection in certain laboratory isolates of that species. References are found in Bates (1998), Lundholm *et al.* (2003), and Bates and Trainer (2006).

SEM and TEM procedures are time consuming and costly. Therefore, a suite of other techniques has been developed in an attempt to expedite the process (Scholin *et al.*, 2003). These are based on designing molecular probes targeting ribosomal RNA sequences that are presumably unique to each species. However, the challenge has been that a probe developed for a *Pseudo-nitzschia* strain from one geographic region is not always able to detect the same species in another part of the world. Thus, it is sometimes necessary to tailor the probes for the diatom strains at the location studied.

Taxonomy of *Pseudo-nitzschia* Species

The taxonomy (i.e., the assigning of scientific names) of phytoplankton sometimes undergoes changes as new information about an organism's morphology or molecular characteristics is discovered. This has certainly been the case with the diatoms associated with DA production, and it can be a source of confusion. The first DA-producing diatom discovered was originally known as *Nitzschia pungens* forma *multiseries* (Bates *et al.*, 1989). However, based on morphological and molecular evidence, it became apparent that taxonomic revision was necessary. As a result, the species with overlapping cell ends were separated from the other species in the genus *Nitzschia* and moved to the genus *Pseudo-nitzschia* (first described in 1900 for cells that produced chains with overlapping ends) (Bates, 2000). Thus, all DA-producing diatom species, except for *Amphora coffeaeformis* (whose toxin-producing ability is disputed) and the newly discovered *Nitzschia navis-varingica*, are now in the genus, *Pseudo-nitzschia* (Bates, 2000). Nevertheless, because of advances in molecular techniques, we are now undergoing a period when some species of *Pseudo-nitzschia* are being reclassified, as some species are split and new species are being discovered (e.g., Lundholm *et al.*, 2003).

Geographic Distribution of Toxigenic *Pseudo-nitzschia* Species

Soon after the 1987 poisoning event in Canada, other species of *Pseudo-nitzschia* became recognized as toxin-producers in other parts of the world. Most *Pseudo-nitzschia* species proven to be capable of producing DA (i.e., toxigenic) in laboratory culture (Table 12-1) are thought to be cosmopolitan, although some are restricted to certain latitudes (Hasle, 2002). Figure 12-6 shows the locations around the world where toxigenic species of *Pseudo-nitzschia* have been identified. Although many locations are shown, only a few have actually reported problems with DA in marine shellfish, finfish (summarized in Table 12-2), birds, or mammals, as indicated by the shading in Figure 12-6. This

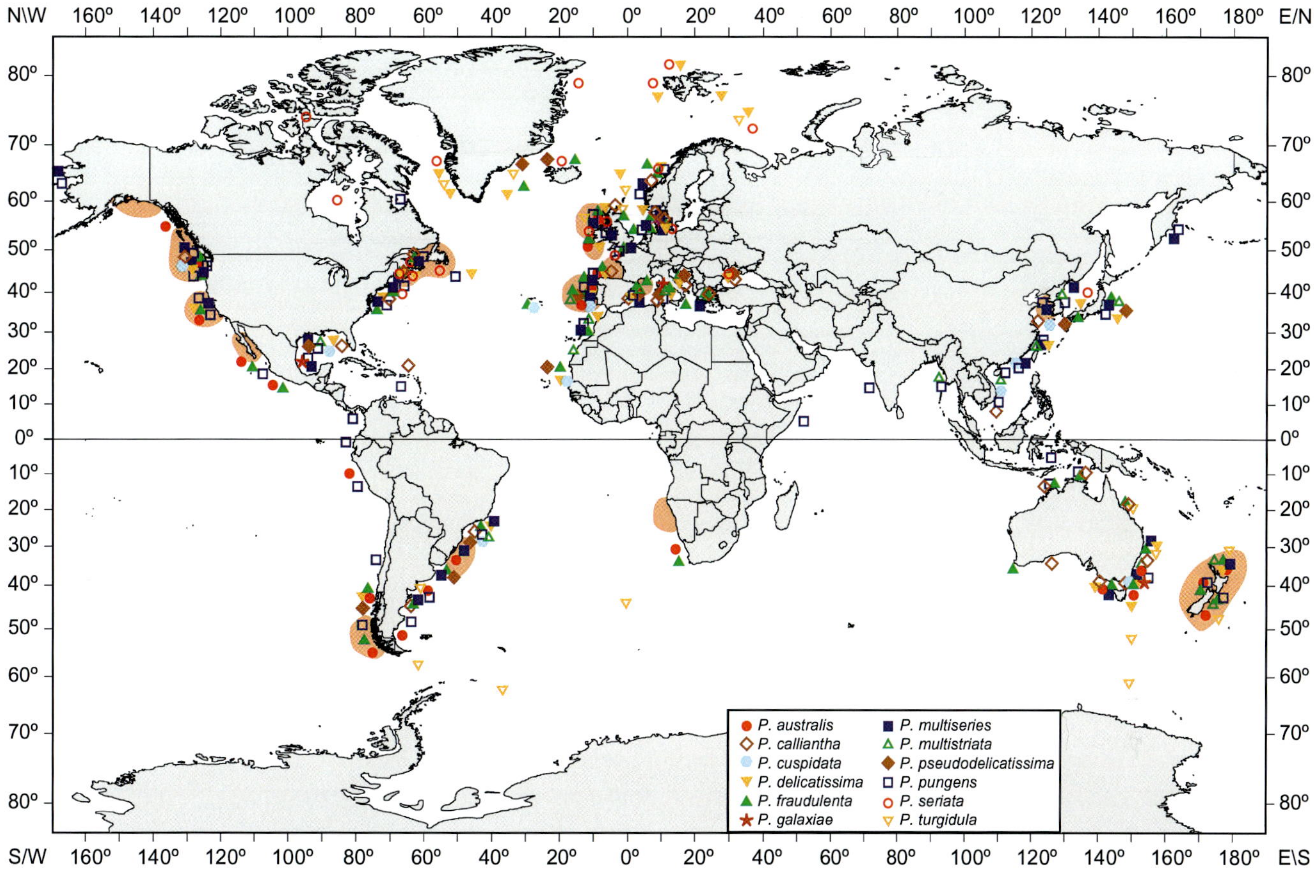

FIGURE 12-6. Worldwide recorded distribution of *Pseudo-nitzschia* species that have been shown to produce domoic acid in laboratory cultures; some strains do not always produce the toxin at detectable levels. Those regions that have experienced domoic acid contamination in marine zooplankton, fish, birds, molluscan shellfish, or mammals are shaded in red. Redrawn and updated from Fehling (2004).

may be because (1) strains of some toxigenic *Pseudo-nitzschia* species may not produce DA at detectable levels; (2) conditions may not be appropriate to trigger the pathway for DA production; (3) the cellular concentration of DA may be low (Table 12-1); (4) the relative abundance of the species may not be great enough, even if it does produce high amounts of DA, for the toxin to accumulate in the animals; and (5) shellfish retain or depurate (eliminate) DA at different rates. In some locations (e.g., coastal Mexico, Namibia, Newfoundland, Alaska), the *Pseudo-nitzschia* species that are the source of the DA contamination have not yet been identified with certainty.

Interestingly, toxigenic species are found mostly in coastal waters, thriving on high nutrient concentrations. Coastal locations are those most prone to high nutrient inputs, from terrestrial sources (Glibert *et al.*, 2005b) or upwelling events, and this could stimulate HABs. However, *Pseudo-nitzschia* species from midocean waters (e.g., Station P, eastern north Pacific) have also been shown to produce low but detectable levels of DA, as measured by the enzyme-linked immunosorbent assay (ELISA). To date, most ASP and DAP events have occurred in temperate regions, although most of the toxic or potentially toxic species can also be found in tropical or subtropical regions, and some of them have been involved in HAB events (Fig. 12-6). Although *Pseudo-nitzschia* species, including known toxigenic species, are found in both Arctic and Antarctic waters, there has been no testing for DA in these areas.

The presence of toxigenic *Pseudo-nitzschia* species in coastal waters around the world often overlaps with those areas that provide food, via aquaculture activities and wild harvests, for human consumption. To address this concern, many countries have established programs to monitor for the presence of phycotoxins in the flesh of molluscan shellfish and finfish destined for human consumption (Andersen *et al.*, 2003). In some countries, this is supplemented with counts of toxic phytoplankton in order to provide an early warning of impending toxicity and to direct further sampling of the food product (Todd, 2003). Trainer and Suddleson (2005) provide an example of such a program for DA events on the U.S. Pacific Northwest coast. The number of areas demonstrated to be affected by DA appears to be on the

TABLE 12-2. Chronology of events, showing the detection of domoic acid (DA) in various marine animals used for human consumption and the physical oceanographic regime in which the event occurred (WBS = Western Boundary System; EBS = Eastern Boundary System).

Location	Year	Affected Species: Common Name	Affected Species: Scientific Name	*Pseudo-nitzschia* Species Implicated	Oceanographic Regime
Prince Edward Island, Canada	1987	Blue mussel	*Mytilus edulis*	*P. multiseries*	WBS, Shallow bay
Bay of Fundy, Canada	1988	Soft-shell clam Blue mussel Horse mussel Sea scallop	*Mya arenaria* *Mytilus edulis* *Volsella modiolus* *Placopecten magellanicus*	*P. pseudodelicatissima* or *P. calliantha*	WBS, Estuary
Washington and Oregon coasts, United States	1991	Razor clam Dungeness crab	*Siliqua patula* *Cancer magister*	*P. australis*	EBS, Upwelling
Monterey Bay, California, United States	1991	Northern anchovy	*Engraulis mordax*	Not directly linked	EBS, Upwelling, Bay
Pacific coast of the United States	1991 to 1993	Blue crab Rock crab Stone crab Spiny lobster	*Cancer spidus* *Cancer pagurus* *Menippe adina* *Palinurus elephas*	Not directly linked	EBS, Upwelling
Coastal New Zealand	1993 to 1997	Maori scallop Greenshell mussel Pacific oyster New Zealand cockle Chilean oyster Tuata surf clam	*Pecten novaezealandiae* *Perna canaliculus* *Crassostrea gigans* *Austrovenus stutchburyi* *Tiostrea chilensis* *Paphies subtriangulata*	*P. australis*, *P. pungens*	WBS, Upwelling
Galicia, NW Spain	1994	Mediterranean mussel	*Mytilus galloprovincialis*	*P. australis*	EBS, Upwelling
Georges, German and Browns Banks, Gulf of Maine	1995	Sea scallop	*Placopecten magellanicus*	*P. seriata* (likely)	WBS, Banks
Baja California peninsula, Mexico	1995	Pacific mackerel	*Scomber japonicus*	*Pseudo-nitzschia* spp.	EBS, Upwelling
Offshore Portugal	1996	Blue mussel Common cockle Peppery furrow shell clam Pullet carpet shell European oyster Razor clam Clam	*Mytilus edulis* *Cerastoderma edule* *Scrobicularia plana* *Venerupis pullastra* *Ostrea edulis* *Ensis* spp. *Ruditapes decussata*	*P. australis* (likely)	EBS, Upwelling
Chinhae Bay, South Korea	1998	Various shellfish	Not specified	*P. multiseries*	WBS, Shallow bay
Washington and Oregon coasts, United States	1991 to 2005	Razor clam	*Siliqua patula*	*P. pseudodelicatissima*, *P. australis*	EBS, Upwelling
Central coast, California, United States	1998	Northern anchovy	*Engraulis mordax*	*P. australis*	EBS, Upwelling
Offshore Scotland	1999 to 2000	King scallop	*Pecten maximus*	*P. australis*, *P. seriata*	EBS, Tidal, Downwelling
Offshore Ireland	1999	King scallop	*Pecten maximus*	*P. australis*	EBS, Tidal, Downwelling
Western Brittany, France	1999	Wedge shell clam	*Donax trunculus*	*P. multiseries*	EBS, Upwelling
Monterey Bay, California, United States	2000	Pacific mackerel Albacore tuna Northern anchovy Pacific sardine Market squid	*Scomber japonicus* *Thunnus alalunga* *Engraulis mordax* *Sardinops sagax* *Loligo opalescens*	*P. australis*	EBS, Upwelling

(*continued*)

TABLE 12-2. *(continued)*

Location	Year	Affected Species: Common Name	Affected Species: Scientific Name	*Pseudo-nitzschia* Species Implicated	Oceanographic Regime
Offshore Portugal	2000 to 2001	European sardine European anchovy Blue mussel Pacific sardine Common cockle Pullet carpet shell Clam Oyster Razor clam	*Sardina pilchardus* *Engraulis enchrasicolus* *Mytilus edulis* *Sardinops sagax* *Cerastoderma edule* *Venerupis pullastra* *Ruditapes decussate* *Crassostrea japonica* *Ensis* spp., *Solen* spp.	Not determined	EBS, Upwelling
Southern Gulf of St. Lawrence, Canada	2002	Blue mussel	*Mytilus edulis*	*P. seriata*	WBS, Deep estuary
Offshore Portugal	2002	Swimming crab	*Polybius henslowii*	Not directly linked	EBS, Upwelling
Offshore Portugal	2003	Common octopus Common cuttlefish	*Octopus vulgaris* *Sepia officinalis*	Not directly linked	EBS, Upwelling
Puget Sound, Washington, United States	2003	Blue mussel	*Mytilus edulis*	*P. australis*	Deep estuary
Southern California, United States	2003 to 2004	Red crab Pacific mackerel Jack mackerel Pacific sanddab Longspine combfish	*Pleuroncodes planipes* *Scomber japonicus* *Trachurus symmetricus* *Citharichthys sordidus* *Zaniolepis latipinnis*	*P. australis* and *P. multiseries*	EBS, Upwelling
Monterey Bay, California, United States	2003 to 2004	Rex sole Dover sole English sole Curlfin turbot	*Errex zachirus* *Microstomus pacifcus* *Pleuronectes vetulus* *Pleuronectes decurrens*	*P. australis* (likely)	EBS, Upwelling
Santa Cruz wharf, California, United States	2004	White croaker Staghorn sculpin	*Genyonemus lineatus* *Gymnocanthus tricuspis*	*P. australis* (likely)	EBS, Upwelling

The *Pseudo-nitzschia* species implicated may have been fed on either directly or indirectly by the animals. The table does not include marine zooplankton, birds, and mammals that have also been affected by DA. References are found in Bates *et al.* (1998) and Bates and Trainer (2006).

increase since the original 1987 ASP outbreak in eastern Canada. This is probably because toxigenic *Pseudo-nitzschia* species are ubiquitous, and more events are being detected as more countries establish regulatory programs to monitor for the presence of DA in food products from the sea.

OCEANOGRAPHY AND TOXIC DIATOM BLOOMS

Toxic blooms may arise under several different oceanographic settings, and the challenge is to tease out which controlling factors are most important. In spite of intense research on the biological and chemical influences on the bloom formation of HABs, the details of bloom initiation and termination and the species composition of a bloom remain elusive. Why does one species of diatom (e.g., toxic *Pseudo-nitzschia multiseries*) begin to grow and become dominant at a particular location and time? The given species must, of course, be present, but then certain biological factors (such as grazing by zooplankton and filter-feeding molluscs, infection by fungi and viruses, and inherent physiological properties) must also exert an important influence. In the case of *Pseudo-nitzschia* species, there is evidence that they are more lightly silicified than other coastal diatoms (Marchetti *et al.*, 2004) and may therefore have a competitive growth advantage at low silicon concentrations. Some work has indicated that toxigenic *Pseudo-nitzschia* species may also have unique capabilities to acquire trace metals such as iron and copper (e.g., Wells *et al.*, 2005), thereby giving them a competitive growth advantage over other phytoplankton.

Ocean circulation and seawater properties exert an additional important influence in the development of toxic diatom blooms; the details are largely unknown, but this is a subject of intense research. Even though a particular physical cause may be invoked for a specific bloom, relationships have been difficult to generalize with any certainty. On the other hand, research has been successful at determining the role of ocean circulation in transporting HABs along the continental shelf (e.g., Adams *et al.*, 2006; Anderson *et al.*, 2006). This circulation can spread the toxin over a greater area; it may also move a bloom onshore to beaches (e.g., Trainer *et al.*, 2002), or export it off the shelf. In areas with certain types of bottom topography (such as banks and submarine canyons), circulation likely plays a role in retaining blooms in a region, thus allowing diatom densities and toxins to increase to dangerous levels (Anderson *et al.*, 2006). Water properties such as stratification may play a role in bloom development by inhibiting turbulence. Because diatoms tend to sink, stratification may allow them to be retained in layers exposed to higher light levels by preventing their sinking across the pycnocline. Macronutrients (such as nitrate and silicate) and micronutrients (such as iron and copper) play an important role in bloom development and the control of DA production.

The Major Coastal Current Systems

Blooms of toxic diatoms have been observed somewhat more frequently worldwide on eastern continental boundaries (where the ocean is bounded by the land on its eastern side) than on western boundaries (Fig. 12-6). Differences between these types of current systems, especially as they might pertain to HABs, are discussed briefly next.

Eastern boundary systems (EBS) support four out of the five most productive upwelling systems on the planet (Hill *et al.*, 1998). For example, the Peruvian/Chilean system produces 15% of all fish landed worldwide. The most productive areas on eastern boundaries are generally those in mid- to lower-, often semi-arid latitudes, where winds blow persistently toward the equator, causing upwelling of deeper, nutrient-rich water masses near the coast. The near-continuous addition of nutrients fuels dense phytoplankton blooms in these regions. Continental shelves on eastern boundaries are generally narrower (~10 to 100 km) than those on western boundaries (~10 to 200 km) (Fig. 12-7). The narrow width results in much greater movement and exchange of suspended material (such as phytoplankton) between the inner shelf (roughly bottom depths shallower than 30 m) and the outer shelf and slope than occurs in a western boundary system (Boicourt *et al.*, 1998; Loder *et al.*, 1998).

The preceding characteristics may be one reason for the more frequent occurrence of toxic diatom impacts on coastal beaches on eastern than western boundaries, especially along U.S. and Canadian coastlines. In general, in EBS upwelling systems (Fig. 12-7, panels at left), currents flow toward the equator at speeds of 10 to 20 km/day over the upper depths of the continental shelf and slope during the spring to summer seasons. At more poleward latitudes, the mean flow reverses to poleward in the fall and winter, reflecting the reversal in along-coast wind direction. Seaward of the continental slope to ~1000 km offshore, flow is generally equatorward throughout the year; this broad, sluggish portion of the "eastern boundary current" (e.g., the California Current) constitutes the eastern limb of the whole ocean basin circulation.

Variability in water properties and currents in an upwelling EBS are dominated by changes in the along-coast component of wind stress (Hickey, 1998). Seasonal variation is generally much greater than day-to-day variability, although the latter may be more critical for moving near-surface phytoplankton onto coastal beaches or offshore. This onshore-offshore movement occurs because the wind dragging on the sea surface generates near-surface currents that move to the right of the wind stress in the northern hemisphere (i.e., on the U.S. West Coast, offshore when winds are blowing toward the south, and onshore when winds are blowing toward the north) (Fig. 12-7, lower-left panel). Currents at deeper depths respond to the near-surface flow, moving in the opposite direction to compensate for the loss of mass in the upper layers (near-bottom dashed arrows in Fig. 12-7); upwelling of deeper, colder, nutrient-rich waters thus occurs to close the loop. Seasonally varying wind stress results in upwelling of nutrient-rich deeper water from the continental slope onto the shelf throughout the spring to fall period. At some locations, upwelling occurs all season long. Episodic wind events infuse nutrients into the coastal zone at several-day intervals, resulting in the renewal of existing phytoplankton blooms or the development of new blooms over the inner to midshelf.

Another important feature of an EBS, with possible relevance to HABs, is the existence of poleward-flowing currents beneath the equatorward surface currents over the continental slope (Fig. 12-7, lower-left panel). This flow, denoted an "undercurrent," is generally concentrated just below the shelf break (~150 to 300 m depths) and has a relatively narrow (~10 to 30 km) high-speed (10 to 20 km/day) core. The undercurrent can be continuous for hundreds or even thousands of kilometers along the coast (Hill *et al.*, 1998). Much of the water that is seasonally upwelled comes from the undercurrent. Also, the undercurrent is thought to play an important role in the poleward transport of larvae or phytoplankton. The existence of an undercurrent may provide a mechanism for phytoplankton, such as toxigenic diatoms, to spread along a coast counter to the prevailing surface currents. We note that evidence for this spreading mechanism has not been documented.

In shallow regions with modest stratification, coastal currents tend to follow the orientation of isobaths (i.e., contour lines mapping areas of identical depth). However, in regions where isobaths change direction abruptly, such as when the continental shelf narrows near a coastal cape, coastal cur-

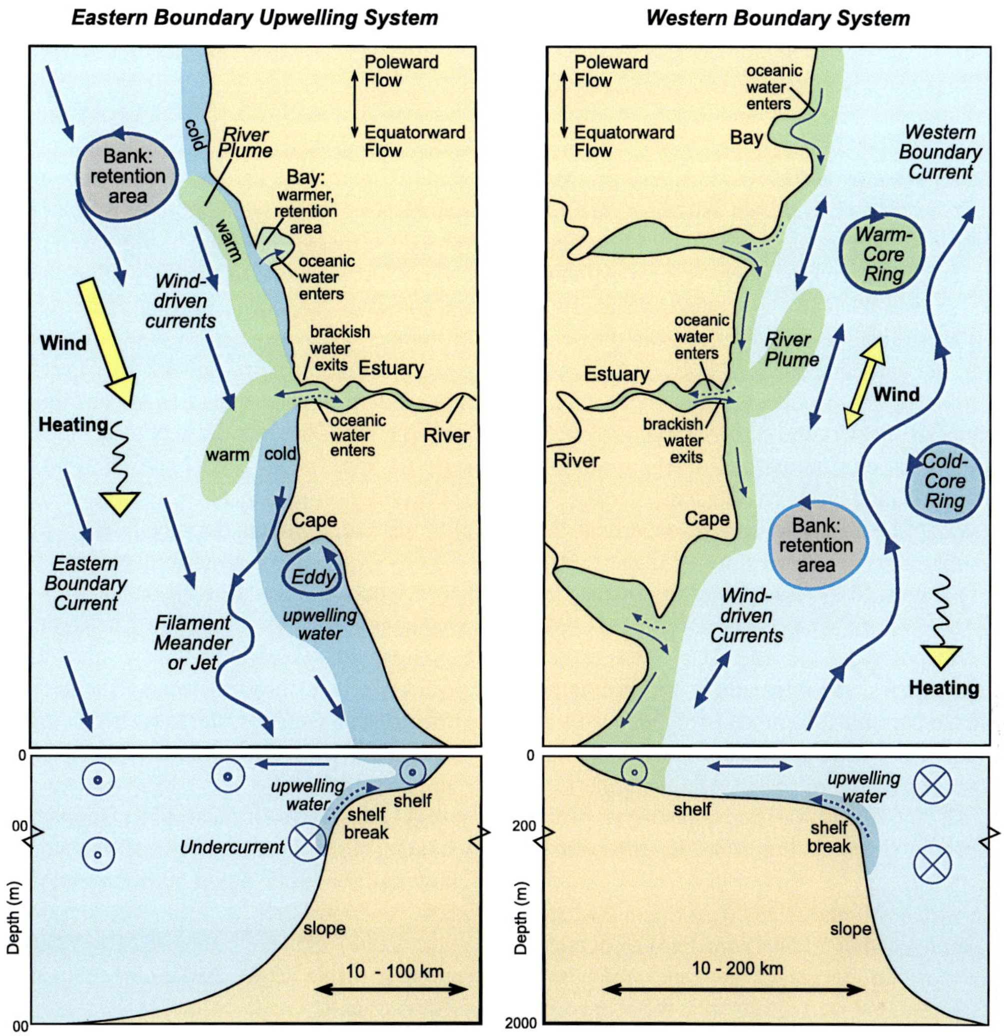

FIGURE 12-7. Schematic depicting major oceanographic features in Eastern (left) and Western (right) Boundary Current Systems (EBS and WBS, respectively) in the northern hemisphere. Upper panels show plan view, lower panels show depth versus distance offshore. The left panels are applicable to the U.S. West Coast; the right panels to the U.S. East Coast. For the U.S. West Coast, the EBS is the California Current; the WBS on the East Coast is the Gulf Stream. Symbols indicate currents (blue arrows; solid for near surface, dashed for beneath the surface), wind (yellow arrows), and heat flux (yellow arrows with tail). Shallow banks (which generally have more retentive circulation patterns) are shown in gray. Upwelled water is shown in dark blue; river plumes, bays, and estuaries are generally warmer (light green). Filaments, meanders, eddies, and jets that are frequently associated with features on an EBS coast, such as capes, are depicted, as well as rings (eddies) and meanders related to the WBS. Flow direction on the lower panels is indicated with a circle with a dot in the center (coming toward you) or a circle with a cross (away from you).

rents may be forced off the shelf by the rapid change in the direction of ocean bottom isobaths. This happens frequently in an EBS, such as the U.S. West Coast (Fig. 12-7, upper-left panel), so that the region offshore of the shelf is dominated by meandering filaments (thin features that stream from a site of origin), jets and eddies, many of which originate on the shelf near the coast (Hill, 1998; Hickey, 1998). The implication for HABs in general, and for toxigenic diatoms in particular, is that cells would be moved offshore and would therefore be less likely to reach the coast in regions where such features occur.

Major physical disturbances to an EBS generally arrive from outside the system (i.e., they travel in the ocean along the coastal boundary from "remote" locations). In particular, the greatest disturbances are caused by El Niño phenomena (see Chapter 1), whose origin is along the equator (Hill, 1998). El Niño is associated with warmer waters and enhanced poleward flow. In the northern hemisphere, nutrients and phytoplankton concentrations are reduced during an El Niño event, and more southern types of plankton and fish are carried poleward in the enhanced poleward undercurrent. Water properties in an EBS can also be impacted by changes in the amount of water entering the system from poleward locations (e.g., enhanced amounts of subarctic water in the Pacific Ocean). In contrast to an El Niño, these intrusions bring enhanced nutrients and result in higher standing stocks of phytoplankton.

Western boundary systems (WBSs) are usually (but not always) characterized by much wider shelves than an EBS (>100 km) (Fig. 12-7) (Loder *et al.*, 1998). This is particularly true for the northwestern and southwestern Atlantic region. Moreover, the mechanisms that control water properties and currents on the shelves often differ from those on eastern boundaries. For example, on the U.S. East Coast, the inner shelf is dominated by the effects of brackish water exiting from the myriad of rivers and estuaries that generally occur on western boundaries at midlatitudes, where rainfall is not inhibited by mountains as on the west coast. On the U.S. East Coast, an equatorward current occurs on the inner continental shelf, driven by freshwater that extends from Greenland to the mid-U.S. East Coast, a distance of more than 2000 km. On the outer continental shelf and slope, currents are poleward on the U.S. East Coast because of the Gulf Stream, and fluctuations are dominated by rings and eddies or currents associated with the meandering the Gulf Stream. The western boundary current (WBC) constitutes the western limb of basin scale circulation. Unlike the sluggish and broad eastern boundary current (such as the California Current), the WBC is narrow and swift.

Because of the wide shelf and fronts (i.e., regions where density changes rapidly within a few kilometers) associated with the freshwater-driven current on the inner shelf, the inner shelf and outer shelf/slope regions in a WBS are frequently somewhat isolated from each other (Boicourt *et al.*, 1998). Along-coast winds can move surface waters onshore and offshore, just as in an EBS, as well as alongshore. However, the winds are generally weaker and less persistently in an upwelling-favorable direction in a WBS (e.g., to the north on the U.S. East Coast) than in an EBS in the spring/summer growing season.

Therefore, upwelling of deep nutrients by along-coast winds, which dominates and controls water properties (including nutrients) on many eastern boundaries, is rarely important on western boundaries. Rather, the dominant source of nutrients on western boundary inner shelves is freshwater runoff that passes through estuaries to the ocean. Much of the nutrient supply is of anthropogenic origin (i.e., resulting from human activities) on many western boundaries. Farther offshore, nutrients are supplied from the continental slope via interactions with the WBC and its meanders and frontal eddies. Tidal mixing can also contribute to the nutrient supply. Major disruptions in a WBS are usually caused by local atmospheric disturbances such as hurricanes and tropical storms, rather than by large-scale oceanic features such as El Niño as on eastern boundaries.

Topographic Features with Importance to HABs

Smaller-scale (~10 to 50 km) features (such as capes, banks and submarine canyons) can locally modify the flow patterns described earlier. These features can also affect stratification in ways that may accelerate phytoplankton growth, perhaps contribute to the onset of toxicity, or influence HAB transport. In regions with equatorward flow over a submarine canyon, upwelling of nutrient-rich water may be enhanced (Hickey, 1997). As well, a retentive circulation pattern forms over the canyon. However, the circulation is modified by local stratification, so that this eddy-like feature is frequently confined to depths below the euphotic zone (i.e., the upper water layer exposed to sufficient sunlight for photosynthesis to occur), where it would not impact local phytoplankton blooms directly.

Eddy-like current patterns form downstream of coastal capes (such as Cape Blanco on the U.S. West Coast) and over banks (such as offshore of the Strait of Juan de Fuca, a feature that separates the western U.S. and western Canada)—for example, Heceta Bank (offshore of central Oregon) and Georges Bank (offshore of the central U.S. East Coast). These features tend to be retentive (i.e., they retain phytoplankton for longer periods than at an open, straight coastline). Because of the longer time spent in that location, phytoplankton may accumulate to high densities. They may also draw down nutrients and thereby experience stress, a contributor to toxin production (e.g., Anderson *et al.*, 2006). Several examples from U.S. coastal regions (given in Table 12-2) are described in more detail in the following text.

A map showing DA levels along the U.S. West Coast during the summer of 1998 shows high DA levels only near features that are known to be retentive: the Juan de Fuca eddy, Heceta Bank, the Farrallon Islands, Monterey Bay, and the Santa Barbara Channel (Trainer *et al.*, 2001; Hickey and Banas, 2003; Fig. 12-8). The Juan de Fuca eddy region has now been shown to sustain toxic *Pseudo-nitzschia* blooms in the summer of almost every year. Blooms appear to be ejected from the eddy under upwelling-favorable wind conditions (Mac-Fadyen *et al.*, 2005). They travel equatorward in the coastal currents on the middle to outer continental shelf and slope. However, brief reversals in wind conditions can transport these toxic patches to the coast, where they toxify razor clams and other shellfish (Trainer *et al.*, 2002).

In Monterey Bay (California), another retentive coastal region off the West Coast of the United States. (Fig. 12-8), toxic *Pseudo-nitzschia* blooms were first identified in 1991, when Brandt's cormorants and brown pelicans were found dying, diving into windows, and otherwise exhibiting behavior indicating that their nervous systems had been impacted (Bates *et al.*, 1998). High levels of DA were found in the sardines and anchovies (phytoplankton-feeding fish) on

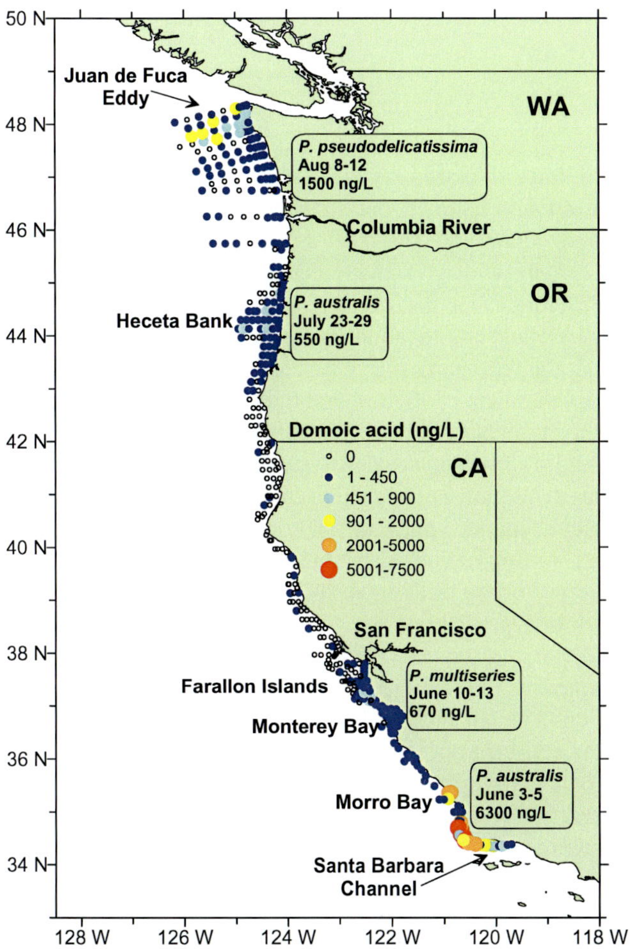

FIGURE 12-8. Particulate domoic acid (ng/L) in surface seawater during cruises off the U.S. West Coast in 1998. The timing and approximate locations of each cruise, as well as the species of *Pseudo-nitzschia* responsible for maximum toxin levels measured during each cruise, are shown in the boxes. Redrawn from Trainer *et al.* (2001); see also Hickey and Banas (2003).

which these birds had fed. A bloom of *Pseudo-nitzschia australis*, identified for the first time as a species producing high levels of DA, was observed in Monterey Bay during this event. However, scientists believe that such marine bird poisonings had also occurred in prior years. In August 1961, in the small central Californian town of Capitola near Santa Cruz, seabirds collided with inanimate objects and attacked police cars and people. This is thought to have inspired Alfred Hitchcock's 1963 movie *The Birds,* based on Daphne du Maurier's short story. Scientists have speculated that this strange behavior was because the birds had consumed anchovies contaminated with DA, much like the 1991 event.

In May and June 1998, more than 400 California sea lions were found dead, and many others, as well as sea otters, displayed signs of neurological dysfunction on beaches along the central California coast, including Monterey Bay and Morro Bay (Scholin *et al.*, 2000). The poisoning behavior in these marine mammals (including scratching, tremors, and seizures) was remarkably similar to that seen in mice used for DA bioassays. DA was found in their tissues, and some contained *P. australis* frustules in their gut. In addition, several affected sea lions showed histological lesions evidenced by nerve cell death in the CA3 region of the hippocampus and dentate gyrus. As in the case of the seabirds, these sea lions had fed upon sardines and anchovies that, in turn, had fed upon toxic *P. australis* cells.

Estuaries and Bays

Coastal estuaries and shallow bays are sometimes sites for toxic diatom blooms. The two environments have significant differences (Fig. 12-7). In particular, an estuary is fed by a river, or several rivers providing fresh water, so that its waters are brackish. Surface and deeper currents generally move in opposite directions in an estuary, with less saline waters flowing out of the estuary at the surface and more saline waters into the estuary at or near the bottom (Hickey and Banas, 2003), sometimes resulting in a density-based stratification of the two layers (see below). In contrast, waters in a bay are usually entirely of oceanic origin and circulation is likely more dominated by currents flowing in the same direction at all depths. Note that the designation "bay" in a name does not necessarily mean that the body of water is a "bay" in the physical sense defined here. Even within these two categories, large differences occur. Estuaries may be deep (e.g., the Strait of Juan de Fuca, Puget Sound or the Gulf of St. Lawrence) or shallow (e.g., Chesapeake Bay or Cardigan Bay).

From the point of view of HABs, estuaries in an EBS and a WBS may have even greater differences. On eastern boundaries, estuaries tend to be rapidly flushed by tidal currents. This means that phytoplankton (including HABs) and nutrients in EBS estuaries tend to be imported from coastal waters. On western boundaries, estuaries tend to have indigenous ecosystems; therefore, nutrients are supplied by runoff from land, and ocean intrusions and tidal flushing are much reduced (Hickey and Banas, 2003).

Bays and estuaries do have one feature in common that derives from their small size and relatively retentive nature: they tend to be warmer than the adjacent ocean in the spring-summer growing season. This leads to enhanced stratification, and in some cases, weaker turbulence. Shallow bays are ideal locations for the aquaculture of molluscan shellfish because they are sheltered and, if well flushed, have a constant source of phytoplankton needed as food for the shellfish. However, they are also vulnerable to HABs, which often occur at these same locations. Several examples of toxic diatom blooms in estuaries and bays are described next (also, Table 12-2); in these examples, Cardigan Bay and Penn Cove are "estuaries," and East Sound is a "bay."

Bays and estuaries within the Gulf of St. Lawrence, a large estuary in eastern Canada, have experienced periodic blooms of toxic *Pseudo-nitzschia*, resulting in closures of molluscan shellfish harvesting. The original bloom in the fall of 1987 occurred in Cardigan Bay, eastern Prince Edward Island (PEI), and it recurred there during the fall for the next 2 years (Bates *et al.*, 1998). In subsequent years (i.e., 1991, 1992, 1994, 2000, and 2001), the toxic blooms occurred in three different bays on northern PEI. Aquacultured blue mussels (*Mytilus edulis*) were the primary vector for the DA, reaching 790 μg DA/g in the whole animal in the 1987 incident, considerably above the 20 μg DA/g limit. In each case, the blooms were composed of DA-producing *P. multiseries*. These blooms likely originated within the bays, as evidenced by lower cell concentrations outside the bays.

In the southern Gulf of St. Lawrence, the mussel and oyster aquaculture industries were caught off guard in 2002 by high levels of DA that forced the closure of shellfish harvesting for the first time during the spring (March to May). This event involved a different species of *Pseudo-nitzschia*: *P. seriata*, a cold-water diatom that is found only at northern latitudes of the Atlantic Ocean (Hasle, 2002). In contrast to the *P. multiseries* blooms, these originated offshore; they were then drawn into the bays by the tides. Because the source organism bloomed throughout the Gulf of St. Lawrence at the same time, it caused a wide shellfish bed closure for most of the southern Gulf.

Record levels of DA in October 2005 resulted in the closure of an important commercial mussel fishery in Penn Cove (Washington State), considered an "estuary" because of its proximity to the Skagit River. DA levels in Dungeness crabs measured 28 μg/g, just below the closure level of 30 μg/g in this commercial and recreational fishery. Studies in Penn Cove (Trainer *et al.*, 2007) have documented blooms of *Pseudo-nitzschia* after periods of strong freshwater discharge from the nearby Skagit River. Stratification was facilitated by weak winds, sunshine, and a freshwater lens at the mouth of the cove. It has been suggested that because *Pseudo-nitzschia* cells have an efficient nutrient uptake capability, they can survive in macro- and micronutrient depleted environments (Wells *et al.*, 2005) such as stratified systems, where freshwater is layered over denser, saltier water, and there are no sustained inputs of nutrients.

Pseudo-nitzschia cells have been found to be a major component of chlorophyll maxima or "thin layers" found at depth within some bays (e.g., in East Sound, Orcas Island in Puget Sound, considered a "bay" because of the lack of freshwater input) on the West Coast of the United States (Rines *et al.*, 2002). During the 1999 bloom observed in East Sound, sustained winds forced a less saline, lighter plume of water into the Sound, displacing the surface bloom of *P. fraudulenta* to a greater depth, where it formed a thin layer. Although high numbers of *P. fraudulenta* were found within the thin layer in East Sound (>1 million cells per L), this species is not always toxic, so shellfish were not toxic at this site.

Plumes from Rivers

A brackish plume develops in the ocean where each river and estuary empties onto the coast (Hill, 1998). These plumes affect both the local current patterns and the local ecosystem, with greater effects from larger rivers such as the Columbia or the Mississippi. Plumes have associated large vertical stratification, and hence decreased turbulence. Thus, they are particularly susceptible to heating, and also to development of high phytoplankton densities, especially if the nutrient source is sustained by upward mixing of nutrient-rich water in deeper layers or by other processes. Some river plumes on western boundaries, such as the Mississippi, support massive phytoplankton blooms off the coast, a result of the terrigenous (i.e., derived from the land) nutrient loading. Sinking of material from these blooms can result in hypoxic or anoxic (low or no oxygen, respectively) conditions on the ocean floor.

Currents may be accelerated within a plume relative to a region with no plume. Hence, plumes can provide rapid along-coast transport to new regions, facilitating the spread of toxic or toxigenic phytoplankton. Evidence has shown the interaction of the Columbia River plume and toxic *Pseudo-nitzschia* in Washington State coastal waters (Adams *et al.*, 2006). Plumes may even transport blooms in directions opposite to the ambient local currents. If a plume hugs the coast, as it frequently does on both eastern and western boundaries, its seaward front may provide a barrier to onshore transport of toxic blooms that develop offshore of the plume. Hickey *et al.* (2005) hypothesized that the plume from the Columbia River frequently prevents toxigenic *Pseudo-nitzschia* blooms in the Juan de Fuca eddy from reaching the central and southern Washington coast.

Impacts of Offshore HABs

Not all molluscan shellfish are harvested along coastlines; there are also deep-water harvests of wild bivalve shellfish. One such fishery is on Georges and nearby banks, retentive features in the Gulf of Maine. Sea scallops (*Placopecten magellanicus*) are harvested as a "roe-on" product (i.e., sold with roe still attached to the adductor muscle). In May 1995, extremely high levels of DA were found in these sea scallops. The most toxic sample showed the following tissue distribution: digestive gland (3400 μg/g wet weight), roe (55 μg/g), gills plus mantle (19 μg/g), and adductor muscle (0.62 μg/g). This event essentially shut down the lucrative wild harvest of "roe-on" scallops for that year; no contaminated product reached the market. As in other cases of DA in scallops, the adductor muscle had DA levels well below the safety guideline.

SUMMARY AND CONCLUSIONS

In summary, HABs composed of diatoms of the genus *Pseudo-nitzschia* can produce the neurotoxin DA and lead to the risk of ASP in many different coastal environments. It is important that medical professionals be educated about the symptoms of ASP. Although there is presently no antidote to ASP other than supportive care (e.g., respiratory support and control of convulsions), it is essential that a rapid diagnosis be made. Ultimately, early warning of HAB events will minimize exposure of humans to this toxin. However, researchers and coastal managers face a great challenge when designing early warning systems to protect humans from the neurotoxic effects of eating shellfish contaminated with DA while also sustaining the harvest of shellfish on our coasts. The forecasting of these HABs must be a dynamic process that satisfies the unique biological, chemical, and physical characteristics of each coastal area. For example, an offshore HAB in the Juan de Fuca eddy will be detected using very different sensors than a HAB in a bay or estuary such as East Sound or the Gulf of St. Lawrence. To this end, remote, automated, and sensitive ocean observing systems, such as the Environmental Sample Processor, ocean gliders, and sensor arrays, are gradually being used in coastal regions to detect toxins, toxic cells, and the environmental conditions associated with HAB development and transport (Babin *et al.*, 2005).

Acknowledgments

This publication was supported by a grant to the West Coast Center for Oceans and Human Health (WCCOHH) to V.L.T. and B.M.H. as part of the National Oceanic and Atmospheric Administration (NOAA) Oceans and Human Health Initiative, WCCOHH publication no. 15. The WCCOHH is part of the National Marine Fisheries Service's Northwest Fisheries Science Center, Seattle, WA. Support was also provided to V.L.T and B.M.H. from the Ecology and Oceanography in the Pacific Northwest (ECOHAB PNW) program (publications no. 12 for ECOHAB PNW and 232 for the national ECOHAB program) by grants from NOAA (NA17OP2789) and NSF (OCE 0234587) and to B.M.H. by the Pacific Northwest Center for Human Health and Ocean Sciences (NIH/National Institute of Environmental Health Sciences [NIEHS] [P50 ES012762] and NSF [OCE-0434087]).

We thank Rita Horner, Nina Lundholm, Andy Tasker, and Cynthia Brown for reviewing an earlier version of this manuscript. We thank Maureen Auro and William Cochlan for providing unpublished data about *P. cuspidata* toxin production, Johanna Fehling for permission to update her map on *Pseudo-nitzschia* distribution, and Luiz Mafra (Brazil) and Martha Ferrario (Argentina) for information about the distribution of *Pseudo-nitzschia* species in their waters. We thank Sheryl Day for creating Fig. 12-3.

The statements, findings, conclusions, and recommendations are those of the authors and do not necessarily reflect the views of NOAA, the Department of Commerce, or the National Science Foundation.

References

Adams, N.G., MacFadyen, A., Hickey, B.M., Trainer, V.L., 2006. The nearshore advection of a toxigenic *Pseudo-nitzschia* bloom and subsequent domoic acid contamination of intertidal bivalves. Afr. J. Mar. Sci. 28, 271–276.

Andersen, P., Enevoldsen, H., Anderson, D., 2003. Harmful algal monitoring programme and action plan design. In Hallegraeff, G.M., Anderson, D.M., and Cembella A.D. (eds.), Manual on Harmful Marine Microalgae, pp. 627–647. Paris, UNESCO.

Anderson, C.R., Brzezinski, M.A., Washburn, L., Kudela, R., 2006. Circulation and environmental conditions during a toxigenic *Pseudo-nitzschia* bloom in the Santa Barbara Channel, California. Mar. Ecol. Prog. Ser. 327, 119–133.

Babin, M., Cullen, J.J., Roesler, C.S., Donaghay, P.L., Doucette, G.J., Kahru, M., Lewis, M.R., Scholin, C.A., Sieracki, M.E., Sosik, M.M., 2005. New approaches and technologies for observing harmful algal blooms. Oceanography 18, 210–227.

Bates, S.S., 1998. Ecophysiology and metabolism of ASP toxin production. In Anderson, D.M., Cembella, A.D., and Hallegraeff, G.M. (eds.), Physiological Ecology of Harmful Algal Blooms, pp. 405–426. Heidelberg, Springer-Verlag.

Bates, S.S., 2000. Domoic-acid-producing diatoms: Another genus added! J. Phycol. 36, 978–983.

Bates, S.S., Bird, C.J., de Freitas, A.S.W., Foxall, R., Gilgan, M., Hanic, L.A., Johnson, G.R., McCulloch, A.W., Odense, P., Pocklington, R., Quilliam, M.A., Sim, P.G., Smith, J.C., Subba Rao, D.V., Todd, E.C.D., Walter, J.A., Wright. J.L.C., 1989. Pennate diatom *Nitzschia pungens* as the primary source of domoic acid, a toxin in shellfish from eastern Prince Edward Island, Canada. Can. J. Fish. Aquat. Sci. 46, 1203–1215.

Bates, S.S., Garrison, D.L., Horner, R.A., 1998. Bloom dynamics and physiology of domoic-acid-producing *Pseudo-nitzschia* species. In Anderson, D.M., Cembella, A.D., and Hallegraeff, G.M. (eds.), Physiological Ecology of Harmful Algal Blooms, pp. 267–292. Heidelberg, Springer-Verlag.

Bates, S.S., Trainer, V.L., 2006. The ecology of harmful diatoms. In Granéli, E., and Turner, J. (eds.), Ecology of Harmful Algae, pp. 81–93. Heidelberg, Springer-Verlag.

Blaylock, R.L., 1994. Excitotoxins: The Taste That Kills. Santa Fe, NM, Health Press.

Boicourt, W.C., Wiseman, W.J., Jr., Valle-Levinson A., Atkinson L.P., 1998. Continental shelf of the southeastern United States and Gulf of Mexico: In the shadow of the western boundary current. In Brink K.H., and Robinson, A.R. (eds.), The Sea, vol. 11, The Global Coastal Ocean: Regional Studies and Syntheses, pp. 135–182. New York, Wiley & Sons.

Bradbury, J., 2004. Nature's nanotechnologists: Unveiling the secrets of diatoms. PLoS Biol. 2: e306.

Cembella, A.D., Doucette, G.J., Garthwaite, I., 2003. *In vitro* assays for phycotoxins. In Hallegraeff, G.M., Anderson, D.M., and Cembella, A. D. (eds.), Manual on Harmful Marine Microalgae, pp. 297–345. Paris, UNESCO.

Davidovich, N.A., Bates, S.S., 1998. Sexual reproduction in the pennate diatoms *Pseudo-nitzschia multiseries* and *P. pseudodelicatissima* (Bacillariophyceae). J. Phycol. 34, 126–137.

Doble, A., 2000. Pharmacology of domoic acid. In Botana, L.M. (ed.), Seafood and Freshwater Toxins; Pharmacology, Physiology and Detection, pp. 359–372. New York, Marcel Dekker.

Fehling, J., 2004. Diversity, ecology and domoic acid production of *Pseudo-nitzschia* spp. in Scottish waters. Ph.D. thesis, The Open University, Milton Keynes, UK.

Field, C.B., Behrenfeld, M.J., Randerson, J.T., Falkowski, P., 1998. Primary production of the biosphere: Integrating terrestrial and oceanic components. Science 281, 237–240.

Friedman, M.A., Levin, B.E., 2005. Neurobehavioral effects of harmful algal bloom (HAB) toxins: A critical review. J. Internat. Neuropsychol. Soc. 11, 331–338.

Fryxell, G.A., Hasle, G.R., 2003. Taxonomy of harmful diatoms. In Hallegraeff, G.M., Anderson, D.M., and Cembella, A.D. (eds.), Manual on Harmful Marine Microalgae, pp. 465–509. Paris, UNESCO.

Glibert, P.M., Anderson, D.M., Gentien, P., Granéli, E., Sellner, K.G., 2005a. The global, complex phenomena of harmful algal blooms. Oceanography 18, 136–147.

Glibert, P.M., Seitzinger, S., Heil, C.A., Burkholder, J.M., Parrow, M.W., Codispoti, L.A., Kelly, V., 2005b. The role of eutrophication in the global proliferation of harmful algal blooms. Oceanography 18, 198–209.

Grattan, L.M., Lesoing, M., King, A., Silbergeld, E., Morris, J.G., 2003. Human health effects of domoic acid in the Pacific Northwest: A preliminary study. Behavioral Toxicology Society Abstract, 22nd Annual Meeting, June 21–23, Philadelphia.

Hasle, G.R., 2002. Are most of the domoic acid-producing species of the diatom genus *Pseudo-nitzschia* cosmopolites? Harmful Algae 1, 137–146.

Hasle, G.R., Syvertsen, E.E., 1997. Marine diatoms. In Tomas, C.R. (ed.), Identifying Marine Diatoms and Dinoflagellates, pp. 5–385. New York, Academic Press.

Hickey, B.M., 1997. Response of a narrow submarine canyon to strong wind forcing. J. Phys. Oceanogr. 27, 697–726.

Hickey, B.M., 1998. Coastal oceanography of western North America from the tip of Baja California to Vancouver Island. In Brink, K.H., and Robinson, A.R. (eds.), The Global Coastal Ocean: Regional Studies and Syntheses, The Sea, vol. 11, pp. 345–393. New York, Wiley & Sons.

Hickey, B.M., Banas, N.S., 2003. Oceanography of the U.S. Pacific northwest coastal ocean and estuaries with application to coastal ecology. Estuaries 26, 1010–1031.

Hickey, B.M., Geier, S., Kachel, N., MacFadyen A., 2005. A bi-directional river plume: The Columbia in summer. Cont. Shelf Res. 25, 1631–1636.

Hill, A.E., 1998. Buoyancy effects in coastal and shelf seas. In Brink, K.H., and Robinson, A.R. (eds.), The Global Coastal Ocean: Processes and Methods, The Sea, vol. 10, pp. 21–62. New York, Wiley & Sons.

Hill, E.D., Hickey, B.M., Shillington, F.A., Strub, P.T., Barton, E.D., Brink, K., 1998. Eastern Boundary Current Systems of the world. In Brink, K.H., and Robinson, A.R. (eds.), The Global Coastal Ocean: Regional Studies and Syntheses, The Sea, vol. 11, pp. 29–67. New York, Wiley & Sons.

Jeffery, B., Barlow, T., Moizer, K., Paul, S., Boyle C., 2004. Amnesic shellfish poison. Food Chem. Toxicol. 42, 545–577.

Landsberg, J., Van Dolah, F., Doucette, G., 2005. Marine and estuarine harmful algal blooms: Implications on human and animal health. In Belkin, S., and Colwell, R.R. (eds.), Oceans and Human Health: Pathogens in the Marine Environment, pp. 165–215. New York, Springer.

Lebeau, T., Robert, J.-M., 2003. Diatom cultivation and biotechnologically relevant products. Part II: Current and putative products. Appl. Microbiol. Biotechnol. 60, 624–632.

Loder, J.W., Boicourt, W.C., Simpson, J.H., 1998. Western Ocean Boundary shelves. In Brink, K.H., and Robinson, A.R. (eds.), The Global Coastal Ocean: Regional Studies and Syntheses The Sea, vol. 11, pp. 3–28. New York, Wiley & Sons.

Lundholm, N., Moestrup, Ø., Hasle, G.R., Hoef-Emden K., 2003. A study of the *P. pseudodelicatissima/cuspidata* complex (Bacillariophyceae): What is *P. pseudodelicatissima*? J. Phycol. 39, 797–813.

MacFadyen, A., Hickey, B.M., Foreman, M.G.G., 2005. Transport of surface waters from the Juan de Fuca eddy region to the Washington coast: Implications for HABs. Cont. Shelf Res. 25, 2008–2021.

MacPhee, D.J., L.A., Hanic, D.L., Friesen, Sims, D.E., 1992. Morphology of the toxin-producing diatom *Nitzschia pungens* Grunow forma *multiseries* Hasle. Can. J. Fish. Aquat. Sci. 49, 303–311.

Mann, D.G., 1999. The species concept in diatoms. Phycologia 38, 437–495.

Marchetti, A., Trainer, V.L., Harrison, P.J., 2004. Environmental conditions and phytoplankton dynamics associated with *Pseudo-nitzschia* abundance and domoic acid in the Juan de Fuca eddy. Mar. Ecol. Prog. Ser. 281, 1–12.

Perl, T.M., Bédard, L., Kosatsky, T., Hockin, J.C., Todd, E., Remis, R.S., 1990. An outbreak of toxic encephalopathy caused by eating mussels contaminated with domoic acid. N. Engl. J. Med. 322, 1775–1780.

Pollanen, M.S., 1997. Forensic diatomology and drowning. Elsevier Science B.V., Amsterdam, The Netherlands, 170 pp.

Quilliam, M.A., 2003. Chemical methods for domoic acid, the amnesic shellfish poisoning (ASP) toxin. In Hallegraeff, G.M., Anderson, D.M., and Cembella A.D. (eds.), Manual on Harmful Marine Microalgae, pp. 247–266. Paris, UNESCO.

Quilliam, M.A., Wright, J.L.C., 1989. The amnesic shellfish poisoning mystery. Anal. Chem. 61, 1053A-1060A.

Ramsdell, J.S., 2007. The molecular and integrative basis to domoic acid toxicity. In Botana, L. (ed.), Phycotoxins: Chemistry and Biochemistry, pp. 223–250. Cambridge, MA, Blackwell Professional.

Rines, J.E.B., Donaghay, P.L., Dekshenieks, M.M., Sullivan, J.M., Twardowski, M.S., 2002. Thin layers and camouflage: Hidden *Pseudo-nitzschia* spp. (Bacillariophyceae) populations in a fjord in the San Juan Islands, Washington, United States. Mar. Ecol. Prog. Ser. 225, 123–137.

Round, F.E., Crawford, R.M., Mann, D.G. (eds.), 1990. The Diatoms: Biology and Morphology of the Genera. New York, Cambridge University Press.

Sarthou, G., Timmermans, K.R., Blain, S., Tréguer P., 2005. Growth physiology and fate of diatoms in the ocean: A review. J. Sea Res. 53, 25–42.

Scholin, C., Vrieling, E., Peperzak, L., Rhodes, L., Rublee, P., 2003. Detection of HAB species using lectin, antibody and DNA probes. In Hallegraeff, G.M., Anderson, D.M., and Cembella, A.D. (eds.), Manual on Harmful Marine Microalgae, pp. 131–164. Paris, UNESCO.

Scholin, C.A., Gulland, F., Doucette, G.J., Benson, S., Busman, M., Chavez, F.P., Cordaro, J., DeLong, R., De Vogelaere, A., Harvey, J., Haulena, M., Lefebvre, K., Lipscomb, T., Loscutoff, S., Lowenstine, L.J., Marin, R.I., Miller, P.E., McLellan, W.A., Moeller, P.D.R., Powell, C.L., Rowles, T., Silvagni, P., Silver, M., Spraker, T., Trainer, V., Van Dolah, F.M., 2000. Mortality of sea lions along the central California coast linked to a toxic diatom bloom. Nature 403, 80–84.

Stoermer, E.F., Smol, J.P., (eds.) 1999. The Diatoms: Applications for the Environmental and Earth Sciences. Cambridge, Cambridge University Press.

Tasker, R.A.R., 2002. Domoic acid. In Waring, R.H., Steventon G.B., and Mitchell S.C. (eds.), Molecules of Death, pp. 36–59. London, Imperial College Press.

Teitelbaum, J.S., Zatorre, R.J., Carpenter, S., Gendron, D., Evans, A.C., Gjedde, A., Cashman N.R., 1990. Neurologic sequelae of domoic acid intoxication due to the ingestion of contaminated mussels. N. Engl. J. Med. 322, 1781–1787.

Todd, K., 2003. Role of phytoplankton monitoring in marine biotoxin programmes. In Hallegraeff, G.M., Anderson, D.M., and Cembella, A.D. (eds.), Manual on Harmful Marine Microalgae, pp. 649–655. Paris, UNESCO.

Trainer, V.L., Adams, N.G., Wekell, J.C., 2001. Domoic acid producing *Pseudo-nitzschia* species off the U.S. west coast associated with toxification events. In Hallegraeff, G.M., Blackburn, S.I., Bolch, C.J., and Lewis, R.J. (eds.), Harmful Algal Blooms 2000, pp. 46–49. Intergov. Oceanogr. Comm., Paris, UNESCO.

Trainer, V.L., Bill, B.D., 2004. Characterization of a domoic acid binding site from Pacific razor clam. Aquat. Toxicol. 69, 125–132.

Trainer, V.L., Suddleson, M., 2005. Monitoring approaches for early warning of domoic acid events in Washington State. Oceanography 18, 228–237.

Trainer, V.L., Cochlan, W.P., Erickson, A., Bill, B.D., Cox, F.H., Borchert, J.A., Lefebvre, K.A., 2007. Recent domoic acid closures of shellfish harvest areas in Washington State inland waterways. Harmful Algae 6, 449–459.

Trainer, V.L., Hickey, B.M., Horner, R.A., 2002. Biological and physical dynamics of domoic acid production off the Washington U.S.A. coast. Limnol. Oceanogr. 47, 1438–1446.

Wells, M.L., Trick, C.G., Cochlan, W.P., Hughes, M.P., Trainer, V.L., 2005. Domoic acid: The synergy of iron, copper, and the toxicity of diatoms. Limnol. Oceanogr. 50, 1908–1917.

STUDY QUESTIONS

1. What potential impacts would there be on the ecosystem and on human health if Arctic and Antarctic species of *Pseudo-nitzschia* were found to be high producers of domoic acid?
2. In what ways would knowledge of the molecular biology of *Pseudo-nitzschia* species help to reduce the impact of these toxic species on human health?
3. What is the relationship between the incidence of toxic blooms of *Pseudo-nitzschia* species and (a) eutrophication; (b) aquaculture activities?
4. What measures can be taken to mitigate the harmful effects of toxic *Pseudo-nitzschia* blooms?
5. Describe what is known about sexual reproduction in pennate diatoms. Why is sex necessary for these diatoms?
6. If you were asked to develop an ocean observing system for *Pseudo-nitzschia* blooms that might impact the U.S. West Coast, where would you place arrays of sensors in the ocean (consider both offshore and coastal sites)? What sensors would you place on your moorings to give the most optimal early warning of HAB events?
7. Why are storms thought to bring toxic cells from offshore regions toward the coast in eastern boundary systems? Describe a sequence of events during the summer months that might result in closures of shellfish beds in an eastern boundary system.
8. How might a river plume affect the development and movement of toxic blooms in the coastal ocean?
9. Why are coastal bays more susceptible to toxic diatom blooms than other coastal regions?
10. How can currents below the euphotic layer have any relation to surface diatom blooms in the coastal ocean?

C H A P T E R

13

Toxic Dinoflagellates

KAREN A. STEIDINGER, JAN H. LANDSBERG, LEANNE J. FLEWELLING, AND BARBARA A. KIRKPATRICK

INTRODUCTION

Dinoflagellates are single-celled microorganisms (5 µm to typically less than 1 mm) that live in aquatic environments from the Arctic to the Antarctic. Most of them are free living, but some are symbiotic (e.g., in coral species) and even parasitic on fish or in other dinoflagellates. Half of the described species are primary producers with photosynthetic pigments. This means they contribute to the world's atmospheric oxygen and produce organic carbon that enters the base of the oceans' food webs. Primary production is a critical process that allows the seas to yield crops high in protein for human consumption. Phytoplankton, including dinoflagellates, also produce (during photosynthesis) over 70% of the oxygen we breathe. The other half of dinoflagellates species are "heterotrophic." It is not known how many of the photosynthetic (or "auxotrophic") species are also mixotrophic or heterotrophic, but this mode of nutrition can be advantageous (Stoecker *et al.,* 2006).

At some point in their life cycle, dinoflagellates are motile and can "swim" about 1 meter per hour, and therefore they can move or disperse within the water column. They have a characteristic nucleus (with condensed chromosomes) and other features (e.g., pigments and sterols) that separate them from other phytoplankton and microalgae. They are classified as protists in the Division Dinoflagellata (Fensome *et al.,* 1993), and typically the International Code of Botanical Nomenclature is used for this group of about 1500 to 2000 species (Gomez, 2005; Taylor, 1990). Those that are free living can be "cryptic" and hidden, or they can "bloom" and reach population densities that discolor seawater.

In earlier years, all dinoflagellate blooms that caused animal mortalities and presented human health risks were called "red tides." Today toxic dinoflagellate blooms are more accurately called harmful algal blooms (HABs). What constitutes a HAB is still debated, but Smayda (1997) stated that it should have ecological relevance and its harmful effects should be quantifiable. For this chapter, a bloom of a HAB organism, no matter the species, means a higher-than-normal cell abundance, or above background levels, of a dinoflagellate that can result in animal (e.g., fish, bird, or marine mammal) kills, toxic shellfish, human illnesses or death, or altered habitat and community structure.

Of the known dinoflagellates today, only about 5% of them are toxic (i.e., capable of producing neurotoxic and cytolytic compounds and other bioactive substances). Almost all toxic dinoflagellates are marine, occurring in open seas, continental shelf systems, and even bays and lagoons. Not all isolates of a toxic species produce detectable amounts of toxins, and that capacity can vary with geographical area and environmental conditions. Toxin content per cell varies within a species and between species, but with all shellfish poisonings in humans secondary to HAB toxins, it should be pointed out that the toxins are accumulated and bioconcentrated over time because shellfish are filter feeders.

Most of the toxic dinoflagellates are planktonic, but more benthic species have been identified (Faust and Gulledge, 2002). For the planktonic realm, dinoflagellate HABs follow a sequence in development of initiation, growth, maintenance, and termination (Steidinger, 1975; Steidinger *et al.,* 1998; Taylor, 2001). Initiation of a pelagic bloom is an inoculation of the water column in an area with a seed stock that can be via benthic resting stages, pelagic stages, or even accidentally by humans. That is why knowledge of the life cycles of particular dinoflagellates, as well as current and circulation patterns in an area, is paramount to understanding how a "bloom" initiates or is seeded. The second stage of bloom maturation is growth, when the dinoflagellate population increases, typically at a slow rate of reproduction (<1 division per day). The third and most visible stage is maintenance, where bloom concentrations are maintained for a period of time. The last stage of a HAB is termination or dissipation, the mechanisms of which are little understood.

Economic impacts of specific HAB events over a set period of time are difficult to interpret because much of the needed data are not readily available or do not exist. However, Anderson *et al.* (2000) estimated (for 1987–1992) that the average loss in revenue (in Yr 2000 dollars) for a HAB in the United States was 49 million dollars; for long duration HABs, this is an underestimate. A breakdown in the derivation of the costs reveals that 45% was for public health impacts, 37% for commercial fisheries, 13% for tourism, and 4% for monitoring and management. This breakdown no doubt varies by geographic area and the type of HAB impact experienced.

This chapter on toxic dinoflagellates, their ecology and relationship to the world's oceans, and consequently to human health (e.g., through seafood consumption and recreational exposure), was designed to give an introduction to the topic by concentrating on two of the main human health risks involving phycotoxins produced by dinoflagellates: the saxitoxins and the brevetoxins. The third section briefly addresses other toxins such as okadaic acid and yessotoxins. Finally, the last section involves a discussion on what new information is needed and where research is headed to answer today's questions on causes, consequences, and management of HABs in the seas worldwide.

SAXITOXINS, PARALYTIC SHELLFISH POISONING, AND SAXITOXIN PUFFER FISH POISONING

Blooms

Saxitoxins (also called paralytic shellfish poisons or paralytic shellfish poisoning toxins, both abbreviated "PSPs") are a family of low molecular weight, heat-stable, water-soluble neurotoxins. These toxins occur in both dinoflagellates and cyanobacteria (see Chapters 12 and 15), and their occurrence in eubacteria has been reported but is still debated (Baker *et al.,* 2003). When the toxins pass up the food chain and bioaccumulate in vectors, such as shellfish or puffer fish, they can create a risk for human illness if the contaminated seafood is eaten. Paralytic shellfish poisoning (also abbreviated PSP) occurs worldwide (Fig. 13-1), in boreal to tropical waters and in near-shore bay bottoms to offshore habitats. Mollusks, crustaceans, and even planktivorous fishes can concentrate the saxitoxins in their gastrointestinal tracts and other internal organs.

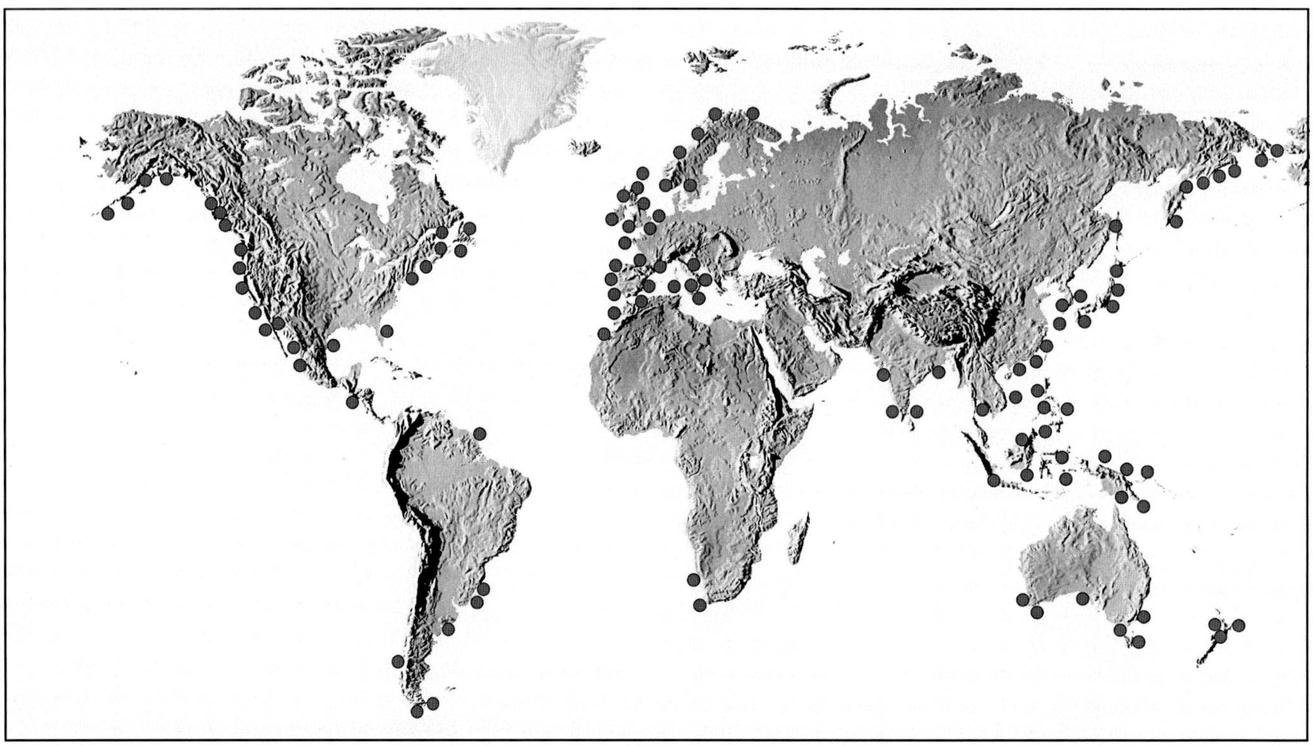

FIGURE 13-1. Incidents of saxitoxin in seafood around the world. Typically the incidents are PSP in bivalve mollusks, but there can be saxitoxins in gastropod mollusks, crustaceans, and in puffer fish (causing saxitoxin puffer fish poisoning or SPFP). From U.S. National Office for Marine Biotoxins and Harmful Algal Blooms.

Dinoflagellates that produce PSPs are the most diverse group of toxic phytoplankton with species in the genera *Alexandrium, Pyrodinium,* and *Gymnodinium.* All of the species associated with recurrent PSP episodes have a life cycle that favors their reoccurrence in the same area, sometimes year after year (e.g., *Alexandrium fundyense, A. tamarense, A. minutum, A. catenella, Gymnodinium catenatum, Pyrodinium bahamense*, and others). In addition to regular binary fission where they replicate by dividing in half, they can also undergo a form of sexual reproduction during which they may form resilient benthic resting cysts or stages. These benthic stages can survive adverse conditions by not being in the water column at certain times of the year. Benthic cyst populations can remain dormant for years and serve as seedbeds. Through reworking or just benthic turbulence of sediments, cysts can be brought to the surface layer, and if conditions are right (e.g., oxygen, light, and other factors), viable cysts can excyst and populate the water column again. In certain cases, cyst dormancy is entrained, and for excystment to occur out of seasonal cycles, the entrainment has to be broken (Anderson, 1998; Anderson *et al.,* 2005; Garcés *et al.,* 2002; Steidinger and Garcés, 2006). Many saxitoxin-producing dinoflagellates with temporary and resting cysts have adaptive strategies to deal with competition, stress, and disturbances (Reynolds and Smayda, 1998).

Two dramatic cases of the introduction of PSP-producing dinoflagellates by currents have been described. In 1972, a Bay of Fundy *Alexandrium* bloom was transported down to the northeast United States beyond Maine (Anderson, 1997), presumably by currents. The bloom was then advected inshore where the population formed benthic cysts and provided a permanent resident population. Federally funded Ecology and Oceanography of Harmful Algal Blooms (ECOHAB) studies in the Gulf of Maine (Anderson *et al.,* 2005) clarified the circulation patterns and transport mechanisms for this inoculation. Was this species there before 1972? Certainly in Maine it was, but it was probably not present in Massachusetts and New Hampshire at population densities that would cause shellfish toxicity, as it does now. Another example described by Hallegraeff and Maclean (1989) was the distribution of *Pyrodinium bahamense*, first evident in Papua New Guinea in 1972, to Brunei and Sabah in 1976, and then to the Phillipines in 1983, associated with meteorological and hydrological events. These cases serve as a reminder that nature itself can influence the distribution of HABs and their consequences.

Toxins

Saxitoxin was the first of the phycotoxin toxins to be purified and structurally defined (Schantz *et al.,* 1957, 1975). It takes its name from *Saxidomus giganteus*, the Alaskan butter clam, one of the organisms from which it was first isolated. Since then, several natural analogs of saxitoxin have been identified with at least 28 saxitoxins described to date, including analogs identified from the cyanobacterium *Lyngbya wollei* (Onodera *et al.,* 1997) and the dinoflagellate *Gymnodinium catenatum* (Llewellyn *et al.,* 2004). The saxitoxin analogs produced by dinoflagellates vary among species, and the profile of toxins produced by a single species can vary geographically. However, although overall toxin production can be affected by environmental conditions, the toxin profile is typically considered to be a stable characteristic of a particular microalgal strain or clone (Kodama, 2000).

All saxitoxins are tetrahydropurines possessing two guanidinium groups, which are essential for their activity (Shimizu, 2000). Saxitoxin analogs are variously substituted at four positions around a basic tricyclic structure and have been divided into groups based on these substitutions. Of the analogs most commonly found in shellfish, the most toxic are the carbamate toxins including saxitoxin (STX), neosaxitoxin (neoSTX), and gonautoxins I-IV (GTX1-4). The decarbamoyl analogs (dcSTX, dcNEO, dcGTX1-4) and the deoxydecarbamoyl analogs (doSTX, doGTX2, doGTX3) are of intermediate toxicity. The least toxic derivatives are the N-sulfocarbamoyl toxins B1 (GTX5), B2 (GTX6) and C1-C4 (Oshima, 1995) (for structures, see www.fao.org/docrep/007/y5486e/y5486e00.htm and Llewellyn, 2006; Shimizu, 2000; Van Egmond *et al.,* 2004).

The polar nature of the molecule makes saxitoxin very water soluble. As a result, it is rapidly absorbed across the gastrointestinal tract (as seen by the rapid onset of symptoms; see Table 1 in Backer *et al.,* 2005) and excreted in the urine (Baden *et al.,* 1995). The neurotoxic effects of saxitoxins are a result of their interaction with voltage-sensitive sodium channels (VSSCs). Voltage-sensitive sodium channels are transmembrane proteins found in neurons and electrically excitable cells that selectively allow the passage of sodium ions into the cell (see Chapter 28). When VSSCs are activated, the influx of sodium ions depolarizes the cell. Proper functioning of VSSCs is a necessary component for the propagation of nerve impulses that ultimately control our higher neurological processes. Sodium channels are complexes of multiple subunits, the largest of which, the alpha subunit, is the toxin-binding component.

Saxitoxin shares a specific binding site (site 1) with tetrodotoxin, a sodium channel-blocking neurotoxin most familiarly associated with puffer fish but produced by many genera. Binding of saxitoxin (or tetrodotoxin) to the VSSC blocks the inward passage of sodium ions, thereby preventing signal transmission in nerves and resulting in paralysis. Multiple genes that encode for the VSSC alpha subunit have been identified. This diversity in VSSC isoforms results in varying affinities for saxitoxins (as well as other toxins that act on VSSCs) and may lead to different clinical manifestations (Llewellyn, 2006).

Animal Health Impacts

Although saxitoxin-producing dinoflagellate blooms have been well documented and their role in human fatalities associated with PSP incidents has been reported globally, the involvement of saxitoxins in animal mortalities has been less well studied or definitively linked. Additionally, saxitoxins may not kill animals directly but may have sublethal effects (i.e., affecting the animal's ability to grow, reproduce, or remain healthy).

Saxitoxins primarily have been associated with causing the death of marine mammals and sea birds. Saxitoxins originating from an *Alexandrium* bloom were implicated in the mortality of 14 humpback whales (*Megaptera novaeangliae*) in Cape Cod Bay, Massachusetts (Geraci *et al.*, 1989), with mackerel identified as the likeliest vector. Saxitoxins believed to have originated from a *G. catenatum* bloom were a key suspect in the mortality of more than 100 highly endangered Mediterranean monk seals (*Monachus monachus*) along the coast of Mauritania (West Africa) during May and June 1997 (Hernández *et al.*, 1998). However, the role of a virulent morbillivirus in the die-off could not be definitively excluded (Osterhaus *et al.*, 1998).

Saxitoxins have been suspected to affect reproduction in endangered North Atlantic right whales (*Eubalaena glacialis*). During the summer, right whales feed on toxic copepods, including *Calanus finmarchius*, which in turn feed on *A. fundyense* in the Bay of Fundy, a common feeding area for the whales. It was estimated that right whales may ingest daily toxin levels via copepods that are similar to minimum doses lethal for humans (Durbin *et al.*, 2002), and that chronic exposure to these levels was sufficient to affect the whale swimming and diving behavior. Such a change in behavior could result in decreased feeding, leading to poor physical condition and ultimately a reduction in the calving rate among the exposed whales (Durbin *et al.*, 2002).

Birds have also been killed by saxitoxins (Landsberg *et al.*, 2007; Shumway *et al.*, 2003). For example, in May 1942, during an *A. catenella* bloom, more than 2000 dead seabirds were observed along the Pacific Northwest. At the same time, six human cases of fatal PSP were reported, and cats and chickens that had consumed razor clam (*Siliqua patula*) viscera also died. Bird species affected included herring gull (*Larus argentatus*), Western gull (*L. occidentalis*), white-winged scoter (*Melanitta fusca*), common murre (*Uria aalge californica*), Pacific loon (*Gavia arctica pacifica*), tufted puffin (*Fratercula cirrhata*), sooty shearwater (*Puffinus griseus*), and black-footed albatross (*Diomedea nigripes*), many of which were found dead with crustaceans and clams in their stomachs. The peak of the mortality occurred about 2 weeks after the crest of the HAB, with the numbers of dead birds on the beaches beginning to decline about 2 weeks after the disappearance of the bloom (McKernan and Scheffer 1942). Landsberg *et al.* (2007) summarized other bird mortality events involving saxitoxins.

Saxitoxins have been associated with fish, shellfish, and plankton mortalities in addition to vertebrate mortalities (Cembella *et al.*, 2002; Landsberg, 2002) and with causing a range of sublethal effects on diverse organisms (Landsberg, 2002; Yan *et al.*, 2003). Whereas the traditional exposure route for saxitoxins is via the food chain from the transfer of intracellular toxins via dinoflagellate cells or inside animal tissues vectored up the food chain, it also appears that dissolved saxitoxins in water could be a potential risk factor. Laboratory larval fish exposed to dissolved saxitoxins have had problems with sensorimotor function, morphological development, and growth and survival (Lefebvre *et al.*, 2004).

Human Health Impacts

Humans become ill with PSP as a result of the consumption of molluscan and crustacean filter feeders (such as oysters, mussels, clams, and crabs) contaminated with saxitoxin. Symptoms usually occur within 2 hours after ingestion of the shellfish. Predominant symptoms are tingling of the mouth and extremities, clumsy or unsteady movement, speech impairment, and muscle paralysis (Baden *et al.*, 1995). Other symptoms can be nausea, vomiting, headache, and dizziness. High doses can lead to the paralysis of the diaphragm, respiratory failure, and, in extreme cases, death (Rodrigue *et al.*, 1990). The toxins associated with PSP are tasteless, odorless, and stable to both heating and freezing. There is no specific antidote for PSP; treatment is supportive including ventilator support for respiratory insufficiency. Evacuation of the stomach may help in removing any remaining toxin-containing shellfish. The acute effects of PSP usually resolve within in 2 to 3 days; the long-term effects in survivors of PSP poisoning are not known.

Worldwide, usually puffer fish poisoning (also known as "fugu" poisoning) occurs after the consumption of certain species of puffer fish contaminated with tetrodotoxins. Tetrodotoxins can cause human fatalities with a presentation similar to PSP poisoning. However, since January 2002, there have been 28 cases of saxitoxin puffer fish poisoning (SPFP) as a result of the consumption of puffer fish from the Indian River Lagoon in Florida. Analysis of the toxins from puffer fish fillet from one of the human cases revealed saxitoxin, not tetrodotoxin (Quilliam *et al.*, 2004). Since that time, the Florida Fish and Wildlife Conservation Commission has banned puffer fish harvesting in that area. The dinoflagellate *Pyrodinium bahamense* is the putative source of the saxitoxin (Landsberg *et al.*, 2006). Previously, this organism was thought to be nontoxic in Florida waters, but isolates from both the east and west coast of Florida have been confirmed to produce saxitoxins (Landsberg *et al.*, 2006).

BREVETOXINS

Blooms

Karenia brevis, formerly known as *Gymnodinium breve* and *Ptychodiscus brevis* (synonyms), has a limited geographic distribution from the Gulf of Mexico, the North Atlantic off Florida and in the Gulf Stream, as well as possibly the Caribbean (e.g., Trinidad and Jamaica) (Lackey, 1956; Steidinger, 1983; E. Ramston, personal communication). It has also been misidentified from the Pacific and many parts of the Atlantic. *K. brevis* has been documented in the Gulf of Mexico since 1946, whereas a related *K. mikimotoi* has been recorded worldwide since 1935. Haywood *et al.* (2004) and others have identified more than 10 new *Karenia* species, at least two of which (*K. selliformis* and *K. papilionacea*) are also found worldwide. All *Karenia* species are unarmored dinoflagellates without cell walls, unlike the PSP-producing *Alexandrium* and *Pyrodinium*, which are armored with cell walls divided into plates. Unarmored species such as *Karenia* are more pleomorphic and fragile.

Brevetoxins were first isolated from *K. brevis* in the 1970s (Baden *et al.,* 1979), and most of our knowledge of these toxins and their effects are based on studies *K. brevis*. In addition to *K. brevis*, brevetoxin production has been confirmed in *K. concordia* isolated from New Zealand waters and is suspected for some other *Karenia* species (Chang *et al.,* 2006; Haywood *et al.,* 2004). Although not known to produce brevetoxin, *K. mikimotoi* and *K. selliformis* produce the fast-acting toxin gymnodimine and are associated with ichthyotoxicity and fish kills. Brevetoxins in shellfish because of dinoflagellates have been recorded from New Zealand and Florida (Fig. 13-2).

Karenia brevis blooms off the west coast of Florida can last up to 2 years, admittedly in different geographic areas at different times, or in patches offshore, but it is still a threat to human health. Among the *Karenia*, only the life cycle of *K. brevis* has been studied. Although it goes through a sexual cycle and produces gametes and planozygotes, hypnozygotes (or resting cysts) have not been produced under the culture conditions used. No benthic resting cysts have been found, yet it is still speculated that the seed population for *K. brevis* could be a benthic stage, or it could be a planktonic source of vegetative cells maintained in circulation features off the west coast of Florida (Steidinger *et al.,* 1998). This species can be entrained in Florida west coast shelf waters and moved great distances by currents (e.g., the Loop Current System) (Tester and Steidinger, 1997).

As with other HAB species, bloom sequences are initiation (in this case in shelf waters offshore) followed by the

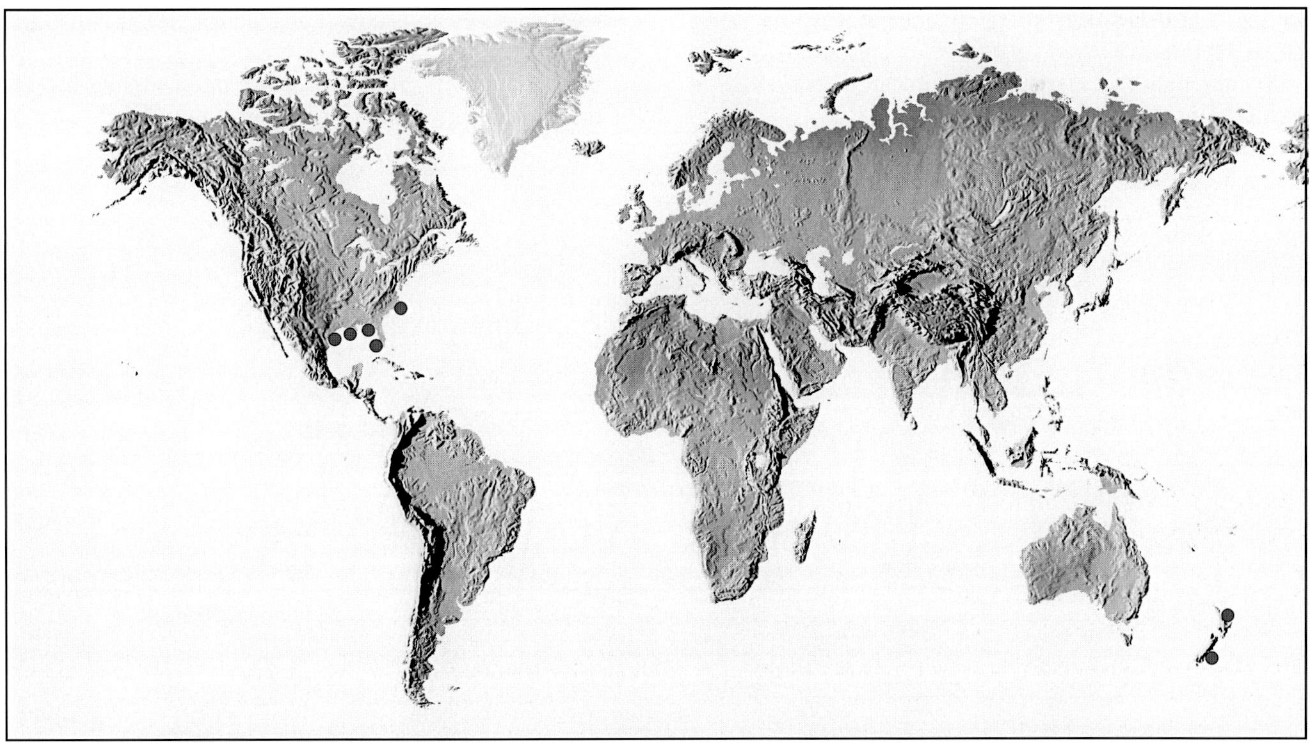

FIGURE 13-2. Incidence of brevetoxins in seafood around the world. The incidents represent the presence of the toxin in bivalve mollusks but not necessarily human intoxication cases. From U.S. National Office for Marine Biotoxins and Harmful Algal Blooms.

growth phase whereby the population increases in biomass as it is moved by currents and winds cross shelf. The next phase, maintenance, is typically inshore but also can be offshore. Eventually the *K. brevis* bloom terminates or is moved out of an area. The highest human health risks are when the bloom is in its maintenance phase inshore near human populations, recreational activities, and shellfish harvesting. This phase is the most obvious because of possible seawater discoloration depending on cell concentrations, dead and dying fish in recreational waters or on the beaches, and the *K. brevis* red tide aerosol experienced by people at the beach or anywhere there is aerosolized seawater with *K. brevis* red tide organisms.

Walsh *et al.* (2007) summarized much of the results from a federally funded ECOHAB : FL program on *K. brevis* blooms by proposing a model that takes *K. brevis* from deep offshore waters to inshore waters at beaches. As the *K. brevis* population increases, there are diverse sources of inorganic and organic nutrients to support the different stages of the bloom. Offshore nutrients from upwelled water, as well as nutrients from *Trichodesmium* blooms, fuel the initial phases of *K. brevis* HABs. *Trichodesmium* is a nitrogen fixer and able to do well in oligotrophic midshelf waters. But to support the increased biomass inshore, more substantive sources are needed (organics from dying and dead fishes, phosphorus from fossil deposits, etc.). *Karenia brevis* bloom dynamics are a complex system of the organism's biology, nutrient supplies, and physical oceanographic forcing, and it has been going on since at least the 1800s off of Florida. The transfer of brevetoxins from *Karenia brevis* through the marine environment is portrayed in Figure 13-3.

Toxins

Brevetoxins (PbTx) are a family of lipid-soluble toxins composed of *trans*-fused polyether rings. Structurally, all active brevetoxins are relatively linear with a lactone function on the A-ring, a rigid tail region formed by the last four rings and a spacer region separating the lactone from the rigid region (Baden *et al.,* 2005). They are divided into two groups based on two distinct molecular backbone structures. At present, 14 brevetoxins have been identified from *K. brevis* cultures, although three of these are potentially artifacts of the extraction process. The brevetoxin A backbone group includes PbTx-1, PbTx-7, and PbTx-10; the remainder (plus PbTx-tbm) are members of the brevetoxin B backbone group (Baden *et al.,* 2005). Although brevetoxin A backbone toxins are more potent (Baden 1989), the brevetoxin B backbone toxins are the most abundant (Baden, 1983) (for structures see Baden *et al.,* 2005, or Van Egmond *et al.,* 2004).

Like saxitoxins, brevetoxins exert their neurotoxic effects by acting on VSSCs; however, the result is persistent sodium channel *activation*, not blocking. Brevetoxins (as well as ciguatoxins, see Chapter 14) bind to a site designated as site 5 on the alpha subunit of VSSCs (Poli *et al.,* 1986), with half-maximal binding in the nanomolar concentration range. Binding results in prolonged opening of sodium channels at normal resting potential and inhibits inactivation, leading to membrane depolarization and repetitive firing in nerves (Baden *et al.,* 1995; Catterall and Gainer, 1985; Poli *et al.,* 1986).

Brevetoxins are nonpolar, noncharged toxins that are both heat and acid stable. In shellfish, some brevetoxins are

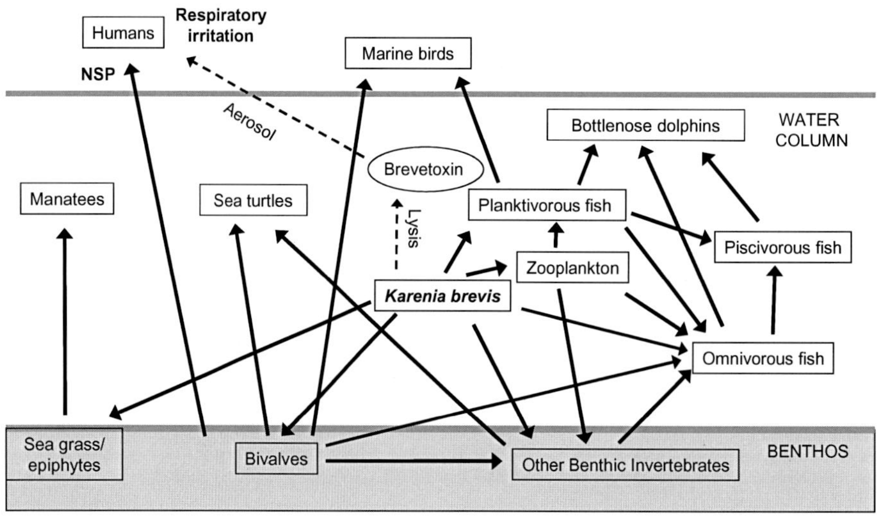

FIGURE 13-3. Common routes of transfer of brevetoxins in the marine environment that lead to harmful impacts on humans, marine mammals, sea turtles, and marine birds, principally through the food web and sea spray. Other impacts such as fish kills resulting from acute exposure in water are not part of this transfer model.

metabolized to both nonpolar and polar compounds. The metabolites produced vary among species of bivalve, and the extent to which they contribute to neurotoxic shellfish poisoing (NSP) appears to vary as well (Ishida *et al.,* 2004; Nozawa *et al.,* 2003; Plakas *et al.,* 2004). The combination of parent toxins and metabolites of different, and in many cases undetermined, potencies presents a complicated problem for shellfish management programs. NSP has only been recorded from the southeastern United States and New Zealand (Fig. 13-2).

Animal Health Impacts

In the Gulf of Mexico, *Karenia brevis* red tides producing brevetoxins have caused wide-scale animal mortalities that have been reported since the mid-1840s. Red tides annually cause mass mortalities of hundreds of thousands of fish and severely impact marine mammals (particularly manatees and bottlenose dolphins), turtles, birds, and invertebrates (Galstoff, 1948; Gunter *et al.,* 1947, 1948; Leverone *et al.,* 2006; Roberts *et al.,* 1979; Smith, 1976; Steidinger *et al.,* 1973; Summerson and Peterson, 1990). Brevetoxins are potent ichthyotoxins, responsible for the deaths of billions of fish; mortalities are common, widespread, and affect hundreds of species (Landsberg, 2002; Steidinger *et al.,* 1973). Signs of intoxication in fish include violent twisting and corkscrew swimming, defecation and regurgitation, pectoral fin paralysis, caudal fin curvature, loss of equilibrium, quiescence, vasodilation, and convulsions, culminating in death from respiratory failure (Baden, 1989; Quick and Henderson, 1974, 1975; Steidinger *et al.,* 1973).

Brevetoxins can be present in water and sediments, are adsorbed to particles, and can be transferred up the food chain. Their persistence in the food web is pervasive (Landsberg, 2002; Tester *et al.,* 2000). Because *K. brevis* cells are fragile, when the cells break apart they release brevetoxin into the water and (with crashing waves) into the marine aerosol. This means that fish and other gill-breathing or filter feeding organisms can be exposed to toxins not only from feeding directly on cells containing intact toxin but also from brevetoxins dissolved in the water (Fig. 13-3).

Florida's red tides have contributed to significant economic losses, causing declines in economically valued fisheries resources, threatening endangered animal species, and impacting local tourist businesses. Since 1996, four major springtime red tide events in southwest Florida have contributed to endangered manatee (*Trichechus manatus latirostris*) mortalities representing more than 10% of the local population (Florida Fish and Wildlife Commission [FWC], unpublished data). In 1999–2000 and in 2004 (see Human Health Impacts), an unusual red tide in the Florida panhandle was responsible for several hundred bottlenose dolphin (*Tursiops truncatus*) deaths (Flewelling *et al.,* 2005; Van Dolah *et al.,* 2003) and for increased turtle strandings (FWC, unpublished data). In 2003, possibly for the first time, domestic dogs from the beach area of Little Gasparilla Island in southwest Florida were admitted to local veterinary clinics after being reportedly affected by exposure to brevetoxins during a highly concentrated red tide event (Flewelling *et al.,* unpublished data).

Birds are also affected by brevetoxins, especially those that consume toxic shellfish or fish (Forrester *et al.,* 1977; Kreuder *et al.,* 2002; Quick and Henderson, 1974). Substantial numbers of sick and dying cormorants co-occurred with red tide outbreaks along the west Florida coast during 1995–1999 (Kreuder *et al.,* 2002); in the spring of 2002, a mortality of more than 20 lesser scaup was also attributed to "brevetoxicosis" in southwest Florida; and annually hundreds of birds are admitted to bird rehabilitation centers (Landsberg *et al.,* 2007).

Research has also demonstrated how animals are at risk from exposure to brevetoxins, even well after the bloom has dissipated. Brevetoxins are fairly stable in the environment and can be present in a range of species providing a source of toxin for animals in the food web. A major die-off of manatees in 2002 was attributed to their being exposed to brevetoxins associated with sea grass up to several months after a red tide bloom. Residual toxin was present in sufficient concentrations to be lethal to manatees feeding on sea grass in their common feeding grounds (Flewelling *et al.,* 2005).

Although acute exposure to lethal doses of brevetoxins results in massive animal mortalities, effects from low-level exposure are harder to interpret. In southwest Florida, manatees are routinely exposed to brevetoxins from *Karenia brevis* red tides. Bossart *et al.* (1998) postulated that chronic exposure to aerosolized brevetoxins could also affect the cellular immune system of manatees, compromising their response to disease or toxin exposure. At this stage it is unclear what impacts, if any, brevetoxins may have on survival of young manatees and how manatees may be more susceptible to a range of health impacts because of routine exposure to brevetoxins in a red tide endemic area.

Human Health Impacts

Brevetoxins are somewhat unique among the dinoflagellate toxins because they have two routes of exposure for human (and animal) impacts. The first route of exposure, ingestion, is similar to that of other toxins produced by dinoflagellates when toxins are transferred through the food chain, particularly through the consumption of contaminated filter feeders such as clams, oysters, and mussels. The second route of exposure occurs by inhalation of toxic aerosols. Toxins are released into seawater, possibly through cell lysis or after cell death, and then the toxins are incorporated into the marine aerosols (Kirkpatrick *et al.,* 2004; Pierce *et al.,* 2003).

With ingestion of the toxin and subsequent NSP, the poisoning is a milder form of intoxication, compared to PSP, with no reported deaths. Symptoms are similar to PSP with tingling of the mouth and lips, dizziness, and gastrointestinal symptoms of diarrhea, nausea, and vomiting (see Table 1 in Backer *et al.,* 2005). One symptom that is particular to NSP (and ciguatera fish poisoning) is the reversal of hot and cold sensations. If a person suffering from severe NSP holds an ice cube in his or her hand, the person may experience a hot sensation. As in PSP, respiratory insufficiency can occur and ventilator support may be required while the toxin is cleared through the body. Removal of any remaining toxic shellfish from the stomach may speed recovery. The acute effects of NSP reportedly resolve within 2 to 3 days; the long-term effects from NSP poisoning are not known.

In 2004, planktivorous menhaden (*Brevoortia* sp.) containing extremely high levels of brevetoxins in their viscera were implicated in a large-scale bottlenose dolphin mortality in the Florida Panhandle. In association with this event, Flewelling *et al.* (2005) measured brevetoxins in the viscera and tissue of fish caught live in the area where the mortality occurred. Brevetoxin concentrations in the edible flesh of these fish as well as several others subsequently collected during *K. brevis* blooms did not exceed the comparable action limit for brevetoxins in shellfish (Flewelling *et al.,* 2005; Naar *et al.,* 2005). The long-term possibility of repeated exposure to low levels of brevetoxins from ingestion of fish (and even shellfish) during and following *K. brevis* blooms is an unknown public health threat to humans as well as animals. These findings raise new public health questions regarding the possibility of repeated exposure to low levels of brevetoxins from ingestion of fish and shellfish harvested during and following Florida red tide blooms. Possible effects of chronic low-level exposure to brevetoxins have not been investigated.

Respiratory irritation from inhalation of the toxic aerosols was first associated with *K. brevis* red tide blooms in 1947 (Woodcock, 1948). Until recently, it was believed that symptoms (such as cough, eye irritation, and tearing) were temporary and would subside after people left the beach and discontinued exposure to the toxic aerosols. Research currently under way indicates that this may not be true for all people. It should be noted that concentrations of brevetoxins in air samples are substantially less that the amount in contaminated shellfish, and respiratory symptoms are induced at concentrations that are several orders of magnitude lower than those that induce illness from ingestion. Furthermore, only 3% to 7 % of the airborne toxins are in the particle size range that allows for deposition in the lung, substantially reducing the actual dose (Cheng *et al.,* 2005). However, in spite of these extremely small doses, Fleming *et al.* (2005) have documented a change in reported symptoms and spirometry after asthmatics took a 1-hour walk on the beach during a *Karenia* red tide. In addition, in a follow-up study, asthmatics were asked to record symptoms and peak flow for 5 days following the 1-hour beach exposure. Study participants reported increased symptoms continuing 5 days following the 1-hour beach exposure (Kirkpatrick *et al.,* 2007). Much more work lies ahead to identify further sensitive populations at risk from these inhaled toxins, chronic effects, and long-term effects.

OTHER TOXINS

Blooms

Diarrheic shellfish poisoning (DSP) with gastrointestinal symptoms was first recognized in the 1970s with human health cases from Japanese harvested shellfish (Van Dolah, 2000). Because of the generality of the symptoms, it could have occurred earlier but been confused with other seafood-associated illnesses. The identification of *Dinophysis* species, and later some *Prorocentrum* species, as producers of okadaic acid (OA) and its analogs helped to indicate the sources and led to the establishment for monitoring programs to detect these organisms, particularly in aquaculture farms.

The most common OA-producing species in temperate waters is *D. acuminata*, whereas *D. caudata* is the most common species in tropical waters. Another OA producer is *P. lima*, a benthic, tychoplanktonic species that can attach to living and dead substrate, and presents a threat to shellfish-growing areas or aquaculture (Van Dolah, 2000) as much as the planktonic and pelagic *Dinophysis*. Blooms of toxic *Dinophysis* are above background numbers of cells, but these numbers are low compared to blooms of *K. brevis*, which can reach millions of cells per liter. *Dinophysis* blooms of several thousand cells per liter occur in Europe, whereas in Scandinavia, a bloom can be hundreds of cells per liter. The most common OA producers in Scandinavia are *D. acuta, D. acuminata, D. norvegica*, and *Phalacroma rotundatum* (Edvardsen *et al.,* 2003). There are hundreds of species of *Dinophysis*, including *Phalacroma* species (see Evardsen *et al.,* 2003, for the recent evaluation of *Dinophysis* and *Phalacroma* as separate genera). Some of the species represent morphological variation, whereas others can represent smaller reproductive cells (Reguera and González-Gil 2001; Reguera *et al.,* 2004). It is, however, important to differentiate species and their life forms because closure of shellfish beds by cell counts varies with the species in Norway, with closures from *D. acuta* at 500 cells per liter, *D. acuminata* at 900 cells per liter, and *D. norvegica* at 2000 cells per liter.

As with other dinoflagellate blooms and resulting human illnesses, physical oceanographic features and dinoflagellate life histories play significant roles. Moita *et al.* (2006) reported that *D. acuta* bloomed (50,000 cells per liter) in a thin layer off northwest Portugal in 2003 and that this bloom also contained *D. dens*, a smaller, possibly gametic form of *D. acuta*. The authors speculated that stratification mechanisms (such as pycnoclines) provided an enhanced environment for chance encounters of gametes and emphasized what Donaghay and Osborne (1997) postulated, that dinoflagellate HABs in thin layers could be the interaction between physical features (such as current sheer) and biological factors (such as swimming behavior). Other physical features such as upwelling systems have been shown to be instrumental in transport of *Dinophysis* populations in South Africa and Spanish rías (drowned river valleys in northern Spain) (Figueiras *et al.*, 2006). Additionally, there are other dinoflagellate species that produce the toxins yessotoxins, azaspiracids, and spirolides, among others, that can co-occur with OA-producing species.

There is a diverse benthic assemblage of dinoflagellates that can be interstitial, attached to macroalgae or grasses, attached to sand grains or rocks, or reside on the bottom and vertically migrate during daylight hours. Several species of the benthic dinoflagellate genus *Ostreopsis* have been found to produce palytoxin analogs and are believe to be one source of palytoxin, the most potent marine toxin known (Lenoir *et al.*, 2006). These toxins can accumulate up the food chain, and present a risk to human health. In some areas, palytoxin is suspected to have contributed to Clupeidae poisonings (i.e., soft-finned schooling food fishes such as menhaden of shallow waters of northern seas). *Ostreopsis* and other toxic benthic species (e.g., *Prorocentrum, Gambierdiscus, Coolia*, and others) can be seasonally abundant. At times, they can be in the water column individually, in mucous strings, or clumps. Whether a dinoflagellate is benthic or planktonic, it is important to know its environmental tolerances and requirements (e.g., temperature, salinity, light, nutrients), life cycle, mechanisms of dispersal, and the timing of bloom sequences in order to forecast events and assess human risk.

Toxins

Often grouped together as the "DSP toxins" are okadaic acid (and its derivatives, the dinophysistoxins), pectenotoxins, and yessotoxins. Although they are all lipophilic and can co-occur in shellfish, they differ in chemical structure, mode of action, and the degree of risk they pose to human health. The dominant compounds involved in DSP incidents vary geographically, and reflect the presence of different toxin-producing organisms (Aune and Yndestad, 1993).

Okadaic acid (OA), a polyether derivative of a C38 fatty acid, was the first DSP toxin to be identified. It was initially isolated from the marine sponges *Halichondria melanodocia* and *H. okadaii*, from which it takes its name (Tachibana and Scheuer, 1981). Dinophysistoxins (DTX-1 through -6), OA diol esters, and other related compounds co-occur with OA in culture and in shellfish (Fernández *et al.*, 2003; Suarez-Gomez *et al.*, 2001; Vieytes *et al.*, 2000) (for structures, see Van Egmond *et al.*, 2004). Okadaic acid is a potent inhibitor of protein phosphatase 1 and 2A, enzymes in mammalian cells that dephosphorylate serine and threonine residues in proteins (Cohen, 1997). These enzymes help regulate many important cellular processes, including control of the cell cycle and transcription of oncogenes (Dounay and Forsyth, 2002). In laboratory studies, OA has also been found to be a potent tumor promoter when applied to mouse skin that had been previously exposed to the carcinogen DMBA [7,12-dimethylbenz(*a*)anthracene] (Suganuma *et al.*, 1988).

Pectenotoxins (PTXs) are an expanding family of cyclic polyether macrolide compounds. They are produced by the same organisms that produce OA and DTXs and therefore co-occur in shellfish. Although PTXs do not contribute to symptoms typically associated with DSP, they are lethal to mice by intraperitoneal (i.p.) injection, and have been shown to be extremely hepatotoxic (Draisci *et al.*, 2000) and are thus generally accepted to represent a human health risk.

Like PTXs, yessotoxins (YTXs) can also co-occur with OA and DTXs but are produced by different organisms. To date, YTXs have been detected in two dinoflagellates, *Protoceratium reticulatum* and *Lingulodinium polyedrum* (Paz *et al.*, 2004; Satake *et al.*, 1997); however, the likelihood of additional dinoflagellate sources has been hypothesized (Ciminiello *et al.*, 2003). Yessotoxins are sulfated transfused polyether compounds, and the number of known naturally occurring analogs is rapidly increasing (Ciminiello *et al.*, 2003; Miles *et al.*, 2005). Many *in vitro* studies have demonstrated toxic effects of yessotoxins (e.g., Franchini *et al.*, 2004a, 2004b; Perez-Gomez *et al.*, 2006; Vinagre *et al.*, 2003), but the specific threat of YTX to human health remains unclear. Although more toxic to mice by i.p. injection than either PTX or OA and its derivatives (Draisci *et al.*, 2000), YTX is nonlethal to adult mice when administered orally, even at dosages calculated to represent 100 times more than the potential human daily intake (Tubaro *et al.*, 2003).

A new shellfish poisoning syndrome, azaspiracid poisoning or AZP, was discovered following a 1995 illness outbreak in northern Europe associated with the consumption of mussels (*Mytilus edulis*) harvested in Ireland (James *et al.*, 2004). Because the symptoms of AZP are similar to DSP symptoms, azaspiracids are sometimes included in discussions of DSP toxins. Azaspiracids are polyether toxins, as are many of the other dinoflagellate toxins discussed here, but they have many unique structural characteristics (Satake *et al.*, 1998). Substitutions on four side chains of the molecule result in at least 11 different analogs. *In vitro* studies have demonstrated that azaspiracids are cytotoxic and that they can induce cytoskeletal, morphological, and growth changes in cells (Twiner *et al.,* 2005; Vilarino *et al.,* 2006); however, the precise mode of action is not yet known.

Cyclic imines are a class of nitrogen-containing shellfish toxins often called "fast-acting toxins" based on the rapid toxic effects they produce in mouse bioassays (James *et al.,* 2000). They include prorocentrolides, gymnodimines, and spirolides, which are known to have dinoflagellate sources (Hu *et al.,* 2001; James *et al.,* 2000), and also pinnatoxins, whose sources has not been identified. Although all are acutely toxic to mice by i.p. injection, only pinnatoxin has been implicated in actual cases of human intoxication (Hu *et al.,* 2001; James *et al.,* 2000).

A multitude of other bioactive compounds from dinoflagellates including ichthyotoxins, hemolytic compounds, and antibacterial and antifungal compounds have also been reported and warrant investigations into potential human health impacts.

Animal Health Impacts

The okadaic acid family of polyether toxins is unique among the algal toxins in that it targets serine/threonine protein phosphatases involved ubiquitously in intracellular signaling mechanisms that are frequently altered in tumor formation. Okadaic acid has been shown to induce skin papillomas and carcinomas in mice, and adenocarcinomas in rats (Fujiki and Suganuma, 1993; Fujiki *et al.,* 1988; Suganuma *et al.,* 1988, 1990). However, studies on the chronic effects of exposure to OA or dinophysistoxins have not been conducted in humans. Correlation between exposure to dinoflagellates producing OA and the occurrence of fibropapillomatosis in marine turtles has been demonstrated, but a causal linkage has not been made (Landsberg *et al.,* 1999).

After oral exposure to mice, azaspiracid was shown to cause lung tumors and necrosis of the T and B lymphocytes (Ito *et al.,* 2000, 2002), while *in vitro* studies showed a range of intracellular effects (Roman *et al.,* 2002). However, the exact mode of action of azaspiracid is still unclear and, currently, there are no documented adverse impacts on animals in the wild.

As each new toxin is described and characterized for its impacts on human health, it is likely that we can anticipate documented effects on many animal species. Thus far, much information usually follows from unexplained mortality events that are subsequently linked with newly recognized toxic species and where analytical methods reveal previously undetected toxins. Many of the toxins described in the previous discussion are only known for their experimental effects; however, impacts on animals in the wild are anticipated.

Human Health Impacts

Diarrheic shellfish poisoning (DSP) occurs after ingestion of okadaic acid and associated dinophysistoxins (Backer *et al.,* 2005). The onset of symptoms, depending on the dose of toxin ingested, may be as soon as 30 minutes to 2 to 3 hours, with symptoms of the illness lasting as long as 2 to 3 days. Symptoms are similar to the other shellfish poisonings, nausea, vomiting, diarrhea as well as chills, headache, and fever. The diarrheic effects of OA may result from phosphorylation of proteins in intestinal cells that control sodium secretion or permeability of cell membranes, resulting in the inability to maintain water balance (Aune and Yndestad, 1993). The illness is generally not life threatening, and recovery from acute DSP is reportedly complete with no aftereffects. The foods associated with DSP are bivalve shellfish such as clams, oysters, and mussels. As with other shellfish poisonings, diagnosis is based entirely on observed symptoms and recent dietary history. There is concern regarding long-term effects that the toxin may be a carcinogen as okadaic acid is a powerful tumor promoter (Backer *et al.,* 2005). An epidemiological study of digestive-tract cancer mortality in relation to the distribution of DSP was conducted in France; although there appeared to be a tentative positive association between the two, Cordier *et al.* (2000) recognized the need for more extensive surveys and in-depth research before any link could be definitively proven.

Azaspiracid shellfish poisoning is one of the lesser known, and more recently characterized, shellfish poisoning syndromes. Exposure is through the consumption of shellfish contaminated with azapiracids produced by the heterotrophic dinoflagellate *Protoperidinium* (James *et al.,* 2004). The original cases of AZP were reported in the Netherlands in 1995 (McMahon and Silke, 1996) and were caused by mussels harvested in Ireland. Since then, cases of AZP have been reported throughout Europe. Symptoms are similar to the gastrointestinal symptoms of DSP (i.e., nausea, vomiting, diarrhea, and stomach cramps) although there is some information to suggest that AZP may be associated with liver damage. As with the other shellfish poisonings, diagnosis is based entirely on observed symptoms and recent dietary history and is easily confused with DSP. There are no known deaths associated with AZP.

TOMORROW'S CHALLENGES

The 2005 U.S. National Plan for Algal Toxins and Harmful Algal Blooms, HARRNESS (Harmful Algal

Research and Response National Environmental Science Strategy 2005–2015) (HARRNESS, 2005) presented a plan to address current issues on harmful algae and their toxic effects. The plan's recommendations for research, monitoring, and management are pertinent to toxic dinoflagellates in the world's oceans, their impacts on ecosystems, and the risks that they pose to human and living resources health. Major recommendations included the development of methodologies and approaches for rapid field-based detection of HABs and toxins for early warning; food web models (including man as one end consumer) for the fate and effect of toxins; and methods for the prevention, control, and mitigation of HABs.

The HARRNESS plan ended with a vision for 2015 that addressed some of the major issues facing us in regard to the world's oceans, the sustained effects of HABs, and even other environmental events. For example, will our wild fisheries be sustained? If not, will aquaculture be able to provide sufficient diversity in protein? How will global HABs affect aquaculture production and selection of facility locations? Can nutrient inputs into the sea be decreased enough to minimize the blooms of nuisance algae that often occur in enclosed bays or lagoons? Will new technologies in the form of handheld sensors or even remotely operated large area sensors be useful for early detection or monitoring? Will new kits be available for rapid and accurate testing of toxins, HAB species, and other variables to provide analyses for resource management, aquaculture, seafood industries, and human health clinics? Will large-scale, networked, and automated observing systems be available with nanotechnological application that can measure toxins and species as well as oceanic variables? Will there be accurate prediction models that natural resource and public health managers can use to prepare for and manage blooms? It will take new initiatives, new technologies, and integrated resources to reach the goals projected in that report. As an example, minute-to-minute geographic data storage, retrieval, interpretation, and management will certainly be a big issue because of the amount and frequency of HAB data being generated worldwide.

The impacts from toxic dinoflagellates span many disciplines that involve research and monitoring and that cover multiple fields including natural resources, oceanography, public health, and seafood safety. In recent years, the NOAA's and NSF-NIEHS Oceans and Human Health Initiatives have identified the need for multidisciplinary work to address the issues associated with HABs; this emphasis is likely to increase in future years. Young investigators should anticipate working with a diversity of scientists specializing in marine biology, marine chemistry, botany, physical oceanography, fisheries, public health, veterinary medicine, aquaculture science, epidemiology, geography, electronics and electrical engineering, and information management.

Blooms

Technological advances have gone beyond the traditional microscopic enumeration of toxic dinoflagellates to include automated techniques that provide for rapid sampling, species identification, and accurate analyses to quantify composition and abundance of HAB dinoflagellates. However, there is much work to be done, both in new design and in the implementation of existing sensors. As the U.S. Commission on Ocean Policy (2004) cited, Integrated Ocean Observing Systems (IOOS) that provides real time, continuous data from a variety of ocean based platforms are needed (see Chapter 1, Case Study 1). Robust, real-time biological sensors (such as those that identify and quantify toxic dinoflagellates) need to be incorporated into an IOOS as part of the future of HAB management. In addition to monitoring for known toxic dinoflagellates in a specific region, sensors are needed that can scan for new HAB species that may arise in the phytoplankton community. Changes from eutrophication, pollution, or physical actions (e.g., release of ballast water or changes in ocean circulation) may create new ecological niches for toxic species that did not previously occur in specific regions or local areas.

As the aquaculture industry continues to grow worldwide and the consumption of seafood in the United States continues to increase, the need to safeguard operations and protect this industry from HAB impacts is paramount. It will be essential to accurately identify early bloom development and to forecast and track bloom movement. Several HAB bulletins have been developed or are being developed that forecast bloom movement based on near real-time data including satellite-measured chlorophyll or pigment levels, known distribution of toxic dinoflagellates, wind direction and speed, and current trajectories. In addition, mitigation and control strategies and techniques are needed to safely protect aquaculture products with minimal ecological impacts.

Some of the challenges specifically for dinoflagellate HABs and their impacts are determining the underlying reasons for why blooms occur. How does one species have an advantage over another in the same habitat and geographic area? Is the advantage conferred by their life cycle that allows them to become dormant and then to reappear in the same area under certain conditions, or in their ability to exploit currents, or in their preference for specific nutrients? Adaptive life cycles (the ability to form temporary or resting stages) and broad adaptations (the ability to tolerate specific or wide ranges in temperature, salinity, light, and nutrient conditions) are key to the success of species either in their regular environment or in their ability to colonize new environments. How vulnerable are the world's oceans, countries' territorial waters, and popular tourist areas to new HAB events? Once a bloom species of dinoflagellate is established, how does it outcompete the other phytoplankton in the water column or pelagic zone? Many dinoflagellates may avoid predation or competition by producing allelopathic chemicals (such as toxins and other bioactive compounds), but in general we do not know why the dinoflagellates bloom or why they produce these compounds.

There were approximately 80 toxic or harmful dinoflagellates listed in 2002 (Landsberg 2002); since then, at least 20 additional species have been identified. The systematics (a science discipline that uses taxonomy as a primary tool in understanding organisms) of these organisms is advancing rapidly and is being supplemented by genetic markers (e.g., large-subunit [LSU] rRNA, small-subunit [SSU], rRNA, rRNA internal transcribed space [ITS], and mitochondrial Cytochrome B gene [cyto b]), specific genes, unique pigments and sterols, and other markers to differentiate species. Molecular technology is paralleling and advancing classical taxonomical research in its application to species detection and quantification. Classical taxonomists must work together with molecular biologists so that they both can reach some common ground for species identifications that can discern meaningful morphological descriptors in parallel with molecular information.

Toxins

How are toxic dinoflagellates different from nontoxic ones? The answer is toxins! One interesting point about toxic dinoflagellates is that different isolates of the same species can either be "nontoxic" (or low in toxicity), whereas other isolates are highly toxic. The genetics of toxicity, environmental factors, and the on-off regulators need to be determined for species, particularly in situations where they were previously considered to be harmless or nontoxic.

As the consumption of seafood in the United States continues to increase, as aquaculture facilities around the world continue to expand, and as global HAB events continue to increase, quick and effective toxin detection methods are becoming vital components for the successful and safe production of cultured seafood products. The need for toxin detection assays that are near real time with simple, yet accurate control methods that can be used in the field or remotely deployed are essential. Progress has been made, but although the number and diversity of both structurally or functionally based rapid screening methods and quantitative analytical methods (LC-MS and LC-MS/MS) for toxin detection are rapidly increasing, the validation and implementation of these methods in management programs tasked with protecting public health has been slow. The mouse bioassay remains the standard for seafood safety for public consumption by ensuring that toxin concentrations remain below federal guidelines. In large part, this is because the application of chemical assays and analyses rather than the use of whole animal bioassays requires a full understanding of exactly which toxins present a threat or are the best indicators of toxicity.

Accurately assessing the risk dinoflagellate toxins pose to human health is not straightforward for many reasons. Families of related toxins (e.g., the brevetoxins) or different toxins that act on the same sites (e.g., saxitoxins and tetrodotoxins) can have different affinities for the same binding sites with more or less extreme consequences. In addition to the multiple toxins that may be present, the metabolism or biological transformation of these toxins in shellfish may render them more or less toxic than the initial toxins produced by the dinoflagellate; toxicity assessments are lacking for many of these recently identified toxin metabolites. Differences in routes of exposure, mechanisms of uptake, transport, storage, transformation, and elimination of the toxins in the body can also affect toxicity. Other bioactive compounds produced by dinoflagellates may also inhibit or counter the effects of the toxins. For example, the brevenals (recently discovered nontoxic compounds found in seawater and cultures of *Karenia brevis*) have been shown to protect fish and laboratory sheep from the neurotoxic effects of brevetoxins and may also modulate the toxicity of brevetoxins to humans (Abraham *et al.,* 2005; Bourdelais *et al.,* 2004).

Many of the toxins discussed here have multiple receptors in the body that have only recently been discovered. Therefore, the full potential mode of toxicity and the impact of many dinoflagellate toxins have not yet been realized. For example, research has identified the effects of saxitoxins on certain types of calcium and potassium channels (Llewellyn, 2006), as opposed to the traditionally known effects on the sodium channel. Similarly, because brevetoxins induce respiratory effects at concentrations that are two to three orders of magnitude lower than have been found in neuronal studies, an additional pulmonary receptor has been hypothesized (Abraham *et al.,* 2005). Brevetoxins have also been shown to inhibit cathepsins (key enzymes in the formation of antigenic determinants), which may result in immunosuppression (Benson *et al.,* 2005). In addition, thus far, because HAB toxins are widely responsible for rapid poisoning events, most studies have focused on the acute effects of toxins. Little or no information is available for the chronic effects of acute exposure or the effects of chronic, low-level exposure to common HAB toxins.

A question often asked and as yet unanswered is this: Why do these organisms produce toxins? The identification of biosynthetic pathways and the genetic origins of toxin production are important areas of study revealing new information that may shed some light on the function, if any, these toxins serve. Most notably, the genes responsible for saxitoxin synthesis in cyanobacteria have been identified (Kellman *et al.,* 2006). These findings may lead to similar discoveries in saxitoxin-producing marine dinoflagellates.

As toxin detection methods improve through new technological advances, issues of toxin transfer through the food web will also need investigation. As new vectors of toxin transmission are identified, concurrent epidemiological studies to identify populations at risk for illness, as well as the rates of illness, will also be needed.

Animal Health Impacts

By their very presence, HABs can affect water quality, by producing toxins or simply because of their physical presence. For example, decomposing and respiring HAB cells can cause a reduction of oxygen (hypoxia) or lead to a total absence of oxygen (anoxia). They can produce ammonia or other nitrogenous by-products or lead to the formation of toxic sulfides. Such poor water quality alone can cause animal or plant mortalities, contribute to weakness, or cause an increased susceptibility of animals to disease. HABs can simply lead to a species avoidance of an area, an impact that can influence the local productivity, which in turn may be critical for ecologically or economically impoverished geographical areas. Ultimately, HABs can cause major ecosystem shifts, the severity of which increases with the length of the bloom and the area covered. These effects, however, are only just being recognized. Broader-scale ecosystem monitoring and modeling is required to gather the information needed to monitor and assess these indirect but potentially large-scale and lasting consequences of HABs. Therefore, not only does the HAB need to be monitored, but so do the affected species and their ecosystems.

Most algal toxin effects on humans and animals are recognized largely because they cause seafood poisoning incidents or mass mortality events associated with their acute toxic activity. In general, the effects of sublethal, or chronic, or repeated low level exposures to dinoflagellate toxins are, however, poorly documented. Preliminary evidence indicates that algal toxins may affect animal development, cause immunosuppression and may even promote tumor formation; but it has been difficult to demonstrate these effects in the field or to discern how much impact chronic exposure may have on the health of animal populations (Landsberg, 2002). The study of the potential chronic effects of phycotoxins can be assessed by using animals as sentinels or biomarkers for possible activity. Such studies could be applied and interpreted for their effects on humans and would thus be of benefit to human health.

In addition to causing sublethal effects or potentially affecting growth and reproduction of animals, HABs are also being considered as possible vectors or transport mechanisms for aquatic animal and human pathogens. This area has not been well researched, but it poses a number of questions for the way in which many disease outbreaks appear to parallel HAB incidents. For example, outbreaks of cholera (caused by the bacterium *Vibrio cholerae*) have been linked to HAB incidents in Bangladesh (Epstein, 1993). Similarly, animal disease syndromes have been postulated to be linked to pathogenic vectors associated with HABs (Landsberg *et al.,* 2005), but once more definitive evidence is lacking. Such circumstantial indications again support the need for holistic multidisciplinary wide-scale approaches embracing numerous fields that can demonstrate the all-reaching effects of HABs.

Human Health

Although the acute effects from shellfish poisonings have been known for some time, a critical public health need in the context of human exposure to dinoflagellates and their toxins is for accurate, rapid diagnostic tools for documenting exposure and effect. Currently, the diagnosis of seafood poisoning is definitively made through the testing of the suspected seafood. The human response to toxic seafood, particularly in mild cases, involves a suite of general gastrointestinal symptoms that can potentially be caused by a myriad of illnesses. Definitive diagnosis may therefore be missed if the health care professional interviewing the patient does not conduct a thorough interview regarding food consumption history. Although the treatment for marine HAB toxins is primarily supportive, misdiagnoses of incidents mean that there is general under-reporting of these illnesses. In addition, global travel complicates the diagnosis as there may be a spatial and temporal delay from exposure to the effect (illness), and a lack of universal knowledge regarding HAB exposures, the nature of public health incidents, and the information needed for accurate diagnoses particularly in nonendemic areas.

Another example of the need for diagnostic tools is in the case of the toxic marine aerosols caused by *Karenia brevis.* Although the suite of brevetoxins is the same as in NSP, the dose of the airborne toxins is several orders of magnitude less (picogram to nanogram range compared to micrograms in ingested shellfish). Therefore, accurate detection requires an even more sensitive diagnostic tool. For both routes of exposure, the ideal would be a noninvasive, rapid analysis tool that would allow medical professionals to quickly identify the causative agent of the symptoms.

Much epidemiological work is needed with the exposure to toxins from dinoflagellates regarding possible long-term effects from acute high dose and chronic low dose exposures. As the population in the United States ages, concern over this susceptible population, especially in retirement areas prone to HAB events, will increase. Other potentially vulnerable populations include fetuses and young children, particularly with exposure to potential neurotoxins and carcinogens.

Some areas of the United States have made improvements in their reporting systems for marine toxin exposures. The Marine and Freshwater Toxin hotline (888-232-8635) based at the South Florida Poison Information Center in Miami, Florida, is an example. Although the system has demonstrated increased reporting of HAB associated illnesses, it remains a passive reporting system and thus still only reports a fraction of cases. Active surveillance (such as the real-time analysis of data collection by poison information centers and in areas frequently impacted by HAB events) would provide the public health community accurate data to quantify the public health impacts of these HAB illnesses. This would require outreach and education of health care providers and public health officials on HAB associated illnesses, their diagnosis, treatment, and reporting requirements.

The study of toxic dinoflagellate blooms is a rapidly evolving science where advancements are made almost daily. With progress comes change, so that scientific methods, approaches, interpretation, and applications advance and change. This progress can leave a murky trail and what appears to be conflicting data; but when taken as a slice of progress over time, this can be sorted out. The vigilance of updating all aspects of the data on each species, toxin, and geographic HAB is a challenge for today as well as the future.

References

Abraham, W.M., Bourdelais, A.J., Ahmed, A., Serebriakov, I., Baden, D.G., 2005. Effects of inhaled brevetoxins in allergic airways: Toxin-allergen interactions and pharmacologic intervention. Environ. Health Perspect. 113, 632–637.

Anderson, D., 1998. Physiology and bloom dynamics of toxic *Alexandrium* species, with emphasis on life cycle transitions. In Anderson, D.M., Cembella, A.D., and Hallegraeff, G.M. (eds.), Physiological Ecology of Harmful Algal Blooms, pp. 29–48. Heidelberg, Springer-Verlag.

Anderson, D.M., 1997. Bloom dynamics of toxic *Alexandrium* species in the northeastern U.S. Limnol. Oceanogr. 42, 1009–1022.

Anderson, D.M., Hoagland, P., Kaoru, Y., White, A.W., 2000. Estimated annual economic impacts resulting from harmful algae blooms (HABs) in the United States. Woods Hole Oceanographic Institute Technical Report, WHOI-200-11, Woods Hole, MA.

Anderson, D.M., Townsend, D.W., McGillicuddy, D.J., Jr., Turner, J.T., 2005. The ecology and oceanography of toxic *Alexandrium fundyense* blooms in the Gulf of Maine. Deep-Sea Res. II 52, 2365–2876.

Aune, T., Yndestad, M., 1993. Diarrhetic shellfish poisoning. In Falconer, I.R. (ed.), Algal Toxins in Seafood and Drinking Water, pp. 87–104. London, Academic Press Limited.

Backer, L.C., Schurz-Rogers, H., Fleming, L.E., Kirkpatrick, B., Benson, J., 2005. Marine phycotoxins in seafood. In Dabrowski, W.M., and Sikorski, Z.E. (eds.), Toxins in Food, pp. 155–189. Boca Raton, FL, CRC Press.

Baden, D.G., 1983. Marine food-borne dinoflagellate toxins. Int. Rev. Cytol. 82, 99–149.

Baden, D.G., 1989. Brevetoxins: Unique polyether dinoflagellate toxins. FASEB J. 3, 1807–1817.

Baden, D.G., Bourdelais, A.J., Jacocks, H., Michelliza, S., Naar, J., 2005. Natural and derivative brevetoxins: Historical background, multiplicity, and effects. Environ. Health Perspect. 113, 621–625.

Baden, D.G., Fleming, L.E., Bean, J.A., 1995. Marine toxins. In DeWolf, F.A. (ed.), Handbook of Clinical Neurology: Intoxications of the Nervous System. Part H. Natural Toxins and Drugs, pp. 141–175. Amsterdam, Elsevier Press.

Baden, D.G., Mende, T.J., Block, R.E., 1979. Two similar toxins isolated from *Gymnodinium breve*. In Taylor, D.L., and Seliger, H.H. (eds.), Toxic Dinoflagellate Blooms, pp. 327–334. New York, Elsevier/North-Holland.

Baker, T.R., Doucette, G.J., Powell, C.L., Boyer, G.L., Plumley, F.G., 2003. GTX(4) imposters: Characterization of fluorescent compounds synthesized by *Pseudomonas stutzeri* SF/PS and *Pseudomonas/Alteromonas* PTB-1, symbionts of saxitoxin-producing *Alexandrium* spp. Toxicon 41, 339–347.

Benson, J., Hahn, F.F., March, T.H., McDonald, J.D., Gomez, A.P., Sopori, M.J., Bourdelais, A.J., Naar, J., Zaias, J., Bossart, G.D., Baden, D.G., 2005. Inhalation toxicity of brevetoxin 3 in rats exposed for twenty-two days. Environ. Health Perspect. 113, 626–631.

Bossart, G.D., Baden, D.G., Ewing, R.Y., Roberts, B., Wright, S.D., 1998. Brevetoxicosis in manatees (*Trichechus manatus latirostris*) from the 1996 epizootic: Gross, histologic, and immunohistochemical features. Toxicol. Pathol. 26, 276–282.

Bourdelais, A.J., Campbell, S., Jacocks, H., Naar, J., Wright, J.L.C., Carsi, J., Baden, D.G., 2004. Brevenal is a natural inhibitor of brevetoxin action in sodium channel receptor binding assays. Cell. Molec. Neurobiol. 24, 553–563.

Catterall, W.A., Gainer, M., 1985. Interaction of brevetoxin A with a new receptor site on the sodium channel. Toxicon 23, 497–504.

Cembella, A.D., Quilliam, M.A., Lewis, N.I., Bauder, A.G., Dell'Aversano, C., Thomas, K., Jellett, J., Cusack, R.R., 2002. The toxigenic marine dinoflagellate *Alexandrium tamarense* as the probable cause of mortality of caged salmon in Nova Scotia. Harmful Algae 1, 313–315.

Chang, F.H., Bourdelais, A., Baden, D.G., Gall, M., Hulston, D., Webb, V., 2006. *Karenia concordia* (Dinophyceae) as a brevetoxin-producer and comparison with two closely related species *K. brevisulcata* and *K. mikimotoi*. In 12th International Conference on Harmful Algae, pp. 51–52, Abstract. Copenhagen, Denmark.

Cheng, Y.S., Zhou, Y., Irvin, C.M., Pierce, R., Naar, J., Backer, L.C., Fleming, L.E., Kirkpatrick, B., Baden, D.G., 2005. Characterization of marine aerosol for assessment of human exposure to brevetoxins. Environ. Health Perspect. 113, 638–643.

Ciminiello, P., Dell'Aversano, C., Fattorusso, E., Forino, M., Magno, S., Guerrini, F., Pistocchi, R., Boni, L., 2003. Complex yessotoxins profile in *Protoceratium reticulatum* from north-western Adriatic sea revealed by LC-MS analysis. Toxicon 42, 7–14.

Cohen, P.T., 1997. Novel protein serine/threonine phosphatases: Variety is the spice of life. Trends Biochem. Sci. 22, 245–251.

Cordier, S., Monfort, C., Miossec, L., Richardson, S., Belin, C., 2000. Ecological analysis of digestive cancer mortality related to contamination by diarrhetic shellfish poisoning toxins along the coasts of France. Environmental Research 84, 145–150.

Donaghay, P.L., Osborn, T.R., 1997. Toward a theory of biological-physical control of harmful algal bloom dynamics and impacts. Limnol. Oceanogr. 42, 1283–1296.

Dounay, A.B., Forsyth, C.J., 2002. Okadaic acid: The archetypal serine/threonine protein phosphatase inhibitor. Curr. Med. Chem. 9, 1939–1980.

Draisci, R., Lucentini, L., Mascioni, A., 2000. Pectenotoxins and yessotoxins: Chemistry, toxicology, pharmacology, and analysis. In Botana, L.M. (ed.), Seafood and Freshwater Toxins: Pharmacology, Physiology, and Detection, pp. 289–324. New York, Marcel Dekker.

Durbin, E.G., Teegarden, G.J., Campbell, R.G., Cembella, A.D., Baumgartner, M.F., Mate, B.R., 2002. North Atlantic right whales, *Eubalaena glacialis*, exposed to paralytic shellfish poisoning (PSP) toxins via a zooplankton vector, *Calanus finmarchicus*. Harmful Algae 1, 243–251.

Edvardsen, B., Shalchian-Tabrizi, K., Jakobsen, K.S., Medlin, L.K., Dahl, E., Brubak, S., Paasche, E., 2003. Genetic variability and molecular phylogeny of *Dinophysis* species (Dinophyceae) from Norwegian waters inferred from single cell analyses of rDNA. J. Phycol. 39, 395–408.

Epstein, P.R., 1993. Algal blooms in the spread and persistence of cholera. Biosystems 31, 209–221.

Faust, M.A., Gulledge, R.A., 2002. Identifying harmful marine dinoflagellates. Contributions from the United States Herbarium, Smithsonian Institution 42, 1–144.

Fensome, R.A., Taylor, F.J.R., Norris, G., Sarjeant, W.A.S., Wharton, D. I., Williams, G.L., 1993. A classification of living and fossil dinoflagellates. Micropaleontology, Special Publication Number 7.

Fernández, J.J., Suarez-Gomez, B., Souto, M.L., Norte, M., 2003. Identification of new okadaic acid derivatives from laboratory cultures of *Prorocentrum lima*. J. Nat. Prod. 66, 1294–1296.

Figueiras, F.G., Pitcher, G.C., Estrada, M., 2006. Harmful algal bloom dynamics in relation to physical processes. In Granéli, E., and Turner, J.T. (eds.), Ecology of Harmful Algae, pp. 127–138. Berlin, Springer-Verlag.

Fleming, L.E., Kirkpatrick, B., Backer, L.C., Bean, J.A., Wanner, A., Dalpra, D., Tamer, R., Zaias, J., Cheng, Y.S., Pierce, R., Naar, J., Abraham, W., Clark, R., Zhou, Y., Henry, M., Johnson, D., Van de Bogart, G., Bossart, G.D., Harrington, M., Baden, D.G., 2005. Initial evaluation of the effects of aerosolized Florida red tide toxins (brevetoxins) in persons with asthma. Environ. Health Perspect. 113, 650–657.

Flewelling, L.J., Naar, J.P., Abbott, J.P., Baden, D.G., Barros, N.B., Bossart, G.D., Bottein, M.-Y.D., Hammond, D.G., Haubold, E.M., Heil, C.A., Henry, M.S., Jacocks, H.M., Leighfield, T.A., Pierce, R.H., Pitchford, T.D., Rommel, S.A., Scott, P.S., Steidinger, K.A., Truby, E.W., Dolah, F.M.V., Landsberg, J.H., 2005. Red tides and marine mammal mortalities. Nature 435, 755–756.

Forrester, D.J., Gaskin, J.M., White, F.H., Thompson, N.P., Quick, J.A., Jr., Henderson, G.E., Woodard, J.C., Robertson, W.D., 1977. An epizootic of waterfowl associated with a red tide episode in Florida. J. Wildl. Dis. 13, 160–167.

Franchini, A., Marchesini, E., Poletti, R., Ottaviani, E., 2004a. Acute toxic effect of the algal yessotoxin on Purkinje cells from the cerebellum of Swiss CD1 mice. Toxicon 43, 347–352.

Franchini, A., Marchesini, E., Poletti, R., Ottaviani, E., 2004b. Lethal and sub-lethal yessotoxin dose-induced morpho-functional alterations in intraperitoneal injected Swiss CD1 mice. Toxicon 44, 83–90.

Fujiki, H., Suganuma, M., 1993. Tumor promotion by inhibitors of protein phosphatase-1 and phosphatase-2a—the okadaic acid class of compounds. Adv. Cancer Res. 61, 143–194.

Fujiki, H., Suganuma, M., Suguri, H., Yoshizawa, S., Takagi, K., Uda, N., Wakamatsu, K., Yamada, K., Murata, M., Yasumoto, T., Sugimura, T., 1988. Diarrhetic shellfish toxin, dinophysistoxin-1, is a potent tumor promoter on mouse skin. Jpn. J. Cancer Res. 79, 1089–1093.

Galstoff, P.S., 1948. Red tide. Progress report on the investigations of the cause of the mortality of fish along the west coast of Florida conducted by the U.S. Fish and Wildlife Service and cooperating organizations. US Fish Wildl. Serv. Spec. Sci. Rep. No. 46 1–44.

Garcés, E., Zingone, A., Montresor, M., Reguera, B., Dale, B., 2002. LIFEHAB: Life histories of microalgal species causing harmful blooms, www.icm.csic.es/bio/projects/lifehab/LIFEHAB.pdf.

Geraci, J.R., Anderson, D.M., Timperi, R.J., St. Aubin, D.J., Early, G.A., Prescott, J.H., Mayo, C.A., 1989. Humpback whales (*Megaptera novaeangliae*) fatally poisoned by dinoflagellate toxin. Can. J. Fish. Aquat. Sci. 46, 1895–1898.

Gomez, F., 2005. A list of free-living dinoflagellate species in the world's oceans. Acta Bot. Croat. 64, 129–212.

Gunter, G., Smith, F.G.W., Williams, R.H., 1947. Mass mortality of marine animals on the lower west coast of Florida, November 1946–January 1947. Science 105, 256–257.

Gunter, G., Williams, R.H., Davis, C.C., Smith, F.G.W., 1948. Catastrophic mass mortality of marine animals and coincident phytoplankton bloom on the west coast of Florida, November 1946 to August 1947. Ecol. Monogr. 18, 309–324.

Hallegraeff, G.M., Maclean, J.L., 1989. Biology, epidemiology and management of *Pyrodinium* red tides. Proceedings of the Management and Training Workshop, Bandar Seri Begawan, Brunei Darussalam, May 23–30, 1989. Fisheries Department, Ministry of Development and International Center for Living Aquatic Resources Management, Manila, Philippines.

HARRNESS, 2005. Harmful Algal Research and Response: A National Environmental Science Strategy 2005–2015, Ramsdell, J.S., Anderson, D.M., and Glibert, P.M. (eds.). Washington, DC, Ecological Society of America.

Haywood, A.J., Steidinger, K.A., Truby, E.W., Bergquist, P.R., Bergquist, P.L., Adamson, J., Mackenzie, L., 2004. Comparative morphology and molecular phylogenetic analysis of three new species of the genus *Karenia* (Dinophyceae) from New Zealand. J. Phycol. 40, 165–179.

Hernández, M., Robinson, I., Aguilar, A., Gonzalez, L.M., Lopez-Jurado, L.F., Reyero, M.I., Cacho, E., Franco, J., Lopez-Rodas, V., Costas, E., 1998. Did algal toxins cause monk seal mortality? Nature 393, 28–29.

Hu, T., Burton, I.W., Cembella, A.D., Curtis, J.M., Quilliam, M.A., Walter, J.A., Wright, J.L., 2001. Characterization of spirolides a, c, and 13-desmethyl c, new marine toxins isolated from toxic plankton and contaminated shellfish. J. Nat. Prod. 64, 308–312.

Ishida, H., Nozawa, A., Nukaya, H., Tsuji, K., 2004. Comparative concentrations of brevetoxins PbTx-2, PbTx-3, BTX-B1 and BTX-B5 in cockle, *Austrovenus stutchburyi*, greenshell mussel, *Perna canaliculus*, and Pacific oyster, *Crassostrea gigas*, involved in neurotoxic shellfish poisoning in New Zealand. Toxicon 43, 779–789.

Ito, E., Satake, M., Ofuji, K., Higashi, M., Harigaya, K., McMahon, T., Yasumoto, T., 2002. Chronic effects in mice caused by oral administration of sublethal doses of azaspiracid, a new marine toxin isolated from mussels. Toxicon 40, 193–203.

Ito, E., Satake, M., Ofuji, K., Kurita, N., McMahon, T., James, K., Yasumoto, T., 2000. Multiple organ damage caused by a new toxin azaspiracid, isolated from mussels produced in Ireland. Toxicon 38, 917–930.

James, K.J., Bishop, A.G., Furey, A., 2000. New toxins on the horizon. In Botana, L.M. (ed.), Seafood and Freshwater Toxins: Pharmacology, Physiology, and Detection, pp. 693–714. New York, Marcel Dekker.

James, K.J., Saez, M.J.F., Furey, A., Lehane, M., 2004. Azasparacid poisoning, the food-borne illness associated with shellfish consumption. Food Add. Cont. 21, 879–892.

Kellman, R., Jeon, Y.J., Mihali, T.K., Cavaliere, R., Neilan, B.A., 2006. The genetic basis for the biosynthesis of PSTs in cyanobacteria and algae. In 12th International Conference on Harmful Algae, pp. 52–53, Abstract, Copenhagen, Denmark.

Kirkpatrick, B.A., Bean, J.A., Fleming, L.E., Backer, L.C., Clark, R., Abraham, W., Benson, J., Cheng, Y.S., Johnson, D., Pierce, R., Zaias, J., Bossart, G.D., Baden, D.G., 2007. Aerosolized red tide toxins (brevetoxins) and asthma: A 10 day follow up after 1 hour beach exposure. In Moestrup, Ø. (ed.) Harmful Algae 2006, International Society for the Study of Harmful Algae and Intergovernmental Oceanographic Commission of UNESCO, submitted.

Kirkpatrick, B.A., Fleming, L.E., Squicciarini, D., Backer, L.C., Clark, R., Abraham, W., Benson, J., Cheng, Y.S., Johnson, D., Pierce, R., Zaias, J., Bossart, G.D., Baden, D.G., 2004. Literature review of Florida red tide: Implications for human health effects. Harmful Algae 3, 99–115.

Kodama, M., 2000. Ecobiology, classification, and origin. In Botana, L.M. (ed.), Seafood and Freshwater Toxins: Pharmacology, Physiology, and Detection, pp. 125–149. New York, Marcel Dekker.

Kreuder, C., Mazet, J., Bossart, G.D., Carpenter, T., Holyoak, M., Elie, M., Wright, S.D., 2002. Clinicopathologic features of suspected brevetoxicosis in double-crested cormorants (*Phalacrocorax auritus*) along the Florida Gulf coast. J. Zoo Wildl. Med. 33, 8–15.

Lackey, J.B., 1956. Known geographic range of *Gymnodinium brevis* Davis. Q. J. Fla. Acad. Sci. 19, 71.

Landsberg, J.H., 2002. The effects of harmful algal blooms on aquatic organisms. Rev. Fish. Sci. 10, 113–390.

Landsberg, J.H., Balazs, G.H., Steidinger, K.A., Baden, D.G., Work, T.M., Russell, D.J., 1999. The potential role of natural tumor promoters in marine turtle fibropapillomatosis. J. Aquat. Anim. Health 11, 199–210.

Landsberg, J.H., Hall, S., Johannessen, J.N., White, K.D., Conrad, S.M., Abbott, J.P., Flewelling, L.J., Richardson, R.W., Dickey, R.W., Jester, E.L.E., Etheridge, S.M., Deeds, J.R., Van Dolah, F.M., Leighfield, T.A., Zou, Y.L., Beaudry, C.G., Benner, R.A., Rogers, P.L., Scott, P.S., Kawabata, K., Wolny, J.L., Steidinger, K.A., 2006. Saxitoxin puffer fish poisoning in the United States, with the first report of *Pyrodinium bahamense* as the putative toxin source. Environ. Health Perspect. 114, 1502–1507.

Landsberg, J.H., Van Dolah, F., Doucette, G., 2005. Marine and estuarine harmful algal blooms: Impacts on human and animal health. In Belkin, S., and Colwell, R. (eds.), Oceans and Health: Pathogens in the Marine Environment, pp. 165–215. New York, Springer.

Landsberg, J.L., Vargo, G.A., Flewelling, L.J., Wiley, F., 2007. Algal biotoxins. In Thomas, N.J., Hunter, D.B., and Atkinson, C.T. (eds.), Infectious Diseases of Wild Birds, pp. 431–455. Ames, Iowa, Blackwell.

Lefebvre, K.A., Trainer, V.L., Scholz, N.L., 2004. Morphological abnormalities and sensorimotor deficits in larval fish exposed to dissolved saxitoxin. Aquat. Toxicol. 66, 159–170.

Lenoir, S., Ten-Hage, L., Turquet, J., Quod, J.P., Hennion, M.C., 2006. Characterisation of new analogues of palytoxin isolated from an *Ostreopsis mascarenensis* bloom in the south-western Indian Ocean. Afr. J. Mar. Sci. 28, 389–391.

Leverone, J.R., Blake, N.J., Pierce, R.H., Shumway, S.E., 2006. Effects of the dinoflagellate *Karenia brevis* on larval development in three species of bivalve mollusc from Florida. Toxicon 48, 75–84.

Llewellyn, L., Negri, A., Quilliam, M., 2004. High affinity for the rat brain sodium channel of newly discovered hydroxybenzoate saxitoxin analogues from the dinoflagellate *Gymnodinium catenatum*. Toxicon 43, 101–104.

Llewellyn, L.E., 2006. Saxitoxin, a toxic marine natural product that targets a multitude of receptors. Nat. Prod. Rep. 23, 200–222.

McKernan, D.L., Scheffer, V.B., 1942. Unusual numbers of dead birds on the Washington coast. The Condor 44, 264–266.

McMahon, T., Silke, J., 1996. Winter toxicity of unknown aetiology in mussels. Harmful Algae News 14, 2.

Miles, C.O., Samdal, I.A., Aasen, J.A.G., Jensen, D.J., Quilliam, M.A., Petersen, D., Briggs, L.M., Wilkins, A.L., Rise, F., Cooney, J.M., MacKenzie, A.L., 2005. Evidence for numerous analogs of yessotoxin in *Protoceratium reticulatum*. Harmful Algae 4, 1075–1091.

Moita, M.T., Sobrinho-Goncalves, L., Oliveira, P.B., Palma, S., Falcao, M., 2006. A bloom of *Dinophysis acuta* in a thin layer off north-west Portugal. Afr. J. Mar. Sci. 28, 265–269.

Naar, J., Flewelling, L.J., Baden, D.G., Jacocks, H.M., Steidinger, K.A., Landsberg, J.L., 2005. Brevetoxins, like ciguatoxins, are potent ichthyotoxic neurotoxins that accumulate in fish. In: 3rd Symposium on Harmful Algae in the U.S., p. 44, Abstract, Pacific Grove, CA.

Nozawa, A., Tsuji, K., Ishida, H., 2003. Implication of brevetoxin B1 and PbTX-3 in neurotoxic shellfish poisoning in New Zealand by isolation and quantitative determination with liquid chromatography-tandem mass spectrometry. Toxicon 42, 91–103.

Onodera, H., Satake, M., Oshima, Y., Yasumoto, T., Carmichael, W.W., 1997. New saxitoxin analogues from the freshwater filamentous cyanobacterium *Lyngbya wollei*. Nat. Toxins 5, 146–151.

Oshima, Y., 1995. Postcolumn derivation liquid chromatographic method for paralytic shellfish toxins. J. AOAC Int. 78, 528–532.

Osterhaus, A., van de Bildt, M., Vedder, L., B., M., Niesters, H., Vos, J., van Egmond, H., Liem, D., Baumann, R., Androukaki, E., Kotomatas, S., Komnenou, A., Abou Sidi, B., Jiddou, A.B., Barham, M.E., 1998. Monk seal mortality: Virus or toxin? Vaccine 16, 979–981.

Paz, B., Riobo, P., Fernandez, M.L., Fraga, S., Franco, J.M., 2004. Production and release of yessotoxins by the dinoflagellates *Protoceratium reticulatum* and *Lingulodinium polyedrum* in culture. Toxicon 44, 251–258.

Perez-Gomez, A., Ferrero-Gutierrez, A., Novelli, A., Franco, J.M., Paz, B., Fernandez-Sanchez, M.T., 2006. Potent neurotoxic action of the shellfish biotoxin yessotoxin on cultured cerebellar neurons. Toxicol. Sci. 90, 168–177.

Pierce, R.H., Henry, M.S., Blum, P.C., Lyons, J., Cheng, Y.S., Yazzie, D., Zhou, Y., 2003. Brevetoxin concentrations in marine aerosol: Human exposure levels during a *Karenia brevis* harmful algal bloom. Bull. Environ. Contam. Toxicol. 70, 161–165.

Plakas, S.M., Wang, Z., Said, K.R.E., Jester, E.L.E., Granade, H.R., Flewelling, L., Scott, P., Dickey, R.W., 2004. Brevetoxin metabolism and elimination in the Eastern oyster (*Crassostrea virginica*) after controlled exposures to *Karenia brevis*. Toxicon 44, 677–685.

Poli, M.A., Mende, T.J., Baden, D.G., 1986. Brevetoxins, unique activators of voltage-sensitive sodium channels, bind to specific sites in rat brain synaptosomes. Mol. Pharmacol. 30, 129–135.

Quick, J.A., Jr., Henderson, G.E., 1974. Effects of *Gymnodinium breve* red tides on fishes and birds: A preliminary report on behavior, anatomy, hematology, and histopathology. In Amborski, R.L., Hood, M.A., and Miller, R.R. (eds.), Proceedings of the Gulf Coast Regional Symposium on Diseases of Aquatic Animals, Louisiana Sea Grant: Louisiana State University, pp. 85–113.

Quick, J.A., Jr., Henderson, G.E., 1975. Evidence of new ichthyointoxicative phenomena in *Gymnodinium breve* red tides. In LoCicero, V.R. (ed.), Proceedings of the First International Conference on Toxic Dinoflagellate Blooms, Massachusetts Science and Technology Foundation, Wakefield, MA, pp. 413–422.

Quilliam, M., Wechsler, D., Marcus, S., Ruck, B., Wekell, M., Hawryluk, T., 2004. Detection and identification of paralytic shellfish poisoning toxins in Florida pufferfish responsible for incidents of neurologic Illness. In Steidinger, K.A., Landsberg, J.H., Tomas, C.R., and Vargo, G.A. (eds.), Harmful Algae 2002, Florida Fish and Wildlife Conservation Commission, Florida Institute of Oceanography, and Intergovernmental Oceanographic Commission of UNESCO, St. Petersburg, FL, pp. 116–118.

Reguera, B., González-Gil, S., 2001. Small cell and intermediate cell formation in species of *Dinophysis* (Dinophyceae, Dinophysiales). J. Phycol. 37, 318–333.

Reguera, B., González-Gil, S., Delgado, M., 2004. Formation of *Dinophysis dens* Pavillard and *D. diegensis* Kofoid from laboratory incubations of *Dinophysis acuta* Ehrenberg and *D. caudata* Saville-Kent. In Steidinger, K.A., Landsberg, J.H., Tomas, C.R., and Vargo, G.A. (eds.), Harmful Algae 2002, Florida Fish and Wildlife Conservation Commission, Florida Institute of Oceanography, and Intergovernmental Oceanographic Commission of UNESCO, St. Petersburg, FL, pp. 440–442.

Reynolds, C.S., Smayda, T.J., 1998. Principles of species selection and community assembly in the phytoplankton: Further explorations of the Mandala. In Reguera, B., Blanco, J., Fernandez, M.L., and Wyatt, T. (eds.), Harmful Algae, Xunta de Galicia and Intergovernmental Oceanographic Commission of UNESCO, pp. 8–10.

Roberts, B.S., Henderson, G.E., Medlyn, R.A., 1979. The effect of *Gymnodinium breve* toxin(s) on selected mollusks and crustaceans. In Taylor, D.L., and Seliger, H.H. (eds.), Toxic Dinoflagellate Blooms, pp. 419–424. New York, Elsevier North Holland,

Rodrigue, D.C., Etzel, R.A., Hall, S., de Porras, E., Velasquez, O.H., Tauxe, R.V., Kilbourne, E.M., Blake, P.A., 1990. Lethal Paralytic Shellfish Poisoning in Guatemala. Am. J. Trop. Med. Hyg. 42, 267–271.

Roman, Y., Alfonso, A., Louzao, M.C., de la Rosa, L.A., Leira, F., Vieites, J.M., Vieytes, M.R., Ofuji, K., Satake, M., Yasumoto, T., Botana, L.M., 2002. Azaspiracid-1, a potent, nonapoptotic new phycotoxin with several cell targets. Cellular Signaling 14, 703–716.

Satake, M., MacKenzie, L., Yasumoto, T., 1997. Identification of *Protoceratium reticulatum* as the biogenetic origin of yessotoxin. Nat. Toxins 5, 164–167.

Satake, M., Ofuji, K., Naoki, H., James, K.J., Furey, A., McMahon, T., Silke, J., Yasumoto, T., 1998. Azaspiracid, a new marine toxin having unique spiro ring assemblies, isolated from Irish mussels, *Mytilus edulis*. J. Am. Chem. Soc. 120, 9967–9968.

Schantz, E.J., Ghazarossian, V.E., Schnoes, H.K., Strong, F.M., Springer, J.P., Pezzanite, J.O., Clardy, J., 1975. The structure of saxitoxin. J. Am. Chem. Soc. 97, 1238–1239.

Schantz, E.J., Mold, J.D., Stanger, D.W., Shavel, J., Riel, F.J., Bowden, J.P., Lynch, J.M., Wyler, R.S., Riegel, B., Sommer, H., 1957. Paralytic shellfish poison. VI. A procedure for the isolation and purification of the poison from toxic clam and mussel tissues. J. Am. Chem. Soc. 79, 5230–5235.

Shimizu, Y., 2000. Chemistry and mechanism of action. In Botana, L.M. (ed.), Seafood and Freshwater Toxins: Pharmacology, Physiology, and Detection, pp. 151–172. New York, Marcel Dekker.

Shumway, S.E., Allen, S.M., Boersma, P.D., 2003. Marine birds and harmful algal blooms: sporadic victims or under-reported events? Harmful Algae 2, 1–17.

Smayda, T.J., 1997. What is a bloom? A commentary. Limnol. Oceanogr. 42, 1132–1136.

Smith, G.B., 1976. The impact of fish-killing phyoplankton blooms upon mideastern Gulf of Mexico reef fish communities. In Bullis, H.R., Jr., and Jones, A.C. (eds.), Proceedings of the Colloquium on Snapper-Grouper Fishery Resources of the Western Central Atlantic Ocean, Florida Sea Grant Report, pp. 185–190.

Steidinger, K.A., 1975. Basic factors influencing red tides. In LoCicero, V.R. (ed.), Proceedings of the First International Conference on Toxic Dinoflagellate Blooms, Massachusetts Science and Technology Foundation, Wakefield, MA, pp. 153–162.

Steidinger, K.A., 1983. A re-evaluation of toxic dinoflagellate biology and ecology. In Round, F.E., and Chapman, D.T. (eds.), Progress in Psychological Research, vol. 2, pp. 147–188. Amsterdam, Elsevier, North Holland.

Steidinger, K.A., Burklew, M.A., Ingle, R.M., 1973. The effects of *Gymnodinium breve* toxin on estuarine animals. In Martin, D.F., and Padilla, G.M. (eds.), Marine Pharmacognosy, pp. 179–202. New York, Academic Press.

Steidinger, K.A., Garcés, E., 2006. Importance of life cycles in the ecology of harmful microalgae. In Granéli, E., and Turner, J.T. (eds.), Ecology of Harmful Algae, pp. 37–49. Berlin, Springer-Verlag.

Steidinger, K.A., Vargo, G.A., Tester, P.A., Tomas, C.R., 1998. Bloom dynamics and physiology of *Gymnodinium breve* with emphasis on the Gulf of Mexico. In Anderson, D.M., Cembella, A.D., and Hallegraeff, G.M. (eds.), Physiological Ecology of Harmful Algal Blooms, pp. 133–153. Heidelberg, Springer-Verlag.

Stoecker, D.K., Tillman, U., Granéli, E., 2006. Phagotrophy in harmful algae. In Granéli, E., and Turner, J.T. (eds.), Ecology of Harmful Algae, pp. 177–187. Berlin, Springer-Verlag.

Suarez-Gomez, B., Souto, M.L., Norte, M., Fernandez, J.J., 2001. Isolation and structural determination of DTX-6, a new okadaic acid derivative. J. Nat. Prod. 64, 1363–1364.

Suganuma, K., Suttajit, M., Suguri, H.F., Ojika, M., Yamada, K., Fujiki, H., 1990. Specific binding of okadaic acid, a new tumor promotor in mouse skin. FEBS Letters 250, 615–618.

Suganuma, M., Fujiki, H., Suguri, H., Yoshizawa, S., Hirota, M., Nakayasu, M., Ojika, M., Wakamatsu, K., Yamada, K., Sugimura, T., 1988. Okadaic acid: An additional non-phorbol-12–tetradecanoate-13–acetate-type tumor promoter. Proceedings of the National Academy of Sciences of the United States of America 85, 1768–1771.

Summerson, H.C., Peterson, C.H., 1990. Recruitment failure of the bay scallop, *Argopecten irradians concentricus*, during the first red tide, *Ptychodiscus brevis*, outbreaks recorded in North Carolina. Estuar. 13, 322–331.

Tachibana, K., Scheuer, P.J., 1981. Okadaic acid, a ctyotoxic polyether from two marine sponges of the genus *Halichondria*. J. Am. Chem. Soc. 103, 2469–2471.

Taylor, F.J.R., 1990. Phylum Dinoflagellata. In: Margulis, L., Corliss, J.O., Melkonian, M., Chapman, D.J. (eds.), Handbook of Protoctista, pp. 419–437, Jones and Bartlett, Boston.

Taylor, F.J.R., 2001. Harmful algal bloom studies enter the new millenium. In Hallegraeff, G.M., Blackburn, S.I., Bolch, C.J., and Lewis, R.J. (eds.), Harmful Algal Blooms 2000, pp. 3–7. Paris, International Oceanographic Commission of UNESCO.

Tester, P.A., Steidinger, K.A., 1997. *Gymnodinium breve* red tide blooms: Initiation, transport, and consequences of surface circulation. Limnol. Oceanogr. 42, 1039–1051.

Tester, P.A., Turner, J.T., Shea, D., 2000. Vectorial transport of toxins from the dinoflagellate *Gymnodinium breve* through copepods to fish. J. Plankton Res. 22, 47–61.

Tubaro, A., Sosa, S., Carbonatto, M., Altinier, G., Vita, F., Melato, M., Satake, M., Yasumoto, T., 2003. Oral and intraperitoneal acute toxicity studies of yessotoxin and homoyessotoxins in mice. Toxicon 41, 783–792.

Twiner, M.J., Hess, P., Dechraoui, M.Y., McMahon, T., Samons, M.S., Satake, M., Yasumoto, T., Ramsdell, J.S., Doucette, G.J., 2005. Cytotoxic and cytoskeletal effects of azaspiracid-1 on mammalian cell lines. Toxicon 45, 891–900.

U.S. Commission on Ocean Policy, 2004. An Ocean Blueprint for the 21st Century. Final Report. Washington, DC.

Van Dolah, F.M., 2000. Marine algal toxins: Origins, health effects, and their increased occurrence. Environ. Health Perspect. 108 (S1), 133–141.

Van Dolah, F.M., Doucette, G.J., Gulland, F.M.D., Rowles, T.L., Bossart, G.D., 2003. Impacts of algal toxins on marine mammals. In Vos, J.G., Bossart, G.D., Fournier, M., and O'Shea, T.J. (eds.), Toxicology of Marine Mammals, pp. 247–269. New York, Taylor & Francis.

Van Egmond, H.P., Van Apeldoorn, M.E., Speijers, G.J.A., 2004. Marine Biotoxins. Food and Agriculture Organization of the United Nations, Rome, www.fao.org/docrep/007/y5486e/y5486e00.htm.

Vieytes, M.R., Louzao, M.C., Alfonso, A., Cabado, A.G., Botana, L.M., 2000. Mechanism of action and toxicology. In Botana, L.M. (ed.), Seafood and Freshwater Toxins: Pharmacology, Physiology, and Detection, pp. 239–256. New York, Marcel Dekker.

Vilarino, N., Nicolaou, K.C., Frederick, M.O., Cagide, E., Ares, I.R., Louzao, M.C., Vieytes, M.R., Botana, L.M., 2006. Cell growth inhibition and actin cytoskeleton disorganization induced by azaspiracid-1 structure-activity studies. Chem. Res. Toxicol. 19, 1459–1466

Vinagre, T.M., Alciati, J.C., Regoli, F., Bocchetti, R., Yunes, J.S., Bianchini, A., Monserrat, J.M., 2003. Effect of microcystin on ion regulation and antioxidant system in gills of the estuarine crab *Chasmagnathus granulatus* (Decapoda, Grapsidae). Comp. Biochem. Physiol. C Toxicol. Pharmacol. 135, 67–75.

Walsh, J.J., Jolliff, J.K., Darrow, B.P., Lenes, J.M., Milroy, S.P., .Remsen, A., Dieterle, D.A., Carder, K.L., Chen, F.R., Vargo, G.A., Weisberg, R.H., Fanning, K.A., Muller-Karger, F.E., Shinn, E., Steidinger, K.A., Heil, C.A., Tomas, C.R., Prospero, J.S., Lee, T.N., Kirkpatrick, G.J., Whitledge, T.E., Stockwell, D.A., Villareal, T.A., Jochens, A.E., Bontempi, P.S., 2006. Red tides in the Gulf of Mexico: Where, when, and why? J. Geophys. Res. 111, CI 1003; doi: 10.1029/2004JC002813.

Woodcock, A.H., 1948. Note concerning human respiratory irritation associated with high concentrations of plankton and mass mortality of marine organisms. J. Mar. Res. 7, 56–62.

Yan, T., Zhou, M., Fu, M., Yu, R., Y., W., Li, J., 2003. Effects of the dinoflagellate *Alexandrium tamarense* on early development of the scallop *Argopecten irradians concentricus*. Aquacult. 217, 167–178.

STUDY QUESTIONS

1. What is a dinoflagellate HAB bloom? Give examples.
2. Discuss the importance of dinoflagellate life cycles in the occurrence of HABs.
3. What are the four developmental stages of blooms, and why do you think it is important to know the difference?
4. Of all the dinoflagellate toxins associated with HABs, which ones cause marine mortality? Which ones cause human mortality?
5. Of the phycotoxins discussed in this chapter, what is the most common underlying chemical structure? Why is this significant?
6. How do brevetoxins work at the cellular level, and how does this relate to the ways they are toxic at the whole organism level?
7. What is known and unknown about the effects of the dinoflagellate toxins on human health?
8. New technology is providing new tools for detection of HAB species as well as HAB toxins. What are some of these tools?
9. List four future needs or directions in HAB research, and discuss their importance.
10. New dinoflagellate toxins are being discovered, species that were previously thought to be nontoxic are being shown to produce toxins, and even new species (i.e., previously undescribed) are being described that produce known existing toxins. Does this represent new technology, more thorough investigations, and increased sampling, or does this represent a real increase in toxic dinoflagellate events?

C H A P T E R

14

Ciguatera Fish Poisoning

A Synopsis from Ecology to Toxicity

P. K. BIENFANG, M. L. PARSONS, R. R. BIDIGARE, E. A. LAWS, AND P. D. R. MOELLER

INTRODUCTION

Ciguatera fish poisoning (CFP) is a worldwide health problem associated with the consumption of seafood. Humans acquire ciguatoxin (CTX), the causative poison, by eating reef fish that have accumulated the toxin via the marine food web. Ciguatoxin originates in dinoflagellates *Gambierdiscus* spp. (Fig. 14-1) that are found in association with various macroalgae in coral reef ecosystems. These dinoflagellates are consumed by herbivorous fish, initiating the processes of bioaccumulation, biomagnification, and biomodification in the reef ecosystem food web, as the herbivores are consumed by carnivores and, ultimately, by humans. The ecological complexities, biochemistry, and molecular biology associated with the production of ciguatoxin remain poorly understood, as do the etiology and epidemiology of the disease as presented in fish-consuming human populations. This overview addresses the ecological, biochemical, and human health aspects of this fascinating and important disease phenomenon.

CFP: THE HUMAN HEALTH PORTION

History

Ciguatera fish poisoning (CFP) has affected coastal populations for centuries. The term, first coined in the Caribbean by the Spanish in the 17th century, is derived from "cigua," the name used by indigenous populations in the Spanish Antilles for a marine turban snail (Bagnis, 1994; Banner, 1976; Halstead, 1967; Helfrich and Banner, 1963). In the Pacific, ships' records from the early 1600s in the New Hebrides, and Captain James Cook in 1774 from New Caledonia (Helfrich, 1964) indicated that ships' crews displayed clinical symptoms that today are associated with CFP. Evidence exists for CFP in all tropical ocean areas of the world. Outbreaks of fish poisoning occurred throughout the Pacific before, during, and after World War II and became quite a serious problem for military troops stationed in various island locales where CFP was endemic (Hokama and Yoshikawa-Ebesu, 2001).

In part, it was those experiences that focused scientific attention on CFP by Randall (1958), Banner *et al.* (1960), and Cooper (1964). Randall (1958) hypothesized that the toxin, ciguatoxin (CTX), was derived from food consumed by the fish, and Helfrich (1964) and Banner (1974, 1976) showed that toxicity could be transferred to/induced in non-toxic fish via consumption of toxic fish. These contributions evolved into the "food web concept" for CFP that is held to this day. The food chain hypothesis held that fish acquired the CTX toxin in their diet, that this acquisition occurred without marked detrimental effects to the fish, and that the ingested toxin was stored in the body without degradation (Randall, 1958).

For decades, there has been especially rich and active ecological research regarding CFP in Australia and French Polynesia (e.g., Bagnis 1977; Bagnis *et al.*, 1985; Chinain *et al.*, 1999b; Holmes *et al.*, 1991; Holmes and Lewis, 1994; Lewis *et al.*, 1994; Yasumoto *et al.*, 1977, 1979b) and in the Caribbean (e.g., Ballantine *et al.*, 1985; Bomber *et al.*, 1988; Carlson 1984; Pottier *et al.*, 2002; Tindall *et al.*, 1984). Scheuer and colleagues (Nukina *et al.*, 1984, 1986; Scheuer *et al.*, 1967; Tacnibana *et al.*, 1980, 1987) first determined the molecular structure of CTX derived from moray eels. The etiology of CFP was advanced when work in the Gambier Islands of French Polynesia by Yasumoto *et al.* (1977, 1979a, 1980) revealed that the guts of toxic (herbivorous) fish contained significant numbers of a dinoflagellate that came to be known as *Gambierdiscus toxicus* (Adachi

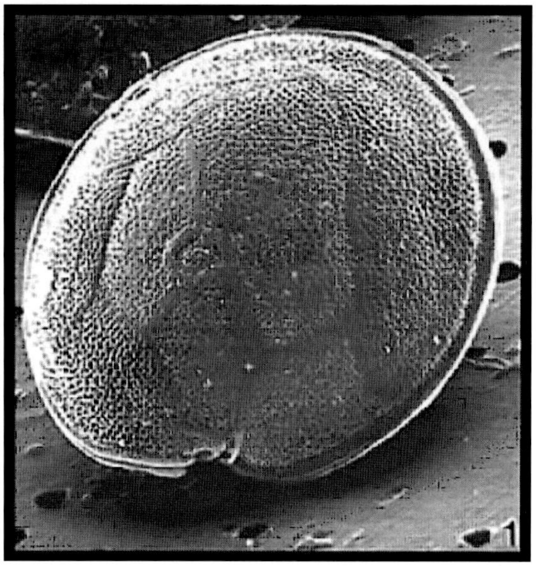

FIGURE 14-1. *Gambierdiscus toxicus.*

and Fukuyo, 1979). Numerous contributions to CTX detection methods followed (Hokama *et al.*, 1977, 1985, 1987, 1998; Lewis *et al.*, 1999a and 1999b; Lewis and Jones, 1997; Pottier *et al.*, 2002). Legrand *et al.* (1992) chemically characterized ciguatoxins from a variety of different fish species. Following the evolution of robust chemical methodologies, significant advances were made toward synthesis of the CTX molecule (Inoue, 2004; Inoue *et al.*, 2006).

Regional Occurrence and Underreporting

There are at least 50,000 reported cases of CFP per year (Lewis, 2001; Quod and Turquer, 1996), but because of the high degree of misdiagnosis and underreporting, it is estimated that the actual frequency of CFP cases is closer to 500,000 per year (Arena *et al.*, 2004; Pearn, 2001). Because of this inaccuracy, it is not possible to ascertain whether CFP incidence is increasing over time. Increased awareness of CFP has given rise to improved identification and increased reporting within both the general population and the medical profession in certain endemic areas.

A CDC report for a period in the 1970s indicted that reported CFP incidences accounted for 25% of all food-borne outbreaks, which was five times the reported incidence for paralytic shellfish poisoning and neurological shellfish poisoning combined (Steidinger and Baden, 1985). Various studies for the Caribbean and Pacific have indicated frequencies of reported CFP incidence ranging from 0.1% to 3% of the population per year (Vernoux, 1988). It is believed that reported values must be increased by a factor of at least 2 to 20 × to account for underreporting (Arena *et al.*, 2004; Fleming *et al.*, 2000; Lawrence *et al.*, 1980). The CDC estimates that in the United States, only 2% to 10% of CFP cases are actually reported. It is estimated that 3% of the population in the U.S. Virgin Islands and French West Indies suffer CFP annually; for St. Thomas and Puerto Rico, these figures were estimated to be 4.4% and 7%, respectively (Escalona de Motta and De la Noceda, 1985). An Australian study estimated that 1.8% to 2.5% of the total population from two Pacific island communities experienced a CFP intoxication at some time in their life (Gillespie *et al.*, 1985).

One study presented a classic case of CFP symptoms to 36 physicians in Florida, an endemic area for CFP because of local and regional coral reefs. About 68% of the physicians correctly diagnosed CFP, but only about 17% correctly recommended mannitol therapy (discussed later) as the acute treatment of choice. Importantly, only about 47% of the physicians knew that CFP was a reportable disease. This study illustrates that CFP is an underdiagnosed, inadequately treated, and underreported disease especially among U.S.-born and U.S.-trained physicians even in endemic areas (McKee *et al.*, 2001). In Mexico, CFP is often misdiagnosed as some form of food poisoning from traditional sanitary microbial contamination, illustrating another way that CFP is a neglected disease.

Symptoms and Effects

Arena *et al.* (2004) summarized the frequency of clinical symptoms of CFP at the time of diagnosis from a number of works. CFP symptomatology has been addressed in considerable detail in numerous publications (Bagnis, 1968, 1990, Bagnis *et al.*, 1979; Calvert, 1990; Nicholson and Lewis, 2006; Steidinger and Baden, 1985; Withers, 1982; Yasumoto *et al.*, 1984). Because of similarities in symptomatology, the diagnostic differentiation of CFP from neurological shellfish poisoning or paralytic shellfish poisoning commonly relies simply on the history of fish versus shellfish consumption as the principal guide. CFP also has symptoms in common with type E botulism, scombroid poisoning, eosinophilic meningitis, and organophosphate pesticide poisoning.

CTX produces gastrointestinal, neurological, and cardiovascular symptoms. These normally develop within 12 to 24 hours of eating the contaminated fish. Gastrointestinal effects usually disappear within 1 to 4 days. The normal progression of symptoms is (1) gastrointestinal symptoms (e.g., diarrhea, abdominal pain, nausea, and vomiting) followed by (2) neurological symptoms (e.g., numbness and tingling of hands and feet, dizziness, altered hot/cold perception, muscle aches, low heart rates and low blood pressure). A pathoneumonic symptom involves the neurological paresthesia of the reversal of hot/cold sensation. Symptoms may persist in some form for weeks, months, or even years (Arena *et al.*, 2004; Benoit *et al.*, 2000; Cameron *et al.*, 1991; Kodama, 1990). Generally, feelings of weakness last

~1 week, and neurosensory manifestations or paresthesias (e.g., muscle and joint aches, itching, tingling extremities, and thermal reversals) commonly represent the most prolonged discomfort. Fortunately, death is rare (i.e., <0.1%) and is most commonly the result of respiratory failure resulting from cardiovascular shock induced by severe dehydration during the initial onset of effects (Bagnis, 1993; Withers, 1982). In the Pacific, this is normally associated with eating the most toxic portions of the fish (e.g., liver, viscera, roe).

Chronic CFP is often misdiagnosed as a psychiatric disorder of general malaise, depression, headaches, and peculiar feelings in the extremities (Chan and Wang, 1993). Interestingly, it appears that acutely gastrointestinal symptoms dominate CFP cases in the Caribbean, whereas neurological symptoms seem to be more prominent acutely in CFP cases in the Pacific (Nicholson and Lewis, 2006). It has been speculated that this difference is the result of different toxin composition among the Caribbean and Pacific *G. toxicus* strains or differences in the species of *Gambierdiscus* contributing to the toxicology.

An interesting feature is that CFP intoxication does not confer any immunity in its victims; on the contrary, it frequently results in heightened sensitivity to CTX or fish products generally. It has been suggested (Nicholson and Lewis, 2006) that, because ciguatoxins may be sequestered in adipose tissue, the reoccurrence of symptoms may be exacerbated during periods of physical stress that catabolize fats (e.g., exercise or weight loss) or ingestion of alcohol and high-protein foods (such as fish, chicken, or nuts, and foods with caffeine).

One well-documented case illustrates the progression of CFP symptoms. Thirty French citizens who consumed portions of a 10– to 12–pound barracuda (muscle tissue only) in Mexico developed gastrointestinal disorders in 4 to 6 hours and were given antidiarrheal drugs before their flight home. During the flight, neurosensory disorders with diffuse pain developed, and on day 5 they were admitted to a poison treatment center. To make an objective assessment of the severity of the clinical features, the center scored each victim upon admission using a quantification method applied in Tahiti that gives points for each of the following: paraesthesiae, pain, tiredness, and cardiovascular and gastrointestinal signs (De Haro, 1997). This study concluded that the severity of intoxication was proportional to the amount of fish eaten.

Pharmacology

The complex suite of symptoms associated with CFP is caused by the ciguatoxin's capability to increase Na^+ permeability through the Na channels that open at normal resting membrane potentials; this enhanced excitability of the membrane in turn affects Na^+–Ca^{+2} exchange and mobilizes intracellular Ca^{+2}. The primary receptor site of the CTX action is the fifth domain of the Na^+ channel where it causes increased sodium ion permeability and depolarization of the resting membrane; the intense depolarization of nerve cells is believed to cause the suite of sensory discomfort symptoms associated with CFP (Arias, 2006; Cameron *et al.*, 1991; Hokama and Yoshikawa-Ebesu, 2001).

Prevention

The heat stability of the ciguatoxin molecule means that normal/proper food preparation (i.e., heating) will not mitigate or eliminate toxicity from fish tissue. This property, together with the fact that exceptionally small (i.e., picogram levels) amounts of CTX toxin can confer toxicity, further reduces any reasonable likelihood for reliable detection/prevention during fish handling/cooking. Therefore, there is no way to "decontaminate" a fish having CTX. There is no way to reliably distinguish a contaminated fish by smell or appearance. In addition to the amount of contaminated fish consumed, intoxication seems to be most closely associated with the *size* and *type* of fish that was consumed. Generally, people are advised to avoid the viscera of reef fish or the consumption of large amounts of large predatory fish (discussed later). Also, many fish are misidentified, they have little or no traceability, or there is a lack of knowledge of the CFP-propensity of the particular fish at the time of their sale. Improvements in these areas are needed to aid prevention.

Tropical islanders have used numerous folk techniques in the attempt to determine if candidate fish for consumption are ciguatoxic: (1) cooking the fish with a silver object to see if discoloration results, (2) seeing if the fish repels flies or ants, (3) avoiding sliced fish that fails to reflect a rainbow when held to the Sun, and (4) rubbing one's gums with the liver to see if a tingling results (Lobel, 1979). Giving a sample of the questionable fish to a household pet or even an elderly relative as a bioassay is still practiced in some island communities.

Clinical Responses/Treatments

By far the most effective therapy for CFP has been mannitol infusions, administered at 0.5 to 1.0 g/kg body weight and given within 48 hours of ingestion of toxic fish (Palafox *et al.*, 1988). The mechanism of mannitol's effect is partly understood. CTX opens sodium channels at the nodes of Ranvier in myelinated neurons, causing the cleft to swell, eliminating saltatory conductance. Mannitol relieves the swelling at the nodes, presumably by an osmotic effect. IV mannitol is the only CFP treatment evaluated by a randomized blinded trial (Bagnis *et al.*, 1992; Schnorf *et al.*, 2002). Although one of these randomized trials (Schnorf *et al.*, 2002) did not find mannitol to be superior to saline, that study included patients who received mannitol up to 28 days after symptom onset and not within the recommended 2 to 3 days from consumption, which may have prevented the identification of benefits in patients who received it within the recommended period (Blythe *et al.*, 1992; Friedman *et al.*, in press).

Many other treatments have been tried with variable results and without appropriate randomized trial testing (Watters, 2007). Other clinical treatments, targeting various elements of the long list of CFP symptoms, have involved a variety of agents (e.g., vitamins, antihistamines, anticholinesterases, steroids, and antidepressants). Treatments frequently focus on the most demonstrative symptoms and thus may involve injections of steroids, nonrespiratory depressants, antihistamines, antidiarrhetics, and vitamins. Gut emptying and decontamination with charcoal have also used, but ongoing vomiting and diarrhea often prevent this. Atropine has been used for bradycardia, and dopamine or calcium gluconate for shock.

Folk Remedies

Folk remedies involving extracts of *Argusia argentea* leaves or *Davalliea solida* rhizomes have been used traditionally in New Caledonia (Benoit, 2000). Baden *et al.* (1995) and others (Blythe *et al.*, 2001; Fleming *et al.*, 2004) reported that there are 64 different local remedies including medicinal teas that are used in both the Indo-Pacific and West Indies regions. Folk remedies frequently involve inducing vomiting; it is generally accepted that vomiting and diarrhea should not be suppressed because it aids the body in voiding the poison.

Hazard Management

CFP ranks as a significant hazard from seafood consumption. A risk assessment tool of 10 seafood hazard/product combinations using data from Australia showed CFP to be among the most hazardous incidents. Examples of hazard/product pairs with a lower ranking (i.e., less hazardous) included mercury poisoning, *Clostridium botulinum* in canned or cold-smoked fish, parasites in sushi/sashimi, viruses in shellfish from uncontaminated waters, enteric bacteria in imported cooked shrimp, *Vibrio parahaemolyticus* or *V. cholerae* in cooked prawns, *Listeria monocytogenes* in cold-smoked seafoods, scombroid, and *V. vulnificus* in oysters. The only elements deemed more hazardous than ciguatera were *L. monocytogenes* in susceptible populations and enteric bacteria in imported cooked shrimp eaten by vulnerable consumers (Sumner and Ross, 2002).

As noted previously, CTX in fishery products is odorless, tasteless, and generally undetectable by simple chemical tests. Assurances that susceptible foods are safe to eat will likely come from the marketplace in the form of screening, isolation of high-risk products, and (where feasible) prediction of potentially hazardous harvesting areas. Effective screening methods must be easy to use/interpret; able to test large numbers of samples quickly, accurate, low-cost, readily available; and capable of identifying the causative CFP toxins. CFP incidence does not reflect on the quality of food handling, storage, preparation, or procurement; no known method of cooking, boiling, baking, or frying can destroy the toxin (Bagnis, 1993).

Risk Assessment

A study in French Polynesia simultaneously evaluated three indices of CFP hazard. These included CFP cases per 1000 residents (CIR), percentage of toxic fish in a group (PCI), and the abundance of *G. toxicus* per gram algae (GTD). During the 20–year study period 24,000 cases were reported (CI = 8%), caused by ~100 fish species. Results indicated that *G. toxicus* abundance data *in situ* and the respective ciguatoxicity of significant samples of microphagous herbivorous fishes and ichthyophagous carnivorous fish provided a reasonable estimate of the CFP potential at any given time of the year (Bagnis *et al.*, 1985). In Hawaii, a general correlation is also apparent seasonally between the frequency of reported CFP incidents and the abundance of *Gambierdiscus* spp.; the highest values of both being observed during the warmest period at the end of the summer.

Fish to Avoid

The fish most commonly involved in CFP incidents include barracudas (*Sphyraena* spp.), groupers (*Epinephelus* spp.), jacks (*Caranx* spp.), snappers (*Lutjanus* spp.), surgeonfish (*Ctenochaetus* spp.), parrot fish (*Scarus* spp.), and moray eels such as *Gymnothorax* spp. (Ebesu, 1998; Lewis *et al.*, 1999a; Tosteson *et al.*, 1998; Vernoux and Lejeune, 1993). In the Pacific, the detritivorous grazer, *Ctenochaetus striatus* (a surgeonfish), is frequently ciguatoxic and is thought to be a key vector sending CTX up the food chain. Surgeonfish and parrotfish are dominant families by weight on many coral reefs and the most common prey of large piscivores. High levels of CTX and other Gambierdiscus-associated toxins in biodetritus are believed to account for frequency of ciguatoxicity in *Ctenochaetus striatus*, and in some areas, mullet (*Mugil* spp.) because both have detritivorous grazing behaviors (Steidinger and Baden, 1985). Studies of seasonality of CFP in different regions have frequently shown temporal variability, but often at different times of the year in different locations.

A study of toxic barracuda in the Caribbean concluded that CTX was accumulated/retained in barracuda tissue for extended periods of time. Fish flesh toxicity was shown to be inducible. Snapper fed CTX fish became toxic in six months and retained potency for 30 months (Banner *et al.*, 1966; Helfrich and Banner, 1968). Toxin chemistry results suggest that CTX precursors from dinoflagellates may be oxidatively biotransformed to numerous congeners within herbivorous and carnivorous fish (Legrand, 1998).

Diagnostic Kits

Development of simplified procedures for the assessment of CTX and related polyethers in ciguateric fish has been actively pursued. The first tested was the radioimmunoassay (RIA) reported by Hokama *et al.* (1977) in which the antibody was radiolabeled with ^{125}I. Although the RIA was effective, its complexity and need for a radioactivity counter demonstrated that a simpler, more cost-effective method was needed (Hokama and Yoshikawa-Ebesu, 2001). Next, Hokama's group developed the stick-enzyme immunoassay (S-EIA) using sheep anti-CTX and MAb-CTX coupled to horseradish peroxidase (Hokama, 1985). A UV monitor was used to detect when the MAb-CTX-horseradish complex bound to the putative CTX molecule. Herein lies a nontrivial diagnostic challenge for such kits (i.e., the fact that multiple CTX congeners having different molecular structures may both be present and may present similar symptomology). Adding to the diagnostic challenge and ranging degrees of efficacy from various therapies, there may also be a number of other toxins (such as okadaic acid, derived from *Ostreopsis* spp.) associated with ciguateric fish. Although this method was used extensively through the mid-1980s and 1990s, the presence of ciguatoxin still could not be directly detected (a UV detector was needed).

The latest developments have focused on the membrane immunobead assay (MIA). The MIA utilizes a plastic stick for the solid phase and a synthetic hydrophobic membrane laminated onto one end that serves as the solid-phase receptor for the binding of methanol-extracted CTX or its related polyethers from fish using detection via colored polystyrene beads coated with MAb- to CTX (Hokama *et al.*, 1998). The MIA was deemed successful enough to merit commercial production in the late 1990s through Oceanit Test Systems as the Cigua-Check Test Kit. The level of satisfaction with clarity, accuracy, or cost of commercially available CTX kits is far from universal, however (Dickey *et al.*, 1994).

Societal Impacts

Considerable negative impacts are associated with CFP. These include (1) the obvious public health impacts of the intoxications, (2) constraints on an important protein source for sub/tropical island communities globally, (3) constraints to the development of small-scale tropical fish resources for export, and (4) potentially immobilizing hazards to personnel involved in operations (military, construction, etc.) in tropical areas. In response to endemic ciguatera from the consumption of local fish, people in the Marshall Islands reportedly changed their eating preferences to include more imported and canned fish because of increased CFP incidence in locally caught fish (Tebana, 1992). Negative ecological impacts of various forms of fish and shellfish poisonings have included mass deaths of shellfish, seagulls, dolphins, and turtles (Ochoa *et al.*, 1996).

CFP is believed to be the leading cause of nonbacterial food poisoning in the United States, and at an estimated incidence rate of 500,000 cases annually it is the most frequent seafood-toxin illness worldwide (Arena *et al.*, 2004; Lewis, 2001; Pearn, 2001; Quod and Turquet, 1996). For example, a study of fish poisonings in a location in the Indian Ocean concluded that ~80% were caused by CTX (Quod and Turquor, 1996). Studies have variously estimated the approximate cost to be \$5000 to \$9000 per CFP incident, particularly in nonendemic areas (Arena *et al.*, 2004; Pohland *et al.*, 1990). It has also been suggested that by any latitudinal extension of warmer water (e.g., because of global warming) will similarly increase the incidence of CFP (De Sylva, 1992).

CFP: THE OCEAN PRODUCTION PORTION

It is now generally understood that the dinoflagellate *Gambierdiscus toxicus*, the putative producer of the toxins associated with *Gamberdiscus* (also known as "gambiertoxins") that are transformed to CTX, grows epiphytically on macroalgae together with a complex consortium of other symbiotic microflora. This microbial biomass is ingested by herbivorous fish, which accumulate the toxin and in turn are grazed by carnivorous fish that bioamplify and chemically modify the toxin. The simplicity of this schema belies a great deal of uncertainty about virtually every facet of the ecological or biochemical processes that lead to the manifestation of CFP in humans (Fig. 14-2). Difficulties in resolving unknowns in the CFP process arise from the ephemeral character of CTX production by *Gambierdiscus* spp. and the need for extremely robust analytical methods because of the CTX molecular structure, its extreme toxicity, and its common association with other phytotoxins. What follows is our best understanding of the currently understood CFP paradigm.

Ciguatoxin

CTX is a polar, lipid-soluble, highly oxygenated polyether molecule. It is an oxygenated long chain fatty acid with cyclic oxoether linkages (Fig. 14-3). Derived from polyketide biosynthetic pathways, CTX is soluble in methanol, ethanol, acetone, and 2–propanol, but not benzene or

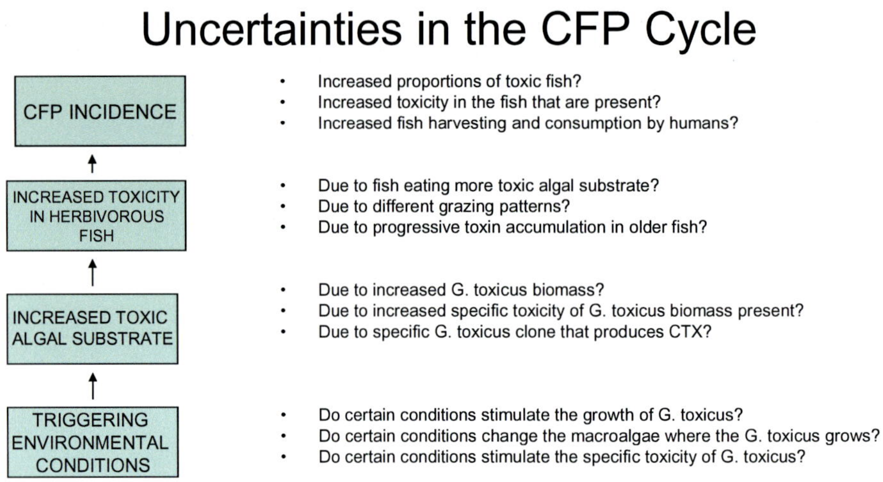

FIGURE 14-2. Uncertainties in the CFP Cycle (after Lewis, 2001).

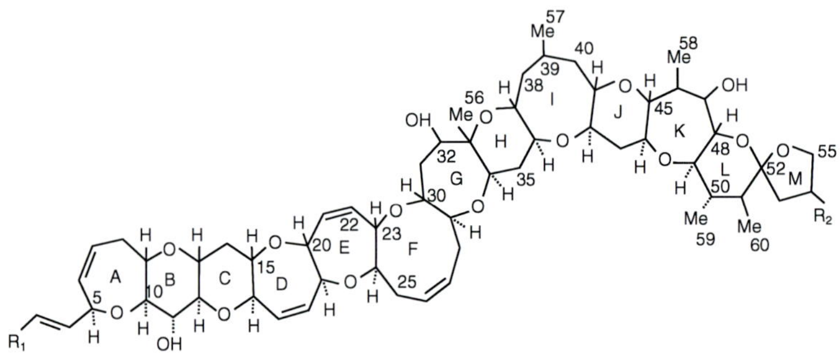

CTX R_1 = $HOCH_2CH(OH)-$ (from moray eel liver)
R_2 = OH

CTX4B R_1 = $CH_2{=}CH-$ (from Gambierdiscus toxicus)
R_2 = H

FIGURE 14-3. Ciguatoxin molecule.

water. Toxin isolated from different regions or organisms has been shown to exist in a number of forms that have different molecular weight, chemical structure, and toxicity. Recognition of these "congeners" necessitated new nomenclature, and in the more recent literature an I, P, or C prefix refers to the ocean of origin (i.e., Indian, Pacific or Caribbean) and a number following refers to a specific congener/form of CTX. Thus, P-CTX-1 would refer to CTX congener 1 isolated from the Pacific Ocean. The advent of more robust analytical methods has lead to important new details emerging regarding the chemical structure of CTX and its congeners. For example, the molecular weight of various P-CTX congeners from moray eels were found to range from 1023 to 1112 amu (Yasumoto and Murata, 1993), and P-CTX in fish from Hawaii were recently found to be 993 amu (P. Moeller, personal communication). Ciguatoxin specifically may also be present with another lipid-soluble toxin (scaritoxin) and the larger (3422 amu), water soluble maitotoxin, named after the Tahitian name for *Ctenochaetus striatus*. The molecular formulae and weights of these and other phytotoxins are summarized in Pohland *et al.* (1990).

Ciguatoxin and maitotoxin are two of the most lethal natural substances known. In mice CTX and MTX are lethal at injected (i.p.) levels of 0.45 ug/kg and 0.13 ug/kg (Anderson and Lobel 1987; Bagnis *et al.*, 1987; Tachibana, 1980; Yasumoto, 1985). Intake of picogram amounts of CTX can cause illness in human adults. Because of the prevalent use of bioassays in lieu of chemistry, the MU (i.e., mouse unit) is frequently encountered in the literature. A MU is defined as either the amount of toxin required to kill a 20–gm mouse in 15 minutes, or the lethal dose for 50% of the population (LD50) for 24 hours (Steidinger and Baden, 1985).

CTX Assay Methods

Analytical difficulties arising from the complexity of the CTX molecule cannot be overemphasized. Interest circumventing the complex analytical chemistry required has led to the development/application of a number of bioassays. None of these assays has proven to be completely satisfactory for routine laboratory testing, field samples, clinical studies, or specific investigations because of individual limitations (Luckas, 2000).

The mouse bioassay is an historical method for determining algal toxins, though limited by low sensitivity and increasing societal aversion to animal testing. A number of other assay organisms have been used in CTX bioassay trials over the years; these have included mongoose, rat, Artemia, cell bioassays, Diptera larvae, shrimp, rabbits, guinea pigs, cats, chicks, and mosquitoes (Labrousse, 1991; Calvert, 1990, and references cited within). A MAb (monocolonal antibody)-based assay has been used as an alternative assay to the mouse bioassay for routine monitoring (Hokama *et al.*, 1998), and radioimmune (RIA) or enzyme linked immunosorbent (ELISA) assays are used for detecting CTX in fish (Hokama and Yoshikawa-Ebesu, 2001; Labrousse, 1991). Cell-based assays for CTX using mouse neuroblastoma cells have proven to be especially functional and sensitive since development of refinements that enable the assessment of sodium-channel activity specifically (Manger *et al.*, 1995). Currently, the best methodology for CTX assessment involves LC/MS/MS/NMR technologies. Though significantly expensive to purchase, maintain, and operate, this suite of applications is by far the most sensitive and informative. The National Science Foundation (NSF) and National Institute of Environmental Health Sciences (NIEHS) Center for Oceans and Human Health at the University of Hawaii uses a two-tiered CTX assessment strategy. The Na-channel neuroblastoma assay (tier 1) is used to screen samples to confirm sodium channel sensitivity, and positive samples are then analyzed via LC/MS/MS/NMR procedures (personal communication, authors).

Gambierdiscus Toxicus

One of six species of *Gambierdiscus*, the dinoflagellate, *G. toxicus*, is roughly 93 × 83 um (Chinain *et al.*, 1999a). It is named after the Gambier Islands in French Polynesia (Adachi and Fukuyo, 1979) where it was found in high abundance and association with CFP (Yasumoto *et al.*, 1977). There is considerable uncertainty about the growth rates or relative toxicity of clones, strains, or Gambierdiscus species from different locales (e.g., Holmes *et al.*, 1991; Legrand, 1998), but several Gambierdiscus species are reported to produce gambiertoxins/ciguatoxins (Chinain *et al.*, 1999a). Examination of the 5.8S + ITS rDNA and the LSU rDNA D8–D10 regions in eleven Polynesian isolates proved to be a reliable biogeographic identifier for the various isolates (Babinchak *et al.*, 1996). Isozyme analysis of intraspecific variation among 19 isolates of G. toxicus from French Polynesia, New Caledonia, and French West Indies revealed biochemically distinct strains. Growth rates and cell sizes varied considerably among the clones; however, no relationship was found between the electrophoretic profiles of these isolates and their capacity to produce CTX compounds (Sako *et al.*, 1996). Parsons (personal communiation, 2006) isolated what appear to be two new *Gambierdiscus* species from Hawaiian waters. Other dinoflagellate genera frequently found in the epiphytic assemblages with *Gambierdiscus* spp. include *Ostreopsis, Coolia, Amphidinium, and Prorocentrum* (Parsons and Preskitt, in press; Steidinger, 1993).

In Vitro Growth

Like other dinoflagellates, *Gambierdiscus* spp. are sensitive to excessive agitation, abrupt changes in temperature, salinity, light, and high silicate and metal levels, particularly copper (Bomber *et al.*, 1988; Durand-Clement, 1986; Guillard and Keller, 1984). The modified K-media (Keller and Guillard, 1985) and Harrison *et al.*'s (1980) ESNW are widely used; the latter has higher levels of chelation preventing Fe-deficiency in the medium. Culture is best at 32 to 35 ppt salinity, and growth is optimal within 26 to 30°C though possible from 19.5 to 34°C. Strong evidence points to *G. toxicus* as a shade-loving species (Villareal and Morton, 2002); it prefers low light (<10% surface intensity) and a 12 : 12 to 14 : 10 light to day (L : D) cycle. *G. toxicus* is a slow growing species (i.e., maximum growth rates ~0.3 to 0.5/d; Bomber *et al.*, 1988).

G. toxicus produces mucoid sheaths that aid their epiphytic life history but complicate *in vitro* culture. The degree of toxicity has frequently been shown to be variable. Toxicity has been seen to increase with increasing temperature and light, and in culture, toxicity has frequently been observed to be highest in mid to late log phase. There is some evidence for an inverse relationship between toxicity and chlorophyll content. Because of the slow growth and the fact that CTX production may well be out-of-phase from optimal growth conditions, it may take considerable time (e.g., weeks) for the actual toxin production to be normalized to a given set of environmental conditions under evaluation. Large-scale culture has been performed by several investigators (Babinchak *et al.*, 1994), and *G. toxicus* has continued to produce CTX in culture. It has also been regularly reported to stop producing toxin (presumably CTX) under *in vitro* culture conditions.

It is unclear why dinoflagellates produce potent toxins, how these secondary metabolites are synthesized, or why *G. toxicus* may cease production of CTX when grown *in vitro*. Although this may in part reflect the limitations of the chemistry employed, it may also be that the provision of optimal growth conditions *in vitro* eliminates the stress/austerity in the growth environment that triggers toxin production. The challenges of maintaining *Gambierdiscus* spp. cultures, the limited concentrations of these targeted secondary metabolites, their ephemeral production *in vitro,* and possibly the role of viable but noncultivable bacteria have limited significant CTX production from culture facilities.

In Vivo Growth

Gambierdiscus spp. are found in coastal tropical and subtropical waters within the 35°N to 35°S band of all oceans. The preferred habitat is typically not a high-energy area but rather a quiet, high-salinity leeward area with minimum freshwater runoff and at depths where cells can chromatically adapt to low light levels (Steidinger and Baden, 1985; Taylor, 1985). Application of pulse amplitude modulated (PAM) fluorometry to *in vivo* cells was used to show that *G. toxicus* exploits the three-dimensional structure of the macroalgal host to achieve shading necessary to permit these high-light intolerant organisms to thrive in shallow, high-light tropical coastal waters (Villareal and Morton, 2002). *G. toxicus* shows strong association with macroalgae but may also be found in detritus or on dead coral surfaces. Normal cell densities range from 1 to 50 cells/g algae; densities >1000 cells/g algae are considered "high," and the maximum abundance ever recorded (~450,000 cells/g algae) was off Gambier Island (Bagnis *et al.*, 1985). It is far from clear whether increased amounts of toxic fish that may occur from time to time are the result of increased *G. toxicus* biomass or increased specific toxicity of the biomass (i.e., higher CTX content or composition).

Among the hypothesized causes for the observed spatial or temporal variations in toxicity are variations in *G. toxicus* abundance, specific toxicity, clonal participants, or some combination of these. There exists some direct and indirect evidence for each. It is not known whether the increased toxicity in stationary phase is the result of increased toxin production, increased toxin flux to the food web, or merely a conversion of a toxin congener to more potent congeners. Studies in the Virgin Islands showed no seasonality in either the number of CFP cases (Morris *et al.*, 1982) or the abundance of *G. toxicus* (Bomber, 1985), whereas coincident seasonality in both CFP cases and *G. toxicus* abundance appear evident in Hawaii (Nakata, personal communication, 2006; Parsons, personal communication, 2006). In the Florida Keys, seasonal minima in *G. toxicus* abundance correlated with minimum winter water temperatures (Bomber *et al.*, 1988). Weekly sampling off Tahiti from 1993 to 1997 showed cell densities frequently >1000 cells/g, and attaining maximum abundance at the beginning and end of the hot season. A noticeable increase in peak densities (~5000 cells/g) was preceded by unusually high water temperatures severe enough to cause a serious coral-bleaching episode (Chinain *et al.*, 1999b). No correlation was found between the toxicity of these *G. toxicus* blooms and their biomass or the seasonal temperature patterns. In an impressively extensive database from 1993–2001, Chateau-Degat *et al.* (2005) correlated *G. toxicus* densities, seawater temperatures, and reported CFP cases, and concluded that peak cell densities lagged peak temperatures by 13 to 17 months, and reported CFP cases lagged peak cell densities by 3 months, at least in French Polynesia.

Early CFP incidents in environmentally disturbed areas led to speculation that anthropogenic disturbances (shipwrecks, dredging, etc.) or natural disturbances (e.g., cyclones, hurricanes, tsunamis, coral bleaching) triggered CTX incidence (Bagnis, 1994; Randall, 1958; Steidinger and Baden, 1985). Despite a low correlation rate between such disturbances and the onset of CFP, this remains a common perception, and for that reason is addressed here (Banner, 1974; Kohler and Kohler, 1992). The rationale for this causation theory is that disturbances that create new surfaces create new substrate for colonization by algal turf that could theoretically provide additional macroalgal colonization, and thus associated epiphytes, which may grow on such algae (i.e., Randall's "new surface theory") (Quod *et al.*, 2001; Randall, 1958). In general, available analyses of areas affected by natural or manmade perturbations have not shown casual linkages to changes in CFP incidence that could be isolated from natural, periodic or stochastic processes (Brusle *et al.*, 1998).

An intriguing paradigm introduced to the literature (Cruz-Rivera and Villareal, 2006) focuses attention on survival strategies of the macroalgal symbionts of *Gambierdiscus* spp. to explain the persistence of these slow growing species and the stochastic nature of increases in cell abundances or CFP incidence. Various algal hosts may tolerate/manage the high rates of herbivory that are characteristic on reef environments by growing fast, having poor nutritional quality, or using chemical or structural defenses. The macroalgal hosts exhibiting these various strategies likewise represent drastically different fluxes of gambiertoxins into herbivorous grazers and subsequently to the food web. Similarly, the relatively inedible algal species are seen to serve as refuges for the slow-growing *Gambierdiscus* spp., which may well attain elevated densities, but serve primarily to reseed the algae that are more heavily grazed. In this way the CTX flux potential to the food web is enhanced. By requiring that the ecology of both *Gambierdiscus* spp. and their host algae are essential to causing patterns of CFP, this paradigm provides an explanation for the lack of correlation between *Gambierdiscus* spp. abundances, frequencies of CTX-positive fish, and CFP in humans.

Symbiotic Relationships

G. toxicus have been associated with foliose or filamentous red, brown, green, and even blue-green algae, especially those genera with structural interstices (e.g., *Ceramium, Chondria, Wrangelia, Gelidium, Amphiroa, Sargassum, Padina, Turbinaria, Caulerpa, Pseudobryopsis, Hypnea, Gracilaria, Jania, Halimeda, Colpomenia, Spyridea, Cladophora sp*., benthic detritus, *Heterosiphonia,* and *Acanthophora*) (Bomber *et al.*, 1988; Carlson and Tindall, 1988; Cruz-Rivera and Villareal, 2006; Gillespie *et al.*, 1985). There is some evidence that various macroalgae may supply important growth factors (Carlson *et al.*, 1984; Nakahara *et al.*, 1996; Steidinger and Baden, 1985). Nonetheless, the occurrence of *G. toxicus* on many macroalgae species suggests opportunism in regard to its macroalgal substrate and a low likelihood that a particular algal metabolite regulates its abundance.

Macroalgae are leaky and release a variety of substances from polysaccharides to complex enzymes, and the derivation of chelating agents from the macroalgae has been suggested as a strategy of *G. toxicus* to mitigate its hypersensitivity to Cu and other metals (Bomber, 1987; Bomber *et al.*, 1988). The apparent benefits to the epiphytic *Gambierdiscus* that are received from the association with macrobiota are location fixation, higher nutrient availability, shading from direct sunlight, protection from turbulence, and access to organic compounds within the thallisphere (Villareal and Morton, 2002). For slowly growing organisms that cannot compete via reproductive rates and are relatively intolerant of variations in physico-chemical conditions, these benefits from the symbiosis seem to be significant. Although allelopathic functions in the thallisphere that are CTX-specific are largely unknown, the capability for extracts of epiphytic dinoflagellates to suppress growth of other dinoflagellates (Fistarol *et al.*, 2004; Kubanek *et al.*, 2005; Sakamoto *et al.*, 1996) has led to the suggestion that the diversity of toxins synthesized by dinoflagellates plays a role in competition among algal species of the thallisphere.

The complex chemical structure of CTX suggests that multiple biochemical steps are required for its synthesis (Murata *et al.*, 1989; Plumley, 1997; Snyder *et al.*, 2003). This, together with the fact that the CTX molecule resembles a polyketide, a class of secondary metabolites known to be produced by other microorganisms (e.g., bacteria, filamentous fungi), has led to speculation that other members of the symbiotic microbial consortium may participate in the production of CTX. Some species of dinoflagellates have been shown to contain bacteria in the cytoplasm or nucleus (Ashton *et al.*, 2003), and it has been suggested that dinoflagellate toxins may originate from bacteria (or viral recombinant RNA). In bacteria, the polyethers often occur as a complex of closely related compounds, as is the case with dinoflagellate polyethers such as CTX. Efforts to evaluate toxin production by bacteria associated with dinoflagellate cultures have been done, but the meaning of negative data is not clear since it could simply mean that the proper environment for toxigenesis was not provided. This hypothesis has been supported by knowledge that a very potent nonproteinaceous marine toxin (palytoxin) is actually produced by an endocellular bacterium associated with zoanthid polyps (Moore *et al.*, 1982; Sakami *et al.*, 1999; Steidinger and Baden, 1985). The fact that only a fraction of marine bacteria are cultivable and that CTX production has been seen to cease over time in dinoflagellate cultures has been taken to support this microbial symbiont hypothesis. However, the failure of TEM analysis to show intracellular bacteria in a toxic dinoflagellate strain, or good correlation between toxicity and presence of intracellular bacteria in multiple samples, leaves this intriguing bacteria theory lacking empirical support from well-designed experiments.

Though not prominent in the literature, it is not unreasonable to suspect endocellular viruses of participating/inducing the production of toxic secondary metabolites in toxic dinoflagellates (Plumley, 1997). Only in 1977 did studies show that pelagic bacteria were abundant, and a decade later that viral densities 10X bacterial densities were common (Azam and Worden, 2004). Viral studies have shown species specificity, host density-dependence, and a variety of behaviors potentially influential to the CTX dynamics in the thallisphere (Brussaard *et al.*, 2004; Tarutani *et al.*, 2000). At present, there are only a few phytoplankton viruses in culture, and most of these infect HAB species. The primitive origin of both dinoflagellates and viruses implies the potential for a coevolution relevant to CTX dynamics. Just as viruses aid in the mitigation of pelagic bloom "hot spots," it has been speculated that they may play a role in controlling hot spots in the sessile, epiphytic realm. Studies where viruses were shown to reduce the density of bloom-forming phytoplankton have also shown that the surviving cells were resistant to most of the virus clones. This implies a change in the properties of the dominant cells pre– and post–hot spot in time. Thus, viruses' capability to quantitatively and qualitatively affect pelagic phytoplankton populations has suggested that they may similarly influence temporal or spatial changes in toxicity of epiphytic dinoflagellates.

ISSUES AND QUESTIONS

Because of the molecular complexity of the ciguatoxins, it seems unlikely that these secondary metabolites are merely waste products that provide no benefit to the dinoflagellates. Clearly, the biosynthesis of CTX did not evolve as a means to harm humans. The bacterial literature suggests that such compounds may be allelochemic agents directed against closely competing or pathogenic organisms. Many such algal, dinoflagellate, and bacterial compounds have been shown to influence the growth of organisms in their vicinity, including inhibiting the growth of diatoms. Their role in allelopathic competition for space in the thallisphere seems a reasonable speculation but awaits improved understanding of the eco-physiological benefits of toxin production to the dinoflagellates.

Given the extreme potency of CTX and its precursors, the negligibly low incidence of fatality in humans is notable. Although there have been reports of abnormal fish behavior associated with CFP fish (Davin *et al.,* 1988), the toxins produced by *G. toxicus* have shown limited evidence of ichythiyotoxicity (Capra *et al.*, 1988; Coleman *et al.*, 2004; Lewis, 1992). One wonders whether an ecological factor such as enhanced predation on heavily intoxicated fish constrains the accumulation of CTX in fish so that the toxin dose within a reasonably sized portion of fish consumed by humans is below sublethal levels.

In the future, prevention of ciguatera fish poisoning is most likely to result from the application of new knowledge at the fish-human interface. Improved assays to expeditiously assess marine toxins are needed by clinicians, fishermen, seafood distributors, restaurateurs, and food safety regulators, and they appear to be tractable, based on advances of technologies now under development.

Improved understanding of how and why environmental factors control toxin production will require good data on the effects of environmental conditions on the growth and toxin production by *Gambierdiscus* spp., improved information on the genetics and biochemical pathways of toxin biosynthesis, and development of a molecular understanding of the genes involved in toxin synthesis.

Marine organisms are a rich, largely untapped source of a wide variety of biologically important secondary metabolites (Faulkner, 2002; Snyder *et al.*, 2003). The ability to manipulate the pathways for secondary metabolites has stimulated interest in marine organisms for novel biosynthetic functions. Extracts from cultured CTX-producing dinoflagellates are potential sources of biomedically useful compounds. These polyethers have proven antifungal and antineoplastic properties. As a group, the polyether class of antibiotics is known to improve food conversion in ruminant animals and to have usefulness in molecular probes for the study of essential ligand-receptor interactions in living systems. In the biomedical research area, these dinoflagellate bioproducts could be used to enhance or inhibit nerve conduction through the voltage-sensitive sodium channel, as well as other biomedical research venues.

Acknowledgments

This chapter was made possible through the Centers for Oceans and Human Health (COHH) program of the National Institute of Environmental Health Sciences (P50ES012740), the National Institutes of Health, and the National Science Foundation (OCE04–32479).

References

Adachi, R., Fukuyo, Y., 1979. The thecal structure of a marine toxic dinoflagellate *Gambierdiscus toxicus* gen. et sp. nov. collected in a ciguatera endemic area. Bull. Jpn. Soc. Sci. Fish. 45, 67–71.

Anderson, D.M., Lobel, P.S., 1987. The continuing enigma of ciguatera. Biol. Bull. 172, 89–107.

Arena, P., Levin, B., Fleming, L.E., Friedman, M.A., Blythe, D.G., 2004. A pilot study of the cognitive and psychological correlatyes of chronic ciguatera poisoning. Harmful Algae 3, 51–60.

Arias, H.R., 2006. Marine toxins targeting ion channels. Mar. Drugs 4, 37–69.

Ashton, M., Rosado, W., Govind, N.S., Tosteson, T.R., 2003. Culturable and nonculturable bacterial symbionts in the toxic benthic dinoflagellate *Ostreopsis lenticularis*. Toxicon 42, 419–424.

Azam, F., Worden, A.Z., 2004. Microbes, molecules and marine ecosystems. Science 303, 1622–1624.

Babinchak, J.A., Doucette, G.J., Ball, R.M., 1996. Partial characterization of the LSU rRNA gene from the ciguatoxic dinoflagellate, *Gambierdiscus toxicus*. Harmful and Toxic Algal Blooms, UNESCO: Paris, France. pp. 459–462.

Babinchak, J.A., Moeller, P.D.R., Van Dolah, F.M., Eyo, P.B., Ramsdell, J.S., 1994. Production of ciguatoxins in cultured Gambierdiscus toxicus. Mem. Queensl. Mus. 34, 447–453.

Baden, D., Fleming, L.E., Bean, J.A., 1995. Marine toxins. In deWolff, F.A. (ed.), Handbook of Clinical Neurology: Intoxications of the Nervous System Part II. Natural Toxins and Drugs, 21(65), 141–175. Amsterdam, Elsevier Press.

Bagnis, R., 1968. Clinical aspects of ciguatera (fish poisoning) in French Polynesia. Hawaii Med. J. 28, 25–28.

Bagnis, R., 1977. Modalites evolutives et biogenese de la ciguatera en Polynesie francaise. Ph.D. dissertation, Universite de Bordeaux (France).

Bagnis, R., 1993. Ciguatera fish poisoning. In Falconer, I.R. (ed.), Algal. Toxins in Seafood and Drinking Water, pp. 105–115. New York, Academic Press.

Bagnis, R., 1994. Natural versus anthropogenic disturbances to coral reefs: Comparison in epidemiological patterns of ciguatera. Mem. Queensld. Mus. Brisbane 34(3), 455–460.

Bagnis, R., Bennett, J., Prieur, C., Legrand, A.M., 1985. The dynamics of three toxic benthic dinoflagellates and the toxicity of ciguateric surgeonfish in French Polynesia. In Anderson, D.M., White, A.W., and Baden, D.G. (eds.), Toxic Dinoflagellates, pp. 177–182. New York, Elsevier Science.

Bagnis, R., Kugerski, T., Laugier, S., 1979. Clinical observations on 3,009 cases of ciguatera (fish poisoning) in the South Pacific. Am. J. Trop. Med. Hyg. 28, 1067–1073.

Bagnis, R., Legrand, A.M., Inoue, A., 1990. Follow-up of a bloom of the toxic dinoflagellate Gambierdiscus toxicus on a fringing reef of Tahiti. In Graneli, E., Sundström, B., Anderson, D.M. (eds.), Toxic Marine Phytoplankton., pp. 98–103. New York, Elsevier.

Bagnis, R., Spiegel, A., Boutin, J.P., Burucoa, C., Nguyen, L., Cartel, J.L., Capdevielle, P., Imbert, P., Prigent, D., Gras, C., Roux, J., 1992. Evaluation de l'efficacité du mannitol dans le traitement de la ciguatera en Polynésie Française. Médecine Tropicale 52, 67–73.

Bagnis, R.A., Barsinas, M., Prieur, C., Pompon, A., Chungue, F., Legrand, A.M., 1987. The use of the mosquito bioassay for determining the toxicity to man of ciguateric fish. Biol. Bull. 172, 137–143.

Ballantine, D.L., Bardales, A.T., Tosteson, T.E., 1985. Seasonal abundance of *Gambierdiscus toxicus* and *Ostreopsis* sp. in coastal waters of southwest Puerto Rico. In Delesalle, B., Galzin, R., and Salvat, B. (eds.), Proceedings of the Fifth International Coral Reef Congress, Tahiti, Antenne Museum-EPHE, Moorea (French Polynesia), pp. 417–422.

Banner, A.H., 1974. The biological origin and transmission of ciguatoxin. In Hum, H.J., and Lance, C.E. (eds.), Bioactive Products from the Sea, pp. 15–36. New York, Marcel Dekker.

Banner, A.H., 1976. Ciguatera: A disease from coral reef fish. In Hum, H.J., and Lance, C.E. (eds.), Bioactive Products from the Sea, pp. 177–215. New York, Marcel Dekker.

Banner, A.H., Helfrich, P., Piyakarnchana, T., 1966. Retention of ciguatera toxin by the red snapper, *Lutjanus bohar*. Copeia 2, 297–301.

Banner, A.H., Sasaki, S., Helfrich, P., Alender. C., Scheuer. P., 1960. A new bioassay method for toxins from fish. Nature. 189, 229–230.

Benoit, E., Laurent, D., Mattei, C., Legrand, A.M., Molgo, J., 2000. Reversal of Pacific ciguatoxin-1B effects on myelinated axons by agents used in ciguatera treatment. Premieres Rencontres de I'Ichtyologie en France 24, 33–40.

Blythe, D.G., De Sylva, D.P., Fleming, L.E., Ayyar, R.A., Baden, D.G., Shrank, K., 1992. Clinical experience with IV mannitol in the treatment of ciguatera. Bulletin de la Sociét́é de Pathologie Exotique, 85, 425–426.

Blythe, D.G., Hack, E., Washington, G., Fleming, L.E., 2001. The Medical Management of Seafood Poisoning. Hui, Kits, Stanfield. Seafood and Environmental Toxins, pp. 311–319. New York, Marcel Dekker.

Bomber, J.W., 1985. Ecological studies of benthic dinoflagellates associated with ciguatera from the Florida Keys. M.S. thesis, Florida Institute of Technology, Melbourne, Florida, 104 pp.

Bomber, J.W., 1987. Ecology, genetic variability and physiology of the ciguatera-causing dinoflagellate *Gambierdiscus toxicus* Adachi and Fukuyo. Ph.D. Dissertation, Florida Institute of Technology, Melbourne, Florida, 147 pp.

Bomber, J.W., Guillard, R.R.L., Nelson, W.G., 1988. Roles of temperature, salinity, and light in seasonality, growth, and toxicity of ciguatera-causing *Gambierdiscus toxicus* Adachi et Fukuyo (Dinophyceae). J. Exp. Mar. Biol. Ecol. 115, 53–65.

Brusle, J., Brusle-Sicard, S., Fourcault, B., 1998. Gonads of groupers (Serranidae from Mururoa and Hao, French Polynesia) used as indicators of the quality of the coral reef environment: Histological and ultrastructural aspects. Cybium 22, 371–382.

Brussaard, C.P.D., Short, S.M., Frederickson, C.M., Suttle, C.A., 2004. Isolation and phylogenetic analysis of novel viruses infecting the phytoplankton *Phaeocystis globosa* (Prymnesiophyceae). Appl. Environ. Microbiol. 70, 3700–3705.

Calvert, G.M., 1990. The recognition and management of ciguatera fish poisoning. In Miller, D.M. (ed.), Ciguatera Seafood Toxins, pp. 1–10. New York, Elsevier.

Cameron, J., Flowers, A.E., Capra, M., 1991. Electrophysiological studies on ciguatera fish poisoning in men (Part II). J. Neurol. Sci. 101, 93–97.

Capra, M.F., Cameron, J., Flowers, A.E., Coombe, I.F., Blanton, C.G., Hahn, S.T., 1988. The effects of ciguatoxin on teleosts. Proceedings of the 6th International Coral Reef Symposium, Australia 1988, vol. 3., pp. 37–41.

Carlson, R.D., 1984. Distribution, periodicity, and culture of benthic/epiphytic dinoflagellates in a ciguatera endemic region of the Caribbean. Ph.D. thesis. Southern Illinois University, Carbondale, IL.

Carlson, R.D., Morey-Gaines, G., Tindall, D.R., Dickey, R.W., 1984. Ecology of toxic dinoflagellates from the Caribbean Sea: Effects of macroalgal extracts on growth in culture. In Ragelis, E.P. (ed.), Seafood Toxins, pp. 272–287, Washington, DC, Amer. Chem. Soc.

Carlson, R.D., Tindall, D.R., 1988. Distribution and periodicity of toxic dinoflagellates in the Virgin Islands. In Anderson, D.B. (ed.), Toxic Dinoflagellates, 1985, Proc. 3rd Intl Conf. Toxic Dinoflagellates, New Brunswick.

Chan, T.Y.K., Wang, A.Y.M., 1993. Life-threatening bradycardia and hypotension in a patient with ciguatera fish poisoning. Trans. R. Soc Trop. Med. Hyg. 87, 71–7.

Chateau-Degat, M.L., Chinain, M., Cerf, N., Gingras, S., Hubert, B., Dewailly, E., 2005. Seawater temperature, *Gambierdiscus* spp. variability and incidence of ciguatera poisoning in French Polynesia. Harmful Algae 4, 1053–1062.

Chinain, M., Faust, M.A., Pauillac, S., 1999a. Morphology and molecular analyses of three toxic species of *Gambierdiscus* (Dinophyceae): *G. pacificus*, sp. nov., *G. australes*, sp. nov., and *G. polynesiensis*, sp. nov. J. Phycol. 35, 1282–1296.

Chinain, M., Germain, M., Deparis, X., Pauillac, S., Legrand, A.M., 1999b. Seasonal abundance and toxicity of the dinoflagellate *Gambierdiscus* spp. (Dinophyceae), the causative agent of ciguatera in Tahiti, French Polynesia. Mar. Biol. 135, 259–267.

Colman, J.R., Dechraoui, M.-Y. B., Dickey, R.W., Ramsdell, J.S., 2004. Characterization of the developmental toxicity of Caribbean ciguatoxins in finfish embryos. Toxicon 44, 59–66.

Cooper, M.J., 1964. Ciguatera and other marine poisoning in the Gilbert Islands. Pac. Sci. 18, 411–440.

Cruz-Rivera, E.,Villareal, T.A., 2006. Macroalgal palatability and the flux of ciguatera toxins through marine food webs. Harmful Algae 5, 497–525.

Davin, W.T., Jr., Kolher, C.C., Tindall, D.R., 1988. Ciguatera toxins adversely affect piscivorous fishes. Trans. Am. Fish Soc. 117, 374–384.

De Haro, L., Hayek-Lanthois, M., Jooson, F., Castanier, L., Jouglard, J., 1997. Mediacal management in the Marseilles Poison Centre of a ciguatera poisoning collective case after a meal of one barracuda in Mexico. Toxicon. 35, 810.

de Sylva, D.P., 1992. Distribution and ecology of ciguatera fish poisoning in Florida, with emphasis on the Florida Keys. Bulletin of Marine Science 54, 944–954.

Dickey, R.W., Granade, H.R., McClure, F.D., 1994. Evaluation of a solid-phase immunobead assay for detection of ciguatera-related biotoxins in Caribbean finfish. Memoirs of the Queensland Museum, Brisbane 34, 481–488.

Durand-Clement, M., 1986. A study of toxin production by *Gambierdiscus toxicus* in culture. Toxicon 24, 1153–1157.

Ebesu, J.S.M., 1998. Isolation and Characterization of Novel Ciguateric Compounds from *Acanthurus triostegus* (Manini). Ph.D. dissertation, University of Hawaii, Honolulu.

Escalona de Motta, G., De la Noceda, G.G., 1985. Epidemiological, clinical and experimental aspects of ciguatera in Puerto Rico. Proc. 5th Intl Coral Reef Congress, Tahiti (French Polynesia). vol. 4, pp. 415–416.

Faulkner, D.J., 2002. Marine natural products. Nat. Prod. Rep. 19, 1–48.

Fistarol, G.O., Legrand, C., Selander, E., Hummert, C., Stolte, W., Graneli, E., 2004. Allelopathy in *Alexandrium* spp.: Effect on a natural plankton community and on algal monocultures. Aquat. Microb. Ecol. 35, 45–56.

Fleming, L., Dewailly, E., Baden, D.G., 2000. The epidemiologic of marine harmful algal blooms, Epidemiology 11, 143.

Friedman, M.A., Arena, P., Levin, B., Fleming, L.E., Weisman, R., Bernstein, J., Schrank, K., Blythe, D., Backer, L., Reich, A, in press. Neuropsychological study of ciguatera fish poisoning: A longitudinal case-control study. Arch. Clin. Neurol.

Gillespie, N.C., Holmes, M.J., Burke, J., Doley, J., 1985. Distribution and periodicity of Gambierdiscus toxicus in Queensland, Australia. In Anderson, D.M., White, A., and Baden, D.G. (eds.), Toxic Dinoflagellate, pp. 183–188. London, Elsevier.

Guillard, R.R.L., Keller, M.D., 1984. Culturing dinoflagellates. In Dinoflagellates, pp. 391–441. New York, Academic Press.

Halstead, B.W., 1967. Poisonous and Venomous Marine Animals of the World, Vol. II. Washington, DC, U.S. Government Printing Office.

Harrison, P.J., Waters, S.E., Taylor, F.J.R., 1980. A broad spectrum artificial seawater medium for coastal and open ocean phytoplankton. J. Phychol. 16, 28–35.

Helfrich, P., 1964. Fish Poisoning in Hawaii. Haw. Med J. 22, 361–372.

Helfrich, P., Banner, A.H., 1963. Experimental induction of ciguatera toxicity in fish through diet. Nature 197, 1025–1026.

Helfrich, P., Banner, A.H., 1968. Ciguatera fish poisoning. 2. General patterns of development in the Pacific. Occasional Papers of the Bernice P. Bishop Museum 23, 371–382.

Hokama, Y., 1985. A rapid, simplified enzyme immunoassay stick test for the detection of ciguatoxin and related polyethers from fish tissues. Toxicon 23, 939–946.

Hokama, Y., Banner, A.H., Boylan, D.B., 1977. A radioimmunoassay for the detection of ciguatoxin. Toxicon 15, 317–325.

Hokama, Y., Nishimura, K., Takenaka, W., Ebesu, J.M., 1998. Simplified solid-phase membrane immunobead assay (MIA) with monoclonal anti-ciguatoxin antibody (MAb-CTX) for detection of ciguatoxin and related polyether toxins. J. Nat. Toxins 7, 1–22.

Hokama, Y., Shirai, L.K., Kurosawa, M.N., Iwamoto, L.M., Goto, C.S., Osugi, A.M., 1987. Assessment of a rapid enzyme immunoassay sitck test for the detection of ciguatoxin and related polyether toxins in fish tissues. Biol. Bull. 172, 144–153.

Hokama, Y., Yoshikawa-Ebesu, J.S.M., 2001. Ciguatera fish poisoning: A foodborne disease. J. Toxicol.—Toxin Reviews 20(2), 85–139.

Holmes, M.J., Lewis, R.J., 1994. The origin of ciguatera. Mem. Queensld. Mus. 34, 497–504.

Holmes, M.J., Lewis, R.J., Poli, M.A., Gillespie, N.C., 1991. Strain dependent production of ciguatoxin precursors (gambiertoxins) by *Gambierdiscus toxicus* (Dinophycean) in culture. Toxicon 29, 761–775.

Inoue, M., 2004. Convergent synthesis of polycyclic ethers. illustrations of the utility of acetal-linked intermediates. Org. Biomol. Chem. 2, 1811–1817.

Inoue, M., Ishihara, Y., Tamashita, S., Masahiro, M., 2006. Convergent assembly of polycyclic ethers via acyl radical addition to unactivated enol ether. Org. Lett. 8, 5801–5804.

Keller, M.D., Guillard, R.L.L., 1985. Factors significant to marine dinoflagellate culture. In Anderson, D.M., White, A.W., and Baden, D.G. (eds.), Toxic Dinoflagellates, pp. 113–116. New York, Elsevier.

Kodama, M., 1990. Production of paralytic shellfish toxins by a bacterium *Moraxellasp.* isolated from *Protogonyaulax tamerensis*. Toxicon 28, 707–714.

Kohler, S.T., Kohler, C.C., 1992. Dead bleached coral provides new surface for dinoflagellates implicated in ciguatera fish poisonings. Environ. Biol. Fish. 35, 413–416.

Kubanek, J., Hicks, M.K., Naar, J., Villareal, T.A., 2005. Does the red tide dinoflagellate *Karenia brevis* use allelopathy to outcompete other phytoplankton? Limnol. Oceanogr. 50, 883–895.

Labrousse, H., Pauillac, S., Jehl-Marinex, C., Legrand, A.M., Avrameas, L., 1991. In vivo and in vitro techniques for the detection of ciguatoxin. Oceanic 18, 182–245.

Lawrence, D.N., Enriquez, M.B., Lumish, R.M., 1980. Ciguatera fish poisoning in Miami. J. Am. Med. Assoc. 244, 254–258.

Legrand, A.M., Fukui, M., Cruchet, P., Yasumoto, T., 1992. Progress on chemical knowledge of ciguatoxins. Bull. Soc. Pathol. Exot. 85, 467–469.

Legrand, A.-M., 1998. Ciguatera toxins: Origin, transfer through the food chain and toxicity to humans. In Reguera, B., Blanco, J., Fernández, M. L., and Wyatt, T. (eds.), Harmful Algae, pp. 39–43. Vigo, Spain Xunta de Galicia and IOC.

Lewis, R.J., 1992. Ciguatoxins are potent ichthyotoxins. Toxicon 30, 207–211.

Lewis, R.J., 2001. The changing face of ciguatera. Toxicon 39, 97–106.

Lewis, R.J., Holmes, M.J., Sellin, M., 1994. Invertebrates implicated in the transfer of gambiertoxins to the benthic carnivore *Pomadasys maculates*. Memoirs of the Queensland Museum 34, 561–564.

Lewis, R.J., Jones, A., 1997. Characterization of ciguatoxins and ciguatoxin congeners present in ciguateric fish by gradient reverse-phase high-performance liquid chromatography/mass spectrometry. Toxicon 35, 159–168.

Lewis, R.J., Jones, A., Vernoux, J.P., 1999a. HPLC/Mass spectrometry and HPLC/MS/MS methods for the detection of multiple ciguatoxins. Proceedings of the 5th Indo-Pacific Fish conference, Noumea–New Caledonia, pp. 739–744.

Lewis, R.J., Jones, A., Vernoux, J.P., 1999b. Societe Francaise d' Ichtyologie, Paris (France), and Institut de Recherche pour le Developpement, Paris (France). HPLC/Mass spectrometry and HPLC/MS/MS methods for the detection of multiple ciguatoxins. 5. Indo - Pacific fish conference, Noumea (New Caledonia), 3–8 Nov 1997, Seret, B., and Sire, J. Y. (eds.). Paris (France): Societe Francaise d' Ichtyologie.

Lobel, P.S., 1979. Folk remedies for tropical fish poisoning in the Pacific. Sea Frontiers 25, 239–245.

Luckas, B., 2000. Phycotoxins in the marine environment: Control of marine organisms for contamination with algal toxins. Int. J. Environ. Pollut. 13, 148–172.

Manger, R.L., Leja, L.S., Lee, S.Y., Hungerford, J.M., Hokama, Y., Dickey, R.W., Grande, H.R., Lewis, R., Yasumoto, T., Wekell, M.M., 1995. Detection of sodium channel toxins: directed cytotoxicity assays of purified ciguatoxins, brevetoxins, saxitoxins, and seafood extracts. J. of AOAC Intl. 78, 521–527.

McKee, D.B., Fleming, L.E., Tamer, R., Weisman, R., Blythe, D., 2001. Physician diagnosis and reporting of ciguatera fish poisoning in an endemic area. Harmful Algal Blooms 2000, 451–453.

Moore, R.E., Bartolini, G., Barchi, J., Bother-By, A.A., Dodok, J., Ford, J., 1982. Absolute stereochemistry of palytoxin. J. Amer. Chem. Soc. 104, 3776–3778.

Morris, H.G., Lewin, P., Smith, W.C., Blake, P.A., Schneider, R., 1982. Ciguatera fish poisoning: Epidemiology of the disease on Saint Thomas, U.S. Virgin Islands. AM. H. Trop. Med. Hyg. 31, 574–578.

Murata, M., Legrand, A.M., Ishibashi, Y., Yasumoto, T., 1989. Excretion of ciguatoxin from moray eels (Muraenidae) of structure of ciguatoxin and its congener. J. Am. Chem. Soc. 111, 8929–8931.

Nakahara, H., Sakami, T., Chinain, M., Ishida, Y., 1996. The role of macroalgae in epiphytism of the toxic dinoflagellate *Gambierdiscus toxicus* (Dinophyceae). Psychol. Res. 44, 113–117.

Nicholson, G.M., Lewis, R.J., 2006. Ciguatoxins: Cyclic polyether modulators of voltage-gated ion channel function. Marine Drugs 4, 82–118.

Nukina, M., Koyanagi, I.M., Scheur, P.J., 1984. Two interchangeable forms of ciguatoxin. Toxicon 22, 169–176

Nukina, M., Tachibana, K., Scheurer, P.J., 1986. Recent developments in ciguatera research. In J.B. White (ed.), Natural Toxins, Animal, Plant and Microbial, pp. 45–55. Oxford, Clarendon Press.

Ochoa, J., Sanchez-Paz, L.A., Crus-Villacorta, A., Nunez-Vazquez, E., 1996. Toxic events in the Northwest Pacific coastline of Mexico during 1992–1995. Hydrobiologia 352, 195–200.

Palafox, N.A., Jain, L.G., Piezano, A.Z., 1988. Successful treatment of ciguatera fish poisoning with intravenous mannitol. JAMA 259, 2740–2742.

Parsons, M.L., Preskitt, L.B., in press. A survey of epiphytic dinoflagellates from the coastal waters of the island of Hawai'i. October.

Pearn, J., 2001. Neurology of ciguatera. J. Neurol. 70, 4–8.

Plumley, F.G., 1997. Marine algal toxins: Biochemistry, genetics, and molecular biology. Limnol. Oceanogr., 42, 1252–1264.

Pohland, A.E., Dowell, V.R., Richard, J.L. (eds.), 1990. Microbial Toxins in Foods and Feeds: Cellular and Molecular Modes of Action. New York, Plenum Press.

Pottier, I., Vernoux, J.P., Jones, A., Lewis, R.J., 2002. Characterization of multiple Caribbean ciguatoxins and congeners in individual specimens of horse-eye jack (*Caranx latus*) by high-performance liquid chromatography/mass spectrometry. Toxicon 40, 929–939.

Quod, J.-P., Turquer, J., 1996. Ciguatera in Reunion Island (SW Indian Ocean): Epidemiology and clinical patterns. Toxicon 34, 779–785.

Quod, J-P., Turquer, J., Conejero, S., Ralijaona, C., 2001. Ciguatera risk assessment in the Indian Ocean following the 1998 coral bleaching event, pp. 166–168. CORDIO. SARREC. Mar. Sci. Prgm. Status Report. D. Souter, D. Obura, and O. Linden (eds.).

Randall, J.E., 1958. A review of ciguatera tropical fish poisoning with a tentative explanation of its cause. Bull. Mar. Sci. Gulf Carib. 8, 236–267.

Sakami, H., Nakahara, M., Chinain, M., Ishida, Y., 1999. Effects of epiphytic bacteria on the growth of the toxic dinoflagellate *Gambierdiscus toxicus* (Dinophyceae). L. Exp Mar. Biol. Ecol. 233, 231–246.

Sakamoto, B., Nagai, H., Hokama, Y., 1996. Stimulators of *Gambierdiscus toxicus* (Dinophycease) growth: The possible role of gambieric acid-A as an endogenous growth enhancer. Phycologia 35, 350–353.

Sako, Y., Chinain, M., Legreand, A., Nakahara, H., Kuruawa, T.S., Uchida, A., Ishida, Y., Inouye, A., 1996. Phylogentic relationship of ciguatera-causing dinoflagellate *Gambierdiscus toxicus* with 18s ribosomal DNA sequence comparison. Harmful and Toxic Algal Blooms, UNESCO: Paris, France. pp. 443–446.

Scheuer, P.J., Takahashi, W., Tsutsumi, J., Yoshida, T., 1967. Ciguatoxin, isolation, and chemical nature. Science 155, 1267–1268.

Schnorf, H., Taurarii, M., Cundy, T., 2002. Ciguatera fish poisoning: A double-blind randomized trial of mannitol therapy. Neurology 58, 873–880.

Snyder, R.V., Gibbs, P.D., Palacios, A., Abiv, L., Dickey, R., Lopez, J.V., Rein, K.S., 2003. Polyketide synthase genes from marine dinoflagellates. Mar. Biotechnol. (NY). 5, 1–12.

Steidinger, K.A., 1993. Some taxonomic and biologic aspects of toxic dinoflagellates. Algal Toxins in Seafood and Drinking Water, pp. 1–28. Orlando, FL, Academic Press.

Steidinger, K.A., Baden, D.G., 1985. Toxic marine dinoflagellates. In Spector, D.L. (ed.), Dinoflagellates, pp. 210–261. Orlando, FL, Academic Press.

Sumner, J., Ross. T., 2002. A semi-quantitative seafood risk assessment. Int. J. Food Microbiol. 77, 55–59.

Tachibana, K., 1980. Structural Studies in Marine Toxins. Ph.D. dissertation, University of Hawaii, Honolulu.

Tachibana, K., Nukina, M., Joh, Y-G., Scheuer, P.J., 1987. Recent developments in the molecular structure of ciguatoxin. Biol. Bull. Mar. Biol. Lab. Woods Hole 172, 122–127.

Tarutani, K., Nagasaki, K., Yamaguchi, M., 2000. Viral impacts on total abundance and clonal composition of the harmful bloom-forming phytoplankton *Heterosigma akashiwo*. Appl. Envir. Microbiol. 66, 4916–4920.

Taylor, F.J.R., 1985. The distribution of the dinoflagellate *Gambierdiscus toxicus* in the eastern Caribbean. In Proceedings of the Fifth International Coral Reef Congress, Delesalle, B., Galzin, B., Repsilon, D., and Salvat, B., (eds.), vol. 4, pp. 423–428. Antenne Museum-EPHE, Moorea (French Polynesia).

Tebana, T., 1992. Ciguatera fish poisoning and reef disturbance in South Tarawa, Kiribati. Ciguatera Info. Bull. 2, 7.

Tindall, D.R., Dickey, R.W., Carlson, R.D., Morey-Gaines, G., 1984. Ciguatoxic dinoflagellates from the Caribbean Sea. In Ragelis, E.P. (ed.), Seafood Toxins, pp. 225–240. Washington, DC, American Chemical Society Symposium Series.

Tosteson, T.R., Ballantine, D.L., Winter, A., 1998. Sea surface temperature, benthic dinoflagellate toxicity and toxin transmission in the ciguatera food web. In Reguera, B., Blanco, J., Fernández, M.L., Wyatt, T. (eds.), Harmful Algae, Xunta de Galicia and IOC, pp. 48–49.

Vernoux, J.P., 1988. Ciguatera fish poisoning: Epidemiology, toxicology, and prevention of the illness on Saint-Barthelemy Island, French West Indies. Oceanologica acta. 11, 37–46.

Vernoux, J-P., Lejeune, J., 1993. Ciguatera in the French West Indies. Mem. Queensl. Mus. 34, 631–638.

Villareal, T.A., Morton, S.L., 2002. Use of cell-specific PAM-fluorometry to characterize host shading in the epiphytic dinoflagellate *Gambierdiscus toxicus*. Mar. Ecol. 23, 127–140.

Watters, M.R., 2007. Marine neurotoxins as a starting point to drugs. In Botana, L.M. (ed.), Seafood and Freshwater Toxins. Pharmacology, Physiology and Detection, 2nd ed. Taylor & Francis (CRC Press: New York).

Withers, N.W., 1982. Ciguatera fish poisoning. Annu. Rev. Med. 33, 97–111.

Yasumoto, T., 1985. Recent progress in the chemistry of dinoflagellate toxins. In Anderson, D.M., White, A.W., and Baden, D.G. (eds.), Toxic Dinoflagellates, pp. 259–270. New York, Elsevier.

Yasumoto, T., Inoue, A., Bagnis, R., 1979a. Ecological survey of a toxic dinoflagellate associated with ciguatera. In Taylor, D.L., and Seliger, H. (eds.), Toxic Dinoflagellate Blooms, pp. 221–224. New York, Elsevier.

Yasumoto, T., Murata, M., 1993. Marine toxins. Chem. Rev. 93, 1897–1907.

Yasumoto, T., Nakajima, I., Bagnis, R., Adachi, R., 1977. Finding of a dinoflagellate as a likely culprit of ciguatera. Bull. Jpn. Soc. Sci. Fish. 43, 1021–1026.

Yasumoto, T., Nakajima, I., Oshima, Y., Bagnis, R., 1979b. A new toxic dinoflagellate found in association with ciguatera. In Taylor, D.L., and Seliger, H.H. (eds.), Toxic Dinoflagellate Blooms, pp. 65–70. New York, Elsevier.

Yasumoto, T., Oshima, Y., Murakami, Y., Nakajima, I., Bagnis, R., Fukuyo, Y., 1980. Toxicity of benthic dinoflagellates. Bull. Jap. Soc. Fish. 46, 327–331.

Yasumoto, T., Raj, U., Bagnis, R., 1984. Seafood poisoning in tropical regions. Lab. Food Sci. Hyg., Faculty of Agriculture, Tohoku, Japan, 1–74.

STUDY QUESTIONS

1. Because of the molecular complexity of the ciguatoxins (CTX), it seems unlikely that these secondary metabolites are merely waste products that provide no benefit to the dinoflagellates. Clearly, the biosynthesis of CTX did not evolve as a means to harm humans. What might be the benefit to the dinoflagellates for the production of these CTX compounds? Allelochemic agents directed against closely competing organisms (i.e., to influence the growth of organisms and compete for space in the thallisphere)? To chelate metals? For protection from oxidation in the supersaturated oxygen microenvironment of the thallisphere? Others? How would you go about setting up an experiment(s) to test your hypothesis?
2. The dinophyceae and cyanobacteria both produce toxins that are harmful to humans Do you think there is a relationship between the production of toxins and the

fact that both groups represent primitive organisms that are low on the phylogenic scheme?
3. Discuss a number of ways that *Gambierdiscus toxicus* might benefit from its symbiotic relationship with the macroalgae.
4. Given the extreme potency of CTX and its precursors, what do you think is the reason for the apparently negligibly low incidence of fatality in humans?
5. If you contract ciguatera fish poisoning (CFP) after eating in a restaurant, do you think you should sue the restaurant, given the state of diagnostic capability for this malady? If so, present the rationale for your case; if not, state your reasons.
6. Your education has given you the intellectual acuity to suspect that your friend has recently become intoxicated with CFP. Describe the information you sought, gleaned, and acquired to come to that conclusion/diagnosis, and express what you would suggest to the attending physician as possible treatment(s).
7. What do you think are currently the most effective means of reducing the incidence of CFP throughout the most susceptible geographic regions? What are the most promising developments that would improve that situation?

C H A P T E R

15

Cyanobacteria and Cyanobacterial Toxins

IAN STEWART AND IAN R. FALCONER

INTRODUCTION

Cyanobacteria are arguably the most successful group of microorganisms on Earth. They are the most genetically diverse; they occupy a broad range of habitats across all latitudes, widespread in freshwater, marine, and terrestrial ecosystems; and they are found in extreme niches such as hot springs, salt works, and hypersaline bays. Photoautotrophic, oxygen-producing cyanobacteria created the conditions in the planet's early atmosphere that directed the evolution of aerobic metabolism and eukaryotic photosynthesis. Cyanobacteria fulfill vital ecological functions in the world's oceans, being important contributors to global carbon and nitrogen budgets. Cyanobacteria have come to the attention of public health workers because they can produce a suite of potent natural toxins that have killed animals (including humans) in acute intoxication and mass poisoning incidents. Some cyanotoxins are thought to be carcinogenic, whereas some cyanobacteria and their toxins may have the capacity to act as allergens. Most reports of acute human and animal cyanobacteria-related poisoning arise through the consumption of untreated drinking water and water treatment failure episodes that allow cyanotoxins to contaminate potable water supplies. Chronic exposure to low levels of contaminant cyanotoxins in drinking water is also the focus of attention with regard to the carcinogenic potential of cyanobacteria.

The freshwater cyanotoxins have been studied more widely by researchers in toxicology and epidemiology to date. Therefore, this chapter includes an overview of the common freshwater cyanobacteria and their toxins. This chapter also reviews marine cyanobacteria, which, by contrast, have received less attention by public health workers. Some coastal and open-ocean cyanobacteria are known to produce specific toxins that are also very potent. Some marine cyanotoxins are, like many freshwater cyanotoxins, well characterized and their chemical structures have been fully elucidated, whereas other marine cyanobacteria and cyanotoxins are less well understood regarding source species, toxicology, and epidemiology. Marine cyanobacteria are being actively and intensively researched as rich sources of structurally diverse, biologically active compounds; some hold significant potential as pharmaceutically active products (see Section IIA).

BIOLOGY AND ECOLOGY OF TOXIC CYANOBACTERIA

The cyanobacteria were earlier called "blue-green algae" because of their presence in lakes and rivers and the blue-green scum that formed when they dried on the shoreline. As they are without a nucleus or mitochondria and their DNA is free in the cytoplasm, they are classified as prokaryotic life, as are the bacteria. They are not necessarily blue-green in appearance; a common species in the marine environment is *Trichodesmium*, which appears as reddish-brown "sawdust" in the sea.

One of the distinguishing features of the cyanobacteria is the presence of particular photosynthetic pigments, the phycocyanins (blue) and the phycoerythrins (red), which do not occur in eukaryotic algae or higher plants. These pigments absorb light at shorter wavelengths than chlorophyll and thus increase the ability of the cells to utilize solar radiation in the green-yellow region. As a consequence, the cyanobacteria can grow at low light intensities and in clear waters at considerable depths. Chlorophyll-a is the common photosynthetic pigment in all cyanobacteria, algae, and plants. Cyanobacteria carry out photosynthesis by a similar mechanism to higher plants, using the energy of light to activate electron transfer. The coenzyme nicotinamide adenine dinucleotide phosphate (NADP) is reduced, which is then utilized in the reduction of carboxylic acid to aldehyde, leading to sugars. Oxygen gas is liberated as a result of water molecules providing the source of the transferred electrons (Heldt, 2004).

The cyanobacteria are distributed worldwide, with considerable populations in the polar oceans, in soils, freshwaters, on desert rocks, and even in volcanic hot springs. Fossil remains of cyanobacteria occur in rocks dating back 3500 million years. The characteristic rock domes called "stromatolites," which are found in present-day shallow marine waters are formed by cyanobacteria, and examples from the geological record have been dated at 3500 million years old in Australia, 2150 in Canada, 1650 in Australia, and 900 million years old in Russia (Schopf, 2000). In the present-day oceans, they occur in hypersaline bays where the salinity suppresses growth of marine organisms that graze the cyanobacteria. A well-known location is Shark Bay in Western Australia, where a shallow bay with seagrass growing at the entrance slows the exchange of water with the ocean and high temperatures evaporate the seawater to high salinity (Fig. 15-1).

The cyanobacteria first appeared in ancient seas completely lacking in free oxygen, and through photosynthetic oxygen liberation, they gradually caused oxidation of their environment. The seas were locally rich in ferrous iron and in carbon dioxide, and the oxygen gas from photosynthesis oxidized the ferrous iron, resulting in the precipitation of ferric iron. This formed deposits of iron ore on the ocean floor. These banded iron deposits were most frequently formed about 2000 million years ago, indicating a large cyanobacterial population at that time living in an anaerobic sea. Gradually, the availability of oxygen from cyanobacterial photosynthesis rose to the level at which free oxygen occurred in the seas and in the atmosphere, and evolution could proceed to multicellular plant and animal life using aerobic respiration requiring oxygen gas.

FIGURE 15-1. Stromatolites at Hamelin Pool, Shark Bay, Western Australia. Photograph by author I.R.F.

The morphology of the ancient cyanobacteria is essentially identical to those growing today, and their biochemical processes are similar to other oxygen-producing photosynthetic organisms. It has been proposed that the chloroplasts present in plants' cells may have evolved from symbiotic cyanobacteria. One interesting aspect is that plant chloroplasts have their own DNA, coding for photosynthetic components.

There are many examples of present-day symbiosis between eukaryotic organisms and cyanobacteria, among which are fungi and cyanobacteria forming lichens, ferns and cyanobacteria (in the water fern *Azolla*) and cycad plants, which contain cyanobacteria. Of interest in the marine context is the symbiosis of dinoflagellates with cyanobacteria. In most examples of cyanobacterial symbiosis, the apparent gain to the host is the nitrogen-fixing ability of the cyanobacteria, rather than the ability to fix carbon dioxide; however, in the dinoflagellates, carbon fixation may be the benefit (Adams, 2000).

ECOLOGY OF CYANOBACTERIA

The majority of cyanobacteria are as follows:

- *Autotrophic*. They do not require organic nutrients and can grow on entirely inorganic materials.
- *Nitrogen-fixing*. They can "fix" atmospheric nitrogen into organic compounds and hence do not require organic or inorganic nitrogenous compounds as nutrients.
- *Photosynthetic*. They can "fix" carbon dioxide into organic molecules using light as the energy source. Like plants, the cyanobacteria release oxygen gas as a by-product of photosynthesis.

Nutrients

The nutrient requirements of cyanobacteria include phosphorus, nitrogen, sulfur, and metals and trace elements (usually in the form of salts). Because of the wide availability of water, sunlight, and carbon dioxide on this planet, the growth of cyanobacteria is normally limited by nutrients. In the oceans, cyanobacteria flourish in surface waters that are nutrient rich, the nutrients coming from ocean upwelling or from terrestrial runoff. The limiting nutrients are usually phosphate and nitrate. Enclosed marine waters such as the Baltic Sea, which has large nutrient loads derived from rivers emptying into the sea, are particularly vulnerable to growth of abundant cyanobacteria. The toxic marine/estuarine cyanobacterium, *Nodularia spumigena*, forms dense masses of filaments on the Baltic Sea in the summer, which is a potential hazard to both swimmers and to the consumers of fish and shellfish, as discussed later. The floating scums of cyanobacteria are sufficiently large to be seen from the air. Similar nutrient enrichment occurs worldwide in bays and estuaries that have low water exchange with the open ocean (Figs. 15-2 and 15-3).

Phosphorus

Phosphorus concentration is frequently the most limiting for cyanobacterial growth, and marine and freshwaters that have very low phosphorus concentrations of 10 μg/L or less do not support substantial growth. At higher phosphorus concentrations, the cyanobacteria can become so prolific that they form water blooms, coloring the water blue-green or red depending on the species. Cell concentrations of 100,000 cells/mL are common in water blooms, and because some species are buoyant, they can form scums that accumulate on shorelines and beaches.

FIGURE 15-2. Aerial photograph of a ship's wake in the Baltic Sea, traveling through a *Nodularia* bloom near Bornholm. Photograph by author I.R.F.

FIGURE 15-3. Mixed bloom of *Microcystis* and *Anabaena*, St Johns River, Florida. Photograph courtesy John W. Burns, Jr.

Cyanobacteria have several strategies that give them a competitive advantage against eukaryotic algae. One is the ability to concentrate and store phosphate in their cells in low phosphate environments. This stored polyphosphate is sufficient for several cell divisions. Another is to use buoyancy to move up and down the water column in order to obtain phosphate close to the sediments and in the anaerobic zones where phosphate is solubilized, and then to rise up into the photosynthetic zone. The ability to use sediment phosphorus is an advantage to both the planktonic cyanobacteria, which are free in the water column, and to the benthic filamentous species that grow on the bottom sediments in shallow waters. One outcome of this phosphate dependence is the ability of cyanobacteria to form dense bands deep in clear lakes, which may be 10 m or more below the surface at the intersection of warm aerobic and cold anaerobic water (Mur *et al.*, 1999).

Nitrogen

Atmospheric nitrogen can be "fixed" by particular cells in some species of cyanobacteria. These cells are known as heterocysts and differ from normal cells in lacking the photosynthetic apparatus, which gives them a clear appearance under the microscope. The heterocysts have a unique ability to convert nitrogen gas into amino groups by an enzymic process of reduction. This is an anaerobic process, and the cells carrying it out have cell walls that minimize oxygen entry. The heterocysts appear to produce excess nitrogenous organic compounds, which are transferred to adjoining photosynthetic cells. Presumably the exchange is reciprocal with photosynthetic products moving into heterocysts to provide the energy for nitrogen fixation, as these cells have no photosynthetic capacity. Microscopic examination of the region adjacent to these heterocysts shows clusters of bacteria on the cell surface, gaining an advantage from the extra nutrients leaking from the cells. Some other cyanobacterial species that do not have visible heterocysts can also fix nitrogen, for example, the marine species *Trichodesmium*, which is discussed later. In this case, the bundles of filaments appear to fix nitrogen toward the center of the bundle.

Several toxic species that provide a potential hazard to human and animal health do not possess heterocysts, and depend on nitrate, nitrite, or ammonia for growth. There is a cyanobacterial preference for ammonia as a nitrogen source, which requires less energy to incorporate into amino acids. Soluble forms of nitrogen in freshwaters are relatively abundant compared to phosphate, often in milligrams/L rather than micrograms, and largely do not limit growth. Under conditions of very low nitrogen availability, the nitrogen-fixing species of cyanobacteria have a competitive advantage over other cyanobacteria and phytoplankton. The relationship between nutrient availability and toxin production is complex, and in laboratory cultures of organisms producing the toxin microcystin, nitrogen limitation appears to reduce toxin content (Falconer, 2005).

Although cyanobacteria can exist in temperatures ranging from those of the polar oceans to volcanic hot springs at 74°C, the majority thrive at 20 to 30°C. In temperate climates, cyanobacterial blooms in marine and fresh waters occur more commonly in late summer when the water temperature is higher than in the winter. In lakes, there may be a sequence of phytoplankton each season, with diatoms preceding cyanobacteria and different cyanobacterial species achieving dominance as the water warms. The main recreational hazards from toxic cyanobacteria result from the coincidence of warmer water temperatures and water sports on sea beaches and inland waters, when cyanobacterial populations rapidly increase. The majority of cyanobacterial species overwinter at low population density and rebuild their numbers as light availability and temperature increase. As discussed later, the annual cycle of many cyanobacterial species includes the formation of "akinetes" or spores in the autumn, which rest on the sediments in winter and germinate in spring as the water temperature rises.

The effect of light intensity and color on cyanobacterial growth is considerable, but is very species specific. Some species are capable of growth at very low light intensities, deep in the water, under ice or sediment, or in turbid water, with other species growing on the surface of rocks or exposed sediments. Because of the widespread occurrence of cyanobacteria and their long evolutionary history, they have adapted to almost every lighted environment (Whitton and Potts, 2000).

Reproduction and Morphology

Cyanobacteria largely multiply by the growth and division of vegetative cells, forming longer filaments that break up or larger colonies that break off smaller colonies. In warm water with adequate nutrition, some cyanobacterial species can double their cell numbers in a few days, eventually forming water blooms of hundreds of thousands of cell/mL of water. Depending on the species, these may be seen as a red stain in inshore marine water or as a blue-green scum on freshwater lakes. Water blooms naturally terminate, normally as water temperatures fall in autumn and the wind causes mixing and dispersion. As conditions become more unfavorable for cyanobacteria, a new, larger cell type develops in filaments, called an "akinete." These resistant cells survive the death of the mass water bloom and are capable of resting over winter in water or sediments or on surfaces when water bodies dry up. Not all species have been observed with akinetes; in particular, the majority of species that have spherical cells in colonies appear to survive over the winter in their absence. Most filamentous species that form heterocysts also form akinetes.

In actively growing cyanobacterial suspensions of filamentous species (planktonic organisms) or surface mats of filamentous cyanobacteria (benthic organisms), there is a novel means of dispersal. This is the formation of motile short filaments, which actively swim and disperse from the parent. These are termed "hormogonia," and they grow at the end of a filament from which they detach and migrate. In some species, the hormogonia can aggregate and form a colony. In others, a single hormogonium can form a colony (Whitton and Potts, 2000).

Identification and Classification

Cyanobacteria have a great diversity of forms and habits of growth. These characters were used as the basis for genus and species identification until recently, when genetic techniques have come to the fore. Cyanobacteria can initially be divided into filamentous forms and nonfilamentous forms. There are two orders within the nonfilamentous group, Chroococcales and Pleurocapsales. Both orders are of spherical cells in unicellular or aggregate form, in which a gel-like matrix holds them together. Chroococcales reproduce by binary division in one, two, or three planes or by budding, and they may form symmetrical colonies or randomly distributed colonies of cells. An example is the common and toxic species *Microcystis aeruginosa*, which is discussed at length here because of its impact on human and animal health (Fig. 15-4).

Pleurocapsales reproduce by multiple fission forming daughter cells smaller than the parent cells, or with multiple and binary fission. They also have the capacity to form

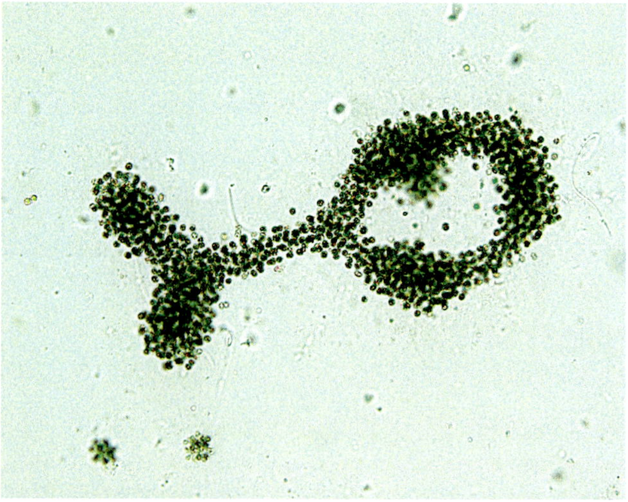

FIGURE 15-4. *Microcystis aeruginosa* colony. Micrograph courtesy Glenn McGregor, Queensland Department of Natural Resources and Water.

motile cells. No toxic species have so far been described, reflecting the relative lack of knowledge of this order.

The filamentous cyanobacteria form three orders. Oscillatoriales divide in one plane to form simple filaments (trichomes), without heterocysts or akinetes. The toxic genus *Planktothrix*, which occurs abundantly in cold freshwaters, is in this order (Fig. 15-5). Other examples are the toxic and dermonecrotic genus *Lyngbya*, which grows on rocks and macroalgae in shallow marine, estuarine, and freshwaters, and the toxic marine bloom-forming genus *Trichodesmium* (Figs. 15-6 and 15-7).

Cyanobacteria of the order Nostocales also divide in one plane to form simple filaments, but they differ from Oscillatoriales as they form heterocysts when grown under low nitrogen conditions. Some species form akinetes. This order includes many of the most abundant cyanobacterial species, including several toxic species that have adverse health effects, which will be discussed later. Examples are the freshwater neurotoxic species *Anabaena circinalis* and the toxic brackish water *Aphanizomenon ovalisporum* and *Nodularia spumigena* (Figs. 15-8, 15-9, and 15-10).

The third order is the Stigonematales, which can divide in more than one plane, forming clumps of branched filaments on surfaces or in free water suspension. Only one toxic species has presently been identified in this order, but this may be a result of more abundant free-floating planktonic species of cyanobacteria receiving the most attention. This

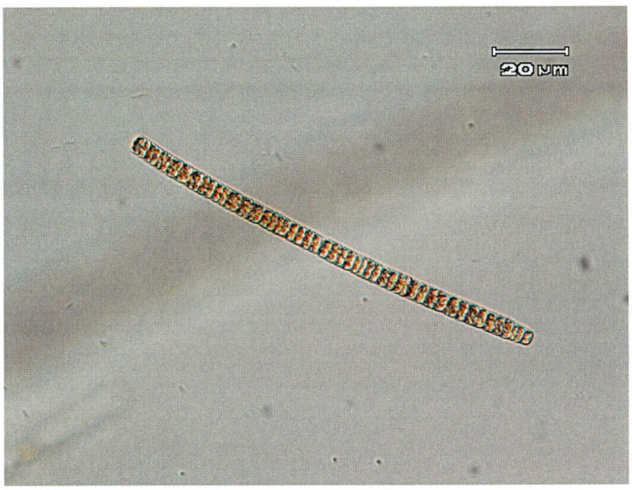

FIGURE 15-5. *Planktothrix agardhii.* Micrograph courtesy Glenn McGregor, Queensland Department of Natural Resources and Water.

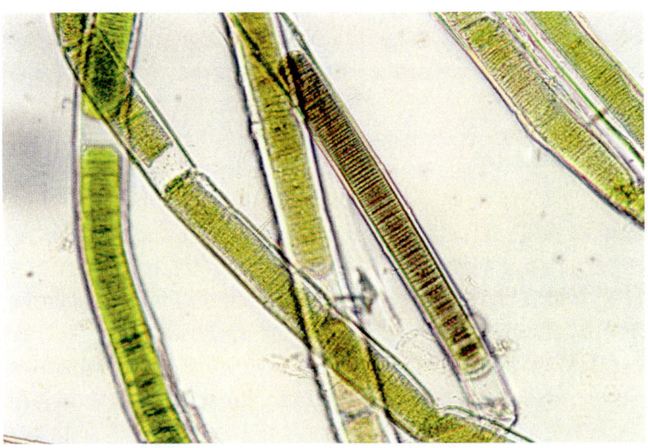

FIGURE 15-6. *Lyngbya majuscula.* Mcrograph courtesy Glenn McGregor, Queensland Department of Natural Resources and Water.

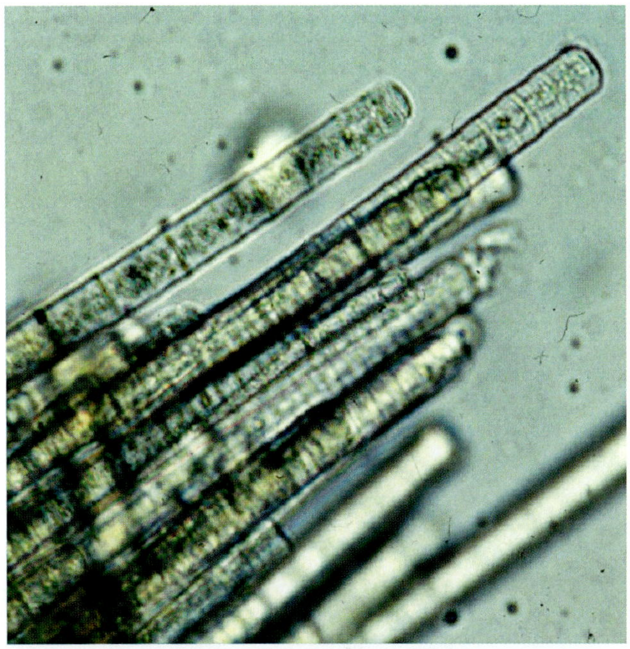

FIGURE 15-7. *Trichodesmium erythraeum.* Micrograph courtesy Glenn McGregor, Queensland Department of Natural Resources and Water.

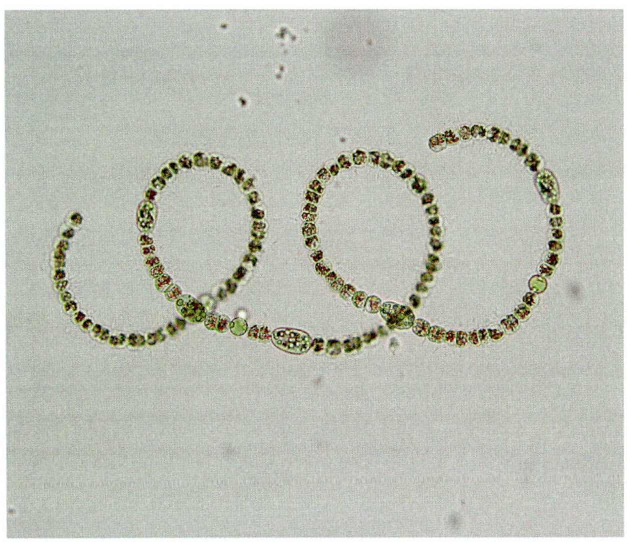

FIGURE 15-8. *Anabaena circinalis* trichome. Micrograph courtesy Glenn McGregor, Queensland Department of Natural Resources and Water.

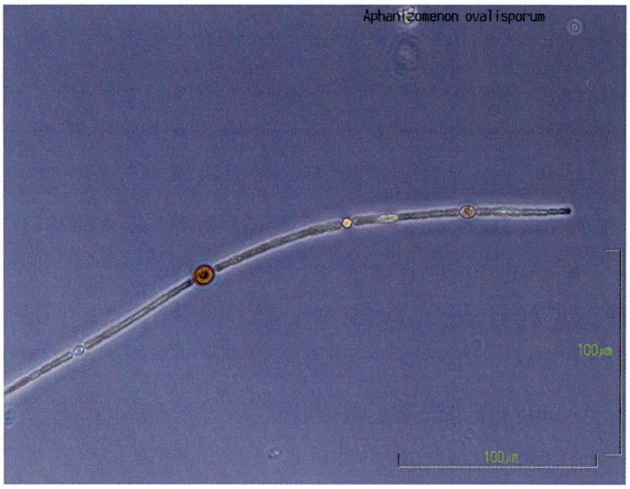

FIGURE 15-9. *Aphanizomenon ovalisporum* trichome. Micrograph courtesy of Phycology Unit, Queensland Health Forensic and Scientific Services.

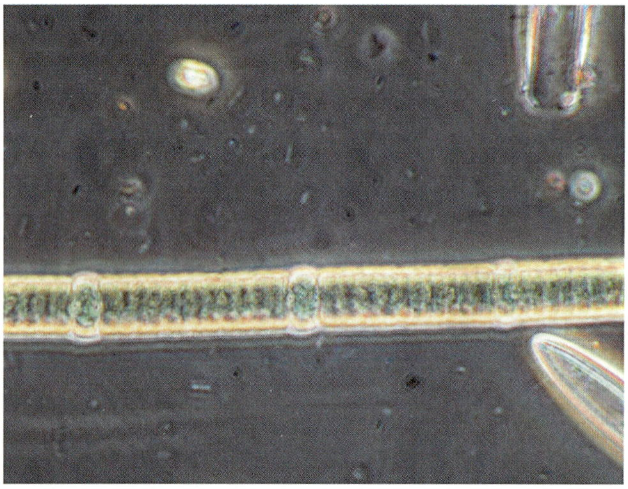

FIGURE 15-10. Detail of *Nodularia* trichome. Micrograph courtesy of Phycology Unit, Queensland Health Forensic and Scientific Services.

species, *Umezakia natans*, was identified in a Japanese lake and shown to be toxic to mice (Watanabe, 1987). The toxin has been characterized and is discussed later (Fig. 15-11).

Far less is known about the toxicity of benthic cyanobacterial species than about the planktonic species. The benthic species are of great importance in soil fertility, particularly in rice cultivation where they provide nitrogen for plant growth. Only if animal poisoning or human injury results from the consumption of these organisms or from contact with the filaments would their toxicity be investigated.

Genetic Approaches to Identification

Several molecular approaches have been successfully applied to the classification of cyanobacterial species. As the

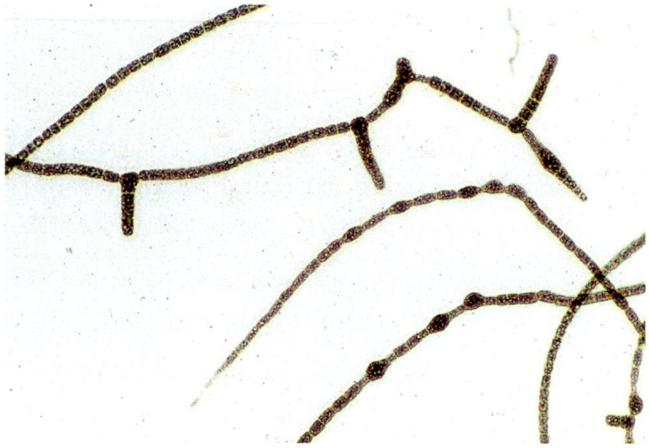

FIGURE 15-11. Umezakia natans. Micrograph by permission of The National Museum of Nature and Science of Japan and M. Watanabe.

genetic information for phycocyanin biosynthesis is highly conserved, in that the base sequences are effectively identical between species, these regions will identify cyanobacteria but are not helpful in classification within the family. However, linking regions, which do not code for proteins and lie between the highly conserved regions, offer a location for study of genus and species differences as they will have mutated as the species evolved. One genetic region that has been studied is the intergenic spacer between the two bilin subunit proteins. DNA "fingerprinting" has been used, employing the polymerase chain reaction to generate DNA replicates, which are then cleaved using endonuclease enzymes to produce characteristic fragments (Neilan *et al.*, 1995). DNA sequencing has further developed this approach, demonstrating that the classification based on morphology was basically correct and that the genetic techniques can be applied to field samples (Baker *et al.*, 2001). However, within the spherical Chroococcales there is much genetic diversity, with less morphological variation, so that the ultimate classification may depend on genetic differences.

Another molecular genetic approach is the sequencing of the 16S ribosomal subunit RNA from cyanobacteria. This method is widely used in strain identification of microbial pathogens and is equally applicable to cyanobacterial identification. Extensive phylogenetic trees have been constructed showing the relationships between strains, species, and genera of cyanobacteria, which provide considerable information of the origins of toxic strains. One organism that has been investigated in detail by this method is *Cylindrospermopsis raciborskii*, found in North America, South America, Europe, and Australasia. This species appears to produce three different toxins: a neurotoxin in South America, a general cytotoxin in Australasia and North America, and an unidentified toxin in Europe. DNA sequence analysis of 16S rRNA from these strains showed three different groupings (the Americas, Europe, and Australasia), which were not the same as the toxin groups (Neilan *et al.*, 2003).

As further genetic analysis is carried out on the cyanobacteria, it will be possible to follow the evolution of the organisms and their distribution around the world. Climate change will affect the species that are present in particular locations, as warmer water species move outward from the tropics into temperate regions, providing new challenges to public health.

TOXINS OF CYANOBACTERIA, CHEMISTRY, BIOSYNTHESIS, AND GENETICS

That such an ancient group of organisms as cyanobacteria produce a diversity of toxins may reflect their need for defense mechanisms against eukaryotic life forms. Grazing invertebrates in marine environments have reduced the abundance of stromatolites to hypersaline locations in the present seas, whereas they have a wide distribution throughout early geological strata, demonstrating successful competition by a more recent eukaryote. Chemical defense is widespread in plants, though the biological function of the plant alkaloids is as disputed as the function of cyanobacterial toxins. Other hypotheses for the presence of toxins, and plant alkaloids, have included their value as storage molecules or chelating agents.

There are three major groups of cyanobacterial toxins: the cyclic peptides, the alkaloids, and the polyketide/polycyclic derivatives. The most abundant toxins are the peptide group, named after the genera in which they were first isolated. They are contained largely within the cells in which they were synthesized and hence have been regarded as endotoxins. This is misleading, as the toxins are soluble and leak out of the cells, with some alkaloid toxins being largely in the free water in which the cyanobacteria are growing (Falconer, 2005). The word *endotoxin* is best reserved for the cell membrane toxins of gram-negative bacteria.

Cyclic Peptide Toxins-Microcystins and Nodularins

These toxins have the general structure illustrated in Figure 15-12 and are cyclic peptides linked through a unique β-linked amino acid (ADDA). The structure of microcystin, from *Microcystis* sp., is a cyclic heptapeptide. A family of these toxins has now been identified, and the first completely characterized had the L-amino acids leucine and alanine, together with five other D-form or unusual amino acids. The sequence of the peptide ring is γ-linked D-glutamic acid, N-methyldehydroalanine, D-alanine, L-alanine, β-linked erythro-β-methylaspartic acid, L-leucine, and the completely novel β-amino acid, abbreviated to ADDA (3-amino-9-methoxy-10-phenyl-2,6,8,-trimethyldeca-4,6-dienoic acid). The molecular weight of this microcystin variant, described as microcystin-LA after the two L-form amino acids, is 909 daltons. Other variants have a variety of alternative L-amino acids and substitutions with molecular weights up to 1115 (Sivonen and Jones, 1999).

The other related family of cyclic peptide toxins are the nodularins from the marine cyanobacterium *Nodularia*, which are cyclic pentapeptides also containing ADDA. The sequence in the peptide ring is γ-linked D-glutamic acid, 2-(methylamino)-2-dehydrobutyric acid, β-linked erythro-β-

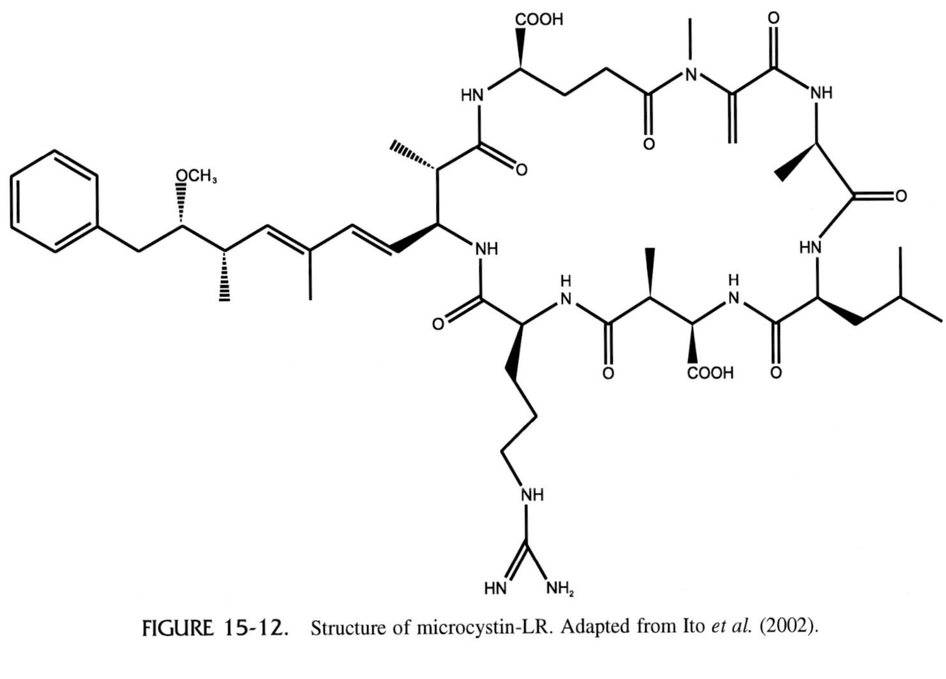

FIGURE 15-12. Structure of microcystin-LR. Adapted from Ito *et al.* (2002).

methylaspartic acid, L-arginine, ADDA. The molecular weight is 824. Minor chemical variants have also been identified.

Since the original identification of the structure of microcystin-LA, more than 90 molecular variants have been identified. The most abundant, and the variant regarded as the type against which the others are compared, is microcystin-LR, having L-leucine and L-arginine in the two positions occupied by L-form amino acids. The different variants are six to ten times less toxic than microcystin-LR. In general, the more hydrophobic variants (for example, those with L-tyrosine, L-alanine, or L-methionine) are more toxic than those with two L-arginine residues (Sivonen and Jones, 1999).

The ADDA component of the molecule is essential for toxicity, and even a stereochemical change in the side chain reduces toxicity greatly. The molecule binds firmly to the active site of protein phosphatase in eukaryotic cells, as discussed later. Nodularins have comparable toxicity to microcystin-LR.

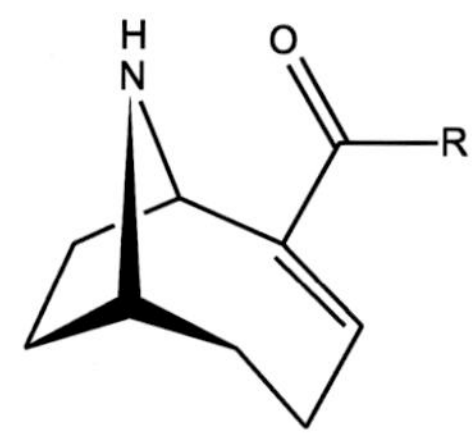

FIGURE 15-13. Structure of anatoxin-a (R = CH_3) and homoanatoxin-a (R = CH_2CH_3). Adapted from Wonnacott and Gallagher (2006).

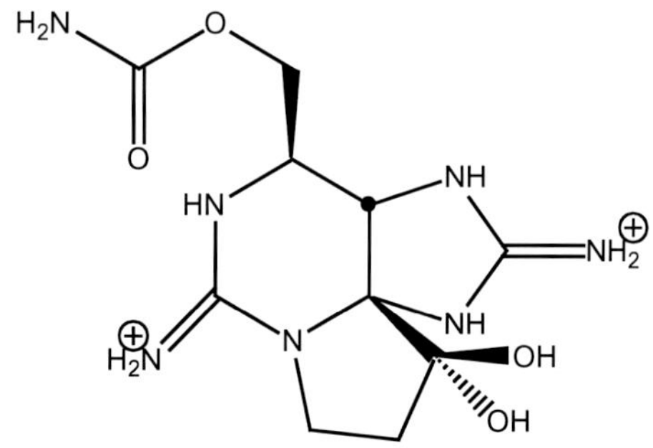

FIGURE 15-14. Structure of saxitoxin. Adapted from Kellmann and Neilan (2007).

Biosynthesis

Since the initial isolation of microcystin from the colonial unicellular *Microcystis*, it has been identified in many other common cyanobacterial genera, including the filamentous *Planktothrix, Anabaena*, and *Nostoc*. Genetic analysis of the potential biosynthetic mechanism has been successful. The microcystin molecule resembles the peptide antibiotic gramicidin synthesized by a Gram-positive bacterium, which had earlier been shown to be produced by a nonribosomal peptide synthetase complex. Using DNA sequence information for the microbial enzyme, it was possible to locate the genetic elements in the cyanobacterium *Microcystis aeruginosa* responsible for microcystin synthesis. Two quite separate pathways are involved. One is the nonribosomal peptide synthetase pathway, which is responsible for the amino acids of the ring, with the exception of ADDA. This unusual amino acid is synthesized by polyketide synthase, which is a sequence of enzymes similar to those involved in fatty acid synthesis. For a detailed discussion, see Borner and Dittmann (2005) and Falconer (2005).

Alkaloid Toxins

Cyanobacteria, like higher plants, synthesize a range of alkaloid toxins. The smallest toxic alkaloid so far characterized is a neurotoxin called anatoxin-a, isolated first from the cyanobacterium *Anabaena flos-aquae*. This is a secondary amine 2-acetyl-9-azabicyclo(4-2-1)non-2-ene (Fig. 15-13). The structure is similar to cocaine and acetylcholine and blocks synaptic transmission. The compound is quite stable and has a half-life in water of 14 days in laboratory tests (Sivonen and Jones, 1999).

Anatoxin-a, and the closely related homoanatoxin-a were identified as a result of animal poisonings. These alkaloids are produced in the genera *Anabaena, Phormidium, Raphidiopsis, Aphanizomenon*, and *Oscillatoria*, both in planktonic strains and in benthic mats of cyanobacteria. All these species are filamentous and occur widely in freshwater. Not all strains of these cyanobacteria produce anatoxins, and even within a single lake, some samples may be toxic and others not so. The pathway for anatoxin-a biosynthesis is under investigation, commencing with glutamic acid and acetate. The general scheme is that of polyketide synthesis, which has been the subject of genetic analysis, with the final step being decarboxylation (Selwood *et al.*, 2007). Similar mechanisms can be expected to produce homoanatoxin-a, though in different species of cyanobacteria.

Saxitoxins, illustrated in Figure 15-14, are a family of tricyclic alkaloids that are produced by both marine dinoflagellates and freshwater cyanobacteria. As discussed in Chapter 13, their main identified human health impact has been as the cause of paralytic shellfish poisoning (PSP) because of the accumulation of saxitoxins within filter-feeding marine and estuarine shellfish.

In freshwaters, there are a number of species of cyanobacteria that produce saxitoxin and its derivatives, including *Aphanizomenon flos aquae* (United States, Canada, Portugal), *Anabaena circinalis* (Australia), *Anabaena lemmermannii* (Scandinavia), *Lyngbya wollei* (United States), *Cylindrospermopsis raciborskii* (Brazil), and *Planktothrix* sp. (Italy). It can be expected that further species and genera will be identified in the future. The concentration of saxitoxin in water during a cyanobacterial bloom can be sufficiently high to be a potential health hazard, without shellfish accumulation. Levels as high as 1 mg/L have been recorded in Finland (Rapala *et al.*, 2005).

There are a series of variants of the basic saxitoxin molecule; the mechanism of toxicity is well understood. Saxitoxin is the most toxic congener by weight; the sulfated C-toxins are the least toxic. Saxitoxins exert pathological effects by blocking the movement of sodium ions through voltage-gated sodium channels in motor nerves, resulting in a conduction block that manifests as acute respiratory insufficiency from partial paralysis of the diaphragm and accessory respiratory muscles. Severe intoxications can result in hypoxia and death by asphyxiation. The genetic basis of saxitoxin biosynthesis is currently under investigation (Pomati *et al.*, 2006).

Cylindrospermopsins

These alkaloid toxins are illustrated in Figure 15-15 and constitute a tricyclic ring structure containing a guanido group, with a sulfate and a methyl group attached. The ring is bridged to hydroxymethyl uracil, which can occur in keto and enol forms at neutral pH. As a result of the negatively charged sulfate group and the positively charged guanido group, the molecule is a zwitterion with high water solubility (Ohtani *et al.*, 1992). The molecule can be bridged by the hydroxymethyl in α or β configuration, which does not appear to affect toxicity. Loss of the oxygen of the hydroxy group reduces but does not abolish toxicity. All three variants of the basic molecule are found naturally in cyanobacteria.

The molecule has been chemically synthesized, with considerable difficulty. Biosynthesis has been studied using genetic and isotopic techniques. The genetic approach uses an examination of nucleotide sequences on the DNA of the synthesizing organism, and comparing these to known sequences for enzymes that are likely to be involved. One of the important pathways for bioactive compound production is the enzyme complex polyketide synthase, which has been extensively studied in fungal antibiotic biosynthesis.

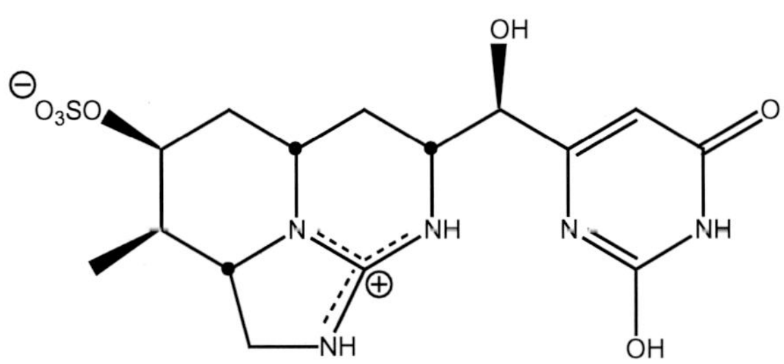

FIGURE 15-15. Structure of cylindrospermopsin. Adapted from Heintzelman *et al.* (2002).

Comparative examination of cyanobacterial species that synthesize cylindrospermopsin and those that do not has demonstrated that polyketide synthase was present in those producing the toxin and absent in those that do not (Falconer, 2005). This is not proof, but highly indicative of the importance of this enzyme system. Final proof will need a "knock-out" insertion into the enzyme complex genes; if this mutation blocks synthesis, then these are crucial genes.

Isotopic studies using stable isotopes that are rare in nature (such as ^{2}H, ^{12}C, ^{15}N, and ^{18}O) have shown the origins of the atoms in the cylindrospermopsin molecule. ^{13}C acetate formed the carbon backbone, with guanidinoacetic acid the starting molecule. This is consistent with the function of a polyketide synthase (for a more detailed discussion, see Falconer, 2005).

HEALTH EFFECTS OF HEPATOTOXIC FRESHWATER SPECIES

Peptide Toxins: Nodularins and Microcystins

Cyanobacterial toxicity was first identified as a result of livestock poisoning in Australia in 1878. Various farm animals were drinking from a brackish estuarine lake, which had a green scum of the filamentous cyanobacterium *Nodularia spumigena* along the margins. The government analyst, George Francis, was sent to investigate the deaths of horses, sheep, pigs, and dogs on lakeshore farms. He found that a "thick scum like green oil paint" had drifted onto the lee shores of the lake, which had a low water level and was unusually warm (74°F; 23°C). He described the deaths of animals drinking the scum as "stupor and unconsciousness, falling and remaining quiet, as if asleep, unless touched when convulsions come on, with head and neck drawn back in rigid spasm, which subsides before death" (Francis, 1878). The interval between drinking and death was relatively short; the longest time was seen in horses, which died between 8 and 13 hours after consuming the scum. He dosed a sheep that received 30 oz (840 mL) of scum and died 15 hours later. He reported that there were no changes in lungs, liver, kidneys, or brain (Francis, 1878). This differs from current pathological reports on poisoning from cyanobacterial hepatotoxins (discussed later), but no one has since dosed sheep with *Nodularia*. The toxin present, nodularin, has been well characterized since the original event (discussed previously). *Nodularia* still forms water blooms in this lake, which cause the shutdown of drinking water supplies from this source when scums approach the water intakes.

Dog deaths from this organism are more common than deaths of other farm animals, particularly on the shores of the Baltic Sea, where scums of *Nodularia* frequently occur during summer (Edler *et al.*, 1985). These deaths are probably the result of dogs habitually licking off material attaching to their coats after swimming, resulting in their swallowing scum. No human poisoning has been reported from *Nodularia* consumption, though nodularin does occur in shellfish (such as mussels) growing in waters containing the organism. Shellfish harvesting is banned when bloom conditions occur near shellfish beds.

A much more frequent source of livestock poisoning is *Microcystis aeruginosa*, which forms scums in late summer in eutrophic freshwater bodies. These organisms are buoyant and move up and down the water column (Falconer and Humpage, 2006) becoming trapped on the surface. When a slight wind drifts the cells onto the shoreline, concentrated scums may occur which are highly toxic. Sheep and cattle deaths have been widely reported, in some cases with thousands of animals dying (Harding and Paxton, 2001). Postmortem examinations of sheep poisoned by this organism show massive liver damage and small hemorrhages throughout the abdominal cavity with accumulation of yellow fluid. Deaths are the result of liver failure and may occur for some weeks after the poisoning, though most mortality is within 24 hours.

Human poisoning by identified microcystins has been investigated in only three cases. There have been numerous other accounts of gastroenteritis and liver dysfunction associated with drinking water contaminated with cyanobacteria, but there has been no identification of the actual toxin(s) (Falconer, 2005). The earliest cases of human injury were in a rural town (Armidale, NSW, Australia), which obtained its drinking water supply from a reservoir that intermittently had large *Microcystis* blooms in late summer. During one of these blooms, scum samples were collected near the drinking water off-take and shown to be toxic to mice. The predominant toxin was later shown to be microcystin-YM (L-tyrosine, L-methionine). As a result of complaints from the population of taste and odor in the tap water, the drinking water utility dosed the reservoir with copper sulfate, which killed the cyanobacterial cells ending the bloom but liberated the toxin into the free water. A retrospective analysis of data was carried out from blood samples collected before, during and just after, and well after the bloom. Elevated hepatic enzymes in blood samples collected from residents indicated liver malfunction. It was shown that the indicator enzyme for toxic liver injury—gamma glutamyl transferase—increased *only* in the blood of residents collected during the bloom, compared with before and after the bloom, and also with blood from people living in adjacent towns at the same time (Falconer *et al.*, 1983). These results caused an international assessment of drinking water safety when cyanobacterial blooms were present in drinking water supplies, ultimately leading to the World Health Organization (WHO) setting a guideline value for microcystin in drinking water (WHO, 2006).

The second group of human cases were recruits undergoing army training in the United Kingdom. Their exercises, including canoe rolls and swimming, took part in a lake with a heavy bloom of *Microcystis*. Within 4 to 5 days, two recruits were admitted to the medical center with blistering round the mouth, sore throat, vomiting, abdominal pain, dry cough, and pneumonia. Eight other recruits were found to have similar symptoms and diarrhea. Microcystin-LR was identified in bloom samples. The inhalation toxicity of microcystin was later shown to be 10 to 50 times more than oral toxicity, and it is likely that in these cases the individuals had both swallowed and inhaled the toxic organisms (Turner *et al.*, 1990).

The third group of human cases were far more severe. These were the poisoning of 131 patients during hemodialysis for renal impairment at a clinic in Brazil. Almost immediately after dialysis, the patients had visual disturbances, nausea and vomiting, headache, muscle weakness, and abdominal pain. One hundred patients developed liver failure, and 76 died in the following months. Fifty-two of the deaths were directly attributable to liver failure. Brazilian authorities in association with the U.S. Centers for Disease Control and Prevention investigated these cases. Two cyanobacterial toxins were found in the water filters of the dialysis unit: microcystin and cylindrospermopsin, both of which cause liver injury (Carmichael *et al.*, 2001; Jochimsen *et al.*, 1998; Pouria *et al.*, 1998). Microcystin was found in the liver and blood of patients, and the histopathology of the liver showed extensive damage. It was concluded that the patients had been killed by dialysis with water containing microcystin, causing severe liver injury and failure.

The origin of the toxins was a drinking water reservoir supplying the municipal water treatment plant; the reservoir was frequently affected by cyanobacterial blooms. The clinic was not connected to the main piped water supply and obtained its water by tanker. The exact location from which the water was collected by the tanker could not be identified, but it was before treatment. No monitoring for cyanobacteria in the reservoir had taken place for some time. Kidney patients are particularly at risk from contaminated water used in dialysis, as the process is equivalent to passing 120 to 150 liters of water across the patient's blood supply. Soluble chemicals of lower molecular weight can pass across the dialysis membrane in both directions.

The Brazilian clinic had ion exchange and activated carbon filters in its in-house water purification system, which were the filters from which the toxins were identified. Standard drinking water treatment will remove intact cyanobacterial cells from the incoming water but will not reliably remove toxins during a cyanobacterial bloom. More sophisticated treatment with activated carbon filtration or ozonation will effectively eliminate toxins from drinking water (Falconer, 2005). Later examination of the cyanobacterial population of the water supply reservoir from which the clinic obtained its water did not identify the toxic species.

A number of cyanobacterial species produce microcystins. The most common in Australia and South Africa are in the *Microcystis* genus; in Europe the most important are in the *Planktothrix* genus, while some *Anabaena, Nostoc, Anabaenopsis*, and *Snowella* strains also produce these toxins (Sivonen and Jones, 1999). From the point of view of public health, any substantial cyanobacterial population in a water source must be viewed with caution as potentially poisonous.

Mechanisms of Action

There has been extensive experimental investigation of *Microcystis* poisoning, and of the toxicity of the predominant peptide hepatotoxin microcystin-LR. By intraperitoneal injection the LD_{50} (i.e., the dose that will kill 50% of the test animals within 24 hours) is about 50 μg/kg bodyweight in mice. Other variants of the microcystins with amino acids of lower hydrophobicity have LD_{50} up to 600 μg/kg (Sivonen and Jones, 1999). This level of toxicity for the pure compounds demonstrates that these substances are highly toxic compared to potassium cyanide with an LD_{50} of about 6 mg/kg in mice.

The biochemical sites in liver cells that are inhibited by microcystins and nodularins are located in protein phosphatase enzymes. These enzymes remove phosphate from the hydroxyl groups of the amino acids serine and threonine, which are constituents of particular proteins. In living cells, sensitive control mechanisms occur for metabolic processes, which often depend on reversible changes in molecules. One of these reversible changes is addition and removal of phosphate from enzymes. If the removal of phosphate groups is inhibited by microcystins or nodularins, the proteins that are the targets of this reversible process will accumulate excess phosphate. In the case of the cytokeratins, which make up the intermediate filaments of the cell cytoskeleton, accumulation of phosphate groups leads to disaggregation of the filaments (Falconer and Yeung, 1992).

Exposed to microcystin the hepatocytes deform and the structure of the liver breaks up. In mice, blood pools in the liver to the extent that the animal rapidly dies of shock. The effects of microcystin poisoning are not limited to acute death by shock or liver failure. Protein phosphatase enzymes 1 and 2A are inhibited, and in turn regulate the activity of a wide range of other enzymes. Cell cycle control is affected at very low concentrations of microcystin, and microcystin has been shown to act as a promoter of tumor growth.

There is increasing evidence that microcystins and nodularins are possible carcinogens resulting in the World Health Organization's International Agency for Research in Cancer (IARC) establishing in 2006 an Expert Group to evaluate the scientific literature on cyanotoxins. This evaluation investigated human epidemiological evidence that microcystins or nodularins in water supplies may be associated with human cancer rates, animal trials in which these toxins were injected or fed to rodents may cause tumor growth, and the molecular mechanisms by which microcystins or nodularins may cause or promote cancers. The overall IARC evaluation concluded that microcystin-LR is a "possible carcinogen" and that insufficient evidence is available for evaluation of other microcystins and nodularin (Grosse *et al.*, 2006). The whole evaluation is to be published in IARC Monograph 94.

Alkaloid Hepatotoxins: Cylindrospermopsins

These toxins came to attention as a result of human poisoning, which started the investigations into the nature of their toxicity. A population of indigenous Australians living on a tropical island off the east coast of Australia called Palm Island suffered a severe outbreak of hepatoenteritis among the children. Only a few adults were affected. The symptoms were unusual: abdominal pain, swollen livers, constipation, and vomiting. Later, the children had profuse bloody diarrhea. Analysis of urine showed glucose, protein, ketones, and some blood, demonstrating considerable kidney damage (Byth, 1980). This acute renal injury was life threatening, and 82% of the 138 children treated had acidosis and hypokalemia (low blood potassium). Of these, the majority were treated by intravenous replacement of fluids and electrolytes, whereas the most severely poisoned also received plasma proteins in an intensive care unit. Within 26 days, all the children had recovered.

Investigation of the cause of the outbreak initially focused on microbial pathogens, but none were detected. The water supply was identified as the source of the problem, as individual families that did not receive their water from the main treated supply but from their own wells did not get sick. The reticulated water supply was drawn from a small storage reservoir, which was at a low level, and complaints had been made about taste and odors in the drinking water before the outbreak. The reservoir was found to have a water bloom of cyanobacteria, so the authority dosed the storage with copper sulfate to kill the organisms. Within days, the children began to get ill. The drinking water was supplied through a standard treatment plant with filtration and chlorination but no advanced treatment (such as activated carbon filtration or ozonation).

Samples of the cyanobacteria were collected from the reservoir and found to be a known tropical species, *Cylindrospermopsis raciborskii.* These organisms were cultured, and the cells tested for toxicity in mice. This cyanobacterial species was found to be highly toxic, causing liver and kidney damage as well as injuring a range of other organs (Hawkins *et al.*, 1985). Later the toxic alkaloid was purified and characterized and named cylindrospermopsin (Ohtani *et al.*, 1992).

Since the initial identification of cylindrospermopsin in cyanobacteria from the Palm Island drinking water supply, analytical methods have been developed for its identification. The most accurate is mass spectrometry, which is also the most costly to use. The molecule can also be identified by high performance liquid chromatography (HPLC), using a known sample as a standard, and by enzyme-linked immunoassay (ELISA). By employing these methods, cylindrospermopsin has been identified in a number of other species of cyanobacteria, including *Aphanizomenon ovalisporum* (Israel), *Umezakia natans* (Japan), *Anabaena bergii* and *Lyngbya wollei* (Australia), and *Aphanizomenon flos-aquae* (Germany). Water bodies in both North and South America have been shown to contain cylindrospermopsin, so it is clear that it is a worldwide contaminant of freshwaters (Falconer, 2005). It is likely that more species producing this toxin await identification.

Cylindrospermopsin is under further investigation to clarify the mechanism of toxicity. A primary component of the poisoning by cylindrospermopsin is the inhibition of protein synthesis, which is the mode of action of several other potent toxins, including the fungal poison amanitin, the protein poison ricin, and the diphtheria toxin. The protein synthesis inhibition has been shown in a cell-free system to be because of the unaltered molecule. These toxins do not rapidly kill poisoned individuals (unlike neurotoxins or respiratory poisons which kill within minutes or a few hours). Protein synthesis inhibitors cause death over days through progressive organ failure, particularly of the liver and kidneys.

One of the puzzles over cylindrospermopsin poisoning is that high doses in mice kill rapidly ($LD_{50\ 24hours}$ 2 mg/kg intraperitoneal), whereas lower doses (0.2 mg/kg) cause death over a week. This effect is likely the result of two independent mechanisms of action, with the more rapid action resulting from a toxic metabolite of the cylindrospermopsin molecule. The toxic metabolite may also be responsible for the DNA damage and possible carcinogenic actions resulting from dosing with cylindrospermopsin (Falconer and Humpage, 2006).

It appears likely that cylindrospermopsin is carcinogenic at low doses in mammals, but the definitive research has not yet been done because of the high cost of the toxin and the large amount needed to do a long-term carcinogenicity trial. This issue is currently under consideration by the U.S. Environmental Protection Agency (EPA). As detailed investigation of this toxin has only just begun, much is currently unknown. The possibility of teratogenic effects on embryos is under active study, and experiments are being planned to look at cancer formation in animals supplied with the toxin in drinking water.

CONCLUSIONS ON THE KNOWN HEALTH IMPACTS OF CYANOBACTERIA AND ASSOCIATED HEALTH REGULATIONS

Cyanobacteria are present in soil, freshwater and marine environments. Their proliferation depends on the availability of nutrients, which can come from rocks or from biological sources. With our increasing human population, nutrients arising from agriculture or aquaculture and from animal and human waste are accumulating in fresh and marine waters. As a consequence, cyanobacterial blooms are becoming more widely distributed and at higher densities. Climate change may exacerbate bloom growth by increasing summer water temperatures, thus providing a longer growing season for cyanobacteria.

Because of the risks of human injury from both recreational and drinking water exposure to cyanobacterial toxins, health authorities in many countries now specify guideline levels for cyanobacteria and for microcystin. In Europe and in Australia, the guidelines for recreational exposure to cyanobacteria are set on the basis of cell counts in the water. Although it is the responsibility of local authorities to legislate safety levels for cyanobacteria, the European Union, the WHO, and national authorities in Australia, Brazil, and other countries have already provided legislation or guidance on standards for exposure to cyanobacteria and microcystins (Chorus, 2005).

For recreational exposure, the WHO recommends that a level of 20,000 cyanobacterial cells/mL of water sample provides a mild/low probability of adverse health effects, and it is sufficient to notify visitors to bathing sites by warning signs of this risk. The guidelines also recommend increased surveillance of the site. At cell concentrations up to 100,000 cells/mL, a moderate risk of short-term adverse health effects such as skin rash and gastrointestinal illness occurs, with the possibility if the water is swallowed of long-term illness (WHO, 2003).

If the local concentration of cyanobacterial cells reaches scum proportions, there is a high probability of adverse health effects from body contact, inhalation, or ingestion. The potential for acute poisoning occurs, and cases of pneumonia from inhaled water have been reported. Immediate action to control human contact with scums, with possible prohibition of swimming and other body contact water sports is recommended (WHO, 2003). In practice, most authorities controlling bathing areas close them for public use if there is a cyanobacterial scum on the beach.

In a similar process, the WHO set drinking water standards through published guidelines. Because of the presence of toxic cyanobacteria in many drinking water storage reservoirs at concentrations ranging from negligible to potentially dangerous, WHO has set a guideline value of 1 μg/L for microcystin in drinking water (WHO, 2006). This was based largely on oral toxicity testing in mice, supported by oral toxicity in pigs, and the guideline value has been adopted with small variations in a number of countries. No long-term carcinogenesis study has yet been undertaken and no similar reference dose has been set for the United States at the time of writing.

MARINE CYANOBACTERIA

Trichodesmium

The genus *Trichodesmium* is the most widely studied open-ocean cyanobacterium. It is a colonial filamentous type, taxonomically assigned to the order Oscillatoriales. The trichomes form buoyant macroscopic spherical or fusiform aggregates. *Trichodesmium* sp. are widely distributed across oligotrophic pelagic waters in the tropics and subtropics and can form massive surface blooms that appear red or brownish-yellow because of the presence of the cyanobacteria-specific photosynthetic pigment phycoerythrin and photoprotective carotenoids. *Trichodesmium* blooms have been observed for centuries: *Trichodesmium erythraeum* was first described in 1830 from bloom material collected from the Red Sea, with the accompanying suggestion that the Red Sea was named after such blood-red *Trichodesmium* blooms. The botanist Joseph Banks in 1770 described "sea sawdust" surface blooms in South Pacific waters, subsequently attributed to *Trichodesmium*, and Charles Darwin made detailed reports of large *Trichodesmium* blooms in pelagic and coastal southern Atlantic Ocean waters (Capone *et al.*, 1997; Capone and Carpenter, 1999; Paerl, 2000; Walsby, 1992). *Trichodesmium* can form very extensive surface blooms, with some individual bloom surface areas estimated at 7000 km^2 (Gulf of Thailand), 52,000 km^2 (Great Barrier Reef area, north-east Australia), 90,000 km^2 (New Caledonia-Vanuatu, South Pacific), and 1×10^6 km^2 (Arabian Sea) (Capone *et al.*, 1997; Carpenter and Capone, 1992; Suvapepun, 1992).

Trichodesmium does not develop heterocysts in its trichomes but is nevertheless an efficient nitrogen-fixing cyanobacterium, and makes a significant contribution to oceanic nitrogen budgets. The ability to convert atmospheric nitrogen into ammonia is particularly important in open ocean waters that are chronically nitrogen deficient; ammonia in such waters is a new source of nitrogen that surmounts N-limited growth and thus supports the growth of heterotrophic bacteria, eukaryotic algae, and animals. The nitrogen-fixing mechanisms adopted by *Trichodesmium* appear to be unusual in comparison to the strategies adopted by other N-fixing cyanobacteria insofar as *Trichodesmium* fixes nitrogen (N_2) during daylight (i.e., concurrent with photosynthetic oxygenesis).

Other diazotrophic (diazotrophic: capable of reducing ["fixing"] atmospheric nitrogen into available forms, mainly ammonia) cyanobacteria have strategies to protect nitrogenase, the enzyme complex that catalyses the reduction of N_2 to NH_3 from oxygen, which inactivates nitrogenase. Heterocystous cyanobacteria have specialized cells, "heterocysts," that facilitate N_2-fixation in microanaerobic conditions; other nonheterocystous cyanobacteria fix nitrogen in oxygen-deplete niches (like sediments, mudflats, reefs, and fouling layers). Some coccoid and filamentous forms adopt a temporal separation of N_2 and O_2, with distinct diel cycles of photosynthesis during the day and peak nitrogen fixation at night.

The conundrum presented by *Trichodesmium* nitrogen fixation (where growth of the organism occurs in oxygen-rich surface waters, and no temporal separation of photosynthesis and N_2 fixation occurs to protect nitrogenase from intracellular O_2) has resulted in *Trichodesmium* being the most widely studied oceanic cyanobacterium. The answers are complex and may in part be related to the propensity of *Trichodesmium* trichomes to form dense macroaggregates in field conditions in calm ocean waters. Partitioning of N_2 and CO_2 fixation is suggested by optically dense, oxygen deplete core regions of aggregates, where efficient N_2 fixation occurs, and diffuse, projecting, photosynthetically active external trichomes. Spatial segregation of nitrogen fixation within trichomes appears to be accomplished by morphologically distinct cell zones (diazocytes) in some trichomes, and some (not all) genes responsible for early stages of heterocyst differentiation in heterocystous cyanobacteria are found in *Trichodesmium.* Contributing to the complexity of this mechanism are observations from laboratory studies that demonstrate single trichomes of *Trichodesmium* are able to both photosynthesize and fix nitrogen at low levels of irradiance (i.e., cellular pO_2 remains below the threshold at which nitrogenase is inactivated). *Trichodesmium* can form symbiotic associations with a variety of microflora, which may promote the nitrogen-fixing efficiency of the cyanobacterium through localized oxygen consumption (El-Shehawy *et al.*, 2003; Gallon and Stal, 1992; Paerl, 1999, 2000; Zehr, 1992).

Trichodesmium has been shown to produce toxic substances, but some toxins remain unidentified and uncharacterized, and the confirmatory identification of *Trichodesmium* as the source organism of identified cyanotoxins remains outstanding. A neurotoxic factor caused increased respiratory rate, convulsions and death by acute respiratory failure in laboratory mice at 2 to 30 minutes after exposure; invertebrate assays (the brine shrimp *Artemia salina* and freshwater fleas *Daphnia pulex* and *Daphnia magna*) and the Microtox assay supported the *in vivo* toxicity findings and evidence of differential toxicity across several Caribbean *Trichodesmium* bloom samples, with *T. thiebautii* samples demonstrating toxic effects and *T. erythraeum* being nontoxic (Hawser *et al.*, 1991; Hawser and Codd, 1992). The same *Trichodesmium* species demonstrated differential toxicity to various copepods, with some species affected by *T. thiebautii*, whereas specialized grazers were resistant to toxic effects (Hawser *et al.*, 1992). Healthy *Trichodesmium* cells from a bloom near North Carolina were nontoxic to the copepod *Arcatia tonsa*, but exposure to senescent or lysed cells caused high levels of mortality, which suggests the activity of an intracellular toxin. Affected copepods displayed unusual and extreme responses to the putative toxin, with distension and prolapsing eversion of the gut (Guo and Tester, 1994).

The toxicity of *Trichodesmium* has been discussed in reviews by Sellner (1997) and Codd (1999): early reports from 1944 and 1967 suggested that toxins may have harmed fish, birds, and invertebrates, as well as the physical effects of bloom material on fish gills. *Trichodesmium* blooms have been temporally associated with mass deaths of fish, crabs, sea urchins, and mollusks near Ceylon (Chacko, 1942), avoidance behavior of tuna in the Arabian Sea (Nagabhushanam, 1967), and mass mortalities of fish and shrimp in aquaculture ponds on the east coast of Thailand, with significant economic losses, although it was unclear whether these incidents in 1983 were caused by toxic materials or local oxygen depletion (Suvapepun, 1992). High mortalities in farmed pearl oysters were temporally associated with severe *T. erythraeum* blooms off Western Australia in 1996, with affected oysters displaying pathological signs in digestive gland tissues. Low levels of saxitoxins were detected by HPLC in some affected oysters, but the toxin was not found in *Trichodesmium* extracts (Negri *et al.*, 2004).

Known cyanotoxins are reportedly associated with *Trichodesmium*. Bloom material from the Caribbean was found to contain n-sulfocarbamoyl toxins (C toxins) and gonyautoxin-3. These toxins belong to the saxitoxin group, although saxitoxin and neosaxitoxin were not detected (Jackson *et al.*, 2001). Microcystins were identified by HPLC with photodiode array detection and confirmatory immunoassay from *Trichodesmium* bloom samples collected from the Canary Islands (Ramos *et al.*, 2005). *Trichodesmium* was not unequivocally confirmed as the source organism in either of these cases; such confirmation requires the isolation and culture of *Trichodesmium* and production of toxin in a laboratory setting.

Trichodesmium blooms function as a habitat for a diverse range of organisms, including bacteria, other cyanobacteria, eukaryotic microalgae, protozoa, fungi, hydrozoans, and copepods (Capone *et al.*, 1997; Ramos *et al.*, 2005), so the question must be asked as to whether epiphytic dinoflagellates might have been present in sufficient numbers in the Caribbean *Trichodesmium* bloom to produce the detected C toxins and gonyautoxin. Similarly, other cyanobacteria may have been the source of microcystins found in the Canary Islands bloom material, though the current absence of confirmatory investigations does not detract from the high degree of suspicion that *Trichodesmium* is indeed capable of producing saxitoxins and microcystins. *T. erythraeum* in culture has produced an intracellular cyclic peptide, an undecapeptide named trichamide, which is reportedly the first novel compound isolated from *Trichodesmium*. The toxic properties (or lack of them) of trichamide remain to be elucidated (Sudek *et al.*, 2006).

The observation that *Trichodesmium* blooms constitute a complex microhabitat warrants further investigation into the potential for oceanic and coastal blooms to act as a reservoir for *Vibrio cholerae*, the bacterium that causes cholera. Cholera epidemics are seasonally linked to coastal blooms off Bangladesh, and algal blooms, including the cyanobacterium *Anabaena variabilis*, are thought to function as an environmental reservoir for *Vibrio*. Copepods are also strongly suspected to be carriers of cholera, as *Vibrio* and other enteric pathogens have been shown to preferentially attach to live and molted zooplankton exoskeletons (Colwell, 1996; Epstein, 1993; Islam *et al.*, 1999).

The only report (to our knowledge) of human health impacts from *Trichodesmium* is that of Satô *et al.* (1963 and 1964). This report details phychological investigations into a *Trichodesmium* bloom that occurred regularly off the coast of Recife in northeast Brazil. Anecdotal reports of an acute febrile illness arose from the region of Tamandaré, 120 km south of Recife. The illness is characterized by respiratory distress, debility, myalgia, arthralgia, and postorbital headache, sometimes accompanied by skin rash affecting the upper body, and is associated by local inhabitants with the seasonal (summer) appearance of a "red tide." The disease reportedly affected "almost all the population," and mass mortality of fish accompanied the red tide. The illness appears to have assumed legendary status among the local population, who named it "Tingui," and a Brazilian physician who investigated the disease in 1943 coined the term "Tamandaré Fever." The team of Satô *et al.* identified and investigated a *Trichodesmium* bloom in the region in 1963 (the investigators did not report any illness within their team, who presumably would have experienced significant exposure to the bloom). Reports of illness dating back to the 1940s, apparently related to unidentified algal blooms (which may have been caused by eukaryotic microalgae), must remain in the realm of anecdote, particularly if there are no contemporary reports of similar disease outbreaks to initiate detailed epidemiological investigations (Satô *et al.*, 1963/1964).

Trichodesmium blooms off the Australian coast are regularly associated with skin rashes and irritation among bathers and are the subject of health warnings on the beaches. A sample of *Trichodesmium* collected on the Western Australian coast was demonstrated to cause a skin rash on experimental application to skin (personal observation, author IRF).

The implications for human health because of toxin ingestion relate primarily to the capacity for trophic transfer of toxins produced by *Trichodesmium* into fish and shellfish. Unlike the issue of toxins produced by freshwater cyanobacteria, acute and chronic exposure to cyanotoxins in drinking water is not a consideration here, although the potential for direct exposure to *Trichodesmium* cells and toxins by inhalation or cutaneous exposure may be realized for individuals engaged in recreational activities when blooms are present in coastal waters. Water-soluble cyanotoxins are known to accumulate in the tissues of seafood and shellfish from both freshwater and coastal marine systems (many references are available, including Magalhães *et al.*, 2001; Negri and Jones, 1995; Sipiä *et al.*, 2002; Xie *et al.*, 2005).

Toxic lipid-soluble polyether compounds were extracted from *T. erythraeum* bloom material and oysters, a marine snail and a molluscivorous fish from Great Barrier Reef waters in Queensland (Australia), with inferences drawn on a possible relationship between *Trichodesmium* and toxins in ciguateric fish (Hahn and Capra, 1992). Other researchers working on *T. erythraeum* bloom materials from the same region found water-soluble and lipid-soluble toxins from bloom material that were also apparently found in the flesh of Spanish mackerel (*Scomberomorus commersoni*), with suggestions again made about the potential for *Trichodesmium* to be implicated in the production of some ciguatera-like fish toxins (Endean *et al.*, 1993). Questions about the source of such toxins must again be raised when considering these reports, although Endean *et al.* (1993) undertook detailed steps in their methodology to identify dinoflagellates associated with *Trichodesmium* filaments; dinoflagellates were not observed in tested material. These two Australian studies somewhat predate the more recent understanding of the biotransformation of precursor toxins (gambiertoxins) produced by the dinoflagellate *Gambierdiscus toxicus* into ciguatoxins through several trophic levels (Lehane and Lewis, 2000). However, there would seem to be good reason to investigate further the capacity for *Trichodesmium* to be involved either directly in the production of other lipophilic compounds with toxic properties, or indirectly as a participant in the complex ecological dynamics of ciguatoxin production. On the latter topic, a review by Cruz-Rivera and Villareal (2006) discussed the food web relationships involving herbivory on macroalgae, defensive strategies by same, and the potential for grazing invertebrates to function as ciguatoxin vectors. *G. toxicus* is epiphytic on many eukaryotic macroalgae and has also been found attaching to filamentous cyanobacteria; these host organisms may function as either ciguatoxin vectors or sinks, depending on grazing pressure (Cruz-Rivera and Villareal, 2006). The question is open as to whether or not tropical *Trichodesmium* blooms, in their capacity as a complex habitat for epiphytes and grazers, might play a part in ciguatera toxin dynamics.

Ultimately, systematic field and laboratory investigations of *Trichodesmium* spp. are necessary and as-yet unexplored research endeavors that are required to allow isolation, identification, characterization, and understanding of the toxicology of related toxins.

OTHER PELAGIC CYANOBACTERIA

The oceanic picoplankton are another important component of the open ocean phytoplankton. *Synechococcus* (order: Chroococcales) are unicellular phycoerythrin-containing coccoid cyanobacteria, and the principal component of the cyanobacterial picoplankton (the picoplankton comprises organisms that are less than 5 μm in diameter), and cyanobacterial picoplankton constitute some 30% to more than 50% of the phytoplankton biomass. *Synechocystis* and *Prochlorococcus* are cyanobacterial genera in the order Chroococcales that are also found in the picoplanktonic size range. Essentially overlooked by researchers before the advent of fluorescence microscopy and electron microscopy, the picoplankton are now seen to be significant contributors to oceanic carbon budgets. Oceanic primary production is thought to constitute about half of global productivity, so the cyanobacterial picoplankton perform important ecological functions. *Synechococcus*, *Synechocystis*, and *Prochlorococcus* are solitary, though essentially ubiquitous, organisms in open ocean waters; these cyanobacteria can, however, form blooms in hypersaline waters. Their small size (and thus high surface area to volume ratio) and superior light harvesting capacity allow them to occupy a niche in nutrient-rich deeper ocean waters (Carr and Mann, 1994; Paerl, 1999, 2000). Most forms were until recently thought not to be diazotrophic, but unicellular cyanobacteria and bacterioplankton are now considered to make a significant contribution to oceanic nitrogen budgets. Unicellular cyanobacteria in the pico- and nanoplankton size range may exceed the nitrogen-fixing capacity of *Trichodesmium* and *Richelia* (Falcón *et al.*, 2004; Montoya *et al.*, 2004; Zehr *et al.*, 2001).

Most of the cyanobacteria literature describes the marine picoplankton as nontoxic; however, there are some reports of toxic effects by *Synechococcus* and *Synechocystis*. Martins *et al.* (2005) isolated littoral zone cyanobacteria from various locations in Portugal; extracts from several strains of *Synechococcus* and *Synechocystis* caused nonspecific neurotoxic or hepatotoxic signs and histopathology in mice exposed by intraperitoneal injection. In a separate study, extracts of isolated strains of *Synechocystis* and *Synechococcus* were found to exert toxic effects on brine shrimp (*Artemia salina*) and to inhibit embryonic and larval development in sea urchins (*Paracentrotus lividus*) and marine mussels (*Mytilus galloprovincialis*) (Martins *et al.*, 2007). Five novel compounds with cytotoxic activity against several human and animal cell lines were isolated from a *Synechocystis* sp. mat overgrowing a Japanese coral reef (Nagle and Gerwick, 1995). Freshwater *Synechocystis* strains from Morocco were found to produce several microcystin variants by ELISA and HPLC with photodiode array detection (Oudra *et al.*, 2000, 2002). A sulfolipid that was toxic to minnows at a 20 mg/L LC_{50} and inhibited the growth of a human lymphoma cell line was isolated from a freshwater *Synechococcus* strain cultured from a Japanese lake (Kaya *et al.*, 1993). *Synechococcus* strains isolated and cultured from the Salton Sea, a brackish to hypersaline inland water body in California, were found to produce microcystins by ELISA and LC/MS. One microcystin-producing *Synechococcus* strain was subjected to 16S rRNA PCR analysis and was shown to be more closely related to marine *Synechococcus* than to freshwater strains, which raises the strong possibility that some marine picoplankton may be capable of microcystin synthesis (Carmichael and Li, 2006).

Other surface bloom-forming pelagic cyanobacteria are *Richelia*, a filamentous nitrogen-fixing (heterocystous) form in the order Nostocales. *Richelia intracellularis* is occasionally seen as a free-living organism but most often is found as an endosymbiont (an organism that lives within another organism, forming a mutually beneficial relationship) of the diatom genera *Rhizosolenia*, *Hemiaulus*, and *Chaetocereros* in tropical, subtropical, and temperate oceans. *Katagnymene* (order: Oscillatoriales) is a free-living nitrogen-fixer that can be found in high concentrations in the open ocean, although this cyanobacterium is less well understood than for *Trichodesmium* (Capone and Carpenter, 1999; Carr and Mann, 1994; Paerl, 2000).

Studies into the ecology and ecophysiology of marine planktonic cyanobacteria have revealed the important function of these organisms in oceanic primary production and nitrogen cycling. However, some researchers have discussed knowledge gaps that show there is still much to learn about the role of cyanobacteria in the open oceans. Carr and Mann (1994) noted that the diversity of pelagic cyanobacteria is much lower than seen in other habitats, particularly freshwater and soil, but also in coastal and littoral marine systems. The open ocean is inhabited by a restricted cyanobacterial species assemblage in numbers sufficient to perform a measurable ecological function, typically the picoplanktonic *Synechococcus*, the nonheterocystous diazotroph *Trichodesmium*, and *Richelia*, the heterocystous symbiont of some marine diatom genera. Paerl (2000) noted that nitrogen-limited oligotrophic ocean waters present a significant but essentially unoccupied niche for various cyanobacterial genera that can employ different physiological strategies to fix nitrogen. Carr and Mann (1994) suggested that the volume of the oceans acts as a significant buffer in relation to variables such as metal ion flux, salinity, anthropogenic manipulation, and natural organic input; therefore, there is less selection pressure to evolve diversifying adaptive responses than in less stable environments such as freshwater systems. Continually limited supply of several essential nutrients such as phosphate and iron in the open ocean may manifest in little selective advantage for evolving strategies to acquire and store individual nutrients.

This difference in cyanobacterial diversity between freshwater and open ocean systems leads the authors of this chapter to question whether a similar dichotomy may apply regarding the production of toxins by cyanobacteria in oceanic and freshwater systems. As discussed earlier, the impetus for researching cyanotoxins began and has essentially continued mainly on cyanobacteria from freshwater systems because of the potential (often realized in the case of livestock and wild animals) for acute and chronic morbidity and mortality from cyanotoxin poisoning in humans and animals by intoxication through drinking water supplies. But does the relative dearth of information on marine cyanotoxin production and fate reflect the bias toward research effort into freshwater cyanotoxins, or is there a genuine paucity of toxin production by open ocean cyanobacteria? If the latter, might the structural and functional diversity of cyanotoxins in freshwater systems be another indication that cyanotoxin production confers a competitive advantage on toxin-producing cyanobacteria in highly variable habitats, rather than toxins being viewed solely as secondary metabolites? Is toxin production in the open ocean limited because the relatively homeostatic environment does not select for toxin production?

Research into trophic transfer of cyanotoxins and related regulatory monitoring of seafood and shellfish should continue in order to facilitate health risk assessment of cyanotoxins in coastal marine waters. The issue of toxin production by *Trichodesmium* and the associated potential for trophic transfer of toxins into the marine food web is unresolved and clearly underresearched, and studies into the potential for toxin production by marine picoplanktonic and bloom-forming cyanobacteria would complement understanding of the ecological function of these organisms.

Filamentous Cyanobacteria in Coastal Habitats

In contrast to open ocean systems, coastal environments display much greater heterogeneity in environmental parameters and niches, so coastal habitats typically support a more diverse array of inhabitants, including cyanobacteria (Golubic *et al.*, 1999). Filamentous cyanobacteria in the order Oscillatoriales (mainly *Lyngbya*, *Symploca*, *Oscillatoria*, and *Schizothrix*) are the main focus of research reports into toxic and bioactive compounds produced by marine cyanobacteria. The benthic and epiphytic coastal species *Lyngbya majuscula* produces three toxins, aplysiatoxin, debromoaplysiatoxin, and lyngbyatoxin A, which have harmed humans by oral and cutaneous exposure routes and will be discussed later. Debromoaplysiatoxin has also been isolated from a mixed assemblage of *Schizothrix* and *Oscillatoria* (Mynderse *et al.*, 1977). Another important toxic bloom-forming cyanobacterium that is not seen in open-ocean habitats is the filamentous form *Nodularia spumigena* (order: Nostocales). *Nodularia* forms blooms in eutrophic, semi-enclosed brackish waters, most notably the Baltic Sea and some Australian estuaries and estuarine lakes. The cyclic peptide toxin produced by *Nodularia*, nodularin, has been discussed earlier in this chapter, and interested readers are referred to reviews of *Nodularia*, for example, Sellner (1997) and Stal *et al.* (2003). The remaining discussion on coastal cyanobacteria will focus on the benthic and epiphytic *Lyngbya majuscula*, some of the toxins it produces, and the implications of exposure to these toxins by various exposure routes.

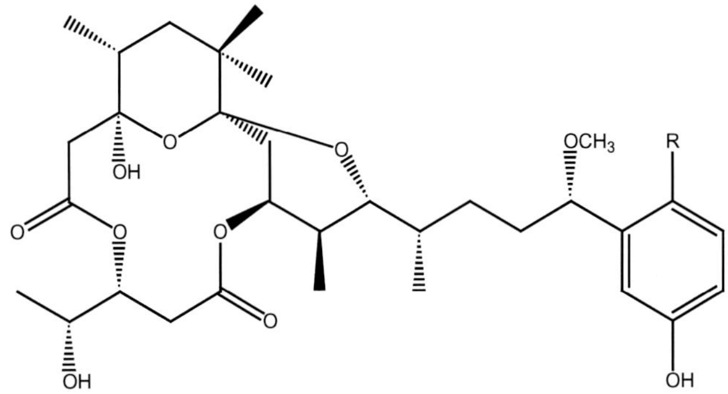

FIGURE 15-16. Structure of aplysiatoxin (R = Br) and debromoaplysiatoxin (R = H). Adapted from Yotsu-Yamashita *et al.* (2004).

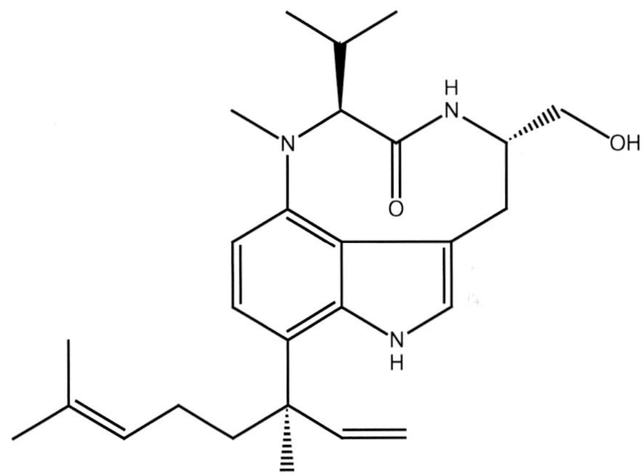

FIGURE 15-17. Structure of lyngbyatoxin A. Adapted from Osborne *et al.* (2001).

Lyngbya

Lyngbya is a genus of filamentous, nonheterocystous cyanobacteria in the order Oscillatoriales. Some forms are capable of fixing nitrogen, for example, *Lyngbya aestuaraii*, which employs a dual strategy of diel cycling, in which peak N_2 production occurs at night when oxygenic photosynthesis ceases, and spatial partitioning of nitrogen fixation and photosynthesis within individual filaments (Paerl, 2000). *Lyngbya* spp. are (like surface bloom-forming cyanobacteria) capable of forming large mass developments, although they typically coalesce into macroscopic aggregates that form floating mats or dense benthic assemblages that can smother the coastal sea floor and its macrophytes. The most prodigious bloom-forming species is *Lyngbya majuscula*, also capable of fixing nitrogen, and this cyanobacterium is the most widely studied of the *Lyngbya* genus because of its production of toxins, associated human health risks and investigation for biologically active metabolites.

L. majuscula (Fig. 15-6) is a common tropical and subtropical cyanobacterium with extensive and toxic blooms reported mainly from coastal Pacific, Indian Ocean, Caribbean, and Mediterranean sites. The public health concerns relate to a group of well-characterized toxins that affect the skin and mucous membranes, principally through cutaneous exposure, but the respiratory and oral exposure routes must also be considered when assessing the human health risks from this cyanobacterium. Whereas numerous metabolites produced by *L. majuscula* have been shown to exert cytotoxic properties *in vitro* (and will be discussed below), the three most widely studied Lyngbya-related toxins are the phenolic bis-lactones aplysiatoxin (AT) and debromoaplysiatoxin (DAT) (Fig. 15-16), and lyngbyatoxin A (LA), an indole alkaloid (Fig. 15-17). Aplysiatoxin differs structurally from DAT only in that the phenol structure is substituted with a single bromine atom. These toxins resemble the structurally unrelated croton oil phorbol ester compound 12-O-tetradecanoylphorbol-13-acetate (TPA) in several important effects: they induce an acute inflammatory reaction on the skin; they are potent tumor promoters; they activate protein kinase C; and they stimulate similar *in vitro* innate immune reactions (e.g., arachidonic acid release and production of prostaglandins, IFN-γ and IL-2, and suppression of IL-1α production) (Levine and Watanabe, 1991; Moore, 1984; Osborne *et al.*, 2001; Watanabe *et al.*, 1995; Yip *et al.*, 1983).

L. majuscula first came to the attention of public health workers in the late 1950s, when more than 100 swimmers at a Hawaiian beach complained of dermatitis (mainly affecting genital, perineal, and perianal areas) that was attributed to exposure to *L. majuscula.* Subsequent skin testing on animals, human volunteers, and auto-experiments demonstrated acute irritant reactions to *L. majuscula* extracts and preparations; in later experiments the isolated compounds LA, AT, and DAT were shown to be the toxic agents capable of initiating an acute contact irritant dermatitis (Hashimoto *et al.*, 1976; Osborne *et al.*, 2001; Solomon and

Stoughton, 1978). Additional mass outbreaks of so-called seaweed dermatitis have occurred among swimmers at coastal beaches in Japan and Hawaii, and anecdotal reports of acute cutaneous reactions in recreational and occupational settings that are temporally related to *L. majuscula* blooms have prompted beach closures and recreational exposure warnings in southeast Queensland. Affected individuals report burning sensations, pruritus, and erythema, followed by blister formation and deep desquamation (peeling of the skin in layers).

Figure 15-18 illustrates the enormous biomass that can be generated by *L. majuscula* when conditions are suitable for bloom growth. Red Beach is one of the Deception Bay beaches (Bribie Island, Queensland, Australia) from which an estimated 2300 tonnes of decomposing material was removed in 2004 (Dennison *et al.*, 1999; Osborne *et al.*, 2001, 2007). Government and research agencies in Queensland have established a telephone hotline and disseminated information and alerts in response to the regular occurrence of Lyngbya blooms in southeast Queensland coastal waters (Marine Botany, 2002; Moore *et al.*, 2006).

The epidemiology of *L. majuscula* and *Lyngbya*-related toxins is limited, and the current knowledge base is informed by a series of outbreak-initiated retrospective studies and anecdotal reports. Osborne *et al.* (2001) reviewed the topic. This same research group subsequently conducted a questionnaire-based cross-sectional study of residents of an Australian coastal area subject to regular *L. majuscula* bloom events. The authors found increased reporting of skin and eye symptoms associated with an increased level of coastal water exposure (exposure levels being determined by the type of recreational pursuit they undertook, with, for example, wading or shore fishing categorized as "low exposure" and swimming or diving being categorized as "high exposure" activities). However, only a small proportion of symptoms were reportedly of serious extent, which the authors thought indicative of limited exposure to *Lyngbya*-related toxins across study subjects (Osborne *et al.*, 2007).

FIGURE 15-18. *Lyngbya majuscula* bloom biomass on Red Beach, Bribie Island, Queensland, Australia, summer of 2004–2005. Photograph courtesy Caboolture Shire Council, Caboolture, Queensland.

This finding highlights one of the principal challenges when conducting retrospective studies in this particular field of environmental epidemiology—that of reliably determining exposure. Cyanobacterial blooms and the production of specific cyanotoxins during bloom events are subject to considerable spatial and temporal variability, such that prospective epidemiological study designs, though generally more reliable than retrospective studies for determining causation of acute, short-duration illnesses, may still suffer from exposure misclassification (Stewart *et al.*, 2006b, 2006c) (see Chapter 11). Yet the potential remains for serious acute morbidity from exposure to *Lyngbya*-related toxins, and the possible hazards of chronic exposure to these potent tumor promoters are worthy of research attention. There are intriguing anecdotal and case reports of human injury and mortality linked to consumption of *L. majuscula* and *Lyngbya*-related toxins either directly or from trophic transfer *via* dietary fish or turtle flesh (see the review by Osborne *et al.*, 2001). *Lyngbya majuscula* biomass, and presumably toxins if present, may aerosolize on beaches when desiccated; government agencies in southeast Queensland now recommend the use of sealed respirators to protect against inhalation for workers disposing of such material (Moore *et al.*, 2006).

A high proportion of individuals exposed to *L. majuscula* or *Lyngbya*-related toxins by cutaneous or oral exposure while engaging in recreational activities reportedly experienced symptoms (Marshall and Vogt, 1998; Osborne *et al.*, 2001). This observation supports associated laboratory investigations into isolated and purified *Lyngbya*-related toxins that describe these toxins as "irritants." Dermatologists work from the premise that exposure to irritant concentrations of a particular chemical agent will cause signs and symptoms in a high proportion of unsensitized individuals, whereas substantially subirritant concentrations may produce allergic reactions in a smaller proportion of people who have experienced prior exposure and sensitization.

The allergic potential of *Lyngbya* remains essentially unexplored, and the allergenicity of the cyanobacteria in general is a significantly underresearched topic. Although an early research paper into *L. majuscula*-related dermatitis concluded that "hypersensitivity as mediated through an immune mechanism was not involved" (Moikeha and Chu, 1971, p. 12), such a statement might be interpreted in terms of eliminating a hypersensitivity reaction as the fall-back explanation for an *in vivo* lesion that was shown to be an acute inflammatory reaction in unsensitized laboratory animals. These and other laboratory investigations that have conclusively demonstrated the irritant potential of DAT,

AT, and LA do not preclude the possibility that these toxins may also have the capacity to function as allergens. The experimental finding that *in vivo* subcutaneous exposure to AT causes mast cell degranulation and histamine release (Ohuchi *et al.*, 1987) should make this and related toxins candidates for investigation into their potential to cause anaphylactic reactions; chemicals that activate mast cells or basophils can provoke anaphylaxis (Kemp and Lockey, 2002).

The freshwater *Lyngbya* species *L. wollei* has the capacity to produce toxins: samples from Alabama were shown to produce neurotoxins in the saxitoxin group (Carmichael *et al.*, 1997), and an Australian strain from southeast Queensland produces cylindrospermopsin and deoxycylindrospermopsin (Seifert *et al.*, 2007). Dense assemblages of *L. wollei* are widely distributed in lakes, reservoirs, and freshwater spring systems in the southeastern United States, less so currently in Australia. There are anecdotal reports from Florida of acute water-contact-related skin and respiratory illnesses, some requiring ambulance retrieval and hospitalization, from spring systems coincidentally affected by mass developments of *L. wollei*. However, some of these springs are popular, highly visited recreational sites, so the incidence of severe reactions would initially appear to be infrequent. Formal epidemiological investigations to establish exposure, outcome, and the presence or absence of known cyanotoxins would be valuable in order to assess the human health risk potential. No published reports of LA, DAT, or AT production by freshwater *Lyngbya* could be found in the literature, but systematic screening, including a sampling regime to encompass possible seasonal variability of toxin production, would also be a valuable exercise.

MARINE CYANOBACTERIA AS SOURCES OF NOVEL BIOACTIVE COMPOUNDS

In contrast to the public health risks of exposure to *Lyngbya* and related cyanotoxins, which remain significantly underresearched, the marine species *Lyngbya majuscula* has received considerable scrutiny by biochemists for the presence of novel biologically active compounds. *L. majuscula* collected from numerous locations worldwide has been found to be a rich source of structurally diverse bioactive metabolites, with considerable potential for application as pharmaceutical agents. With a particular focus on compounds with anticancer and antiviral properties, this significant research effort has identified many new structures with cytotoxic, neuroactive, antibacterial, and antifungal activities. Many publications on novel compounds from *L. majuscula* can be found in the literature (see, for example, Jiménez and Scheuer, 2001; Luesch *et al.*, 2001; Milligan *et al.*, 2000; Wu *et al.*, 2000; and a review in Shimizu, 2003).

Many more marine, freshwater, and terrestrial cyanobacteria are the subject of research attention for their potential pharmaceutical applications: the marine genus *Symploca* produces compounds with cytotoxic properties and is therefore being investigated for its antitumor potential (e.g., Horgen *et al.*, 2002; Williams *et al.*, 2004); novel tridecapeptides with antifungal properties have been isolated from the terrestrial cyanobacterium *Tolypothrix byssoidea* collected from a granite block in Nepal (Jaki *et al.*, 2001); and a novel linear tetrapeptide with antialgal activity was isolated from a strain of the freshwater *Microcystis aeruginosa* (Ishida and Murakami, 2000). Again, many more examples of novel bioactive compounds produced by a broad range of cyanobacteria can be found in the biochemistry and related literature. Interested readers are referred to reviews on novel cyanobacterial metabolites (Bickel *et al.*, 2001; Forchert *et al.*, 2001; Singh *et al.*, 2005); toxins from marine cyanobacteria (Long and Carmichael, 2003); preclinical and clinical pharmacotherapy trials of marine natural products (Newman and Cragg, 2004); and antiviral (with an emphasis on anti-HIV) compounds extracted from cyanobacteria (Schaeffer and Krylov, 2000). A review of cytotoxic compounds isolated from marine sea hares proposes that many of these compounds are dietary in origin and originate from *Lyngbya* and *Symploca*; novel compounds isolated from carnivorous mollusks and marine sponges are also thought to have a cyanobacterial origin (Luesch *et al.*, 2002). Tan (2007) compiled an extensive review of the literature between 2001 and 2006 on novel compounds produced by marine cyanobacteria. For an earlier comprehensive review, see Burja *et al.* (2001). Marine sponges are a particularly interesting example of symbiotic relationships involving cyanobacteria; in some so-called cyanobacteriosponges, cyanobacteria constitute more than half of their tissues. These organisms are highly competitive, which highlights the mutualistic benefit to both sponge and cyanosymbiont. Density-gradient or fluorescence-based cell separating techniques have allowed the cellular origin of sponge-derived novel compounds to be determined; for example, in one study sponge cells were found to produce terpenes and cyanobacterial cells produced peptide compounds (König *et al.*, 2006).

EDIBLE CYANOBACTERIA AND MACROALGAE: CONTAMINATION BY CYANOTOXINS

Brief mention should be made of cyanobacteria that are used as food and medicine for humans and animals. The genus *Spirulina* (now classified as *Arthrospira*, order: Oscillatoriales) has a long history of use as a seasonal food source in Africa and Central America. *Arthrospira* spp. can colonize a wide variety of habitats in marine, brackish water, freshwater, and terrestrial environments. Surface blooms of *A. fusiformis* and *A. maxima*, which occur as virtual monocultures in certain lakes in Africa and Mexico, have been enthusiastically harvested, prepared, and traded as food staples and supplements; such practices were observed and documented in the 16th century by the Spanish invaders of the Aztec lands. These cyanobacteria are now grown commercially in mass culture in countries such as India, Thailand, and the United States and sold principally as dietary supplements, with reputed benefits as a high-quality source of protein and micronutrients.

A significant body of literature has, with few exceptions, shown that consumption of *Arthrospira* is not harmful (Ciferri, 1983; Stewart *et al.*, 2006a). However, the potential for nontoxic cyanobacteria to acquire toxin-producing genes is suggested by investigations into mass poisonings of flamingos in African Rift Valley lakes. *Arthrospira* is the principal food source of Lesser Flamingos inhabiting these lakes; studies initiated by the deaths of tens of thousands of birds have found microcystins and anatoxin-a in tissue samples. Although these studies are incomplete and continuing, one interesting finding has been to show that *Arthrospira* strains collected from these soda lakes can produce microcystin-YR and anatoxin-a in laboratory cultures. Toxic cyanobacterial genera such as *Microcystis* and *Anabaena* coexist with *Arthrospira* in Rift valley alkaline lakes, and it seems that *Arthrospira* has the capacity to acquire toxin-producing genes (Ballot *et al.*, 2004, 2005; Codd *et al.*, 2003; Krienitz *et al.*, 2003; Ndetei and Muhandiki, 2005). *Microcystis* and *Anabaena* are also found in Lake Chad, where the Kanenbou people, who constitute the main ethnic group to the east and northeast of the lake, harvest *Arthrospira* bloom material. This is then dried to produce biscuits that are later reconstituted in water and added to sauces that supplement the standard millet meal (Ciferri, 1983).

So the possibility of exposure to cyanotoxins from direct consumption of dietary *Arthrospira* must be considered if wild-harvested material has grown in association with sufficient biomass of toxigenic cyanobacteria such as *Microcystis* or *Anabaena*. The potential for *Arthrospira* itself to produce cyanotoxins, although essentially unexplored in terms of ecological capacity and dynamics, suggests that public health research into commercially grown *Arthrospira* strains and products may be necessary. This is the case with other commercially produced cyanobacteria, particularly *Aphanizomenon flos-aquae* harvested from Lake Klamath in Oregon. This cyanobacterium is processed and sold worldwide as a nutritional supplement, and microcystin toxins have been found to contaminate some batches of the retailed product. The source of these cyanotoxins is presumed to be from contaminating growth of *Microcystis* in the lake. In 1996, a bloom of *M. aeruginosa* affected Upper Lake Klamath, which initiated intensive investigations into dietary *A. flos-aquae* products by the Oregon departments of agriculture and health. Oregon adopted a regulatory standard of 1 μg microcystins per gram dry weight of commercial product as a result of these studies (Gilroy *et al.*, 2000). In addition, *Arthrospira* tablets and capsules marketed in Italy as health supplements were found to contain degradation products of anatoxin-a and homoanatoxin-a (Draisci *et al.*, 2001).

A related but unusual oral exposure route relates to the propensity for filamentous cyanobacteria to grow epiphytically on edible marine macroalgae. Consumption of seaweed is common in many Pacific Island and Pacific Rim countries. At a picnic in Hawaii in 1994, all seven individuals who ate a seaweed dish became ill between 15 and 90 minutes after consuming the dish. Six individuals experienced burning sensations in the mouth and throat; five of the seven who consumed larger portions of the seaweed suffered gastrointestinal symptoms (all had nausea and diarrhea, three of the five reported vomiting). The seaweed was identified as the red macroalga, *Gracilaria coronopifolia*, with cyanobacteria attached and interwoven among seaweed stems. While the cyanobacterium was not formally identified, it was presumed to be *L. majuscula*. Extracts of the seaweed were toxic to mice, causing diarrhea and death. Debromoaplysiatoxin and aplysiatoxin were isolated from a *Gracilaria*/cyanobacteria extract. It is thought that that epiphytic *Lyngbya* was the true source of aplysiatoxin and debromoaplysiatoxin associated with this *Gracilaria*-related poisoning incident (Hanne *et al.*, 1995; Marshall and Vogt, 1998; Nagai *et al.*, 1996).

A related and more serious matter concerns the macrolide toxin polycavernoside-A, which is responsible for several food poisoning outbreaks in Guam and the Philippines; 11 fatalities have been reported. Polycavernoside-A poisoning is associated with consumption of certain species of red macroalga, *Gracilaria edulis* and *Acanthophora specifera*, but the true origin of the toxin is thought to be cyanobacterial (Yasumoto, 2001, 2005; Yotsu-Yamashita *et al.*, 1993, 2004). As polycavernoside-A has been confirmed as the causative agent in fatal human poisonings, a research program to identify the presumptive cyanobacterial source species would appear to be warranted.

CONCLUDING REMARKS

Marine cyanobacteria are important participants in oceanic carbon and nitrogen budgets, and as these large oceanic nutrient cycles are significant contributors to global systems, cyanobacteria should be considered to be important determinants of ecosystem health. This observation alone should designate the marine cyanobacteria as significant, though indirect, determinants of human health. Such a statement is not intended to be tokenistic; human health and well-being are intimately and inextricably bound to functioning ecosystems. Readers interested in the macrocosmic theme of environmental public health are referred to the work of Anthony McMichael, particularly his book *Human Frontiers, Environments and Disease*, which is an excellent source manual on the topic (McMichael, 2001). Regarding oceanic cyanobacteria (in conjunction with eukaryotic microalgae) on this matter, large-scale anthropogenic manipulation of influencing factors (such as temperature and acidity, through the enhanced greenhouse effect, overfishing in upper trophic levels, or eutrophication) may affect oceanic cyanobacteria and microalgae in essentially unknown and probably unexpected ways. Such perturbations in oceanic primary productivity and bloom dynamics are unlikely in most instances to result in a net overall benefit to ecosystem health, and thus to human health.

More directly, cyanobacteria are a rich source of novel biologically active compounds, and several marine genera, particularly *Lyngbya* and *Symploca*, are the focus of growing research attention to assess their potential as pharmacotherapies. Compounds with potent antitumor, antiviral, and enzyme-blocking effects have been discovered. Some cyanobacteria have a long history as sources of food and traditional medicine; the potential for large-scale production of edible cyanobacteria for use as animal and human dietary protein and nutritional supplements would appear to be significant in the face of expanding human populations that place ever-increasing demands on the availability and productive capacity of agricultural lands. However, the potential for commercial cyanobacterial products to contain harmful levels of cyanotoxins, acquired through wild-harvested contaminant toxigenic cyanobacteria growing among edible cyanobacteria or by the acquisition of toxin-producing genes by edible strains, will necessitate research and monitoring by food safety authorities.

Finally, and the main topic of this chapter, cyanobacteria present actual and potential hazards to humans (and many other animal species) through their capacity to produce potent toxins. Cyanobacteria can produce a variety of structurally diverse toxic compounds that harm different organ systems, depending on the particular toxin the individual is exposed to. Confirmed mortalities in humans have occurred through exposure to cyanotoxins in an outbreak at a hemodialysis clinic when a water treatment failure allowed cyanotoxins to enter the dialysate. The intravenous exposure route involved in this incident may be somewhat unusual, but the potential for acute illness or death by exposure to cyanotoxins through various natural exposures is real and thus worthy of vigilance by researchers and public health workers. Acute exposure via the oral route may occur through ingestion of cyanotoxin-contaminated water in recreational settings or via untreated drinking water, via trophic transfer from contaminated fish and shellfish from marine or freshwater sources, from edible cyanobacteria contaminated by cyanotoxins, or potentially through the consumption of market garden produce irrigated by cyanotoxin-contaminated water. Inhalational exposure to cyanotoxins may also occur in recreational and occupational settings, and this exposure route is underresearched with respect to cyanobacteria. Some freshwater cyanotoxins are tumor promoters and suspect carcinogens, so chronic exposure to subacute doses via drinking water supplies drawn from surface reservoirs invites further research attention. There are many well-documented reports and investigations into mass mortalities of wild animals and livestock caused and implicated by exposure to cyanotoxins in drinking water; such observations further emphasize the potential hazards to humans.

Freshwater cyanobacterial toxins have received more research attention than cyanotoxins of marine origin, probably because of the concerns relating to drinking water exposures that do not apply to marine cyanotoxins. However, there seems to be growing evidence that the marine bloom-forming genus *Trichodesmium* may have the capacity to produce toxins, which then leads to concerns regarding the potential for bioaccumulation in fish and shellfish and economic losses to aquaculture industries. Such concerns already exist in the case of some eutrophic marine waters such as the Baltic Sea that are subject to blooms of the toxin-producing cyanobacterium *Nodularia*. And the filamentous genus *Lyngbya* is capable of growing to dense and prolific bloom proportions in tropical and subtropical regions in coastal and inland waters. Freshwater *Lyngbya* can produce toxins in both the saxitoxin and cylindrospermopsin groups, and marine *Lyngbya* can generate several potent inflammatory and tumor-promoting toxins that affect the skin of bathers and fishers and are likely to cause harm by inhalation and oral exposure.

Although many of the major cyanotoxins are well characterized with regard to elucidation of chemical structures and to important mechanisms of toxicity, much remains to be learned about these potent toxins. The epidemiology of cyanobacteria and cyanotoxins is limited, and difficult to study because of the sporadic and ephemeral nature of bloom events. Inhalational exposures require more investigation in both field and laboratory settings, and the allergic potential of cyanobacterial products is also underresearched. Long-term *in vivo* exposure studies are needed in order to investigate the possible carcinogenicity of cyanotoxins.

References

Adams, D.G., 2000. Symbiotic interactions. In Whitton, B.A., and Potts, M. (eds.), The Ecology of Cyanobacteria: Their Diversity in Time and Space, pp. 523–561. Dordrecht, Kluwer Academic.

Baker, J.A., Neilan, B.A., Entsch, B., McKay, D.B., 2001. Identification of cyanobacteria and their toxigenicity in environmental samples by rapid molecular analysis. Environ. Toxicol. 16, 472–482.

Ballot, A., Krienitz, L., Kotut, K., Wiegand, C., Metcalf, J.S., Codd, G.A., Pflugmacher, S., 2004. Cyanobacteria and cyanobacterial toxins in three alkaline Rift Valley lakes of Kenya–Lakes Bogoria, Nakuru and Elmenteita. J. Plankton Res. 26, 925–935.

Ballot, A., Krienitz, L., Kotut, K., Wiegand, C., Pflugmacher, S., 2005. Cyanobacteria and cyanobacterial toxins in the alkaline crater lakes Sonachi and Simbi, Kenya. Harmful Algae 4, 139–150.

Bickel, H., Neumann, U., Weckesser, J., 2001. Peptides and depsipeptides produced by cyanobacteria. In Chorus, I. (ed.), Cyanotoxins: Occurrence, Causes, Consequences, pp. 281–286. Berlin, Springer-Verlag.

Borner, T., Dittmann, E., 2005. Molecular biology of cyanobacterial toxins: Genetic basis of microcystin production. In Huisman, J., Matthijs, H.C.P., and Visser, P.M. (eds.), Harmful Cyanobacteria. Aquatic Ecology Series, vol. 3, pp. 25–40. Dordrecht, Springer.

Burja, A.M., Banaigs, B., Abou-Mansour, E., Burgess, J.G., Wright, P.C., 2001. Marine cyanobacteria: A prolific source of natural products. Tetrahedron 57, 9347–9377.

Byth, S., 1980. Palm Island mystery disease. Med. J. Aust. 2, 40, 42.

Capone, D.G., Carpenter, E.J., 1999. Nitrogen fixation by marine cyanobacteria: Historical and global perspectives. In Charpy, L., and Larkum, A.W.D. (eds.), Marine Cyanobacteria, pp. 235–256. Monaco City, Musée océanographique.

Capone, D.G., Zehr, J.P., Paerl, H.W., Bergman, B., Carpenter, E.J., 1997. *Trichodesmium*, a globally significant marine cyanobacterium. Science 276, 1221–1229.

Carmichael, W.W., Azevedo, S.M., An, J.S., Molica, R.J.R., Jochimsen, E.M., Lau, S., Rinehart, K.L., Shaw, G.R., Eaglesham, G.K., 2001. Human fatalities from cyanobacteria: Chemical and biological evidence for cyanotoxins. Environ. Health Perspect. 109, 663–668.

Carmichael, W.W., Evans, W.R., Yin, Q.Q., Bell, P., Moczydlowski, E., 1997. Evidence for paralytic shellfish poisons in the freshwater cyanobacterium Lyngbya wollei (Farlow ex Gomont) comb. nov. Appl. Environ. Microbiol. 63, 3104–3110.

Carmichael, W.W., Li, R.H., 2006. Cyanobacteria toxins in the Salton Sea. Saline Systems 2, 5

Carpenter, E.J., Capone, D.G., 1992. Nitrogen fixation in *Trichodesmium* blooms. In Carpenter, E.J., Capone, D.G., and Rueter, J.G. (eds.), Marine Pelagic Cyanobacteria: *Trichodesmium* and Other Diazotrophs, pp. 211–217. Dordrecht, Kluwer Academic.

Carr, N.G., Mann, N.H., 1994. The oceanic cyanobacterial picoplankton. In Bryant, D.A. (ed.), The Molecular Biology of Cyanobacteria, pp. 27–48. Dordrecht, Kluwer Academic.

Chacko, P.I., 1942. An unusual incidence of mortality of marine fauna. Curr. Sci. 11, 404.

Chorus, I. (ed.), 2005. Current approaches to cyanotoxin risk assessment, risk management and regulations in different countries. Berlin, Umweltbundesamt (Federal Environmental Agency), www.umweltdaten.de/publikationen/fpdf-l/2910.pdf.

Ciferri, O., 1983. Spirulina, the edible microorganism. Microbiol. Rev. 47, 551–578.

Codd, G.A., 1999. Cyanobacterial toxins: Their occurrence in aquatic environments and significance to health. In Charpy, L., and Larkum, A.W.D. (eds.), Marine Cyanobacteria, pp. 483–500. Monaco City, Musée océanographique.

Codd, G.A., Metcalf, J.S., Morrison, L.F., Krienitz, L., Ballot, A., Pflugmacher, S., Wiegand, C., Kotut, K., 2003. Susceptibility of flamingos to cyanobacterial toxins via feeding. Vet. Rec. 152, 722–723.

Colwell, R.R., 1996. Global climate and infectious disease: The cholera paradigm. Science 274, 2025–2031.

Cruz-Rivera, E., Villareal, T.A., 2006. Macroalgal palatability and the flux of ciguatera toxins through marine food webs. Harmful Algae 5, 497–525.

Dennison, W.C., O'Neil, J.M., Duffy, E.J., Oliver, P.E., Shaw, G.R., 1999. Blooms of the cyanobacterium *Lyngbya majuscula* in coastal waters of Queensland, Australia. In Charpy, L., and Larkum, A.W.D. (eds.), Marine Cyanobacteria, pp. 501–506. Monaco City, Musée océanographique.

Draisci, R., Ferretti, E., Palleschi, L., Marchiafava, C., 2001. Identification of anatoxins in blue-green algae food supplements using liquid chromatography-tandem mass spectrometry. Food Addit. Contam. 18, 525–531.

Edler, L., Fernö, S., Lind, M.G., Lundberg, R., Nilsson, P.O., 1985. Mortality of dogs associated with a bloom of the cyanobacterium *Nodularia spumigena* in the Baltic Sea. Ophelia 24, 103–109.

El-Shehawy, R., Lugomela, C., Ernst, A., Bergman, B., 2003. Diurnal expression of *hetR* and diazocyte development in the filamentous non-heterocystous cyanobacterium *Trichodesmium erythraeum*. Microbiology 149, 1139–1146.

Endean, R., Monks, S.A., Griffith, J.K., Llewellyn, L.E., 1993. Apparent relationships between toxins elaborated by the cyanobacterium *Trichodesmium erythraeum* and those present in the flesh of the narrow-barred Spanish mackerel *Scomberomorus commersoni*. Toxicon 31, 1155–1165.

Epstein, P.R., 1993. Algal blooms in the spread and persistence of cholera. Biosystems 31, 209–221.

Falcón, L.I., Carpenter, E.J., Cipriano, F., Bergman, B., Capone, D.G., 2004. N_2 fixation by unicellular bacterioplankton from the Atlantic and Pacific Oceans: Phylogeny and in situ rates. Appl. Environ. Microbiol. 70, 765–770.

Falconer, I.R., 2005. Cyanobacterial Toxins of Drinking Water Supplies: Cylindrospermopsins and Microcystins. Boca Raton, FL, CRC Press.

Falconer, I.R., Beresford, A.M., Runnegar, M.T., 1983. Evidence of liver damage by toxin from a bloom of the blue-green alga, *Microcystis aeruginosa*. Med. J. Aust. 1, 511–514.

Falconer, I.R., Humpage, A.R., 2006. Cyanobacterial (blue-green algal) toxins in water supplies: Cylindrospermopsins. Environ. Toxicol. 21, 299–304.

Falconer, I.R., Yeung, D.S., 1992. Cytoskeletal changes in hepatocytes induced by *Microcystis* toxins and their relation to hyperphosphorylation of cell proteins. Chem. Biol. Interact. 81, 181–196.

Forchert, A., Neumann, U., Papendorf, O., 2001. New cyanobacterial substances with bioactive properties. In Chorus, I. (ed.), Cyanotoxins: Occurrence, Causes, Consequences. Springer-Verlag, Berlin, pp. 295–315.

Francis, G., 1878. Poisonous Australian lake. Nature 18, 11–12.

Gallon, J.R., Stal, L.J., 1992. N_2 fixation in non-heterocystous cyanobacteria: An overview. In Carpenter, E.J., Capone, D.G., and Rueter, J.G. (eds.), Marine Pelagic Cyanobacteria: *Trichodesmium* and Other Diazotrophs, pp. 115–139. Dordrecht, Kluwer Academic.

Gilroy, D.J., Kauffman, K.W., Hall, R.A., Huang, X., Chu, F.S., 2000. Assessing potential health risks from microcystin toxins in blue-green algae dietary supplements. Environ. Health Perspect. 108, 435–439.

Golubic, S., Le Campion-Alsumard, T., Campbell, S.E., 1999. Diversity of marine cyanobacteria. In Charpy, L., and Larkum, A.W.D. (eds.), Marine Cyanobacteria, pp. 53–76. Monaco City, Musée océanographique.

Grosse, Y., Baan, R., Straif, K., Secretan, B., El Ghissassi, F., Cogliano, V; WHO International Agency for Research on Cancer Monograph Working Group, 2006. Carcinogenicity of nitrate, nitrite, and cyanobacterial peptide toxins. Lancet Oncol. 7, 628–629.

Guo, C., Tester, P.A., 1994. Toxic effect of the bloom-forming *Trichodesmium* sp. (Cyanophyta) to the copepod *Arcatia tonsa*. Nat. Toxins 2, 222–227.

Hahn, S.T., Capra, M.F., 1992. The cyanobacterium *Oscillatoria erythraea*: A potential source of toxin in the ciguatera food-chain. Food Addit. Contam. 9, 351–355.

Hanne, M., Matsubayashi, H., Vogt, R., Wakida, C., Hau, S., Nagai, H., Hokama, Y., Solorzano, L., 1995. Outbreak of gastrointestinal illness associated with consumption of seaweed: Hawaii, 1994. MMWR Morb. Mortal. Wkly. Rep. 44, 724–727.

Harding, W.R., Paxton, B.R., 2001. Cyanobacteria in South Africa: A review. Pretoria, Water Research Commission.

Hashimoto, Y., Kamiya, H., Yamazato, K., Nozawa, K., 1976. Occurrence of a toxic blue-green alga inducing skin dermatitis in Okinawa. In Ohsaka, A., Hayashi, K., and Sawai, Y. (eds.), Animal, Plant, and Microbial Toxins, vol. 1: Biochemistry, pp. 333–338. New York, Plenum.

Hawkins, P.R., Runnegar, M.T., Jackson, A.R., Falconer, I.R., 1985. Severe hepatotoxicity caused by the tropical cyanobacterium (blue-green alga) *Cylindrospermopsis raciborskii* (Woloszynska) Seenaya and Subba Raju isolated from a domestic water supply reservoir. Appl. Environ. Microbiol. 50, 1292–1295.

Hawser, S.P., Codd, G.A., 1992. The toxicity of *Trichodesmium* blooms from Caribbean waters. In Carpenter, E.J., Capone, D.G., and Rueter, J.G. (eds.), Marine Pelagic Cyanobacteria: *Trichodesmium* and Other Diazotrophs, pp. 319–329. Dordrecht, Kluwer Academic.

Hawser, S.P., Codd, G.A., Capone, D.G., Carpenter, E.J., 1991. A neurotoxic factor associated with the bloom-forming cyanobacterium *Trichodesmium*. Toxicon 29, 277–278.

Hawser, S.P., O'Neil, J.M., Roman, M.R., Codd, G.A., 1992. Toxicity of blooms of the cyanobacterium *Trichodesmium* to zooplankton. J. Appl. Phycol. 4, 79–86.

Heintzelman, G.R., Fang, W.K., Keen, S.P., Wallace, G.A., Weinreb, S.M., 2002. Stereoselective total syntheses and reassignment of stereochemistry of the freshwater cyanobacterial hepatotoxins cylindrospermopsin and 7–epicylindrospermopsin. J. Am. Chem. Soc. 124, 3939–3945.

Heldt, H.-W.E., 2004. Plant biochemistry, 3rd ed. New York, Academic Press.

Horgen, F.D., Kazmierski, E.B., Westenburg, H.E., Yoshida, W.Y., Scheuer, P.J., 2002. Malevamide D: Isolation and structure determination of an isodolastatin H analogue from the marine cyanobacterium *Symploca hydnoides*. J. Nat. Prod. 65, 487–491.

Ishida, K., Murakami, M., 2000. Kasumigamide, an antialgal peptide from the cyanobacterium *Microcystis aeruginosa*. J. Org. Chem. 65, 5898–5900.

Islam, M.S., Rahim, Z., Alam, M.J., Begum, S., Moniruzzaman, S.M., Umeda, A., Amako, K., Albert, M.J., Sack, R.B., Huq, A., Colwell, R.R., 1999. Association of *Vibrio cholerae* 01 with the cyanobacterium, *Anabaena* sp., elucidated by polymerase chain reaction and transmission electron microscopy. Trans. R. Soc. Trop. Med. Hyg. 93, 36–40.

Ito, E., Takai, A., Kondo, F., Masui, H., Imanishi, S., Harada, K., 2002. Comparison of protein phosphatase inhibitory activity and apparent toxicity of microcystins and related compounds. Toxicon 40, 1017–1025.

Jackson, W.C., Jr., Carmichael, W.W., Carpenter, E.J., 2001. The paralytic shellfish toxins, n-sulfocarbamoyl and gonyautoxin, are the major neurotoxic factors associated with bloom-forming *Trichodesmium* spp. [abstract]. Abstracts of the 5th international conference on toxic cyanobacteria, July 16–20, 2001, Noosa, Queensland, Australia.

Jaki, B., Zerbe, O., Heilmann, J., Sticher, O., 2001. Two novel cyclic peptides with antifungal activity from the cyanobacterium *Tolypothrix byssoidea* (EAWAG 195). J. Nat. Prod. 64, 154–158.

Jiménez, J.I., Scheuer, P.J., 2001. New lipopeptides from the Caribbean cyanobacterium *Lyngbya majuscula*. J. Nat. Prod. 64, 200–203.

Jochimsen, E.M., Carmichael, W.W., An, J.S., Cardo, D.M., Cookson, S.T., Holmes, C.E.M., Antunes, M.B.C., de Melo Filho, D.A., Lyra, T.M., Barreto, V.S.T., Azevedo, S.M.F.O., Jarvis, W.R., 1998. Liver failure and death after exposure to microcystins at a hemodialysis center in Brazil. N. Engl. J. Med. 338, 873–878.

Kaya, K., Sano, T., Watanabe, M.M., Shiraishi, F., Ito, H., 1993. Thioic *O*-acid ester in sulfolipid isolated from freshwater picoplankton cyanobacterium, *Synechococcus* sp. Biochim. Biophys. Acta 1169, 39–45.

Kellmann, R., Neilan, B.A., 2007. Biochemical characterization of paralytic shellfish toxin biosynthesis *in vitro*. J. Phycol. 43, 497–508.

Kemp, S.F., Lockey, R.F., 2002. Anaphylaxis: A review of causes and mechanisms. J. Allergy Clin. Immunol. 110, 341–348.

König, G.M., Kehraus, S., Seibert, S.F., Abdel-Lateff, A., Müller, D., 2006. Natural products from marine organisms and their associated microbes. Chembiochem 7, 229–238.

Krienitz, L., Ballot, A., Kotut, K., Wiegand, C., Pütz, S., Metcalf, J.S., Codd, G.A., Pflugmacher, S., 2003. Contribution of hot spring cyanobacteria to the mysterious deaths of Lesser Flamingos at Lake Bogoria, Kenya. FEMS Microbiol. Ecol. 43, 141–148.

Lehane, L., Lewis, R.J., 2000. Ciguatera: Recent advances but the risk remains. Int. J. Food Microbiol. 61, 91–125.

Levine, L., Watanabe, M., 1991. Amplification of prostaglandin I_2 production by thapsigargin. Prostaglandins 42, 291–302.

Long, B.M., Carmichael, W.W., 2003. Marine cyanobacterial toxins. In Hallegraeff, G.M., Anderson, D.M., and Cembella, A.D. (eds.), Manual on Harmful Marine Microalgae, pp. 279–296. Paris, UNESCO.

Luesch, H., Harrigan, G.G., Goetz, G., Horgen, F.D., 2002. The cyanobacterial origin of potent anticancer agents originally isolated from sea hares. Curr. Med. Chem. 9, 1791–1806.

Luesch, H., Pangilinan, R., Yoshida, W.Y., Moore, R.E., Paul, V.J., 2001. Pitipeptolides A and B, new cyclodepsipeptides from the marine cyanobacterium *Lyngbya majuscula*. J. Nat. Prod. 64, 304–307.

Magalhães, V.F., Soares, R.M., Azevedo, S.M.F.O., 2001. Microcystin contamination in fish from the Jacarepaguá Lagoon (Rio de Janeiro, Brazil): Ecological implication and human health risk. Toxicon 39, 1077–1085.

Marine Botany, University of Queensland, and Moreton Bay Waterways and Catchments Partnership, 2002. *Lyngbya majuscula* blooms: Cyanobacteria impacting our coastline. Brisbane, www.marine.uq.edu.au/marbot/publications/pdffiles/lyngbyablooms.pdf.

Marshall, K.L.E., Vogt, R.L., 1998. Illness associated with eating seaweed, Hawaii, 1994. West J. Med. 169, 293–295.

Martins, R., Fernandez, N., Beiras, R., Vasconcelos, V., 2007. Toxicity assessment of crude and partially purified extracts of marine *Synechocystis* and *Synechococcus* cyanobacterial strains in marine invertebrates. Toxicon 50, 791–799.

Martins, R., Pereira, P., Welker, M., Fastner, J., Vasconcelos, V.M., 2005. Toxicity of culturable cyanobacteria strains isolated from the Portuguese coast. Toxicon 46, 454–464.

McMichael, A.J., 2001. Human Frontiers, Environments and Disease: Past Patterns, Uncertain Futures. Cambridge, Cambridge University Press.

Milligan, K.E., Marquez, B.L., Williamson, R.T., Gerwick, W.H., 2000. Lyngbyabellin B, a toxic and antifungal secondary metabolite from the marine cyanobacterium *Lyngbya majuscula*. J. Nat. Prod. 63, 1440–1443.

Moikeha, S.N., Chu, G.W., 1971. Dermatitis-producing alga *Lyngbya majuscula* Gomont in Hawaii. II. Biological properties of the toxic factor. J. Phycol. 7, 8–13.

Montoya, J.P., Holl, C.M., Zehr, J.P., Hansen, A., Villareal, T.A., Capone, D.G., 2004. High rates of N_2 fixation by unicellular diazotrophs in the oligotrophic Pacific Ocean. Nature 430, 1027–1031.

Moore, N., Yeates, M., Esdaile, J., Beumer, J., Dobos, S., Clements, G., Milne, P., 2006. Guidelines for contingency response to coastal algal blooms. Moreton Bay Waterways and Catchments Partnership, Brisbane, www.healthywaterways.org/FileLibrary/algalblooms_smaller.pdf.

Moore, R.E., 1984. Public health and toxins from marine blue-green algae. In Ragelis, E.P. (ed.), Seafood Toxins. ACS Symposium Series No. 262, pp. 369–376. Washington, DC, American Chemical Society.

Mur, L.R., Skulberg, O.M., Utkilen, H., 1999. Cyanobacteria in the environment. In Chorus, I., and Bartram, J. (eds.), Toxic Cyanobacteria in Water: A Guide to Their Public Health Consequences, Monitoring and Management. E & FN Spon on behalf of WHO, London, pp. 15–40, www.who.int/water_sanitation_health/resourcesquality/toxcyanchap2.pdf.

Mynderse, J.S., Moore, R.E., Kashiwagi, M., Norton, T.R., 1977. Antileukemia activity in the Oscillatoriaceae: Isolation of debromoaplysiatoxin from *Lyngbya*. Science 196, 538–540.

Nagabhushanam, A.K., 1967. On an unusually dense phytoplankton "bloom" around Minicoy Island (Arabian Sea), and its effect on the local tuna fisheries. Curr. Sci. 36, 611–612.

Nagai, H., Yasumoto, T., Hokama, Y., 1996. Aplysiatoxin and debromoaplysiatoxin as the causative agents of a red alga *Gracilaria coronopifolia* poisoning in Hawaii. Toxicon 34, 753–761.

Nagle, D.G., Gerwick, W.H., 1995. Nakienones A-C and nakitriol, new cytotoxic cyclic C_{11} metabolites from an Okinawan cyanobacterial (*Synechocystis* sp.) overgrowth of coral. Tetrahedron Lett. 36, 849–852.

Ndetei, R., Muhandiki, V.S., 2005. Mortalities of lesser flamingos in Kenyan Rift Valley saline lakes and the implications for sustainable management of the lakes. Lakes Reserv. Res. Manage. 10, 51–58.

Negri, A.P., Bunter, O., Jones, B., Llewellyn, L.E., 2004. Effects of the bloom-forming alga *Trichodesmium erythraeum* on the pearl oyster *Pinctada maxima*. Aquaculture 232, 91–102.

Negri, A.P., Jones, G.J., 1995. Bioaccumulation of paralytic shellfish poisoning (PSP) toxins from the cyanobacterium *Anabaena circinalis* by the freshwater mussel *Alathyria condola*. Toxicon 33, 667–678.

Neilan, B.A., Jacobs, D., Goodman, A.E., 1995. Genetic diversity and phylogeny of toxic cyanobacteria determined by DNA polymorphisms within the phycocyanin locus. Appl. Environ. Microbiol. 61, 3875–3883.

Neilan, B.A., Saker, M.L., Fastner, J., Torokne, A., Burns, B.P., 2003. Phylogeography of the invasive cyanobacterium *Cylindrospermopsis raciborskii*. Mol. Ecol. 12, 133–140.

Newman, D.J., Cragg, G.M., 2004. Marine natural products and related compounds in clinical and advanced preclinical trials. J. Nat. Prod. 67, 1216–1238.

Ohtani, I., Moore, R.E., Runnegar, M.T.C., 1992. Cylindrospermopsin: a potent hepatotoxin from the blue-green alga *Cylindrospermopsis raciborskii*. J. Am. Chem. Soc. 114, 7941–7942.

Ohuchi, K., Watanabe, M., Takahashi, C., Hayashi, Y., Hirasawa, N., Tsurufuji, S., Fujiki, H., Sugimura, T., 1987. Analysis of tumor-promoter-induced inflammation in rats: Participation of histamine and prostaglandin E2. Biochim. Biophys. Act. 925, 156–163.

Osborne, N.J., Shaw, G.R., Webb, P.M., 2007. Health effects of recreational exposure to Moreton Bay, Australia waters during a *Lyngbya majuscula* bloom. Environ. Int. 33, 309–314.

Osborne, N.J.T., Webb, P.M., Shaw, G.R., 2001. The toxins of *Lyngbya majuscula* and their human and ecological effects. Environ. Int. 27, 381–392.

Oudra, B., El Andaloussi, M., Franca, S., Barros, P., Martins, R., Oufdou, K., Sbiyyaa, B., Loudiki, M., Mezrioui, N., Vasconcelos, V., 2000. Harmful cyanobacterial toxic blooms in waste stabilization ponds. Water Sci. Technol. 42, 179–186.

Oudra, B., Loudiki, M., Vasconcelos, V., Sabour, B., Sbiyyaa, B., Oufdou, Kh., Mezrioui, N., 2002. Detection and quantification of microcystins from cyanobacteria strains isolated from reservoirs and ponds in Morocco. Environ. Toxicol. 17, 32–39.

Paerl, H.W., 1999. Physical-chemical constraints on cyanobacterial growth in the oceans. In Charpy, L., and Larkum, A.W.D. (eds.), Marine Cyanobacteria, pp. 319–349. Monaco City, Musée océanographique.

Paerl, H.W., 2000. Marine plankton. In Whitton, B.A., and Potts, M. (eds.), The Ecology of Cyanobacteria: Their Diversity in Time and Space, pp. 121–148. Dordrecht, Kluwer Academic.

Pomati, F., Kellmann, R., Cavalieri, R., Burns, B.P., Neilan, B.A., 2006. Comparative gene expression of PSP-toxin producing and non-toxic *Anabaena circinalis* strains. Environ. Int. 32, 743–748.

Pouria, S., de Andrade, A., Barbosa, J., Cavalcanti, R.L., Barreto, V.T., Ward, C.J., Preiser, W., Poon, G.K., Neild, G.H., Codd, G.A., 1998. Fatal microcystin intoxication in haemodialysis unit in Caruaru, Brazil. Lancet 352, 21–26.

Ramos, A.G., Martel, A., Codd, G.A., Soler, E., Coca, J., Redondo, A., Morrison, L.F., Metcalf, J.S., Ojeda, A., Suárez, S., Petit, M., 2005. Bloom of the marine diazotrophic cyanobacterium *Trichodesmium erythraeum* in the northwest African upwelling. Mar. Ecol. Prog. Ser. 301, 303–305.

Rapala, J., Robertson, A., Negri, A.P., Berg, K.A., Tuomi, P., Lyra, C., Erkomaa, K., Lahti, K., Hoppu, K., Lepisto, L., 2005. First report of saxitoxin in Finnish lakes and possible associated effects on human health. Environ. Toxicol. 20, 331–340.

Satô, S., Paranaguá, M.N., Eskinazi, E., 1963/1964. On the mechanism of red tide of *Trichodesmium* in Recife northeastern Brazil, with some considerations of the relation to the human disease, "Tamandaré Fever." Trab-s Inst. Oceanogr. Univ. Recife. 5/6, 7–49.

Schaeffer, D.J., Krylov, V.S., 2000. Anti-HIV activity of extracts and compounds from algae and cyanobacteria. Ecotoxicol. Environ. Saf. 45, 208–227.

Schopf, J.W., 2000. The fossil record: Tracing the roots of the cyanobacterial lineage. In Whitton, B.A., and Potts, M. (eds.), The Ecology of Cyanobacteria: Their Diversity in Time and Space, pp. 13–35. Dordrecht, Kluwer Academic.

Seifert, M., McGregor, G., Eaglesham, G., Wickramasinghe, W., Shaw, G., 2007. First evidence for the production of cylindrospermopsin and deoxy-cylindrospermopsin by the freshwater benthic cyanobacterium, *Lyngbya wollei* (Farlow ex Gomont) Speziale and Dyck. Harmful Algae 6, 73–80.

Sellner, K.G., 1997. Physiology, ecology, and toxic properties of marine cyanobacteria blooms. Limnol. Oceanogr. 42, 1089–1104.

Selwood, A.I., Holland, P.T., Wood, S.A., Smith, K.F., McNabb, P.S., 2007. Production of anatoxin-a and a novel biosynthetic precursor by the cyanobacterium *Aphanizomenon issatschenkoi*. Environ. Sci. Technol. 41, 506–510.

Shimizu, Y., 2003. Microalgal metabolites. Curr. Opin. Microbiol. 6, 236–243.

Singh, S., Kate, B.N., Banerjee, U.C., 2005. Bioactive compounds from cyanobacteria and microalgae: An overview. Crit. Rev. Biotechnol. 25, 73–95.

Sipiä, V.O., Kankaapää, H.T., Pflugmacher, S., Flinkman, J., Furey, A., James, K.J., 2002. Bioaccumulation and detoxication of nodularin in tissues of flounder (*Platichthys flesus*), mussels (*Mytilus edulis*, *Dreissena polymorpha*), and clams (*Macoma balthica*) from the northern Baltic Sea. Ecotoxicol. Environ. Saf. 53, 305–311.

Sivonen, K., Jones, G., 1999. Cyanobacterial toxins. In Chorus, I., and Bartram, J. (eds.), Toxic Cyanobacteria in Water: A Guide to Their Public Health Consequences, Monitoring and Management. E & FN Spon on behalf of WHO, pp. 41–111. London, www.who.int/water_sanitation_health/resourcesquality/toxcyanchap3.pdf.

Solomon, A.E., Stoughton, R.B., 1978. Dermatitis from purified sea algae toxin (debromoaplysiatoxin). Arch. Dermatol. 114, 1333–1335.

Stal, L.J., Albertano, P., Bergman, B., von Bröckel, K., Gallon, J.R., Hayes, P.K., Sivonen, K., Walsby, A.E., 2003. BASIC: Baltic Sea cyanobacteria. An investigation of the structure and dynamics of water blooms of cyanobacteria in the Baltic Sea: Responses to a changing environment. Cont. Shelf. Res. 23, 1695–1714.

Stewart, I., Schluter, P.J., Shaw, G.R., 2006a. Cyanobacterial lipopolysaccharides and human health: A review. Environ. Health 5, 7.

Stewart, I., Webb, P.M., Schluter, P.J., Fleming, L.E., Burns, J.W., Jr., Gantar, M., Backer, L.C., Shaw, G.R., 2006b. Epidemiology of recreational exposure to freshwater cyanobacteria: An international prospective cohort study. BMC Public Health 6, 93.

Stewart, I., Webb, P.M., Schluter, P.J., Shaw, G.R., 2006c. Recreational and occupational field exposure to freshwater cyanobacteria: A review of anecdotal and case reports, epidemiological studies and the challenges for epidemiologic assessment. Environ. Health 5, 6.

Sudek, S., Haygood, M.G., Youssef, D.T.A., Schmidt, E.W., 2006. Structure of trichamide, a cyclic peptide from the bloom-forming cyanobacterium *Trichodesmium erythraeum*, predicted from the genome sequence. Appl. Environ. Microbiol. 72, 4382–4387.

Suvapepun, S., 1992. *Trichodesmium* blooms in the Gulf of Thailand. In Carpenter, E.J., Capone, D.G., and Rueter, J.G. (eds.), Marine Pelagic Cyanobacteria: *Trichodesmium* and Other Diazotrophs, pp. 343–348. Dordrecht, Kluwer Academic.

Tan, L.T., 2007. Bioactive natural products from marine cyanobacteria for drug discovery. Phytochemistry 68, 954–979.

Turner, P.C., Gammie, A.J., Hollinrake, K., Codd, G.A., 1990. Pneumonia associated with contact with cyanobacteria. BMJ 300, 1440–1441.

Walsby, A.E., 1992. The gas vesicles and buoyancy of *Trichodesmium*. In Carpenter, E.J., Capone, D.G., and Rueter, J.G. (eds.), Marine Pelagic Cyanobacteria: *Trichodesmium* and Other Diazotrophs, pp. 141–161. Dordrecht, Kluwer Academic.

Watanabe, M., 1987. Studies on the planktonic blue-green algae 2. *Umezakia natans* gen. et sp. nov. (Stigonemataceae) from the Mikata Lake, Fukui Prefecture. Bull. Natl. Sci. Mus. Ser. B. Bot. 13, 81–88.

Watanabe, M., Kawase, Y., Tanabe, J., Min, K.R., Mue, S., Ohuchi, K., 1995. Suppression of interleukin-1 alpha production by protein kinase C activators in human vascular endothelial cells. J. Pharmacol. Exp. Ther. 272, 808–814.

Whitton, B.A., Potts, M., 2000. Introduction to the cyanobacteria. In Whitton, B.A., and Potts, M. (eds.), The Ecology of Cyanobacteria: Their Diversity in Time and Space, pp. 3–11. Dordrecht, Kluwer Academic.

Williams, P.G., Yoshida, W.Y., Moore, R.E., Paul, V.J., 2004. Micromide and guamamide: Cytotoxic alkaloids from a species of the marine cyanobacterium *Symploca*. J. Nat. Prod. 67, 49–53.

Wonnacott, S., Gallagher, T., 2006. The chemistry and pharmacology of anatoxin-a and related homotropanes with respect to nicotinic acetylcholine receptors. Mar. Drugs 4, 228–254.

World Health Organization (WHO), 2003. Guidelines for safe recreational water environments, volume 1: Coastal and fresh waters. World Health Organization, Geneva, www.who.int/water_sanitation_health/bathing/srwg1.pdf.

World Health Organization (WHO), 2006. Guidelines for drinking-water quality. First addendum to third edition, volume 1: Recommendations. World Health Organization, Geneva, www.who.int/water_sanitation_health/dwg/gdwq0506.pdf.

Wu, M., Okino, T., Nogle, L.M., Marquez, B.L., Williamson, R.T., Sitachitta, N., Berman, F.W., Murray, T.F., McGough, K., Jacobs, R., Colsen, K., Asano, T., Yokokawa, F., Shioiri, T., Gerwick, W.H., 2000. Structure, synthesis and biological properties of kalkitoxin, a novel neurotoxin from the marine cyanobacterium *Lyngbya majuscula*. J. Am. Chem. Soc. 122, 12041–12042.

Xie, L., Xie, P., Guo, L., Li, L., Miyabara, Y., Park, H., 2005. Organ distribution and bioaccumulation of microcystins in freshwater fish at different trophic levels from the eutrophic Lake Chaohu, China. Environ. Toxicol. 20, 293–300.

Yasumoto, T., 2001. The chemistry and biological function of natural marine toxins. Chem. Rec. 1, 228–242.

Yasumoto, T., 2005. Chemistry, etiology, and food chain dynamics of marine toxins. Proc. Jpn. Acad. Ser. B. Phys. Biol. Sci. 81, 43–51.

Yip, Y.K., Kelker, H.C., Stone-Wolff, D.S., Pearlstein, K., Urban, C., Vilcek, J., 1983. Stimulation of lymphokine production by teleocidin, aplysiatoxin, and debromoaplysiatoxin. Cell Immunol. 79, 389–395.

Yotsu-Yamashita, M., Haddock, R.L., Yasumoto, T., 1993. Polycavernoside A: A novel glycosidic macrolide from the red alga *Polycavernosa tsudai* (*Gracilaria edulis*). J. Am. Chem. Soc. 115, 1147–1148.

Yotsu-Yamashita, M., Yasumoto, T., Yamada, S., Bajarias, F.F.A., Formeloza, M.A., Romero, M.L., Fukuyo, Y., 2004. Identification of polycavernoside A as the causative agent of the fatal food poisoning resulting from ingestion of the red alga *Gracilaria edulis* in the Philippines. Chem. Res. Toxicol. 17, 1265–1271.

Zehr, J.P., 1992. Molecular biology of nitrogen fixation in natural populations of marine cyanobacteria. In Carpenter, E.J., Capone, D.G., and Rueter, J.G. (eds.), Marine Pelagic Cyanobacteria: *Trichodesmium* and Other Diazotrophs, pp. 249–264. Dordrecht, Kluwer Academic.

Zehr, J.P., Waterbury, J.B., Turner, P.J., Montoya, J.P., Omoregie, E., Steward, G.F., Hansen, A., Karl, D.M., 2001. Unicellular cyanobacteria fix N_2 in the subtropical North Pacific Ocean. Nature 412, 635–638.

STUDY QUESTIONS

1. Cyanobacteria are microorganisms, yet cyanobacteria can often be seen with the naked eye. What characteristics of cyanobacteria allow them to be seen without a microscope?
2. When did cyanobacteria originate, and what part did they play in the evolution of eukaryotic life?
3. Cyanobacteria play a major part in the nutrient supply of ocean life. How does this come about?
4. Many cyanobacterial strains and species are capable of producing potent toxins, and some novel cyanobacterial compounds express cytotoxic effects on mammalian cell lines. How are the genetic origins of these toxicities examined?
5. The algae are a very diverse group, including (for example) the multicellular forms *Macrocystis pyrifera* (giant bladder kelp), *Ulva lactula* (sea lettuce), and the (mostly) microscopic and unicellular diatoms (Class: Bacillariophyceae). Although "algae" is not a current taxonomic group, what characteristics are common to these three kinds of algae?
6. Cyanobacteria are often known as blue-green algae. Are *cyanobacteria* and *blue-green algae* completely synonymous terms? Why or why not? What features do cyanobacteria/blue-green algae share with algae? What features separate them?
7. The marine cyanobacterium *Richelia* forms a symbiotic relationship with certain diatoms. What benefits might accrue to both organisms from this relationship? Are there other symbiotic interactions involving cyanobacteria?

8. Human injury has occurred through ingestion of cyanobacteria. Under what circumstances has this occurred? How can risks from cyanobacterial injury be minimized?
9. Design a study to evaluate the acute and chronic health effects of cyanobacteria or their toxins. What human population(s) would you pick and why? What exposure routes would you evaluate? What cyanobacteria or their toxins would you focus on? What health effects would you study and why?

C H A P T E R

16

Pfiesteria

WOLFGANG K. VOGELBEIN, VINCENT J. LOVKO, AND KIMBERLY S. REECE

INTRODUCTION

During the 1990s, dinoflagellates of the genus *Pfiesteria* were considered a serious emerging fish and human health problem in mid-Atlantic U.S. estuaries, primarily the Chesapeake Bay (in Maryland and Virginia) and the Pamlico Sound (in North Carolina). At the time, these dinoflagellates were implicated as the causative agents of massive fish kills and fish lesion events, primarily in Atlantic menhaden (*Brevoortia tyrannus*). They were also implicated in human health effects (including gastrointestinal, respiratory, and skin abnormalities, as well as neurocognitive deficits). These adverse health effects in fishes and humans were purportedly caused by a potent *Pfiesteria* toxin, the production of which was intimately tied to life-cycle transformations and the specific environmental cues that regulated them. The complex life history of *Pfiesteria* was reported to consist of at least 24 stages, several of which were thought to produce toxins.

Pfiesteria piscicida was initially thought to be the primary small heterotrophic dinoflagellate present in the environment. It was subsequently recognized that *P. piscicida* co-occurred with a diverse group of morphologically similar *Pfiesteria*-like dinoflagellates (referred to in this chapter as "PLDs"), including a second *Pfiesteria* species. As research intensified, many aspects of *Pfiesteria* biology (including its life cycle, toxigenicity and its purported role in fish kill and lesion events, and human illness) became disputed and highly controversial.

In this chapter, we use the term "*Pfiesteria*" to include *Pfiesteria piscicida* and *Pseudopfiesteria shumwayae*, the latter initially described as *Pfiesteria shumwayae* but reclassified based on morphological and genetic differences. The majority of early *Pfiesteria* literature considered the biology, behavior, and toxicity of these two species to be identical. We distinguish *P. piscicida* and *P. shumwayae* from the other PLDs because only these two species are reported to produce potent toxins implicated in major fish and human health impacts. Although PLDs will be referred to occasionally throughout this chapter, the primary focus is on *Pfiesteria piscicida* and *Pseudopfiesteria shumwayae*.

The goal of this chapter is to objectively present the body of scientific research on *Pfiesteria* so that the reader may better understand the biology of these interesting organisms. This chapter also aims to provide the reader with an understanding of the controversial aspects of *Pfiesteria* science, so that the evidence regarding the role that *Pfiesteria*-like dinoflagellates may play in human illness and fish death and disease can be effectively evaluated.

WHAT IS *PFIESTERIA*?

Description

Pfiesteria and *Pfiesteria*-like dinoflagellates are small (~10 to 20 μm) heterotrophic, peridinoid protists belonging to the Family Pfiesteriaceae (class Dinophyceae, division Dinoflagellata, order Peridiniales). Although often referred to as "algae," *Pfiesteria* spp. are microzooplankters that do not synthesize chlorophyll or other photopigments. The primary nutritional strategy of these dinoflagellates is heterotrophic feeding using a peduncle. This tubelike structure emerges from a longitudinal groove called the "sulcus" and is used to attach to and withdraw the contents of prey cells by a process called "myzocytosis" (Steidinger *et al.*, 2001). However, they reportedly can supplement their nutritional requirements using chloroplasts captured from algal prey and sequestered in food vacuoles, a process called "kleptochloroplasty" (Burkholder and Glasgow, 1995, 1997a;

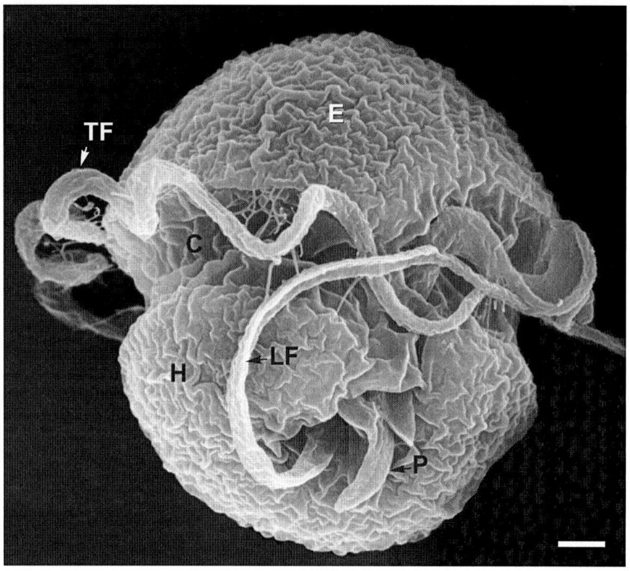

FIGURE 16-1. Scanning electron micrograph of *Pseudopfiesteria shumwayae* showing basic morphology of a typical flagellated zoospore: E, epitheca; H, hypotheca; C, cingulum; TF, transverse flagellum; LF, longitudinal flagellum; P, peduncle. Scale bar, 1 µm.

Lewitus *et al.*, 1999; Steidinger *et al.*, 1996). Like other dinoflagellates, *Pfiesteria* uses two flagella for locomotion: a transverse flagellum encircling the cell within an equatorial groove called the cingulum (Fig. 16-1) and a longitudinal flagellum oriented perpendicular to the transverse flagellum and arising from a longitudinal groove called the sulcus. A typical dinokaryotic nucleus, containing permanently condensed chromosomes, is located in the hypotheca (Steidinger *et al.*, 1996).

Identification

Identification and taxonomic placement of dinoflagellates is based on morphological criteria. However, genetic identification methods have recently been developed to supplement the morphological observations and will be discussed in a later section. Lightly armored dinoflagellates, such as *Pfiesteria*, contain thin cellulose plates, or "theca," within vesicles called "amphiesma" located immediately under the outer cell membrane ("plasmalemma"). Dinoflagellate species are assigned to a genus based on the number, shape, and arrangement of these thecal plates (Kofoid, 1909). Visualization of thecal plates by light microscopy is difficult because they are very thin and may be obscured by the membranes of the amphiesma and the plasmalemma. For these lightly armored species, scanning electron microscopy (SEM) is the preferred method of visualizing thecal plate arrangement. Before imaging, either the plasmalemma and amphiesmal membranes must be removed by chemical or mechanical stripping, or the cells must be osmotically swollen so that the sutures between the plates become visible through these membranes (Mason *et al.*, 2003; Steidinger *et al.*, 1996). The plate tabulation for the genus *Pfiesteria* includes an apical pore complex (APC) made up of a pore plate (Po), a closing plate (cp), and a canal plate (X); an apical series comprised of four apical plates (4′); one three-sided, or triangular, anterior intercalary plate (1a); five precingular plates (5″); six cingular plates (6c); five or more sulcal plates (5 + s); and an antapical series composed of five postcingular plates (5‴) and two antapical plates (2⁗) (Steidinger *et al.*, 1996) (Fig. 16-2). The plate tabulation for the genus *Pseudopfiesteria* differs in having six precingulars instead of five and a four-sided, or diamond-shaped, anterior intercalary plate instead of a three-sided one (AlgaeBase, www.algaebase.org; Burkholder *et al.*, 2001a, 2001b; Glasgow *et al.*, 2001b; Litaker *et al.*, 2005) (Fig. 16-3).

Abundance and Distribution

Pfiesteria piscicida and PLDs are members of the plankton community in shallow, eutrophic estuaries along the mid-Atlantic and southeastern U.S. coasts. They have been detected in samples collected from New York, Delaware Bay, Chesapeake Bay, the Albemarle-Pamlico Estuarine System in North Carolina, South Carolina, the Atlantic and Gulf coasts of Florida and Mobile Bay, Alabama (Burkholder *et al.*, 1995a; Lin *et al.*, 2006). The development of species-specific molecular assays for detection of *Pfiesteria* has broadened the distribution to several other locations around the world (Bowers *et al.*, 2000; Lin *et al.*, 2006; Rhodes *et al.*, 2006; Rublee *et al.*, 2005). Using light microscopy, the abundance of *Pfiesteria* spp. was reported to be high in North Carolina estuaries during fish kill events, averaging ~1000 to 5000 cells ml^{-1} (Burkholder and Glasgow, 1997a). These "presumptive cell counts," however, can be misleading, as light microscopy cannot distinguish between species. At the time these early counts were made, the diversity of PLDs was not appreciated and many cells were likely misidentified as *Pfiesteria*, leading to significant overestimations of the actual *Pfiesteria* concentrations in samples. More recently, lower abundances (<20 cells ml^{-1}) of *Pfiesteria*-like dinoflagellates have been reported in tidal creeks in Virginia and South Carolina, using light microscopy (Lewitus *et al.*, 2002; Marshall *et al.*, 1999). Quantitative molecular assays have shown that *Pfiesteria* concentrations in estuaries measured worldwide are typically less than 1 to 2 cells ml^{-1} and seldom exceed 5 cells ml^{-1} (Coyne *et al.*, 2001; Lin *et al.*, 2006; Litaker *et al.*, 2003; Reece *et al.*, 2002, 2005). These concentrations are several orders of magnitude lower than those reported to harm fish (Burkholder and Glasgow, 1997a).

Although euryhaline, *Pfiesteria* and PLDs are most commonly found at salinities between 5 and 20 psu, with optimal

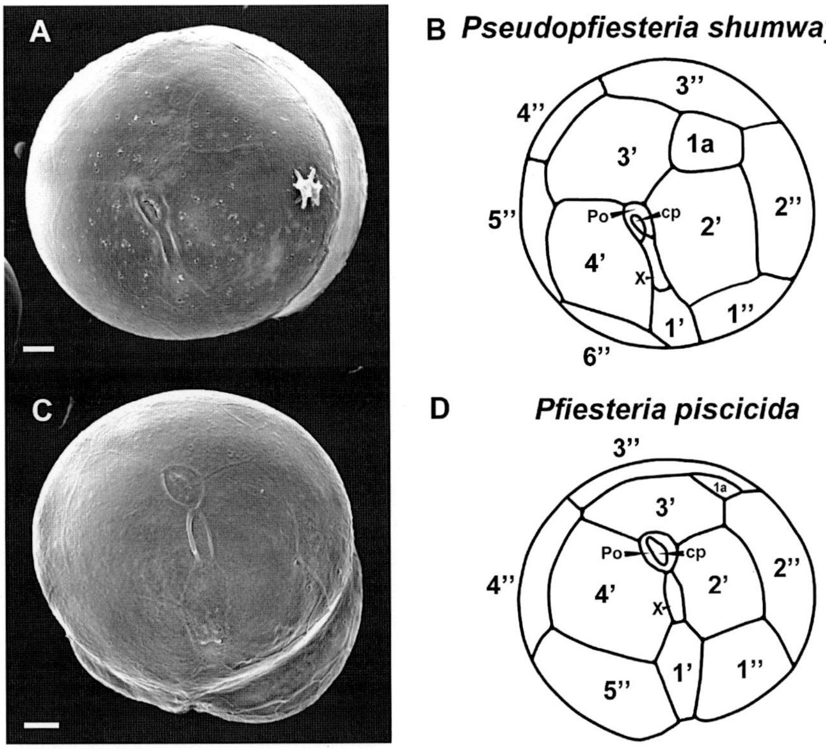

FIGURE 16-2. Comparison of the plate tabulations for *Pseudopfiesteria shumwayae* and *Pfiesteria piscicida.* (A) Membrane stripped SEM preparation showing apical view of a typical *P. shumwayae* zoospore. (B) Diagrammatic representation of plate tabulation for *P. shumwayae*, apical view. Note the rectangular shaped anterior intercalary plate (1a). (C) Membrane stripped SEM preparation showing apical view of a typical *P. piscicida* zoospore. (D) Diagrammatic representation of plate tabulation for *P. piscicida*, apical view. Note the triangular shaped anterior intercalary plate (1a). Scale bars, 1 μm. Reproduced by permission of the Phycological Society of America, from Litaker *et al.* (2005).

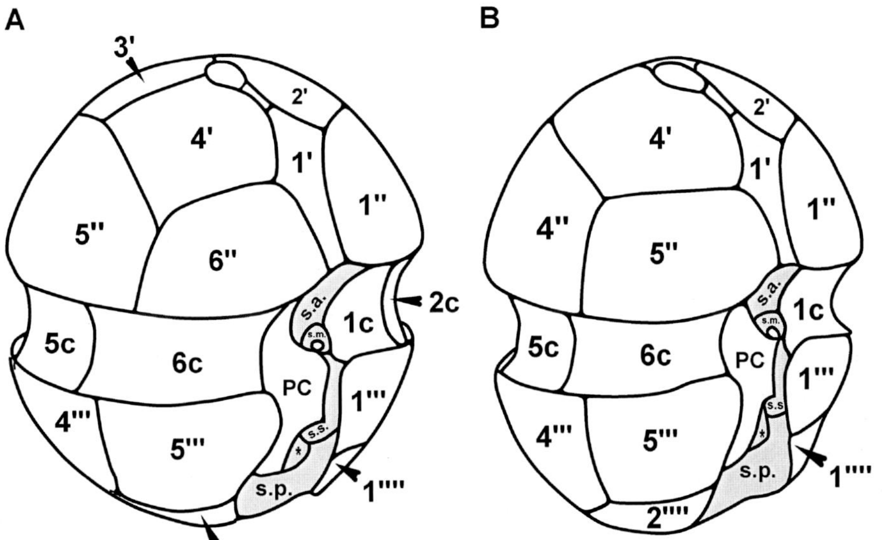

FIGURE 16-3. Diagrammatic representation of the plate tabulation of (A) *Pseudopfiesteria shumwayae* and (B) *Pfiesteria piscicida* illustrating some of the differences in plate morphology. Note the six precingular plates in *P. shumwayae* versus five precingulars in *P. piscicida*. Sinistral-ventral view showing the sulcus. Reproduced by permission of the Phycological Society of America, from Litaker *et al.* (2005).

growth at ~15 psu (Burkholder *et al.*, 1995a; Lin *et al.*, 2006). They tolerate a broad temperature range of 6 to 31°C but demonstrate greatest abundance and activity at warmer temperatures (15 to 30°C) (Burkholder *et al.*, 1995a; Lin *et al.*, 2006) and in nutrient-rich waters (Burkholder *et al.*, 1992, 2001a). A light intensity optimum for growth of *Pfiesteria* has not been reported (Burkholder *et al.*, 1995a; Eriksen *et al.*, 2002). Growth of flagellated stages has been stimulated with inorganic phosphorous (50 to 400 μg PO_4 L^{-1}) but not with nitrogen (Burkholder *et al.*, 1992). Although cryptophytes (small flagellated phytoplankton) are the preferred prey of *Pfiesteria* (Burkholder and Glasgow, 1997a), they can also survive on a variety of microbial and animal prey (Burkholder *et al.*, 2001a). Several species of protists have been shown to prey on *Pfiesteria*, including ciliates (*Strombidium* sp., *Mesodinium pulex*) (Stoecker *et al.*, 2000), rotifers (*Brachionus plicatilis*) (Stoecker and Gustafson, 2002), and calanoid copepods (*Acartia tonsa*) (Mallin *et al.*, 1995), suggesting that *Pfiesteria* is readily consumed by grazers.

THE EARLY PFIESTERIA PARADIGM

Discovery

Pfiesteria piscicida was discovered in 1988 at the North Carolina State University, School of Veterinary Medicine, when cultured tilapia (*Oreochromis aureus*) died after exposure to water from the Pamlico River, North Carolina (Burkholder *et al.*, 1992; Noga *et al.*, 1993). Water from the affected tanks contained a small dinoflagellate that increased in number before the fish mortalities then decreased rapidly when live fish were no longer present (Burkholder *et al.*, 1992). Affected fish displayed shallow, rapid ventilation, dull skin, and "neurological signs" (including depression of normal activity, hyperexcitability, and loss of equilibrium) (Noga *et al.*, 1993). Histopathological analysis revealed epidermal edema and necrosis, as well as focal to widespread epithelial erosion of the skin (Noga *et al.*, 1996).

Bioassays using juvenile tilapia were conducted in 9-L aquaria inoculated with dinoflagellates (initial densities <100 cells L^{-1}) from the original fish-killing tank (Noga *et al.*, 1993). Exposed fish demonstrated clinical signs identical to those observed in the original tanks. Fish mortalities of 80% to 100% were associated with increasing dinoflagellate densities (500 to 10,000 cells ml^{-1}), suggesting a cause and effect relationship. Sublethally exposed fish transferred to clean water, subsequently developed skin ulcers that became secondarily colonized by bacteria and oomycete water moulds (Noga *et al.*, 1996). Tilapia exposed to water that had been filtered (0.22 μm) to remove the dinoflagellates, displayed similar clinical signs as well as 60% mortality after 48 hours (Burkholder *et al.*, 1992; Noga *et al.*, 1993). No mortalities occurred in fish exposed to filtrate from control tanks. Researchers concluded that the disease and mortality in the tilapia, as well as the adverse effects observed in 33 additional finfish and shellfish species, were caused by a potent neurotoxin secreted by the dinoflagellate (Burkholder *et al.*, 1992, 1995a; Noga *et al.*, 1993; Steidinger *et al.*, 1996).

Based on results of these studies and the fact that water used in the original fish-killing tanks derived from the Pamlico River (North Carolina), it was hypothesized that *Pfiesteria* could have been involved in numerous fish kills occurring in the Albemarle-Pamlico Estuarine system during the late 1980s and early 1990s. A massive fish kill event involving more than a million fish, primarily Atlantic menhaden (*Brevoortia tyrannus*), was investigated while in progress in the Pamlico River during May 1991 (Burkholder *et al.*, 1992; Noga *et al.*, 1993). Dinoflagellates resembling the original fish-killing aquarium contaminant were abundant in the fish kill zone and cultures derived from these samples killed juvenile tilapia in laboratory challenges (Burkholder *et al.*, 1992, 1995a, 1995b).

To further define the role of *Pfiesteria* species in estuarine fish kills, a standardized approach to investigating these events was developed. First, field-collected water samples from an in-progress fish kill were evaluated by light microscopy for the presence of *Pfiesteria* using the presumptive cell count method, which, as discussed earlier, did not distinguish between *Pfiesteria* and other similar dinoflagellates (Burkholder *et al.*, 1992, 1995a). Samples with cell counts of ≥300 cells ml^{-1} were considered indicative of potential *Pfiesteria* activity and were subsequently used in standard fish bioassays in which juvenile tilapia were exposed in 9-L aquaria to the environmental water sample (Burkholder *et al.*, 1992, 1995a). Controls consisted of fish in tanks with artificial seawater, but without dinoflagellates. If fish mortality occurred (which in many assays took several weeks or even months), tank water was evaluated microscopically for the presence of *Pfiesteria* (Burkholder *et al.*, 1995a). If present, dinoflagellates were enumerated, clonally isolated, and cultured for a second tier of fish bioassays. If mortalities occurred in the second set of bioassays and if the clonal culture was positively identified as *Pfiesteria* using SEM, then *Pfiesteria* would be implicated as the causative agent of the fish kill (Burkholder *et al.*, 2001a, 2001c; Glasgow *et al.*, 2001a).

This fish bioassay approach was claimed to fulfill the Henle-Koch postulates modified for toxic, rather than infectious organisms and thus purportedly confirmed involvement of toxic *Pfiesteria* in the fish kill (Burkholder and Glasgow, 1997a; Burkholder *et al.*, 1992, 1995a, 2001a, 2001c). Using this approach, *Pfiesteria* was implicated in ~50% of the major (involving ≥1000 fish per event) annual fish kill events, involving predominantly menhaden, in the Neuse and Pamlico estuaries (North Carolina) between 1991

FIGURE 16-4. Gross appearance of ulcerative mycosis in Atlantic menhaden, *Brevoortia tyrannus*. These ulcers were commonly attributed to *Pfiesteria* spp. during the 1990s but are now known to be caused by the oomycete, *Aphanomyces invadans*. Ulcers are commonly located perianally and penetrate deeply into underlying muscle. Photographs reproduced from Vogelbein *et al.* (2001).

and 1993 (Burkholder *et al.*, 1995a). In many of these events, fish exhibited ulcerative skin lesions often penetrating into the musculature and visceral organs (Burkholder *et al.*, 1995a) (Fig. 16-4). Thus, *Pfiesteria* was presumed to play a causative role in these fish kill events and in the development of ulcerative lesions in natural fish populations in the Pamlico and Neuse estuaries of North Carolina and was postulated to be involved in other unexplained kills and disease events in similar nutrient rich and poorly flushed estuaries in the mid-Atlantic (Burkholder *et al.*, 1992, 1995a; Noga *et al.*, 1996).

In fact, the association between *Pfiesteria* and ulcerous lesions in fishes was considered strong enough to use lesioned fish as an indicator of actively toxic *Pfiesteria* in the environment (Burkholder and Glasgow, 1997a; Burkholder *et al.*, 2001a, 2001c; Glasgow *et al.*, 2001b; Magnien, 2001; Noga *et al.*, 1996). Additionally, a causative link also was made between *Pfiesteria* and human health, with reports of adverse effects in humans that had contact with *Pfiesteria*, either through laboratory exposure while working with *Pfiesteria* (Glasgow *et al.*, 1995) or by recreational or occupational exposure to waterways with purportedly toxic *Pfiesteria* (Grattan *et al.*, 1998b). Involvement of *Pfiesteria* in fish and human health will be discussed in greater detail later in this chapter.

Life History

Under the early paradigm, *Pfiesteria* was reported to have a complex, multiphasic life cycle involving numerous flagellated, amoeboid, and encysted forms occupying various niches in the water column and sediments (Burkholder *et al.*, 1995b; Burkholder and Glasgow, 1997a, 1997b). At least 24 different life stages were proposed, encompassing these three basic morphological forms, with sizes ranging from 5 to >400 μm. The proposed life cycle included lobose, rhizopodial, and filose amoebae; free-swimming toxic and nontoxic zoospores (biflagellated cells); planozygotes; gametes; and distinct cysts types, including smooth-walled and ornamented cysts exhibiting a covering of scales with spiny bracts that resembled marine chrysophytes (Fig. 16-5) (Burkholder *et al.*, 1995b; Burkholder and Glasgow, 1995, 1997a, 1997b).

Pfiesteria was described as an "ambush predator" with ephemeral flagellated forms referred to in the literature as "phantom" existing for only a few hours following exposure to live fish (Burkholder *et al.*, 1995a, 1995b; Burkholder and Glasgow, 1995, 1997b). According to this proposed life cycle, large reservoirs of *P. piscicida* exist in the sediment as encysted or amoeboid forms. Unidentified substances produced by fish are believed to stimulate these sediment-dwelling forms to excyst or transform from amoeboid to flagellated stages, referred to as toxic zoospores (TZs). These stages exhibit directed chemotaxis toward fish and purportedly secret a potent neurotoxin that stuns fish and causes the epidermis to slough off, which the dinoflagellates then consume phagotrophically (Burkholder *et al.*, 1992, 1995a, 1995b; Burkholder and Glasgow, 1997b).

It was proposed that flagellated cells reproduce either by simple binary fission (asexual reproduction) to produce additional flagellated cells or by concentrating in a gelatinous mass and undergoing repeated cytoplasmic divisions generating multiple "male" or "female" gametes from each parent cell (Burkholder *et al.*, 1995b; Burkholder and Glasgow, 1997b). Fusion of gametes (sexual reproduction) to form a larger, triflagellated planozygote was reported to occur only in the presence of live fish, with the planozygotes eventually undergoing division to produce four biflagellated cells (Burkholder *et al.*, 1995b; Burkholder and Glasgow, 1995).

In the absence of live fish, toxic cells and planozygotes reportedly decrease in abundance as they encyst or revert to amoeboid forms, whereas some flagellated cells may revert to nontoxic forms (NTZs) and remain in the water column if there is an abundant alternative prey source (Fig. 16-6) (Burkholder *et al.*, 1995b; Burkholder and Glasgow, 1995, 1997a, 1997b). Flagellated cells may also attach to fish carcasses and transform into lobose amoebae that continue to feed on the fish remains (Burkholder and Glasgow, 1995, 1997a). Without live fish, gametes that have not fused may

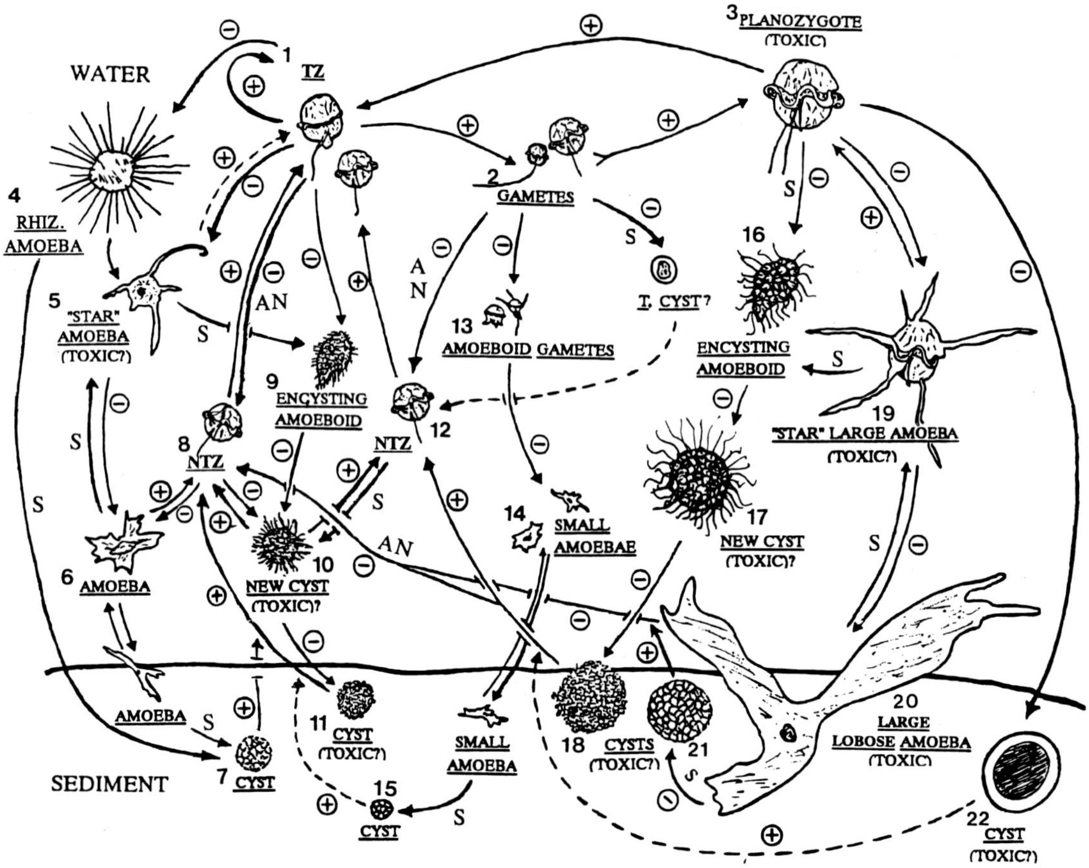

FIGURE 16-5. Diagrammatic representation of a complex, multiphasic life cycle initially proposed for *Pfiesteria*. Twenty-four distinct life history stages including flagellated, amoeboid, and cyst forms were initially considered to comprise the life cycle. Copyright 1997 by the American Society of Limnology and Oceanography, Inc. Burkholder and Glasgow (1997a), Figure 1, p. 1053.

form temporary cysts or transform into small lobose amoeba, reported to be present in large numbers in the water column and sediment surface at fish kills (Burkholder *et al.*, 1995b). Planozygotes that have not yet divided may instead form a scaled hypnozygote cyst or they may increase two- to three-fold in size and transform into large, benthic lobose amoebae (≥250 μm) with flexible pseudopodia, rigid, tapering extensions, and a fibrous, reticulate external covering (Burkholder *et al.*, 1995b).

This complex life cycle purportedly provides a mechanism for the "ambush predator" lifestyle wherein a large reservoir of cells in the sediment and water column can be rapidly mobilized to cause massive fish disease and mortality events (Fig. 16-6) (Burkholder and Glasgow, 1997a).

Functional Types and the Toxic *Pfiesteria* Complex

Much of the early literature referred to a second species that is distinct from, yet similar to *P. piscicida* (Burkholder *et al.*, 1995a, 1995b; Burkholder and Glasgow, 1997a; Steidinger *et al.*, 1996). This species, *Pfiesteria shumwayae*, was formally described in 2001 and, although morphologically and genetically distinct, was considered identical to *P. piscicida* with regard to toxicity and life history (Glasgow *et al.*, 2001b). Because of the similarities, these two species are collectively referred to as the "Toxic *Pfiesteria* Complex" (TPC), and are described as having three distinct functional types with respect to toxin production (Burkholder *et al.*, 2001b).

Noninducible strains (NON-IND functional type) are incapable of producing toxin under any circumstances (Burkholder *et al.*, 2001b, 2001c; Glasgow *et al.*, 2001a, 2001b) and are defined as being "unable to cause fish distress, disease or death." Temporarily nontoxic strains (TOX-B functional type) have lost their toxicity after extended culture in the absence of fish (Burkholder *et al.*, 2001a, 2001b). They can, however, regain toxicity when again cultured with fish. The longer the period away from fish, the more time required to restore toxicity. After a sufficient duration (weeks to months), TOX-B strains permanently lose their ability to kill fish and become NON-IND (Burkholder

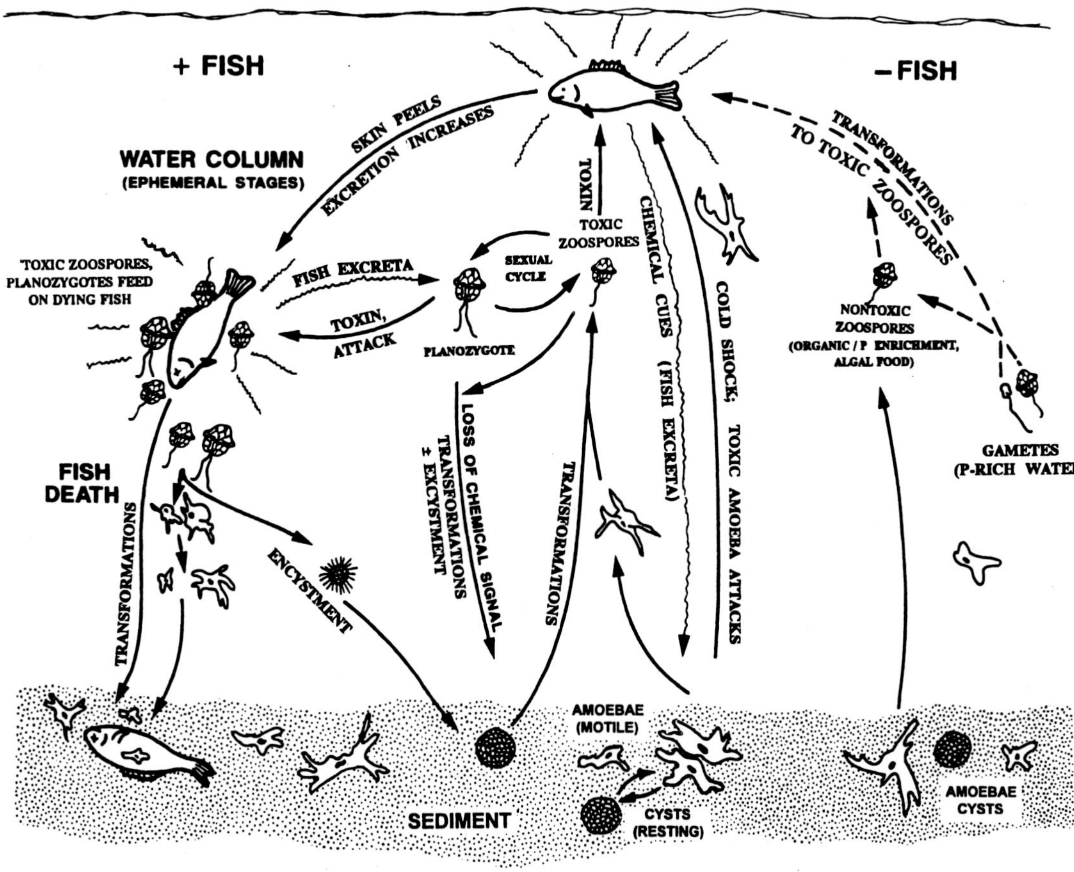

FIGURE 16-6. Diagrammatic representation of a complex, multiphasic life cycle illustrating the "ambush predator" hypothesis wherein *Pfiesteria* spp. become actively toxic in the presence of fish secreta or excreta. The potent exotoxin stuns fish and causes their tissues to slough off. Zoospores were then believed to feed on the sloughed fish tissues. Copyright 1997 by the American Society of Limnology and Oceanography, Inc. Burkholder and Glasgow (1997a), Figure 2, p. 1054.

et al., 2001b, 2001c; Glasgow *et al.*, 2001a, 2001b). Actively toxic strains (TOX-A functional type) demonstrate a strong attraction to fish and fish tissues and cause rapid fish death (hours to days) in bioassays (Burkholder *et al.*, 1992, 2001a, 2001b). Fish exposed to TOX-A strains exhibit narcosis, hyperexcitability, ataxia, diffuse skin hemorrhage, scale loss, and death (Burkholder *et al.*, 1992; Burkholder and Glasgow, 1997a, 1997b; Noga *et al.*, 1996). Filtrate from fish-reared cultures that were actively killing fish has been reported to have similar effects on fish, supporting the hypothesis that a toxin was responsible for the observed disease signs (Burkholder and Glasgow, 1997a, 1997b).

CONTROVERSIAL ISSUES

Life History

The life cycle of *Pfiesteria piscicida* as originally described was of note because it was reported as "the first known free-living estuarine or marine dinoflagellate with a complex life cycle that involves rapid transformations among at least 24 stages that include flagellates, amoebae, and cysts" (p. 1056, Burkholder and Glasgow, 1997a). These various life stages were of importance because they were claimed to be instrumental in the "ambush predator hypothesis" described earlier in this chapter (Burkholder, 1999; Burkholder *et al.*, 1992). As mentioned previously, many of these stage transformations are dependent on the presence of fish (Burkholder and Glasgow, 1997b; Burkholder *et al.*, 2001b). Of particular importance were numerous amoeboid stages, able to transform to produce large numbers of toxic zoospores (TZ) when exposed to fish excreta (Burkholder *et al.*, 1995b). It was reported that in winter ">90% of the *P. piscicida* population existed as large lobose amoebae that exhibited similar behavior as TZs during warmer seasons, with apparent attraction to fish followed by prey mortality" (p. 1056, Burkholder and Glasgow, 1997a). Current evidence, however, indicates that *P. piscicida* exhibits a typical dinoflagellate life cycle and that the amoeboid and chrysophyte-like cyst stages originally reported belong to other benign organisms (Litaker *et al.*, 2002).

A possible reason for the mistaken complexity of the original life cycle was that material was collected directly from aquaria where extremely high densities of *Pfiesteria* were purportedly killing fish by toxin (Burkholder *et al.*, 1992). No single-cell isolates for development of clonal cultures were produced for this study. Molecular fingerprinting subsequently has revealed that aquaria containing *P. piscicida* and fish also contain a diverse microbial community including numerous bacteria, amoebae, algae, and fungi (Drgon *et al.*, 2005). These organisms are introduced from the fish epithelia, the estuarine water, or sediment samples used to initiate the experiment. Many of these introduced species have overlapping morphologies and life histories, making unambiguous identification of life cycle stages of any given organism virtually impossible. Even when single cell isolations (clonal cultures) are used, unambiguous identification of the life cycle stages can be difficult. Attempted isolations of pure *Pfiesteria* cultures, for example, often produced mixed amoebae/*P. piscicida* zoospore isolates (Litaker *et al.*, 2002). This result suggested that this association was an artifact caused by bactiverous amoebae adhering to the surface of the *P. piscicida* during isolation. Subsequent serial single cell isolations always produced cultures that were uniformly either amoebae or *Pfiesteria* zoospores. These isolates never demonstrated any of the reported transformations of zoospores to amoebae or amoebae to zoospores.

Although papers published after Litaker *et al.* (2002) mentioned the existence of amoebae in the *Pfiesteria* life history, they failed to present any data confirming their existence, and instead described typical dinoflagellate life cycles (Parrow *et al.*, 2002; Parrow and Burkholder, 2003). The inability of other researchers to find *Pfiesteria* amoebae was ascribed either to using NON-IND strains that lacked the ability to produce amoebae, or to incorrect culture conditions (Burkholder and Glasgow, 2002). A study by the same research group, however, showed no genetic differences between strains based on the sequences analyzed (Tengs *et al.*, 2003).

Because *Pfiesteria* amoebae were reported as easily obtained from estuarine sediments (Burkholder *et al.*, 2001a, 2001b), investigations were undertaken to establish single-cell (clonal) amoebae isolates from Maryland estuaries where fish kills had been attributed to *Pfiesteria*. These isolates were then analyzed using SEM in combination with ribosomal sequencing (Peglar *et al.*, 2003, 2004). Contrary to the claim that most dinoflagellates have amoeboid life cycle stages (Burkholder *et al.*, 1998, 2001a, 2001b), no dinoflagellate amoebae were obtained in the Peglar *et al.* (2003, 2004) studies. Several of the amoebae observed, however, were morphologically identical to those originally described as being *Pfiesteria* life cycle stages, and the results were instead consistent with previous taxonomic studies indicating that free-living dinoflagellates lack amoeboid life cycle stages. The isolated amoebae exhibited morphological features characteristic of true amoebae (Peglar *et al.*, 2004) that had been previously described for purported *Pfiesteria* amoebae (Burkholder, 2002; Burkholder and Glasgow, 1995, 2001; Glasgow *et al.*, 2001b).

Sequencing the small subunit (SSU) ribosomal RNA genes from the amoebae and zoospore cultures confirmed that all the amoebae were true sarcodinian amoebae, and that the zoospores were uniformly *P. piscicida* (Litaker *et al.*, 2002). The only molecular studies claiming to support the existence of *Pfiesteria* amoebae (Burkholder *et al.*, 2001b; Rublee *et al.*, 1999) used molecular probes that emitted light within the same wavelengths at which preserved amoebae autofluoresce, making it impossible to identify *Pfiesteria*-specific hybridization. Furthermore, this study did not use amoebae-specific probes or PCR assays that could have disproved that the amoebae were part of the *Pfiesteria* life cycle. Hybridization with molecular probes specific to *P. piscicida* and those that targeted conserved amoeba-specific sequences (which fluoresced at different wavelengths) demonstrated that all amoeba were hybridization negative with the *P. piscicida* probe, but hybridized to the amoeba probe (Litaker *et al.*, 2002). Cumulatively these data, plus the unavailability of purported *Pfiesteria* amoebae cultures to the greater research community, lead to the conclusion by many researchers that there are no amoeboid stages in the life cycle of *P. piscicida* (Litaker *et al.*, 2002; Peglar *et al.*, 2003, 2004).

The life cycles of *P. piscicida* and *Pseudopfiesteria shumwayae* are similar to each other and appear to be typical of other dinoflagellates, consisting of swimming flagellated stages (zoospores, gametes, and planozygotes) and mitotic, meiotic, and resting cyst stages (Fig. 16-7). During mitosis, free-swimming zoospores shed their flagella and form cysts, which sink to the bottom and subsequently excyst, releasing rapidly swimming zoospores. The only remaining uncertainty concerning the life cycle of *P. piscicida* involved a few steps during meiosis. There is general agreement that at the start of meiosis, two isogamous, biflagellated gametes fuse to form a free-swimming triflagellated planozygote (Litaker *et al.*, 2002; Parrow *et al.*, 2002). The planozygotes eventually shed their flagella, forming nonmotile hypnozygotes or "reproductive cysts." During this process, the separate nuclei fuse and replicate producing a 4n nucleus. At this point, the data suggest that the hypnozygote can differentiate along two different pathways. In the first, the hypnozygote undergoes meiosis I, forming two secondary cysts within the primary cyst wall (Fig. 16-7). Completion of meiosis II results in two closely opposed cells within each secondary cyst. These secondary cysts then excyst independently, releasing two zoospores each, resulting in the release of a

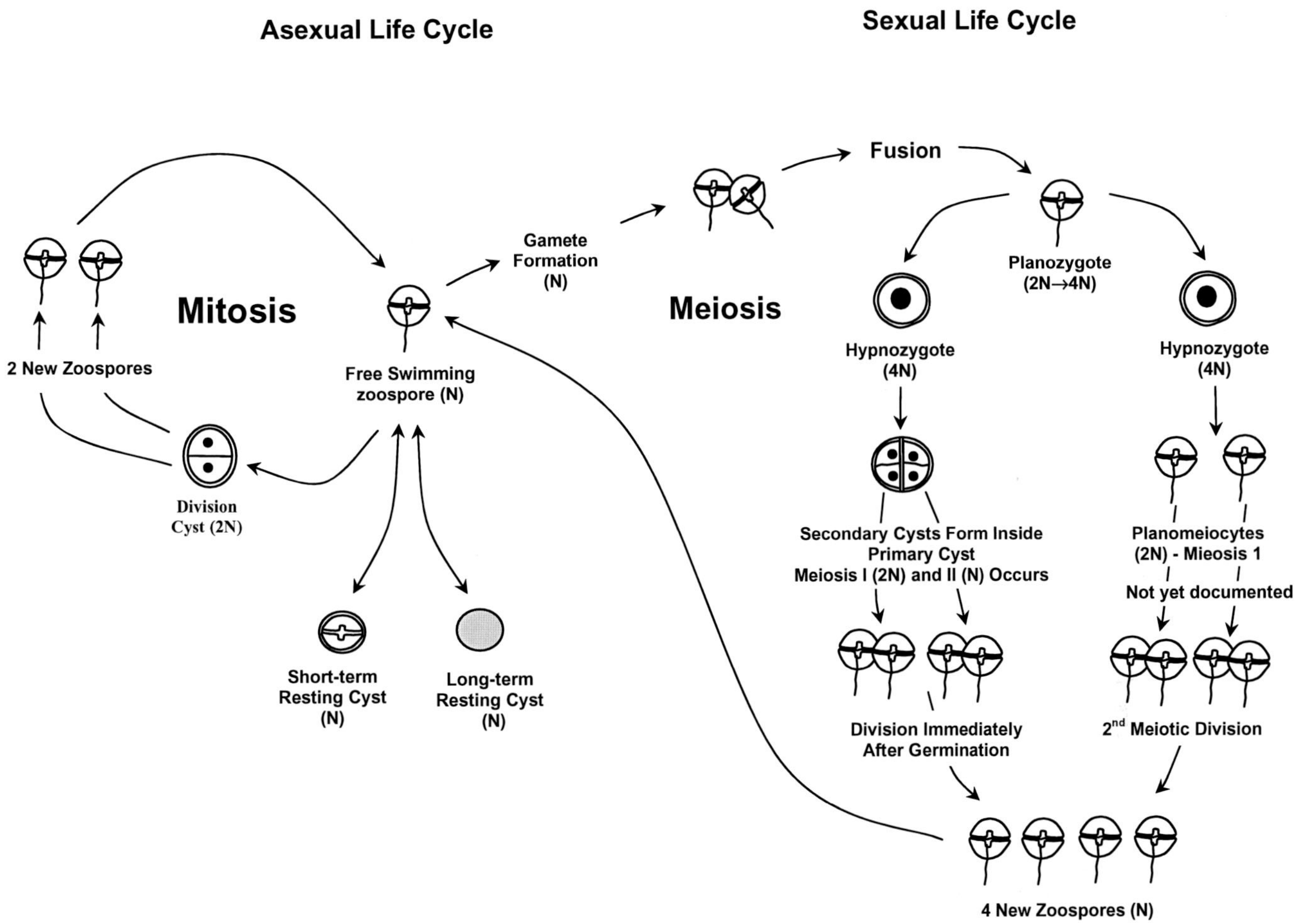

FIGURE 16-7. The simple life cycle of *Pfiesteria piscicida* as proposed in Litaker *et al.* (2002).

total of four haploid zoospores from each hypnozygote. This pathway is observed regardless of whether the cells are fed algae or fish (Litaker *et al.*, 2002), indicating that food source does not affect which life cycle stages are produced, as was hypothesized by Burkholder and Glasgow (1997b).

Parrow *et al.* (2002) used nuclear cyclosis, a rapid swirling motion of the chromosomes in the nucleus, as both an indicator of meiosis and to identify hypnozygotes. They observed that some hypnozygotes developed along the first pathway described earlier, but that others did not form secondary cysts. Instead, they underwent a single division within the primary cyst (hypnozygote), followed by the release of only two free-swimming cells. Though not directly documented, these newly germinated cells were presumed to be diploid planomieocytes, each of which subsequently divided in the water column forming two haploid zoospores. In contrast, the hypnozygotes of *P. shumwayae* can divide mitotically multiple times producing two, four, or more secondary cysts within the thin outer membrane of the primary cyst depending on the amount of food consumed (Parrow and Burkholder, 2003). In this species, four or more free-swimming cells are ultimately produced from a single hypnozygote.

Another distinct cyst type has been observed in the *P. piscicida* life cycle. Haploid resting cysts can be induced in response to environmental stressors. Cells disturbed by salinity changes, turbulence, and so on may respond within seconds by shedding their flagella and forming thin walled temporary cysts that may persist for minutes to several days. Alternatively, if the cells are starved or severely stressed, they can form thicker walled resting cysts that can remain dormant for months. This provides a mechanism whereby *P. piscicida* can survive adverse conditions.

In summary, *P. pisicicda* and *P. shumwayae* do not exhibit the complex, multiphasic life cycle that had been previously proposed. Rather, they have a typical dinoflagellate life cycle, well adapted to life in estuaries where food abundances and environmental conditions are highly variable. The life cycle is composed of free-swimming flagellated stages, reproductive cysts, and resting cysts that allow for dormancy periods from minutes to years, sufficient to cope with most environmental fluctuations (Litaker *et al.*, 2002;

Parrow and Burkholder, 2002). The chrysophyte and sarcodinian amoebae species initially ascribed to the *Pfiesteria* life cycle are, in fact, common benign fish tank contaminants. The lack of these amoeboid or chrysophyte-like stages brings into question the ambush predator hypothesis put forward to account for how *Pfiesteria* kills fish.

Taxonomy and Phylogeny

Dinoflagellates fall into a larger group of organisms referred to as the Alveolates (Cavalier-Smith, 1993), which includes ciliates, apicomplexans, and the *Perkinsus* species, a genus of parasites infecting marine and estuarine molluscan hosts around the world (Villalba *et al.*, 2004). Over the years, there has been considerable confusion over the taxonomic affinities of the family Pfiesteriaceae. Based on the reported multiphasic life cycle with flagellated, amoeboid, and cyst stages, the family was initially assigned to the order Dinamoebales. As discussed previously, however, subsequent life cycle studies have failed to consistently identify an amoeboid stage (Litaker *et al.*, 2002; Parrow *et al.*, 2002; Parrow and Burkholder, 2003), and therefore, this assignment is now considered invalid. The family was, therefore, placed into the order Peridiniales (Litaker *et al.*, 1999, 2002; Parrow *et al.*, 2002), and this assignment has been supported by subsequent studies (Litaker *et al.*, 2005; Marshall *et al.*, 2006).

Currently, there are four or five described genera in the family Pfiesteriaceae, depending on whether *Pfiesteria piscicida* and *Pseudopfiesteria shumwayae* are considered to be in the same genus. Three genera constitute the PLDs: *Stoeckeria*, *Cryptoperidiniopsis*, and *Luciella* (Jeong *et al.*, 2005; Mason *et al.*, 2007; Steidinger *et al.*, 2006). *Pseudopfiesteria shumwayae* was initially described as *Pfiesteria shumwayae* (Glasgow *et al.*, 2001b). It was later argued, however, that there are important differences in Kofoidian plate tabulations between the two species. In particular, *P. shumwayae* has six versus the five precingular plates in *P. piscicida*, and a four-sided, rather than three-sided, apical intercalary plate. Following historical convention, these differences indicated that *P. shumwayae* should be in a different genus from *P. piscicida* (Litaker *et al.*, 2005). In addition, molecular phylogenetic analyses that included only one species from each clade supported *P. piscicida* and *P. shumwayae* to be as genetically distinct from each other as from species in the genera *Luciella* and *Cryptoperidiniopsis* (Litaker *et al.*, 2005), supporting the placement of *P. shumwayae* into a different genus.

The appropriate designation for *P. shumwayae*, however, is still debated, and not all investigators agree on its taxonomic position. Marshall *et al.* (2006) contended that *P. shumwayae* should remain in the genus *Pfiesteria* because the plate differences were not great enough to support the reassignment, and the molecular phylogenies tended to group the two species, albeit generally with low statistical support values. The morphological criteria used to separate *Pfiesteria* and *Pseudopfiesteria*, however, are no different than those used to separate *Luciella, Stoeckeria*, and *Cryptoperidiniopsis* (Mason *et al.*, 2007; Steidinger *et al.*, 2006), which are all accepted as valid genera in the same family. The phylogenetic association cited in Marshall *et al.* (2006) between *P. piscicida* and *P. shumwayae* is possibly an artifact that is inherent to phylogenetic programs designed to place species in a bifurcating tree. Additional molecular data from the species in this group, along with the addition of sequences from new morphologically defined *Pfiesteria* and *Pseudopfiesteria* species, will minimize any analysis artifacts and help resolve the question of whether *P. piscicida* and *P. shumwayae* should be in the same or different genera. All the molecular phylogenies, however, agree that the *Pfiesteria* species and PLDs form a monophyletic group with high statistical support (Jakobsen *et al.*, 2002; Litaker *et al.*, 1999, 2005; Marshall *et al.*, 2006; Oldach *et al.*, 2000; Rublee *et al.*, 2005; Zhang and Lin, 2005).

Detection Techniques and Environmental Monitoring

Because of uncertainties in presumptive cell counts commonly used to associate *Pfiesteria* to fish lesion events, development of more specific and rapid molecular methods became necessary. Definitive detection and identification of *Pfiesteria* and PLD species in environmental samples initially depended on slow and labor-intensive procedures, involving identification of morphological features using SEM. This approach also required cumbersome clonal isolation and culture of the *Pfiesteria* cells. It was impossible to obtain rapid, accurate, and objective assessments of the prevalence, distribution, and concentration of these organisms in the environment, especially given that several of these closely related species may be present at varying concentrations in the same environment (Reece *et al.*, 2003, 2005; Rublee *et al.*, 2005). Therefore, it is possible that many of the early studies using presumptive cell counts may have significantly overestimated the number of *Pfiesteria* cells present in some environmental samples.

To respond to these difficulties, rapid molecular detection techniques have been developed for *Pfiesteria* spp. and the related PLDs (for review see Rublee *et al.*, 2005). Most assays are based on the polymerase chain reaction (PCR), which targets species-specific gene sequences. Numerous so called standard PCR assays have been developed (Litaker *et al.*, 2003; Oldach *et al.*, 2000; Rublee *et al.*, 1999; Saito *et al.*, 2002; Vogelbein *et al.*, 2001, 2002; Zhang and Lin, 2002), as well as some real-time PCR assays that allow quantification of particular species (Bowers *et al.*, 2000; Lin *et al.*, 2006; Reece *et al.*, 2002, 2003, 2005). There are also nucleic acid, protein nucleic acid (PNA), and immunofluorescent molecular probes available for *in situ* hybridization studies that allow for the visualization and identification of whole cells using light microscopy (Glasgow *et al.*, 2001a; Lin and Zang, 2003; Litaker *et al.*, 2002; Reece *et al.*, 2002, 2003, 2005; Rublee *et al.*, 1999). Hybridization protocols are not as rapid as PCR, so they cannot readily be applied to large-scale screening of environmental samples. They do provide, however, the ability to confirm the presence and identification of live cells in cultures and environmental samples, as well as demonstrate the association of these cells with fish and shellfish tissues in histopathological sections (Litaker *et al.*, 2002; Reece *et al.*, 2002, 2003, 2005).

Additional molecular techniques include the heteroduplex mobility assay (HMA) (Oldach *et al.*, 2000) and denaturing gradient gel electrophoresis (DGGE) (Coyne *et al.*, 2001). These combine PCR with gel electrophoresis of DNA fragments to separate these fragments based on structural or sequence differences. The HMA assay, however, is only appropriate for cultures and not environmental samples. With DGGE analyses, it can be difficult to determine whether a sample contains multiple variant strains of a species or if the observed variations occur in a single strain, as these molecular assays typically target multicopy regions of the genome where there may be differences among sequence copies in a single genome.

Development of these accurate and rapid methods for detecting *Pfiesteria* spp. and PLD species markedly augmented the capacity of ongoing field monitoring programs conducted over various time spans between 1998 and 2006 in the states of Delaware, New Jersey, Maryland, Virginia, North Carolina, South Carolina, and Florida (Lin *et al.*, 2006; Magnien, 2001; Reece *et al.*, 2003, 2005). In addition, molecular detection assays have facilitated studies examining the effects of various environmental parameters on growth and behavior, as well as determining geographic distributions (Rublee *et al.*, 2005). Both *Pfiesteria* spp. and PLDs have been detected along the east and Gulf coasts of the United States from Maine to Texas (Rublee *et al.*, 1999; Bowers *et al.*, 2000; Lin *et al.*, 2006), and either one or both species have been reported from waters of several other regions including northern Europe (Jakobsen *et al.*, 2002; Rublee *et al.*, 2005), New Zealand (Rhodes *et al.*, 2002, 2006), and South America (Lin *et al.*, 2006). These studies universally indicate very low abundance of these organisms (Lin *et al.*, 2006; Rublee *et al.*, 2005). Lin *et al.* (2006), for example, used real-time PCR to quantify *P. piscicida* abundance in areas along the East and Pacific coasts of the United States. *Pfiesteria piscicida* was detected in all locations along the U.S. Atlantic coast and in Chile at concentrations ranging from <1 to 1.5 cell ml^{-1} (Lin *et al.*, 2006). Interestingly, no strong correlation between abundance and particular environmental factors was found in the Lin *et al.* (2006) study.

A likely explanation for low *Pfiesteria* abundances comes from grazing and prey preference studies. Numerous protists including ciliates (*Strombidium* sp., *Mesodinium pulex*) (Stoecker *et al.*, 2000) and rotifers (*Brachionus plicatilis*) (Stoecker and Gustafson, 2002), as well as calanoid copepods (*Acartia tonsa*) (Mallin *et al.*, 1995), readily prey on *Pfiesteria* spp. Prey preference studies have also shown that *Pfiesteria* prefers cryptophytes (such as *Rhodomonas*) and can consume on average 10 or more cells per day. This rate represents an extremely high food requirement and suggests that the abundance of appropriate prey may be an important factor regulating *P. piscicida* abundance in nature (Lin and Zhang, 2003; Lin *et al.*, 2004). Because populations of *Pfiesteria* are likely to be limited by both grazing pressure and food supply, extensive *Pfiesteria* blooms of are unlikely events (Coyne *et al.*, 2001; Lin *et al.*, 2006; Litaker *et al.*, 2003; Reece *et al.*, 2002, 2005).

Fish Kills and Fish Lesions

As detailed in the preceding section, massive menhaden kills and associated ulcerative lesions in dead and dying fish were attributed during the 1990s to the activity of a *Pfiesteria* toxin (Burkholder, 1999; Burkholder and Glasgow, 1997a, 1997b; Burkholder *et al.*, 2001a; Magnien, 2001). This presumed cause and effect relationship became the basis for using fish mortality, fish lesions, presumptive counts of *Pfiesteria*-like cells in water samples from a fish kill or lesion event, and positive fish bioassays to implicate toxic *Pfiesteria* as the causative agent (Burkholder *et al.*, 1995a, 2001a; Glasgow *et al.*, 2001a). However, more recent studies have disputed the assertion that massive fish kills of Atlantic menhaden, or deep, penetrating cutaneous ulcers frequently found on menhaden, are indicative of the presence of *Pfiesteria* or the action of a *Pfiesteria* toxin. This section offers a detailed examination of this body of research and considers a variety of alternative causes for fish mortality and disease events.

Fish Kills

Approximately 80% to 85% of the fish kills in the Pamlico Sound were attributed to oxygen depletion (hypoxia/anoxia) of bottom waters caused by eutrophication, nutrient-stimulated phytoplankton blooms and resultant high organic loading (Paerl *et al.*, 1998). Although this work was formally challenged (Burkholder *et al.*, 1999), Paerl *et al.* (1999) provided a plausible mechanism whereby surface-dwelling menhaden might succumb to O_2-depleted bottom waters by a physical process called "seiching." In this scenario, winds blowing across a river push surface water toward the downwind shore, where it then downwells. A compensating flow of bottom water in the opposite direction causes rapid upwelling of oxygen-depleted water along the upwind shore

FIGURE 16-8. Large fish kill event in Naragansett Bay (RI) attributed to an anoxia/upwelling event. Photo from Thomas Ardito, Narragansett Bay Estuary Program, www.nbep.org.

(Luettich *et al.*, 2000). Although fish may be able to swim away from weak upwelling events, a large pocket of O_2-depleted water resulting from a strong near-shore upwelling event may rapidly engulf fish before they can escape, resulting in massive fish kills (Fig. 16-8). By the time a fish kill is reported and investigators are on scene, hours to days may have passed. Local conditions may no longer reflect conditions that preceded the fish kill, or fish may have been transported long distances away from the initial location of the event. Thus, dissolved oxygen levels measured in the presence of floating dead fish may no longer be relevant to the conditions that were present when the kill occurred.

Other researchers have demonstrated that the use of observational data linking the presence and activity of *Pfiesteria* with estuarine fish kills is not adequate for statistical determination of a cause and effect relationship (Stow, 1999; Stow and Borsuk, 2003). Applying probability models to published data on *Pfiesteria* and fish kills, Stow and Borsuk (2003) determined the assertion that *Pfiesteria* causes fish kills to be a statistical improbability but that the reverse situation (that fish kills cause the proliferation of *Pfiesteria*) is statistically more plausible. However, a scenario involving some other initiating event that causes both the fish kill and the presence of *Pfiesteria*, was considered equally probable.

Thus, it has become clear that this association, if it exists, is not one of demonstrated cause and effect.

Fish Lesions

One of the hallmarks of *Pfiesteria*-induced fish kills was said to be the presence of deeply penetrating skin lesions (Burkholder and Glasgow, 1997a; Burkholder *et al.*, 2001a, 2001c; Glasgow *et al.*, 2001a). Skin ulceration in fishes, however, is a well-recognized response to polluted or otherwise stressed aquatic environments (Noga, 2000; Sinderman, 1990). It has many different etiologies including infectious agents, toxins, physical causes, immunological causes, and nutritional and metabolic perturbations (Kane *et al.*, 1998; Law, 2001; Noga, 2000; Sindermann, 1998). Additionally, environmental stressors can directly or indirectly modulate expression of disease.

Since the early 1980s, estuarine fishes, primarily Atlantic menhaden, have experienced seasonal outbreaks of cutaneous ulcer disease in eastern U.S. estuaries (Dykstra *et al.*, 1986; Noga, 1993). This epizootic disease is characterized by the presence of deeply penetrating ulcers exhibiting severe dermatitis, underlying granulomatous myositis and tissue destruction associated with fungus-like hyphae (Dykstra *et al.*, 1986; Noga and Dykstra, 1986). Additionally, lesions typically exhibit a diverse mixture of microbial contaminants colonizing the outermost necrotic layer (Blazer *et al.*, 1999; Dykstra *et al.*, 1986, 1989; Levine *et al.*, 1990a; Noga *et al.*, 1988, 1993). First recognized in North Carolina estuaries in 1984, this disease has since been documented in menhaden and other fish species from New York to Florida (Noga, 1993). The disease etiology was actively investigated during the late 1980s. Based on pathology, microbial characterization and laboratory challenges, the condition was ascribed to infection by oomycetes (water molds) and called ulcerative mycosis (UM) (Dykstra *et al.*, 1986, 1989; Levine *et al.*, 1990a, 1990b; Noga *et al.*, 1988, 1993; Noga and Dykstra, 1986). Oomycetes (including *Aphanomyces* sp. and *Saprolegnia* sp.) were associated with >90% of the lesions (Noga and Dykstra, 1986). However, laboratory challenges of menhaden with a strain of *Aphanomyces* sp. (ATCC 62427) isolated from a diseased menhaden were unable to reproduce the typical lesions seen in wild fish (Noga, 1993). This finding suggested that the oomycetes were opportunistic secondary invaders, and that other, as yet unknown, biological or environmental factors were playing a prominent role in disease expression (Dykstra *et al.*, 1989).

Thus, these early studies failed to resolve whether the mixture of microbial agents, including the deeply invasive oomycete found in the ulcers, were opportunists colonizing a superficial lesion caused by unknown environmental stressors or biological insults (such as a purported toxin produced by *Pfiesteria*), or whether one of the microbes represented a primary pathogen capable of causing UM directly without other stressors playing a role (Kane *et al.*, 1998; Noga, 2000; Noga *et al.*, 1996). Noga *et al.* (1996) and Sindermann (1998) proposed that ulcerative lesions in fishes are multifactorial in their etiology and reasonably hypothesized a primary role for toxic *Pfiesteria* in initiating early epidermal damage by causing superficial skin erosion and mild ulceration, thereby providing a portal of entry for secondary microbial invaders including bacteria and water mold fungi such as *Aphanomyces* spp.

The attribution of menhaden UM to *Pfiesteria* in the 1990s became controversial, however, because a similar disease also characterized by deeply penetrating skin ulcers had been reported in more than 100 species of wild and cultured freshwater and estuarine fishes from the Indo-Pacific region (Callinan, 1994). This disease was attributed to a highly pathogenic oomycete called *Aphanomyces invadans* and was referred to respectively as "epizootic ulcerative syndrome" (EUS) in Asia, "red spot disease" (RSD) in Australia, and "mycotic granulomatosis" (MG) in Japan (Callinan *et al.*, 1995; Frazer *et al.*, 1992; Hatai, 1980; Hatai *et al.*, 1977, 1980; Lilley *et al.*, 1998; Willoughby *et al.*, 1995). In contrast to the unsuccessful early laboratory exposures of menhaden to strain ATCC-62427, typical ulcerative lesions in Indo-Pacific fishes were experimentally induced with *A. invadans* through multiple exposure routes. These included intramuscular injection of fish with hyphae or secondary zoospores (Catap and Munday, 1998; Hatai *et al.*, 1977; Kahn *et al.*, 1998; Lilly and Roberts, 1997; Roberts *et al.*, 1993; Wada *et al.*, 1996), cohabitation of healthy with infected fish (Cruz-Lacierda and Shariff, 1995), and aqueous exposures of healthy fish to secondary zoospores, the infective stage of this oomycete (Callinan *et al.*, 1996). This disease has devastated major commercial fisheries, both wild and farmed, and is now collectively referred to as "epizootic ulcerative syndrome" (EUS) (Lilley and Roberts, 1997).

Blazer *et al.* (1999) investigated menhaden mortality and lesion events attributed to *Pfiesteria* in Maryland and Virginia portions of Chesapeake Bay. They found that the lesions in these fish were identical to those reported as menhaden UM from eastern U.S. estuaries during the 1980s (e.g., Noga and Dykstra, 1986). Additionally, based on growth characteristics, molecular assays (including immunoassays, polymerase chain reaction [PCR] and fluorescent *in situ* hybridization [FISH]), the oomycete isolates obtained from Chesapeake Bay menhaden were definitively identified as *A. invadans* (Blazer *et al.*, 2002). Molecular analyses also identified *A. invadans* in ulcerative lesions of menhaden from North Carolina estuaries (Vandersea *et al.*, 2006). In addition, the strain ATCC-62427 described earlier (which did not cause typical ulcers in laboratory challenged menhaden) was shown to be a different nonpathogenic oomycete species (Blazer *et al.*, 2002; Lilly *et al.*, 2003).

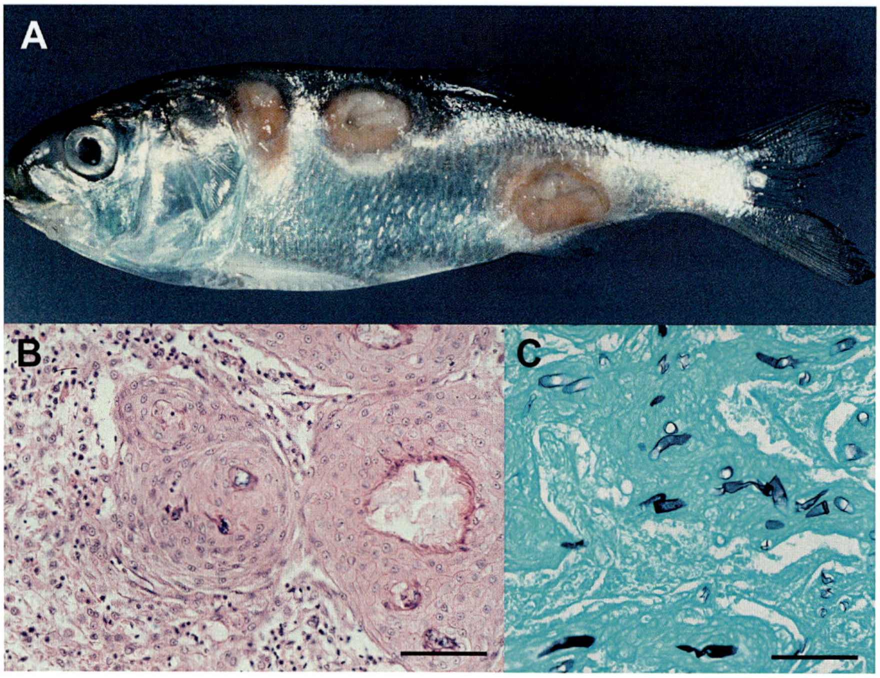

FIGURE 16-9. Gross and histological features of ulcerative mycosis experimentally induced in Atlantic menhaden, *Brevoortia tyrannus*. (A) Gross appearance of a fish 18 days after experimental aqueous exposure to secondary zoospores of *Aphanomyces invadans*. (B) Histopathology showing chronic granulomatous inflammation. G, granulomas consisting of host defensive cells called macrophages encapsulating fungal hyphae. (C) Grocott's silver stain for fungi showing black staining fungal hyphae within the affected tissues. Scale bars (for figures B and C) are 45 μm. Panel (A) is reproduced by permission of the Journal Diseases of Aquatic Organisms, from Kiryu *et al.* (2003). Panels (B) and (C) are reproduced from Vogelbein *et al.* (2001).

Laboratory studies also support *A. invadans* as the primary pathogen in menhaden. Kiryu *et al.* (2002, 2003) challenged menhaden with a Chesapeake Bay isolate called WIC (Blazer *et al.*, 2002), an Indo-Pacific isolate (PA7), and strain ATCC-62427. Fish were challenged by subcutaneous injection or aqueous exposure, with or without trauma-inducing treatments. Isolates WIC and PA7 were highly infectious and pathogenic by both exposure routes, with or without mild skin trauma. Grossly and histologically characteristic ulcerative lesions developed in exposed menhaden (Fig. 16-9), and mortality rates ranged from 24% to 100%, with highest rates in groups that had their skin mildly abraded. In some cases, a single zoospore/fish could induce the ulcerous lesions, and ultimately cause death.

However, strain ATCC-62427 (not *A. invadans*) did not cause lesions or mortality. Although there have been other species designations for this pathogen (e.g., *Aphanomyces piscicida*; Hatai, 1980), it is now recognized as a single species and listed in the Index of Fungi as *Aphanomyces invadans* (IMI, 1997). Most researchers currently believe that *A. invadans* is the intrinsic causative agent of EUS/UM and that it occurs in all outbreaks. In some events, the presence of *A. invadans* appears to be the only biological factor required for disease to occur (Lilley *et al.*, 1998).

Other observations increasingly led to questions regarding the purported role of *Pfiesteria* in fish kill and lesion events (Dykstra and Kane, 2000; Vogelbein *et al.*, 2001). *Pfiesteria* lesions occurring in laboratory fish bioassays and attributed to toxin production were described as forming rapidly in as little as several hours (Noga *et al.*, 1996). In contrast, based on the granulomatous inflammatory response elicited by *A. invadans* (Fig. 16-9B), ulcers in wild menhaden are chronic lesions (days to weeks old) and therefore could not have formed rapidly in response to acute toxin exposure. In addition, menhaden are mobile, thus the presence of a chronic lesion days to weeks old may have no bearing on current local conditions where fish are collected. Finally, presumptive cell counts of *Pfiesteria*-like dinoflagellates are not accurate in identifying *Pfiesteria*, and it was impossible to determine the exact *Pfiesteria* cell counts in the water when fish were collected by the methods that were being used at that time.

Vogelbein *et al.* (2001) distinguished menhaden UM from laboratory-induced "pfiesteriosis" in tilapia (the early model fish species used in bioassays) based on histopathology. Tilapia (*Oreochromis* sp.) experimentally exposed to *P. shumwayae* using the standard fish bioassay approach available at the time (Burkholder et al, 1995a; Burkholder *et al.*, 2001a, 2001b) exhibited widespread loss of the epidermis, mucus, and scales (Vogelbein *et al.*, 2001). This diffuse superficial pathology, affecting large portions of the body, was clearly distinct from the deeply penetrating, single chronic ulcers in wild-caught menhaden. The diffuse and superficial epidermal damage reported in laboratory-exposed fish has never been observed in wild menhaden from fish kill or lesion events (Law, 2001) and was essentially identical to that previously reported by Noga *et al.* (1996) as an effect of *Pfiesteria piscicida* toxin.

Therefore, the preponderance of laboratory and field evidence indicates that *A. invadans* is a primary pathogen rather than a secondary invader and is the principal causative agent of menhaden UM. There is no direct evidence that *Pfiesteria* spp. play any role whatsoever in menhaden UM, although some investigators continue to claim a marginally statistically significant association between the presence of *Pfiesteria* and menhaden lesions (Tango *et al.*, 2006). *Aphanomyces invadans* must clearly be distinguished from saprophytic oomycetes, including other *Aphanomyces* spp. that secondarily colonize the ulcers, but are not themselves pathogenic (e.g., ATCC-62427). Although superficial skin damage may augment infectivity of *A. invadans* (Kiryu *et al.*, 2002), it is not a necessary precursor to infection. Thus, UM in menhaden and in other North American fish species is essentially identical to EUS in the Indo-Pacific region. This conclusion is consistent with the fact that the same disease is seen in fish worldwide, including in many environments where *Pfiesteria* has not been observed.

Fish Bioassays and Micropredation

As previously mentioned, much of the reported biology of *Pfiesteria* is based on observations from the "standard" fish bioassay developed specifically to test *Pfiesteria* toxicity (Burkholder, 2002; Burkholder *et al.*, 1995a, 2001a, 2001c; Glasgow *et al.*, 2001a). In this system, juvenile tilapia are exposed in aquaria (9 to 38 liters) to *Pfiesteria* from sediment samples, water samples, or laboratory cultures. Aquaria are maintained for weeks to months until fish mortalities occur. As fish die, they are removed and replaced with live fish without changing the water. During the incubation, *Pfiesteria* and bacterial populations increase to densities that greatly exceed those observed in the field. Because of the complex community of microorganisms and degradation of water quality that develop over time, concern has been raised regarding the validity of these assays.

Furthermore, the "standard" fish bioassay developed specifically to test *Pfiesteria* toxicity suffers from poor reproducibility and difficulties with assigning causality to fish mortality (Drgon *et al.*, 2005; Lovko *et al.*, 2003; Vogelbein *et al.*, 2002). Microbial contaminants in these systems include amoebae, bacteria, chrysophytes, ciliates, diatoms, rotifers, and other protists (Burkholder *et al.*, 1995a, 2001a, 2001c; Burkholder and Glasgow, 1997a, 1997b; Drgon *et al.*, 2005; Peglar *et al.*, 2004; Vogelbein *et al.*, 2001). Replicating the standard aquarium-format bioassay Drgon *et al.* (2005) found bacteria (including pathogenic *Vibrio* sp., and *Aeromonas* sp.) among the vast assemblage of microorganisms that developed. The ability of bacteria commonly associated with harmful algal species to produce toxins, or contribute to their production, has been documented (Doucette, 1995). This finding raises the possibility that toxins attributed to *Pfiesteria* may have originated from tank contaminants, some known to be toxin-producing pathogens.

Alternative, small-scale bioassays were developed to address many of the problems associated with the standard bioassays (Lovko *et al.*, 2003; Quesenberry *et al.*, 2002). These studies exposed larval or juvenile fishes to *Pfiesteria* in tissue culture plates or small culture flasks. The smaller format system and shorter assay duration eliminated many of the difficulties associated with the larger tank-format assays. Additionally, with the ability to view the flasks under the microscope, these assays allowed better elucidation of the mechanisms and biological factors playing a role in fish killing by *Pfiesteria*.

Studies mentioned earlier in this chapter demonstrating epidermal erosions in *P. shumwayae*-exposed tilapia, also reported a direct physical association between *Pfiesteria* zoospores and the eroded surface epithelia of exposed fish (Vogelbein *et al.*, 2001). Histologically, zoospores were attached to damaged skin, oral mucosa, gills, and deep within the lateral line canal and olfactory organs (Fig. 16-10). An alternative mechanism of fish pathogenicity was hypothesized whereby the zoospores attach to and feed on the fish epidermis (Vogelbein *et al.*, 2001). In contrast, earlier studies held that *Pfiesteria* toxins were responsible for adverse effects in fish and invertebrates, rather than any direct physical association with, or ingestion of, *Pfiesteria* (Burkholder *et al.*, 1998, 1999, 2001a, 2001b; Kane *et al.*, 1998; Noga 2000; Noga *et al.*, 1996; Samet *et al.*, 2001; Silbergeld *et al.*, 2000). In fact, the original paper describing the effects *P. piscicida* specifically stated, "the alga has not been observed to attack fish directly" (Burkholder *et al.*, 1992). However, intense chemoattraction to fish and heterotrophic feeding on single celled algae, protozoans, and isolated vertebrate cells (such as human red blood cells) was described previously (Burkholder *et al.*, 1995a; Burholder and Glasgow, 1997a).

To help resolve this issue, experiments were performed to assess the relative importance of toxin production versus direct feeding by *Pfiesteria* in fish mortality. Using the 96-hour larval fish bioassay of Lovko *et al.* (2003) (Fig. 16-11), Vogelbein *et al.* (2002) demonstrated that physical attachment and myzocytotic feeding on fish skin by *P. shumwayae* resulted in epidermal damage and fish mortality identical to what had previously been reported as a toxin effect (Burkholder *et al.*, 2001a; Noga *et al.*, 1996). These studies used

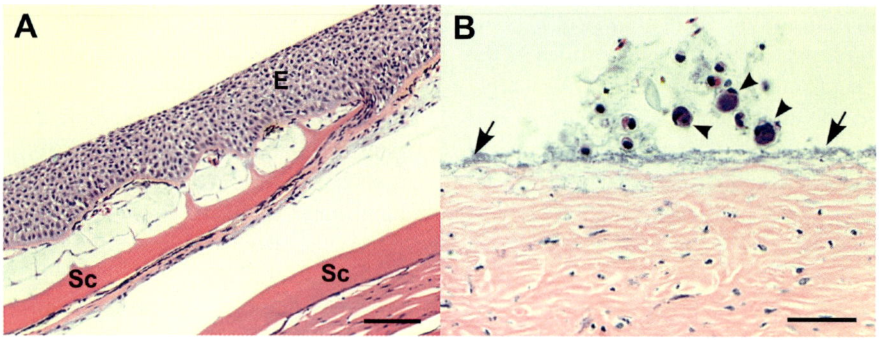

FIGURE 16-10. Histological sections of the skin of tilapia (*Oreochromis* sp.) before and after exposure to *Pseudopfiesteria shumwayae* in the standard fish bioassay format. (A) Unexposed tilapia (Control) showing intact healthy skin. Note the intact multicellular layer of epidermal cells (E), scales (Sc). (B) Exposed tilapia showing complete erosion of the epidermis of the skin and bacterial colonization of the exposed dermis (arrows). *P. shumwayae* zoospores are seen to be attached to the eroded skin surface (arrowheads). Scale bars: A, 100 µm; B, 50 µm. Panel (A) reproduced from Vogelbein *et al.* (2001).

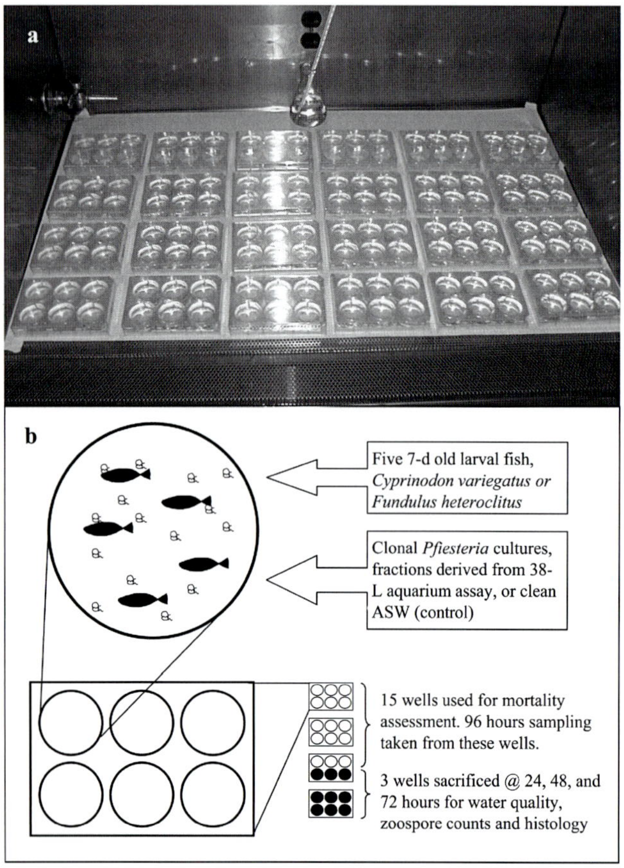

FIGURE 16-11. Laboratory setup and general experimental design for conducting the 96-hour larval fish bioassay. Figure reproduced by permission of the Phycological Society of America, from Lovko *et al.* (2003).

6-well tissue culture plates fitted with semipermeable membrane inserts that physically separated the dinoflagellates from the fish but permitted bacteria and soluble substances to cross (Fig. 16-12). Fish mortality (100%) occurred only in treatments where fish and *P. shumwayae* zoospores were in physical contact. No mortalities occurred in treatments

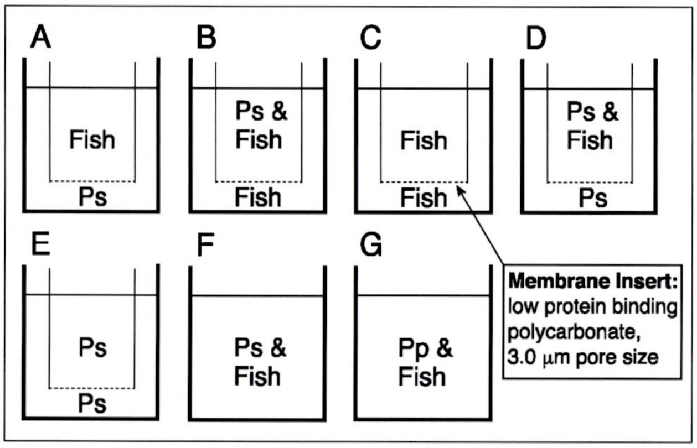

FIGURE 16-12. Experimental design for laboratory investigations of the mechanisms of fish killing by *Pseudopfiesteria shumwayae* (Ps) sp. piscicida and *Pfiesteria piscicida* (Pp). Modification of the larval fish bioassay in which membrane inserts are used to prevent physical contact between larval fish and *Pfiesteria* spp., one of several approaches used to determine if *Pfiesteria* spp. secrete potent exotoxins or kill fish by alternate mechanisms. In this experimental design, fish and dinoflagellates were variously combined to determine the mechanisms of killing of *P. piscicida*. Figure reprinted from: Vogelbein *et al.* (2002).

where the membrane prevented physical contact (Fig. 16-13). Differential centrifugation and filtration of water from a fish-killing aquarium was used to separate soluble, bacterial, and dinoflagellate fractions so that they could be tested independently. Mortalities (60% to 100%) occurred only in fractions that contained live *Pfiesteria* zoospores, whereas no mortalities occurred in cell-free (soluble) or bacteria-enriched fractions or controls. Videomicrography and SEM confirmed that zoospores were highly chemotactic toward fish larvae, attached to and actively fed on the skin, rapidly denuding fish of their epidermis (Fig. 16-14). Two videos illustrating chemoattraction, attachment, and myzocytotic feeding can be found at the Nature Web site (www.nature.com/nature/journal/v418/n6901/suppinfo/nature01008.html).

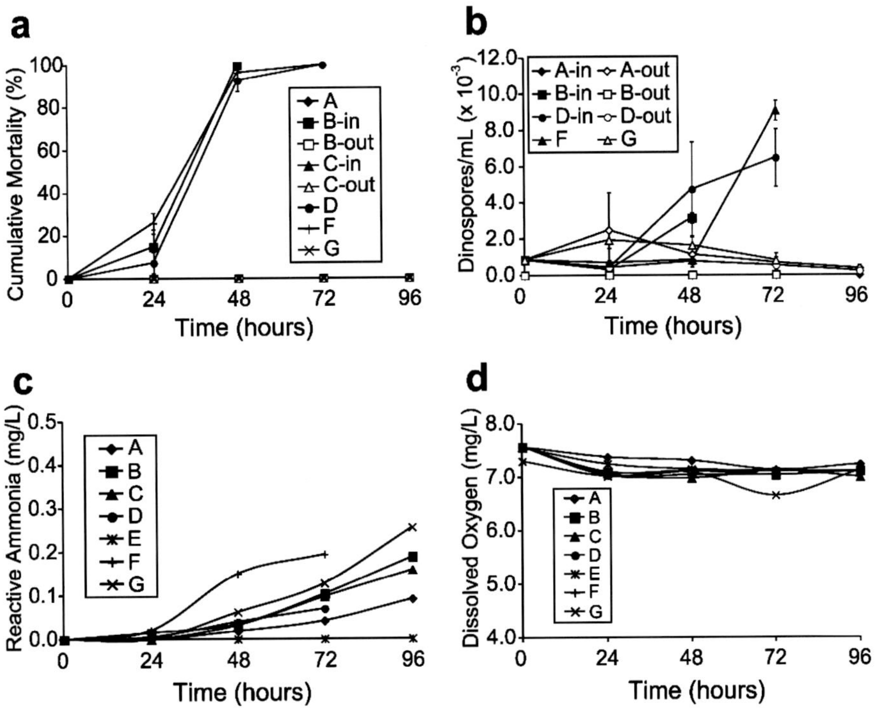

FIGURE 16-13. Results of the membrane insert studies. A. Cumulative fish mortality in the various treatments. Note that only treatments affording direct physical contact (treatments B, D, F: refer to Figure 16-12) killed fish. In treatments where physical contact was prevented by the membrane insert, no mortalities were observed. Pore size of the polycarbonate membrane used in these studies was 3 μm. Any diffusible toxin being produced by dinoflagellates should have diffused across the semipermeable membrane and killed the fish. Reprinted from Vogelbein *et al.* (2002).

These findings were further supported by exposure studies using cell-free supernatants of fish killing cultures, lysates of whole *Pfiesteria* cells and organic extracts of lyophilized *Pfiesteria* cells, all of which failed to kill larval fish (Berry *et al.*, 2002). These findings helped confirm that mortality was due primarily to direct epidermal damage associated with a feeding process described as "micropredation" (Vogelbein *et al.*, 2001, 2002).

Micropredation by *Pfiesteria* spp. was rapidly confirmed in studies with larval and juvenile fish (*Cyprinodon variegatus, Oreochromis* sp.) (Burkholder *et al.* 2001a, 2001b, 2001c; Gordon *et al.*, 2002), bay scallops (*Argopecten irradians*), and eastern oysters (*Crassostrea virginica*) (Springer *et al.*, 2002). However, the concept of micropredation as the primary cause of fish mortality in these assays was largely discounted, and the adverse effects on finfish and shellfish continued to be attributed to *Pfiesteria* toxin. It is important to note that in the application of the standard fish bioassay, fish death was consistently used to identify a culture as "toxin producing," whereas strains failing to cause fish distress, disease, or death were consistently referred to as non-toxic, or NON-IND (Burkholder *et al.*, 2001a, 2001b; Glasgow *et al.*, 2001a, 2001b). Variations in mortality rate were viewed mainly as a function of differences in toxin production among strains, whereas the critically important role of *Pfiesteria* cell density and its effect on mortality was not considered.

This interpretation is problematic because cell numbers in these systems can change drastically over time. With an adequate food source (such as fish), dinoflagellate populations can dramatically increase over a relatively short time period (24 to 48 hours) (Vogelbein *et al.*, 2002; Lovko *et al.*, 2003). Higher cell densities in these assays resulted in increased feeding, greater tissue damage and shorter times to fish death. Lovko *et al.* (2003) demonstrated a clear dose response in larval fish exposed to a range of *P. shumwayae* zoospore concentrations (Fig. 16-15).

The relationship between fish mortality and cell density was subsequently confirmed for *P. piscicida* using a fish tank bioassay system (Drgon *et al.*, 2005). Ichthyocidal (fish-killing) activity of *P. piscicida* was due mostly to micropredation, rather than to a soluble toxin. Cages with varying mesh sizes were used to demonstrate that fish mortality was directly proportional to how well the mesh allowed the passage of intact *Pfiesteria* cells into the cage. Only 3% to 10% total mortality was observed in tanks with mesh sizes that excluded *Pfiesteria* cells, suggesting the presence of a soluble factor making a minor contribution to overall fish

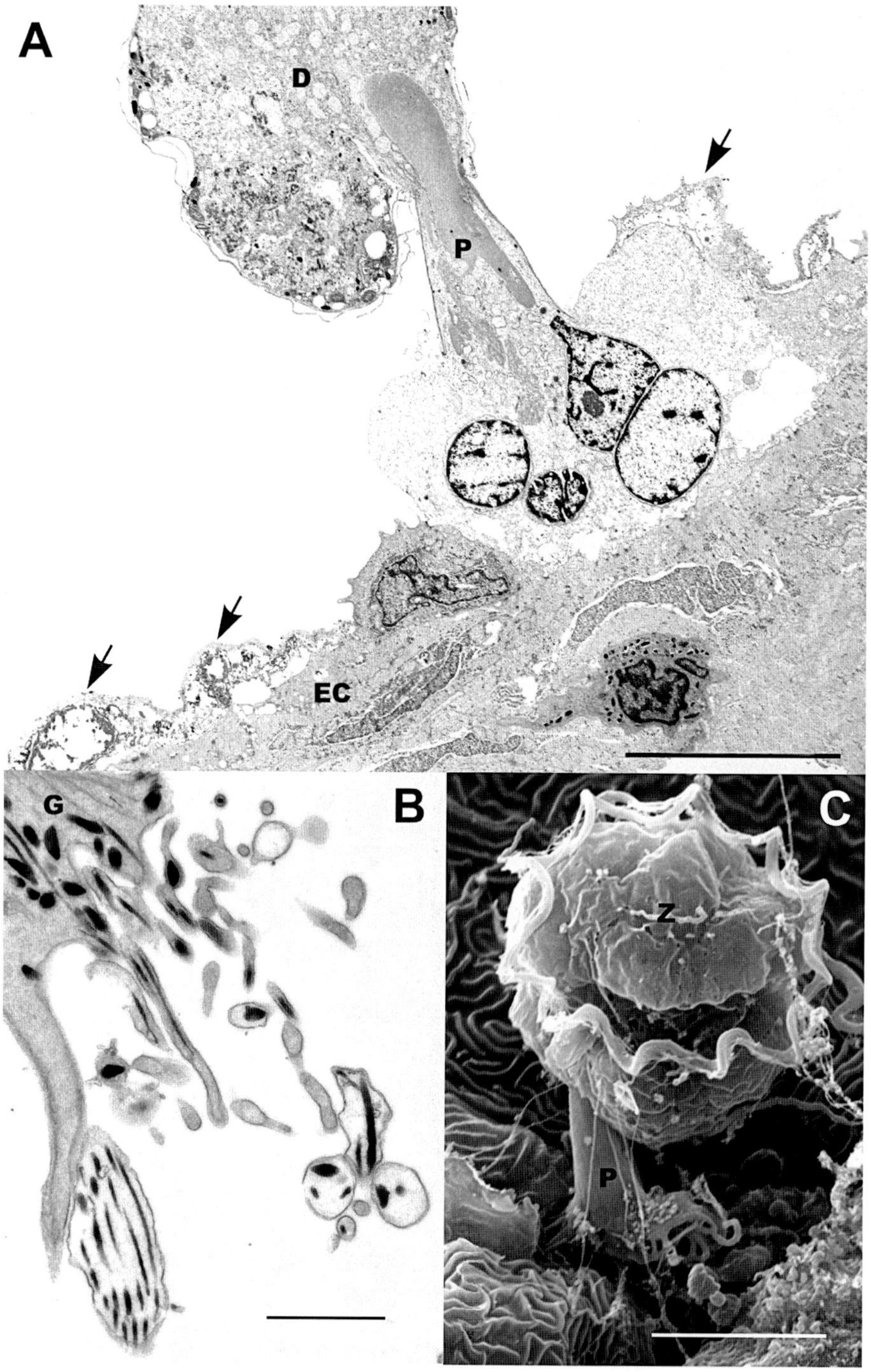

FIGURE 16-14. Scanning and transmission electron micrographs of the direct physical attachment of *Pseudopfiesteria shumwayae* to the epidermis of lightly anesthetized larval (7-day-old) sheepshead minnow, *Cyprinodon variegatus*. (A) Transmission electron micrograph showing the dinoflagellate/fish skin interface and attachment site. The dinoflagellate (D) attaches to the fish epidermal cells (EC) with the peduncle (P), seen here withdrawing the contents of several dead cells. Note the extensive damage to adjacent epidermal cells (arrows) because of this active feeding event. (B) Transmission electron micrograph of an as yet unattached *P. shumwayae* peduncle exhibiting numerous electron dense granules and rodlets (G), not apparent once the dinoflagellate has attached and commenced feeding (see panel A). (C) Scanning electron micrograph of a *P. shumwayae* zoospore (Z) attached to the fish skin via a peduncle (P) and actively feeding. Scale bars: A, C, 10 μm; B, 5 μm. Reprinted from Vogelbein *et al.* (2002).

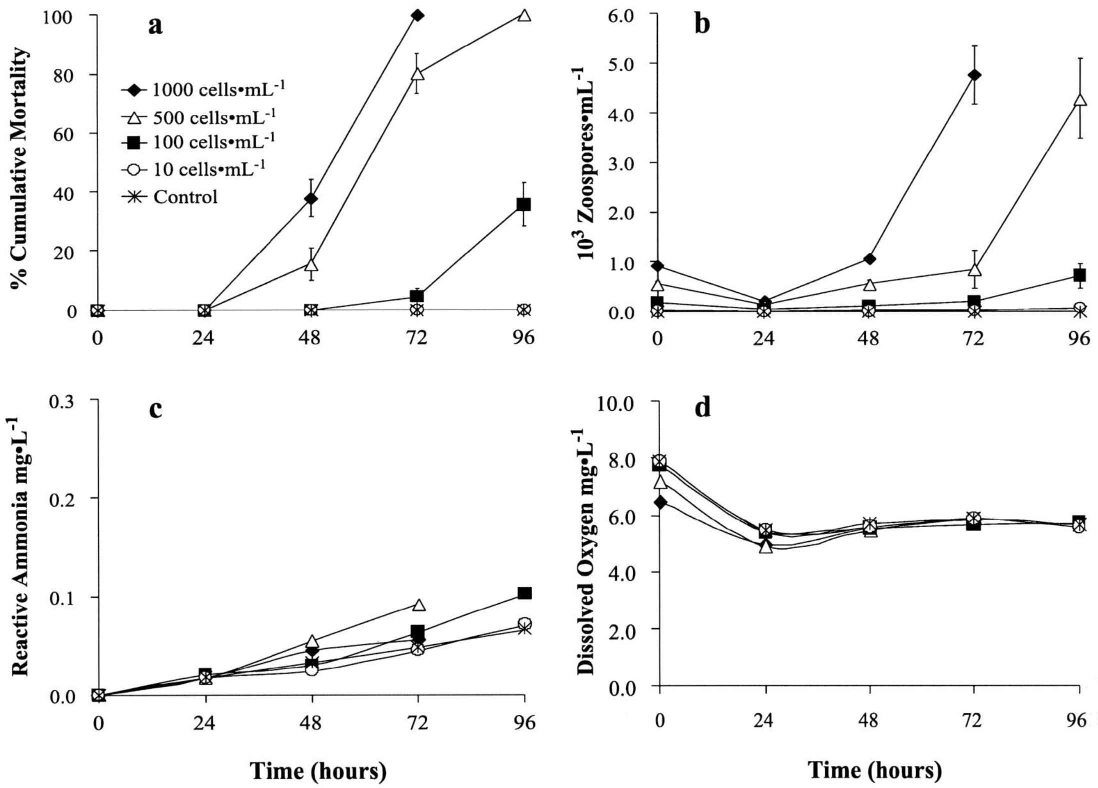

FIGURE 16-15. Dose-response study for *P. shumwayae* in the larval fish assay: (a) cumulative mortality over 96 hours with initial zoospore densities of 0, 10, 100, 500, and 1000 cells • mL^{-1} and standard errors shown (n = 45), (b) change in zoospore density in each treatment over 96 hours (n = 3), (c) mean reactive ammonia from each treatment (n = 3), and (d) mean dissolved oxygen from each treatment (n = 3). Reproduced by permission of the Phycological Society of America, from Lovko *et al.* (2003).

mortality. However, the investigators could not attribute these mortalities to *Pfiesteria* directly, as the tanks contained a rich microbial flora, including bacteria, fungi, and protists, some of which are known fish pathogens and may also produce soluble toxic factors (Drgon *et al.*, 2005).

Studies have further confirmed micropredation as the primary mechanism of fish killing by *P. shumwayae* in both finfish and shellfish. Gordon and Dyer (2005) confirmed micropredation as the dominant and most consistent cause of fish death but attributed low levels of fish death (6% to 19%) in containers that prevented physical contact between fish and the zoospores to a soluble toxin. However, a possible role of the complex microbial community present in their systems was acknowledged. Shumway *et al.* (2006), adopting the experimental methods of Lovko *et al.* (2003) and Vogelbein *et al.* (2002), attributed mortality of larval eastern oysters, northern quahogs, and bay scallops largely to micropredation by *P. shumwayae* zoospores, and only secondarily to a toxin effect. Greatly reduced mortality in mollusk larvae occurred when physical contact was prevented by a semipermeable barrier.

Human Health

Human exposure to most known algal toxins typically occurs by ingestion of contaminated finfish (e.g., ciguatera food poisoning) or shellfish (e.g., amnesic, diarrhetic, paralytic, and neurotoxic shellfish poisoning). Filter-feeding finfish, shellfish, and zooplankton can bioaccumulate toxins, acting as vectors of toxin directly to humans (e.g., consumption of contaminated shellfish) or indirectly through transfer through higher trophic levels of the food web (Friedman and Levin, 2005; Van Dolah, 2000). Exposure can also occur through skin contact or aerosol inhalation (e.g., brevetoxin) or through drinking water (cyanobacterial toxins). In humans, a major target of many of these toxins is the central nervous system (Friedman and Levin, 2005). Patients with acute exposure usually present with gastrointestinal disorders (nausea, vomiting, diarrhea), cardiac or respiratory distress, and neurological symptoms that may persist for days to years (Van Dolah, 2000). Neurological complaints may include memory and learning deficits, abnormal sensations in the extremities (paresthesia), headaches, dizziness, fatigue, mood disorders, seizures, and hallucinations (e.g., Baden *et al.*, 1995; Landsberg, 2002). Some of the reported properties of *Pfiesteria* toxin(s), namely the lack of bioaccumulation in finfish and shellfish (McClellan-Green *et al.*, 1998) and rapid degradation (Moeller *et al.*, 2001, 2007), differ substantially from those of most other known dinoflagellate toxins. Thus, the primary routes of human exposure to *Pfiesteria* toxin were thought to be through inhalation of aerosols from, or skin contact with, contaminated estuarine waters, rather than through the ingestion of contaminated seafood.

Pfiesteria Exposure in Laboratory Personnel

The initial concept of a human clinical syndrome was based on acute exposure of laboratory workers to *Pfiesteria* cultures and bioassays. In one case, prolonged exposure resulted when air from a *Pfiesteria* laboratory was inadvertently vented into an adjacent office space. These exposures resulted in multisystem health complaints, including learning and memory deficits in several laboratory workers (Glasgow *et al.*, 1995). Of note, this first study was self-reported by the individual scientists who experienced the exposures (Fleming *et al.*, 1999). Skin and aerosol contact was postulated as the exposure route. Symptoms included mucus membrane and skin irritation, cognitive problems, fatigue, paresthesias, and gastrointestinal complaints. One individual reported peripheral sensory disturbances, but a neurological exam and electromyography/nerve conduction test results were normal. Another technician experienced burning eyes, nausea, cramping, and disorientation; however, this individual did not follow up by contacting a medical professional.

The third and most severely affected individual had worked with *Pfiesteria* cultures for about 3 years and underwent a medical exam 2 weeks after the onset of symptoms and exposure cessation. Schmechel and Koltai (2001) provide a more detailed report of this case. Physical examination of this patient identified numerous subjective complaints and 2- to 4-week-old healing skin sores around the wrists and dorsum. Routine medical screening tests (blood work, liver enzymes, renal panel), neurophysiological exam and neuroimaging analyses (MRI, PET scan) were largely normal, but with mild asymmetry of the hippocampi indicated by MRI. Neuropsychological evaluation confirmed a syndrome with memory deficits of amnestic proportion, impaired auditory attention, slowed calculation skills, and mild alexia. A follow-up neuropsychological screen 6 weeks later indicated the resolution of the observed deficits with only mild residual evidence of reduced cognitive functions. Diagnostic considerations suggested occupational exposure to *P. piscicida* cultures as the likely cause of the syndrome. Although inappropriate ventilation was suggested as a contributing factor to exposure in the case of this individual, no formal industrial hygiene evaluation of the workplace was conducted (Swinker *et al.*, 2002b).

Several other occupational and civilian cases of suspected *Pfiesteria* exposure were also reported (Schmechel and Koltai, 2001). However, confounding medical and psychological factors were common in these patients, as was the possibility of unconscious modeling (power of suggestion), attributed to the high publicity of *Pfiesteria* in the lay and scientific press at the time (Fleming *et al.*, 1999). Thus, the evidence for human health effects caused by laboratory *Pfiesteria* exposure appears to rest largely on the one compelling clinical case and two others that are less so. To our knowledge, there are no other published reports detailing this type of exposure.

Possible Estuary-Associated Syndrome (PEAS)

In 1998, on a directive from the U.S. Congress and in response to the concern over *Pfiesteria*, the Centers for Disease Control and Prevention (CDC) funded a multistate surveillance program comprised of three prospective human cohort studies in North Carolina, Virginia, and Maryland to investigate what was termed "possible estuary-associated syndrome" (PEAS), so called because of the uncertainty concerning the involvement of *Pfiesteria* (CDC, 1997; Moe *et al.*, 2001; *Morbidity and Mortality Weekly Report* [MMWR], 1999). The key symptom and exposure criteria (CDC, 1997, 1999) that indicated PEAS were (1) exposure to estuarine water with a fish kill or fish with lesions consistent with *Pfiesteria* spp.; (2) symptoms of memory loss and confusion or three or more of the following symptoms: skin rash/sensation of burning skin at the site of water contact, headache, eye irritation, upper respiratory irritation, muscle cramps and gastrointestinal symptoms, any of which persisted for more than 2 weeks; (3) symptoms developing within 2 weeks of exposure to estuarine water and persisting for 2 weeks or longer; and (4) inability of a health care provider to identify another cause for the symptoms.

In addition to an early report indicating that no PEAS was found during year 1 (1998) of the Virginia cohort study (Turf *et al.*, 1999), investigators have published the results of the entire 4-year cohort study that focused on Maryland waters (Morris *et al.*, 2006). They found no correlation between PEAS symptomology and environmental exposure of human subjects to water containing *Pfiesteria* spp. As fish kill and severe lesion events were not recorded during this study, sensitive PCR-based detection methods (see section on Detection Techniques and Environmental Monitoring) were used to specifically identify *Pfiesteria* spp. and related dinoflagellates in water. Thus, the investigators concluded that although high-level or outbreak-associated exposures may have had adverse human health effects in the past as in Grattan *et al.* (1998a, 1998b), recurrent exposure to waters in which these organisms are routinely found, does not appear to pose a significant health risk.

Occupational and Recreational Exposure

Concurrent studies of occupational and recreational environmental exposures in individuals using Mid-Atlantic estuaries provided contradictory results on human health effects attributed to *Pfiesteria* spp. A retrospective, single-blind, case-controlled clinical study of occupational exposures in watermen was conducted in Maryland estuaries following fish kills attributed to *Pfiesteria* spp. in the Pocomoke River and nearby tributaries during 1997 (Grattan *et al.*, 1998a, 1998b). Twenty-four persons reporting direct contact with lesioned fish or waters during an active fish kill were evaluated medically. The most commonly reported symptoms were confusion, forgetfulness, headache, skin lesions, and paresthesia. Other than skin lesions, no abnormalities were found on physical exam or in routine clinical laboratory studies. However, 19 subjects that underwent neuropsychological testing exhibited unexplained and significantly decreased performance on tests that measured new learning and selected and divided attention. Five of 13 subjects that underwent a thorough dermatological exam exhibited nonspecific skin lesions not readily attributed to any one cause (Grattan *et al.*, 1998b; Lowitt and Kauffman, 1998). The investigators believed that the available evidence supported the involvement of *Pfiesteria*, however, it was acknowledged that this clinical syndrome could not be attributed definitively to *Pfiesteria* toxins and did not constitute a cause and effect relationship.

Friedman and Levin (2005) considered the retrospective human health evaluation conducted in the Pocomoke River, Maryland (Grattan *et al.*, 1998b), to be limited by small sample sizes and a lack of evidence that *Pfiesteria* spp. were actually present in waters at the time the subjects were exposed, and they suggested that the "impairment" observed in the neuropsychological exams was likely the result of chance. Moe *et al.* (2001) voiced additional cautions stating that the study by Grattan *et al.* (1998b) was limited by (1) unknown preexposure cognitive test performances of subjects, (2) a small number of subjects that were self-selected, (3) exposure status that was self-reported and may have been affected by recall bias, and (4) a medical team that was not blinded to exposure status of subjects during evaluations.

To date, this is the only study of environmental exposure that has shown a convincing adverse human health effect associated with exposure to an estuarine environment, although that effect could not be directly attributed to *Pfiesteria* spp. In contrast, similar studies in North Carolina and Virginia found no clear association between human health and occupational exposures to estuaries in which fish kills and lesion events attributed to *Pfiesteria* spp. had occurred (Griffith *et al.*, 1998; Morris, 1996; Smith and Music, 1998; Turf *et al.*, 1999).

The remaining environmental studies were equivocal and failed to present convincing evidence for significant adverse human health effects as a result of exposure to *Pfiesteria* toxin. For example, a suite of nonspecific symptoms similar to those described previously (Grattan *et al.*, 1998b) was reported in 5 persons with occupational exposure (Shoemaker and Hudnell, 2001) and 77 individuals having recreational or residential exposure to Maryland estuaries in which fish lesion events and fish death attributed to *Pfiesteria* spp. had occurred during 1997 (Shoemaker, 2001). These studies used visual contrast sensitivity (VCS), a measure of visual pattern detection ability, as a proxy for neuropsychological testing and reported sharply reduced VCS in the exposed group. Hudnell *et al.* (2001) and Hudnell (2005) evaluated 22 exposed (exhibiting symptoms meeting PEAS criteria) and 20 unexposed persons from North Carolina estuaries and found significant VCS deficits in the exposed group.

Although VCS is considered a sensitive indicator of neurotoxicity, the test is nonspecific, with deficits observed following eye disorders attributed to chronic sunlight exposure, anticonvulsant medications, alcohol use and occupational exposure to solvents, petrochemical fuels, heavy metals, and combustion products. Abnormal VCS is not considered diagnostic for any particular disease or exposure, although some neurotoxicants may produce a characteristic deficit pattern in VCS (Swinker *et al.*, 2001a, 2001b). Because VCS deficits are considered to be nonspecific, the aforementioned studies were highly controversial and heatedly debated (Hudnell and Shoemaker, 2002, 2003a, 2003b; Swinker and Burke, 2002; Swinker, 2003a, 2003b). Additionally, the studies have been criticized for their omission of neuropsychological evaluation and their sole reliance on VCS to determine the presence of cognitive deficits (Friedman and Levin, 2005; Morris, 2001; Swinker *et al.*, 2002).

Swinker *et al.* (2001a, 2001b) conducted two cross-sectional case-controlled studies on the long-term chronic human health and neuropsychological effects of possible *Pfiesteria* exposure in North Carolina 3 months after a purported *Pfiesteria* fish kill. One study (Swinker *et al.*, 2001b) examined the effects of occupational exposure in individuals recruited from a roster of commercial fishermen (22 cases and 21 controls). The other study examined the effects of occupational and recreational exposure among individuals calling a *Pfiesteria* hotline (11 cases and 11 controls). In both studies, the surrogate index for toxin exposure was exposure to waters associated with *Pfiesteria* fish kills, disease/distressed fish, or presence of *Pfiesteria* spp. itself. Both studies used standardized neuropsychological test batteries assessing multiple cognitive domains, VCS analysis, and medical and neurological examinations. Neither study identified an increased likelihood of PEAS or evidence of persistent health or neuropsychological effects attributable to *Pfiesteria* toxin exposure, even despite the potential for sampling bias among the hotline callers.

Friedman and Levin (2005), in a critical review of the neurobehavioral investigations conducted on reported exposure to the toxins of harmful algae, correctly pointed out that research has neither identified a specific toxin associated with *Pfiesteria* spp. nor confirmed that exposure to it leads to human illness. Despite this lack of scientific evidence, neuropsychological research on PEAS has resulted in broad sweeping impacts on public attitude, policy, and regulations. In the case of *Pfiesteria*, this climate led to the rapid closure of Maryland waterways thought to be impacted, enactment of new legislation to regulate nutrient inputs to protect water quality, and appropriation of millions of federal dollars to fund research and manage the perceived health and environmental threats posed by *Pfiesteria* spp. (Magnien, 2001). These changes have impacted farming, fishing, and seafood industries, with an estimated $43 million dollar loss of sales in Maryland alone in 1997 (Lipton, 1998).

In summary, the most significant documented medical consequence of exposure to abnormally high levels of *Pfiesteria* in a laboratory setting was a learning deficit that appeared to resolve within several months in 1 individual (Glasgow *et al.*, 1995). No deaths or prolonged illnesses have been attributed to *Pfiesteria*. The early studies claiming to show human health effects have been questioned based on significant methodological issues. All subsequent studies have failed to substantiate any significant health effects resulting from exposure to estuarine water containing *Pfiesteria*. Only one study showed mild adverse health effects following environmental exposure to water where fish were dying presumably because of *Pfiesteria* (Grattan *et al.*, 1998b). However, as stated previously, others have questioned the conclusions of this study.

Rodent Model Studies

Several studies have used rodents as models to study the effects of the purported *Pfiesteria* toxin on cognitive impairment in humans (Duncan *et al.*, 2005; Levin, 2001; Levin *et al.*, 2003; Rezvani *et al.*, 2001). Many of these studies used a baited radial arm maze (RAM) to measure the ability of rats to accurately find bait randomly placed in several arms of the maze. Errors (such as entering an arm of the maze that does not contain bait or entering the same arm twice) were calculated. A comparison of *Pfiesteria*-treated and control groups over repeated trials permitted assessment of memory and spatial learning. Before testing, rats were injected subcutaneously either with water samples taken directly from an aquarium containing purportedly toxic *Pfiesteria* (fish actively dying), a filtrate (0.2 μm) of this same water, or water from aquaria containing fish, but no dinoflagellates (or less commonly, saline) as a control.

Studies by Levin (2001) demonstrated a mild, yet statistically significant learning deficit in rats that were injected with *Pfiesteria* versus controls. There was no consistent *Pfiesteria* effect when rats were first exposed to the maze, but during repeated sessions, when the rats had gained more experience with the maze and improved their performance, *Pfiesteria* injected rats improved more slowly than control-injected rats. This effect persisted for up to 10 weeks (Levin *et al.*, 1997). There was, however, no consistent effect of dose (Levin *et al.*, 1999), *Pfiesteria* strain (Duncan *et al.*, 2005; Levin *et al.*, 2000) or culture filtrate versus whole-cell culture (Levin *et al.*, 1999). Levin *et al.* (2003) also examined the effect of *Pfiesteria* toxin on learning impairment by directly injecting the purported toxin (extracted from large-volume algal-fed clonal cultures of *P. piscicida* demonstrated to rapidly kill fish in bioassays) into the hippocampus of rats before RAM testing. Although this study purported to "clearly" show a *Pfiesteria*-induced learning deficit in rats, the data demonstrated a statistically significant effect, relative to controls, in only 50% of the sessions in one trial of a five-trial series.

A figure-8 maze, widely used in behavioral toxicology studies (Crofton *et al.*, 1991, as cited in Levin *et al.*, 1999) was used in several studies to test locomotor activity of rats. The maze consisted of a continuous alley in the form of a figure-8, with photobeams positioned at various points to track differences in rat movement patterns between treatments and controls. In all treatment groups, rat activity was typically high following introduction to the maze with a marked decrease in activity over time. Differences in the rate of activity reduction between control rats (injected with clean water) and rats injected with *P. piscicida* were compared. In one study, initial activity in rats injected with high doses of *P. piscicida* was higher than controls (Levin *et al.*, 1999). Although activity in all groups decreased over time, there was a statistical difference in the rate at which this activity reduction occurred between the high-dose *Pfiesteria* treatment and controls because of the higher initial activity in the high-dose treatment. Another study failed to demonstrate a convincing effect of strain (TOX-A versus TOX-B) and dose (35,600 versus 106,800 cells/kg) or any effect on juvenile rats (Levin *et al.*, 2000).

Additionally, many of these studies also incorporated a series of observations and tests used to assess neurological integrity, referred to as a functional observational battery (FOB) (Levin *et al.*, 1997, 1999, 2000). These included measurements of abnormal motor movements, number of rearing responses, arousal level, excretion activity, and reactions to such stimuli as the sound of a metal clicker, approach of an object, or a pinch near the end of the tail. No significant differences in FOB were observed in adults (Levin *et al.*, 1997) or juvenile rats (Levin *et al.*, 2000), except for reduction of arousal and rearing in adults. Further, no abnormalities were observed in blood cell counts, clinical state, or histological analyses of spleen, brain, liver, lungs, or kidneys in *Pfiesteria* treated rats (Levin *et al.*, 1997).

To date, studies on neurocognitive behavior in mammals using a rodent model fall short of providing conclusive, compelling evidence for the existence of a potent *Pfiesteria* toxin that can affect human health. These studies demonstrate evidence of only minor and inconsistent learning deficits in *Pfiesteria*-treated rats. Furthermore, it is not evident that the effects described in the rodent studies were the result of a *Pfiesteria* toxin. These studies do not rule out the possibility that other substances in these fish bioassay systems (i.e., metabolites produced by bacteria or the decay of dead fish, etc.) could be responsible for the observed effects. One study (Rezvani *et al.*, 2001) reported bacterial densities in the fish-killing *Pfiesteria* tanks up to 15 times higher than the densities in their control tanks. Additionally, most of these studies used water samples containing *Pfiesteria* cells that were frozen, then thawed before injecting into rats. This process would likely result in the disruption of the *Pfiesteria* cells, liberating numerous proteins and other cell components that could potentially elicit a response in these studies. These possible confounding effects were not accounted for in the reported studies.

The *Pfiesteria* Toxin

Perhaps the most controversial aspect of *Pfiesteria* biology has been the source and structure of the purported toxin. It was claimed during the initial 15 years of research that *Pfiesteria* requires the presence of fish to secrete a potent exotoxin and that algal-grown cultures rapidly became noninducible and incapable of causing fish distress, disease and death (e.g., the NON-IND functional type) (Burkholder and Glasgow 1997a; Burkholder *et al.*, 2001c; Gordon and Dyer, 2005). Thus, methods to detect toxicity have, until recently, relied exclusively on the standardized fish bioassay (Burkholder *et al.*, 2001c), the pitfalls of which were described previously (also reviewed in Miller and Belas, 2003; Drgon *et al.*, 2005).

Early research suggested that a toxin could be partially purified from fish bioassays in which fish were dying in the presence of *Pfiesteria* (Moeller *et al.*, 2001). A molecular assay was developed in which rat pituitary cells carrying a *c-fos* luciferase gene would produce light following activation (Fairey *et al.*, 1999). Assays of water from fish bioassays actively killing fish exhibited elevated (41% above control water) luciferase activity in cells, indicating the presence of a substance that induced *c-fos* gene expression. This substance was suggested to be *Pfiesteria* toxin (pPfTx).

Several polar, methanol-soluble fractions of water derived from *P. piscicida* bioassays actively killing fish were partially purified and tested for biological activity using four distinct bioassay systems. These included brine shrimp (*Artemia salina*) and larval fish (*Cyprinodon variegatus*) bioassays, a cell cytotoxicity assay, and the *c-fos*-luciferase assay described above. However, lipid soluble fractions that killed fish and *Artemia*, and activated the *c-fos* luciferase gene contained a phthalate ester as the principal active ingredient (Moeller *et al.*, 2001). Phthalates are common contaminants resulting from the plastic polymerization process, and in this study, they were thought to be in the Instant Ocean artificial sea salts used to make the water for the fish bioassays (Moeller *et al.*, 2001). Several water-soluble fractions also killed larval *Artemia* and fish, exhibited cell cytotoxicity and *c-fos* luciferase activity; however, no compounds were identified from these fractions, nor were toxin concentrations provided. Furthermore, the authors acknowledged that neither the lipophilic nor the hydrophilic fractions caused formation of skin lesions in the fish bioassay. A partial structure of a potential *Pfiesteria* toxin (pPfTx) was subsequently presented at the 10th International Conference on Harmful Algal Blooms (HABs) in St. Petersburg, FL (Moeller *et al.*, 2002), and summarized as a "glycoside, a molecule that's half sugar, half some other chemical group that hasn't been identified" (p. 946, Kaiser, 2002). The complete structure of this toxin was never described.

As detailed in a previous section, studies showed micropredation rather than exotoxin to be the predominant cause of fish death by *P. shumwayae* and *P. piscicida* (Berry *et al.*, 2002; Drgon *et al.*, 2005; Gordon and Dyer, 2005; Lovko *et al.*, 2003; Shumway *et al.*, 2006; Vogelbein *et al.*, 2002). Additionally, Berry *et al.* (2002) showed that (1) centrifugation, with or without sonication, was sufficient to "detoxify" water from actively fish-killing *P. shumwayae* (CCMP-2089) cultures; (2) organic extracts of lyophilized cells from this culture were not toxic; and (3) *P. shumwayae* most likely lacked the polyketide synthases needed to produce polyketides, a class of toxins produced by most other toxic dinoflagellates (Berry *et al.*, 2002). This same fish-killing culture (CCMP 2089) was subsequently reported to be "toxic" based on the "standard" fish bioassay method (Burkholder *et al.*, 2005) described earlier and criticized by other investigators (summarized by Gordon and Dyer, 2005).

Recently, pPfTx was reported to be a ligated copper compound with numerous congeners (Moeller *et al.*, 2007). It was hypothesize that "rapid, free-radical-mediated toxicity of *Pfiesteria* toxins may occur via production of a redox-cycling metal center and free radical(s) that can lead to specific reactions with 'pro-toxins' which, in turn, can produce more active toxic species" (p. 1170, Moeller *et al.*, 2007). This toxic free radical cascade was suggested to work as follows: "Light exposure could initiate redox cycling of the metal ion(s) resulting in free radical formation and release of the toxin species" (p. 1170, Moeller *et al.*, 2007). Unfortunately, no direct evidence was presented that free radical formation can actually take place in estuarine waters. Furthermore, incomplete structural information was presented that primarily focused on the copper reaction center.

Additionally, this toxin was obtained from *Pfiesteria* grown on algae, which, by definitions put forward in previous publications, should have represented a NON-IND strain, unable to produce toxin and cause fish disease and death (Burkholder *et al.*, 2001a, 2005).

It should be noted that, although grown on algae, these cultures are acknowledged to be nonaxenic (contaminated by microbial organisms) (Moeller *et al.*, 2007). In addition to *Pfiesteria,* they contain abundant algal prey and other microbial contaminants, possibly including bacteria, fungi, and protozoans. Unfortunately, the microbial community associated with these cultures was not characterized or quantified in this study, making definitive attribution of any associated "biological activity" in these cultures to *Pfiesteria* problematic.

Most important, however, the environmental relevance of the newly reported mechanism of toxicity was not clarified. Concentrations of *Pfiesteria* cells used in laboratory studies (300 to 10,000 cells/ml) are far higher than those typically found in nature (<20 cells/ml) (Coyne *et al.*, 2001; Lin *et al.*, 2006; Litaker *et al.*, 2003; Reece *et al.*, 2002, 2005). Additionally, reports regarding toxin stability are conflicting. Moeller *et al.* (2007) reported that the toxin was highly unstable and rapidly degraded, yet earlier studies by Gordon *et al.* (2002, 2005) indicated that the toxin was quite stable.

Micropredation by *Pfiesteria* spp. is now recognized as the predominant cause of fish mortality in laboratory studies (Drgon *et al.*, 2005; Lovko *et al.*, 2003; Vogelbein *et al.*, 2001, 2002). Other investigators have confirmed this observation, but continue to attribute a minor component of fish mortality to a toxin (Drgon *et al.*, 2005; Gordon and Dyer, 2005; Shumway *et al.*, 2006). However, it has not been possible for these investigators to attribute this mortality directly to a *Pfiesteria* toxin because of microbial contamination in their assays. Taken as a whole, these observations and uncertainties continue to raise questions concerning the ability of *Pfiesteria* spp. to impact wild fish populations. Further, the hypothesis that the adverse impacts on fishes observed in the laboratory, including micropredation, are artifacts and do not occur in the wild must receive serious consideration and testing.

SUMMARY AND CONCLUSIONS

Mounting evidence suggests that *Pfiesteria* species may be benign microzooplankters, acting as opportunists, feeding on a variety of prey, and playing no significant role in fish or human health. Based on the literature available to date, the following conclusions summarize the current state of knowledge.

- Although widely distributed, analyses indicate a very low abundance of *Pfiesteria* species cells in environmental samples, so that relevance with respect to fish and human health effects is increasingly questioned.
- Ulcerous lesions, particularly those observed during fish kills and attributed to *Pfiesteria* species during the 1990s, are now known to be caused by the highly pathogenic oomycete, *Aphanomyces invadans*. Thus, the cause-and-effect relationship between lesions in wild fish, fish kills, and *Pfiesteria* is tenuous.
- The early evidence of toxicity, based on fish death in bioassays, was likely the result of misinterpretation of micropredatory feeding on captive prey. The occurrence of this pathogenic mechanism in natural systems is unknown.
- The proposed complex life cycle appears to be based on imprecise methodology, allowing unrelated contaminants present in water samples and bioassays to be misidentified as stages of *Pfiesteria.*
- Some investigators continue to provide evidence of bioactivity in certain *Pfiesteria* strains that may be the result of production of bioactive substances by *Pfiesteria.* However, the relevance of such substances to the manifestation of health impacts to fishes or humans is not well supported and likely insignificant.

References

AlgaeBase, www.algaebase.org.

Baden, D.G., Fleming, L.E., Bean, J.A., 1995. Marine toxins. In deWolff, F.A. (ed.), Handbook of Clinical Neurology, 21: Intoxications of the Nervous System, Part II, pp. 141–175. Amsterdam, Elsevier Science BV.

Berry, J.P., Rein, K., Reece, K.S., Haas, L., Shields, J.D., Vogelbein, W.K., Baden, D.G., Gawley, R.E., 2002. Are *Pfiesteria* spp. toxicogenic? Evidence against production of polyketide and other ichthyotoxins by *Pfiesteria shumwayae.* Proc. Nat. Acad. Sci. 99, 10970–10975.

Blazer, V.S., Lilley, J.H., Schill W.B., Kiryu, Y., Densmore C.L., Panyawachira, V., Chinabut, S., 2002. *Aphanomyces invadans* in Atlantic menhaden along the East Coast of the United States. J. Aquat. Anim. Health 14, 1–10.

Blazer, V.S., Vogelbein, W.K., Densmore, C., Zwerner, D.E., May, E.B., 1999. *Aphanomyces* as a cause of ulcerative skin lesions in menhaden from Chesapeake Bay tributaries. J. Aquat. Anim. Health 11, 340–349.

Bowers, H.A., Tengs, T., Glasgow, H.B., Jr., Burkholder, J.M., Rublee, P.A., Oldach, D.W., 2000. Development of real-time PCR assays for rapid detection of *Pfiesteria piscicida* and related dinoflgellates. Appl. Environ. Microbiol. 66, 4641–4648.

Burkholder, J.M., 1999. The lurking perils of *Pfiesteria.* Sci. Amer. 281, 42–49.

Burkholder, J.M., 2002. *Pfiesteria*: The toxic *Pfiesteria* complex. In Bitton, G. (ed.), Encyclopedia of Environmental Microbiology, pp. 2431–2447. New York, Wiley.

Burkholder, J.M., Glasgow, H.B., 1995. Interactions of a toxic estuarine dinoflagellate with microbial predators and prey. Arch. Protistenkd. 145, 177–188.

Burkholder, J.M., Glasgow, H.B., 1997a. *Pfiesteria piscicida* and other *Pfiesteria*-like dinoflagellates: Behavior, impacts, and environmental controls. Limnol. Oceanogr. 42, 1052–1075.

Burkholder, J.M., Glasgow, H.B., 1997b. Trophic controls on stage transformations of a toxic ambush-predator dinoflagellate. J. Euk. Microbiol. 44, 200–205.

Burkholder, J.M., Glasgow, H.B., 2001. History of toxic *Pfiesteria* in North Carolina estuaries from 1991 to the present. Bioscience 51, 827–841.

Burkholder, J.M., Glasgow, H.B., 2002. The life cycle and toxicity of *Pfiesteria piscicida* revisited. J. Phycol. 38, 1261–1267.

Burkholder, J.M., Glasgow, H.B., Deamer-Melia, N.J., 2001a. Overview and present status of the toxic *Pfiesteria* complex. Phycologia. 40, 186–214.

Burkholder, J.M., Glasgow, H.B., Deamer-Melia, N.J., Springer, J., Parrow, M.W., Zhang, C., Cancellieri, P.J., 2001b. Species of the toxic *Pfiesteria* complex and the importance of functional type in data interpretation. Environ. Health Perspect. 109, 667–686.

Burkholder, J.M., Glasgow, H.B., Hobbs, C.W., 1995a. Fish kills linked to a toxic ambush-predator dinoflagellate: Distribution and environmental conditions. Mar. Ecol. Progr. Ser. 124, 43–61.

Burkholder, J.M., Glasgow, H.B., Jr., Lewitus, A.J., 1998. Physiological ecology of *Pfiesteria piscicida* with general comments on "ambush-predator" dinoflagellates. In Anderson, D.M., Cembella, A., and Hallegraeff, G.M. (eds.), Physiological Ecology of Harmful Algae, vol. 41, NATO ASI Sciences. Berlin, Springer-Verlag.

Burkholder, J.M., Glasgow, H.B., Jr., Steidinger, K.A., 1995b. Stage transformations in the complex life cycle of an ichthyotoxic "ambush-predator" dinoflagellate. In Lassus, P., Arzul, G., Erard, E., Gentien, P., and Marcaillou, C. (eds.), Harmful Marine Algal Blooms. Paris, Technique et Documentation–Lavoisier, Intercept.

Burkholder, J.M., Gordon, A.S., Moeller, P.D., Law, J.M., Coyne, K.J., Lewitus, A.J., Ramsdell, J.S., Marshall, H.G., Deamer, N.J., Cary, S.C., Kempton, J.W., Morton, S.L., Rublee, P.A., 2005. Demonstration of toxicity to fish and to mammalian cells by *Pfiesteria* species: Comparison of assay methods and strains. Proc. Natl. Acad. Sci. USA 102, 3471–3476.

Burkholder, J.M., Mallin, M.A., Glasgow, H.B., Jr., 1999. Fish kills, bottom-water hypoxia, and the toxic *Pfiesteria* complex in the Neuse River and Estuary. Mar. Ecol. Prog. Ser. 179, 301–310.

Burkholder, J.M., Marshall, H.G., Glasgow, H.B., Seaborn, D.W., Deamer-Melia, N.J., 2001c. The standardized fish bioassay procedure for detecting and culturing actively toxic *Pfiesteria*, used by two reference laboratories for Atlantic and Gulf Coast states. Environ. Health Perspect. 109, 745–756.

Burkholder, J.M., Noga, E.J., Hobbs, C.H., Glasgow, H.B., Smith, S.A., 1992. New "phantom" dinoflagellate is the causative agent of major estuarine fish kills. Nature 358, 407–10.

Callinan, R.B., 1994. A comparative review of *Aphanomyces* species associated with epizootic ulcerative syndrome, red spot disease and mycotic granulomatosis. In Roberts, R.J., Campbell, B., and MacRae, I.H. (eds.), Proceedings of the ODA Regional Seminar on Epizootic Ulcerative Syndrome, pp. 248–252. Bangkok, Thailand, The Aquatic Animal Health Research Institute.

Callinan R.B., Paclibare, J.O., Bonad-Reantaso, M.G., Chin, J.C., Gogolewski, R.P., 1995. *Aphanomyces* species associated with epizootic ulcerative syndrome (EUS) in the Phillipines and red spot disease (RSD) in Australia: Preliminary comparative studies. Dis. Aquat. Org. 21, 233–238.

Callinan, R.B., Sammut, J., Fraser, G.C., 1996. Epizootic ulcerative syndrome (red spot disease) in estuarine fish—confirmation that exposure to acid sulfate soil runoff and an invasive aquatic fungus, *Aphanomyces* sp., are causative factors. In Associates and NSW Acid Sulfate Soils Management Advisory Committee (ASSMAC), R.J. Smith and H.J. Smith (eds.), Proceedings of the Second National Conference on Acid Sulfate Soil, Australia,pp. 146–151.

Catap, E.S., Mundy, B.L., 1998. Effects of variations of water temperature and dietary lipids on the expression of experimental epizootic ulcerative syndrome (EUS) in sand whiting, *Sillago ciliata*. Fish Path. 33, 327–335.

Cavalier-Smith, T., 1993. Kingdom Protozoa and its 18 phyla. Microbiol. Rev. 57, 953–994.

Centers for Disease and Control and Prevention, 1997. Results of the public health response to *Pfiesteria* workshop. Sept. 29–30, 1997, Atlanta, GA. Mor. Mortal. Wkly. Rep. 46, 951–952.

Centers for Disease and Control and Prevention, 1999. Possible estuary-associated syndrome. Mor. Mortal. Wkly. Rep. 48, 381–383.

Coyne, K.J., Hutchins, D.A., Hare, C.E., Cary, S.C., 2001. Assessing temporal and spatial variability in *Pfiesteria piscicida* distributions using molecular probing techniques. Aquat. Microbial Ecol. 24, 275–285.

Crofton, K.M., Howard, J.L., Moser, V.C., Gill, M.W., Reiter, L.W., Tilson, H.A., MacPhail, R.C., 1991. Interlaboratory comparisons of motor activity experiments: Implications for neurotoxicological assessments. Neurotoxicol. Teratol. 13, 599–609.

Cruz-Lacierda, E.R., Shariff, M., 1995. Experimental transmission of epizootic ulcerative syndrome (EUS) in snakehead *Ophicephalus striatus*. In Shariff, M., Arthur, J.R., and Subasinghe, R. (eds.) Diseases of Asian aquaculture II. Manila, Asian Fisheries Society, Fish Health Section.

David, J.C., Kirk, P.M., IMI (International Mycological Institute). 1997. Index of fungi, volume 6, Part 13 in Ainsworth and Bisby's dictionary of the fungi, 8th edition. CAB International, Wallingford, UK, 706.

Doucette, G.J., 1995. Interactions between bacteria and harmful algae: A review. Natural Toxins 3, 65–74.

Drgon, T., Saito, K., Gillevet, P.M., Sikaroodi, M., Whitaker, B., Krupatkina, D.N., Argemi, F., Vasta, G.R., 2005. Characterization of ichthyocidal activity of *Pfiesteria piscicida*: Dependence on the dinospore cell density. App. Env. Microbiol. 71, 519–529.

Duncan, P.M., Parris, B., Schultz, S., Hones, J., Gordon, A., Dyer, B. Marshall, H., 2005. Behavioral effects and drug vulnerability in rats exposed to *Pfiesteria* toxin. Neurotox. Teratol. 27, 701–710.

Dykstra, M.J., Kane, A.S., 2000. *Pfiesteria piscicida* and ulcerative mycosis of Atlantic menhaden current status of understanding. J. Aquat. Anim. Health 12, 18–25.

Dykstra, M.J, Levine, J.F., Noga, E.J., Hawkins, J.H., Gerdes, P., Hargis, W.J., Jr., Grier, H.J., Strake, T.E., 1989. Ulcerative mycosis: A serious menhaden disease of the southeastern coastal fisheries of the United States. J. Fish Dis. 12, 175–178.

Dykstra, M.J., Noga, E.J., Levine, J.F., Moye, D.W., Hawkins, J.H., 1986. Characterization of the *Aphanomyces* species involved in ulcerative mycosis (UM) in menhaden. Mycologia 78, 664–672.

Eriksen, N.T., Hayes, K.C., Lewitus, A.J., 2002. Growth responses of the mixotrophic dinoflagellates *Cryptoperidiniopsis* sp. and *Pfiesteria piscicida*, to light under prey-saturated conditions. Harmful Algae 1, 191–203.

Fairey, E.R., Edmunds, J.S., Deamer-Melia, N.J., Glasgow, H.B., Jr., Johnson, F.M., Moeller, P.D., Burkholder, J.M., Ramsdell, J.S., 1999. Reporter gene assay for fish killing activity produced by *Pfiesteria piscicida*. Environ. Health. Perspect. 107, 711–714.

Fleming, L.E., Easom, J., Baden, D., Rowan, A., Levin, B., 1999. Emerging harmful algal blooms and human health: *Pfiesteria* and related organisms. Toxicol. Pathol. 27, 573–581.

Fraser, G.C., Callinan, R.B., Calder, L.M., 1992. *Aphanomyces* species associated with red spot disease: An ulcerative disease of estuarine fish from eastern Australia. J. Fish Dis. 15, 173–181.

Friedman, M.A., Levin, B.E., 2005. Neurobehavioral effects of harmful algal bloom (HAB) toxins: A critical review. J. Int. Neurobiol. Soc. 11, 331–338.

Glasgow, H.B., Burkholder, J.M., Mallin, M.A., Deamer-Melia, N.J., Reed, R.E., 2001a. Field ecology of Toxic *Pfiesteria* Complex species and a conservative analysis of their role in estuarine fish kills. Environ. Health Perspect. 109, 715–730.

Glasgow, H.B., Burkholder, J.M., Morton, S.L., Springer, J., 2001b. A second species of ichthyotoxic *Pfiesteria* (Dinamoebales, Dinophyceae). Phycologia 40, 234–245.

Glasgow, H.B., Jr., Burkholder, J.M., Schmechel, D.E., Tester, P.A., Rublee, P.A., 1995. Insidious effects of a toxic estuarine dinoflagellate on fish survival and human health. J. Toxicol. Environ. Health 46, 501–522.

Gordon, A.S., Dyer, B.J., 2005. Relative contribution of exotoxin and micropredation to icthyotoxicity of two strains of *Pfiesteria shumwayae* (Dinophyceae). Harmful Algae 4, 423–431.

Gordon, A.S., Dyer, B.J., Seaborn, D., Marshall, H.G., 2002. Comparative toxicity of *Pfiesteria* spp., prolonging toxicity of *P. piscicida* in culture and evaluation of toxin(s) stability. Harmful Algae 1, 85–94.

Grattan, L., Oldach, D., Tracy, K., Greenberg, D., 1998a. Neurobehavioral complaints of symptomatic persons exposed to *Pfiesteria piscicida* or morphologically related organisms. Maryland Med. J. 47, 127–129.

Grattan, L.M., Oldach, D., Perl, T.M., Lowitt, M.H., Matuszak, D.L., Dickson, C., Parrott, C., Shoemaker, R.C., Kauffman, C.L., Wasserman, M.P., Hebel, J.R., Charache, P., Morris, J.G., Jr., 1998b. Learning and memory difficulties after environmental exposure to waterways containing toxin-producing *Pfiesteria* or *Pfiesteria*-like dinoflagellates. Lancet 352, 532–539.

Griffith, D., Shecter, A., Borre, K., Kelley, V., 1998. Occupational risks of crabbing: A report with special emphasis on the public health threat of *Pfiesteria piscicida*. An exploratory study of possible human health effects of deteriorating water quality among North Carolina crabbers, research on toxic algae: *Pfiesteria*-like organisms. NC Sea Grant Publication No. UNC-56-98-02. Raleigh, NC, North Carolina State University.

Hatai, K., 1980. Studies on pathogenic agents of saprolegniasis in fresh water fishes. Nagasaki Prefectural Institute of Fisheries, Special Report 8, Nagasaki, Japan.

Hatai, K., Egusa, S., Takahashi, S., Ooe, K., 1977. Study on the pathogenic fungus of mycotic granulomatosis. I. Isolations and pathogenecity of the fungus from cultured ayu infected with the disease. J. Fish Path. 12, 129–133 (in Japanese).

Hudnell, H.K., House, D., Schmid, J., Koltai, D., Wilkins, J., Stopford, W., Savitz, D., Swinker, M., Music, S., 2001. Human visual function in the North Carolina clinical study on possible estuary-associated syndrome. J. Toxicol. Environ. Health 62, 575–594.

Hudnell, K.H., 2005. Chronic biotoxin-associated illness: Multiple-system symptoms, a vision deficit, and effective treatment. Neurotox. Teratol. 27, 733–743.

Hudnell, K.H., Shoemaker, R.C., 2002. Visual contrast sensitivity: Response. Environ. Health Perspect. 110, A121-A123.

Hudnell, K.H., Shoemaker, R.C. 2003a. Neuropsychologic testing versus visual contrast sensitivity: Response. Environ. Health Perspect. 111, A14-A15.

Hudnell, K.H., Shoemaker, R.C., 2003b. A letter of comment on "human health effects of exposure to *Pfiesteria piscicida*: A review" by Swinker and colleagues. Microbes Infect. 5, 345–347.

Jakobsen, K.S., Tengs, T., Vatne, A., Bowers, H.A., Oldach, D.W., Burkholder, J.M., Glasgow, H.B., Rublee, P.A., Klaveness, D., 2002. Discovery of the toxic dinoflagellate, *Pfiesteria*, from northern European waters. Proc. Royal. Soc. Lond. (B) 269, 211–214.

Jeong, H.J., Seong, J., Park, J.Y., Kim, J.H., Kim, S., Lee, I., Lee, S.H., Ha, J.H., Yih, W.H., 2005. *Stoeckeria algicida* n. gen., n. sp. (Dinophyceae) from the coastal waters off southern Korea: Morphology and small subunit ribosomal DNA gene sequence. J. Eukaryot. Microbiol. 52, 382–390.

Kaiser, J., 2002a. The science of *Pfiesteria*: Elusive, subtle, and toxic. Science 298, 346–349.

Kaiser, J, 2002b. Is sugary toxin the smoking gun? Science 298, 945–946.

Kane, A.S., Oldach, D., Reimschuessel, R., 1998. Fish lesions in the Chesapeake Bay: *Pfiesteria*-like dinoflagellates and other etiologies. Maryland Med. J. 47, 106–112.

Khan, M.H., Marshall, L., Thompson, K.D., Campbell, R.E., Lilley, J.H., 1998. Susceptibility of five fish species (Nile tilapia, rosy barb, rainbow trout, stickleback and roach) to intramuscular injection with the oomycete fish pathogen, *Aphanomyces invadans*. Bulletin of the European Association of Fish Pathologists 18, 192–197.

Kiryu, Y., Shields, J.D., Vogelbein, W.K., Kator, H., Blazer, V.S., 2003. Infectivity and pathogenicity of the oomycete *Aphanomyces invadans* in Atlantic menhaden *Brevoortia tyrannus*. Dis. Aquat. Org. 54, 135–146.

Kiryu Y., Shields, J.D., Vogelbein, W.K., Zwerner, D.E., Kator, H., 2002. Induction of skin ulcers in Atlantic menhaden by injection and aqueous exposure to zoospores of *Aphanomyces invadans*. J. Aquat. Anim. Health 14, 11–24.

Kofoid, C.A., 1909. On *Peridinium steini* Jörgensen, with a note on the nomenclature of the skeleton of the Peridinidae. Archiv für Protistenkunde. 16, 25–47.

Landsberg, J.H., 2002. The effects of harmful algal blooms on aquatic organisms. Rev. Fisheries Sci. 10, 113–390.

Law, M., 2001. Differential diagnosis of ulcerative lesions in fish. Env. Health Perspect. 109 (suppl 5), 681–686.

Levin, E.D., 2001. A rat model of the cognitive impairment from *Pfiesteria piscicida* exposure. Environ. Health Perspect. 109 (suppl 5), 757–763.

Levin, E.D., Blackwelder, W.P., Glasgow, H.B., Burkholder, J.M., Moeller, P.D., Ramsdell, J.S., 2003. Learning impairment caused by a toxin produced by *Pfiesteria piscicida* infused into the hippocampus of rats. Neurotox. Teratol. 25, 419–426.

Levin, E.D., Rezvani, A.H., Christopher, N.C., Glasgow, H.B., Jr., Deamer-Melia, N.J., Burkholder, J.M., Moser, V.C., Jensen, K., 2000. Rapid neurobehavioral analysis of *Pfiesteria piscicida* effects in juvenile and adult rats. Neurotox. Teratol. 22, 533–540.

Levin, E.D., Schmechel, D.E., Burkholder, J.M., Glagow, H.B., Jr., Deamer-Melia, N.J., Moser, V.C., Harry, G.J., 1997. Environ. Health. Perspect. 105, 1320–1325.

Levin, E.D., Simon, B.B., Schmechel, D.E., Glasgow, H.B., Jr., Deamer-Melia, N.J., Burkholder, J.M., Moser, V.C., Jensen, K., Harry, G.J., 1999. *Pfiesteria* toxin and learning performance. Neurotox. Teratol. 21, 215–221.

Levine, J.F., Hawkins, J.H., Dykstra, M.J., Noga, E.J., Moye, D.W., Cone, R.S., 1990a. Epidemiology of ulcerative mycosis in Atlantic menhaden in the Tar-Pamilico estuary, North Carolina. J. Aquat. Anim. Health 2, 162–171.

Levine, J.F., Hawkins, J.H., Dykstra, M.J., Noga, E.J., Moye, D.W., Cone, R.S., 1990b. Species distribution of ulcerative lesions on finfish in the Tar-Pamlico River Estuary, North Carolina. Dis. Aquat. Org. 8, 1–5.

Lewitus, A.J., Glasgow, H.B., Burholder, J.M., 1999. Kleptoplastidy in the toxic dinoflagellate, *Pfiesteria piscicida*. J. Phycol. 35, 303–312.

Lewitus, A.J., Hayes, K.C., Willis, B.M., Burkholder, J.M., Glasgow, H.B., Jr., Holland, A.F., Maier, P., Rublee, P.A., Magnien, R., 2002. Low abundance of dinoflagellates, *Pfiesteria piscicida*, *P. shumwayae*, and *Cryptoperidiniopsis* spp. in South Carolina tidal creeks and open estuaries. Estuaries 25, 586–597.

Lilley, J.H., Callinan, R.B., Chinabut, S., Kanchanakhan, S., MacRae, I.H., Phillips, M.J., 1998. Epizootic ulcerative syndrome (EUS) technical handbook. Bangkok, Thailand, The Aquatic Animal Health Research Institute.

Lilley, J.H., Hart, D., Panyawachira, V., Kanchanakhan, S., Chinabut, S., Soderhall, K., Cerenius, L., 2003. Molecular characterization of the fish-pathogenic fungus *Aphanomyces invadans*. J. Fish Dis. 26, 263–275.

Lilley, J.H., Roberts, R.J., 1997. Pathogenicity and culture studies comparing the *Aphanomyces* involved in epizootic ulcerative syndrome (EUS) with other similar fungi. J. Fish Dis. 20, 135–144.

Lin, S., Mulholland, M.R., Zhang, H., Feinstein, T.N., Jochemd, F.J., Carpenter, E.J., 2004. Intense grazing and prey-dependent growth of *Pfiesteria piscicida* (Dinophyceae). J. Phycol. 40, 1062–1073.

Lin, S., Zhang, H., 2003. Mitogen-activated protein kinase in *Pfiesteria piscicida* and its growth rate-related expression. Appl. Environ. Microbiol. 69, 343–349.

Lin, S., Zhang, H., Dubois, A., 2006. Low abundance distribution of *Pfiesteria piscicida* in Pacific and Western Atlantic as detected by mtDNA-18S rDNA real-time polymerase chain reaction. J Plank. Res. 28, 667–681.

Lipton, D.W., 1998. *Pfiesteria*'s economic impact on seafood industry sales and recreational fishing. *Pfiesteria*: Where do we go from here? Economics of Policy Options for Nutrient Management and Dinoflagellates Conference. University of Maryland, Department of Agriculture and Natural Resources, College Park.

Litaker, R.W., Steidinger, K.A., Mason, P.L., Landsberg, J.H., Shields, J. D., Reece, K.S., Haas, L.W., Vogelbein, W.K., Vandersea, M.W., Kibler, S.R., Tester, P.A., 2005. The Reclassification of *Pfiesteria shumwayae*: *Pseudopfiesteria*, gen. nov. J. Phycol. 41, 643–651.

Litaker, R.W., Tester, P.A., Haugen, E.M., Colorni, A., Levy, M.G., Noga, E.J., 1999. The phylogenetic relationship of *Pfiesteria piscicida*, a Cryptoperidiniopsoid sp., *Amyloodinium ocellatum* and a *Pfiesteria*-like dinoflagellate to other dinoflagellates and apicomplexans. J. Phycol. 35, 1379–1389.

Litaker, R.W., Vandersea, M.W., Kibler, S.R., Madden, V.J., Noga, E.J., Tester, P.A., 2002. Life cycle of the heterotrophic dinoflagellate *Pfiesteria piscicida* (Dinophyceae). J. Phycol. 38, 442–463.

Litaker, R.W., Vandersea, M.W., Kibler, S.R., Reece, K.S., Stokes, N.A., Steidinger, K., Millie, D.F., Bendis, B.J., Pigg, R.M., Tester, P.A., 2003. Identification of *Pfiesteria piscicida* (Dinophyceae) and *Pfiesteria*-like organisms using ITS-specific PCR assays. J. Phycol. 39, 754–761.

Lovko, V.J., Vogelbein, W.K., Shields, J.D., Haas, L.W., Reece, K.S., 2003. A new larval fish bioassay for testing the pathogenicity of *Pfiesteria* spp. (Dinophyceae). J. Phycol. 39, 600–609.

Lowitt, M., Kauffman, L., 1998. *Pfiesteria* and the skin. Maryland Med. J. 47, 124–126.

Luettich, R.A., McNinch, J.E., Paerl, J.W., Peterson, C.H., Wells, J.T., Alpern, M., Martens, C.S., Pinkney, J.L., 2000. Neuse River Estuary modeling and monitoring project stage 1: Hydrography and circulation, water column nutrients and productivity, sedimentary processes and benthic-pelagic coupling. In Water Resources Research Institute of the University of North Carolina, Raleigh, North Carolina.

Magnien, R.E., 2001. State monitoring activities related to *Pfiesteria*-like organisms. Environ. Health Perspect. 109 (suppl 5), 711–714.

Mallin, M.A., Burkholder, J.M., Larsen, L.M., Glasgow, H.B., Jr., 1995. Response of two zooplankton grazers to an ichthyotoxic estuarine dinoflagellate. J. Plankton Res. 17, 351–363.

Marshall, H.G., Hargraves, P.E., Burkholder, J.M., Parrow, M.W., Elbrächter, M., Allen, E.H., Knowlton, V.M., Rublee, P.A., Hynes, W. L., Egerton, T.A., Remington, D.L., Wyatt, K.B., Coyne, K.J., Lewitus, A.J., Henrich, V.C., 2006. Taxonomy of *Pfiesteria* (Dinophyceae). Harmful Algae 5, 481–496.

Marshall, H.G., Seaborn, D., Wolny, J., 1999. Monitoring results for *Pfiesteria piscicida* and *Pfiesteria*-like organisms from Virginia waters in 1998. Virginia J. Sci. 50, 287–298.

Mason, P.L., Litaker, W., Jeong, H.J., Reece, K.S., Vogelbein, W.K., Steidinger, K., 2007. Description of a new genus of *Pfiesteria*-like dinoflagellate, *Luciella* gen. nov. (Dinophyceae), including two new species: *Luciella masanensis* sp. nov. and *Luciella atlantis* sp. nov. J. Phycol. 43, 799–810.

Mason, P.M., Vogelbein, W.K., Haas, L.W., Shields, J.D., 2003. An improved stripping technique for lightly armored dinoflagellates. J. Phycol. 39, 1–7.

McClellan-Green, P.D., Jaykus, L.A., Green, D.P., 1998. Consumer health risks due to incidental exposure of fish to *Pfiesteria piscicida*, North Carolina. Sea Grant, Publication No. UNC-5G-98-02. Raleigh, North Carolina State University.

Miller, T.R., Belas, R., 2003. *Pfiesteria piscicida*, *P. shumwayae*, and other *Pfiesteria*-like dinoflagellates. Res. Microbiol. 154, 85–90.

Moe, C.L., Turf, E., Oldach, D., Bell, P., Hutton, S., Savitz, D., Koltai, D., Turf, M., Ingsrisawang, L., Hart, R., Ball, J.D., Stutts, M., McCarter, R., Wilson, L., Haselow, D., Grattan, L., Morris, J.G., Weber, D.J., 2001. Cohort studies of health effects among people exposed to estuarine waters: North Carolina, Virginia and Maryland. Environ. Health Perspect. 109 (suppl 5), 781–786.

Moeller, P.D., Beuchesne, K.R., Huncik, K.M., Davis, W.C., Christopher, S.J., Riggs-Gelasco, P., Gelasco, A.K., 2007. Metal complexes and free radical toxins produced by *Pfiesteria piscicida*. Environ. Sci. Technol. 41, 1166–1172.

Moeller, P.D., Morton, S.L., Mitchell, B.A., Sivertsen, S.K., Fairey, E.R., Mikulski, T.M., Glasgow, H.B., Jr., Deamer-Melia, N.J., Burkholder, J.M., Ramsdell, J.S., 2001. Current progress in isolation and characterization of toxins isolated from *Pfiesteria piscicida*. Environ. Health Perspect. 109 (suppl 5), 739–743.

Moeller, P.D.R., Ramsdell, J.S., Morton, S.L., Mitchell, B.A., Eaker, S., Burkholder, J.M., Glasgow, H.B., Deamer-Melia, N.J., 2002. Isolation, characterization and current chemical structural information on a water soluble toxin derived from *Pfiesteria piscicida*. 10th International Conference on Harmful Algae, St. Petersburg, FL.

Morbidity and Mortality Weekly Report (MMWR). May 14, 1999, vol. 48(19), www.cdc.gov/mmwr/PDF/wk/mm4818.pdf.

Morris, J.G., 2001. Human health effects and *Pfiesteria* exposure: A synthesis of available clinical data. Environ. Health Perspect. 109, 787–790.

Morris, J.G., Grattan, L.M., Wilson, L.A., Meyer, W.A., McCarter, R., Bowers, H.A., Hebel, J.R., Matuszak, D.L., Oldach, D.W., 2006. Occupational exposure to *Pfiesteria* species in estuarine waters is not a risk factor for illness. Environ. Health Perspect. 114, 1038–1043.

Morris, P.D., 1996. Acute symptoms reported by persons exposed to fish kills associated with *Pfiesteria piscicida*. Occupational and Environmental Epidemiology Section, Division of Epidemiology, North Carolina State Dept. of the Environment, Health, and Natural Resources, Raleigh.

Noga, E.J., 1993. Fungal diseases of marine and estuarine fishes. In Couch, J., and Fournie, J.W. (eds.), Pathobiology of marine and Estuarine Organisms, pp. 85–109. Boca Raton, FL, CRC Press.

Noga, E.J., 2000. Skin ulcers in fish: *Pfiesteria* and other etiologies. Toxicol. Pathol. 28, 807–823.

Noga, E.J., Dykstra, M.J., 1986. Oomycete fungi associated with ulcerative mycosis in menhaden, *Brevoortia tyrannus* (Latrobe). J. Fish. Dis. 9, 47–53.

Noga, E.J., Khoo, L., Stevens, J.B., Fan, Z., Burkholder, J.M., 1996. Novel toxic dinoflagellate causes epidemic disease in estuarine fish. Mar. Poll. Bull. 32, 219–224.

Noga, E.J., Levine, J.F., Dykstra, M.J., Hawkins, J.H., 1988. Pathology of ulcerative mycosis in Atlantic menhaden. Dis. Aquat. Org. 4, 189–197.

Noga, E.J., Smith, S.A., Burkholder, J.M., Hobbs, C., Bullis, R.A., 1993. A new ichthyotoxic dinoflagellate: Cause of acute mortality in aquarium fishes. Vet. Rec. 133, 96–97.

Oldach, D.W., Delwiche, C.F., Jakobsen, K.S., Tengs, T., Brown, E.G., Kempton, J.W., Schaefer, E.F., Bowers, H.A., Glasgow, H.B., Burkholder, J.M., Steidinger, K.A., Rublee, P.A., 2000. Heteroduplex mobility assay-guided sequence discovery: Elucidation of the small subunit (18S) rDNA sequences of *Pfiesteria piscicida* and related dinoflagellates from complex algal culture and environmental sample DNA pools. Proc. Nat. Acad. Sci. 97, 4303–4308.

Paerl, H.W., Pinckney, J.L., Fear, J.M., Peierls, B.J., 1998. Ecosystem responses to internal and watershed organic matter loading: Consequences for hypoxia in the eutrophying Neuse River Estuary, North Carolina, USA. Mar. Ecol. Prog. Ser. 166, 17–25.

Paerl, H.W., Pinckney, J.L., Fear, J.M., Peierls, B.L., 1999. Fish kills, bottom-water hypoxia, and the toxic *Pfiesteria* complex in the Neuse River and Estuary: Reply to Burkholder *et al.* Mar. Ecol. Prog. Ser. 186, 307–309.

Parrow, M., Burkholder, J.M., Deamer, N.J., Zhang, C., 2002. Vegetative and sexual reproduction in *Pfiesteria* spp. (Dinophyceae) cultured with algal prey, and inferences for their classification. Harmful Algae 1, 5–33.

Parrow, M.W., Burkholder, J.M., 2003. Reproduction and sexuality in *Pfiesteria shumwayae* (Dinophyceae). J. Phycol. 39, 697–711.

Peglar, M.T., Amaral Zettler, L.A., Anderson, O.R., Nerad, T.A., Gillevet, P.M., Mullen, T.E., Frasca, S., Jr., Silberman, J.D., O'Kelly, C.J., Sogin, M.L., 2003. Two new small-subunit ribosomal RNA gene lineages within the subclass Gymnamoebia. J. Euk. Microbiol. 50, 224–232.

Peglar, M.T., Nerad, T.A., Anderson, O.R., Gillevet, P.M., 2004. Identification of amoebae implicated in the life cycle of *Pfiesteria* and *Pfiesteria*-like dinoflagellates. J. Euk. Micro. 51, 542–552.

Quesenberry, M.S., Saito, K., Krupatkina, D.N., Robledo, J.A.F., Drgon, T., Pecher, W.T., O'Leary, N., Alavi, M., Miller, T., Schneider, R.E., Belas, R., Deeds, J.R., Place, A.R., Zohar, Y., Vasta, G.R., 2002. Bioassay for ichthyocidal activity of *Pfiesteria piscicida*: Characterization of a culture flask assay format. J. Appl. Phycol. 14, 241–254.

Reece, K.S., Shields, J., Vogelbein, W., Haas, L., 2005. Annual Report: Toxicity and Life-Cycle Analyses of Species and Strains of *Pfiesteria*, Submitted to the Virginia Department of Health, National Centers for Disease Control and Prevention.

Reece, K.S., Stokes, N.A., Burreson, E.M., 2002. Final Report: ECOHAB: DNA-based molecular diagnostics for *Pfiesteria*-complex organisms in the Chesapeake Bay. Submitted to ECOHAB STAR Grant Program, U.S. Environmental Protection Agency.

Reece, K.S., Vogelbein, W.K, Shields, J.D., Haas, L.W., Hoenig, J., Kator, H., Burreson, E.M., 2003. Final report: Environmental monitoring in support of the CDC human cohort study: Year 2002. Submitted to the Virginia Department of Health, National Centers for Disease Control and Prevention.

Rezvani, A.H., Bushnell, P.J., Burkholder, J.M., Glasgow, H.B., Jr., Levin E.D., 2001. Specificity of cognitive impairment from *Pfiesteria piscicida* exposure in rats: Attention and visual function versus behavioral plasticity. Neurotox. Teratol. 23, 609–616.

Rhodes, L.L., Adamson, J.E., Rublee, P.A., Schaefer, E.F., 2006. Geographic distribution of *Pfiesteria* spp. (Pfiesteriaceae) in Tasman Bay and Canterbury, New Zealand (2002–03). N. Z. J. Mar. Freshwater Res. 40, 211–220.

Rhodes, L.L., Burkholder, J.M., Glasgow, H.B., Jr., Rublee, P.A., Allen, C., Adamson, J.E., 2002. *Pfiesteria shumwayae* (Pfiesteriaceae) in New Zealand. N. Z. J. Mar. Freshwater Res. 36, 621.

Roberts, R.J., Willoughby, L.G., Chiabut, S., 1993. Mycotic aspects of epizootic ulcerative syndrome (EUS) of Asian fishes. J. Fish. Dis. 16, 169–183.

Rublee, P.A., Kempton, J., Schaefer, E., Burkholder, J.M., Glasgow, H.B., Jr., Oldach, D., 1999. PCR and FISH detection extends the range of *Pfiesteria piscicida* in estuarine waters. Va. J. Sci. 50, 325–335.

Rublee, P.A., Remington, D.L., Schaefer, E.F., Marshall, M.M., 2005. Detection of the dinozoans *Pfiesteria piscicida* and *P. shumwayae*: A review of detection methods and geographic distribution. J. Eukaryot. Microbiol. 52, 83–89.

Saito, K., Drgon, T., Robledo, J.A.F., Krupatkina, D.N., Vasta. G.R. 2002. Characterization of the rRNA locus of *Pfiesteria piscicida* and development of standard and quantitative PCR-based detection assays targeted to the nontranscribed spacer. Appl. Environ. Microbiol. 68, 5394–5407.

Samet, J., Bignami, G.S., Feldman, R., Hawkins, W., Neff, J., Smayda, T., 2001. *Pfiesteria*: Review of the science and identification of research gaps. Report for the National Center for Environmental Health, Centers for Disease Control and Prevention. Environ. Health Perspect. 109, 639–659.

Schmechel, D.E., Koltai, D.C., 2001. Potential human health effects associated with laboratory exposures to *Pfiesteria piscicida*. Environ. Health Perspect. 109, 775–779.

Shoemaker, R.C., 2001. Residential and recreational acquisition of possible estuary-associated syndrome: A new approach to successful diagnosis and treatment. Environ. Health Perspect. 109 (suppl 5), 791–796.

Shoemaker, R.C., Hudnell, K.H., 2001. Possible estuary-associated syndrome: Symptoms, vision and treatment. Environ. Health Perspect. 109, 539–545.

Shumway, S.E., Burkholder, J.M., Springer J., 2006. Effects of the estuarine dinoflagellate *Pfiesteria shumwayae* (Dinophyceae) on survival and grazing activity of several shellfish species. Harmful Algae 5, 442–458.

Silbergeld, E.K., Grattan, L., Oldach, D., Morris, J.G., 2000. *Pfiesteria*: Harmful algal blooms as indicators of human: Ecosystem interactions. Envir. Res. Section A 82, 97–105.

Sindermann, C.J., 1990. Principal Diseases of Marine Fish and Shellfish. 2nd ed., vol. 1. New York, Academic Press.

Sindermann, C.J., 1998. External ulcers of fish: Some general considerations. In Jordan, S.J., Sindermann, C.J., Rosenfield, A., and May, E.B. (eds.), Causes and Effects of Ulcerative Lesions in Fish: Proceedings of a Workshop, Easton, MA.

Smith, C.G., Music, S.I., 1998. *Pfiesteria* in North Carolina: The medical inquiry continues. North Carolina Med. J. 59, 216–220.

Springer, J.J., Shumway, S.E., Burkholder, J.M., Glasgow, H.B., 2002. Interactions between the toxic estuarine dinoflagellate *Pfiesteria piscicida* and two species of bivalve molluscs. Mar. Ecol. Prog. Ser. 245, 1–10.

Steidinger, K.A., Burkholder, J.M., Glasgow, H.B., Jr., Hobbs, C.W., Garrett, J.K., Truby, E.W., Noga, E.J., Smith, S.A., 1996. *Pfiesteria piscicida* gen. et sp. nov. (Pfiesteriaceae fam. nov.), a new toxic dinoflagellate with a complex life cycle and behavior. J. Phycol. 32, 157–64.

Steidinger, K.A., Landsberg, J.H., Mason, P.L., Vogelbein, W.K., Tester, P.A., Litaker, R.W., 2006. *Cryptoperidiniopsis brodyi* gen. Et sp. Nov. (Dinophyceae), a small lightly armored dinoflagellate in the Pfiesteriaceae. J. Phycol. 42, 951–961.

Steidinger, K.A., Landsberg, J., Richardson, R.W., Truby, E., Blakesley, B., Scott, P., Tester, P., Tengs, T., Mason, P., Morton, S., Seaborn, D., Litaker, W., Reece, K., Oldach, D., Haas, L., Vasta, G., 2001. Classification and identification of *Pfiesteria* and *Pfiesteria*-like species. Environ. Health Perspect. 109, 661–665.

Stoecker, D.K., Gustafson, D.E., Jr., 2002. Predicting grazing mortality of an estuarine dinoflagellate, *Pfiesteria piscicida*. Mar. Ecol. Prog. Ser. 233, 31–38.

Stoecker, D.K., Stevens, K., Gustafson, D.E., 2000. Grazing on *Pfiesteria piscicida* by microzooplankton grazing. Aquat. Microb. Ecol. 22, 261–270.

Stow, C.A., 1999. Assessing the relationship between *Pfiesteria* and estuarine fishkills. Ecosystems 2, 237–241.

Stow, C.A., Borsuk, M.E., 2003. Enhancing causal assessment of estuarine fishkills using graphical models. Ecosystems 6, 11–19.

Swinker, M., 2003a. Correspondence: neuropsychologic testing versus visual contrast sensitivity in diagnosing PEAS. Environ. Health Perspect. 111, A13–A14.

Swinker, M., 2003b. Response to letter from Drs. Hudnell and Shoemaker. Microbes Infect. 5, 349–350.

Swinker, M., Burke, W.A., 2002. Correspondence: Visual contrast sensitivity as a diagnostic tool. Environ. Health Perspect. 109, A120–A121.

Swinker, M., Koltai, D., Wilkins, J., Hudnell, K., Hall, C., Darcey, D., Robertson, K., Schmechel, D., Stopford, W., Music S., 2001a. Estuary-associated syndrome in North Carolina: An occupational prevalence study. Environ. Health Perspect. 109, 21–26.

Swinker, M., Koltai, D., Wilkins, J., Stopford, W., 2001b. Is there an estuary associated syndrome in North Carolina? North Carolina Med. J. 62, 126–132.

Swinker, M., Tester, P., Koltai Attix, D., Schmechel, D., 2002. Human health effects of exposure to *Pfiesteria piscicida*: A review. Microbes Infect. 4, 751–762.

Tango, P., Magnien, R., Goshorn, D., Bowers, H., Michael, B., Karrh, R., Oldach. D., 2006. Associations between fish health and *Pfiesteria* spp. in Chesapeake Bay and mid-Atlantic estuaries. Harmful Algae 5, 352–362.

Tengs, T., Bowers, H.A., Glasgow, H.B., Jr., Burkholder, J.M., Oldach, D.W., 2003. Identical ribosomal DNA sequence data from *Pfiesteria piscicida* (Dinophyceae) isolates with different toxicity phenotypes. Environ. Res. 93, 88–91.

Turf, E., Ingsrisawang, L., Turf, M., Ball, J.D., Stutts, M., Taylor, J., Jenkins, S., 1999. A cohort study to determine the epidemiology of estuary-associated syndrome. Virginia J Sci 50, 299–310.

Vandersea, M.W., Litaker, R.W., Yonnish, B., Sosa, E., Landsberg, J.H., Pullinger, C., Moon-Butzin, P., Green, J., Morris, J.A., Kator, H., Noga, E.J., Tester, P.A., 2006. Molecular assays for detecting *Aphanomyces invadans* in ulcerative mycotic fish lesions. Appl. Environ. Microbiol. 72, 1551–1557.

Van Dolah, F.M., 2000. Marine algal toxins: Origins, health effects, and their increased occurrence. Environ. Health Perspect. 108 (suppl 1), 133–141.

Villalba, A., Reece, K.S., Ordás, M.C., Casas, S.M., Figueras, A., 2004. Perkinsosis in molluscs: A review. Aquat. Liv. Res. 17, 411–432.

Vogelbein, W.K., Lovko, V.J., Shields, J.D., Reece, K.S., Mason, P.L., Haas, L.W., Walker, C.C., 2002. *Pfiesteria shumwayae* kills fish by myzocytosis not exotoxin secretion. Nature. 418, 967–970.

Vogelbein, W.K., Shields, J.D., Haas, L.W., Reece, K.S., Zwerner, D.E., 2001. Skin ulcers in estuarine fishes: A comparative pathological evaluation of wild and laboratory-exposed fish. Environ. Health Perspect. 109 (suppl 5), 687–693.

Wada, S., Rha, S., Kondoh, T., Suda, H., Hatai, K., Ishi, H., 1996. Histopathological comparison between ayu and carp artificially infected with *Aphanomyces piscicida*. Fish Pathol. 31, 71–80.

Willoughby, L.G., Roberts, R.J., Chinabut, S., 1995. *Aphanomyces invaderis* sp. nov., the fungal pathogen of freshwater tropical fish affected by epizootic ulcerative syndrome. J. Fish Dis. 18, 273–275.

Zhang, H., Lin, S., 2002. Identification and quantification of *Pfiesteria piscicida* by using the mitochondrial cytochrome *b* gene. Appl. Environ. Microbiol. 68, 989–994.

Zhang, H., Lin, S., 2005. Development of a *cob*-18S rDNA real-time PCR assay for quantifying *Pfiesteria shumwayae* in the natural environment. Appl. Environ. Microbiol. 71, 7053–7063.

STUDY QUESTIONS

1. Describe the key evidence that originally led investigators to suggest that *Pfiesteria* secretes a potent toxin. Do you consider this evidence to be sufficient to support this hypothesis? Why or why not? Consider various lines of evidence derived from early *Pfiesteria* studies, human and rodent studies, as well as more recent laboratory investigations.
2. What lines of evidence from field and laboratory studies initially suggested that *Pfiesteria* may be responsible for major fish kill events in North Carolina estuaries? Describe a plausible alternative explanation to the massive fish kills attributed to *Pfiesteria* during the 1990s.
3. During the early 1990s, scientists investigating the causes of ulcerative mycosis in menhaden fish from North Carolina estuaries isolated a species of oomycete commonly found within the skin ulcers and, therefore, thought to be the causative agent of the disease. They designated this isolate (ATCC 62427) as the species associated with the disease. Describe how subsequent experimental studies with this isolate led researchers to consider *Pfiesteria* as a possible contributory factor in the development of the ulcerous lesions in menhaden. How has our understanding of the etiology of these lesions evolved?
4. Compare and contrast the lesions found in wild menhaden and the lesions induced in the laboratory by *Pfiesteria.* What is the predominant mechanism whereby *Pfiesteria* causes pathology and death in laboratory-exposed fishes? Is there sufficient evidence to suggest that *Pfiesteria* causes the ulcerous lesions seen in wild menhaden? Explain.
5. Compare and contrast the proposed complicated *Pfiesteria* life cycle, which involved numerous stages to the more recently proposed simple, typical dinoflagellate life cycle. Explain how these two very different ideas could have developed. Consider how methodology and interpretation of observations may have influenced the competing ideas.
6. Fish bioassays have been important in identifying how *Pfiesteria* kills fish. Describe the different fish bioassays that have been used, and discuss their similarities and differences. How has the interpretation of results from these bioassays contributed to the perception of toxicity of *Pfiesteria*?
7. Consider the purported effects of *Pfiesteria* on human health. Is there sufficient evidence for considering *Pfiesteria* a valid risk to human health? Why or why not?
8. Given the available evidence, do you think it is feasible that *Pfiesteria* has any significant health effects on wild fish and other estuarine organisms? Support your answer with what is currently known about the biology, behavior, and ecology of this dinoflagellate.

CASE STUDY

16

Media Coverage of Environmental Health Issues: Where Morality, Science, and the News Reflect and Depend on Fundamental Philosophical Perspectives

RUBEN RABINSKY

INTRODUCTION

Most of us learn about recent notable environmental issues from the news, and we gain a sense and understanding of the significance of these environmental issues through the various forms of coverage that are publicly disseminated by news organizations worldwide. For specialized knowledge and information, we turn to academic journals and technical or science-based, online Web sites, and programs. For general information, the lay (nonscientific) public will most likely turn to national and international news coverage on major networks and news organizations (the Associated Press, the BBC, etc.) and prominent newspapers such as the *New York Times*. The coverage of environmental issues is (or potentially can be) of great sociopolitical significance, and philosophically interesting in its own right, because we all share a common planet and our survival depends largely on the preservation and well-being of our planet as a whole. Moreover, the funds and resources devoted to environmental preservation and conservation may depend, in large part, on how well informed our citizens are with regard to a variety of environmental issues, including "global warming," levels of pollution throughout the world, and local, potential environmental risks in specific communities throughout the United States and in other nations.

The ability to make informed and rational decisions with regard to the allocation of funds and resources for the preservation of our environment depends, therefore, on the ability of individuals and groups to have relevant, timely, and objective information about the environment. Traditionally, journalists have been expected to provide such objective, accurate, veridical information through their news reports and coverage of newsworthy events. However, because media organizations are largely private enterprises within a profit-driven industry (like any other business), the news reports that journalists produce and disseminate to the public may be (and often are) sensationalized in order to increase the news organizations ratings and overall profitability. Although most of us can readily identify such sensationalized news reports when they are presented in the context of tabloid-style newspapers or in the context of media reports about the private lives of celebrities and public figures, the task of sorting the facts and identifying science from science fiction becomes more difficult when the subject of the distortion and misrepresentation of information is news about the environment itself.

In this case study, we will focus on ways of identifying sensationalized or popularized news reports about the environment, and we will consider how to apply a variety of philosophical principles and concepts (in ethics and epistemology) to gain an appreciation of the nature of the creation and dissemination of objective news and science reports on the environment.

THE CASE OF *PFIESTERIA*: THE "CELL FROM HELL"?

In 1997, a newly discovered and little-known microorganism—a dinoflagellate that was ultimately named "*Pfiesteria piscicida*" became the subject of national media attention as journalists (who began to refer to the organism as "the Cell from Hell" in the mainstream media) began to speculate on the organism's alleged toxicity and health hazards to humans, fish, and the environment. An analysis of media coverage of (what may be described as) the "*Pfiesteria* controversy" suggests that, with few exceptions, the media as a whole did not cover the scientific controversy with much objectivity (Rabinsky and Fleming, 2004).

Although a variety of scientific questions may remain unresolved or open to further research (Berry *et al.*, 2002; Fleming *et al.*, 1999; Litaker *et al.*, 2002; Vogelbein *et al.*, 2002) (e.g., questions relating to whether the organism is toxic, or whether the dinoflagellate has a 24-stage life cycle, as some scientists believe), several fundamental ethical and epistemological issues arose, which all scientists, journalists, and citizens alike should consider and be aware of in order to appreciate the need for objective news reports on the environment and the significance that such reports have for our environmental well-being.

Several factors relating to the historical evolution of the scientific research on *Pfiesteria* are relevant to consider in this case study: (1) the bulk of the research on the organism was conducted by a group of scientists who essentially controlled all of the access to specimens of the microorganism and who did not openly share their data or research materials with the rest of the scientific community that was interested in studying the microorganism; (2) the scientific community was divided by those who regarded themselves as or were regarded by the public as being the "experts on *Pfiesteria*" (that is, the researchers who controlled the specimens of the microorganism) and the "nonexperts" on *Pfiesteria* (everyone else); (3) the distinction between popularized reports on *Pfiesteria* and scientific reports was gradually blurred as mainstream media news coverage became sensationalized and nonobjective when discussing "the Cell from Hell."

Epistemological Issues: Objectivity and the Sharing of Scientific Data

Often in scientific research, the opinions of "experts" in a given field of inquiry are valued and used as hypotheses by which scientific progress can be achieved in terms of guiding research and development. Yet how can one identify an expert in a given field? What constitutes being an expert? And in a dispute between experts, how does one decide which expert opinion to believe or accept in light of their conflicting claims to expertise and knowledge? Who has the logical burden of proof to demonstrate the existence of a hypothesized scientific phenomenon: the researcher who asserts the phenomenon's existence or the researcher who is skeptical of, or denies, the existence of such phenomena?

Scientists and the public at large rely extensively on the testimony of experts to form opinions and ultimately to decide what beliefs to accept. As noted, difficulties in deciding what beliefs to accept may arise when different experts in the same or related fields of inquiry judge that different beliefs should be accepted or rejected. When various (several or many) experts disagree on whether a theory or belief is true, and when there is significant lack of consensus among the experts, then the public—and the experts themselves—may find it unusually difficult to determine which belief or theory to accept (Coady, 1992; Czubaroff, 1997). The lack of scientific consensus, even among the experts, need not preclude the possibility of the critical and analytical efforts to make epistemic progress in determining which beliefs to accept (at least tentatively, fallibly) and which beliefs to reject for lack of good evidence.

In this regard, existing case studies involving research into the nature of scientific controversy may be used to illustrate and understand the nature of the debate surrounding the *Pfiesteria* organism and how this controversy can impact the objective reporting of scientific research in general. Consider, for example, the controversy surrounding the alleged health and therapeutic effects of vitamin C. As philosopher and argumentation theorist Douglas Walton (1997) has noted, the scientific community was initially skeptical of the research conducted by the Nobel laureate biochemist Linus Pauling who defended the claim that consuming large doses of vitamin C is good for human health (especially for preventing "the common cold") and that taking large doses of other vitamins could also enhance longevity and health. In reply to Pauling's claims, various physicians and nutritionists disputed Pauling's conclusions. One physician, Dr. Richard Rivlin, professor of medicine at Cornell University, asserted that vitamins in excess can cause significant damage.

In this type of scientific dispute, in which two experts disagree, the public and other scientists must decide which belief to accept, based on the consideration of alternative expert opinions. In this case, argumentation theorist D. Walton suggests that experts with different (but significantly relevant) scientific backgrounds or areas of expertise may contribute to, and enhance, the quality of the intellectual inquiry as a whole: one should not simply dismiss as false the opinions of experts who are considered by the public or other experts to be outside of the realm of established expertise. Because knowledge is fallible and scientific hypotheses may be either corroborated or falsified (and theories may be revised, accordingly), knowledge should be construed or regarded as being in a continuous stage of evolution: knowledge—specifically, empirical/scientific knowledge—is not static and permanent [as contrasted with timeless or immutable conceptual truths, such as the basic laws of logic—for example, the law of identity (A = A) or the law of noncontradiction (– (A + –A))—which are true independently of empirical evidence].

Accordingly, a precondition for the possibility of the rational resolution to scientific debate is that the participants to the dialogue maintain a genuine effort to avoid the fallacious species of argument known as the "illegitimate appeal to authority"—that is, arguments that are designed to exclude opposing or conflicting points of view by maintaining that the opinions of those who disagree with one's own point of view are biased or erroneous simply because they do not conform to the orthodox, mainstream, or official views and theories. (For example, in the case of the dispute between Pauling and Rivlin on the nature of vitamin C, Walton maintains that Pauling's views on the properties of vitamin C cannot be dismissed simply because Pauling is not a physician, unlike Rivlin who is. Although Pauling is not a physician or nutritionist, he still has expertise in biochemistry, a relevant or related field of inquiry in the debate on the properties of vitamin C.)

Similarly, in the scientific controversy surrounding the properties of *Pfiesteria* (and the alleged risks it poses for humans, fish, and the environment), a precondition for objective scientific knowledge to emerge is that there exist the opportunity for open scientific discussion and debate with a full exchange of contrasting views, theories, and arguments on all sides of the inquiry. Appeal to expert opinion can work to either promote or hinder research, depending on the nature of the expertise and who decides or determines who the experts are (i.e., who gets to join the dialogue and who, if anyone, is excluded from the scientific dialogue).

Unfortunately, a media analysis of the case concerning the controversy surrounding research into *Pfiesteria* suggests that scientific inquiry on the microorganism has largely been controlled by a group of researchers who have claimed to possess unique expertise on the dinoflagellate (such that alternative or competing theories on the microorganism have been readily dismissed as the views of "nonexperts," without substantive consideration) (Burkholder and Glasgow, 1999; Griffith, 1999). However, in order to engage in a reasonable analysis of the competing and conflicting claims of scientists and to try to reach a justifiable conclusion, it is necessary to allow and consider all relevant and critical perspectives on the issue being debated (Rescher, 1997; Walton, 1989, 1997).

Ethical Issues: The Sharing of Scientific Data That Are of Public Concern

One of the key ethical issues arising in the case study of the scientific controversy on *Pfiesteria* is the question of whether scientists and researchers have an ethical duty or moral responsibility to share information and research materials with other scientists (who may disagree with each other's views and conclusions) and to share information and test results with the public as well. Because the alleged public health risks that may arise because *Pfiesteria* might affect humans, fish, and the environment, knowledge of such risks and of the organism itself would seem to be of vital public concern, not just abstract, theoretical, scientific interest. Interestingly, because of the relevance for public and environmental well-being of research into *Pfiesteria*, two of the most influential (and often conflicting) theories of ethics and sociopolitical philosophy—utilitarianism and libertarianism—converge in yielding similar conclusions and recommendations for public policy as presented in this case study.

Utilitarianism is an ethical and sociopolitical philosophy that maintains that standards of right and wrong are determined by the consequences of the action or situation in question: for example, an action or situation is morally good if it promotes (or enhances) the overall happiness and well-being of the greatest number of persons concerned with, or affected by, the given action or situation in question. For example, if providing free public education will increase the happiness and well-being for most citizens, then the morally right action would be for the government (or the majority of citizens) to provide for free public education. Although many people find the theory of utilitarianism to be appealing for various reasons (it is a very egalitarian and democratic philosophy), one traditional criticism of utilitarianism is that an action or situation may be deemed good or morally praiseworthy if it results in the greatest happiness for the greatest number of people, even if this results in a single person (or a few people) being ignored, inconvenienced, or harmed in the process of realizing the greatest happiness for the greatest number.

The preceding criticism of utilitarianism, that the implementation of utilitarianism may lead to the unjust violation of individual rights and minority interests, is often raised by libertarians, who espouse an alternative theory of ethics and sociopolitical philosophy. The philosophy of *libertarianism* maintains that all individuals possess equally natural human rights to life, liberty, health, property, and the pursuit of happiness and that these individual human rights may not be sacrificed or compromised in the quest of maximizing the greatest happiness for the greatest number of persons. Libertarians maintain that an action is morally permissible if it does not violate the individual rights of others, and that all humans possess these natural inalienable rights (the only legitimate purpose of government, for libertarians, is the protection of our natural human rights).

In the context of the debate on *Pfiesteria* and for the goal of promoting objective knowledge, scientific research, and a reliable media coverage of the environmental and scientific issues relating to *Pfiesteria*, *both* the utilitarian and libertarian perspectives would conclude that (1) there is a moral or ethical duty to share the scientific data and research materials among the relevant members of the scientific community that study the *Pfiesteria* organism, and (2) both scientists and journalists have a moral duty to present this information as objectively as possible to other scientists and the public at large.

From a utilitarian view, there is a moral duty for scientists to share samples, information, and materials precisely because of the alleged nature of *Pfiesteria* itself: if *Pfiesteria* does indeed pose the serious biohazards to humans and the environment that some scientists maintain, then the public's well-being will be maximized if scientific progress is enhanced through an optimization of the free-flow of information and data among all scientists (with the tools and relevant credentials) to conduct a balanced inquiry into *Pfiesteria*.

Moreover, from a libertarian perspective, there is a moral duty for scientists to share samples, data, and information on *Pfiesteria* because research into the dinoflagellate has been funded publicly through tax dollars; thus, the rightful owners of information and samples of *Pfiesteria* is the general tax-paying public who has funded the research into *Pfiesteria*, not private individuals, corporations, or academic institutions. If all scientists who participate in the *Pfiesteria* research share their samples, data, and information, the public's opportunity to learn more about *Pfiesteria*'s properties will be enhanced, and scientific inquiry stands a better chance of functioning according to the scientific process—which involves putting out theories and then trying to disprove and prove them in a public forum, where dissenting points of view have the opportunity to be heard.

CONCLUSIONS

Controversy is inherent in the scientific progress (and it is this aspect of the practice of science that the media is perhaps most apt to report), but a precondition for the reasonable resolution of scientific controversy is that all parties to the dispute have access to the same materials, data, and so on (Griffith, 1999). Of course, this precondition depends on, or presupposes, that all parties to a scientific dispute, as well as the media and the public, have a common interest and willingness to revise or examine their beliefs and conclusions in light of all the available and relevant evidence.

References

Berry, J., Reece K., Rein, K., Baden, D., Haas, L., Ribeiro, W., Shields, J., Snyder, R., Vogelbein, W., Gawley, R., 2002. Are *Pfiesteria* species toxicogenic? Evidence against production of ichthyotoxins by *Pfiesteria Shumwayae*. Proc. Natl. Acad. Sci. 99, 10970–10975.

Burkholder, J.M., Glasgow, H., Jr., 1999. Science ethics and its role in early suppression of the *Pfiesteria* issue. Hum. Organ. 58, 443–455

Coady, C.A.J., 1992. Testimony: A Philosophical Study. Oxford, Oxford University Press.

Czubaroff, J., 1997. The public dimension of scientific controversies. Argumentation 11, 51–74

Fleming, L.E., Easom J., Baden, D., Rowan, A., Levin, B., 1999. Emerging harmful algal blooms and human health: *Pfiesteria* and related organisms. Toxicol. Pathol. 27, 573–581.

Griffith, D., 1999. Exaggerating environmental health risk: The case of the toxic dinoflagellate, *Pfiesteria*. Hum. Organ. 58, 119–127.

Litaker, R.W., Vandersea, M.W., Kibler, S.R., Madden, V.J., Noga, E.J., Tester, P.A., 2002. Life cycle of the heterotrophic dinoflagellate *Pfiesteria piscicida* (Dinophyceae). J. Phycol. 38, 442–463.

Rabinsky, R., Fleming, L.E., 2004. Philosophical insights from an analysis of media coverage of the *Pfiesteria* controversy. In Steidinger, K.A., *et al.* (eds.), Harmful Algae 2002, Florida Fish and Wildlife Conservation Commission, Florida Institute of Oceanography, and Intergovernmental Oceanographic Commission of UNESCO.

Rescher, N., 1997, Objectivity: The obligations of impersonal reason. Notre Dame, IN, University of Notre Dame Press.

Vogelbein, W.K., Lovko, V.J., Shields, J.D., Reece, K.S., Mason, P.L., Haas, L.W., Walker, C.C., 2002. *Pfiesteria shumwayae* kills fish by myzocytosis not exotoxin secretion. Nature 418, 967–970.

Walton, D., 1989. Informal Logic: A Handbook for Critical Argumentation. Cambridge, England: Cambridge University Press.

Walton, D., 1997. Appeal to Expert Opinion: Arguments from Authority. Philadelphia, Pennsylvania State University Press.

STUDY QUESTIONS

1. In the case of the scientific study of the *Pfiesteria* dinoflagellate, a variety of philosophical issues are raised concerning the sharing of data on this microorganism. Ethically, there was the question of whether scientists who had obtained information on the organism had a moral or ethical duty to share the information with other scientists and research institutes in order to advance the public's knowledge, understanding, and appreciation of the actual or potential health risks that the *Pfiesteria* organism was suspected of posing to the environment and humans. What would a utilitarian suggest is the ethically right way to address the issue of sharing scientific information, if it is for the public benefit? How would a libertarian reply to the view that there is an ethical duty to share one's information and publicly funded research materials and data? Which of these philosophical perspectives is more plausible in your view? Are there other ethical approaches or solutions that may offer a more plausible way to address the question of how, if at all, to share scientific data that are or can be of public interest or on which the public's health and safety may depend?
2. From the perspective of epistemology (the branch of philosophy that focuses on the study of questions relating to knowledge and the justification of empirical, scientific, and logical claims to knowledge), there is a fundamental question of whether scientists have a professional obligation to share information about their research with other scientists, as well as the public in order to promote and enhance objective research and knowledge. How may knowledge be more objective or accurate if claims to knowledge (research, data, information) are examined from multiple perspectives and subjected to a variety of critical questions? What role does the falsification of scientific theories play in the development of scientific knowledge? Should scientists

only seek to gain positive evidence or support for their theories, or should they also attempt to disprove (falsify) their theories? Why or why not?

3. Who should have a greater ethical duty in providing objective information about the environment, scientists or journalists? Or do scientists and journalists have an equal moral responsibility to provide objective news reports about the environment and other environmental-health issues? Why? Discuss, giving reasons for your view.
4. In your view, does the popular media overall tend to sensationalize news reports about the environment and scientific research in general? Or does the mainstream media tend to provide objective, accurate, veridical news reports? Look through a variety of news magazines, journals, newspapers (both online and hard copy), as well as television programs or radio broadcasts for examples to support of your own viewpoint on this issue. Then discuss how the news reports could have been developed more objectively and whether this would have made them of greater value and why. You may also discuss the question of whether popularized or sensationalized news reports about the environment are intrinsically or inherently unethical in some way, and state the reasons that you hold your view on this issue.

SECTION I

D. Infectious Microbes in Coastal Waters

HELENA SOLO-GABRIELE

Microbes in water can cause human illness through different means (Fig. D-1). First, microbes such as harmful algae or bacteria such as *Clostridium botulinum* can cause illness through the release of toxins. These toxins are neurotoxins, causing neurological disorders and in some cases severe gastrointestinal symptoms. For example, the toxin produced by the bacteria *Clostridium botulinum*, botulin, when ingested orally blocks nerve function, which leads to respiratory and musculoskeletal paralysis. Because of its impact on the musculoskeletal system, botulin is now routinely used as a cosmetic agent through direct injection to facial muscles for wrinkle reduction. Although neurotoxins can cause serious disease when ingested orally, the microbes that cause them are considered to be noninfectious. In other words, these microbes generally do not multiply within humans, and humans do not generally serve as the reservoir; the illness caused by these toxins cannot be transferred from one person to the next. People who become ill must be exposed to the toxin produced by microbe, not to the person who is ill from the toxin.

Infectious microbes represent the second group of microbes causing illness. The subsequent four chapters within this book focus on infectious microbes that can be transmitted through coastal or oceanic waters. As opposed to toxin-producing microbes, infectious microbes multiply within human hosts. Primary reservoirs for these microbes are varied and include humans (e.g., *Shigella* sp. and norovirus), other animals (e.g., *Salmonella sp.* and *Giardia* sp.), and the general environment (*Vibrio* sp. and *Legionella* sp.). Although there is a tendency to group the microbes into one reservoir, in reality, humans can serve as the reservoir in all relevant cases as only those

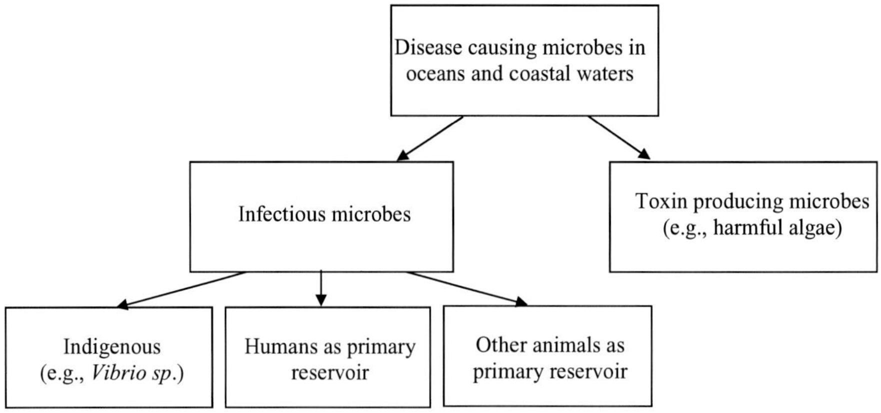

FIGURE D-1. Basic categories of microbes in oceans and coastal waters. The following chapters focus on infectious microbes that can be transmitted through water. An emphasis is placed on those where humans serve as the primary reservoir.

microbes, which can multiply within human hosts are the focus of current regulatory monitoring programs. The four chapters within this section focus on human infectious microbes that can be transmitted through water and in some cases ultimately concentrated in shellfish, as shellfish tend to filter infectious microbes from the water. Of note is that water may not serve as the only vehicle of transmission. For example, many of the gastrointestinal illnesses are transmitted through the feces of infected individuals. In addition to contaminating water, these feces can also directly contaminate food and fomites (objects capable of carrying infectious organisms), so alternative transmission routes exist for many waterborne infectious microbes. These alternative routes should also be considered when evaluating outbreaks associated with waterborne infectious microbes.

The transmission of infectious disease can be considered as a cycle (Fig. D-2). Consider the case of an infectious gastrointestinal microbe that survives in aquatic environments. An ill person becomes the reservoir for that disease and releases large quantities of the infectious agent through the feces, which then becomes part of sanitary sewage. Sanitary sewage is typically treated through wastewater treatment plants. Wastewater treatment systems are generally not effective at inactivating or removing many pathogenic microbes from the water. The water from these wastewater treatment plants is then discharged via ocean outfall and mixes with the ocean water. Under some conditions, this mixture of ocean water and treated sanitary sewage (containing pathogens) can migrate back toward bathing beaches and shellfish harvesting areas. Susceptible populations swimming at the impacted beach can then become exposed to the pathogens through swimming. Alternatively, shellfish harvested from contaminated waters can be consumed resulting in illness. The role of the regulatory community is to eliminate the transmission route by minimizing the transmission of the infectious microbes once released to the environment (e.g., improved treatment of wastewater, more effective design of an ocean outfall) and by issuing impaired water advisories. Either of these precautions, removing the source or eliminating exposures, will break the transmission route. The medical community also plays a key role in breaking the cycle of disease transmission by vaccinating the population to minimize the number

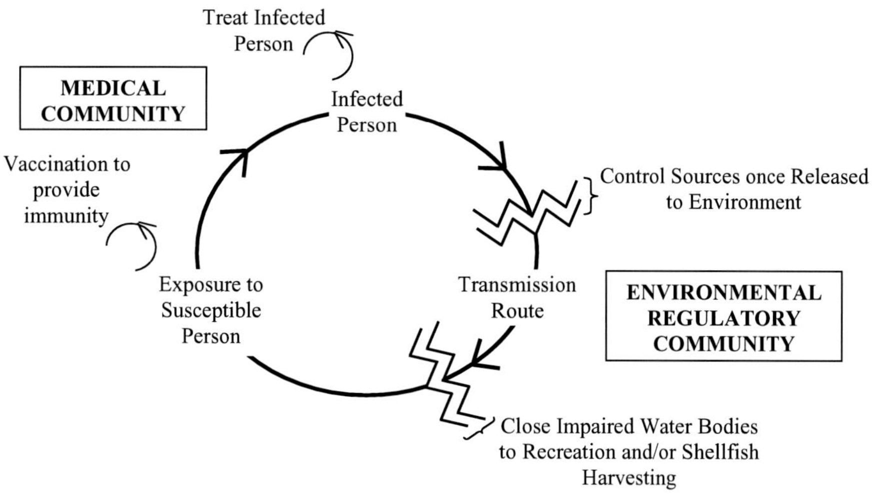

FIGURE D-2. Transmission cycle of infectious diseases and means to minimize the transmission route. The transmission cycle originates with the infected person followed by transmission and then the infection of a susceptible person. The goal of most environmental regulatory programs is to break the transmission route by eliminating the source as it is moves either through the environment or through the closure of beaches or shellfish harvesting areas that are impaired. The medical community also plays an important role in eliminating the transmission route through the treatment of individuals.

of susceptible individuals and by treating infected individuals so they recover quicker, thereby minimizing the release of infectious microbes during human illness.

The particular transmission route as described in the previous paragraph is the classical representation of a "point source" of sewage contamination and represents the basis for current recreational water quality and shellfish harvesting monitoring programs throughout the world. As a result of this traditional view of coastal water contamination, current regulatory programs utilize fecal indicator bacteria (FIB) to evaluate the potential for contamination of coastal waters. FIB are used to indicate the potential presence of sewage or pathogens in the water. In the traditional view of coastal water contamination, which considers only point sources of sewage contamination, the FIB concept works very well, as FIB are found in large numbers within the human gut, whether or not a particular human is ill. Thus, FIB are found in sewage in large numbers (contributed by all humans) regardless of the transient illnesses that may affect the population. Pathogens will only be present in sewage if the humans contributing to the sewage are ill. As many illnesses can affect a human population at any given time, the types of pathogens found in sewage will vary. Furthermore, the number of pathogens in sewage will typically be much lower than the number of FIB as only a subset of the population will be infected at any given time. Also of note is that the indicator utilized is a bacterium. Many bacteria, including FIB, can be easily cultured in the laboratory and, if present in large numbers, require processing relatively small volumes of water for analysis. Many different classes of microbes (e.g., viruses, bacteria, and protozoans) can be pathogenic. Viruses, because they require a living cell in order to replicate, are more difficult to analyze in the laboratory using traditional culture methods. Protozoans transmitted through water are typically in an encysted dormant form, which results in difficulty culturing this class of microbe outside of the human host. The difficulty in measuring viruses and protozoans along with their low numbers in the environment has resulted in the use of the FIB concept for monitoring the safety of recreational waters and shellfish harvesting areas. The use of FIB is effective for tracking sanitary sewage impacts for areas impacted by point sources of sanitary sewage contamination as the number of FIB in impacted waters should be high, and laboratory measurements are relatively simple.

The use of only FIB measurements to establish the safety of recreational waters and shellfish harvesting areas, however, has come under scrutiny. There are many situations during which FIB can provide false negatives (indicates water is safe when actually the water is impaired) and also false positives. FIB have been evaluated for only one specific transmission route of illness: gastrointestinal disease through the ingestion of sewage contaminated waters. As a result of this limitation, there are many situations in which FIB can provide false negative results, as FIB are used to in most cases evaluate the safety of waters regardless of transmission route and type of illness. In addition to gastrointestinal disease, water can also serve as a means to transfer skin ailments, eye infections, and respiratory disease. FIB were established based on a relationship between FIB levels in water and gastrointestinal disease only. Other types of disease are not addressed through measurements of FIB. In addition, indigenous microbes such as *Vibrio* sp. are natural inhabitants of marine systems and under certain environmental conditions may be capable of multiplying in the environment to high levels, which upon exposure to susceptible populations can cause disease. The use of FIB does not address the indigenous infectious microbes (even those associated with gastrointestinal transmission routes) that can be present in the environment as the numbers of FIB would not necessarily relate to the numbers of indigenous infectious microbes. Thus, a negative FIB result does not ensure that the water is safe from a microbiological point of view. It simply states that the likelihood of contracting a gastrointestinal disease from a point source of sewage is low and as a consequence there are situations in which FIB may provide false negatives.

In addition to false negatives, FIB are also prone to false positive results (indicating that the water is impaired when actually the water is safe). Epidemiological studies used to establish regulatory guidelines have been conducted in areas impacted by a point source of sanitary sewage and in nontropical environments. Under these conditions, measurements of FIB have helped communities control and eliminate point sources of contamination through the improved design of wastewater treatment plants and ocean outfalls. Even after eliminating point sources of sewage, communities may find that coastal water bodies continue to fail regulatory guidelines for FIB. For these communities without point sources of sewage, the source of FIB is now shifted toward nonpoint sources. Nonpoint sources of FIB have not been evaluated in the epidemiological studies that were used to establish current regulatory guidelines. Nonpoint sources include animal and environmental sources. The feces from animals are generally considered to be much less infectious to humans as humans do not share as many diseases with other animals. Furthermore, research has shown that FIB multiply in the environment, whereas pathogenic viruses and protozoans are not considered to replicate under environmental conditions and there is a question about whether or not bacterial pathogens are capable of replication outside of the human host. The regrowth of FIB is especially apparent within subtropical and tropical environments because of the warmer wetter climates. Thus, for nonpoint sources there is no clear link between FIB and pathogens. As a consequence, questions have been raised among the research and regulatory communities about the meaning of FIB levels in areas impacted by nonpoint sources of FIB. Should the same regulatory guidelines be used in areas impacted by point sources versus nonpoint sources? To answer this question, epidemiological studies relating public health to microbial water quality are needed in areas characterized by nonpoint sources of pollution, with a special emphasis on areas within the subtropics and tropics as these areas are the most heavily used for recreational purposes and are also most prone to growth of FIB in the environment.

The following four chapters focus on describing waterborne disease and infectious agents that can be transmitted through recreational waters and waters used for shellfish harvesting. A theme that resounds throughout the four chapters within this section is that improvements are needed in current monitoring strategies, in particular within areas impacted by nonpoint sources of FIB. Measurement tools are especially emphasized, and all four chapters address the need to improve detection methods and management strategies for assuring that the public is properly informed. Chapter 17 by Santo Domingo and Hansel provides a comprehensive history with respect to the development of the FIB concept for monitoring water quality. The authors of this chapter describe regulations focusing on general recreational water quality and shellfish sanitation. The chapter focuses on microbes relevant to fecal pollution of water and describes ecological factors that impact survival of fecal pathogens in the environment. A relatively new method (microbial source tracking) is also described in the chapter as a potential tool for identifying human versus nonhuman sources of FIB. Chapter 18 by Girones *et al.* focuses on foodborne infectious diseases, in particular diseases that can be transmitted from shellfish grown in contaminated waters. This chapter describes bacterial, parasitic (protozoan and helminths), and viral contamination and infections from seafood. The chapter provides a comprehensive summary of viral infections and methods to detect, control, and manage potentially contaminated shellfish. The chapter closes with a discussion of consumer safety programs and standards for maintaining seafood quality in the United States and the European Union. Goodwin and Litaker in Chapter 19 describe the state-of-the art with respect to new emerging technologies for monitoring recreational waters for bacteria and viruses. In this chapter, the authors describe common principles of emerging technologies and provide details with respect to new detection methods that are based on antibody, enzyme, or nucleic acid assays. The advent of polymerase chain reaction (PCR), which increases the number of copies of target genetic material, has circumvented the need to culture infectious microbes and has opened the door for the detection of viruses and protozoans that have been difficult to detect in the past using traditional culture or microscopic methods. Goodwin and Litaker describe the basis of different nucleic acid amplification methods and also describe new and innovative ways for detecting the amplified product. Chapter 20 by Palmer *et al.* expands on the technical content of Goodwin and Litaker by describing the potential future of microbial water quality monitoring. The chapter emphasizes the need for improved measurement methods and describes the potential use of the current emerging technologies for managing beach water quality. The chapter also discusses the use of models for forecasting closures and describes the potential impacts of new technologies within the regulatory community. The chapter by Palmer *et al.* closes with a discussion about the public health benefits of improved coastal monitoring and a discussion of the importance of improving coastal water management. In summary, the chapters in this section describe new tools and techniques that may lead to better decisions concerning water quality advisories in coastal areas. Consideration should be given to utilizing these new tools to supplement results obtained from routine monitoring of FIB.

C H A P T E R

17

Waterborne Diseases and Microbial Quality Monitoring for Recreational Water Bodies Using Regulatory Methods

JORGE W. SANTO DOMINGO AND JOEL HANSEL

INTRODUCTION

Coastal zones harbor some of the most important ecosystems on Earth judging by the levels of productivity and biological diversity associated with them. The economic impact of coastal zones is also remarkable, not only because of tourism and shellfishing, but also because of the high percentage of the global population living and working close to marine/oceanic waters. For example, in 1995 it was estimated that 39% of the world's population (2.2 billion people) lived within 100 km (60 miles) of an oceanic coast (Burke *et al.*, 2001). Preserving coastal areas for future generations is therefore one of our most important goals. Unfortunately, coastal zones are undergoing tremendous ecological pressure because of the ever-increasing number of people moving near to coastal areas, a rate that has been estimated at about 30 million people a year worldwide. Anthropogenic impacts are particularly stressful as they tend to concentrate in accessible coastal areas. For example, there are about 620,000 kilometers (372,000 miles) of coastline on Earth, of which only 40% could be considered habitable. As humans continue to relocate closer to the ocean, the number of temporary beach closures, the areas closed to shellfishing, and the degradation of marine habitats are likely to increase.

Fecal pollution is an important stressor to coastal zones as it has a direct impact on human health (Henrickson *et al.*, 2001). Microbial waterborne diseases associated with exposure to polluted marine recreational waters are creeping up to levels that are becoming noticeable to epidemiologists. Unfortunately, monitoring systems for tracking waterborne diseases for marine recreational beaches are lacking. Furthermore, beach-related outbreaks are difficult to recognize because of the low numbers of people that seek medical help as a result of the self-limiting nature of many illnesses (e.g., mild gastroenteritis lasting less than 48 hours). The fact that enteric microorganisms tend to have lower survival rates in marine waters also encourages a perception of lesser urgency and importance. However, the number of beaches impacted by fecal pollution is increasing, which is a good reason to presume that the number of waterborne cases that go undetected is also increasing. The impact on human health is only part of the picture as fecal pollution can also have a negative effect on marine life (Austin and Austin, 1999; Lipp and Griffin, 2004) and ecosystem stability (Kennish, 1997; Koop *et al.*, 2001).

This chapter provides a historical perspective of microbial water quality and monitoring of recreational waters, with special attention to marine environments. It reviews the regulations that are currently in effect in the United States and discusses critical issues regarding microbial water quality and the ecology of fecal pollution indicators and waterborne pathogens.

HISTORICAL BACKGROUND OF FECAL POLLUTION AND WATERBORNE DISEASES

Microbial diseases like plague, pneumonia, and cholera can be traced back several millennia through religious documents like the Book of Exodus and the writings of Homer, Thucydides, and Hippocrates. In his Humoral theory, Hippocrates recognized that it was important for doctors to have clean hands before treating their patients (Table 17-1). Thus, host-to-host contact was recognized as critical to disease transmission, although air, water, and food were also believed to be important vectors. Microbial agents were not established as the specific cause of disease for another 2000 years, sometime after the independent microscopical observations made by Robert Hooke and Anton Von

Table 17-1. Milestones in water microbiology and waterborne diseases.

Circa 400 B.C.: Hippocrates develops the humoral theory and recommends boiling and straining water.

312 B.C.–226 A.D.: The 11 major Roman aqueducts are built.

1668: Francesco Redi's experiments with maggots challenge the theory of spontaneous generation.

1677: Anton Van Leeuwenhoek reports the discovery of microorganisms using a microscope.

1768: Lazzaro Spallanzani designs biogenesis experiments.

1774: Carl Wilhelm Scheele discovers chlorine in Sweden.

1849: William Budd publishes a book titled *Malignant Cholera, Its Mode of Propagation and Its Prevention.*

1850s: John Snow links the London cholera outbreak to a contaminated well on Broad Street.

1859: Louis Pasteur designs experiments disproving the theory of spontaneous generation.

1877–1882: Louis Pasteur develops the germ theory of disease.

1882: Filtration of London drinking water begins.

1890: Robert Koch refines the postulates to identify the causative microbial agent of a particular disease.

1890s: Chlorine is proven an effective disinfectant of drinking water.

1890s: Robert Koch suggests that the low incidence of cholera in Altona was the result of filtration of the water supply.

1896: The Louisville Water Company used coagulation and rapid-sand filtration to remove bacteria from water.

1899: The Refuse Act was established to control pollution discharges into navigable waters.

1902: Belgium implements the first continuous use of chlorine to make drinking water biologically "safe."

1905: Chlorine was added to London's water supply.

1908: First public water supply in Jersey City (NJ) begins a chlorination disinfection program.

1908: Chick-Watson law of microbial disinfection is established.

1912: U.S. Congress passes the Public Health Service Act.

1914: First standards under the Public Health Service Act are established.

1948: U.S. Water Pollution Control Act was passed.

1955: Hepatitis epidemic in New Delhi, India, is caused by inadequate water treatment (1 million people are infected).

1956: Federal Water Pollution Control Act is passed.

1962: U.S. Public Health Service Drinking Water Standards Revision is accepted as minimum standards for all public water suppliers.

1965: Reported cases of polio in the United States decreased from 20,000 in 1955 to 100 because of immunization.

1965: U.S. Water Quality Act is passed.

1969: U.S. Public Health Service Community Water Supply study reveals major deficiencies in the nation's public water supplies.

1972: U.S. Clean Water Act provides for restoring and maintaining all bodies of surface water.

1974: The Safe Drinking Water Act is passed (amended several times, e.g., 1977, 1986, 1996).

1993: Milwaukee *Cryptosporidium* outbreak occurs.

2000: Beaches Environmental Assessment and Coastal Health (BEACH) Act is passed.

Leeuwenhoek. The discovery of these small "animacules" and the development of solid culture media by Hesse allowed scientists to prove that microorganisms indeed were responsible for diseases in plants and animals.

Some of the early descriptions of microbial agents came from the work of Prévost, Bassi, and Berkeley who implicated fungi as plant pathogens. Koch, who was the first to implicate bacteria as human pathogens, along with Pasteur, whose experiments once and for all disproved the theory of spontaneous generation, were primarily responsible for what is called the germ theory of disease. By the end of the 19th century, pathogens causing diseases like anthrax, cholera, malaria, diphtheria, small pox, tetanus, and plague were discovered or described, and microbiology was established as an important field in the public health arena. Simultaneously, scientists began to look for ways to combat microbial diseases and prevent infections by developing processes like pasteurization, by using antiseptics, and by searching for chemicals that killed bacterial pathogens.

Water fecal pollution became of great concern to public health officials shortly after John Snow elegantly linked cholera outbreaks in London to a sewage-impacted water well more than 150 years ago. Since then, drinking water research has been the primary driver for method development and regulations pertaining to microbial water quality. By the mid-1800s, cholera epidemics had also been documented in North America, specifically in Montreal, New York, Saint Louis, Cincinnati, and Chicago. The agent responsible for cholera, *Vibrio cholerae*, was first isolated by Filippo Pacini, although it was not until 1883, when it was again isolated by Koch, that the medical establishment accepted this bacterium as the cause of cholera. Before Snow, William Budd had implicated fecal matter and contaminated well water as the source of typhoid fever in Bristol, England, and even recommended the use of chloride to prevent the spread of diseases. Like Snow, Budd encountered significant opposition as the miasma theory of disease (i.e., bad air containing particles from decomposed matter was the cause of diseases) prevailed among scholars before 1850. It was several years after he first suspected that water was implicated in typhoid transmission that Budd's observations were finally published.

The importance of water sanitation was recognized in ancient civilizations and treatments like boiling, filtration, and exposure to sunlight were practiced in some cities. Sedimentation was also used to decrease turbidity and hence increase the clear appearance of drinking water. In fact, settling tanks were used as part of the Roman aqueducts to remove impurities via sedimentation. However, for centuries no significant developments in water treatment were recorded. In the 1700s, filtration techniques were further developed, and by the early 1800s, slow sand filtration was used with some frequency in Europe, a practice that was later adopted in the United States. These approaches were used chiefly to improve the overall taste and odor of drinking water. Evidence that water filtration was an effective process for the microbiological improvement of water was not recognized until Koch in 1892 noticed that, although the cities of Hamburg and Altonia used the same drinking water source, fewer cases of cholera were reported in the latter city. The main difference was that Altonia filtered its water. Koch suggested that careful management of the filters was important to maintain the safety as well as the quality of water and proposed that to protect public health, source water must have fewer than 100 colony-forming units (CFU) per milliliter (ml). This concept was used not only to evaluate water purity but also to assess the effectiveness of water disinfection treatments.

Other important developments in microbial water quality include the identification of bacteria present in human feces, their use as indicators of fecal pollution, and the use of chlorine as part of disinfection treatment. Enteric bacteria like *Klebsiella pneumonia* and *Escherichia coli* were first identified in the 1880s by Von Fritsch and Escherich, respectively. Because these bacteria were commonly isolated from human feces and sewage, Frankland suggested that their detection could be used to better identify dangerous pollution. Recognizing the sanitary risk of fecal pollution, the development of media for the specific detection of *E. coli* and coliform bacteria became an area of intense research, resulting in their use as indicators of fecal contamination by the beginnings of the 20th century. In 1914, the first water standards based on coliform counts were established by the U.S. Public Health Service (two coliforms per 100 ml) and later revised in 1925 (one coliform per 100 ml). The standards, however, were only enforced against systems that provided drinking water to interstate carriers like ships and trains, although many states recognized their value and adopted them as guidelines. In 1974, the U.S. Congress passed the Safe Drinking Water Act (SDWA) formally establishing enforceable regulations for public drinking water systems. Amendments made to the SDWA in 1996 include actions to protect drinking water sources. This act is relevant to marine recreational waters as freshwater quality has an important impact on the overall pollution of coastal areas.

Chlorine disinfection was suggested by Budd, but it was not routinely applied as part of drinking water treatment until Sims Woodhead in 1897 began to use a "bleach solution" in potable water distribution mains in Maidstone, England. As the spread of diseases was significantly controlled by chlorination treatment, it became common practice in Europe and later in the United States and Canada. In 1908, water utilities in New Jersey and Chicago started using chlorine to disinfect water. At about the same time, Chick (1908) and Watson (1908) developed the first models for microbial inactivation based on the concentration and contact time (Ct) of disinfectant. By the 1930s, the combination of filtration and chlorination proved to have a significant impact on reducing the incidence of waterborne outbreaks (Fig. 17-1). Ozone, first used in Europe and the United States in the early 1900s, has become a popular alternative to chlorine-based treatment since the 1980s as it reduces the production of harmful disinfection by-products.

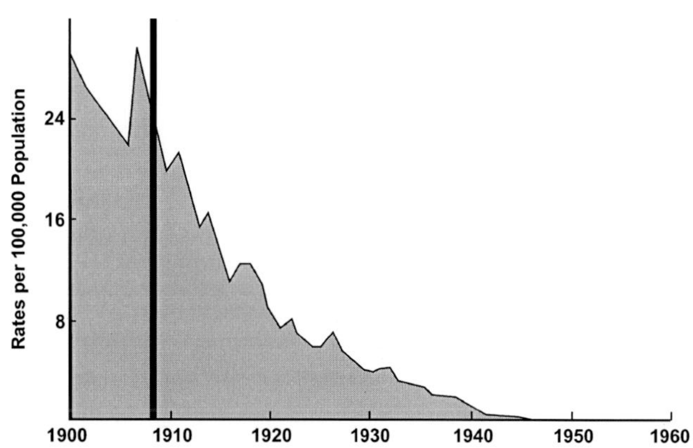

FIGURE 17-1. Number of typhoid fever cases reported in the United States in the first half of the 20th century. The bar indicates the time chlorination was introduced as a disinfection treatment (Center for Disease Control and Prevention).

HISTORICAL PERSPECTIVE ON MONITORING OF RECREATIONAL WATERS

Exposure to fecally polluted recreational waters has long been recognized as a potential hazard to human health. In ancient times, the problems were mostly associated with freshwater, but the relocation of people toward coastal areas has significantly increased the number of illnesses associated with marine waters, to the point that it is now recognized as a global crisis. For example, the World Health Organization (WHO) and the United Nations recently estimated that "250 million clinical cases of mild gastroenteritis and upper respiratory disease are caused every year by bathing in contaminated seawater" (p. 28, Shuval, 1999). Moreover, the same study reported that using the disability-adjusted life-years (DALYs) metric, "gastroenteritis and upper respiratory disease from bathing in polluted seas reduce active life by as many years as diphtheria and leprosy" (Table 17-2).

Several regulations and legislative statutes have been instituted in the United States since the late 1800s to protect against pollution of navigable waters. For example, the Refuse Act of 1899 was established to prohibit the discharge of refuse matter vessels, buildings, structures, or facilities into the nation's navigable bodies of water, or tributaries to such waters, unless a federal permit was obtained to do so. Additionally, the Water Pollution Control Act of 1948, the Federal Water Pollution Control Act of 1956, and the Water Quality Act of 1965 were instituted to develop programs of pollution control and water quality standards for interstate waters. Although by the 1970s all the states had adopted water quality standards, these statutes were not effective at preventing pollution. In fact, criteria for bathing waters were not strictly enforced for another ten years because of the lack of data showing the relationship between water quality and exposure risks. However, earlier reports from the American Public Health Association (APHA) indicated the need for investigating "the extent and prevalence of infections which may be conveyed by means of swimming pools and other bathing places" (Simons *et al.*, 1922). Surveys conducted by the APHA established the importance of bathing places in the transmission of diseases and identified eye, ear, nasal, throat, and skin infections, as well as gastrointestinal illnesses, among the potential diseases associated with bathing waters.

TABLE 17-2. Economic and health impacts associated with pollution of coastal waters.

Disease or Cause	Disability-Adjusted Life-Years (DALY)	Corresponding Economic Losses (Rounded) in U.S. Million Dollars
Disease		
Tuberculosis	38 000 000	115 000
Malaria	31 000 000	95 000
Diabetes	11 000 000	35 000
Trachea, Brachia and Lung cancer	8 800 000	26 000
Stomach cancer	7 700 000	23 000
Intestinal nematodes	5 000 000	15 000
Upper respiratory tract infections	1 300 000	4 000
Trachoma	1 000 000	3 000
Onchocherciasis	900 000	2 700
Dengue fever	750 000	2 200
Japanese encephalitis	740 000	2 200
Chagas disease	660 000	2 000
Leprosy	380 000	1 100
Diphtheria	360 000	1 100
Marine exposures		
Contaminated bathing water	400 000–800 000	1 200–2 400
Contaminated shellfish	3 500 000–7 000 000	10 000–20 000

Adapted from Shuval, 1999.

In 1924, the APHA recommended the first set of standards for bathing waters (Simons *et al.*, 1924): (1) Not more than 10% of samples covering any considerable period of time should exceed 1000 bacteria per ml or 5000 bacteria per ml for any single sample. Enumeration was performed on agar plates after a 2-day incubation at 20°C. (2) Not more than 10% of samples covering any considerable period of time should exceed 100 per ml or 200 bacteria per ml for a single sample when using agar or litmus lactose agar incubated for 24 hours at 37°C. (3) Not more than two out of five samples collected on the same day, or not more than three out of 10 consecutive samples collected on different dates, should be positive for *Bacillus coli* in 10 ml of water using lactose-bile tube or litmus lactose agar. These standards were not mandatory and moreover were specifically recommended only for swimming pools, as natural bathing waters were not considered by many to present a major public health problem even though there was ample evidence signaling the opposite.

The correlation between illness and the microbial quality of surface contaminated waters was first supported experimentally from studies conducted by the U.S. Public Health Services (Stevenson, 1953). Two freshwater sites (one in the Ohio River and one in Lake Michigan) and two marine sites (both in New York) containing different densities of coliform bacteria were selected for the study. Although swimming-associated gastroenteritis was not observed at the marine sites, the data from the latter study demonstrated that swimmers at the Ohio River site were more prone to get ill than nonswimmers when the average coliform counts exceeded 2300 per 100 ml. Other studies followed but did not conclusively show significant increases in health effects

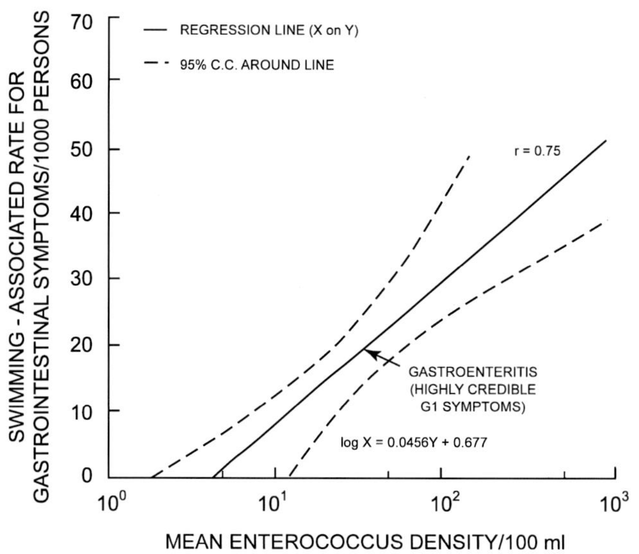

FIGURE 17-2. Correlation between enterococcus densities and the gastroenteritis of swimmers exposed to marine recreational waters.

associated with exposure to recreational waters. In the early 1970s, the Environmental Protection Agency (EPA) initiated a series of epidemiological studies that lasted nearly a decade and were used to establish guidelines for both freshwater and marine waters used for recreational activities in the United States (Fig. 17-2) (Cabelli, 1983; EPA, 1986). Overall, most epidemiological studies since then have confirmed the results obtained in the EPA studies that there is a cause-effect relationship between gastrointestinal symptoms and fecal pollution as measured by bacterial indicator levels. However, problems associated with the precision of the enumeration methods, spatial and seasonal variation of indicator bacteria, and the poor correlation between indicator bacteria and etiological agents (i.e., viruses) can lead to underestimating public health risks associated with swimming in waters that meet standard levels (Pruss, 1998).

The Federal Water Pollution Control Act was enacted in 1972 to control water pollution of surface waters and thereby maintain the chemical, physical, and biological integrity of the nation's waters. In 1977, the name was changed to the Clean Water Act (CWA), which has since been amended and revised several times. Section 402 of the CWA describes the National Pollutant Discharge Elimination System (NPDES) established to require point source dischargers to waters of the United States to obtain a discharge permit. By regulating the direct and indirect discharges of waste, the goal has been to support "the protection and propagation of fish, shellfish, and wildlife and recreation in and on the water." Section 303(d) of the CWA establishes that states have to meet total maximum daily loads (TMDL) for a number of pollutants including fecal bacteria. As defined by the EPA, TMDL is "a calculation of the maximum amount of a pollutant that a water body can receive and still meet water quality standards, and an allocation of that amount to the pollutant's sources." The calculation includes contributions from point sources (e.g., publicly owned treatment works, industrial facilities, and storm water discharges associated with industrial activity) and nonpoint sources (e.g., overland runoff from livestock pasturing farming operations and agricultural activities, leaky septic tanks).

The Marine Protection, Research and Sanctuaries Act of 1972 was a legislative action that specifically addressed the dumping of all types of materials into ocean waters by preventing or strictly limiting the dumping into ocean waters of any material that would adversely affect human health, welfare or amenities, or the marine environment, ecological systems, or economic potentialities. This practice is also no longer accepted or is tightly regulated in some countries as it has been proven to have an impact on the levels of pollution nearby recreational areas (Grimes *et al.*, 1984). However, it is not necessarily a universal practice, as implementation and enforcement of regulatory programs is too costly for some developing countries.

In spite of the emergence of regulatory programs in the United States, the CWA goal of keeping all waters below the regulatory standards (hence "fishable and swimmable") has not been fully met, as 39% of watersheds and streams are considered to be impaired by one or more of the primary pollutants of concern like nutrients, sediments, fecal bacteria, and metals (EPA, 2002). Fecal bacteria are among the most important pollutants in the United States, particularly in rivers, streams, estuaries, and ocean shorelines. Fecal bacterial levels exceeding regulatory standards impact close to 90% of the impaired ocean shorelines in the United States. Domestic sewage and agricultural runoff are the most important stressors contributing to high levels of pollution in coastal zones. Fecal pollution also introduces excessive nutrients into the environment, potentially impacting the growth and survival of indicator organisms and enteric pathogens, as well as perturbing the balance of microbial networks. Excessive nutrients are also responsible for harmful blooms of cyanobacteria containing neurotoxins.

REGULATIONS IN EFFECT IN THE UNITED STATES

Environmental regulations regarding water quality in the U.S. is primarily accomplished through the Clean Water Act (CWA) and overseen by the Environmental Protection Agency. In brief, water quality standards (WQS) consistent with the statutory goals of the CWA must be established. Water bodies are then monitored to determine whether the WQS are met. If all WQS are met, then antidegradation policies and programs are employed to keep the water quality at acceptable levels. Ambient monitoring is also needed to ensure acceptable levels. If the water body is not meeting established standards, a strategy for meeting these standards must be developed. The most common type of strategy is the development of a TMDL to determine what level of pollutant load would be consistent with meeting WQS. As mentioned previously, TMDLs allocate acceptable loads among sources of the relevant pollutants.

Necessary reductions in pollutant loading are achieved by implementing strategies authorized by the CWA, along with any other tools available from federal, state, and local governments and nongovernmental organizations. Key CWA tools include (1) the NPDES permit program, which covers point sources of pollution discharging into a surface water body; (2) section 319, which addresses nonpoint sources of pollution, such as most farming and forestry operations, largely through grants; (3) section 404, which regulates the placement of dredged or fill materials into wetlands and other waters of the United States; (4) section 401, which requires federal agencies to obtain certification from the state, territory, or Indian tribes before issuing permits that would result in increased pollutant loads to a water body; and (5) state revolving funds, which provide large amounts of money in the form of loans for municipal point sources, nonpoint sources, and other activities. After implementation of these strategies, ambient conditions are again measured and compared to ambient water quality standards. If standards are now met, only occasional monitoring is needed. If standards are still not being met, then a revised strategy is developed and implemented, followed by more ambient monitoring. This iterative process must be repeated until standards are met.

Additionally, key provisions of the CWA were modified or added to via the Beaches Environmental Assessment and Coastal Health Act (BEACH Act) of 2000. Specifically, the BEACH Act modified sections 104, 303, 502, and 518 and added section 406. Combined, these new sections gave the EPA the ability to issue grants to states, tribes, and territories to monitor (and notify the public about) the microbiological condition of the nation's coastal recreational waters. Also, the EPA was charged with assuring that state water quality standards are up-to-date with respect to bacterial criteria and to conduct research in developing new indicators and methodologies to improve notification time for the public.

Shellfish Sanitation

On June 19, 1975, the Food and Drug Administration (FDA) proposed National Shellfish Safety Program (NSSP) Regulations in the Federal Register. After evaluation of the comments received as a result of the proposed rules, the FDA determined that promulgating federal regulations would not likely achieve NSSP goals. Subsequently, the FDA decided that revision of the 1965 manual of operations was the best approach for strengthening the NSSP (Federal Register of February 26, 1985, 50 F.R. 7797). During this period, many state shellfish control agencies began questioning the uniformity and effectiveness of shellfish programs in other states. These states and FDA began exploring methods for strengthening the NSSP that would not involve federal regulations. In reviewing other approaches, it was noted that since 1950, the National Conference of Interstate Milk Shippers (NCIMS), a successful voluntary public health program, has been successful in assuring a nationwide safe and wholesome milk supply. The NCIMS was consulted for direction and advice. The success of the NCIMS program prompted state shellfish control officials and FDA to select the NCIMS program as a model for developing a shellfish organization.

In 1982, a delegation of state officials from 22 states met in Annapolis, Maryland, and formed the Interstate Shellfish Sanitation Conference (ISSC). The ISSC was formed to promote shellfish sanitation through the cooperation of state and federal control agencies, the shellfish industry, and the academic community. To achieve this purpose, the ISSC adopts uniform procedures incorporated into an Interstate Shellfish Sanitation Program and implemented by all shellfish control agencies; gives state shellfish programs current and comprehensive sanitation guidelines to regulate the harvesting, processing, and shipping of shellfish; provides a forum for shellfish control agencies, the shellfish industry, and academic community to resolve major issues concerning shellfish sanitation; and informs all interested parties of recent developments in shellfish sanitation and other major issues of concern through the use of news media, publications, regional and national meetings, the Internet, and by working closely with academic institutions and trade associations. One of the foremost goals of the ISSC has been the adoption of a model ordinance that would embody the principles and requirements of the ISSP. Adoption of the model ordinance by each of the ISSC participating states implies commitment by each state to provide the necessary legal authority and resources to implement these regulatory requirements. Adoption also ensures uniformity across state boundaries and enhances public confidence in shellfish product.

Other Countries

The WHO has published guidelines for the classification of recreational waters based on enterococci counts per 100 ml (≤40, 41 to 200, 201 to 500, >500) and the potential presence of human sources impacting the waters in question (very low, low, moderate, high, and very high) (Table 17-3; WHO, 2003). These guidelines also recommend monitoring schedules depending on risk categories based on sanitary inspections performed annually, as well as management practices to prevent pollution. Sanitary inspections should include information on the presence of sewage outfalls,

Table 17-3. WHO classification matrix for fecal pollution of recreational waters.

		Microbial Water Quality Assessment Category (95th Percentile Enterococci/100ml)			
		A ≤40	B 41–200	C 201–500	D >500
Sanitary Inspection Category (Susceptibility to Fecal Influence)	**Very low**	Very good	Very good	Follow up	Follow up
	Low	Very good	Good	Fair	Follow up
	Moderate	Good	Good	Fair	Poor
	High	Good	Fair	Poor	Very poor
	Very high	Follow up	Fair	Poor	Very poor

combined sewer overflows, storm water discharges, riverine discharges, bather shedding, rainfall (duration and quantity), wind (speed and direction), tides and currents or water release (e.g., dam-controlled rivers), and coastal physiography. These guidelines are based on the "Annapolis protocol," which was developed by a panel of experts in 1998 (p. 58, WHO, 1999). The protocol's intent is to "move away from the reliance on numerical values of fecal index bacteria as the sole compliance criterion to the use of a two component qualitative ranking of fecal loading in recreational water environments, supported by direct measurement of appropriate fecal indices."

The European Union (EU) recommends that member countries adopt a series of regulations for the protection of both shellfish and recreational uses. For shellfish uses, the EU program is based on the concentration of fecal coliforms in actual shellfish tissue and not in the surrounding water. It should be noted that if any foreign nation wishes to import raw molluscan shellfish in the U.S. market, a memorandum of understanding must be entered into between the foreign government and the FDA showing that the foreign program complies with the NSSP. For recreational water quality, in 2008, member states of the EU are expected to comply with a program similar to the U.S. BEACH program in that enterococci and *E. coli* will be the determining organisms for compliance.

In the south Pacific, countries like New Zealand and Australia currently use a system similar to the WHO guidelines to determine the suitability for recreation of both freshwaters and marine waters (National Health and Medical Research Council [NHMRC], 2005; New Zealand Ministry for the Environment [NZME], 2003). Grading of waters is achieved by determining the microbiological assessment and sanitary inspection categories for each particular body of water. Enterococci are the preferred microbiological indicator for recreational waters in these countries. In contrast, many Latin American countries have continued to favor the use of fecal coliforms as the primary microbial indicator.

FECAL POLLUTION OF RECREATIONAL WATERS AND PUBLIC HEALTH

Fecal pollution is monitored by enumerating the levels of fecal indicator bacteria. An indicator organism should satisfy the following criteria: it must be present in waters whenever the pathogens of concern are present; it must be present only when the presence of pathogenic organisms is an imminent danger; it should occur in much greater numbers than the pathogens; it must be more resistant to disinfectants and to the aqueous environment than the pathogens; it must grow readily on relatively simple media; it must yield characteristic and simple reactions enabling an unambiguous identification of the group species; it should preferably be randomly distributed in the sample to the tested, or it should be possible to obtain a uniform distribution by simple homogenization procedures; and its growth in artificial media must be largely independent of any other organism present (Bonde, 1977). Additionally, indicators should be absent in clean waters, should not be able to proliferate to any greater extent than pathogens in aquatic environment, and their densities should bear some relation to the degree or extent of pollution.

In the United States, enterococci are recommended to determine the levels of pollution in marine recreational waters, whereas enterococci and *E. coli* are indicators for freshwater systems. Male-specific coliphage *C. perfringens*, and *B. fragilis* have been suggested as alternate indicators. The presumed primary ecosystem of bacteria indicators is the mammalian gut, although some indicators have been isolated from the gut of invertebrates, fish, and reptiles. Nongut habitats have also been identified for some indicator bacteria, for example, phyllosphere, algal mats, beach sand, soil, and pristine waters. Overall, coliforms and enterococci might benefit the host by occupying niches of potential pathogens. Water is considered to be a temporary habitat for indicator organisms and most pathogens, particularly within oligotrophic waters. However, after the host excretes feces, fecal microorganisms could spend a significant amount of time outside of the gut before being ingested by another host. Common bacterial indicators do not cause disease, but their presence in high numbers has been used for decades to indicate the potential presence of harmful human microbial pathogens.

Recreational water quality standards are used to protect bathers and swimmers from potential risks associated with polluted waters. Regulations are based on type of use, for example, primary contact versus secondary contact, although there are also aesthetically based regulations. The risks could be numerous, including skin, eye, ear, and upper respiratory infections, as well as gastrointestinal related illnesses (Cheung *et al.*, 1990; Fleisher *et al.*, 1996). The exposure risks associated with these diseases could vary significantly. For example, minimal exposure might be required for skin infections, whereas significant contact time is necessary for gastroenteritis as ingestion of a significant amount of water is a prerequisite. Some human groups are at higher risk when exposed to waterborne pathogens, such as the elderly, infants younger than 3 months old, immunosuppressed and immunocompromised people. Factors like smoking, history of asthma, allergies, and other chronic illnesses (congestive heart failure, sickle-cell disease) might also contribute to a different rate of exposure effects. Recreational activities like fishing, swimming, surfing, yachting, boating, and shellfish harvesting are all associated with coastal zones, and each has a different level of risk exposure for humans. Different pathogens can be associated with similar symptoms, like gastroenteritis, fever, abdominal discomfort, bloating, cramps, nausea, vomiting, and muscle ache, although certain symptoms are primarily characteristic of some diseases.

Marine waterborne outbreaks are not as numerous as those reported for freshwaters and swimming pools, although they are believed to be much higher than those reported by the Center for Disease Control and Prevention because of the fact that the current monitoring system consists of a voluntary surveillance program (Centers for Disease Control and Prevention [CDC], 2002: Dziuban *et al.*, 2006). Most cases do not require a visit to the doctor or to a medical clinic unless the symptoms last longer than 48 hours and include dehydrating diarrhea, extreme vomiting, and persistent fever. Although endemic levels of waterborne diseases are not known for marine recreational waters, it is reasonable to assume that they are also higher than the number reported. A relatively small number of different fecal pathogens have been reported to be responsible for waterborne outbreaks, primarily because of the limited number of cases in which isolation of the culprit is attempted. In fact, isolation and identification are performed only for extreme cases. Moreover, there are severe limitations associated with the current culture-based approaches used to isolate and identify pathogens, and consequently the number of cases of unknown etiology or that are misdiagnosed is high. The fact that it is necessary to concentrate pathogens from large volumes of water (as they are usually present in very low numbers) is also a significant problem. However, the fact that foodborne outbreaks associated with shellfish and other marine foods is relatively high is evidence that oceans are impacted with fecal pathogens and that fecal pathogens could be implicated in waterborne disease. The fact that some fecal organisms can survive long enough to contaminate marine and estuarine food products suggests that pollution prevention of coastal areas continues to be an important challenge in developed countries. Most waterborne pathogens are not indigenous to aquatic environments but rather transitory inhabitants. On the other hand, *Vibrio*, *Aeromonas*, and *Mycobacterium* are ubiquitous to marine and estuarine waters; however, their levels tend to be relatively low, unless factors like nutrients and temperature stimulate their growth. For example, the environmental levels of *Vibrio* spp. are known to show seasonal trends and have shown linkages to climatological phenomena.

The modes of transmission of marine waterborne pathogens include direct contact, aerosols, ingestion, food handling (in the fishing boats), and contaminated food consumption. With the exception of food-related cases, the overall frequency and importance of mode in the transmission of disease is not known because of the lack of a nationally reportable surveillance program. People that perform recreational activities like surfing, diving, or long-distance swimming are at higher risks because of the prolonged exposure to water and to the amount of water ingested. Children playing at the swash zone (i.e., part of foreshore washed by waves) are also at higher risk because of their underdeveloped immune system and because of the higher densities of indicator bacteria associated with this zone (Koop and Griffiths, 1982; Shibata *et al.*, 2004). Beaches near urban areas and wastewater treatment plants are more likely to be impacted. In countries like the United States, New Zealand, and Australia, recreational areas are intensely used because a significant percentage of the coastal zone is inaccessible increasing exposure risk.

ENUMERATION OF INDICATOR BACTERIA IN RECREATIONAL WATERS

Microbial water quality of recreational waters has historically been measured using fecal bacterial indicators. The EPA 1986 Ambient Water Quality Criteria for Bacteria was established to protect people from gastrointestinal illness in recreational waters. Different microbial water quality criteria can be used to determine the level of pollution of beaches, and this can vary depending on the type of recreational water

TABLE 17-4. Criteria for indicator bacterial densities used for recreational waters.

	Acceptable Swimming Associated Gastroenteritis Rate per 1000 Swimmers	Steady State Geometric Mean Indicator Density	Single Sample Maximum Allowable Density: Designated Beach Area (Upper 75% C.L.)	Single Sample Maximum Allowable Density: Moderate Full Body Contact Recreation (Upper 82% C.L.)	Single Sample Maximum Allowable Density: Lightly Used Full Body Contact Recreation (Upper 90% C.L.)	Single Sample Maximum Allowable Density: Infrequently Used Full Body Contact Recreation (Upper 95% C.L.)
Freshwater enterococci	8	33	61	78	107	151
E. coli	8	126	235	298	409	575
Marine Water enterococci	19	35	104	158	276	501

(i.e., marine water versus freshwater) or on the primary use of the water body (designated bathing beach to infrequent use for bathing). Current standards recommended by the EPA for recreational water are based on enterococci and *E. coli* densities (Table 17-4). Specifically, EPA recommends both a geometric mean and a single sample maximum value, based on the intensity of primary contact use, for each indicator. *E. coli* is recommended only for use in freshwater areas and has a recommended geometric mean value of 126 CFU/100 ml and a range of single sample maxima of 235 to 575 CFU/100 ml based on an acceptable risk of 8 additional gastrointestinal illnesses per 1000 swimmers. Enterococci are recommended for both freshwater and marine water. In freshwater, the recommended values are a geometric mean of 33 CFU/100 ml and a range of single sample maxima of 61 to 151 CFU/100 ml, again based on an acceptable risk of 8 additional illnesses per 1000 swimmers. For marine waters, a geometric mean of 35 CFU/100 ml and a range of single sample maxima of 104 to 501 CFU/100 ml is recommended for enterococci. However, the acceptable risk is an additional 19 gastrointestinal illnesses per 1000 swimmers. This is in part because of the presumed risk that was associated with areas that met the previous fecal coliform recommendation was different for fresh and marine waters.

Local agencies might recommend posting advisory signs depending on levels measured from a single sample or after the levels exceed the recommended geometric mean as calculated by the following formula:

$$G = \sqrt[n]{(X_1)(X_2)(X_3)\ldots(X_n)}$$

where X is the densities calculated for a particular sample, *n* is the number of samples taken for a given period of time, and G is the calculated geometric mean. The results from five samples taken over a 30-day period are used to calculate G for recreational waters.

Although there are federal guidelines on issues like sampling and advisory or closure decision rules, the actual procedures may vary even between neighboring states (EPA, 2003). States often use more stringent values than those recommended by the federal government to protect beach goers (e.g., Hawaii uses seven enterococci CFU/100 ml of water). The use of multiple indicator bacterial groups has also been suggested to confirm the degree of fecal pollution (Shibata *et al.*, 2004). To prevent swimmers from getting ill, public health officials can post Beach Closure signs when there is a known sewage spill or when the bacterial indicator levels exceed the regulatory standards (Fig. 17-3). A Beach Warning sign is often posted when a bacterial standard has been exceeded, without a known source of human sewage. Additionally, Rain Advisory signs can be issued when it rains because it is known from past experience that rainwater carries pollution to the beach. Prohibiting people from swimming in water exceeding regulatory standards, however, is rarely enforced by local authorities.

There are two basic approaches for indicator bacterial enumeration: the membrane filtration technique and the most probable number technique. The following paragraphs provide a general description of the microbial enumeration methods that are accepted for regulatory activities. The reader is encouraged to consult with *Standard Methods for the Examination of Water and Wastewater* for additional details on standard operating procedures (American Public Health Association [APHA], 2005) as well as publications by the WHO on microbial monitoring of recreational waters (Bartram and Rees, 2000; WHO, 2003)

Membrane Filtration (MF)

For membrane filtration (MF) counts, water is filtered onto cellulose acetate membranes, which are then transferred onto the medium of choice. The targeted bacterial indicators are total coliforms, fecal coliforms, *E. coli*, and enterococci. Densities should be estimated from dilutions producing 20 to 80 colonies. It is recommended that different water volumes be used, normally between 1 and 100 ml.

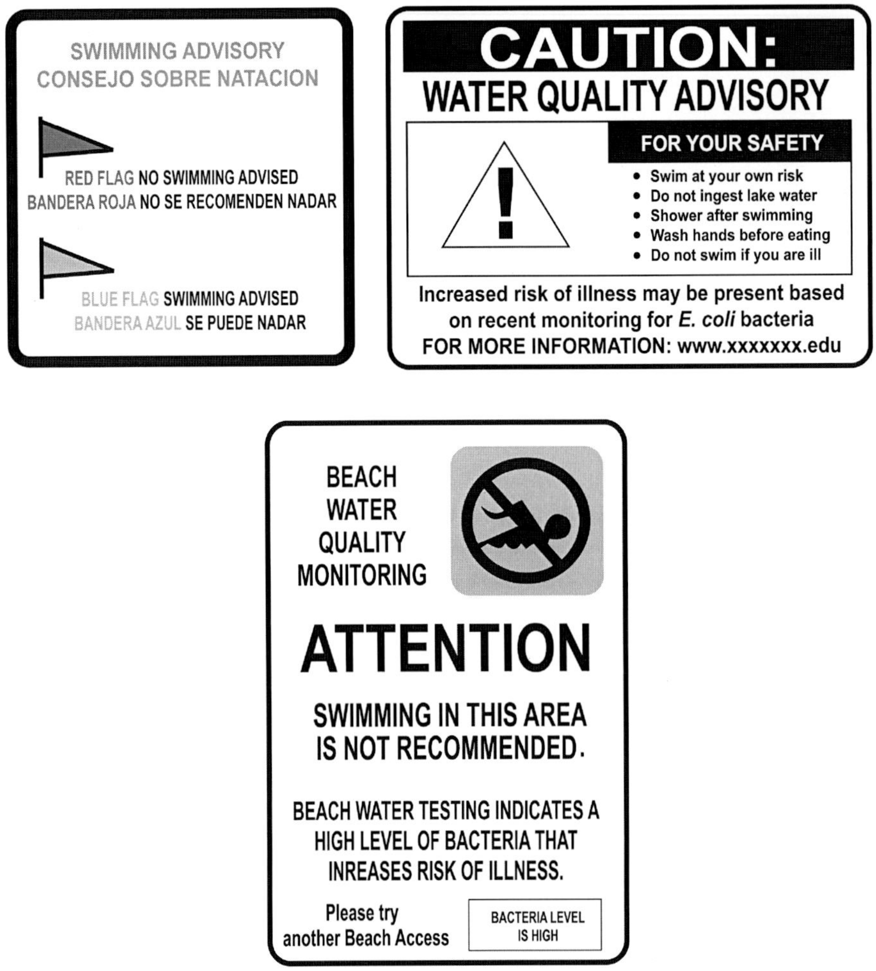

FIGURE 17-3. Examples of beach advisory signs.

Samples can be diluted using phosphate buffer to minimize the interference of turbidity or for high bacterial densities. If necessary, results from multiple volumes to estimate densities could be combined.

Total coliforms are enumerated using the MF technique and mEndo agar as it selects against Gram-positive cocci and endospore-forming bacteria. The plates are incubated at 35°C for 24 hours. Green-sheen colonies are considered members of the total coliform group. Fecal coliforms are enumerated using mFC agar. This is a one-step method that uses rosolic acid and bile salts to inhibit growth of nonfecal coliforms. Aniline blue is added to the medium to differentiate between lactose fermenting bacteria (i.e., acid production of fecal coliforms will turn colonies blue) from other bacterial colonies. Agar plates are incubated at 44.5°C for 22 to 24 hours.

There are several ways *E. coli* densities could be determined using MF techniques. For example, *E. coli* colonies could be counted from mEndo or mFC membranes by using an additional incubation on NA-MUG (nutrient agar with 4-methylumbelliferyl-β-D-glucuronide). Hydrolysis of MUG by *E. coli* cells can be determined using a fluorescence detector. *E. coli* present in freshwater, estuarine, and marine waters can also be enumerated using mTEC medium (EPA method 1103.1) and modified mTEC Medium (EPA method 1603) (APHA, 2005). Once the samples are filtered, the membranes are placed onto the mTEC agar and incubated for 2 hours at 35°C. Culture plates are then transferred to a Whirl-Pack bag and incubated for an additional 24 hours at 44.5°C in a water bath. Although the medium is fairly selective, the membranes need to be placed onto a pad wet with a urea-based solution at room temperature for 20 minutes to discriminate between *E. coli* and non-*E. coli* colonies. *E. coli* densities are estimated by counting yellow, yellow-green, or yellow-brown colonies. Method 1603 uses most of the media components and steps of EPA method 1103.1 with the exception that the urea incubation step is avoided by adding a chromogen (5-bromo-6-chloro-3-indolyl-β-D-glucuronide) to the basic mTEC medium. This substrate is catabolized to glucuronic acid and a red- or magenta-colored metabolite by *E. coli* cells containing the enzyme -D-glucuronidase.

In the United States, enterococci densities can be determined using either of two membrane filtration techniques: EPA method 1106.1 (mE medium) or EPA method 1600 (mEI medium) (APHA, 2005). In other countries, the use of mEnterococcus medium is often used instead of the mE or mEI. Originally developed by Levin *et al.* (1975), mE medium contains sodium azide and cycloheximide to inhibit other microorganisms. Water is processed like other MF methods. After filtration, the membrane is transferred to the mE agar plate and then incubated at 41°C for 48 hours. To confirm enterococci colonies the membrane is transferred to an Esculin Iron Agar plate. Colonies with a black or reddish brown precipitate are considered enterococci. In EPA method 1600, mE agar is amended with nalidixic acid and triphenyltetrazolium chloride. Colonies with a blue halo after a 24-hour incubation period at 41°C are counted as enterococci.

Most Probable Number (MPN)

The most probable number (MPN) procedure is performed using the multiple-tube fermentation (MTF) technique. Different dilutions of the samples are inoculated into sterile selective liquid media in glass tubes. Like the membrane filtration technique, the targeted bacterial group determines the medium used. Lauryl tryptose broth is used for the presumptive detection of total coliforms. This medium is mixed with bromocresol purple to indicate acid production. An inverted tube is added to the test tube to verify for gas production. Aliquots from tubes showing growth, gas, and acid production are inoculated into a confirmatory medium to determine final results. EC medium (at 35°C) or EC-MUG (at 44.5°C) can be used for the detection of fecal coliforms or thermo-tolerant *E. coli*, respectively, using an MPN approach. In most cases, five tubes per dilution are inoculated in the MTF method, but this number could be higher to increase the accuracy of the results. The actual densities can be calculated using an MPN formula or by consulting an MPN table (APHA, 2005). Most environmental laboratories use conventional MPN methods only in special cases, for instance, to increase enumeration accuracy when the contribution of sediments to total contamination is in question. Some U.S. public health laboratories, especially those involved in shellfish harvesting protection, use the MPN method as the definitive regulatory method.

The Quanti-tray method is another variation of the MPN approach. Quanti-tray based methods to target indicator bacteria (i.e., Colilert and Enterolert) are gaining acceptance among scientists in environmental monitoring programs in the U.S. and other countries primarily because of their simplicity and time savings (Ashbolt *et al.*, 2001; Eckner, 1998). Enteroliert measures enterococci. Colilert measures both total coliforms and *E. coli* simultaneously. Some reports suggest, however, that Colilert is prone to errors when analyzing subtropical waters (Chao, 2006; Pisciotta *et al.*, 2002). The steps involve first mixing water with the selective medium containing a fluorescence substrate. The Quanti-tray is then sealed and placed in an incubator for 24 hours. To detect the presence of the indicator organism, a source of ultraviolet light is used to identify fluorescent chambers containing the targeted indicator bacteria. A table provided by the manufacturer is used to estimate bacterial densities within a 1 to 2419 CFU/100 ml range.

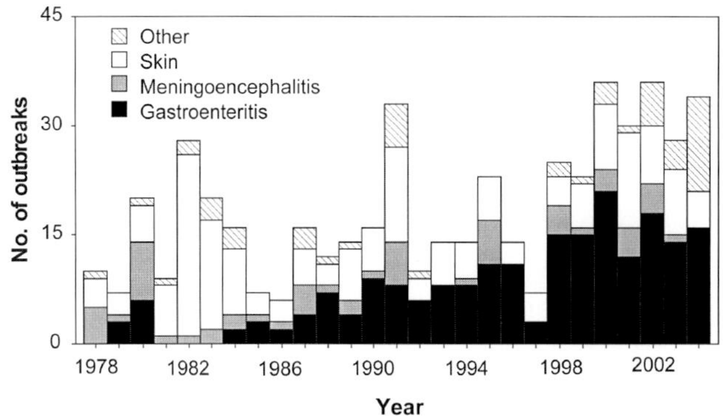

FIGURE 17-4. Waterborne outbreaks associated with recreational waters (Adapted from Yoder *et al.*, 2004).

MICROORGANISMS RELEVANT TO WATER FECAL POLLUTION AND HUMAN HEALTH

Historically, the primary public health concern (related to recreational waters fecal pollution) is the prevention of gastrointestinal illnesses (Fig. 17-4). Several fecal microorganisms have been implicated in waterborne outbreaks, although generally speaking in most gastroenteritis cases conclusive evidence linking marine recreational water with a specific pathogen is lacking. This is the case for a number of pathogens like *Salmonella*, *Shigella*, *Aeromonas*, *Toxoplasma*, *Cryptosporidium*, and many enteric viruses, to mention a few. However, in many instances some of these organisms have been documented as the culprits of contaminated seafood outbreaks; therefore, their potential role in endemic as well as unreported cases deserves some future consideration (Table 17-5). In other cases, water is a secondary vehicle of transmission as contaminated soils might be directly associated with pollution and survival of relevant fecal microorganisms. The following section provides a general description of some microorganisms relevant to microbial water quality and public health.

Bacterial Indicators

Enterococci are Gram-positive facultative anaerobic bacteria that colonize the large intestine of humans. These

Table 17-5. Some of pathogens potentially associated with waterborne diseases.

Bacteria	Protozoa	Viruses	Fungi
Escherichia coli	*Giardia* spp.	Noroviruses	Candida albicans
Shigella	*Cryptosporidum parvum*	Enteroviruses	Aspergillus fumigatus
Klebsiella spp.	Enterocytozoon bieneusi	Adenoviruses	Fusarium solani
Vibrio spp.	Entamoeba histolytica	Rotaviruses	*Penicillium* spp.
Aeromonas spp.	Naegleria fowleri	Reoviruses	
Mycobacterium spp.	Cyclospora cayetanensis	Hepatitis A and E	
Acinetobacter spp.	Toxoplasma gondii		
Campylobacter jejuni			
Nocardia spp.			
Pseudomonas aeruginosa			
Helicobacter pylori			
Listeria monocytogenes			
Plesiomonas shigelliodes			
Legionella pneumophila			
Clostridium spp.			
Acenitobacter spp.			
Burkholderia spp.			
Shewanella algae			

Most pathogens in this list have been associated with freshwater or drinking water outbreaks but not with marine water outbreaks. Some genera, like *Mycobacterium* and *Vibrio,* harbor several pathogenic species (*Mycobacterium avium* complex, *M. maritimum*; *V. cholera*, *V. parahemolyticus*, *V. vulnificus*). Only a small number of *E. coli* strains are pathogenic (e.g., *E. coli* O157:H7). Water can be a temporary reservoir of other pathogenic organisms like *Ascaris* spp. and *Schistosomas* spp., particularly in tropical areas exhibiting low sanitary conditions.

bacteria can be found at densities up to 10^7 CFU/gram (g) of feces. As a group, they were part of the fecal streptococci group until 1984 when the *Enterococcus* genus was proposed. Most strains are catalase negative, are capable of growth at 10°C and 45°C, are resistant to 60°C for 30 min, can grow in the presence of bile salts, show growth at pH 9.6 and at 6.5% NaCl, are able to reduce 0.1% methylene blue, and are homofermentative, without gas production. Glucose fermentation results in production of lactic acid. Some enterococci strains are of clinical importance because of their resistance to vancomycin, whereas others have been used as probiotics. The most isolated enterococci species from water are *E. faecalis*, *E. faecium*, *E. casseiflavus*, *E. gallinarium*, *E. mundtii*, *E. hirae*, and *E. durans*. Most other species do not seem to be associated with fecal pollution, although the fact that some species require unique growth factors might explain why they are rarely isolated from environmental waters. Alternatively, the survival of the latter species might be poor outside of the gut environment.

A normal member of the mammalian gut microbiota, *E. coli* was one of the first fecal indicators to be suggested to estimate microbial quality of surface waters. This organism is a Gram-negative, facultative anaerobic, rod shaped bacterium, member of the Enterobacteriaceae family, known to be one of the earliest colonizers of the human intestines, and is normally found in levels up to 10^8 CFU/g of feces of adult mammals. *E. coli* is capable of degrading simple sugars (typically via mixed acid fermentation pathways) and of synthesizing vitamins beneficial to the host. Cell motility is achieved via peritrichous flagella, which are critical to chemotaxis. As a group, it has been shown that *E. coli* has a broad range of hosts, including reptiles and amphibians, although some have argued that there is adaptation to different ecological niches resulting in the preferential host distribution of some strains.

Bacterial Pathogens

Vibrio spp. are halophilic heterotrophic Gram-negative straight or curved rods, which exhibit motility by means of a single polar flagellum. Species within this genus use glucose as primary carbon source and require 2% to 3% sodium or seawater amendment in order to grow on artificial media. Like most marine bacteria, their survival in marine waters is linked to redox-driven ion pumps (in this case sodium) that are used for metabolic work. Vibrios are among the most studied human waterborne pathogens of nonfecal origin. Some of the most pathogenic species are *V. cholerae, V. vulnificus*, and *V. parahaemolyticus.* Many vibrio species are associated with *shellfish, explaining why they are responsible for many foodborne outbreaks.* Cytotoxin-hemolysin and thermolabile hemolysin are important human toxins within this group. *Some vibrios* exhibit seasonal patterns in temperate waters, with highest densities during the summer. During the winter the culturable densities decrease significantly, although their actual numbers (as determined by microscopical counts) might not vary as much. A viable but nonculturable state has been suggested to explain this seasonal phenomenon for *V. vulnificus* (Oliver *et al.*, 1995).

Aeromonas is a genus within the *Aeromonadaceae* family that contains human and fish pathogens isolated from a wide variety of marine waters and freshwaters (Hazen *et al.*, 1978). *A. hydrophila, A. caviae*, and *A. veronii* are the most important human pathogens of the genus and contain several putative virulence factors including endotoxins, hemolysins, enterotoxins, proteases, and adhesins. Aeromonads are gamma Proteobacteria that share many biochemical characteristics with members of the *Enterobacteriaceae.* Although aeromonads can be isolated from feces and sewage, they are not considered of fecal origin and, therefore, their presence does not always correlate with fecal pollution or with fecal indicators. In fact, although they have been isolated from marine and surface waters, no typical recreational waterborne outbreaks have been reported to this date. Most human illnesses linked to this bacterial group are foodborne related. However, *A. hydrophila, A. jandaei*, and *A. veronii* have been associated with wound infections of people exposed to recreational waters (Joseph *et al.*, 1991; Semel and Trenholme, 1990).

Several members of the *Enterobacteriaceae*, namely, *E. coli, Shigella*, and *Salmonella*, are relevant foodborne and drinking water pathogens. Although most *E. coli* strains are nonpathogenic, there are some like *E.coli O157* (VTEC), and *E. coli* non-O157 enterotoxigenic, enteroaggregative, and enteroinvasive strains that are indeed important. Pathogenic *E. coli* and *Salmonella* strains receive a lot media attention because of food outbreaks, but they have also been implicated in waterborne outbreaks (Ackman *et al.*, 1997; Keene *et al.*, 1994). Although no marine recreational waterborne outbreaks have implicated these bacteria in recent times, they have been detected with some frequency in marine polluted waters.

Plesiomonas shigelloides is another member of the *Enterobactericeae* family (oxidase-positive) that is commonly isolated from freshwater and estuarine waters, as well as from the feces of a variety of animals, including amphibians and reptiles. It is rarely isolated from temperate waters, although one study showed the presence of nine different isolates from six lakes in Sweden (Gonzalez-Rey *et al.*, 2003). Filter feeders can be a reservoir of *P. shigelloides*, particularly along waters impacted by sewage sources (Miller *et al.*, 2006). Many cases in temperate regions are associated with travelers' diarrhea. Most human infections are usually mild and self-limiting cases of diarrhea, the majority registered in tropical and subtropical regions, and in association with consumption of food or to exposure to waters of unsanitary conditions. Asymptomatic people can also be carriers of pathogenic strains.

Campylobacter spp. are Gram-negative, spiral-shaped, microaerophilic bacteria, primarily known as foodborne pathogens responsible for a significant number of cases of diarrhea worldwide. The most important clinical species of this genus are *C. jejuni, C. coli*, and *C. enteritis.* These organisms are not common members of the human gut flora but are often isolated from the gut of wildlife and domesticated animals. Contaminated chicken meat is the primary source of *C. jejuni. Campylobacter* waterborne outbreaks have been recorded in association with drinking water (Andersson *et al.*, 1997; Melby *et al.*, 2000). Although campylobacteriosis outbreaks in marine recreational waters have not been reported, many cases have been reported for divers (Schijven and de Roda Husman, 2006), suggesting that ingestion is critical to infection. These bacteria change from a spiral to coccoid morphology under environmental stress, a feature that has been suggested to be relevant to their survival outside of the gut (Tangwatcharin *et al.*, 2006).

Pseudomonas aeruginosa is a Gram-negative bacterium that has been isolated from several environments, including feces, although it is mostly regarded as a terrestrial organism. Some strains are pathogenic to humans, whereas others are important animal and plant pathogens. *P. aeruginosa* could be considered a versatile opportunistic pathogen as it has been associated with a variety of human infections (e. g., pulmonary tract, urinary tract, septicimea, endocarditis). Physiologically it is also versatile as it can grow at a wide range of temperatures and carbon sources, resist a wide array of antibiotics, exhibit chemostatic behavior, form biofilms, and is genetically promiscuous. Genomic analyses of clinical and environmental strains have indicated that some of the virulence factors are conserved (Wolfgang *et al.*, 2003). This organism has been isolated from marine waters without evidence of fecal pollution. In one study, 50% of marine samples analyzed were positive for *P. aeruginosa* even though many of these same samples were not exceeding bathing standards (Guimarães *et al.*, 1993). Open ocean isolates have developed greater tolerance to sodium chloride than their river and clinical counterparts (Khan *et al.*, 2007). Divers are particularly receptive to skin infections from this pathogen (Ahlen *et al.*, 1998).

Staphylococcus aureus is a Gram positive cocci associated with skin infections. Methicillin-resistant *S. aureus*

(MRSA) is an important pathogen, which in previous years was only associated with nosocomial cases. Recently this pathogen has been isolated from a variety of matrices increasing its public health importance. Isolation of *S. aureus* from tropical marine waters has been reported (Charoenca and Fujioka, 1993), including waters that did not exceed the recommended microbial water quality standards (Charoenca and Kungskulniti, 2001), further suggesting that the current indicators might underestimate risks for nongastroenteritis illness associated with bathing waters.

Mycobacterium spp. are among the most relevant bacteria in public health and environmental microbiology. Mycobacteria are found in a number of different environments, are capable of resisting harsh environmental conditions, survive a wide range of pH and temperatures, and have a broad metabolic capability. Several mycobacteria species are important human pathogens from the standpoint of number of deaths per year worldwide. For example, the death toll associated with *M. tuberculosis* (agent of tuberculosis) is in the millions. *M. bovis* (tuberculosis agent in cattle) and *M. leprae* (agent of leprosy) are other nonwaterborne pathogens within the genus. Relevant to water quality are strains of *M. avium* and *M. intracellulare*, recognized as the *M. avium* complex (Mac) group. Mac-like bacteria are considered opportunistic pathogens primarily associated with immunocompromised or immunosupressed people. Crohn's disease, cervical adenitis, and pulmonary and skin infections are among the diseases associated with Mac. Their resistance to chlorination, association with biofilms, and intracellular lifestyle of many mycobacteria species makes them a particularly difficult group to eliminate from drinking water, hot tubs, and swimming pools (Keinanen-Toivola *et al.*, 2006; Mangione *et al.*, 2001).

Protozoan Pathogens

Giardia, the infectious agent of giardiasis, can be found in contaminated water and soil. This protozoan is a flagellated member of the Metamonada phylum. The species associated with human disease include *Giardia intestinalis* and *G. lamblia*. This parasite infects the gastrointestinal tract and can be transmitted via contaminated water, foods, and fomites. The primary symptoms of giardiasis are diarrhea, gas or flatulence, greasy stools that tend to float, stomach cramps, upset stomach, and nausea. Severe cases can lead to dehydration. The cells are called cysts that become trophozoites in the lumen. The trophozoites then multiply by longitudinal binary fission in the lumen of the proximal small bowel, a place where they can reside for long periods after infection. Outside of the lumen (e.g., water) *Giardia* is found as a cyst. Microscopic coupled with immunological techniques is the traditional approach to detect and identify this parasite. Giardia is often isolated from oysters, providing evidence that the waters are impacted by fecal pollution. This protozoan has also been isolated from marine mammals implicating that they are a reservoir of zoonotic agents (Hughes-Hanks *et al.*, 2005). The role of this and other zoonotic protozoa in marine waterborne outbreaks is poorly understood (Fayer *et al.*, 2004).

Cryptosporidium species are coccodian protozoa that belong to the apicomplexa phylum of nonfree-living parasites. They were first recognized as pathogenic to turkeys and cows. The first cases of human cryptosporidiosis were reported in 1976. These protozoa are of significant importance to public health, particularly to immunocompromised people. *Cryptosporidium* oocysts contain four sporozoites, a feature that can be used to differentiate from other protozoa but that is not exclusive to this parasite. Oocysts have been isolated from the feces of humans, cattle, guinea pigs, turkeys, chickens, and snakes. The taxonomy of *Cryptosporidium* is currently in a state of flux as molecular studies are providing new ways of differentiating among populations that were assigned the same species name (as in the case of *C. parvum*). One important characteristic of cryptosporidia is that they are capable of resisting different harsh environmental conditions, including disinfection treatments.

Viral Pathogens

Discovered in 1968 after an outbreak in Norwalk, Ohio, noroviruses are members of the single-stranded RNA, nonenveloped *Caliciviridae* family (Fong and Lipp, 2005). These viruses are responsible for most of the viral gastroenteritis infections worldwide. Sapovirus is another calicivirus that causes gastroenteritis. Although noroviruses have been observed in human, swine, cattle, cat, dog, and mice hosts, these are all different genogroups and as a result are considered to be host specific. Some of the most common symptoms include nausea, vomiting, diarrhea, and abdominal cramps, although headache, fever, chills, and muscle aches are also possible. These symptoms can begin as early as 12 hours after ingestion of the virus and normally last for up to 3 days. Diagnosis can be performed microscopically, by detecting an increase in specific antibodies, using reverse transcriptase polymerase chain reaction (RT-PCR), or by a series of commonly associated symptoms. No good cultivation method or animal models are available for this viral group. The major mode of transmission is through the fecal-oral route—that is, by ingesting fecally contaminated water or food. Person-to-person contact, air route, and contact with contaminated surfaces have also been implicated in the transmission of noroviruses.

Polyomaviruses and adenoviruses infect a wide variety of vertebrate species and are frequently found in the excreta and urine of humans. Polyomaviruses belong to the *Picornaviridae* family, whereas adenoviruses belong to the *Adenoviridae* family. Their isolation from environmental waters has been suggested as potential indication of human, porcine, and bovine fecal pollution (Fong and Lipp, 2005; Hundesa *et al.*, 2006). Both groups are double-stranded DNA viruses. Polyomaviruses can infect the respiratory system, kidneys, and brain and produce tumors, whereas adenovirises can also produce respiratory diseases, conjunctivitis, and gastroenteritis.

Enteroviruses are small nonenveloped isometric viruses commonly found in association with several significant human and animal illnesses, like polio, conjunctivitis, meningitis, and myocarditis, among many others. Children are very susceptible to enterovirus-related illnesses. Enteroviruses are single stranded RNA viruses that multiply in the gut mucosa and respiratory tract. The main form of transmission is the fecal-oral route. Three different groups are known to date: polioviruses, coxsackie viruses, and echoviruses, many of which have been isolated in coastal waters (Griffin *et al.*, 1999). Their detection in marine waters has been suggested to imply recent fecal contamination events (Wetz *et al.*, 2004). Enteroviral infections are more frequent in the summer and fall in the United States, whereas in tropical countries their incidence is not associated with any particular season.

MICROBIAL SURVIVAL: ECOLOGICAL AND GENETIC FACTORS

Differences in water survival rates have been documented between indicators and pathogens (Anderson *et al.*, 2005; Meschke and Sobsey, 2003). These differences are more striking when survival rates of bacterial indicators are compared to those of protozoa and enteric viruses. The latter two microbial groups tend to resist harsh environmental conditions that are often lethal to bacterial indicators. Laboratory studies have shown that viral particles can survive for weeks, whereas fecal bacterial indicators normally survive several days only. This has been used as an argument against the use of current indicators of fecal pollution and the proposal of coliphages as better surrogates for enteric viruses (Leclerc *et al.*, 2000). Several studies have shown that enterococci survive longer than fecal coliforms in marine waters, which explains the better correlation between illness and enterococci in marine recreational waters observed in epidemiological studies.

Some of the main factors affecting the survival of fecal indicators and pathogens in marine environments are sunlight, salinity, sediments, and nutrient content. Ultraviolet light can severely damage nucleic acids and other microbial macromolecules, directly controlling the densities of fecal microorganisms in surface waters. In contrast, sediments have been shown to increase the survival of *E. coli* and enteroviruses in marine waters (Smith *et al.*, 1978). For example, indicator bacteria have been shown to have longer survival rates in sediments (Davies *et al.*, 1995), perhaps because of the availability of nutrients, suggesting that sediments are a good reservoir for bacterial pathogens. Adsorption of enteric viruses to sediments particles can prevent their degradation.

Changes in global temperature (i.e., global warming) could have significant consequences to the overall abundance of pathogens in coastal areas. For example, freeze/thaw cycles can control abundance of pathogens. On the other hand, prolonged warm seasons and mild winters could have a direct impact on the survival of indicator bacteria and bacterial pathogens in temperate regions. Optimal growth of many fecal bacteria is between 30 to 35°C, and therefore higher temperatures coupled with the transport of nutrients caused by events can favor rapid growth of indicators and pathogens in coastal waters. Indeed, the sudden availability of nutrients can stimulate growth of bacterial indicators in tropical marine waters (Santo Domingo *et al.*, 1989). Weather also has an impact on human activity on or nearby coastal areas and hence the number of human fecal bacteria in coastal zones. For example, the link between precipitation and waterborne diseases has been recognized for years (Curriero *et al.*, 2001). Higher global temperatures are also associated with a higher incidence of intense rainstorms favoring the transport of inland pathogens to coastal zones (Rose *et al.*, 2001). Climatological phenomena like El Niño have been linked to cholera outbreaks in East Africa, Bangladesh, Ecuador, Chile, and Peru, particularly with increases in copepod (zooplankton) that served as the vector for *V. cholerae* (Anyamba *et al.*, 2006; Colwell, 1996; Pascual *et al.*, 2000). Moreover, global warming has also been implicated in the increase of coral disease (i.e., bleaching) in tropical areas (Graham *et al.*, 2006). Although temperature is relevant in marine environments, Smith *et al.* (1994) showed that nutrient availability and not temperature was the limiting factor impacting bacterial activity of some fecal bacteria in cold environments. Indeed, fecal coliform and enterococci have been isolated from water samples collected near the McMurdo Station in Antartica as a result of improper human waste disposal (Lisle *et al.*, 2004).

The genetic makeup of fecal microorganisms is greatly responsible for their fate outside of the gut environment. For example, the presence of flagella and pilli in some fecal bacteria is important from the standpoint of chemotaxis and for their attachment to surfaces. Attachment is important to biofilm formation, which provides a refuge to predators and increases the availability of nutrients (Costerton *et al.*, 1995). Because microorganisms prefer to live as part of biofilms, any type of surfaces including aquatic plants and debris also play a role in microbial survival. Appendages can also promote the attachment of phages, in turn playing a negative role in the survival of bacterial hosts. Additionally, marine waterborne pathogens like *Vibrio* and *Aeromonas* are equipped with proton pumps relevant to withstanding the salty conditions of marine and brackish waters and with the ability to enter a viable but nonculturable state.

ISSUES ASSOCIATED WITH MONITORING FECAL POLLUTION

Several problems have been documented throughout the years with the use of the current indicators of fecal pollution. The problems include the lack of correlation between indicators and pathogens, poor correlation between indicator levels and waterborne illnesses, potential survival and regrowth of bacterial pathogens, the underestimation of injured indicators, representativeness of sampling, and their inability to discriminate between different sources of fecal pollution (Fujioka *et al.*, 2004; Leclerc *et al.*, 2001). Alternate indicators have been suggested, including the coliphages, *C. perfringens*, *Bacteroides fragilis*, *Bifidobacterium* spp., and *Candida albicans*, to mention a few. Unfortunately, an ideal indicator does not exist for every type of water. Most fecal bacterial indicators do not survive well in water with a high salt content, so the number of alternate indicators for marine waters is even more limited.

Two critical issues regarding monitoring recreational waters and prevention of illnesses are the need for real-time data and accurate risk assessment models. Assays used to detect indicator bacteria take at least 24 hours after sampling is performed, and, therefore, the communication of potential exposure risks cannot be performed on a timely basis. Quantitative PCR (Q-PCR) methods have been developed to enumerate indicators in less than 3 hours (Haugland *et al.*, 2005) and more recently to determine microbial risks assessments (Wade *et al.*, 2006). Although such molecular techniques can be useful for rapidly detecting indicators and pathogens, currently most methods are not able to discriminate between dead and alive/active organisms, which is critical to establishing an accurate correlation between fecal indicators and microbial pathogens and hence to predicting real risks. Pathogen enumeration is not a simple task as most pathogens are present in relatively low numbers in surface waters. Preenrichments steps are often required to determine the occurrence and abundance of bacterial pathogens (e.g., *Salmonella*). In these cases, the data are often reported as presence/absence, and as a result it is difficult to establish a good mathematical correlation between indicators and pathogens. Additionally, distribution of pathogens in aquatic environments is not homogenous, which means that comprehensive sampling schemes are needed to increase the chances of detection. Moreover, several pathogenic groups require specific growth factors, whereas others cannot be cultured (e.g., spirochetes, mycoplasmas, viruses) or have yet to be discovered. Molecular methods also need to discriminate between infectious and noninfectious populations. Virulence factors have not been conclusively determined in many pathogens, information that is useful to discriminate between pathogenic and nonpathogenic forms and ultimately to develop pathogen-specific detection methods.

Most techniques and target organisms used to measure fecal pollution in surface waters have been developed in countries of temperate climate. Historically, tropical countries have adopted monitoring regulations developed in temperate regions. There is mounting evidence suggesting that microbial indicators of fecal pollution might not be adequately signaling fecal pollution levels and health risks in tropical waters. For example, several studies have indicated that bacterial indicators can persist for an extended period of time in tropical waters (Carrillo *et al.*, 1985). Moreover, isolation of fecal coliforms from pristine tropical waters (Rivera *et al.*, 1988) and from tropical soils (Byappanahalli and Fujioka, 2004) suggests that conditions typical to tropical environments could favor the survival of indicator bacteria. This finding is in agreement with studies showing that the animal gut might not be the only environment conducive to *E. coli* persistence (Gordon *et al.*, 2002).

Other issues specifically associated with microbial water quality in marine environments can be categorized into three different areas—different indicators for determining water quality acceptability, varied purposes for each of the regulatory programs in the United States, and issues associated with the naturally occurring pathogens.

Varied Indicators

In the United States, two bacterial indicator indices are normally used for regulatory activities in marine environments: fecal coliforms for shellfish regulation and enterococci for recreational water quality. These indicators have been shown to be effective for controlling disease outbreaks for their respective programs; however, the use of two different indicators comes with some unique problems. In marine environments, it does not follow that an exceedence of one indicator is indicative of an exceedence of the other indicator. This is because of a number of factors, including the stability of entercocci in a saline environment (as opposed to the inability of fecal coliforms to maintain cell integrity in the same environment), the intermittent presence of high volumes of freshwater, differences in cell densities in different fecal sources, and the classification of some soil microbes into the fecal coliform group. Second, the control of pollution sources is complicated by the use of these two indicators since they are not equally applicable to all marine waters (i.e., some waters are not designated or intended for use as shellfish harvesting areas and do not have the fecal coliform criteria applied to them). This can create an air of confusion as some sources are regulated for both indicators and others are not.

Varied Purposes

Current bacterial indicators are used in the United States by a number of programs for varied purposes that are compatible on some levels and incompatible on others. For example, fecal coliforms are used in marine areas primarily to determine the acceptability of shellfish for harvest and eventual raw consumption. When harvest areas exceed an acceptable level, they are either temporarily closed to harvest, restricted as to eventual disposition of the harvested shellfish (i.e., they are not immediately sent to the marketplace for raw consumption), or the area is permanently closed. Similarly, enterococci are used in marine areas to inform the public as to the acceptability of swimming areas for primary contact recreation. When a swimming area exceeds the acceptable level, an advisory is issued to notify the public about the increased risk of gastroenteritis with swimming in these areas. Both indicators are used by environmental regulatory programs to determine whether waters are meeting their designated uses and, if not, to what degree remediation efforts as needed to return the water body to its attainable use.

As to incompatibility of these various programs, there are a number of issues. As previously cited, because of the difference in indicators, it is possible for areas to be closed to shellfish harvesting, yet open to primary contact recreation or vice versa. Also, because the primary function of the shellfish and recreational programs is to be conservative in the protection of public health, the responsible environmental agency (usually a state agency) may make a determination that either the water body is not impaired, even though it is closed or has an advisory posting because the frequency of these events is low or that the highest attainable use is one that allows for some frequency of exceedence or total closure to the activity.

Natural Pathogens

Although not unique to marine systems, human environmental pathogens do present their own issues. Environmental pathogens are defined as microorganisms that normally spend a substantial part of their life cycle outside human hosts, but when introduced to humans they cause disease with measurable frequency (Cangelosi *et al.*, 2004). They are carried in the water, soil, air, food, and other parts of the environment and can affect almost every individual on the planet. Under the aforementioned definition, zoonotic agents are also considered environmental pathogens. Risks from environmental pathogens are not controlled by the use of current indicator bacteria (i.e., pathogens and diseases associated with them are expected to occur in waters that fully met the applicable indicator densities). Examples of these pathogens include *V. vulnificus, V. parahaemolyticus, Giardia*, and *Cryptosporidium* spp., and in freshwater environments, *Naegleria fowleri*. Currently, there are no water column levels for these pathogens to be able to inform the public as to the possibility of disease from contact recreation activities.

TRACKING SOURCES OF POLLUTION

Current methods used to monitor levels of fecal pollution cannot discriminate among different fecal sources because bacterial indicators are present in different hosts implicated in fecal pollution (e.g., human, cattle, swine, poultry, wildlife). As a result, environmental scientists have recently relied on microbial source tracking (MST) methods to determine the presence of human and nonhuman fecal pollution in waters. TMDL regulations are forcing state and local agencies to target the sources impacting polluted waters. In some cases the primary sources are obvious, whereas in others they are not; therefore it is necessary to first identify the source(s) in order to implement pollution control practices. Several reviews have been published on the subject (EPA, 2005; Simpson *et al.*, 2002), so this section focuses only on providing a general overview of this research area and its relevance to microbial risk assessment.

Since the mid-1990s, source tracking has normally been performed using two different types of methods: library dependent (LDMs) and library independent methods (LIMs). Library here refers to the use of a collection of phenotypic or genotypic profiles from bacterial strains isolated from known sources (i.e., fecal samples, septic tanks, animal waste lagoons). The most significant problem with LDMs relates to the need to develop large fingerprint databases of bacterial isolates from potential hosts and from environmental samples (including sediments) in order to represent the large genetic diversity of fecal and water isolates (Jenkins *et al.*, 2003). In contrast, LIMs usually target specific genetic markers that have shown some level of host specificity. The latter methods normally do not rely on culturing, are relatively rapid, sensitive, and inexpensive and can be automated to target multiple hosts simultaneously. Some of the current issues associated with LIMs include potential PCR inhibition, low target numbers, and the lack of standard operating procedures. There is also limited information on host specificity, geographic and temporal stability, and survival of host-specific populations.

Source tracking has been performed primarily with indicator bacteria. More recently, fecal anaerobic bacteria have been used as a target of host-specific assays, and in such cases, the 16S rDNA has been the marker of choice. An alternate approach is to directly type pathogens to their sources (Jiang *et al.*, 2005; Martinez-Urtaza[a]Instituto de Acuicultura, Universidad de Santiago de Compostela, Campus Universitario Sur, 15782 Santiago de Compostela, Spain, corresponding author. Tel.: +34 981

528024/563100x16043; fax: +34 981 547165. ucmjmur@ usc.es and Liebana, 2005). This approach is particularly useful with pathogens that exhibit true host specificity, like enteric parasites and enteric viruses. For example, polyomaviruses, enteroviruses, and adenoviruses have been used to show that human and bovine feces are impacting coastal waters (Fong and Lipp, 2005). However, in the latter cases, there has been a poor correlation between identified fecal sources and indicator bacteria. Moreover, the primary drawback of targeting pathogens in source tracking studies is that pathogens are often present in small numbers. Larger volumes of water could be used to increase microbial DNA yields, although careful attention should be given to the removal of PCR inhibitors as they tend to co-precipitate in the nucleic extracts. Preenrichment methods are used to deal with concentration issues when targeting bacterial pathogens for source tracking. Because the original densities are altered during the enrichment process, the data are not amenable to quantification.

Source tracking is relevant to public health because not all fecal pollution events impacting the microbial water quality of coastal zones are directly associated with human feces. Pollution associated with wildlife (e.g., birds and aquatic mammals) can be a major contributor to fecal inputs of coastal waters (Grant *et al.*, 2001). Moreover, fecal inputs from wildlife might not always be related to the direct access of animals to coastal waters as in watersheds. Wildlife pollution could gain entry to coastal areas as many different animals often impact wetlands. Fecal pollution sources from domesticated animals (i.e., cattle, poultry, and swine) can also be important to coastal recreational waters, primarily via storm runoff.

Water quality standards applied to marine recreational waters are based on human health risks, in turn based on epidemiological data obtained from areas primarily impacted by human fecal pollution. Although risks associated with nonhuman sources are believed to be of lesser magnitude (assuming that enteric viruses are associated with most cases of unknown etiology), the fact that zoonotic agents are increasingly important in human health (e.g., salmonellosis, campylobacteriosis, cryptosporidiosis) denotes the need to develop methods to identify when the source is not a wastewater treatment plant but rather pollution from an animal source. Although it is clearly understood that some animals serve as reservoirs for pathogens that can affect humans, the fate and transport of these pathogens as well as the route of exposure all make a clear association difficult to ascertain (Craun *et al.*, 2004). As a matter of policy, the EPA has determined that, until information to the contrary is received, the precautionary principle will be used, and fecal material associated with animals, both domestic and wild, is deemed to carry a risk equal to that of humans. Altogether this suggests the need for assays to detect nonhuman fecal pollution and need for alternate approaches to determine the risks associated with the different sources of pollution (Colford *et al.*, 2007). The latter information is critical to develop accurate risk assessment models and to evaluate best management practices.

FUTURE REGULATIONS AND FUTURE NEEDS

As new science develops, new regulations may call for different acceptable levels of existing indicator organisms or an entirely new suite of organisms. What is certain is that under the Clean Water Act section 304(a)(9)(B), the EPA will review and revise, as necessary, the water quality criteria for primary contact recreation. Although this may mean no change in the criteria for some years to come (especially if new science on the topic is not generated), it still places an investigative burden on the EPA to at least review any new science that is generated.

As mentioned previously, there is a disparity between the various indicators used for different purposes. Future regulations may attempt to harmonize the indicator used for these programs to allow for simplification and still protect public health. Additionally, to support current and any future regulations regarding microbial water quality, a number of critical research areas need to be addressed. First, more rapid methods of detection and enumeration of indicator organisms and pathogens are needed to fulfill mandates in certain statutes and to better inform the public and the regulated community as to the possible increased risks of consuming certain food products or recreating in specific areas. These rapid methods should also lead to a decrease in the amount of waterborne and foodborne (e.g., raw molluscan shellfish) disease. Second, it may be necessary to investigate other methods of deriving new criteria other than through classical epidemiological cohort studies. Although these studies are invaluable and have tremendous power in elucidating associations between risk factors and outcomes, the costs and difficulty of these types of studies makes them less likely to be done in the future. Risk assessment models and other systems will need to be employed to assist in the generation of the next set of pathogen or indicator criteria and to determine the cost benefit associated with implementing new methods using new microbial targets. The latter will ultimately provide the rationale for changing new policies and monitoring standards.

From a method development standpoint, biotechnological advances in robotics, genomics, and bioinformatics will likely contribute to the solution of many issues pertaining public health and pollution of coastal zones. Future microarray-like platforms will help microbiologists, microbial risk assessors, and environmental managers determine the severity of pollution by simultaneously looking at the presence of indicators, pathogens, and their sources (Lemarchand *et al.*, 2004; Sadowsky *et al.*, 2007). Emerging areas like metagenomics and computational biology will provide microbial ecologists the opportunity to look at fecal pollution from a microbial community perspective (Santo Domingo *et al.*, 2007) and engineers the tools to implement and evaluate remediation practices.

References

Ackman, D., Marks, S., Mack, P., Caldwell, M., Root, T., Birkhead, G., 1997. Swimming-associated haemorrhagic colitis due to *Escherichia coli* O157:H7 infection: Evidence of prolonged contamination of a fresh water lake. Epidemiol. Infect. 119, 1–8.

Ahlen, C., Mandal, L.H., Iversen, O.J., 1998. Identification of infectious *Pseudomonas aeruginosa* strains in an occupational saturation diving environment. Occup. Environ. Med. 55, 480–484.

American Public Health Association (APHA), 2005. Standard methods for the examination of water and wastewater, 21st ed. Washington, DC, American Public Health Association.

Anderson, K.L., Whitlock, J.E., Harwood, V.J., 2005. Persistence and differential survival of fecal indicator bacteria in subtropical waters and sediments. Appl. Environ. Microbiol. 71, 3041–3048.

Andersson, Y., de Jong, B., Studahl, A., 1997. Waterborne *Campylobacter* in Sweden: The cost of an outbreak. Water Sci. Technol. 35, 11–14.

Anyamba, A., Chretien, J.P., Small, J., Tucker, C.J., Linthicum, K.J., 2006. Developing global climate anomalies suggest potential disease risks for 2006–2007. Int. J. Health Geogr. 5, 60.

Ashbolt, N.J., Grabow, W.O.K., Snozzi, M., 2001. Indicators of microbial water quality. In Fewtrell, L., and Bartram, J. (eds.), Water Quality: Guidelines, Standards and Health. World Health Organization (WHO). London, IWA.

Austin, D.A., Austin, B., 1999. Bacterial Fish Pathogens: Disease of Farmed and Wild Fish, 3rd ed., p. 457. Praxis, UK, Springer Praxis Books.

Bartram, J., Rees, G., 2000. Monitoring bathing waters: A practical guide to the design and implementation of assessments and monitoring programmes. World Health Organization. London, F & FN Spon.

Bonde, G.J. 1977. Bacterial indicators of water pollution. Adv. Aquatic Microbiol. 1, 273–364.

Burke, L., Kura, Y., Kassem, K., Revenga, C., Spalding, M., McAllister, D., 2001. Pilot Analysis of Global Ecosystems: Coastal Ecosystems. Washington, DC, World Resources Institute.

Byappanahalli, M., Fujioka, R., 2004. Indigenous soil bacteria and low moisture may limit but allow faecal bacteria to multiply and become a minor population in tropical soils. Water Sci. Technol. 50, 27–32.

Cabelli, V.J. 1983. Health effects criteria for marine recreational waters. EPA-600/1-80–031. Cincinnati, OH, U.S. Environmental Protection Agency.

Cangelosi, G.A, Freitag, N.E., Buckley M.R., 2004. From outside to inside: Environmental organisms as human pathogens. Washington, DC, American Academy of Microbiology, American Society for Microbiology.

Carrillo, M., Estrada, E., Hazen, T.C., 1985. Survival and enumeration of the fecal indicators *Bifidobacterium adolescentis* and *Escherichia coli* in a tropical rain forest watershed. Appl. Environ. Microbiol. 50, 468–476.

CDC. 2002. Surveillance for Waterborne-Disease Outbreaks, US, 1999–2000. MMWR 51(SS-8), www.cdc.gov/mmwr/PDF/ss/ss5108.pdf.

Chao, W.L., 2006. Evaluation of Colilert-18 for the detection of coliforms and *Escherichia coli* in tropical fresh water. Lett. Appl. Microbiol. 42, 115–120.

Charoenca, N., Fujioka, R.S., 1993. Assessment of Staphylococcus bacteria in Hawaii's marine recreational waters. Water Sci. Technol. 27, 283–289.

Charoenca, N. Kungskulniti, N., 2001. Staphylococci and fecal indicators recovered from coastal recreational waters in Thailand. Intern. Med. 17, 43–49.

Cheung, W.H.S., Chang, K.C.K., Hung, R.P.S., 1990. Health effects of beach water pollution in Hong Kong. Epidemiol. Infect. 105, 139–162.

Chick, H., 1908. An investigation of the laws of disinfection, J. Hygiene 8, 92–157.

Colford, J.M., Jr, Wade, T.J., Schiff, K.C., Wright, C.C., Griffith, J.F., Sandhu, S.K., Burns, S., Sobsey, M., Lovelace, G., Weisberg, S.B., 2007. Water quality indicators and the risk of illness at beaches with nonpoint sources of fecal contamination. Epidemiology 18, 27–35.

Colwell, R.R., 1996. Global climate and infectious disease: The cholera paradigm. Science 274, 2025–2031.

Costerton, J.W., Lewandowski, Z., Caldwell, D.E., Korber, D.R., Lappin-Scott. H.M., 1995. Microbial biofilms. Annu. Rev. Microbiol. 49, 711–745.

Craun, G.F., Calderon, R.L., Craun, M.F., 2004. Waterborne outbreaks caused by zoonotic pathogens in the USA. pp. 120–135. In Cotruvo, J.A., Dufour, A., Rees, G., Bartrum, J., Carr, R., Cliver, D.O., Craun, G.F., Fayer, R., and Gannon, V.P.J. (eds.), Waterborne Zoonoses: Identification, Causes and Control. London, World Health Organization, IWA.

Curriero, F.C., Patz, J. A., Rose, J.B., Lele, S., 2001. The association between extreme precipitation and waterborne disease outbreaks in the United States, 1948–1994. Am. J. Public Health 91, 1172–1174.

Davies, C.M., Long, J.A., Donald, M., Ashbolt, N.J., 1995. Survival of fecal microorganisms in marine and freshwater sediments. Appl. Environ. Microbiol. 61, 1888–1896.

Dziuban, E.J., Liang, J.L., Craun, G.F., Hill, V., Yu, P.A., Painter, J., Moore, M.R., Calderon, R.L., Roy, S.L., Beach, M.J., 2006. Surveillance for waterborne disease and outbreaks associated with recreational water—United States, 2003–2004. MMWR Surveill. Summ. 55, 1–30.

Eckner, K.F., 1998. Comparison of membrane filtration and multiple-tube fermentation by the colilert and enterolert methods for detection of waterborne coliform bacteria, *Escherichia coli*, and enterococci used in drinking and bathing water quality monitoring in southern Sweden. Appl. Environ. Microbiol. 64, 3079–3083.

Environmental Protection Agency (EPA), 1986. Ambient water quality criteria for bacteria, EPA440/5-84-002. Washington, DC, U.S. Environmental Protection Agency, Office of Water.

Environmental Protection Agency (EPA), 2002. The 2000 national water quality inventory, EPA-841-R-02-001. Washington, DC, U.S. EPA Office of Water.

Environmental Protection Agency (EPA), 2003. Bacterial water quality standards for recreational waters (freshwater and marine waters) status report, EPA-823-R-03-008. Washington, DC, U.S. EPA Office of Water.

Environmental Protection Agency (EPA), 2005. Microbial source tracking guide document, EPA-600/R-05-064. Washington, DC, U.S. EPA Office of Research and Development.

Fayer, R., Dubey, J. P., Lindsay, D.S., 2004 Zoonotic protozoa: From land to sea. Trends Parasitol. 20, 531–536.

Fleisher, J.M., Kay, D., Salmon, R.L., Jones, F., Wyer, M.D., Godfree, A.F., 1996. Marine waters contaminated with domestic sewage: Nonenteric illnesses associated with bather exposure in the United Kingdom. Am. J. Public Health 86, 1228–1234.

Fong, T.T., Lipp E.K., 2005. Enteric viruses of humans and animals in aquatic environments: Health risks, detection, and potential water quality assessment tools. Microbiol. Mol. Biol. Rev. 69, 357–371.

Fujioka, R., Sian-Denton, C., Borja, M., Castro, J., Morphew K., 1999. Soil: The environmental source of *Escherichia coli* and enterococci in Guam's streams. J. Appl. Microbiol. Symposium Supplement, 85, 83S–89S.

Gonzalez-Rey, C., Svenson, S.B., Eriksson, L.M., Ciznar, I., Krovacek, K. 2003. Unexpected finding of the "tropical" bacterial pathogen *Plesiomonas shigelloides* from lake water north of the Polar Circle. Polar Biol. 26, 495–499.

Gordon, D.M., Bauer, S., Johnson J.R., 2002. The genetic structure of *Escherichia coli* populations in primary and secondary habitats. Microbiology 148, 1513–1522.

Graham, N.A., Wilson, S.K., Jennings, S., Polunin, N.V., Bijoux, J.P., Robinson, J., 2006. Dynamic fragility of oceanic coral reef ecosystems. Proc. Nat. Acad. Sci. USA 103, 8425–8429.

Grant, S.B., Sanders, B.F., Boehm, A.B., Redman, J.A., Kim, J.H., Mrse, R.D., Chu, A. K., Gouldin, M., McGee, C.D., Gardiner, N.A., Jones, B.H., Svejkovsky, J., Leipzig, G. V., Brown, A., 2001. Generation of enterococci bacteria in a coastal saltwater marsh and its impact on surf zone water. Environ. Sci. Technol. 35, 2407–2416.

Griffin, D.W., Gibson 3rd, C.J., Lipp, E.K., Riley, K., Paul 3rd, J.H., Rose, J.B., 1999. Detection of viral pathogens by reverse transcriptase PCR and of microbial indicators by standard methods in the canals of the Florida Keys. Appl. Environ. Microbiol. 65, 4118–4125.

Grimes, D.J., Singleton, F.L., Stemmler, J., Palmer, L.M., Brayton, P., Colwell, R.R., 1984. Microbiological effects of wastewater effluent discharge into coastal waters of Puerto Rico. Water Res. 18, 613–619.

Guimarães, V.F., Araújo, M.A.V., Mendonçla-hagler, L.C.S., Hagler, A. N., 1993. *Pseudomonas aeruginosa* and other microbial indicators of pollution in fresh and marine waters of Rio de Janeiro, Brazil. Environ. Toxicol. Water Qual. 8, 313–322

Haugland, R.A., Siefring, S.C. Wymer, L.J. Brenner, K.P., Dufour, A.P., 2005. Comparison of *Enterococcus* measurements in freshwater at two recreational beaches by quantitative polymerase chain reaction and membrane filter culture analysis. Water Res. 39, 559–568.

Hazen, T.C., Fliermans, C.B., Hirsch, R.P., Esch, G.W., 1978. Prevalence and distribution of *Aeromonas hydrophila* in the United States. Appl. Environ. Microbiol. 36, 731–738.

Henrickson, S.E., Wong, T., Allen, P., Ford, T., Epstein, P.R., 2001. Marine swimming-related illness: Implications for monitoring and environmental policy. Environ. Health Perspect. 109, 645–650.

Hughes-Hanks, J.M., Rickard, L.G., Panuska, C., Saucier, J.R., O'Hara, T. M., Dehn, L., Rolland, R.M., 2005. Prevalence of *Cryptosporidium* spp. and *Giardia* spp. in five marine mammal species. J. Parasitol. 91, 1225–1228.

Hundesa, A., Maluquer de Motes, C., Bofill-Mas, S., Albinana-Gimenez, N., Girones, R., 2006. Identification of human and animal adenoviruses and polyomaviruses for determination of sources of fecal contamination in the environment. Appl. Environ. Microbiol. 72, 7886–7893.

Jenkins, M.B., Hartel, P.G., Olexa, T.J., Stuedemann J.A., 2003. Putative temporal variability of *Escherichia coli* ribotypes from young steers. J. Environ. Qual. 32, 305–309.

Jiang, J., Alderisio, K.A., Xiao, L., 2005. Distribution of *Cryptosporidium* genotypes in storm event water samples from three watersheds in New York. Appl. Environ. Microbiol.71, 4446–4454.

Joseph, S.W., Carnahan, A.M., Brayton, P.R., Fanning, G.R., Almazan, R., Drabick, C., Trudo. E.W., Jr, Colwell, R.R., 1991. *Aeromonas jandaei* and *Aeromonas veronii* dual infection of a human wound following aquatic exposure. J. Clin. Microbiol. 29, 565–569.

Keene, W.E., McAnulty, J.M., Hoesly, F.C., Williams, L.P., Jr, Hedberg, K., Oxman, G.L., Barrett, T.J., Pfaller, M.A., Fleming, D.W., 1994. A swimming-associated outbreak of hemorrhagic colitis caused by *Escherichia coli* O157:H7 and *Shigella sonnei*. N. Engl. J. Med. 331, 579–584.

Keinänen-Toivola M.M., Revetta R.P., Santo Domingo, J.W.. 2006. Identification of active bacterial communities in drinking water biofilms using 16S rRNA-targeted clone libraries. Accepted in FEMS Microbiol. Lett. 257, 182–188.

Kennish, M.J., 1998. Pollution Impacts on Marine Biotic Communities, p. 310. Boca Raton, FL: CRC Press.

Khan, N.H., Ishii, Y., Kimata-Kino, N., Esaki, H., Nishino, T., Nishimura, M., Kogure, K., 2007. Isolation of *Pseudomonas aeruginosa* from open ocean and comparison with freshwater, clinical, and animal isolates. Microb Ecol. 53, 173–186.

Koop, K., Booth, D., Broadbent, A., Brodie, J., Bucher, D., Capone, D., Coll, J., Dennison, W., Erdmann, M., Harrison, P., Hoegh-Guldberg, O., Hutchings, P., Jones, G.B., Larkum, A.W.D., O'Neil, J.O., Steven, A., Tentor, E., Ward, S., Williamson, J., Yellowlees, D., 2001. ENCORE: The effect of nutrient enrichment on coral reefs; synthesis of results and conclusions. Mar. Poll. Bull. 42, 91–120.

Koop, K., Griffiths, C. L., 1982. The relative significance of bacteria, meio- and macrofauna on an exposed sandy beach. Mar. Biol. 66, 295–300.

Leclerc, H., Edberg, S., Pierzo, V., Delattre, J.M., 2000. Bacteriophages as indicators of enteric viruses and public health risk in groundwaters. J. Appl. Microbiol. 88, 5–21.

Leclerc, H., Mossel, D.A.A., Edberg, S.C., Struijk, C.B., 2001. Advances in the bacteriology of the coliform group: Their suitability as markers of microbial water safety. Ann. Rev. Microbiol. 55, 201–234.

Lemarchand, K., Masson, L., Brousseau, R., 2004. Molecular biology and DNA microarray technology for microbial quality monitoring of water. Critical Rev. Microbiol. 30, 145–172.

Levin, M.A., Fischer, J.R., Cabelli, V.J., 1975. Membrane filter technique for enumeration of enterococci in marine waters. Appl. Microbiol. 30, 66–71.

Lipp, E.K., Griffin, D.W., 2004. Analysis of coral mucus as an improved medium for detection of enteric microbes and for determining patterns of sewage contamination in reef environments. EcoHealth 1, 317–323.

Lisle, J.T., Smith, J.J., Edwards, D.D., McFeters, G.A., 2004. Occurrence of microbial indicators and *Clostridium perfringens* in wastewater, water column samples, sediments, drinking water, and Weddell seal feces collected at McMurdo Station, Antarctica. Appl. Environ. Microbiol. 70, 7269–7276.

Mangione, E.J., Huitt, G., Lenaway, D., Beebe, J., Bailey, A., Figoski, M., Rau, M.P., Albrecht, K.D., Yakrus, M.A., 2001. Nontuberculous mycobacterial disease following hot tub exposure. Emerging Infectious Diseases. 7, 1039–1041.

Martinez-Urtaza, J., Liebana, E., 2005. Investigation of clonal distribution and persistence of *Salmonella* Senftenberg in the marine environment and identification of potential sources of contamination. FEMS Microbiol. Ecol. 52, 255–263.

Melby, K.K., Svendby, J.G., Eggeb ø, T., Holmen, L.A., Andersen, B.M., Lind, L., Sjøgren, E., Kaijser, B., 2000. Outbreak of *Campylobacter* infection in a subartic community. European J. Clin. Microbiol. Infectious Dis. 19, 542–544.

Meschke, J.S., Sobsey, M.D., 2003. Comparative reduction of Norwalk virus, poliovirus type 1, F+ RNA coliphage MS2 and *Escherichia coli* in miniature soil columns. Water Sci. Technol. 47, 85–90.

Miller, W.A., Miller, M.A., Gardner, I.A., Atwill, E.R., Byrne, B.A., Jang, S., Harris, M., Ames, J., Jessup, D., Paradies, D., Worcester, K., Melli, A., Conrad P.A., 2006. *Salmonella* spp., *Vibrio* spp., *Clostridium perfringens*, and *Plesiomonas shigelloides* in marine and freshwater invertebrates from coastal California ecosystems. Microb. Ecol. 52, 198–206.

National Health and Medical Research Council (NHMRC), 2005. Guidelines for managing risks in recreational water. Canberra, Australia, National Health and Medical Research Council.

New Zealand Ministry for the Environment (NZME), 2003. Microbiological water quality guidelines for marine and freshwater recreational areas. Wellington, New Zealand, New Zealand Ministry for the Environment.

Oliver, J.D., Hite, F., McDougald, D., Andon, N.L., Simpson L.M., 1995. Entry into, and resuscitation from, the viable but nonculturable state by

Vibrio vulnificus in an estuarine environment. Appl. Environ. Microbiol. 61, 2624–2630.

Pascual, M., Rodo, X., Ellner, S.P., Colwell, R., Bouma M.J., 2000. Cholera dynamics and El Nino-Southern Oscillation. Science 289, 1766–1769.

Pisciotta, J.M., Rath, D.F., Stanek, P.A., Flanery, D.M., Harwood V.J., 2002. Marine bacteria cause false-positive results in the Colilert-18 rapid identification test for *Escherichia coli* in Florida waters. Appl. Environ. Microbiol. 68, 539–544.

Pruss, A., 1998. Review of epidemiological studies on health effects from exposure to recreational water. Int. J. Epidemiol. 27, 1–9.

Rivera, S.C., Hazen, T.C., Toranzos, G.A., 1988. Isolation of fecal coliforms from pristine sites in a tropical rain forest. Appl. Environ. Microbiol. 54, 513–517.

Rose, J.B., Epstein, P.R., Lipp, E.K., Sherman, B.H., Bernard, S.M., Patz, J.A., 2001. Climate variability and change in the United States: Potential impacts on water-and foodborne diseases caused by microbiologic agents. Environ. Health Perspect.109, 211–221.

Santo Domingo, J.W., Bambic, D.G., Edge, T. A., Wuertz, S., 2007. Quo vadis source tracking? Towards a unifying framework for environmental monitoring of fecal pollution. Water Res 41, 3529–3552.

Santo Domingo, J.W., Fuentes, F.A., Hazen, T.C., 1989. Survival and activity of *Streptococcus faecalis* and *Escherichia coli* in petroleum-contaminated tropical marine waters. Environ. Pollut. 56, 263–281.

Schijven, J., de Roda Husman, A.M., 2006. A survey of diving behavior and accidental water ingestion among Dutch occupational and sport divers to assess the risk of infection with waterborne pathogenic microorganisms. Environ. Health Perspectives. 114, 712–717.

Semel, J.D., Trenholme, G., 1990. *Aeromonas hydrophila* water-associated traumatic wound infections: A review. J. Trauma 30, 324–327.

Shibata, T., Solo Gabriele, H.M., Fleming, L.E., Elmir, S., 2004. Monitoring marine recreational water quality using multiple microbial indicators in an urban tropical environment. Water Res. 38, 3119–3131.

Shuval, H.I., 1999. Scientific, economic and social aspects of the impact of pollution in the marine environment on human health—a preliminary quantitative estimate of the global disease burden. Report. World Health Organization and GESAMP.

Simons, G.W., Jr., Gillespie, C.G., Gage, S. de M., Ferguson, H., Tisdale, E.S., 1924. The report of the Committee on Bathing Places. Am. J. Public Health 14, 597–602.

Simons, G.W., Jr., Hilscher, R., Ferguson, H. F., Gage. S. De M., 1922. Report of the Committee on Bathing Places. American J. Public Health 12, 121–123.

Simpson, J.M., Santo Domingo, J.W., Reasoner, D.J., 2002. Microbial source tracking: State of the science. Environ. Sci. Technol. 36, 5279–5288.

Smith, E.M., Gerba, C.P., Melnick, J.L., 1978. Role of sediment in the persistence of enteroviruses in the estuarine environment. Appl. Environ. Microbiol. 35, 685–689.

Smith, J.J., Howington, J.P., McFeters, G.A., 1994. Survival, physiological response and recovery of enteric bacteria exposed to a polar marine environment. Appl. Environ. Microbiol. 60, 2977–2984.

Stevenson, A.H., 1953. Studies of bathing water quality and health. Am. J. Pub. Health. 43, 529–538.

Tangwatcharin, P., Chanthachum, S., Khopaibool, P., Griffiths, M.W., 2006. Morphological and physiological responses of *Campylobacter jejuni* to stress. J. Food Prot. 69, 2747–2753.

Wade, T.J., Calderon, R.L., Sams, E., Beach, M., Brenner, K.P., Williams, A.H., Dufour. A.P., 2006. Rapidly measured indicators of recreational water quality are predictive of swimming-associated gastrointestinal illness. Environ. Health. Perspect. 114, 24–28.

Watson, H.E., 1908. A note on the variation of the rate of disinfection with change in the concentration of the disinfectant. J. Hygiene 8, 536–542.

Wetz, J.J., Lipp, E.K., Griffin, D.W., Lukasik, J., Wait, D., Sobsey, M.D., Scott, T.M., Rose, J.B., 2004. Presence, infectivity, and stability of enteric viruses in seawater: Relationship to marine water quality in the Florida Keys. Mar. Pollut. Bull. 48, 698–704.

Wolfgang, M.C., Kulasekara, B.R., Liang, X., Boyd, D., Wu, K., Yang, Q., Miyada, C.G., Lory. S., 2003. Conservation of genome content and virulence determinants among clinical and environmental isolates of *Pseudomonas aeruginosa*. Proc. Natl. Acad. Sci. USA 100, 8484–8489.

World Health Organization (WHO), 2003. Guidelines for safe recreational water environments: Coastal and fresh waters, vol. 1. Geneva, World Health Organization.

Yoder, J.S., Blackburn, B.G., Craun, G.F., Hill, V., Levy, D.A., Chen, N., Lee, S.H., Calderon, R.L., Beach, M.J., 2004. Surveillance for waterborne-disease outbreaks associated with recreational water—United States, 2001-MMWR Surveill. Summ. 53, 1–22.

STUDY QUESTIONS

1. Discuss some of the historical events leading to implicating water as vector of diseases.
2. Discuss the economic and public health impacts of fecal pollution at a global scale.
3. What are some of the specific U.S. regulations that protect waters from fecal pollution?
4. What are the main differences between the U.S. and WHO microbial water quality guidelines for recreational waters?
5. What are the different organisms used as fecal microbial indicators, and what methods are used to enumerate their densities?
6. What different types of illnesses are associated with swimming in fecally polluted waters? What are some of the activities that increase the risks of getting sick from exposure to contaminated waters? Who is at higher risk?
7. What main microbial groups are responsible for waterborne diseases? Provide examples for each group, and discuss their potential implication in marine recreational outbreaks.
8. Why are the numbers of reported outbreaks associated with marine recreational waters relatively low?
9. What are some of the factors influencing the survival of bacterial indicators and pathogens in marine waters?
10. What are the different types of water quality objectives that fecal bacterial indicators are used for? How are indicators used in each case?
11. Why is tracking the source of fecal pollution important to the regulatory community?

CHAPTER

18

Foodborne Infectious Diseases and Monitoring of Marine Food Resources

ROSINA GIRONES, SÍLVIA BOFILL-MAS, M. DOLORES FURONES, AND CHRIS RODGERS

INTRODUCTION

The World Health Organization (WHO) has indicated that foodborne diseases are a widespread public health problem, both in developed and developing countries (Valdimarsson, 2005). In fact, the percentage of people suffering from foodborne diseases in industrialized countries has been reported to be up to 30%, although developing countries bear the brunt of the problem because they have major food safety risks that are less well documented (WHO, 2001).

Essentially, foodborne diseases occur when an individual consumes food contaminated with certain pathogenic microorganisms. However, although microbial contamination is an increasing worldwide problem (Fleming *et al.*, 2006), consumer health risks from marine food resources are usually considered to be low for seafood derived from unpolluted open marine environments. Nevertheless, the risk of foodborne illnesses from products arising from more closed environments (e.g., coastal or brackish estuarine areas) is higher, directly relating to the greater potential for contamination in coastal regions compared to capture fisheries (Reilly and Käferstein, 1997). The higher concentrations of pathogens in more coastal areas are the direct result of land-based human activities, because pollution from urban or rural runoffs leads to nutrient-rich coastal waters that provide ideal conditions for the growth and reproduction of these microorganisms (Marine Conservation Research Institute [MCRI], 2005). Typical point and nonpoint pollution sources include sewer overflows, discharges of inadequately treated wastes from sewage treatment plants, leakage from septic tanks, and storm water runoff. Shellfish are at particular risk of contamination because they feed by filtering large volumes of water per unit time and they can become carriers of potential pathogens if the water is contaminated with certain bacteria, parasites, or viruses.

Infections transmitted through consumption of contaminated seafood are a significant source of human morbidity (Ripabelli *et al.*, 2004). For instance, the U.S. Centers for Disease Control and Prevention (CDC) estimate that seafood is one of the leading causes of foodborne outbreaks in the United States, as seafood has been shown to constitute 15% of all outbreaks with a confirmed source, which is a level greater than that associated with meat and poultry products that are consumed at eight and six times the rate of seafood, respectively (Valdimarsson, 2005). However, the number of individual cases associated with seafood is lower than for meat and poultry, apparently because of the much lower consumption rate (U.S. Accounting Office, 2001). On the other hand, more than 50% of seafood-related outbreaks in the United States are the result of virus infections caused by eating molluscan shellfish (Valdimarsson, 2005). Some of the larger known outbreaks of illness (>400 cases) related to seafood consumption worldwide are detailed in Table 18-1.

Another source of seafood-borne disease can be postharvest food processing, because potential pathogens can either be introduced from infected humans who handle the food, by cross-contamination from raw product being processed at the same time, or through defects in storage conditions after product sale. For example, hepatitis A virus and norovirus can be introduced by the unwashed hands of food handlers who are themselves infected (CDC, 2006a).

Fortunately, the use of good manufacturing practice in conjunction with raised consumer awareness related to risks such as eating raw or partially cooked products that may contain microbiological contamination can effectively control most food safety hazards (Reilly and Käferstein,

TABLE 18–1. Examples of larger outbreaks (>400 cases) of illness caused by seafood consumption.

Year	Country	Product	Causal Agent	Cases	Reference
1924/1925	United States	Oysters	*Salmonella typhi*	"Widespread"	Lumsden *et al.*, 1925
1944	United States	Clams	"Gastroenteritis"	400	Richards, 1985
1955	Sweden	Oysters	Hepatitis A virus	629	Roos, 1956
1961	United States	Clams	Hepatitis A virus	459	Verber, 1984
1976/1977	England	Cockles	SRSVs	Approx. 800	Appleton and Pereira, 1977
1978	Australia	Oysters	Norwalk virus	>2000	Murphy *et al.*, 1979
1978	United States	Boiled shrimp	*Vibrio parahaemolyticus*	1133	Anonymous, 2001
1982	United States	Clams and oysters	"Gastroenteritis"	>400	Richards, 1985
1983	United States	Clams	"Gastroenteritis"	>1000	Richards, 1985
1988	China	Clams	Hepatitis A virus	300,000	Xu *et al.*, 1992
1990	Australia	Oysters	Norwalk virus	446	Bird and Kraa, 1995
1991/1993	Latin America (especially Peru)	Multiple factors incl. crab meat	*Vibrio cholerae* O1	Approx. 1 million	Guthmann, 1995
1996	Italy	Oysters	Hepatitis A virus	>400?	Malfait *et al.*, 1996
1996	Japan	Boiled crabs	*Vibrio parahaemolyticus*	691	Anon, 2001
1997	United States	Oysters	Norwalk virus	>400	CDC, 1997
1998	Japan	Catered meals	*Vibrio parahaemolyticus*	1167	Anonymous, 2001
2003/2004	Mexico	Shrimps	*Vibrio parahaemolyticus*	>1230	Cabanillas-Beltrán *et al.*, 2006
2004	Chile	Shellfish	*Vibrio parahaemolyticus*	1500	González-Escalona *et al.*, 2005
2005	Chile	Shellfish	*Vibrio parahaemolyticus*	10,491	Heitmann *et al.*, 2005

1997). Therefore, concentration on management of all the points related to the contamination risk of food resources is the most effective way to improve food quality and maintain consumer confidence.

The following sections of this chapter focus on bacterial, parasitic, and viral contamination and infections from seafood. Table 18-1 provides a summary of outbreaks associated with seafood consumption, and Table 18-2 provides additional examples of pathogenic agents recovered from seafood to supplement the information included within the text of this chapter. The discussion of viral infections focuses on shellfish, as shellfish tend to concentrate viruses from waters. The discussion of viruses in shellfish includes detection, control, and management. The chapter closes with a discussion of consumer safety programs and standards for maintaining seafood quality in the United States and the European Union.

BACTERIAL INFECTIONS FROM SEAFOOD

Fecal Bacterial Infections

This group of pathogens can be classed as the nonindigenous enteric bacteria that occur in marine or estuarine waters as a result of fecal contamination. They include *Salmonella* spp., pathogenic *E. coli*, *Campylobacter* spp., *Shigella* spp., and *Yersinia enterocolitica*. The contamination is related to agricultural land runoff, large seabird or sea mammal colonies, as well as the occurrence of both point and nonpoint sources of pollution (e.g., sewage discharges, septic tanks, and leisure craft sea toilets). Foodborne outbreaks caused by enterotoxigenic *E. coli* spp. and *Shigella* spp. have also occurred on ships, particularly cruise ships. Factors associated with shipboard outbreaks include inadequate temperature control, infected food handlers, contaminated raw ingredients, cross-contamination, inadequate heat treatment, and onshore excursions (Rooney *et al.*, 2004). Seafood was the most common food vehicle implicated in 50 outbreaks on passenger ships, involving almost 10,000 people, in a retrospective study by Rooney *et al.* (2004).

Salmonella

Salmonella spp. are among the most important causes of gastrointestinal disease worldwide, and many seafood-importing countries restrict the import of raw food products containing these pathogens (Feldhusen, 2000). *Salmonella* spp. have been isolated in the United States from oysters harvested and intended for human consumption (Brands

TABLE 18–2. Additional examples of potential pathogenic agents associated with seafood.

Pathogenic Agent	Isolated From	Country	Comment	Reference
Fecal Bacterial Infection				
Campylobacter spp.	Clams *Mya arenaria*	Canada	Pathogens tended to be less frequently detected in opened, as opposed to closed, harvesting sites	Lévesque *et al.* (2006)
Enteropathogenic (ETEC) *E. coli*	Red snapper *Lutjanus purpureus*	Brazil	Samples collected from a seafood market	Teophilo *et al.* (2002)
	Seabob shrimp *Xiphopenaeus kroyeri*	Brazil	Samples collected from a seafood market	Teophilo *et al.* (2002)
Salmonella spp.	Oysters *Pinctada imbricate*	Venezuela	Two shucked samples suggested that contamination in bivalve growing areas and during the harvesting operation was increased by handling of the product	Villalobos de Bastardo and Elguezabal Aristizabal (2001)
	Crab *Ucides cordatus*	Brazil	Marketed seafood	Vieira *et al.* (2004)
	Crustacean shrimps	Italy	66% of frozen seafood previously harvested in Thailand	Ripabelli *et al.* (2004)
	Clams	Italy	Low level (0.7%) isolation	Legnani *et al.* (2002)
	Clams *Mya arenaria*	Canada	Pathogens tended to be less frequently detected in opened, as opposed to closed, harvesting sites	Lévesque *et al.* (2006)
Salmonella Typhimurium	Tiger shrimp *Penaeus monodon*	India	One sample out of 846 raw, peeled tail-on aquacultured shrimp obtained from a processor operating under HACCP conditions	Mohamed Hatha *et al.* (2003)
Yersinia enterocolitica	Prawns	UK (England)	Low level (2%) found in cooked product obtained at the point of sale to the consumer	Greenwood *et al.* (1985)
Nonfecal Bacterial Infection				
Aeromonas hydrophila	Oysters and fish	Mexico	Raw street-vended seafood found contaminated during May, the warmest month of the year	Estrada-Garcia *et al.* (2005)
Clostridium botulinum	Salmon	Canada	Fermented roe (“stink eggs” or “gink”), a delicacy of the West Coast Aboriginal people in British Columbia, and other traditional foods, such as smoked salmon, have been shown to cause illness and even death	Dawar *et al.* (2002)
	Fish and shrimp	India	Shown to occur in fresh and cured samples in the retail trade	Lalitha and Surendran (2002)
	Tuna	Italy	Home-canned fish	Aureli *et al.* (1984)
	Fish	Japan	A total of 7.7% of samples from different sea foods available in supermarkets	Haq and Sakaguchi (1980)
	Mussels	New Zealand	Outbreak associated with the consumption of traditional Maori Tiroi, a combination of mussels and Puha, the sow thistle *Sonchus asper*	Whyte *et al.* (2001)
	Whitefish	United States (New York)	Seafood-associated disease outbreaks examined from 1980 to 1994 showed three deaths caused by ungutted, dried kapchunka or ribetz prepared and eaten by persons of Russian or Eastern European descent	Wallace *et al.* (1999)
Vibrio spp.	Mussels *Mytilus galloprovincialis*	Italy	Isolated from 48.4% of mussels harvested from approved shellfish waters in the Adriatic Sea	Ripabelli *et al.* (1999)
Vibrio cholerae non-O1/O139	Blue mussels	Norway	Reported in 3% of mussels from various coastal sites	Bauer *et al.* (2006)
	Oysters *Ostrea edulis*	Norway	Reported in 11% of oysters from the midcoast	Bauer *et al.* (2006)

(*continued*)

TABLE 18-2. *(continued)*

Pathogenic Agent	Isolated From	Country	Comment	Reference
V. cholerae	Reef fish	United States (Guam)	A small number of cholera events associated with eating imported fish were reported but the imposition of a narrowly focused ban on the importation of inshore seafood and processed food products from the identified source, led to no additional local or imported cases of cholera being detected	Haddock *et al.* (2002)
V. parahaemolyticus	Crabs *Ucides cordatus*	Brazil	Isolated from marketed product	Vieira *et al.* (2004)
	Short-necked clam, hen clam and rock oyster	Japan	Isolated from 33% of samples, as well as from the sediments of rivers near the coast	Hara-Kudo *et al.* (2003)
	Shellfish, jellyfish, fish intestine, seaweed or seaslug	New Zealand	Seafood consumption privately imported from the Pacific Islands (Tonga, Samoa, and Niue) led to 358 patients with gastroenteritis	Thornton *et al.* (2002)
	Oysters	Spain (Galicia)	A large outbreak (64 cases admitted to a single hospital) associated with raw shellfish consumption	Martínez-Urtaza *et al.* (2004)
V. vulnificus	Razor clams, giant tiger prawns and mantis shrimps	China	Live seafood sampled from markets in coastal cities	Yano *et al.* (2004)
	Blue mussels	Norway	Reported for the first time in 2% of samples from the south coast in August 2003	Bauer *et al.* (2006)
Parasitic Infection				
Protozoa				
Giardia duodenalis	Clams	United States	Cysts present	Graczyk *et al.* (1999)
Toxiplasma gondii	Marine mammals	Canada	50% of Inuit women tested seropositive in northern Canada, which was linked to the consumption of contaminated dried seal meat	Fayer *et al.* (2004)
Cestodes				
Diphyllobothrium pacificum	Fish	Peru	Children diagnosed with infection because raw, mainly marinated, fish is widely present in the diet of coastal communities	Medina Flores *et al.* (2002)

et al., 2005), as well as raw oysters and oyster cocktails in Trinidad (Rampersad *et al.*, 1999). A study on the presence of *Salmonella* spp. in live mollusks in Spain showed a low overall incidence but, although mussels and oysters had a higher incidence than clams or cockles, there was a seasonal pattern; 54% of the isolations were detected from September to November (Martínez-Urtaza *et al.*, 2003). Bivalve mollusks (cockles, mussels, scallops, and oysters) in Northern Ireland, examined according to European Commission (EC) shellfish bed classification regulations, had an 8% level of *Salmonella* serotypes, including a 2% isolation from category A beds (see EU Directive 91/492/EEC, presented later in the chapter, for a description of category criteria), which are nominally suitable for immediate consumption according to *E. coli* indicator counts (Wilson and Moore, 1996).

Escherichia Coli

E. coli is a common contaminant of seafood in the tropics and is often encountered in high numbers. The first report of the presence of Shiga toxin-producing *E. coli* (STEC) and *E. coli* O157:H7 in shellfish from French coastal environments has been published from EC category B and category D prohibited areas, indicating that shellfish can serve as a vehicle for STEC transmission (Gourmelon *et al.*, 2006). This is supported by the isolation of STEC *E. coli* strains from clams in Mangalore, India (Kumar *et al.*, 2004; Sanath Kumar *et al.*, 2001). Enterohaemorrhagic *E. coli* (EHEC) O157:H7 ingested through contaminated salmon roe prepared with rice cakes (Ikura-Sushi) was shown to be the cause of disease outbreaks in Japan (Terajima *et al.*, 1999), and it has been suggested that EHEC *E. coli* might enter the viable but nonculturable state in salted salmon roe (Makino *et al.*, 2000), which would have implications for seafood quality safety programs.

Campylobacter *spp.*

Campylobacter spp. can cause mild to severe diarrhea, although the infective dose depends on the strain, damage to cells from environmental stress, and the susceptibility of the host (Price and Tom, 2003). Thermophilic *Campylobacter* spp. have been reported in 42% of shellfish from Northern Ireland, although they have not been associated with illness, and the findings were used to show that bed classification on the basis of indicator organisms alone is not sufficient to assure the absence of bacterial pathogens (Wilson and Moore, 1996).

Shigella *spp.*

Shigella spp. can cause dysentery and are associated with diarrheal or gastrointestinal illness. *Shigella flexneri* has been isolated from ready-to-eat shrimp, representing four countries of origin, which were obtained from local grocery stores in the United States (Duran and Marshall, 2005). It has also been implicated in a foodborne disease outbreak related to the lack of adequate personnel hygiene facilities in the galley area of a passenger ship, although it is not clear if this was the result of seafood consumption (Rooney *et al.*, 2004).

Yersinia Enterocolitica

Yersinia enterocolitica is a relatively infrequent cause of diarrhea and abdominal pain, with children being infected more often than adults, and it is not usually associated with a seafood vector (Feldhusen, 2000). However, *Y. enterocolitica* is ubiquitous and has been recovered from frozen seafood in Italy where 3% of hen clams purchased from local vendors were shown to be contaminated (Ripabelli *et al.*, 2004).

Nonfecal Bacterial Infections

This group of pathogens can be classed as the indigenous bacteria that naturally inhabit marine or estuarine waters, such as *Aeromonas hydrophila*, *Clostridium botulinum*, and *Listeria monocytogenes*, as well as certain *Vibrio* spp. (particularly *V. cholerae*, *V. parahaemolyticus*, and *V. vulnificus*). They are not necessarily associated with waters that are contaminated with fecal pollution and would normally occur at low levels in freshly produced seafood. Nevertheless, the microflora associated with live or recently harvested product reflects the microbial populations of the harvest water and sediments in which they are living. Subsequently, poor storage or inadequate cooking can lead to food safety hazards when fish and shellfish are eaten (Reilly and Käferstein, 1997). Cross contamination, failures in the processing of seafood or natural pollution can lead to the presence of other potential pathogens, such as *Bacillus cereus* (crab sticks), *Clostridium perfringens* (live mussels and oysters, sea urchins), and *Staphylococcus aureus* (hake fish fingers, raw shrimp), but these are of less importance (Cordoba *et al.*, 2003; Hudson *et al.*, 2001; Mohamed Hatha *et al.*, 2003; Muniain-Mujika *et al.*, 2003; Pinon *et al.*, 2004; Vieira *et al.*, 2001).

Aeromonas *spp.*

Aeromonas infection after ingestion of contaminated food can produce a diarrheal illness with associated fever and vomiting (Butt *et al.*, 2004a). *Aeromonas* spp. have rarely been associated with foodborne illness even though strains are common in aquatic environments. For instance, there were no reported cases related to *Aeromonas* spp. in England and Wales during the 1996 to 2000 period (Adak *et al.*, 2005). Nevertheless, *Aeromonas* spp. can contaminate raw fish and shellfish, especially if the product has had prolonged exposure to elevated temperatures (Fleming *et al.*, 2006). *Aeromonas hydrophila* and *Aeromonas sobria* are the most commonly isolated species from seafood (Butt *et al.*, 2004a). *Aeromonas hydrophila* is primarily a fish pathogen but is occasionally implicated in human illness and has been found from cooked or hot-smoked foods, in conjunction with *A. caviae* and *A. sobria* (Gobat and Jemmi, 1993), associated with recontamination at the slicing and packaging stage (Feldhusen, 2000). It has also been suggested that psychrotrophic *Aeromonas* strains capable of growing at low temperatures could be of public health significance in food products that have an extended shelf life at refrigeration temperature (Feldhusen, 2000).

Clostridium Botulinum

Clostridium botulinum causes botulism, a severe food poisoning, resulting from the ingestion of a botulinal toxin, and there is a high mortality rate (Price and Tom, 2003). The prevalence of *C. botulinum* varies considerably with fish species and geographical area (Hyytiä *et al.*, 1999) but has occasionally been implicated in foodborne infections (Wallace *et al.*, 1999). Botulism has been associated with the consumption of airtight sealed smoked salmon consumed 3 days after the sell by date had expired (Dressler, 2005). Foodborne botulism may constitute a safety hazard in processed lampreys from the Baltic Sea area if packaging and extended shelf lives are used instead of more traditional methods (Merivirta *et al.*, 2006). *C. botulinum* is known to occur in the gastrointestinal tract of fish species such as tilapia (Nol *et al.*, 2004).

The trend toward extended shelf life products is a potential hazard. For example, a study of the raw materials used in refrigerated processed foods of extended durability, manufactured in France, detected *C. botulinum* using PCR-ELISA (polymerase chain reaction–enzyme linked immunosorbent assay) in a total of 8 out of 102 samples of fish and shellfish (Carlin *et al.*, 2004). A similar study reported *C. botulinum* in 214 fresh fish and environmental samples from a coastal area in northern France (Fach *et al.*, 2002).

Listeria Monocytogenes

Of the seven different species of *Listeria*, only *Listeria monocytogenes* causes the disease listeriosis in humans. Illness can range from meningitis in neonatal infection to vomiting and diarrhea in cases of food poisoning (Anonymous, 1999). The consumption of contaminated food is an important route for listeriosis transmission, with the population most at risk being the elderly and the immunocompromised (Anonymous, 1999). *L. monocytogenes* is widely distributed in the general environment, including coastal waters, as well as live fish from these areas (Huss *et al.*, 2000). Contamination or recontamination of seafood may also take place during processing and low levels of *L. monocytogenes* are frequently found on seafood (Huss *et al.*, 2000). Ready-to-eat products represent the greatest threat of listeriosis because they do not require further cooking at home (Price and Tom, 2003), as one of the most important properties of *L. monocytogenes* is its ability to multiply slowly in foods at chill temperatures (Australian Dairy Authorities' Standards Committee [ADASC], 1999). *L. monocytogenes* in raw food that will be cooked before consumption is less of a concern to the food industry because the bacteria are killed during cooking (Price and Tom, 2003). *L. monocytogenes* has been isolated from raw fish, smoked fish, fish roe, minced tuna, seafood, cooked crabs, raw and cooked shrimp, raw crawfish, raw lobster, surimi, smoked mussels, and raw oysters (Anonymous, 1999; Degnan *et al.*, 1994; Handa *et al.*, 2005; Huss *et al.*, 2000; Kwiatek, 2004; Thimothe *et al.*, 2002).

Vibrio *spp.*

Vibrio species are common in estuarine and other aquatic environments, and a number of them are human pathogens (Reilly and Käferstein, 1997). They are also commonly present in or on shellfish and other seafood (Morris, 2003). Additionally, they occur on the surfaces of marine plants and animals, naturally in the intestinal content of marine animals (Anonymous, 2001), and they can be concentrated in species of phytoplankton, zooplankton, and planktonic copepods (Borroto, 1997). There are many different *Vibrio* spp., but only three species, *V. cholerae* (toxigenic strains of serogroups O1 and O139), *V. parahaemolyticus,* and *V. vulnificus*, represent a serious and growing public health hazard (Anonymous, 2001) related to cases of cholera, gastroenteritis, and primary septicemia, respectively. As long as oysters and other shellfish are harvested from warm waters and eaten raw or with minimal cooking, there is risk of infection with *Vibrio* species (Morris, 2003).

Nearly all *Vibrio* infections in the United States are caused by noncholeragenic *Vibrio* species, such as *V. parahaemolyticus*, *V. vulnificus*, non-O1, and non-O139 *V. cholerae* (CDC, 2006a). However, two cases of toxigenic *V. cholerae* O1 infection attributed to consumption of undercooked or contaminated seafood were reported in Louisiana after Hurricanes Katrina and Rita caused flooding of large residential areas in 2005. Seafood samples obtained in seafood markets and supermarkets in Malaysia showed that strains of both O1 and O139 *V. cholerae* serotypes were present regardless of the season (Elhadi *et al.*, 2004). *V. cholerae* O139 Bengal emerged on the Indian subcontinent in late 1992. It was later recognized in Thailand in 1993 during an epidemiological study that included the evaluation of serotype O1. This study showed that risk factors for cholera development included untreated water, consumption of uncooked seafood, and food served at group gatherings (Hoge *et al.*, 1996). A link to the consumption of seafood, including shellfish, mantis shrimps, and crabs, was also shown for an outbreak of cholera, caused by *V. cholerae* O1 biotype El Tor, in Hong Kong during June/July 1994, when contaminated seawater in fish tanks used for keeping the seafood alive was found to be the most likely vehicle of transmission (Kam *et al.*, 1995). A study of the epidemiology and ecology of *V. cholerae* O1 and O139 carried out in two coastal areas of Bangladesh, where cholera occurs seasonally, indicated that cholera was endemic in both regions and that serotypes O1, O139, and non-O1/non-O139 were autochthonous to the aquatic environment (Alam *et al.*, 2006). Street-vended seafood is a potential transmission route of infection, particularly in developing countries, and raw oysters along with fish ("ceviche") have been shown to harbor *V. cholerae* non-O1 and non-O139 (Estrada-García *et al.*, 2005).

V. parahaemolyticus occurs in many different fish, crustaceans, mollusks, and shellfish, such as eels, octopus, squid, sardines, tuna, mackerel, flounder, rockfish, red snapper, pompano, clams, oysters, lobsters scallops, shrimps, and crabs (Anonymous, 2001). The consumption of raw bivalve mollusks has caused outbreaks of food poisoning because of *V. parahaemolyticus* (Normanno *et al.*, 2006), and they are almost exclusively associated with the consumption of raw, improperly cooked, or cooked, recontaminated fish and shellfish, with up to 86% of seafood samples being positive in the United States (Anonymous, 2001). During only the period May–July 2006, New York City, New York State, Oregon, and Washington health departments reported a total of 177 cases of *V. parahaemolyticus* infection. The cases were related separately to restaurants, certain seafood markets and recreational harvesting (CDC, 2006b). In fact, *V. parahaemolyticus* is the leading cause of seafood-associated gastroenteritis in the United States and is typically related to the consumption of raw oysters gathered from warm-water estuaries (McLaughlin *et al.*, 2005). It is also one of most important foodborne pathogens in Asia, causing approximately half of the food poisoning outbreaks in Taiwan, Japan, and Southeast Asian countries (Martínez-Urtaza *et al.*, 2004, citing Joseph *et al.*, 1982), but it is rarely reported in Europe (Anonymous, 2001). The prevalence and density of human pathogenic vibrios in the environment and also in seafood products have been shown to be highly dependent on the ambient temperature, with the largest numbers occurring at higher seawater temperatures in countries such as the United States, United Kingdom, Italy, and Mexico (Anonymous, 2001). The northernmost documented source of oysters causing illness caused by *V. parahaemolyticus* has been Alaska where rising ocean water temperatures were reported to have contributed to one of the largest known outbreaks of *V. parahaemolyticus* in the United States from oysters obtained locally and consumed onboard a cruise ship (McLaughlin *et al.*, 2005). The seasonal nature of the presence of *V. parahaemolyticus* has also been reported in oysters from two Indian estuaries (Deepanjali *et al.*, 2005) and from oysters sampled in the shell from restaurants or oyster bars, retail seafood markets, and wholesale seafood markets in coastal and inland markets throughout the United States (Cook *et al.*, 2002). In this latter study, oysters had been harvested from the Gulf, Pacific, Mid-Atlantic, and North Atlantic coasts of the United States and from Canada. Densities of *V. parahaemolyticus* in market oysters from all harvest regions also followed a seasonal distribution, with highest densities in the summer (Cook *et al.*, 2002). Outbreaks of diarrhea caused by *V. parahaemolyticus* in two remote regions of Chile (Antofagasta in 1998 and Puerto Montt in 2004) associated with shellfish consumption were likely to have been triggered by higher than normal temperatures during the summer months, at least in the southern-most shellfish producing region of Puerto Montt (González-Escalona *et al.*, 2005). In 1998, the warm seawater caused by the El Niño phenomena could have favored the geographic dispersion of the bacterium in Antofagasta (Cordova *et al.*, 2002). Analysis of shellfish and clinical samples during a subsequent outbreak in 2005 (also Puerto Montt) indicated that 53% of shellfish samples from both 2004 and 2005 contained *V. parahaemolyticus* but most isolates corresponded to nonpandemic clones and therefore the causative agent during epidemics was only a minor component of a small but diverse population of *V. parahaemolyticus* in shellfish (Fuenzalida *et al.*, 2006). The spread of this *V. parahaemolyticus* serotype (O3: K6), which is related to shellfish consumption, likely started in 1996 in Calcutta, India, and subsequently achieved pandemic proportions since it appeared quickly in Taiwan, Laos, Japan, Thailand, and Korea, as well as more recently from foodborne outbreaks and from sporadic cases in additional countries such as Bangladesh, Chile, Korea, Mozambique, Russia, and the United States (Balakrish Nair and Hormazábal, 2005). The first outbreak of gastroenteritis in Mexico caused by the pandemic strain of *V. parahaemolyticus* (O3:K6) occurred in 2003–2004 when more than 1230 cases were reported, and they were associated with the consumption of commercialized shrimp (Cabanillas-Beltrán *et al.*, 2006). The majority of Japanese outbreaks have been caused by fish, and although outbreaks can be relatively small, they are frequent (Anonymous, 2001). In Norway, 10% of mussels have been shown to harbor *V. parahaemolyticus*, but only when seawater temperatures were 15°C or higher and not during the cold season (Anonymous, 2001), although a level of 49% was found by concentrating on a "worst case" area where temperature and salinity were thought to be more favorable (Bauer *et al.*, 2006). Strains have also been isolated from water and seafood collected along the Atlantic coast of France (Urdaci *et al.*, 1988), natural samples of seawater and shellfish on the Mediterranean coast of Spain (Macián *et al.*, 2000), and 9.5% of shellfish samples (mussels and clams) from European category B harvesting areas in the Adriatic Sea, Italy. A seasonal trend was apparent in the Adriatic Sea with more frequent isolations during the summer months (Suffredini *et al.*, 2006). A total of 94 *V. parahaemolyticus* strains were isolated in 1999 from seafood imported into France from China, Ecuador, India, Iran, Madagascar, Senegal, Tanzania, Thailand, and Vietnam (Anonymous, 2001). Seafood imported into Europe has also been notified as containing *V. parahaemolyticus* from other countries such as Bangladesh, Belize, Canada, Indonesia, Ivory Coast, Malaysia, Mozambique, Namibia, Nigeria, Sri Lanka, and Turkey (Anonymous, 2001).

Vibrio vulnificus is a naturally occurring estuarine bacterium often associated with disease such as septicemia in humans following consumption of raw and lightly cooked seafood (Yano *et al.*, 2004), especially raw bivalve mollusks (Normanno *et al.*, 2006). Infections caused by *V. vulnificus* (biotype 1) in Europe are rare but usually severe (e.g., rapid septicemia, cutaneous wound infections), and in the United States, most cases are associated with the consumption of raw oysters, although there is a notable seasonality that correlates with water temperature (Anonymous, 2001; Cook *et al.*, 2002). The disease has a high mortality rate of up to 60% in cases involving predisposed persons (Anonymous, 2001). A study of *V. vulnificus* distribution in seawater, sediment, and shellfish along the coast of Japan, as well as the contamination levels in retail fish and shellfish, showed that *V. vulnificus* was isolated from 68.1% of seawater samples, 70.4% of sediment samples, 17.2% of fish samples, and

43.8% of shellfish samples, although it occurred at a lower level in prawns (Fukushima, 2006). It has been observed in intestinal samples of fish feeding on mollusks and crustaceans from the Gulf of Mexico and the United States (De Paola *et al.*, 1994), as well as oysters and crabs in the summer months along the Atlantic coast of the United States (Morris, 1988) and in 16.6% of marine fish from coastal areas of Cochin on the west coast of India, including oil sardine, Indian mackerel, threadfin bream, and white pomfret from a fish market and lizard fish and croaker collected fresh from fishing vessels (Thampuran and Surendran, 1998). The low numbers of *V. vulnificus* found in seawater and shellfish along the Spanish Mediterranean coast may be associated with the higher salinities recorded in this sea (Arias *et al.*, 1999). The number of reported clinical cases in Nordic countries (e.g., Denmark, Norway, and Sweden) is sporadic, likely because of the lower water temperature (Anonymous, 2001).

Other strains of *Vibrio* spp., such as *V. cholerae* (strains other than O1 or O139), *V. fluvialis, V. hollisae,* and *V. mimicus* have been associated with a significant number of infections arising from contaminated seafood, but consumption usually leads to gastroenteritis (Anonymous, 2001) rather than septicemia. Seafood samples obtained in seafood markets and supermarkets in Malaysia resulted in the isolation of eight potentially pathogenic *Vibrio* species, including *V. cholerae* (4.6%), *V. parahaemolyticus* (4.7%), *V. vulnificus* (6.0%), *V. alginolyticus* (11%), *V. metschnikovii* (9.9%), *V. mimicus* (1.3%), *V. damsela* (13%), and *V. fluvialis* (7.6%), which together represented 52% of the overall sample size. The *Vibrio* species were detected in shrimp, squid, and peel mussels, but the incidence was highest (82%) in cockles *Anadara granosa* (Elhadi *et al.*, 2004). The occurrence of various *Vibrio* species in water, sediment, and shrimp samples from multiple shrimp farm environments from the east and west coast of India showed predominantly *V. alginolyticus* (3% to 19%), *V. parahaemolyticus* (2% to 13%), *V. harveyi* (1% to 7%), and *V. vulnificus* (1% to 4%) (Gopal *et al.*, 2005).

PARASITIC INFECTIONS FROM SEAFOOD

Parasites are responsible for a substantial number of seafood-associated infections (Butt *et al.*, 2004b). Parasitic infections can arise following consumption of seafood, generally fish, which has been inadequately cooked or eaten raw if the products harbor the infectious stages of a particular parasite. The parasites in question are usually trematodes (mainly from freshwater cyprinid fish and not seafood), nematodes, or cestodes. Recreational coastal water can also harbor protozoan parasites (e.g., *Cryptosporidium* spp., *Giardia intestinalis, Toxoplasma*). Seafood-borne parasitic infections occur in the United States with sufficient frequency to make preventive controls necessary during the processing of parasite-containing species of fish that are intended for raw consumption (Food and Drug Administration [FDA], 2001).

Products implicated in parasitic human infection include ceviche (fish and spices marinated in lime juice), lomi lomi (salmon marinated in lemon juice, onion and tomato), poisson cru (fish marinated in citrus juice, onion, tomato and coconut milk), herring roe, sashimi (slices of raw fish), sushi (pieces of raw fish with rice and other ingredients), green herring (lightly brined herring), drunken crabs (crabs marinated in wine and pepper), cold-smoked fish, and undercooked grilled fish (Price and Tom, 2003).

Protozoa

The zoonotic protozoa are found worldwide and are among the most frequently reported parasites of humans and animals (Fayer *et al.*, 2004). Although problems of illness related to protozoan infections have not been associated with seafood, commercial and noncommercial oysters (*Crassostrea gigas*) in the Netherlands have been shown to harbor *Cryptosporidium* oocysts or *Giardia* cysts in their intestines, suggesting that consumption of raw oysters may occasionally lead to cases of gastrointestinal illness (Schets *et al.*, 2006). A similar finding has been observed in oysters from Chesapeake Bay in the United States, with the highest percentage found shortly after the greatest rainfall event during a 3-year study, indicating that runoff was the most likely source (Fayer, 2004). *Cryptosporidium* oocysts have also been detected in other shellfish, such as mussels, cockles, and clams (Freire-Santos *et al.*, 2000; Gomez-Bautista *et al.*, 2000; Gómez-Couso *et al.*, 2003) in coastal waters of Canada, Ireland, Italy, Portugal, Spain, and the United States (Fayer, 2004; Fayer *et al.*, 2004). The presence of these parasites in the marine environment has clear implications for the potential development of zoonotic protozoan infections related to the consumption of seafood, and cases may be underreported.

Trematodes

Consumption of fish from brackish water has been known to be involved in the transmission of some trematode infections. Examples include *Heterophyes heterophyes* in mullet (*Mugil* spp.), *Nanophyetus* sp. in salmonids, *Spelotrema brevicaeca* in crustaceans and amphipods, and *Paragonimus westermani* in crabs and shrimps (Reilly and Käferstein, 1997, citing Abdussalem *et al.*, 1995). However, the incidence of these infections seems to be regional and may not be of particular significance on a global scale. Products from aquaculture involved in international trade are generally free from infectious trematodes because they are usually frozen

(Reilly and Käferstein, 1997). Nevertheless, locally traded aquaculture products represent a risk of trematode parasitic infection, particularly in areas such as Asia where consumption of raw or undercooked fish is often a cultural practice or is related to the difficulty of obtaining sufficient fuel for cooking.

Some studies suggest that those who consume raw estuarine fish, such as perch, shad, mullet, redlip mullet, or goby, are at a high risk of infection with heterophyid flukes. For instance, Do Gyun *et al.* (2006) indicated that in Korea the proportion of the population eating raw fish could lead to a high human infection rate.

Nematodes

The risk of nematode parasitic infections is also greatest from the consumption of raw or minimally processed fish that are already infected. Fish and marine invertebrates are usually intermediate hosts in the life cycle of nematodes, whereas the definitive hosts are piscivorous marine mammals (e.g., seals). Nematodes are quite common in wild-caught fish, although it is considered that marine farmed fish (e.g., salmon) fed on artificial feed do not harbor nematodes with potential to infect humans. However, in certain other types of aquaculture systems where cultured marine fish are fed on raw fish there would be a risk of nematode transmission (Reilly and Käferstein, 1997). The risk of infection can be greatly reduced by heating or freezing fishery products above 55°C or below −20°C, respectively.

Human infections can occur from eating raw fish, and examples include the cod worm *Phocanema decipiens,* a parasite found in fish and seals, *Dioctophyme renale* and *Gnathostoma hispidium* larvae (Myers, 1970), *Anisakis* spp., *Pseudoterranova* spp., *Eustrongylides* spp., and *Gnathostoma* spp. (Price and Tom, 2003).

Large outbreaks of human anisakiasis have been reported from countries with a high consumption of raw or undercooked seafood (Butt et al., 2004b). It is caused by the accidental ingestion of *Anisakis simplex* and *Pseudoterranova decipiens* larvae (Price and Tom, 2003). Following penetration of the gastric and intestinal mucosa they may result in violent abdominal pain, nausea and vomiting, or even severe anaphylactic reactions. The adult stages reside in the stomach of marine mammals and their eggs are passed in the feces. Second stage larvae are ingested by crustaceans, resulting in development of third stage larvae that are infective for fish and squid, in which they migrate from the intestine to the peritoneal cavity and then to the muscle tissues. They can be transferred from fish to fish, and humans become infected by eating raw or undercooked marine fish (Price and Tom, 2003). Legislation in the European Union recognizes the growing anisakis problem and requires that fresh fish eaten raw or lightly cooked has to be frozen for 24 hours before consumption (Anonymous, 2004).

Cestodes

Cestode tapeworms use fish and copepods as intermediate hosts, although humans are one of the definitive hosts for some species. Most human diphyllobothriasis infections are asymptomatic but can result in abdominal discomfort, diarrhea, vomiting, and weight loss. It has been reported in Europe, Asia, North America, and South America, although not all incidences involve marine fish products. On a worldwide scale, new infections are reported regularly, especially from Russia and parts of Japan, whereas in South America there has been an increase in reports from fish, particularly salmonids.

There are several recognized species that infect humans, including *Diphyllobothrium latum, D. pacificum, D. dendriticum, D. klebanovskii,* and *D. nihonkaiense* (Mello Sampaio *et al.*, 2005). *D. pacificum* infects only saltwater fish, whereas *D. latum* infects only freshwater fish or species that spend part of their life in fresh water (Mello Sampaio *et al.*, 2005).

Human infections with *Diphyllobothrium pacificum* are related to marine fish and cases have been reported from Chile and Peru, probably because of the consumption of raw fish or ceviche (raw fish with salt and lemon) (Sagua *et al.*, 2000, 2001; Villena *et al.*, 2002). It has also been shown that new cases of human infection by *D. pacificum* were clearly associated with a cyclic manifestation of El Niño on the Chilean Pacific coast during the 1975–2000 period (Sagua *et al.*, 2001). However, South American diphyllobothriasis is not a new phenomenon, as *D. pacificum* eggs were found in coprolites, 4000- to 5000-year-old Chinchorro Chilean mummies, indicating that it is actually an ancient disease (Reinhard and Urban, 2003).

HUMAN VIRUSES AND SHELLFISH

Many viruses transmitted by the fecal-oral route are widely prevalent in the community, and infected individuals can shed millions of viral particles in their feces and also in some cases in urine. Sewage treatment processes, when present, are only partially effective at viral removal (Bofill-Mas *et al.*, 2006; Sorber, 1983), and coastal discharges constantly release human viruses into the marine environment. Bivalve shellfish, in the process of filter feeding, concentrate and retain human pathogens derived from sewage contamination in the shellfish-growing water, and the transmission of diseases by consumption of shellfish harvested in contaminated areas represents the most clearly identified health risk associated with coastal pollution by urban wastewater.

Viruses are obligate intracellular parasites depending entirely on a specific living host cell. Viruses then, unlike bacteria, do not grow or multiply in or on foods, but foods may become contaminated with human viruses that are excreted in feces or urine, or even in vomit, and transmit infection. Consequently, seafood may become contaminated by viral pathogens directly from infected people or through sewage pollution in the seawater.

Not all the viruses present in seawater represent a health risk for humans. In all natural aquatic environments, viral particles have proven to be highly abundant (approximately 10^8 virus-like particles/ml in seawater) (Bergh *et al.*, 1989) and show higher concentrations than bacteria. A wide diversity of natural populations of viruses forms part of aquatic ecosystems and infects the organisms present in these ecosystems; however, only human viruses have ever been associated with illness in seafood consumers.

Many viruses transmitted through the oral route produce subclinical infections in most people, but a symptomatic disease is diagnosed in only a small proportion of the population. However, such diseases may also be life threatening, as occurs in acute hepatitis infections in adults, as well as in severe gastroenteritis in young children and the elderly. The development of disease is related to the infective dose of the viral agent, the age, and the health status of the individual (pregnancy, presence of other infections or diseases), as well as his or her immunological and nutritional status and the availability of health care. Among viruses infecting humans, many different types are excreted in high concentrations in the feces of patients with gastroenteritis or hepatitis and in lower concentrations in the feces or urine of patients with other viral diseases. Moreover, viruses are also present in healthy individuals, and thus high viral loads are detected throughout the year in urban sewage and are regarded as environmental contaminants. Those viruses considered stable in the environment, and that are transmitted via contaminated food or water, lack the lipid envelopes that would render viral particles more fragile to environmental agents. Some viruses, such as human polyomaviruses and some adenovirus strains, infect humans during childhood, thereby establishing persistent infections. In the case of many frequent adenoviral respiratory infections, viral particles may continue to be excreted in feces for months or even years afterward.

A wide variety of studies have shown that even in highly industrialized countries, there is a high prevalence of human viruses in the aquatic environment, posing a threat to public health and being responsible for substantial economic losses. Following shedding into the environment, viruses can survive for weeks or months and can be detected either in the water column or for longer periods by attaching to particulate matter and accumulating in sediments (Bosch *et al.*, 1988; Callahan *et al.*, 1995; Girones *et al.*, 1989).

Viruses have not been associated with seafood spoilage issues and may survive well on, or in, contaminated seafood, especially as the common seafood processing procedures are freezing and icing that, in general, enhance the survival of viruses in the food. Of the various seafood species, only the bivalve mollusks have consistently proven to be an effective vehicle for the transmission of viral diseases. Historically there has been a strong association between the consumption of contaminated bivalve shellfish and outbreaks of enteric illnesses. According to data provided by the Food and Agriculture Organization of the United Nations, bivalve shellfish are consumed by inhabitants of all five continents (Food and Agriculture Organization of the United Nations, 2000). During the past three decades, there has been a constant increase in the number of reports of outbreaks of infection, especially for viral infections, and most reports have involved oysters, followed by clams, mussels, and other types of shellfish.

The detection of viruses in shellfish has been limited by the lack of appropriate methodologies, as most pathogenic viruses producing outbreaks related to shellfish consumption could not be efficiently detected using classical techniques of virus isolation in cell culture. The development of molecular methods, such as PCR-related techniques and quantitative real-time PCR assays in particular, has provided rapid and sensitive analytical tools (see Chapter 19). Their application in the specific detection of widely excreted viruses, as specific detection targets of human and animal viral contamination, provides an effective method for evaluating the microbiological quality of food and water and may be useful for tracing the sources of fecal contamination and for quantifying virus removal efficiency in water treatment and shellfish depuration plants.

PATHOGENIC AND EMERGENT VIRUSES DETECTED IN SHELLFISH

Epidemiological evidence suggests that human enteric viruses are the most common pathogens transmitted by bivalve shellfish (Lees, 2000). Human viruses transmitted by the fecal-oral route have been associated with shellfish either by the isolation of viruses from shellfish samples or through epidemiological studies linking them with diseases in shellfish consumers. Application of molecular methods of analysis has greatly increased the number of identifiable viruses in the environment. Moreover, these techniques permit not only the identification of those viruses causing clinical and subclinical infections in population groups but also the genetic characterization of viruses that cannot be efficiently cultured in cell lines. Such is the case for hepatitis E virus (HEV) and noroviruses. The use of genomic amplification techniques has also generated more comprehensive information on hepatitis A strains and certain adenoviruses present in the environment that remain inefficiently isolated in cell cultures.

Viruses identified in urban sewage and potentially transmitted through contaminated shellfish have been associated with a wide diversity of diseases including meningitis, paralysis, respiratory diseases, diarrhea and vomiting, myocarditis, congenital heart anomalies, acute hepatitis, and ocular infections. The most frequent pathology is gastroenteritis that also represents a cause of mortality not only in countries of emergent economy but also in highly industrialized countries, especially for the elderly, followed by acute hepatitis of high clinical importance. Most of the asymptomatic infections related to many of the excreted viruses represent a source of viral contamination in the environment and a potential source of new infections in the population, which is the case, for example, of most of the infections caused by enterovirus. The human polyomaviruses and some adenoviruses are excreted in urine or feces by healthy individuals over a period of years, and they may cause sporadic clinical syndromes and important diseases in immunocompromised people. This is the case for JCPyV that is the cause of progressive multifocal leucoencephalopathy (PML). Studies suggest a relation between colorectal cancer and the human polyomavirus JCPyV. The entry of oncogenic genes through the gastrointestinal route and their potential role in colon and other type of cancers are not yet elucidated but they deserve the development of further specific studies.

The most important pathogens identified in shellfish are hepatitis A viruses and noroviruses, which are being increasingly detected. HEV has been related to few cases of transmission of hepatitis through shellfish consumption and may also be considered as an emergent pathogen. Emergent and common groups of viruses detected in or transmitted through shellfish are briefly described with more detail next.

Hepatitis A Virus (HAV)

The hepatitis A virus is classified as genus *Hepatovirus* in the family *Picornaviridae* and causes acute hepatitis. HAV infection is the most serious viral infection linked to shellfish consumption, causing a debilitating disease and, occasionally, death.

Detection of the acute-phase antibody response is the mainstay of diagnosis. Preceding the appearance of antibodies, there is the shedding of virus in the feces and a viremia. HAV is excreted in feces for 1 to 2 weeks before the onset of illness. Prolonged shedding may occur, particularly among infants and children. It has been reported that some patients shed HAV in their stools for more than 1 month (up to 3 months) (Yotsuyanagi *et al.*, 1996). Detection of HAV RNA by RT-PCR in feces and serum allows an earlier diagnosis of the infection.

HAV infection occurs worldwide, being highly prevalent in regions where sanitation is limited or absent (Cuthbert, 2001). The transmission of the infection is by the fecal-oral route, mainly by person-to-person contact, food, and water. In developing countries with low standards of sanitation, HAV is endemic and infection occurs during early childhood. In contrast, in developed countries (with improved socioeconomic and environmental conditions), the average age increases, because adults are susceptible to the infection. Outbreaks of hepatitis A are common in crowded situations (such as institutions, e.g., schools, prisons, and in military forces) and have also been frequently associated with the consumption of contaminated shellfish. One impressive example is the outbreak described in Shanghai in 1988, when 250,000 people contracted hepatitis A after eating contaminated clams (Halliday *et al.*, 1991). The introduction in some areas of a vaccination program for preadolescents using an inactivated vaccine against HAV will probably reduce the number of susceptible individuals and the dissemination of the virus in these populations, but this effect has not yet been reflected in the number of cases and outbreaks registered at present.

In outbreak situations, up to 20% of cases are caused by secondary transmission (Koopmans *et al.*, 2002). Investigations of waterborne outbreaks of viral disease have shown that bacteriological indicators of fecal contamination may be inadequate to detect viral contamination. This lack of correlation may be related to the fact that some viruses (e.g., HAV) survive longer than fecal coliform bacteria in the environment (De Serres *et al.*, 1999).

Noroviruses

The genus *Norovirus* (Fig. 18-1) belongs to the *Caliciviridae* family, which is made up of four genetically and antigenetically diverse virus groups that differ in the animal host of preference. In the genus *Vesivirus,* strains such as the *San Miguel Sea Lion virus* serotype 5 (SMSV-5) have exhibited an extraordinary lack of host specificity, and these

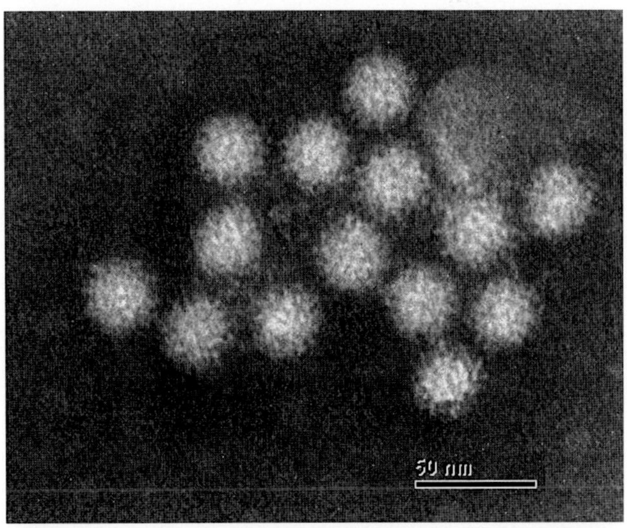

FIGURE 18-1. Norovirus in a gastroenteritis patient observed with the use of electron microscopy. A scale bar (50 nm) is shown. Image kindly donated by the Electron Microscopy Unit of the National Center of Microbiology, Madrid, Spain.

viruses are able to infect a wide diversity of species including fish, marine mammals, pigs, cattle, and also humans, albeit with very low incidence. Human caliciviruses are distributed in two genera, *Norovirus* and *Sapovirus*. Norovirus (previously Norwalk-like viruses) and other enteric caliciviruses cause acute gastroenteritis in humans. Although asymptomatic infections are common, studies have shown that noroviruses are the most frequent cause of gastroenteritis in all age groups, with higher incidences in young children, but they often produce infections in adults as well. This is in contrast to other virus groups such as rotaviruses, astroviruses, and adenoviruses, all of which principally cause gastroenteritis in children. Noroviruses are responsible for sporadic gastroenteritis and outbreaks in schools, hospitals, homes for the elderly, hotels, and on cruise ships. Caliciviruses, especially the noroviruses, have emerged as the predominant cause of gastroenteritis associated with shellfish consumption. Gastrointestinal symptoms in such cases are typically described as relatively mild, often including nausea, vomiting, diarrhea, and abdominal cramps after an incubation period of 30 to 40 hours. The symptoms last for 2 days, followed by complete recovery with no complications (Potasman *et al.*, 2002).

The identification of noroviruses has proved problematic, in addition to the fact that it has not been possible to culture these viruses in cell lines. Moreover, this group features a high degree of genetic diversity, and detection in environmental or food samples requires the use of RT-PCR assays with highly degenerated primers, as well as a confirming step involving hybridization with a diversity of probes or a sequencing analysis. This genus is distributed in two genogroups that are similarly divided into diverse genotypes. Within these genotypes, new variants and recombinant strains have been identified in the epidemiological and surveillance studies currently being undertaken.

Hepatitis E Virus (HEV)

The hepatitis E virus belongs to the newly designated *Hepeviridae* family and is considered an emergent virus in industrialized countries. The *Hepevirus* contain an RNA lineal genome. Only one serotype has been described until now, although the detected strains have been distributed in four genogroups. Although hepatitis E most commonly causes subclinical infections, it is also the principal cause of sporadic acute hepatitis in many countries with low sanitation levels and is a source of severe outbreaks. In the general population, the lethality index associated with liver failure is 1%, a percentage that exceeds HAV and that nears 20% in pregnant woman, especially during the third trimester. Although there is no direct evidence of association between shellfish and transmission of HEV, several studies strongly suggest shellfish consumption as the origin of HEV infections (Koizumi *et al.*, 2004; Mechnik *et al.*, 2001).

One possible source of HEV is terrestrial animals. According to a study undertaken in the United States, 80% of pigs older than 3 months were seropositive (Meng *et al.*, 1997). In addition, data from other countries confirmed that HEV is spread throughout the world's swine population. In Catalonia (Spain), a seroprevalence in pigs measuring approximately 20% has been recorded. Moreover, the analyzed genomic regions of swine strains are closely related, not only to those strains identified in the clinical serum samples of patients with acute hepatitis E, but also to urban sewage samples collected in the same geographic area. Descriptions of humans infected with hepatitis E after consuming wild boar and venison in Japan have also been published. Antibodies against HEV have been identified in rats, chickens (Avian HEV), cows, goats, cats, and dogs, although until now there has been no proof establishing these animals as viral reservoirs of virus that could potentially infect humans.

More than 20 sequences of HEV have been identified in a study analyzing urban sewage samples in industrialized countries, some proving identical to those sequenced strain regions detected in clinical serum samples (Clemente-Casares *et al.*, 2003). The phylogenetic tree generated with these strains, as well as other representative strains, is presented in Figure 18-2. Hepatitis E infections have been traditionally regarded as imported diseases in both the United States and Europe. However, recently described autochthonous strains strongly suggest that some level of endemicity also persists in highly industrialized countries and that there is still a need for sensitive diagnostic kits and HEV environmental contamination controls.

Enterovirus

Enterovirus is a genus of the *Picornaviridae* family, which comprises a large family of RNA viruses. Enteroviruses infect the gastrointestinal tract but they do not often cause the classical gastrointestinal symptoms. However, they could spread to other organs and cause serious diseases, such as poliomyelitis, myocarditis, aseptic meningitis, and other infections. The enterovirus genus currently consists of about 111 serotypes, which are grouped into five species that contain human viruses: poliovirus (3 serotypes), human enterovirus A (12 serotypes), human enterovirus B (36 serotypes), human enterovirus C (11 serotypes), and human enterovirus D (2 serotypes).

Enteroviruses replicate in the gastrointestinal tract and are shed in feces in large numbers (Lees, 2000). Because it is relatively easy to isolate them in cell culture, they have been used in the past as classical viral indicators of human viral contamination. However, the presence of outbreaks of enterovirus in a population related to seafood consumption has not been clearly established.

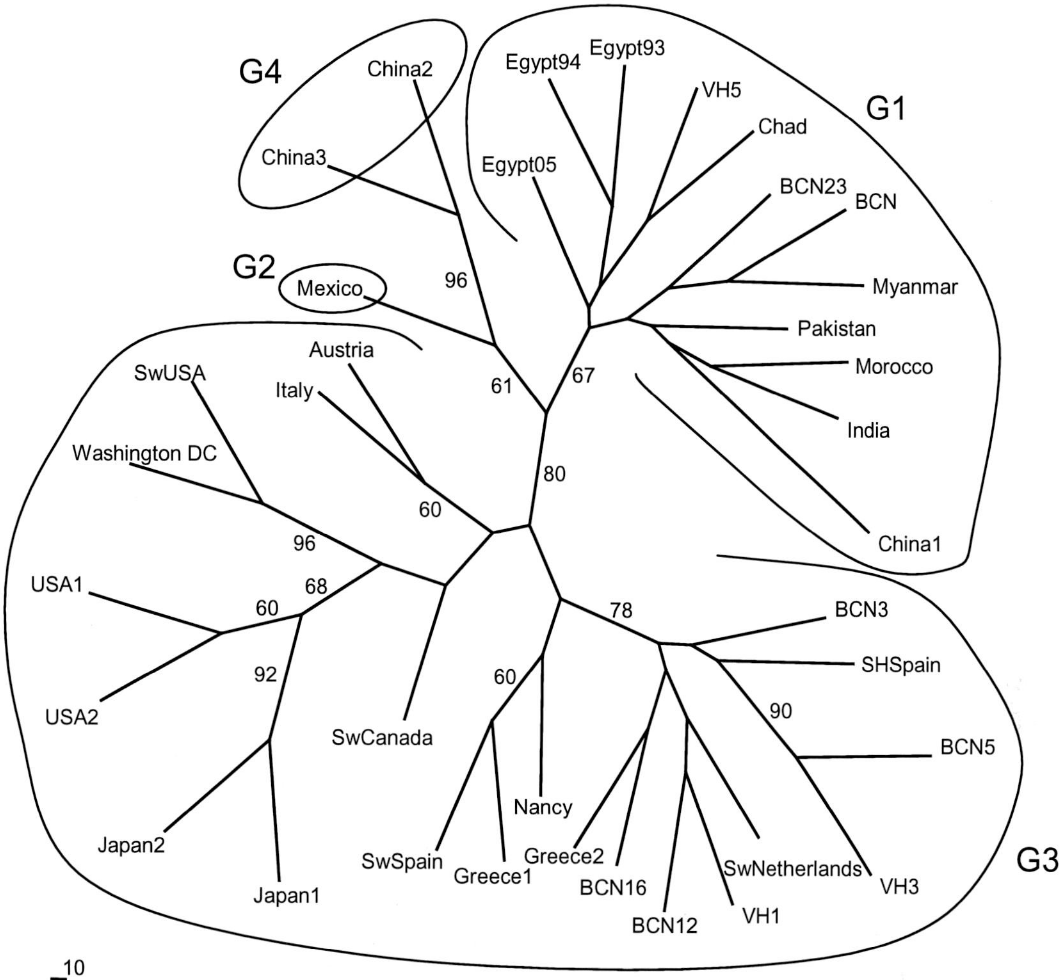

FIGURE 18-2. Unrooted phylogenetic tree depicting the relationship, when comparing 101 nucleotides within ORF2, between HEV strains isolated from sewage (collected in Spain -BCN, BCN3, BCN5, BCN12, BCN16, BCN23-, France –Nancy-, USA -Washington DC-, Egypt –Egypt05-) and other isolates from genotype 1 (China1, Pakistan, India, Myanmar, Chad, Egypt93, Egypt94, Morocco, VH5), genotype 2 (Mexico), genotype 3 (USA1, USA2, Greece1, Greece2, Italy, Austria, Japan1, Japan2, VH1, VH3, SwUSA, SwCanada, SwNetherlands, SwSpain, SHSpain), and genotype 4 (China2, China3). VH1, VH3, and VH5 are strains isolated from clinical cases in Barcelona (Spain). SHSpain is a sequence isolated from sewage generated by a slaughterhouse in the Barcelona area, and strains with "Sw" have a porcine origin. The internal node numbers represent bootstrap values (1000 replicates) expressed as the percentage of all trees. Only values greater than 60 are represented. Tree kindly donated by Dr. Pilar Clemente Casares from the Department of Microbiology, Faculty of Biology, University of Barcelona, Spain.

Adenoviruses

The *Adenoviridae* (Fig. 18-3) family includes a group of icosahedral nonenveloped viruses containing a double-stranded DNA genome. Adenoviruses may be excreted by healthy people during a period of months or even years after an initial infection and may cause different levels of respiratory and gastrointestinal infections. Adenoviruses infect a broad spectrum of species, and 51 human, 6 porcine, and 10 bovine serotypes have been described; there are adenoviruses that also infect other hosts, such as avian or amphibian. Human adenoviruses have been classically included in the genus Mastadenovirus. Methods based on nucleic acid amplification for adenovirus detection in environmental and shellfish samples have been described in the literature (Hernroth *et al.*, 2002; Muniain-Mujika *et al.*, 2003; Pina *et al.*, 1998). In these studies, the detection of adenoviruses has been proposed as a molecular indicator of human viral contamination. These viruses have a higher prevalence in shellfish than other viruses studied.

There are no significant reports on outbreaks of adenoviral infections related to shellfish consumption, but the detection of this group of viruses has been suggested as a useful molecular index for the evaluation and control of shellfish virological quality (Formiga-Cruz *et al.*, 2002).

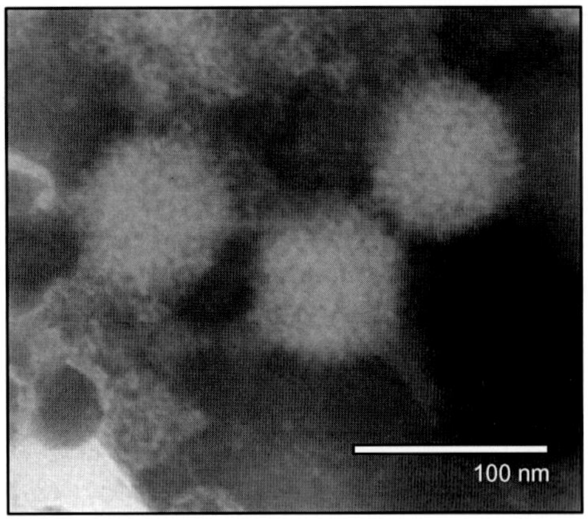

FIGURE 18-3. Electron micrograph of adenovirus viral particles. A scale bar (100 nm) is shown.

Rotaviruses and Astroviruses

Rotaviruses and astroviruses are RNA viruses belonging to two different families, *Reoviridae* and *Astroviridae*, respectively. *Reoviridae* has double-stranded segmented RNA, and *Astroviridae* contains a single-stranded RNA genome. Both groups have been related to gastroenteritis especially in young children. Rotavirus is the most common cause of severe diarrhea among children, resulting in the hospitalization of approximately 55,000 children each year in the United States and the death of more than 600,000 children annually worldwide. Astrovirus infection is generally mild and self-limiting, rarely leading to severe dehydration, hospitalization, or death. A few cases of infection caused by these viruses as a result of the consumption of seafood have been described (Lees, 2000). Water and food is a significant route of transmission for these viruses, mostly in developing countries.

PRESENCE, STABILITY, AND CONTROL OF VIRUSES IN SHELLFISH

Current regulations pertaining to shellfish and their growing waters are based on bacterial standards (fecal coliforms and *Escherichia coli*) and have largely prevented bacterial gastrointestinal infections. However, they are believed to have limited predictive value for viral pathogens such as noroviruses and hepatitis A (Desenclos *et al.*, 1991; Formiga-Cruz *et al.*, 2003; Wanke and Guerrant, 1990). Several studies have revealed the differential rates of reduction of bacteria and viruses in depurating shellfish (Dore and Lees 1995; Formiga-Cruz *et al.*, 2002), and thus shellfish quality should include detection of viruses.

The detection of viral contamination through mollusk production and processing procedures requires the use of highly sensitive and molecular techniques. Diverse studies have evaluated the presence of viruses in shellfish and have identified the presence of adenoviruses, noroviruses, hepatitis A, or enteroviruses. HEV has also been suggested to be transmitted through contaminated shellfish in a few studies (Koizumi *et al.*, 2004; Mechnik *et al.*, 2001). Various research laboratories have been developing and evaluating diverse procedures for the detection of viruses in shellfish, but there is no single procedure that could be considered standard, although there are ongoing efforts to define standard methods for the detection of at least the two principal pathogens associated with contaminated shellfish consumption, noroviruses and hepatitis A. Several distinct methods have been evaluated and applied in a large array of studies (Formiga-Cruz *et al.*, 2003; Jothikumar *et al.*, 2005; Le Guyader *et al.*, 2003; Mullendore *et al.*, 2001).

Harvesting from areas with good water quality is generally the most widely acceptable approach for controlling the contamination of shellfish. Once shellfish is harvested, microbial contamination may be reduced by heat treatment (cooking) or by extending the natural filter-feeding processes in clean seawater to remove microbial contaminants. This can be performed in tanks, termed "depuration," or in the natural environment, termed "relaying" (Lees, 2000), and shellfish will then be commercialized only after the level of the standard fecal indicators are in accordance with the sanitary legislation applicable. Shellfish harvesting may also be prohibited where there are unacceptable levels of pollution.

Virus stability is an important parameter to be considered in conjunction with the treatments applied to shellfish containing even moderate levels of contamination. Hepatitis A virus in shellfish could be inactivated by more than 4 $\log_{10}$ infectious units by raising the internal temperature of shellfish meats to 85 to 90°C for 1 minute (Millard *et al.*, 1987); therefore, it is recommended that commercial heat treatment processes raise the internal temperature of shellfish meats to 90°C for 1.5 minutes (Anonymous, 1993). It seems probable that viral problems associated with home and restaurant cooked shellfish are a consequence of under- or inconsistent cooking.

Numerous problems contribute to shellfish-associated infections, and outbreaks of infection continue to occur despite the control measures highlighting the role of viruses, especially noroviruses and hepatitis A, as contaminants in shellfish. Commercial depuration has been applied to most bivalve shellfish species sold live, including oysters, clams, mussels, and cockles and, when used as a treatment process to reduce microbial contaminants, it is subject to legal control in the European Union, the United States, and other countries. Tank-based depuration is widely practiced in many countries and depuration periods may vary from 1 to 7 days, although 1 or 2 days are the most widely used. The systems used for depuration are diverse and include processes where water is static or changed in batches, systems where seawater is flushed through continually or recycled through a disinfection treatment. Water disinfection processes include ozone, chlorination, UV irradiation, and iodophors (Lees, 2000). In a study described by Muniain-Mujika *et al.* (2002), the depuration of naturally contaminated mussels was monitored using a tank equipped with particle filters where seawater at 17 to 18°C was recirculated through the tank six to eight times per hour, passed through a biological filter and a skimmer, and disinfected by UV irradiation and ozone. The presence of viruses was monitored using nested-PCR and PFU for various groups of bacteriophages. In this study, bacteriophages infecting *B. fragilis* HSP40 were detected by PCR and PFU quantification and the purification curves followed equivalent kinetics, which showed that free viral DNA is not stable under these conditions in shellfish meat. Negative results for the presence of the most prevalent viruses, such as adenoviruses and enteroviruses, were not obtained until the fifth day of treatment, despite the fact that the acceptable *E. coli* levels were observed after the first 24 hours of depuration treatment (Muniain-Mujika *et al.*, 2002).

There is then experimental and epidemiological evidence from various sources suggesting that depuration plants functioning satisfactorily according to the standard fecal coliform criteria may still fail to remove human enteric viruses fully (Grohmann *et al.*, 1981). A study undertaken by various laboratories of the European Union showed that the percentage of shellfish samples with moderate levels of fecal contamination (classified as B according to the European Union standards), which were positive for the human enteric viruses studied (adenoviruses, hepatitis A, enteroviruses, and noroviruses) did not change after depuration of 1 or 2 days, although they complied with current regulations based on the concentration of *E. coli* in the samples (Formiga-Cruz *et al.*, 2002).

Because of uncertainties in the effectiveness of depuration processes, the design, technologies and operation of depuration plants require reevaluation by considering the new methods now currently available for the detection of viruses.

MANAGEMENT AND MONITORING

There are several identifiable stages in fish and shellfish production, from capture or harvesting to the point of consumption, where the product can be susceptible to contamination from a variety of sources, such as fecal pollution, cross-contamination, biotoxins or hazardous substances, and inadequate food packaging. Consequently, there are certain standards, including legislation and quality assurance programs, designed to improve the safety of marine food resources and lead to enhanced consumer protection. The most common and effective approach to identify potential risks is the hazard analysis critical control point (HACCP) concept, successfully applied in other areas of food safety. However, HACCP does not necessarily point to the potential consequences of the occurrence of a hazard or the effectiveness of risk reduction techniques. A more systematic approach to identifying what could go wrong in any procedure is risk analysis and this procedure can additionally assess the likelihood and consequences of a hazardous event occurring and formulate measures to manage any perceived risk.

Nevertheless, in general terms, the best means of minimizing the risk to public health resulting from, for instance, fecal contamination, is to ensure that shellfish are grown in waters free from sewage contamination and that fish are harvested in more open waters. However, the provision or maintenance of clean growing waters can be difficult and is not always possible, particularly in more urban areas. Consequently, harvesting from coastal areas, in particular, should be from carefully controlled, monitored areas and not from grossly polluted waters, unclassified areas or from classified areas with a significant temporary reduction in water quality.

Consumer Safety and Quality Assurance Programs

The objective of a quality assurance program is ultimately to protect the consumer. However, because of a variety of factors, the existing programs cannot completely guarantee that the product is free from fecal pollution. Because mollusks are filter feeders that obtain their nutrition by filtering several liters of water a day, they may also concentrate potentially infectious microorganisms during the same process, whereas fish products are more prone to spoilage organisms or be carriers of zoonotic parasites. In addition, mollusks are more frequently marketed and eaten raw or only partly cooked, thereby increasing the risk of infectious illness by organisms that would normally be killed or inactivated by heat. Therefore, in shellfish programs, for instance, the use of an indicator organism provides presumptive evidence that disease causing fecal pathogens may or may not be present. When pathogenic organisms are present, they will usually be found at numbers lower than the indicator organisms. Thus, when indicator organisms are absent or in low numbers, it is assumed that the water can be safely used to grow shellfish and that the shellfish themselves should be safe to consume. Although the test methods do not guarantee that harvesting area waters or their shellfish are free of bacterial or viral pathogens, they do statistically reduce the potential of shellfish contamination and therefore also reduce shellfish related incidences of disease. Although correctly tested market shellfish are usually of high quality, none of the programs currently in use can provide a zero risk of ingesting a disease producing organism, especially enteric viruses and *Vibrio* spp. Consequently, the aim of management programs is to reduce the risks from consuming contaminated fish and shellfish to an acceptable level.

Standards

The international requirements for microbiological analysis of fish, shellfish, and marine water samples are contained in the Codex Alimentarius, as well as procedures stipulated by the United States and the European Union. The fundamental basis for shellfish programs is a monitoring and classification program, using fecal coliform indicators to show the levels of pollution in harvesting areas. In addition, there are requirements for routine monitoring of marine phytoplankton biotoxins and other potential contaminants, such as hydrocarbons and heavy metals.

Codex Alimentarius

The Codex Committee on Fish and Fishery Products has a joint FAO/WHO (Food and Veterinary Office/World Health Organization) Food Standards Program draft for molluscan shellfish. Essentially, the draft standard applies to live bivalve mollusks intended for direct consumption and quick frozen and canned bivalve mollusks. Product traceability is also an important feature of the draft. The Codex Committee establishes the Codex Code of Practice for Fish and Fishery Products, which provides general advice on the production, processing, storage, and handling of fishery products on board fishing vessels and on shore, as well as their distribution, export, import, and sale (Codex, 2005). The code includes guidelines related to the minimization of product contamination and deterioration in conjunction with the HACCP concept and defect analysis.

U.S. National Shellfish Sanitation Program (NSSP)

The voluntary, tripartite cooperative U.S. NSSP is comprised of two operation manuals that are periodically revised to reflect changing circumstances and improvements in methodology or working practices (www.issc.org/On-Line_docs/onlinedocs.htm). Part I focuses on the sanitation of shellfish growing areas, which includes classification, laboratory procedures, relaying, patrol operations, and marine biotoxins. Part II focuses on sanitation of harvesting, processing, and distribution of shellfish, which includes operating, inspecting and certifying shellfish shippers, processors, and depuration facilities, in addition to controlling interstate shipments of shellfish.

The concept of classification of growing water areas based on a comprehensive sanitary survey and water based classification system are key features of the NSSP. Classification is related to a number of critical factors, including actual and potential pollution sources and their quantification, as well as meteorological and seasonal factors affecting the area. The U.S. NSSP classification categories are defined as approved, conditionally approved, restricted, conditionally restricted, or prohibited.

In addition, the U.S. FDA can use the NSSP as the basis for certifying foreign shellfish sanitation programs by initiating a Memorandum of Understanding with the official agency in the country that, for instance, wants to export shellfish to the United States. The U.S. FDA also has a zero tolerance for *V. vulnificus* in their guidelines for microbial contaminants in seafood (Anonymous, 2001).

EU Directive 91/492/EEC (Placing Shellfish on the Market)

The EU program uses shellfish flesh samples for classification of harvesting areas and stipulates a specified end product standard. This directive requires all member states to classify their shellfish harvesting areas into one of three categories (A, B, or C), according to the degree of fecal indicator bacteria present in samples of shellfish flesh. Shellfish from a category A area can go for direct human consumption if they contain less than 230 *E. coli* (or 300 fecal coliforms)/100 g flesh and no *Salmonella* sp. in 25 g flesh. Shellfish from category B areas must not exceed, in 90% of samples, the limit of 4,600 *E. coli* (or 6000 fecal coliforms)/100 g flesh. Shellfish from category C areas must not exceed the limit of 60,000 fecal coliforms/100 g of flesh. The EU shellfish hygiene controls also apply a common import system for third countries, since the provisions applied to imports of live bivalve mollusks from third countries must be at least equivalent to those governing the production and placing on the market within the European Union itself. The EU system monitors the quality of the end product and the consumer is protected through microbiological testing, incorporated facilities standards and a "paper trail" approach for back tracing any eventual problems. This is the responsibility of the appropriately designated national competent authority.

EU Directive 91/493/EEC (Fishery Products)

This directive deals with the conditions applied to cooked crustacean and molluscan shellfish products but is mainly concerned with fish. It recognizes that freshly caught fishery products are in principle free of contamination with microorganisms, but contamination and subsequent decomposition may occur when handled and treated in an unhygienic fashion. The directive therefore stipulates the essential requirements for correct hygienic handling of fresh and processed products at all stages of production, as well as during storage and transport.

The EU Food and Veterinary Office (FVO) is the group that promotes effective control systems for food safety and quality through inspection checks on compliance with the requirements of EU legislation both within the EU member states and in third countries exporting to the European Union. The FVO makes recommendations to the country's competent authority to deal with any shortcomings revealed during inspections.

European Legislative Changes

The current horizontal EU legislation regulating all water bodies and all food commodities is established by the Water Framework Directive (WFD; Anonymous, 2000) and by the new EU food safety regulations (Anonymous, 2004, 2005), respectively. These are designed to supersede the current legislation regarding shellfish growing waters (Directive 79/923/EEC) and placing shellfish on the market (Directive 91/492/EEC). In the case of live bivalve mollusks, criteria for official controls are developed in Annex II of Regulation (EC) 854/2004 (Anonymous, 2004), which requires classification of production areas and their monitoring, as previously stipulated in Directive 91/492. Microbiological monitoring points have to be representative of the pollution contamination conditions and sampling programs should be conducted on a systematic basis. The process of gathering these data and making an assessment is commonly described as a "sanitary survey," in a similar way to the U.S. NSSP. Of note, present EU directives and decisions regarding fisheries products and live shellfish do not include either standards or guidelines for vibrios (Anonymous, 2001). However, a guide to good practice for microbiological monitoring has been developed to assist competent authorities and scientific institutes responsible for implementing the current requirements (www.crlcefas.org).

Hazard Analysis Critical Control Point (HACCP)

HACCP was originally developed in the United States for large commercial food production operations but is now applied across most aspects of food hygiene, inspection, and safety. The HACCP system for seafood safety control originated in the United States following concerns within industry, government, and consumer groups about the need to improve seafood safety in the 1980s (National Research Council [NRC], 2003). Essentially, it is the sequential inspection of a food preparation process in order to identify the possible effect of failure in any individual step, as a result of error or the inability to reach minimum specified standard parameters, particularly regarding microbiological hazards. The system should include a science-based analysis of potential hazards at any point in the food chain, which is then used to prevent potential problems and take corrective actions if necessary. The U.S. FDA has estimated that the introduction of HACCP by fish processors in the United States averted 20% to 60% of seafood-borne illness cases (FDA, 1994). In general terms, reports from the United States indicate sustained decreases of 15% to 49% in the reported incidences of *Campylobacter, Yersinia, Listeria,* and *Salmonella* infections from 1996 to 2001. These reductions have been attributed to implementation of food safety measures, which include the HACCP concept (CDC, 2002).

Risk Analysis

The risk analysis process is made up of components that include risk assessment, risk communication, and risk management (MacDiarmid, 2001). The essential first step though within the assessment phase can be referred to as hazard identification, and, in the case of food safety, virtually every interested group would agree that the risk to the consumer should be the minimum possible. Unfortunately, one potential problem with a developing science such as risk analysis is the availability of sufficient data that are relevant and in the correct form. For instance, without the necessary statistics on illness related directly to food poisoning following shellfish consumption, it is difficult to interpret the problem. In addition, there is a basic lack of quality information concerning pathogen occurrence or removal parameters.

Control programs are designed to focus on risk management. This system recognizes that a problem exists but that it is "managed" by a combined, integrated stepwise strategy through the use of, in the case of shellfish, sanitary surveys, water quality or shellfish meat testing, classification of harvesting areas, the perceived need for depuration or relaying, and crisis reaction management plans. However, usually the control program does not directly address the cause of the problem. Nevertheless, once certain key management options have been identified, controls can be implemented in order to reduce any potential risk. For instance, *V. parahaemolyticus* and the occurrence of illness following consumption of shellfish is associated with the summer months when the seawater temperature is higher. In this case, control by prohibiting harvesting of shellfish from areas associated with food poisoning outbreaks when the water temperature exceeds a certain level could be considered (Anonymous, 2001).

One of the key elements of an effective program is an enforcement infrastructure capable of preventing the harvest of seafood from compromised areas and subsequently introducing these into the market, although this is more difficult in developing countries that often lack the resources necessary to implement such programs.

SUMMARY

Although foodborne diseases are a known, widespread public health problem, consumer health risks from marine food resources are usually considered to be low for seafood derived from unpolluted open marine environments. The risk of foodborne illnesses from products arising from more closed coastal environments is higher and related to the greater potential for contamination in such regions as opposed to capture fisheries. Illness can also be related to failures in postharvest food processing, cross-contamination, poor storage conditions or inadequate cooking, and the current trend toward extended shelf life products.

Risks associated with contaminated seafood can be reduced through standards based on the HACCP, including regulatory monitoring and quality assurance programs, which have been shown to be effective for reducing the likelihood of illness. New techniques based on molecular methods are also now available for the identification of pathogens in seafood, and the implementation of these techniques may lead to improved monitoring and subsequently result in the improved quality of seafood distributed worldwide.

References

Abdussalem, M., Käferstein, F.K., Mott, K., 1995. Food safety measures for the control of foodborne trematode infections. Food Control, 6, 71–79.

Adak, G.K., Meakins, S.M., Yip, H., Lopman, B.A., O'Brien, S.J., 2005. Disease risks from foods, England and Wales, 1996–2000. Emerg. Infect. Dis. 11, 365–372.

Alam, M., Hasan, N.A., Sadique, A., Bhuiyan, N.A., Ahmed, K.U., Nusrin, S., Nair, G.B., Siddique, A.K., Sack, R.B., Sack. D.A., Huq, A., Colwell, R.R., 2006. Seasonal cholera caused by *Vibrio cholerae* serogroups O1 and O139 in the coastal aquatic environment of Bangladesh. Appl. Environ. Microbiol. 72, 4096–4104.

Anonymous, 1993. Council Decision of 11 December 1992 approving certain heat treatments to inhibit the development of pathogenic microorganisms in bivalve moluscs and marine gastropods (93/25/EEC). Off. J. Eur. Communities. 16, 22–23.

Anonymous, 1999. Opinion of the Scientific Committee on Veterinary Measures Relating to Public Health on *Listeria monocytogenes*. European Commission. Health and Consumer Protection Directorate General.

Anonymous, 2000. Council Directive 2000/60/EC of the European Parliament and of the Council of 23 October 2000 establishing a framework for Community action in the field of water policy. Off. J. Eur. Communities, L 327, 22.12.2000, 1.

Anonymous, 2001. Opinion of the Scientific Committee on Veterinary Measures Relating to Public Health on *Vibrio vulnificus* and *Vibrio parahaemolyticus*. European Commission. Health and Consumer Protection Directorate General.

Anonymous, 2004. Regulation (EC) No 853/2004 of the European Parliament and of the Council of 29 April 2004 laying down specific hygiene rules for food of animal origin. Off. J. Eur. Communities (L226: 22–82).

Anonymous, 2005. Regulation (EC) No 2073/2005 of the European Parliament and of the Council of 15 November 2005 on microbiological criteria for foodstuffs. Off. J. Eur. Communities (L338: 1–26).

Appleton, H., Pereira, M.S., 1977. A possible virus aetiology in outbreaks of food poisoning from cockles. Lancet i, 780–781.

Arias, C.R., Macián, M.C., Aznar, R., Garay, E., Pujalte, M.J., 1999. Low incidence of *Vibrio vulnificus* among *Vibrio* isolates from sea water and shellfish of the western Mediterranean coast. J. Appl. Microbiol. 86, 125–134.

Aureli, P., Fenicia, L., Ferrini, A.M., 1984. Food poisoning caused by *Clostridium botulinum* type E (article in Italian). Boll. Ist. Sieroter. Milan. 63, 552–555.

Australian Dairy Authorities' Standards Committee (ADASC), 1999. Australian manual for control of *Listeria* in the dairy industry.

Balakrish Nair, G., Hormazábal, J.C., 2005, The *Vibrio parahaemolyticus* pandemic Rev. Chil. Infect. 22, 125–130.

Bauer, A., Østensvik, Ø., Florvåg, M., Rørvik, L.M., 2006. *Vibrio parahaemolyticus, V. cholerae* and *V. vulnificus* in Norwegian shellfish. In Molluscan Shellfish Safety. Proceedings of 5th International Conference on Molluscan Shellfish Safety, Galway, Ireland 2004, Shell Viruses and Pathogens, 295–298.

Bergh, O., Borsheim, K. Y., Bratbak, G., Heldal, M., 1989. High abundance of viruses found in aquatic environments. Nature 340, 467–468.

Bird, P.D., Kraa, E., 1995. Overview of the 1990 viral gastro-enteritis outbreak from oysters. In Poggi, R, and Le Gall, J.-Y. (eds.), Shellfish Depuration, IFREMER-ENSP. Proceedings of the 2nd International Conference on Shellfish Purification. Rennes, France, April 1992, 31–36.

Bofill-Mas, S., Albinana-Gimenez, N., Clemente-Casares, P., Hundesa, A., Rodriguez-Manzano, J., Allard, A., Calvo, M., Girones, R., 2006. Quantification and stability of human adenoviruses and polyomavirus JCPyV in wastewater matrices. Appl. Environ. Microbiol. 72, 7294–7296.

Borroto, R.J., 1997. Ecology of *Vibrio cholerae* serogroup 01 in aquatic environments (article in Spanish). Rev. Panam. Salud Publica. 1, 3–8.

Bosch, A., Lucena, F., Girones, R., Jofre, J., 1988. Occurrence of enteroviruses in deep marine sediment. Can. J. Microbiol. 34, 921–924.

Brands, D.A., Inman, A.E., Gerba, C.P., Mare, C.J., Billington, S.J., Saif, L.A., Levine, J.F., Joens, L.A., 2005. Prevalence of *Salmonella* spp. in oysters in the United States. Appl. Environ. Microbiol. 71, 893–897.

Butt, A.A., Aldridge, K.E., Sanders, C.V., 2004a. Infections related to the ingestion of seafood. Part I: Viral and bacterial infections. Lancet Infect. Dis. 4, 201–212.

Butt, A.A., Aldridge, K.E., Sanders, C.V., 2004b. Infections related to the ingestion of seafood. Part II: Viral and bacterial infections. Lancet Infect. Dis. 4, 294–300.

Cabanillas-Beltrán, H., LLausás-Magaña, E., Romero, R., Espinoza, A., García-Gasca, A., Nishibuchi, M., Ishibashi, M., Gomez-Gil, B., 2006. Outbreak of gastroenteritis caused by the pandemic *Vibrio parahaemolyticus* O3:K6 in Mexico. FEMS Microbiol. Lett. 265, 76–80.

Callahan, K.M., Taylor, D.J., Sobsey, M.D., 1995. Comparative survival of hepatitis A virus, poliovirus and indicator viruses in geographically diverse seawaters. Water Sci. Technol. 3, 189–193.

Carlin, F., Broussolle, V., Perelle, S., Litman, S., Fach, P., 2004. Prevalence of *Clostridium botulinum* in food raw materials used in REPFEDs manufactured in France. Int. J. Food Microbiol. 91, 141–145.

Centers for Disease Control and Prevention (CDC), 2002. Preliminary Food Net data on the incidence of food bore illness—selected sites, United States, 2001. Morbid Mortal Weekly Rep. 51, 325–329.

Centers for Disease Control and Prevention (CDC), 2006a. Two cases of toxigenic *Vibrio cholerae* O1 infection after Hurricanes Katrina and Rita—Louisiana, October 2005. Morbid Mortal Weekly Rep. 55, 31–32.

Centers for Disease Control and Prevention (CDC), 2006b. *Vibrio parahaemolyticus* infections associated with consumption of raw shellfish—three states, 2006. Morbid Mortal Weekly Rep. 55, 854–856.

Clemente-Casares, P., Pina, S., Buti, M., Jardi, R., Gironés, R., 2003. Hepatitis E virus epidemiology in industrialized countries. Emerg. Infect. Dis. 9, 448–455.

Codex Alimentarius Commission (CAC), 2005. Code of Practice for Fish and Fishery Products (CAC/RCP 52–2003, Rev. 2–2005).

Cook, D.W., Oleary, P., Hunsucker, J.C., Sloan, E.M., Bowers, J.C., Blodgett, R.J., Depaola, A., 2002. *Vibrio vulnificus* and *Vibrio parahaemolyticus* in U.S. retail shell oysters: A national survey from June 1998 to July 1999. J. Food Prot. 65, 79–87.

Cordoba, M.G., Jordano, R., Aranda, E., Benito, M.J., Cordoba, J.J., 2003. Differentiation of toxigenic *Staphylococcus aureus* in staphylococcal isolates from prepared and frozen foods by combined arbitrarily primed polymerase chain reaction and DNA probe. Nahrung 47, 166–170.

Cordova, J.L., Astorga, J., Silva, W., Riquelme, C., 2002. Characterization by PCR of *Vibrio parahaemolyticus* isolates collected during the 1997–1998 Chilean outbreak. Biol. Res. 35, 433–440.

Cuthbert, J. A., 2001. Hepatitis A: Old and new. Clin. Microbiol. Rev. 14, 38–58.

Dawar, M., Moody, L., Martin, J.D., Fung, C., Isaac-Renton, J., Patrick, D.M., 2002. Two outbreaks of botulism associated with fermented salmon roe: British Columbia, August 2001. Can. Commun. Dis. Rep. 28, 45–49.

Deepanjali, A., Kumar, H.S., Karunasagar, I., Karunasagar, I., 2005. Seasonal variation in abundance of total and pathogenic *Vibrio parahaemolyticus* bacteria in oysters along the southwest coast of India. Appl. Environ. Microbiol. 71, 3575–3580.

Degnan, A.J., Kaspar, C.W., Otwell, W.S., Tamplin, M.L., Luchansky, J.B., 1994. Evaluation of lactic acid bacterium fermentation products and food-grade chemicals to control *Listeria monocytogenes* in blue crab (*Callinectes sapidus*) meat. Appl. Environ. Microbiol. 60, 3198–3203.

De Paola, A., Capers, G.M., Alexander, D., 1994. Densities of *Vibrio vulnificus* in the intestines of fish from the U.S. Gulf Coast. Appl. Eviron. Microbiol. 60, 984–988.

Desenclos, J.C., Klontz, K.C., Wilder, M.H., Nainan, O.V., Margolis, H. S., Gunn, R.A., 1991. A multistate outbreak of hepatitis A caused by the consumption of raw oysters. Am. J. Public Health 81, 1268–1272.

De Serres, G., Cromeans, T.L., Levesque, B., Brassard, N., Barthe, C., Dionne, M., Prud'homme, Paradis, D., Shapiro, C.N., Naiman, O.V., Margolis, H.S., 1999. Molecular confirmation of hepatitis A virus from well water: Epidemiology and public health implications. J. Infect. Dis. 179, 37–43.

Do Gyun, K., Tong-Soo, K., Shin-Hyeong, C., Hyeon-Je, S., Woon-Mok, S., 2006. Heterophyid metacercarial infections in brackish water fishes from Jinju-man (Bay), Kyongsangnam-do, Korea. Korean J. Parasitology, 44, 7–13.

Dore, W.J., Lees, D.N., 1995. Behaviour of *Escherichia coli* and male specific bacteriophage in environmentally contaminated bivalve molluscs before and after depuration. Appl. Environ. Microbiol. 61, 2830–2834.

Dressler, D., 2005. Botulism caused by consumption of smoked salmon (article in German). Nervenarzt 76, 763–766.

Duran, G.M., Marshall, D.L., 2005. Ready-to-eat shrimp as an international vehicle of antibiotic-resistant bacteria. J. Food Prot., 68, 2395–2401.

Elhadi, N., Radu, S., Chen, C.H., Nishibuchi, M., 2004. Prevalence of potentially pathogenic *Vibrio* species in the seafood marketed in Malaysia. J. Food Prot., 67, 1469–1475.

Estrada-Garcia, T., Lopez-Saucedo, C., Arevalo, C., Flores-Romo, L., Luna, O., Perez-Martinez, I., 2005. Street-vended seafood: A risk for foodborne diseases in Mexico. Lancet Infect. Dis., 5, 69–70.

Fach, P., Perelle, S., Dilasser, F., Grout, J., Dargaignaratz, C., Botella, L., Gourreau, J.M., Carlin, F., Popoff, M.R., Broussolle, V., 2002. Detection by PCR-enzyme-linked immunosorbent assay of *Clostridium botulinum* in fish and environmental samples from a coastal area in northern France. Appl. Environ. Microbiol., 68, 5870–5876.

Fayer, R., 2004. *Cryptosporidium*: A water-borne zoonotic parasite. Vet Parasitol., 126, 37–56.

Fayer, R., Dubey, J.P., Lindsay, D.S., 2004. Zoonotic protozoa: From land to sea. Trends in Parasitology, 20, 531–536.

Feldhusen, F., 2000. The role of seafood in bacterial foodborne diseases. Microbes and Infection, 2, 1651–1660.

Fleming, L.E., Broad, K., Clement, A., Dewailly, E., Elmir, S., Knap, A., Pomponi, S.A., Smith, S., Solo-Gabriele, H., Walsh, P., 2006. Oceans and human health: Emerging public health risks in the marine environment. Mar. Pollut. Bull., 53, 545–560.

Food and Agriculture Organization of the United Nations. 2000. Statistical Database. Vol. 2000.

Food and Drug Administration (FDA), 1994. Proposal to establish procedures for the safe processing and importing of fish and fishery products: proposed rule. Dept. of Health and Human Services, 21, CFR Parts 123 and 1240. Federal Register, 59, 4142–4214.

Food and Drug Administration (FDA), 2001. Parasites. Ch. 5. In: Fish and Fishery Products Hazards and Controls Guidance, 3rd ed., p. 65–72. Washington, DC, Food and Drug Administration, Center for Food Safety and Applied Nutrition, Office of Seafood.

Formiga-Cruz, M., Allard A. K., Condin-Hansson, A-C., Henshilwood, K., Hernroth, B.E., Jofre, J., Lees, D.N., Lucena, F., Papapetropoulou, M., Rangdale, R.E., A., Tsibouxi, Vantarakis, A., Girones, R., 2003. Evaluation of potential indicators of viral contamination in shellfish with applicability to diverse geographical areas. Appl. Environ. Microbiol. 69, 1556–1563.

Formiga-Cruz, M., Tofiño-Quesada, G., Bofill-Mas, S., Lees, D.N., Henshilwood, K., Allard A.K., Condin-Hansson, A-C., Hernroth, B.E., Vantarakis, A., Tsibouxi, A., Papapetropoulou, M., Furones, D., Girones, R., 2002. Distribution of human viral contamination in shellfish from different growing areas in Greece, Spain, Sweden and UK. Appl. Environ. Microbiol. 68, 5990–5998.

Freire-Santos, F., Oteiza-López, A.M., Vergara-Castiblanco, C.A., Ares-Mazás, E., Alvarez-Suárez, E., García-Martín, O., 2000. Detection of *Cryptosporidium* oocysts in bivalve molluscs destined for human consumption. J. Parasitol., 86, 853–854.

Fuenzalida, L., Hernandez, C., Toro, J., Rioseco, M.L., Romero, J., Espejo, R.T., 2006. *Vibrio parahaemolyticus* in shellfish and clinical samples during two large epidemics of diarrhoea in southern Chile. Environ. Microbiol., 8, 675–683.

Fukushima, H., 2006. Distribution of *Vibrio vulnificus* along the coastal area of Shimane Prefecture and contamination of retail fish and shellfish with *V. vulnificus* in Shimane Prefecture, Japan (article in Japanese). Kansenshogaku Zasshi, 80, 220–230.

Girones, R., Jofre, J., Bosch, A., 1989. Natural inactivation of enteric viruses in seawater. J. Environ. Quality. 18, 34–39.

Gobat, P.F., Jemmi, T., 1993. Distribution of mesophilic *Aeromonas* species in raw and ready-to-eat fish and meat products in Switzerland. Int. J. Food Microbiol., 20, 117–120.

Gomez-Bautista, M., Ortega-Mora, L.M., Tabares, E., Lopez-Rodas, V., Costas, E., 2000. Detection of infectious *Cryptosporidium parvum* oocysts in mussels (*Mytilus galloprovincialis*) and cockles (*Cerastoderma edule*). App. Environ. Micro. 66, 1866–1870.

Gómez-Couso, H., Freire-Santos, F., Martínez-Urtaza, J., García-Martín, O., Ares-Mazás, M.E., 2003. Contamination of bivalve molluscs by *Cryptosporidium oocysts*: The need for new quality control standards. Int. J. Food Micro. 87, 97–105.

González-Escalona, N., Cachicas, V., Acevedo, C., Rioseco, M.L., Vergara, J.A., Cabello, F., Romero, J., Espejo, R.T., 2005. *Vibrio parahaemolyticus* diarrhea, Chile, 1998 and 2004. Emerg. Infect. Dis., 11, 129–131.

Gopal, S., Otta, S.K., Kumar, S., Karunasagar, I., Nishibuchi, M., Karunasagar, I., 2005. The occurrence of *Vibrio* species in tropical shrimp culture environments; implications for food safety. Int. J. Food Microbiol., 102, 151–159.

Gourmelon, M., Montet, M.P., Lozach, S., Le Mennec, C., Pommepuy, M., Beutin, L., Vernozy-Rozand, C., 2006. First isolation of Shiga toxin 1d producing *Escherichia coli* variant strains in shellfish from coastal areas in France. J. Appl. Microbiol., 100, 85–97.

Graczyk, T.K., Thompson, R.C.A., Fayer, R., Adams, P., Morgan, U.M., Lewis, E.J., 1999. *Giardia duodenalis* cysts of genotype A recovered from clams in the Chesapeake Bay subestuary, Rhode River. Am. J. Trop. Med. Hyg., 61, 526–529.

Greenwood, M.H., Coetzee, E.F., Ford, B.M., Gill, P., Hooper, W.L., Matthews, S.C., Patrick, S., 1985. The microbiology of cooked prawns and shrimps on retail sale. J. Hyg. (Lond), 94, 319–326.

Grohmann, G.S., Murphy, A.M., Christopher, P.J., Auty, E., Greenberg, H.B., 1981. Norwalk virus gastroenteritis in volunteers consuming depurated oysters. Aust. J. Exp. Biol. Med. Sci. 59(Pt 2), 219–228.

Guthmann, J.P., 1995. Epidemic cholera in Latin America: Spread and routes of transmission. J. Trop. Med. Hyg., 98, 419–427.

Haddock, R.L., Truong, L.T., Aguon, T.S., 2002. Cholera control on Guam, 2000. Pac. Health Dialog., 9, 190–192.

Halliday, M.L., Kang, L.Y., Zhou, T.K., Hu, M.D., Pan, Q.C., Fu, T.Y., Huang, Y.S, Hu, S.L., 1991. An epidemic of hepatitis A attributable to the ingestion of raw clams in Shanghai, China. J. Infect. Dis. 164, 852–859.

Handa, S., Kimura, B., Takahashi, H., Koda, T., Hisa, K. and Fujii, T., 2005. Incidence of *Listeria monocytogenes* in raw seafood products in Japanese retail stores. J. Food Prot., 68, 411–415.

Haq, I., Sakaguchi, G., 1980. Prevalence of *Clostridium botulinum* in fishes from markets in Osaka. Jpn. J. Med. Sci. Biol., 33, 1–6.

Hara-Kudo, Y., Sugiyama, K., Nishibuchi, M., Chowdhury, A., Yatsuyanagi, J., Ohtomo, Y., Saito, A., Nagano, H., Nishina, T., Nakagawa, H., Konuma, H., Miyahara, M., Kumagai, S., 2003. Prevalence of pandemic thermostable direct hemolysin-producing *Vibrio parahaemolyticus* O3:K6 in seafood and the coastal environment in Japan. Appl. Environ. Microbiol., 69, 3883–3891.

Heitmann, I., Jofré, L., Hormázabal, J.C., Olea, A., Vallebuona, C., Valdés, C., 2005. Review and guidelines for treatment of diarrhea caused by *Vibrio parahaemolyticus* (article in Spanish). Rev. Chil. Infect., 22, 131–140.

Hernroth, B.E., Conden-Hansson, A.C., Rehnstam-Holm, A.S., Girones, R., Allard, A.K., 2002. Environmental factors influencing human viral pathogens and their potential indicator organisms in the blue mussel, *Mytilus edulis*: The first Scandinavian report. Appl. Environ. Microbiol. 68, 4523–4533.

Hoge, C.W., Bodhidatta, L., Echeverria, P., Deesuwan, M., Kitporka, P., 1996. Epidemiologic study of *Vibrio cholerae* O1 and O139 in Thailand: At the advancing edge of the eighth pandemic. Am. J. Epidemiol., 143, 263–268.

Hudson, J.A., Hasell, S., Whyte, R., Monson, S., 2001. Preliminary microbiological investigation of the preparation of two traditional Maori foods (Kina and Tiroi). J. Appl. Microbiol., 91, 814–821.

Huss, H.H., Jorgensen, L.V., Vogel, B.F., 2000. Control options for *Listeria monocytogenes* in seafoods. Int. J. Food Microbiol., 62, 267–274.

Hyytiä, E., Hielm, S., Björkroth, J., Korkeala, H., 1999. Biodiversity of *Clostridium botulinum* type E strains isolated from fish and fishery products. Appl. Environ. Microbiol., 65, 2057–2064.

Joseph, S.W., Colwell, R.R., Kaper, J.B., 1982. *Vibrio parahaemolyticus* and related halophilic vibrios. Crit. Rev. Microbiol., 10, 77–124.

Jothikumar, N., Lowther, J. A., Henshilwood, K., Lees, D.N., Hill, V.R., Vinjé, J., 2005. Rapid and sensitive transcription-PCR assays and application to naturally contaminated shellfish samples. Appl. Environ. Microbiol. 71, 1870–1875.

Kam, K.M., Leung, T.H., Ho, Y.Y., Ho, N.K., Saw, T.A., 1995. Outbreak of *Vibrio cholerae* 01 in Hong Kong related to contaminated fish tank water. Public Health, 109, 389–395.

Koizumi, Y., Isoda, N., Sato, Y., Iwaki, T., Ono, K., Ido, K., Sugano, K., Takahashi, M., Nishizawa, T., Okamoto, H., 2004. Infection of a Japanese patient by genotype 4 hepatitis E virus while traveling in Vietnam. J. Clin. Microbiol. 42, 3883–3885.

Koopmans, M., von Bonsdorff, C-H., Vinjé, J., de Medici, D., Monroe, S., 2002. Foodborne viruses. FEMS Microb. Rev. 26, 187–205.

Kumar, H.S., Karunasagar, I., Karunasagar, I., Teizou, T., Shima, K., Yamasaki, S., 2004. Characterisation of Shiga toxin-producing *Escherichia coli* (STEC) isolated from seafood and beef. FEMS Microbiol. Lett., 233, 173–178.

Kwiatek, K., 2004. Occurrence of *Listeria monocytogenes* in selected food of animal origin. Bull. Vet. Inst. Pulawy, 48, 269–272.

Lalitha, K.V., Surendran, P.K., 2002. Occurrence of *Clostridium botulinum* in fresh and cured fish in retail trade in Cochin (India). Int. J. Food Microbiol., 72, 169–174.

Lees, D., 2000. Viruses and bivalve shellfish, Int. J. Food Microbiol. 59, 81–116.

Legnani, P.P., Leoni, E., Villa, G.C., 2002. Microbiological monitoring of mussels and clams collected from the shellfish-growing marine areas in Rimini Province (article in Italian). Ann. Ig., 14, 105–113.

Le Guyader, F.S., Neill, F.H., Dubois, E., Bon, F., Loisy, F., Kohli, E., Pommepuy, M., Atmar, R.L., 2003. A semiquantitative approach to estimate Norwalk-like virus. Int. J. Food Microbiol. 87, 107–112.

Lévesque, B., Gagnon, F., Valentin, A., Cartier, J.-F., Chevalier, P., Cardinal, P., Cantin, P., Gingras, S., 2006. Étude de la contamination microbienne des myes (*Mya arenaria*) de la rive nord de l'estuaire maritime du fleuve Saint-Laurent (Québec, Canada) (article in French). Can. J. Microbiol. 52, 984–991.

Lumsden, L.L., Haseltine, H.E., Leak, J.P., Veldee, M.V., 1925. A typhoid fever epidemic caused by oyster-borne infection. Public Health Rep. (suppl.) 50, 1–102.

MacDiarmid, S.C., 2001. Risk analysis in aquatic animal health. In C.J. (ed.), Proceedings of the OIE Conference on Risk Analysis in Aquatic Animal Health, Rodgers, Paris, February 8–11, 2000, 1–6.

Macían, M.C., Arias, C.R., Aznar, R., Garay, E., Pujalte, M.J., 2000. Identification of *Vibrio* spp. (other than *V. vulnificus*) recovered on CPC agar from marine natural samples. Int. Microbiol. 3, 51–53.

Makino, S.I., Kii, T., Asakura, H., Shirahata, T., Ikeda, T., Takeshi, K., Itoh, K., 2000. Does enterohemorrhagic *Escherichia coli* O157:H7 enter the viable but nonculturable state in salted salmon roe? Appl. Environ. Microbiol. 66, 5536–5539.

Malfait, P., Lopalco, P.L., Salmaso, S., Germinario, C., Salamina, G., Quarto, M., Barbuti, S., Cipriati, R., Mundo, A., Pesone, G., 1996. An outbreak of hepatitis A in Puglia, Italy, 1996. Eurosurveillance 1.

Marine Conservation Research Institute (MCRI), 2005. Marine Bacteria and Viruses. In Connecting the Oceans and Human Health, Chapter 23; Ocean Commission Report; February 11, 2005, www.aopmcri.org/com/article/oid/31/ht/fish.

Martínez-Urtaza, J., Lozano-Leon, A., DePaola, A., Ishibashi, M., Shimada, K., Nishibuchi, M., Liébana, E., 2004. Characterization of pathogenic *Vibrio parahaemolyticus* isolates from clinical sources in Spain and comparison with Asian and North American pandemic isolates. J. Clin. Microbiol. 42, 4672–4678.

Martínez-Urtaza, J., Saco, M., Hernandez-Cordova, G., Lozano, A., Garcia-Martin, O., Espinosa, J., 2003. Identification of *Salmonella* serovars isolated from live molluscan shellfish and their significance in the marine environment. J. Food Prot. 66, 226–232.

McLaughlin, J.B., DePaola, A., Bopp, C.A., Martinek, K.A., Napolilli, N. P., Allison, C.G., Murray, S.L., Thompson, E.C., Bird, M.M., Middaugh, J.P., 2005. Outbreak of *Vibrio parahaemolyticus* gastroenteritis associated with Alaskan oysters. N. Engl. J. Med. 353, 1463–1470.

Mechnik, L., Bergman, N., Attali, M., Beergabel, M., Mosenkis, B., Sokolowski, N., Malnick, S., 2001. Acute hepatitis E virus infection presenting as a prolonged cholestatic jaundice. J. Clin. Gastroenterol. 33, 421–422.

Medina Flores, J., Tantaleán Vidaurre, M., León Rivera, M., Cano Rosales, M., 2002. *Diphyllobothrium pacificum* en niños del Perú (in Spanish with English abstract). Diagnóstico. July-August, 41(4), www.fihu-diagnostico.org.pe/revista/numeros/2002/julago02/161–164.html.

Mello Sampaio, J.L., Andrade, V.P., Lucas, M.C., Fung, L., Gagliardi, S.M.B., Santos, S.R.P., Figueiredo Mendes, C.M., Paula Eduardo, M. B., Dick, T., 2005. Diphyllobothriasis, Brazil. Emerg. Infect. Dis., 11, 1598–1600, www.cdc.gov/ncidod/EID/vol11no10/05–0377.htm.

Meng, X.J., Purcell, R.H., Halbur, P.G., Lehman, J.R., Webb, D.M., Tsareva, T.S., Haynes, J.S., Thacker, B.J., Emerson, S.U., 1997. A novel virus in swine is closely related to the human hepatitis E virus. Proc. Natl. Acad. Sci. USA 94, 9860–9865.

Merivirta, L.O., Lindstrom, M., Bjorkroth, K.J., Korkeala, H.J., 2006. The prevalence of *Clostridium botulinum* in European river lamprey (*Lampetra fluviatilis*) in Finland. Int. J. Food Microbiol. 109, 234–237.

Millard, J., Appleton, H., Parry, J.V., 1987. Studies on heat inactivation of hepatitis A virus with special reference to shellfish. Epidemiol. Infect. 98, 397–414.

Mohamed Hatha, A.A., Maqbool, T.K., Suresh Kumar, S., 2003. Microbial quality of shrimp products of export trade produced from aquacultured shrimp. Int. J. Food Microbiol. 82(3), 213–221.

Morris, J.G., 1988. *Vibrio vulnificus*: A new monster of the deep? Ann. Intern. Med. 109(4), 261–263.

Morris, J.G., Jr., 2003. Cholera and other types of vibriosis: A story of human pandemics and oysters on the half shell. Clin. Infect. Dis. 37, 272–280.

Mullendore, J.L., Sobsey, M.D., Shieh, Y.C., 2001. Improved method for the recovery of hepatitis A virus from oysters. J. Virol. Methods. 94, 25–35.

Muniain-Mujika, I., Calvo, M., Lucena, F., Girones, R., 2003. Comparative analysis of viral pathogens and potential indicators in shellfish. Int. J. Food Microbiol. 83, 75–85.

Muniain-Mujika, I., Girones, R., Tofiño-Quesada, G., Calvo, M., Lucena, F., 2002. Depuration dynamics of viruses in shellfish. Int. J. Food Microbiol. 77, 125–133.

Murphy, A.M., Grohmann, G.S., Christopher, R.J., Lopez, W.A., Davey, G. R., Millsom, R.H., 1979. An Australian-wide outbreak of gastroenteritis from oysters caused by Norwalk virus. Med. J. Australia 2, 329–333.

Myers B.J., 1970. Nematodes transmitted to man by fish and aquatic mammals. J. Wildl. Dis. 6, 266–271.

National Research Council (NRC)., 2003. Scientific Criteria to Ensure Safe Food. Chapter 5: Scientific criteria and performance standards to control hazards in seafoods. Committee on the Review of the Use of Scientific Criteria and Performance Standards for Safe Food.

Nol, P., Williamson, J.L., Rocke, T.E., Yuill, T.M., 2004. Detection of *Clostridium botulinum* type C cells in the gastrointestinal tracts of Mozambique Tilapia (*Oreochromis mossambicus*) by polymerase chain reaction. J. Wildl. Dis. 40, 749–753.

Normanno, G., Parisi, A., Addante, N., Quaglia, N.C., Dambrosio, A., Montagna, C., Chiocco, D., 2006. *Vibrio parahaemolyticus, Vibrio vulnificus* and microorganisms of fecal origin in mussels (*Mytilus galloprovincialis*) sold in the Puglia region (Italy). Int. J. Food Microbiol. 106, 219–222.

Pina, S., Puig, M., Lucena, F., Jofre, J., Girones, R., 1998. Viral pollution in the environment and in shellfish; detection of human adenovirus by PCR as an index of human viruses. Appl. Environ. Microbiol. 64, 3376–3382.

Pinon, A., Zwietering, M., Perrier, L., Membre, J.M., Leporq, B., Mettler, E., Thuault, D., Coroller, L., Stahl, V., Vialette, M., 2004. Development and validation of experimental protocols for use of cardinal models for prediction of microorganism growth in food products. Appl. Environ. Microbiol. 70, 1081–1087.

Potasman, I., Paz, A., Odeh, M., 2002. Infectious outbreaks associated with bivalve shellfish consumption: A worldwide perspective. CID 35, 921–928.

Price, R.J., Tom, P.D., 2003. Compendium of Fish and Fishery Product Processing Methods, Hazards and Controls. National Seafood HACCP Alliance for Training and Education. University of California, Davis. Updates, http://seafood.ucdavis.edu/haccp/compendium/compend.htm.

Rampersad, F.S., Laloo, S., La Borde, A., Maharaj, K., Sookhai, L., Teelucksingh, J., Reid, S., McDougall, L., Adesiyun, A.A., 1999. Microbial quality of oysters sold in Western Trinidad and potential health risk to consumers. Epidemiol. Infect. 123, 241–250.

Reilly, A., Käferstein, F., 1997. Food safety and the application of the principles of the hazard analysis and critical control point (HACCP) system for their control in aquaculture production. Aquaculture Res. 28, 735–752.

Reinhard, K., Urban, O., 2003. Diagnosing ancient diphyllobothriasis from Chinchorro mummies. Mem. Inst. Oswaldo Cruz. 98(suppl 1), 191–193.

Richards, G.P., 1985. Outbreaks of shellfish-associated enteric viruses illness in the United States: requisite for development of viral guidelines. J. Food Protec. 48, 815–823.

Ripabelli, G., Sammarco, M.L., Fanelli, I., Grasso, G.M., 2004. Detection of *Salmonella*, *Listeria* spp., *Vibrio* spp., and *Yersinia enterocolitica* in frozen seafood and comparison with enumeration for faecal indicators: implication for public health (article in Italian). Ann. Ig. 16, 531–539.

Ripabelli, G., Sammarco, M.L., Grasso, G.M., Fanelli, I., Caprioli, A., Luzzi, I., 1999. Occurrence of *Vibrio* and other pathogenic bacteria in *Mytilus galloprovincialis* (mussels) harvested from Adriatic Sea, Italy. Int. J. Food Microbiol. 49, 43–48.

Rooney, R.M., Cramer, E.H., Mantha, S., Nichols, G., Bartram, J.K., Farber, J.M., Benembarek, P.K., 2004. A review of outbreaks of foodborne disease associated with passenger ships: Evidence for risk management. Public Health Rep. 119, 427–434.

Roos, B., 1956. Hepatitis epidemic conveyed by oysters. Sven Lakartidningen. 53, 989–1003.

Sagua F.H., Aliaga R.P., Neira C.I., Araya R.J., González C.J., 2000. *Diphyllobothrium pacificum*: Infection in a 3 year old boy in Antofagasta (in Spanish with abstract in English). Rev. chil. pediatr. 71, 427–429, http://scielo-test.conicyt.cl/scielo.php?script = sci_arttextandpid = S0370–41062000000500009andlng = enandnrm = iso.

Sagua, H., Neira, I., Araya, J., González. J., 2001. New cases of *Diphyllobothrium pacificum* (Nybelin, 1931) Margolis, 1956 human infection in North of Chile, probably related with the El Niño phenomenon, 1975–2000. Bol. Chil. Parasitol. 56, 22–25.

Sanath Kumar, H., Otta, S.K., Karunasagar, I., Karunasagar, I., 2001. Detection of Shiga-toxigenic *Escherichia coli* (STEC) in fresh seafood and meat marketed in Mangalore, India by PCR. Lett. Appl. Microbiol. 33, 334–338.

Schets, F.M., van den Berg, H.H., Engels, G.B., Lodder, W.J., de Roda Husman, A.M., 2006. *Cryptosporidium* and *Giardia* in commercial and non-commercial oysters (*Crassostrea gigas*) and water from the Oosterschelde, the Netherlands. Int. J. Food Microbiol. (E-publication, in press, doi:10.1016/j.ijfoodmicro. 2006.06.031).

Sorber, C.A., 1983. Removal of viruses from wastewater and effluent by treatment processes. In Berg, G. (ed.), Viral Pollution of the Environment, pp. 39–52. Boca Raton, FL, CRC Press.

Stoner, G.L., Jobes, D.V., Fernandez Cobo, M., Agostini, H.T., Chima, S.C., Ryschkewitsch, C.F., 2000. JC virus as a marker of human migration to the Americas. Microbes Infect. 2, 1905–1911.

Suffredini, E., Di Pasquale, S., Alessi, E., Cozzi, L., Croci, L., Toti, L., 2006. In: Molluscan Shellfish Safety. Proceedings of 5th International Conference on Molluscan Shellfish Safety, Galway, Ireland 2004. Shell viruses and pathogens, pp. 299–303.

Teophilo, G.N., dos Fernandes Vieira, R.H., dos Prazeres Rodrigues, D., Menezes, F.G., 2002. *Escherichia coli* isolated from seafood: Toxicity and plasmid profiles. Int. Microbiol. Mar. 5, 11–14.

Terajima, J., Izumiya, H., Iyoda, S., Tamura, K., Watanabe, H., 1999. Detection of a multi-prefectural *E. coli* O157:H7 outbreak caused by contaminated Ikura-Sushi ingestion. Jpn. J. Infect. Dis. 52, 52–53.

Thampuran, N., Surendran, P.K., 1998. Occurrence and distribution of *Vibrio vulnificus* in tropical fish and shellfish from Cochin (India). Lett. Appl. Microbiol. 26, 110–112.

Thimothe, J., Walker, J., Suvanich, V., Gall, K.L., Moody, M.W., Wiedmann, M., 2002. Detection of *Listeria* in crawfish processing plants and in raw, whole crawfish and processed crawfish (*Procambarus* spp.). J. Food Prot. 65, 1735–1739.

Thornton, V., Hazell, W., Simmons, G., 2002. Acute gastroenteritis associated with seafood privately imported from the Pacific Islands. N.Z. Med. J. 115, 234–236.

United States Accounting Office, 2001. Food Safety: Federal oversight of seafood does not sufficiently protect consumers. GAO-01–204.

Urdaci, M.C., Marchand, M., Grimont, P.A., 1988. Species of the genus *Vibrio* associated with marine products from Arachon Bay (article in French). Ann. Inst. Pasteur Microbiol. 139, 351–362.

Valdimarsson, G., 2005. Evolving inspection regimes: An FAO perspective. Fifth World Fish Inspection and Quality Control Congress, The Hague, Netherlands, 20–22 October 2003. Ryder, J. and Ababouch, L. (eds.) FAO Fisheries Proceedings, www.fao.org/docrep/008/y5929e/y5929e06.htm.

Vieira, R.H., Lima, E.A., Sousa, D.B., Reis, E.F., Costa, R.G., Rodrigues D.P., 2004. *Vibrio* spp. and *Salmonella* spp., presence and susceptibility in crabs *Ucides cordatus*. Rev. Inst. Med. Trop. Sao Paulo. 46, 179–182.

Vieira, R.H., Rodrigues, D.P., Goncalves, F.A., Menezes, F.G., Aragao, J.S., Sousa OV., 2001. Microbicidal effect of medicinal plant extracts (*Psidium guajava* Linn. and *Carica papaya* Linn.) upon bacteria isolated from fish muscle and known to induce diarrhea in children. Rev. Inst. Med. Trop. Sao Paulo. 43, 145–148.

Villalobos de Bastardo, L.B., Elguezabal Aristizabal, L., 2001. Microbiological quality of the bivalve *Pinctada imbricata* commercialized in Cumana, Venezuela. Acta Cient. Venez. 52, 55–61.

Villena, O., Kianman, W., Núñez, J., Alcántara, A., Villena, A., 2002. Pancreatitis aguda por *Diphyllobothrium pacificum*. Enfermedades del Aparato Digestivo. 5, 40–41, http://sisbib.unmsm.edu.pe/BVrevistas/Enfermedades_digestivo/v05_n1/pdf/pancreatitis_aguda.pdf#search=%22%22Diphyllobothrium%20pacificum%22%22.

Wallace, B.J., Guzewich, J.J., Cambridge, M., Altekruse, S., Morse, D.L., 1999. Seafood-associated disease outbreaks in New York, 1980–1994. Am. J. Prev. Med. 17, 48–54.

Wanke, C.A., Guerrant, R.L., 1990. Viral hepatitis and gastro-enteritis transmitted by shellfish and water. Infect. Dis. Clin. N. Am. 19, 435–438.

Whyte, R., Hudson, J.A., Hasell, S., Gray, M., O'Reilly, R., 2001. Traditional Maori food preparation methods and food safety. Int. J. Food Microbiol. 69, 183–190.

Wilson, I.G., Moore, J.E., 1996. Presence of *Salmonella* spp. and *Campylobacter* spp. in shellfish. Epidemiol. Infect. 116, 147–153.

World Health Organization (WHO), 2001. Food safety and food borne illness. Fact Sheet 237. http://www.who.int/mediacentre/factsheets/fs237/en/index.html.

Xu, Z.-Y., Li, Z.-H., Wang, J.-X., Xiao, Z.-P., Dong, D.-X., 1992. Ecology and prevention of a shellfish-associated hepatitis A epidemic in Shanghai, China. Vaccine 10 (suppl. 1), 67–68.

Yano, Y., Yokoyama, M., Satomi, M., Oikawa, H., Chen, S.S., 2004. Occurrence of *Vibrio vulnificus* in fish and shellfish available from markets in China. J. Food Prot. 67, 1617–1623.

Yotsuyanagi, H., Koike, K., Yasuda, K., Moriya, K., Shintani, Y., Fujie, H., Kurokawa, K., Iino, S., 1996. Prolonged fecal excretion of hepatitis A virus in adult patients with hepatitis A as determined by polymerase chain reaction. Hepatology 24, 10–13.

STUDY QUESTIONS

1. Describe the sources of pathogens related to seafood-borne diseases. May pathogens be regular inhabitants of aquatic ecosystems?
2. Explain the definition of nonindigenous enteric bacteria causing infections in seafood.
3. Explain why molluscan shellfish are the more frequent seafood product implicated in outbreaks related to viral diseases.
4. Explain the factors that are associated with shipboard outbreaks of seafood-related pathogens.
5. How could the important pathogen *E. coli* O157:H7 be transmitted through seafood? Also, show some examples of reported outbreaks. Is it distributed homogeneously through distinct geographical areas?
6. Explain which species of *Vibrio* represent a serious public health hazard and how seafood becomes contaminated.
7. Suggest an efficient process for reducing the risk of infection by nematodes in fishery products, and show an example of frequent human infection by nematodes associated with consumption of raw or undercooked seafood.
8. Explain the general characteristics of the methods applied for the detection of human viruses in seafood and the most significant pathologies related to viral infections.
9. Outline the available techniques and methods used for reducing the likelihood of seafood-borne illness.

CHAPTER

19

Emerging Technologies for Monitoring Recreational Waters for Bacteria and Viruses

KELLY D. GOODWIN AND R. WAYNE LITAKER

INTRODUCTION

Microbial contamination of coastal waters is detrimental to the health of humans, marine mammals, and ecosystems with corresponding negative feedback on coastal economies (Dwight *et al.*, 2005). Environmental managers need quick and accurate assessments of water quality to restrict human access to contaminated water and seafood products and to formulate remediation strategies (Ocean.US, 2006b). Current monitoring techniques for fecal contamination are slow and provide limited information, making management decisions, ecological study, and assessment of control measures difficult. Emerging technologies seek to rapidly quantify microbial contaminants and to provide information about pollution sources and the presence of human pathogens.

Emerging technologies for water quality testing are science-driven innovations that have yet to be fully exploited in the marketplace. There are numerous examples of technologies poised to meet the needs of water quality testing (Clark *et al.*, 2001; Deisingh and Thompson, 2004; Homola, 2006; Kerman *et al.*, 2004; Lemarchand *et al.*, 2004; Liron *et al.*, 2001; Narayanaswamy, 2004; Noble and Weisberg, 2005; Thompson, 2006). Given the pace of research and engineering, it is not possible to provide an exhaustive discussion or one that will remain current. In an attempt to circumvent this limitation, this chapter highlights the underlying fundamentals common to numerous detection strategies. An understanding of these fundamentals should act as a primer for comprehending additional advanced technologies not discussed here. That said, the researchers who take a novel turn away from these fundamentals may be the ones to overcome the challenges facing environmental biosensing and bring the needed technologies to the field.

THE CURRENT STATE OF WATER QUALITY TESTING

Feces containing pathogenic microbes (bacteria, viruses, protozoa) may enter coastal ecosystems through discharges from wastewater treatment plants, septic systems, and runoff. Coastal waters polluted by human feces can have negative consequences to human health and the economy because of the acquisition of waterborne illness and the closing of recreational or seafood harvesting waters (Cabelli *et al.*, 1979, 1982; Dufour, 1984b; Kay *et al.*, 1994; Leclerc *et al.*, 2002). Coastal waters are routinely monitored for the presence of fecal pollution by incubating samples in selective media and monitoring for either the growth of indicator species or diagnostic enzymes produced by the indicator species. Indicator species, which are relatively abundant, are employed because the actual pathogens that cause waterborne illness are generally present at low concentrations. In addition, measuring all potential pathogens is technically and financially unfeasible. Therefore, the concentration of fecal indicator bacteria (FIB) is typically used to monitor and manage coastal water quality.

The fecal indicator itself is not a pathogen per se, but rather an organism whose presence is used to indicate or infer the presence of sewage-associated pathogens. A fecal indicator should (1) be present when pathogens are present, (2) not reproduce in the environment, (3) be easy and cost-effective to grow and identify in the laboratory from environmental samples, (4) share a similar fate and transport to pathogens (including being at least equally resistant as the pathogen to disinfection and to environmental factors), and (5) be uniformly distributed in the sample (National Research Council, 2004).

As discussed in Chapter 17, Environmental Protection Agency (EPA) guidelines recommend that states test marine recreational waters for enterococci bacteria and freshwater recreational waters for *Escherichia coli* or enterococci (EPA, 2003). Shellfish harvesting waters are regulated by the concentration of total or fecal coliform (Food and Drug Administration/Interstate Shellfish Sanitation Conference [FDA/ISSC], 2003). The concentration of fecal indicator bacteria correlates to the risk of human gastrointestinal (GI) illness when subjects are exposed to waters impacted by point sources of human fecal pollution (Cabelli *et al.*, 1979; Dufour, 1984a; Wade *et al.*, 2003). A concentration of fecal indicator bacteria in marine recreational waters higher than the EPA recommendation (5-day geometric mean >35 CFU enterococci/100 ml; single sample standard ≥104 CFU enterococci/100 ml) is thought to indicate that the risks of contracting gastrointestinal illness exceeds a threshold level, and the water is unsafe for public use (EPA, 2003).

Current Enumeration Practices

The concentration of fecal indicator bacteria in a water sample is usually determined by either the membrane-filtration (MF) technique or by the most-probable-number (MPN) technique (Rompré *et al.*, 2002). In the membrane filtration technique, a water sample is vacuum filtered onto a 0.45-μm filter, the filter is placed face-up on selective and indicating agar media, it is incubated at the proper temperature, the colony-forming units (CFU) that develop are recorded, and the concentration of bacteria in the original sample is calculated and reported in terms of CFU/100 ml water. In the MPN method, water samples are diluted into a selective growth medium, put into a series of tubes or wells, and incubated. The tubes/wells that are positive for growth are then counted. Results are compared to a MPN table to determine the concentration of bacteria (MPN/100 ml) in the original sample.

Fecal indicators have traditionally been identified functionally rather than by a specific molecular signature. For example, fecal coliform are operationally defined by the ability to grow and ferment lactose at 44°C, and growth media are designed to select for these characteristics. Traditional identification requires a series of biochemical tests that are time consuming and provide variable results.

Detection schemes that have become the "new traditional" methods identify fecal indicators based on characteristic enzymes. These detection schemes utilize chromogenic or fluorogenic enzyme substrates to identify the target bacteria, in combination with selective culturing procedures. The enzyme substrates consist of a chromophore or fluorophore linked to the normal substrate of the target enzyme, such as a carbohydrate, amino acid, or phosphate. The target enzyme cleaves the chromogenic or fluorogenic substrate, and the free chromophore or fluorophore, respectively, produces a color change or fluorescence that can be detected (Manafi *et al.*, 1991). For example, most coliform bacteria produce the enzyme β-galactosidase. To detect this enzyme, water samples are cultured under selective conditions in the presence of the chromogenic substrate o-nitrophenol-β-galactopyranoside (ONPG) or with chlorophenol red-β-D-galactopyranoside (CPRG). Coliform bacteria hydrolyze the ONPG, turning the colonies or solution (for MPN studies) yellow. If CPRG is used, the colonies or solution turn red or magenta.

The concentration of *E. coli* is considered in many cases to serve as a better indicator of human health risk than is the concentration of total or fecal coliform (Doyle and Erickson, 2006). *E. coli* strains produce β-glucuronidase that hydrolyzes the fluorogenic substrate methylumbilliferyl-β-glucuronide (MUG) and produces fluorescence under UV light (366 nm) (Munn, 2004). For marine waters, the concentration of *Enterococcus* species is usually considered the best predictor of human health risk compared to the other indicators tested to date (Cabelli *et al.*, 1982, 1983). The enterococci are Gram-positive bacteria that produce β-D-glucosidase. Enterococci colonies have a blue halo when grown on agar medium containing the chromogenic substrate indoxyl-β-D-glucoside and under culture conditions that inhibit Gram-negative bacteria and select for enterococci (e.g., sodium azide, cycloheximide, esculin, 41°C). Enterococci are also detected in liquid medium by production of fluorescence through β-glucuronidase cleavage of 4-methylumbilliferyl-β-D-glucoside. This is the foundation of the Enterolert MPN test, an American Society for Testing and Materials method (ASTM #D6503-99) for enumerating enterococci in water and wastewater.

CURRENT NEEDS OF WATER QUALITY TESTING

Rapid Detection and High Throughput

Despite the simplicity of selectively growing bacteria as a means for identification, the standard culture-based methods have a variety of drawbacks. A primary concern is the time (>18 hours) between sample collection and result reporting. In the event that the water is contaminated, this lag time creates a risk that humans will be exposed. Alternatively, in the event that the water is not contaminated, the lag time creates a risk of false positive reporting (posting that waters are contaminated when clean) because quickly fluctuating conditions (Boehm *et al.*, 2002; Leecaster and Weisberg, 2001) give rise to indicator concentrations that are poorly correlated between the sampling and the reporting day (Kim and Grant, 2006; Whitman and Nevers, 2003). At the present time, many researchers are looking to achieve results within 4 to 6 hours of collection (ACT, 2006). In addition, high through put is desired. Throughput describes the number of samples that can be processed in a given time. Many monitoring programs need to process hundreds of samples a day; therefore, high throughput is as critical as, if not more so than, rapid detection.

Alternate Indicators

There is growing consensus that alternate indicators are needed (Griffin *et al.*, 2001; Henrickson *et al.*, 2001). Some data suggest that the ecology, prevalence, survival, and distribution of indicators in aquatic environments might differ significantly from the group of pathogens for which they are a proxy (Noble and Fuhrman, 2001). One possible reason for the lack of correlation between indicators and pathogens is the inability of culture-based methods to detect viable, but not culturable (VBNC) indicator species. Another possibility is that traditional indicators such as *Enterococcus* spp. and *E. coli* may persist or grow in sediment and sand environments (Alm *et al.*, 2006; Anderson *et al.*, 2005; Desmarais *et al.*, 2002; Ferguson *et al.*, 2005; Lee *et al.*, 2006; Whitman and Nevers, 2003; Whitman *et al.*, 2003), thereby creating a source of indicators to nearshore waters. Regrowth violates axiom 2 of indicator theory (see the earlier discussion). Regardless of the indicator, momentum is growing for inclusion of sand in the analysis of recreational water quality (Clean Beaches Council, 2005).

Bacteroides spp. are one of the suggested alternative indicators (Allsop and Stickler, 1985; Bernhard and Field, 2000a, 2000b; Fiksdal *et al.*, 1985; Kreader, 1995). These bacteria are anaerobic and do not form spores and thus should not survive long outside of the host. Molecular methods primarily have been used for evaluation of this indicator because culturing requires maintenance of anaerobic conditions. Bacteriophages, viruses that infect bacteria, are another suggested indicator because they may better mimic the fate and transport of human pathogenic viruses (Gantzer *et al.*, 1998; Jiang *et al.*, 2001; Paul *et al.*, 1997).

Source Tracking

Source tracking is a method to identify the origin of fecal contamination. Normally the term refers to determining whether the fecal contamination is from human or animal origin; however, it can also denote spatial tracking of the source to determine the physical origin of contamination. Identifying the origin of fecal pollution in aquatic ecosystems is a requirement for taking logical actions to remedy the problem (Scott *et al.*, 2002). The lengthy turnaround time of the culture-based indicator methods is not compatible with source tracking. The most commonly used tracking approach is to look for differential bacterial concentrations at the convergence of upstream tributaries (Scott *et al.*, 2002). Unfortunately, the fecal contamination signal may dissipate or disperse while the samples that would trigger an investigation are being processed, making it difficult to source track. When tracking is initiated, the slow processing time requires that many locations be examined simultaneously. If more rapid, and field-based methods were available, tracking would benefit by allowing a spatially sequential sampling approach to be pursued. The culture-based indicator methods also lack the species resolving power required to differentiate between human and animal sources of fecal contamination. In contrast, molecular approaches have been used successfully to track human versus animal sources of fecal pollution (Bernhard and Field, 2000b; Bonjoch *et al.*, 2004; Scott *et al.*, 2002; Simpson *et al.*, 2002; Stewart *et al.*, 2003; Stoeckel, 2005).

Multiplexed Detection

Overall, no single indicator or pathogen is likely to monitor all exposure routes adequately (National Research Council, 2004). Therefore, a suite of indicators may provide a better approach than single-species analysis (Harwood *et al.*, 2005). Rapid detection of multiple species could yield a "fingerprint" of water quality that would be a useful addition to fecal indicator enumeration (Baums *et al.*, 2007). Such a multitiered approach (Boehm *et al.*, 2003; Noble *et al.*, 2006) could yield more information about the source of contamination, the potential health risk, and the best strategy for remediation. This approach could benefit environmental research, epidemiological studies, and routine water quality analysis. Investigators have increasingly turned toward molecular biotechnologies to meet the need for rapid, multiplexed, species-level detection that also gives information about fecal contamination sources.

Affordability and Usability

The technology application (i.e., the market sector) and the competition (the other technologies already in the marketplace) help determine the maximum price for a technology without inhibiting its entrance into the marketplace. For example, a competitive cost estimate for sensor units in the realm of real-time oceanographic detection (discussed further below) is in the range of US$1000 per unit (ACT, 2006). Costs for recreational water quality monitoring are usually tabulated on a per sample basis. Prices vary but commercial prices run approximately US$40 to $100 per sample for standard fecal indicator enumeration, and US$100 to $600 per sample for analysis or viruses, pathogens, or source tracking markers. To move into markets supporting routine monitoring and regulation (versus research), emerging technologies need to meet such prices or provide additional value over current practices. In addition, they must contend with other drivers—for example, EPA or FDA

(Food and Drug Administration) regulations, ease of use, and the availability of personnel able to perform or be trained to perform the analyses.

COMMON PRINCIPLES UNDERLYING EMERGING TECHNOLOGIES

Emerging technologies are based on advanced biological and engineering protocols. Yet many of these technologies incorporate underlying concepts that are relatively simple (Fig. 19-1). For example, a conversational knowledge of emerging technologies can be obtained if one understands that most solution-based technologies utilize the following steps:

- Design primers and probes specific for the molecular target.
- Concentrate the organism of interest from the environment.
- Isolate the molecular target from the organism of interest.
- Amplify, label, and detect the target molecule.

In addition, surface-based technologies include the following steps before target detection:

- Immobilize the capture probe.
- Capture the target and wash away unwanted constituents.

Design Probes and Primers Specific to the Molecular Target

A probe is a molecule designed to specifically bind to the molecular target of interest. A probe may be labeled with a detectable molecule such as fluorescein. Selecting or designing a molecule that is specific for a target must take into account the structure of the target molecule; for example, whether the target is a protein, double-stranded DNA, or single-stranded RNA (Litaker and Tester, 2002). When the target molecule is a protein, the probe is typically an antibody or antigen. When the target molecule is DNA or RNA, the probe is generally an oligonucleotide or a peptide nucleic acid (PNA). An oligonucleotide is a short nucleic acid molecule, typically <20 to 25 base pairs (bp). A PNA is a synthetic oligonucleotide that uses peptide N-(2-

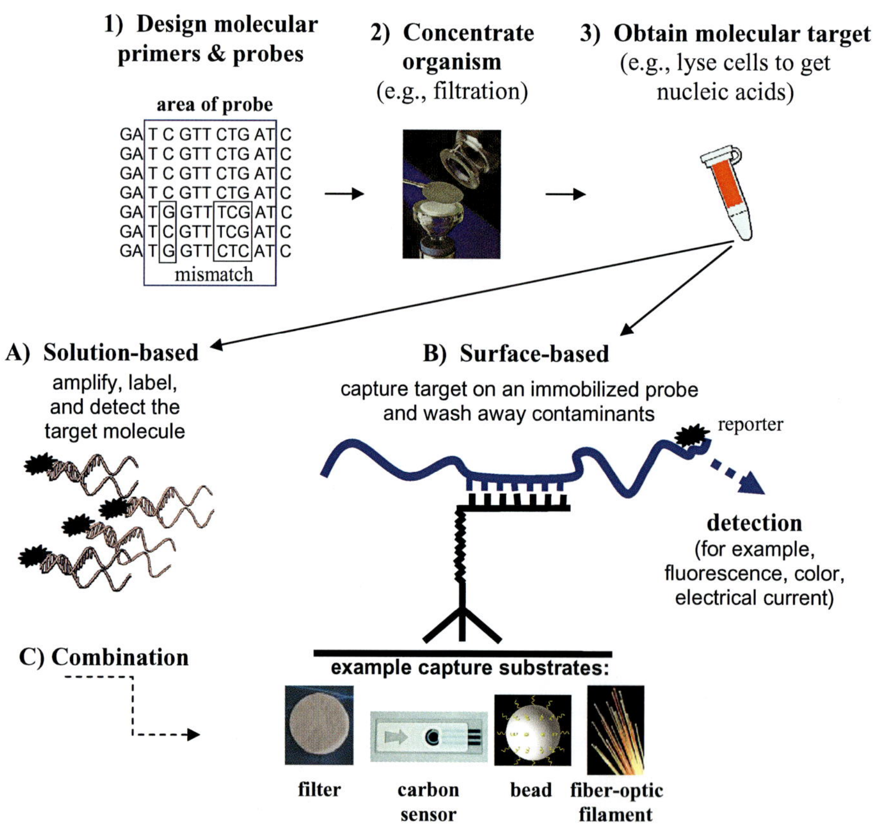

FIGURE 19-1. Generalization of sample processing (steps 1–3) and approaches to biological sensing by (A) solution-based and (B) surface-based emerging technologies. In addition, some technologies (C) combine solution-based amplification and labeling steps (e.g., PCR, for amplification and labeling) with surface-based capture and detection steps.

aminotheyl)-glycine as the backbone rather than a deoxyribose or ribose sugar. The lack of charged phosphate groups in the backbone provides PNA molecules some advantages with regard to hybridization stability and specificity (Pellestor and Paulasova, 2004).

Bioinformatics software and sequences from databases, such as GenBank (www.ncbi.nlm.nih.gov), as well as sequences obtained from organisms of interest not yet deposited in GenBank, are used to identify diagnostic nucleic acid or protein sequences. Typically, the sequence of interest is aligned with a series of related sequences and the alignment is examined for a region that uniquely identifies the organism of interest. Primers and probes are then designed to specifically bind a unique molecular signature. The specificity of the probe is important because the organism of interest is often closely related to other organisms—sometimes as little as one base pair in the region of the probe will differentiate the organism of concern from non-target organisms. More detailed discussions of primer and probe design are included in the individual sections that follow.

Concentrate the Pathogen or Indicator Organism from the Environment

To detect the target molecule (e.g., nucleic acid, protein, antigen) the organism of interest must be collected and concentrated from the environment. A concentration step is necessary because the organism of interest is normally present at concentrations that are too dilute to be measured directly. For this reason, some "dipstick" or "point of care" detection methods that work for clinical samples (Herron *et al.*, 2006) cannot be directly transferred to environmental samples. Nonetheless, these dilute concentrations still may pose a risk to human and ecosystem health; therefore, procedures to concentrate the sample are used in order to satisfy the detection limit of the molecular assay.

Size exclusion is the most common way to concentrate microorganisms from environmental samples. Vacuum filtration onto a membrane filter is commonly used to remove microorganisms from bulk water based on size. Large particles are sometimes removed with screens or large pore size filters to reduce clogging of downstream filters. However, coastal waters contain particles similar in size to the pores of the membrane filters used to filter bacteria (0.45 or 0.2 μm), so clogging remains problematic. Also, the organisms of interest are often attached to larger particles, so removing the larger particles may be undesired. Tangential flow filtration and dialysis separation are methods sometimes employed to overcome clogging and to reduce the volume of the sample when large volumes need to be processed. However, these methods concentrate substances such as humic materials that interfere with downstream molecular methods, particularly the polymerase chain reaction (PCR). Because of the small size of viruses, turbid coastal waters would quickly clog membrane filters small enough to capture them; therefore, viruses are usually removed from bulk water based on charge rather than size (Haramoto *et al.*, 2005; Katayama *et al.*, 2002).

Obtain the Target Molecule from the Organism of Interest

In general, once the organism of interest has been concentrated away from the bulk water or sediment, the next step is to isolate the molecular target from the organism. In some cases the target molecule is on the surface of the organism (e.g., antigen) and the whole cell is captured, but in many cases the molecular target is a cellular component (e.g., DNA, RNA, internal protein) that must be released from the organism before detection. Release generally involves a lysis step to disrupt the cell membrane or viral coat and may be accomplished by enzymatic or chemical methods, heating, freeze-thaw cycles, or sonication. A combination of physical and chemical disruption also may be used depending on the organism and the matrix of the sample.

Many investigators use commercially available kits to perform cell lysis, nucleic acid extraction, and nucleic acid purification. After cells are lysed, the free molecular targets must be protected from enzymes that will degrade them. The buffers used in these protocols contain enzymes or inhibitors that prevent degradation. To isolate the molecular target, most of the commercial kits alter the chemical nature of the target molecule to selectively immobilize it to a column of solid substrate. The column is washed with a solution of appropriate ionic strength that rinses away contaminants but allows the target to remain immobilized. The purified target is eluted from the column by rinsing with a solution whose chemistry disrupts the bond between the target molecules and the solid substrate.

Most commercial kits are designed for concentrated bacterial cultures or clinical specimens, although some manufacturers do carry products aimed at environmental markets. These products focus on removing chemicals such as humic acids and polysaccharides, which are compounds that can inhibit subsequent isolation and/or amplification of the intended target. Other investigators use older organic extraction protocols (Sambrook *et al.*, 1989) and some bypass purification all together and instead obtain the molecular target directly from the crude cell lysate (Haugland *et al.*, 2005; LaGier *et al.*, 2005; LaGier *et al.*, 2007). Efficient recovery of the molecular target is an important issue, as discussed in later sections.

GENERAL PRINCIPLES OF SOLUTION-BASED TECHNOLOGIES

Label and Detect Target Molecules

There are a variety of methods to detect molecular targets in solution. The method most commonly employed to detect nucleic acids is the polymerase chain reaction (PCR). PCR is a technique that replicates DNA molecules in an exponential fashion so that a small amount of starting material can be amplified by about 10^6–fold or more (Gerba *et al.*, 2001). PCR is the endpoint of many detection schemes. In these cases, the presence of the organism of interest may be indicated by visualizing the PCR product (amplicons) on an electrophoresis gel. The resulting band is compared to a ladder that indicates the size of a band in base pairs (bp); if the band is of the correct size a positive result is indicated (Sambrook *et al.*, 1989). In addition, PCR is the starting point for many of the advanced detection technologies discussed below. For some of these technologies, PCR represents both amplification and a labeling step (Figs. 19-1B and C). Weakness or failure at the PCR step thus translates to failure at the detection stage. Therefore, the underlying mechanisms of the PCR technique will be discussed in some detail here.

The PCR Reaction

To understand PCR reactions it is necessary to recall that DNA is double stranded. The paired strands of the double helix run in opposite directions (i.e., the forward strand runs 5′ to 3′ from left to right, whereas the reverse or complementary strand runs 5′ to 3′ from right to left (Fig. 19-2). The PCR reaction consists of three steps, which are repeated multiple times: denature, anneal, and extend. In the first step, the double-stranded DNA is denatured by heating so that the strands come apart (typical melting temperatures are 93 to 95°C). In the second step, the reaction is cooled to a lower temperature, typically 50 to 72°C, such that short pieces of single stranded DNA (the oligonucleotide primers) anneal to specific regions of the target. Lastly, the reaction is heated to 72 to 75°C to allow the DNA polymerase to extend the primers such that a copy of the target region is synthesized. Typically, this three-step sequence of denature, anneal, and extend is cycled 25 to 35 times, although protocols with 40 cycles are not uncommon for amplification of pathogens.

A PCR reaction mix for amplifying DNA consists of (1) the template DNA, (2) a forward and reverse primer, (3) dNTPs (deoxyribonucleoside triphosphates), (4) a buffer containing salts, and (5) a thermostable DNA polymerase. The dNTPs (the standard ones are dATP, dCTP, dGTP, and dTTP) provide the nucleotides that are incorporated into the growing DNA chain. In some advanced detection schemes, dNTPs are manufactured with a label such as biotin or a fluorescent molecule and used to create labeled PCR products (Fig. 19-1C). The DNA polymerase is the enzyme that links the dNTPs onto the 3′ end of the growing chain; therefore the DNA is synthesized in a 5′ to 3′ direction (Fig. 19-2). The polymerase requires a terminal 3′-OH group in order to incorporate a nucleotide. Primers initiate the PCR reaction by supplying this group to the polymerase (after the initial nucleotide incorporation, the dNTPs themselves supply the hydroxyl group needed for the addition of subsequent nucleotides). Two primers (a primer set) are needed—one for each strand of DNA. Each primer lies down 5′ to 3′ relative to its strand of template DNA (Fig. 19-2), and the two primers flank the region to be amplified. The nucleotide added to the 3′ end is determined by the complementary nucleotide on the template strand—that is, an adenine (A) in the template strand dictates a thymine (T) be added to the growing chain and vice versa, and a guanine (G) directs a cytosine (C) be added and vice versa. The DNA polymerase requires a divalent cation to function (usually Mg^{++}), and this is typically included in the PCR buffer.

PCR Primer Design

PCR primers are designed for specificity (i.e., to bind only to specific DNA sequences within the genome of an organism). This specificity can be achieved because there is a 1 in 4 chance that a particular nucleotide will occupy any individual position in a DNA strand. This means that the probability of encountering a given sequence randomly is $(1/4)^n$, where n is the number of consecutive nucleotides. Consequently, oligonucleotides of 17 to 20 bp will normally hybridize to a unique site on a DNA, even for genomes as

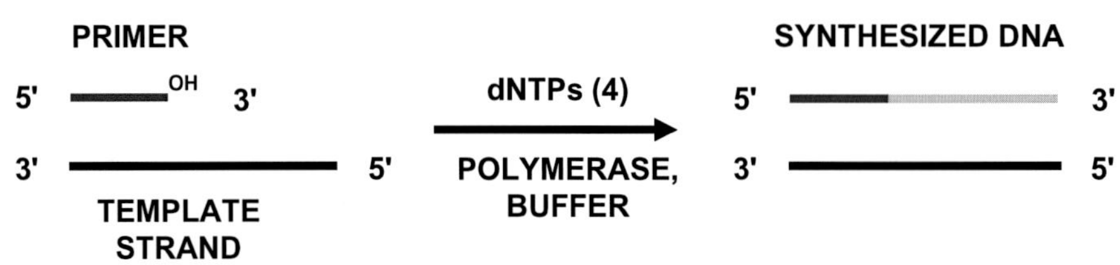

FIGURE 19-2. Simple depiction of a PCR reaction used to amplify a nucleic acid target. Principal components of the reaction include the template, a primer designed to bind specifically to a portion of the template, dNTPs, polymerase, and the reaction buffer. The newly formed DNA strand (indicated in gray) is synthesized in a 5′ to 3′ direction. For amplification of a DNA target, a pair of forward and a reverse primers are used—one primer for each strand of DNA.

large as that of human beings (3×10^9 bp). Primers are typically 18 to 25 bp long, usually providing adequate specificity. However, DNA bases are not randomly distributed—different species do share the same or similar genetic code; therefore, the uniqueness of the primer site must be checked. This can be done by comparing the sequence of the target to the sequences available in the GenBank database found at the National Center for Biotechnology Information Web site (www.ncbi.nlm.nih.gov). Submission of a sequence for comparison to this database is called "BLASTing" (BLAST = basic local alignment search tool). The BLAST program compares and aligns the submitted sequence, showing other similar sequences. If sufficient unique sequences are available, the primers will amplify only the DNA or RNA from a particular species. If unique sequences are not available, the DNA of related organisms (or functional genes) will be amplified; in fact, sometimes amplification of a species group rather than a species is the goal.

Good primer design is critical to the success of PCR. Well-designed primers should not contain hairpin loops—a sequence of complimentary bases that causes the primer to fold onto itself forming a secondary structure that prevents proper annealing. In addition, the primers should not contain sequence that allows the primers to hybridize to themselves or to each other (i.e., a "primer dimer"). Sequence repeats and runs of more than three bases of a single nucleotide are also best avoided because they can misprime or compromise specificity by annealing to unintended repeat regions in the genome. Furthermore, the two primers should have similar melting temperatures (within 5°C of each other). Otherwise, the primer with the higher temperature may bind nonspecifically during the low temperature conditions required by the other oligonucleotide. Primer melting temperatures around 52 to 58°C are typical. The melting temperature (T_m) is a useful parameter for characterizing the binding process. It represents the temperature at which, on average, half of the strands are hybridized and the remaining half are denatured. In practice, it suggests the temperature to use for specific annealing of primers and probes. The annealing temperature must be empirically tested, but it is usually about 5°C lower than the melting temperature. The optimum annealing temperature (T_a Opt) can be calculated (but still must be empirically tested) by the following equation: T_a Opt = 0.3 x(T_m primer) + 0.7 x(T_m product) − 25, where T_m primer is the lowest melting temperature of the primer-template pair and T_m product is the melting temperature of the PCR product (PREMIER BioSoft, 2007).

In addition, the GC content of the primers (the percentage of guanine and cytosine) should be kept between 40% to 60%. Under normal PCR conditions, each adenosine thymidine (AT) pair contributes ~2°C to the annealing temperature, but the G-C bond is more stable so each GC pair contributes ~4°C. A high GC content results in high annealing temperatures, and an additive to facilitate DNA denaturation (e.g., glycerol, dimethyl sulfoxide [DMSO], formamide) will be needed to achieve reasonable reaction temperatures (50 to 72°C). Finally, the primer should have a GC clamp and yet have relative instability at the 3′ end, meaning that the primer should have a G or a C at the 3′ end to provide a strong bond, but no more than two of the last five base pairs should be a G or C to avoid nonspecific priming. Computer programs are available that assist in the process of primer design.

Optimization of the PCR Reaction

Many factors influence the fidelity of a PCR reaction, particularly when environmental samples are the source of the nucleic acids. PCR reaction conditions vary for different targets and primer sets. In fact, reactions vary between thermal cycling instruments and types of tubes (e.g., regular versus thin wall). PCR optimization is usually required to eliminate nonspecific and inefficient amplification. Unfortunately, the optimum conditions have to be determined empirically. Among the variables that can be changed to achieve optimization are (1) thermocycling conditions, (2) type of polymerase, (3) primer concentration, (4) amount of DNA template, (5) concentration of divalent cation in the buffer (commonly Mg^{++}), and (6) the presence or absence of additives such as DMSO, formamide, or single stranded binding proteins. The constituents of the PCR buffer, such as the Mg^{++} concentration and other additives, control the stringency of nucleotide incorporation; this concept is discussed further in the surface-based technologies section regarding hybridization and wash buffers.

The reaction temperature is the easiest parameter to change during PCR optimization. The annealing temperature must be low enough to get sufficient annealing to start the PCR reaction efficiently but high enough to prevent nonspecific priming. Some thermal cycling instruments are able to generate a temperature gradient to simplify PCR optimization. Typically the three-step sequence of PCR (denature, anneal, extend) is repeated for 25 to 35 cycles. Additional cycles may provide more amplification, but this will often generate nonspecific products; therefore, it is best to use as few cycles as needed.

Hot start and proofreading polymerases provide more specific and error-free PCR products. Hot start polymerases are chemically engineered to remain inactive until after the first high-temperature step. This aids fidelity by limiting extension until after the primers have properly annealed during the first temperature drop. This is crucial because PCR is an exponential process and what happens in the first 1 to 2 rounds of amplification greatly affects the outcome of the later rounds.

The lowest primer concentration that generates a good yield of the desired product should be used. Usually about 100 to 500 ng of each primer per 50 μl reaction will work in most cases. Higher concentrations may be needed if the primers are degenerate (a primer mix synthesized so that some of the bases vary to ensure hybridization to DNA of variable composition). In that case, the concentration of the actual primer with the correct binding sequence may be low, so more overall primer is used.

The amount of original DNA template should be approximately 20 to 100 ng per 50 μl reaction. Too little template results in poor initial priming and thus low product yield. Conversely, too much template causes inefficient amplification, which can reduce or eliminate product yield. Excess DNA consumes too many dNTPs in the first few rounds of priming because the polymerase continues to transcribe new DNA far beyond where the reverse primer site is located. In addition, too much DNA can increase the chances of nonspecific binding between the primers and template, resulting in the production of nonspecific amplification products.

PCR Controls

Incorporation of adequate controls is an essential part of environmental detection systems. Both positive and negative controls are required when performing PCR reactions. A positive control contains a given quantity of template DNA isolated from a known culture of organism. This control ensures that all the reagents are working properly. A negative control contains all of the reaction reagents, but an aliquot of sterile water replaces the addition of DNA. This control checks for contamination of the PCR reagents.

In addition, assays of environmental samples should include inhibition controls, which assess whether the PCR reactions were inhibited by contaminants, such as phenolic compounds or carbohydrates carried over during the isolation process. Inhibition results in decreased sensitivity, and for quantitative analysis (discussed further later), an underestimate of the amount of template DNA present in the starting sample. Sometimes diluting the template DNA slightly with water or buffer will alleviate most of the inhibition. However, in some cases alternative extraction procedures are needed. Although there are a variety of control strategies (Shepley and Wolk, 2004), one way of incorporating an inhibition control into a PCR assay is to spike replicate reactions with a small amount of positive control DNA. The amount of amplicon in the inhibition control is compared to that in the positive control. The production of significantly less DNA in the spiked sample relative to the positive control means that inhibitors reduced the PCR efficiency. For logistical reasons, some investigators check only a subset of samples (e.g., one out of five) and rerun any with initial negative results. Other investigators use an internal control, as discussed in the reverse transcriptase PCR section.

Finally, experiments using field material should incorporate controls of the extraction procedure. An extraction blank consists of processing samples using exactly the same extraction reagents, forceps, filters, and so on except that the actual sample is omitted. If these controls are PCR positive, then all results are suspect. In addition, the efficiency of nucleic acid recovery should be assessed to establish operational detection limits (in contrast to analytical detection limits based on diluted genomic DNA, without the extraction step). Recovery is tested by spiking samples with known amounts of bacteria or viral particles, extracting the sample, and determining the efficiency of nucleic acid recovery. If quantitative recovery can be achieved, the quantitative amplification methods discussed later in this chapter can be employed.

A variety of procedures are available to prevent cross-contamination to ensure that negative controls and blank samples remain as such. For example, contamination by aerosols generated during pipetting and centrifugation is controlled by use of plugged pipette tips and by allowing centrifuged samples to settle before opening. Instruments used to process samples (e.g., forceps) may be disposable or are cleaned with bleach or other products designed to render trace amounts of DNA unamplifiable before sterilization. PCR reaction mixtures are made in hoods designed to protect the samples, and the template DNA is added last to the PCR mix. Importantly, PCR products are kept physically separated from the part of the laboratory in which PCR reaction mixtures are set up.

ADDITIONAL PRINCIPLES FOR SURFACE-BASED TECHNOLOGIES

General Formats for Surface-Based Technologies

For many surface-based technologies, capture probes are immobilized onto a solid surface, the target molecule is captured, and unwanted constituents are washed away before detection (Fig. 19-1B; Fig. 19-3). The captured target is linked to a reporter molecule that emits a signal that can be detected. The reporter probe usually binds to the target molecule at a site near to but distinct from the site of the capture probe. A configuration where the target molecule is bound between the capture and reporter molecules is typically referred to as a sandwich hybridization assay. In this type of assay, hybridization conditions appropriate for both the capture and reporter molecules are needed.

The basic sandwich hybridization format used to detect proteins is known as an enzyme linked immunosorbent assay (ELISA). In this case, the capture and reporter probes are antibodies specific to the target molecule (Figs. 19-3A and B). In some cases (direct detection schemes), the signal molecule is directly linked to the reporter molecule (Fig. 19-3A), whereas in other cases (secondary detection), the

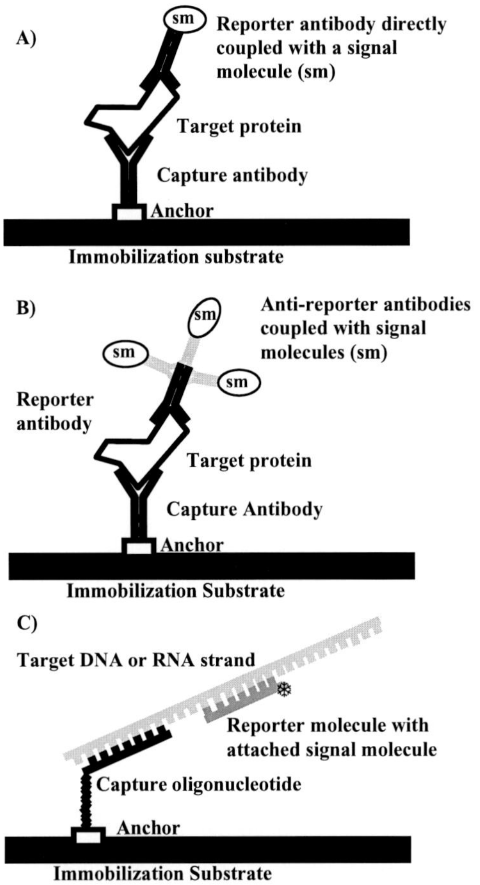

FIGURE 19-3. Various formats for surface-based technologies. (A) Direct detection ELISA format, in which the reporter molecule is directly coupled with a signal molecule. (B) Secondary detection ELISA format, which employs labeled antibodies that specifically bind the reporter antibody. (C) Standard sandwich hybridization format for detection of nucleic acids. The capture molecule typically incorporates modifications to help lift it from the surface (C) and may be anchored by a variety of methods including direct adsorption, photoimmobilization, or biotin-avidin chemistry.

signal molecule is linked to the reporter via an additional antibody (Fig. 19-3B). This additional antibody is designed to specifically bind to the reporter probe without disrupting binding to the target molecule. All antibodies have areas that are highly conserved; therefore, antibodies are designed against these generic regions. For example, if the original reporter antibody was produced in mice, an antimouse antibody (produced in goats, for example) designed to a conserved region in the mouse genome will recognize all mouse antibodies. This strategy is used to economically manufacture a small number of antibodies and label them with a variety of signal molecules that can then be employed in a variety of assays. The secondary antibody is often coupled to an enzyme and the product of the enzymatic reaction is what is actually detected. Multiple secondary molecules can bind a single reporter molecule (Fig. 19-3B), and this can provide secondary detection systems with increased sensitivity compared to reporter antibodies directly linked to the signaling molecule.

There are also RNA and DNA versions of the sandwich hybridization assay (Fig. 19-3C). When single-stranded RNA is used, the detection scheme is analogous to that described for proteins (Figs. 19-3A and B), except that appropriate oligonucleotides are used in place of antibodies. For typical DNA assays, the protocol accounts for the double stranded nature of the target molecule by adding a denaturation step or by capturing double stranded DNA that was labeled previously; for example, during PCR. If PCR labeling was used, this variation of the sandwich hybridization assay is actually a combination of solution-based and surface-based approaches (Fig. 19-1C).

Capture Probe Design

The methods described previously for extracting bulk DNA, RNA, or proteins from an organism are not specific. Capture molecules are needed to retain specific proteins or nucleic acid sequences. It is generally desirable to have a capture molecule that recognizes a region of the target that is unique and also distinct from the region that subsequently will be used to bind a reporter molecule.

In the case of proteins, the capture and reporter molecules usually are antibodies (Figs. 19-3A and B). Most microbial species have at least one antigen that can be used to uniquely identify that organism, and sequence alignments are used to help identify useful proteins or protein fragments. If a specific antigen can be purified and antibodies generated, these molecules can provide effective diagnostic tools. Typically, an internal protein is the target and cell lysis is required to free it; however, surface proteins also may be used. Immunoseparation using surface proteins (Porter and Pickup, 1999) is theoretically the method of choice to process environmental samples because organism concentration and target detection steps could be combined. However, cost and lack of good antibodies have hampered the use of immunoseparation for many bacteria and viruses important to water quality monitoring. Part of the problem is that many organisms have a variety of surface antigens (serotypes). For example, *E. coli* has approximately 160 serotypes, thus 160 different capture antibodies would be needed to capture all *E. coli*. Another issue is that bacterial antigens seldom are expressed uniformly in different growth phases and this affects the ability to quantify the result.

Nucleic acid molecules are being used increasingly in diagnostic tests. Hybridization is the term used to describe the binding of DNA or RNA to a molecule with complementary base pair sequence. During hybridization, a single-stranded nucleic acid is joined or annealed to a complementary capture molecule to form a double-stranded molecule. For nucleic acid targets, the capture molecule is generally an oligonucleotide. Stringency describes the level of base-pair mismatch tolerated by the hybridization conditions (Li and Hanna, 2004). For example, high stringency describes hybridization conditions that allow little to no mismatch between the base-pair sequence of the captured molecule and the capture probe. Important considerations in the design of target-specific oligonucleotides include the length of the probe, the number of mismatched bases between the probe and the target sequence, and the location of the mismatch (e.g., in the middle of the probe versus the end) because such factors affect the specificity and stringency of target capture. In general, highly specific and stringent capture is desired, particularly in environmental biosensing, because the rich molecular diversity present in natural samples (discussed further later) increases the risk of cross-hybridization to the nucleic acids of nontarget organisms. However, high stringency means high per base binding efficiency, which results in high melting temperatures. To employ practical hybridization conditions, probes are kept short or chemicals are used in the hybridization buffer (discussed later) to decrease the T_m. However, these conditions, in turn, can reduce probe specificity because of cross-hybridization to nontarget sequences. Therefore, testing with laboratory cultures and environmental samples is always required to validate the design of capture probes.

Capture Probe Immobilization

There are a variety of chemistries used to immobilize capture molecules onto surfaces (Fig. 19-1C, Fig. 19-3). The binding may be hydrophobic, ionic, or covalent. Immobilization protocols often involve simple incubation and drying steps. This simplicity allows laboratory technicians to custom-coat substrates with capture molecules specific to their needs. Adsorption and photoimmobilization are common methods to apply capture probes. Biotin-avidin chemistry is another common strategy to anchor molecules to a substrate (Fig. 19-3) or bind molecules to each other (e.g., reporter to signal molecules). Avidin and relatives such as streptavidin or neutroavidin are proteins that tightly (but noncovalently) bind a small molecule called biotin. Therefore, a substrate coated with or a molecule labeled with streptavidin will bind to any compound containing biotin (Fig. 19-3).

A wide range of immobilization substrates are currently employed in emerging technologies. These include microplate wells, microspheres (e.g., fluorescent polystyrene or paramagnetic beads), glass slides, membrane filters, wave guides (planar or optical fibers), and carbon electrodes (Fig. 19-1C). Capture probes may coat a surface or be physically isolated as spots or strips (Ligler and Taitt, 2006).

Target Capture and Wash

The target is incubated with the immobilized capture probe and allowed to bind. Careful design of the capture probe ensures specific target capture; however, nontarget molecules must be removed by a series of wash steps before detection. Hybridization and wash solutions contain a blend of ingredients designed to remove unbound and nonspecifically bound constituents from the capture substrate without disrupting specific binding of the target. For typical nucleic acid assays, a nonstringent wash is first used to remove nonspecifically bound nucleic acids, and a second (or sometimes third) wash is performed under conditions of higher stringency so that only targets with near perfect matches to the capture probe are left.

The stringency of hybridization is adjusted by changing the chemical characteristics of the hybridization solution and wash buffers (similar to optimizing binding during PCR). Parameters that affect the stringency of hybridization include temperature, ionic strength, base composition, and viscosity (Hames and Higgins, 1985). Increasing the temperature or lowering the salt concentration are common methods to increase stringency. A lower salt concentration increases stringency by increasing the electrostatic repulsion between nucleic acid strands; therefore, only homologous bases are likely to pair. Denaturing and chaotropic agents effect hybridization by destabilizing the hydrogen bonds and hydrophobic interactions of double stranded DNA structures. Chemicals such as formamide (Chakrabarti and Schutt, 2001), guanidine thiocyanate, and tetramethylammonium chloride (TMAC) are used to (1) reduce secondary and tertiary structures that interfere with hybridization, (2) effectively lower the melting temperature (T_m) so that hybridization can be performed at convenient laboratory temperatures and protect the DNA (e.g., 40°C rather than 80°C), and (3) normalize for differences in base composition (GC base pairs are more stable than AT base pairs). Components such as ethylenediaminetetraacetic acid (EDTA) bind metals and therefore protect DNA by inactivating nucleases—enzymes that destroy nucleic acids. Inert polymers such as dextran sulphate increase the rate of hybridization by pushing DNA out of bulk solution and into contact with immobilized capture probes, thereby enhancing the kinetics of the capture reaction by increasing the effective concentration of the target relative to the probe.

Use Reporter Molecules to Detect the Captured Target

After the capture and wash steps are completed, the next step is to bind the reporter molecule, as summarized earlier (Fig. 19-3). The reporter molecule often consists of two parts: a reporter probe that binds to the captured target and a signal molecule that can be quantified by the detector. Numerous signal molecules are available (Caruana, 2004; Didenko, 2006; Li and Hanna, 2004; Thompson, 2006). Common types of signal molecules include chromophores, flurophore, electrochemical, and electrochemiluminescent (ECL) molecules. The reporter molecule is activated (for example, by light of the appropriate wavelength or addition of enzyme substrate), and the detectable signal is quantified to determine how much target was captured. The amount of signal produced is correlated to the amount of bound reporter, which in turn is correlated to the amount of target bound. For quantitative assays, standard curves using known amounts of the target molecule are run so that the signal in each sample can be converted to an estimate of the amount of target present.

Detection Instrumentation

A biosensor is a system that combines a biological recognition event (i.e., binding, hybridization) with detection ("sensing") of the event. The detector ultimately measures some property brought about by the reporter molecule, such as fluorescence, emitted light, or change in electrical current. Change in weight or change in refractive index can also be measured in detector systems employing microcantilever (Hansen and Thundat, 2005; Sepaniak *et al.*, 2002) or surface plasmon resonance (SPR) and total internal reflection fluorescence (TIRF) sensors (Herron *et al.*, 2006; Homola, 2006), but these (and other sensor types) are not discussed here. Spectrophotometers, luminometers, photodetectors, flow cytometers, confocal microscopes, and color-capture device (CCD) cameras are examples of detectors used in advanced technologies. In the most mature state, the biosensor is integrated into an advanced detection system. As a system, the detector platform may carry reagents (using macro or microfluidics), separate constituents (microspheres, analytes, sequencing reactions, etc.), excite labels, transfer the resulting signal to the detector, and translate the signal into an electrical output. The instrumentation interfaces with a computer and computer software to process and organize the data output. In addition, some detector systems miniaturize or automate the detection protocols (Chang *et al.*, 2006; Greenfield *et al.*, 2006). Production of a detector system requires an interdisciplinary workforce with molecular biologists and engineers working together to integrate molecular protocols into a detection platform.

EXAMPLES OF ADVANCED DETECTION TECHNOLOGIES

An inclusive discussion of emerging technologies for water quality applications is not feasible, especially in light of the momentum (and investment) created by the clinical and biosecurity sectors. Instead, the underlying principals outlined previously will serve as the platform for the discussion of advanced technologies to be discussed later. The interested reader can consult a number of books and review articles for more examples of detection technologies (Clark *et al.*, 2001; Cooper and Cass, 2004; Hansen and Thundat, 2005; Liron *et al.*, 2001; Narayanaswamy, 2004; Noble and Weisberg, 2005; Thompson, 2006; Wang, 2000).

The examples that follow address or have the potential to address many of the current needs of water quality testing that were discussed in the beginning of this chapter. One technology, quantitative PCR, is rapidly transitioning into an established technology for the environmental sector. It is becoming widely used in research settings and is being seriously considered as a technology appropriate for monitoring programs. One of the obstacles to implementation is that there is not a one-to-one correlation between molecular enumeration and culture-based enumeration, and this poses an obstacle to simply replacing the current methods on which regulations are based. One issue is that molecular protocols do not distinguish between viable and nonviable organisms, unlike traditional culture based methods (EPA, 2004). In addition, molecular- and culture-based targets do not necessarily measure the same thing; for example, a molecular protocol may measure one species of enterococci whereas the culture method will grow numerous species from the genus. The most direct way to address the situation (albeit expensive) is to incorporate molecular assays such as qPCR into epidemiology studies and derive risk-based guidelines based on those results. Steps in this direction have begun (Colford, Jr. *et al.*, 2005; Wade *et al.*, 2006).

Solution-Based Technologies

Enzyme-Based Assays

Strides have been made to increase the convenience of current culture-based methods that identify fecal indicators based on characteristic enzymes (see the description of chromogenic and fluorogenic enzyme substrates in the Current Enumeration Practices section). Assay systems such as Colilert and Enterolert (IDEXX Corp., Westbrook, Maine) are becoming commonly used detection systems in the environmental sector. In addition, rapid and field deployable instruments that utilize enzymatic assays as the technological basis have been designed (Anglès d'Auriac *et al.*, 2000; Davies and Apte, 1999; Noble and Weisberg, 2005; Sartory *et al.*, 2001). However, the specificity of these enzymatic methods with environmental samples is dependent on selective enrichment, and this appears to be a limiting factor in further reducing the turnaround time of these assays (Tallon *et al.*, 2005).

Quantitative PCR Assays

PCR reactions follow a sigmoidal curve—production of amplicon is exponential after an initial lag and then plateaus as the reaction components are consumed and the polymerase loses activity. Standard PCR (Fig. 19-1A) and technologies that utilize standard PCR (Fig. 19-1C) often quantify the amount of amplicon present. However, standard PCR is not quantitative in the sense of being able to determine the amount of organism that was in the original sample. The amount of amplicon could be related back to the original sample if the PCR reaction did not plateau and if it were 100% efficient—that is, if (the number of copies at cycle $n + 1$)/(number of copies at cycle n) = 2. However, standard PCR only samples the end product of the reaction, so one is unable to determine if plateau was reached. In addition, the reaction is not completely efficient and efficiency varies from sample to sample (Shepley and Wolk, 2004; Wittwer and Kusukawa, 2004). In contrast to endpoint PCR, quantitative PCR (qPCR) measures the rate of DNA synthesis over the entire reaction and provides cycle-by-cycle output ("real time" measurements) to ensure that the measurements are collected within the exponential phase of the reaction. The amount of starting target can be quantified because the more that was present initially, the faster (earlier cycle) the fluorescence signal moves pass a set threshold value (C_t) and into the exponential phase (Fig. 19-4).

In qPCR, the progress of the PCR reaction can be followed cycle by cycle because the newly formed DNA molecules are fluorescently labeled. The fluorescence is measured over time by a specialized thermocycler equipped with an excitation light source that can illuminate each reaction tube and a detector system for measuring fluorescence light emission. Sample quantification is achieved by use of a standard curve, which plots the C_t value of known amounts of target DNA against the log of the starting DNA concentration (Fig. 19-5). The amount of target DNA in the sample can be estimated by plotting the sample C_t values onto the standard curve. The cell concentration in the original sample (reported in units of copy number or genome equivalents) is calculated by estimating the amount of DNA in each bacterial cell or viral particle. The controls discussed for standard PCR are applicable to qPCR; however, the emphasis on controls is greater (Shepley and Wolk, 2004) because proper use is critical to the precision and accuracy of the method.

The simplest means of monitoring DNA formation is by adding a DNA-intercalating, fluorescent dye (e.g., SYBR Green) into the PCR reaction mix. The dye binds to all double-stranded DNA molecules, whereupon it transforms from a nonfluorescent to a fluorescent state. As a result, this method cannot distinguish legitimate amplicon from nonspecific PCR products or primer dimers (see the PCR section above). If these products are present, the number of organisms in the initial sample will be overestimated. A melting curve analysis is performed to ensure that this is not the case. The analysis is performed by rapidly heating an amplified sample. As DNA is heated, it has an unusual property. The two DNA strands remain together then suddenly denature as the T_m is reached, rather than coming apart gradually as the temperature rises. The denatured, single-stranded DNA does not fluoresce, and shorter double-stranded DNA molecules have lower melting temperatures than do longer amplicons. As a result, if only one PCR product is formed, the signal will decline uniformly to a low level. In contrast, if multiple products are present the signal will decrease in a series of plateaus as each double stranded DNA product reaches the temperature at which it denatures. Although SYBR Green is easier to use and more economical that other qPCR approaches, a melting curve analysis with multiple peaks indicates that the PCR reaction must be optimized further and/or a different type of qPCR probe is needed.

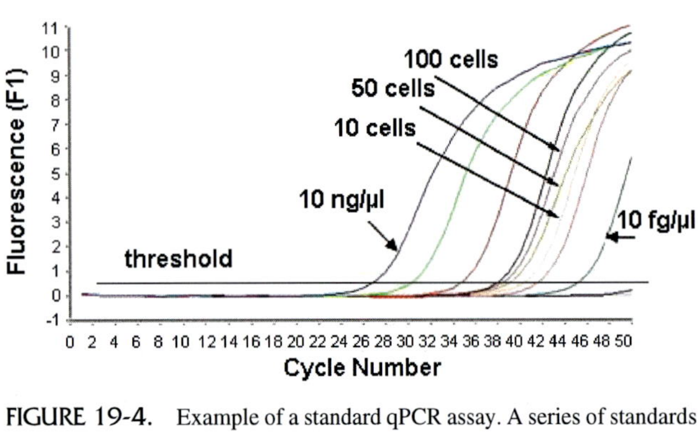

FIGURE 19-4. Example of a standard qPCR assay. A series of standards containing 10 fg/µl to 10 ng/µL target DNA was run along with the genome equivalents from 10, 50, and 100 cells. Note that the higher the initial target concentration, the earlier the cycle number at which the signal (generated from SYBR green, molecular beacon, or other probe) rises above the background threshold.

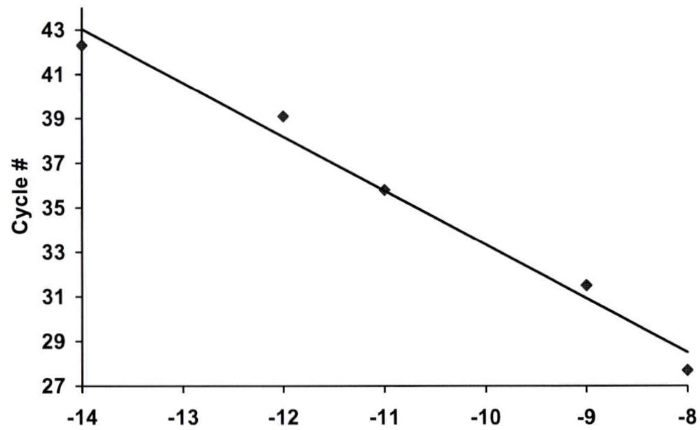

FIGURE 19-5. Example of a standard curve ranging from 10 fg/µl to 10 ng/µl versus the threshold cycle number (C_t) for amplification obtained from 10, 50, and 100 bacterial cells. In practice, the standard curve and the sample C_t value are used to estimate the number of organisms in the original sample.

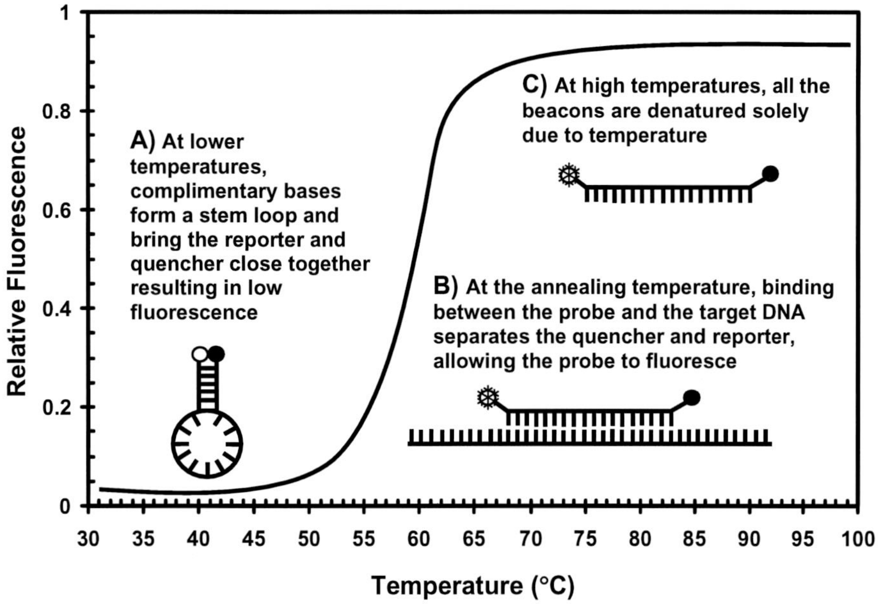

FIGURE 19-6. Diagram showing the effect of temperature and target hybridization on molecular beacon fluorescence. (A) Lower temperatures produce low fluorescence because of maintenance of the stem loop structure. (B) At the annealing temperature, the binding between the probe and the complementary bases in the template is stronger than within the stem loop, and probe hybridization separates the quencher and reporter, allowing it to fluoresce. Fluorescence is measured during the same phase of each annealing step. (C) At denaturation temperatures, all the beacons are disassociated and fluorescent. ○ indicates the fluorescent reporter molecule is being quenched, • represents the quencher molecule, and ❁ represents an actively fluorescing reporter molecule.

Molecular beacons and other fluorescent probes overcome the limitation of SYBR Green because the probe binds to a specific DNA site located within the amplicon. Primer dimers and nonspecific amplification products should not contain this binding site, thus no fluorescence should be produced by nonspecific products. A common molecular beacon is the 5′-nuclease probe, such as that used in TaqMan assays (Applied Biosystems, Foster City, California). The probes are usually constructed with a fluorescent molecule attached to the 5′ end and a quenching molecule at the 3′ end (Fig. 19-6). This configuration holds the reporter and quencher molecules in close enough proximity that most of the light emitted by the fluorescent reporter molecule is absorbed by the quencher molecule, resulting in low background fluorescence (Fig. 19-6). This probe binds during the annealing phases of PCR amplification. During extension, the polymerase moves in a 5′ to 3′ direction, and when it reaches the 5′ fluorescent molecule of the molecular beacon, the 5′-exonuclease activity of the polymerase cleaves the probe and releases it free into solution (Fig. 19-7). Once released, the fluorescent molecule is no longer in close proximity to the quenching molecule and can fluoresce freely when excited. The DNA polymerase then continues down the chain, displacing the remainder of the beacon and synthesizing new DNA. This process repeats every cycle and the amount of fluorescence increases in correlation to the number of new DNA strands.

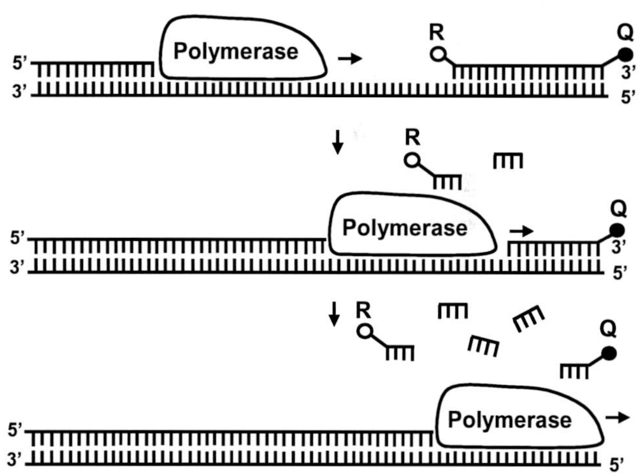

FIGURE 19-7. Illustration of the 5′ exonuclease activity of DNA polymerase, which degrades the molecular beacon and liberates the signal molecule of the reporter probe (R). The signal molecule is free to fluoresce when no longer in close proximity to the quencher (Q).

The primer design parameters described for standard PCR (discussed earlier) should be incorporated into quantitative PCR. In addition, empirical experience has shown that qPCR works best when the assays are designed to amplify targets between 50 and 150 bp long. Short amplification products also decrease the time of the qPCR assay by reducing the extension time required to synthesize the amplicon.

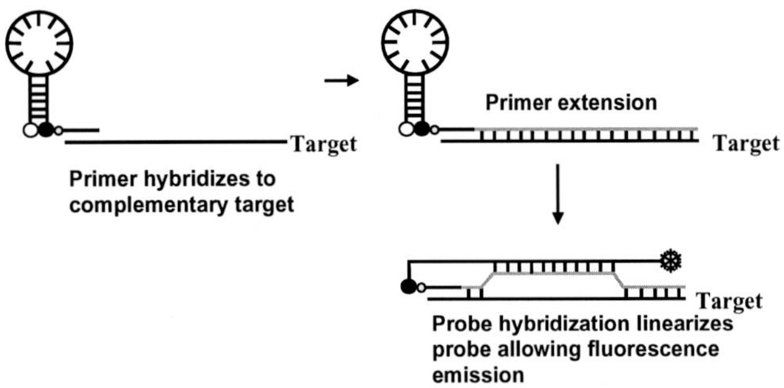

FIGURE 19-8. Diagram showing the function of a Scorpion primer. The key component is that the stem loop of the primer is complementary to the resulting strand of synthesized DNA. Binding of the primer to the amplicon strand separates the quencher and fluorophore, allowing fluorescence.

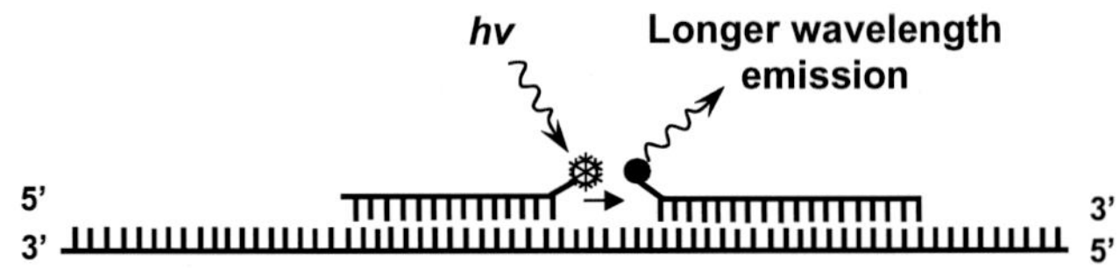

FIGURE 19-9. Diagram illustrating fluorescence resonance energy transfer (FRET) between adjacent probes. Probe binding brings the fluro-phores close together, enabling energy transfer from the donor to the acceptor when the sample is excited with the appropriate wavelength of light. By using the appropriate filter sets, it is possible to measure only the fluorescence emitted by the second dye.

In a molecular beacon assay, the T_m of each primer should be the same and ideally within the range of 58 to 60°C. The T_m of the molecular beacon itself should be 10°C higher than the primers and the GC content should be 30% to 80%. Another common modification of a molecular beacon such as the TaqMan probe involves attaching the quencher to the central portion of the probe and adding a blocking molecule to the 3′ end.

Scorpion primers (PREMIER Biosoft International, Palo Alto, California) are another type of molecular beacon, but in this case a PCR primer itself is covalently linked to a fluorophore and a quencher. In the absence of target DNA, the quencher and reporter are held next to each other in a stem loop structure and fluorescence is not emitted. Successful amplification physically separates the reporter and the quencher, and the increase in fluorescence intensity is correlated to the number of new DNA molecules. The separation occurs as follows: the annealed Scorpion primer maintains the stem-loop structure until after the polymerase has extended the new strand. The Scorpion is designed to hybridize to this DNA, which separates the reporter from the quencher and allows emission of fluorescence (Fig. 19-8).

Another qPCR probe type is the adjacent probe, such as that used in the LightCycler (Roche, Indianapolis, Indiana). In this design, two flurophores are used in which the emission spectrum of the donor overlaps with the excitation spectrum of the acceptor. The donor fluorophore is linked to the 3′ end of a probe and the acceptor fluorophore to the 5′ end of a second probe (Fig. 19-9). The probes are designed to bind to the amplicon with a gap of 3 to 5 bp between them. This gap allows an excited donor fluorophore to transfer energy to the acceptor fluorophore in a process termed fluorescence resonance energy transfer (FRET) (Marras, 2006). Appropriate filters are used so that only the wavelength of the acceptor fluorophore is measured. No energy transfer should occur unless both strands are hybridized; thus, the increase in fluorescence can be correlated to the amount of DNA synthesized, given appropriate standards and controls.

Traditionally, a drawback of qPCR was that multiplexing was not available (i.e., only one target could be detected at a time). However, great progress has been made at increasing the number of different fluorescent signals that can be followed. This capacity allows multiple reactions to be run simultaneously using reporter dyes that fluoresce at different wavelengths. These multiplex assays, however, are difficult to design so that cross-reactivity between the probes does not occur and that the various reactions do not directly compete for nucleotides during the PCR reaction. If the later occurs, a form of PCR bias can result in which the dominant template reaction causes the other reactions to fail because they lack sufficient nucleotides to carry out an efficient amplification.

Limits to Quantification

The term *quantitative* in qPCR is somewhat of a misnomer. Experts agree that that there is no way to quantify PCR reactions with full accuracy, although incorporation of certain controls helps improve precision and accuracy (Shepley and Wolk, 2004). One reason is that the amount of product is dependent on stochastic processes (random or probalistic, but with some direction) that occur in the first and second round of PCR, particularly when target concentrations are low. Small differences in the starting reaction are exponentially compounded, which can lead to large differences in the yield of final product. Even for samples that are not inhibited, PCR reaction efficiency varies within and between samples, with the DNA increase typically ranging between 1.8 and 2.0 fold each cycle and variability ranging from 10 to 20% in replicate samples (Shepley and Wolk, 2004). The result is that replicate samples will often vary by ± 0.5 C_t or greater. As an example, a C_t of 30 might correspond to 8000 cell equivalents, a C_t of 31 to 16,000 cells, and a C_t of 32 to 32,000 cells. Given the variability in replicates, a sample with a C_t value of 31.5 is expected to contain about 20,000 cell equivalents, but the actual number may range from 17,300 to 30,200. The difference between 17,300 and 30,200 cells in the original sample, however, may be significant in terms of a threat to human health. In addition, quantitative results at the level of the PCR reaction assume that the DNA itself was extracted from the environmental samples in a quantitative manner. As discussed later, extraction issues are often the greatest impediment to performing a quantitative assay. Despite these limitations, progress is being made at successfully incorporating qPCR into monitoring of recreational waters (Haugland *et al.*, 2005; Noble *et al.*, 2006).

Quantitative Reverse Transcriptase PCR

Quantitative reverse transcriptase PCR (qRT-PCR) uses RNA as the template molecule. This technique is commonly used for detecting RNA viruses. The procedure is essentially the same as that described for qPCR, except that a reverse transcription (RT) step must first be performed to quantitatively convert the RNA to complementary DNA (cDNA). The RT step is necessary because the DNA polymerases of qPCR reactions do not amplify RNA. The conversion to cDNA is accomplished by enzymes such as the avian myeloblastosis reverse transcriptase (AMV-RT), the Moloney murine leukemia virus (MMLV) RT (Gerard *et al.*, 1997), or other reverse transcriptases that have been genetically engineered to better handle secondary structures and to improve fidelity.

In nature, reverse transcriptases function to convert single stranded RNA genomes into DNA, an essential step in the life cycle of some viruses. Unlike some RNA polymerases, reverse transcriptases do require primers (Snyder and Champness, 1997). Detection of RNA viruses in water samples is a logical application of the RT technique because certain RT enzymes were cloned from RNA viruses. Bacterial targets can also be quantified by qRT-PCR, but less effectively because RNA levels in bacteria vary by 10- to 100-fold, depending on growth stage. Thus, unlike viral genomes, which tend to contain a fixed amount of RNA, the quantitation of bacteria using this approach can be off by one to two orders of magnitude simply because of the growth state of the cells.

Before the RT step, DNA should be removed from the sample to ensure that only RNA is amplified, otherwise quantification will be compromised. Therefore, the sample is treated with DNAase and then heated to inactivate the DNAase. Care must be taken to ensure that the DNAase is truly inactivated because carryover will degrade the cDNA produced in the RT reaction, leading to a gross underestimate of the amount of target in the original sample. A typical RT reaction occurs as follows: An aliquot of RNA is heated to reduce secondary structure, which will hinder the ability of the RT to synthesize cDNA. The sample is rapidly cooled and added to a standard qPCR mixture also containing RT. The reaction is incubated at 37°C for 20 minutes to 2 hours to allow the RT reaction to proceed. Afterward, the sample is processed by qPCR.

The controls described for PCR and qPCR apply to qRT-PCR reactions (i.e., positive and negative controls for the extraction and PCR steps and an estimate of nucleic acid recovery are needed). In addition, controls are included to account for the RT reaction. The RNA used to create the standard curve should be treated exactly like the samples themselves in order to account for the variability in the RT step. In addition, no RT controls are used to alert against DNA contamination. These controls are processed like the samples, except that the RT is omitted from the reaction. If these controls amplify, genomic DNA was present from insufficient DNAase treatment or from cross-contamination, leading to an over-estimate of the amount of target present in the original sample.

A competitive internal positive control (CIPC) (Noble *et al.*, 2006) can be used as an inhibition control and to normalize results from one experiment to the next. This control consists of spiking each sample with a synthetic target before the RT step of the assay. The CIPC is distinguished from the target by using a different fluorescent reporter molecule. It is important to add as little of the CIPC to each sample as possible to obtain a consistent signal, but not so much that this primer competes for the pool of dNTPs. If the CIPC concentration is sufficiently low, detection of the primary target over a wide range of concentrations is possible. PCR inhibition is indicated by a higher C_t value for the CIPC in the sample relative to the control. An inhibited sample can be diluted and rerun or, over a certain range of inhibition, the CIPC data can be used to estimate the degree of inhibition. These data allow a calculation of the genome equivalents that would have been detected if no inhibition had occurred (Noble *et al.*, 2006). As described for qPCR, there are limits to the degree to which this procedure can be considered quantitative.

Nucleic Acid Sequence-Based Amplification (NASBA)

Besides PCR, other methods are available to amplify nucleic acids (Hayden, 2004; Shepley and Wolk, 2004). For example, Nucleic Acid Sequence-Based Amplification (NASBA) is an isothermal method (i.e., no temperature ramping is required) designed to amplify RNA from either an RNA or DNA template, although it is most commonly used to amplify RNA targets (Cook, 2003).

NASBA amplifications use three enzymes: a reverse transcriptase, a RNA polymerase, and a ribonuclease. The reverse transcriptase is AMV-RT, which can synthesize a complementary DNA strand from either RNA (to produce cDNA) or from single-stranded DNA. In either case, the enzyme requires a primer to begin synthesis. The second enzyme is a RNA polymerase, which is used to transcribe DNA into RNA. Unlike a DNA polymerase, this RNA polymerase does not require primers. Instead, it employs a specific area of double-stranded DNA called the promoter region to initiate transcription. Initially, the RNA polymerase binds to the promoter and separates the DNA strands, exposing the bases. A ribonucleoside triphosphate (NTP) then binds to the template strand of the DNA (or cDNA in the case of NASBA), and the RNA polymerase builds the RNA chain by linking NTPs to the 3′ end (Snyder and Champness, 1997). The resulting RNA transcript is a reverse complement or is "antisense" to the template strand of DNA (Fig. 19-10). In nature, RNA transcription ends at a transcription terminator. In NASBA, the transcription ends when the polymerase comes to the end of the cDNA, releasing the RNA transcript into solution and leaving the cDNA intact so that more RNA can be made. The third enzyme used is RNAse H, which is a ribonuclease used to degrade the RNA of the RNA-cDNA duplex.

A typical NASBA reaction contains the following: (1) RNA template; (2) two primers, one containing a T7 polymerase binding site; (3) AMV-RT; (4) RNase H; (5) T7 polymerase; (6) dNTPs; (7) NTPs; and (8) buffer. As depicted in Figure 19-10, the NASBA reaction begins when a primer containing the T7 polymerase binding site (primer 1) specifically binds to the target RNA. This primer contains a T7 RNA polymerase binding site at the 5′ end. Once the primer is bound, AMV-RT synthesizes a strand of cDNA, creating a RNA-cDNA duplex. The RNase H enzyme subsequently degrades the RNA of the RNA-cDNA duplex, releasing single stranded cDNA into solution. The second primer (primer 2) is designed to anneal to that cDNA, which allows the AMV-RT to synthesize a second strand of cDNA that also contains a T7 binding site. The result is a double stranded cDNA containing the promoter for the T7 RNA polymerase. The promoter allows the T7 polymerase to bind and synthesize numerous (up to 700) RNA molecules from each double stranded cDNA molecule. The sequence of the new RNA is a reverse complement to the template strand of cDNA and is also "antisense" to the initial RNA strand (Fig. 19-10). At this point, the NASBA reaction enters an amplification phase. The reagents in the reaction mix convert each of the new RNAs into double stranded cDNA molecules, which in turn are converted into more copies of the antisense RNA, allowing the cycle to repeat (Fig. 19-10). Unlike PCR, only one copy of product is made per cycle, so the amplification is linear rather than exponential. Nonetheless, more than 10^{14} RNA molecules can be generated in a NASBA reaction, with demonstrated sensitivities similar to that of qRT-PCR (Rutjes *et al.*, 2005). The primer concentrations are on the order of 10^{12} so that high template loads (e.g., on the order 10^5-10^6 virus particles) do not cause the primers to be rate limiting (Baudart *et al.*, 2002). Therefore, accumulation of RNA is linear with time and thus quantifiable given appropriate standards and controls (Casper *et al.*, 2005). The production of a single-stranded product simplifies the hybridization of reporter molecules. Approaches using fluorescent and electrochemiluminescent reporter molecules are

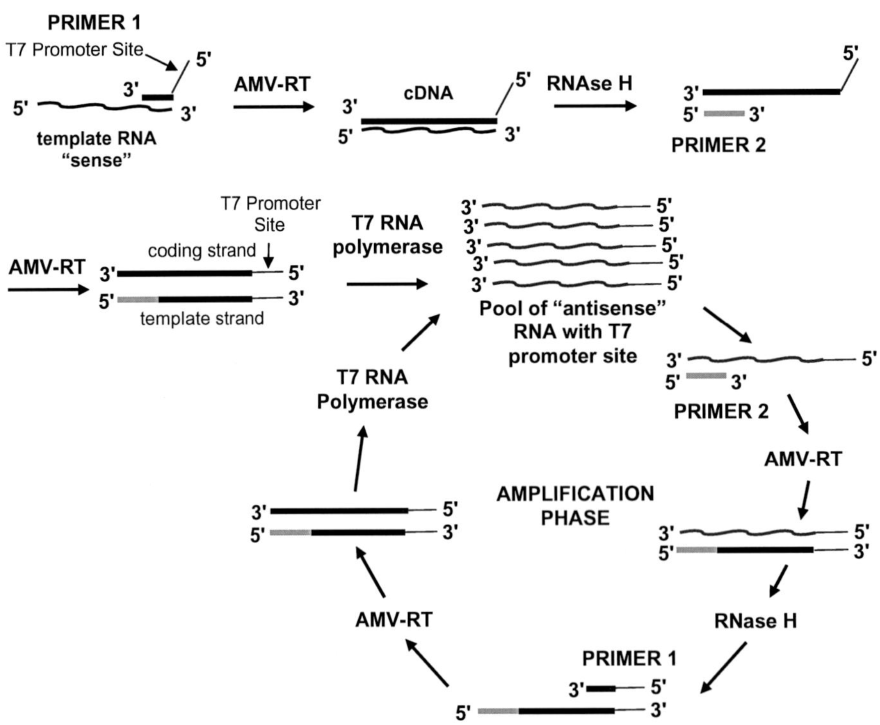

FIGURE 19-10. Diagram of a NASBA reaction. Reaction components include a primer that contains a promoter site for the T7 polymerase (represented by small blue line), which enables the amplification phase of the reaction, as described in the text.

used to provide real-time detection and product quantification. NASBA chemistry is expensive and the additional enzyme adds complexity; however, the isothermal chemistry (i.e., no temperature ramping required) may be more conducive to advanced detection technologies designed for field deployment than protocols using PCR amplification (Patterson *et al.*, 2006).

Surface-Based Technologies

Scale is the major difference between some detection technologies. For example, a "macroarray" may be in the format of a 96-well microtiter plate or it may be a filter with tens to hundreds of spots of capture probes. In contrast, a microarray may contain tens of thousands of spots, each as small as the head of a pin (on the order of 100 μm diameter). Although there are significant differences in detail, all of these technologies can be understood within the general framework discussed for surface-based technologies.

Microarrays

A microarray (or gene chip) contains thousands to tens of thousands of capture molecules precisely spotted onto a substrate such as a glass microscope slide or a nylon membrane. Traditionally, microarrays have been used primarily in clinical applications to look at gene expression or gene mutations, for example, to study human diseases such as cancer. The capture molecules are usually derived from various genes of a particular organism, such as a mouse, human, or yeast. In practice, two samples are compared—for example, tissue from a benign versus a cancerous tumor. In a two-color array, the targets (RNA, cDNA, or proteins) in each sample are labeled differently (e.g., Cy3 versus Cy5 fluorescent labeling), and both samples are added to the same array and allowed to compete for the capture probe. Special scanning instrumentation and image-processing software is used to determine the ratio of Cy3 to Cy5. After washing away unbound probe, the flurophores are excited. Spots bound predominately with Cy3 targets will appear green, those with Cy5 will appear red, and the genes that were not differentially expressed will be a mix of red and green, and thus will appear yellow. The pattern is interpreted in terms of genes being "up-regulated" (induced) or "down-regulated" (repressed), and this information can be used, for example, to help design new drugs. Alternatively, the target is labeled with biotin and the two samples are added to two different chips and the images are compared. Similar experimental designs have been used with environmental samples to study sample toxicity. However, microarrays for organism detection in environmental samples do not require differential comparison. Instead, the capture probes correspond to different species, strains, serovars, pathotypes, or virulence factors (Lemarchand *et al.*, 2004; Sharkey *et al.*, 2004). In this way, microarrays and gene chips are a prime example of advanced technologies representing miniaturization and/or automation of standard molecular techniques (such as Northern blots). The use of microarrays for coastal water quality applications is currently being explored (J. Rose, Michigan State University, personal communication).

Suspension Arrays

In suspension arrays, capture molecules are immobilized to microspheres (beads) rather than to spots on a slide (Nolan and Sklar, 2002). Binding or hybridization of the target is carried out in a suspension contained in the well of a microtiter plate. Instead of a single capture molecule being in each well like in a traditional ELISA format, many (10s to 100) coexist in the same well. Each type of capture molecule is immobilized onto beads of different "color" (varying ratios of flurochromes). The beads are separated by flow cytometry, which control the flow of the microspheres through the path of two lasers. The excitation from one laser determines the bead color and the other laser excites the label of the target; in other words, one laser determines the identity of the capture probe and one registers whether or not the bead has captured a target. A main benefit of a suspension array over a microarray is that chip fabrication is not required, and this allows increased flexibility and cost savings (Dunbar, 2006). Suspension arrays provide rapid, high-throughput detection of multiple targets and are sold commercially (Luminex Corporation). The potential to use them with coastal samples is under investigation (Baums *et al.*, 2007).

Waveguide Technology

With evanescent wave biosensors, capture molecules (e.g., nucleic acids or antibodies) are immobilized onto the surface of a structure that propagates laser light along its length (the waveguide). In some cases, the waveguide is a fiber-optic cable (Ahn *et al.*, 2006; Harwood *et al.*, 2004; Lim, 2003; Simpson and Lim, 2005), but other geometries, such as planar waveguides (Herron *et al.*, 2006; Martinez *et al.*, 2005), have been employed. The laser light is mostly contained within the waveguide, but some light escapes (the evanescent field), and this light is used to directly excite the fluorophore label of the captured target. The light emitted from the flurophores is collected and focused back down the waveguide. The signal intensity registered by the photodetector is correlated to the amount of captured target. Evanescent wave biosensors are sold commercially (RAPTOR, Research International, Woodinville, Washington), and researchers are trying to adapt them to water quality applications (Anderson and Rowe-Taitt, 2001), for example, to detect enterococci in marine waters (D. Lim, University of South Florida, personal communication).

Electrochemistry

Handheld electrochemical sensors are commonly used to measure certain blood products (e.g., over-the-counter glucose sensors). Numerous approaches have been formulated to detect nucleic acids by electrochemical means (LaGier *et al.*, 2005, 2007; Noble and Weisberg, 2005), some showing amazing analytical sensitivity (Kerman *et al.*, 2004; Wang, 2000). Many electrochemical formats adopt standard capture and reporter configurations (Fig. 19-3), and some commercial formats include battery operated, handheld instruments (Alderon Biosciences, Inc., Research Triangle Park, North Carolina).

An electrochemical biosensor for nucleic acids applies a controlled potential and monitors the electrical current resulting from the hybridization of the target and probe molecules. For example, the reporter molecule generates an electrical current, rather than color or fluorescence, upon addition of an appropriate enzyme substrate and in the presence of a charged electrode surface. Some formats hybridize PCR products and thus are a combination of solution- and surface-based approaches (Fig. 19-1C). For example, during PCR one strand of DNA may be labeled with biotin to create a target for capture and one strand labeled with fluorescein to serve as a signal molecule (LaGier *et al.*, 2007). Neutroavidin-coated electrodes capture the biotinylated amplicons, and unbound constituents are washed away. An antifluorescein antibody links the fluorescein label to horseradish peroxidase (HRP). Addition of enzyme substrate (hydrogen peroxide) oxidizes the HRP [Equation (1)]. The HRP is regenerated upon donation of an electron to 3,3′,5,5′-tetramethylbenzidine (TMB) [Equation (2)], and the TMB is regenerated by transferring an electron to the charged sensor surface, thus completing the catalytic cycle. The charged sensor surface is generated by applying pulses of electrical potential (various methods are used). The electrical current produced is correlated to the amount of nucleic acid hybridized:

$$H_2O_2 + 2H^+ + HRP_{red} \rightarrow HRP_{ox} + 2H_2O \quad (1)$$

$$2TMB + HRP_{ox} \rightarrow 2TMB+ + HRP_{red} \quad (2)$$

$$2TMB+ + 2e- \rightarrow 2TMB \quad (3)$$

AUTOMATED SENSORS AND REAL-TIME DATA RELAY

The ultimate goal of many emerging technologies is to deploy and maintain sensors in the field (*in situ* biosensing) and to provide the data in near real time. The objective is to provide a rapid, automated, and cost-effective means to make decisions about the closings of fisheries and recreational waters. The ultimate goal is to provide predictive capacity in order to eliminate human contact with contaminated waters and seafood products. For example, the goal of the Integrated Ocean Observing System (IOOS) is to integrate physical, biological, chemical, and geological observations into data products that allow prediction of hazardous coastal and ocean water conditions and the impact of such conditions (see Chapter 1). The IOOS design relies on three interdependent subsystems: observing (global and coastal components), data management and communication (DMAC), and modeling and analysis. IOOS-derived models and information are designed to improve management and response capabilities to threats such as harmful algal blooms (HABs), invasive species, water quality and habitat degradation, and organisms that pose a risk to human and ecosystem health (Ocean.US, 2006a, 2006b).

The development and integration of various technologies and platforms will be required if IOOS is to provide *in situ*, long-term observation of coastal ecosystems. Example technologies include advanced acoustic, optic, chemical, physical, and biological sensors. Example platforms include autonomous underwater vehicles (AUVs), buoys, ships, handheld devices, satellites, aircraft, and unmanned airborne vehicles. To achieve *in situ* biosensing, the automated devices will need to concentrate and prepare samples for detection of the molecular targets. Prototype devices to meet such needs have been deployed and continue to be developed (Alliance for Coastal Technology [ACT], 2006; Scholin *et al.*, 2007). Examples include the environmental sampling processor (ESP) developed by the Monterey Bay Aquarium Research Institute (MBARI) (www.mbari.org/microbial/esp) (Greenfield *et al.*, 2006) and the Autonomous Microbial Genosensor (AMG) developed at the University of South Florida (www.marine.usf.edu/microbiology/genosensor.shtml).

CHALLENGES FACING EMERGING TECHNOLOGIES

The underlying premise of many emerging technologies for water quality research and monitoring is that sensitive and rapid molecular technologies designed for clinical applications can be transferred to environmental applications. There are a number of promising technologies on the horizon to meet the needs of biodetection (Deisingh and Thompson, 2004), and many emerging technologies are being tried in the arena of recreational water quality (Baums *et al.*, 2007; Gersberg *et al.*, 2006; Haugland *et al.*, 2005; LaGier *et al.*, 2007; Noble and Weisberg, 2005; Noble *et al.*, 2006; Tallon *et al.*, 2005; Wade *et al.*, 2006). Despite the great promise of molecular technologies for water quality analysis, environmental samples offer several challenges, which are discussed next. These challenges include the choice of appropriate targets, the rich molecular diversity inherent in environmental samples, and the difficulty of delivering adequate amounts of clean nucleic acids to a detection system.

Choosing Appropriate Targets

Despite attempts to engineer a robust design, emerging technologies depend on the details associated with the organism to be detected, including its biology, ecology, and molecular sequence. Therefore, uncertainty regarding what is appropriate to measure (Griffin *et al.*, 2001) presents a challenge to the technology developer. Presently, the choice of appropriate targets for water quality analysis is guided by research that includes the molecular microbiology of fecal indicators (Haugland *et al.*, 2005), alternative indicators, and source tracking markers (Bernhard *et al.*, 2003; Carson *et al.*, 2003; Noble *et al.*, 2003; Scott *et al.*, 2005), pathogens, and markers of pathogenicity (Blanch *et al.*, 2003; Fuhrman *et al.*, 2005; Garcia-Aljaro *et al.*, 2004; Shetab *et al.*, 1998). Given the uncertainty surrounding appropriate target choice, a combination of targets may provide a more useful measure of water quality. In this case, technologies offering multiplex detection in a flexible format are desired (Baums *et al.*, 2007). Ultimately, epidemiology studies are needed to establish the correlation between health effects and a given determinant of microbiological water quality (National Research Council, 2004). Indeed, the emerging technology itself would best be tested within the context of an epidemiology study (Noble and Weisberg, 2005). This is especially true if the technology is to compete in the environmental market, which is controlled by regulatory drivers (e.g., EPA guidelines) in addition to financial ones. However, epidemiology studies are not widely available because of the expense.

Overcoming Molecular Diversity

Rich molecular diversity is a fundamental difference between environmental and clinical samples. A clinical sample such as blood or urine will contain little or no other organisms besides the one causing illness. In contrast, the organism used to indicate environmental pollution will represent a tiny fraction of the natural microbial population. To further complicate matters, coastal waters contain a wealth of molecular diversity, much of which has not yet been cataloged to date. Indeed, many DNA and protein sequences obtained from coastal samples are closely related to, but not exact matches to, sequences available in GenBank (Baums *et al.*, 2007). This has become a particularly acute issue given our current ability to obtain hundreds of thousands of unique sequences from the organisms present in single environmental sample (Venter *et al.*, 2004). Despite this remarkable ability, a majority of sequences have yet to be obtained. Because probe specificity is based *in-silico*, underrepresentation of environmental sequences in the databases provides the opportunity to design probes that will inadvertently cross-hybridize with nonintended organisms. Therefore, even under the most specific and stringent DNA amplification and hybridization conditions, environmental molecular diversity makes it difficult to prove correspondence of detected and intended targets. In other words, we know so little about the microbiology and molecular microbiology of the environment that just because a sequence obtained from the environment is found in feces does not prove that the sequence came from feces.

In the future, protein analysis may provide another approach to more specific detection. Fundamentally, bacterial species differ from one another because of variability at the DNA sequence level. This sequence variability confers differences at the protein level, which further translates into differences at the physiological level. Identification of these unique proteins may lead to the development of diagnostic tests. There are databases available that catalog proteins thought to be unique to certain organisms. One such database is the Core and Unique Protein Identification System (CUPID), which can provide a list of proteins specific to a genus, species, or strain (Mazumder *et al.*, 2005). The gene sequences of these proteins could be mined for primer and probe sequences to be used in advanced detection technologies.

The Need to Improve Concentration and Nucleic Acid Extraction Methods

The ability to efficiently concentrate and extract nucleic acids from environmental samples constrains the utility of any molecular technology for water quality research and management, including the ones described here. One problem is that molecular protocols utilize small volumes (Noble and Weisberg, 2005). For example, a typical PCR reaction uses only 1 to 5 μl of template, so even when several liters of water have been processed down to 50 μl of DNA (a typical final volume for a standard DNA extraction spin kit), only 1/10th to 1/50th of the sample is amplified in the PCR reaction. In addition, low extraction efficiencies and PCR inhibition (Wilson, 1997) create a gap between the analytical sensitivity achieved with diluted genomic DNA and what can be achieved with environmental samples (Table 19-1). Variable and low (0 to ~40%) DNA extraction efficiencies from soil have been observed with standard spin kits and bead-beat lysate protocols (Baums *et al.*, 2007). Table 19-1 illustrates how small volumes, inefficient recovery, and PCR inhibition work to undermine the theoretical detection limit achieved with diluted genomic DNA and translate into a relatively large number of cells required in the starting material. However, simply increasing the volume of sample processed does not compensate for inefficient nucleic acid extraction, because concentration of PCR inhibitors results in diminishing returns—increasing the level of

TABLE 19-1. Table illustrating the effect of extraction protocol, PCR inhibition, and the efficiency of DNA recovery on the starting material needed to meet laboratory-determined detection limits for a PCR reaction.

PCR Detection Limit (Genome Equivalent)[a]	Eluant Vol (μl)	Volume Added to PCR rx (μl)[b]	Dilution Needed to Overcome PCR Inhibition	Percentage Recovery[c]	~Number of Cells on Filter to Deliver PCR Detection Limit
3	50	5	1	10	300
3	50	5	1	2	1500
3	600	5	5	40	4500
10	600	5	5	40	15000

[a]Analogous to cells; example detection limits based on dilution of genomic bacterial DNA rather than extraction of cells from a membrane filter.
[b]Typical volumes for crude lysate protocols (600 μl) (Haugland *et al.*, 2005) versus commercial spin kits (50 μl).
[c]Example values observed in laboratory experiments.

concentration often reduces the likelihood of achieving detection. Furthermore, the risk of false-negative reporting increases for rare targets. The probability that no template will be placed into a reaction tube follows a Poisson distribution, such that if an average of two copies of target are present in each PCR reaction, the risk that no template will be in the reaction is 10%. The risk is 1% for an average of five copies per reaction (Wittwer and Kusukawa, 2004). Progress in the area of upstream sample processing is needed to realize the full potential of emerging technologies (Noble and Weisberg, 2005), particularly as molecular assays attempt to move away from indicators to rare targets such as human pathogens.

SUMMARY

Current water quality monitoring practices require improvement with regard to rapidity, throughput, multiplexed detection, and the ability to detect pathogens and source tracking markers. Emerging technologies in the environmental sector seek to establish themselves by improving current monitoring practices. Approaches used in the clinical sector are often explored for water quality applications because investments in clinical applications drive research, development, and technology transfer. Many of these advanced detection schemes share underlying principles that can be understood within a relatively simple conceptual framework. Common steps include designing primers and probes for the molecular target, concentrating the organism from the environment and obtaining the molecular target from the organism. Solution-based technologies such as PCR and qPCR amplify, label, and detect the target molecule. In contrast, surface-based technologies use molecular probes to capture the target onto some type of solid surface such as a filter, bead, electrode, or waveguide. Unwanted constituents are washed away and the molecular target is labeled and detected, such as in an ELISA assay. Some detection schemes vary this approach slightly by amplifying or labeling the molecular target before the capture, wash, and detection steps. For example, a suspension array may amplify and label molecular targets by PCR and then employee capture probes that are immobilized onto beads to retain specific amplicons for detection. Overall, most detection schemes utilize a label that emits color or fluorescence, although other technologies measure attributes such as electrical signal, mass, or refractive index. All molecular detection technologies, particularly those seeking quantitative detection, must utilize a variety of control measures. These include appropriate blanks, positive and negative controls, standards, and extraction controls. As work continues to overcome the challenges that face environmental biosensing, advanced detection technologies hold promise to improve current monitoring practices and to advance the protection of health by helping environmental managers to minimize human contact with harmful microbes and contaminated seafood products.

References

Ahn, S., Kulis, D.M., Erdner, D.L., Anderson, D.M., Walt, D.R., 2006. Fiber-optic microarray for simultaneous detection of multiple harmful algal bloom species. Appl. and Environ. Microbiol. 72, 5742–5749.

Alliance for Coastal Technology (ACT), 2006. Genetic sensors for environmental water quality. [UMCES]CBL 05–069; Act-05–01. Alliance for Coastal Technologies.

Allsop, K., Stickler, D.J., 1985. An assessment of *Bacteroides fragilis* group organisms as indicators of human faecal pollution. J. Appl. Bacteriol. 58, 95–99.

Alm, E.W., Burke, J., Hagan, E., 2006. Persistence and potential growth of the fecal indicator bacteria, *Escherichia coli* in shoreline sand at Lake Huron. J. Great Lakes Res. 32, 401–405.

Anderson, G.P., Rowe-Taitt, C.A., 2001. Water quality monitoring using an automated portable fiber optic biosensor: Raptor. Proceedings of SPIE: The International Society for Optical Engineering 4206, 58–63.

Anderson, K.L., Whitlock, J.E., Harwood, V.J., 2005. Persistence and differential survival of fecal indicator bacteria in subtropical waters and sediments. Appl. Environ. Microbiol. 71, 3041–3048.

Anglès D'auriac, M.B., Roberts, M.F., Shaw, T., Sirevåg, R., Hermansen, L.F., Berg, J.D., 2000. Field evaluation of a semiautomated method for rapid and simple analysis of recreational water microbiological quality. Appl. Environ. Microbiol. 66, 4401–4407.

Baudart, J., Coallier, J., Laurent, P., Prevost, M., 2002. Rapid and sensitive enumeration of viable diluted cells of members of the family Enterobacteriaceae in freshwater and drinking water. Appl. Environ. Microbiol. 68, 5057–5063.

Baums, I.B., Goodwin, K.D., Kiesling, T., Wanless, D., Fell, J.W., 2007. Luminex® detection of fecal indicators in river samples, marine recreational water, and beach sand. Mar. Pollut. Bull. 54, 521–536.

Bernhard, A.E., Field, K.G., 2000a. Identification of nonpoint sources of fecal pollution in coastal waters by using host-specific 16S ribosomal DNA genetic markers from fecal anaerobes. Appl. Environ. Microbiol. 66, 1587–1594.

Bernhard, A.E., Field, K.G., 2000b. A PCR assay to discriminate human and ruminant feces on the basis of host differences in *Bacteroides-Prevotella* genes encoding 16S rRNA. Appl. Environ. Microbiol. 66, 4571–4574.

Bernhard, A.E., Goyard, T., Simonich, M.T., Field, K.G., 2003. Application of a rapid method for identifying fecal pollution sources in a multi-use estuary. Water Res. 37, 909–913.

Blanch, A.R., Caplin, J.L., Iversen, A., Kuhn, I., Manero, A., Taylor, H.D., Vilanova, X., 2003. Comparison of enterococcal populations related to urban and hospital wastewater in various climatic and geographic European regions. J. Appl. Microbiol. 94, 994–1002.

Boehm, A.B., Furhman, J.A., Mrse, R.D., Grant, S.B., 2003. Tiered approach for identification of a human fecal pollution source at a recreational beach: Case study at Avalon Bay, Catalina Island, CA. Environ. Sci. Technol. 37, 673–680.

Boehm, A.B., Grant, S.B., Kim, J.H., Mowbray, S.Lo., Mcgee, C.D., Clark, C.D., Foley, D.M., Wellman, D.E., 2002. Decadal and shorter period variability of surf zone water quality at Huntington Beach, California. Environ. Sci. Technol. 36, 3885–3892.

Bonjoch, X., Balleste, E., Blanch, A.R., 2004. Multiplex PCR with 16S rRNA gene-targeted primers of *Bifidobacterium* spp. to identify sources of fecal pollution. Appl. Environ. Microbiol. 70, 3171–3175.

Cabelli, V.J., Dufour, A.P., Levin, M.A., Mccabe, J., Haberman, P.W., 1979. Relationship of microbial indicators to health effects at bathing beaches. Am. J. Public Health 69, 690–696.

Cabelli, V.J., Dufour, A.P., Mccabe, L.J., Levin, M.A., 1982. Swimming-associated gastroenteritis and water quality. Am. J. Epidemiol. 115, 606–616.

Cabelli, V.J., Dufour, A.P., Mccabe, L.J., Levin, M.A., 1983. A marine recreational water quality criterion consistent with indicator concepts and risk analysis. J. Water Pollut. Control Fed. 55, 1306–1314.

Carson, C.A., Shear, B.L., Ellersieck, M.R., Schnell, J.D., 2003. Comparison of ribotyping and repetitive extragenic palindromic-PCR for identification of fecal *Escherichia coli* from humans and animals. Appl. Environ. Microbiol. 69, 1836–1839.

Caruana, D.J., 2004. Hybridization at oligonucleotide sensitive electrodes. In Cooper, J., and Cass, T. (eds.), Biosensors, 2nd ed. New York, Oxford University Press.

Casper, E.T., Patterson, S.S., Smith, M.C., Paul, J.H., 2005. Development and evaluation of a method to detect and quantify enteroviruses using NASBA and internal control RNA (IC-NASBA). J. Virol. Methods 124, 149–155.

Chakrabarti, R., Schutt, C.E., 2001. The enhancement of PCR amplification by low molecular weight amides. Nucleic Acids Res. 29, 2377–2381.

Chang, H.N., Hatch, A.V., Hawkins, K.R., Yager, P. 2006. Lab-on-a-chip and fluorescence sensing on the microscale. In Thompson, R.B. (ed.), Fluorescence Sensors and Biosensors, pp. 233–262. Boca Raton, FL, CRC Press.

Clark, S.A., Thompson, K.C., Keevil, C.W., Smith, M.S., 2001. Rapid Detection Assays for Food and Water. Cambridge, The Royal Society Of Chemistry.

Clean Beaches Council, 2005. 2005 State of the beach report: Bacteria and sand: A national call to action.

Colford, J.M., Jr., Wade, T.J., Schiff, K., Wright, C., Griffith, J.F., Sandhu, S.K., Weisberg, S.B., 2005. Recreational water contact and illness in Mission Bay, California. Southern California Coastal Water Research Project. Technical Report 449.

Cook, N., 2003. The use of NASBA for the detection of microbial pathogens in food and environmental samples. J. Microbiol. Methods 53, 165–174.

Cooper, J., Cass, T., 2004. Biosensors, 2nd ed. New York, Oxford University Press.

Davies, C.M., Apte, S.C., 1999. Field evaluation of a rapid portable test for monitoring fecal coliforms in coastal waters. Environ. Toxicol. 14, 355–359.

Deisingh, A.K., Thompson, M., 2004. Biosensors for the detection of bacteria. Can. J. Microbiol. 50, 69–77.

Desmarais, T.R., Solo-Gabriele, H.M., Palmer, C.J., 2002. Influence of soil on fecal indicator organisms in a tidally influenced subtropical environment. Appl. Environ. Microbiol. 68, 1165–1172.

Didenko, V.V., 2006. Fluorescent Energy Transfer Nucleic Acid Probes Designs and Protocols. Totowa, NJ, Humana Press.

Doyle, M.P., Erickson, M.C., 2006. Closing the door on the fecal coliform assay. Microbe 1, 162–163.

Dufour, A.P., 1984a. Bacterial indicators of recreational water quality. Can. J. Public Health 75, 49–56.

Dufour, A.P., 1984b. Health effects criteria for fresh recreational waters. Research and Development EPA-600/1-84/004. US EPA, Health Effects Research Laboratory, NC.

Dunbar, S.A., 2006. Applications of Luminex[R] Xmap[TM] technology for rapid, high-throughput multiplexed nucleic acid detection. Clini. Chim. Acta 363, 71–82.

Dwight, R.H., Fernandez, L.M., Baker, D.B., Semenza, J.C., Olson, B.H., 2005. Estimating the economic burden from illnesses associated with recreational coastal water pollution: A case study in Orange County, California. J. Environ. Manage. 76, 95–103.

Environmental Protection Agency (EPA), 2003. Guidelines establishing test procedures for the analysis of pollutants; Analytical methods for biological pollutants in ambient water; final rule. Federal Register V68, No. 139 40 CFR Part 136, 43272–43283.

Environmental Protection Agency (EPA), 2004. Water quality standards for coastal and Great Lakes recreation waters; final rule. Federal Register V69, No. 220 40 CFR Part 131, 67218–67243.

Ferguson, D.M., Moore, D.F., Getrich, M.A., Zhowandai, M.H., 2005. Enumeration and speciation of enterococci found in marine and intertidal sediments and coastal water in Southern California. J. Appl. Microbiol. 99, 598–608.

Fiksdal, L., Maki, J.S., Lacroix, S.J., Staley, J.T., 1985. Survival and detection of *Bacteroides* spp., prospective indicator bacteria. Appl. Environ. Microbiol. 49, 148–150.

Fuhrman, J.A., Liang, X., Noble, R.T., 2005. Rapid detection of enteroviruses in small volumes of natural waters by real-time quantitative reverse transcriptase PCR. Appl. Environ. Microbiol. 71, 4523–4530.

Gantzer, C., Maul, A., Audic, J.M., Schwartzbrod, L., 1998. Detection of infectious enteroviruses, enterovirus genomes, somatic coliphages, and *Bacteroides fragilis* phages in treated wastewater. Appl. Environ. Microbiol. 64, 4307–4312.

Garcia-Aljaro, C., Muniesa, M., Jofre, J., Blanch, A.R., 2004. Prevalence of the *Stx2* gene in coliform populations from aquatic environments. Appl. Environ. Microbiol. 70, 3535–3540.

Gerard, G.F., Fox, D.K., Nathan, M., D'alessio, J.M., 1997. Reverse transcriptase: The use of cloned moloney murine leukemia virus reverse transcriptase to synthesize DNA from RNA. Mol. Biotechnol. 8, 61–77.

Gerba, C.P., Reynolds, K.A., Dowd, S.E., Pepper, I.L. 2001. Polymerase chain reaction for the detection of parasites and viruses. In Clark, S.A., Thompson, K.C., Keevil, C.W., and Smith, M.S. (eds.), Rapid Detection Assays for Food and Water. Cambridge, The Royal Society of Chemistry.

Gersberg, R.M., Rose, M.A., Robles-Sikisaka, R., Dhar, A.K., 2006. Quantitative detection of Hepatitis A virus and enteroviruses near the United States–Mexico border and correlation with levels of fecal indicator bacteria. Appl. Environ. Microbiol. 72, 7438–7444.

Greenfield, D.I., Marin III, R., Jensen, S., Massion, E., Roman, B., Feldman, J., Scholin, C., 2006. Application of the environmental sample processor (ESP) methodology for quantifying *Pseudo-Nitzschia australis* using ribosomal RNA-targeted probes in sandwich and fluorescent *in situ* hybridization. Limnology Oceanography Methods 4, 426–435.

Griffin, D.W., Lipp, E.K., Mclaughlin, M.R., Rose, J.B., 2001. Marine recreation and public health microbiology: Quest for the ideal indicator. Bioscience 51, 817–825.

Hames, B.D., Higgins, S.J., 1985. Nucleic Acid Hybridization: A Practical Approach. Washington, DC, IRL Press.

Hansen, K.M., Thundat, T., 2005. Microcantilever biosensors. Methods 37, 57–64.

Haramoto, E., Katayama, H., Oguma, K., Ohgaki, S., 2005. Application of cation-coated filter method to detection of noroviruses, enteroviruses, adenoviruses, and Torque teno viruses in the Tamagawa River in Japan. Appl. Environ. Microbiol. 71, 2403–2411.

Harwood, V.J., Delahoya, N.C., Ulrich, R.M., Kramer, M.F., Whitlock, J.E., Garey, J.R., Lim, D.V., 2004. Molecular confirmation of *Enterococcus faecalis* and *E. faecium* from clinical, faecal and environmental sources. Lett. Appl. Microbiol. 38, 476–482.

Harwood, V.J., Levine, A.D., Scott, T.M., Chivukula, V., Lukasik, J., Farrah, S.R., Rose, J.B., 2005. Validity of the indicator organism paradigm for pathogen reduction in reclaimed water and public health protection. Appl. Environ. Microbiol. 71, 3163–3170.

Haugland, R.A., Siefring, S.C., Wymer, L.J., Brenner, K.P., Dufour, A.P., 2005. Comparison of *Enterococcus* measurements in freshwater at two recreational beaches by quantitative polymerase chain reaction and membrane filter culture analysis. Water Res. 39, 559–568.

Hayden, R.T. 2004. *In vitro* nucleic acid amplification techniques. In Persing, D.H., Tenover, F.C., Smith, T.F., White, T.J. (eds.), Molecular Microbiology: Diagnostic Principles and Practice, pp. 43–69. Washington, DC, ASM Press.

Henrickson, S.E., Wong, T., Allen, P., Ford, T., Epstein, P.R., 2001. Marine swimming-related illness: Implications for monitoring and environmental policy. Environ. Health Perspect. 109, 645–650.

Herron, J.N., Wang, H.-K., Tan, L., Brown, S.Z., Terry, A.H., Tolley, S.E., Durtschi, J.D., Simon, E.M., Astill, M.E., Smith, R.S., Christensen, D.A. 2006. Planar waveguide biosensors for point-of-care clinical and molecular diagnostics. In Thompson, R.B. (ed.), Fluorescence Sensors and Biosensors, pp. 263–281. Boca Raton, FL, CRC Press.

Homola, J., 2006. Surface Plasmon Resonance Based Sensors. Berlin, Heidelberg, New York, Springer-Verlag.

Jiang, S., Noble, R., Chu, W., 2001. Human adenoviruses and coliphages in urban runoff-impacted coastal waters of Southern California. Appl. Environ. Microbiol. 67, 179–184.

Katayama, H., Shimasaki, A., Ohgaki, S., 2002. Development of a virus concentration method and its application to detection of enterovirus and Norwalk virus from coastal seawater. Appl. Environ. Microbiol. 68, 1033–1039.

Kay, D., Jones, F., Wyer, M.D., Fleisher, J.M., Salmon, R.L., Godfree, A.F., Zelenauch-Jacquotte, A., Shore, R., 1994. Predicting likelihood of gastroenteritis from sea bathing: Results from randomised exposure. Lancet 344, 905–909.

Kerman, K., Kobayashi, M., Tamiya, E., 2004. Recent trends in electrochemical DNA biosensing technology. Meas. Sci. Technol. 15, R1–R11.

Kim, J.H., Grant, S.B., 2006. Public mis-notification of coastal water quality: A probabilistic evaluation of posting errors at Huntington Beach, California. Environ. Sci. Technol. 38, 2497–2503.

Kreader, C.A., 1995. Design and evaluation of *Bacteroides* DNA probes for the specific detection of human fecal pollution. Appl. Environ. Microbiol. 61, 1171–1179.

LaGier, M.J., Fell, J.W., Goodwin, K.D., 2007. Electrochemical detection of harmful algae and other microbial contaminants in coastal waters using hand-held biosensors. Mar. Pollut. Bull. 54, 757–770.

LaGier, M.J., Scholin, C.A., Fell, J.W., Wang, J., Goodwin, K.D., 2005. An electrochemical rna hybridization assay for detection of the fecal indicator bacterium *Escherichia coli*. Mar. Pollut. Bull. 50, 1251–1261.

Leclerc, H., Schwarzbrod, L., Dei-Cas, E., 2002. Microbial agents associated with waterborne diseases. Crit. Rev. Microbiol. 28, 371–409.

Lee, C.M., Lin, T.Y., Lin, C.C., Kohbodi, G.A., Bhatt, A., Lee, R., Jay, J.A., 2006. Persistence of fecal indicator bacteria in Santa Monica Bay beach sediments. Water Res. 40, 2593–2602.

Leecaster, M., Weisberg, S.B., 2001. Effect of sampling frequency on shoreline microbiology assessments. Mar. Pollut. Bull. 42, 1150–1154.

Lemarchand, K., Masson, L., Brousseau, R., 2004. Molecular biology and DNA microarray technology for microbial quality monitoring of water. Crit. Rev. Microbiol. 30, 145–172.

Li, J., Hanna, B.A. 2004. DNA probes for culture confirmation and direct detection of bacterial infections: A review of the technology. In Persing, D.H., Tenover, F.C., Smith, T.F., White, T.J. (eds.), Molecular Microbiology: Diagnostic Principles and Practice, pp. 19–26. Washington, DC, ASM Press.

Ligler, F.S., Taitt, C.R. 2006. The array biosensors. In Thompson, R.B. (ed.), Fluorescence Sensors and Biosensors, pp. 263–281. Boca Raton, Fl, CRC Press.

Lim, D.V., 2003. Detection of microorganisms and toxins with evanescent wave fiber-optic biosensors. Proceedings of the IEEE 91, 902–907.

Liron, Z., Bromberg, A., Fisher, M., 2001. Novel Approaches in Biosensors and Rapid Diagnostic Assays. New York, Kluwer Academic/Plenum.

Litaker, R.W., Tester, P.A., 2002. Molecular methods for detecting and characterizing harmful phytoplankton. In Hurst, C.J., Crawford, R.L., Knudsen, G.R., Mcinerney, M.J., and Stetzenbach, L.D. (eds.), Manual of Environmental Microbiology, 2nd ed., pp. 342–353. Washington, DC, ASM Press.

Manafi, M., Kneifel, W., Bascomb, S., 1991. Flurogenic and chromogenic substrates used in bacterial diagnostics. Microbiol. Rev. 55, 335–348.

Marras, S.A.E. 2006. Selection of fluorophore and quencher pairs for fluorescent nucleic acid hybridization probes. In Didenko, V.V. (ed.), Fluorescent Energy Transfer Nucleic Acid Probes Designs and Protocols. Totowa, NJ, Humana Press.

Martinez, J.S., Grace, W.K., Grace, K.M., Hartman, N., Swanson, B.I., 2005. Pathogen detection using single mode planar optical waveguides. J. Mater. Chem. 15, 4639–4647.

Mazumder, R., Natale, D.A., Murthy, S., Thiagaranjan, R., Wu, C.H., 2005. Computational identification of strain-, species-, and genus-specific proteins. BMC Bioinformatics 6, 279–287.

Munn, C.B., 2004. Marine Microbiology. New York, Garland Science/Bios Scientific.

Narayanaswamy, R., 2004. Optical Sensors Industrial, Environmental and Diagnostic Applications. Berlin, Heidelberg, New York, Springer-Verlag.

National Research Council, 2004. Indicators for waterborne pathogens. committee on indicators for waterborne pathogens. Washington, DC, National Academy of Sciences.

National Shellfish Sanitation Program (NSSP), 2005. National shellfish program guide for the control of molluscan shellfish 2005. U.S. Food and Drug Administration and the Interstate Shellfish Sanitation Conference.

Noble, R.T., Fuhrman, J.A., 2001. Enteroviruses detected by reverse transcriptase polymerase chain reaction from the coastal waters of Santa Monica Bay, California: Low correlation to bacterial indicator levels. Hydrobiologia 460, 175–184.

Noble, R.T., Griffith, J.F., Blackwood, A.D., Fuhrman, J.A., Gregory, J.B., Hernandez, X., Liang, X., Bera, A.A., Schiff, K., 2006. Multitiered approach using quantitative PCR to track sources of fecal pollution affecting Santa Monica Bay, California. Appl. Environ. Microbiol. 72, 1604–1612.

Noble, R.T., Weisberg, S.B., 2005. A review of technologies being developed for rapid detection of bacteria in recreational waters. J. Water Health 3, 381–392.

Noble, R.T., Weisberg, S.B., Leecaster, M.K., McGee, C.D., Ritter, K., Walker, K.O., Vainik, P.M., 2003. Comparison of beach bacterial water quality indicator measurement methods. Environ. Monit. Assess. 81, 301–312.

Nolan, J.P., Sklar, L.A., 2002. Suspension array technology: Evolution of the flat-array paradigm. Trends Biotechnol. 20, 9–12.

Ocean.US, 2006a. The first U.S. Integrated Ocean Observing System (IOOS) development plan. Publication No. 9. Arlington, VA.

Ocean.US, 2006b. Public health risks: Coastal observations for decision-making. Publication No. 15. Arlington, VA.

Patterson, S.S., Smith, M.W., Casper, E.T., Huffman, D., Stark, L., Fries, D., Paul, J.H., 2006. A nucleic acid sequence-based amplification assay for real-time detection of norovirus genogroup II. J. Appl. Microbiol. 101, 956–963.

Paul, J.H., Rose, J.B., Jiang, S.C., London, P., Xhou, X., Kellogg, C., 1997. Coliphage and indigenous phage in Mamala Bay, Oahu, Hawaii. Appl. Environ. Microbiol. 63, 133–138.

Pellestor, F., Paulasova, P., 2004. The peptide nucleic acids, efficient tools for molecular diagnosis (review). Int. J. Epidem. 13, 521–525.

Porter, J., Pickup, R. 1999. Magnetic particle-based separation techniques for monitoring bacteria from natural environments. In Edwards, C. (ed.), Environmental Monitoring of Bacteria, pp. 75–96. Totowa, NJ, Humana Press.

Premier Biosoft, 2007. PCR Primer Design Guidelines, www.Premierbiosoft.Com/Tech_Notes/Pcr_Primer_Design.html.

Rompré, A., Servais, P., Baudart, J., De-Roubin, M.R., Laurent, P., 2002. Detection and enumeration of coliforms in drinking water: Current methods and emerging approaches. J. Microbiol. Methods 49, 31–54.

Rutjes, S.A., Italiaander, R., Van Den Berg, H.H.J.L., Lodder, W.J., De Roda Husman, A.M., 2005. Isolation and detection of enterovirus RNA from large-volume water samples by using the Nuclisens Minimag system and real-time nucleic acid sequence-based amplification. Appl. Environ. Microbiol. 71, 3734–3740.

Sambrook, J., Fritsch, E.F., Maniatis, T., 1989. Molecular Cloning: A Laboratory Manual, 2nd ed. Cold Spring Harbor, NY, Cold Spring Harbor Laboratory.

Sartory, D.P., Parton, A., Rackstraw, C. 2001. Rapid same-day microcolony enumeration technique for *E. Coli* in drinking water. In Clark, S.A., Thompson, K.C., Keevil, C.W., and Smith, M.S. (eds.), Rapid Detection Assays for Food and Water. Cambridge, The Royal Society of Chemistry.

Scholin, C.A., Doucette, G., Cembella, A.D. 2007. Prospects for developing automated systems for *in situ* detection of harmful algae and their toxins. In Babin, M., Roesler, C., and Cullen, J. (eds.), Monographs on Oceanographic Methodology. Paris, Unesco.

Scott, T.M., Jenkins, T.M., Lukasik, J., Rose, J.B., 2005. Potential use of a host associated molecular marker in *Enterococcus faecium* as an index of human fecal pollution. Environ. Sci. Technol. 39, 283–287.

Scott, T.M., Rose, J.B., Jenkins, T.M., Farrah, S.R., Lukasik, J., 2002. Microbial source tracking: Current methodology and future directions. Appl. Environ. Microbiol. 68, 5796–5803.

Sepaniak, M., Datskos, P., Lavrik, N., Tipple, C., 2002. Microcantilever transducers: A new approach in sensor technology. Anal. Chem. 74, 569a-575a.

Sharkey, F.H., Banat, I.M., Marchant, R., 2004. Detection and quantification of gene expression in environmental bacteriology. Appl. Environ. Microbiol. 70, 3795–3806.

Shepley, D.P., Wolk, D.M., 2004. Quantitative molecular methods: Result standardization, interpretation, and laboratory quality control. In Persing, D.H., Tenover, F.C., Smith, T.F., White, T.J. (eds.), Molecular Microbiology: Diagnostic Principles and Practice, pp. 95–129. Washington, DC, ASM Press.

Shetab, R., Cohen, S.H., Prindiville, T., Tang, Y.J., Cantrell, M., Rahmani, D., Silva, J., Jr., 1998. Detection of *Bacteroides fragilis* enterotoxin gene by PCR. J. Clin. Microbiol. 36, 1729–1732.

Simpson, J.M., Lim, D.V., 2005. Rapid PCR confirmation of *E. coli* O157: H7 after evanescent wave fiber optic biosensor detection. Biosensors Bioelectronics 21, 881–887.

Simpson, J.M., Santo Domingo, J.W., Reasoner, D.J., 2002. Microbial source tracking: State of the science. Environ. Sci. Technol. 36, 5279–5288.

Snyder, L., Champness, W., 1997. Molecular Genetics of Bacteria. Washington, DC, American Society of Microbiology.

Stewart, J.R., Ellender, R.D., Gooch, J.A., Jiang, S., Myoda, S.P., Weisberg, S.B., 2003. Recommendations for microbial source tracking: Lessons from a methods comparison study. J. Water Health 1, 225–231.

Stoeckel, D.M., 2005. Selection and application of microbial source tracking tools for water-quality investigations. In U.S. Geological Survey Techniques and Methods, Book 2, 1–43. http://pubs.usgs.gov/tm/2005/tm2a3

Tallon, P., Magajna, B., Lofranco, C., Leung, K.T., 2005. Microbial Indicators of faecal contamination in water: A current perspective. Water Air Soil Pollut. 166, 139–166.

Thompson, R.B., 2006. Fluorescence Sensors and Biosensors. Boca Raton, FL, CRC Press.

Venter J.C., Remington K., Heidelberg, J.F., Halpern, A.L., Rusch, D., Eisen, J.A., Wu, D., Paulsen, I., Nelson, K.E., Nelson, W., Fouts, D.E., Levy, S., Knap, A.H., Lomas, M.W., Nealson, K., White, O., Peterson, J., Hoffman, J., Parsons, R., Baden-Tillson, H., Pfannkoch, C., Rogers, Y.H., Smith, H.O., 2004. Environmental genome shotgun sequencing of the Sargasso Sea. Science 304, 66–74.

Wade, T.J., Calderon, R.L., Sams, E., Beach, M.J., Brenner, K.P., Williams, A.H., Dufour, A.P., 2006. Rapidly measured indicators of recreational water quality are predictive of swimming associated gastrointestinal illness. Environ. Health Perspect. 114, 24–28.

Wade, T.J., Pai, N., Eisenberg, J.N.S., Colford, J.M., Jr., 2003. Do U.S. Environmental Protection Agency water quality guidelines for recreational waters prevent gastrointestinal illness? A systematic review and meta-analysis. Environ. Health Perspect. 111, 1102–1109.

Wang, J., 2000. Analytical Electrochemistry, 2nd ed. New York, John Wiley & Sons.

Whitman, R.L., Nevers, M.B., 2003. Foreshore sand as a source of *Escherichia coli* in nearshore water of a Lake Michigan Beach. Appl. Environ. Microbiol. 69, 5555–5562.

Whitman, R.L., Shively, D.A., Pawlik, H., Nevers, M.B., Byappanahalli, M.N., 2003. Occurrence of *Escherichia coli* and enterococci in Cladophora (Chlorophyta) in nearshore water and beach sand of Lake Michigan. Appl. Environ. Microbiol. 69, 4714–4719.

Wilson, I.G., 1997. Inhibition and facilitation of nucleic acid amplification. Appl. Environ. Microbiol. 63, 3741–3751.

Wittwer, C.T., Kusukawa, N., 2004. Real-time PCR. In Persing, D.H., Tenover, F.C., Smith, T.F., White, T.J. (eds.), Molecular Microbiology: Diagnostic Principles and Practice, pp. 71–84. Washington, DC, ASM Press.

STUDY QUESTIONS

1. Discuss the drawbacks of current monitoring practices and the needs that emerging technologies attempt to meet.
2. What are the concerns regarding the current fecal indicators? Discuss which alternative indicators might be better and why.
3. Describe sample concentration and extraction of target nucleic acids or proteins, and discuss why these steps are a primary factor limiting the development of rapid molecular detection techniques for environmental samples.
4. Assuming a diagnostic PCR reaction is producing nonspecific amplification products, detail what might be going wrong with the reaction and what tests might be done to determine how best to correct the problem. Include what controls you would use and why.
5. Assuming a diagnostic PCR reaction is producing little to no amplification products, detail what might be going wrong with the reaction and what tests might be done to determine how best to correct the problem. Include what controls you would use and why.
6. You want to develop a detection method for measuring low concentrations of pathogenic RNA virus in estuarine waters. Which molecular detection methods would you choose? Explain the strengths and weaknesses of the technique you selected and the limitations you might expect to encounter. What necessary controls should be included?
7. Given three unique DNA sequences for a new bacterial pathogen within 250 base pairs of each other, describe how you could construct a solid phase detection system for that pathogen. Include controls that you would need to include. What alternative methods might be considered, and what would be the advantages and disadvantages relative to the assay you have designed?
8. Explain the major similarities and differences in a quantitative PCR assay and a quantitative reverse transcriptase PCR assay for an environmental pathogen.
9. What are the advantages and disadvantages of using a nucleic acid sequence based amplification (NASBA) type assay compared to quantitative PCR?
10. What would be the main differences in applying/developing the detection methods outline in this chapter to bacteria and viruses versus the species that generate HABs as outlined earlier in the book?

C H A P T E R

20

Future of Microbial Ocean Water Quality Monitoring

CAROL J. PALMER, J. ALFREDO BONILLA, TONYA D. BONILLA, KELLY D. GOODWIN, SAMIR M. ELMIR, AMIR M. ABDELZAHER, AND HELENA M. SOLO-GABRIELE

INTRODUCTION

Increased use of our coastal waters for swimming, recreational sports, and shellfishing has focused attention on the need to develop a reliable monitoring strategy to ensure that coastal waters do not pose a public health danger. Disease-causing microorganisms enter coastal waters through various pathways (Table 20-1), and their ability to survive in coastal waters is related to nearshore coastal water conditions. Offshore marine environments are fairly stable with regard to salinity, oxygen concentration, temperature, and other physical parameters. Characteristics of nearshore coastal water, on the other hand, can fluctuate dramatically in response to rainfall events, land/soil runoff, and direct impact from human coastal activities, especially in areas where there are high-density swimming beaches, boating, and shellfish-growing activities. Coastal water quality is critical to human and ecosystem health and has an impact on fishing, aquaculture, and tourism. Common coastal water problems include blooms of harmful algae, bacterial, and viral contaminants from sewage and chemical runoff from agriculture or other nearshore land use. Virtually every coastal state is threatened by these problems.

Beaches serve an important role in the U.S. economy. Coastal recreation is estimated to contribute approximately 85% of all U.S. tourist revenues (National Resources Defense Council [NRDC], 2006). This revenue, however, depends on the availability of coastal areas that are safe for recreational purposes. According to the latest surveillance of the U.S. Centers for Disease Control and Prevention (CDC, 2006), the largest number of recreational water-associated disease cases occurred between 2003 and 2004. During this period, there were 62 outbreaks causing illness among an estimated 2698 persons. Among these cases, there were 58 hospitalizations and one death. The U.S. Environmental Protection Agency (U.S. EPA, 2006) reports that during 2005, U.S. beaches had 27,177 beach closing and advisory days, the highest in 17 years since they started reporting this data. In 2005, 85% of the total closings and advisories were issued because water quality measurements exceeded the recommended bacterial indicator standards. In a majority of cases, point sources of contamination were not identified. The inability to identify sources of pollution, in particular when point sources of pollution are not obvious or not present, has made it difficult to remediate and prevent the impacts to beaches.

This information illustrates the compelling reasons for improving our beach monitoring efforts. Improved monitoring is needed so that we can close beaches based on solid evidence of the potential for disease transmission and prevent unwarranted beach closures when possible. Pathogens from human and animal waste may enter coastal ecosystems through discharges from wastewater treatment plants, septic systems, and runoff. Swimmers have a higher risk for infections and gastrointestinal illness, and swimming near polluted beaches poses a greater risk to health. Sewage pathogens also pose a threat to health through contamination of shellfish. In recognition of the risks posed by contaminated coastal waters, 25 states and six territories have adopted marine water quality standards.

Historically, growing public concern and awareness about the problem of water pollution in the 1970s led to the enactment of the Federal Water Pollution Control Act of 1972, later amended in 1977, and commonly known as the Clean Water Act (www.epa.gov/watertrain/cwa). This law established the basic structure for regulating discharges of pollutants into U.S. waterways and gave the U.S. EPA the authority to implement pollution control programs for waterways. This act is the cornerstone of today's water quality programs. Initially, the approach to monitor ocean water was

TABLE 20-1. Pathways by which microorganisms enter coastal water

1. Urban and storm drain runoff
2. Agricultural runoff
3. Municipal sewage discharge and accidental or storm event sewage spills
4. Leaking septic tanks
5. Boats illegally discharging human waste
6. Contaminated beach sand and soil
7. Human and animal swimming and subsequently shedding pathogens in coastal water

based on the approach used to monitoring drinking water—that is, we tested for coliforms assuming that, similar to freshwater, the presence of high levels of coliform bacteria may suggest that the water is contaminated by sewage and thus may potentially contain pathogenic microorganisms. Further research in this area over the years showed that coliforms (1) did not survive well in coastal water (Carlucci and Pramer, 1960); (2) did not correlate well with the presence or absence of pathogenic microorganisms (Joseph *et al.*, 1982); (3) did not correlate to high numbers of indigenous pathogenic marine bacteria such as *Vibrio* spp. (Joseph *et al.*, 1982); (4) did not correlate to the presence of enteric viruses (Goyal, 1983); and (5) can originate from nonfecal sources such as soil, sand, vegetation, and certain industrial wastes (Cabelli, 1978; Desmarais *et al.*, 2002; Hardina and Fujioka, 1991; Hendricks, 1978; Solo-Gabriele *et al.*, 2000). Additionally, in tropical and subtropical waters, *E. coli* may actually multiply in soil embankments and wash into coastal waters falsely elevating the concentration of coliforms. Because these coliforms are not of fecal origin, they would not indicate the potential presence of sewage contamination with a resultant potential health threat (Solo-Gabriele *et al.*, 2000).

In the 1980s, enterococci was proposed as a replacement indicator organism that could serve as a better sentinel organism with which to judge ocean water quality. The U.S. Environmental Protection Agency recommended (U.S. EPA, 1986) that states utilize the indicator microbes enterococci or *Escherichia coli* to determine whether health advisories or closures should be issued for recreational coastal waters. *E. coli* was recommended for freshwaters and enterococci was recommended for both fresh and marine waters. The recommendation of enterococci as the indicator organism in marine waters was based on epidemiological studies by Cabelli *et al.*, (1982) that showed a positive correlation between illness and the presence of enterococci in beach bathing waters impacted by point sources of sewage. Although problems with the epidemiological study that formed the basis of suggesting this organism as an alternative indicator were reported to exist (Fleisher, 1990, 1991), the original studies plus more recent studies using molecular methods, have shown that enterococci remains a valid indicator organism as this organism was consistently shown to correlate with increased health risk at swimming beaches (Bonilla *et al.*, 2006; Wade *et al.*, 2006). Other microorganisms, such as *Clostridium perfringens*, *Bacteroides*, and coliphage, have also been suggested as a replacement or supplement to coliform or enterococci indicator organisms (Bettancourt and Fujioka, 2006; Colford *et al.*, 2007; Fujioka 2001). However, to date, these organisms have not gained regulatory acceptance.

The use of fecal indicator bacteria to monitor and regulate the recreational use of coastal waters continues to raise questions, particularly in the tropical and subtropical marine environments. Specifically, the *USEPA's Action Plan for Beaches and Recreational Water* (p. 7, 1999) states the following:

> Currently recommended fecal indicators may not be suitable for assessing human health risks in the tropics. Studies have suggested that at tropical locales such as Puerto Rico, Hawaii, and Guam, *E. coli* and enterococci can be detected in waters where there is no apparent warm-blooded animal source of contamination. If this phenomenon is widespread under tropical conditions, additional research should be conducted to modify approaches for monitoring, or to develop new tropics-specific indicators.

To make matters even more complicated, there have been documented cases where coastal waters monitored for both sets of fecal indicator bacteria (fecal coliforms and enterococci) have passed regulatory limits for enterococci and not for fecal coliforms (Shibata *et al.*, 2004). So a regulator is left with a perplexing situation where it is not clear which indicator microbe(s) should be utilized, and once the data are obtained, how these data should be interpreted.

Similar to beach water quality concerns, there is also a need to improve methods for detection of microorganisms in seafood. In the United States, there are between 7.6 million and 14 million cases of foodborne illness caused by eating contaminated seafood each year (Butt *et al.*, 2004). Most often, seafood-associated foodborne illness is caused by consumption of raw or undercooked seafood. Whereas viruses are the most common cause of infection, most hospitalizations and morbidity and mortality are caused by bacteria agents, such as *Vibrio vulnificus*, an indigenous marine bacteria. Mollusks may pose the greatest risk because they are filter feeders and, as a result, are ideally suited to bioconcentrate large numbers of microorganisms. Indicator organisms may not necessarily correlate to large numbers of indigenous marine pathogenic bacteria such as vibrios, thus direct pathogen detection in seafood may provide a valuable tool with which to assess the seafood safety. It is also crucial to provide better public health information on seafood consumption to consumers, alerting them to the dangers of eating raw and undercooked seafood items.

In addition to pathogenic bacteria, viruses, and protozoans, harmful algal blooms (HABs) threaten many coastal states and can impact large geographic areas. Since the 1980s, harmful algal blooms have been increasing in frequency, intensity, and distribution. Blooms of toxic dinoflagellates are associated with fish and marine mammal kills, neurotoxic shellfish poisoning, and eye and respiratory irritation in marine animals and humans. Harmful algae have been spread through aquaculture activities and ballast water releases. Microscopic identification of HABs requires considerable taxonomic expertise that is not broadly available. Many HAB species are delicate and difficult to identify when preserved. Although blooms are usually considered to be monospecific, this is an oversimplification, and extensive experience is necessary to distinguish closely related species. In addition, sampling is often restricted spatially and temporally, giving a limited view of the extent and dynamics of a bloom. Conversely, limited resources can require managers to impose blanket closures on shellfish harvesting. Closures have caused millions of dollars in economic losses per year.

LIMITATIONS OF CURRENT MONITORING PROTOCOLS AND METHODS

Water quality assays are based on traditional, culture-based methods using bacterial indicators of fecal contamination. However, traditional microbiological assays suffer from serious drawbacks. Weaknesses in culture methods are evident; the assays are labor and supply intensive, samples must be transported at 1 to 4°C, samples must be processed within 6 hours, and results are not available for approximately 24 hours. Moreover, traditional assays fail to differentiate between human and animal-derived wastes. Fecal-indicator bacteria also have a limited ability to accurately model the behavior of actual human pathogens in seawater and do not monitor for bloom forming algae, cyanobacteria, or indigenous vibrios.

Public health agencies are responsible for assuring that drinking water, recreational water, and seafood are safe. Fisheries and beaches impacted by HABs and sewage contamination are closed to protect human health. Aquaculture industries lose million of dollars annually to closures. However, it is usually reports from citizens and physicians that alert public health agencies to a problem. This is in part because public agencies oversee large areas of coast with limited funds, and management decisions are based on time-consuming and labor-intensive assays. Monitoring programs that rely on microscopy (e.g., for HABs) require taxonomic expertise (Millie *et al.*, 1997), are time and labor intensive, and are prone to error (Culverhouse *et al.*, 2003). Monitoring programs that rely on traditional culture techniques (e.g., fecal indicators) are slow and do not return information on human pathogens or source tracking markers (Bower *et al.*, 2005; Griffin, 2001; Scott *et al.*, 2002). As a result, resource management, ecological study, and planning and assessment of remediation efforts are difficult.

FUTURE RECOVERY OF MICROORGANISMS

The field of water quality monitoring is moving toward more precise detection methods that will eliminate the need for lengthy culture times and specialized media for indicator bacterial species. In all probability, the future of water quality monitoring will also include rapid methods for pathogen specific detection using quantitative polymerase chain reaction (qPCR) as initial studies using this method have shown promising results (Wade *et al.*, 2006). This is especially important for nonbacterial pathogens, as bacteria are fairly easy to culture from ocean water but viruses and protozoan parasites, such as *Giardia* and *Cryptosporidium,* are much more challenging to recover and enumerate using currently recommended methodologies. Traditional methods for virus detection from coastal water are burdensome. Viruses are usually present in smaller numbers than bacteria, so large volumes of water must be filtered using conventional adsorption to and elution from microporous filters followed by cell culture methods. It can take several weeks to propagate a virus in cell culture, plus the sensitivity of detection by cell culture is low and the method is labor intensive and tedious. Many viruses of public health importance, such as norovirus and hepatitis A virus, are difficult to culture even under optimum conditions. Detection of *Giardia* and *Cryptosporidum* also require the filtration of large amounts of ocean water, highly specialized equipment and labor-intensive recovery steps. To quickly evaluate water for viruses and protozoan parasites, future methodologies will most likely include genetic detection using a variety of molecular technologies. Molecular methods (Luminex, nucleic acid analysis and qPCR/microarrays) require concentration of large volumes of water, which present challenges in removing inhibitors (e.g., humic/fulvic acids) to the enzymes utilized in the procedures, but will yield more accurate and precise results.

There is still ongoing debate over the attributes of one indicator organism over another with no real consensus in the scientific community on which is the better indicator organism with which to monitor ocean water quality. However, as we look to the future of ocean water quality monitoring, there are some exciting developments that may play a pivotal role in evaluating the quality of our coastal waters. Molecular methods such as qPCR hold the promise of rapid evaluation of ocean water within hours. A study by Wade *et al.* (2006) presented the first data demonstrating the ability of rapid indicator methods to predict health effects. Measuring levels of *Enterococcus* by qPCR, they were able to predict gastrointestinal illness after swimming in fecally contaminated fresh water. Applying these results to beach water suggests that collecting and analyzing samples in the morning could allow beach managers to assess the microbiological safety of the beach before most beachgoers are exposed. These authors concluded that incorporation of rapid molecular measurements into a regulatory framework has the potential to improve beach management decisions and protect swimmers' health.

Improved assays are needed to better protect human health and economic interests. Molecular methods and rapid enzyme detection technology can provide highly specific detection in real time without the need for microscopic or culturing expertise. Samples undergoing molecular analysis can be frozen, allowing easier sample storage. Nucleic-acid based methods have the potential to provide rapid, high-throughput detection, allowing more widespread screening of waters. With the development and refinement of new and more sensitive testing methods, we have the potential to thoroughly and intensively evaluate water samples for a host of organisms simultaneously. Chapter 19 by Goodwin and Litaker provides an overview of a number of molecular technologies in detail, including some of the technological challenges that molecular methods face. This current chapter assumes that science will meet those challenges and explores some of the ways in which these new monitoring tools would impact the future of water quality testing in the 21st century. The sections in this chapter include a discussion of water-quality models and their potential role in providing forecasts for beach closures. The description of an existing relatively new commercial rapid enzyme test follows. New technologies for processing samples using nucleic acid based methods are first introduced by describing sample concentration methods as a first step in processing samples for analysis. The chapter describes the potential use of high throughput quantitative PCR, oligonucleotide microarrays, and Luminex xMAP technologies for pathogen detection. The chapter also describes how the incorporation of such technologies would impact the regulatory community. The chapter closes with a description of the benefits of utilizing new technologies and emphasizes the need to supplement current monitoring programs with new methods.

MODELS FOR MANAGEMENT OF RECREATIONAL WATER QUALITY

Water quality models can play an important role in establishing recreational safety. Water quality models range from simple regression type models (Kelsey *et al.*, 2004; Maillard and Pinheiro Santos, in press; Nevers and Whitman, 2005; Siewicki *et al.*, 2007; Wickham *et al.*, 1989) to complex two- and three-dimensional hydrodynamic-water quality models, which incorporate functions that account for microbial sources, dieoff, deposition, regrowth (Liu *et al.*, 2006; Jin *et al.*, 2003), and in some cases impacts to shellfish (Riou *et al.*, 2007). Intermediate models in their complexity include stream models (Hyer and Moyer, 2004; Irvine *et al.*, 2005; Scarlatos, 2001) and lumped parameter models, which have a physical basis but do not simulate microbe levels to the same level of detail as the more complex hydrodynamic models (Elshorbagy *et al.*, 2006). Also, innovative approaches including the use of neural networks have been incorporated into modeling efforts (Montgomery Brion and Lingireddy, 1999). Common factors among all models are their ability to provide estimates for fecal indicator given environmental conditions and algorithms of the suspected sources. All models require measurements of fecal indicators for calibration and verification. Although models require a considerable amount of data to develop, once developed they can provide insights with respect to factors that influence fecal indicator levels within recreational water bodies.

Models are useful for management purposes for two primary reasons. First models provide managers with the ability to assess the impacts of different sources on fecal indicator levels within receiving water bodies. Managers can use models to evaluate the impact of suspected sources by simply removing those sources from the model (Canale *et al.*, 1993). Such a concept is manifested within the regulatory framework by the establishment of total maximum daily loads (TMDL) for a particular watershed. TMDLs represent maximum limits (mass or number per unit time) for contaminants within watersheds. The TMDLs provide allowances for different sources of contamination and provide a means by which these different sources can be managed in a way to minimize contaminant loads (Benham *et al.*, 2006). Second, models are valuable tools for management purposes because they can be used in a predictive mode to forecast beach closures. As mentioned in earlier sections of this chapter, an emphasis has been placed on the reduction of sample analysis times. Currently, traditional methods of analysis based on culture techniques result in a 24–hour turnaround time to obtain results. Newer qPCR based methods can reduce the analysis time to 2 to 4 hours, providing for earlier postings of advisories more consistent with the time frame during which samples were collected. The lag time between sample collection and analysis results in a contradiction as beaches could remain open for a significant period of time as impaired water samples are analyzed. They will also remain closed during the time lag needed to measure samples that are of acceptable quality. Models are capable of accomplishing what monitoring could never accomplish even in the ideal scenario of real-time measurements. Models could be used to forecast beach advisories. Forecasts are especially useful as the public can be warned ahead of time. The use of predictive models thus eliminates the problems associated with time lags between sample collection and analysis. Routine sampling and analysis should continue, nevertheless, as such data should be used to continuously calibrate and verify the model.

Predictive models have been used in various management scenarios. For example, two sets of regression expressions were developed for southern Michigan (Nevers and Whitman, 2005), one for times dominated by onshore winds and one for times dominated by offshore winds. These relationships provide estimates of a fecal indicator (*Escherichia coli*) given parameters that can be measured in real-time including measurements of wave height, wave period, lake chlorophyll, lake turbidity, river turbidity, and precipitation. Output from the model with respect to fecal indicator levels was converted to probability of beach closure because of exceedences. Beach managers were given this information and this information was used along with routine monitoring to make daily determinations of swimming advisories. Comparing fecal indicator measurements with model forecasts showed that the model incorrectly predicted open/close determinations only 2% of the time for onshore winds and only 3% of the time for offshore winds. The same was true for a hydrodynamic water quality model that was developed for forecasting beach closures for the south shore of Lake Pontchartrain, Louisiana (McCorquodale *et al.*, 2004). Input to the hydrodynamic portion of the model (the portion of the model that simulates currents) included tidal and wind conditions, salinity, and water temperature. The source of fecal indicators was simulated through inputs from a series of canals for which pumping records were available or estimated. The model then advects the indicator bacteria from the canals by the lake currents, and specific functions were used to simulate bacterial decay. The decay function was dependent on salinity, temperature, turbidity, light, and sedimentation. The model results were shown to accurately simulate fecal indicator dilution and decay within reasonable confidence limits, given the uncertainties associated with the concentrations observed at the canal discharge sites. Given that the run time of the model was relatively rapid, the authors recommend the use of the model to forecast beach closures.

Given the ability of models to assimilate data, it is anticipated that their use will increase in the future for forecasting beach closures. However, limitations exist in the development of such models. Regression type models are limited for forecasting indicator levels for situations observed within the dataset used to establish the model. Regression models should not be used to forecast fecal indicator levels for extreme events nor for changing watershed characteristics. Also care should be taken to eliminate bias associated with the processing of data, which may be correlated in time series (Ge and Frick, 2006). Limitations of hydrodynamic modeling efforts include the need for large data sets, limitations in the current state of science with respect to factors that control fecal indicator deposition, adsorption, and regrowth (Bai and Lung, 2005; Jamieson *et al.*, 2004; Pachepsky *et al.*, 2006), and the need to simulate shoreline shear and rip currents (Clarke *et al.*, 2007; Grant *et al.*, 2005), which can result in contaminants generally hugging the shoreline upon release.

COMMERCIAL RAPID ENZYME TESTING METHODS FOR ENTEROCOCCUS

Traditional detection and enumeration of enterococci is completed using the membrane filtration (MF) method followed by incubation on differential medium. A most probable number method (MPN) may also be utilized. Details of these techniques are described in previous chapters. Both MF and MPN methods are cumbersome, labor intensive, require specialized laboratory equipment that must be cleaned and sterilized between water samples, and it can take 48 to 72 hours or longer for final confirmatory identification of the target organism.

Several commercial products have been introduced that eliminate several of the drawbacks of traditional detection methods. One such product, Enterolert (IDEXX Laboratories, Inc., Westbrook, ME), is a defined substrate technology MPN method for detection and enumeration of enterococci. This product is semiautomated and utilizes a nutrient indicator substrate, 4-methylumbelliferone-B-D-glucoside, that fluoresces when metabolized by enterococci present in the water sample. This is the only defined substrate technology currently approved by the three major agencies/oversight groups for water quality testing: the EPA (Federal Register, 2003, www.epa.gov/fedrgstr/EPA-WATER/2003/July/Day-21/w18155.pdf), American Society for Testing Materials (American Society for Testing and Materials [ASTM], 1999), and Standard Methods for the Examination of Water and Wastewater (American Public Health Association [APHA], 2005).

The test procedure is simple and straightforward, requiring little more equipment than an incubator to complete. A package of powdered Enterolert reagent is mixed with the water sample under evaluation. The mixture is added to one of two types of disposable trays. Trays come in either a 51 or a 97 well format, with the 97 well format providing for a larger range of analyses. The trays are mechanically sealed and incubated for 24 hours at 41°C. Test results are read under a 365 nm UV light. Fluorescence in a well is considered positive, indicating the presence of enterococci in the water sample. MPN tables are utilized to enumerate the enterococci present in the water sample based on the number of positive wells.

A study by Budnick *et al.* (1996), comparing the Enterolert Quanti-Tray test with the MF method, showed that Enterolert more effectively and accurately recovered enterococci from recreational bathing waters and may provide an improved method for recovery of injured or stressed organisms. Time studies indicated that the Enterolert test required significantly less time per sample for setup, interpretation, and recording of results. Further studies by Abbott *et al.* (1998) concurred with the Budnick *et al.* (1996) findings that Enterolert showed increased sensitivity and specificity when compared to the MF method. These authors concluded that this method provides for a more user friendly and less expensive method to quickly analyze beach water quality to protect public health. Moreover, they found that the Enterolert was superior to the MF method in detection of target organisms in turbid waters, a common problem in environmental water samples.

There has been one negative report on the use of Enterolert and that was by a group using the test on fresh water. Kinzelman *et al.* (2003) evaluated the product in freshwater Lake Michigan beaches and found that they were unable to verify the presence of enterococci in Quanti-Tray wells exhibiting fluorescence. As the test has been shown to work well detecting enterococci in marine water samples, additional studies on the use of Enterolert in freshwater systems may be of value to further investigate this finding. It could be that another fresh water organism could be multiplying faster than the enterococci in the well and masking their presence on isolation efforts or some component in the Lake Michigan freshwater beach sample may have caused interference in the test.

The studies performed on marine waters show the Enterolert test to be a valuable tool with which to monitor coastal water quality. Commercial test products such as Enterolert are simple to perform with easy-to-interpret results. Another advantage is that it takes minimal training and minimal laboratory equipment. Because the test results are available in 24 hours instead of the 48 to 72 hours required for standard MPN and MF methods, it allows for faster monitoring of coastal waters and allows regulators to more quickly respond to beach water quality issues, especially after storm events.

CONCENTRATION METHODS

Direct monitoring for pathogens of concern may be the direction taken by regulators and researchers in assessing microbial water quality in the near future. However, one of the main obstacles in pathogen monitoring is concentrating water samples for the pathogens of interest. Usually pathogens, and especially viruses, are found in small numbers in water. Thus, large volumes of water (up to 1000 L) need to be concentrated to a small volume (a few milliliters) to allow for detection via culturing or molecular methods. The reason we are concerned about one pathogen in a large volume of water is because sometimes it only takes one pathogenic organism to cause illness to a bather or person coming in contact directly or indirectly with the ocean.

Marine pathogens can be split into three main classes: protozoa, bacteria, and viruses. The different classes can differ significantly in size and characteristics. Most waterborne pathogens range in size from 0.01 to 100 μm (Gerba, 1996). For example, the size of an enteric virus such as Norovirus is only 10 nanometers, whereas the size of an enteric protozoan such as *Cryptosporidium parvum* is 4 micrometers (Fig. 20-1). By analogy, these relative sizes are like comparing the size of a football player and a football field, respectively. Many of the traditional methods used in drinking water analysis have been used in ocean water to concentrate various pathogens. However, given the differences in the chemical characteristics of the two water matrices (as well as the difference in risk from having a certain concentration of a pathogen in drinking water that is directly ingested verses having that same concentration of pathogen in ocean water that is not intentionally swallowed), the detection methods and acceptable levels of microorganisms can be very different.

Two main approaches are used in pathogen concentration: size exclusion and membrane adsorption. In the size exclusion approach, large volumes of water are pushed through filters, which have a pore size that is usually one order of magnitude smaller than the pathogens of concern. After filtration, the pathogens are usually left to float in a fluid suspension above the filter and subsequently removed for analysis. Alternatively, the pathogens attach to the filter and must then be eluted with a surfactant in order to transfer the pathogen into a liquid matrix for analysis. In the membrane adsorption approach, the membranes used have pore sizes that are not necessarily smaller than the pathogen size. However, through electrostatic forces the pathogen is attracted to the membrane. This method is usually used for viruses, which tend to attach to membranes given the appropriate water chemistry and the composition of the membrane itself.

Ideally, regulators and researchers may want to concentrate all three classes of pathogens (bacteria, protozoa, and viruses) simultaneously and efficiently in a standardized

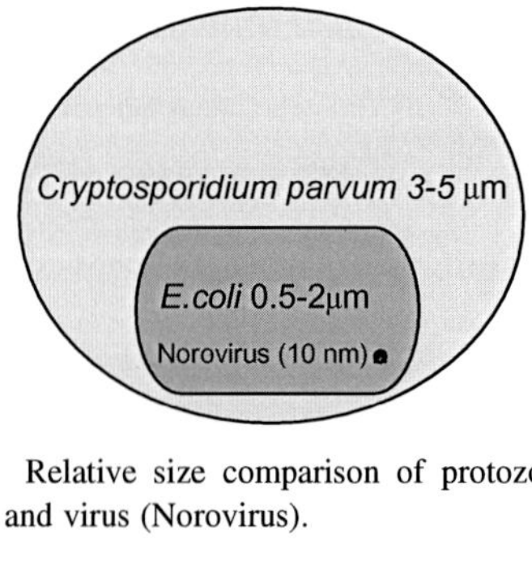

FIGURE 20-1. Relative size comparison of protozoans (*C. parvum*) bacteria (*E. coli*) and virus (Norovirus).

technique to avoid time and cost delays as illness may result from any member of the three classes as mentioned earlier. Both the size exclusion and membrane adsorption approaches have been used for this task and both have advantages and disadvantages. Methods based on size exclusion have been shown to be a promising option for concentrating all three classes of organisms (Olszewski *et al.*, 2005). Some of the advantages of this approach are that high recoveries can be obtained, because as long as the filter used is intact, most pathogens will not pass through as the pores tend to be much smaller than the pathogens. However, the small pore size of the membranes, e.g., 100 kDA, may cause the membrane to clog prematurely as such filters will also concentrate a considerable amount of naturally occurring dissolved materials (e.g., humic and fulvic acids) and additional debris in environmental water samples. Moreover, these materials are known to cause inhibition in subsequent molecular analysis of the water sample (Jiang *et al.*, 2001) and must be eliminated before analysis for microorganisms.

Membrane adsorption has also shown promising results with respect to the concentration of viruses (Fuhrman *et al.*, 2005; Katayama *et al.*, 2002) and may also prove useful to concentrate bacteria and viruses simultaneously (Abdelzaher *et al.*, 2007; Rolland and Block, 1980). In this technology, viruses attach to charged membranes with large pore sizes through electrostatic forces. Potential inhibitors to molecular analysis, as well as debris that may clog the membrane, do not attach to the membrane and thus do not concentrate with the viruses. The viruses can then be eluted from the membrane into a relatively clean solution by altering the charge. However, the differences in the chemistry of the water filtered (e.g., pH and presence of cations) and the charge of the membrane cause significantly varied recovery levels in adsorption to and elution off the membrane (Lukasik *et al.*, 2000; Sobsey and Hickey, 1985).

New technologies are starting to tackle some of the disadvantages of previous systems. New methods based on size exclusion (ultrafiltration and tangential flow filtration) are being used to concentrate all three classes of pathogens simultaneously. Some of the limitations of size exclusion methods are being remedied by precoating the membranes so that pathogens can easily be eluted off once attached to membranes as well as prefiltering to avoid membrane clogging (Hill *et al.*, 2005).

One example of a newly proposed system that combines advantages of both the size exclusion and membrane adsorption approaches is the bilayer membrane approach. This technique may be used to concentrate bacteria and viruses, and it may also be suitable for protozoa (Fig. 20-2). The technique can also allow for the simultaneous concentration of different classes of pathogens and indicators from one water sample while separating bacteria on one membrane and viruses on another. This system uses commercially available 90–mm diameter 0.45 μm pore size membranes.

FIGURE 20-2. The bilayer concentration unit.

The top PVDF (Millipore Corp) membrane is used to filter bacteria by physical straining while the bottom mixed cellulose esters (Millipore Corp) membrane retains viruses through adsorption. The technique is still in a research phase but has shown potential when concentrating *Enterococci faecalis* as a representative bacterial organism (87 ± 16% recovery) and F+ coliphage as a representative virus (82 ± 8% recovery) (Abdelzaher *et al.*, 2007).

There are a few limitations in the bilayer method. Because viruses are attached via charge on the bottom membrane, water chemistry can greatly affect recovery. Also, concentrations of the different classes of microbes may need different dilutions or filtration volumes, especially if the membranes are placed on media for culturing, because overgrowth and hence the inability to quantify the organism may occur. This is also a drawback of MPN and MF bacterial detection methods as well. In the bilayer approach, this situation may be corrected by continuously replacing the top membrane to prevent potential overgrowth, allowing for larger effective volumes to pass through the bottom membrane.

The bilayer method is also limited in the volume of water that can be filtered because the system utilizes flat filters. As in any detection scheme, risk must be quantified in order to determine which volume is adequate for the filtration. For example, is the risk from pathogen X so high that 1 organism in 100 L poses an unacceptably high risk to bathers, or is this risk negligible? The answer to this question would affect the volume of sample that should be processed to establish the acceptable risk level.

The aim of this method, as well as other novel concentration approaches under investigation, is to simplify and develop a more efficient detection method for regulatory agencies, public health laboratories, or research facilities to monitor beaches and oceans directly for pathogens and thereby more accurately assess the public health threat from microbes.

NUCLEIC ACID ANALYSIS IN WATER QUALITY MONITORING

There are many established culture-based assays for the detection of microorganisms (indicator and pathogenic) in recreational water samples. Generally, pure-culture methods are used to detect bacteria, and cell culture techniques are used to detect some viruses. Microscopic methods are often used to detect protozoa in environmental samples. As described previously, these methods of water quality analyses have worked relatively well but do have some limitations in their usefulness. The major limitations of culture methods are that target microorganisms may not grow in culture media because of injury from exposure to environmental stressors such as disinfectants used during water treatment, and culture methods to detect several important enteric viruses and protozoa have not been developed. In addition, the number of pathogens in wastewater and sludge is usually lower than the number of nonpathogens present, and cumbersome concentration methods of large volumes of water are often necessary. Lastly, culture-based methods for bacteria and viruses and microscopic analysis for protozoa can take several days to weeks to complete, resulting in obvious difficulties when trying to protect the public from exposure to a pollution event.

Since the 1980s, environmental microbiologists have developed and refined molecular methods (or nucleic acid-based methods) to rapidly detect the presence of fecal indicators and fecal pathogens in environmental water samples (Coupe *et al.*, 2006; Fong and Lipp, 2005; Guy *et al.*, 2003; Mayer and Palmer, 1996; Palmer *et al.*, 1995; Rochelle *et al.*, 1997; Tsai *et al.*, 1994). Nucleic acid-based methods are able to detect specific microorganisms with a high degree of sensitivity and specificity without the need for complex cultivation. Consequently, these methods allow the detection of pathogenic microorganisms (culturable or not) within a few hours, instead of the days normally required by culture-based assays. For many nucleic acid-based methods, sample collection and processing are often still necessary, but the length of time in which one acquires the results is substantially shorter. One important limitation of most nucleic acid-based detection assays is that they do not discriminate between viable and nonviable microorganisms. Despite these limitations, molecular detection of specific pathogens has proven useful in identifying the presence of certain human pathogens in a sensitive and timely manner (Albinana-Gimenez *et al.*, 2006; Behets *et al.*, 2007; Betancourt and Fujioka, 2006; Thompson *et al.*, 2006). Moreover, advances in high-density DNA analysis are providing platforms to analyze water samples for a large number of pathogens simultaneously, possibly circumventing the issue of viability (Noble and Weisberg, 2005).

Chapter 19 by Goodwin and Litaker provides a thorough description of the scientific and technical details of many nucleic acid-based technologies emerging in environmental microbiology. The sections that follow discuss the potential of high-density DNA analysis, both quantitative polymerase chain reaction [qPCR (or "real time" PCR)] and DNA microarray technology, in determining the hygienic quality of recreational watersheds and their implementation into a city or state monitoring system.

HIGH-THROUGHPUT QUANTITATIVE PCR AND OLIGONUCLEOTIDE MICROARRAYS FOR PATHOGEN DETECTION

Microbiological and engineering advances are providing platforms for running 96 and 384 qPCR assays simultaneously on a single plate. Liquid-handling robotic systems for preparing and loading the samples are available as are methods for lyophilizing reagents in wells. An exciting technology being used in biomedical research is Applied Biosystem's MicroFluidic card. Currently, this card is produced by ABI for simultaneous gene expression analysis of up to 384 human or mouse targets. The customer provides the experimental design of the card by selecting gene targets from its TaqMan Assay-on-Demand system, and ABI produces a batch of cards preloaded with the primer and probe sets manufactured into the card. The user simply adds the sample into a port (there are eight ports per card), seals the card, and places it into a qPCR instrument for temperature cycling and data acquisition.

This system demonstrates the direction of multiple qPCR analysis and would be amenable to the analysis of many environmental microorganisms by designing appropriate primer and probe sets to be included in the card. The use of multiple ports to insert sample is convenient, as some of the ports would be reserved for the addition of a mixture of synthesized plasmids that contain known concentrations of the gene targets necessary to generate a standard curve. Robotic systems of preparing 384 well plates for qPCR analysis are also a direction that may prove useful in environmental monitoring for a high number of target microorganisms. Together, they represent the evolution of engineering technologies capable of providing standardized and quality-controlled preparations of the reagents needed for multiple qPCR analyses.

There are several advantages to using a 384–well qPCR analysis of a particular water sample, including the following:

1. Many pathogens and indicators can be quantitatively assayed simultaneously in a single water sample.
2. The inclusion of multiple gene targets for a single microorganism would dramatically increase the specificity of detection.
3. A complete profile of the microbiological status of the sample could more accurately predict a health risk to exposure.
4. Microbial source tracking can be performed if the appropriate gene targets are included.

Microarray technology is also an extremely powerful technique to analyze an environmental sample for many gene targets simultaneously. With thousands of oligonucleotide probes immobilized onto a microarray slide, the analysis of multiple gene targets is feasible in environmental samples (Bodrossy and Sessitsch, 2004, Desantis *et al.*, 2007). PCR amplification of bacterial 16s rRNA genes using universal primers followed by hybridization onto microarray slides will provide a powerful method of determining the bacterial content of a particular water sample.

There is little doubt that a multiple-microorganism approach to assaying the hygienic quality of an environmental water samples would be more informative than strictly using an indicator-microorganism approach. For these technologies to find their way into the study of environmental water quality, several important issues will need to be addressed including epidemiological studies to correlate the microbiological profiles of water samples with a health risk to exposure. If multiple (>20) gene targets related to various pathogenic and indicator organisms are included, it may be possible to circumvent the question of viability because the presence of certain concentrations of targets present in conjunction with the combinations of the targets present can be used in epidemiological studies to assess the relative risk to exposure to a particular microbiota in the water sample. This type of analysis could be a highly sensitive and specific method of evaluating the relative risk to human health. Additional research is needed to move these technologies from a proof-of-concept scenario to actual studies in water quality monitoring.

LUMINEX xMAP SUSPENSION ARRAY

A multitiered approach toward coastal water quality monitoring (Boehm *et al.*, 2003; Noble *et al.*, 2006) includes analysis of a variety of targets in addition to enumeration of traditional fecal indicators. A multitiered approach can yield more information about the source of contamination, health risk, and the best approach for remediation. Identification of a matrix of species could provide new dimensions to the examination of water quality. A single indicator or pathogen is not adequate for monitoring all exposure routes (National Research Council, 2004); therefore, a suite of targets may provide a better "fingerprint" of water quality and risk to human health. Such an approach would give environmental managers comprehensive information on which to base decisions in order to protect human health from toxins and pathogens in food, fish, shellfish, and recreational waters.

A multitiered approach could benefit from a technology that can deliver rapid, multiplexed, species-level detection. PCR is a powerful tool for identification of a few molecular targets of interest. However, problems arise when there is a requirement to identify a large numbers of species. The need to coordinate PCR stringency conditions for a variety of primers places unrealistic requirements on the design of a multiplex reaction. A more versatile approach is the use of hybridization assays. The Luminex xMAP hybridization assay system is a suspension array that has primarily been used for clinical applications (Dunbar, 2006; Dunbar *et al.*, 2003). Research development includes detection methods for pathogenic fungi in which the capability of detecting molecular targets that differ by a single nucleotide mismatch has been demonstrated (Diaz and Fell, 2004, 2005; Diaz *et al.*, 2006). More recently, the system has been investigated for coastal water quality applications (Baums *et al.*, 2007) and marine microplanktonic dynamics (Ellison and Burton, 2005).

Luminex is essentially a flow cytometer equipped with two lasers, one that identifies a color-coded bead (100 are available) and the other that registers whether or not the capture probe has captured a target (e.g., DNA, RNA, protein, antigen). If the target is DNA, the DNA is first isolated from the sample and amplified via PCR with biotin-labeled primers (see Chapter 19). The amplified DNA is then hybridized to capture probes that have been conjugated to microspheres (Fig. 20-3). These beads contain a varying ratio of red and infrared fluorophores, which imparts a unique "color" to each set. The biotinylated DNA that has been captured is coupled to a reporter molecule (streptavidin R-phycoerythrin) to generate fluorescence. Microfluidics control the flow of the microspheres though the path of two lasers. The red laser (636 nm) identifies the spectral address of the color-coded beads, and the green laser (532 nm) registers whether or not the probe has captured a target and quantifies the fluorescence (which is proportional to the amount of DNA that has been captured). The 100 available microsphere colors allow detection of many targets in a single sample well. Hybridization time is approximately 1

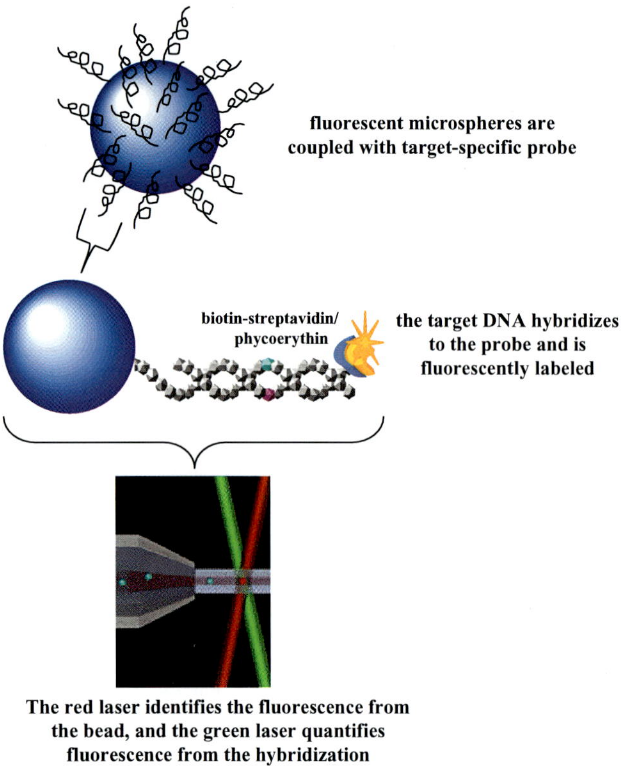

FIGURE 20-3. Illustration of the Luminex xMap detection system. Microspheres are interrogated individually in a fast-flowing fluid as they pass by two separate laser beams. Signal processing classifies the fluorescence of the bead, thus identifying the covalently bound probe. The hybridized DNA (labeled with biotin during amplification) is quantified based on emission from the fluorochrome phycoerythrin.

hour, and each well of a 96-well microtiter plate is assayed in approximately 0.47 second; thus, the Luminex system has the potential to provide rapid, high-throughput detection of multiple targets.

A major benefit of this technology is the versatility imparted by the use of beads, allowing probes to be added or subtracted for specific studies. In comparison to microarrays, suspension arrays can offer increased flexibility, cost effectiveness, statistical power, and faster hybridization kinetics (Dunbar, 2006). The use of molecular probes allows for specific identification and can alleviate problems associated with some morphological, physiological, and biochemical techniques. In addition, the Luminex suspension array can be used to detect a wide variety of compounds such as toxins, proteins, oligonucleotides, enzymes, antibodies, and antigens (Bellisario *et al.*, 2000, 2001; Fulton et al, 1997).

Market sectors for a technology such as Luminex include coastal zone management, risk management, environmental science, and aquaculture. High-throughput sample processing would allow more widespread screening of coastal waters, helping to protect health while avoiding the economic burden of blanket closures. Public health agencies could benefit from prescreening large sample sets, rapidly identifying only those requiring enumeration. Presently, routine monitoring work relies on tedious microscopic analysis or extensive bacterial culturing. For microscopic analysis, a highly trained person is required to accurately identify the species in the sample. Such training is no longer routinely provided in education curricula, resulting in sample bottlenecks. In contrast, molecular biology training has become widely available, with introduction now given in many high schools. In addition to monitoring applications, a variety of ecological studies would benefit from rapid and accurate species identification. Examples include investigations of bloom dynamics, species biogeography, introduction of invasive species, monitoring of ballast water release, and microbial source tracking.

Aquaculture management could also benefit by the ability to rapidly screen products before distribution. Molecular methods could offer competitive advantages to those industries using advanced product testing. Aquaculture is a growing industry worldwide with sales in the United States of nearly $1 billion during 1998. The market for aquatic organisms has grown while sustainable fisheries and habitats have dwindled. Aquaculture has been heralded as a critical solution to many coastal and estuarine problems. However, aquaculture can be severely affected by algal and bacterial contaminants, leading to livestock and human disease. In addition, aquaculture also has introduced foreign species, including toxic dinoflagellates into previously uncontaminated environments. Closures of shellfish harvesting costs local business millions of dollars annually.

Some end users may opt for centralized testing to defray the capital cost of the Luminex platform. Such users would likely prefer water quality test kits that could be purchased commercially. Probes (e.g., for source tracking, pathogen detection, fecal indicators) could be marketed individually so that users could design assays to their specific needs. The market potential for suspension array technology could be expanded, particularly in sectors of risk management and resource management fields, if the assays could be designed for quantitative testing rather than presence/absence testing. Of note, entry into these sectors also requires regulatory verification and acceptance.

A primary technical obstacle for this and other molecular biological technologies arises from working with environmental or food samples. As discussed in previous chapters, challenges to molecular microbial ocean water quality monitoring include the need to filter large volumes of water, the presence of PCR inhibitors, the inherent patchiness of target organisms, and the rarity of microbial contaminants in comparison to the rich microbial background. However, molecular approaches can offer improvements over current methods. A commitment to work through these challenges will allow those improvements to be made available to the coastal research and management communities.

IMPACTS OF NEW TECHNOLOGIES: A REGULATORY PERSPECTIVE

Integrating new technologies into a regulatory framework presents a means of potentially improving decision making when issuing beach advisories and warnings. However, the integration of these technologies within regulations is challenging. First, the measurements taken by these new technologies must be clearly linked to human health. The general public will ask, "Is the water safe for swimming?" Rarely will the public ask about the levels of particular contaminants. The regulator's responsibility is thus to bridge the gap between measurements and public health perception. Bridging this gap is difficult, as it requires quantifiable relationships between environmental measurements and human health.

The linkage between measurements and human health is generally best established through epidemiological studies that associate illness by human subjects (which are usually self-reported) and an exposure to a recreational water body. Although straightforward in concept, many complexities are associated with epidemiological studies, making them difficult to design, execute, and interpret (Eisenberg *et al.*, 2002). These complexities include differences in self-reporting of illness (Fleisher and Kay, 2006) and confounding factors from other exposures. Quantification of these exposures also represents a challenge. In the case of water ingestion during swimming, there are uncertainties with the amount of water ingested and the quality of that water. Measurements can be highly variable in space and time within a water body (Boehm *et al.*, 2002; Shibata *et al.*, 2004; Solo-Gabriele *et al.*, 2000); collecting a representative water sample for comparison with human health reports is very difficult. Epidemiological studies are expensive and can require from thousands to tens of thousands of human participants to establish meaningful associations between exposure and illness. In the absence of epidemiological studies, regulators may rely on risk assessments, which relate dose (amount of contaminant ingested) to response (illness) through a quantitative analysis utilizing many assumptions. For example, when evaluating water ingestion as the route of exposure, the assumptions may include the amount of water ingested by age, the concentration of contaminant in the water at the time of ingestion, the susceptibility of the individual, length exposure (intermittent, lifetime), and so on. Because of the number of assumptions, the relationship between water quality characteristics and human illness in this kind of approach may be subject to a considerable amount of uncertainty.

Once established relationships have been made between environmental conditions and human health, several additional criteria must be met before a new measurement technique can be implemented within the regulatory framework. These include incorporation of new criteria or techniques into regulatory language, financial impact analyses, and basic logistical issues associated with implementation. Incorporating new criteria into regulatory language is a long process frequently requiring a sponsor within the regulatory authority. Language is circulated among the regulated community and public hearings are held. Throughout this process the language is discussed, modified, and hopefully, adopted as part of the regulations. With respect to recreational water quality, the U.S. EPA recommends nonenforceable guidelines at the federal level. Only the states have the power to adopt and enforce the U.S. EPA guidelines or other guidelines. Most states adopt the EPA values. An exception is Hawaii, which adopted stricter standards and also includes provisions for monitoring an alternative fecal indicator (Fujioka and Shizumura, 1985). New criteria would require adoption of appropriate regulatory language at the state level. For larger scale implementation, new criteria would need adoption at the federal level as many states establish their standards based on federal guidelines.

The cost of using new technologies also plays a significant factor with implementation. Newer, more innovative technologies tend to be more expensive than using traditional and well-established methods. Most of the current beaches monitoring programs are underfunded, relying exclusively on the BEACH ACT grant (i.e., no local funding is available to supplement this grant money). As mentioned in previous chapters, the U.S. EPA transitioned its fecal indicator guideline from total and fecal coliform toward *E. coli* and enterococci during 1986 (U.S. EPA, 1986). Methods used for analyses of both groups of fecal indicators are similar; conversion would be relatively simple. Even so, many programs did not convert to fecal indicators (recommended in 1986 guideline) until the implementation of the BEACH Act of 2000, which provided state funding specifically for *E. coli* and enterococci measurements. Thus financial backing is critical for implementation of new methods of analysis. If new methodologies are different relative to current techniques, a considerable amount of resources may be required for successful implementation. If techniques shift from culture-based methods toward molecular-based methods, considerable funds will be needed for capital equipment, training, and hiring scientists with specialized skills.

Inclusion of new measurements within regulatory standards will also require a considerable amount of logistical changes on the part of the agencies. Before a new measurement can be adopted, there must be assurances that the new method can be utilized at most laboratories. Meeting quality control/quality assurance requirements such as reproducibility may be an issue. There is not a wealth of experience and training in applying emerging new technologies and testing methods among regulatory agencies staff, beach managers, and laboratory technicians. In fact, even with existing methods, there have been cases where different laboratories consistently reported different results for the same method, rendering the data sharing and comparisons invalid and impractical. Reproducibility of results among a large number of laboratories is critical for nationwide monitoring of recreational water quality. Assuring reproducibility will typically require "round robin" testing where blind samples are sent to a subset of laboratories to evaluate method comparisons. Once the technique has been tested and fine-tuned, a broader scale rollout would be required where additional laboratories are to be certified. This process requires a labor force capable of conducting the new analysis and a training program for new personnel. The amount of training will depend on the complexity of the new method. For example, the same workforce could likely be used when converting from a culture-based method for measuring fecal coliform to a culture-based method for measuring enterococci; however, conversion to a molecular-based method will require retraining and perhaps require a new workforce with the appropriate skills. Such a change would put a considerable burden on regulatory agencies from a human resources point of view. Implementation of a new measurement will also require that each laboratory be fitted with the appropriate equipment and supplies with new quality assurance protocols established. Although straightforward, implementation will require considerable planning and resources. Existing regulatory facilities are also frequently required to assess other types of samples including drinking water, wastewater, and clinical samples. Facilities, resources, and labor force available for measuring these other sample types may be leveraged assuming that the new measurement technique for recreational waters is similar to those used for these other samples. Of note is that routine monitoring of drinking water and wastewater samples is based on measurements of traditional fecal indicators. Only for specific studies are molecular techniques utilized for monitoring drinking water and wastewater. Clinical samples in many cases are analyzed using molecular methods; leveraging their equipment and expertise may be beneficial if new measurement techniques for recreational waters require the incorporation of similar methods.

PUBLIC HEALTH BENEFITS OF IMPROVED COASTAL MONITORING

New measurement techniques would assist regulators in evaluating the safety of recreational waters if the technique shortens the time frame between sample collection and analysis, provides information about additional exposure routes, identifies potential sources, and provides information directly about agents that cause disease. Currently, the time period between sample collection and provision of results is on the order of 18 to 24 hours using traditional culture-based methods. New qPCR-based methods can reduce the analysis time down to 2 to 4 hours. Results of samples collected in the morning would be available by late morning to early afternoon as the usage of the beach begins to increase. A beach advisory or warning could be issued the same day the sample is collected in the event that the measurement indicates poor water quality.

Information concerning additional exposure routes associated with human illness would also be advantageous. Current standards for measuring recreational water quality are based on risks from gastrointestinal disease. New measurements may be able to provide insights into health risks associated with direct contact with the water (e.g., skin, eye and ear infections) and inhalation of the water (e.g., respiratory illness), in addition to gastrointestinal illness associated with inadvertent ingestion. Furthermore, water at a beach may not be the only potential vector of exposure. Research has shown that fecal indicator bacteria can be found in beach sand and sediments (Alm *et al.*, 2006; Desmarais *et al.*, 2002; Fujioka *et al.*, 1999; Lee *et al.*, 2006); potential health effects from exposure to these media are not clear. Ideally, new measurement techniques would provide more effective monitoring of additional vectors and various exposure routes.

Newer methods, in particular microbial source tracking (MST) technologies, would be helpful in identifying sources of fecal indicator bacteria (Simpson *et al.*, 2002; MST Guide Document, 2005). This technique is moving into the application phase within applied research, public health, and legal investigations. Because MST methods are suited for identifying if fecal indicators are of human or nonhuman origin, MST has been used as a tool to develop the total maximum daily loads (TMDL) for surface water systems mainly impacted by nonpoint fecal sources such as storm water runoff, animal waste, and other environmental sources. This is of significance as human health risks are generally considered to be less if the source of the fecal indicators is from nonhuman sources.

New measurements may also provide information about direct agents of disease. Many have argued that fecal indicators may not be adequate surrogates for disease, in particular when from nonpoint sources of pollution (Colford *et al.*, 2005). Adding direct pathogen evaluations to the suite of measurements used to assess the quality of the beach would allow regulators to make more informed decisions. For example, in 2000, a 95 million liter spill of untreated sewage into Biscayne Bay, Florida, resulted in immediate beach closures. This spill was caused by the rupture of a 137-cm wastewater force main. During repairs, the sewage was chlorinated and discharged via ocean outfall. Upon rerouting the sewage to the outfall, fecal indicator levels at the beaches returned to acceptable levels. However, because of the limited treatment of the sewage through only chlorination, the Miami-Dade Health Department requested that the utility test the beach waters for enteroviruses (Hepatitis A, Norwalk, and Rotavirus) and protozoans (*Cryptosporidium* and

Giardia) in addition to testing for the regulatory bacteria indicators fecal coliforms and enterococci. Measurements for pathogens were negative, and the beaches were reopened. In this case, there was reason to suspect that a negative fecal indicator reading may not have been protective; these measurements were supplemented with pathogens testing. Fecal indicators and pathogen results supported regulators' decision to reopen the beaches.

There have, however, been documented cases where results from fecal indicator measurements did not correlate with pathogen measurements (Griffin *et al.*, 1999; Harwood *et al.*, 2005; Lipp *et al.*, 2001). Fecal bacteria are known to persist and even regrow (Desmarais *et al.*, 2002) in the environment. There is a question about whether these environmental sources of fecal bacteria are correlated with the presence of pathogens. Persistence and regrowth of fecal organisms in the environment would result in potential false positives and result in unnecessary beach closures, which have negative economic impacts. This phenomenon is likely emphasized within tropical and subtropical environments because of the warmer and wetter climate (Fujioka and Roll, 1997). Evidence of this phenomenon has also been observed within temperate regions (Whitman *et al.*, 2003). Direct measurements of pathogens would be useful when there is reason to suspect that the fecal indicators may be providing false positive results.

In summary, new methods and technologies will lead to improvements in bridging the gap between environmental measurements and the identification of "safe" water quality. Potential advantages of these technologies are a decrease in time between sample collection and results; greater information about potential sources of fecal indicator microbes; additional insights on relationship between exposure and illness; and, ultimately, direct measurements of disease causing agents to supplement existing measurements of sewage surrogates. Such information will be of great value to regulators as they interpret results from environmental measurements, make decisions on beach safety and health, and inform the community of potential risks.

DISCUSSION

The rapid growth of human populations and industrial output has profoundly impacted coastal water quality. Each year, miles of coastal water areas are temporarily closed to the public when indicator organism levels escalate, indicating potential contamination from pathogenic microorganisms. This not only impacts recreational use of coastal waters but can adversely impact the economy since tourism is also affected by beach closures. In addition, when shellfish-growing areas are closed because of high levels of indicator organisms, this action can impact the nation's food supply. Thus, the development and implementation of better coastal water quality analysis methods is crucial to support the many facets of coastal water usage.

Coliforms were the indicator organism of choice for water quality for most of the 20th century for monitoring both fresh and marine waters. In 1986, the EPA put forth new microbiological water quality guidelines for recreational marine water, recommending enterococci as a better choice of indicator organism for possible fecal contamination. Many coastal areas have since adopted this organism for use in coastal water quality monitoring. However, there is still much room for improvement in coastal water monitoring. Although the current system is fairly effective at monitoring beach water quality, there are several deficiencies in the existing monitoring programs. Should we change to direct pathogen monitoring using the new technologies? Another issue in ocean monitoring is that all coastal states do not all monitor for the same organisms. In fact, even within a state, different beaches may be monitored for different organisms, depending on local preferences. We need to develop a national policy for ocean monitoring with all states in agreement to monitoring choices.

We also need to address the issue of indigenous marine pathogens. What should be the regulatory stand on indigenous pathogenic marine bacteria such as *Vibrio vulnificus*, which can enter wounds and lead to rapid necrosis and even death or *Vibrio parahaemolyticus* in shellfish, which can cause severe/debilitating diarrhea. Would we even want to monitor coastal water for these organisms as they are always present, and on what basis would we decide how many indigenous organisms constitute risk? The epidemiological studies that may provide answers to the risk from both pathogens entering our coast from land-derived sources and blooms of naturally occurring pathogenic marine organisms are expensive to complete and will take significant time for regulators and other oversight agencies to decide whether change is necessary in our coastal monitoring programs. However, the tools to do such studies are available today.

New methods provide timely (within 2 to 4 hours) data, provide additional information concerning alternative exposure routes, elucidate fecal pollution sources (point and non-point sources), and provide direct measurements of agents that cause disease. Such information would supply the data necessary to design and implement the best management practices to reduce or eliminate the source in a sustainable and consistent approach. New technologies will also likely minimize the issuance of unnecessary beach closings and advisories, resulting in a significant positive impact on tourism.

Although the use of these new technologies and testing methods are slowly entering the application phase, their use is still limited to research and legal investigations. This integration may continue for some years allowing the necessary time for development of technologies and testing methods followed by adoption by the regulatory communities. Historically, regulatory agencies need to allocate time and funds to assist in passing the necessary laws and write new regulations to adopt new technologies and methods as this process is complex and can be lengthy. According to the NRDC 2006 Testing the Waters report, the EPA will not be ready to revise the standards and establish new methods for better characterizing the public health risk until 2011. Given the lag between enacting regulations and implementation, the inclusion of new techniques within the regulatory community will likely be seen at best within the next decade.

The methods discussed in this chapter have the potential to radically change the way we examine marine water and should lead to improved public health protection. The ability to directly detect pathogenic microorganisms or more rapidly detect indicator organisms will lead to more educated decisions by regulators as to when to limit swimming, fishing, seafood harvesting, and other recreational activities. Monitoring efforts that can target multiple microorganisms in a real time schedule will allow for beach closures based on presence of pathogens (the disease causing agent) rather than elevated indicator organisms. The future of water quality monitoring in the 21st century, utilizing the technological advances discussed in this chapter, provides the promise of cleaner and safer beaches and improved public health outcomes for all those who enjoy our coastal environment.

References

Abbott, S., Caughley, B., Scott, G., 1998. Evaluation of enterolert for the enumeration of enterococci in the marine environment. New Zealand J. Mar. Freshwater Res. 32, 505–513.

Abdelzaher, A, Solo-Gabriele, H., Palmer, C., Wright, M., 2007. Sequential Concentration of Bacteria and Viruses from Marine Waters Using a Layered Membrane System. Proceedings of the 106th General Meeting of the American Society for Microbiology, Abstract number Q-451.

Albinana-Gimenez, N., Clemente-Casares, P., Bofill-Mas, S., Hundesa, A., Ribas F., Girones, R., 2006. Distribution of human polyomaviruses, adenoviruses, and hepatitis E virus in the environment and in a drinking-water treatment plant. Environ. Sci. Technol. 40, 7416–7422.

Alm, E.W., Burke, J., Hagan, E., 2006. Persistence and potential growth of the fecal indicator bacteria, Escherichia coli, in shoreline sand at Lake Huron. J. Great Lakes Res., 32, 401–405.

American Public Health Association (APHA), 2005. Standard Methods for the Examination of Water and Wastewater, 21st ed., sec. 9223. Washington, DC, American Public Health Association.

American Society for Testing and Materials (ASTM). 1999. Standard Test Method for Enterococci in Water Using Enterolert™, D6503–99. West Conshohocken, PA, American Society Testing and Materials.

Bai, S., Lung, W-S., 2005. Modeling sediment impact on the transport of fecal bacteria. Water Res., 39, 5232–5240.

Baums, I.B., Goodwin, K.D., Kiesling, T., Wanless, D., Fell, J.W., 2007. Luminex detection of fecal indicators in river samples, marine recreational water, and beach sand. Mar. Pollut. Bull. 54, 521–536.

Behets, J., Declerck, P., Delaedt, Y., Verest, L., Ollevier, F., 2007. A duplex real-time PCR assay for the quantitative detection of Naegleria fowleri in water samples. Water Res. 41, 118–126.

Bellisario, R., Colinas, R.J., Pass, K.A., 2000. Simultaneous measurement of thyroxine (T4) and thyrotropin (TSH) from newborn dried blood-spot specimens using a multiplexed fluorescent microsphere immunoassay. Clin. Chem. 46, 1422–1424.

Bellisario, R., Colinas, R.J., Pass, K.A., 2001. Simultaneous measurement of antibodies to three HIV-1 antigens in newborn dried blood-spot specimens using a multiplexed microsphere-based immunoassay. E. Hum. Develop. 64, 21–25.

Benham, B.L., Baffaut, C., Zeckoski, R.W., Mankin, K.R., Pachepsky, Y.A., Sadeghi, A.A., Brannan, K.M., Soupir, A.L., Habersack, M.J., 2006. Modeling bacteria fate and transport in watersheds to support TMDLs. Transactions of the ASABE, 49, 987–1002.

Betancourt W.Q., Fujioka R.S., 2006. *Bacteroides spp.* As reliable marker of sewage contamination in Hawaii's environmental waters using molecular techniques. Water Sci. Technol. 54, 101–107.

Bodrossy, L. Sessitsch, A., 2004. Oligonucleotide microarrays in microbial diagnostics. Curr. Opin. Microbiol. 7, 245–254.

Boehm, A.B., Furhman, J.A., Morse, R.D., Grant, S.B., 2003. Tiered approach for identification of a human fecal pollution source at a recreational beach: Case study at Avalon Bay, Catalina Island, CA. Environ.l Sci. Technol. 37, 673–680.

Boehm, A.B., Grant, S.B., Kim, J.H., Mowbray, S.L., McGee, C.D., Clark, C.D., Foley, D.M., Wellman, D.E., 2002. Decadal and shorter period variability of surf zone water quality at Huntington Beach, California. Environ. Sci. Technol., 36, 3885–3892.

Bonilla, T.D., Nowosielski, K., Esiobu, N., McCorquodale, D.S., Rogerson, A., 2006. Species assemblages of Enterococcus indicate potential sources of fecal bacteria at a south Florida recreational beach. Mar. Pollut. Bull. 52, 807–810.

Bower, P.A., Scopel, C.O., Jensen, E.T., Depas, M.M., McLellan, S.L., 2005. Detection of genetic markers of fecal indicator bacteria in Lake Michigan and determination of their relationship to *Escherichia coli* densities using standard microbiological methods. Appl. Environ. Microbiol. 71, 8305–8313.

Budnick, G.E., Howard, R.T., Mayo, D.R., 1996. Evaluation of enterolert for enumeration of entrococci in recreational waters. Appl Env. Microbiol. 62, 3881–3884.

Butt, A.A., Aldridge, K.E., Sanders, C.V., 2004. Infections related to the ingestion of seafood Part I: Viral and bacterial infections. Lancet Infect. Dis 4, 201–212.

Cabelli, V., 1978. Water Pollution Microbiology, vol. 2. Mitchell, R. (ed.). New York, John Wiley & Sons.

Cabelli, V.J., Dufour, A.P., McCabe, L.J., Levin, M.A., 1982. Swimming associated gastroenteritis and water quality. Am. J. Epidemiol. 115, 606–616.

Canale, R.P., Auer, M.T., Owns, E.M., Heidtke, T.M., Effler, S.W., 1993. Modeling fecal coliform bacteria–II: Model development and application. Water Res. 27, 703–714.

Carlucci, A.F., Pramer, D. 1960. An evaluation of factors affecting the survival of *E. coli* in seawater: Salinity, pH and nutrients. Appl. Environ. Microbiol. 8, 243–247.

Centers for Disease Control and Prevention (CDC), 2006. Surveillance for waterborne disease and outbreaks associated with recreational water–United States 2003–2004. MMWR Surveill. Summ. December 22, 2006 / 55(SS12), 1–30.

Clarke, L.B., Ackerman, D., Largier, J., 2007. Dye dispersion in the surf zone: Measurements and simple models. Continental Shelf Res. 27, 650–669.

Colford, C.M., Jr., Wade, T.J., Schiff, K.C., Griffith, J.F., Sandhu, S.K., Weisberg, S.B., 2005. Recreational Water Contact and Illness in Mission Bay, California. Southern California Coastal Water Research Project, Technical Report 449.

Colford, J.M., Jr., Wade, T.J., Schiff, K.C., Wright, C.C., Griffith, J.F., Sandhu, S.K., Burns, S., Sobsey, M., Lovelace, G., Weisberg, S.B.,

2007. Water quality indicators and the risk of illness at beaches with nonpoint sources of fecal contamination. Epidemiology 18, 27–35.

Coupe, S., Delabre, K., Pouillot, R., Houdart, S., Santillana-Hayat M., Derouin, F., 2006. Detection of cryptosporidium, Giardia and Enterocytozoon bieneusi in surface water, including recreational areas: A one-year prospective study. FEMS Immunol. Med. Microbiol. 47, 351–359.

Culverhouse, P.F., Williams, R., Reguera, B., Herry, V., González-Gil, S., 2003. Do experts make mistakes? A comparison of human and machine identification of dinoflagellates. Mar. Ecol. Prog. Ser. 247, 17–25.

Desantis, T. Z., Brodie, E.L., Moberg, J.P., Zubieta, I.X., Piceno, Y.M., Andersen, G.L., 2007. High-density universal 16S rRNA microarray analysis reveals broader diversity than typical clone library when sampling the environment. Microbial. Ecol. 53, 371–383.

Desmarais, T.R., Solo-Gabriele, H.M., Palmer, C.J., 2002. Influence of soil on fecal indicator organisms in a tidally influenced subtropical environment. Appl. Environ. Microbiol. 68, 1165–1172.

Diaz, M.R., Boekhout, T., Theelen, B., Bovers, M., Cabanes, F.J., Fell, J.W., 2006. Microcoding and flow cytometry as a high-throughput fungal identification system for *Malassezia* species. J. Med. Microbiol. 55, 1197–1209.

Diaz, M.R., Fell, J.W., 2004. High through-put detection of pathogenic yeasts in the genus *Trichosporon*. J. Clin. Microbiol. 42, 3696–3706.

Diaz, M.R., Fell, J.W., 2005. Use of a suspension array for rapid identification of the varieties and genotypes of the *Cryptococcus neoformans* species complex. J. Clin. Microbiol. 43, 3662–3672.

Dufour, A.P., 2006. Rapidly measured indicators of recreational water quality are predictive of swimming-associated gastrointestinal illness. Environ. Health Perspect. 114, 24–28.

Dunbar, S.A., Vander Zee, C.A., Oliver, K.G., Karem, K.L., Jacoboson, J. W., 2003. Quantitative, multiplexed detection of bacterial pathogens: DNA and protein application of the Luminex LabMAP™ system. J. Microbiol. Meth. 53, 245–252.

Eisenberg, J.N.S., Brookhart, M.A., Rice, G., Brown, M., Colford, J.M., Jr., 2002. Disease transmission models for public health decision making: Analysis of epidemic and endemic conditions caused by waterborne pathogens. Environ. Health Perspect., 110, 783–790.

Ellison, C.K., Burton, R.S., 2005. Application of bead array technology to community dynamics of marine phytoplankton. Mar. Ecol. Prog. Ser. 288, 75–85.

Elshorbagy, A., Teegavarapu, R.S.V., Ormsbee, L., 2006. Assessment of pathogen pollution in watersheds using object-oriented modeling and probabilistic analysis. J. Hydroinformatics, 8, 51–63.

Environmental Protection Agency (EPA), 2003. Guidelines establishing test procedures for the analysis of pollutants; Analytical methods for biological pollutants in ambient water; Final Rule. Federal Register V68, No. 139 40 CFR Part 136, 43272–43283.

Fleisher, J.M., 1990. The effects of measurement error on previously reported mathematical relationships between indicator organism density and swimming-associated illness: A quantitative estimate of the resulting bias. Int. J. Epidemiol. 19, 1100–1106.

Fleisher, J.M., 1991. A reanalysis of data supporting U.S. federal bacteriological water quality criteria governing marine recreational waters. Res. J. WPCP 63, 259.

Fleisher, J.M., Kay, D., 2006. Risk perception bias, self-reporting of illness, and the validity of reported results in an epidemiologic study of recreational water associated illnesses. Mar. Pollut. Bull. 52, 264–268.

Fong, T.T. Lipp, E.K., 2005. Enteric viruses of humans and animals in aquatic environments: Health risks, detection, and potential water quality assessment tools. Microbiol. Mol. Biol. Rev. 69, 357–371.

Fuhrman, J.A., Liang, X., Noble, R., 2005. Rapid detection of enteroviruses in small volumes of natural waters by real-time quantitative reverse transcriptase pcr. Appl. Environ. Microbiol. 71, 4523–4530.

Fujioka, R., Shizumura L.K., 1985. *Clostridium perfringens*, a reliable indicator of stream water quality. J. Water Pollut. Control Fed. 57, 986–992.

Fujioka, R.S., 2001. Monitoring coastal marine waters for spore-forming bacteria of faecal and soil origin to determine point from non-point source pollution. Water Sci. Technol. 44, 181–188.

Fujioka, R.S., Roll, B.M., 1997. Sources of fecal indicator bacteria in a brackish, tropical stream and their impact on recreational water quality. Water Sci. Technol. 35, 179–186.

Fujioka, R.S., Sian-Denton, C., Borja, M., Castro, J., Morphew, K., 1999. Soil: The environmental source of *Escherichia coli* and *Enterococci* in Guam's stream. J. Appl. Microbiol. Sympos. (suppl.) 85, 83S–89S.

Fulton, R., McDade, R., Smith, P., Kienker, L., Kettman, J., 1997. Advanced multiplexed analysis with the FlowMetrix system. Clin. Chem. 43, 1749–1756.

Ge, Z., Frick, W.E., 2006. Some statistical issues related to multiple linear regression modeling of beach bacteria concentrations. Environ. Res. 103, 358–364.

Gerba, C.P., 1996. Pathogens in the environment. In Pepper, I.L., Gerba, C.P., and Brusseau, M.L. (eds.), Pollution Science. New York, Academic Press.

Goyal, S.M., 1983. Viral Pollution of the Environment. G. Gerg (ed.). Boca Raton, FL, CRC Press.

Grant, S.B., Kim, J.H., Jones, B.H., Jenkins, S.A., Wasyl, J., Cudaback, C., 2005. Surf zone entrainment, along-shore transport, and human health implications of pollution from tidal outlets. J. Geophys. Res.110, C10025, doi:10.1029/2004JC002401.

Griffin, D.W., 2001. Microbial public health indicators in the marine environment. In Methods in Microbiology, 541–558. New York: Academic Press.

Griffin, D.W., Gibson III, C.J., Lipp, E.K., Riley, K., Paul III, J.H., Rose, J.B., 1999. Detection of viral pathogens by reverse transcriptase PCR and of microbial indicators by standard methods in the canals of the Florida Keys. Appl. Environ. Microbiol. 65, 4118–4125.

Guy, R.A., Payment, P., Krull, U.J., Horgen, P.A., 2003. Real-time PCR for quantification of *Giardia* and *Cryptosporidium* in environmental water samples and sewage. Appl. Environ. Microbiol. 69, 5178–5185.

Hardina, C.M., Fujioka, R.S., 1991. Soil: The environmental source of *E. coli* and enterococci in Hawaii's streams. Environ. Toxicol. Water Quality 6, 185–195.

Harwood, V.J., Levine, A.D., Scott, T.M., Chivulkula, V., Lukasik, J., Farrah, S.R., Rose, J.B., 2005. Validity of the indicator organism paradigm for pathogen reduction in reclaimed water and public health protection. Appl. Environ. Microbiol. 71, 3163–3170.

Hendricks, C.W., 1978. Indicators of Virus in Water and Food. G. Berh (ed.), Ann Arbor, MI, Ann Arbor Science.

Hill, V.R., Polaczyk, A., Hahn, D., Narayanan, J., Cromeans, R., Roberts, J., Amburgey, J., 2005. Development of a rapid method for simultaneous recovery of diverse microbes in drinking water by ultrafiltration with sodium polyphosphate and surfactants. Appl. Environ. Microbiol. 71, 6878–6884.

Hyer, K.E., Moyer, D.L., 2004. Enhancing fecal coliform total maximum daily load models through bacterial source tracking. J. Am. Water Resources Assoc. December, 1511–1526.

Irvine, K.N., Perrelli, M.F., McCorkhill, G., Caruso, J., 2005. Sampling and modeling approaches to assess water quality impacts of combined sewer overflows: The importance of a watershed perspective. J. Great Lakes Res. 31, 105–115.

Jamieson, R., Gordon, R., Joy, D., Lee, H., 2004. Assessing microbial pollution of rural surface waters, a review of current watershed scale modeling approaches. Agricultural Water Manage. 70, 1–17.

Jiang, S., Noble, R., Chui., W.P., 2001. Human adenoviruses and coliphages in urban runoff-impacted coastal waters of Southern California. Appl. Environ. Microbiol. 67, 179–184.

Jin, G., Englande, A.J., Jr., Liu, A., 2003. A preliminary study on coastal water quality monitoring and modeling. Journal of Environmental Science and Health, Part A–Toxic/Hazardous Substances and Environmental Engineering, A38, 493–509.

Joseph , S.W., Colwell, R.R., Kaper, J.B., 1982. *Vibrio parahaemolyticus* and related halophilic vibrios. Crit. Rev. Microbiol. 10, 77–124.

Katayama, H., Shimasaki, A., Ohgaki, S., 2002. Development of a virus concentration method and its application to detection of enterovirus and Norwalk virus from coastal seawater. Appl. Environ. Microbiol. 68, 1033–1039.

Kelsey, H., Porter, D.E., Scott, G., Neet, M., White, D., 2004. Using geographic information systems and regression analysis to evaluate relationships between land use and fecal coliform bacterial pollution. J. Exp. Mar. Biol. Ecol. 298, 197–209.

Kinzelman, J., Ng, C, Jackson, E., Gradus, S., Bagley, R., 2003. Enterococci as indicators of Lake Michigan recreational water quality: Comparison of two methodologies and their impacts on public health regulatory events. Appl. Environ. Microbiol. 69, 92–96.

Lee, C.M., Lin, T.Y., Lin, C-C, Kohbodi, G.A., Bhatt, A., Lee, R., Jay, J.A., 2006. Persistence of fecal indicator bacteria in Santa Monica Bay beach sediments. Water Res.40, 2593–2602.

Lipp, E.K., Farrah, S.A., Rose, J.B., 2001. Assessment and impact of microbial fecal pollution and human enteric pathogens in a coastal community. Mar. Pollut. Bull. 42, 286–293.

Liu, L., Phanikumar, M.S., Molloy, S.L., Whitman, R.I., Shively, D.A., Nevers, M.B., Schwab, D.J., Rose, J.B., 2006. Modeling the transport and inactivation of E. coli and enterococci in the near-shore region of Lake Michigan. Environ. Sci. Technol. 40, 5022–5028.

Lukasik, J., Scott, T., Andryshak, D., Farrah, S., 2000. Influence of salts on virus adsorption to microporous filters. Appl. Environ. Microbiol. 66, 2914–2920.

Maillard, P., Pinheiro Santos, N.A., in press. A spatial-statistical approach for modeling the effect of non-point pollution on different water quality parameters in the Velhas river watershed—Brazil. J. Environ. Manage. doi:10.1016/j.jenvman.2006.12.009.

Mayer, C. L. Palmer, C.J., 1996. Evaluation of PCR, nested PCR, and fluorescent antibodies for detection of *Giardia* and *Cryptosporidium* species in wastewater. Appl. Environ. Microbiol. 62, 2081–2085.

McCorquodale, J.A., Georgiou, I., Carnelos, S., Englande, A.J., 2004. Modeling coliforms in storm water plumes. J. Environ. Eng. Sci. 3, 419–431.

Microbial Source Tracking Guide Document, EPA, National Risk Management Research laboratory, Office of Research and development, Cincinnati, OH, EPA/600/R-05/064, June 2005.

Millie, D.F., Schofield, O.M., Vinyard, B.T., 1997. Detection of harmful algal blooms using photopigments and absorption signatures: A case study of the Florida red tide dinoflagellate, *Gymnodinium breve*. Limnol. Oceanogr. 42, 1240–1251.

Montgomery Brion, G., Lingireddy, S., 1999. A neural network approach to identifying non-point sources of microbial contamination. Water Res. 33, 3099–3106.

National Research Council, 2004. Indicators for Waterborne Pathogens. Committee on Indicators for Waterborne Pathogens. Washington, DC, National Academy of Sciences.

National Resources Defense Council (NRDC), Testing the Waters 2006 and 2005 annual reports.

Nevers, M.B., Whitman, R.L., 2005. Nowcast modeling of Escherichia coli concentrations at multiple urban beaches of southern Lake Michigan. Water Res. 39, 5250–5260.

Noble, R.T., Weisberg, S.B., 2005. A review of technologies for rapid detection of bacteria in recreational waters. J. Water Health 3, 381–392.

Noble, R.T., Griffith, J.F., Blackwood, A.D., Fuhrman, J.A., Gregory, J.B., Hernandez, X., Liang, X., Bera, A.A., Schiff, K., 2006. Multitiered approach using quantitative PCR to track sources of fecal pollution affecting Santa Monica Bay, California. Appl. Environ. Microbiol. 72, 1604–1612.

Olszewski, J., Winona, L., Oshima, K., 2005. Comparison of 2 ultrafiltration systems for the concentration of seeded viruses from environmental waters. Can. J. Microbiol. 51, 295–303.

Pachepsky, Y.A., Sadeghi, A.M., Bradford, S.A., Shelton, D.R., Guber, A.K., Dao, T., 2006. Transport and fate of manure-borne pathogens: Modeling perspective. Agric. Water Manage. 86, 81–92.

Palmer, C. J., Lee, M.H., Bonilla, G.F., Javier, B.J., Siwak, E.B., Tsai, Y.L., 1995. Analysis of sewage effluent for human immunodeficiency virus (HIV) using infectivity assay and reverse transcriptase polymerase chain reaction. Can. J. Microbiol. 41, 809–815.

Riou, P., Le Saux, J.C., Dumas, F., Caprais, M.P., Le Guyader, S.F., Pommepuy, M., in press. Microbial impact of small tributaries on water and shellfish quality in shallow coastal areas. Water Res. 41, 2774–2786.

Rochelle, P.A., De Leon, R., Stewart, M.,Wolfe, R., 1997. Comparison of primers and optimization of PCR conditions for detection of Cryptosporidium parvum and Giardia lamblia in water. Appl. Environ. Microbiol. 63, 106–114.

Rolland, D., Block, J.C., 1980. Simultaneous concentration of *Salmonella* and enteroviruses from surface water by using micro-fiber glass filters. Appl. Environ. Microbiol. 39, 659–661.

Scarlatos, P.D., 2001. Computer modeling of fecal coliform contamination of an urban estuarine system. Water Sci. Technol. 44, 9–16.

Scott, T.M., Rose, J.B., Jenkins, T.M., Farrah, S.R., Lukasik, J., 2002. Microbial source tracking: Current methodology and future directions. Appl. Environ. Microbiol. 68, 5796–5803.

Shibata, T., Solo-Gabriele, H.M., Fleming, L., Elmir, S., 2004. Monitoring marine recreational water quality using multiple microbial indicators in an urban tropical environment. Water Res. 38, 3119–3131.

Siewicki, T.C., Pullaro, T., Pan, W., McDaniel, S., Glenn, R., Stewart, J., 2007. Models of total and presumed wildlife sources of fecal coliform bacteria in coastal ponds. J. Environ. Manage. 82, 120–132.

Simpson, J.M., Santo Domingo, J.W., Reasoner D.J., 2002. Microbial source tracking: State of the science. Environ. Sci. Technol. 36, 5279–5288.

Sobsey, M.D., Hickey, A.R., 1985. Effects of humic and fulvic acids on poliovirus concentration from water by microporous filtration. Appl. Environ. Microbiol. 49, 259–264.

Solo-Gabriele, H.M., Wolfert, M.A., Desmarais, T.R. Palmer, C.J. 2000. Sources of *Escherichia coli* in a coastal subtropical environment. Appl. Environ. Microbiol. 66, 230–237.

Thompson, D.E., Rajal, V.B., De Batz, S., Wuertz, S., 2006. Detection of *Salmonella spp.* in water using magnetic capture hybridization combined with PCR or real-time PCR. J. Water Health 4, 67–75.

Tsai, Y.L., Tran, B., Sangermano, L.R., Palmer, C.J., 1994. Detection of poliovirus, hepatitis A virus, and rotavirus from sewage and ocean water by triplex reverse transcriptase PCR. Appl. Environ. Microbiol. 60, 2400–2407.

U.S. Environmental Protection Agency (U.S. EPA), 1986. Ambient Water Quality Criteria for Bacteria—1986. U.S Environmental Protection Agency, Office of Water, Washington, DC, EPA 440/5–84–002.

U.S. Environmental Protection Agency (U.S. EPA), 1999. USEPA's Action Plan for Beaches and Recreational Water, EPA/600/R-98/079.

U.S. Environmental Protection Agency (U.S. EPA), 2006. EPA's BEACH Report: 2005 Swimming Season, EPA 823–F-06–010.

Wade, T.J., Calderon, R.L., Sames, E., Beach, M., Brenner, K.P., Williams, A.H., Dufour, A.P., 2006. Rapidly measured indicators of recreational water quality are predictive of swimming-associated gastrointestinal illness. Environ. Health Perspect. 114, 24–28.

Whitman, R.L., Nevers, M.B., 2003. Foreshore sand as a source of Escherichia coli in nearshore water of Lake Michigan Beach. Appl. Environ. Microbiol. 69(9), 5555–5562.

Wickham, J.D., Nash, M.S., Wade, T.G., Currey, L., 1989. Statewide empirical modeling of bacterial contamination of surface waters. J. Am. Water Resources Assoc. June, 583–591.

STUDY QUESTIONS

1. Name four ways in which pathogenic microorganisms enter coastal water.
2. Which microorganism gained favor in the 1980s as the best indicator organism for use in beach water quality analysis?
3. Name two indigenous pathogenic microorganisms found in coastal water regardless of sewage or other contamination input.
4. Why are better detection methods needed to evaluate beach water samples?
5. List some of the areas of environmental research and management that might utilize the technologies discussed.
6. List some strengths of one of these new technologies that will help it move into the environmental marketplace. List some of the weaknesses or obstacles for moving this technology into the marketplace.
7. What is the impact of new technology in assessing the microbial or sanitary recreational water quality on regulatory agencies including beach managers or operators?
8. Name several public health benefits of improving/updating water quality monitoring technologies.
9. In what year did the Clean Water Act become law?

SECTION II

REMEDIES

A. Pharmaceuticals and Other Natural Products

CHAPTER

21

Marine Remedies

WILLIAM GERWICK

Adaptations to the unique environmental features of a watery world underlie much of the unusual chemical and biochemical adaptations of marine organisms. From the intense interspecies competition for food and space in the sea, to competition for nutrients, coping with a submerged lifestyle in which microbial pathogens have intimate and direct contact with potential hosts, the presence of and access to an abundance of halogen salts are just a few of the factors that contribute to these unique adaptations. Because the oceans are so vast and contain so many species, each with its own distinctive adjustment to the sea environment, it has been articulated that one could search among marine creatures and find just about any adaptation one desired to study (G. Somero, Stanford University, personal communication). Indeed, the marine realm has provided a wealth of model organisms for the study of specific physiological and biochemical systems, and this has given much insight into human health as well as the creatures on which we depend for life.

This chapter was written while the author participated in an expedition to the north coast of New Britain, Papua New Guinea, aimed at collecting new species of marine algae and cyanobacteria for their anticancer potential, and it provides a poignant example of the essential point made in the introduction, which is that marine life is a rich source of fundamentally new adaptive natural product chemicals that have great biomedical potential. One species under continuing investigation by the author's laboratory is what we call "the cobweb Lyngbya," known scientifically as *Lyngbya bouillonii*. This is a fascinating cyanobacterium that forms densely woven veils of red filaments that cover small holes in the coral reef at about a 50–foot water depth. As they are collected, it is common for small shrimp to make aggressive snaps at the fingers of the collector, for these shrimp live beneath the protective veil of *Lyngbya* filaments and, in fact, play a role in "stitching" the cobweb to the reef to provide firm anchorage. What protects both the cyanobacterium and the associated shrimp from the profuse predation so typical of these reefs is a rich assortment of natural products that have properties that are powerfully toxic to potential predators. One of these compounds is the complex lipopeptide known as apratoxin A, a molecule we and others have worked with to advance through early stages of cancer drug development (Luesch *et al.*, 2001). Indeed, the amazing adaptations of marine life to a watery existence in which both predators and pathogens have direct access to the tissues of delicate benthic organisms such as cyanobacteria are yielding an exciting array of molecules for drug and biotechnological development.

For a number of reasons, marine natural products have been examined in only a limited number of specific therapeutic or biotechnological areas. This prominently includes the search and discovery of small molecules with anticancer or anti-infectious disease properties, peptides with antipain properties, and proteins with diverse applications in biotechnological research. In part, these focus areas have arisen because of society's need; for instance, following the call for a war on cancer in 1971 by then U.S. President Richard Nixon, the National Cancer Institute at the National Institutes of Health has been a major source of biomedical research funding for new cancer drug discovery, including from diverse marine life. In Chapter 22, Simmons and Gerwick present an overview of the results of these investigations. A focus by the U.S. Department of Defense as well as the National Institute of Allergy and Infectious Disease (NIAID) on the discovery of countermeasures to bioterrorism agents has funded a resurgence of effort in this area, made especially important given the virtual absence of anti-infectious disease programs in the modern pharmaceutical industry. The search of marine life for antibiotic natural products is the focus of Chapter 23 by Carter.

However, additional reasons why these particular therapeutic areas have received so much attention by marine natural products researchers is a reflection of prevalent activity themes in the chemistry of marine organisms themselves. For example, marine organisms are relatively well known for their diverse toxins and poisons, and this explains why they have been so aggressively studied for compounds that might be toxic to unwanted cell types, such as cancer cells or bacteria. Indeed, it is a fundamental concept in therapeutic drug application that the difference between a drug having a toxic versus a beneficial property is simply a matter of dose. A corollary to this principle is that the difference between the beneficial and toxic dose represents the "therapeutic window," or a measure of a drug's safety.

Other observational data have provided the impetus to study and derive value from the unique protein adaptations of marine life. For example, discovery of deep sea bacteria living in the effluvia of superheated hydrothermal vents led to discovery of their unique DNA polymerase enzymes. Observation of luminescent organs in marine fish led to fundamental studies of its biochemical cause and subsequently to diverse biotechnological applications. A similar observation and development path led to the highly useful fluorescent proteins, such as green fluorescent protein (GFP), as well as the many development candidates as discussed by Wiedenmann in Chapter 24.

The discovery of the therapeutic potential of cone snail toxins has a slightly different history. Because the prey animals of some cone snails are highly motile fish, the slower-moving cone snail must ensure the almost immediate incapacitation of fish following a predatory attack. This insight led to detailed studies of the pharmacology of cone snail toxins, and by serendipitous events (an undergraduate is reported to have injected into mice the individual toxins from a high-performance liquid chromatography [HPLC] chromatogram of crude cone snail toxin and then observed highly distinctive neurological behaviors resulting from each), it was recognized that these toxins recognized specific neurochemical receptor subtypes that were previously unknown. Hence, they immediately became a highly useful set of molecular tools, and more recently a novel class of pain therapeutic. This is richly detailed by Teichert and Olivera in Chapter 25.

A significant shortcoming of the field of marine natural products chemistry and drug development has been the issue of "supply" of material sufficient for drug development. Although it is reasonable to collect small amounts of a relatively rare seaweed or sponge and examine its extract in the laboratory for a novel cancer cell toxin or antimicrobial agent, gaining access to enough of the new compound to allow animal or human trials is, in most cases, extremely difficult. Obviously, it is dissatisfying to all involved to abandon exciting drug leads simply because of a lack of supply of the compound. The supply need can be addressed in a number of ways, no one of which is appropriate or realistic in all cases, and they include (1) harvest of the organism from natural stocks, (2) aquaculture of the source organism in pond or net culture, (3) chemical synthesis, (4) semisynthesis from a natural material obtained economically and in good yield from another source, and (5) biosynthesis using the source organism's enzymes transplanted into an easily grown bacterium, such as *E. coli*. In Chapter 26, Udwary *et al.* discuss the fundamental underpinning knowledge of the biosynthetic pathways, enzymes, and genes to enable this latter approach. Capture and harness of a biosynthetic pathway to produce compounds of biomedical utility can be accomplished in a variety of ways, and these future technologies are examined in some intriguing and insightful depth in Chapter 26. Indeed, genomic information and technologies are being applied with great power to the search for new therapeutics among marine microorganisms.

How do researchers go about the process of discovering and isolating new anticancer- or anti-infective types of natural products from marine organisms? In general, researchers in this area focus on a class or a few classes of marine organisms so as to be able to develop deeper and more profound insight and understanding of the organisms under study, including their taxonomy, chemistry, biochemistry, physiology, and ecology. For example, several marine natural products scientists focus on marine bacteria, or sponges, or algae, or corals. Fieldwork to make high-quality collections occurs worldwide, but it requires considerable effort to gain permission from the host country to make the collections and to have in place a proper benefit-sharing document. The Natural Products Branch of the National Cancer Institute under the leadership of Gordon Cragg has provided worldwide leadership in developing fair and equitable templates for governance of these multinational investigations, and these were developed in accordance with the 1992 Convention on Biological Diversity.

At the current time, most researchers make collections of the organism either frozen or pickled in alcohol solvent as well as prepare cultures of the associated microorganisms for subsequent culture. Additionally, voucher samples are prepared at the time of collection, as well as field notes and photographic records. Subsequently in the laboratory, extracts are produced by a variety of techniques, and generally use organic solvents rather than aqueous extraction procedures. Most investigators process the crude extracts in some fashion so as to produce a series of derivative fractions, each of which is not a pure substance but rather a reduced complexity mixture. These have advantages over crude extracts in biological screening programs because of their reduced complexity, the increase in the relative concentration of minor constituents in the derivative fractions, the segregation of nuisance compounds into discreet fractions, and the ability to utilize high-resolution separation methods immediately upon obtaining activity in a particular fraction. Bioassays are diverse and depend on an investigator's collaborations and interests. Some employ a strategy of screening to isolated proteins in assay wells of 96- or 384-well plates. This so-called mechanism-based screening is attractive in its high level of focus on targets of importance to a particular disease condition. However, as an approach, mechanism-based screening suffers in that many potential mechanisms by which to treat the target disease are not evaluated at all. Other approaches use cells or complex systems, such as zebra fish embryos, with molecular readouts that indicate that an extract, fraction, or compound has impacted a general feature or pathway of interest to a given disease state. These "mechanism-rich targets" are proving highly effective in screening diverse biomaterials; however, they do require subsequent dissection of the pathway in order to precisely understand how the agent has impacted the cellular pathway under study.

Active materials are subject to finer and finer levels of separation, most usually employing HPLC as a final step. In a process known as "bioassay-directed fractionation," after each chromatographic step the derivative fractions are re-valuated in the relevant bioassay, and the results of these assays are then used to direct which material is chosen for further chromatography. The desired goal of this process, indeed the ultimate "reductionist" aspiration, is the isolation of a single compound of high biological potency in the assay of interest. At this point, the investigation turns to spectroscopic techniques, such as nuclear magnetic resonance spectroscopy, mass spectrometry, infrared and ultraviolet spectroscopy to piece various parts of the molecule together and formulate a working structure. The logic used to develop such a molecular structure from these various spectroscopic methods is much like that used to solve a Sudoku puzzle in that one deduces small features from reiteratively considering the clues given by each method. This is subject to additional spectroscopic analysis as well as probing with chemical reagents to gain additional proof of the structural hypothesis. Ultimate proof of structure comes from increasingly focused spectroscopic studies, X-ray diffraction analysis, or chemical synthesis of the candidate structure. Determination of the complete three-dimensional chemical structure of a new and bioactive compound is a tremendous accomplishment indeed. However, in many respects, this is really just the starting place for many additional studies, such as the pharmacological mechanism by which the agent works, what role it plays in nature (= chemical ecology), or how the producing organism makes the compound from simple and ubiquitous precursors (= biosynthesis).

However, the principal area of continued investigation upon discovery and structural description of a new and bioactive natural product focuses on development of its potential utility to treat human disease. In this regard, a given bioactive lead molecule is progressively advanced through more and more rigorous and advanced models of the disease condition under study. This usually involves animal testing to evaluate efficacy; for example, in the discovery of new anticancer agents, human cancers are implanted into special mice that do not reject these foreign tissues and give initial insight as to whether the experimental agent is stable and effective *in vivo.* A large number of compounds must be studied in these initial levels of evaluation, retrospectively estimated at 10,000 to 15,000 in cancer drug discovery, in order for even a dozen compounds to be successfully advanced into early stage trials in humans (phase I trials). In these initial human studies, the goal is not so much efficacy but rather to learn how and at what level the new agent should be applied. The scaling factors for predicting dose in humans from experiments in mice, rats, or even dogs are imprecise at best. The next stage of human evaluation, phase II trials, are larger in scope and have the intent of evaluating the clinical efficacy of the new treatment, often in comparison with established therapies for that disease. Drugs showing signs of benefit to patients in phase II trial are advanced to phase III trials, which involve many patients at many different hospitals and represent the true evaluation of whether the new agent will be advanced into general clinical utility. On average, of the 15,000 substances initially evaluated, only a single new agent enters the marketplace as drug approved by the Food and Drug Administration. Indeed, the odds against the successful discovery and development of a new anticancer agent are daunting, and this result provides an important rationale for why society should utilize a variety of very different or orthogonal approaches to new cancer drug discovery, including structure-based design, synthetic medicinal chemistry, combinatorial chemistry, and natural products chemistry.

Marine life forms are an exciting and productive group of organisms from which to prospect for new pharmaceuticals. Pioneering studies in the 1960s and 1970s quickly established that diverse marine life forms were rich in natural products and that many of these had potentially useful biological properties. No one knows with certainty the number of species in the sea, and much of this uncertainty is due to the great diversity of microorganisms that are still largely unknown; however, estimates range up to several million. And while the sea may be vast, its coastal fringe is in fact very limited, thus causing a crowded accumulation of species that compete for surface area, light, and nutrients. Hence, sessile marine life forms rely heavily on physical and chemical defenses in their fight for survival and reproduction. Defensive chemicals of this type are secondary metabolites in that they are not involved in the primary metabolic life processes (physical cell walls, energy metabolism, reproduction), but rather, have an adaptive role that enhances their competitiveness. The secondary metabolites of marine organisms are distinctively different from those of terrestrial organisms due to their abundant utilization of chlorine and bromine in covalent linkage with organic molecules. This may be due in part to their ready access to these elements as seawater possesses some 19,000 mg/L of chlorine and 67 mg/L of bromine. A variety of unique biochemical systems have evolved in marine life by which to incorporate halogen atoms, including haloperoxidases, halogenases, and more recently, radical halogenases that utilize extremely high energy biochemical catalysts (Vaillancourt *et al.*, 2006).

As a result of the strength of the evolutionary pressures, the length of time that some of these organisms have had to develop their secondary metabolite pathways (e.g., cyanobacteria arose some 3.5 billion years ago), and the diversity of species present in the oceans, the marine realm is an extraordinarily rich source of novel natural product structures, currently tallied at more than 18,000 distinct molecular entities (MarinLit Database). Moreover, these molecules are made by biological systems as mediators of biological interactions; they are inherently biologically relevant, and hence, a great starting place for the discovery of new pharmaceutical lead compounds. This contrasts sharply with libraries of synthetic molecules that have little inherent relevance to biology, unless their structures are patterned after some feature of a natural product.

Are there specific groups of organisms that are richer than others in their production of biologically active secondary metabolites? Are their regions in the world that are more productive from a drug discovery and bioprospecting perspective? Historically, it has been the perception that sponges and seaweeds are the richest sources of new chemistry, and those from the tropics are especially plentiful. As relatively large, sessile creatures with considerable occurrence in nature, sponges and algae are especially vulnerable to predation (as well as collection by marine natural products chemists!). In tropical reef systems wherein there are large numbers of species co-existing and competing for resources, algae and sponges are particularly rich in secondary metabolites. However, even marine life from the cold waters of Antarctica have yielded exciting drug-like molecules, such as the anticancer lead compound known as palmerolide A from the tunicate *Synoicum adareanum* (Diyabalanage *et al.*, 2006).

However, our understanding of the sources of these exciting drug-like molecules has been undergoing a revolution in the last few years. It is now recognized that a majority of the most potent and novel natural products from invertebrates, such as sponges and tunicates, are actually the metabolic products of bacteria that live in association with these creatures. Similarly, an increasing number of natural products are being reported from the fungi that live in association with seaweeds, a direct parallel to the recent recognition that many higher plant natural products, such as the anticancer agents camptothecin and podophyllotoxin, are actually produced by endophytic fungi (Eyberger *et al.*, 2006; Puri *et al.*, 2005). As discussed in greater detail in Chapter 22, one of the more exciting frontiers in marine drug discovery is recognition of the role of microorganisms in the production of bioactive natural products. Ultimately, that microorganisms are responsible for the production of so many useful compounds is fortunate for this implies that familiar technologies of fermentation can be used to produce these substances in mass. Moreover, bacterial systems are more amenable to genetic manipulation wherein the capacity to produce these exotic molecules can be harnessed and manipulated to makes new compounds of utility (see Chapter 26).

The creatures of the oceans have given us a bounty of unique genetic adaptations to their environment, and efforts to explore and examine these features for useful biotechnological and biomedical applications are fully under way. However, it is important to note that these are truly initial efforts as it has only been since the 1980s or so that our technological abilities were up to the challenges that the oceans present. The structural complexity of the natural products produced by marine life is extraordinary, and many times they are made in only small quantities, presumably because of their ultrahigh potency as adaptive chemicals. Bioassays since the 1980s have also reached a level of sophistication and disease relevance such that truly useful molecules are being isolated and examined in great detail. Now, some 15 to 20 years following the initial discovery of a number of these, we are starting to see the fruits of these efforts in the form of new pharmaceutical and biotechnological products reaching the marketplace. However, in this respect only the lowest hanging fruit has so far been examined in sufficient depth to recognize their applications. The coming era is likely to be highly productive as we use greater sophistication in the analytical chemistry area, and this is matched by robust and disease-relevant biological assays. Indeed, natural products in general and marine natural products specifically have been neglected as a source of useful lead substances in many disease areas, including inflammation, allergy, diabetes, obesity, and the neurosciences. Future ocean scientists will hopefully be emboldened by the initial successes in cancer, infectious disease, and pain, as well as marine proteins useful in biotechnology, and will examine these other therapeutic areas in thoughtful, creative, and sophisticated ways. As the chapters that follow highlight, there are many great opportunities remaining or emerging for students of oceans and human health who wish to pursue the remedy side of the equation.

References

Diyabalanage, T., Amsler, C.D., McClintock, J.B., Baker, B.J., 2006. Palmerolide A, a cytotoxic macrolide from the Antarctic tunicate *Synoicum adareanum*. J. Am. Chem. Soc. 128, 5630–5631.

Eyberger, A.L., Dondapati, R., Porter, J.R., 2006. Endophyte fungal isolates from *Podophyllum peltatum* produce podophyllotoxin. J. Nat. Prod. 69, 1121–1124.

Flatt, P.M., Gautschi, J.T., Thacker, R.W., Crews, P., Gerwick, W.H., 2005. Identification of the cellular site of polychlorinated peptide biosynthesis in the sponge *Dysidea* (*Lamellodysidea*) *herbacea* and symbiotic cyanobacterium *Oscillatoria spongeliae* by CARD-FISH analysis. Mar. Biol. 147, 761–774.

Luesch, H., Yoshida, W.Y., Moore, R.E., Paul, V.J., Corbett, T.H., 2001. Total structure determination of apratoxin A, a potent novel cytotoxin from the marine cyanobacterium *Lyngbya majuscula*. J. Am. Chem. Soc. 123, 5418–5423.

Puri, S.C., Verma, V., Amna, T., Qazi, G.N., Spiteller, M., 2005. An Endophytic Fungus from *Nothapodytes foetida* that Produces Camptothecin. J. Nat. Prod. 68, 1717–1719.

Vaillancourt, F.H., Yeh, E., Vosburg, D.A., Garneau-Tsodikova, S., Walsh, C.T., 2006. Nature's inventory of halogenation catalysts: oxidative strategies predominate. Chem. Rev. 106, 3364–3378.

C H A P T E R

22

Anticancer Drugs of Marine Origin

T. LUKE SIMMONS AND WILLIAM H. GERWICK

INTRODUCTION

The notion that exotic marine organisms contain secondary metabolites that can be therapeutically useful inherently captures our interest and imagination. That a toxin produced by a marine invertebrate or microbe for its own chemical defense could also be useful in fighting human disease is remarkable. This process, the systematic evaluation of natural products from diverse life forms to discover new drug leads, is, in fact, how humanity has discovered many of the drugs currently in use worldwide. The origins of our effective anticancer drugs have been analyzed in some detail; 65% of the 175 agents used between 1940 and June 2006 have, in some sense, come from nature (Newman and Cragg, 2007). It is not necessarily that a natural product has been extracted from a plant, bacterium, or marine organism and then used directly as a drug (although about 25 have); rather, compounds with potentially useful, but not perfect, anticancer properties have been obtained from natural sources, and these have become the chemical idea around which synthetic analogs have been generated to create an effective pharmaceutical. In this latter variety, nearly 90 agents (51%) used to treat cancer are natural product derivatives or synthetic drugs that are patterned after features of the natural product. Of the agents not derived from a small molecule natural product origin, 11% are biologics (e.g., proteins) or vaccines, with only 24% being of a completely independent synthetic origin. As of the writing of this chapter, the field of marine natural products is poised to make a major contribution to our arsenal of anticancer agents with 20 such substances in (or recently in) various phases of clinical trial (Table 22-1).

Cancer is not a single disease but a family of perhaps 200 diseases with diverse underlying biochemical causes. For a cancer to develop, several genetic mutations are required to escape the body's normal and redundant control mechanisms that govern cellular proliferation. Generally, these mutations are found in genes encoding for proteins that normally stimulate a cell's growth or division (= oncogenes); however, gene mutations that contribute to cancer are also found in genes encoding proteins that inform cells to stop growing and dividing or to undergo normal cell death (= tumor suppressor genes) (Fesik, 2005; Mitelman *et al.*, 2007). Despite the heterogeneity of cancer causes, there are some shared features between different cancers. For example, all cancers involve an abnormal proliferation of cells that does not respond to the body's normal "stop dividing" signals, a loss of normal cellular morphology and hence biochemical utility as a particular cell type, and an ability to cross membrane barriers within the body to invade adjacent tissues. These features lead to a growth that is essentially parasitic upon the host, and through a cancer's increased use of the body's resources as well as production of chemical signals that weaken general health they can render patients subject to serious infection by microorganisms. As a consequence, many cancer patients ultimately succumb to an opportunistic infection (approximately 30% to 40% of cancer deaths) (Klastersky and Aoun, 2004). The other key characteristic of cancer that leads to patient mortality is the ability of tumor cells to cross basement membranes, structures that normally constrain and define tissues. This ability of cancer cells to cross these barriers allows their invasion into adjacent tissues, as well as entry to the body's circulatory system. In a process known as metastasis, cancer cells can travel to distant portions of the body where they then propagate new tumors. Metastatic tumors in the brain, lung, and other critical tissues account for an appreciable percentage of cancer deaths (about 30%).

Current anticancer drug therapy is largely based on the strategy of outright killing cancer cells, known as cytotoxic

TABLE 22-1. Relevant marine natural products and their current clinical status.

Compound Name	Source Organism and Source of Material for Clinical Trial	Molecular Target	Current Status
Ara-C (Cytarabine; **18**)	*Cryptotethya crypta* (sponge)	Nucleotide mimic	Clinically available; Phase I/II
Ecteinascidin 743 (Yondelis; **14**)	*Ecteinascidia turbinata* (tunicate)	Tubulin	Phase III (registered)
Ǽ-941 (Neovastat)	Shark cartilage	VEGF	Phase II/III (appears withdrawn March 2007)
Bryostatin 1 (**29**)	*Bugula neritina* (bryozoan)	PKC	Phase I/II
Cemadotin (LU103793; Dolastatin 15 derivative)	*Dolabella auricularia/Symploca* sp. (mollusk/cyanobacterium; synthetic analog)	Tubulin	Phase I/II (Discontinued 2004)
Synthadotin (ILX651, Dolastatin 15 derivative)	*Dolabella auricularia* (synthetic analog)	Tubulin	Phase II
Kahalalide F (**2**)	*Elysia rufescens/Bryopsis* sp. (mollusk/green alga, synthetic)	Lysosomes/erbB pathway	Phase II
Squalamine (**32**) (plus trodusquemine, a simple derivative of squalamine)	*Squalus acanthias* (shark)	Phosopholipid bilayer	Phase II
Dehydrodidemnin B (Aplidine; **13**)	*Trididemnum solidum* (tunicate, synthetic)	Ornithine decarboxylase	Phase II
Soblidotin (TZT-1027, Dolastatin 10 derivative; **4**)	*Dolabella auricularia* (synthetic analog)	Tubulin	Phase I/II
E7389 (Halichondrin B derivative; **22**) (plus eribulin mesylate, a simple derivative)	*Halichondria okadai* (sponge, synthetic)	Tubulin	Phase III
NVP-LAQ824 (Psammaplin derivative; **27**)	*Psammaplysilla* sp. (sponge, synthetic)	HDAC/DNMT	Phase I
Discodermolide (**20**)	*Discodermia dissoluta* (sponge)	Tubulin	Phase I (Discontinued 2005)
HTI-286 (taltobulin, hemiasterlin derivative; **24**)	*Cymbastella* sp. (synthetic analog)	Tubulin	Phase I (Discontinued 2005)
LAF-389 (Bengamide B derivative)	*Jaspis digonoxea* (sponge, synthetic)	Methionine aminopeptidase	Phase I (Discontinued 2002)
KRN-7000 (Agelasphin derivative; **28**)	*Agelas mauritianus* (sponge, synthetic)	Vα24 + NKT cell activation	Phase I
E7974 (hemiasterlin derivative; **25**)	*Hemiastrella minor* (semisynthetic analog)	Tubulin	Phase I
Zalypsis (Jorumycin derivative; **15**)	*Jorunna funebris* (mollusk; total synthesis)	DNA-binder	Phase I
Salinosporamide A (**1**)	*Salinospora* sp. (marine bacterium)	20S proteasome	Phase I
NPI-2358 (Halimide derivative)	*Aspergillus* sp. (fungus; semi-synthetic)	Tubulin	Phase I

drug therapy. Because cancer cells and normal cells share more points of commonality than points of difference, cytotoxic drug therapy typically has many side effects, including nausea, appetite loss, diarrhea, hair loss, inability to defend against invading pathogenic microorganisms, and decreased production of various classes of blood cells (e.g., thrombocytes and leukocytes). The specific mechanisms of cytotoxic drug therapy vary; however, many involve disruption at some level in the functioning of DNA. For example, the antimetabolites mimic intermediates in DNA subunit biosynthesis, thus either disrupting the crucial balance in their metabolic pool sizes or being incorporated into DNA with a resultant malfunction during replication or transcription/translation into RNA/proteins. Another class of anticancer drug is known as the alkylating agents, and they function in large part by reacting selectively with basic sites in the purine bases of intact DNA to then lead to its malfunctioning either by strand cleavage or misreading of the genetic code.

Another series of agents bind to DNA by virtue of their flat planar structure (the "intercalators" slip in between the stacked nucleotide bases in DNA) and there interfere with various enzymes, such as topoisomerases, which are critical for unwinding DNA so it can be transcribed into mRNA and hence translated into protein. The antimitotics are another class of cytotoxic agent that works downstream of these molecular events; this class includes such well-known drugs as Taxol and the Vinca alkaloids. These kill cancer cells by interfering with the proteins charged with coordinating the ordered separation of chromosomal DNA during mitosis. In the past few years, discovery efforts have centered on non-cytotoxic drug therapy approaches and build on the identification of specific biochemical or molecular targets that allow cancer cells to escape controls on proliferation. An example of a drug working on a non-DNA target but instead on a protein that distinguishes cancer cells from normal cells is Gleevec, an effective treatment against chronic myelogenous leukemia, which specifically inhibits the aberrant Abelson tyrosine kinase that underlies this disease.

Is there a strong rationale for why we should prospect among marine organisms as sources of new anticancer agents? Yes, there are compelling arguments for why marine organisms should be examined in a thoughtful, sophisticated, and comprehensive manner for new classes of therapeutics, including those effective in the treatment of cancer. First, marine life forms have been little studied for their unique natural products, with the earliest pioneering studies dating just back to the 1960s and 1970s. Some marine environments remain completely uncharacterized in these regards, and from some habitats even the fundamental species composition is fragmentary (e.g., the deep sea). Much remains to be discovered! From the species studied to date, it is clear that marine organisms have been subject to unique adaptive pressures and utilize rather different strategies for producing secondary metabolites compared to their terrestrial counterparts. In some cases, seasoned organic chemists look at the structures of metabolites produced by marine life and characterize them as bizarre, unlike anything found from the land environment. Alternatively, some marine metabolites are of exceptional complexity, representing true milestones of human achievement in the characterization of their convoluted, multicyclic, and three-dimensional structures, such as maitotoxin (Nonomura *et al.*, 1996; Sasaki *et al.*, 1996). Coupled to the uniqueness of their physical structure are their biological properties, which can be exquisitely potent against some cellular targets. Indeed, some of the most potent natural toxins on the planet derive from marine life (once again, maitotoxin is an extreme example). Perhaps even more important than potency is the fact that some of these marine metabolites exert their pharmacological activities through interaction at novel drug sites, such as enzymes or receptors not targeted by any current pharmaceutical agent. Hence, the real possibility exists that entirely new drug classes will be discovered that have novel structures and new sites of action, and this is very exciting indeed.

In this chapter we review a majority of the marine natural products and their derivatives that are in (or were recently in) stage I, II, or III of clinical trial in human cancer patients (Table 22-1), or in a few cases, such compounds in late stage preclinical evaluation (Table 22-2). At first glance, the original biological sources of these agents appear dispersed among microorganisms, especially the eubacteria, and macroorganisms, in particular the sponges and ascidians (Fig. 22-1A). However, it is becoming increasingly apparent that many of the organic molecules ascribed to "sponge" or "ascidian" metabolism are actually produced by the metabolic activities of symbiotic bacteria that live in association with these sessile invertebrates. Although such speculations have been abundant in the literature for many years, largely based on structural relationships between the compounds isolated from sponges and those isolable from free-living bacteria, especially the cyanobacteria, it has been remarkably difficult to obtain experimental proof of this phenomenon. In part, the difficulty has resulted from the near absolute failure to culture the microorganisms found in symbiosis with invertebrates separately from their hosts, and thus the chemical and biochemical relationships between hosts and symbionts remain vague and uncertain. Some partial success has been obtained through the isolation of bacterial and eukaryotic host cells by cell separation techniques followed by chemical profiling of the resultant cell types. This approach, however, suffers the criticism that compounds could be excreted from one cell type and absorbed by another, resulting in misleading or conflicting outcomes. We used a powerful genetic basis to unequivocally demonstrate that a cyanobacterial symbiont, *Oscillatoria spongelae*, is the site of biosynthesis of a series of unique chlorinated peptides that had previously been isolated from the host sponge *Dysidea herbaceae* (Flatt *et al.*, 2005). This technique, known as CARD-FISH analysis, involved developing gene probes that were complementary to the genes encoding the unique halogenase involved in chlorinated peptide biosynthesis. These gene probes were then labeled with fluorescent signatures that allowed microscopic visualization of their location in thin sections of the sponge/cyanobacterial tissue. The gene probes only bound to the cyanobacterial cells, thereby demonstrating that these cells possessed the messenger RNA encoding the unique halogenase enzyme. If one makes reasonable speculations based on distinctive chemical motifs in sponge and ascidian natural products and their relationship to microbial metabolites, then a majority of the marine anticancer agents in clinical trial derive from marine microorganisms (Fig. 22-1B). From work with new early stage anticancer leads not yet in clinical trial, this trend is continuing, and it can be expected that there will be general recognition that the amazing chemical resource in

TABLE 22-2. A selection of marine natural products showing promise in preclinical anticancer studies.

Compound	Source Organism	Molecular Target
Curacin A (**6**)	*Lyngbya majuscula* (cyanobacterium)	Tubulin
Desmethoxymajusculamide C (DMMC)	*Lyngbya majuscula* (cyanobacterium)	Tubulin
Laulimalide	*Cacospongia mycofijiensis* (sponge)	Tubulin
Iejimalide A (**5**)	*Eudistoma rigida/Lyngbya* sp. (tunicate/cyanobacterium)	Vo-ATPase
Vitilevuamide	*Didemnin cucliferum/Polysyncration lithostrotum* (tunicates)	Tubulin
Diazonamide	*Diazona angulata* (tunicate)	Tubulin
Eleutherobin	*Eleutherobia* sp./*Erythropodium caribaeorum* (soft corals)	Tubulin
Sarcodictyin	*Sarcodictyon roseum* (sponge)	Tubulin
Peloruside A	*Mycale hentscheli* (sponge)	Tubulin
Salicylihalimides A and B	*Haliclona* sp. (sponge)	Vo-ATPase
Thiocoraline	*Micromonospora marina* (bacterium)	DNA-polymerase
Ascididemin	*Didemnum* sp. (sponge)	Caspase-2/mitochondria
Variolins	*Kirkpatrickia variolosa* (sponge)	Cdk
Sarcophytol A (**29**)	*Sarcophyton glaucum* (soft coral)	Inhibition of oxidative stress and DNA strand breaks
Lamellarin D	*Lamellaria* sp. (mollusk and various soft corals)	Topoisomerase I/mitochondria
Dictyodendrins	*Dictyodendrilla verongiformis* (sponge)	Telomerase
ES-285 (Spisulosine; **31**)	*Mactromeris polynyma* (mollusk)	Rho (GTP-bp)
Avrainvillamide (**7**)	*Aspergillus* sp. (fungus)	LN-Cap
Thyrsiferyl 23-acetate (**11**)	*Laurencia thyrsifera* (marine alga)	PP2A
Amphidinolide N (**9**)	*Amphidinium* sp. (dinoflagellate)	Unknown

the oceans is largely the result of their rich diversity of microorganisms (Fig. 22-1B).

The agents presented in this chapter are arranged along taxonomic lines considering the originally collected source, be it microbial, invertebrate, or vertebrate. In consideration of space limitations and when faced with multiple relevant examples from each taxonomic source, we have elected to present the agent that has advanced to the furthest degree in clinical trial or the agent that illustrates a particularly intriguing aspect of marine anticancer drug discovery. The chapter begins with examples in which it is clear that the source organism is a microorganism (i.e., cultured bacterial species or field collected cyanobacterium). These examples are followed by a discussion of anticancer agents in clinical trial or late stage preclinical evaluation, which were isolated from various classes of invertebrates. The chapter concludes with an analysis of an anticancer compound that derives from the primary tissues of a marine vertebrate species. For each, we have briefly placed the discovery and development of the new agent in its appropriate biological context as well as given a sense of the unexpected and often fruitful events that occurred during the discovery and development of the agents. We have not discussed the often monumental tasks of structure elucidation, total organic synthesis, or the process of determining the molecular pharmacological mechanism of action of new compounds. We have, however, given references to the primary literature; the interested reader should consult these papers for greater detail. Finally, many excellent reviews on the subject of anticancer drug discovery from marine organisms exist, and several of these were utilized in the construction of this chapter (Cragg *et al.*, 2006; Mayer and Gustafson, 2006; Newman and Cragg, 2006; Simmons *et al.*, 2005).

ANTICANCER AGENTS FROM MARINE MICROORGANISMS

Heterotrophic Bacteria

Salinosporamide A (1)

Advances in the cultivation of obligate marine actinomycete bacteria (Fig. 22-2A) have yielded some exciting new

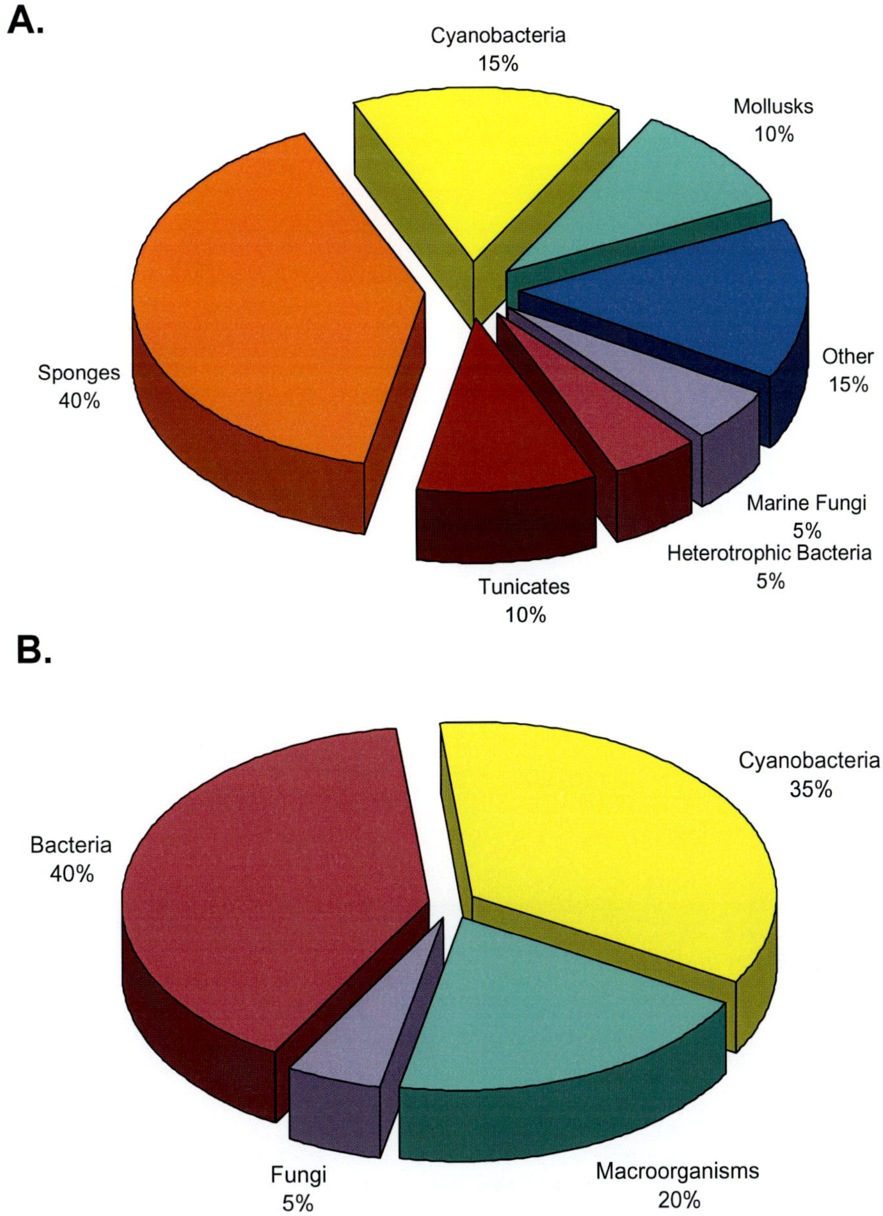

FIGURE 22-1. (A) Pie chart of the reported sources for 20 marine-derived anticancer agents in clinical trial or recently in clinical trial (data from Table 22-1). (B) Pie chart of the predicted metabolic sources of the 20 marine-derived anticancer agents in clinical trial or recently in clinical trial.

small molecules with antibacterial and anticancer properties (Fenical and Jensen, 2006; Mincer *et al.*, 2002). In 2003, researchers from the Scripps Institution of Oceanography published their discovery of salinosporamide A (**1**), a fused γ-lactam-β-lactone bicyclic compound, from a newly discovered marine actinomycete *Salinispora tropica* (Feling *et al.*, 2003) (Fig. 22-3). Bioassay-guided fractionation of the fermentation broth produced a colorless crystalline solid with a 7-mg/L yield. The absolute stereostructure of **1** was solved by extensive NMR analysis and completed by a single-crystal X-ray diffraction study. Evaluation of the biological activity of pure salinosporamide A against the HCT-116 human colon carcinoma cell line indicated it was exquisitely cytotoxic with an IC_{50} = 11 ng/mL. Subsequent mechanism-based and co-crystallization studies have shown that salinosporamide A irreversibly binds within the yeast 20 S proteasome core, an enzyme complex that is responsible for normal protein degradation. The reaction mechanism involves β-lactone hydrolysis with concomitant ester bond formation with the proteasome active site threonine residue (Groll *et al.*, 2006). This reaction is essentially irreversible as the newly produced hydroxyl group within the drug dis-

FIGURE 22-2. (A) Photograph of new marine actinomycete bacterium *Salinospora tropica*, source of the anticancer agent salinosporamide (a) (**1**). (B) Photomicrograph of filaments of the marine cyanobacterium *Lyngbya majuscula*, a source of many anticancer lead molecules such as curacin A (**6**) and DMMC (Table 22-2). (C) Underwater photograph of the sponge *Psammaplysina*, a source of the anticancer lead compound psammaplin (**26**). (D) Underwater photograph of the nurse shark *Ginglymostoma cirratum*, a source of the anticancer lead squalamine (**33**). Photo in part (C) is by P. Crews.

places chloride to form a tetrahydrofuran ring, a group that excludes water from the active site, thereby protecting the ester bond between drug and enzyme from hydrolysis. This is an exceptionally clear case wherein the presence of a halogen atom in a drug increases its biological potency (other halogenated natural products of therapeutic interest include vancomycin, rebeccamycin, chlortetracycline, and chloramphenicol). Salinosporamide A is currently undergoing phase I clinical trial (Nereus Pharmaceuticals) in relapsed or refractory multiple myeloma patients (Chauhan *et al.*, 2006).

Kahalalide F (2; PM02734)

Kahalalide F (**2**) is one of many important compounds discovered in the laboratory of the late Professor Paul Scheuer during his prolific 50–year career at the University of Hawaii. Isolated in 1993 from tissue extracts of the herbivorous marine mollusks *Elysia rufescens* and *E. degeneri*, as well as from the green alga on which they feed (*Bryopsis* sp.), kahalalide F (**2**) is a cyclic depsipeptide containing a reactive dehydroaminobutyric acid residue and, consequently, displays potent biological activity (Hamann and

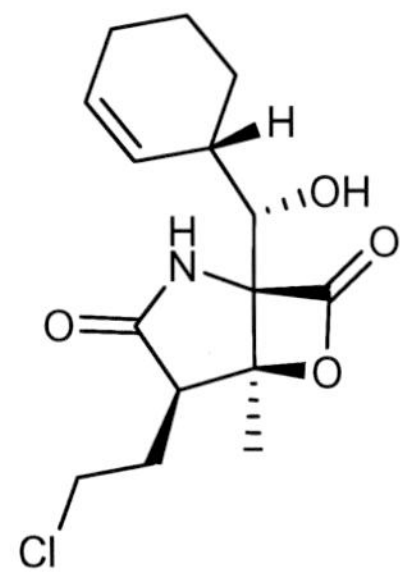

Salinosporamide A (**1**)

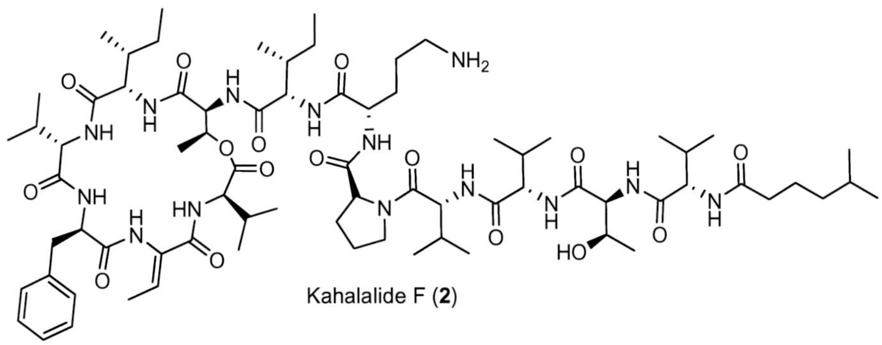

Kahalalide F (**2**)

FIGURE 22-3. Structures of anticancer leads derived from marine heterotrophic bacteria.

Scheuer, 1993) (Fig. 22-3). It is well understood that many sacoglossan mollusks have the ability to acquire "defensive" molecules from their diets and then sequester these into peripheral tissues as chemical deterrents to their own predation. Hence, compounds of biological interest are often found in both herbivores and their respective diets. In this case, however, the situation appears even more complex as a report identifies a *Vibrio* bacterium living on the surface of the alga and within the tissues of the nudibranch as the ultimate source of kahalalide F (Hill *et al.*, 2005). Kahalalide F (**2**) was shown to be the largest and most potent of the diverse depsipeptides isolated from either of the two mollusks or the alga. The biological activity of **2** has stimulated considerable excitement with its selectivity against solid tumor cell lines and IC_{50} values against A-549, HT-29, and LOVO cell lines below the μg/mL range (Hamann and Scheuer, 1993). Driven in part by these *in vitro* results, kahalalide F entered phase I clinical trails (PharmaMar, a Spanish pharmaceutical company) in 2005 for patients with advanced androgen refractory prostate cancer. The results of this study indicated a maximum tolerated dose of 930 μg/m^2/day with the dose limiting toxicity being reversible (Rademaker-Lakhai *et al.*, 2005). Although its mechanism of action is not well understood, kahalalide F has been shown to induce cancer cell death via a necrosis-like process (Janmaat *et al.*, 2005). The National Cancer Institute's COMPARE analysis places **2** within a group of agents that interact with the Erb/Her-neu pathway and that thus selectively down-regulates ErbB3 expression (Jimeno *et al.*, 2006). Currently, kahalalide F (**2**) is in phase II clinical trials for liver cancer, melanoma, and nonsmall cell lung cancer (Jimeno *et al.*, 2006).

Cyanobacteria

Dolastatin 10 (3) and TZT-1027 (4)

Sea slugs are slow-moving marine gastropods (see Chapter 28) that lack obvious means of defense. Nevertheless, they move alone on the seafloor, in an unconcerned manner, secure in their knowledge that no predator will find them tasty! Indeed, extracts or secretions of the skin and organs of sea hares are highly toxic and have played a nefarious role in ancient history. Legend holds that Agrippina, the mother of Nero, used secretions from a sea hare to kill her son's opponents in his quest to become emperor of Rome some 2000 years ago. In less dissolute contexts, the extracts of sea hares can be toxic to cancer cells and have yielded some important drug lead compounds. In 1972, the Pettit group in Arizona collected several thousand *Dolabella auricularia* sea hares from the Indian Ocean; biological evaluation at the National Cancer Institute showed their organic extract to increase life span in the P388 lymphocytic leukemia mouse model by 100%. The molecular basis for this anticancer activity was not characterized until 1987; after 15 years of intense effort, a group of potent toxins known as the dolastatins were isolated and their structures determined (Pettit *et al.*, 1987) (Fig. 22-4). These are present in the sea hare in infinitesimal quantities, which made their characterization extremely challenging. Dolastatin 10 (**3**) and its analogs have generated much excitement because of their potent *in vivo* anticancer properties. Dolastatin 10 has been shown to act via disruption of cancer cell microtubule networks, thus disturbing the normal cell division process (i.e., is antimitotic). Although this natural product progressed into phase II clinical trials, it was ultimately dropped, as a single agent, because of undesired peripheral toxicity. Chemical modification efforts to reduce toxicity resulted in the synthesis of TZT-1027 (**4**, Auristatin; Soblidotin), which recently completed a phase II clinical trial in patients with advanced or metastatic soft-tissue sarcomas. The authors of this latter study indicated that "TZT-1027 was found to be safe and well tolerated" (Patel *et al.*, 2006, p. 2881).

Some have questioned why *D. auricularia* possesses such small quantities of these intricate and highly active

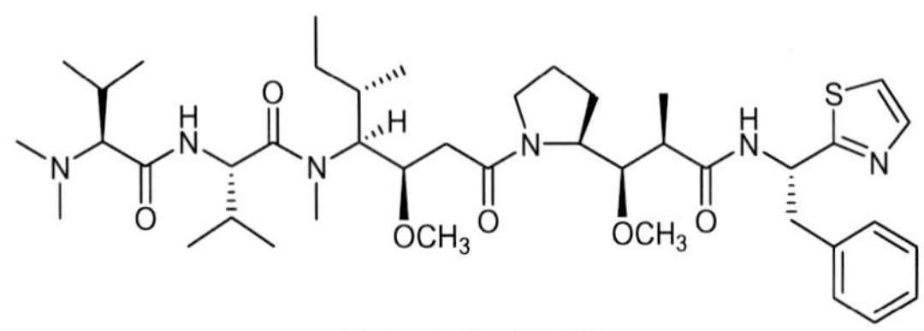

Dolastatin 10 (**3**)

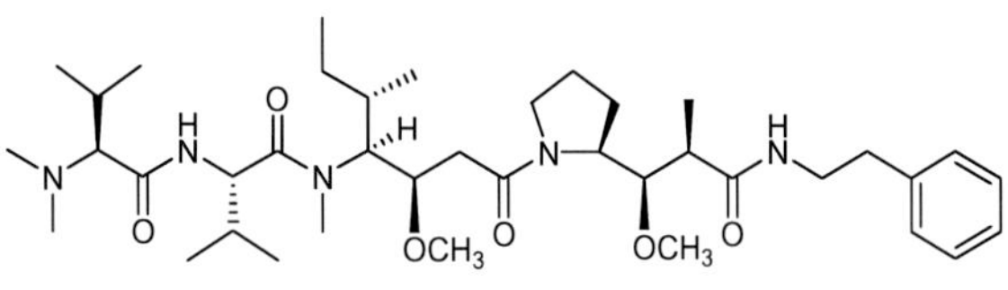

TZT-1027 (**4**)

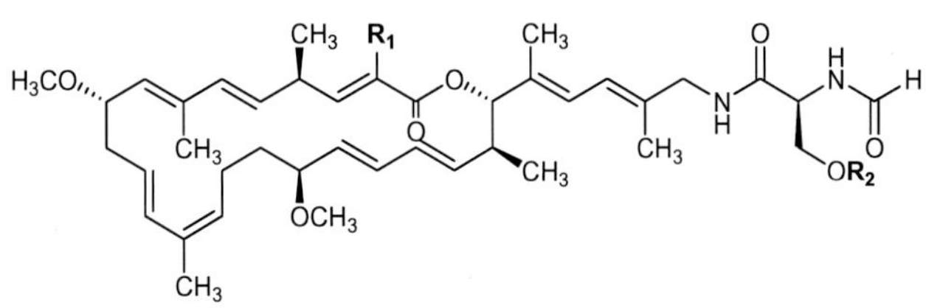

Iejimalide A (**5**) : $R_1 = H$; $R_2 = H$
Iejimalide B : $R_1 = H$; $R_2 = SO_3Na$
Iejimalide C : $R_1 = CH_3$; $R_2 = H$
Iejimalide D : $R_1 = CH_3$; $R_2 = SO_3Na$

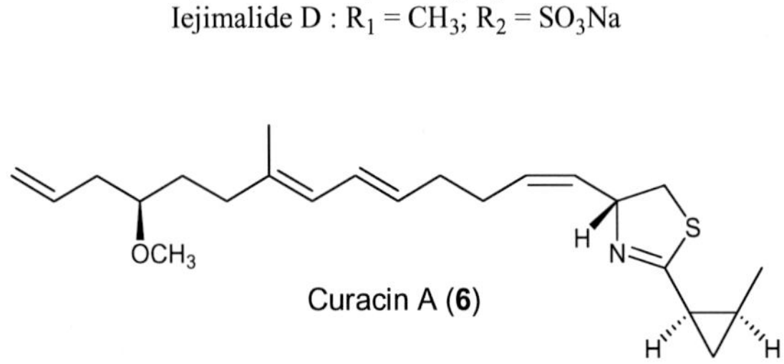

Curacin A (**6**)

FIGURE 22-4. Structures of anticancer leads derived from marine cyanobacteria.

toxins. Insight here developed from isolation of closely similar natural products from the cyanobacterial diet of these sea hares. Various species of tropical cyanobacteria grow as tufts or mats in shallow water environments and sometimes exist in enormous biomass, thus providing a rich food source for these specialist feeders. Ultimately, dolastatin 10 (**3**) itself was isolated directly from the benthic cyanobacterium *Symploca* sp. (Luesch *et al.*, 2001).

Iejimalide A (5)

Roughly 3 miles northwest from the Motobu Peninsula of Okinawa, Japan, lies a small island to which the potent anticancer compound iejimalide A owes its name. A place of myth and war, Ie is a small agricultural island with thriving coral reefs on its southeastern coast. These reefs are home to the beautiful purple ascidian *Eudistoma rigida*, a colonial tunicate comprised of many individual sea squirts (see Chapter 30) and a vast community of associated microorganisms. *E. rigida*, however, is remarkable for another reason. Extracts of this mini biosphere are highly toxic to cancer cells when tested *in vitro*. The active compounds were discovered by the Kobayashi group as a series of 24-membered macrolide polyketide synthase (PKS)/nonribosomal peptide synthetase (NRPS) hybrid molecules named the iejimalides A-D (**5**) (Kikuchi *et al.*, 1991; Kobayashi *et al.*, 1988) (Fig. 22-4). The iejimalides were subsequently isolated from another shallow water tunicate from Ie Island, *Cystodytes* sp., suggesting that a microorganism associated with both tunicates might be responsible for iejimalide biosynthesis (Kazami *et al.*, 2006). Further clarification of this came from our direct isolation of iejimalide A from a Papua New Guinea collection of the filamentous marine cyanobacterium *Lyngbya* sp. (Simmons and Gerwick, unpublished observations). Investigations of the Okinawan ascidians for iejimalide-producing symbiotic cyanobacteria are ongoing in several laboratories.

Although the iejimalides are not yet in clinical trials, much effort is being focused on their development as anticancer drugs. Initial antitumor assay data indicated 120% and 150% life-span increase in mice inoculated with P388 leukemia cells in their intraperitoneal cavities and treated with iejimalides A/B and C/D, respectively. When compared with known anticancer agents for their spectrum of activity against a 39-human cancer cell line panel, there were no correlations to any of the standard agents. These data provide evidence that the iejimalides are effective against cancer models and that they most likely possess a novel mode of action compared to existing anticancer therapies. Work by the Osada group has provided some insight into the molecular target of the iejimalides. Molecular pharmacological studies identified the iejimalides as potent osteoclast inhibitors with specific activity against cellular V-ATPases. Interestingly, the iejimalides have been found to be effective inhibitors of both mammalian and yeast V-ATPases. Yeast strains resistant to bafilomycin (another V-ATPase inhibitor) are also resistant to the iejimalides, thereby suggesting similar binding sites for both compound series (Kazami *et al.*, 2006). Several total chemical syntheses have been published for these exciting lead compounds, although none is yet cost and yield effective. As a recurring theme in marine natural product drug discovery, obtaining a reliable and reasonably priced source of this promising natural product, iejimalide A, is fundamental to its further clinical evaluation and development.

Curacin A (6)

The filamentous marine cyanobacterium *Lyngbya majuscula* has been shown to produce exciting and complex molecular structures that are finding their way into a variety of clinical and biotechnological applications (Ramaswamy *et al.*, 2006) (Fig. 22-2B). Intriguingly, it appears that there are many different chemotypes of this species, each with its own capacity to produce novel natural products with powerful biological properties. This cyanobacterium has gigantic disk-shaped cells that stack into long filaments encased in a highly resistant polysaccharide sheath. Because they resemble human hair in size and come in many colors from red to green to brown, this organism is known by the common name of "mermaid's hair." Collections of this mermaid's hair from Curaçao (Netherlands Antilles) in the southern Caribbean yielded an organic extract that was initially found to be antiviral; however, a more careful look showed it to be highly toxic to the cell line used to grow the experimental virus. Isolation and structure elucidation of the cell toxin was achieved in the author's laboratory in 1993 and created considerable surprise and interest among scientists in the cancer drug discovery area. Curacin A (**6**) possesses a relatively simple structure that contains an interesting juxtaposition of cyclopropyl (three-membered) and thiazoline (five-membered, containing sulfur and nitrogen) rings, a constellation never previously observed in natural products (Fig. 22-4). Curacin A (**6**) displays potent antiproliferative and antimitotic activity with IC_{50} values ranging from 7 to 200 nM against various cell lines from the National Cancer Institute's 60 cell line assay (Gerwick *et al.*, 1994). Follow-up on its anticancer potential has been assisted by chemical synthesis of the natural product as well as many analogs that have better druglike properties (e.g., improved stability and water solubility). It is hoped that by continued exploration of the unique molecular architecture of curacin A and its analogs, in conjunction with their biological properties, a new drug for the treatment of human cancers will be devised.

Also inspired by curacin A's novel structure (**6**), considerable investigation has probed how this marine cyanobacterium creates such an exotic molecule. Early biosynthetic investigations used feeding experiments (i.e., providing ^{13}C-labeled amino acids and acetate) to growing cultures of *L. majuscula*. More recent investigations have pioneered the use of molecular genetic approaches to find the cluster of genes encoding the desired biosynthetic enzymes and then to study in a specific manner how the chemical substrates are manipulated and assembled into the structure of curacin A. Ultimately, it is hoped that these biosynthetic genes and enzymes can be harnessed to provide a steady supply of curacin A (**6**) and its analogs.

Marine Fungi

Avrainvillamide (7)

Work in recent decades has yielded reports of fungal strains being isolated from the marine environment, which raises some interesting questions. For example, to what extent are they responsible for the natural products isolated from marine invertebrates, such as sponges and tunicates? And should they really be considered marine organisms at all? What really constitutes a marine organism? Case in point, several fungal and bacterial strains isolated from the marine environment produce similar or identical compounds to those obtained from their terrestrial counterparts. Can we really call microbial cells that wash into the sea from the land but do not naturally reside or reproduce in the ocean a "marine microorganism"? One further example is the exceptional finding and cultivation of the fungus *Aspergillus* from the surface of the common marine green alga *Avrainvillea* sp. growing in the Bahamas. Fungi belonging to the genus *Aspergillus* are ubiquitous in terrestrial environments. Fermentation of this "marine" fungus gave a strongly bioactive extract, and subsequent biological assay guided fractionation led to the isolation of a novel organic molecule, avrainvillamide (**7**) (Fenical *et al.*, 2000) (Fig. 22-5). Avrainvillamide is composed of a bicyclo-[2.2.2.]diazaoctane ring system that is likely derived from the amino acid tryptophan and represents an alkaloid structure class common to terrestrial fungal secondary metabolites. Thus, it came as little surprise when the following year a group at Pfizer identified avrainvillamide from *Aspergillus ochraceus*, which had been isolated and cultured from a Venezuelan soil sample (Sugie *et al.*, 2001). Avrainvillamide and related indole alkaloids such as stephacidin B possess interesting molecular structures that are slowly succumbing to total organic synthesis (Baran *et al.*, 2006; von Nussbaum, 2003). Avrainvillamide (**7**) shows strong cytotoxicity to breast and melanoma cancer cells in early stage testing, and in ongoing preclinical trials compound **7** displays selective inhibition against LN-Cap progesterone-dependent prostate cancer (www.cancer.ucsd.edu/summaries/wfenical.asp).

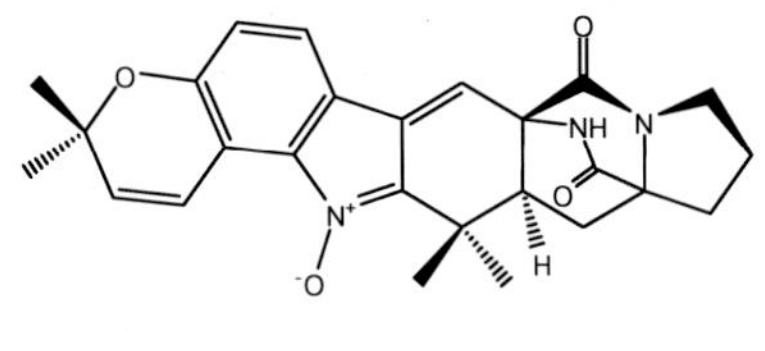

Avrainvillamide (**7**)

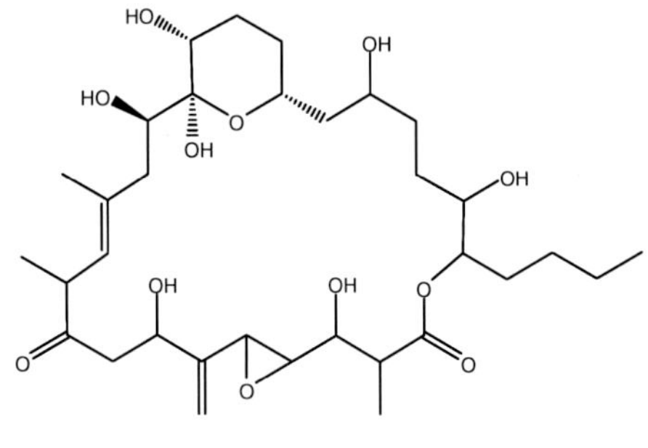

AmphidinolideN (**9**)

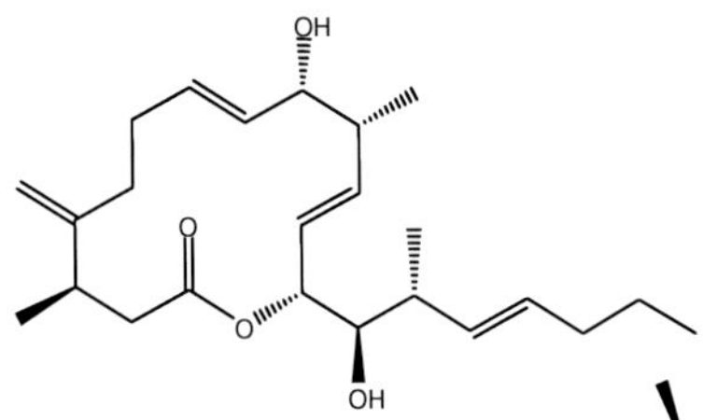

Amphidinolide R (**8**)

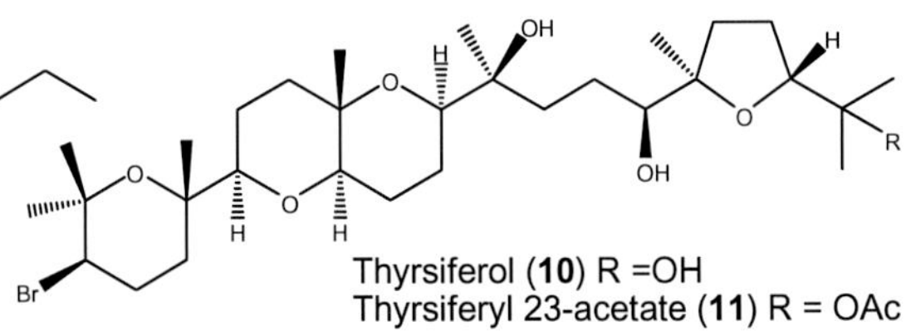

Thyrsiferol (**10**) R =OH
Thyrsiferyl 23-acetate (**11**) R = OAc

FIGURE 22-5. Structures of anticancer leads derived from marine fungi, microalgae, and macroalgae.

ANTICANCER AGENTS FROM MARINE MACROORGANISMS (OR ARE THEY?)

Marine Algae

Amphidinolides (8–9)

Crawling among the tropical seaweeds and corals of Okinawa, Japan, are a number of small, highly colored flatworms belonging to the genus *Amphiscolops*. Living in the inner tissues of these flatworms are symbiotic marine dinoflagellates of the genus *Amphidinium*. The Kobayashi group has been studying the unique natural products of these dinoflagellates by culturing them, in flasks, after isolation from their host worms. *Amphidinium* has been shown to produce a series of highly bioactive macrolides called the amphidinolides, which posses a cyclic core structure of variable carbon number (Fig. 22-5). A consequence of the diverse core ring size is a dramatic effect on biological activity; for example, amphidinolide R (**8**) is among the smaller possessing only 12 atoms in its macrolide core and displays only mild toxicity to L1210 murine lymphoma cells ($IC_{50} > 6$ µg/mL). In contrast, amphidinolide N (**9**) is considerably larger with a 26–membered core and exhibits an impressive $IC_{50} = 0.00005$ µg/mL against the same cells (Kobayashi *et al.*, 2004). The high level of toxicity to cancer cells and novel structural framework of some of the amphidinolides has propelled these molecules to the forefront of early stage anticancer drug discovery; however, once again a lack of supply is impeding these efforts. The dinoflagellate grows slowly and to limited density, and other methods of production would be costly (e.g., total synthesis). Therefore, there is interest to study how these dinoflagellates make such impressive molecules (e.g., their biosynthesis) with the hopes of one day harnessing this metabolic capacity.

The amphidinolides ultimately derive from the repetitive condensations of acetate units (= polyketides); however, they possess interesting structural features not observed in most simple polyketide/macrolide motifs, including many with an odd-numbered lactone ring (macrolides usually have an even-numbered ring), and most also contain at least one exo-methylene unit as well as other unusual carbon branches. To shed light on these unusual features, the Kobayashi group performed biosynthetic studies by supplying various ^{13}C-labeled acetate precursors to the growing dinoflagellates and monitoring incorporation patterns by ^{13}C-NMR. What emerged was completely unexpected. Although there were regions that conformed to the regular acetate polymerization model, some sections of the amphidinolides derived almost exclusively from the methyl group of acetate (Kobayashi and Tsuda, 2004). A satisfactory explanation for this highly unusual labeling/incorporation pattern has not yet been proposed, but hints that entirely new secondary biosynthetic architectures still await discovery and description.

Thyrsiferol (10) and Thyrsiferyl 23-Acetate (11)

For the illustration of structural diversity, terpeniods offer some good examples with a range of ring sizes, levels of oxygenation, and a variety of oxidation states. In particular, the squalene-derived polyether triterpeniods isolated from marine (macro-) algal species have been well characterized. Of these, the most important structural class is the dioxabicyclo[4.4.0]decanes containing thrysiferol (**10**) and its analogs, isolated by the Blunt and Munro groups from the New Zealand red alga *Laurencia thyrsifera* (Blunt *et al.*, 1978) (Fig. 22-5). Although initial biological evaluation of **10** revealed little activity against several common pathogenic microbes (e.g., *Bacillus subtilis, Staphylococcus aureus, Pseudomonas aeruginosa*), later studies revealed profound activity to a panel of human cancer cells. To date, thyrsiferyl 23–acetate (**11**) has demonstrated its greatest drug potential by inhibiting P388 leukemia cells with an IC_{50} = 0.3 ng/mL (Suzuki *et al.*, 1985). Subsequently, thyrsiferyl 23–acetate was shown to selectively inhibit serine/threonine protein phosphatase 2A (PP2A) with an $IC_{50} \sim 4\ \mu M$. Interestingly, this activity was seen for PP2A only and was not observed in the homologs PP1, PP2B, PP2C, or the protein tyrosine phosphatase (Matasuzawa *et al.*, 1994). The exceptional potency of thyrsiferyl 23–acetate (**11**) and its unique molecular structure identify it as a prime candidate for molecular pharmacological studies and further preclinical development.

Tunicates

Didemnin B (12)/Aplidine (13; Dehydrodidemnin B)

As described in the section on the iejimalides, the Urochordates (also known as ascidians or "sea squirts") are primitive animals that can bear a superficial resemblance to sponges. Like sponges, they harbor rich communities of microbial symbionts (e.g., cyanobacteria of the genus *Prochloron*). Members of the class Ascidiacea can be identified by their thick polysaccharide tunic, which often is the physical location of their cyanobacterial symbionts. There has been particular interest in members of the family Didemnidae following the discovery of the potent anticancer and antiviral compound didemnin B (**12**) from *Trididemnum solidum* (Fig. 22-6). Reported in 1981 by the late Professor Ken Rinehart and colleagues at the University of Illinois, didemnin B, displays high potency against L1210 leukemia cells ($LD_{50} = 0.0011\ \mu g/mL$) (Rinehart *et al.*, 1981a, 1981b). Didemnin B (**12**) entered phase I/II clinical trials in the late 1980s but was dropped because of its unpredictable toxicity in patients. However, the structural analog aplidine (dehydrodidemnin B, **13**), isolated from the related tunicate *Aplidium albicans*, continues to show promise for the clinical treatment of cancer. Aplidine has been shown to act by inhibiting DNA and protein synthesis, interfering with signal transduction events in proliferating cancer cells, and by blocking cell cycle progression in G1 phase. Phase I trials show aplidine (**13**) to be "well tolerated with few severe adverse events" (Maroun *et al.*, 2006, p. 1371). Phase II trials are under way with promising results thus far in advanced melanoma and multiple myeloma patients previ-

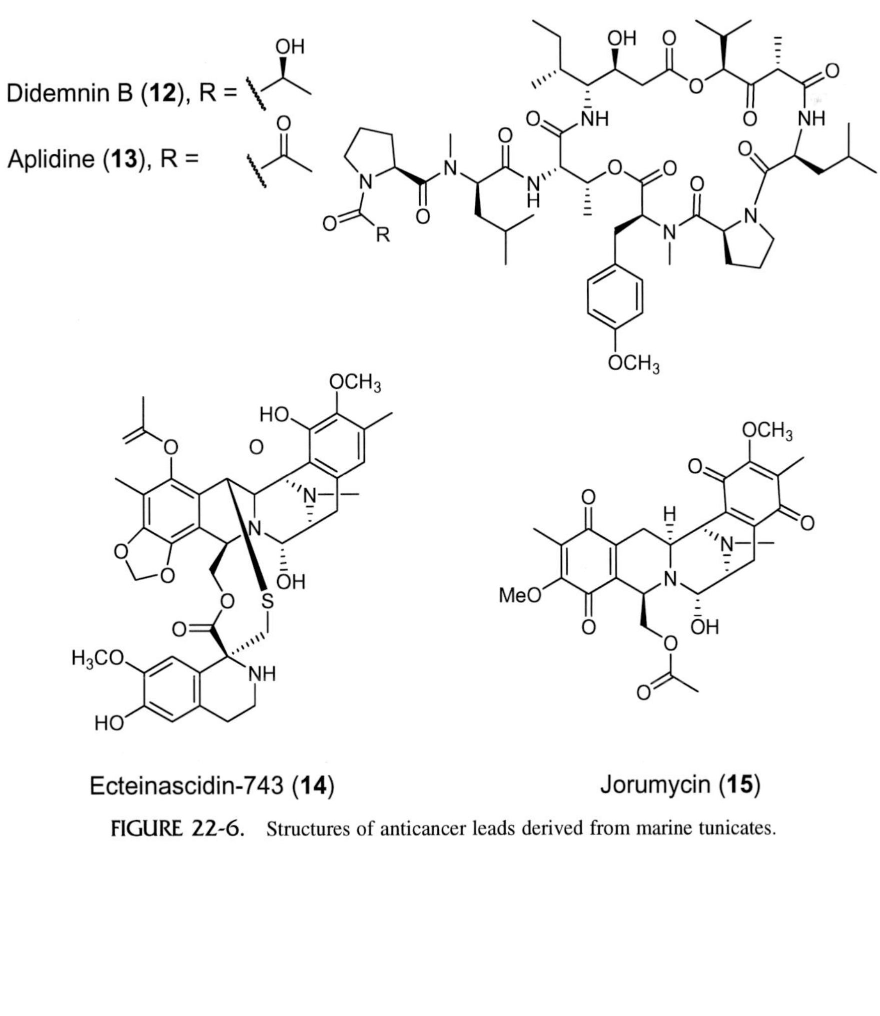

Ecteinascidin-743 (**14**) Jorumycin (**15**)

FIGURE 22-6. Structures of anticancer leads derived from marine tunicates.

ously treated with other anticancer agents (Straight *et al.*, 2006). As in the case of sponges, it has been proposed that didemnin B and related peptides are actually produced by the symbiotic cyanobacteria living in association with the tunicates; however, demonstration of this has not yet been shown experimentally. Many fascinating questions on the metabolic source, biosynthetic pathway, and relationships of host and symbionts remain.

Ecteinascidin-743 (14; ET-743, Yondelis) and Jorumycin (15)

The southern coast of Puerto Rico is fringed by thick mangrove forests, impenetrable except through a maze of small channels of tepid water. In a constant battle for space, these mangroves develop colonizing root structures from the outermost reach of their branches. Such root structures on the edges of channels are thickly encrusted by diverse colorful red and purple sponges, various macroalgae, cyanobacteria filaments, and an abundance of light yellow colored sacklike creatures that hang like bunches of grapes. This latter creature, the tunicate *Ecteinascidia turbinata*, was first reported to contain an anticancer substance by medical researcher Sigel in the late 1960s, but not until the early 1990s was the structure of this potent anticancer agent simultaneously solved by the late Professor Ken Rinehart and one of his former students, Harbor Branch Oceanographic Institution Scientist Amy Wright (Rinehart *et al.*, 1990; Sigel *et al.*, 1969; Wright *et al.*, 1990). The most active compound, named ecteinascidin 743 (**14**; ET-743), is a *tris* (tetrahydroisoquinoline) alkaloid biosynthetically derived from the condensation of two dihydroxyphenylalanine- (DOPA-) derived moieties to form the diketopiperazine core (Fig. 22-6). Preclinical evaluation of these compounds was initiated at the National Cancer Institute and revealed good activity in various cancer models (e.g., B16 melanoma in the mouse model).

To advance promising anticancer leads through the drug development pipeline, many hundreds to thousands of milligrams of the pure compound are required. The inability to produce anticancer lead compounds in these quantities is a recurrent dilemma in natural products drug discovery and development, as noted earlier. This was precisely the problem in the development of ET-743. Five potential solutions considered were to (1) recollect vast quantities of the producing organism from nature; (2) resource intensive aquaculture of the producing organism, either in the sea or in ponds excavated on land; (3) produce an analog by microbial fermentation and convert this to the desired drug; (4) clone and heterologously express the biosynthetic gene cluster in a fermentable organism (e.g., *E. coli*); and (5) achieve total chemical synthesis of the natural product or natural product analog. Although chemical synthesis of ET-743 has been accomplished several times (Chen *et al.*, 2006; Martinez *et al.*, 1999), because of the low yield and high cost of these total syntheses, the second and third approaches outlined earlier have been employed by the Spanish pharmaceutical company PharmaMar to supply **14** for clinical trials. Ecteinascidin 743 has been produced both by aquaculture of one of the original source organisms, the tunicate *Ecteinascidia turbinata*, and by semisynthesis from cyanosafracin B, an alkaloid present in large quantity from fermentation of the bacterium *Pseudomonas fluorescens* (Cuevas *et al.*, 2000).

Of the ecteinascidin alkaloids, ecteinascidin 743 (**14**) is the most promising clinically and is currently undergoing phase II/III trials for pretreated sarcoma, breast and ovarian cancer. The strand-specific DNA binding properties of this chemotype, possessing the distinctive carbinolamine functionality, have been studied extensively. Using advanced 2D NMR techniques, the Hurley group in Arizona was able to show the specific binding of ET-743 to the N-2 position of guanine in the minor groove of DNA. A covalent adduct is formed through catalytic protonation and dehydration of the carbinolamine with hydrogen bond donation from adjacent base pairs. Binding of ET-743 in the minor groove causes the DNA double helix to bend toward the major groove, introducing a distortion of tertiary structure, which in turn interferes with gene transcription and leads to apoptosis. Although clinical evaluations have shown resistance to ET-743, phase II and III trials are currently recruiting patients with advanced prostate and ovarian cancers, respectively (Beumer *et al.*, 2006; www.clinicaltrials.gov).

Jorumycin (**15**) represents a recent addition to this "gifted" molecular class. Isolated from the mantle and mucus of the Pacific marine nudibranch *Jorunna funebris*, this dimeric tetrahydroisoquinoline alkaloid displays 100% growth inhibition against NIH 3T3 (mouse fibroblast) tumor cells at 50 ng/mL and to human cancer cell lines at as low as 12.5 ng/mL (Fontana *et al.*, 2000) (Fig. 22-6). Although less potent than ecteinascidin 743 (**14**), the activity profiles were considered promising enough for PharmaMar to initiate a phase I clinical evaluation under the trade name Zalypsis. Jorumycin is being evaluated in standard dose escalating protocols in patients with solid tumors or lymphomas. Although isolated from a nudibranch, this dimeric alkaloid (as well as ecteinascidin 743 from the tunicate) almost certainly derives from metabolism of associated microorganisms.

Sponges

Spongouridine (16), Spongothymidine (17), (18; Ara-C, Cytarabin), and Ara-A (19; Vidarabine)

Spongouridine (**16**) and spongthymidine (**17**) are quite possibly the most important marine natural products obtained to date. These nucleosides led directly to the development

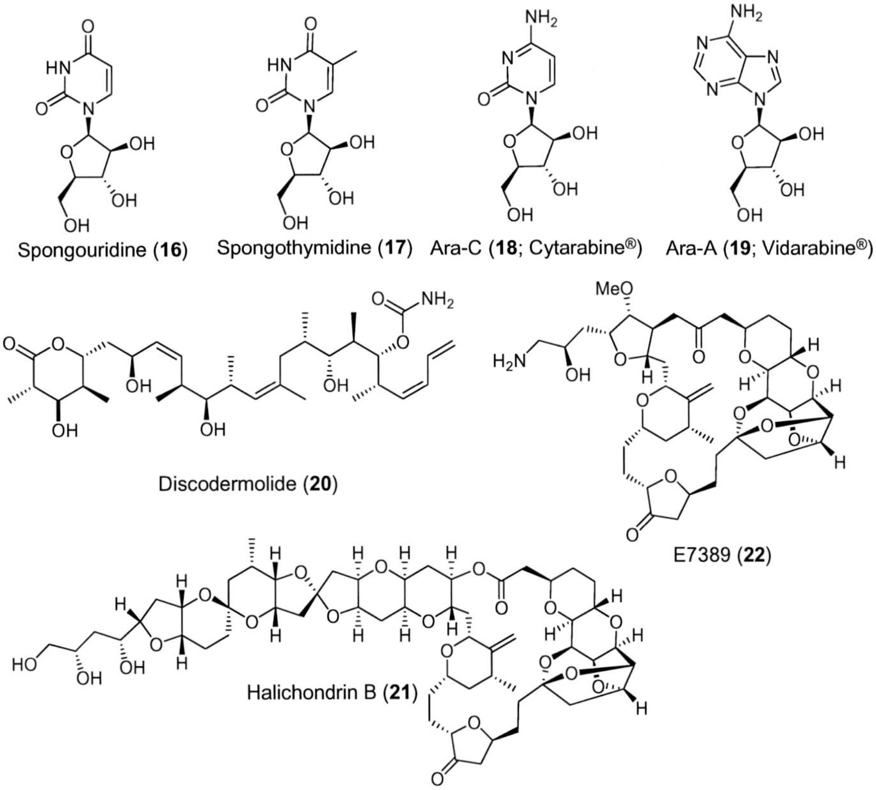

FIGURE 22-7. Structures of anticancer leads **16–22** derived from marine sponges.

of an anticancer agent of proven clinical utility (Fig. 22-7). Discovered from the Caribbean sponge *Cryptotethya crypta* in 1950 by Bergmann and Feeney, these unusual nucleosides inspired the chemical synthesis of structural analogs Ara-C (**18**, Cytarabine) and Ara-A (**19**; Vidarabine), which are used as anticancer and antiviral agents, respectively. In fact, Ara-C remains the only marine natural product-derived or inspired compound that is marketed today with FDA approval as an anticancer drug (Bergmann and Feeney, 1950). The importance of these molecules stems not only from their clinical application but from the momentum they gave to the emerging field of marine natural products drug discovery. In fact, these were some of the first sponge-derived compounds ever isolated; Bergmann began collecting *Cryptotethya crypta* from the shallow waters of Elliot Key, Florida, in 1945. Despite Ara-C having entered the market a number of years ago, there are still nearly 150 clinical trials currently recruiting acute myeloid or lymphoblastic leukemia patients for phase I and II trials; its clinical utility is still being explored and expanded (clinicaltrials.gov). The mode of action for these structurally simple molecules is based on their forming structural mimics of the normal DNA and RNA building blocks. As such, they are remarkably effective at shutting down aberrant cancer cell proliferation in the S-phase by disrupting chromosomal replication. Both Ara-A and Ara-C are actually "prodrugs" in that they require metabolic activation to the corresponding triphosphates before they can exert their disruptive effects on DNA or RNA (Kufe *et al.*, 1980). When integrated into DNA, Ara-C can both inhibit chain elongation and induce DNA chain termination. Studies suggest that "lesions" created by the insertion of Ara-C into chromosomal DNA act as position-specific topoisomerase II inhibitors, thereby stimulating DNA cleavage and ultimately apoptotic cell death. Ara-C (**16**) may also act synergistically with other anticancer agents or other modalities of cancer treatment (e.g., radiation).

Discodermolide (20)

The Harbor Branch Oceanographic Institution in Fort Pierce, Florida, is unique in its exploration of deep sea organisms as a source of new potential drug molecules. To collect samples from depths below those typically reached by SCUBA, they employ a fleet of manned deep sea submersibles, notable for their distinctive Plexiglas bubble design, which allows superb underwater visibility (www.hboi.edu/gallery/photoarchive/subs_gallery_1.html). Using SCUBA initially and deep sea submersibles subsequently, the marine sponge *Discodermia dissoluta* was collected from Lucay, Grand Bahama Island. A bioassay directed investigation of the cancer cell toxicity noted in the crude extract of the sponge led to the isolation of a polyhydroxylated lactone named discodermolide in low yield (Gunasekera *et al.*, 1990) (Fig. 22-7). Moreover, discodermolide (**20**) displayed activity against fungi and in a two-way mixed-lymphocyte culture assay; in the latter assay, antiproliferative responses of both murine splenocytes and human peripheral blood lymphocytes were observed at 0.5 and 5.0 μg/mL, respectively. Moreover, after treatment greater than 85% of the murine splenocytes were still viable, thus indicating selective immunosuppressive activity without overt cell toxicity.

The mechanism of cancer cell toxicity of discodermolide (**20**) has been studied in some detail. In many regards, the activity of this sponge compound is similar to that of taxol (Paclitaxel), the microtubule stabilizing anticancer drug isolated from bark of the Pacific Yew tree. Discodermolide binds to β-tubulin and leads to microtubule stabilization, like taxol. At low concentrations, **20** interferes with the subtle yet critical dynamics of tubulin polymerization at the tips of the microtubules, thereby interfering with chromosome separation during mitosis. In turn, this results in a blockade of the cell cycle at the anaphase-metaphase transition of cells (Sánchez-Pedregal *et al.*, 2006). Discodermolide (**20**) has been taken into phase I clinical trials, but because of patient toxicity it was withdrawn. Nevertheless, it is believed that with other dosing schedules and routes of administration, discodermolide can be applied safely and effectively to treat cancer.

Biosynthetically, discodermolide is another example of a polyketide natural product produced from the polymerization of acetate units with variable levels of reduction after each acetate addition. Polyketides of this sort are created by an assembly line of enzymes (polyketide synthases) that progressively build the molecule one acetate unit at a time and, hence, are attractive targets for genetic engineering approaches. Although the natural biosynthetic pathway for discodermolide (**20**) has not yet been obtained from the sponge, a totally synthetic gene approach has been attempted, and intermediates along the route to discodermolide have been produced (Burlingame *et al.*, 2006). Polyketides of the type represented by discodermolide are most often produced by microorganisms, and it is once again suspected that a bacterium living in association with the sponge is responsible for discodermolide (**20**) biosynthesis.

Halichondrin B (21) and E7389 (22)

In a shallow bay to the south of Tokyo, Japan, the midnight black sponge *Halichondria okadai* is found attached to the rocky substrate in large quantities. In fact, this sponge is common in many parts of the world. Japanese researchers in the mid-1980s, headed by Professor Daisuke Uemura, collected this sponge by SCUBA and evaluated its extract for cancer cell toxicity. Although it was very active, it was also present in the sponge in extraordinarily small quantities and required the collection of 600 kg of sponge to yield just 12.5 mg of halichondrin B (**21**), as well as small quantities of several related metabolites (Fig. 22-7). Halichondrin B (**21**) is one of the most structurally complex and biologically potent marine natural products ever discovered. This polyether macrolide (**21**) is remarkably bioactive with IC_{50} values less than 100 pg/mL against the B16 mouse melanoma *in vitro* cancer cell line (Hirata and Uemura, 1986). Soon after its original discovery, workers at the National Cancer Institute (NCI) in the United States showed that the potent anticancer activity of halichondrin B was due, in whole or in part, to its ability to noncompetitively bind to β-tubulin (Bai *et al.*, 1991). Tubulin binding of this nature results in disruption of the intracellular microtubule networks present in rapidly dividing cancer cells and ultimately leads to apoptotic cell death. Halichondrin B (**21**) showed across the board potent cancer cell toxicity when evaluated in the NCI's 60 cell line panel with most IC_{50} values less than 100 pM. Fostad and coworkers subsequently showed that halichondrin B has excellent activity in several animal tumor models (Fodstad *et al.*, 1996). Based on these early and highly promising results, halichondrin was anticipated as the next blockbuster anticancer drug. However, with the low abundance from nature (often <0.0001% wet weight from the sponge), compound supply seemed an insurmountable obstacle to taking this compound into the clinic. Remarkably, the Kishi group at Harvard University was able to complete a total chemical synthesis of halichondrin B, which led to subsequent syntheses of simplified as well as stabilized forms of the drug (E7389, **22**) by a collaborating pharmaceutical company (Eisai Pharmaceutical) (Aicher *et al.*, 1992) (Fig. 22-7). The simplified version of halichondrin B, E7389 (**22**), was tested in phase II clinical trials for women with advanced breast cancer previously treated with other agents; almost 20% of these patients showed partial responses, some for more than a year following the trial. Phase III trials are ongoing to determine the efficacy of **22** in patients with locally recurrent or metastatic breast cancer. The development of E7389 (**22**) is a remarkable story of persistence and ingenuity in organic synthesis and easily represents the most complex synthetic drug yet taken into the clinic. Where does such a complex natural product come from? Although speculative, it seems highly likely that the halichondrin family of molecules also derives from bacteria living in symbiosis with the sponge.

Hemiasterlin (23), HTI-286 (24; taltobulin), and E7974 (25)

On occasion, researchers come across organisms that either produce or otherwise contain a multitude of biologically exciting secondary metabolites; the South African sponge *Hemiastrella minor* is a prime example of this. From this organism the Kashman group in Israel reported the bioactive yet previously described cyclic peptide jasplakinolide (jaspamide) as well as two new compounds, geodiamolide TA and hemiasterlin (**23**) (Talpir *et al.*, 1994). The later molecule (**23**) generated a particularly high level of excitement because of its therapeutic efficacy against a variety of tumor types (Fig. 22-8). Hemiasterlin (**23**) has been shown to interact with tubulin and, by binding noncompetitively within the "vinca domain" of the β-subunit, results in cell cycle arrest at the G_2/M interface. By doing so, **23** and compounds that bind in a similar manner [e.g., the cryptophycins and dolastatin 10 (**3**)] modulate dynamic instability of cellular microtubules, disrupt spindle microtubules, inhibit microtubule assembly, and induce the self-association of tubulin dimers into single protofilament rings and spirals, all of which culminates in the inhibition of tubulin assembly and cell death. Although a variety of molecules have been shown to disrupt tubulin polymerization in a similar manner, few do it at picomolar or low nanomolar concentrations and with as high specificity as hemiasterlin (**23**). Motivated by the very potent anticancer affect of hemiasterlin, the Andersen group in British Columbia developed the synthetic analog HTI-286 (**24**) with improved activity and pharmaceutical properties (Nieman *et al.*, 2003). In 2003, **24** entered phase I clinical trials as taltobulin (Wyeth) for the treatment of advanced malignant solid tumors where it demonstrated potent activity *in vitro* against several tumor types including both taxane-sensitive and taxane-resistant tumors (Ratain *et al.*, 2003). Subsequently, additional hemiasterlin analogs were synthesized by chemists at Eisai Pharmaceuticals and resulted in E-7974 (**25**), which is currently undergoing three phase I clinical trials for patients with advanced solid malignancies (Campagna *et al.*, 2006).

Psammaplin A (26) and NVP-LAQ824 (27)

As noted elsewhere in this chapter, marine sponges are actually complex ecosystems supporting a diverse community of both macro- and microorganisms. Consideration of the molecules currently undergoing clinical or preclinical investigation identifies a significant number that have been isolated from marine sponge or ascidian species (Tables 22-1 and 22-2). Both of these classes of invertebrates are exceptionally rich in associated microorganisms. For example, a cross section of the sponge *Theonella swinhoei* reveals a diverse microbial community assuming a stratified organization; unicellular cyanobacteria live in the outer ectosomal layer with smaller unicellular bacteria and filamentous bacteria inhabiting the inner sponge cell matrix (Bewley and Faulkner, 1998). Moreover, a number of important compounds isolated from a particular sponge or ascidian have subsequently been isolated from a variety of other unrelated sponges or ascidians, again supporting the hypothesis that microorganisms are the actually producers of these molecules (Fig. 22-1B). Psammaplin A (**26**), isolated from the sponge *Psammaplysilla* sp., provides an excellent example of this phenomenon (Fig. 22-2C). Reported simultaneously by Crews, Schmitz, and Scheuer, psammaplin A (**26**) is a symmetrical oxime-containing bromotyrosine metabolite that exhibits potent cytotoxicity against P388 cells (IC_{50} = 0.3 ng/mL) (Arabshahi and Schmitz, 1987; Quiñoá and Crews, 1987; Rodriguez *et al.*, 1987) (Fig. 22-8). Much

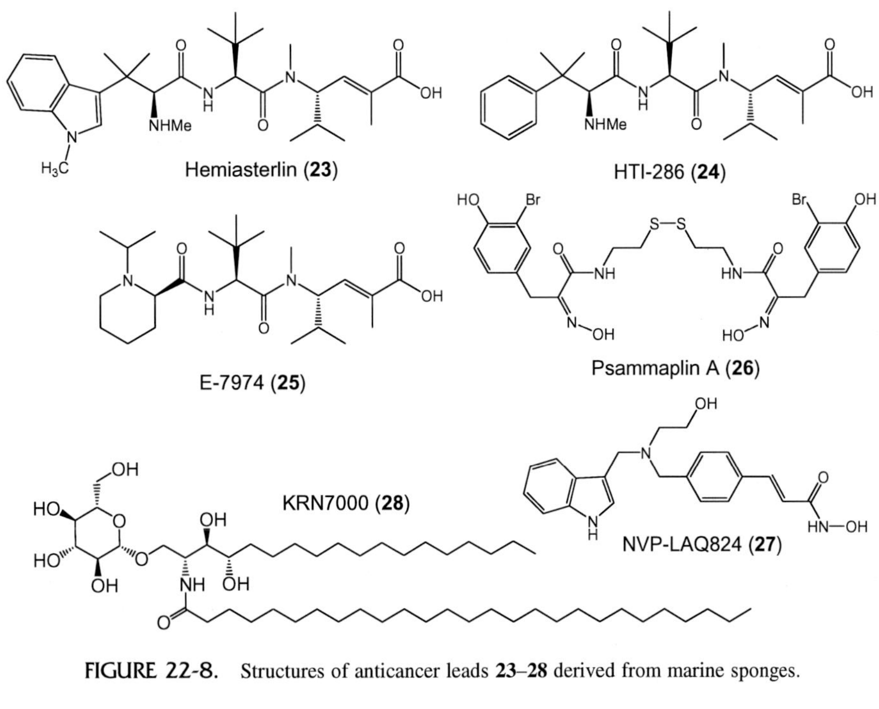

FIGURE 22-8. Structures of anticancer leads **23–28** derived from marine sponges.

work has been done on this promising compound, both in terms of its wide spectrum of biological activity and total organic synthesis (Godert *et al.*, 2006). Notable here is the ability of **26** to inhibit the function of several important enzymes in an array of prokaryotic and eukaryotic systems, including those involved in the epigenetic regulation of gene expression. Psammaplin A (**26**) has also been shown to affect processes relevant to potential cancer therapies such as DNA replication and angiogenesis. The antiproliferative effects of **26** were shown to occur through the inhibition of mammalian aminopeptidase N (IC_{50} = 18 μM), and as a consequence, to suppress tumor cell invasion (Shim *et al.*, 2004). More recently, researchers from the University of Mississippi have shown that **26** is able to activate PPARγ, a ligand-activated transcription factor known to inhibit growth, promote terminal differentiation, and induce apoptosis in human breast tumor cells (Mora *et al.*, 2006). Using a cell-based screen, scientists at Novartis Pharmaceuticals found three natural products (trapoxin B, trichostatin A, and **26**) that induce expression of the cyclin dependent kinase inhibitor $p21^{waf1}$ via inhibition of histone deacetylase (HDAC). These three compounds were investigated for their respective structural features that impart HDAC-inhibitory activity (i.e., their pharmacophores); what emerged was a composite synthetic compound that combined features of all three substances into NVP-LAQ824 (**27**) (Griffin and Anderson, 2003). Compared to the many other analogs produced, drug candidate **27** had several desirable features including its patient tolerability, efficacy to multiple myeloma xenografts, and potency as an HDAC inhibitor (Catley *et al.*, 2006; Remiszewski, 2003) (Fig. 22-8). Researchers found that psammaplin A (**26**) may be a prodrug with the active form of the molecule being the reduced sulfhydryl (Kim *et al.*, 2007). NVP-LAQ (**27**) is currently undergoing phase I clinical trials (Novartis Pharmaceuticals) for a variety of solid tumors and as a combination therapy against malignant melanoma (Kato *et al.*, 2007).

KRN7000 (28)

The biological response induced by many pharmaceuticals, including natural products, is often the result of "biomimicry" in that the applied drug closely resembles an endogenous substance (i.e., a molecule required for day-to-day metabolic function of the organism). In some cases, the foreign compound can be so similar to the endogenous substance that it easily enters cells and localizes in specific subcellular regions but binds abnormally and with higher affinity than the native ligand, thereby causing a downstream inhibitory effect. One of the most studied biomimetic small molecule natural products is KRN7000 (**28**), an α-galactosylceramide molecule that closely resembles numerous monoglycosylated ceramides that are used by nearly all living cells for cell-cell recognition and other cellular regulation processes (e.g., neuronal regulation, protein kinase C activity, hormone receptor function) (Fig. 22-8).

KRN7000 (**28**) was first isolated by researchers employed by the Kirin Brewery Company in Takasaki, Japan, as a novel immunomodulator and antitumor compound from the marine sponge *Agelas mauritianus*. The group observed potent antitumor activity in B16 melanoma bearing mouse models with a noted stimulation of lymphocyte proliferation in an allogeneic mixed leukocyte reaction (MLR). To confirm whether KRN7000 was, in fact, regulating an immunological response, the authors demonstrated the ability of **28** to stimulate natural killer cell responses both *in vitro* and *in vivo* (Kobayashi *et al.*, 1995). Having shown the life-extending properties of KRN7000 (**28**), a series of medicinal chemical studies were undertaken to attempt enhancement of efficacy and to generate sufficient supply for clinical trials. During the course of derivative synthesis and biological testing it became apparent that nature, over the course of eons, had already optimized **28** for stimulation of lymphocyte proliferation. This latter point was shown for both MLR and human peripheral blood cells (Morita *et al.*, 1995). Clinical evaluation indicates that KRN7000 (**28**) activates human Vα24 natural killer T cells, which then "show a strong antitumor activity against various malignant tumors" and activate other antitumor effector cells by enhanced production of cytokines such as IFN-γ and interleukin-4 (Motohashi *et al.*, 2006, p. 6079).

Miscellaneous Marine Organisms

Bryostatin (29)

Bryozoans are sedentary colonial marine organisms in which the individual zooids feed on microalgae using a feathery ciliated structure known as a lophophore. Some types become calcified, and, in general, bryozoans are not preyed upon by nudibranch mollusks or other marine animals. Only a few of the nearly 4000 living species of bryozoans have been studied for their natural products chemistry, but from these a number of highly bioactive and interesting chemical structures have been obtained. For example, the bryozoan *Bugula neritina* was initially collected from the Gulf of California in 1968, and its two-propanol extract showed exciting antitumor results (168% to 200% increase in mouse survival) in an *in vivo* lymphocytic leukemia assay developed by the National Cancer Institute (Pettit *et al.*, 1970). Nearly 14 years of effort resulted in the isolation and structure determination by X-ray crystallography of the first of several anticancer active compounds from this source, known as bryostatin 1 (**29**) (Fig. 22-9). Bryostatin 1 is a potent antineoplastic agent that exerts its toxicity to cancer cells through modulation of the protein kinase C (PKC) cell signaling pathway. Although phase II clinical trials have not shown much promise for bryostatin when

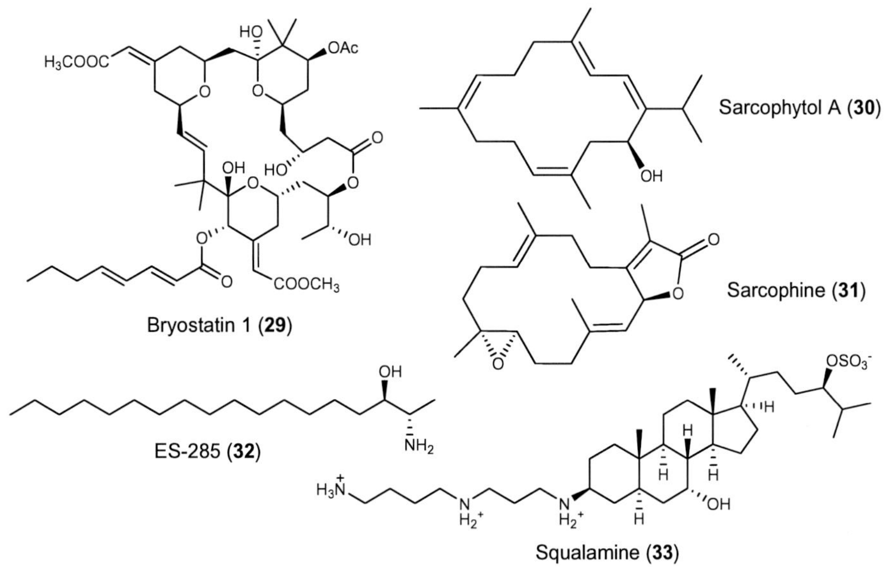

FIGURE 22-9. Structures of anticancer leads derived from miscellaneous marine invertebrates and vertebrates.

used as a single agent, there is hope for better efficacy when used in combination with other cytotoxic drug therapies, such as vincristine. As of the writing of this chapter, the National Cancer Institute is recruiting patients for one phase I and two phase II clinical trials investigating **29** in combination therapies. All three trials follow the rationale that bryostatin 1 (**29**) will enhance tumor cell sensitivity to the other therapeutic agents (temsirolimus, rituximab, and vincristine, respectively). The phase I trial focuses on developing more effective strategies to treat kidney cancer and melanoma; the phase II trials have similar goals with lymphoma (www.clinicaltrials.gov).

Detailed molecular ecological studies by the Haygood laboratory, then at Scripps Institution of Oceanography, revealed that a new bacterium, *Endobugula sertula*, is found in association with mature colonies of *B. neritina*, as well as from free-living larvae, and is the biosynthetic source of the bryostatins. The putative gene cluster encoding bryostatin biosynthesis was obtained from two strains of this bacterium (Sudek *et al.*, 2007). It shows a number of unique architectural features not seen in other biosynthetic pathways and represents a future possibility for functional expression of this gene cluster (e.g., heterologous production) in another more easily cultured microorganism (*E. sertula* has not yet been cultured free of its host). A fermentation source of bryostatin 1 (**29**) would help to insure a plentiful and economic source of this promising agent for future clinical evaluation and use.

Sarcophytol A (30) and Sarcophine (31)

Sarcophyton sp. soft corals are popular with marine reef tank hobbyists largely because of their fast growth rates and easy maintenance. These robust organisms are capable of surviving in varied and competitive environments, yet they have no motility or physical defenses against potential predators. Instead, they have developed the biosynthetic capacity to produce a diverse series of highly bioactive terpenoid metabolites. The most common are the 14–membered macrocyclic cembranoids, which display a range of antibiotic, neuroprotective, and antitumor properties (Pham *et al.*, 2002). In particular, *Sarcophyton glaucum* has been thoroughly studied for its rich complement of bioactive cembranoids, including sarcophytol A (**30**) and sarcophine (**31**), which it produces in surprisingly high concentrations (up to 3% of the dry weight) (Bernstein *et al.*, 1974) (Fig. 22-9). These cembranoids exhibit a wide range of biological activities including neuroprotective, antimicrobial, antitumor, and cancer chemopreventive properties. These latter activities of **30** and **31** have been well studied with the goal of optimizing biological activity through medicinal chemical approaches (El Sayed *et al.*, 1998; Sawant *et al.*, 2004). More recently, El Sayed and coworkers have investigated the biocatalytic transformation (by fermentation of **31** with strains of the fungi *Rhizopus stolonifer* and/or *Absidia spinosa*) to yield novel structures and examine their antimetastatic activities (Sawant *et al.*, 2006). Several cancer prevention mechanism-of-action studies have been reported for **31** and have shown that it has selective inhibitory effects against vital enzymes such as cholinesterase and phosphofructokinase. Moreover, the synthetic derivative sarcotriol and a sarcophine-diol derivative have been shown to induce apoptosis, decrease COX-2 levels, and exhibit overall cancer chemopreventive effects in mouse skin cancer models (Kundoor *et al.*, 2006; Zhang *et al.*, 2007). Although formal clinical trials have not yet begun for these promising cancer chemopreventative metabolites, it seems likely to occur soon.

ES-285 (32)

The marine mollusk *Spisula polynyma* (also known as Simpson's surf clam) has been the source of an exceptionally simple compound, which is nevertheless an important anticancer lead under current clinical evaluation. *S. polynyma* has a geographic distribution throughout the cold ocean waters of the northern hemisphere, including the Arctic, North Atlantic, and North Pacific. It is commonly found near the low-tide line down to a depth of nearly 100 meters and can reach 80 mm in shell diameter in its 13- to 14-year life span. The active anticancer compound, spisulosine or (2*S*,3*R*)-2–amino-3–hydroxyoctadecane (**32**), has a molecular mass of 285 and no UV chromophore, and it was isolated as one of a series of related compounds each of which displayed a unique toxicity against L1210 murine leukemia cells (Rinehart *et al.*, 2000) (Fig. 22-9). Soon following its discovery, **32** was shown to display antiproliferative activity through inhibition of actin fiber assembly. This results in a two-phase morphological effect; first the cells adopt a fusiform morphology followed by an apoptotic rounded-shape phase with an absence of substrate adhesion points (Cuadros *et al.*, 2000). More recently, ES-285 (**32**) has been shown to trigger a highly atypical cell death pathway that involves caspases 3 and 12 activation and p53 phosphorylation. Interestingly, this occurs without affecting other pathways typically associated with cell survival and apoptosis (e.g., JNK, Erks, or Akt), thus suggesting a nonspingosine-dependent apoptotic pathway (Salcedo *et al.*, 2007). ES-285 (**32**) is undergoing preclinical development (PharmaMar) for the treatment of solid cancers.

Shark Cartilage and Squalamine (33)

Observations in mammals that cartilage is a tissue naturally lacking blood vascularization subsequently revealed that cartilage contains substances that potently inhibit new blood vessel growth (i.e., antiangiogenesis). As mammalian-derived cartilage is in relatively short supply, sharks were examined because they lack bones but are rich in cartilage (Fig. 22-2D). Two shark-derived products have reached clinical trial for their ability to inhibit the vascularization associated with tumor growth (Cho and Kim, 2002). First, a small molecule, squalamine (**33**), an amino-steroid derivative from shark cartilage, has strong antiangiogenic effects, is well tolerated by oral administration, and is effective as an antitumor agent when used in combination with other traditional chemotherapies, such as platinum-based anticancer agents (Fig. 22-9). A second product, Neovastat (AE-941), is a concentrate of biologically active water-soluble components extracted from shark cartilage, and it contains peptides of 10,000 and 14,000 Da in size, which also inhibit angiogenesis. Despite some controversy over variable results with shark cartilage preparations, phase III clinical trials were conducted with this latter preparation in combination with traditional cytotoxic drug therapy for the treatment of metastatic lung tumors. Investigations now focus on the genes that encode some of these antiangiogenic factors, and this may be a fruitful approach to both the discovery of additional compounds with desirable properties as well as a method for producing these compounds without relying on the further harvest of animals from nature. Rather, through transfer of these genes into microorganisms, it is hoped that products such as these shark-derived antiangiogenic peptides can be produced through industrial fermentation processes.

CONCLUSIONS

This chapter has showcased some of the most promising and interesting examples of marine natural products that are advancing through the cancer drug development process as of early 2007. Most of these are at relatively early stage clinical evaluation, and, unfortunately, drug development statistics indicate that most will fail as useful agents because of either unacceptable toxicity or lack of efficacy. Nevertheless, with 20 agents currently or recently in clinical trial (Table 22-1) and many more at late stage preclinical evaluation (Table 22-2), the chances are good that a few of these will be successful and contribute to our arsenal of agents for treating this complex family of diseases. Indeed, there will soon be widespread recognition that marine life forms are valuable contributors to our medicine chest of useful pharmaceuticals.

Although in principle these drug leads derive from a wide variety of marine life, from simple bacteria to complex invertebrates, there is a growing body of evidence that most of these are, in fact, the result of metabolic processes present in marine microorganisms, including heterotrophic bacteria, cyanobacteria, and fungi (Fig. 22-1B). The deeper we investigate the complex relationships that are present in such organisms as sponges, tunicates, and corals, the clearer it becomes that we should conceive of these macroscopic life forms as complex assemblies of species and not as individuals per se. There is much to study and learn about these communities of organisms and their complex chemical interactions. Microorganisms are much better suited to large-scale fermentation than are macroorganisms; hence, we can ultimately hope to develop reliable and economic sources for many marine-derived drug leads. Moreover, as discussed later in this text in the chapter by Moore *et al.* (Chapter 26), the pathways utilized by prokaryotic microorganisms are better suited for excision from their original sources and expression in heterologous hosts that grow well in fermentation culture. However, to realize the tangible outcome of these technologies, much underlying knowledge is required on the nature of marine symbiotic relationships and marine microbiology, as well as on the molecular genetics and mechanistic biochemistry of secondary metabolite biosynthesis. Students of oceans and human health have before them large and exciting challenges in both the basic and applied sciences.

Acknowledgments

Anticancer drug discovery research in the author's laboratory has been supported by the following NIH grants: CA100851, CA52955, and TW006634.

References

Aicher, T.D., Buszek, K.R., Fang, F.G., Forsyth, C.J., Jung, S.H., Kishi, Y., Matelich, M.C., Scola, P.M., Spero, D.M., Yoon, S.K., 1992. Total synthesis of halichondrin B and norhalichondrin B. J. Am. Chem. Soc. 114, 3162–3164.

Arabshahi, L., Schmitz, F.J., 1987. Brominated tyrosine metabolites from an unidentified sponge. J. Org. Chem. 52, 3584–3586.

Bai, R., Paull, K.D., Herald, C.L., Malspeis, L., Pettit, G.R., Hamel, E., 1991. Halichondrin B and homohalichondrin B, marine natural products binding in the Vinca domain of tubulin. J. Biol. Chem. 266, 15882–15889.

Baran, P.S., Hafensteiner, B.D., Ambhaikar, N.B., Guerro, C.A., Gallagher, J.D., 2006. Enantioselective total synthesis of avrainvillamide and the stephacidins. J. Am. Chem. Soc. 128, 8678–8693.

Beck, Z.Q., Aldrich, C.C., Magarvey, N.A., Georg, G.I., Sherman, D.H., 2005. Synthesis of cryptophycin/grenastatin natural products, Biochemistry 44, 13457–13466.

Bergmann, W., Feeney, R.J., 1950. The isolation of a new thymine pentoside from sponges. J. Am. Chem. Soc. 72, 2809–2819.

Bernstein, J., Shmeuli, U., Zadock, E., Kashman, Y., Neeman, I., 1974. Sarcophine, a new epoxy cembranolide from marine origin. Tetrahedron 30, 2817–2824.

Beumer, J.H. Buckle, T., Ouwehand, M., Franke, N.E.F., Lopez-Lazaro, L., Schellens, J.H.M., Beijnen, J.H., van Tellingen, O., 2006. Trabectedin (ET-743, Yondelis™) is a substrate for P-glycoprotein, but only high expression of P-glycoprotein confers the multidrug resistance phenotype. Invest. New Drugs 25, 1–7.

Bewley, C.A., Faulkner, D.J., 1998. Lithistid sponges: Star performers or hosts to the stars. Angew. Chem. Int. Ed. 37, 2162–2178.

Blunt, J.W., Hartshorn, M.P., McLennan, T.J., Munro, M.H.G., Robinson, W.T., Yorke, S.C., 1978. Thyrsiferol: A squalene-derived metabolite of *Laurencia thyrsifera*. Tetrahedron Lett. 19, 69–72.

Burlingame, M.A., Mendoza, E., Ashley, G.W., Myles, D.C., 2006. Synthesis of discodermolide intermediates from engineered polyketides. Tetrahedron Lett. 47, 1209–1211.

Campagna, S., Pop, M.M. (Eisai Co., Ltd., Japan), 2006. Unsolvated and host-quest solvated crystalline forms of (2E,4S)-[(N-{[(2R)-1–isopropylpiperdin-2–yl]-carbonyl}-3–methyl-L-valyl)(methyl)amino]-2,5–dimethylhex-2–enoic acid and their pharmaceutical uses. PCT Int. Appl. WO 2006121857 A2 20061116.

Catley, L., Weisberg, E., Tai, Y-T., Atadja, P., Remiszewski, S., Hideshima, T., Mitsiadea, N., Shringarpure, R., LeBlanc, R., Chauhan, D., Munshi, N.C., Schlossman, R., Richardson, P., Chauhan, D., Hideshima, T., Anderson, K.C., 2006. A novel proteasome inhibitor NPI-0052 as an anticancer therapy. Br. J. Cancer 95, 961–965.

Chauhan, D., Hideshima, T., Anderson, K.C., 2006. A novel proteasome inhibitor NPI-0052 as an anticancer therapy. Br. J. Cancer. 95, 961–965.

Chen, X., Chen, J., Zhu, J. 2006. Synthetic studies on ecteinascidin 743 (Et 743): Asymmetric synthesis of a highly oxygenated tetrahydroisoquinoline via a key phenolic mannich reaction. Synthesis 23, 4081–4086.

Cho, J., Kim, Y., 2002. Sharks: A potential source of antiangiogenic factors and tumor treatments. Mar. Biotech 4, 521–525.

Cragg, G.M., Newman, D.J., Yang, S.S., 2006. Natural product extracts of plant and marine origin having antileukemia potential. The NCI experience. J. Nat. Prod. 69, 488–498.

Cuadros, R., Montejo de Garcini, E., Wandosell, F., Faircloth, G., Fernandez-Sousa, J.M., Avila, J., 2000. The marine compound spisulosine, an inhibitor of cell proliferation, promotes the disassembly of actin stress fibers. Cancer Lett. 152, 23–29.

Cuevas, C., Perez, M., Martin, M.J., Chicharro, J.L., Fernandez-Rivas, C., Flores, M., Francesch, A., Gallego, P., Zarzuelo, M., de la Calle, F., Garcia, J., Polanco, C., Rodriguez, I., Manzanares, I., 2000. Synthesis of ecteinascidin ET-743 and phthalascidin Pt-650 from cyanosafracin B. Org. Lett. 2, 2545–2548.

El Sayed, K.A., Hamann, M.T., Waddling, C.A., Jensen, C., 1998. Structurally novel bioconversion products of the marine natural product sarcophine effectively inhibit JB6 cell transformation. J. Org. Chem. 63, 7449–7455.

Feling, R.H., Buchanan, G.O., Mincer, T.J., Kauffman, C.A., Jensen, P.R., Fenical, W., 2003. Salinosporamide A: A highly cytotoxic proteasome inhibitor from a novel microbial source, a marine bacterium of the new genus *Salinospora*. Angew. Chem. Int. Ed. 42, 355–357.

Fenical, W., Jensen, P.R., 2006. Developing a new resource for drug discovery: Marine actinomycete bacteria. Nature Chem. Biol. 2, 666–673.

Fenical, W., Jensen, P.R., Cheng, X.C., 2000. Avrainvillamide, a cytotoxic marine natural product, and derivatives thereof. United States Patent # 6066635.

Fesik, S.W., 2005. Promoting apoptosis as a strategy for cancer drug discovery. Nature Rev. Cancer 5, 876–885.

Flatt, P.M., Gautschi, J.T., Thacker, R.W., Crews, P., Gerwick W.H., 2005. Identification of the cellular site of polychlorinated peptide biosynthesis in the sponge *Dysidea (Lamellodysidea) herbacea* and symbiotic cyanobacterium *Oscillatoria spongeliae* by CARD-FISH analysis. Mar. Biol. 147, 761–774.

Fodstad, O., Breistol, K., Pettit, G.R., Shoemaker, R.H., Boyd, M.R., 1996. Comparative antitumor activities of halichondrins and vinblastine against human tumor xenografts. J. Exp. Ther. Oncol. 1, 119–125.

Fontana, A., Cavaliere, P., Wahidulla, S., Naik, C.G., Cimino, G., 2000. A New antitumor isoquinoline alkaloid from the marine nudibranch *Jorunna funebris*. Tetrahedron 56, 7305–7308.

Gerwick, W.H., Proteau, P.J., Nagle, D.G., Hamel, E., Blokhin, A., Slate, D.L., 1994. Structure of curacin A, a novel antimitotic, antiproliferative, and brine shrimp toxic natural product from the marine cyanobacterium *Lyngbya majuscula*. J. Org. Chem. 59, 1243–1245.

Godert, A.M., Angelino, N., Woloszynska-Read, A., S.R., James, S.R., Karpf, A.R., Sufrin, J.R., 2006. An improved synthesis of psammaplin A. Bioorg. Med. Chem. Lett. 16, 3330–3333.

Griffin, J., Anderson, K.C., 2003. NVP-LAQ824 is a potent novel histone deacetylase inhibitor with significant activity against multiple myeloma. Blood 102, 2615–2622.

Groll, M., Huber, R., Potts, B.C.M., 2006. Crystal structures of salinosporamide A (NPI-0052) and B (NPI-0047) in complex with the 20 S proteasome reveal important consequences of β-lactone ring opening and a mechanism for irreversible binding. J. Am. Chem. Soc. 128, 5136–5141.

Gunasekera, S.P., Gunasekera, M., Longley, R.E., Schulte, G.K., 1990. Discodermolide: A new bioactive polyhydroxylated lactone from the marine sponge *Discodermia dissoluta*. J. Org. Chem. 55, 4912–4915.

Hamann, M.T., Scheuer, P.J., 1993. Kahalalide F: A bioactive depsipeptide from the sacoglossan mollusk *Elysia rufescens* and the green alga *Bryopsis* sp. J. Am. Chem. Soc. 115, 5825–5826.

Hill, R.T., Hamann, M.T., Enticknap, J., Rao, K.V., 2005. Kahalalide -producing bacteria and methods of identifying kahalalide -producing bacteria and preparing kahalalides. PCT Int. Appl. WO 2005042720

Hirata, Y., Uemura, D., 1986. Halichondrins-antitumor polyether macrolides from a marine sponge. Pure Appl. Chem. 58, 701–710.

Janmaat, M.L., Rodriguez, J.A., Jimeno, J., Kruyt, F.A.E., Giaccone, G., 2005. Kahalalide F induces necrosis-like cell death that involves depletion of ErbB3 and inhibition of Akt signaling. Mol. Phamacol. 68, 502–510.

Jimeno, J., Aracil, M., Tercero, J.C., 2006. Adding pharmacogenomics to the development of new marine-derived anticancer agents. J. Trans. Med. 4, 1–9.

Kato, Y., Salumbides, B.C., Wang, X-F., Qian, D.Z., Williams, S., Wei, Y., Sanni, T.B., Atadja, P., Pili, R., 2007. Antitumor effect of the histone deacetylase inhibitor LAQ824 in combination with 13–cis-retinoic acid in human malignant melanoma. Mol. Cancer Ther. 6, 70–81.

Kazami, S., Muroi, M., Kawatani, M., Kubota, T., Usui, T., Kobayashi, J., Osada, H., 2006. Iejimalides show anti-osteoclast activity via V-ATPase inhibition. Biosci. Biotechnol. Biochem. 70, 1364–1370.

Kikuchi, Y., Ishibashi, M., Sasaki, T., Kobayashi, J., 1991. Iejimalides C and D, new antineoplastic 24–membered macrolide sulfates from the Okinawan marine tunicate *Eudistoma rigida*. Tetrahedron Lett. 6, 797–798.

Kim, D.H., Shin, J., Kwon, H.J., 2007. Psammaplin A is a natural prodrug that inhibits class I histone deacetylase. Exp. Mol. Med. 39, 47–55.

Klastersky, J., Aoun, M., 2004. Opportunistic infections in patients with cancer. Ann. Oncol. 15(S4), iv329–iv335.

Kobayashi, E., Motoki, K., Takeshi, U., Fukushima, H., Koezuka, Y., 1995. KRN7000, a novel immunomodulator, and its antitumor activities. Oncol. Res. 7, 529–534.

Kobayashi, J., Cheng, J., Ohta, T., Nakamura, H., Nozoe, S., Hirata, Y., Ohizumi, Y., Sasaki, T., 1988. Iejimalides A and B, novel 24–membered macrolides with potent antileukemic activity from the Okinawan tunicate *Eudistoma cf. rigida*. J. Org. Chem. 53, 6147–6150.

Kobayashi, J., Tsuda, M., 2004. Amphidinolides, bioactive macrolides from symbiotic marine dinoflagellates. Nat. Prod. Rep. 21, 77–93.

Kufe, D.W., Major, P.P., Egan, M., Beardsley, G.P., 1980. Correlation of cytotoxicity with incorporation of ara-C into DNA. J. Biol. Chem. 255, 8997–9000.

Kundoor, V., Zhang, X., Khalifa, S., Fahmy, H., Dwivedi, C., 2006. A possible mechanism of action of the chemopreventive effects on skin tumor development in CD-1 mice. Mar. Drugs 4, 274–285.

Luesch, H., Moore, R.E., Paul, V.J., Mooberry, S.L., Corbett, T.H., 2001. Isolation of dolastatin 10 from the marine cyanobacterium *Symploca* species VP642 and total stereochemistry and biological evaluation of its analogue symplostatin 1. J. Nat. Prod. 64, 907–910.

Maroun, J.A., Belanger, K., Seymour, L., Matthews, S., Roach, J., Dionne, J., Soulieres, D., Stewart, D., Goel, R., Charpentier, D., Goss, G., Tomiak, E., Yau, J., Jimeno, J., Chiritescu, G., 2006. Phase I study of aplidine in a daily x5 one-hour infusion every 3 weeks in patients with solid tumors refractory to standard therapy. A National Cancer Institute of Canada Clinical Trials Group Study: NCIC CTG IND 115. Ann. Oncol. 17, 1371–1378.

Martinez, E.J., Owa, T., Schreiber, S.L., Corey, E.J., 1999. Phthalascidin, a synthetic antitumor agent with potency and mode of action comparable to ecteinascidin 743. Proc. Natl. Acad. Sci. 96, 3496–3501.

Matasuzawa, S., Suzuki, T., Suzuki, M., Matsuda, A., Kawamura, T., Mizuno, Y., Kikuchi, K., 1994. Thyrsiferyl 23–acetate is a novel specific inhibitor of protein phosphatase PP2A. FEBS Lett. 356, 272–274.

Mayer, A.M.S., Gustafson, K.R., 2006. Marine pharmacology in 2003–2004: Anti-tumor and cytotoxic compounds. Eur. J. Cancer, 42, 2241–2270.

Mincer, T.J., Jensen, P.R., Kauffman, C.A., Fenical, W., 2002. Widespread and persistent populations of a major new marine actinomycete taxon in ocean sediments. Appl. Environ. Microbiol. 68, 5005–5011.

Mitelman, F., Johansson, B., Mertens, F., 2007. The impact of translocations and gene fusions on cancer causation. Nature Rev. Cancer 7, 233–245.

Mora, F.D., Jones, D.K., Desai, P.V., Patny, A., Avery, M.A., Feller, D.R., Smillie, T., Zhou, Y-D., Nagle, D.G., 2006. Bioassay for the identification of natural product-based activators of peroxisome proliferator-activated receptor-γ (PPAR-γ): The marine sponge metabolite psammaplin A activates PPARγ and induces apoptosis in human breast tumor cells. J. Nat. Prod. 69, 547–552.

Morita, M., Motoki, K., Akimoto, K., Natori, T., Sakai, T., Sawa, E., Yamaji, K., Koezuka, Y., Kobayashi, E., Fukushima, H., 1995. Structure-activity relationship of α-galactosylceramides against B16–bearing mice. J. Med. Chem. 38, 2176–2187.

Motohashi, S., Ishikawa, A., Ishikawa, E., Otsuji, M., Lizasa, T., Hanaoka, H., Shimizu, N., Horiguchi, S., Okamoto, Y., Fujii, S-I., Taniguchi, M., Fujisawa, T., Nakayama, T., 2006. A phase I study of *in vitro* expanded natural killer T cells in patients with advanced and recurrent non-small cell lung cancer. Clin. Cancer Res. 12, 6079–6086.

Newman, D.J., Cragg, G.M., 2006. Natural products from marine invertebrates and microbes as modulators of antitumor targets. Curr. Drug Targets 7, 279–304.

Newman, D.J., Cragg, G.M., 2007. Natural products as sources of new drugs over the last 25 years. J. Nat. Prod. 70, 461–477.

Nieman, J.A., Coleman, J.E., Wallace, D.J., Piers, E., Lim, L.Y., Roberge, M., Andersen, R.J., 2003. Synthesis and antimitotic/cytotoxic activity of hemiasterlin analogs. J. Nat. Prod. 66, 183–199.

Nonomura, T., Sasaki, M., Matsumori, N., Murata, M., Tachibana, K., Yasumoto, T., 1996. The complete structure of maitotoxin. Part II. Configuration of the C135–C142 side chain and absolute configuration of the entire molecule. Angew. Chem. Int. Ed. 35, 1675–1678.

Patel, S., Keohan, M.L., Saif, M.W., Rushing, D., Baez, L., Feit, K., DeJager, R., Anderson, S., 2006. Phase II study of intravenous TZT-1027 in patients with advanced or metastatic soft-tissue sarcomas with prior exposure to anthracycline-based chemotherapy. Cancer 107, 2881–2887.

Pettit, G.R., Day, J.F., Hartwell, J.L., Wood, H.B., 1970. Antineoplastic components of marine animals. Nature 227, 962–963.

Pettit, G.R., Kamano, Y., Herald, C.L., Tuinman, A.A., Boettner, F.E., Kizu., H., Schmidt, J.M., Baczynskyj, L., Tomer, K.B., Bontems, R.J., 1987. The isolation and structure of a remarkable marine animal antineoplastic constituent: Dolastatin 10. J. Am. Chem. Soc. 109, 6883–6885.

Pham, N-B., Butler, M.S., Quinn, R.J., 2002. Naturally occurring cembranes from an Australian Sarcophyton Species. J. Nat. Prod. 65, 1147–1150.

Quiñoá, E., Crews, P. 1987. Phenolic constituents of Psammaplysilla. Tetrahedron Lett. 28, 3229–3232.

Rademaker-Lakhai, J.M., Horenblas, S., Meinhardt, W., Stokvis, E., de Reijke, T.M., Jimeno, J.M., Lopez-Lazaro, L., Lopez Martin, J.A., Beijnen, J.H., Schellens, J.H.M., 2005. Phase I clinical and pharmacokinetic study of kahalalide F in patients with advanced androgen refractory prostate cancer. Clin. Cancer Res. 11, 1854–1862.

Ramaswamy, A.V., Flatt, P.M., Edwards, D.J., Simmons, T.L., Han, B., Gerwick, W.H., 2006. The secondary metabolites and biosynthetic gene clusters of marine cyanobacteria. Applications in Biotechnology. In Proksch, P., and Müller, W.E.G. (eds.), Frontiers in Marine Biotechnology, pp. 175–224. Norfolk, England, Horizon Bioscience.

Ratain, M.J., Undevia, S., Janisch, L., Roman, S., Mayer, P., Buckwalter, M., Foss, D., Hamilton, B.L., Fischer, J., Bukowski, R.M., 2003. Phase I and pharmacological study of HTI-286, a novel antimicrotubule agent: Correlation of neutropenia with time above a threshold serum concentration. Proc. Am. Soc. Clin. Oncol. 22, 129.

Remiszewski, S.W., 2003. The discovery of NVP-LAQ824: From concept to clinic. Curr. Med. Chem. 10, 2392–2402.

Rinehart, K.L., Gloer, J.B., Cook, J.C., 1981a. Structures of the didemnins, antiviral and cytotoxic depsipeptides from a Caribbean tunicate. J. Am. Chem. Soc. 103, 1857–1859.

Rinehart, K.L., Gloer, J.B., Hughes, R.G., Renis, H.E., McGovren, J.P., Swynenberg, E.B., Stringfellow, D.A., Kuentzel, S.L., Li, H.L., 1981b. Didemnins: Antiviral and antitumor depsipeptides from a Caribbean tunicate. Science 212, 933–935.

Rinehart, K.L., Holt, T.G., Fregeau, N.L., Stroh, J.G., Keifer, P.A., Sun. F., Li, H.L., Martin, D.G., 1990. Ecteinascidins 729, 743, 745, 759A, 759B, and 770: Potent antitumor agents from the Caribbean tunicate *Ecteinascidia turbinate*. J. Org. Chem. 55, 4512–4515.

Rinehart, K.L., Warwick, R.A., Freqeau, N.L., 2000. Spisulosine compounds. United States Patent # 6107520.

Rodriguez, A.D., Akee, R.K., Scheuer, P.J., 1987. Two bromotyrosine-cysteine derived metabolites from a sponge. Tetrahedron Lett. 42, 4989–4992.

Salcedo, M., Cuevas, C., Alonso, J.L., Otero, G., Faircloth, G., Fernandez-Sousa, J.M., Avila, J., Wandosell, F., 2007. The marine sphingolipid-derived compound ES 285 triggers an atypical cell death pathway. Apoptosis 12, 395–409.

Sánchez-Pedregal, V.M., Kubicek, K., Meiler, J., Lyothier, I., Paterson, I., Carlomagno, T., 2006. The tubulin-bound conformation of discodermolide derived by NMR studies in solution supports a common pharmacophore model for epothilone and discodermolide. Angew. Chem. Int. Ed. 45, 7388–7394.

Sasaki, M., Matsumori, N., Maruyama, T., Nonomura, T., Murata, M., Tachibana, K., Yasumoto, T., 1996. The complete structure of maitotoxin. Part I. Configuration of the C1–C14 side chain. Angew. Chem. Int. Ed. 35, 1672–1675.

Sawant, S.S., Sylvester, P.W., Avery, M.A., Desai, P., Youssef, D.T.A., El Sayed, K.A., 2004. Bioactive rearranged and halogenated semisynthetic derivatives of the marine natural product sarcophine. J. Nat. Prod. 67, 2017–2023.

Sawnat, S.S., Youssef, D.T.A., Reiland, J., Ferniz, M., Marchetti, D., El Sayed, K.A., 2006. Biocatalytic and antimetastatic studies of the arine cembranoids sarcophine and 2–epi-16–deoxysarcophine. J. Nat. Prod. 69, 1010–1013.

Shim, J.S., Lee, H-S., Shin, J., Kwon, H.J. 2004. Psammaplin A, a marine natural product, inhibits aminopeptidase N and suppresses angiogenesis *in vitro*. Cancer Lett. 203, 163–169.

Sigel, M.M., Wellham, L.L., Lichter, W., Dudeck, L.E., Gargus, J.L. Lucas, L.H., 1970. In food-drugs from the sea. In Youngken, H.W., (ed.), Proceedings, 1969, pp. 281–295. Washington, DC, Marine Technology Society.

Simmons, T.L., Andrianasolo, E., McPhail, K., Flatt, P. Gerwick, W.H., 2005. Marine natural products as anticancer drugs. Mol. Cancer Ther. 4, 333–342.

Straight, A.M., Oakley, K., Moores, R., Bauer, A.J., Patel, A., Tuttle, R.M., Jimeno, J., Francis, G.L., 2006. Aplidin reduces growth of anaplastic thyroid cancer xenografts and the expression of several angiogenic genes. Cancer Chemother. Pharmacol. 57, 7–14.

Sudek, S., Lopanik, N.B., Waggoner, L.E., Hildebrand, M., Anderson, C., Liu, H., Patel, A., Sherman, D.H., Haygood, M.G., 2007. Identification of the putative bryostatin polyketide synthase gene cluster from "*Candidatus EndoBugula sertula*," the uncultivated microbial symbiont of the marine bryozoan *Bugula neritina*. J. Nat. Prod. 70, 67–74.

Sugie, Y., Hirai, H., Inagaki, T., Ishiguro, M., Kim, Y-J., Koojima, Y., Sakakibara, T., Sakemi, S., Sugiura, A., Suzuki, Y., Brennan, L., Duignan, J., Huang, L.H., Sutcliffe, J., Nakao, K., 2001. A new antibiotic CJ-17665 from *Aspergillus ochraceus*. J. Antibiot. 54, 911–913.

Suzuki, T., Suzuki, M., Furusaki, A., Matsumoto, T., Kato, A., Imanaka, Y., Kurosawa, E., 1985. Teurilene and thyrsiferyl 23–acetate, meso and remarkably cytotoxic compounds from the marine red alga *Laurencia obtuse* (Hudson) Lamourroux. Tetrahedron Lett. 26, 1329–1332.

Talpir, R., Benayahu, Y., Kashman, Y., Pannell, L., Schleyer, M., 1994. Hemiasterlin and geodiamolide TA: Two new cytotoxic peptides from the marine sponge *Hemiasterella minor* (Kirkpatrick). Tetrahedron Lett. 35, 4453–4456.

von Nussbaum, F., 2003. Stephacidin B: A new stage of complexity within prenylated indole alkaloids from fungi. Angew. Chem. Int. Ed. 42, 3068–3071.

Wright, A.E., Forleo, D.A., Gunawardana, G.P., Gunasekera, S.P., Koehn, F.E., McConnell, O.J., 1990. Antitumor tetrahydroisoquinoline alkaloids from the colonial ascidian *Ecteinascidia turbinate*. J. Org. Chem. 55, 4508–4512.

Zhang, X., Kundoor, V., Khalifa, S., Zeman, D., Fahmy, H., Dwivedi, C., 2007. Chemopreventive effects of sarcophine-diol on skin tumor development in CD-1 mice. Cancer Lett. doi:10.1016/j.canlet.2007.01.009.

STUDY QUESTIONS

1. As discussed in the chapter, many of the anticancer agents isolated from marine invertebrates are believed to derive from associated or symbiotic microorganisms. Suppose you had the good fortune to discover such a useful compound from a tropical sponge but were unable to isolate or culture the drug producing bacterium. How would you demonstrate which organism is, in fact, the biosynthetic source of your new anticancer agent?
2. Cryptophycin 1, shown here, is produced by cyanobacterium *Nostoc* sp., and although not discussed in the chapter, it is one of the most potent antimitotic natural products yet discovered.

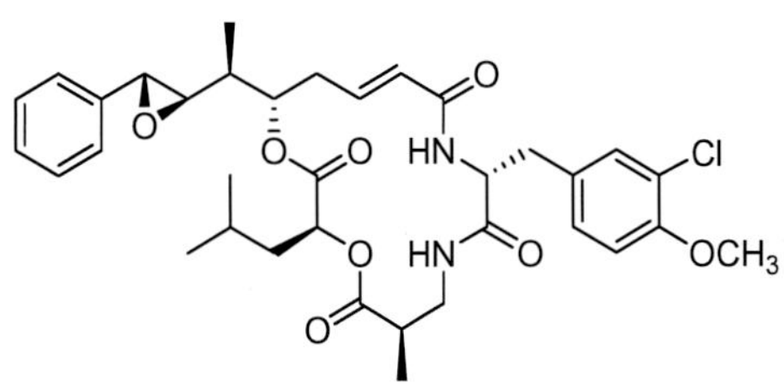

Based on its molecular structure, propose a biosynthetic pathway for cryptophycin 1. [For an answer to this question, see Beck *et al.* (2005).]

3. The chapter presented the material along taxonomic lines of the source from which the anticancer-type compounds were isolated. However, there are many other concepts upon which to organize this chapter. How else could you have arranged this material, and why?
4. Where do you feel the greatest opportunity for scientific advance exists within the field of marine natural products chemistry? What about the greatest opportunity for a contribution of long lasting societal value?

CHAPTER

23

Discovering Anti-infectives from the Sea

GUY T. CARTER

INTRODUCTION

Infectious diseases are among the most common ailments affecting the health, well-being, and productivity in the world today. A number of different classes of organisms may cause infections, including bacteria, fungi, helminthes (worms), protozoa, and viruses. Those infections caused by helminthes and protozoans are often referred to as parasitic diseases. Most parasitic diseases have received scant attention as subjects for new drug discovery, although they are common, especially in the southern hemisphere in underdeveloped regions. The resurgence of malaria (see Chapter 2), however, has resulted in greater efforts aimed at developing new medicinal agents for this devastating parasitic disease caused by protozoan parasites of the genus *Plasmodium*. Bacterial and fungal infections have shown the greatest susceptibility to chemotherapy, whereas viral infections are more difficult to treat and often subjected to immunological approaches. Human immunodeficiency virus (HIV) infections are the obvious exception because of the difficulty in developing effective vaccines. Although considerable effort has been aimed at the discovery of marine natural products effective against HIV infections, the clinically useful agents to date derive from synthetic medicinal chemistry.

This discussion on anti-infectives mainly focuses on bacterial infections, owing to the preponderance of research in this area. General principles related to the discovery of new antibiotics from the sea also hold for antifungal, antiviral, or other types of anti-infective agents. A historically applied definition of an antibiotic is any compound produced by a microorganism that is toxic to, or inhibitory to the growth of, other microbes. Under this definition, all antibiotics are natural products, as distinguished from the more general class of antimicrobial agents, which can be synthetically derived. These terms have become more difficult to strictly apply because of the large number of semisynthetic derivatives of antibiotics that have been developed. In the context of this chapter, however, the term *antibiotic* is used as originally intended and only references natural products—although the organism that produces the compound may be other than a microbe.

During the golden age of antibiotic discovery that spanned from the 1940s through the 1960s, most major pharmaceutical companies around the world were engaged in the discovery of new antibiotics. These antibiotics were discovered by cultivation (fermentation) of microorganisms, followed by the isolation and purification of products from the resulting fermentation broths. Microbial sources for these efforts were derived from the environment, mainly through the isolation of bacteria from soil samples. Soil is an incredibly rich source of microbes and microbial diversity (Handlesman, 2004). The most productive group of microbes was found to be the actinomycetes, a highly diversified division of the bacterial family tree. It was from these actinomycetes, primarily of the genus *Streptomyces*, that most of the progenitor antibiotics that led to the world's first real antibacterial pharmacopeia were derived. Structure diagrams of representative examples of clinically significant antibiotics discovered during this time are drawn in Figure 23-1. Many of these agents remain in use today, particularly as their semisynthetic derivatives, despite the alarming increase in resistance developed by many pathogenic strains of bacteria.

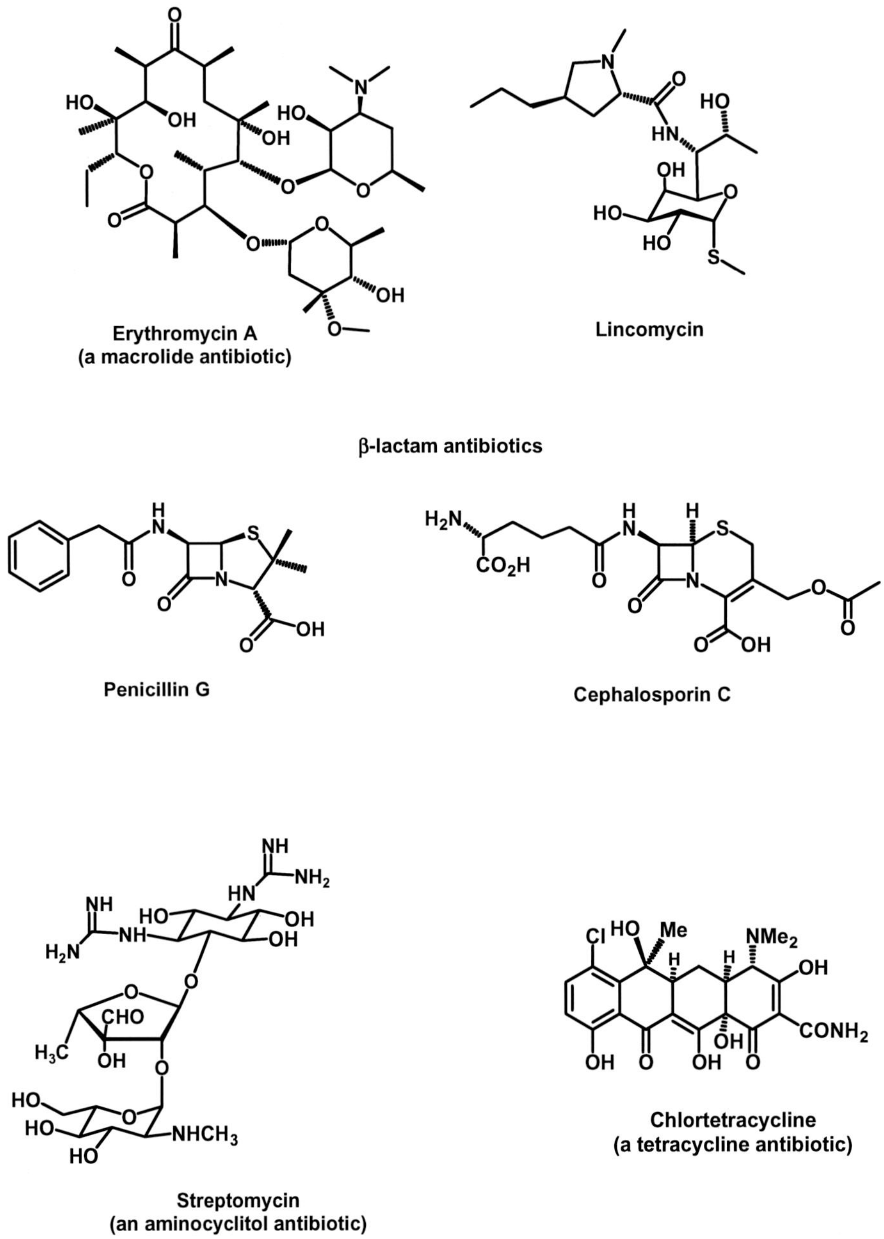

FIGURE 23-1. Examples of antibiotics derived from soil bacteria during the golden age of antibiotic discovery.

The widespread and sometimes indiscriminate use of antibiotics has resulted in the development of resistance mechanisms by pathogenic bacteria that render many antibiotics ineffective. This is particularly serious in hospital settings, where the use of antibiotics is constant. In response to this high antibiotic "pressure," many organisms have developed resistance to more than one antibiotic chemotype. Multiply resistant *Staphylococcus aureus* is one such highly resistant pathogen that is often targeted in antibiotic discovery programs. Some *S. aureus* strains can now only be effectively controlled with the class of glycopeptide antibiotics exemplified by vancomycin. Vancomycin has recently been proclaimed to be the last line of defense against resistant bacteria, and there are reports of emerging resistance to it as well (Singh and Barrett, 2006). The reader is referred to a review by Wright that offers a perspective on the evolutionary development of antibiotic resistance (Wright, 2007). The structure for vancomycin is as follows:

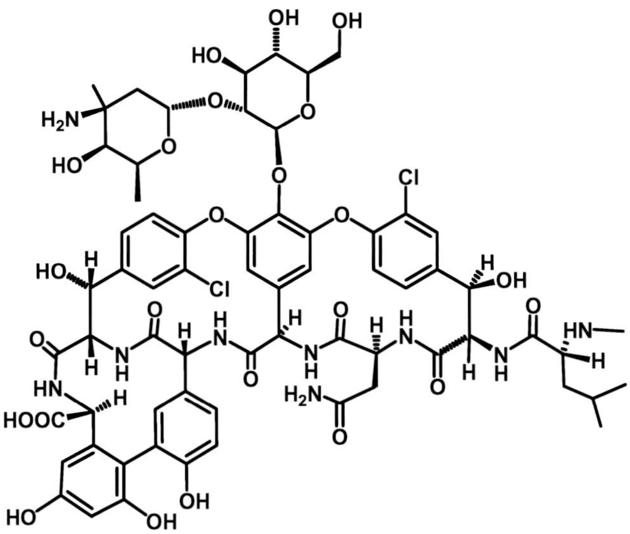

Vancomycin

The continued development of resistance to all agents, including glycopeptides, is an ongoing process. There is no question that pathogenic bacteria will ultimately become resistant to vancomycin; the only unknown parameter is how rapidly this resistance will spread. As long as vancomycin usage escalates, the rate of development and spread of resistance will also increase. Without the development of new chemotypes that lack cross-resistance to existing antibiotics, we are in jeopardy of returning to the preantibiotic era where effective treatment of serious infections is increasingly difficult. The Infectious Diseases Society of America (IDS) estimates that about 2 million hospitalized patients will contract an infection in the hospital annually and that as many as 90,000 of them will die as a result of this infection (IDS, 2004). Resistance to existing agents is a critically important factor that should encourage renewed efforts in the discovery of novel antibiotics.

One shocking result of the development of bacterial resistance is the reemergence of tuberculosis (TB) as a major threat to global human health. This disease was managed by treatment with one of a few standard agents, such as rifamycin or its semisynthetic analogs, throughout the 1960s and 1970s. By the mid-1980s, however, highly resistant *Mycobacterium tuberculosis* strains had emerged whose control required more rigorous chemotherapy. The course of treatment for resistant TB is arduous, requiring long-term dosing (6 to 9 months) of multiple agents to ensure that the infection has been reduced to the point where a healthy immune system can maintain control of bacterial growth and thereby mitigate the progression of disease. Failure to complete the course of therapy increases the likelihood that an individual will retain highly resistant forms of the bacterium, which may then be spread to others. The other major complicating factor is the prevalence of the disease in immunocompromised patients who are co-infected with HIV. Given the logistical obstacles to delivering therapy, particularly in underdeveloped countries, it is highly unlikely that the disease will be managed effectively with the current drug regimen. Coupled with the poor prognosis of recovery in immunocompromised patients, this means that the spread of resistant TB will proceed largely unchecked until a new effective therapy is available. Under the current scenario it has been estimated that 30 million people will succumb to TB infections by 2020 (De Souza, 2006). The structure of rifamycin S is as follows:

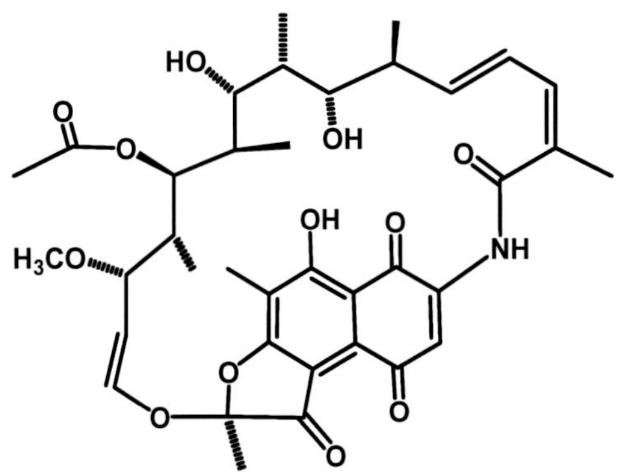

Rifamycin S

Despite the clear indication of current and escalating unmet medical need, there have been relatively few new additions into the antibiotic arsenal since 1970. Most significantly, there have been only three new chemotypes approved for marketing during this period (Butler and Buss, 2006). The reasons for this lack of productivity are manifold. Unquestionably, one factor has been the difficulty of discovering new antibiotic chemotypes. Continued screening of the usual soil-derived microbes for generalized antibiotic activity against a panel of susceptible bacteria most often leads to well-known classes of antibiotics. One of the limiting factors to this conventional approach is the lack of innovative methods to capture the microbial diversity present in the soil. Only those organisms that are amenable to cultivation in the laboratory have been evaluated for the production of antibiotics. Various estimates suggest that those cultivable organisms represent less than 1% of the viable microbes in the soil (Torsvik *et al.*, 1996), and the genomic evidence indicates that the remaining 99% is highly diverse (Torsvik *et al.*, 1990). It is intriguing to speculate as to why so few of the bacterial genera that reside in the soil have been cultivated in the laboratory. Is it due to a lack of effort? Or is it a lack of fundamental knowledge of the physiology and growth requirements of these organisms? This untapped biodiversity may have the biosynthetic potential to produce the new antibiotics of the future; however, significant research effort remains in order to render that biosynthetic capability accessible (Clardy *et al.*, 2006). Therefore, although the soil has continued promise, alternate sources of chemical diversity whose potential is more readily realized are highly desirable.

MARINE NATURAL PRODUCTS

Natural product chemists have been fully engaged in the discovery of novel secondary metabolites from marine organisms since about 1970 when the effort gained significant momentum in several academic laboratories around the world. Initially these efforts were focused on readily available organisms, particularly invertebrates, whose chemistry was intriguing. Many of the secondary metabolites derived from these investigations were cytotoxic, and some appeared to have potential as anti-infective agents. As the search for meaningful biological activities became more focused, bioassays were employed to differentiate generally cytotoxic compounds from those with more specific and selective activities (Rinehart *et al.*, 1981). Some early examples of marine natural products that exhibited antimicrobial activity are illustrated in Figure 23-2. Manoalide, a sesterterpene containing an unusual unsaturated lactol functionality, was extracted from the sponge *Luffariella variabilis* collected in Palau (De Silva and Scheuer, 1980). Another highly modified sesterterpene, variabilin, was obtained from the sponge *Ircinia variabilis* collected in the Gulf of California (Faulkner, 1973). Of particular interest during this period of investigation were halogenated compounds, such as the so-called bromotyrosine metabolites from sponges (e.g., aerothionin; Fatturosso *et al.*, 1970) and the polyhalogenated terpenes, such as elatol, extracted from the red algae *Laurencia elata* (Sims *et al.*, 1974). Naturally occurring peroxides are relatively rare in nature. The marine environment has yielded a number of such compounds; one of the earliest was plakortin, obtained from the sponge *Plakortis halichondriodes*, collected in the Caribbean Sea along the coast of Panama (Higgs and Faulkner, 1978). Each of these compounds has structural features that are rare or unknown in terrestrial products and therefore supported the hypothesis that new biodiversity would translate into new chemical diversity.

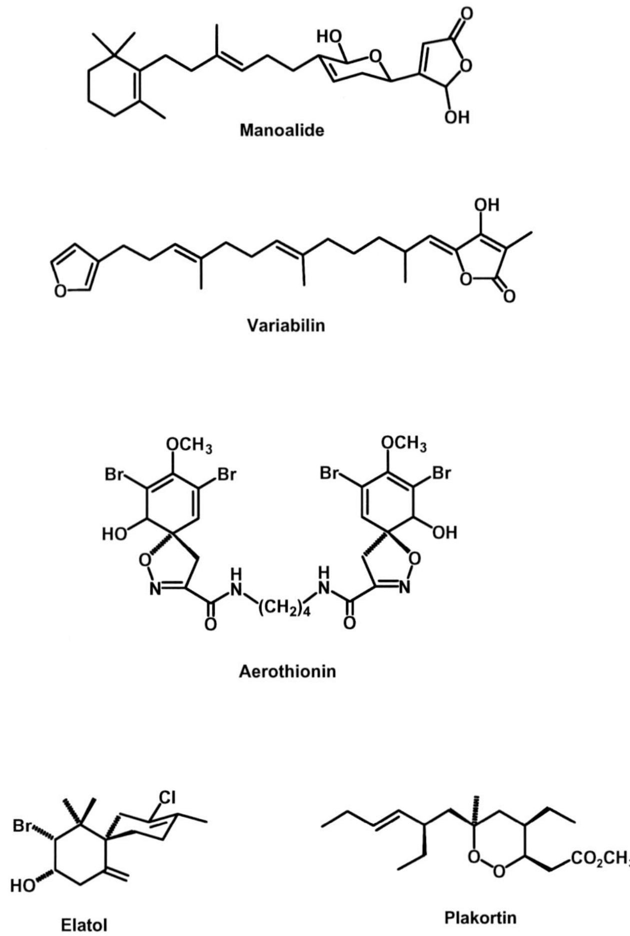

FIGURE 23-2. Examples of antibacterial marine natural products isolated in the 1970s.

THE SUPPLY ISSUE

All of the compounds shown in Figure 23-2 were extracted from invertebrates or macroalgae. Generally, enough material (tens of milligrams) could be obtained from a small-scale collection of the marine organism for antibiotic susceptibility testing as well as preliminary laboratory experiments aimed at defining a mechanism of action. Given favorable potency and spectrum of activity, the next phase of experimentation was testing the compound for *in vivo* activity in an animal model of infection. These animal models represent the most stringent tests for a new compound undergoing evaluation as a potential new clinical agent during the earliest stages of drug development. The ability of an agent to mitigate the course of an induced infection in a live animal, often a rodent, is the first indication that the compound has pharmaceutical potential. Successful protection of the host animal is both an indication of efficacy (bioavailability and access to the infectious agent) and safety (lack of acute toxicity to the host). Material requirements for these *in vivo* experiments (hundreds of milligrams) are such that either a significant recollection of the source organism is required or the material needs to be prepared synthetically. This matter is commonly referred to as the "supply issue." The supply issue becomes increasingly critical as the candidate compound moves through a drug development cycle, as material requirements escalate from milligrams to grams to kilograms. For an invertebrate-derived compound, the most practical method of production is through total synthesis. Although this is often daunting because of the exquisite structural complexity of many of these metabolites, there are successful examples of moderately large-scale syntheses of invertebrate-derived marine natural products. Preparation of hemiasterlin analogs (Zask *et al.*, 2005) illustrates a process whereby the individual amino acid components can be synthesized and then coupled, leading to excellent yields of final products. Similarly, discodermolide was prepared in a convergent synthetic procedure that provided sufficient material for clinical testing (Mickel *et al.*, 2004). In these examples total synthesis overcomes the supply issue. Given the continued advancement of synthetic technologies, it is expected that synthetic processes could supply virtually any natural product with commercial potential (Wilson and Danishefsky, 2006). The structures of hemiasterlin and discodermolide are as follows:

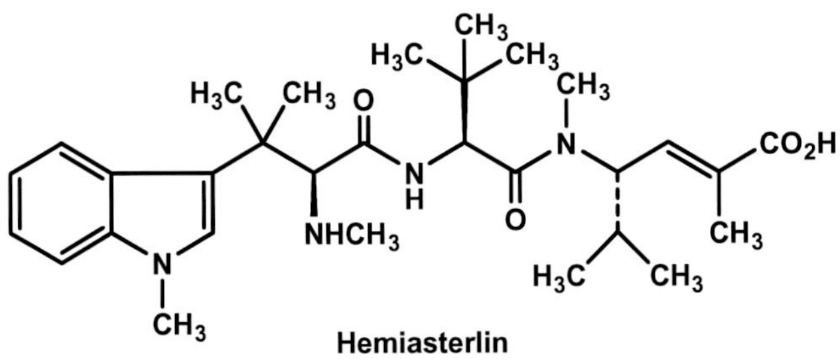

Hemiasterlin

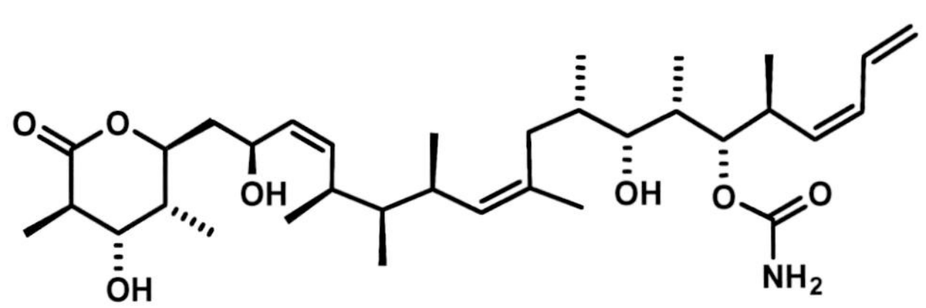

Discodermolide

MARINE NATURAL PRODUCTS AS ANTI-INFECTIVES

Early surveys of the biological activities found in extracts of various classes of marine organisms revealed that a strikingly high percentage of these organisms, particularly the sponges, contained compounds that had some type of antimicrobial activity (Rinehart *et al.*, 1981; Thompson *et al.*, 1985). Many of the secondary metabolites isolated from marine organisms were able to inhibit the growth of bacteria or fungi in classical "zone of inhibition" assays. In these assays, the test organisms are seeded onto Petri dishes containing agar-based media and in the absence of inhibitors will grow into a confluent "lawn." Compounds can be evaluated for their ability to inhibit the growth of the seeded organism by applying known quantities on the agar surface before incubation. The potency of the compound is typically reflected in the size of the zone (diameter) of clearing (i.e., the lack of a lawn) around the applied samples. An example of this technique is shown in Figure 23-3. The assay can be used as a method of detection for antibacterial activity, as shown in Figure 23-3a, where test samples are applied to determine whether or not antibiotic activity is present. It is also highly valuable as a semiquantitative method as indicated by the dilution series shown in Figures 23-3b and c. Tests like the agar plate assay have been exceptionally useful in guiding the isolation of active natural products from crude extracts through sequential rounds of fractionation leading to pure compounds, as illustrated in Figure 23-4. This figure depicts the process that natural products chemists use to obtain purified, biologically active compounds. The process is highly experimental with the choice of fractionation methods relying on experience. Once biologically active fractions are obtained they are assessed for homogeneity (purity) before combining those fractions that might contain the same compound. Once homogeneous, the compound is subjected to spectroscopic analysis to determine its structure. Most of the marine "antibiotics" isolated in the 1970s, such as those shown in Figure 23-2, lacked the potency of the bona fide antibiotics of Figure 23-1. The relatively weak antibacterial activity of these isolated

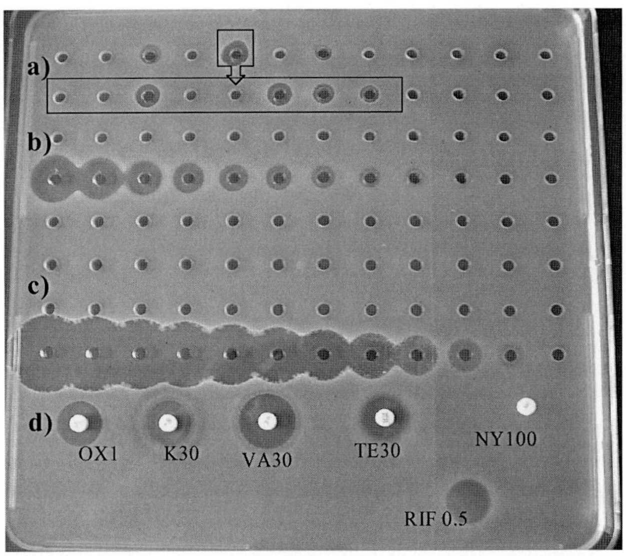

FIGURE 23-3. Zone of inhibition assay on an agar plate seeded with *Bacillus subtilis*. (a) Top row wells contain test samples of broth from newly isolated actinomycetes. Boxed arrow indicates a culture that when cultivated in eight separate media gave the results shown in the second row. (b) Serial twofold dilutions of rifampicin (starting amount 1 μg/well). (c) Serial twofold dilutions of penicillin G (starting amount 2 μg/well). (d) Standards: OX1 is oxacillin (1 unit on a paper disc); K30 is kanamycin (30 μg/disc); VA30 is vancomycin (30 μg/disc); TE30 is tetracycline (30 μg/disc); NY100 is nystatin (100 μg/disc); RIF 0.5 is 0.5 μg of rifampicin applied directly to the agar surface.

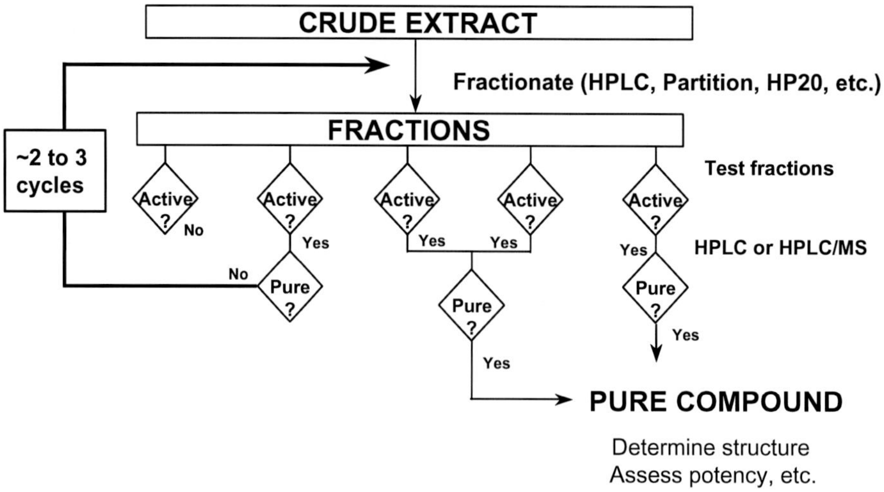

FIGURE 23-4. Bioassay-guided fractionation of natural products.

compounds was disappointing and somewhat surprising given the apparent antibacterial potency found for some of the crude extracts (Rinehart *et al.*, 1981; Thompson *et al.*, 1985).

With the advent of powerful separation technologies, such as high-pressure liquid chromatography (HPLC) and sensitive two-dimensional NMR techniques for structure determination, the field of marine natural products chemistry began to flourish, beginning around 1980. There was a stunning transition from discoveries of relatively abundant and structurally simple compounds, such as those shown in Figure 23-2, to highly potent and complex metabolites. Several of these metabolites are covered in the following sections. A review by Faulkner further highlights the uniqueness of structural features found in marine natural products (Faulkner, 2000).

MARINE MICROBES AS SOURCES FOR NEW ANTIBIOTICS

Marine microbes, as sources for new secondary metabolites, were "discovered" by natural products chemists in the early 1990s. In part this move into marine microbiology was driven by the desire to explore new sources of natural products, owing to the extensive examination of marine algae, sponges, and tunicates during the 1970s and 1980s. Scientists were beginning to find the same compounds repeatedly in their chemical and biological investigations of marine macroorganisms. To find novel metabolites from these common sources, "dereplication" of the materials extracted had become necessary. Dereplication is a process originally defined in the search for antibiotics from soil bacteria in which the products derived from newly isolated bacteria had to be compared against well-known antibiotics to preclude rediscovery of compounds. The earliest dereplication methods for antibiotics employed paper chromatography to separate the components of interest in comparison with expected reference compounds. The development of directly coupled HPLC and mass spectrometry (HPLC/MS) revolutionized the dereplication process, making the identification of known compounds more definitive and eliminating the need for collections of reference compounds (Nielsen and Smedsgaard, 2003). A schematic representation of HPLC/MS-based dereplication is illustrated in Figure 23-5.

Figure 23-5 illustrates the power of coupling a separation method (HPLC) with information-rich detection methods. In this case, the PDA component is a photodiode array ultraviolet absorption detector that records full UV-visible absorption spectra across the entire chromatogram. The effluent stream from the HPLC column is split after the PDA, the majority of which is collected in a bioassay plate and the remainder is directed to an electrospray interface into a mass spectrometer. Analogous to the PDA, the mass spectrometer records mass spectra across the entire chromatographic run. Once the bioassay data pinpoint the active components by correlation with retention times, the data systems are queried to retrieve both UV-visible and mass spectrometric data. These data are then matched against known compounds catalogued in various databases, allowing one to determine whether the compound has been reported previously. In the hypothetical example shown, the active peak has a molecular weight of 612 amu and UV absorption bands around 240 nm, which leads to a prediction of nemadectin α as a possible match.

An equally motivating factor driving the exploration of microbial sources was that several marine natural products derived from invertebrates were closely related in chemical structure to compounds previously known from terrestrial bacteria. One telling example is illustrated in Figure 23-6, which shows the structures of ecteinascidin 743, renieramycin A, and saframycin A. The saframycins were the first of

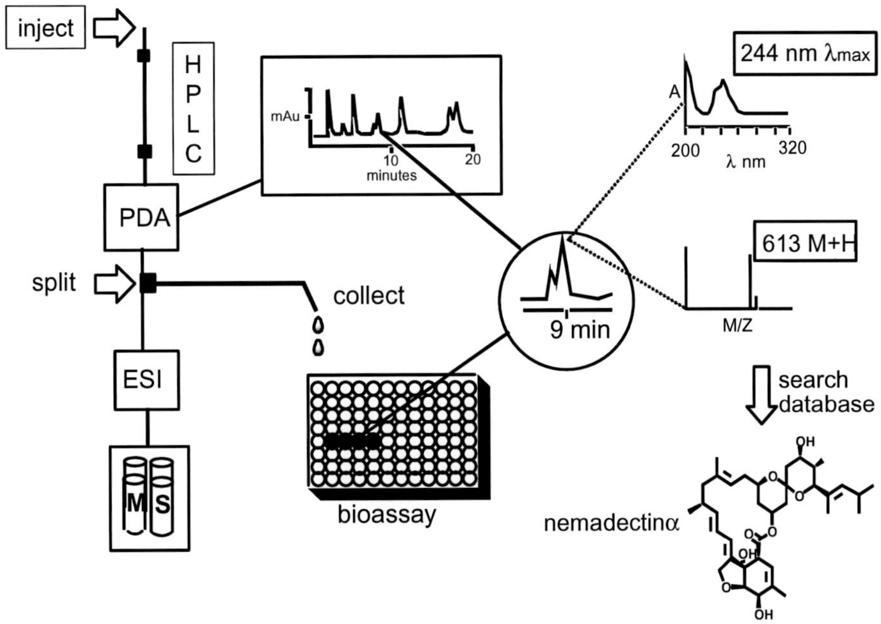

FIGURE 23-5. Schematic of HPLC/UV/MS system for dereplication of natural products.

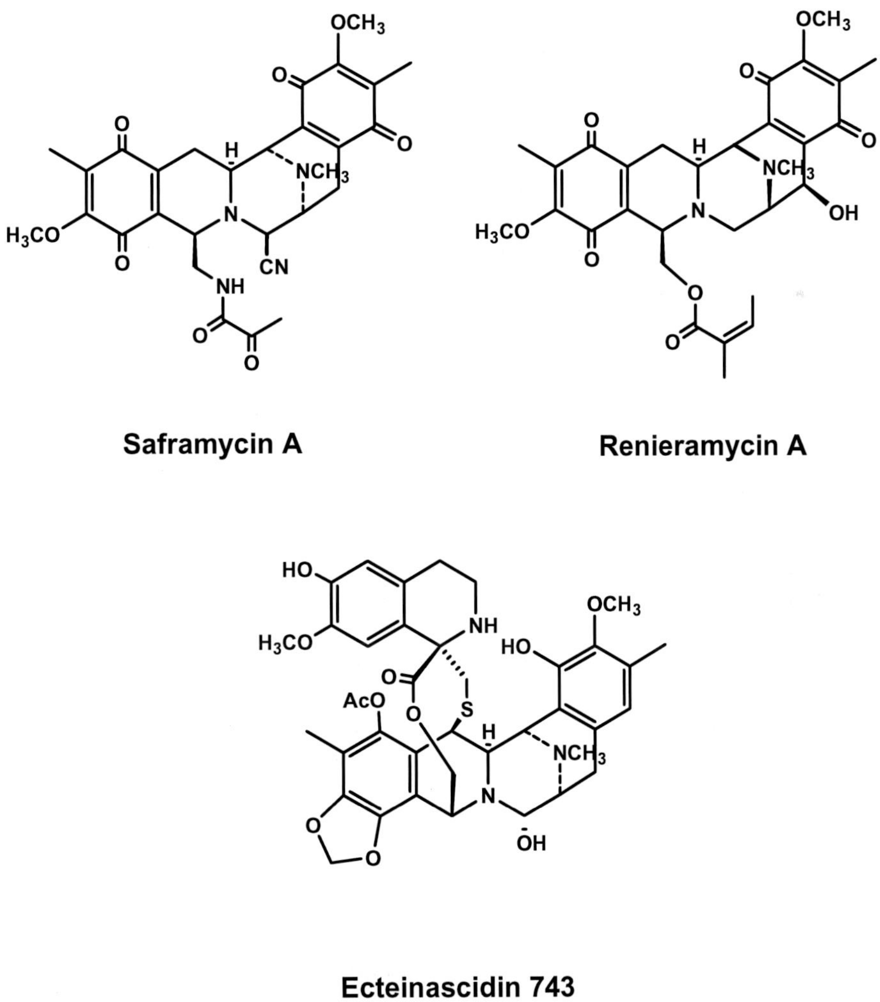

FIGURE 23-6. Structures of saframycin A, renieramycin A and ecteinascidin 743.

these compounds to be reported, having been isolated from the terrestrial actinomycete *Streptomyces levendulae* in the late 1970s (Arai *et al.*, 1980). The renieramycins (as well as saframycins A, B, and C) were described as constituents of the marine sponge *Reniera* sp. collected off the coast of Mexico (Frincke and Faulkner, 1982) in 1982. The ecteinascidins (Rinehart *et al.*, 1990) were isolated in 1986 from the colonial tunicate *Ecteinascidia turbinata* collected in the Caribbean. These compounds not only share the same core structural framework, but their reported biological properties, including antibiotic and antitumor activities, coincide as well. Do biosynthetic enzymes encoded by the genomes of the invertebrates produce the ecteinascidins and renieramycins (and saframycins)? Or are commensal bacteria involved in some or all of the biosynthetic production of these compounds? If this were an isolated example, the argument for common biosynthetic pathways between bacteria and marine invertebrates might have sufficed; however, such was not the case, and much discussion ensued. The speculative nature of this debate is captured quite well in the following quote from Faulkner (Faulkner *et al.*, 2000).

> In recent years, it has been fashionable for marine natural products chemists to propose that certain compounds and classes of chemicals are produced by symbiotic bacteria. While they may indeed be correct in some cases, the arguments used to support their proposals are often simplistic and require closer examination. Is it prudent to assume that a sponge metabolite was produced by a bacterial symbiont just because the structure of the sponge metabolite resembles that of a metabolite from a cultured bacterium, which has usually been isolated from a terrestrial source? While these observations might be helpful in designing an experimental program, they are by no means sufficient to define a bacterial source for a sponge metabolite. Other tests must be applied before assigning the source of a natural product to a symbiont.

In 1996, a discovery was reported that shifted the momentum of the debate. An antibiotic that bore the unmistakable signature of a microbial product was reported from a marine invertebrate. Namenamicin, an enediyne-containing metabolite (Fig. 23-7), was isolated from the marine ascidian *Poly-*

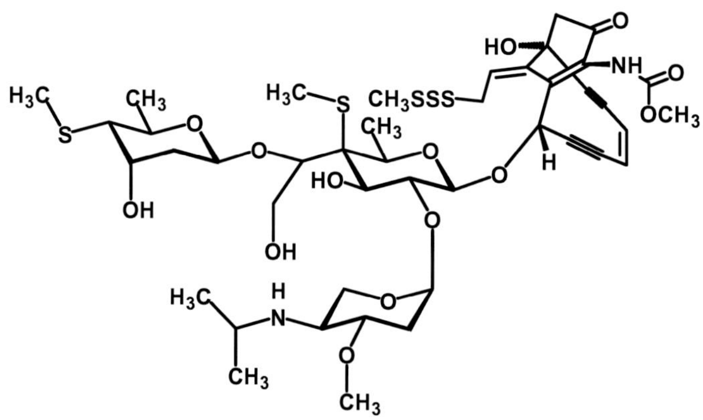

Namenamicin

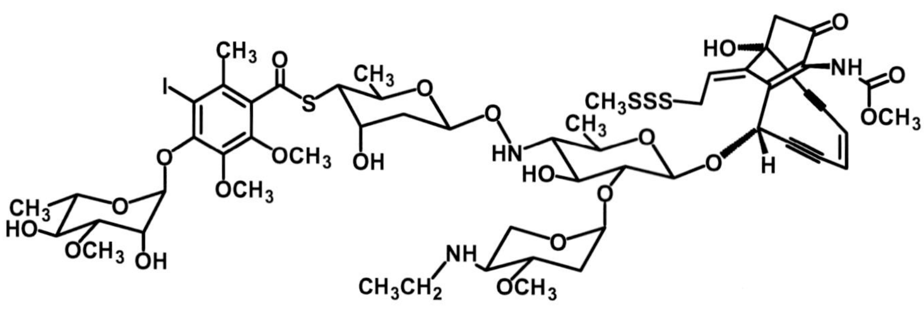

Calicheamicin

FIGURE 23-7. The enediyne antibiotics namenamicin and calicheamicin.

syncraton lithostrotum collected in Fiji (McDonald *et al.*, 1996). Enediynes, as exemplified by calicheamicin (Fig. 23-7), were previously uniquely derived from actinomycetes. The biosynthetic pathways described for the enediynes, including genes for the polyketide-derived aglycone that forms the reactive "warhead" of these compounds, have the hallmarks of bacterial origin (Ahlert *et al.*, 2002). This case, as well as a growing body of additional examples, has greatly strengthened the hypothesis that symbiotic bacteria were at least partially responsible for the production of many secondary metabolites isolated from their hosts. The clearest evidence in favor of this hypothesis would be the isolation of a bacterium from the host invertebrate that, when subjected to cultivation in the laboratory, produces the host metabolite in question. Surprisingly, this hypothesis has proved difficult to support in that there are only a few examples in which a microbe has been isolated from the macroorganism and demonstrated to produce the compound of interest. Some scientists have argued that the inability to successfully cultivate the bacteria responsible for the biosynthesis of secondary metabolites in their invertebrate hosts is not surprising because true symbionts will not be readily cultivable. Genomic approaches are beginning to provide some insights as researchers are able to predict the biosynthetic capacities of microorganisms by cataloging their genomes (Piel *et al.*, 2005). Given the ability to recognize a gene cluster that has features common to a particular biosynthetic pathway, it is possible to predict that an organism has the potential to produce the metabolite. The enediyne biosynthetic pathway has several unique genetic features that allow its ready detection by genomic bioinformatics. Remarkably, the presence of enediyne biosynthetic capability is much more widespread in bacteria than was previously recognized; in one study amounting to approximately 15% of the strains from randomly isolated soil actinomycetes (Zazopoulus *et al.*, 2003). It has become abundantly clear that much of the biosynthetic capabilities of a bacterium are latent. The challenge remains to demonstrate in the laboratory that a pathway is productive, either by inducing the microbe to express the pathway or by transferring it to a capable surrogate host.

One such example where the collaborative synthesis of secondary metabolites between bacteria and invertebrates has been demonstrated is found in the work of Hamann and Hill on the production of manzamine A. The manzamines (Fig. 23-8) are β-carboline alkaloids that have been isolated from a number of different genera of marine sponges (Hu *et al.*, 2003). Manzamine can be derived independently by culturing an actinomycete of the genus *Micromonospora* that was isolated directly from the sponge tissue (R. T. Hill, personnel communication). Once the microbial production of manzamines has been optimized through medium and strain improvement strategies, the practical production of multigram quantities of the manzamine will be available

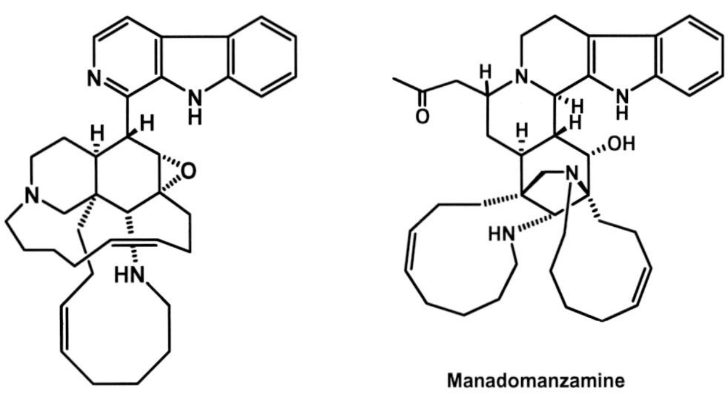

Manzamine B

Manadomanzamine

FIGURE 23-8. Manzamine B and manadomanzamine.

to facilitate their further development as anti-infective agents.

Although there has been limited success in isolating the putative microbial producers of these invertebrate-derived compounds, what has resulted from these experiments is the realization that invertebrate-associated microbes represent a valuable new source for the production of novel secondary metabolites (Donia *et al.*, 2006; Hill, 2004). In a study conducted by the Wyeth group, a host of highly diverse microbes was isolated from *L. patella*, the ascidian from which nemanamicin was isolated. These results are illustrated in Figure 23-9. Representative examples of microbes isolated from *L. patella* and their antibacterial products are shown. Four families of known compounds were revealed from laboratory cultivation of four highly diverse microorganisms. *Pelagiobacter variabilis*, a unique marine organism, was shown to produce new phenazine derivatives, subsequently reported by Imamura (Imamura *et al.*, 1997). Notably, an *Agrobacterium* species was also isolated from the acsidian, and it was shown to produce the structurally intriguing antibiotic thiotropocin. Rifamycin S was isolated from cultures of a third organism that was classified as a *Micromonospora* species. One fungal isolate from the experiment that was not taxonomically characterized was shown to produce a series of small aromatic compounds exemplified by chlorogentisyl alcohol; these were found to possess modest antibacterial activity. The most remarkable antibiotics discovered in these experiments are the lomaivitacins, which were derived from an organism initially identified as *Micromonospora lomaivitiensis* (He *et al.*, 2001). The lomaiviticins are symmetrical dimeric compounds that possess a number of highly unusual structural features including diazo groups. The lomaiviticins are both potent antibiotics and mammalian cell cytotoxins owing to their DNA-damaging activity.

During this period of decreasing accessibility of biodiversity through collection of macroorganisms, Fenical and Jensen at the Scripps Institution of Oceanography pioneered the exploration of marine sediments for productive microbes. A logical extension of the terrestrial realm (i.e., the soil), marine sediments have proved to be a rich source of new

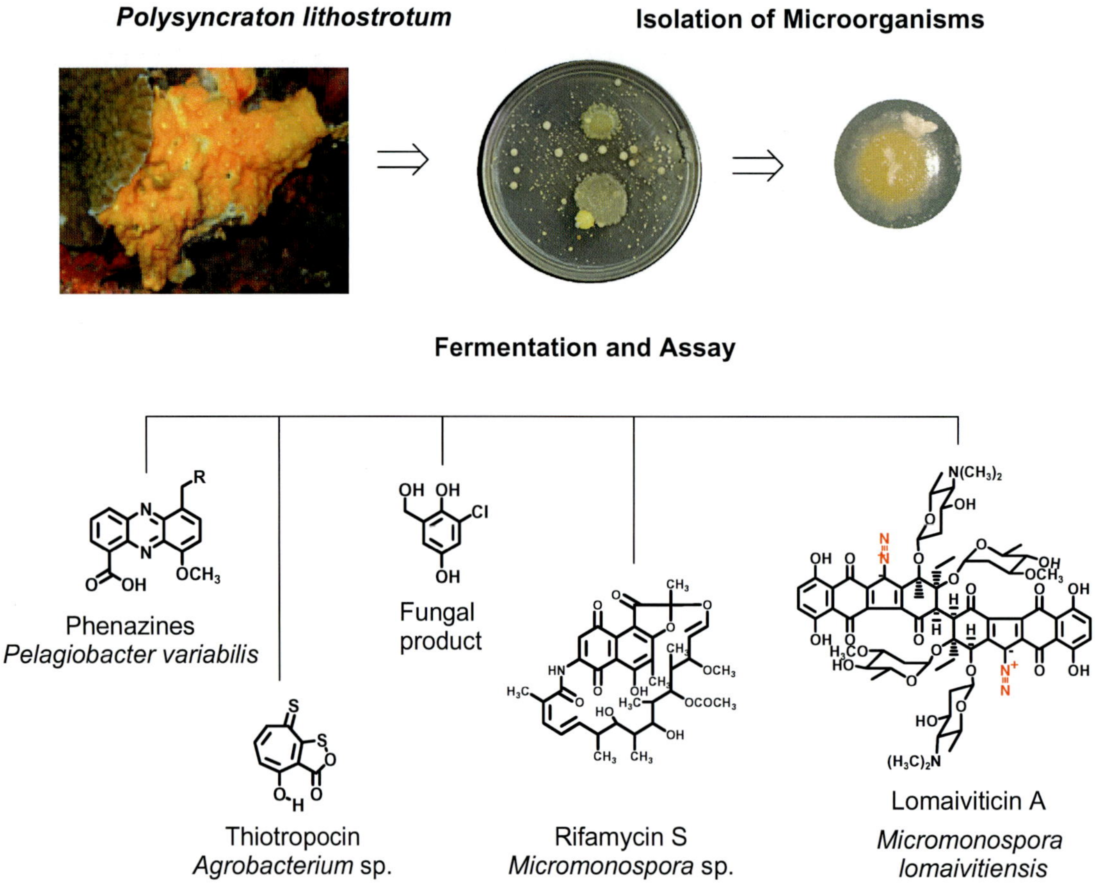

FIGURE 23-9. Microbial inhabitants of *Polysyncraton lithostrotum* and examples of antibiotics derived from laboratory cultivation.

marine biodiversity that had previously not been systematically studied (Fenical and Jensen, 2006). Their work has revealed the existence of several new genera of actinomycetes, particularly in deeper ocean sediments. Of those that have been cultured and tested for biological activity, the majority has yielded active metabolites. Significant new chemical diversity is being derived from these newly uncovered sources (Lam, 2006). One example of a new structure class isolated from a sediment-derived bacterium is the marinomycins. Marinomycin A is a potent antibiotic with activity against antibiotic-resistant strains of bacteria. The compound was isolated from laboratory-scale cultures of a new bacterial strain of the recently described genus designated *Marinospora*, obtained from a sediment sample collected off the coast of southern California (Kwon *et al.*, 2005). Structurally, marinomycin A combines a number of unusual features in its macrodiolide framework, which makes it distinct from the terrestrial counterparts. The structure of marinomycin A is as follows:

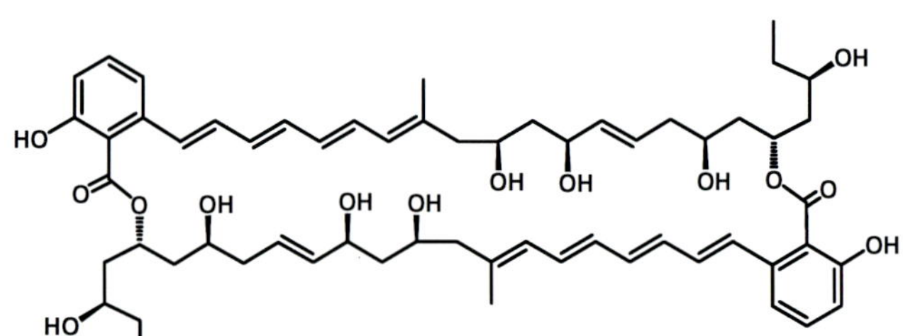

marinomycin A

It is important to note that marine microbial diversity is subject to the same limitations in accessibility as previously noted for terrestrial organisms. That is, to date only a small percentage (<1.0 %) has been successfully cultivated in the laboratory (Torsvik *et al.*, 1996). So although it is clear that some of the bacteria being discovered can be grown in the laboratory, there is much more biosynthetic potential among those that are not cultured, and their fruitful examination will require alternative strategies.

STRATEGIES FOR THE DISCOVERY OF NEW ANTIBIOTICS

How does one discover new metabolites with antibiotic or other specific anti-infective properties? As was touched on previously, when novel source organisms were first examined, such as the actinomycetes in the 1940s or marine macroorganisms in the 1970s, many of the metabolites were novel. In such cases, empirical biological assays, such as the agar plate-based growth inhibition assay shown in Figure 23-3, are adequate because many previously unrecognized metabolites are present. Once a novel compound is isolated, its mechanism of action will be determined by various means including chemical genetics (Spring, 2005). If a large number of new metabolites are available, then the prospect of finding one or more with the desired selective effects on bacteria, fungi, or other infectious organisms versus mammalian cells is reasonable.

The drive to access new biodiversity for drug discovery has motivated scientists to explore the ocean to uncover these hidden resources, and it does appear that marine microbes offer such promise. However, as was found with marine macroorganisms, it is clear that these unique marine microbes also produce well-known antibiotics, as was illustrated in the study depicted in Figure 23-9. For this reason it is desirable to have a screening assay that is more discriminating than one of simple bacterial growth inhibition. This selectivity can be accomplished by using panels of organisms with different susceptibilities, multiply resistant target organisms, or through highly focused assays in which the molecular target of interest is challenged. A successful targeted assay will allow the discovery of compounds with the desired mechanism of action in the presence of other antibiotics with various alternative mechanisms. The most effective anti-infective targets are those that focus on the unique physiology of the pathogen (bacteria, fungi, parasite, etc.) compared to mammalian cells.

The discovery of potent and selective agents active against antibiotic-resistant bacteria continues to be one of the most challenging research objectives of our time. As we alluded to previously, a large proportion of the marine natural products described to date can be shown to inhibit the growth of bacteria *in vitro*, albeit at impractically high concentrations. In the early work reported in the 1970s and 1980s, compounds were described as antibiotics that had marginal potency for the inhibition of bacterial growth. Most of these studies were focused on the unusual chemical characteristics of the marine products rather than details of their biological properties. More recently, as the critical need for new antibacterial agents has come into focus, researchers have been much more attentive to the pivotal properties of antibacterial potency and selectivity. Ideally, selectivity of action is achieved through choice of biochemical targets that are unique to the infecting organisms. However, there are cases in which an agent that does not have the desired specificity with respect to its target can also be useful. One alternative is through selective delivery of the antibiotic. A familiar example of the latter approach is the use of antibacterial ointments for topical use. These products are highly effective antibacterial agents that can be applied to skin to speed healing of wounds. The compounds used in such mixtures typically would not be effective if given systemically owing to poor absorption, distribution, or metabolic properties, or because of adverse effects on host tissues.

An excellent example of an approach for the discovery of a selective antibiotic from the marine environment is captured in the finding of abyssomicin (Fig. 23-10). In this work, the goal was to find compounds that inhibited bacterial growth through inhibition of p-aminobenzoic acid biosynthesis (Bister *et al.*, 2004). p-Aminobenzoic acid is a component of tetrahydrofolate, a key cofactor in amino acid metabolism, and is essential in bacteria but absent in mammals. Mammals require folate in their diet, and thus it is considered a vitamin. The synthetically derived sulfonamides, such as sulfanilamide, act by inhibition of this pathway in bacteria and were among the first antibacterial agents introduced into clinical practice. In the course of the research leading to the discovery of abyssomicin, Fiedler and coworkers screened the extracts of more than 200 organisms to identify a rare actinomycete of the genus *Verrucosispora* that had the desired p-aminobenzoate-reversible biological activity. Their *Verrucosispora* strain was derived from a deep-water sediment sample obtained in the Sea of Japan. The abyssomicins are structurally unique antibiotics that have already elicited great enthusiasm from synthetic chemists around the world (Nicolaou and Harrison, 2007). Abyssomicin C is responsible for the inhibitory activity demonstrated in the assay (abyssomicin B and D are essentially inactive). It is noteworthy that although numerous synthetic antibacterial agents have been prepared that interfere with p-aminobenzoic acid biosynthesis, abyssomicin C is the only known natural inhibitor.

Increasingly, marine natural products researchers have been focusing their efforts on the discovery of new antibiotics effective in the treatment of tuberculosis. Among the previously known families of marine natural products that have engendered renewed interest because of their anti-TB activity are the manzamines (Fig. 23-8). Hamann and coworkers have been actively exploring this class of marine alkaloids to enhance their potential as anti-tuberculosis agents. The manadomanzamines (Peng *et al.*, 2003), isolated from an Indonesian sponge of the genus *Acathostrongylophora*, collected in Manado Bay, show relatively potent activity *in vitro* against *Mycobacterium tuberculosis* strains. These compounds are presumably derived from the

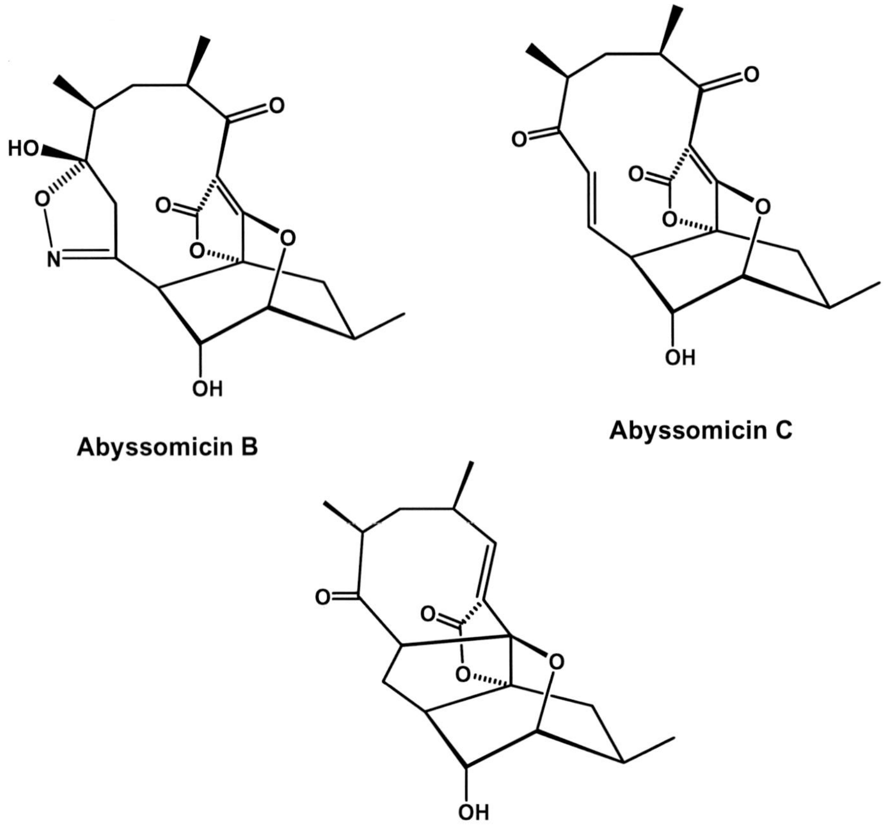

Abyssomicin D

FIGURE 23-10. Structure of abyssomicins B, C, and D.

manzamines, perhaps through the cooperation of symbiotic actinomycetes. As a part of their investigation of this class of compounds, Hamann and coworkers have isolated eight actinomycetes from the source sponge. Owing to the previous demonstration of the biosynthetic role of a species of *Micromonospora* in the biosynthesis of the manzamines, these researchers are actively exploring the biosynthetic potential of the newly isolated bacteria.

ANTIMALARIAL MARINE NATURAL PRODUCTS

There has also been significant renewed interest in the discovery of antimalarial agents as the threat of infection from resistant organisms continues to grow. A number of reports have attributed significant antimalarial activity to classes of previously known marine natural products. One such example is the crambescidin family of marine alkaloids first isolated by the Rinehart group (Jares-Erijman *et al.*, 1991). Originally discovered in a program searching for antiviral activity, this family of compounds has been used as a model for the total synthesis of analogs (Fig. 23-11) that show significant activity *in vitro* against the malaria parasite *Plasmodium falciparum* (Lazaro *et al.*, 2006).

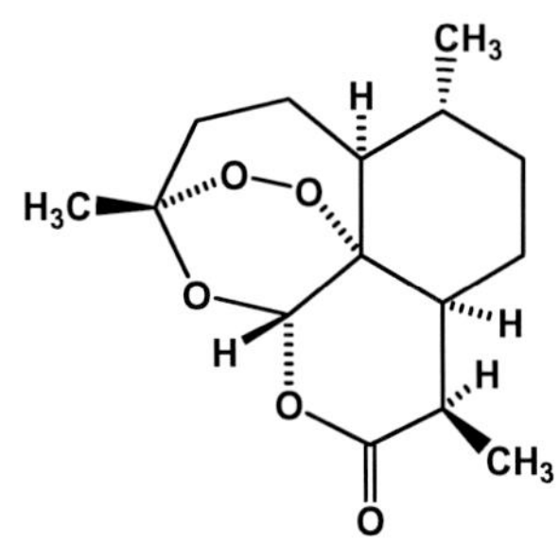

Artemisinin

The effective use of cycloperoxide-containing artemisinin derivatives in the treatment of chloroquine-resistant malaria prompted the reinvestigation of plakortin metabolites for such activity (Fattorosso *et al.*, 2006). Plakortin itself (Fig. 23-2) shows micromolar activity in *in vitro* tests for antimalarial activity. A limited investigation of structure-activity relationships of simple plakortin analogs and synthetic derivatives revealed the intrinsic activity of this

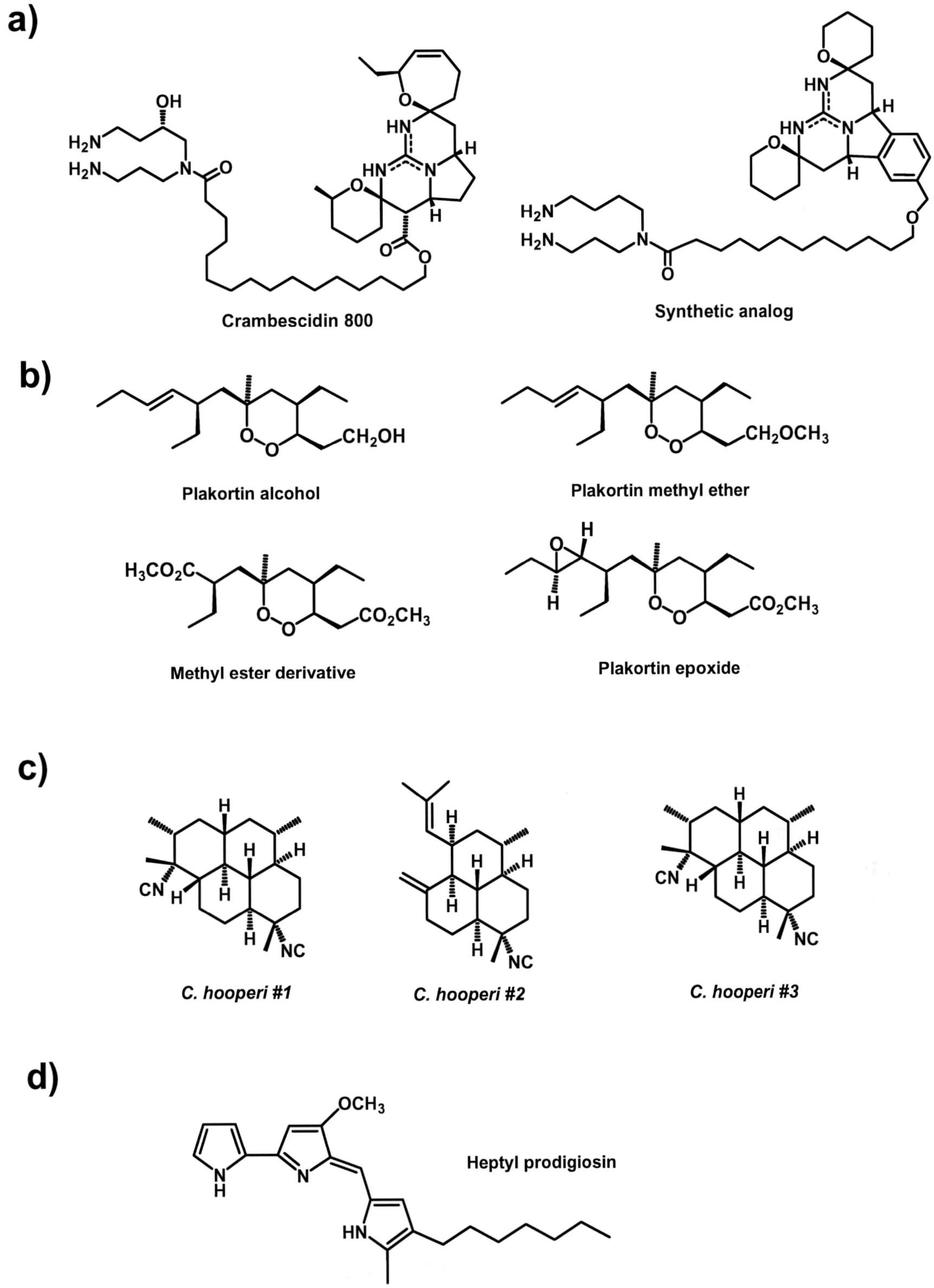

FIGURE 23-11. Antimalarial natural products. (a) Crambescidin and synthetic analog. (b) Plakortin derivatives. (c) Isonitrile-containing terpenes from *Cimbastella hooperi*. (d) Heptyl prodigiosin.

structure class. Shown in Figure 23-11 are some of the plakortin analogs reported with similar potencies in *in vitro* tests against *P. falciparum*. Unsurprisingly, the study demonstrated that the most important structural feature in the plakortin series for conferring antimalarial activity is the endoperoxide function. Analogs lacking the peroxide were not active in the assay.

An example of a well-known class of antibiotics originally isolated from cultures of terrestrial bacteria that has been "rediscovered" from the marine environment are the prodigiosins. These tripyrrole pigments have a history of being evaluated as treatment for various kinds of infectious diseases. The prodigiosins are characteristically bright red in color and as such have been attractive targets for isolation. One study (Lazaro *et al.*, 2002) reports the production of heptyl prodigiosin (Fig. 23-11) by an α-proteobacterium that was derived from a marine tunicate collected in the Philippines. Heptylprodigiosin is a rare congener in the family, and little was known about its specific biological properties. In this study, the researchers showed that heptylprodigiosin could have a protective effect when administered to mice infected with a model of malaria. Although there were issues related to adverse side effects, the researchers were nevertheless able to take the significant step of evaluating the compound for its druglike behavior.

A remarkable characteristic of marine natural products that has caught the imagination of chemists is the presence of unusual functional groups. One such example is the isonitrile (-NC) group. A series of compounds that contain the isonitrile group or a derivative thereof were isolated from the sponge *Cymbastella hooperi* and shown to have antimalarial activity (Wright *et al.*, 1996). In this research, the authors were cognizant of safety concerns and therefore evaluated the selective toxicity of the compounds to the malaria parasite compared to a mammalian cell line. The most selective compounds that retained reasonable antimalarial potency are shown in Figure 23-11.

THE FUTURE OF ANTI-INFECTIVE DISCOVERY FROM THE SEA

The growing need for novel agents to combat life-threatening infections from highly resistant strains of bacteria and other organisms will continue to provide the motivation for exploring the natural products of marine organisms. In the course of the preceding discussion, a number of questions were raised that indicate some of the new directions for anti-infective discovery. Chemical synthesis will continue to play an important role in the production of anti-infective agents. It can be anticipated that synthetic efforts will reveal the essential features of a complex molecule that are required for biological activity. This "pharmacophore" may well be amenable to practical, large-scale synthesis and thereby enable the production of a version of the natural compound that is a viable commercial entity.

Understanding the complex relationships between marine macroorganisms and their microbial neighbors represents one of the key frontiers in marine natural products research. How the production of secondary metabolites is coordinated between host and microbe, to what extent each plays a role in the biosynthesis and distribution of their natural products, and how to harness this biosynthetic potential remain as goals for future investigation. The reader is referred to the work of Haygood (Sudek *et al.*, 2007) regarding the symbiotic production of bryostatins in *Bugula neritina* as a prime example of current research. Continued investigations of bacterial symbionts also should lead to the harnessing of biosynthetic capabilities for the practical production of some natural metabolites.

The exploration of the vast resource of biosynthetic potential represented by the 99% of uncultivable organisms found in marine sediments or that are commensal with marine invertebrates remains as a major challenge. It seems likely that one path to new and potent antibiotics will be found through unlocking the biosynthetic potential of these microbes. In this pursuit there are enormous opportunities for traditional microbiology as well as genetic engineering and metagenomic approaches (Newman and Hill, 1996).

References

Ahlert, J., Shepard, E., Lomovskaya, N., Zazopoulos, E., Staffa, A., Bachmann, B.O., Huang, K., Fonstein, L., Czisny, A., Whitwam, R.E., Farnet, C.M., Thorson, J.S., 2002. The calicheamicin gene cluster and its iterative type I enediyne PKS. Science 297, 1173–1176.

Arai, T., Takahashi, K., Nakahara, S., Kubo, A., 1980. The structure of a novel antitumor antibiotic, saframycin A. Experientia 36, 1025–1027.

Bister, B., Bischoff, D., Stroebele, M., Riedlinger, J., Reicke, A., Wolter, F., Bull, A.T., Zaehner, H., Fiedler, H.-P., Suessmuth, R.D., 2004. Abyssomicin C: a polycyclic antibiotic from a marine Verrucosispora strain as an inhibitor of the p-aminobenzoic acid/tetrahydrofolate biosynthesis pathway. Angewandte Chemie, Int. ed., 43, 2574–2576.

Butler, M.S., Buss, A.D., 2006. Natural products: The future scaffolds for novel antibiotics? Biochem. Pharmacol. 71, 919–929.

Clardy, J., Fischbach, M.A., Walsh, C.T., 2006. New antibiotics from bacterial natural products. Nature Biotechnol. 24, 1541–1550.

De Silva, E.D., Scheuer, P.J., 1980. Manoalide, an antibiotic sesterterpenoid from the marine sponge *Luffariella variabilis* (Polejaeff). Tetrahedron Lett. 21, 1611–1614.

De Souza, M.V.N., 2006. Marine natural products against tuberculosis. Sci. World J. 6, 847–861.

Donia, M.S., Hathaway, B.J., Sudek, S., Haygood, M.G., Rosovitz, M.J., Ravel, J., Schmidt, E.W., 2006. Natural combinatorial peptide libraries in cyanobacterial symbionts of marine ascidians. Nat. Chem. Biol. 2, 729–735.

Fattorusso, C., Campiani, G., Catalanotti, B., Persico, M., Basilico, N., Parapini, S., Taramelli, D., Campagnuolo, C., Fattorusso, E., Romano, A., Taglialatela-Scafati, O., 2006. Endoperoxide derivatives from marine organisms: 1, 2–dioxanes of the plakortin family as novel antimalarial agents. J. Med. Chem. 49, 7088–7094.

Fattorusso, E., Minale, L., Sodano, G., Moody, K., Thomson, R.H., 1970. Aerothionin, a tetrabromo-compound from *Aplysina aerophoba* and *Verongia thiona*. J. Chem. Soc., Chem. Comm. 12, 752–753.

Faulkner, D.J., 1973. Variabilin, an antibiotic from the sponge, *Ircinia variabilis*. Tetrahedron Lett. 39, 3821–3822.

Faulkner, D. J., 2000, Highlights of marine natural products chemistry (1972–1999). Nat. Prod. Reports 17, 1–6.

Faulkner, D.J., Harper, M.K., Haygood, M.G., Salomon, C.E., Schmidt, E.W., 2000. Symbiotic bacteria in sponges: Sources of bioactive substances. In Fusetani, N. (ed.), Drugs from the Sea, p. 108. Basel, Switzerland, Karger.

Fenical, W., Jensen, P.R., 2006. Developing a new resource for drug discovery: Marine actinomycete bacteria. Nat. Chem. Biol. 2, 666–673.

Frincke, J.M., Faulkner, D.J., 1982. Antimicrobial metabolites of the sponge *Reniera* sp. J. Am. Chem. Soc. 104, 265–269.

Handelsman, J., 2004. Soils-the metagenomics approach. In Bull, A.T. (ed.), Microbial Diversity and Bioprospecting, pp. 109–119. Washington, DC, ASM Press.

He, H., Ding, W.-D., Bernan, V.S., Richardson, A.D., Ireland, C.M., Greenstein, M., Ellestad, G.A., Carter, G.T., 2001. Lomaiviticins A and B, potent antitumor antibiotics from *Micromonospora lomaivitiensis*. J. Am. Chem. Soc. 123, 5362–5363.

Higgs, M.D., Faulkner, D.J., 1978. Plakortin, an antibiotic from *Plakortis halichondrioides*. J. Org. Chem. 43, 3454–3457.

Hill, R.T., 2004. Microbes from marine sponges: A treasure trove of biodiversity for natural products discovery. In Bull, A.T. (ed.), Microbial Diversity and Bioprospecting, pp. 177–190. Washington, DC, ASM Press.

Hu, J.-F., Hamann, M.T., Hill, R., Kelly, M., 2003. The manzamine alkaloids. Alkaloids (San Diego, CA, United States) 60, 207–285.

Imamura, N., Nishijima, M., Takadera, T., Adachi, K., Sakai, M., Sano, H., 1997. New anticancer antibiotics pelagiomicins, produced by a new marine bacterium *Pelagiobacter variabilis*. J. Antibiotics 50, 8–12.

Infectious Diseases Society of America (IDS), 2004. Bad Bugs, No Drugs. Alexandria, VA, www.idsociety.org.

Jares-Erijman, E.A., Sakai, R., Rinehart, K.L., 1991. Crambescidins: New antiviral and cytotoxic compounds from the sponge *Crambe crambe*. J. Org. Chem. 56, 5712–5715.

Kwon, H.C., Kauffman, C.A., Jensen, P.R., Fenical, W., 2005. Marinimycins A-D, antitumor antibiotics of a new structure class from a marine actinomycete of the recently discovered genus "*Marinospora*." J. Am. Chem. Soc. 128, 1622–1632.

Lam, K.S., 2006. Discovery of novel metabolites from marine actinomycetes. Curr. Opin. Microbiol. 9, 245–251.

Lazaro, J., Enrico, H., Nitcheu, J., Mahmoudi, N., Ibana, J.A., Mangalindan, G.C., Black, G.P., Howard-Jones, A.G., Moore, C.G., Thomas, D.A., Mazier, D., Ireland, C.M., Concepcion, G.P., Murphy, P.J., Diquet, B., 2006. Antimalarial activity of crambescidin 800 and synthetic analogues against liver and blood stage of *Plasmodium* sp. J. Antibiotics 59, 583–590.

Lazaro, J., Enrico H., Nitcheu, J., Predicala, R.Z., Mangalindan, G.C., Nesslany, F., Marzin, D., Concepcion, G.P., Diquet, B., 2002. Heptyl prodigiosin, a bacterial metabolite, is antimalarial *in vivo* and non-mutagenic *in vitro*. J. Nat. Toxins 11, 367–377.

McDonald, L.A., Capson, T.L., Krishnamurthy, G., Ding, W-D., Ellestad, G.A., Bernan, V.S., Maiese, W.M., Lassota, P., Discafani, C., Kramer, R.A., Ireland, C.M., 1996. Namenamicin, a new enediyne antitumor antibiotic from the marine ascidian *Polysyncraton lithostrotum*. J. Am. Chem. Soc. 118, 10898–10899.

Mickel, S.J., Niederer, D., Daeffler, R., Osmani, A., Kuesters, E., Schmid, E., Schaer, K., Gamboni, R., Chen, W., Loeser, E., Kinder, F.R., Jr., Konigsberger, K., Prasad, K., Ramsey, T.M., Repic, O., Wang, R.-M., Florence, G., Lyothier, I., Paterson, I., 2004. Large-scale synthesis of the anti-cancer marine natural product (+)-discodermolide. Part 5: Linkage of fragments c1–6 and c7–24 and finale. Org. Process Res. Dev. 8, 122–130.

Newman, D.J., Hill, R.T., 2006. New drugs from marine microbes: The tide is turning. J. Indust. Microbio. Biotech. 33, 539–544.

Nicolaou, K.C., Harrison, S.T., 2007. Total synthesis of abyssomicin C, atrop- abyssomicin C, and abyssomicin D: Implications for natural origins of atrop- abyssomicin. C. J. Am. Chem. Soc. 129, 429–440.

Nielsen, K.F., Smedsgaard, J., 2003. Fungal metabolite screening: Database of 474 mycotoxins and fungal metabolites for dereplication by standardized liquid chromatography-UV-mass spectrometry methodology. J. Chromatography, A 1002, 111–136.

Peng, J., Hu, J.-F., Kazi, A.B., Li, Z., Avery, M., Peraud, O., Hill, R.T., Franzblau, S.G., Zhang, F., Schinazi, R.F., Wirtz, S.S., Tharnish, P., Kelly, M., Wahyuono, S., Hamann, M.T., 2003. Manadomanzamines A and B: A novel alkaloid ring system with potent activity against mycobacteria and HIV-1. J. Am. Chem. Soc. 125, 13382–13386.

Piel, J., Butzke, D., Fusetani, N., Hui, D., Platzer, M., Wen, G., Matsunaga, S., 2005. Exploring the chemistry of uncultivated bacterial symbionts: Antitumor polyketides of the pederin family. J. Nat. Prod. 68, 472–479.

Rinehart, K.L., Holt, T.G., Fregeau, N.L., Stroh, J.G., Keifer, P.A., Sun, F., Li, L.H., Martin, D.G., 1990. Ecteinascidins 729, 743, 745, 759A, 759B, and 770: Potent antitumor agents from the Caribbean tunicate *Ecteinascidia turbinata*. J. Org. Chem. 55, 4512–4515.

Rinehart, K.L., Jr., Shaw, P.D., Shield, L.S., Gloer, J.B., Harbour, G.C., Koker, M.E.S., Samain, D., Schwartz, R.E., Tymiak, A.A., Weller, D. L., Carter, G.T., Munro, M. H.G., Hughes, R.G., Jr., Renis, H.E., Swynenberg, E.B., Stringfellow, D.A., Vavra, J.J., Coats, J.H., Zurenko, G.E., Kuentzel, S.L., Li, L.H., Bakus, G.J., Brusca, R.C., Craft, L.L., Young, D.N., Connor, J.L., 1981. Marine natural products as sources of antiviral, antimicrobial, and antineoplastic agents. Pure Appl. Chem. 53, 795–817.

Sims, J.J., Lin, G.H.Y., Wing, R.M., 1974. Marine natural products: X. Elatol, a halogenated sesquiterpene alcohol from the red alga *Laurencia elata*. Tetrahedron Lett. 39, 3487–3490.

Singh, S.B., Barrett, J.F., 2006. Empirical antibacterial drug discovery: Foundation in natural products. Biochem. Pharmacol. 71, 1006–1015.

Spring, D.R., 2005. Chemical genetics to chemical genomics: Small molecules offer big insights. Chem. Soc. Rev. 34, 472–482.

Sudek, S., Lopanik, N.B., Waggoner, L.E., Hildebrand, M., Anderson, C., Liu, H., Patel, A., Sherman, D.H., Haygood, M.G., 2007. Identification of the putative bryostatin polyketide synthase gene cluster from "*Candidatus Endobugula sertula*," the uncultivated microbial symbiont of the marine bryozoan *Bugula neritina*. J. Nat. Prod. 70, 67–74.

Thompson, J.E., Walker, R.P., Faulkner, D.J., 1985. Screening and bioassays for biologically-active substances from forty marine sponge species from San Diego, California, USA. Mar. Biol. 88, 11–21.

Torsvik, V., Goksøyr, J., Daae, F.L., 1990. High diversity in DNA of soil bacteria. Appl. Environ. Microbiol. 56, 782–787.

Torsvik, V., Sorheim, R., Goksøyr, J., 1996. Total bacterial diversity in soil and sediment communities: A review. J. Ind. Microbiol. 17, 170–178.

Wilson, R.M., Danishefsky, S.J., 2006. Small molecule natural products in the discovery of therapeutic agents: The synthesis connection. J. Org. Chem. 71, 8329–8351.

Wright, A.D., Koenig, G.M., Angerhofer, C.K., Greenidge, P., Linden, A., Desqueyroux-Faundez, R., 1996. Antimalarial activity: The search for marine-derived natural products with selective antimalarial activity. J. Nat. Prod. 59, 710–716.

Wright, G.D., 2007. The antibiotic resistome: The nexus of chemical and genetic diversity. Nat. Rev. Microbiol. 5, 175–186.

Zask, A., Kaplan, J., Musto, S., Loganzo, F., 2005. Hybrids of the hemiasterlin analogue taltobulin and the dolastatins are potent antimicrotubule agents. J. Am. Chem. Soc. 127, 17667–17671.

Zazopoulus, E., Huang, K., Staffa, A., Liu, W., Bachmann, B.O., Nonaka, K., Thorson, J.S., Shen, B., Farnet, C.M., 2003. A genomics-guided approach for discovering and expressing cryptic metabolic pathways, Nat. Biotech. 21, 187–190.

STUDY QUESTIONS

1. Why is it that microorganisms, such as Streptomycetes, which are found ubiquitously in the environment, have the biosynthetic capacity to make several different chemical classes of antibiotics?
2. Resistance to antibiotics is often cited as the driving force for the discovery of new agents. Why is the development of resistance considered inevitable?
3. The first anti-infective marine natural products isolated during the 1970s lacked the potency of compounds discovered in more recent times. Why were the more potent compounds not discovered earlier?
4. Some of the most potent anti-infective (and cytotoxic) agents isolated from marine macroorganisms have structural features characteristic of microbial products. For many years, scientists have questioned whether the macroorganism, some associated microorganism, or both produce these. What experiments can be employed to resolve this question?
5. Why is it that the chemical constituents of the same species of sponge collected in different locations are often significantly different?
6. The causative agent of tuberculosis, *Mycobacterium tuberculosis,* is extremely slow growing for a bacterium. How does this affect the treatment of the disease?

CHAPTER

24

Marine Proteins

JÖRG WIEDENMANN

INTRODUCTION

Thinking about marine proteins, pictures of golden fried fish and tasty *frutti di mare* emerge in the mind. Indeed, the oceans contribute a significant component of proteins to the nutrition of humankind. However, proteins derived from the marine realm have more to offer. Frequently they have evolved fascinating, unique structural and biochemical features as adaptations of marine organisms to diverse and sometimes extreme environmental conditions. A number of marine proteins have become indispensable tools, thereby enabling rapid progress in life sciences research. Basic studies seeking to understand the mechanisms underlying diseases such as cancer and HIV infections as well as the work of drug developers are equally dependent on the these protein tools. Polymerases from heat-loving deep sea bacteria ensure the precise amplification of DNA fragments, the light produced by enzymes from glowing marine creatures extends the knowledge about gene activity, and unprecedented insights into live cells are provided by fluorescent proteins isolated from cnidarians. The luminescent systems derived from the oceans are also helpful in sensing the toxicity of compounds of everyday use or of wastewater. Moreover, assays for the detections of bacterial contamination are based on the complex interaction of enzymes from horseshoe crabs. On the other hand, many marine organisms produce poisonous cocktails that can be deadly for humans. However, the toxic mixtures of substances found in cnidarians, fishes, and snails contain peptides and proteins that show high potential as pharmaceuticals; some of these are already in use, for instance, as painkillers (see Chapter 25). Indeed, even after their degradation, marine proteins are useful to farmers as excellent fertilizers (e.g., the guano or droppings of birds). This chapter introduces a selection of useful marine proteins.

LIMULUS PROTEINS MARK BACTERIAL CONTAMINATIONS

A large number of medical products, such as injectable drugs and solutions or artificial heart valves, require absolute sterility. Proteins from the blood of the horseshoe crab *Limulus polyphemus* ensure the detection of traces of bacterial contaminations. Since 1983, the U.S. Food and Drug Administration has accepted the so-called *Limulus* amoebozyte lysate (LAL) test as the standard test for bacterial endotoxins, and it has replaced the rabbit test in many areas of biomedical research and clinical routine work. Moreover, the LAL test can help diagnose diseases like meningitis or infections of the urinary tract that can be caused by Gram-negative bacteria (Matsumoto *et al.*, 1991; Nachum *et al.*, 1973; Nachum and Shanbrom, 1981; Obayashi *et al.*, 1985).

Horseshoe crabs belong to the group of xiphosurans, which together with the sea spiders (Pantopoda) form the extinct group Eurypterida and the arachnids the taxon Chelicerata. One species of horseshoe crabs (*Limulus polyphemus*) inhabits the Atlantic coast of North America, whereas three more species (*Tachypleus gigas, Tachypleus tridentus, Carcinoscorpius rotundicauda*) can be found on the coasts of South and Southeast Asia. These four species survived a rich variety of xiphosurans blossoming in former times of Earth's history (Stormer, 1952). However, since then, their physical and constitutional characteristics have not altered much, and 140 million-year-old Jurassic fossils look essentially the same as recent species, giving them the status of "living fossils."

As for any other organism, the horseshoe crab has to efficiently close wounds to prevent the loss of hemolymph and defend themselves against invasion by pathogenic bacteria. Xiphosurans developed a form of innate immunity that relies on a complex system of enzymes and clotting proteins that are useful for both purposes (Iwananga, 2002; Iwananga and Kawabata, 1998; Iwananga and Lee, 2005). An enzymatic cascade, induced by traces of bacterial endotoxins (lipopolysaccharides) or β-1,3 glucan, results in the clotting of hemolymph. Lipopolysaccharides (LPS) and β-1,3 glucan are components of the cell walls of Gram-negative bacteria and fungi, respectively. The major players of the coagulation cascade are localized in the larger granules of particular hemolymph cells, the amebocytes. If the animals are wounded, amebocytes get into contact with LPS or β-1,3 glucan at the injured parts, the components of the coagulation system are released via secretion, and the serine protease zymogens factors C (LPS sensor) and G (β-1,3 glucan sensor) are autocatalytically activated. The active factor C converts factor B into its active form, which in turn activates the proclotting enzyme to become the clotting enzyme. In contrast, the active factor G directly mediates the activation of proclotting enzyme. The active factor B and clotting enzymes also act as serine proteases. The enzymatic action of clotting enzyme subsequently excises the 27 amino acid-long peptide C from coagulogen to form the active Coagulin monomer (Bergner *et al.*, 1996; Iwananga, 2002; Iwananga and Kawabata, 1998; Iwananga and Lee, 2005). Finally, the Coagulin monomers form head-to-tail homopolymers and produce a gel that can close the wound. Gram-negative bacteria or fungi that enter the animals are engulfed by the protein-gel, immobilized and made accessible to other components of the immune system of the animals. The crabs can regulate the clotting cascade through proteins termed *Limulus* intracellular coagulation inhibitors (LICIs).

Knowledge of the xiphosuran immune system and the development of the LAL test are based on the studies by Howell on the coagulation of *Limulus* blood in 1885 and the seminal work of Bang (1953), who discovered that the presence of bacteria induces clot formation of *Limulus* blood (Bang, 1953; Howell, 1885). Together with his colleague Levin, he found amebocytes to be the effective component in blood clotting and realized the role of bacterial endotoxins in stimulating blood coagulation (Levin and Bang, 1964a, 1964b, 1967) (Figure 24-1).

Practical Aspects of the LAL Test

Although much is known about the biochemical mechanisms that result in the clotting of xiphosuran hemolymph and some of the participating proteins have been cloned, it is not yet possible to combine recombinant proteins to recreate the functional coagulation cascade (Iwananga and Kawabata, 1998; Iwananga and Lee; 2005; Obayashi *et al.*, 1985). Therefore, the LAL test still relies on horseshoe crabs as the source of the required components. The hemolymph is harvested from the crabs with sterile stainless steel syringe needles. The amount of extracted hemolymph is kept small enough to ensure high survival rates (~90%) of the animals and aim for a sustainable exploitation of this natural resource. The amebocytes are separated from the hemolymph by centrifugation, lysed and lyophilized, and then stored under vacuum. The proteins are reconstituted with water and the samples are added as aqueous solutions. The reconstitution step can be combined with the addition of the test sample. The test should be always performed with positive and negative controls. In the presence of bacterial endotoxins, the solution forms a gel at the bottom of the reaction tube. If the gel stays intact upon inverting the tube by 180°, the test can be considered positive. Chromogenic variants of the LAL test were developed that allow a quantitative analysis of endotoxins (Gorny *et al.*, 1999; Iwananga *et al.*, 1978; Lindsay *et al.*, 1989, Nakamura *et al.*, 1977). For example, endotoxin-activated protease activity is exploited to cleave p-nitroaniline from a colorless synthetic peptide substrate. The reaction velocity can be spectrophotometrically monitored by the resulting color development, thereby allowing calculation of the quantity of endotoxins present in the sample.

FROM LIGHTLESS DEPTHS TO THE LABORATORIES OF THE WORLD: THERMOSTABLE POLYMERASES

In the current time, biotechnology, biomedical research and clinical routine work simply would not be imaginable were it not for the discovery of the polymerase chain reaction (PCR) (Kleppe *et al.*, 1971; Mullis *et al.*, 1986; Saiki *et al.*, 1985). Applications of this technology span uses in forensic laboratories to identify offender or victims and their traces, diagnosis of pathogens infecting humans, the elucidation of molecular basis of diseases including cancer and HIV infections, the development of new therapeutic approaches such as gene therapy, and many more. The PCR technique allows the amplification of DNA fragments by the cyclic repetition of a denaturing step that separates the DNA strands at 98°C, one annealing step during which oligonucleotide primers are allowed to bind to the single DNA strands at lower temperatures and an elongation step that allows a DNA polymerase to synthesize new strands. The denaturation step requires temperatures that destroy polymerases from both vertebrates and invertebrates adapted to life at ambient temperatures. The development of the PCR method was therefore promoted by the discovery of a thermostable polymerase, the *Taq*-polymerase, isolated from the heat-loving bacterium *Thermus aquaticus* from a hot spring in Yellowstone National Park (Brock, 1997; Saiki *et al.*, 1988). *Taq*-polymerase has proven extremely useful in many PCR uses and became one of the enzymes most used in molecular biological laboratories. However, its relatively high error rate (1 error per 10,000–50,000 synthetisized base

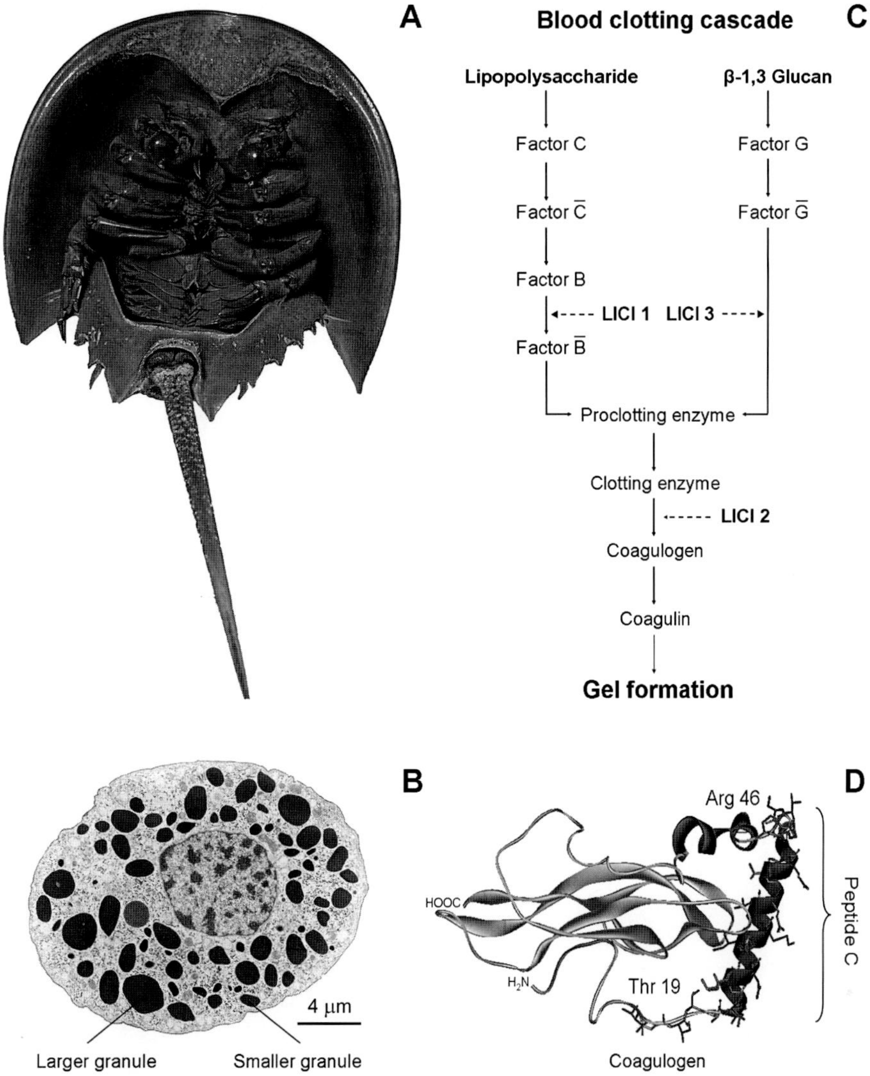

FIGURE 24-1. Xiphosurans and their blood clotting system. (A) A ventral view on the horseshoe crab *Limulus polyphemus*. (B) An electron micrograph of a horseshoe crab amebocyte shows larger and smaller granules. (C) Scheme of the xiphosuran blood clotting cascade depicts the enzymatic activation steps upon stimulation by lipopolysaccharides and β-1,3 glucan. The cascade can be regulated by LICIs (limulus intracellular coagulation inhibitors). The major proteins of the blood clotting system are localized in the larger granules of the amebocytes. (D) Schematic model of the three dimensional structure of the protein coagulogen from *Tachypleus tridentatus* (Bergner *et al.*, 1996). The model was constructed using the coordinates deposited under the PDB accession code 1AOC. The activation is achieved by the excision of peptide C. Peptide C, extending from residue 19 (Thr) to 46 (Arg) is highlighted by the display of the amino acids as stick models. Panel (B) was modified from Iwanaga (2002).

pairs) has identified a need for more accurate polymerases. Therefore, bioprospectors screened diverse hot springs for alternative thermostable polymerases (Adams, 1993; Deming, 1998).

Until the middle of the 19th century, it was assumed that the lightless depths of the ocean would be a cold, hostile environment in which no or only a few living organisms could exist. This view began to change after an overseas telegraph cable was rescued from the deep sea after spending several years on the sea floor. Surprisingly, it was found to be covered by a number of encrusting species (Beebe, 1935). The development of submersible equipment such as deep sea submarines and remotely operated vehicles (ROVs) enabled the discovery of a rich variety of species adapted to life down to even hadal depths. Maybe one of the most notable discoveries of our time was the finding that whole ecosystems exist in the absence of sunlight. Primary production in these deep sea communities is not dependent on plants and photosynthesis, but rather on chemoautotroph bacteria that use H_2S, hydrogen or methane to produce energy and assimilate carbohydrates (Fig. 24-2). Such communities are found in sites with tectonic activity, often in the proximity of hydrothermal vents (Ballard and Grassle, 1979; Londsdale, 1977). Here, the volcanic activity results in the release of chemical compounds necessary to power bacterial life. Some bacteria have adapted to life close to the hydrothermal vents and grow optimally at temperatures around 100°C (Fiala and Stetter, 1986; Gonzalez *et al.*, 1998; Postec *et al.*, 2005). Life under such extreme conditions requires enzymes that function best at high temperatures. Consequently, bioprospectors were proven right in their hypothesis that vent bacteria might contain thermostable polymerases useful for biotechnological applications (Cline *et al.*, 1996; Kong *et al.*, 1993; Lundberg *et al.*, 1991). In particular, the archaeal taxa *Pyrococcus* and *Thermococcus* provided a number of thermostable polymerases that show up to 50 times greater accuracy in DNA replication when compared to *Taq* polymerase, primarily because of their 3′ to 5′ exonuclease proofreading activity (André *et al.*, 1997; Cline *et al.*, 1996; Greag *et al.*, 1999; Moore, 2005). However, at the same time this beneficial activity reduces the processivity of the enzymes in comparison to *Taq* polymerase. One way to improve the extension capacity of proofreading polymerases was found by the addition of elongation factors, protein homologs to the proliferating cell nuclear antigen (PCNA), or so-called sliding clamps (Henneke *et al.*, 2000; Kitabayashi *et al.*, 2002; Moore, 2005). Again, promising representatives of this protein family with molecular masses around 28.2 kDa were isolated from the marine bacterium *Thermococcus* (= *Pyrococcus*) *kodakarensis*.

Aside from PCR enzymes, thermophilic marine bacteria have yielded additional protein tools useful in biotechnological applications. For example, a thermostable alkaline phosphatase and a thermostable proteinase have been described (Dib *et al.*, 1998; Zappa *et al.*, 2001). Most likely, additional heat-resistant enzymes with a high potential for biotechnological application remain to be discovered.

DEADLY BUT USEFUL: TOXIC PEPTIDES AND PROTEINS

Marine organisms produce a vast number of poisons. Many of these deadly cocktails contain peptides and proteins as the active ingredients. On the one hand, these toxins represent a severe threat to humans; on the other, they contain substances with a high potential for application as pharmaceuticals (Pushkar *et al.*, 1986).

In benthic communities, a fierce competition for settlement space rages between various taxa of highly diverse origin although sometimes even between conspecifics. The struggle for space is probably hardest on coral reefs, structures of geological significance though of exclusive biogenic origin. Sessile organisms are constantly in danger of being overgrown. They also represent an easy prey for predators. Therefore, many animals build up hard skeletons, as, for example, scleractinian corals, or they incorporate sharp sclerites in tissues, as in leather corals or sponges, in order to hamper predation. Another common strategy is the evolution of various toxic or distasteful substances that prevent grazers from feeding. This chemical defense makes sponges

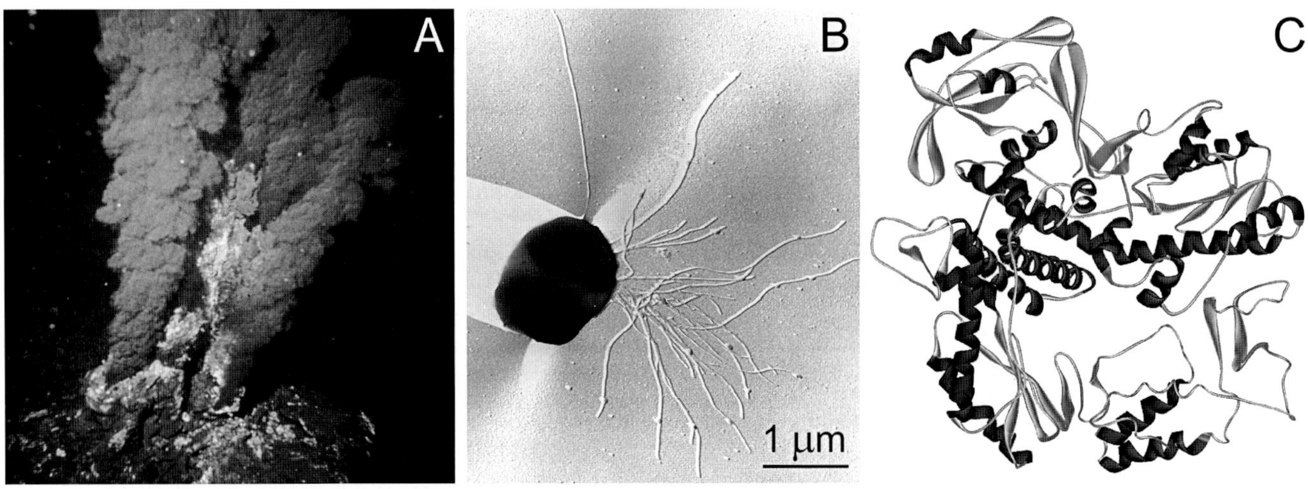

FIGURE 24-2. Heat-loving bacteria and their polymerases. (A) A hydrothermal vent at a midocean ridge. Their classification as "black smokers" is inspired by the dark color of the plume of hot water caused by precipitated minerals. (B) Micrography of the heat-loving bacterium *Pyrococcus furiosus*. (C) Schematic model of the crystal structure of DNA polymerase from the hyperthermophilic archaeon *Pyrococcus (= Thermococcus) kodakaraensis* (Hashimoto et al., 2001). The model was constructed using the coordinates deposited under the PDB accession code 1WNS. Photograph (A) by P. Rona, NOAA Photo Library. Photograph (B) courtesy of R. Rachel and K.O. Stetter, University of Regensburg.

particularly rewarding targets for the isolation of bioactive compounds for pharmaceutical applications. However, highly potent defensive toxins are not only found among invertebrates but also in fish such as scorpionfish (Scorpaeninae), weever fish (e.g., *Echiichthys vipera*), lionfish (*Dendrochirus* sp.), and stonefish (Synanceinae). Members of these taxa can be also dangerous for humans. In particular, the stonefishes represent a considerable threat as the fish usually do not swim away to escape from a person wading in shallow water but rather rely on their perfect camouflage and their toxicity, thereby increasing the risk of stepping on the poisonous spines of their dorsal fins (Khoo, 2002). The potentially deadly poison of stonefish shows the activity of several enzymes, including hyaluronidase, alkaline phosphomonoesterase, 5′ nucleotidase, arginine amidase, arginine esterase, and proteinase. Major lethal effects such as hemolysis, vascular permeability, platelet aggregation, edema induction, and endothelium-dependent vasorelaxation can be attributed to a 148 kDa protein known as "Stonustoxin."

In addition to defense against macroscopic enemies, marine organisms have to counteract invading microbes. Therefore, many sessile species such as sponges, tunicates, and bryozoans are promising sources for antibiotic and antiviral compounds (Thakur *et al.*, 2003).

Chemical weapons are also widely applied offensively for the capture of prey. Cnidarians were named after the effective structure of their stinging cells, the so-called nematocysts. Upon contact with prey, or potential enemies including humans, the nematocysts explode and inject poison through a thin tubule into the tissue of the prey or the attacker (Lotan *et al.*, 1996). The animals do not rely on a single toxin but rather apply complex mixtures of compounds including poisonous peptides and proteins. Proteolytic and hemolytic enzymes support the action of the toxins. The stings of cnidarians can cause intense pain, inflammation, edema, and tissue necrosis (Nagai *et al.*, 2002; Williamson and Burnett, 1995). The poisons can have various effects on the nervous and circulation system that can eventually result in heart or respiratory failure and death (Williamson *et al.*, 1988). Probably most infamous for endangering the life of swimmers are jellyfish as the cubozoan *Chironex fleckeri* or the siphonophoran *Physalia utriculus* (Burnett and Gable, 1989; Currie *et al.*, 1992). Less well known is the fact that other cnidarians are also highly toxic (Fig. 24-3). For instance, the most potent poison of the animal kingdom, palytoxin, is found in the crust anemone *Palythoa toxica*. Similar toxins are also used by other *Palythoa* species that are quite frequently kept in aquaria by seawater hobbyists, and cases of severe intoxication have been reported. In one case, apparently the inhalation of aerosols generated by brushing off crust anemones from a rock under running hot water was enough to provoke an intoxication that required intensive care hospitalization for several days. Also, many sea anemones contain well-studied toxic peptides and proteins, for instance, pore-forming proteins that kill the affected cells by destabilizing membranes (Anderluh *et al.*, 2003). Their toxic peptides often affect nerve cells by targeting voltage-gated ion channels (Messerli and Greenberg, 2006).

One motivation for studying poisonous peptides or proteins is to allow their recombinant production that would be helpful in developing antisera that could be applied after accidents with marine stingers. Moreover, the high bioactivity of the toxic compounds makes them promising lead structures for drug development. Examples with high application potential as pharmaceuticals can be found among the toxic peptides of cone shells, the conotoxins (see Chapter 25). The snails of the genus *Conus* hunt down their prey of fish, other mollusks, or polychaete worms and then shoot them with a poisoned "arrow." This structure evolved from the teeth of the tongue of the mollusks, the so-called radula, which is used for scraping algae off the substrate by other snail species. In particular, for fish-hunting species it is important that the injected poison instantaneously paralyzes the highly motile prey. Therefore, the toxic cocktails contain remarkably active compounds that are also dangerous for humans. A number of human deaths are reported because of intoxication with cone shell poison. The peptide omega-Conotoxin MVIIA served as a template for the development of a synthetic variant known as Ziconotide (Kohno *et al.*, 1995; Wallace, 2006). The peptide was approved for the treatment of severe and chronic pain. The substance selectively blocks neuronal-type voltage sensitive calcium channels, possibly resulting in decreased synaptic transmission of pain signals. A recombinant variant of Ziconotide biotechnically produced in *Escherichia coli* was reported to be 800-fold stronger than morphine (Xia *et al.*, 2006). One further benefit of Ziconotide is that it acts on patients refractory or intolerant to other treatments, for instance, to opiates.

DIVERSE BIOACTIVE PROTEINS WAIT FOR DISCOVERY

The oceans contain an impressive diversity of habitats ranging from ecosystems depending on the activity of hydrothermal vents to communities living in, on, and close to the polar ice. Thus, metabolic processes have evolved to function at temperatures reaching from boiling hot to freezing cold. Life exists under glistening sunlight, exposing intertidal creatures to harmful doses of short wavelength irradiation, down to the lightless depths of the deep sea. Sessile organisms fight an eternal battle for space, whereas other species form inseparable alliances known as symbioses. For this purpose, they use various chemicals to communicate, to poison each other, or to attract mating partners. Marine

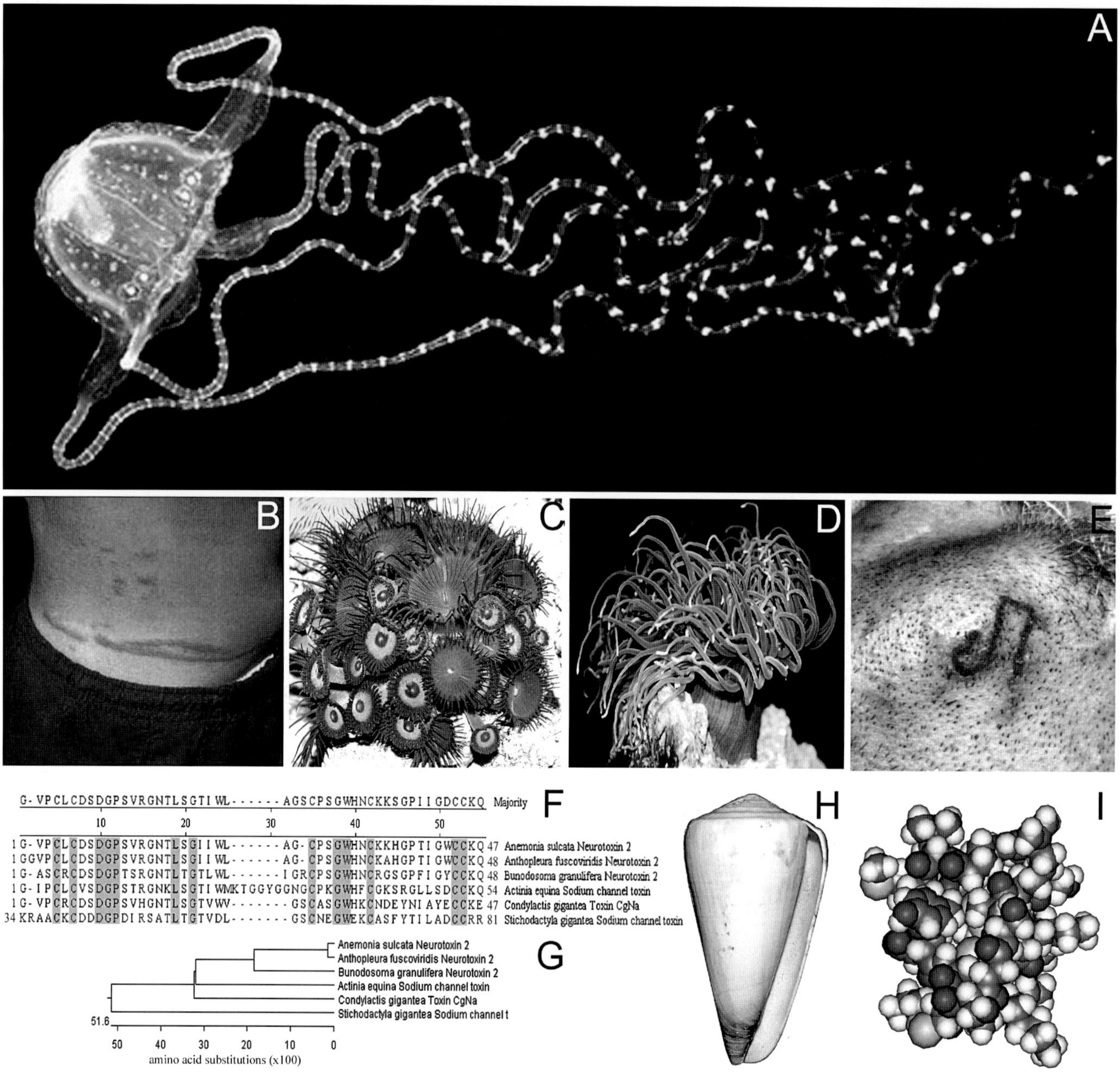

FIGURE 24-3. Poisonous marine invertebrates and some of their toxins. (A, B) Box jellyfishes (Cubozoa) belong to the cnidarians most dangerous for humans. (A) *Carukia barnesi.* (B) Skin injuries in the hip region of a person after contact with tentacles of *Carybdea marsupialis.* (C) Crust anemones (Zoantharia) contain among different toxic proteins and peptides the compound palytoxin, one of the most potent poisons. (D) The sea anemone *Anemonia sulcata*, on of the most abundant sea anemones of the Mediterranean Sea can severely nettle humans by applying its poison of the stinging cells. However, the nematocyst can only penetrate thin parts of the skin. (E) An single tentacle of *Anemonia sulcata* left an imprint of necrotic tissue on the chin of the author. The picture was taken 12 hours after the tentacle stuck for approximately 10 minutes on the skin during collection of anemones. During the first minutes, the stings remain painless. Later, severe itching and moderate pain results from the action of the poisons. (F) Multiple sequence alignment of the amino acid sequences of neurotoxins from various sea anemones. Strictly conserved residues are underlayed in gray and highlight also the unusual abundance of cysteine residues. (G) Representation of the multiple sequence alignment of sea anemone neurotoxins (F) as phylogenetic tree shows the high similarity of neurotoxin 2 of *Anemonia sulcata* and *Anthopleura elegantissima.* (H) Cone shells (*Conus* sp.) obtained their name from the characteristic shape of their shells. The poisonous cocktail used by the mollusks for hunting their prey consists of numerous compounds, including various peptides. (I) Three-dimensional model of of the analgetic peptide Ziconotide from *Conus magus* with the atoms displayed as Van-der-Waals-spheres. The model was constructed using the coordinates deposited under the PDB accession code 1OMG. Photographs in parts (A) and (B) are a courtesy of Griselda Ávila.

organisms build up sophisticated skeletons that might change the landscape, such as in the case of scleractinian corals, or they bore themselves burrows into the substrate by excreting acids, an adaptation of certain sponges. Others, for example, the clam *Mytilus edulis*, build protective shells and attach themselves to the substrate with proteinaceous fibers (the byssus) of surprising durability. Finally, all of these creatures share their habitats with bacteria. Marine microbes are amazing for their ability to colonize all available niches and utilize nearly every organic compound to generate energy, including oil spills that pollute the shores. All of these processes are more or less directly dependent of the action of enzymes or peptides, offering tremendous potential for biotechnology and biomedical applications (Cooper, 2004). The benefits of heat-resistant polymerases were already highlighted. However, some industrial processes require enzymes that work at low temperatures, which has the added benefit of lowering energy costs for biotechnological production. On the other hand, new biocompatible materials might be engineered using marine enzymes. For instance, enzymes building silica fibers in sponges promise interesting biomedical applications (Mueller *et al.*, 2004). Summarizing, there remains many opportunities for exciting discoveries and innovative applications with marine proteins.

Let There Be Light: Bioluminescence Enlightens Biomedical Research

Although Pliny the Elder reported the luminescence of jellyfish in the 1st century A.D. (Morin, 1974), still today anyone who has ever witnessed a bloom of bioluminescent dinoflagellates will never forget the mysterious glow of the water provoked by the movements of a swimmer or the action of the waves. Such events of massive bioluminescence are rather rare, and only a few places exhibit the spectacle on a somewhat regular basis (e.g., some bays on the south coast of Puerto Rico). However, if divers or snorkelers switch off their dive lights in the ocean at night and then wave their hand in front of their eyes, they will almost certainly observe some glow of unicellular algae as *Pyrocystis* or *Noctiluca*. Aside from dinoflagellates, bacteria can cause other major bioluminescent events in the ocean (Miller *et al.*, 2005). Bioluminescence means the active production of light by a living organism (Fig. 24-4). On land, a comparably small number of species produce light (Hastings and Morin, 1998). Among them are bacteria, some insects including the famous fireflies, some oligochaete and nematode worms, particular snails, and, finally, certain fungi. Even fewer freshwater organisms are able to produce light, an example being the limpet *Latia neretoides*. In contrast, bioluminescence in the marine realm is widespread, and luminescent representatives can be found in many divergent phyla such as bacteria and vertebrates (Campbell and Herring, 1990; Haddock and Case, 1999; Hastings and Morin, 1998; Wilson and Hastings, 1998). Although glowing organisms can be found in all water depths, the deep sea is particularly rich in luminescent creatures, and below certain depths, they are the only significant source of light (Haddock and Case, 1999; Reynolds and Lutz, 2001). Most bioluminescent organisms use the enzyme luciferase to process the substrate, luciferin, to produce photons. Usually, an oxida-

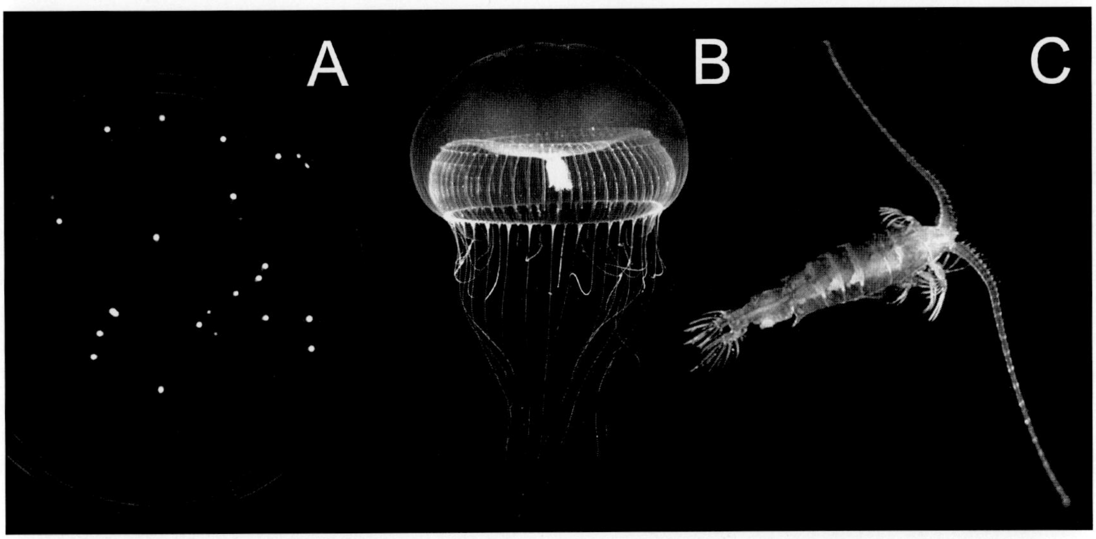

FIGURE 24-4. Bioluminescent organisms of significance for their biotechnological applications. (A) Colonies of the bacterium *Vibrio* sp. isolated from the Mediterranean sponge *Tethya aurantium* growing on an agar plate photographed by their own light. (B) The hydrozan jellyfish *Aequorea victoria* yielded the photoprotein aequorin useful for determination of intracellular calcium levels as well as the green fluorescent protein (GFP). (C) A brightly glowing luciferin/luciferase system was isolated from the copepod *Gaussia* sp. Photographs in panels (B) and (C) are a courtesy of Steve Haddock, Monterey Bay Aquarium Research Station.

tion step is involved at a certain stage of the reaction. However, luciferin and luciferase are generic terms, and they do not indicate that both the enzymes and the substrates are phylogenetically related. On the contrary, the biochemical diversity of the participating compounds clearly indicates that bioluminescence independently evolved several times during Earth's history. It has been proposed that the first bioluminescent systems might have served to detoxify oxygen when earth's atmosphere was becoming aerobic through photosynthesis (Rees *et al.*, 1998).

From the marine environment, the bioluminescent systems of bacteria, cnidarians, and dinoflagellates were understood first; however, since the 1990s several other light reactions have been studied in detail as well. The bacterial luciferases are heterodimers with an approximate molecular weight of 80 kDa (Fisher *et al.*, 1995; Meighen, 1993; Wilson and Hastings, 1998). Two compounds, reduced flavin mononucleotide (FMN) and a long-chain aldehyde, act together to form the excited luciferin. The subunits of the luciferase and additional proteins involved in aldehyde and riboflavin synthesis are encoded by two operons. The bioluminescence is controlled by an autoinducer substance that is also produced as result of the activity of these operons and subsequently secreted by the bacterial cells. Therefore, the inductive internal concentrations are only reached if the cells are sufficiently concentrated. This implies that a single bacterium in dilute solution will not show luminescence.

Bioluminescent cnidarians utilize an imidazolopyrazine derivative named coelenterazine as their common luciferin (Campbell and Herring, 1990; Haddock *et al.*, 2001; Rees *et al.*, 1998; Thomson *et al.*, 1997). In these animals, light is produced by specialized photocytes upon mechanical stimulation. In the hydrozoan jellyfish *Aequorea victoria*, the coelenterazin is bound to the luciferase, forming the photoprotein aequorin (Head *et al.*, 2000; Shimomura, 1995). The luciferase shows structural similarity to calmodulin, a calcium binding protein. The hydroperoxide-compound breaks down upon binding of Ca^{2+} ions to luciferase and light is emitted.

The luciferase of dinoflagellates is a roughly 140-kDa protein that consists of three tandem repeat domains consisting of 375 to 377 amino acids each (Li *et al.*, 1997). The luciferase is localized in specialized organelles together with a luciferin binding protein (LBP) that prevents uncontrolled reaction with the luciferin (Desjardins and Morse, 1993). The luciferin itself is a tetrapyrrole similar to chlorophyll. The luminescent flash produced by dinoflagellates upon mechanical stimulation is triggered by a lowering of the pH from 8 to 6, which results in a release of the bound luciferin from the LBP (Hastings and Morin, 1998).

The color of marine bioluminescence often ranges between blue and green with species-specific differences in the positions of the emission peaks. However, red light emitting animals as the fish *Malacosteus niger* are also known (Douglas *et al.*, 1998). Both bacteria and cnidarians use fluorescent proteins as accessory emitter molecules to shift the emission toward longer wavelengths. Bacteria can push the emission maximum of the bioluminescence spectra up to 540 nm using a yellow fluorescent protein (Petushkov *et al.*, 1995; Wilson and Hastings, 1998). Cnidarians, such as *Aequorea victoria*, *Renilla reniformis* or *Obelia geniculata*, use green fluorescent proteins as acceptor molecules for the energy generated in the luminescent reaction (Hart *et al.*, 1979; Hastings and Morin, 1998, Morin, 1974). It is important to note that this is a true nonradiative energy transfer and not an absorption and reemission of the photon generated in the luminescence reaction. Consequently, the donor and acceptor molecules are found in close proximity in the photocytes (distances of less than 100 Å). The *in vitro* reactions generate blue light that are distinguished by a broad emission spectra with maxima around 460 nm (*A. victoria*), 472 nm (*Obelia*), and 486 nm (*Renilla*). In contrast, the *in vivo* luminescence is characterized by narrow peaks at 508 nm. Exceptions are found in the fluorescent proteins from *Phialidium sp.* where both blue-shifted (498 nm) and red-shifted (537 nm) emission maxima were observed and in *Halistaura* GFP, which has also a blue-shifted (497 nm) emission maximum (Levine and Ward, 1982; Shagin *et al.*, 2004; Ward, 1998). As previously mentioned for the latter species, light is often generated by distinct cells or organs. However, in some animals such as the ostracod *Vargula* (= *Cypridina*) or the copepod *Gaussia* luciferins and luciferases are expelled into the water where they produce a luminescent cloud (Bowlby and Case, 1991; Thompson *et al.*, 1989).

Another fact worth mentioning is that not all animals generate their light on their own. Several species—for example, the flashlight fish *Anomalops katoptron* or the cephalopod *Euprymna scolopes*—host luminescent bacteria of the genus *Vibrio* in highly developed light organs (Haygood and Distel, 1993; Ruby and Lee, 1998; Watson *et al.*, 1978). As described previously, the nature of the bacterial luminescence results in a constant glow that cannot be switched on or off directly. Therefore, anomalopid fishes have developed opaque shutters to cover the light source, or in some cases they rotate the entire organ in order to control light emission (Johnson and Rosenblatt, 1988).

Bioluminescence fulfills diverse biological functions (Buck, 1978). Many organisms use the light for defensive purposes. One strategy of potential prey organisms is to dazzle potential predators by a sudden light flash and then escape into the dark. Crustaceans such as *Vargula* or *Gaussia* expel a luminescent cloud when they escape, and this may confuse the predator as to the actual location of the prey.

However, it is hard to imagine that nudibranchs can be prevented from feeding on a hydrozoan colony by light flashes. Therefore, the light emission may be used to attract larger predators, which subsequently attack the smaller predators of the cnidarians. This scenario may be operational for pelagic light emitting species including dinoflagellates. But why do cnidarians use GFPs as secondary emitters? In the case of *Renilla* luminescence, the answer appears to be straightforward. Light emission via GFP increases the quantum yield of the chemoluminescence reaction and thereby promotes its effect (Ward and Cormier, 1978). Many photoreceptors have a high sensitivity in the green spectral region (Partridge and Hilder, 1990; Partridge and Cummings, 1999). Therefore, a focused emission of green light might increase the perceptibility of the emitted light as compared to the broadbanded bluish light generated in the chemiluminescent reaction. Moreover, the optical properties of coastal waters allow penetration of green-yellow light most efficiently (Jerlov, 1976), and this may have promoted the evolution of the secondary green emitters in order to increase the efficiency of luminescence.

In other organisms, such as the fish *Porichthys notatus*, bioluminescence is produced for camouflage purposes in a strategy called counterillumination (McFall-Ngai, 1990). These animals use ventral photophores to brighten their undersides, making them less visible for predators watching from below against the downwelling light (Harper and Case, 1999; McAllister, 1967). In contrast, fish such as *Malacosteus niger* use the light to hunt their prey (Douglas *et al.*, 1998). Finally, luminescence is used for intraspecific communication such as in cephalopods (Johnsen *et al.*, 1999). However, in many cases the biological meaning is still not fully understood. For example, it is not known why the boring clam *Pholas dactylus* glows deeply with emitted light even though it is hidden deep in the substrate (Dunstan *et al.*, 2000; Henry *et al.*, 1978).

The bioluminescence of ostracods was exploited by the Japanese Army during World War II. The dried crustaceans were ground to a powder by hand and then wetted to reactivate the luciferase. The procedure resulted in a faint diffuse glow on the individual's hands that provided enough light, for instance, to read maps at night without risk of being seen over long distances. During a visit at the Monterey Bay Aquarium Research Institute, Steve Haddock successfully demonstrated the method using an original 60–year-old batch of ostracods.

Bioluminescent creatures have provided a number of tremendously useful tools employed in biomedical research. For example, the luciferases from ostracods such as *Vargula* are used to measure the activity of genes (Thompson *et al.*, 1989). If the cDNA coding for the luciferase is combined with a promoter responsive to certain cellular signals of interest, the activity of the reporter gene can be monitored by luminescence once transfected into suitable cells. However, the exogenous addition of the coelenterazine is also required. The luciferase systems of other crustaceans, such as the copepods *Gaussia* or *Metridia*, can also be applied in reporter gene assays (Markova *et al.*, 2004; Tannous *et al.*, 2005; Verhaegen and Christopoulos, 2002). In this context, the exceptional brightness of the *Gaussia* luciferase reaction increases the sensitivity of the assay. Another beneficial property of the crustacean luciferases is their endogenous excretion signal. In this case, the luciferase is released into the medium in recombinant cellular assays and light emission can be measured directly after addition of the luciferin without the need to lyse the cells. Another well-known luciferase from the cnidarian *Renilla reniformis* is also used as reporter protein in biomedical research, typically in combination with firefly luciferase (Lorenz *et al.*, 1991; McNabb *et al.*, 2005; Parsons *et al.*, 2000). In one application, the light generated by the firefly luciferase serves as a reporter of gene activity whereas the luminescence produced by the *Renilla* luciferase is used to quantify the amount of the firefly luciferase construct present in the experimental cells. The parallel use of the two luciferase systems is possible because the biochemistry of the bioluminescence reactions as well as the substrates are fundamentally different and therefore provide ways in which to selectively block one of the reactions. Used as reporters of gene activity, luciferases have promoted the understanding of disease-related cellular mechanisms, such as the development of cancer. As luminescent reporters, they have also been applied in model organisms such as mice (Bhaumik and Gambhir, 2002). Pharmaceutical drug discovery research widely applies luminescent reporters to screen for the activity of potential drugs (Deo and Daunert, 2001).

Calcium plays an important role in inter- and intracellular signaling. Hence, the understanding of many fundamental biological processes is dependent on the ability to measure changes in the cellular calcium levels *in vivo*. As Ca2+ triggers the luminescence of the photoprotein aequorin from *Aequorea victoria*, aequorin has enabled many applications such as the reporting of intracellular calcium level. Initial studies relied on the microinjection of the purified reporter protein into living cells (Blinks *et al.*, 1982; Ridgeway and Ashley, 1968). However, after the cDNA coding for apoaequorin was isolated (Inouye *et al.*, 1985), the reporter protein could be directly expressed in target cells. Upon addition of coelenterazine, the functional aequorin is formed and can be used to measure free cytosolic calcium (Tanahashi *et al.*, 1990).

Another application of aequorin is together with the green fluorescent protein (GFP) as a "molecular ruler" to study the interaction of any two proteins of interest (Deo *et al.*, 2005). For this purpose, one of the proteins under examination is fused to the luciferase, and the other is linked to GFP. Being expressed in cells, the luciferase and GFP will be brought into intimate contact if the attached proteins interact on the molecular level. If the two tags are closer than a distance of ~100 Å, the energy generated in the bioluminescence reaction is directly transferred to the GFP chromophore resulting in green instead of blue emitted light. Interestingly, this bioluminescence resonance energy transfer (BRET) assay directly mimics the process that transforms the light color from blue to green in *Aequorea* itself.

The luminescence of marine bacteria is also exploited to report gene activity (Chatterjee and Meighen, 1995). In contrast to the cnidarian luciferases, the more complex bacterial luminescence system requires that several genes are transferred to the cells under study. The long chain aldehyde component has to be added exogenously, whereas reduced flavin mononucleotide (FMN) required to form the functional luciferin is already present in cells. Highly sensitive reporter assays for toxic compounds were developed on the basis of the bacterial luminescence (Alexander *et al.*, 2000; Gu, 2005). Certain luminescent bacteria, including *Photobacterium phosphoreum* and *Vibrio fisheri*, respond specifically and rapidly to the presence of toxins with a decrease in luminescence intensity (Thomulka *et al.*, 1993, 1996). This property makes them suitable as test organisms to monitor, for instance, the quality of wastewater, thereby helping to reduce pollution. Lastly, the photoprotein pholasin that was isolated from the clam *Pholas dactylus* (Dunstan *et al.*, 2000; Michelson, 1978) offers potential as reporter of reactive oxygen species.

The preceding discussion presents only a selection of applications exploiting luciferases from marine organisms. Most likely, new applications will as well be facilitated by a better understanding of yet unexplored luciferin—luciferases reactions. Summing up, the future looks bright for marine bioluminescence.

FLUORESCENT PROTEINS: GLOWING TOOLS FROM THE OCEANS

Green Fluorescent Protein

Since the mid-1990s, the green fluorescent protein isolated from the hydrozoan jellyfish *Aequorea victoria* has become one of the most important tools for biomedical research and gave rise to the era of life cell imaging. However, the first observations of the protein had been communicated nearly half a century before. In 1955, Davenport and Nicol reported yellow-green fluorescence from the luminescent tissue of *Aequorea* and *Halistaura* (Davenport and Nicol, 1955). During the isolation and characterization of the photoprotein aequorin, Shimomura *et al.* (1962) identified the green fluorescent pigment of *Aequorea victoria* as a protein. The emission spectrum showed a sharp peak at 508 nm, consistent with the visual perception of green fluorescence color, whereas the excitation spectrum was characterized by maxima at 270, 390, und 460 nm (Johnson *et al.*, 1962). Von Titschack applied contact print photography and fluorescence microscopy on the sea pen *Veretillium cynomorium* to demonstrate that green fluorescence is emitted from the photogenic tissue areas (von Titschack, 1964).

Further green fluorescent pigments were found in the hydroid *Obelia* and the pennatulacean *Renilla* (Hastings and Morin, 1969; Morin, 1974; Morin and Hastings, 1971). The pigments were also shown to be proteins and therefore named for the first time "green fluorescent protein (GFP)" (Hastings and Morin, 1969). In the following text, the abreviation avGFP will be used when the GFP of *Aequorea victoria* is addressed specifically.

In all cases initially examined, the GFPs were exclusively found in the photogenic cells of bioluminescent Cnidaria (Morin 1974; Morin and Reynolds, 1970; von Titschack, 1964). These studies supported the common belief that GFPs, in general, act as secondary emitters in the bioluminescence reaction of marine cnidarians. In 1979, Shimomura suggested an imidazolone structure as the chromophore of avGFP, validated later by Cody *et al.* (Cody *et al.*, 1993; Shimomura, 1979). They showed that the chromophore is formed by a cyclization reaction within the avGFP amino acid sequence. Between 1970 and 1990, numerous studies elucidated biochemical and photophysical properties of GFPs (Ward, 1998).

Finally, Prasher *et al.* (1992) reported the primary structure of avGFP as the final result of screening of an *Aequorea victoria*-cDNA library with oligonucleotides designed on the basis of partial amino acid sequences from purified GFP.

Surprisingly, avGFP showed the same fluorescence properties as native avGFP when recombinantly expressed in the bacterium *Escherichia coli* and the nematode *Caenorhabditis elegans* (Chalfie *et al.*, 1994). This result suggested an autocatalytic formation of the chromophore in the absence of additional cofactors or substrates except for molecular oxygen (Cubitt *et al.*, 1995; Heim *et al.*, 1994; Inouye and Tsuji, 1994). Because of these properties, avGFP became the first genetically encoded marker introduced into life science research that can readily be visualized without the addition of further enzymes, substrates, or cofactors.

The usefulness of avGFP for studying cellular events *in vivo* led to an avalanche of publications. The majority of these papers described applications of avGFP as a marker of gene expression or as a fusion marker to study protein localization. Laborious, large-scale collections of the jellyfish *Aequorea victoria* suddenly became obsolete because of the virtually unlimited amounts of avGFP from recombinant bacteria. This development stimulated further biochemical and photophysical studies and made avGFP research independent from the natural source. An outstanding result of those studies was the elucidation of the crystal structure of avGFP (Ormö *et al.*, 1996; Yang *et al.*, 1996). The chromophore was found to interrupt a central helix near the geometric center of an 11–stranded β-barrel, the so-called β-can. Additional studies aimed to improve avGFP for *in vivo* applications by altering the coding DNA sequence. A special focus was the generation of variants with different spectral properties for multicolor labeling and fluorescence resonance energy transfer (FRET) applications (Heim *et al.*, 1994; Tsien, 1998). These efforts yielded blue, cyan, and yellowish fluorescent variants. However, a particular goal of this work was to develop a red fluorescent variant of avGFP. Red-emitting fluorescent proteins are highly desirable for fluorescent marker applications because of reduced cellular autofluorescence in the red spectral range, the possibility to apply less cytotoxic, longer-wavelength excitation light, and their use in multicolor labeling or FRET experiments. However, the generation of a stably red fluorescent variant of avGFP has not yet been achieved.

Several reports on green and red fluorescent pigments in nonbioluminescent cnidarians can be found in the literature (Catala, 1958, 1959, 1960; Kawaguti, 1944; Marden, 1956). As early as 1927, Phillips demonstrated that the green pigment of a sea anemone collected from a rock pool in Great Britain showed fluorescence upon irradiation with UV light (Phillips, 1927). However, until 1997 these were not considered as possible candidates for marker gene applications for two major reasons: (1) there was the general belief that GFP-like proteins occur only as green secondary emitters in bioluminescent cnidarians, and (2) both green and red fluorescent pigments of reef corals were suspected to be related to biliproteins such as phycoerythrin. These fluorescent proteins contain tetrapyrrole chromophores attached to heteromultimeric apoproteins. Therefore, they did not appear suitable to be used easily as genetically encoded markers.

The author became interested in cnidarian pigments while studying the color morphs of the Mediterranean sea anemone *Anemonia sulcata*. Four types of colored host pigments could be extracted in aqueous buffers from the tentacle tissue of *A. sulcata*: two green fluorescent pigments with excitation/emission maxima at 480/499 nm and 511/522 nm, respectively, and a red fluorescent pigment which could be maximally excited at 574 nm (Fig. 24-5) (the latter pigment showed an emission spectrum peaking at 595 nm); additionally, a nonfluorescent, pinkish pigment with an absorption maximum at 562 nm was isolated from the tentacle tips (Wiedenmann *et al.*, 1999). The color/fluorescence showed a remarkable stability under harsh conditions as demonstrated by exposure to 2% SDS, 6 M urea, temperatures up to 60°C, pH extremes, proteases, or β-mercaptoethanol. The green fluorescent pigment had a molecular mass of ~28 kDa (Wiedenmann, 1997). Similar properties had been already found for the GFPs of *Aequorea victoria* and *Renilla reniformis* and later were explained by the rigid β-can fold that shelters the chromophore in the molecular center (Bokman and Ward 1981; Levine and Ward, 1982; Ormö *et al.*, 1996; Surpin and Ward, 1989; Tsien, 1998; Ward, 1981; Ward and Cormier 1979; Ward *et al.*, 1982; Yang *et al.*, 1996). Therefore, the author concluded that the green fluorescent, orange fluorescent, and nonfluorescent pink proteins from *A. sulcata* represented GFP-homologs from nonbioluminescent anthozoa (Wiedenmann, 1997). Based on the reports on multicolored fluorescent pigments in other anthozoans, a widespread occurrence of nongreen GFP-like proteins appeared likely (Wiedenmann, 1997). The demand for red fluorescent marker proteins motivated the effort to isolate the genes coding for the GFP-like proteins of *A. sulcata*. A cDNA library was constructed from the morph *rufescens*, because this variety expresses all four types of colored pro-

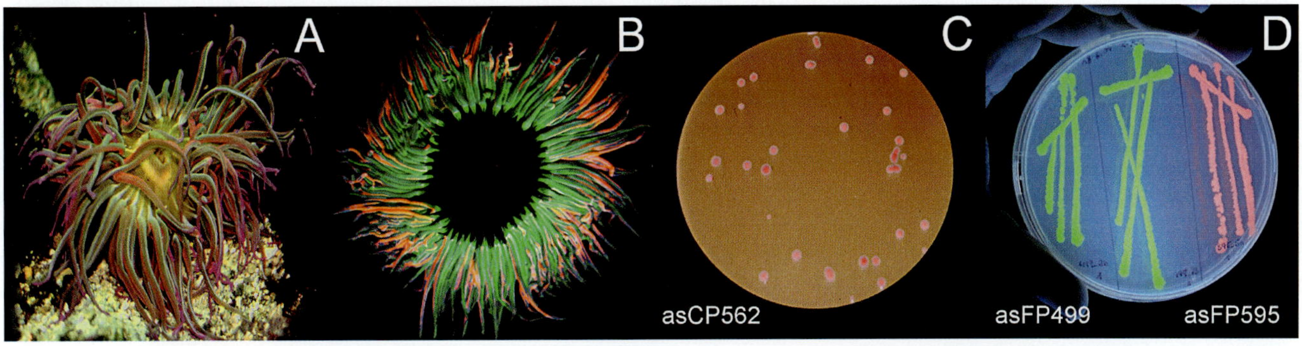

FIGURE 24-5. The Mediterranean sea anemone *Anemonia sulcata* var *rufescens* and its pigments. (A) A daylight picture of *A. sulcata* var. *rufescens*. The animals contain four types of pigment in its tentacles. (B) Under UV excitation (366 nm), the distribution of the pigments becomes visible. Two green fluorescent pigments are localized in the upper side of the tentacles, whereas the underside shows orange fluorescence. The pink pigment of the tentacle tips (A) is nonfluorescent (B) (Wiedenmann, 1997; Wiedenmann *et al.*, 1999). (C) The pink color of *E. coli* colonies under daylight indicates the expression of cDNA, isolated from *A. sulcata* coding for the nonfluorescent chromoprotein asCP562. (D) The green fluorescent protein asFP499 and the red fluorescent protein asFP595/1 can be also functionally expressed in *E. coli* as shown by the fluorescerce of bacterial streaks under 366 nm light (Gundel and Wiedenmann, unpublished; Wiedenmann *et al.*, 2000). Panels (A), (B), and (C) were modified from Wiedenmann *et al.*, 2000.

teins in its tentacles (Fig. 24-5). The phage library was transformed into a phagemid library and the cDNAs were expressed in *E. coli* grown on agar plates. The cloning strategy was based on the assumption that the chromophores of the target proteins would form in a similar autocatalytic manner as in avGFP. Therefore, it was anticipated that positive colonies could be selected by their color or fluorescence. Indeed, an average of one clone out of 700 showed a bright green fluorescence under UV light (Wiedenmann *et al.*, 2000). Out of 10^6 colonies, only 17 appeared a nonfluorescent, pink color. All green fluorescent colonies expressed a protein fluorescing with an excitation/emission maximum at 480/499 nm whereas the pink chromoprotein showed an absorption maximum at 562 nm, the same spectral properties as observed in the native proteins. Sequencing of the cDNAs revealed that the novel proteins shared ~18% identical amino acids with avGFP. The truncated form of the β-can initially proposed for the red proteins, however, turned out not to be valid but rather a misinterpretation of a postranslational cleavage of the peptide backbone, which occurs during maturation of the chromophore (Wiedenmann *et al.*, 2002a). The novel green fluorescent protein from *A. sulcata* was successfully expressed both in insect and plant cells, indicating a universal applicability of GFP-like proteins from nonbioluminescent anthozoans. The red fluorescent protein of *A. sulcata* was missed in the first screen for unknown reasons. Both the green fluorescent and the red fluorescent proteins can be found in comparable amounts in the tissue, suggesting the presence of similar mRNA levels. Later approaches identified the red fluorescent protein as a highly fluorescent isoform of the pink chromoprotein asCP562 (Gundel and Wiedenmann, unpublished) (Fig. 24-5).

Independent from the author's studies at the University of Ulm, Misha Matz and coworkers also worked on the development of a red fluorescent marker protein in the group of Sergey Lukyanov at the Institute of Bioorganic Chemistry at the Russian Academy of Science in Moscow. They synthesized cDNAs of nonbioluminescent anthozoa obtained from a private aquarium. Using degenerate primers based on the sequence of avGFP, they succeeded in cloning six GFP-like proteins from *Anemonia majano (Actiniaria), Discosoma* sp. (Corallimorpharia), *Zoanthus* sp. (Zoanthidea), and *Clavularia* sp. (Stolonifera) (Matz *et al.*, 1999). The emission colors ranged from cyan (483 to 486 nm, *Discosoma striata*, *A. majano*), green (506 nm, *Zoanthus* sp.), yellow (538 nm, *Zoanthus* sp.) to red (583 nm, *Discosoma* sp.) (Fig. 24-6). The expression of both green and red fluorescent proteins in *Xenopus* embryos demonstrated the potential application of multicolored anthozoa FPs as *in vivo* markers. In the following years, further evidence was found in support of the assumption (Wiedenmann, 1997) that GFP-like proteins with different fluorescence colors are widespread among nonbioluminescent cnidaria. For example, the sea anemone *Entacmaea quadricolor* has yielded the most red-shifted natural emitter of the GFP-family to date, eqFP611, with an emission maximum at 611 nm (Wiedenmann *et al.*, 2002b). Numerous GFP-like FPs were identified from scleractinian corals (Ando *et al.*, 2002; Dove *et al.*, 2001; Karasawa *et al.*, 2003, 2004; Kelmanson and Matz, 2003; Labas *et al.*, 2002; Mazel *et al.*, 2003; Oswald *et al.*, 2007; Wiedenmann *et al.*, 2004b), octocorals (Shagin *et al.*, 2004), and ceriantharians (Ip *et al.*, 2004; Wiedenmann *et al.*, 2004a). Furthermore, nonfluorescent, GFP-like chromoproteins were shown to occur frequently among different groups of Cnidaria, such as actiniaria, scleractinia, alcyonaria, and even in hydromedusae (Beddoe *et al.*, 2003; Dove *et al.*, 2001; Gurskaya *et al.*, 2001, 2003; Labas *et al.*, 2002; Lukyanov *et al.*, 2000; Martynov *et al.*, 2001, 2003; Shagin *et al.*, 2004; Wiedenmann *et al.*, 2000; Wiedenmann *et al.*, 2004a).

Nonfluorescent pink, purple, and blue pigments with absorption maxima ranging from 560 to 590 nm were isolated from pocilloporid and acroporid corals (Dove *et al.*, 1995). The water-soluble compounds were identified as proteins with a native molecular mass of 54 kDa, or in the case of *Acropora formosa*, of 86 kDa. Both types of chromoproteins were found to consist of subunits of 28 kDa. In light of studies, these proteins, initially described as "pocilloporins," must also be considered GFP-like chromoproteins (Dove *et al.*, 2001).

The great majority of GFP-like proteins from nonbioluminescent anthozoa were found in species living in a light-dependent symbiosis with zooxanthellae. In the tissue parts expressing these proteins, they contribute up to 14% to the total soluble cellular proteins (Leutenegger *et al.*, 2007; Oswald *et al.*, 2007). Therefore, their distribution is restricted to the photic zone of the oceans. However, green and red fluorescent proteins and nonfluorescent chromoproteins can be found also in azooxanthellate cnidaria, such as *Cerianthus sp., Calliactis parasitica*, and *Adamsia palliata* (Wiedenmann *et al.*, 2004a). This observation indicates that potential sources for GFP-like proteins can also be hidden in lightless habitats such as caves and the deep sea.

Most interestingly, the distribution of GFP-like proteins is not restricted to the phylum containing cnidarians. Six green fluorescent proteins sharing greater than 60% amino acid identify with avGFP were cloned from marine copepods (Crustacea, Calanoida, Pontellidae) (Shagin *et al.*, 2004) (Fig. 24-6). These latter proteins show emission maxima between 500–511 nm and have an excitation spectral peak between 480 and 491 nm.

Despite the impressive variability of the colors among both fluorescent and nonfluorescent GFP homologs, the avGFP is highly conserved (Nienhaus *et al.*, 2005, 2006a, 2006b; Prescott *et al.*, 2003; Remington *et al.*, 2005; Wall *et al.*, 2000; Wiedenmann *et al.*, 2005; Yarbrough, *et al.*, 2001). The rigid β-can fold most likely contributes to the long half-lives of coral FPs, which can persist up to three

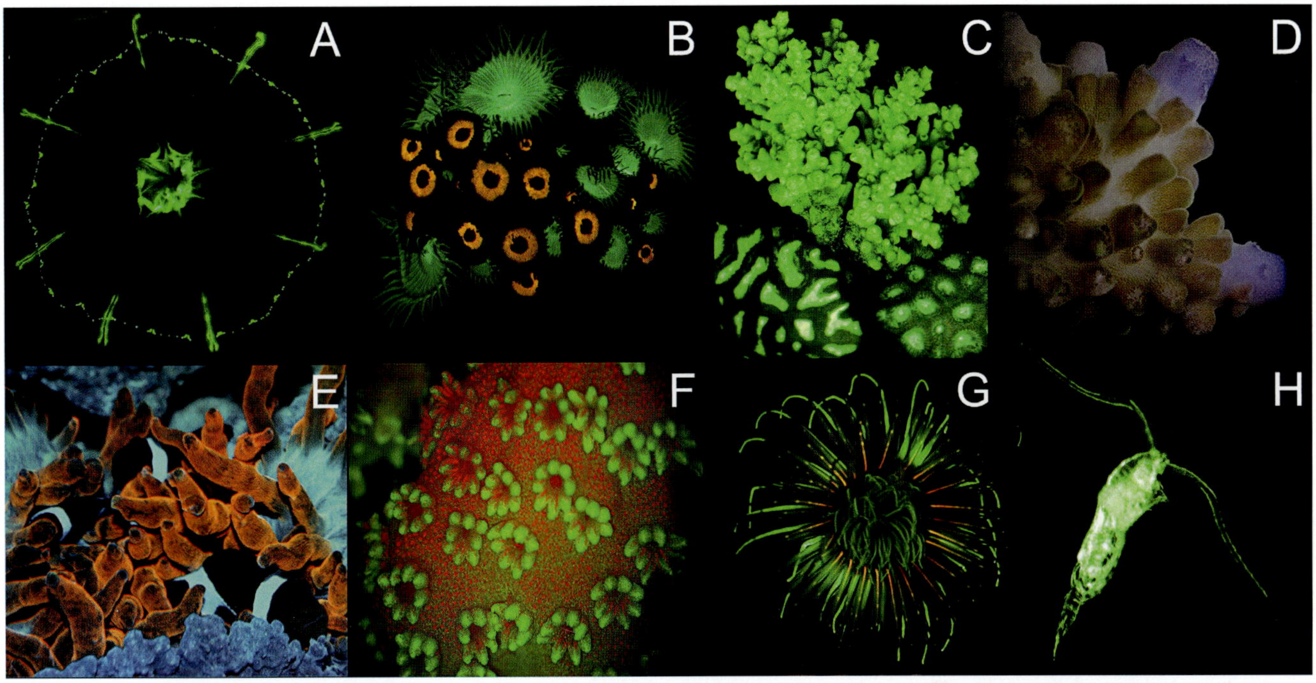

FIGURE 24-6. GFP-like protein pigments in marine invertebrates. (A) GFP fluorescence in a hydrozoan jellyfish. The light-producing photocytes are arranged pearl-chain like at the margin of the umbrella. (B) Green and red fluorescent proteins in the oral discs or tentacles of a mixed colony of *Palythoa/Zoanthus* species. A daylight image of the animals is shown in Figure 24-3B. (C) Green fluorescent corals from the Heron island reef flat: *Acropora* sp. (upper colony), *Platygyra pini* (lower left colony), *Favites flexuosa* (lower right colony). (D) Acroporid corals often contain bluish, nonfluorescent GFP-like protein in the axial polyps. (E) Orange-red protein fluorescence of tentacles of the sea anemone *Entacmaea quadricolor*. The nonfluorescent orange carotinoid pigments of clownfishes (*Amphiprion percula*) living among the tentacles appear black under the blue excitation light. (F) Green fluorescence of the tentacles of a leather coral (Alcyonaria). The red fluorescence is derived from chlorophyll of the symbiotic algae. (G) Green and red FPs in the tentacles of the tube anemone *Cerianthus membranaceus*. (H) A brightly green fluorescent copepod from the plankton of the Mediterranean Sea.

weeks as measured both *in vivo* and *in situ* (Leutenegger *et al.*, 2007). One striking structural feature of GFP-like proteins from nonbioluminescent anthozoans is the widespread formation of homotetramers. In contrast, GFP from *Renilla reniformis* exists as a dimer and avGFP forms dimers only at concentrations greater than 1mg/mL. In contrast, the GFPs isolated from copepods are predominantely monomeric (Shagin *et al.*, 2004).

It is worth mentioning that the β-can fold is not only found among pigments of marine organisms but also in the G2 fragment of the mammalian extracellular matrix protein nidogen, which contains a domain representing a structural homolog of avGFP (Hopf *et al.*, 2001). Although the β-barrel domain of the G2 fragment shares less than 10% sequence identity with avGFP, the tertiary structure is almost perfectly conserved. The GFP-like domain, however, lacks the intrinsic chromophore and is consequently completely colorless. This domain binds to the protein perlecan via a large surface patch conserved among metazoan nidogens. Based on these findings, a potential distribution of GFP-like proteins throughout bilateria is possible.

The color variability of marine GFP-like pigments depends on the structure of the chromophores. The 4–(p-hydroxybenzylidene)-5–imidazolinone structure of the GFP chromophore appears to be universally involved in the development of chromogenic properties in both fluorescent and nonfluorescent GFP-homologs (Figs. 24-7, 24-8, and 24-9). The strict conservation of the second and third chromophore-forming amino acids, tyrosine and glycine, and residues corresponding to Arg96 and Glu222 in avGFP suggest that the basic mechanisms of chromophore formation are all similar. However, significant differences exist concerning the subsequent modification of the GFP-type chromophore and its surroundings so that the structures responsible for cyan, green, yellow, and red fluorescence or nonfluorescent pink to blue colors are produced.

Red fluorescence requires an extension of the conjugated π-electron system. In proteins such as dsRed or eqFP611, the extension of the GFP-type chromophore along the peptide backbone is realized via an alcylimine bond generated in a second oxidation step. Another GFP-like protein with an alcylimine-containing chromophore is the blue

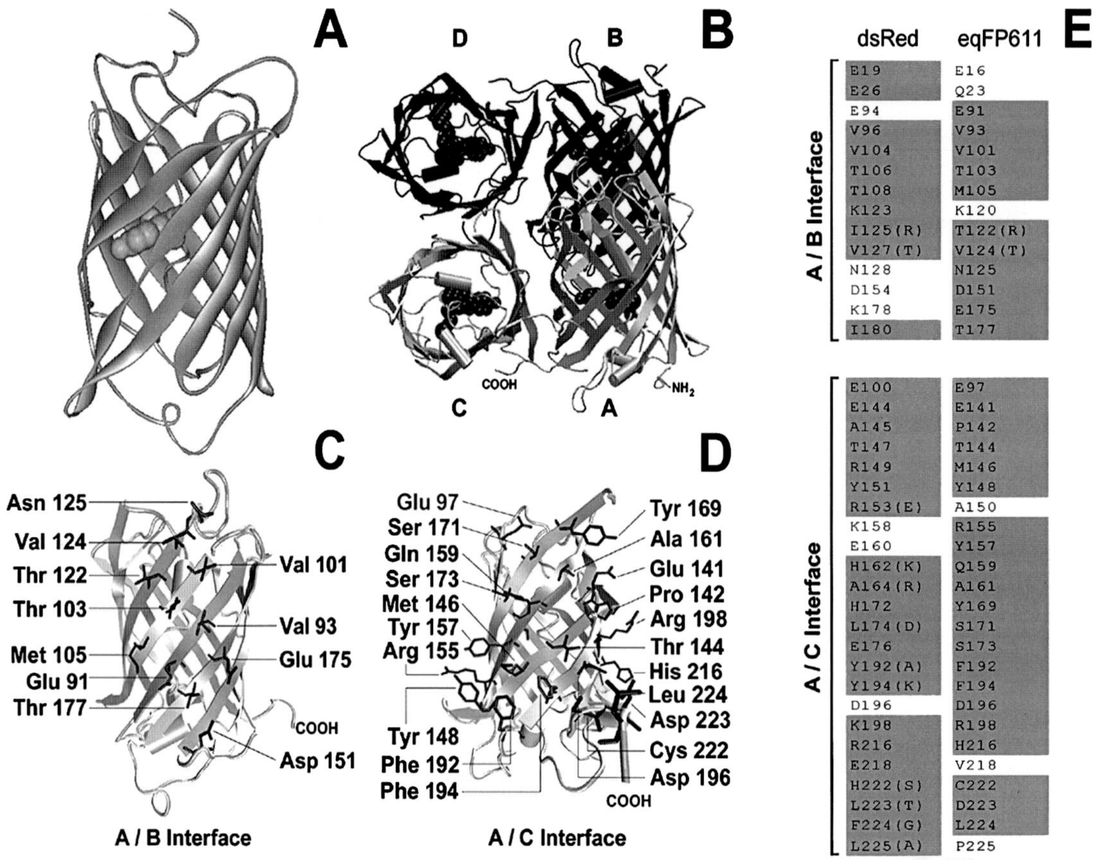

FIGURE 24-7. Secondary, tertiary, and quaternary structures of GFP and eqFP611 shown as ribbon diagrams. The β-can structure of the green fluorescent protein from *Aequorea victoria* consists of 11 antiparallel β-sheets connected by loop regions. The central helix is interrupted by the chromophore displayed as van-der-Waals spheres (A). The β-can fold is conserved in GFP-like proteins such as the red fluorescent protein eqFP611 from *Entacmaea quadricolor*. In contrast to avGFP, these proteins form densely packed tetramers (B). The tetramers are stabilized by interactions mediated by two interfaces per subunit. For eqFP611, the key amino acids (stick representation) are highlighted by their names in three-letter code and their number in the sequence (C, D). Alignment of interface residues involved in the formation of tetramers of the red fluorescent proteins dsRed and eqFP611 (E). The numbering of the amino acids equals the position in the amino acid sequence specific for each protein. Interacting residues in the respective protein are identified by a gray background. Data for dsRed are taken from Verkusha and Lukyanov (2004). For eqFP11, residues within a distance of 3.6 Å to the adjacent monomer are assumed to participate in the tetrameric interactions. Figure adapted from Wiedenmann *et al.* (2005) and Ivanchenko *et al.* (2005).

nonfluorescent protein RTMS5 (Prescott *et al.*, 2003). As in eqFP611, the hydroxyphenyl group assumes a *trans* configuration. However, the chromophore shows a noncoplanar conformation, which might explain the lack of fluorescence.

Interestingly, the extension of the conjugated π-electron system includes also light-driven processes that change the conformation of the chromophore as in the nonfluorescent purple chromoprotein asCP (Lukyanov *et al.*, 2000) or result in a posttranslational modification of the peptide backbone as in the photoconvertible proteins Kaede and EosFP (Ando *et al.*, 2002; Wiedenmann *et al.*, 2004b). These proteins switch irreversibly from a bright green to a bright red fluorescent state upon irradiation of near UV light with maximum efficiency around 390 nm (Mizuno *et al.*, 2003; Wiedenmann *et al.*, 2004b). Structure analysis of EosFP in both the green and the red fluorescent states showed that photoconversion involves a break of the N_α-C_α bond of His62 (Nienhaus *et al.*, 2005; Wiedenmann and Nienhaus, 2006).

Still another type of chromophore was proposed for the yellow fluorescent protein zFP538 from *Zoanthus* sp. (Zoanthidae). This protein contains a three-ring chromophore that is derived by a transimination reaction from a transient dsRed-type chromophore (Remington *et al.*, 2005). The structural analysis of GFP-like proteins has already revealed seven distinct chromophores, and, thus, it is likely that continuing research will yield new types and conformations.

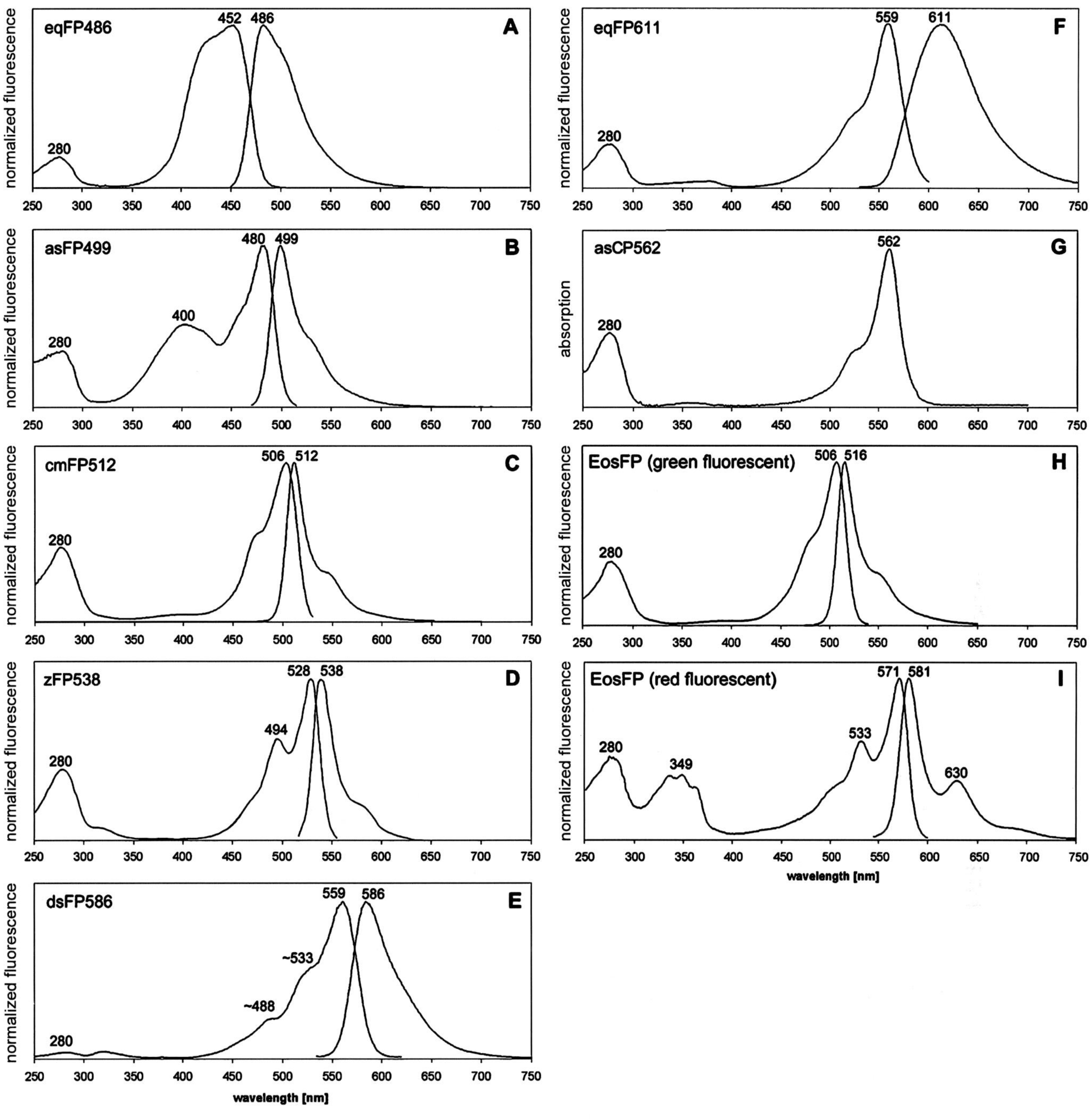

FIGURE 24-8. Spectral properties of major color classes of GFP-like proteins. Panels A through I show the excitation and emission spectra of fluorescent proteins or the absorption spectrum in the case of the nonfluorescent chromoprotein asCP562. (A) CFP (eqFP486) from *Entacmaea quadricolor*. (B) GFP (asFP499) from *Anemonia sulcata*. (C) GFP (cmFP512) from *Cerianthus membranaceus*. (D) YFP (zFP538) from *Zoanthus* sp. (E) RFP (dsFP586) from *Discosoma* sp. (F) Far-red fluorescent protein (eqFP611) from *Entacmaea quadricolor*. (G) Pink chromoprotein asCP562 from *Anemonia sulcata*. (H) Green-to-red photoconverting protein EosFP from *Lobophyllia hemprichii* before and (I) after conversion. Spectral data of zFP538 (Matz *et al.*, 1999) in (D) are a courtesy of Mikhail V. Matz, University of Texas; all others adapted from Wiedenmann *et al.* (2000, 2002b, 2004a, 2004b, and unpublished).

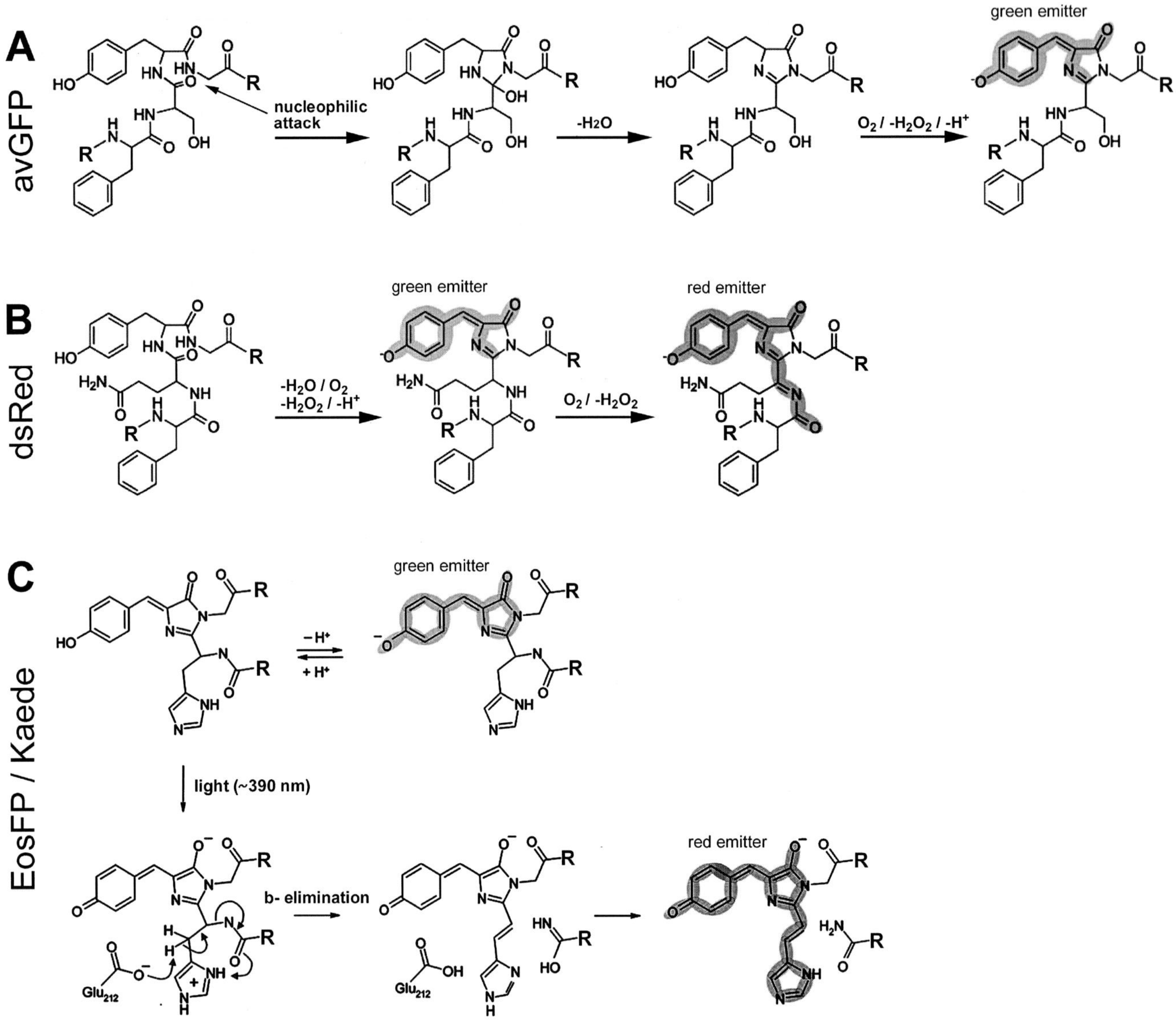

FIGURE 24-9. Chromophore structures of representatives of the GFP family and proposed mechanisms for their formation. (A) The formation of the 4–(p-hydroxybenzylidene)-5–imidazolinone chromophore of GFP from *Aequorea victoria* proceeds via a dehydration and a dehydrogenation reaction. (B) The conjugated π-system of the dsRed chromophore is extended by an alcylimine bond as a result of a second oxidation step. (C) In EosFP or Kaede, the extension of the π-system is achieved by a light driven β-elimination reaction resulting in a cleavage of the peptide backbone. Figure modified from Mizuno *et al.* (2003) and Nienhaus *et al.* (2005).

Applications of GFP

The green fluorescent protein from the bioluminescent jellyfish *Aequorea victoria* (avGFP) has revolutionized life science research as a label of specific proteins, marker of gene expression, and reporter of pH and Ca^{2+} levels in living cells and tissues (Chalfie *et al.*, 1994; Griesbeck, 2004; Llopis *et al.*, 1998; Tsien, 1998; Zhang *et al.*, 2002). GFP and its blue and yellowish emitting variants derive their popularity from the fact that the fluorophore forms in an autocatalytic reaction; no additional cofactor is needed (Tsien, 1998). This makes them superior over other reporter systems (e.g., luciferase based assays). Further advantages of FPs are that they can be designed to respond to a large variety of biological events and signals, targeted to subcellular compartments, and used in a wide variety of tissues and organisms. Biomedical research uses FPs in various applications, such as for tracking metastases, monitoring of aberrant gene regulation during tumor genesis or in studies of viral infections (Condeelis *et al.*, 2000; Hoffman, 1999; Kafri *et al.*, 2000; Lee *et al.*, 1997; Sandman *et al.*, 1999; Srivastava *et al.*, 1998; Stevenson *et al.*, 2000; Tsien, 1998; Zacharias *et al.*, 2000; Zhang *et al.*, 2002).

Pharmaceutical companies have embraced fluorescent protein (FP) technology in their research (e.g., high-throughput and high-content screening) as well as in the development of pharmaceutically relevant lead compounds and validation of production cell lines for biopharmaceuticals (Deo and Daunert, 2001; Wolf *et al.*, 2005, 2006). Some molecular properties of avGFP and its variants are, however, disadvantageous for wide application and require further improvement. In particular, photostability, the excitation by cytotoxic light, and the restricted range of emission wavelengths pose severe limitations for many applications. For cellular studies, red fluorescent proteins are particularly desirable because the background fluorescence of cellular tissue, cellular components, media, cultureware, and certain chemical compounds is markedly reduced in the red spectral region. Thus, the FP emission can be more easily distinguished from the background. The longer wavelengths of red light cause less scattering, and thus the light is transmitted more efficiently through tissue, resulting in clearer images. Low-energy red light is also less cytotoxic, and finally, semiconductor detection systems are more sensitive in the red spectral region. Another attractive aspect is the application of red FPs in multicolor labeling or fluorescence resonance energy transfer (FRET) experiments.

Applications of Multicolored GFP-Like Proteins

The discovery of GFP-like proteins in azooxanthellate anthozoa and even in crustaceans has greatly extended the range of available marker proteins, with colors ranging from cyan to red (Matz *et al.*, 1999; Shagin *et al.*, 2004; Wiedenmann, 1997). The novel red fluorescent proteins (RFPs) show spectral features that promise to overcome the disadvantages of the GFP technology (Fig. 24-10). Sometimes, the obligate formation of tetramers associated with the tendency to form high molecular weight aggregates can hamper the use of the wild-type proteins as markers for protein localization. Moreover, some of these proteins show maturation times of up to ~100 hours to achieve the fully fluorescent state, which restricts the applicability as marker for gene expression. For instance the use of dsRed in double labeling experiments along with GFP was complicated by the presence of a green fluorescent state of the red fluorescent marker (Baird *et al.*, 2000; Jakobs *et al.*, 2000).

Bioprospecting efforts have yielded proteins with the most red-shifted unmodified fluorescent protein of the GFP family, eqFP611 from *Entacmaea quadricolor* (Schenk *et al.*, 2004; Wiedenmann *et al.*, 2002b). The far-red emission at 611 nm and the large Stokes shift of 52 nm makes this protein particularly well suited for cellular applications (Wiedenmann *et al.*, 2002b). In addition, the protein is characterized by a fast and essentially complete maturation. However, also in this case, wide application as a marker protein is restricted by the tetramerization at high concentrations and the lack of folding at 37°C.

FP-based systems were developed that enable optical marking on the subcellular level under physiological conditions by localized irradiation with visible or near-ultraviolet light. Light-induced marking was first performed with avGFP that undergoes photoconversion to a red-emitting form when excited with intense blue light under low oxygen conditions (Elowitz *et al.*, 1997). Irradiation of the GFP T203H mutant (PA-GFP) with intense blue light (413 nm) induced a 100-fold increase of the green emission upon excitation at 488 nm (Patterson and Lippincott-Schwartz, 2002, 2003). The number of available photoactivatable proteins was recently greatly extended. Bright green light was shown to increase by 30-fold the red fluorescence of the kindling fluorescent protein mutant asCP A148G (Chudakov *et al.*, 2003). Another protein that appears useful for localized optical marking is the FP Kaede isolated from *Trachyphyllia geoffroyii* (Ando *et al.*, 2002). The emission color of this protein can be changed irreversibly from green to red upon irradiation with approximately 400-nm light. The author's laboratory introduced another photoswitchable FP, EosFP, which was cloned from the stony coral *Lobophyllia hemprichii* (Ivanchenko *et al.*, 2005; Wiedenmann *et al.*, 2004b). Similar to Kaede, the initially green emitting protein can be converted into a red emitting form by exposure to approximately 390-nm light, and this has proved useful as a fusion marker and for tracking molecules in both live cells and tissues (Fig. 24-10). The proteins also offer potential applications in photolithography and data storage on the molecular level (Wiedenmann *et al.*, 2005b).

Progress in engineering novel fluorescent marker proteins has been swift. These efforts have yielded variants with altered spectral properties, reduced aggregation tendency, and faster and better maturation properties (Bevis and Glick, 2002; Campbell *et al.*, 2002; Gurskaya *et al.*, 2001; Wiedenmann *et al.*, 2005; Yanushevich *et al.*, 2002). Dimeric and monomeric variants of GFP-like proteins were generated from hcRed, dsRed, eqFP611, EosFP, and further FPs form reef corals (Campbell *et al.*, 2002; Gurskaya *et al.*, 2001, 2006; Karasawa *et al.*, 2003, 2004; Shaner *et al.*, 2004; Wiedenmann *et al.*, 2004b, 2005). An *in vivo* mutagenesis system was introduced as valuable tool for the directed evolution of FPs (Wang *et al.*, 2004). A mutant of the dsRed derivative mRFP1, named mPlum, has an excitation maximum at 590 nm while the emission was shifted from 612 to 649 nm.

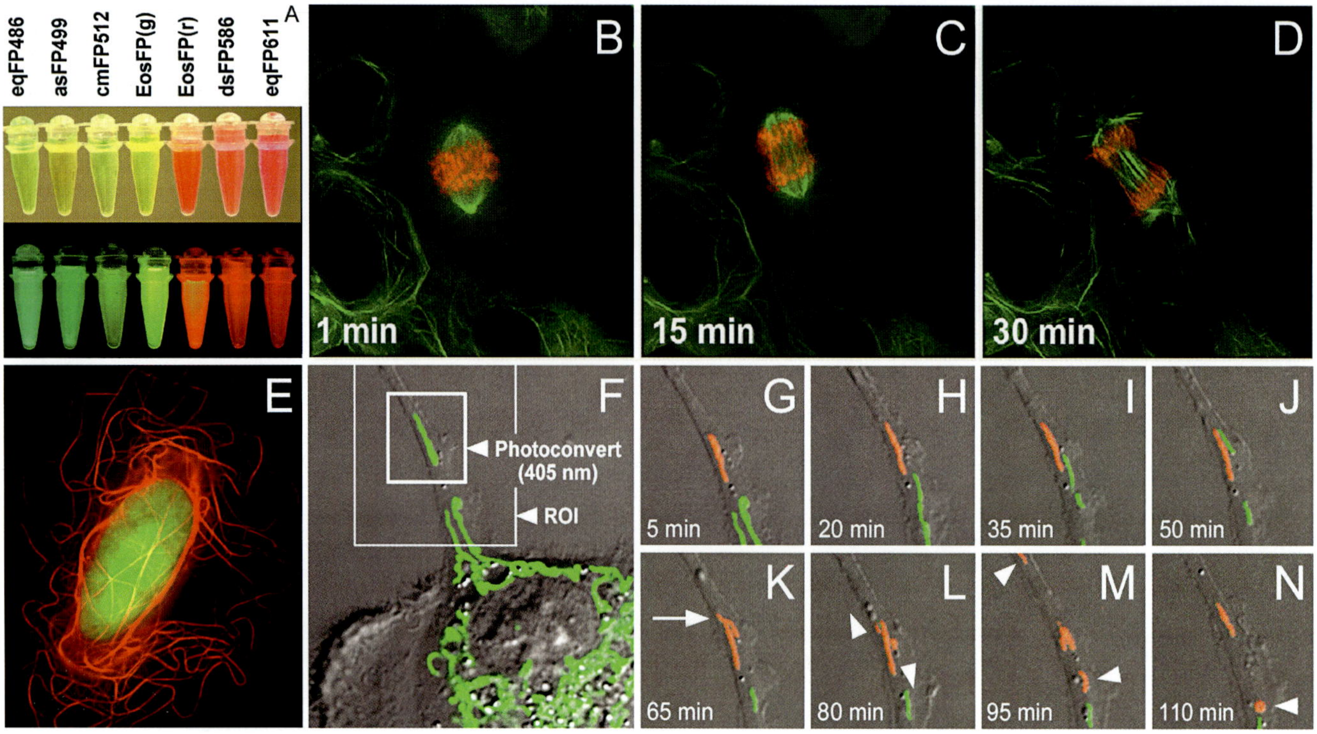

FIGURE 24-10. Applications of GFP-like Proteins. (A) The visual appearance of purified solutions of recombinant GFP-like proteins under visible light (upper row) and under UV light (366 nm; lower row). The proteins were cloned from the sea anemones *Entacmaea quadricolor* (eqFP486; eqFP611) and *Anemonia sulcata* (asFP499), the tube anemone *Cerianthus membranaceus* (cmFP512), the corallimorpharian disc anemone *Discosoma* sp. (dsFP586) and the scleractinian coral *Lobophyllia hemprichii* (EosFP). EosFP solution is shown in the initial green fluorescent state (EosFP(g)) and in the red state after photoconversion (EosFP(r)). (B, C, and D) Mitosis of a HEK293 cell co-transfected with cDNA of the tubulin-binding protein K49 fused to EGFP (enhanced *Aequorea* GFP) and the chromatin binding protein RBP-2N fused to a red fluorescent protein (EosFP after photoconversion). Chromatin association of RBP-2N is clearly visible whereas the EGFP fusion highlights the tubulin fibers. Chromosome segregation and tubulin dynamics during mitosis can be monitored. (E) A monomeric variant of the red fluorescent protein eqFP611 obtained by protein engineering highlights the cytoskeleton of a single HEK293 in fusion with the tubulin-binding protein K49. This fusion is not possible with the tetrameric wildtype protein (S. Kredel, F. Oswald, and J. Wiedenmann, unpublished). The nucleus shows green fluorescence of EGFP fused to the intracellular domain of the Notch receptor protein. (F–N) Application of the green-to-red photoconvertible protein EosFP for regional optical marking. The images show a microscopic brightfield image of a rabbit kidney cell overlaid with fluorescence images. EosFP was targeted to the mitochondria by a targeting peptide fused to the N-terminus of the marker protein. A single mitochondrion was photoconverted by a light pulse of a 400 nm laser. The photolabeled mitochondrion can be tracked over the time by its red fluorescence. A fusion event between two mitochondria can be observed in panel (K) by the color change of the unconverted mitochondrion. Later, fission generates three independently moving fragments. Panels (F) through (N) are a courtesy of Michael W. Davidson, National High Magnetic Field Laboratory, the Florida State University. Panels B–D are modified from Nienhaus *et al.* (2005).

Future Prospects for GFP-like Proteins

The focus of life science research is shifting from the analysis of genomes to the mechanistic study of gene products. Advances in light microscopy techniques allow imaging at the nanoscale and promise to increase the resolution to the molecular level (Betzig *et al.*, 2006; Hofmann *et al.*, 2005). The family of GFP-like proteins offers particularly great potential to satisfy the increasing demand for optical techniques to pursue mechanistic experimentation. Moreover, the relative ease with which these proteins can be modified for special purposes makes them important lead structures for the development of sophisticated marker systems. This is clearly demonstrated by the large number of discoveries from the field of GFP-like proteins that have been converted into viable assays for biomedical research (Tsien, 1998; Wiedenmann and Nienhaus, 2006; Zhang *et al.*, 2002). The discovery of GFP-like proteins in anthozoa species has greatly expanded the range of emission colors to the red spectral region. However, the potential of red fluorescent and photoactivatable proteins have not yet been fully exploited. Further research is required to improve FP technology by a systematic and thorough exploration of the relationship between structure and function of fluorescent proteins. The combination of bioprospecting and subsequent protein engineering promises synergisms that will further extend the set of useful fluorescent tools for life science applications and facilitate a rational design of fluorescent marker proteins for special uses.

Phycobiliproteins

Phycobiliproteins constitute another group of fluorescent proteins that lends the reddish color to cyanobacteria, cryptomonad algae, and red algae (Fig. 24-11). These proteins serve as accessory photosynthetic pigments to exploit the energy of light otherwise not accessible to the chlorophylls (Glazer, 1982, 1989). Prominent representatives are phycoerythrin (PE), phycocyanin (PC), and allophycocyanin (APC). These proteins are hetero-oligomers consisting of a type-dependent number of subunits. The proteins are often organized in accurately ordered, high-molecular-weight complexes called "phycobilisomes." Under daylight, the pigments show red (PE) to blue (PC, APC) coloration. Irradiated with a UV light source, PE fluoresces brightly orange, whereas APC exhibits a deep red glow. Color and fluorescence derive from linear tetrapyrroles attached as chromophores to the apoproteins. It is obvious from the differences in the structure of both the apoprotein and the attached chromophores that phycobiliproteins are not phylogenetically related to GFP-like proteins (Contreras-Martel *et al.*, 2001). Numerous genes act together to form the protein and the chromophore and thereby prevent, at least at present time, the use of phycobiliproteins as genetically encoded markers analogous to GFP and its homologs. However, the phycobilin fluorescence shows an exceptional brightness, which can be explained by the fact that each protein molecule binds several chromophores. The bright fluorescence is exploited for numerous applications in biomedical research, for instance, in the identification of cells (Kronick, 1986). For this purpose, the protein part of PE or APC is cross-linked to antibodies that specifically recognize epitopes characteristic of certain cell types. Consequently, the respective cells can be identified by the fluorescence of the phycobiliprotein via the bound antibody. Such specifically labeled cells can be automatically sorted out from a pool of different cell types by a method called fluorescence-activated cell sorting (FACS) (Sohn and Sautter, 1991; Zola *et al.*, 1990). Further potential applications of phycobiliproteins are in photodynamic cancer therapy (Huang *et al.*, 2002) or as antioxidant/anti-inflammatory agents (Romay *et al.*, 2003). Finally, phycocyanins are used as cosmetic colorants or food additives. Because of the complicated synthesis of the functional pigments, a recombinant production of the functional pigment is not possible. Therefore, the dyes are still purified by chromatographic methods, such as ion exchange and size exclusion chromatography, from natural sources. For example, the red alga *Corallina officinalis*, the

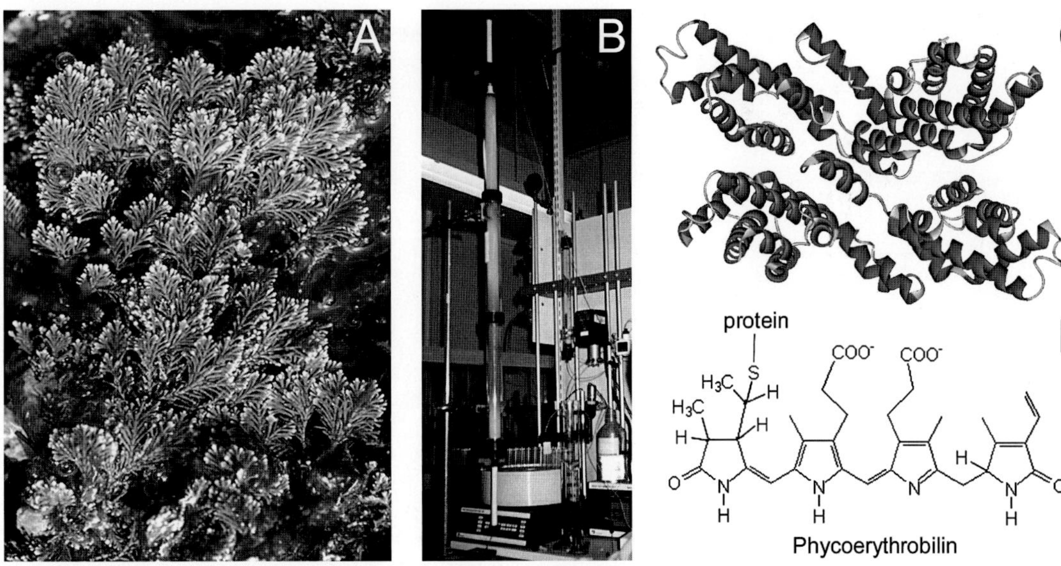

FIGURE 24-11. Phycobiliproteins. (A) The orange-red color of the calcareous alga *Corallina officinals* is derived from phycoerythrin and other phyocobiliproteins. The purified phycobiliproteins required for labeling purposes are produced by chromatographic techniques. (B) Separation of phycobiliproteins with a preparative size exclusion chromatography column. The colorful proteins are visible in the lower third of the column. (C) Schematic model of the crystal structure of R-phycoerythrin from *Gracilaria chilensis* (Contreras-Martel *et al.*, 2001) is characterized by abundant helical regions. The model was constructed using the coordinates deposited under the PDB accession code 1EYX. (D) Phycoerythrins owe their color to several linear tetrapyrrole chromophores including the phycoerythrobilin chromophore shown in the figure. The chromophores are covalently bound to the protein. Photograph (A) by M.D. Guiry, http://www.algaebase.org, was reproduced with kind permission of the photographers. Photograph (B) courtesy of the Haematological Malignancy Diagnostic Service Department, Leeds; www.hmds.org.uk/fluorochrome.html#RPE).

cyanobacteria *Arthrospira* (*Spirulina*) and *Anabaena flos-aquae*, and cryptomonad algae have been used for this purpose. These organisms are either collected from the ocean or grown in aquaculture (Materassi *et al.*, 1984; Tredici *et al.*, 1986).

A NOTE ON THE LEFTOVERS: GUANO

"Guano" refers to the excrements of seabirds that accumulate on their nesting islands in arid climates (Fig. 24-12). Being an end product of the catabolism of marine proteins, guano has had a considerable impact on human health in former times, and it is therefore to briefly discuss this topic. Guano islands can be found off the coasts of Africa, Australia, and South America. The most famous guano islands are located on the Peruvian coast. Upwellings enrich these waters with nutrients and support large populations of anchovy (*Engraulis* sp.). These fish are the major diet of several bird species including pelicans (*Pelecanus thagus*), cormorants (*Phalacrocorax bougainvillii*), and boobies (*Sula variegata*) that inhabit these guano islands (Coker, 1919; Muck and Pauly, 1987). Fish proteins contain high amounts of nitrogen as part of peptide backbone as well as the side chains of the amino acids lysine, arginine, histidine, tryptophan, asparagine, and glutamine. During catabolism of the proteins, the α-amino group is removed from the component amino acids as ammonia. Because ammonia is toxic for birds, they transform it into uric acid as a waste product. Uric acid is excreted in a crystalline-like form as part of the bird droppings. Under the extremely dry conditions of the guano islands, the leftovers accumulate in layers many meters thick over the course of several hundreds of years. The action of bacteria transforms uric acid and other parts of the guano into various nitrogen containing compounds, including nitrate. The high content of both nitrogen and phosphate in chemical forms available to plants makes guano an excellent fertilizer. The application of guano in agriculture has been a long lasting tradition among the South American natives. Mainly in the 19th century to the beginning of the 20th century, guano was massively exploited and

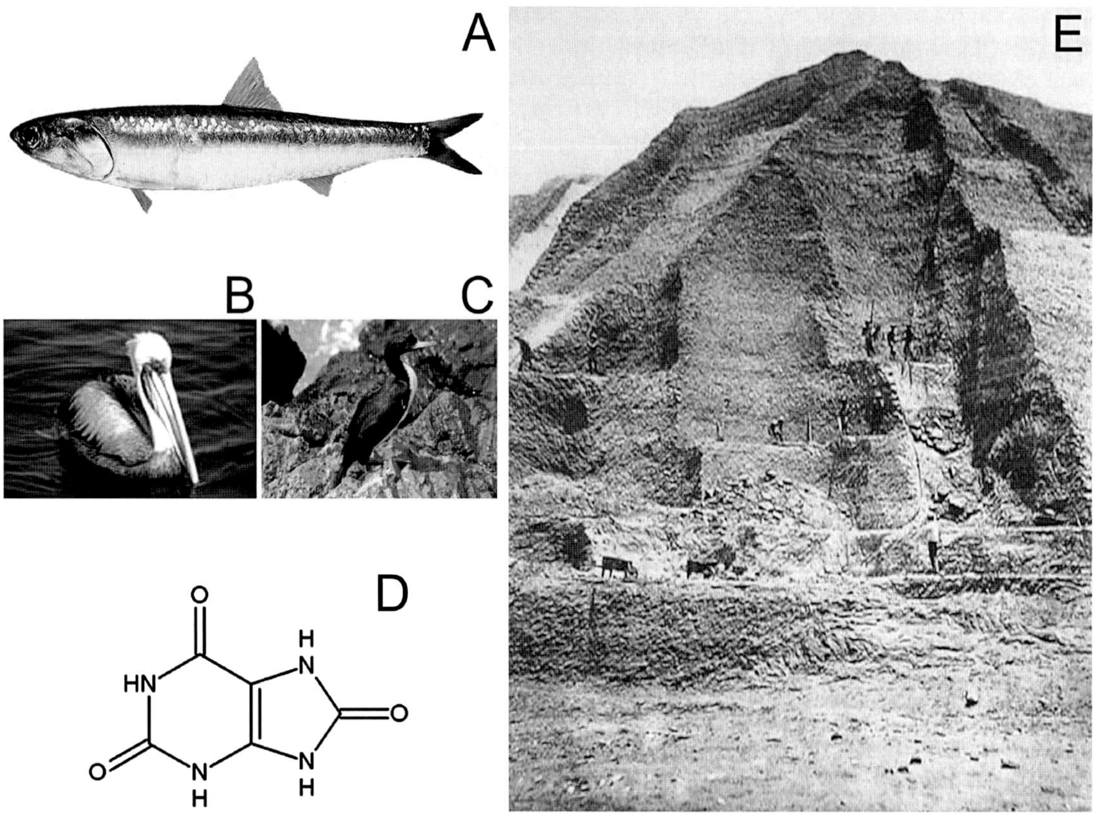

FIGURE 24-12. Guano production. In the Peruvian upwelling ecosystem, anchovies such as *Engraulis ringens* (A) are one major part of the diet of guano-producing birds such as pelicans (*Pelecanus thagus*) (B) or cormorants (*Phalacrocorax bougainvillii*) (C). Nitrogen from the protein rich diet is excreted in the form of uric acid (D). Guano mining on the Central Chinchua Islands, ca. 1860 (E). Photograph (A) is a courtsey of Philippe Béarez, CNRS/Muséum National d'Histoire Naturelle, Paris. Photographs (B–C) are from Wikipedia. Photograph (D) Negative No. 311830 of the American Museum of Natural History. From Wikipedia.

exported to the United States and Europe (Hunt, 1985; Mathew, 1970, 1977). The conditions for the workers mining the guano were miserable, and the constant exposure to guano dust had strongly negative effects on their health. Aside from its use as fertilizer, guano has also been a major source of nitrates for the production of gunpowder. So the nitrogenous remnants of marine proteins has had ambivalent effects on human health worldwide depending on whether one had to dig the excrements, or if one was benefiting from a richer harvest from fertilized soils, or if one was standing in front of a gun muzzle or behind it. The industrial production of nitrates, enabled by the invention of the Haber-Bosch procedure, ended the extensive use of guano. Currently, guano is appreciated as a natural fertilizer by organic farmers and potted plant growers.

CONCLUSION

Marine proteins have already provided a number of tools without which the current progress of life science research would not have been possible. Tremendous potential lies ahead for the discovery of new valuable applications. The usefulness of marine compounds, including peptides and proteins, for pharmaceutical purposes is still largely unexplored. Also in this context, the discovery of bioactive compounds will be facilitated by high throughput and high-content assays using marine protein tools such as luciferases and fluorescent proteins. Continued progress in developing marine proteins will be promoted by new biotechnological and molecular biological methods, such as high-end mass spectroscopy, high-throughput sequencing of genomes, microarray techniques, automated microscopy, and advanced computer-based data analysis, as well as by modern marine instrumentation including submarines and remotely operated vehicles. However, it is imperative to keep in mind that the vast space left for new discoveries is dependent on the presence and diversity of marine creatures that have adapted to different habitats over Earth's history. This diversity is threatened by human activities on a global scale. Whole ecosystems are endangered by pollution and eutrophication. Global warming has negative impacts on the oceans, and with the increasing number of mass coral mortality events, the coral reef community and its pharmaceutical treasure chest is in danger of being lost. Overfishing brings species close to extinction. Most important, not only the directly affected species suffer, but entire food webs can be destabilized by the removal of key components such as the top predators (e.g., sharks and other large fish). Increasingly, modern scientific research including the analysis and application of marine proteins are carried out at the nanoscale. However, this does not imply that science should lose sight of the large-scale connections between the marine realm and humankind. The health of humans is dependent on healthy oceans.

Acknowledgments

I thank Professor Ulrich Nienhaus and Dr. Franz Oswald for the long-lasting and fruitful collaboration. Moreover, I am particularly grateful Dr. Cecilia D'Angelo for her valuable support in the preparation of this chapter.

References

Adams, M.W., 1993. Enzymes and proteins from organisms that grow near and above 100 degrees C. Annu. Rev. Microbiol. 47, 627–658.

Alexander, D.C., Costanzo, M.A., Guzzo, J., Cai, J., Charoensri, N., Diorio, C., Dubow, M.S., 2000. Blazing towards the next millennium: Luciferase fusions to identify genes responsive to environmental stress. Water Air Soil Pollut. 123, 81–94.

Anderluh, G., Dalla Serra, M., Viero, G., Guella, G., Macek, P., Menestrina, G., 2003. Pore formation by equinatoxin ii, a eukaryotic protein toxin, occurs by induction of nonlamellar lipid structures. J. Biol. Chem. 278, 45216–45223.

Ando, R., Hama, H., Yamamoto-Hino, M., Mizuno, H., Miyawaki A., 2002. An optical marker based on the UV-induced green-to-red photoconversion of a fluorescent protein. Proc. Natl. Acad. Sci. USA 99, 12651–12656.

André, P., Kim A., Khrapko, K., Thilly, W.G., 1997. Fidelity and mutational spectrum of Pfu DNA polymerase on a human mitochondrial DNA sequence. Genome Res. 7, 843–852.

Baird, G.S., Zacharias, D.A., Tsien, R.Y., 2000. Biochemistry, mutagenesis, and oligomerization of DsRed, a red fluorescent protein from coral. Proc. Natl. Acad. Sci. USA 97, 11984–11989.

Ballard, R.D., Grassle, J.F., 1979. Return to oases of the deep. National Geographic. 156, 680–705.

Bang, F.B., 1953. The toxic effect of a marine bacterium on *Limulus* and the formation of blood clots. Biol Bull. 105, 637–362.

Beddoe, T., Ling, M., Dove, S., Hoegh-Guldberg, O., Devenish, R.J., Prescott, M., Rossjohn, J., 2003. The production, purification and crystallization of a pocilloporin pigment from a reef-forming coral. Acta Crystallogr. D 59, 597–599.

Beebe, W., 1935. Half Mile Down. London, John Lane.

Bergner, A., Oganessyan, V., Muta, T., Iwanaga, S., Typke, D., Huber, R., Bode, W., 1996. Crystal structure of a coagulogen, the clotting protein from horseshoe crab: A structural homologue of nerve growth factor. EMBO J. 24, 6789–6797.

Betzig, E., Patterson, G.H., Sougrat, R., Lindwasser, O.W., Olenych, S., Bonifacino, J.S., Davidson, M.W., Lippincott-Schwartz, J., Hess, H.F., 2006. Imaging intracellular fluorescent proteins at nanometer resolution. Science 313, 1642–1645.

Bevis, B. Glick, B., 2002. Rapidly maturing variants of Discosoma red fluorescent protein (DsRed). Nat. Biotechnol., 20, 83–86.

Bhaumik, S., Gambhir, S.S., 2002. Optical imaging of *Renilla* luciferase reporter gene expression in living mice. Proc. Natl. Acad. Sci. USA 99, 377–382.

Blinks, J.R., Wier, W.G., Hess, P., Prendergast, F.G., 1982. Measurement of Ca^{2+} concentrations in living cells. Prog. Biophys. Mol. Biol. 40, 1–114.

Bokman, S.H., Ward, W.W., 1981. Renaturation of *Aequorea* green-fluorescent protein. Biochem. Biophys. Res. Commun. 101, 1372–1380.

Bowlby, M.R., Case, J.E., 1991. Flash kinetics and spatial patterns of bioluminescence in the copepod *Gaussia princes*. Mar. Biol. 110, 329–336.

Brock, T.D., 1997. The value of basic research: Discovery of *Thermus aquaticus* and other extreme thermophiles. Genetics 146, 1207–1210.

Burnett, J.W., Gable, W.D., 1989. A fatal jellyfish envenomation by the Portuguese man-o'-war. Toxicon 27, 823–824.

Campbell, A.K., Herring, P.J., 1990. Imidazolopyrazine bioluminescence in copepods and other marine organisms. Mar. Biol. 104, 219–225.

Campbell, R.E., Tour, O., Palmer, A.E., Steinbach, P.A., Baird, G.S., Zacharias, D.A., Tsien, R.Y., 2002. A monomeric red fluorescent protein. Proc. Natl. Acad. Sci. USA 99, 7877–7882.

Catala, R., 1958. Effets de fluorescence provoque sur des coraux par l'action des rayons ultra-violets. C. R. Acad. des Sciences 247, 1678–1679.

Catala, R., 1959. Fluorescence effects from corals irradiated with ultraviolet rays. Nature 183, 949.

Catala, R., 1960. Nouveaux organismes marins presentant des effets de fluorescence par l'action des rayons ultra-violets. C. R. Acad. des Sciences 250, 1128.

Chalfie, M., Tu, Y., Euskirchen, G., Ward, W.W., Prasher, D.C., 1994. Green fluorescent proteins as a marker for gene expression. Science 263, 802–805.

Chatterjee, J., Meighen, E.A., 1995. Biotechnological applications of bacterial bioluminescence (lux) lux genes. Photochem. Photobiol. 62, 651–650.

Chudakov, D.M., Belousov, V.V., Zaraisky, A.G., Novoselov, V.V., Staroverov, D.B., Zorov, D.B., Lukyanov, S., Lukyanov, K.A., 2003. Kindling fluorescent proteins for precise in vivo photolabeling. Nat. Biotechnol. 21, 191–194.

Cline, J., Braman, J.C., Hogrefe, H.H., 1996. PCR fidelity of Pfu DNA polymerase and other thermostable DNA polymerases. Nucleic Acids Res. 24, 3546–3551.

Cody, C.W., Prasher, D.C., Westler, W.M., Prendergast, F.G., Ward, W.W., 1993. Chemical structure of the hexapeptide chromophore of the *Aequorea* green-fluorescent protein. Biochemistry 32, 1212–1218.

Coker, R.E., 1919. Habits and economic relations of the guano birds of Peru. Charge Scientific Inquiry, United States Bureau of Fisheries. Proc. U.S. Nat. Mus. 56, 449–511.

Condeelis, J.S., Wyckoff, J., Segall, J.E., 2000. Imaging of cancer invasion and metastasis using green fluorescent protein. Eur. J. Cancer. 36, 1671–1680.

Cooper, E.L., 2004 Commentary on *traditional and modern biomedical prospecting: Part II—the benefits*. In Mueller, W.E.G., Schroder, H.C., Wiens, M., Perovic-Ottstadt, S., Batel, R., Muller, I.M. Evid. Based Complement Alternat. Med. 1, 207–209.

Cubitt, A.B., Heim, R., Adams, S.R., Boyd, A.B., Gross, L.A., Tsien R.Y., 1995. Understanding, improving and using green fluorescent proteins. TIBS 20, 448–455.

Currie, B., Khanh, D.M., Alderslade, P., Williamson, J., Fenner, P., 1992. Jellyfish envenomation in the Northern Territory of Australia. Toxicon 30, 501.

Davenport, D., Nicol, J.A.C., 1955. Luminescence in hydromedusae. Proc. R. Soc. London Ser. B Biol. Sci. 144, 399–411.

Deming, J.W., 1998. Deep ocean environmental biotechnology. Curr. Opin. Biotechnol. 9, 283–287.

Deo, S.K., Daunert, S., 2001. Special issue paper: Luminescent proteins from *Aequorea victoria*: Applications in drug discovery and in high throughput analysis. Fresenius J. Anal. Chem. 369, 258–266.

Deo, S.K., Mirasoli, M., Daunert, S., 2005. Bioluminescence resonance energy transfer from aequorin to a fluorophore: an artificial jellyfish for applications in multianalyte detection. Anal. Bioanal. Chem. 381, 1387–1394.

Desjardins, M., Morse, D., 1993. The polypeptide components of scintillons, the bioluminescence organelles of the dinoflagellate *Gonyaulax polyedra*. Biochem. Cell. Biol. 71, 176–182.

Dib, R., Chobert, J.M., Dalgalarrondo, M., Barbier, G., Haertle, T., 1998. Purification, molecular properties and specificity of a thermoactive and thermostable proteinase from *Pyrococcus abyssi*, strain st 549, hyperthermophilic archaea from deep-sea hydrothermal ecosystem. FEBS Lett. 431, 279–284.

Douglas, R.H., Partridge, J.C., Dulai, K., Hunt, D., Mullineaux, C.W., Tauber, A.Y., Hynninen, P.H., 1998. Dragon fish see using chlorophyll. Nature 393, 423–424.

Dove, S.G., Hoegh-Guldberg, O., Ranganathan, S., 2001. Major colour patterns of reef-building are due to a family of GFP-like protein. Coral Reefs 19, 197–204.

Dove, S.G., Takabayashi, M., Hoegh-Guldberg, O., 1995. Isolation and partial characterization of the pink and blue pigments of Pocilloporid and Acroporid corals. Biol. Bull. 189, 288–297.

Dunstan, S.L., Sala-Newby, G.B., Fajardo, A.B., Taylor, K.M., Campbell, A.K., 2000. Cloning and expression of the bioluminescent photoprotein pholasin from the bivalve mollusc *Pholas dactylus*. J. Biol. Chem. 275, 9403–9409.

Elowitz, M.B., Surette, M.G., Wolf, P.-E., Stock, J., Leibler, S., 1997. Photoactivation turns green fluorescent protein red. Curr. Biol. 7, 809–812.

Fiala, G. Stetter, K.O., 1986. Pyrococcus furiosus sp. nov. represents a novel genus of marine heterotrophic archaebacteria growing optimally at 100°C. Arch. Microbiol. 145, 56–61.

Fisher, A.J., Raushel, F.M., Baldwin, T.O., Rayment, I., 1995. Three-dimensional structure of bacterial luciferase from *Vibrio harveyi* at 2.4 A resolution. Biochemistry 34, 6581–6586.

Glazer, A.N., 1982. Phycobilisomes: Structures and dynamics. Ann. Rev. Microbiol. 36, 173–198.

Glazer, A.N., 1989. Light guides. Directional energy transfer in a photosynthetic antenna. J. Biol. Chem. 264, 1–4.

González, J.M., Masuchi, Y., Robb, F.T., Ammerman, J.W., Maeder, D.L., Yanagibayashi, M., T amaoka, J., Kato, C., 1998. *Pyrococcus horikoshii* sp. nov., a hyperthermophilic archaeon isolated from a hydrothermal vent at the Okinawa Trough. Extremophiles 2, 123–130.

Gorny, R.L., Douwes, J., Versloot, P., Heederik, D., Dutkiewicz, J., 1999. Application of the classic *Limulus* test and the quantitative kinetic chromogenic LAL method for evaluation of endotoxin concentration in indoor air. Ann. Agric. Environ. Med. 6, 45–51.

Greagg, M.A., Fogg, M.J., Panayotou, G. Evans, S.J., Connolly, B.A., Pearl, L.H., 1999. A read-ahead function in archaeal DNA polymerases detects promutagenic template-strand uracil. Proc. Natl. Acad. Sci. U. S.A. 96, 9045–9050.

Griesbeck, O., 2004. Fluorescent proteins as sensors for cellular functions. Curr. Opin. Neurobiol. 14, 636–641.

Gurskaya, N.G., Fradkov, A.F., Terskikh, A., Matz, M.V., Labas, Y.A., Martynov, V.I., Yanushevich, Y.G., Lukyanov, K.A., Lukyanov, S.A., 2001. GFP-like chromoproteins as a source of far-red fluorescent proteins. FEBS Lett. 507, 16–20.

Gurskaya, N.G., Verkhusha, V.V., Shcheglov, A.S., Staroverov, D.B., Chepurnykh, T.V., Fradkov, A.F., Lukyanov, S., Lukyanov, K.A., 2006. Engineering of a monomeric green-to-red photoactivatable fluorescent protein induced by blue light. Nat. Biotechnol. 24, 461–465.

Haddock, S.H.D., Case, J.F., 1999. Bioluminescence spectra of shallow and deep-sea gelatinous zooplankton: Ctenophores, medusae and siphonophores. Mar. Biol. 133: 571–582.

Haddock, S.H.D., Rivers, T.J., Robison, B.H., 2001. Can coelenterates make coelenterazine? Dietary requirement for luciferin in cnidarian bioluminescence. Proc. Nat. Acad. Sci. USA 98, 11148–11151.

Harper, R.D., Case, J.F., 1999. Disruptive counterillumination and its antipredatory value in the plainfin midshipman *Porichthys notatus*. Mar. Biol. 134, 529–540.

Hart, R.C., Matthews, J.C., Hori, K., Cormier, M.J., 1979. Renilla reniformis bioluminescence, luciferase-catalyzed production of nonradiating excited state species involved in the energy transfer to *Renilla* green fluorescent protein. Biochemistry 18, 2204–2210.

Hashimoto, H., Nishioka, M., Fujiwara, S., Takagi, M., Imanaka, T., Inoue, T., Kai, Y.J., 2001. Crystal structure of DNA polymerase from hyperthermophilic archaeon *Pyrococcus kodakaraensis* KOD1. Mol. Biol. 306, 469–477.

Hastings, J.W., Morin, J.G., 1969. Calcium-triggered light emission in Renilla: A unitary biochemical scheme for coelenterate bioluminescence. Biochem. Biophys. Res. Commun. 37, 493–498.

Hastings, J.W., Morin, J.G., 1998. Photons for reporting molecular events: Green fluorescent protein and four luciferase systems. In Chalfie, M., and Kain, S. (eds.), Green Fluorescent Protein: Properties, Applications and Protocols, pp. 17–41. New York, Wiley-Liss.

Haygood, M.G., Distel, D.L., 1993. Bioluminescent symbionts of flashlight fishes and deep-sea anglerfishes form unique lineages related to the genus *Vibrio*. Nature 363,154–156

Head, J.F., Inouye, S., Teranishi, K., Shimomura, O., 2000. The crystal structure of the photoprotein aequorin at 2.3 Å resolution. Nature 405, 372–376.

Heim, R., Prasher, D.C., Tsien, R.Y., 1994. Wavelength mutations and posttranslational autoxidation of green fluorescent protein. Proc. Natl. Acad. Sci. USA 91, 12501–12504.

Henneke, G., Raffin, J.P., Ferrari, E., Jonsson, Z.O., Dietrich, J., Hubscher, U., 2000. The PCNA from *Thermococcus fumicolans* functionally interacts with DNA polymerase delta. Biochem. Biophys. Res. Commun. 276, 600–606.

Henry, J.P., Monny, C., Michelson, A.M., 1978. Characterization and properties of *Pholas* luciferase as a metalloglycoprotein. Biochemistry 14, 3458–3466.

Hoffman, R.M., 1999. Orthotopic metastatic mouse models for anticancer drug discovery and evaluation: A bridge to the clinic. Invest. New Drugs 17, 343–359.

Hofmann, M., Eggeling, C., Jakobs, S., Hell, S.W., 2005. Breaking the diffraction barrier in fluorescence microscopy at low light intensities by using reversibly photoswitchable proteins. Proc. Natl. Acad. Sci. USA 102, 17565–17569.

Hopf, M., Göhring, W., Ries, A., Timpl, R., Hohenester, E., 2001. Crystal structure and mutational analysis of a perlecan-binding fragment of nidogen-1. Nat. Struct. Biol. 8, 634–640.

Howell, W.H. 1885. Observations upon the chemical composition and coagualtion of the blood of *Limulus polyphemus*, *Callinectes hastatus*, and *Cucumaris* sp. Johns Hopkins University Circular 43, 4–5.

Huang, B., Wang, G.C., Zeng, C.K., Li, Z.G., 2002. The experimental research of R-phycoerythrin subunits on cancer treatment: A new photosensitizer in PDT. Cancer Biother. Radiopharm. 17, 35–42.

Hunt, S., 1985. Growth and guano in nineteenth-century Peru. In Conde, R.C., and Hunt, S.J. (eds.), The Latin American economies: Growth and the export sector, 1880–1930. New York, Holmes & Meier.

Inouye, S., Noguchi, M., Sakaki, Y., Takagi, Y., Miyata, T., Iwanaga, S., Miyata, T., Tsuji, F.I., 1985. Cloning and sequence analysis of cDNA for the luminescent protein aequorin. Proc. Natl. Acad. Sci. USA 82, 3154–3158.

Inouye, S., Tsuji, F.I., 1994. *Aequorea* green fluorescent protein: Expression of the gene and fluorescent characteristics of the recombinant protein. FEBS Lett. 341, 277–280.

Ip, D.T., Chan, S.H., Allen, M.D., Bycroft, M., Wan, D.C., Wong, K.B., 2004. Crystallization and preliminary crystallographic analysis of a novel orange fluorescent protein from the Cnidaria tube anemone *Cerianthus* sp. Acta Crystallogr. D 60, 340–341.

Ivanchenko, S., Röcker, C., Oswald, F., Wiedenmann, J., Nienhaus, G.U., 2005. Targeted green-red photoconversion of EosFP, a fluorescent marker protein. J. Biol. Phys. 31, 249–259.

Iwanaga, S., 2002. The molecular basis of innate immunity in the horseshoe crab. Curr. Opin. Immunol. 14, 87–95.

Iwanaga, S., Kawabata, S., 1998. Evolution and phylogeny of defence molecules associated with innate immunity in horseshoe crab. Front. Biosci. 3, 973–984.

Iwanaga, S., Lee, B.L., 2005. Recent advances in the innate immunity of invertebrate animals. J. Biochem. Mol. Biol. 38, 128–150.

Iwanaga, S., Morita, T., Harada, T., Niwa, M., Takada, K., Kimura, T., Sakakibara, S., 1978. Chromogenic substrates for horseshoe crab clotting enzyme. Its application for the assay of bacterial endotoxins. Haemostasis 7, 183–188.

Jakobs, S., Subramaniam, V., Schonle, A., Jovin, T.M., Hell, S.W., 2000. EFGP and DsRed expressing cultures of Escherichia coli imaged by confocal, two-photon and fluorescence lifetime microscopy. FEBS Lett. 18, 131–135.

Jerlov, N.G., 1976. Marine Optics. Amsterdam, Elsevier.

Johnsen, S., Balser, E.J., Widder, E.A., 1999. Light-emitting suckers in an octopus. Nature 398, 113–114.

Johnson, F.H., Shimomura, O., Saiga, Y., Gershman, G., Reynolds, G.T., Waters, J.R., 1962. Quantum efficiency of *Cypridina* luminescence with a note on that of *Aequorea*. J. Cell. Comp. Physiol. 60, 85–103.

Johnson, G.D., Rosenblatt, R.H., 1988. Mechanisms of light organ occlusion in flashlight fishes, family Anomalopidae (Teleotei: Beryciformes), and the evolution of the group. Ann. Linneun. Sot. Land. 94, 65–96.

Kafri, T., van Praag, H., Gage, F.H., Verma, I.M., 2000. Lentiviral vectors: Regulated gene expression. Mol. Ther. 1, 516–521.

Karasawa, S., Araki, T., Nagai, T., Mizuno, H., Miyawaki, A., 2004. Cyan-emitting and orange-emitting fluorescent proteins as a donor/acceptor pair for fluorescence resonance energy transfer. Biochem. J. 381, 307–312.

Karasawa, S., Araki, T., Yamamoto-Hino, M., Miyawaki, A., 2003. A green-emitting fluorescent protein from Galaxeidae coral and its monomeric version for use in fluorescent labeling. J. Biol. Chem. 278, 34167–34171.

Kawaguti, S., 1944. On the physiology of reef corals VI. Study on the pigments. Palao Trop. Biol. Stn. Stud. 2, 617–674.

Kelmanson, I.V., Matz, M.V., 2003. Molecular basis and evolutionary origins of color diversity in great star coral *Montastraea cavernosa* (Scleractinia: Faviida). Mol. Biol. Evol., 20, 1125–1133.

Khoo, H.E., 2002. Bioactive proteins from stonefish venom. Clin. Exp. Pharmacol. Physiol. 29, 802–806.

Kitabayashi, M., Nishiya, Y., Esaka, M., Itakura, M., Imanaka, T., 2002. Gene cloning and polymerase chain reaction with proliferating cell nuclear antigen from *Thermococcus kodakaraensis* KOD1. Biosci. Biotechnol. Biochem. 66, 2194–2200.

Kleppe, K., Ohtsuka, E., Kleppe, R., Molineux, I., Khorana, H.G.J., 1971. Studies on polynucleotides. XCVI. Repair replications of short synthetic DNA's as catalyzed by DNA polymerases. Mol. Biol. Mar. 56, 341–361.

Kohno, T., Kim, J.I., Kobayashi, K., Kodera, Y., Maeda, T. Sato, K., 1995. Three-dimensional structure in solution of the calcium channel blocker omega-conotoxin MVIIA. Biochemistry 34, 10256–10265.

Kong, H., Kucera, R.B., Jack, W.E., 1993. Characterization of a DNA polymerase from the hyperthermophile archaea *Thermococcus litoralis*. Vent DNA polymerase, steady state kinetics, thermal stability, processivity, strand displacement, and exonuclease activities. J. Biol. Chem. 268, 1965–1975.

Kronick, M.N., 1986. The use of phycobiliproteins as fluorescent labels in immunoassay. J. Imm. Meth. 92, 1–13.

Labas, Y.A., Gurskaya, N.G., Yanushevich, Y.G., Fradkov, A.F., Lukyanov, K.A., Lukyanov, S.A., Matz, M.V., 2002. Diversity and evolution of the green fluorescent protein family. Proc. Natl. Acad. Sci. USA 99, 4256–4261.

Lee, A.H., Han, J.M., Sung, Y.C., 1997. Generation of the replication-competent human immunodeficiency virus type 1 which expresses a jellyfish green fluorescent protein. Biochem. Biophys. Res. Commun. 233, 288–292.

Leutenegger, A., D'Angelo, C., Matz, M.V., Denzel, A., Oswald, F., Salih, A., Nienhaus, G.U., Wiedenmann, J, 2007. It's cheap to be colorful: Anthozoans show a slow turnover of GFP-like proteins. FEBS J., in press.

Levin, J., Bang, F.B., 1964a. A description of cellular coagulation in the *Limulus*. Bull. Johns Hopkins Hosp.115, 337–345.

Levin, J., Bang, F.B., 1964b. The role of endotoxin in the extracellular coagulation of *Limulus* blood. Bull. Johns Hopkins Hosp. 115, 265–274.

Levin, J., Bang, F.B., 1967. Clottable protein in Limulus: Its localization and kinetics of its coagulation by endotoxin. Thromb. Diath. Haernorrh., 19, 186–197.

Levine, L.D., Ward, W.W., 1982. Isolation and characterization of a photoprotein, "phialidin," and a spectrally unique green-fluorescent protein from the bioluminescent jellyfish Phialidium gregarium. Comp. Biochem. Physiol. 72B, 77–85.

Li, L., Hong, R., Hastings, J.W., 1997. Three functional luciferase domains in a single polypeptide chain. Proc. Natl. Acad. Sci. USA 94, 8954–8958.

Lindsay, G.K., Roslansky, P.F., Novitsky, T., 1989. Single-step chromogenic limulus amebocyte lysate assay for endotoxin. J. Clinic. Microbiol. 27, 947–951.

Llopis, J., McCaffery, J.M., Miyawaki, A., Farquhar, M.G., Tsien, R.Y., 1998. Measurement of cytosolic, mitochondrial, and Golgi pH in single living cells with green fluorescent proteins. Proc. Natl. Acad. Sci. USA 95, 6803–6808.

Lonsdale, P., 1977. Clustering of suspension-feeding macrobenthos near abyssal hydrothermal vents at oceanic spreading centers. Deep-Sea Res. 24, 857–863.

Lorenz, W.W., McCann, R.O., Longiaru, M., Cormier, M.J., 1991. Isolation and expression of a cDNA encoding *Renilla reniformis* luciferase. Proc. Natl. Acad. Sci. USA 88, 4438–4442.

Lotan, A., Fishman, L., Zlotkin, E., 1996. Toxin compartimentation and delivery in the cnidaria. The nematocyst's tubule as a multiheaded poisonous arrow. J. Exp. Zool. 275, 444–451.

Lukyanov, K.A., Fradkov, A.F., Gurskaya, N.G., Matz, M.V., Labas, Y.A., Savitsky, A.P., Markelov, M.L., Zaraisky, A.G., Zhao, X., Fang, Y., Tan, W., Lukyanov, S.A., 2000. Natural animal coloration can be determined by a nonfluorescent green fluorescent protein homolog. J. Biol. Chem. 275, 25879–25882.

Lundberg, K.S., Shoemaker, D.D., Adams, M.W., Short, J.M., Sorge, J.A., Mathur, E.J., 1991. High-fidelity amplification using a thermostable DNA polymerase isolated from Pyrococcus furiosus. Gene 108, 1–6.

Marden, L., 1956. Camera under the sea. National Geographic 109, 162–200.

Markova, S., Golz, S., Frank, L.A., Kalthof, B., Vysotski, E.S., 2004. Cloning and expression of cDNA for a luciferase from the marine copepod *Metridia longa*. J. Biol. Chem. 279, 3212–3217.

Martynov, V.I., Maksimov, B.I., Martynova, N.Y., Pakhomov, A.A., Gurskaya, N.G., Lukyanov, S.A., 2003. A purple-blue chromoprotein from *Goniopora tenuidens* belongs to the DsRed subfamily of GFP-like proteins. J. Biol. Chem. 278, 46288–46292.

Martynov, V.I., Savitsky, A.P., Martynova, N.Y., Savitsky, P.A., Lukyanov, K.A., Lukyanov, S.A., 2001. Alternative cyclization in GFP-like proteins family. The formation and structure of the chromophore of a purple chromoprotein from *Anemonia sulcata*. J. Biol. Chem. 276, 21012–21016.

Materassi, R., Tredici, M.R., Milicia, F., Sili, C., Pelosi, E., Vincenzini, M., Torzillo, G., Balloni, W., Florenzano, G., Wagener, K., 1984. Development of a production size system for the mass culture of marine microalgae. In Palz, W., and Pirrwitz, D. (eds.), Energy from Biomass, series E, vol. 5, pp. 150–158. Dordrecht, Reidel.

Mathew, W.M., 1970. Peru and the British Guano Market, 1840–1870. The Economic History Review, New Series, 23, 112–128.

Mathew, W.M., 1977. A primitive export sector: Guano production in mid-nineteenth-century Peru. *J. Lat. Am. Stud.* 9, 35–57.

Matsumoto, T., Tanaka, M., Ogata, N., Mizunoe, Y., Takahashi, K., Kumazawa, J., 1991. Significance of urinary endotoxin concentration in patients with urinary tract infection. Urol. Res., 19, 293–295.

Matz, M.V., Fradkov, A.F., Labas, Y.A., Savitsky, A.P., Zaraisky, A.G., Markelov, M.L., Lukyanov, S.A., 1999. Fluorescent proteins from nonbioluminescent Anthozoa species. Nature Biotechnol. 17, 969–973.

Mazel, C.H., Lesser, M.P., Gorbunov, M.Y., Barry, T.M., Farrell, J.H., Wyman, K.D., Falkowski, P.G., 2003. Green-fluorescent proteins in Caribbean corals. Limnol. Oceanogr. 48, 402–411.

McAllister, D.E., 1967. The significance of ventral bioluminescence in fishes. J. Fish. Res. Board Canada 24, 537–554.

McNabb, D.S., Reed, R., Marciniak, R.A., 2005. Dual luciferase assay system for rapid assessment of gene expression in *Saccharomyces cerevisiae*. Eukaryot. Cell. 4, 1539–1549.

Meighen, E.A., 1993. Bacterial bioluminescence: Organization, regulation, and application of the lux genes. FASEB J. 11, 1016–1022.

Messerli, S.M., Greenberg, R.M., 2006. Cnidarian toxins acting on voltage-gated ion channels. Mar. Drugs 4, 1–12.

Mizuno, H., Mal, T.K., Tong, K.I., Ando, R., Furuta, T., Ikura, M., Miyawaki, A., 2003. Photo-induced peptide cleavage in the green-to-red conversion of a fluorescence protein. Mol. Cell. 12, 1051–1058.

Moore P., 2005. Hot from the vent. Nature 435, 235.

Morin, J.G., 1974. Coelenterate bioluminescence. In Muscatine, L., and Lenhoff, H.M. (eds.), Coelenterate Biology: Reviews and New Perspectives, pp. 397–348. New York, Academic Press.

Morin, J.G., Hastings, J.W., 1971. Biochemistry of the bioluminescence of colonial hydroids and other coelenterates. J. Cell. Physiol. 77, 305–311.

Morin, J.G., Reynolds, G.T., 1970. Luminescence and related fluorescence in coelenterates. Biol. Bull. 139, 430–431.

Muck, P., Pauly, D., 1987. Monthly anchoveta consumption of guano birds. The Peruvian anchoveta and its upwelling ecosystem: Three decades of changes, pp. 219–233. In Pauly, D., and Tsukayama, I. (eds.), ICLARM Studies and Reviews 15, 1953 to 1982. International Center for Living Aquatic Resources Management: Makato, Manila, Philippines.

Mueller, W.E.G., Schroder, H.C., Wiens, M., Perovic-Ottstadt, S., Batel, R., Muller, I.M., 2004. Traditional and modern biomedical prospecting: Part II—the benefits. Approaches for a sustainable exploitation of biodiversity (secondary metabolites and biomaterials from sponges). Evid. Based Complement. Alternat. Med.1:133–144.

Mullis, K., Faloona, F., Scharf, S., Saiki, R., Horn, G., Erlich, H., 1986. Specific enzymatic amplification of DNA in vitro: The polymerase chain reaction. Cold Spring Harb. Symp. Quant. Biol. 51, 263–273.

Nachum, R., Lipsey, A.L., Siegel, S.E., 1973. Rapid detection of Gram-negative bacterial meningitis by the *Limulus* lysate test. N. Engl. J. Med. 289, 931–934.

Nachum, R., Shanbrom E., 1981. Rapid detection of Gram-negative bacteriuria by limulus amoebocyte lysate assay. J. Clinic. Microbiol. 13, 158–162.

Nagai, H., Takuwa-Kuroda, K., Nakao, M., Oshiro, N., Iwanaga, S., Nakajima, T., 2002. A novel protein toxin from the deadly box jellyfish (Sea Wasp, Habu-kurage) *Chiropsalmus quadraticus*. Biosci. Biotechnol. Biochem. 66, 97–102.

Nakamura, S.T., Morita, S., Iwanaga, M., Miwa, Takahashi, K., 1977. A sensitive substrate for the clotting enzyme in horseshoe crab hemocytes. J. Biochem. 81, 1567–1569.

Nienhaus, K., Nienhaus, G.U., Wiedenmann, J., Nar, H., 2005. Structural basis for photo-induced protein cleavage and green-to-red conversion of fluorescent protein EosFP. Proc. Natl. Acad. Sci. USA 102, 9156–9159.

Nienhaus, K., Renzi, F., Vallone, B., Wiedenmann, J., Nienhaus, G.U., 2006a. Chromophore-protein interactions in the anthozoan green fluorescent protein asFP499. Biophys. J. 91, 4210–4220.

Nienhaus, K., Renzi, F., Vallone, B., Wiedenmann, J., Nienhaus, G.U., 2006b. Exploring chromophore-protein interactions in fluorescent protein cmFP512 from *Cerianthus membranaceus*: X-ray structure analysis and optical spectroscopy. Biochemistry, 45, 12942–12953.

Obayashi, T., Tamura, H., Tanaka, S., Ohki, M., Takahashi, M., Arai, M., Matsuda, M., Kawai, T., 1985. A new chromogenic endotoxin-specific assay using recombined *Limulus* coagulation enzyme and its clinical application. Clin. Chim. Acta. 149, 55–65.

Ormö, M., Cubitt, A.B., Kallio, K., Gross, L.A., Tsien, R.Y., Remington, S.J., 1996. Crystal structure of the *Aequorea victoria* green fluorescent protein. Science 273, 1392–1395.

Oswald, F., Schmitt, F.M., Leutenegger, A., Ivanchenko, S., D'Angelo, C., Salih, A., Maslakova, S., Bulina, M., Schirmbeck, R., Nienhaus, G.U., Matz, M.V., Wiedenmann, J., 2007. Contributions of host and symbiont pigments to the coloration of reef corals. FEBS J. 274, 1102–1122.

Parsons, S.J.W., Rhodes, S.A., Connor, H.E., Rees, S., Brown, J., Giles, H., 2000. Use of a dual firefly and renilla luciferase reporter gene assay to simultaneously determine drug selectivity at human corticotrophin releasing hormone 1 and 2 receptors. Anal. Biochem. 281, 187–192.

Partridge, J.C., Cummings, M.E., 1999. Adaptation of visual pigments to the aquatic environment. In Archer, S.N., Djamgoz, M.B.A., Loew, E.R., Partridge, J.C., and Vallerga, S. (eds.), Adaptive Mechanisms in the Ecology of Vision, pp. 251–283. Dordrecht, Kluwer.

Partridge, J.C., Hilder, P.E., 1990. The colour sensitivity and vision of fishes. In Herring, P.J., Campbell, A.K., Whitfield, M., Maddock, L. (eds.), Light and Life in the Sea, pp. 167–184. Cambridge, Cambridge University Press.

Patterson, G.H., Lippincott-Schwartz, J., 2002. A photoactivatable GFP for selective photolabeling of proteins and cells. Science 297, 1873–1877.

Patterson, G.H., Lippincott-Schwartz, J., 2003. Development and use of fluorescent protein markers in living cells. Science 300, 87–91.

Petushkov, V.N., Gibson, B.G., Lee J., 1995. The yellow bioluminescence bacterium, *Vibrio fischeri* Y1, contains a bioluminescence active riboflavin protein in addition to the yellow fluorescence FMN protein. Biochem. Biophys. Res. Commun. 211, 774–779.

Phillips, C.E.S., 1927. Fluorescence of sea anemones. Nature 119, 747.

Postec, A., Urios, L., Lesongeur, F., Ollivier, B., Querellou, J., Godfroy, A., 2005. Continuous enrichment culture and molecular monitoring to investigate the microbial diversity of thermophiles inhabiting deep-sea hydrothermal ecosystems. Curr. Microbiol. 50, 138–144.

Prasher, D.C., Eckenrode, V.K., Ward, W.W., Prendergast, F.G., Cormier, M.J., 1992. Primary structure of the *Aeqourea victoria* green fluorescent protein. Gene 111, 229–233.

Prescott, M., Ling, M., Beddoe, T., Oakley, A.J., Dove, S., Hoegh-Guldberg, O., Devenish, R.J., Rossjohn J., 2003. The 2.2 A crystal structure of a pocilloporin pigment reveals a nonplanar chromophore conformation. Structure 11, 275–284.

Pushkar, N., Kaul, P.N., Daftari, P., 1986. Marine pharmacology: Bioactive molecules from the sea. Ann. Rev. Pharmacol. Toxicol. 26, 117–142.

Rees, J.-F., De Wergifosse, B., Noiset, O., Dubuisson, M., Janssens, B., Thompson, E.M., 1998. The origins of marine bioluminescence: Turning oxygen defense mechanisms into deep-sea communication tools. J. Exp. Biol., 201, 1211–1221.

Remington, S.J., Wachter, R.M., Yarbrough, D.K., Branchaud, B., Anderson, D.C., Kallio, K., Lukyanov, K.A., 2005. zFP538, a yellow-fluorescent protein from *Zoanthus*, contains a novel three-ring chromophore. Biochemistry 44, 202–212.

Reynolds, G.T., Lutz, R.A., 2001. Sources of light in the deep ocean. Rev. Geophys. 39, 123–136.

Ridgeway, E.B., Ashley, C.C., 1968, Simultaneous recording of membrane potential, calcium transient and tension in single muscle fibres. Nature (London) 219, 1168–1169.

Romay, C., Gonzalez, R., Ledon, N., Remirez, D., Rimbau, V., 2003. C-phycocyanin: A biliprotein with antioxidant, anti-inflammatory and neuroprotective effects. Curr. Protein Pept. Sci. 4, 207–216.

Ruby, E.G., Lee, K.-H., 1998. The vibrio fischeri-euprymna scolopes light organ association: Current ecological paradigms. Appl. Environ. Microbiol. 64, 805–812.

Saiki, R.K., Gelfand, D.H., Stoffel, S., Scharf, S.J., Higuchi, R., Horn, G. T., Mullis, K.B., Erlich, H.A., 1988. Primer-directed enzymatic amplification of DNA with a thermostable DNA polymerase. Science 239, 487–491.

Saiki, R.K., Scharf, S., Faloona, F., Mullis, K.B., Horn, G.T., Erlich, H.A., Arnheim, N., 1985. Enzymatic amplification of beta-globin genomic sequences and restriction site analysis for diagnosis of sickle cell anemia. Science 230, 1350–1354.

Sandman, K.E., Marla, S.S., Zlokarnik, G., Lippard, S.J., 1999. Rapid fluorescence-based reporter-gene assays to evaluate the cytotoxicity and antitumor drug potential of platinum complexes. Chem. Biol. 6, 541–551.

Schenk, A., Ivanchenko, S., Röcker, C., Wiedenmann, J., Nienhaus, G.U., 2004. Photodynamics of red fluorescent proteins studied by fluorescence correlation spectroscopy. Biophys. J. 86, 384–394.

Shagin, D.A., Barsova, E.V., Yanushevich, Y.G., Fradkov, A.F., Lukyanov, K.A., Labas, Y.A., Semenova, T.N., Ugalde, J.A., Meyers, A., Nunez, J.M., Widder, E.A., Lukyanov, S.A., Matz, M.V., 2004. GFP-like proteins as ubiquitous metazoan superfamily: Evolution of functional features and structural complexity. Mol. Biol. Evol. 5, 841–850.

Shaner, N.C., Campbell, R.E., Steinbach, P.A., Giepmans, B.N., Palmer, A.E., Tsien, R.Y., 2004. Improved monomeric red, orange and yellow fluorescent proteins derived from *Discosoma* sp. red fluorescent protein. Nat. Biotechnol. 22, 1567–1572.

Shimomura, O., 1979. Structure of the chromophore of *Aequorea* green fluorescent protein. FEBS Lett. 104, 220–222.

Shimomura, O., 1995. A short story of aequorin. Biol. Bull. 189, 1–5.

Shimomura, O., Johnson, F.H., Saiga, Y., 1962. Extraction, purification and properties of aequorin, a bioluminescent protein from the luminous hydromedusan, *Aequorea*. J. Cell. Comp. Physiol. 59, 223–239.

Sohn, G., Sautter C.J., 1991. R-phycoerythrin as a fluorescent label for immunolocalization of bound atrazine residues. Histochem. Cytochem. 39, 921–926.

Srivastava, R., Deb, D.K., Srivastava, K.K., Locht, C., Srivastava B.S., 1998. Green fluorescent protein as a reporter in rapid screening of antituberculosis compounds in vitro and in macrophages. Biochem. Biophys. Res. Commun. 18, 431–436.

Stevenson, A.J., Frolova-Jones, E., Hall, K.T., Kinsey, S.E., Markham, A.F., Whitehouse, A., Meredith, D.M., 2000. A herpes virus saimiri-based gene therapy vector with potential for use in cancer immunotherapy. Cancer Gene Ther. 7, 1077–1085.

Stormer, L., 1952. Phylogeny and taxonomy of fossil horseshoe crab. J. Paleontol. 26, 630–639.

Surpin, M.A., Ward, W.W., 1989. Reversible denaturation of Aequorea green-fluorescent protein-thiol requirement. Photochem. Photobiol. 49, 65.

Tanahashi, H., Ito, T., Inouye, S., Tsuji, F.I., Sakaki, Y., 1990. Photoprotein aequorin: Use as a reporter enzyme in studying gene expression in mammalian cells. Gene 96, 249–255.

Tannous, B.A., Kim, D.-E., Fernandez, J.L., Weissleder, R., Breakefield, X.O., 2005. Codon-optimized *Gaussia* luciferase cDNA for mammalian gene expression in culture and *in vivo*. Mol. Ther. 11, 435–443.

Thakur, N.L., Hentschel, U., Krasko, A., Pabel, C.T., Anil, A.C., Müller, W.E.G., 2003. Antibacterial activity of the sponge Suberites domuncula and its primmorphs: Potential basis for epibacterial chemical defense. Aquat. Microb. Ecol. 31, 77–83.

Thompson, E.M., Nagata, S., Tsuji, F.I., 1989. Cloning and expression of cDNA for the luciferase from the marine ostracod *Vargula hilgendorfii*. *Proc. Nat. Acad. Sci. USA* 86, 6567–6571.

Thomson, C.M, Herring, P.J., Campbell, A.K., 1997. The widespread occurrence and tissue distribution of the imidazolopyrazine luciferins. J. Biolum. Chemilum. 12, 87–91.

Thomulka, K.W., Abbas, C.G., Young, D.A., Lange, J.H., 1996. Evaluating median effective concentrations of chemicals with bioluminescent bacteria. Bull. Environ. Contam. Toxicol. 56, 446–452.

Thomulka, K.W., Kriebel, J.A., Schroeder, J.K., Lange, J.H., 1993. Toxicity of various chemicals in a sand and water mixture using the marine bacterium *Vibrio harveyi* in a direct bioluminescence reduction bioassay. J. Clean Tech. Environ. Sci. 3, 217–226.

Tredici, M.R., Papuzzo, T., Tomaselli, L., 1986. Outdoor mass culture of *Spirulina maxima* in sea-water. Appl. Microbiol. Biotechnol. 24, 47–50.

Tsien, R.Y., 1998. The green fluorescent protein. Annu. Rev. Biochem. 67, 509–544.

Verhaegen, M., Christopoulos, T.K., 2002. Recombinant *Gaussia* luciferase: Overexpression, purification and analytical application of a bioluminescent reporter for DNA hybridization. Anal. Chem. 74, 4378–4385.

von Titschack, H., 1964. Untersuchungen über das Leuchten der Seefeder *Veretillum cynomorium*. Vie Milieu 15, 547–563.

Wall, M.A., Socolich, M., Ranganathan R., 2000. The structural basis for red fluorescence in the tetrameric GFP homolog DsRed. Nat. Struct. Biol. 12, 1133–1138.

Wallace, M.S., 2006. Ziconotide: A new nonopioid intrathecal analgesic for the treatment of chronic pain. Expert Rev. Neurother. 6, 1423–1428.

Wang, L., Jackson, W.C., Steinbach, P.A., Tsien, R.Y., 2004. Evolution of new nonantibody proteins via iterative somatic hypermutation. Proc. Natl. Acad. Sci. USA 101, 16745–16749.

Ward, W.W., 1981. Properties of the coelenterate green fluorescent proteins. In DeLuca, M., and McElroy, D.W. (eds.), Bioluminescence and Chemiluminescence: Basic Chemistry and Analytical Applications, pp. 235–242. New York, Academic.

Ward, W.W., 1998. Biochemical and physical properties of green fluorescent protein. In Chalfie, M., and Kain, S. (eds.), Green Fluorescent Protein: Properties, Applications, and Protocols, pp. 45–75. New York, Wiley-Liss.

Ward, W.W., Cormier, M.J., 1979. An energy transfer protein in coelenterate bioluminescence: Characterization of the *Renilla* green-fluorescent protein (GFP). J. Biol. Chem. 254, 781–788.

Ward, W.W., Prentice, H.J., Roth, A.F., Cody, C.W., Reeves, S.C., 1982. Spectral perturbations of the *Aequorea* green-fluorescent protein. Photochem. Photobiol. 35, 803–808.

Watson, M., Thurston, E.L., Nicol, J.A.C., 1978. Reflectors in the light organ of anomalops (Anomalopidae, Teleostei). Proc. R. Soc. London. Ser. B., 202, 339–351.

Wiedenmann, J., 1997. Die Anwendung eines orange fluoreszierenden Proteins und weiterer farbiger Proteine und der zugehörenden Gene aus der Artengruppe *Anemonia* sp. (sulcata) Pennant, (Cnidaria, Anthozoa, Actinaria) in Gentechnologie und Molekularbiologie. Offenlegungsschrift DE 197 18 640 A1. Deutsches Patent- und Markenamt, 1–18.

Wiedenmann, J., Elke, C., Spindler, K-D., Funke W., 2000. Cracks in the β-can: Fluorescent proteins from *Anemonia sulcata*. Proc. Natl. Acad. Sci USA 97, 14091–14096.

Wiedenmann, J., Elke, C., Spindler, K-D., Funke W., 2002a. Correction for vol. 97, p.14091–14096, Proc. Natl. Acad. Sci. USA 99, 13357b.

Wiedenmann, J, Ivanchenko, S, Oswald, F., Nienhaus G.U., 2004a. Identification of GFP-like proteins in non-bioluminescent, azooxanthellate Anthozoa opens new perspectives for bioprospecting. Mar. Biotechnol. 6, 270–277.

Wiedenmann, J., Ivanchenko, S., Oswald, F., Schmitt, F., Röcker, C., Salih, A., Spindler, K.D., Nienhaus, G.U., 2004b. EosFP, a fluorescent marker protein with UV-inducible green-to-red fluorescence conversion. Proc. Natl. Acad. Sci. USA 101, 15905–15910.

Wiedenmann, J., Nienhaus, G.U., 2006. Live-cell imaging with EosFP and other photoactivatable marker proteins of the GFP family. Expert Rev. Proteomics 3, 361–374.

Wiedenmann, J., Röcker C., Funke, W., 1999. The morphs of *Anemonia* aff. *sulcata* (Cnidaria, Anthozoa) in particular consideration of the ectodermal pigments. In Pfadenhauer, J. (ed.), Verhandlungen der Gesellschaft für Ökologie, Band 29, 497–503. Heidelberg, Berlin: Spektrum Akademischer Verlag.

Wiedenmann, J., Schenk, A., Röcker, C., Girod, A., Spindler, K.-D., Nienhaus, G.U., 2002b. A far-red fluorescent protein with fast maturation and reduced oligomerization tendency from *Entacmaea quadricolor* (Cnidaria, Anthozoa, Actinaria). Proc. Natl. Acad. Sci. USA 99, 11646–11651.

Wiedenmann, J., Vallone, B., Renzi, F., Nienhaus, K., Ivanchenko, S., Röcker, C., Nienhaus, G.U., 2005. The red fluorescent protein eqFP611 and its genetically engineered dimeric variants. J. Biomed. Opt. 10, 1–7.

Williamson, J., Burnett, J., 1995. Clinical toxicology of marine coelenterated injuries. In Meier, J., and White, J., (eds.), Handbook of Clinical Toxicology and Animal Venoms and Poisons, pp. 89–116. Boca Raton, FL, CRC Press.

Williamson, J.A., Burnett, J.W., Fenner, P.J., Hach-Wunderle, V., Hoe, L.Y., Adiga, K.M., 1988. Acute regional vascular insufficiency after jellyfish envenomation. Med. J. Aust. 149, 698–701.

Wilson, T., Hastings, W.J., 1998. Bioluminescence. Annu. Rev. Cell Dev. Biol. 14, 197–230.

Wolff, M., Haasen, D., Merk, S., Kroner, M., Maier, U., Bordel, S., Wiedenmann, J., Nienhaus, G.U., Valler, M., Heilker, R., 2005. Automated high content screening for phosphoinositide kinase 3 inhibition using an AKT1 redistribution assay. Comb. Chem. High Throughput Screen. 9, 339–350.

Wolff, M., Wiedenmann, J., Nienhaus, G.U., Valler, M., Heilker, R., 2006. Novel fluorescent proteins for high content screening. Drug Discov. Today 11, 1054–1060.

Xia, Z., Chen, Y., Zhu, Y., Wang, F., Xu, X., Zhan, J., 2006. Recombinant omega-conotoxin MVIIA possesses strong analgesic activity. BioDrugs 20, 275–281.

Yang, F., Moss, L.G., Phillips, G.N., Jr., 1996. The molecular structure of green fluorescent protein. Nature Biotechnol. 14, 1246–1256.

Yanushevich, Y.G., Staroverov, D.B., Savitsky, A.P., Fradkov, A.F., Gurskaya, N.G., Bulina, M.E., Lukyanov, K.A., Lukyanov, S.A., 2002. A strategy for the generation of non-aggregating mutants of Anthozoa fluorescent proteins. FEBS Lett. 511, 11–14.

Yarbrough, D., Wachter, R.M., Kallio, K., Matz, M.V., Remington, S.J., 2001. Refined crystal structure of DsRed, a red fluorescent protein from coral, at 2.0–A resolution. Proc. Natl. Acad. Sci. USA 98, 462–467.

Zacharias, D.A., Baird, G.S., Tsien, R.Y., 2000. Recent advances in technology for measuring and manipulating cell signals. Curr. Opin. Neurobiol. 10, 416–421.

Zappa, S., Rolland, J.L., Flament, D., Gueguen, Y., Boudrant, J., Dietrich, J., 2001. Characterization of a highly thermostable alkaline phosphatase from the euryarchaeon Pyrococcus abyssi. Appl. Environ. Microbiol. 67, 4504–4511.

Zhang, J., Campbell, R.E., Ting, A.Y., Tsien, R.Y., 2002. Creating new fluorescent probes for cell biology. Nat. Rev. Mol. Cell. Biol. 12, 906–918.

Zola, H., Neoh, S.H., Mantzioris, B.X., Webster, J., Loughnan, M.S., 1990. Detection by immuno fluorescence of surface molecules present in low copy numbers. High sensitivity staining and calibration of flow cytometer. J. Immunol. Methods. 135, 247–255.

STUDY QUESTIONS

1. List examples of proteins or peptides from marine organisms that became useful tools in biomedical research.
2. Give reasons why marine creatures are rich in compounds that are potentially useful for biotechnological or pharmaceutical applications.
3. Explain the basic principle of the *Limulus* Amebocyte Lysate (LAL) test.
4. What are the advantages and disadvantages of thermostable polymerases isolated from hydrothermal vents?
5. Name examples of organisms that yielded bioluminescence system useful for applications in reporter assays. What makes luciferases from crustaceans distinctive?
6. What is a major advantage of the application of GFP-like proteins as reporter protein compared to luciferase- or phycobiliprotein-based assays.
7. Name major applications of fluorescent proteins.
8. Explain the uses of photoactivatable GFP-like proteins.

C H A P T E R

25

Novel Pain Therapies from Marine Toxins

RUSSELL W. TEICHERT AND BALDOMERO M. OLIVERA

INTRODUCTION

Modern pharmacology has its historical roots in the discoveries of natural products used in the ancient world for treating medical disorders. One example of an ancient natural-product discovery highly relevant to modern medical practice is opium, which has been used as an analgesic drug for thousands of years. Even with the many advancements of modern medicine, opiate drugs, such as morphine, remain the mainstay of treatment for severe acute pain in the 21st century. Although some may consider such pharmacological discoveries serendipitous events of history and not drug development informed by rigorous scientific methods, a summary of new pharmacologically active substances introduced between 1981 and 2002 demonstrates that the majority were either natural products, derived from natural products, or began with natural products as lead compounds (Newman *et al.*, 2003).

Since the 1980s, the major pharmaceutical companies mostly abandoned natural products and committed to combinatorial chemistry to generate large libraries of compounds to test against various drug targets in high-throughput screens. Originally proclaimed as the key to accelerating the discovery and approval of new drugs, this approach has not lived up to expectations. The number of new drugs approved per year has consistently declined over the past decade (Frantz, 2006). The brute-force approach of testing libraries of synthetic compounds against potential drug targets appeals to the desire to screen more compounds more quickly. However, starting with natural products may provide the following advantages over a brute-force combinatorial-chemistry approach: (1) Nature has refined biological molecules over eons of evolution for both high affinity and specificity for particular targets. (2) Although it is now possible to efficiently screen for compounds that are active at a particular target, many synthetic organic compounds may functionally affect unrelated targets in other physiological circuits; such compounds are probably selected again by evolution. (3) Even a large library of 1 million synthetic compounds is a relatively small number to adequately explore chemical-diversity space; each natural product has evolved over innumerable generations and has probably resulted from the exploration of a larger sector of chemical diversity space. These potential advantages of natural products over combinatorial chemistry make it reasonable to hypothesize that a part of the drug-pipeline problem may be the result of abandoning natural products as potential therapeutic drugs and drug leads.

Current drug development arguably involves significantly more random discovery (e.g., high-throughput screening) than rational design. Although medicinal chemists are capable of modifying drug leads from initial screens for further testing, it is not generally possible currently to predict *a priori* which chemical modifications will produce greater specificity for a target or avoid undesirable side effects in clinical trials. Because natural products have biological origins, related organisms often produce related products for homologous physiological ends. This gives drug discovery programs that begin with natural products the potential to more systematically elucidate the chemical and structural features that confer both high affinity and specificity. Such biochemical insights should contribute to efforts aimed at rationally designing and optimizing therapeutic drugs, rather than relying mainly on random discovery.

Although natural products have formed the historical cornerstone of pharmacology, until recently there were no native marine natural products approved as drugs. This changed in December 2004, when the United States Food and Drug Administration (FDA) approved a peptide originally discovered in the venom of the marine cone snail,

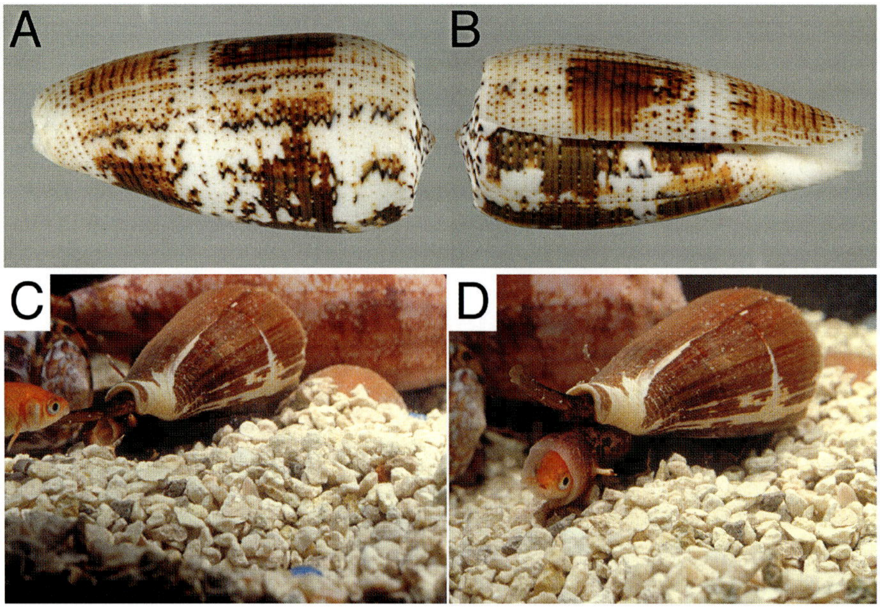

FIGURE 25-1. Photographs of the Indo-pacific cone-snail, *Conus magus*. The drug Prialt (ω-conotoxin MVIIA) was obtained from the venom of this marine mollusk. (A) Dorsal and (B) ventral views of a *Conus magus* shell. (C) *Conus magus* stinging its fish prey and (D) engulfing the same fish.

Conus magus, a fish-hunting marine mollusk (Fig. 25-1). This peptide, ω-conotoxin MVIIA (ω-MVIIA), is marketed as Prialt (Ziconitide) by Elan Corporation, for use in treating intractable chronic pain. Prialt is biochemically identical to the peptide obtained from the cone snail venom; the primary amino acid sequence of ω-MVIIA is the following (with cysteine residues, which form disulfide bonds, in bold font):

CKGKGAK**C**SRLMYD**CC**TGS**C**RSGK**C**

Its unprecedented specificity for a single type of voltage-gated calcium channel (N-type, $Ca_V2.2$) allowed Prialt to be utilized as a powerful analgesic for intractable neuropathic pain. As an illustration of the evolutionary refinement of biological molecules, many analogs of ω-MVIIA were synthesized for preclinical testing in the hope of finding a higher-affinity ligand with improved specificity. However, in the end, the decision was made to develop the natural peptide as a drug without any modifications (Miljanich, 2004).

OVERVIEW OF *CONUS* VENOM PEPTIDES

Cone snails comprise a large genus (*Conus*) of predatory marine snails (~700 species) that each specialize in a particular prey, generally other mollusks, polychaete worms, or fish. The cone snails inject venom into their prey with a hollow, radular tooth that is harpoon-like in its appearance (Fig. 25-2). The venom injected through the tooth is produced in a venom duct, where specialized epithelial cells express and secrete a large array of peptide toxins. The peptides from cone-snail venoms are a particularly interesting source of natural-product drug leads because they provide advantages of both natural products and combinatorial chemistry: they have evolved with high affinity and specificity for particular targets in the nervous system, but they also represent a natural combinatorial library of compounds.

A sufficient number of venom peptides have now been identified from diverse *Conus* species to group subsets of these peptides into structurally similar (genetically related) superfamilies and functionally similar (pharmacologically related) families (Table 25-1). Although related species of cone snails have similar toxins in their venoms, the accelerated evolution of genes encoding these peptides has resulted in each species possessing its own unique set of toxin peptides with almost no overlap, even between closely related species. Because each species has between 50 and 200 unique components in its venom, the approximately 700 species of cone snails may encode for more than 100,000 unique peptide sequences in their venom ducts (Olivera, 1997, 2006; Terlau and Olivera, 2004; Terlau *et al.*, 1996). Additionally, within some of the more extensively characterized conotoxin families (e.g., α-conotoxins), the techniques of combinatorial chemistry are now being applied (e.g., amino-acid substitutions), which in some cases may alter or refine the pharmacological specificity on structurally similar scaffolds.

The evolution of as many as 200 different components in a single species' venom may be explained in part by the fact

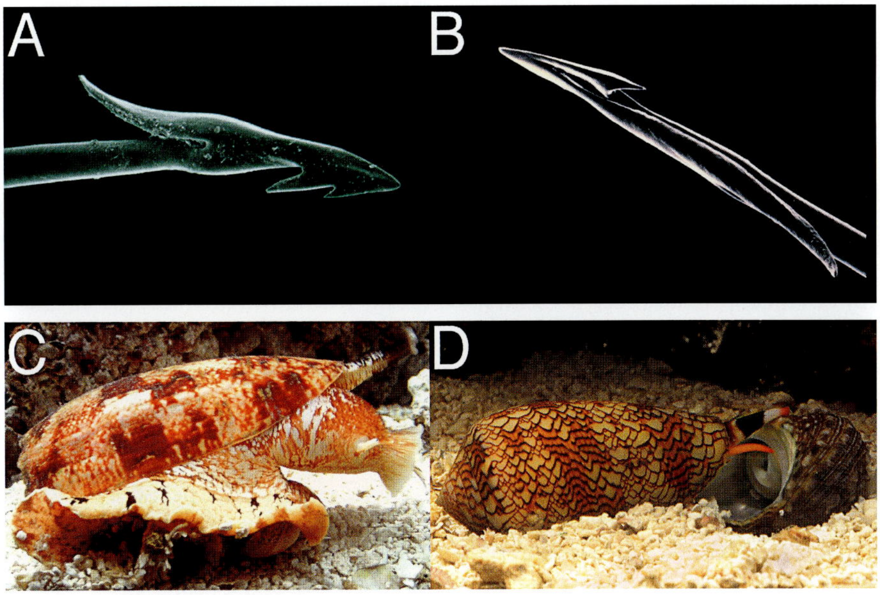

FIGURE 25-2. Predation of cone snails. (A) The radular tooth of *Conus obscurus*, used to inject venom in fish prey. (B) The radular tooth of *Conus purpurascens*. (C) *Conus geographus* engulfing a fish. (D) *Conus textile* stinging its mollusk prey.

TABLE 25-1. Genetically defined *Conus* peptide superfamilies and pharmacologically defined families.

Superfamily	Cys Pattern	Pharmacological Family	Molecular Target
A (Santos *et al.*, 2004)	CC-C-C	**α**	**nAChRs**
		r	
(Hopkins *et al.*, 1995)	CC-C-C-C-C	αA	nAChRs
(Craig *et al.*, 1998)		κA	K^+ channels
M (Corpuz *et al.*, 2005)	CC-C-C-C	μ	Na^+ channels
		κM	K^+ channels
		ψ	nAChRs
O (Terlau and Olivera, 2004)	C-C-CC-C-C	**ω**	**Ca^{2+} channels**
		κ	K^+ channels
		δ	Na^+ channels
		μO	**Na^+ channels**
T (McIntosh *et al.*, 2000)	CC-CC	Undefined	Unknown
T (Walker *et al.*, 1999)	CC-CXOC	**χ**	**NT**
S (England *et al.*, 1998)	C-C-C-C-C-C-C-C-C-C	σ	SR ($5HT_3$)
(Teichert *et al.*, 2005)		αS	nAChRs
P (Lirazan *et al.*, 2000)	C-C-C-C-C-C	Undefined	Unknown
I_1/I_2 (Buczek *et al.*, 2005)	C-C-CC-CC-C-C	Undefined	Na^+ channels
J (Imperial *et al.*, 2006)	C-C-C-C	Undefined	K^+ channels
Undefined (McIntosh *et al.*, 1984)	No disulfide bonds	**Conantokin**	**NMDA-R**
Undefined (Craig *et al.*, 1999)	No disulfide bonds	**Contulakin**	**Neurotensin-R**

Adapted from Terlau and Olivera (2004) and Olivera (2006). At least one *Conus* peptide has reached clinical development as an analgesic drug from each family highlighted in bold text and underlined (preclinical development only for μO-conotoxins). Note that the six conotoxin families in bold text have six different molecular targets. The Cys pattern identifies the position of the cysteine residues relative to each other in the primary amino-acid sequence. A dash represents a variable number of amino acids between Cys residues. Abbreviations used in the table are as follows: nAChR, nicotinic acetylcholine receptor; NT, norepinephrine transporter; SR, serotonin receptor; NMDA-R, NMDA receptor; Neurotensin-R, neurotensin receptor.

that *Conus* venom is used for predation on disparate prey types, defense against myriad predators, and for numerous competitive interactions. Additionally, the relatively slow-moving cone snails are at a disadvantage as predators, particularly those that hunt relatively faster-moving prey (such as fish). The fish-hunting adaptation, from the purportedly ancestral worm-hunting cone snails, may have required the evolution of numerous toxins that could immobilize vertebrate prey quickly and with long-lasting effects. Such effects are achieved by an array of venom peptides, each possessing unique pharmacological properties.

For example, diverse peptides in the venoms of fish-hunting cone snails elicit a nearly-immediate, rigid immobilization of fish-prey by causing repetitive firing of nerves. These peptides antagonize voltage-gated potassium channels, delay inactivation of voltage-gated sodium channels, and possibly activate sodium channels or lower the threshold of firing for certain sodium channels (see Chapter 28). The physiological result elicited is a type of excitotoxic shock, much like electrocution or the shock of electric current passing through the body.

A separate group of peptides collectively cause a delayed, but long-lasting, flaccid neuromuscular paralysis by antagonizing voltage-gated calcium channels in motor neurons, voltage-gated sodium channels in muscle, and nicotinic acetylcholine receptors at the neuromuscular synapses. Groups of conotoxins that elicit a single physiological end point have been termed "cabals" (after secret societies that seek to overthrow existing governments). Hence, the peptides causing excitotoxic shock have been termed the lighting-strike cabal, whereas the peptides causing neuromuscular paralysis have been called the motor cabal. At least one additional cabal appears to function within some fish-hunting cone snails, called the nirvana cabal, which sedates the fish prey, apparently to prevent them from struggling to free themselves during prey capture. Components of the nirvana cabal may include serotonin receptor antagonists and N-methyl-D-aspartic acid (NMDA) receptor antagonists, among others (Olivera, 1997; Terlau and Olivera, 2004; Terlau *et al.*, 1996).

As cone snails manufacture hundreds of different components in their venoms, there is a limited quantity of any particular component in a single venom strike. Consequently it appears that each component has been refined by evolution for both high affinity and specificity for individual targets. However, many conotoxin families remain unexplored and uncharacterized. Thus, new discoveries of pharmacological specificity continue at a steady pace. Conotoxin research has really just begun, if one considers the enormous potential for discovery.

As the prey of cone snails include polychaete worms, marine mollusks, and fish, one may reasonably wonder whether the evolutionarily derived affinity and specificity for molecular targets in these prey types are conserved in humans. Further, this is especially relevant to the goal of finding therapeutic applications in medicine. In some cases, particularly with the fish-hunting cone snails, it appears that some molecular targets are remarkably conserved across all vertebrates, from fish to humans. For example, a subset of the μ-conotoxins appears to cause muscular paralysis in both fish and mammals by targeting highly conserved skeletal-muscle sodium channels. However, in the case of ω-MVIIA (Prialt), the voltage-gated calcium channels targeted are expressed in motor neurons in the fish prey, but not in humans. Consequently, ω-MVIIA paralyzes fish but not mammals, and thus it has clinical application as an analgesic, targeting the homologous calcium channel in the human central nervous system. Similarly, though some α-conotoxins may have evolved to produce muscular paralysis in polychaete worms, these peptides interact with neuronal nAChRs in mammals rather than the mammalian muscle nAChR. Such species differences may account for the therapeutic potential of some conotoxins.

CONUS PEPTIDES, OPIATES, AND NSAIDS

Because the targets of most *Conus* venom peptides are in the nervous system, it is not surprising that some of these peptides are being developed as pain therapeutics. It is, however, surprising that six *Conus* peptides have reached human clinical trials as analgesic drugs with five distinct mechanisms of action, as will be discussed in detail later. In Figure 25-3, we have provided a simplified overview of nociceptive signaling relevant to *Conus* peptides as a conceptual framework for rationalizing disparate mechanisms of analgesia. We have included probable points of intervention in nociceptive pathways by *Conus* peptides that explain their analgesic properties. These disparate mechanisms include pharmacological intervention at various points in ascending nociceptive pathways, from the peripheral nervous system to the central nervous system. They also include intervention in descending antinociceptive signaling (e.g., inhibition of norepinephrine uptake), which is believed to function primarily at the dorsal horn of the spinal cord, a key point of integration for nociceptive and antinociceptive signaling.

At present, the most widely used analgesic drugs are opiates (e.g., morphine) and nonsteroidal anti-inflammatory drugs (NSAIDs, e.g., aspirin and ibuprofen). These act primarily through modulatory effects of nociceptive signaling (via opiates) or by inhibiting a process known as sensitization (via NSAIDs). Sensitization is a reduction in the threshold for firing action potentials in nociceptive neurons, resulting in hypersensitivity to pain signals. Sensitization may occur in either the peripheral or central nervous systems, and this occurs through distinct mechanisms of action.

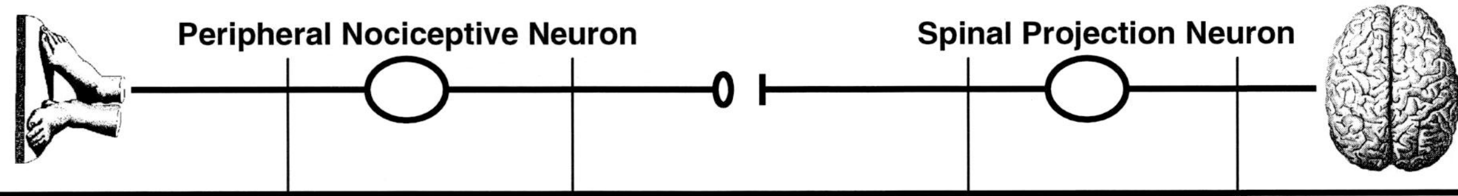

	Peripheral Tissue	Dorsal Root Ganglia (DRG)	Dorsal Horn of Spinal Cord	Spinal Cord	Brain
Functional Neuroanatomy	Free endings of nociceptive neurons detect noxious stimuli or tissue damage.	Cell bodies of peripheral nociceptive neurons reside in DRG, with axons projecting to dorsal horn.	**Integration & Relay:** Peripheral nociceptive neurons synapse on local interneurons and ascending projection neurons. Descending neurons from brain project to dorsal horn with antinociceptive input.	Both ascending (nociceptive) and descending (antinociceptive) pathways run through spinal cord.	Ascending projection neurons synapse primarily in reticular formation of brainstem and thalamus. Descending projections originate from amygdala.
Select Receptors & Ion Channels (Select Drugs for Analgesic Intervention)	Mechanoreceptors Vanilloid receptors Various ligand-gated and voltage-gated ion channels. **nAChRs** (α-Vc1.1 & Rg1A) **Na$_V$1.8** (μO-MrVIB) **ASICs** (APETx2) **Cyclooxygenase** (NSAIDs)	Voltage-gated $Na^+/K^+/Ca^{2+}$ channels **Na$_V$1.7** (?) **Na$_V$1.8** (μO-MrVIB) **ASICs** (APETx2)	Numerous ligand-gated and voltage-gated ion channels. **NMDA receptors** (Conantokin-G) **Ca$_V$2.2** (ω-MVIIA or CVID) **Neurotensin receptor** (Contulakin-G) **Norepinephrine reuptake** (χ-MrIA derivative) **Opiod receptors** (Opiate drugs)	Voltage-gated $Na^+/K^+/Ca^{2+}$ channels. **Ca$_V$2.2** (ω-MVIIA or CVID) **Neurotensin receptor** (Contulakin-G) **Opiod receptors** (Opiate drugs)	Numerous ligand-gated and voltage-gated ion channels. **NMDA receptors** (Conantokin-G) **Ca$_V$2.2** (ω-MVIIA or CVID) **Neurotensin receptor** (Contulakin-G) **Opiod receptors** (Opiate drugs)
Sensitization or Modulation	Reduced firing threshold from arachidonic-acid metabolites, e.g. prostaglandins and leukotrienes, causes peripheral sensitization.		Central sensitization is mediated by NMDA receptors. Norepinephrine acts on α_2-adrenergic receptor to decrease pain signaling. Endogenous opioid peptides are released from descending antinociceptive pathways to reduce pain signaling.		

FIGURE 25-3. Simplified pain signaling and potential points of intervention. A simplified sketch of an ascending pain pathway is overlaid with brief summaries of basic functional neuroanatomy, potential points of pharmacological intervention at key receptors/ion channels, and mechanisms of sensitization or modulation. In general, analgesic intervention either inhibits pain signaling at key points along nociceptive pathways ascending to the brain or promotes antinociceptive signaling in pathways descending from the brain.

NSAIDs effect sensitization by antagonizing the enzyme cyclooxygenase, thereby inhibiting the production of arachidonic-acid metabolites, which in turn decreases the inflammatory response associated with peripheral sensitization.

In contrast, central sensitization appears to be an NMDA-receptor mediated increase in the excitability of dorsal-horn neurons that is believed to play a significant role in chronic, neuropathic pain (pain that persists after tissue damage has healed). Neither NSAIDs nor opiates are effective in the treatment of chronic, neuropathic pain. Although patients are able to use opiate drugs, such as morphine, to relieve severe acute pain, they develop tolerance to these drugs over time, meaning that higher dosages are required to obtain equivalent pain relief. In addition, opiates are associated with several undesirable side effects, such as respiratory depression, constipation, sedation, confusion, and addiction.

Opiate drugs modulate pain signaling by acting as agonists of opioid receptors (G-protein coupled receptors) in the descending antinociceptive pathways. The endogenous ligands of the opioid receptors are neuropeptides, which mediate the body's natural antinociceptive signaling (Nestler *et al.*, 2001). Although some *Conus* peptides, like NSAIDs and opiates, appear to inhibit sensitization or modulate nociceptive signaling, additional *Conus* peptides intervene more directly in nociceptive signal propagation.

POTENTIAL *CONUS* PEPTIDE THERAPEUTICS FOR PAIN: MOVING BEYOND PRIALT

As mentioned previously, Prialt is an ω-conotoxin that acts directly on nociceptive signal propagation in the central nervous system (CNS). Before the discovery of Prialt's analgesic properties, the N-type calcium channel had not been identified as an analgesic drug target, underscoring the importance and utility of using natural products as lead compounds. This novel mechanism of action has provided pain relief to patients with severe neuropathic pain without the tolerance or addiction associated with opiate analgesics. Because the peptide will not cross the blood-brain barrier, the potential market for Prialt is necessarily limited by the requirement to deliver the drug directly to the CNS via intrathecal administration. It also has some undesirable side effects at therapeutic dosages, including confusion, hypotension, and sedation (Hogg, 2006). A similar peptide, ω-conotoxin CVID (AM336 from Zenyth Therapeutics, formerly Amrad), like Prialt, targets N-type voltage-gated Ca^{2+} channels ($Ca_V2.2$). It is of interest because it is reported to have a higher therapeutic index than Prialt in neuropathic and inflammatory pain models in rats, and thus it may have reduced side effects (Scott *et al.*, 2002; Smith *et al.*, 2002). However, it will likely share Prialt's requirement for intrathecal administration in order to deliver the peptide directly to the spinal-cord neurons.

Just as the specificity of some ω-conotoxins for a single subtype of Ca^{2+}-channels provides a favorable therapeutic index for pain management, the selectivity of various α-conotoxins for particular subtypes of nicotinic acetylcholine receptors (nAChRs) has allowed some of these peptides to advance in preclinical and clinical development as analgesic drugs. Significantly, the α-conotoxins are presently the most extensively characterized family of *Conus* peptides, and many nAChR-subtype selective peptides have been identified within this family. The identification of subtype-selective ligands within the α-conotoxin family suggests that many target-selective ligands will also be found in the less-characterized conotoxin families. This is a research area under active investigation.

Various neuronal nicotinic acetylcholine receptors (nAChRs) are expressed throughout the nervous system, including nociceptive neurons. Additionally, nAChRs are expressed in macrophages and lymphocytes, and they may play a role in the inflammatory aspects of pain stimuli (peripheral sensitization). α-conotoxin Vc1.1 (α-Vc1.1) (ACV-1 from Metabolic Pharmaceuticals) has reached phase II clinical trials for neuropathic pain after demonstrating potent analgesia in rat pain models (Livett *et al.*, 2006; Sandall *et al.*, 2003; Satkunanathan *et al.*, 2005). A proposed hypothesis for its analgesic effects is via antagonism of α3β4 or α3α5β4 nAChRs expressed in sensory nerves (Satkunanathan *et al.*, 2005).

However, a recent paper demonstrated that α-conotoxin Vc1.1 has significantly greater affinity for the α9α10 nAChR than for other neuronal nAChR subtypes (Vinkler *et al.*, 2006). These authors also demonstrated that α-conotoxin RgIA (α-RgIA) is a high-affinity antagonist of the α9α10 nAChR and exhibits analgesic effects in rat nerve pain models. While antagonism of α9α10 nAChRs in sensory neurons may account for the analgesic effects of both α-Vc1.1 and RgIA, α-RgIA also inhibited migration of lymphocytes, macrophages, and acetylcholine-producing cells at a site of injury (Vinkler *et al.*, 2006), suggesting that analgesic effects may be the result of a reduction of the inflammatory response at the site of injury. As analgesic drugs, these peptides may have an advantage over the ω-conotoxins because they could potentially be delivered to the peripheral nervous system to produce analgesia, rather than to the CNS through intrathecal administration, which is required for delivery of ω-conotoxins to N-type Ca^{2+}-channels in the spinal cord.

Although antagonism of N-type Ca^{2+}-channels or neuronal nAChRs attenuates nociceptive signaling, it is important to understand that all nociceptive signaling propagates as action potentials along neuronal axons via the opening of voltage-gated Na^+-channels. In humans, there are nine known Na^+-channels, and these exhibit complex patterns of expression. However, if we hypothesize that all nociceptive signaling depends on a single Na^+-channel subtype, then that particular subtype would be an important drug target for analgesia, if not the key drug target for analgesia, because the disparate sources of pain signaling in the peripheral nervous system would all converge on a single receptor subtype. This hypothetical scenario now appears to be plausible. One report has identified humans with very rare homozygous nonsense mutations in the $Na_V1.7$ Na^+-channel subtype, which results in a complete loss of function of this particular Na^+-channel subtype. The $Na_V1.7$ Na^+-channel is normally expressed at high levels in dorsal-root ganglia neurons (peripheral nervous system) and at lower levels in the spinal cord and brain. The subjects lacking the $Na_V1.7$ Na^+-channel were able to experience normal sensations of temperature, touch, and pressure, but remarkably they had never experienced the sensation of pain at any time in any part of their bodies, implying that all pain signaling depends upon $Na_V1.7$, whereas other sensations do not (Cox *et al.*, 2006).

In light of this significant discovery, the *Conus* peptides targeting Na^+-channels are of particular interest in the development of analgesic drugs. Although no *Conus* peptide demonstrating $Na_V1.7$ subtype selectivity has been identified yet, there are many uncharacterized peptides belonging to the μ- and μO-conotoxin families, which are known collectively to target sodium channels. Understandably, the pursuit of conotoxins selective for particular Na^+-channel subtypes is an area of active investigation. In addition to $Na_V1.7$, the $Na_V1.8$ and 1.9 Na^+-channel subtypes are also expressed in nociceptive neurons and appear to have roles in the propagation of pain signaling in the peripheral nervous system. An inhibitor of $Na_V1.8$ that has analgesic activity in preclinical animal models of pain is μO-conotoxin MrVIB (μO-MrVIB) (CGX-1002 from Cognetix) (Bulaj *et al.*, 2006). Perhaps the most promising group of *Conus* peptides for the discovery of subtype selective Na^+-channel inhibitors is the μ-conotoxin family, which includes peptides that target tetrodotoxin (TTX)-sensitive and TTX-resistant sodium channels. The μ-conotoxins are an extensive group of peptides founds in the venoms of most *Conus* species.

Even though Na^+-channels propagate all pain signals along axons in ascending pain pathways, an important point of integration for both ascending nociceptive pathways and descending antinociceptive pathways is at the dorsal horn of the spinal cord. Peripheral nociceptive neurons make synaptic connections in the dorsal horn with interneurons and with projection neurons ascending to the brain. Glutamate functions as a neurotransmitter in these dorsal-horn synapses. One class of glutamate receptors known as N-methyl-D-aspartic acid (NMDA) receptors are believed to be involved in neuropathic pain by increasing excitatory signaling via a process known as central sensitization (Woolf and Thompson, 1991). However, NMDA receptors are expressed in various locations within primary afferent neurons, the spinal cord, and the brain (reviewed in Bleakman *et al.*, 2006), complicating the potential mechanisms of analgesia mediated by NMDA receptor antagonists.

A family of *Conus* peptides that acts on NMDA receptors is known as conantokins. Conantokin-G (CGX-1007 from Cognetix), originally purified from the venom of *Conus geographus*, has reached phase I clinical trials for pain and epilepsy (Malmberg *et al.*, 2003). Conantokin-G is a 17 amino-acid peptide and contains five residues of the unusual posttranslationally modified amino acid gamma-carboxy-glutamate (McIntosh *et al.*, 1984). Conantokin-G is a selective inhibitor of NMDA receptors containing NR2B subunits. Selective NR2B antagonists have been a focal point for analgesic drug development because of their efficacy and potential for reduced psychomotor effects (reviewed by Bleakman *et al.*, 2006).

Several *Conus* peptide families have molecular targets other than voltage-gated and ligand-gated ion channels. Some of those peptides have analgesic properties as well (McIntosh *et al.*, 2000). The χ-conotoxins inhibit the reuptake of norepinephrine in the spinal cord (Sharpe *et al.*, 2001). Norepinephrine is released from descending neurons in the spinal cord to modulate nociceptive signaling (Furst, 1999). Thus, inhibition of norepinephrine uptake amplifies the antinociceptive signal. A derivative of χ-conotoxin, MrIA, from *Conus marmoreus* (Xen2174 from Xenome), has advanced to phase I clinical trials after demonstrating antiallodynic properties in rat models of spinal nerve injury (Nielsen *et al.*, 2005).

Similar to the analgesic effects of norepinephrine in descending antinociceptive pathways, an endogenous neuropeptide, neurotensin, appears to have analgesic properties in the CNS. A glycosylated *Conus* peptide that is a neurotensin analog was purified from the venom of *Conus geographus* and named contulakin-G. This peptide (CGX-1160 from Cognetix) has advanced to phase I clinical trials after demonstrating agonist activity on the neurotensin receptor, NTSR1 (Craig *et al.*, 1999), and demonstrating analgesic efficacy in several animal models of pain. Efficacy was demonstrated in a few cases of human spinal cord injury.

AN ANALGESIC PEPTIDE FROM A SEA ANEMONE

Sensory nerve endings and dorsal root ganglia express an acid-sensing ion channel known as ASIC3. This channel is involved in responses to nociceptive stimuli such as heat and acid, and it has been implicated in nociceptive signaling involving acidosis in ischemic and inflamed tissue (Chagot *et al.*, 2005; Price *et al.*, 2001). The activity of ASIC3 increases substantially when lactate is produced under ischemic conditions, and the ion channel is known to transmit pain signals associated with myocardial ischemia (Immke and McClesky, 2001). A recently isolated 42–amino acid peptide from the sea anemone *Anthopleura elegantissima*, named APETx2, selectively inhibits the ASIC3 subtype of ASIC ion channels. The peptide potently inhibits homomeric ASIC3 ion channels and also inhibits (with variable affinity) most heteromeric acid-sensing ion channels containing an ASIC3 subunit. These experimental results suggest that the peptide may have novel analgesic properties (Diochot *et al.*, 2004).

TETRODOTOXIN AS AN ANALGESIC DRUG

Among the marine toxins that may have useful analgesic properties is tetrodotoxin, found in many organs of the puffer fish (*fugu*). Although tetrodotoxin is known to be deadly, causing paralysis via antagonism of voltage-gated sodium channels at the neuromuscular junction and in motor neurons, low dosages may be administered systemically to produce analgesia in rodent models of inflammatory and neuropathic pain. At proper dosages, the analgesic effects were not accompanied by any obvious adverse events (Marcil *et al.*, 2006).

AN OCEAN OF OPPORTUNITY

Together with the many analgesic *Conus* peptides, the discoveries of APETx2 and the analgesic efficacy of tetrodotoxin collectively highlight the potential of the oceans as a vast and largely untapped resource for the discovery of analgesic drugs with novel mechanisms of action. As the cornerstone of modern pharmacology, natural products remain a critical source of therapeutic drugs and lead compounds. Successes in the clinic with natural-product based drugs, such as the FDA approval of Prialt for intractable, chronic pain, should underscore the continued importance of natural products in pharmaceutical development. The *Conus* peptides have been studied extensively and intensively over a period of decades, and this has allowed many of these compounds to advance to clinical trials as therapeutic drugs. Remarkably, six *Conus* peptides have advanced to human clinical trails as pain therapeutics, with five different mechanisms of action. Nevertheless, the cone snails represent just one genus of the incredibly biodiverse marine world; much of that biodiversity remains to be explored. The ongoing discoveries of marine natural products with novel analgesic mechanisms, from a variety of sources, demonstrate that the oceans contain incredible potential for therapeutic drug discovery.

Although the oceans provide great opportunity for drug discovery, many natural products, including peptides, do have some limitations as therapeutic drugs. The limitations of peptides as drugs include difficulty in delivery to the CNS, enzymatic degradation when taken orally, and potential immunogenicity. However, it now appears that some of these limitations can be overcome. For example, the necessity of intrathecal administration for analgesic, CNS-targeted *Conus* peptides, such as Prialt, has been a market-limiting factor and consequently has limited interest in developing additional peptides as drugs. However, as described earlier, some *Conus* peptides should produce analgesia upon administration to the peripheral nervous system alone. For example, the cone-snail and sea-anemone peptides that target sodium channels (e.g., $Na_V1.7–1.9$), neuronal nAChRs (e.g., $\alpha9\alpha10$ nAChRs), and ASIC3 in the peripheral nervous system should produce analgesia without the necessity of CNS drug delivery. Furthermore, there have been successes with peptide delivery to the brain via nasal administration (i.e., nasal sprays), which may ultimately provide a more attractive drug-delivery method for the mass market with CNS-targeting peptides (Li *et al.*, 2007; Vyas *et al.*, 2005). Nasal drug delivery eventually may provide a method to bypass intrathecal administration to the CNS, and intravenous administration to the peripheral nervous system, with the convenience of oral administration. In addition, analogs of conotoxins have been synthesized to improve their bioavailability (e.g., improved ability to cross intestinal mucosal membranes and the blood-brain barrier) and to improve their stability (e.g., resistance to proteolytic degradation) (Blanchfield *et al.*, 2003; Hogg, 2006). As scientific and technological solutions arise to address the limitations of peptide-based drugs, these natural products should play a more significant role in the therapeutic tool kit in the years to come.

References

Blanchfield, J.T., Dutton, J.L., Hogg, R.C., Gallagher, O.P., Craik, D.J., Jones, A., Adams, D.J., Lewis, R.J., Alewood, P.F., Toth, I., 2003. Synthesis, structure elucidation, *in vitro* biological activity, toxicity and Caco-2 cell permeability of lipophilic analogues of α-conotoxin MII. J. Med. Chem. 46, 1266–1272.

Bleakman, D., Alt, A., Nisenbaum, E.S., 2006. Glutamate receptors and pain. Semin. Cell Dev. Biol. Oct 26 [Epub ahead of print].

Buczek, O., Yoshikami, D., Watkins, M., Bulaj, G., Jimenez, E.C., Olivera, B.M., 2005. Characterization of D-amino-acid-containing excitatory conotoxins and redefinition of the I-conotoxin superfamily. FEBS Lett. 272, 4178–4188.

Bulaj, G., Zhang, M.M., Green, B.R., Fiedler, B., Layer, R.T., Wei, S., Nielsen, J.S., Low, S.J., Klein, B.D., Wagstaff, J.D., Chicoine, L., Harty, T.P., Terlau, H., Yoshikami, D., Olivera, B.M., 2006. Synthetic muO-conotoxin MrVIB blocks TTX-resistant sodium channel NaV1.8 and has a long-lasting analgesic activity. Biochemistry 45, 7404–7414.

Chagot, B., Escoubas, P., Diochot, S., Bernard, C., Lazdunski, M., Darbon, H., 2005. Solution structure of APETx2, a specific peptide inhibitor of ASIC3 proton-gated channels. Protein Sci. 14, 2003–2010.

Corpuz, G.P., Jacobsen, R.B., Jimenez, E.C., Watkins, M., Walker, C., Colledge, C., Garrett, J.E., McDougal, O., Li, W., Gray, W.R., Hillyard, D.R., Rivier, J., McIntosh, J.M., Cruz, L.J., Olivera, B.M., 2005. Definition of the M-conotoxin superfamily: Characterization of novel peptides from molluscivorous *Conus* venoms. Biochemistry 44, 8176–8186.

Cox, J.J., Reimann, F., Nicholas, A.K., Thornton, G., Roberts, E., Springell, K., Karbani, G., Jafri, H., Mannan, J., Raashid, Y., Al-Gazali, L., Hamamy, H., Valente, E.M., Gorman, S., Williams, R., McHale, D.P., Wood, J.N., Gribble, F.M., Woods, C.G., 2006. An SCN9A channelopathy causes congenital inability to experience pain. Nature 444, 894–898.

Craig, A.G., Norberg, T., Griffin, D., Hoeger, C., Akhtar, M., Schmidt, K., Low, W., Dykert, J., Richelson, E., Navarro, V., Macella, J., Watkins, M., Hillyard, D., Imperial, J., Cruz, L.J., Olivera, B.M., 1999. Contulakin-G, an O-glycosylated invertebrate neurotensin. J. Biol. Chem. 274, 13752–13759.

Craig, A.G., Zafaralla, G., Cruz, L.J., Santos, A.D., Hillyard, D.R., Dykert, J., Rivier, J.E., Gray, W.R., Imperial, J., DelaCruz, R.G., Sporning, A., Terlau, H., West, P.J., Yoshikami, D., Olivera, B.M., 1998. An O-glycosylated neuroexcitatory *Conus* peptide. Biochemistry 37, 16019–16025.

Diochot, S., Baron, A., Rash, L.D., Deval, E., Escoubas, P., Scarzello, S., Salinas, M., Lazdunski, M., 2004. A new sea anemone peptide, APETx2, inhibits ASIC3, a major acid-sensitive channel in sensory neurons. EMBO J. 23, 1516–1525.

England, L.J., Imperial, J., Jacobsen, R., Craig, A.G., Gulyas, J., Akhtar, M., Rivier, J., Julius, D., Olivera, B.M., 1998. Inactivation of a serotonin-gated ion channel by a polypeptide toxin from marine snails. Science 281, 575–578.

Frantz, S., 2006. Pipeline problems are increasing the urge to merge. Nat. Rev. Drug Disc. 5, 977–979.

Furst, S., 1999. Transmitters involved in antinociception in the spinal cord. Brain Res. Bull. 48, 129–141.

Hogg, R.C., 2006. Novel approaches to pain relief using venom-derived peptides. Curr. Med. Chem. 13, 3191–3201.

Hopkins, C., Grilley, M., Miller, C., Shon, K., Cruz, L.J., Gray, W.R., Dykert, J., Rivier, J., Yoshikami, D., Olivera, B.M., 1995. A new family of *Conus* peptides targeted to the nicotinic acetylcholine receptor. J. Biol. Chem. 270, 22361–22367.

Immke, D.C., McClesky, E.W., 2001. Lactate enhances the acid-sensing Na^+-channel on ischemia-sensing neurons. Nat. Neurosci. 4, 869–870.

Imperial, J.S., Bansal, P.S., Alewood, P.F., Daly, N.L., Craik, D.J., Sporning, A., Terlau, H., Lopez-Vera, E., Bandyopadhyay, P.K., Olivera, B.M., 2006. A novel conotoxin inhibitor of Kv1.6 channel and nAChR subtypes defines a new superfamily of conotoxins. Biochemistry 45, 8331–8340.

Li, F., Feng, J., Cheng, Q., Zhu, W., Jin, Y., 2007. Delivery of (125)I-Cobrotoxin after intranasal administration to the brain: A microdialysis study in freely moving rats. Int. J. Pharm. 328, 161–167.

Lirazan, M.B., Hooper, D., Corpuz, G.P., Ramilo, C.A., Bandyopadhyay, P., Cruz, L.J., Olivera, B.M., 2000. The spasmodic peptide defines a new conotoxin superfamily. Biochemistry 39, 1583–1588.

Livett, B.G., Sandall, D.W., Keays, D., Gayler, K.R., Satkunanathan, N., Khalil, Z., 2006. Therapeutic applications of conotoxins that target the neuronal nicotinic acetylcholine receptor. Toxicon 48, 810–829.

Malmberg, A.B., Gilbert, H., McCabe, R.T., Basbaum, A.I., 2003. Powerful antinociceptive effects of the cone snail venom-derived subtype-selective NMDA receptor antagonists conantokins G and T. Pain 101, 101–116.

Marcil, J., Walczak, J.S., Guindon, J., Ngoc, A.H., Lu, S., Beaulieu, P., 2006. Antinociceptive effects of tetrodotoxin (TTX) in rodents. Br. J. Anaesth. 96, 761–768.

McIntosh, J.M., Corpuz, G.P., Layer, R.T., Garrett, J.E., Wagstaff, J.D., Bulaj, G., Vyazovkina, A., Yoshikami, D., Cruz, L.J., Olivera, B.M., 2000. Isolation and characerization of a novel *Conus* peptide with apparent antinociceptive activity. J. Biol. Chem. 275, 32391–32397.

McIntosh, J.M., Olivera, B.M., Cruz, L.J., Gray, W.R., 1984. γ-Carboxyglutamate in a neuroactive toxin. J. Biol. Chem. 259, 14343–14346.

Miljanich, G.P., 2004. Ziconotide: Neuronal calcium channel blocker for treating severe chronic pain. Curr. Med. Chem. 11, 3029–3040.

Nestler, E.J., Hyman, S.E., Malenka, R.C., 2001. Molecular Neuropharmacology: A Foundation for Clinical Neuroscience, 433–452. New York: McGraw-Hill.

Newman, D.J., Cragg, G.M., Snader, K.M., 2003. Natural products as sources of new drugs over the period 1981–2002. J. Nat. Prod. 66, 1022–1037.

Nielsen, C.K., Lewis, R.J., Alewood, D., Drinkwater, R., Palant, E., Patterson, M., Yaksh, T.L., McCumber, D., Smith, M.T., 2005. Anti-allodynic efficacy of the chi-conopeptide, Xen2174, in rats with neuropathic pain. Pain 118, 112–124.

Olivera, B.M., 1997. *Conus* venom peptides, receptor and ion channel targets and drug design: 50 million years of neuropharmacology (E.E. Just Lecture, 1996). Mol. Biol. Cell 8, 2101–2109.

Olivera, B.M., 2006. *Conus* peptides: Biodiversity-based discovery and exogenomics. J. Biol. Chem. 281, 31173–31177.

Price, M.P., McIlwrath, S.L., Xie, J., Cheng, C., Qiao, J., Tarr, D.E., Sluka, K.A., Brennan, T.J., Lewin, G.R., Welsh, M.J., 2001. The DRASIC cation channel contributes to the detection of cutaneous touch and acid stimuli in mice. Neuron 32, 1071–1083.

Sandall, D.W., Satkunanathan, N., Keays, D.A., Polidano, M.A., Liping, X., Pham, V., Down, J.G., Khalil, Z., Livett, B.G., Gayler, K.R., 2003. A novel alpha-conotoxin identified by gene sequencing is active in suppressing the vascular response to selective stimulation of sensory nerves *in vivo*. Biochemistry 42, 6904–6911.

Santos, A.D., McIntosh, J.M., Hillyard, D.R., Cruz, L.J., Olivera, B.M., 2004. The A-superfamily of conotoxins: Structural and functional divergence. J. Biol. Chem. 279, 17596–17606.

Satkunanathan, N., Livett, B., Gayler, K., Sandall, D., Down, J., Khalil, Z., 2005. Alpha-conotoxin Vc1.1 alleviates neuropathic pain and accelerates functional recovery of injured neurones. Brain Res. 1059, 149–158.

Scott, D.A., Wright, C.E., Angus, J.A., 2002. Actions of intrathecal omega-conotoxins CVID, GVIA, MVIIA, and morphine in acute and neuropathic pain in the rat. Eur. J. Pharmacol. 451, 279–286.

Sharpe, I.A., Gehrmann, J., Loughnan, M.L., Thomas, L., Adams, D.A., Atkins, A., Palant, E., Craik, D.J., Adams, D.J., Alewood, P.F., Lewis, R.J., 2001. Two new classes of conopeptides inhibit the α1-adrenoceptor and noradrenaline transporter. Nat. Neurosci. 4, 902–907.

Smith, M.T., Cabot, P.J., Ross, F.B., Robertson, A.D., Lewis, R.J., 2002. The novel N-type calcium channel blocker, AM336, produces potent dose-dependent antinociception after intrathecal dosing in rats and inhibits substance P release in rat spinal cord slices. Pain 96, 119–127.

Teichert, R.W., Jimenez, E.C., Olivera, B.M., 2005. αS-Conotoxin RVIIIA: A structurally unique conotoxin that broadly targets nicotinic acetylcholine receptors. Biochemistry 44, 7897–7902.

Terlau, H., Olivera, B.M., 2004. *Conus* venoms: A rich source of novel ion channel-targeted peptides. Physiol. Rev. 84, 41–68.

Terlau, H., Shon, K., Grilley, M., Stocker, M., Stühmer, W., Olivera, B. M., 1996. Strategy for rapid immobilization of prey by a fish-hunting cone snail. Nature 381, 148–151.

Vinkler, M., Wittenauer, S., Parker, R., Ellison, M., Olivera, B.M., McIntosh, J.M., 2006. Molecular mechanism for analgesia involving specific antagonism of alpha9alpha10 nicotinic acetylcholine receptors. Proc. Natl. Acad. Sci. USA 103, 17880–17884.

Vyas, T.K., Shahiwala, A., Marathe, S., Misra, A., 2005. Intranasal drug delivery for brain targeting. Curr. Drug Deliv. 2, 165–175.

Walker, C., Steel, D., Jacobsen, R.B., Lirazan, M.B., Cruz, L.J., Hooper, D., Shetty, R., DelaCruz, R.C., Nielsen, J.S., Zhou, L., Bandyopadhyay,

P., Craig, A., Olivera, B.M., 1999. The T-superfamily of conotoxins. J. Biol. Chem. 274, 30664–30671.

Woolf, C.J., Thompson, S.W., 1991. The induction and maintenance of central sensitization is dependent on N-methyl-D-aspartic acid receptor activation; implication for the treatment of post-injury pain hypersensitivity states. Pain 44, 293–299.

STUDY QUESTIONS

1. What are the potential advantages and disadvantages of peptides or other natural products as therapeutic drugs?
2. List the molecular targets (e.g., receptor and ion channel subtypes) of the *Conus* peptides identified as potential analgesic drugs. How can the targeting of just one of such a diverse array of biological molecules produce analgesia?
3. The venom from any particular species of cone snail contains approximately 50 to 200 unique components, primarily peptides. How do the different venom components described in this chapter produce both rapid and long-lasting immobilization of prey? What are the molecular mechanisms of action that produce the motor cabal versus the lightning-strike cabal?
4. What are the molecular targets of opiate drugs and NSAIDs? Are their targets ion channels? Compare and contrast the various mechanisms of action for opiate drugs, NSAIDs, and analgesic *Conus* peptides.
5. Explain why Prialt (ω-conotoxin MVIIA) must be administered intrathecally (directly to the spinal cord) to produce analgesic effects. Would the α- or μ-conotoxins described in the chapter likely require intrathecal administration to produce analgesia? Use specific examples in explaining your answer.
6. What is central sensitization versus peripheral sensitization?
7. Would an antagonist of a voltage-gated potassium channel be a likely candidate as an analgesic drug? Why or why not?
8. Why are opiate drugs ineffective in the treatment of chronic, neuropathic pain?

CHAPTER

26

Emerging Marine Biotechnologies

Cloning of Marine Biosynthetic Gene Clusters

DANIEL W. UDWARY, JOHN A. KALAITZIS, AND BRADLEY S. MOORE

INTRODUCTION

The turn of the century signaled the beginning of a new era in the field of marine natural products research with the inclusion of molecular genetic techniques to the natural product chemical repertoire. Whereas feeding experiments with isotopically labeled precursors had been the main analytical tool to elucidate biosynthetic pathways in laboratory cultured marine microorganisms, the addition of modern molecular techniques involving bioinformatics and recombinant technology has been instrumental in deepening our understanding and appreciation for complex marine microbial pathways (Moore, 2005, 2006). Several biosynthetic pathways to complex polyketides, peptides, and hybrids thereof were elucidated for the first time at the molecular level in marine bacteria and cyanobacteria, thus opening the door to new research directions involving metabolic engineering and biocatalysis. Furthermore, recombinant technology provided new opportunities to interrogate uncultured marine microbes associated with eukaryotic hosts such as sponges, tunicates, and bryozoans and to probe their biosynthetic abilities in producing natural chemicals advantageous to their host that could have human health benefits.

Genes associated with the biosynthesis, regulation, and resistance of natural products are typically clustered in prokaryotic genomes. This convenient assemblage of genes into so-called biosynthetic gene clusters, which can range in size from a few kilobases (kb) to well over 100 kb, imparts tremendous information regarding how, when, and sometimes why a natural product is produced. As a consequence, its cloning provides distinct opportunities in the drug discovery and development process. Such opportunities include the expression and production of the natural product in a model heterologous (surrogate) host such as *Escherichia coli*, the reengineering of the biosynthetic pathway to yield "unnatural" natural products that may have altered biological activities, the elimination of metabolic processes for the streamlined (over)production of a desired natural product, the discovery of new enzymes that may have applied therapeutic or biocatalytic properties themselves, and so on.

Natural products such as polyketides, nonribosomal peptides, and hybrids thereof represent an important group of marine chemicals with important druglike properties. These molecules are biosynthesized by large, multifunctional enzymatic complexes that in an assembly line process sequentially assemble small carboxylic acid and amino acid building blocks into their products (Fischbach and Walsh, 2006). In the case of modular polyketide synthase (PKS) and nonribosomal peptide synthetase (NRPS) systems, the repetitive domain structures associated with these megasynthases (Fig. 26-1) generally follow a colinearity rule that not only allows for their precise genetic reprogramming but also for their annotation to provide predictive chemical structures for unknown products. PKS modules contain at a minimum three enzymatic domains, namely the acyltransferase (AT) for the selection of acyl-coenzyme A thioester substrates, the acyl carrier protein (ACP) for the covalent tethering of substrates and resulting products during the assembly line process, and the ketosynthase (KS) for catalyzing the condensation of two ACP-bound substrates. This fatty acid synthase-related process also involves up to three additional catalytic domains per PKS module that act on the newly synthesized β-keto group and convert it to either a hydroxyl group by the NADPH-dependent ketoreductase (KR), an α,β-unsaturated olefin by the KR and dehydratase (DH), or a saturated methylene by the KR, DH, and the NADPH-dependent enoylreductase (ER). NRPSs operate in an analogous fashion for the condensation of amino acid precursors

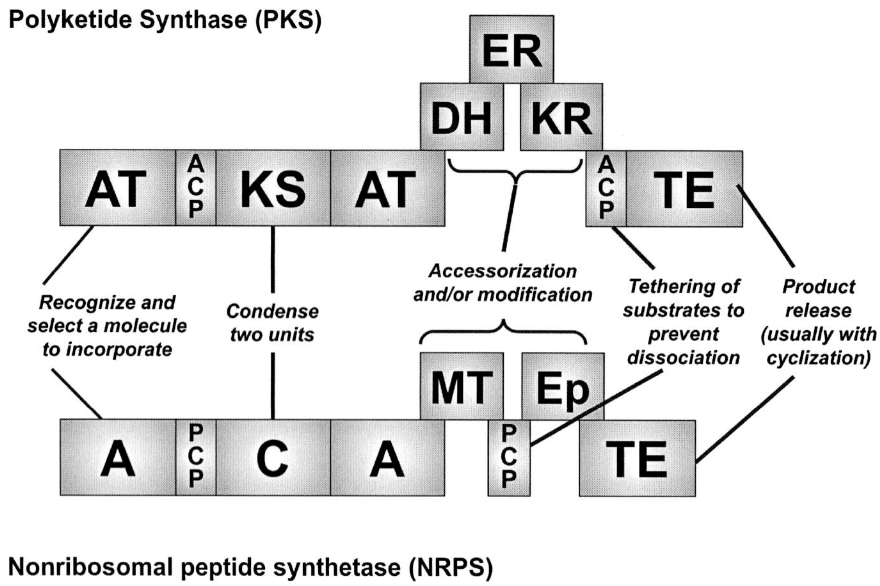

FIGURE 26-1. Comparative modular organization of type I polyketide synthase and nonribosomal peptide synthetase enzymes.

and likewise minimally employ three enzyme domains, namely the ATP-dependent adenylation (A) domain for the activation of an amino acid to its AMP analog, the ACP related peptidyl carrier protein (PCP), and the condensation domain (C), which facilitates the formation of amide bonds between PCP-bound amino acid substrates. NRPS accessory domains include the epimerase (Ep) for the in trans conversion of L- to D-amino acid substrates and the methyltransferase (MT) for catalyzing the *N*-methylation of amide linkages. PKS and NRPS megasynthases can harbor over a dozen such modules for the construction of their products; once constructed, they are released from their associated carrier proteins by the C-terminal thioesterase (TE) as either linear carboxylic acids or cyclized to large, cyclic esters (i.e., macrolides or depsipeptides) or amides (i.e., cyclic peptides). The amalgamation of these related mechanisms in hybrid PKS–NRPS megasynthases results in remarkably versatile enzymatic machines that employ carrier protein-based biosynthetic logic to yield varied natural products derived from carboxylic and amino acid substrates.

This chapter focuses on the progress of cloning natural product biosynthetic gene clusters from marine organisms and their utilization in addressing fundamental questions and applied goals in the drug discovery and development process. Select examples will be provided in the following section from a growing number of cloned marine natural product biosynthetic pathways from cultured and uncultured marine microorganisms (Table 26-1). Advances and emerging technology associated with genomics and metagenomics are highlighted in the second half of the chapter, which points to new trends in biotechnology and their utilization in the discovery of new marine natural products.

CLONING OF MARINE BIOSYNTHETIC GENE CLUSTERS

Marine Bacteria

The search for new anti-infective, anticancer, and other pharmaceuticals agents has led researchers, particularly natural products chemists, to investigate actinomycetes from the marine environment as a potential source of new chemical entities (Fenical and Jensen, 2006). Although soil-dwelling bacteria are notable for their production of natural products, including more than 60% of all natural antibiotics, the marine environment for the most part has remained a largely untapped microbial resource with respect to drug discovery. The isolation of marine actinomycetes through advances in culturing techniques has resulted in the discovery of new chemistry and drug leads as well as the classification of new bacterial taxa (Lam, 2006). The characterization of novel marine bacterial agents has fueled the field of marine biotechnology by providing opportunities not only in drug discovery and development but also in the discovery of new enzymatic processes with applied potential in biocatalysis and therapeutics. This section highlights genetic approaches used to clone marine bacterial biosynthetic gene clusters and in some cases genetically reengineer these pathways to provide "unnatural" natural products.

The 21.3 kilobase (kb) enterocin biosynthetic gene cluster (*enc*) from *Streptomyces maritimus* was the first fully characterized gene cluster encoding a polyketide synthase (PKS) from a marine microbe (Piel *et al.*, 2000a). In addition to genes encoding biosynthetic enzymes, the 20 open reading frame (ORF) containing *enc* gene cluster codes for transport

TABLE 26-1. Cloned natural product biosynthetic gene clusters from cultured and uncultured marine prokaryotes.

Gene Cluster	Source Organism	~Size (kb)	Biosynthetic Origin
Enterocin (*enc*)	*S. maritimus* (bacterium)	21	Polyketide
Napyradiomycin (*nap*)	*S. aculeolatus* NRRL 18422 and CNQ525 (bacteria)	43	Polyketide/terpenoid
Salinosporamide (*sal*)	*S. tropica* (bacterium)	41	Polyketide/peptide
Griseorhodin (*grh*)	Unidentified streptomycete (bacterium)	34	Polyketide
Eicosapentaenoic acid	*P. profundum* SS9 (bacterium)	17	Polyketide
Docosahexeanoic acid	*M. marina* MP-1 (bacterium)	>20	Polyketide
Barbamide (*bar*)	*L. majuscula* (cyanobacterium)	26	Polyketide/peptide
Curacin (*cur*)	*L. majuscula* (cyanobacterium)	64	Polyketide/peptide
Jamaicamide (*jam*)	*L. majuscula* (cyanobacterium)	58	Polyketide/peptide
Lyngbyatoxin (*ltx*)	*L. majuscula* (cyanobacterium)	11	Peptide/terpenoid
Microcystin (*mcy*)	*M. aeruginosa* (cyanobacterium)	55	Polyketide/peptide
Nodularin (*nda*)	*N. spumigena* (cyanobacterium)	48	Polyketide/peptide
Patellamide (*pat*)	*L. patella* (tunicate)	11	Peptide
Onnamide (*onn*)	*T. swinhoei* (sponge)	>36	Polyketide/peptide
Bryostatin (*bry*)	*B. neritina* (bryozoan)	65	Polyketide

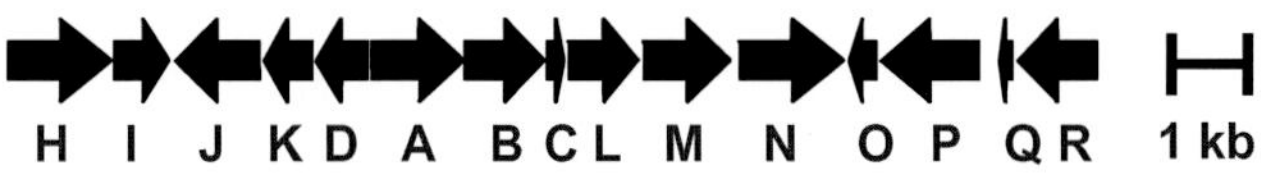

FIGURE 26-2. Partial map of the enterocin biosynthetic gene cluster (*enc*) in *S. maritimus*. Each vector represents the direction of transcription of an open reading frame. The functions of the depicted *enc* genes are as follows: *encA–encB* (ketosynthase αβ subunits), *encC* (acyl carrier protein), *encD* (ketoreductase), *encH–encJ* (starter unit biosynthesis genes), *encK* (*O*-methyltransferase), *encL* (acyl transferase homolog), *encM* (FAD-dependent oxygenase), *encN* (benzoate : ACP ligase), *encO* (unknown), *encP* (phenylalanine ammonia lyase), *encQ* (ferredoxin), and *encR* (cytochrome P_{450} hydroxylase).

and resistance proteins associated with enterocin production (Fig. 26-2). The sediment-derived isolate from Hawaii produces a structurally diverse series of broad-spectrum bacteriostatic polyketides that include the major product enterocin together with a series of related molecules including 5-deoxyenterocin and the wailupemycins (Piel *et al.*, 2000b). These polyketides are biosynthesized by a novel iterative type II PKS and are derived from a single biosynthetic pathway with numerous metabolic options for creating molecular diversity (Fig. 26-3).

The enterocin PKS complex is composed of three proteins—two ketosynthase subunits EncA and EncB and the acyl carrier protein (ACP) EncC—that are required for polyketide chain assembly from benzoyl-coenzyme A and seven malonate molecules (Hertweck *et al.*, 2004). Investigations into the biosynthesis of the benzoate building block revealed that this rare bacterial metabolite is derived from the common amino acid L-phenylalanine via a plant-like β-oxidative pathway through cinnamic acid (Xiang *et al.*, 2002; Xiang and Moore, 2003). Genetic experiments identified a novel prokaryotic phenylalanine ammonia-lyase encoding gene *encP* in *S. maritimus*, which codes for the first enzyme in the pathway to the enterocin PKS primer unit benzoyl-CoA. Disruption of this gene completely inhibited the production of cinnamic acid and as a consequence enterocin itself. Restoration of enterocin and wailupemycin biosynthesis in the knockout mutant with cinnamic and benzoic acids opened the door for the mutasynthesis of a series of unnatural polyketide analogs in which the natural background of the benzoyl-CoA starter unit was eliminated and replaced with unnatural aromatic acids (Kalaitzis *et al.*, 2003). The further combination of benzoyl-CoA biosynthesis genes from the enterocin pathway with PKS genes from other biosynthetic pathways such as for the macrolide antibiotic erythromycin allowed for the production of novel aryl-primed polyketides (Garcia-Bernardo *et al.*, 2004). These sets of experiments validated the use of biosynthetic genes from marine microbes in the combinatorial biosynthesis of new chemical entities.

The enterocin biosynthetic gene cluster also provided clues regarding the unprecedented oxidative rearrangement reaction that uniquely characterizes this biosynthetic pathway and allowed for the discovery of a novel flavoprotein called EncM (Xiang *et al.*, 2004). *In vivo* characterization of the gene *encM* through mutagenesis and heterologous biosynthesis demonstrated that its product is solely responsible for the oxidative carbon–carbon rearrangement of the polyketide backbone as well as the aldol condensation and

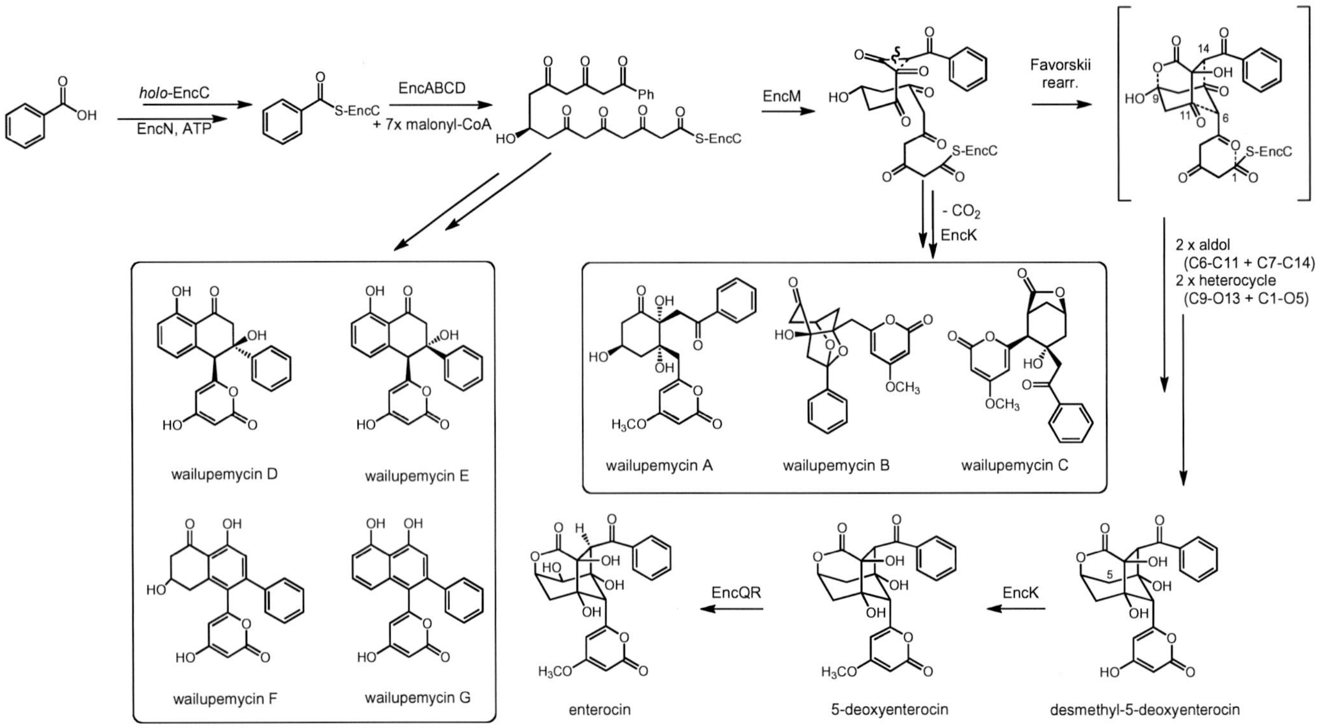

FIGURE 26-3. Proposed biosynthesis of the benzoate-primed polyketides enterocin and the wailupemycins. The acyl carrier protein (EncC) is primed with benzoate by EncN (in the presence of ATP and Mg^{2+}) before undergoing a series of malonate extensions as controlled by the ketosynthase αβ subunits (EncA/EncB). The growing polyketide chain then undergoes a series of tailoring reactions, cyclizations, and, in the case of enterocin and similar polyketides, a Favorskiiase rearrangement (facilitated by EncM) to form the final products. Wailupemycins D–G are formed via spontaneous cyclization of the C-9 reduced poly-β-ketide intermediate. The biosynthesis of benzoic acid from phenylalanine is not shown on this abbreviated scheme.

heterocycle forming reactions. As a result of this phenomenal activity, five chiral centers and four rings are generated by this multifaceted flavoprotein. The enterocin rearrangement reaction may serve as a model for other such biosynthetic reactions, most notably for dinoflagellate polyketides, such as the toxin okadaic acid, which have been postulated to originate through related mechanisms (Wright *et al.*, 1996). Given its remarkable reactivity and rarity in nature, the biosynthetic enzyme EncM has potential utility in combination with other PKS systems as a recombinant biocatalyst to engineer novel polyketide products and perhaps potential drug leads.

The salinosporamide series of potent anticancer agents from the marine bacterium *Salinispora tropica* isolated from sediments in the Bahamas (Feling *et al.*, 2003; Williams *et al.*, 2005) represents another example in which the cloning of the biosynthetic gene cluster has provided opportunities to impact how the drug candidate salinosporamide A (Chauhan *et al.*, 2005) and fermentation-based chemical variants are produced (Fig. 26-4). Stable isotope experiments initially laid the foundation for salinosporamide biosynthesis in which the precursor building blocks acetyl-CoA, a substituted malonyl-CoA, and β-hydroxycyclohexenylalanine are assembled by a hybrid PKS–NRPS to generate the unusual bicyclic γ-lactam-β-lactone nucleus (Beer and Moore, 2007). The biochemical knowledge of the pathway was instrumental in the identification of the 41 kb biosynthetic gene cluster *sal* spanning 29 ORFs in *S. tropica* CNB-440 whose 5,183,331 bp circular genome was sequenced (Udwary *et al.*, 2007). The salinosporamide hybrid PKS–NRPS pathway involves new enzymatic mechanisms in biological chlorination and β-lactone synthesis as well as 20S proteasome resistance. The cloning and sequencing of biosynthetic gene clusters associated with other marine actinomycete-derived antibiotic natural products such as griseorhodin A (Li and Piel, 2002) and napyradiomycin (Winter *et al.*, 2007) further revealed new metabolic processes associated with natural product biosynthesis involving remarkable oxidative and halogenation biochemistry (Fig. 26-4). The ongoing characterization of these novel biochemical processes not only positively impacts the drug discovery and development process but also expands our basic knowledge of how enzymes catalyze diverse chemical reactions in nature.

The nutraceutical long-chain polyunsaturated fatty acids (PUFAs), eicosapentaenoic acid (EPA) and

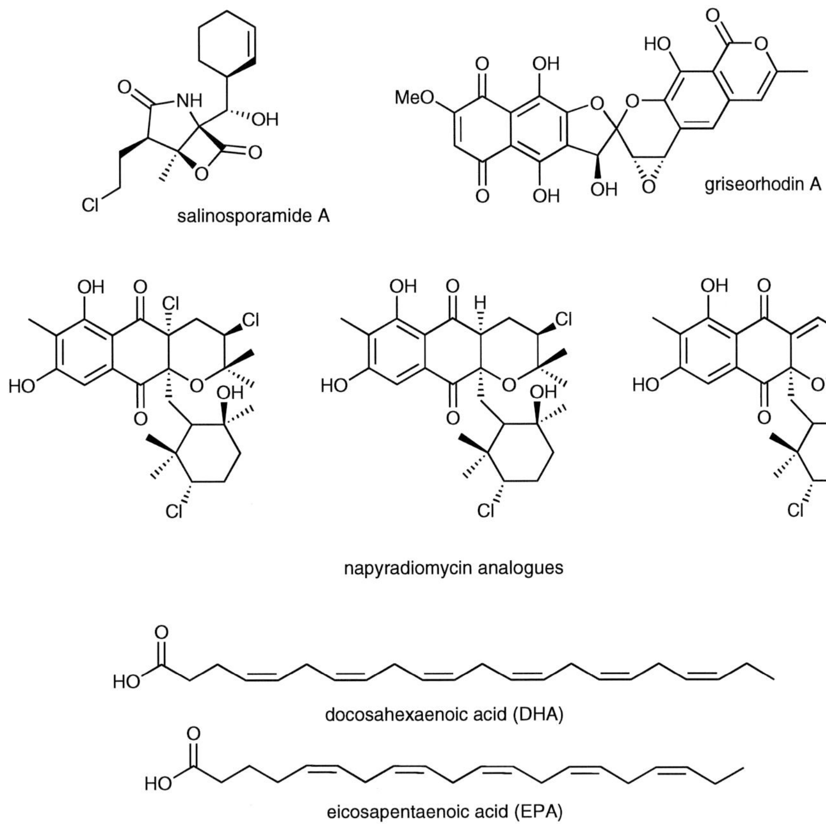

FIGURE 26-4. Natural products derived from gene clusters cloned from marine bacteria.

docosahexaenoic acid (DHA), were once believed to be eukaryotic products until they were discovered in psychrophilic (low temperature) marine bacteria where they are biosynthesized via an anaerobic pathway rather than the more common aerobic pathway in plants and animals (Metz *et al.*, 2001). Investigations into PUFA biosynthesis by marine bacteria unexpectedly revealed that these essential metabolites are constructed in a similar manner to polyketide natural products rather than the commonly accepted fatty acid synthase (FAS) pathway by which PUFAs are typically biosynthesized. Sequence analysis of a 38 kb genomic fragment from the marine bacterium *Shewanella pneumatophori* strain SCRC-2738 led to the identification of five ORFs related to PKSs that were required for synthesis of EPA in *Escherichia coli* (Orikasa *et al.*, 2004). Bioinformatic analysis revealed 11 putative enzyme domains within the five ORFs, eight of which appeared to be more closely related to PKS proteins than FAS proteins (Metz *et al.*, 2001). The PKS domain organization differs from typical microbial iterative type I PKSs, and the novelty of this system will provide new mechanistic details on PKS biochemistry. Furthermore, some of these new enzymes may even be useful in engineering new chemical entities. Genes with homology to the *S. pneumatophori* EPA gene cluster have also been found in the marine protist *Schizochytrium* sp. The deep sea bacterium *Photobacterium profundum* strain SS9 was also shown to construct EPA in a similar manner to *S. pneumatophori* (Allen and Bartlett, 2002). Four genes (*pfaA–pfaD*) from the strain *P. profundum* SS9 required for EPA synthesis were identified and found to span a region of approximately 17 kb. Comparison of these enzyme domains with those of *S. pneumatophori* SCRC-2738 and *Moritella marina* strain MP-1 revealed high degrees of similarity and identity. It is interesting to note that *M. marina* MP-1 produces DHA, whereas *S. pneumatophori* SCRC-2738 and *P. profundum* SS9 produce EPA. *In vivo* recombination of *pfaA–pfaD* with a fifth gene *pfaE* in *E. coli* has been shown to be an efficient method of synthesizing PUFAs. Co-expression of *pfaA–pfaD* and *pfaE* from the DHA producing strain *M. marina* MP-1 yielded DHA as expected, and this approach represented the first report of a recombinant biosynthesis of a PUFA via this new polyketide-like pathway (Orikasa *et al.*, 2006). EPA was produced in a similar manner by co-expressing *pfaA–pfaD* from *S. pneumatophori* SCRC-2738 and *pfaE* from *M. marina* MP-1 and thus demonstrated that genes from different organisms can be successfully co-expressed. Even though the PKS domain organization between the eukaryotic protist and the bacterial organisms is different, homology between the genes suggests that the PUFA PKS has possibly undergone lateral gene transfer.

However, the lack of conserved sequence in the regions flanking the *pfa* gene clusters, and the absence of flanking genes that could facilitate such horizontal transfer, does not support this notion (Allen and Bartlett, 2002).

Marine Cyanobacteria

Cyanobacteria, which are often referred to as blue-green algae, are commonly associated with toxic blooms and the accompanying production of a structurally diverse array of harmful neurotoxins and hepatotoxins. This phenomena alerted researchers to the potential of these organisms as prolific producers of bioactive metabolites that may have utility as drug candidates (Burja *et al.*, 2001; Gerwick *et al.*, 2001; Moore *et al.*, 1996). The portfolio of cyanobacterial natural products includes many mixed polyketide–peptide molecules derived from hybrid PKS–NRPS pathways, and several of these molecules are highlighted in the following section. Some of the most significant cyanobacterial natural products with respect to drug discovery and human health are the potent anticancer peptolides known as the dolastatins (Poncet, 1999). Although the dolastatins were originally isolated from the sea hare (mollusk) *Dolabella auricularia*, they are produced by marine cyanobacteria belonging to the genera *Lyngbya* and *Symploca* (Luesch *et al.*, 2002) and sequestered as chemical defensive agents through grazing by the sea hare. Given the exceedingly low isolation yields of dolastatins from the mollusk, cyanobacterial fermentation and subsequent bioengineering provide new opportunities to supply these promising anticancer agents and analogs thereof for drug discovery and development.

The biosynthesis of the cyclic heptapeptide microcystin-LR, the major hepatotoxin of the toxic bloom-forming alga *Microcystis aeruginosa*, was shown through feeding and preliminary genetic studies to originate from a hybrid PKS–NRPS (Fig. 26-5). The microcystin biosynthetic gene cluster was first cloned and sequenced from two *M. aeruginosa* strains (Nishizawa *et al.*, 1999, 2000; Tillet *et al.*, 2000) and more recently from strains of the genera *Planktothrix* (Christiansen *et al.*, 2003) and *Anabaena* (Rouhiainen *et al.*, 2004). The *M. aeruginosa* PCC7806 gene set, which represents the first complete characterization of a complex cyanobacterial secondary metabolic pathway, spans 55 kb and is composed of 10 bidirectionally transcribed ORFs arranged in two operons (*mcyABC* and *mcyDEFGHIJ*) (Kaebernick *et al.*, 2002; Tillet *et al.*, 2000). The microcystin synthetase

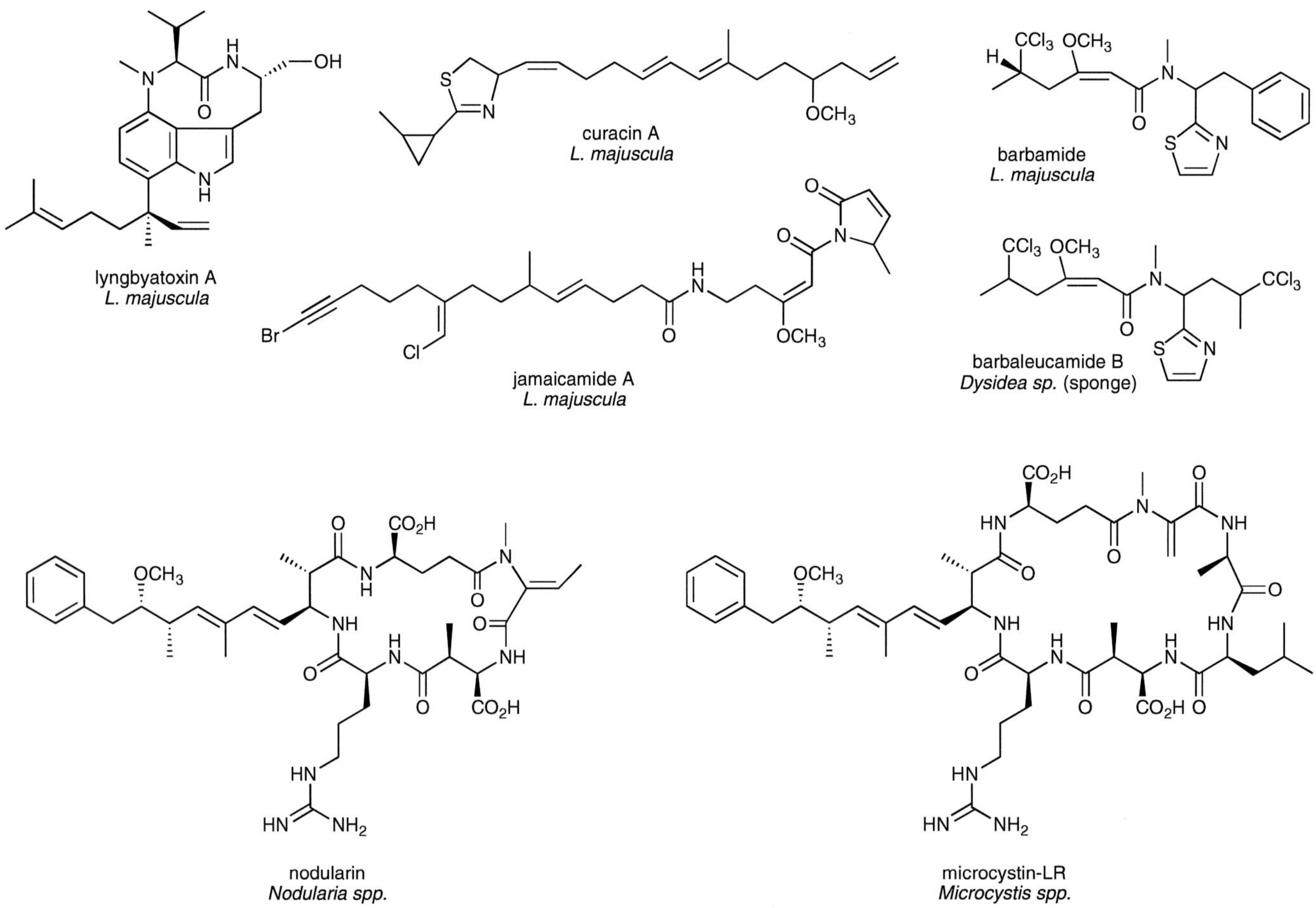

FIGURE 26-5. Natural products derived from gene clusters cloned from marine cyanobacteria.

activates and loads a starter unit derived from L-phenylalanine for subsequent extension by four malonate and six amino acid residues to a protein-bound linear microcystin precursor before it is hydrolyzed and cyclized to microcystin-LR by the C-terminal thioesterase (TE) of the NRPS McyC. The genetic characterization of the microcystin synthetase genes allowed the elucidation of genetic variants that correlate to different microcystin isoforms and the study of the regulation of the biosynthetic pathway (Kurmayer *et al.*, 2002; Mikalsen *et al.*, 2003). The biosynthesis of the structurally related cyclic pentapeptide nodularin, produced by toxic strains of the cyanobacterial genus *Nodularia*, proceeds via a similar pathway in *Nodularia spumigena* and provides valuable clues about the evolution of these related biosynthetic gene clusters (Moffitt and Neilan, 2004).

The filamentous marine cyanobacterium *Lyngbya majuscula*, on the other hand, is a remarkably prolific producer of structurally diverse natural products possessing broad ranges of biological activities from environmental toxins to promising anticancer agents. In many cases, *L. majuscula* products possess chemical features rarely encountered in nature. The cloning and sequencing of *L. majuscula* biosynthetic gene clusters associated with the molluscicidal chlorinated lipopeptide barbamide (Chang *et al.*, 2002), the antitubulin agent curacin A (Chang *et al.*, 2004), the mixed polyketide–peptide neurotoxin jamaicamide A (Edwards *et al.*, 2004), and the potent skin irritant lyngbyatoxin A (Edwards and Gerwick, 2004) have identified a multitude of novel biosynthetic processes that shed light on how these fascinating molecules (Fig. 26-5) are transformed from common precursors of the primary metabolic pool. For instance, precursor incorporation studies revealed that barbamide A is derived from one molecule each of L-leucine, acetate, L-phenylalanine, and L-cysteine together with two methyl groups from *S*-adenosyl-L-methionine (Sitachitta *et al.*, 2000). The novelty of this compound relates to the incorporation of the unusual 5,5,5-trichloroleucine moiety and the manner in which the unactivated *pro-R* methyl group of leucine is multiply halogenated. A multidisciplinary approach involving synthetic organic chemistry (Flatt *et al.*, 2006), enzymology (Galonic *et al.*, 2006), and molecular biology (Chang *et al.*, 2002) was instrumental in unveiling a new route for the halogenation of unactivated carbon centers in natural products by the none-heme iron halogenases BarB1 and BarB2. Further sequence analysis of the 12 ORF barbamide biosynthetic gene cluster extending 26 kb revealed a colinear genetic arrangement of the barbamide cluster in which the hybrid PKS–NRPS megasynthetase assembles a linear protein-bound diketide dipeptide intermediate that undergoes an unusual thiazole ring forming reaction catalyzed by the *C*-terminal thioesterase domain of BarG (Fig. 26-6). Biosynthetic studies with this laboratory cultured *L. majuscula* cyanobacterium were instrumental in further exploring the origin of related compounds such as barbaleucamide B from a Philippine *Dysidea* sp. sponge where *bar* gene probes were employed to provide strong support to the presumption that many *Dysidea* natural products are of cyanobacterial origin (Flatt *et al.*, 2005; Harrigan *et al.*, 2001).

Marine Tunicates

Two of the most promising drug candidates derived from marine invertebrates are the antitumor agents dehydrodidemnin B (trade name Aplidine) and ecteinascidin-743 (Et-743, trade name Yondelis) isolated from tunicate (ascidian) species of *Aplidium* and *Ecteinascidia*, respectively

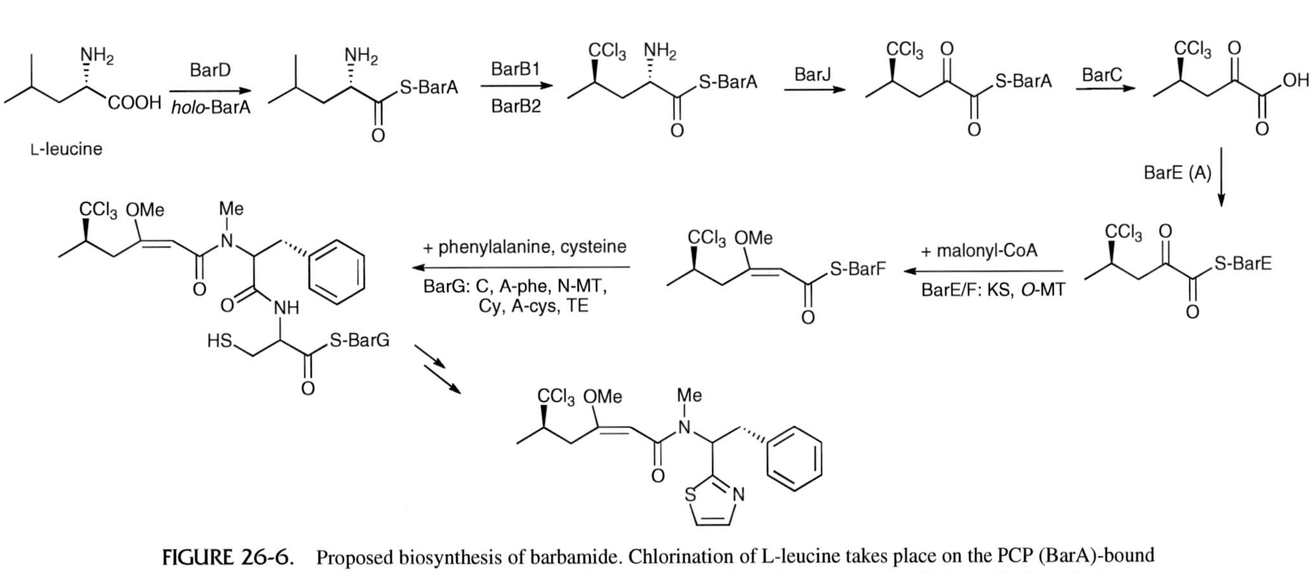

FIGURE 26-6. Proposed biosynthesis of barbamide. Chlorination of L-leucine takes place on the PCP (BarA)-bound substrate leucyl-BarA by the halogenase enzymes BarB1 and BarB2. Further processing of this trichlorinated substrate and its transfer to the PCP domain of BarE initiates a hybrid PKS–NRPS pathway involving the BarE/BarF/BarG modular enzymes to give the natural product.

(Fig. 26-7) (Rinehart, 2000). Et-743 and all of the ecteinascidin family of compounds were isolated from the mangrove tunicate *Ecteinascidia turbinata*. Raw antitumor activities displayed by the extracts of *E. turbinata* were first observed in 1969; however, because of the small amounts of the ecteinascidins available from the natural source, the isolation and structure elucidation were elusive. Modern spectroscopic techniques allowed the unambiguous assignment of the structure in 1990—more than 20 years after its biological activity was first observed (Rinehart *et al.*, 1990). Although there is no evidence to support that Et-743 is produced by a microbial symbiont, the structurally related saframycin family of molecules is produced by the terrestrial myxobacterium *Myxococcus xanthus* (Pospiech *et al.*, 1995). Like Et-743, these molecules have also shown promising anticancer activities (Spencer *et al.*, 2006). A gene cluster spanning approximately 58 kb has been identified for the biosynthesis of saframycin Mx1 (Pospiech *et al.*, 1995) that may provide clues regarding Et-743 biosynthesis in the tunicate.

The patellamide family of bioactive cyclic peptides (Fig. 26-7) from the tropical ascidian *Lissoclinum patella* is related in structure to compounds synthesized by cultured cyanobacteria and also has long been suspected to be of microbial origin. *Prochloron* spp. are obligate cyanobacterial symbionts of many didemnid family ascidians (Salomon and Faulkner, 2002). However, preliminary studies involving cellular localization were inconclusive as the peptides were distributed throughout the ascidian tunic and not in the cyanobacterial symbiont *Prochloron didemni* itself. These data suggested that either the host is the biosynthetic producer or the symbionts, having an active transport mechanism, are the producer. An additional genetic study examined the biosynthetic potential of *L. patella*-associated *Prochloron* spp. and demonstrated that the uncultured symbiont contained genes for NRPS biochemistry that could be associated with patellamide biosynthesis (Schmidt *et al.*, 2004). Although an NRPS-containing gene cluster was indeed cloned and sequenced, its predicted function was not consistent with patellamide biosynthesis. As part of a project to sequence the genome of *P. didemni*, the true patellamide biosynthesis gene cluster was identified as an ~11 kb gene cluster comprising the genes *patA–patG* (Schmidt *et al.*, 2005). Unexpectedly, the bioinformatics analysis suggested that the patellamides are synthesized ribosomally and that its precursor peptides undergo extensive posttranslational modifications involving multiple heterocycle-forming reactions to yield cytotoxic cyclic peptides. Heterologous expression of fosmid DNA containing the complete gene set in *E. coli* and production of patellamide A confirmed unambiguously that the *patA–patG* gene cluster was indeed responsible for the biosynthesis in *P. didemni*. This observation represented the first example of a genetics-based identification, transfer, and expression of a complete biosynthetic pathway from a marine microbial symbiont and opened the

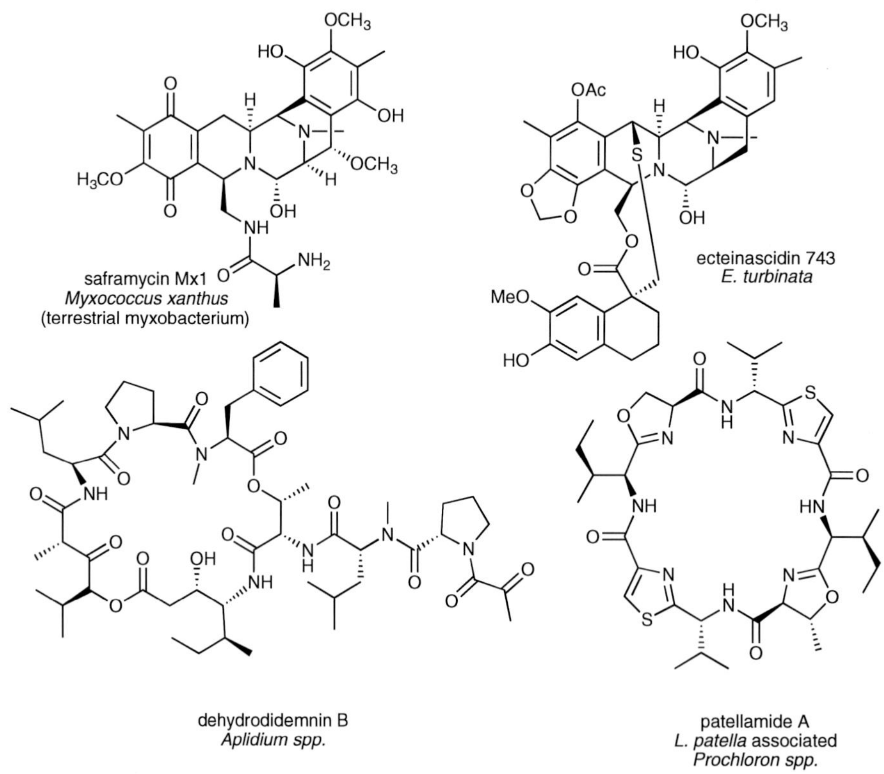

FIGURE 26-7. Tunicate metabolites and related bacterial products.

door for the genetic engineering of combinatorial peptide libraries (Donia *et al.*, 2006) as well as genome mining of new cyclic peptides (Sudek *et al.*, 2006). Working independently, a similar conclusion was achieved by expressing randomly cloned *Prochloron* sp. DNA in bacterial artificial chromosomes in *E. coli* for the production of patellamide D (Long *et al.*, 2005). These studies nicely illustrate the power and validate the use of genomics from highly elusive and largely unculturable symbiotic microorganisms in the natural product drug discovery process.

Marine Sponges

Whereas sponges provide the majority of marine natural products currently in clinical and preclinical trials, their associated microflora, which can account for over half of the sponge biomass, are thought in many cases to synthesize "sponge" natural products. Unlike the patellamide example from the ascidian *L. patella* that harbors the major cyanobacterial symbiont *P. didemni*, sponges typically host complex microbial communities that can greatly complicate locating a natural product-producing member.

A metagenomics study with the marine sponge *Theonella swinhoei* from Japan provided the first biosynthetic insight of an uncultivated symbiotic bacterium associated with a marine sponge for the production of the antitumor polyketide onnamide A (Piel *et al.*, 2004). This molecule shares structural similarities to a number of other bioactive natural products such as theopederin A also from the sponge *T. swinhoei*, mycalamide A from the sponge *Mycale* sp., and pederin from *Paederus* beetles (Fig. 26-8) and, hence, was suggested and later confirmed to be of microbial origin (Piel *et al.*, 2005). The cloning of the putative onnamide biosynthetic gene cluster (*onn*) was achieved by using a PCR-based approach in which PKS genes were amplified from the *T. swinhoei* metagenome and phylogenetically characterized. Biosynthetic investigations of the natural product pederin from the terrestrial beetle *Paederus fuscipes* demonstrated that the pederin biosynthetic gene cluster (*ped*) resided on three isolated gene fragments in the genome of an uncultured bacterial symbiont related to *Pseudomonas aeruginosa.* A detailed bioinformatics analysis of the *ped* gene cluster revealed a number of novel features, which were important not only for elucidating the biochemistry of the pederin biosynthetic pathway but also for targeting its related biosynthetic pathway in the unrelated sponge microbial community. The pederin megasynthetase is a hybrid PKS–NRPS whose encoding genes are arranged co-linearly with respect to the putative order of the biosynthetic assembly of the natural product. Interestingly, the *ped* gene cluster appears to code for the biosynthesis of a much larger polyketide than pederin itself with the addition of the gene *pedH*. The hypothetically extended pederin-based structure, which is remarkably related to onnamide, likely undergoes an oxidative cleavage reaction by the FAD-dependent PedG oxygenase to yield pederin (Piel, 2002). With this information in hand, *T. swinhoei* PKS genes were analyzed from a 60,000-member clone library representing the large and diverse sponge metagenome to yield a single *ped*-related cosmid that was sequenced. Unlike the *ped* cluster that was distributed on three regions in the *Paederus* symbiont genome, the putative onnamide gene cluster (*onn*) was clustered on just one genomic region. Although the cloned *onn* gene cluster is unfortunately incomplete, its architecture

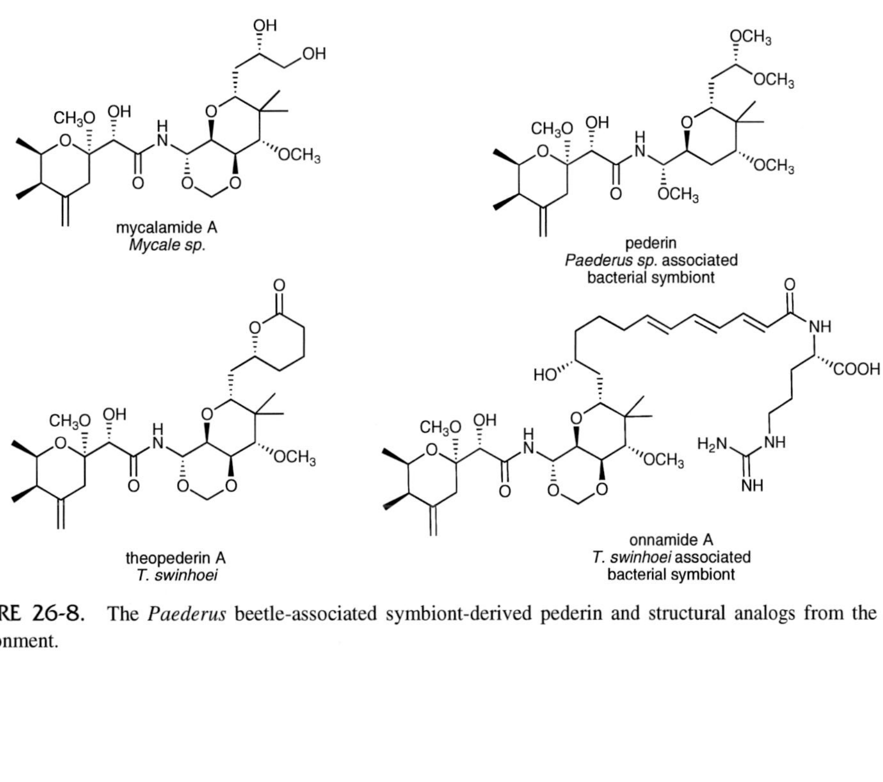

FIGURE 26-8. The *Paederus* beetle-associated symbiont-derived pederin and structural analogs from the marine environment.

mirrors that of the *ped* system to a high degree. The cloned *onn* genes correspond to the region of the molecule that is largely identical with that of pederin and support the claim that complex sponge-derived polyketides are produced by associated microbes.

Marine Bryozoans

The bryostatin family of cytotoxic macrolides is found in the marine bryozoan *Bugula neritina*, a common fouling organism of temperate and tropical waters. Bryostatins, including the potent anticancer agent bryostatin 1 (Fig. 26-9), are proposed metabolites of the *B. neritina* bacterial symbiont *Candidatus* Endobugula sertula. Lending strong support to this proposal is the fact that treatment of the bryozoan with antibiotics greatly diminished the production of bryostatin 1, suggesting that the producing organism (in this case, *Candidatus* Endobugula sertula) had been nullified (Davidson *et al.*, 2001; Lopanik *et al.*, 2004). Furthermore, the bryostatins resemble bacterial modular PKS products, thus supporting speculation that these macrolides are derived from a microbial symbiont much like in the case of tunicate-derived patellamide and sponge-derived onnamide. Attempts to culture the bacterial symbiont have been unsuccessful, thus fermentation of the microbe in an attempt to produce bryostatin 1 for human clinical trials not feasible. Consequently, heterologous expression of the biosynthetic gene cluster is seen as another avenue to address the supply of this anticancer agent. To identify the bryostatin biosynthetic gene cluster and establish the true producer of the bryostatins, a combined PCR and *in situ* hybridization approach was implemented (Sudek *et al.*, 2007).

A 300 bp ketosynthase gene fragment (KSa) was cloned from a total DNA preparation of the bryozoan by PCR and shown to be present in all bryostatin-containing bryozoans (Davidson *et al.*, 2001). It was further determined by *in situ* hybridization studies that KSa transcripts were located and expressed in *Candidatus* E. sertula cells in *B. neritina* larvae,

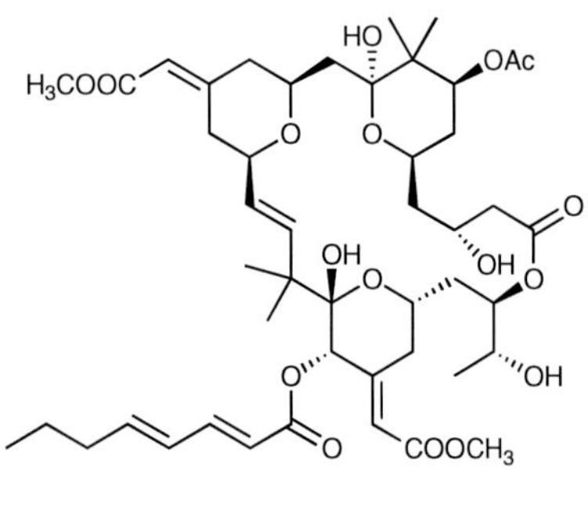

bryostatin 1

FIGURE 26-9. The chemical structure of bryostatin 1, a putative product of the bacterial symbiont *Candidatus* Endobugula sertula of the bryozoan *Bugula neritina*.

yet only those that had not been treated with antibiotics (Hildebrand *et al.*, 2004). Further confirmation of the origin of the bryostatins was given by cloning a homologous PKS fragment from the closely related gamma-proteobacterium *Candidatus* Endobugula glebosa, which is the larval symbiont of the Northern Atlantic *Bugula simplex* that produces bryostatin-related compounds (Lim and Haygood, 2004). The gene fragment KSa was used as a probe to clone the putative 80 kb bryostatin biosynthetic gene cluster (*bry*) from deep and shallow species of *B. neritina*. Despite being a type I PKS, the organization of the putative gene cluster does not strictly obey the co-linearity rule for polyketide biosynthesis, which has complicated its characterization. Two clinical modular PKS products, avermectin from *Streptomyces avermitilis* (Ikeda *et al.*, 1999) and rapamycin from *Streptomyces hygroscopicus* (Schwecke *et al.*, 1995), also share this architectural anomaly. Sequence analysis of the centrally located gene *bryX* encoding one of the five bryostatin modular PKSs does not correlate with bryostatin polyketide assembly and may be largely nonfunctional (Sudek *et al.*, 2007). With the putative bryostatin biosynthetic gene cluster *bry* now in hand, its heterologous expression in a suitable surrogate host provides an exciting opportunity to produce this important anticancer agent from an uncultured marine symbiont. Numerous technical challenges, however, stand in the way of this achievement involving practical metagenomic analyses and expression of large megasynthases in unrelated heterologous hosts. Overcoming these challenges will provide many more opportunities to develop drug leads for those working in the field of marine biotechnology.

FUTURE TARGETS OF OPPORTUNITY

Microbial Genomics and Biosynthesis

From the previous section, we see that coupling chemical isolation of bioactive molecules to the biosynthetic machinery encoded in DNA has clear benefits to our understanding of how microorganisms synthesize molecules and use them. What if it were possible to look at all of the biosynthetic machinery of an organism, from both primary and secondary metabolism, all at the same time? Once we know what to look for, this can, of course, be achieved through analysis of the complete genome sequence of an organism.

The sequencing and assembly of the complete genomic DNA of numerous organisms has affected the field of biology more than any other single advance in technology since the 1990s. The study of microbial biosynthesis is no exception, and sequencing of well beyond 1000 microbial genomes is complete or in progress, a number that has increased rapidly since the completion of the first bacterial sequence of *Haemophilus influenzae* in 1995 (Fig. 26-10) (Fleischmann

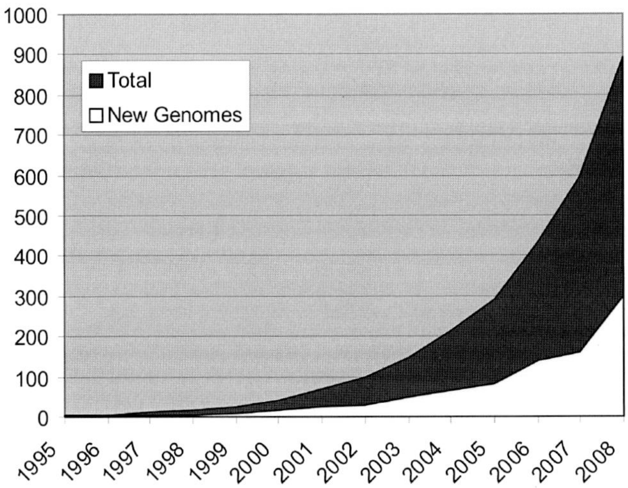

FIGURE 26-10. Numbers of microbial genomes sequenced by year. Numbers for 2007 and 2008 are conservative predictions based on genomes currently reported in sequencing pipelines. *(White)* Number of complete, reported microbial sequences released in each year. *(Black)* Total number of microbial genome sequences completed.

et al., 1995). Microbial genomics has grown in importance to biosynthesis research because, traditionally, identification and sequencing of a biosynthetic gene cluster has required construction of a cosmid library from genomic DNA, followed by screening and sequencing of one (or more likely, several) clones. This process can be labor intensive and the materials expensive. Automated DNA sequencing technology has steadily advanced since the 1990s to the point that in the near future it will be more cost-effective to sequence the entire genome of an organism of interest, especially when one takes into account the astounding degree of additional information provided. For this reason, future researchers must be prepared to interpret and utilize the exponentially growing volume of publicly available information encoded in DNA sequence.

The soil-dwelling actinomycetes of the genus *Streptomyces* are a major source of biologically active natural products, and one organism in particular, *Streptomyces coelicolor*, has served as the model organism for genomics-related biosynthesis studies. Analysis of its complete genome sequence, which was reported in 2002, revealed that many more biosynthetic clusters were identified than molecules isolated from this bacterium following decades of research (Bentley *et al.*, 2002). These unidentified biosynthetic pathways are often referred to as encoded by "cryptic" or "orphan" gene clusters. Orphan clusters are found in most genomes, including the closely related *Streptomyces avermitilis* (which shares few secondary metabolic pathways with *S. coelicolor*), the mycobacteria (which contain many cryptic PKS clusters likely to synthesize components of the mycobacterial lipid coat), and *Nocardia farcinica* and *Rhodococcus* RHA1 (each with numerous cryptic NRPS pathways). Marine microorganisms are no exception, and completed genomes with cryptic clusters include several species of cyanobacteria and, more recently, the obligate marine actinomycetes *Salinispora tropica* and *Salinispora arenicola* (see www.ncbi.nlm.nih.gov/genomes/lproks.cgi).

The 5.2-Mb genome of the marine bacterium *S. tropica* (previously discussed as producer of the anticancer agent salinosporamide A) shows a wide array of secondary metabolite biosynthetic gene clusters (Udwary *et al.*, 2007). In addition to salinosporamide, biosynthetic pathways were identified for the lymphocyte kinase inhibitor lymphostin, a large polyene macrolide named salinilactam, and the enediyne-derived sporolide polyketides (Fig. 26-11). Several cryptic pathways were also detected and are anticipated to produce numerous siderophore-like molecules, nonribosomal peptides, aromatic polyketides, and terpene-derived molecules of novel composition. Preliminary analysis of the related 5.8-Mb *S. arenicola* genome shows an even greater number of biosynthetic gene clusters, and the difference in size of the two organisms is largely accounted for by additional DNA devoted to secondary metabolism. Many clusters, from both organisms, appear to have been accumulated by horizontal gene transfer, which further indicates that the sea floor is a rich source of secondary metabolite-producing microorganisms.

The obvious advantage of having access to genomic information has been the ability to predict the products of pathways to guide researchers in the isolation and identification of druglike molecules. However, the prediction of pathways from DNA alone is often a detective game, requiring broad knowledge of bioinformatic analysis, chemical enzymology, primary metabolism, and the ecology of the organism. Analysis of genes dedicated to secondary metabolism suggests that they are often specially adapted versions of genes copied from primary metabolism, so a thorough understanding of the enzymology of the primary gene often yields insights into the function of the secondary copy. That said, generalized identification of PKS- or NRPS-based biosynthetic pathways is often trivial—modular PKS and NRPS genes are normally the largest open reading frames to be found in a genome, and catalytic domains are generally highly conserved and easily identified by BLAST or other homology searching. Unfortunately, automated methods are not routinely used to determine modular protein domain structure, so genes are typically found in completed genome sequences annotated only as "polyketide synthase" or "nonribosomal peptide synthetase." Predictive methodology currently exists for determining the degree of oxidation of polyketide backbones (Staunton and Weissman, 2001), the stereospecificity of ketoreductase product alcohol groups (Siskos *et al.*, 2005), and the nature of PKS (Reeves *et al.*, 2001) and NRPS (Challis *et al.*, 2000; Stachelhaus *et al.*, 1999) based building blocks (extender units). As more

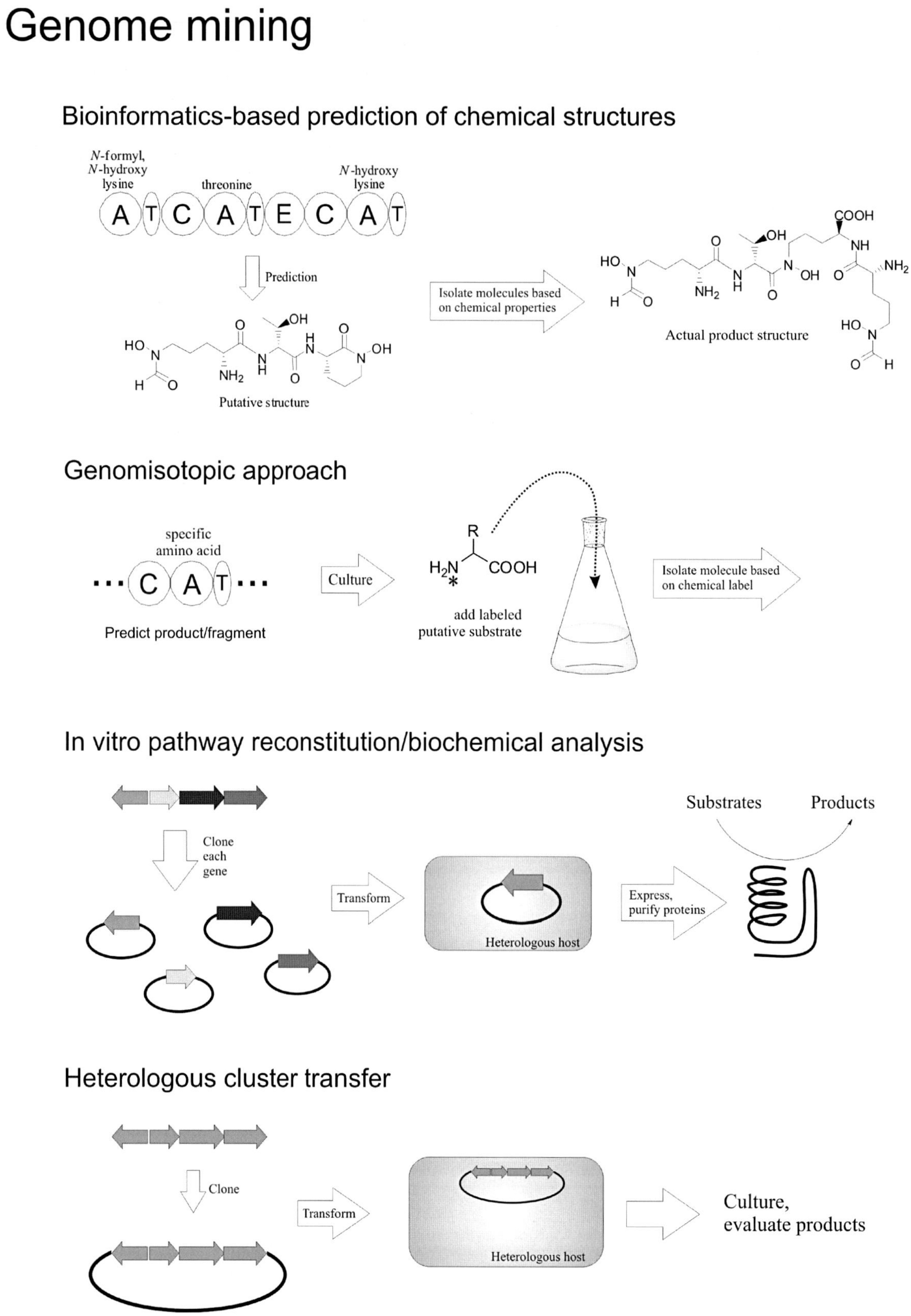

FIGURE 26-11. Genome mining approaches involving bioinformatics, stable isotopes, and expression studies.

sophisticated analysis routines are developed specifically for secondary metabolic systems, the accuracy of predictions of their product molecules will improve.

Mining Marine Microbial Genomes for Natural Product Drug Discovery

Exploitation of the information encoded in biosynthetic gene clusters for the purpose of isolation or identification of novel bioactive molecules or biosynthetic mechanisms is now frequently referred to as "genome mining," although similar research has been conducted in biosynthetic circles for many years without the benefit of genomic information. Genome mining studies can take several forms (Fig. 26-12), depending on resources and the circumstances of what is known about a given pathway (Correl and Challis, 2007).

Because of the types of studies outlined in the previous section, in many cases it is possible to predict chemical properties or ecological utility of a secondary metabolite by examining the sequence homology and domain structure of PKS, NRPS, and accessory genes. The first example that can be found in the literature is the structure prediction for the siderophore coelichelin, which was based upon analysis of a cryptic NRPS gene cluster in *S. coelicolor* (Challis and Ravel, 2000). Whereas the exact structural prediction proved only partially correct, prediction of general chemical properties of the molecule allowed for its isolation and identification and revealed a novel biosynthetic mechanism (Lautru *et al.*, 2005). More recently, initial evaluation of the *Salinispora tropica* genome was done almost entirely by predictive methods, which allowed for the isolation or confirmation of several previously un-isolated secondary metabolites (Udwary *et al.*, 2007). Because current methodology is imperfect (and will be for the foreseeable future), predictive biosynthesis is at its most effective when coupled with rigorous chemical analysis, which allows proof and refinement of these predictions.

An improvement to this methodology, recently reported as the "genomisotopic approach," couples this predictive methodology with the administration of labeled substrate compounds (i.e., ^{15}N-labeled amino acids) to the fermentation. If incorporated into a natural product, the isotopic label can be followed even at low concentrations by highly sensitive analytical chemistry methods such as high-field nuclear magnetic resonance and mass spectroscopy to aid in the isolation process. This procedure was recently reported for the isolation and characterization of a bioactive nonribosomal peptide from a cryptic gene cluster observed in the

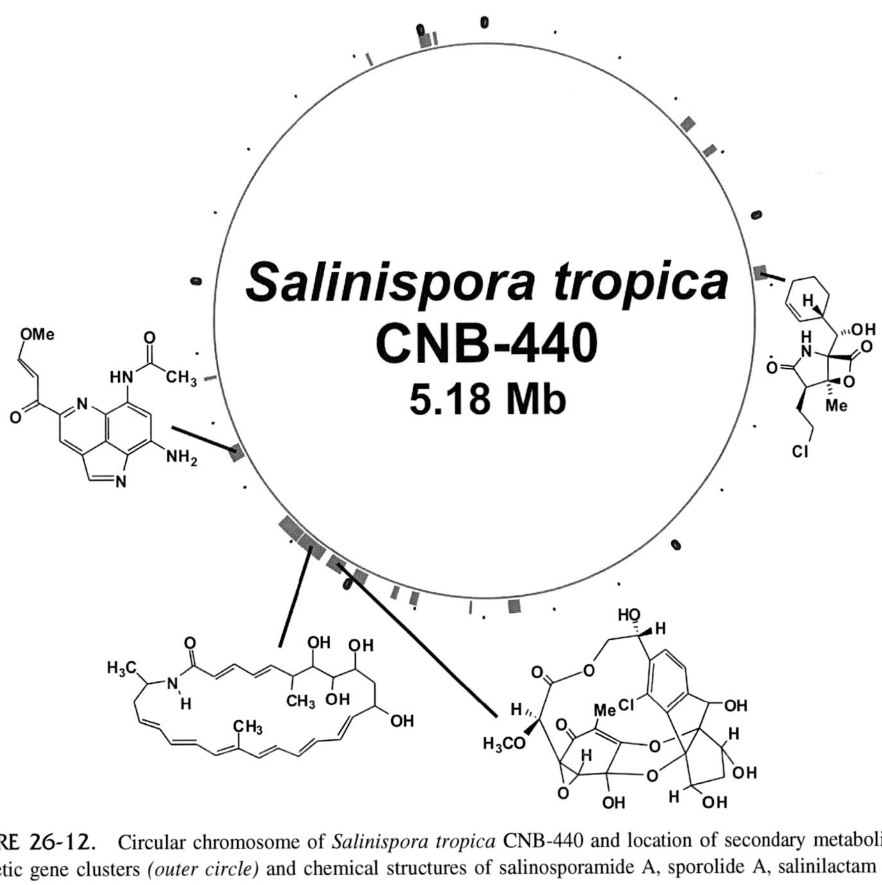

FIGURE 26-12. Circular chromosome of *Salinispora tropica* CNB-440 and location of secondary metabolite biosynthetic gene clusters *(outer circle)* and chemical structures of salinosporamide A, sporolide A, salinilactam A, and lymphostin *(clockwise from far right).*

Pseudomonas fluorescens Pf-5 genome (Gross *et al.*, 2007).

Perhaps still most common is the cloning and heterologous expression of individual biosynthetic genes for the purpose of *in vitro* biochemical characterization or reconstitution of pathways. Of course, biochemical characterization of enzymes is not new, but ready availability of entire gene clusters through genome sequencing, as well as information about an organism's primary metabolism, now often allows for reevaluation of the substrates and products of enzymes in a biosynthetic cluster based on homology and context. There are occasional technical difficulties with heterologous expression because of incompatible codon usage, or protein stability or toxicity, but many smaller bacterial proteins can be expressed without difficulty in *E. coli* using conventional molecular biology techniques, and in a few cases it has been possible to reconstruct entire biosynthetic pathways *in vitro.*

Large enzymes, such as PKSs and NRPSs, are notoriously difficult to work with *in vitro,* as the large repetitive DNA fragments are often difficult to amplify and clone, and traditional affinity chromatography is not conducive to intact purification of very large functional proteins. In such cases it has been helpful to express specific domains of larger modular proteins, allowing study of their specific enzymatic reactions outside of the context of the larger protein. *In vitro* analysis is particularly useful in studying the early stages of a biosynthetic pathway in which primary metabolites or simple molecules are taken up and committed to a pathway, simply because of commercial availability of putative enzyme substrates. Evaluation of enzymes involved in later stages of a pathway may require the chemical synthesis of much more specific, elaborated intermediates.

Similarly, another option is the heterologous expression of entire pathways. It is sometimes possible to transfer the DNA encoding a biosynthetic gene cluster (or a portion thereof) to a host organism, which may then confer the ability to produce the product of that cluster. Because many biosynthetic gene clusters show evidence of species-to-species horizontal transfer (and, thus, probably did not originate in the organism from which they were found), this technique is not conceptually difficult to imagine. Transfer of the entire gene cluster is often necessary for functional expression, and thus genomics or complete sequencing of the cluster may be necessary to determine the cluster's extremities. Heterologous cluster transfer typically works best when transferring DNA between closely related organisms, as metabolic flux and regulatory mechanisms between the originator and host must be similar for the biosynthesis to be efficient enough to be detectable, and often requires sensitive analytical instrumentation to perform accurate comparative metabolic profiling. For this reason, heterologous pathway expression has been most commonly achieved using streptomycete DNA in a well-defined laboratory strain with a manipulable genetics system, such as *S. lividans*, as a host. This technique has been used to heterologously produce the anticancer agent epothilone, the antibiotics erythromycin and tetracenomycin, as well as marine microbial agents such as enterocin and napradiomycin (Bode and Muller, 2005).

Marine Invertebrate Metagenomics and Symbiosis

For many years in studies of natural products from the marine environment, sponges were the kings of unique, active metabolites, and they continue to be important today. However, it is increasingly recognized that sponges and many other multicellular marine and terrestrial organisms often maintain a close symbiotic relationship with complex assemblages of microorganisms and that these microbes are often the actual producers of the druglike molecules isolated from these communities (e.g., see Chapter 22). Effort over the years to culture these associated microbes has been met with varying degrees of success, and large-scale DNA sequencing of the uncultured community is now playing a major role in this field.

The sequencing of "libraries" of DNA isolated from communities of organisms is referred to as metagenomics, and it is likely to have important ramifications on studies of secondary metabolism. Availability of metagenomic sequence similarly allows one to pursue many of the genome mining techniques described previously for identification of the products of clusters, or to study biosynthesis, without necessarily having access to a cultured organism, or even identifying it.

One of the earliest metagenomic studies of prolific marine secondary metabolite producing communities was conducted on two phylogenetically distinct sponges collected from different geographic regions (Hentschel *et al.*, 2002). A 16S ribosomal DNA library was sequenced from whole sponge extracts of *T. swinhoei* and *Aplysina aerophoba*, and it was found that each sponge bore a distinguishable microbial community, sharing very few species. One could, therefore, expect that unique microbial communities (sponge or otherwise) will tend to produce unique secondary metabolites. This work demonstrates the need to examine potentially diverse communities of natural product-producing microbes in a more comprehensive manner.

There are currently several marine metagenomics projects completed or under way. One of the first large community sequencing efforts was J. Craig Venter and colleagues' sequencing of the water column of the Sargasso Sea (Venter *et al.*, 2004). Intended primarily as proof that shotgun sequencing of large DNA samples could be assembled into analyzable fragments of DNA, the experiment was a success. Sequences from an estimated 1800 species—including an entire *Prochlorococcus* genome—were assembled, as well as several novel plasmids of varying size. It was further estimated that more than 1.2 million unknown genes were sequenced, and analysis of 782 novel rhodopsin genes gave insight into photobiology in the Sargasso Sea water column. A massive expansion of this project is now under way, and some initial results were recently reported from the Global Ocean Sequencing Expedition from the J. Craig Venter Institute, which is by far the largest ever metagenomics effort (see http://collections.plos.org/plosbiology/gos-2007.php). Their first study focused on analysis of the more than 6 million novel genes sequenced, with comparison to the existing 3 million gene sequences found in current databases. Most intriguing, several unique protein families were identified, demonstrating that much of the world's diversity is yet to be discovered. A more specific comparative analysis of protein kinase families using these data was also simultaneously published, demonstrating that such data can be of great utility to even well-studied enzymes and proteins (Kannan *et al.*, 2007).

Such large-scale metagenomic sequencing projects are currently rare, requiring resources beyond the grasp of most researchers. When looking for specific genes or gene types from a community, it is typically much more efficient to first screen a large community-derived library to select for these specific genes, a process called "enrichment." Examination of enriched metagenomic libraries is now used with increasing frequency in secondary metabolism studies. Analysis of PKS ketosynthase sequences found in soil samples has shown just how rich polyketide diversity is (Courtois *et al.*, 2003), and it is only a matter of time before such analyses are conducted for the water column or sea bed.

There are clear benefits of a metagenomics approach to the sequencing of secondary metabolic gene clusters. PKS and NRPS sequences can be found in almost every searchable metagenomic sequence database, attesting to how widespread these systems are and how much more needs to be investigated. It is theoretically possible to sequence and identify biosynthetic clusters from unculturable organisms, potentially allowing for heterologous expression and fermentative production, rather than large-scale isolation from the wild or difficult chemical synthesis. With the world's oceans and its denizens already under enormous pressure from destructive human influence, perhaps it makes sense to collect only a few grams of sponge tissue to produce a metagenomic library and examine it, rather than to collect the kilograms of (possibly rare) sponge tissue often required for a thorough chemical isolation study.

A cautionary tale of metagenomics comes in a study of the sponge *Discodermia dissoluta,* which harbors the promising antitumor agent discodermolide. The structure of discodermolide strongly suggests that it is produced by an associated bacterium via a type I modular PKS. Despite the fact that discodermolide is the most plentiful polyketide isolated chemically from *D. dissoluta*, genetic screening of a >4 Gb DNA library with ketosynthase probes failed to uncover PKS clones specifically associated with discodermolide production from the several hundred PKS-containing clones sequenced from the library (Schirmer *et al.*, 2005). This Herculean effort to locate the discodermolide biosynthetic machinery in order to produce the compound heterologously rather than by total synthesis to address long-term supply issues of this drug candidate was unfortunately unsuccessful and represents the challenges in this rapidly evolving field of science.

Metagenomics currently has several drawbacks and technical hurdles that remain to be crossed to provide more benefits to natural products research. First is the cost associated with this approach. Sequencing of metagenomes can be much more expensive than whole genome sequencing, depending on the size of the library and desired coverage. Rich sources, while potentially the most interesting to study, will also require the most sequencing. Targeted screening and selection before sequencing brings the cost down somewhat, but these are also time-consuming and costly processes. Again, the costs of DNA sequencing are expected to continue to drop as technology develops, but such large-scale sequencing efforts are currently beyond the reach of most academic researchers.

Construction of the metagenomic library can be a perilous process as well. Library construction requires harvesting of cells, uniform lysis of all cells, purification of genomic DNA of widely varying sizes, uniform randomized digestion of the DNA into fragments of a specified size, and, finally, cloning of fragments into a cosmid or fosmid, which must be taken up and maintained by a host organism. However, bias may be accrued at any of these stages because of unusual biology in the large numbers of unstudied microbes present in a community sample. The end result may be a library that is nonrepresentative of the sample, and one should take care before drawing too many conclusions from specific population numbers. This is of particular concern to those searching for biosynthetic gene clusters, as under the right conditions dominating amounts of secondary metabolites may be produced from a small, inaccessible population, as seems to be the case with discodermolide.

Although metagenomic sequencing does potentially provide access to DNA of unculturable organisms, the inability to do further work with an isolated organism can be a limitation. In natural products research, this can be a significant problem, as one would not have access to cultures that produce the product of the gene clusters. Thus, it may be difficult to directly tie observed chemistry to putative biosynthetic gene clusters without functional expression in a heterologous host, a process that has its own technical limitations. Furthermore, if only a portion of a gene cluster of interest is isolated, as in the case of the onnamide gene set from the sponge *T. swinhoei*, it may be difficult to locate the remainder among a population of thousands of unique species.

Nonetheless, with these limitations come opportunities for the discovery and development of new methodology as in the case of isolating rare clones from complex DNA libraries by PCR analysis of liquid gel pools (Hrvatin and Piel, 2007). The new marriage of genomics and biosynthesis with marine natural product discovery is beginning to mature into a formidable multidisciplinary approach to expand the limits of marine biotechnology into new arenas that impact human health.

References

Allen, E.E., Bartlett, D.H., 2002. Structure and regulation of the omega-3 polyunsaturated fatty acid synthase genes from the deep-sea bacterium *Photobacterium profundum* strain SS9. Microbiology 148,1903–1913.

Beer, L.L., Moore, B.S., 2007. Biosynthetic convergence of salinosporamides A and B in the marine actinomycete *Salinispora tropica*. Org. Lett. 9, 845–848.

Bentley, S.D., Chater, K.F., Cerdeño-Tárraga, A.-M., Challis, G.L., Thomson, N.R., James, K.D., Harris, D.E., Quail, M.A., Kieser, H., Harper, D., Bateman, A., Brown, S., Chandra, G., Chen, C.W., Collins, M., Cronin, A., Fraser, A., Goble, A., Hidalgo, J., Hornsby, T., Howarth, S., Huang, C.-H., Kieser, T., Larke, L., Murphy, L., Oliver, K., O'Niel, S., Rabbinowitsch, E., Rajandream, M.-A., Rutherford, K., Rutter, S., Seeger, K., Saunders, D., Sharp, S., Squares, R., Squares, S., Taylor, K., Warren, T., Wietzorrek, A., Woodward, J., Barrell, B.G., Parkhill, J., Hopwood, D.A., 2002. Complete genome sequence of the model actinomycete *Streptomyces coelicolor* A3(2). Nature 417, 141–147.

Bode, H.B., Muller, R., 2005. The impact of bacterial genomics on natural product research. Angew. Chem. Int. Ed. 44, 6828–6846.

Burja, A.M., Banaigs, B., Abou-Mansour, E., Burgess, J.G., Wright, P.C., 2001. Marine cyanobacteria: A prolific source of natural products. Tetrahedron 57, 9347–9377.

Challis, G.L., Ravel, J., 2000. Coelichelin, a new peptide siderophore encoded by the *Streptomyces coelicolor* genome: Structure prediction from the sequence of its non-ribosomal peptide synthetase. FEMS Microbiol. Lett. 187, 111–114.

Challis, G.L., Ravel, J., Townsend, C.A., 2000. Predictive, structure-based model of amino acid recognition by nonribosomal peptide synthetase adenylation domains. Chem. Biol. 7, 211–224.

Chang, Z., Flatt, P., Gerwick, W.H., Nguyen, V.-A., Willis, C.L., Sherman, D.H., 2002. The barbamide biosynthetic gene cluster: A novel marine cyanobacterial system of mixed polyketide synthase (PKS)–non-ribosomal peptide synthetase (NRPS) origin involving an unusual trichloroleucyl starter unit. Gene 296, 235–247.

Chang, Z.X., Sitachitta, N., Rossi, J.V., Roberts, M.A., Flatt, P.M., Jia, J. Y., Sherman, D.H., Gerwick, W.H., 2004. Biosynthetic pathway and gene cluster analysis of curacin A, an antitubulin natural product from the tropical marine cyanobacterium *Lyngbya majuscula*. J. Nat. Prod. 67, 1356–1367.

Chauhan, D., Catley, L., Li, G.L., Podar, K., Hideshima, T., Velankar, M., Mitsiades, C., Mitsiades, N., Yasui, H., Letai, A., Ovaa, H., Berkers, C., Nicholson, B., Chao, T.H., Neuteboom, S.T.C., Richardson, P., Palladino, M.A., Anderson, K.C., 2005. A novel orally active proteasome inhibitor induces apoptosis in multiple myeloma cells with mechanisms distinct from Bortezomib. Cancer Cell 8, 407–419.

Christiansen, G., Fastner, J., Erhard, M., Börner, T., Dittmann, E., 2003. Microcystin biosynthesis in *Planktothrix*: Genes, evolution, and manipulation. J. Bacteriol. 185, 564–572.

Correl, C., Challis, G.L., 2007. Heavy tools for genome mining. Chem. Biol. 14, 7–9.

Courtois, S., Cappellano, C.M., Ball, M., Francou, F.X., Normand, P., Helynck, G., Martinez, A., Kolvek, S.J., Hopke, J., Osburne, M.S., August, P.R., Nalin, R., Guerineau, M., Jeannin, P., Simonet, P., Pernodet, J.L., 2003. Recombinant environmental libraries provide access to microbial diversity for drug discovery from natural products. Appl. Environ. Microbiol. 69, 49–55.

Davidson, S.K., Allen, S.W., Lim, G.E., Anderson, C.M., Haygood, M.G., 2001. Evidence for the biosynthesis of bryostatins by the bacterial symbiont "*Candidatus* Endobugula sertula" of the bryozoan *Bugula neritina*. Appl. Environ. Microbiol. 67, 4531–4537.

Donia, M.S., Hathaway, B.J., Sudek, S., Haygood, M.G., Rosovitz, M.J., Ravel, J., Schmidt, E.W., 2006. Natural combinatorial peptide libraries in cyanobacterial symbionts of marine ascidians. Nat. Chem. Biol. 2, 729–735.

Edwards, D.J., Gerwick, W.H., 2004. Lyngbyatoxin biosynthesis: Sequence of biosynthetic gene cluster and identification of a novel aromatic prenyltransferase. J. Am. Chem. Soc. 126, 11432–11433.

Edwards, D.J., Marquez, B.L., Nogle, L.M., McPhail, K., Goeger, D.E., Roberts, M.A., Gerwick, W.H., 2004. Structure and biosynthesis of the jamaicamides, new mixed polyketide-peptide neurotoxins from the marine cyanobacterium *Lyngbya majuscula*. Chem. Biol. 11, 817–833.

Feling, R.H., Buchanan, G.O., Mincer, T.J., Kauffman, C.A., Jensen, P.R., Fenical, W., 2003. Salinosporamide A, an antitumor proteasome inhibitor from a novel microbial source, a marine bacterium of the new genus *Salinospora*. Angew. Chem. Int. Ed. 115, 369–371.

Fenical, W., Jensen, P.R., 2006. Developing a new resource for drug discovery: Marine actinomycete bacteria. Nat. Chem. Biol. 2, 666–673.

Fischbach, M.A., Walsh, C.T., 2006. Assembly-line enzymology for polyketide and nonribosomal peptide antibiotics: Logic, machinery, and mechanisms. Chem. Rev. 106, 3468–3496.

Flatt, P., Gautschi, J., Thacker, R., Musafija-Girt, M., Crews, P., Gerwick, W.H., 2005. Identification of the cellular site of polychlorinated peptide biosynthesis in the marine sponge *Dysidea* (Lamellodysidea) *herbacea* and symbiotic cyanobacterium *Oscillatoria spongeliae* by CARD-FISH analysis. Mar. Biol. 147, 761–774.

Flatt, P., O'Connell, S.J., McPhail, K.L., Zeller, G., Willis, C.L., Sherman, D.H., Gerwick, W.H., 2006. Characterization of the initial enzymatic steps of barbamide biosynthesis. J. Nat. Prod. 69, 938–944.

Fleischmann, R.D., Adams, M.D., White, O., Clayton, R.A., Kirkness, E.F., Kerlavage, A.R., Bult, C.J., Tomb, J.F., Dougherty, B.A., Merrick, J.M., McKenney, K., Sutton, G., Fitzhugh, W., Fields, C., Gocayne, J.D., Scott, J., Shirley, R., Liu, L.I., Glodek, A., Kelley, J.M., Weidman, J.F., Phillips, C.A., Spriggs, T., Hedblom, E., Cotton, M.D., Utterback, T.R., Hanna, M.C., Nguyen, D.T., Saudek, D.M., Brandon, R.C., Fine, L.D., Fritchman, J.L., Fuhrmann, J.L., Geoghagen, N.S.M., Gnehm, C.L., McDonald, L.A., Small, K.V., Fraser, C.M., Smith, H.O., Venter, J.C., 1995. Whole-genome random sequencing and assembly of *Haemophilus influenzae* RD. Science 269, 496–512.

Galonic, D.P., Vaillancourt, F.H., Walsh, C.T., 2006. Halogenation of unactivated carbon centers in natural product biosynthesis: Trichlorination of leucine during barbamide biosynthesis. J. Am. Chem. Soc. 128, 3900–3901.

Garcia-Bernardo, J., Xiang, L., Hong, H., Moore, B.S., Leadlay, P.F., 2004. Engineered biosynthesis of phenyl-substituted polyketides. ChemBioChem 5, 1129–1131.

Gerwick, W.H., Tan L.T., Sitachitta, N., 2001. Nitrogen-containing metabolites from marine cyanobacteria. In G. Cordell (ed.), The Alkaloids, vol. 57, pp. 75–184. San Diego, Academic Press.

Gross, H., Stockwell, V.O., Henkels, M.D., Nowak-Thompson, B., Loper, J.E., Gerwick, W.H., 2007. The genomisotopic approach: A systematic method to isolate products of orphan biosynthetic gene clusters. Chem. Biol. 14, 53–63.

Harrigan, G.G., Goetz, G., Luesch, H., Yang, S., Likos, J., 2001. Dysideaprolines A-F and barbaleucamides A-B, novel polychlorinated compounds from a *Dysidea* species. J. Nat. Prod. 64, 1133–1138.

Hentschel, U., Hopke, J., Horn, M., Friedrich, A.B., Wagner, M., Hacker, J., Moore, B.S., 2002. Molecular evidence for a uniform microbial community in sponges from different oceans. Appl. Environ. Microbiol. 68, 4431–4440.

Hertweck, C., Xiang, L., Kalaitzis, J.A., Cheng, Q., Palzer, M., Moore, B.S., 2004. Context-dependent behavior of the enterocin iterative polyketide synthase: A new model for ketoreduction. Chem. Biol. 11, 461–468.

Hildebrand, M., Waggoner, L.E., Liu, H.B., Sudek, S., Allen, S., Anderson, C., Sherman, D.H., Haygood, M., 2004. *bryA*: An unusual modular polyketide synthase gene from the uncultivated bacterial symbiont of the marine bryozoan *Bugula neritina*. Chem. Biol. 11, 1543–1552.

Hrvatin, S., Piel, J., 2007. Rapid isolation of rare clones from highly complex DNA libraries by PCR analysis of liquid gel pools. J. Microbiol. Meth. 68, 434–436.

Ikeda, H., Nonomiya, T., Usami, M., Ohta, T., Omura, S., 1999. Organization of the biosynthetic gene cluster for the polyketide anthelmintic macrolide avermectin in *Streptomyces avermitilis*. Proc. Natl. Acad. Sci. USA 96, 9509–9514.

Kaebernick, M., Dittmann, E., Börner, T., Neilan, B.A., 2002. Multiple alternate transcripts direct the biosynthesis of microcystin, a cyanobacterial nonribosomal peptide. Appl. Environ. Microbiol. 68, 449–455.

Kalaitzis, J.A., Izumikawa, M., Xiang, L., Hertweck, C., Moore, B.S., 2003. Mutasynthesis of enterocin and wailupemycin analogues. J. Am. Chem. Soc. 125, 9290–9291.

Kannan, N., Taylor, S.S., Zhai, Y., Venter, J.C., Manning, G., 2007. Structural and functional diversity of the microbial kinome. PLoS Biol. 5, 467–478.

Kurmayer, R., Dittmann, E., Fastner, J., Chorus, I., 2002. Diversity of microcystin genes within a population of the toxic cyanobacterium *Microcystis* spp. in Lake Wannsee (Berlin, Germany). Microb. Ecol. 43, 107–118.

Lam, K.S., 2006. Discovery of novel metabolites from marine acrinomycetes. Curr. Opin. Microbiol. 9, 245–251.

Lautru, S., Deeth, R.J., Bailey, L.M., Challis, G.L., 2005. Discovery of a new peptide natural product by Streptomyces coelicolor genome mining. Nat. Chem. Biol. 1, 244–245.

Li, A.Y., Piel, J., 2002. A gene cluster from a marine *Streptomyces* encoding the biosynthesis of the aromatic spiroketal polyketide griseorhodin A. Chem. Biol. 9, 1017–1026.

Lim, G.E., Haygood, M.G., 2004. "*Candidatus endobugula glebosa*," a specific bacterial symbiont of the marine bryozoan *Bugula simplex*. Appl. Environ. Microbiol. 70, 4921–4929.

Long, P.F., Dunlap, W.C., Battershill, C.N., Jaspars, M., 2005. Shotgun cloning and heterologous expression of the patellamide gene cluster as a strategy to achieving sustained metabolite production. ChemBioChem 6, 1760–1765.

Lopanik, N., Lindquist, N., Targett, N., 2004. Potent cytotoxins produced by a microbial symbiont protect host larvae from predation. Oecologia 139, 131–139.

Luesch, H., Harrigan, G.G., Goetz, G., Horgen, F.D., 2002. The cyanobacterial origin of potent anticancer agents originally isolated from sea hares. Curr. Med. Chem. 9, 1791–1806.

Metz, J.G., Roessler, P., Facciotti, D., Levering, C., Dittrich, F., Lassner, M., Valentine, R., Lardizabal, K., Domergue, F., Yamada, A., Yazawa, K., Knauf, V., Browse, J., 2001. Production of polyunsaturated fatty acids by polyketide synthases in both prokaryotes and eukaryotes. Science 293, 290–293.

Mikalsen, B., Boison, G., Skulberg, O.M., Fastner, J., Davies, W., Gabrielsen, T.M., Rudi, K., Jakobsen, K.S., 2003. Natural variation in the microcystin synthetase operon *mcyABC* and impact on microcystin production in *Microcystis* strains. J. Bacteriol. 185, 2774–2784.

Moffitt, M.C., Neilan, B.A., 2004. Characterization of the nodularin synthetase gene cluster and proposed theory of the evolution of cyanobacterial hepatotoxins. Appl. Environ. Microbiol. 70, 6353–6362.

Moore, B.S., 2005. Biosynthesis of marine natural products: Microorganisms (part A). Nat. Prod. Rep. 22, 580–593.

Moore, B.S., 2006. Biosynthesis of marine natural products: Macroorganisms (part B). Nat. Prod. Rep. 22, 615–629.

Moore, R.E., Corbett, T.H., Patterson, G.M.L., Valeriote, F.A., 1996. The search for new antitumor drugs from blue-green algae. Curr. Pharm. Des. 2, 317–330.

Nishizawa, T., Asayama, M., Fujii, K., Harada, K., Shirai, M., 1999. Genetic analysis of the peptide synthetase genes for a cyclic heptapeptide microcystin in *Microcystis* spp. J. Biochem. 126, 520–529.

Nishizawa, T., Ueda, A., Asayama, M., Fujii, K., Harada, K., Ochi, K., Shirai, M., 2000. Polyketide synthase gene coupled to the peptide synthetase module involved in the biosynthesis of the cyclic heptapeptide microcystin. J. Biochem. 127, 779–789.

Orikasa, Y., Nishida, T., Yamada, A., Yu, R., Watanabe, K., Hase, A., Morita, N., Okuyama, H., 2006. Recombinant production of docosahexaenoic acid in a polyketide biosynthesis mode in *Escherichia coli*. Biotechnol. Lett. 28, 1841–1847.

Orikasa, Y., Yamada, A., Yu, R., Ito, Y., Nishida, T., Yumoto, I., Watanabe, K., Okuyama, H., 2004. Characterization of the eicosapentaenoic acid biosynthesis gene cluster from *Shewanella* sp strain SCRC-2738. Cellu. Molec. Biol. 50, 625–630.

Piel, J., 2002. A polyketide synthase-peptide synthetase gene cluster from an uncultured bacterial symbiont of *Paederus* beetles. Proc. Natl. Acad. Sci. USA 99, 14002–14007.

Piel, J., Butzke, D., Fusetani, N., Hui, D.Q., Platzer, M., Wen, G.P., Matsunaga, S., 2005. Exploring the chemistry of uncultivated bacterial symbionts: Antitumor polyketides of the pederin family. J. Nat. Prod. 68, 472–479.

Piel, J., Hertweck, C., Shipley, P.R., Hunt, D.M., Newman, M.S., Moore, B.S., 2000a. Cloning, sequencing and analysis of the enterocin biosynthesis gene cluster from the marine isolate "Steptomyces maritimus": Evidence for the derailment of an aromatic polyketide synthase. Chem. Biol. 7, 943–955.

Piel, J., Hoang, K., Moore, B.S., 2000b. Metabolic diversity encoded by the enterocin biosynthesis gene cluster. J. Am. Chem. Soc. 122, 5415–5416.

Piel, J., Hui, D.Q., Wen, G.P., Butzke, D., Platzer, M., Fusetani, N., Matsunaga, S., 2004. Antitumor polyketide biosynthesis by an uncultivated bacterial symbiont of the marine sponge *Theonella swinhoei*. Proc. Natl. Acad. Sci. USA 101, 16222–16227.

Poncet, J., 1999. The dolastatins, a family of promising antineoplastic agents. Curr. Pharm. Design 5, 139–162.

Pospiech, A., Cluzel, B., Bietenhader, J., Schupp, T., 1995. A new *Myxococcus xanthus* gene cluster for the biosynthesis of the antibiotic saframycin MX1 encoding a peptide synthetase. Microbiology 141, 1793–1803.

Reeves, C.D., Murli, S., Ashley, G.W., Piagentini, M., Hutchinson, C.R., McDaniel, R., 2001. Alteration of the substrate specificity of a modular polyketide synthase acyltransferase domain through site-specific mutations. Biochemistry 40, 15464–15470.

Rinehart, K.L., 2000. Antitumor compounds from tunicates. Med. Res. Rev. 20, 1–27.

Rinehart, K.L., Holt, T.G., Fregeau, N.L., Stroh, J.G., Keifer, P.A., Sun, F., Li, L.H., Martin, D.G., 1990. Ecteinascidins 729, 743, 745, 759A,

759B, and 770: Potent antitumor agents from the Caribbean tunicate *Ecteinascidia turbinata*. J. Org. Chem. 55, 4512–4515.

Rouhiainen, L., Vakkilainen, T., Siemer, B.L., Buikema, W., Haselkorn, R., Sivonen, K., 2004. Genes coding for hepatotoxic heptapeptides (microcystins) in the cyanobacterium *Anabaena* strain 90. Appl. Environ. Microbiol. 70, 686–692.

Salomon, C.E., Faulkner, D.J., 2002. Localization studies of bioactive cyclic peptides in the ascidian *Lissoclinum patella*. J. Nat. Prod. 65, 689–692.

Schirmer, A., Gadkari, R., Reeves, C.D., Ibrahim, F., DeLong, E.F., Hutchinson, C.R., 2005. Metagenomic analysis reveals diverse polyketide synthase gene clusters in microorganisms associated with the marine sponge *Discodermia dissoluta*. Appl. Environ. Microbiol. 71, 4840–4849.

Schmidt, E.W., Nelson, J.T., Rasko, D.A., Sudek, S., Eisen, J.A., Haygood, M.G., Ravel, J., 2005. Patellamide A and C biosynthesis by a microcin-like pathway in *Prochloron didemni*, the cyanobacterial symbiont of *Lissoclinum patella*. Proc. Natl. Acad. Sci. USA 102, 7315–7320.

Schmidt, E.W., Sudek, S., Haygood, M.G., 2004. Genetic evidence supports secondary metabolic diversity in *Prochloron* spp., the cyanobacterial symbiont of a tropical ascidian. J. Nat. Prod. 67, 1341–1345.

Schwecke, T., Aparicio, J.F., Molnar, I., König, A., Khaw, L.E., Haydock, S.F., Oliynyk, M., Caffrey, P., Cortes, J., Lester, J.B., Böhm, G.A., Staunton, J., Leadlay, P.F., 1995. The biosynthetic gene cluster for the polyketide immunosuppressant rapamycin. Proc. Nat. Acad. Sci. USA 92, 7839–7843.

Siskos, A.P., Baerga-Ortiz, A., Bali, S., Stein, V., Mamdani, H., Spiteller, D., Popovic, B., Spencer, J.B., Staunton, J., Weissman, K.J., Leadlay, P.F., 2005. Molecular basis of Celmer's rules: Stereochemistry of catalysis by isolated ketoreductase domains from modular polyketide synthases. Chem. Biol. 12, 1145–1153.

Sitachitta, N., Márquez, B.L., Williamson, R.T., Rossi, J., Roberts, M.A., Gerwick, W.H., Nguyen, V.-A., Willis, C.L., 2000. Biosynthetic pathway and origin of the chlorinated methyl group in barbamide and dechlorobarbamide, metabolites from the marine cyanobacterium *Lyngbya majuscula*. Tetrahedron 56, 9103–9113.

Spencer, J.R., Sendzik, M., Oeh, J., Sabbatini, P., Dalrymple, S.A., Magill, C., Kim, H.M., Zhang, P.L., Squires, N., Moss, K.G., Sukbentherng, J., Graupe, D., Eksterowicz, J., Young, P.R., Myers, A.G., Green, M.J., 2006. Evaluation of antitumor properties of novel saframycin analogs *in vitro* and *in vivo*. Bioorg. Med. Chem. 16, 4884–4888.

Stachelhaus, T., Mootz, H.D., Marahiel, M.A., 1999. The specificity-conferring code of adenylation domains in nonribosomal peptide synthetases. Chem. Biol. 6, 493–505.

Staunton, J., Weissman, K.J., 2001. Polyketide biosynthesis: A millennium review. Nat. Prod. Rep. 18, 380–416.

Sudek, S., Haygood, M.G., Youssef, D.T.A., Schmidt, E.W., 2006. Structure of trichamide, a cyclic peptide from the bloom-forming cyanobacterium *Trichodesmium erythraeum*, predicted from the genome sequence. Appl. Environ. Microbiol. 72, 4382–4387.

Sudek, S., Lopanik, N.B., Waggoner, L.E., Hildebrand, M., Anderson, C., Liu, H.B., Patel, A., Sherman, D.H., Haygood, M.G., 2007. Identification of the putative bryostatin polyketide synthase gene cluster from "*Candidatus* endobugula sertula," the uncultivated microbial symbiont of the marine bryozoan *Bugula neritina*. J. Nat. Prod. 70, 67–74.

Tillet, D., Dittmann, E., Erhard, M., von Döhren, H., Börner, T., Neilan, B.A., 2000. Structural organization of microcystin biosynthesis in *Microcystis aeruginosa* PCC7806: An integrated peptide-polyketide synthetase system. Chem. Biol. 7, 753–764.

Udwary, D.W., Zeigler, L., Asolkar, R.N., Singan, V., Lapidus, A., Fenical, W., Jensen, P.R., Moore, B.S., 2007. Genome sequencing reveals complex secondary metabolome in the marine actinomycete *Salinispora tropica*. Proc. Nat. Acad. Sci. USA, PNAS 104, 10376–10381.

Venter, J.C., Remington, K., Heidelberg, J.F., Halpern, A.L., Rusch, D., Eisen, J.A., Wu, D., Paulsen, I., Nelson, K.E., Nelson, W., Fouts, D.E., Levy, S., Knap, A.H., Lomas, M.W., Nealson, K., White, O., Peterson, J., Hoffman, J., Parsons, R., Baden-Tillson, H., Pfannkoch, C., Rogers, Y.H., Smith, H.O., 2004. Environmental genome shotgun sequencing of the Sargasso Sea. Science 304, 66–74.

Williams, P.G., Buchanan, G.O., Feling, R.H., Kauffman, C.A., Jensen, P. R., Fenical, W., 2005. New cytotoxic salinosporamides from the marine actinomycete *Salinispora tropica*. J. Org. Chem. 70, 6196–6203.

Winter, J.M., Moffitt, M.C., Zazopoulos, E., McAlpine, J.B., Dorrestein, P.C., Moore, B.S., 2007. Molecular basis for chloronium-mediated meroterpene cyclization: Cloning, sequencing, and heterologous expression of the napyradiomycin biosynthetic gene cluster. J. Biol. Chem. 282, 16362–16368.

Wright, J.L.C., Hu, T., McLachlan, J.L., Needham, J., Walter, J.A., 1996. Biosynthesis of DTX-4: Confirmation of a polyketide pathway, proof of a Baeyer-Villiger oxidation step, and evidence for an unusual carbon deletion process. J. Am. Chem. Soc. 118, 8757–8758.

Xiang, L., Kalaitzis, J.A., Moore, B.S., 2004. EncM, a versatile enterocin biosynthetic enzyme involved in Favorskii oxidative rearrangement, aldol condensation, and heterocycle-forming reactions. Proc. Natl. Acad. Sci. USA 101, 15609–15614.

Xiang, L., Kalaitzis, J.A., Nilsen, G., Chen, L., Moore, B.S., 2002. Mutational analysis of the enterocin Favorskii biosynthetic rearrangement. Org. Lett. 4, 957–960.

Xiang, L., Moore, B.S., 2003. Characterization of benzoyl coenzyme A biosynthesis genes in the enterocin-producing bacterium "*Streptomyces maritimus*." J. Bacteriol. 185, 399–404.

STUDY QUESTIONS

1. Bacterial biosynthetic gene clusters can be manipulated to generate "unnatural" natural products. Describe some genetic approaches that have been employed in an effort to generate novel natural product-like compounds.
2. Expression of biosynthesis gene clusters in heterologous hosts is seen as another avenue to the production of natural products. Discuss (a) the advantages of this *in vivo* approach and (b) the technical difficulties that may be encountered.
3. A greater understanding toward the biosynthesis of a natural product can be gained from the nature of biosynthetic genes present in its associated gene cluster. What other information besides biosynthesis can be deduced?
4. Several types of macroorganisms were mentioned as plentiful sources of microbial biosynthetic gene clusters, and thus they could be good targets of metagenomic sequencing and analysis. What other marine organisms or systems not mentioned might be good targets of study? Why?
5. As discussed, the onnamide and pederin biosynthetic gene clusters are related, despite the fact that one is found in a sponge and the other in a beetle. How might this be possible? Why might similar chemical compounds be advantageous to both a beetle and a sponge?

SECTION II

B. Aquatic Animal Models of Human Health

C H A P T E R

27

Aquatic Animal Models of Human Health

PATRICK J. WALSH AND CHRISTER HOGSTRAND

INTRODUCTION

Much of the rapidly expanding interest in the benefits of using aquatic animal models for biomedical research in the context of human health stems from the parent disciplines of comparative biochemistry and physiology and comparative pathology/toxicology. These disciplines seek to unify our understanding of common physiological (osmoregulation, respiration, etc.) and pathological (infection, carcinogenesis, etc.) processes across many taxa (see, e.g., Somero, 2000). In this unification, it is not surprising that knowledge arises from a so-called model species that is relevant to the understanding of basic human physiology and pathology. In this quest for unification, however, it has often been the discovery of *natural exceptions* to various rules that lead to a species becoming a key subject for study. These valuable exceptions present themselves in at least two ways. First, selected animal species are capable of withstanding extremes of environmental circumstances, be it a natural variable, such as temperature, salinity, or oxygen, or an unnatural/anthropogenic substance. Study of these "champions" or "supermodels" of survival can often give us great insight into why humans might be especially susceptible to a disease and how we might treat or prevent the disease. The flip side of this coin is also informative: often within the animal kingdom, aquatic species exist that are *more* sensitive than mammals to toxicants, and their study can give great insights into mechanisms of pathology.

In a second sense, however, animals can be extremely useful in biomedical research simply because of the ease of studying a particular process or phenomenon purely within an experimental context. Comparative biochemistry and physiology in particular gave birth to an experimental approach that has come to be known as the August Krogh principle: "For many problems, there is an animal on which it can be most conveniently studied." Although this quote is attributed to the Danish physiologist August Krogh, it is clear that the Krogh principle was applied even earlier by physiologists preceding Krogh (Jørgenson, 2001). Among his many discoveries, indeed it was Krogh, using amphibians, who discovered the basic mechanisms of how blood capillaries function (Krogh, 1922), for which he was awarded a Nobel Prize. Regardless of the principle's exact origin, we believe that the approach applies today in the choice of aquatic animal models for the efficient and insightful study of human disease.

GENERAL FEATURES OF AQUATIC ANIMAL MODELS FOR HUMAN HEALTH

There are numerous instances where aquatic animals have contributed to our basic understanding of cell biology, physiology, biochemistry, and molecular biology, as well as directly contributing to better insights into human disease states. An historical review can be found in the 1999 NAS Report "Monsoons to Microbes" (Fenical *et al.*, 1999), with the chapters that follow offering detailed insights into a selected few of these models. Some of the concepts discussed in this chapter were discussed in Grosell and Walsh (2006), and these also include focus on sentinel species (species that warn of environmental degradation because of their susceptibility to environmental change). Before turning to specific examples, however, we look at attributes that apply generally to aquatic organisms, or at least to many aquatic species, making them suitable subjects for

TABLE 27-1. Some general attributes of many aquatic species that make them attractive experimental models.

Trait	Advantage
High species diversity	Limitless choice of native environments and susceptibility or resistance to toxicants/ environmental variables
High fecundity/external fertilization	Ultimate application of (large scale) breeding in captivity, potentially lowering costs compared to mammalian models and reducing "supply issues" and environmental impacts of scientific collection
Rapid development/ short generation times	Shorter duration of experiments involving development or genetics
Large egg size/external fertilization	Facilitates genetic manipulation and production of transgenics
Transparent eggs/embryos	Development readily observed and manipulated
Nonkeratinized skin, gills/water breathing	Simple natural intensive exposure system where organism can be bathed directly in toxicants or other test solutions; "washout kinetics" readily applied
Variable body temperature	Allows use of temperature as a realistic experimental variable; colder temperatures especially can "slow down" the progression of a disease or phenomenon to enhance ability to observe
Nonmammalian	Greater social acceptance; in the case of invertebrates in the United States; no current animal protocol requirements

experimental study (Table 27-1). The first and most obvious advantage to the use of aquatic animals as experimental models for human disease is the nearly limitless palette of species from which to choose. There are more than 25,000 described species of finfish and hundreds of thousands of described aquatic invertebrate species, many inhabiting environments as extreme as Antarctica, deep ocean hydrothermal vents, hypersaline brine pools, and extremely polluted estuaries. Thus, myriad adaptational stories are known, and many more are waiting to be told for these thousands of species.

A second important consideration is that many aquatic species produce large numbers of eggs, and many either have natural external fertilization or external fertilization can be readily adopted in the laboratory. Thus, typically eggs and sperm can be harvested at will (although sometimes hormonal treatment or photoperiod preconditioning is needed to induce spawning), or in some cases gametes can be stored for use at a later date (see Chapter 33). External fertilization, of course, allows external development, and therefore the subsequent culture of the organism can be typically in a medium that is no more complicated than fresh or saltwater (perhaps with antibiotics or a few other simple additives). Thus, individual investigators or large-scale facilities serving many investigators can readily raise aquatic animals for research. This process yields several distinct advantages. (1) Animals can be raised under known and controlled conditions (relating to, for example, nutrition, light/photoperiod, salinity, temperature, oxygen, or diet) to yield an extremely consistent research subject. Thus, the use of feral aquatic species can be minimized once the life cycle can be duplicated in captivity (unless, of course, the examination of particular natural populations is the focus of the study). (2) Animal supply can be continuous and year round, which is especially important when species do not naturally spawn year round in the wild. (3) Some studies require thousands to tens of thousands of individuals, so captive breeding also ensures that investigators will not contribute to the decline of a species in the wild because of overharvesting. This is important not only from the supply issue aspects raised in the prior section, but also from the standpoint of sound environmental practices and stewardship. (4) The simplified culture and feeding conditions for aquatic organisms typically means that they have much lower per diem costs than mammalian models. In some cases (see, e.g., the rainbow trout model of carcinogenesis, Chapter 32), the cost differential between fish and mammals is so pronounced that a mammalian study is practically not possible. Notably, for nearly all of the examples in the chapters that follow, aquaculture efforts for the species are either well developed or in development. In many cases, national resources for these species are supported by the National Institutes of Health's National Center for Research Resources, allowing multiple investigators to have access to these organisms (www.ncrr.nih.gov/comparative_medicine).

In addition to the obvious advantages of conducting research on a relatively less expensive and consistent model, there are other generic advantages of many aquatic species. Because many possess large (and often transparent) eggs, these eggs lend themselves to ready manipulation during both prefertilization and early embryonic stages (e.g., one- and two-celled stages). This trait means that the developmental fate of specific cell types can be easily followed and manipulated by, for example, removal of specific cells, *and* that transgenic organisms can be easily created by injection (or other methods of introduction) of foreign DNA. Many aquatic embryos and larvae are also often transparent until rather late in development, so that the developmental process can be easily observed and followed to near completion in selected species.

Especially important from the standpoint of toxicology or even in the administration of drugs or compounds to manipulate a process in an experimental sense, aquatic organisms are easy to dose. Typically, they readily exchange substances with their medium because of their lack of keratinized skin and their need to "breath" via gills (with a high surface area) or skin. Furthermore, because the effective concentration of oxygen in aqueous media is substantially lower than air, water-breathing species must convect large volumes of water per unit time to extract the same amount of oxygen as an air-breather (the ratio being typically 30:1). Thus, where an intraperitoneal or intravenous injection might be necessary in a mammalian model, the introduction of the test chemical can simply be via the water. "Batch" treatment of many embryos or small organisms at once is also possible and should minimize potential experimental variation. Furthermore, while in a mammalian study the organism must clear the substance via hepatic and renal pathways, at times a slow process, aquatic organisms can often be rapidly purged of a test substance simply by putting them in water free of the substance. Thus, "washout kinetics" can be more easily studied.

Most aquatic nonmammalian species are "ectotherms," meaning that their body temperature is largely determined by their external environment. Many species are also "eurythermic," which means they are able to tolerate a wide temperature range. Thus, aquatic species can realistically be used in experiments where temperature is needed as a variable. Where a process might happen relatively rapidly in a mammalian model (where body temperature is typically 37°C), an investigator might be able to slow down a particular biological process in an ectothermic aquatic model, making the phenomenon unfold more slowly and thus easier to describe and investigate.

Importantly, in addition to the general advantages of aquatic models discussed previously and the specific experimental advantages of individual species discussed in the chapters that follow, the use of aquatic model species appears to have wider social acceptance compared to mammalian counterparts. For better or worse, society appears to be more comfortable with an experiment on a fish or sea hare than on a mouse or primate. Nonetheless, animal experimentation with *any* vertebrate species (and in some countries including either all invertebrates or sentient invertebrates such as the octopus) is closely regulated; animal research protocols must adhere to strict federal guidelines to prevent/minimize pain and suffering, and research must be preapproved by ethical committees and regulatory bodies. Interestingly, the subject of whether or not lower vertebrates sense pain has itself become an important area for research (Newby *et al.*, 2007; Sneddon, 2004). Investigators using animal models of any sort seek to observe the three R's: reduce, refine, and replace—that is, to reduce the numbers of animals used, to refine animal protocols, and to attempt to replace animal models with less sentient species or *in vitro* preparations and computer simulations where possible (Russell and Burch, 1959). Aquatic animal models readily contribute to this strategy, in particular in terms of use of lower vertebrates and invertebrates as replacements for animal testing on mammals.

SOME SPECIFIC EXAMPLES

The chapters that follow present several examples of aquatic species that have become or are emerging as key biomedical models. In advance of these chapters, we would like to introduce these models and highlight some of the reasons for their success in research and for being featured in this book. Chapter 28 examines the use of aquatic animal models for neurosciences research. One consistent theme in that chapter is that size and simplicity matter. Because invertebrate nerves do not possess the myelination seen in vertebrate nerves, when rapid conduction was adaptive in a species or nerve, evolution acted to increase the diameter of the nerve fiber. Thus, many invertebrate nerves are large (by nerve cell standards) and thus more easily studied with microelectrodes and other approaches. Furthermore, many invertebrate "central nervous systems" are in fact organized into several ganglia with relatively few large cells. Thus, investigators can readily identify and, in fact, reidentify over time or between individuals the same cells, and they can also determine readily which behaviors are linked to these cells. Chapter 28 also emphasizes the theme that the properties of the aquatic medium itself (e.g., how sound propagates in water relative to air) also lead to interesting adaptations in sensory systems. Given these characteristics, it is little wonder that several Nobel Prizes have been awarded for discoveries using nervous systems in aquatic invertebrates, and these discoveries have in turn underpinned our basic understanding of how all brains, including the human brain, function.

The sensory and metabolic systems of a particular group of fish, the toadfish and midshipmen (family Batrachoididae), are examined in Chapter 29. Here, the reliance of this group on sound production as an important part of their reproductive ecology has led to their use in a strictly Kroghian experimental sense (their lateral line and vestibulary system). However, their champion abilities to tolerate ammonia, which also appears linked to the sound production important in their reproduction, has led to their use as a model for the disease hepatic encephalopathy. Specific aquaculture efforts are also discussed for this group.

Developmental biology takes center stage in Chapter 30, in which arguably the prototypical "developmental" model (sea urchins) and a more recently developed model (tunicates) are featured. Especially in the case of the sea urchin, the experimental simplicity of an externally fertilized, rapidly developing species has revealed the rudiments of development for all animals over an experimental history of nearly two centuries. From a anthropocentric standpoint, both groups are from critical branch points in the tree of animal evolution, and the sequencing of their genomes has revealed insights not only into how networks of genes regulate the process of development but also into the evolution of vertebrates and the invertebrate/vertebrate transition.

Rapid external development in a transparent embryo and larvae are also the case for the zebrafish. Chapter 31 emphasizes the power of genetic manipulation in a vertebrate model. Several mutants in zebrafish mimic specific human diseases, and studies of this organism in both cancer and abnormalities of hematopoiesis are presented. The theme of carcinogenesis continues in Chapter 32, where two powerful finfish models are presented. The *Xiphophorus* (platyfish and swordtails) model of melanoma is an example where years of painstaking identification of genetic lines and genetic mapping of chromosomal markers has led to a powerful system with which to study environment/gene interactions in the causation of cancer. The rainbow trout model is arguably an aquatic system that has shown the most direct connection to human health: observations of aflatoxin-induced hepatic carcinogenesis in trout decades ago have led to trout-based studies of chemoprevention and then to clinical trials on the use of chlorophyllin in the prevention of human aflatoxin-induced cancers. The rainbow trout model also underscores the cost factor advantage in animal models where, quite simply, low-dose effect studies could not be carried out in mammalian models because of cost.

We end this section, and indeed the book, with a forward-looking chapter on cell preservation. Chapter 33 examines the issues surrounding cryopreservation and "desiccation preservation" from the viewpoint of lineage/gamete conservation for biomedical research, for conservation of species diversity, and for preservation of mammalian, even human, cell types. In that chapter, it is fitting that we close with both the themes of water and extreme adaptation. Our book opened with the transport of water on a planetary scale; indeed, the transport of water on a microscale, across the cell membrane through aquaporin protein channels, appears to play a key role in our abilities to cryopreserve cells. In parallel, understanding of the mechanisms underpinning the extreme hypometabolic states achieved during desiccation by brine shrimp (*Artemia*) are leading to developments in the preservation of mammalian cell types. It is likely that a large percentage of the readers of this book will have experimented with growing these "sea monkeys" as children. Thus, *Artemia* are a simple reminder that aquatic animals germane to human health can come from any quarter.

One of the difficult choices faced in selecting specific models to cover in this textbook was how to limit ourselves to a manageable number of chapters. Clearly, we have left out many important models. A more complete listing of past and current models (dogfish sharks, horseshoe crabs, damselfish, medaka, etc.) can be found in Fenical *et al.* (1999), in Schmale *et al.* (2007), and in the many texts on biochemical and physiological adaptation (e.g., Hochachka and Somero, 2002; Moyes and Schulte, 2005).

EPILOGUE: THE AUGUST KROGH PRINCIPLE IN THE MODERN GENOMIC ERA

The sequencing of full genomes of aquatic species is rapidly accelerating, providing even greater utility of these models for human health. Can the August Krogh principle be applied even to these new molecular genome- and proteome-wide approaches to biology? The answer is undoubtedly yes. In fact, knowingly or not, researchers who identified the fugu (Japanese puffer fish) as one of the first vertebrates to have its genome sequenced applied the August Krogh principle. The reason is that the puffer fish genome is only one-eighth the size of the human genome, although it retains an almost identical number of genes (current estimates of number of genes are 21,667 for human and 21,161 for puffer fish). The reason for the much more compact size of the puffer fish genome is that these organisms have somehow cut out unnecessary chunks of DNA that contain neither coding information for protein synthesis nor sequences that contribute to regulation of gene expression. Thus, when sequencing costs and effort were much higher during initial attempts at full genome sequencing, it made Kroghian sense to first pick an organism with an exceptionally information-rich genome to begin to answer questions such as what constitutes the beginning and end of a gene, how many genes are there in vertebrates, and what genes are on which chromosomes? The findings that the puffer fish has 90% of the genes found in humans and that, vice versa, 90% of the genes in puffer fish can be identified in humans have further increased the usefulness of the puffer fish genome in terms of biomedical research. For example, genome organization is ideally studied in the puffer fish as are searches for regulatory noncoding DNA sequences, such as transcription factor binding sites. The discoveries made with the fugu genome (see, e.g., Watabe *et al.*, 2006, and articles within that issue) enabled a less laborious and much faster approach to the sequencing of other genomes that undoubtedly is leading to incredible advances in human health. Indeed, genome projects are now completed, in process, or planned for many of the species discussed in the chapters that follow.

Lastly, the costs for sequencing genomes and especially expressed sequence tags (ESTs) (i.e., partial sequences of the mRNA/cDNA that are expressed in a given organism and tissue) are rapidly falling, and the availability of the technology to small laboratories is increasing, such that genome and EST projects and related microarray studies of gene expression are now underway for all manner of aquatic species, health models included (for a review, see, e.g., Cossins and Crawford, 2005; Cossins and Somero, 2007, and articles contained within). It is also likely that not too far behind, proteomic approaches (simultaneous study of all proteins expressed in a cell/tissue/organism) will proliferate for these species. Because of the unifying conceptual aspects of gene and protein sequences and mechanisms of changes in gene expression, we predict that the lines between research on aquatic animal health models and direct human health impacts will blur even more favorably. In fact, the zebrafish is already becoming increasingly popular among pharmaceutical industries as an early tier screening organism to sort out drugs with potential benefits from those with likely low efficacy or severe side effects. In some laboratories, such studies involve experiments where a gene or suite of genes, whose expression levels are quantified and impacted in zebrafish, are used to predict target genes to examine in human studies. The zebrafish might be the most commonly used model fish species at the moment, but it is certainly far from the only useful and molecularly accessible aquatic model organism. Other models might be selected based on their specific properties to solve a molecular biomedical problem. For example, genes that are affected in a particular human disease with a component of environmental susceptibility could be used to guide choices for particular model species (and genes) for environmental monitoring and experimental biology. We are on the cusp of an even greater appreciation of the role of aquatic organisms in affecting and improving human health.

References

Cossins A.R., Crawford, D.L., 2005. Fish as models for environmental genomics. Nat. Rev. Genet. 6, 324–333.

Cossins, A.R., Somero, G.N., 2007. Guest editor's introduction. Special issue on "Post-genomic and system approaches to comparative and integrative physiology." J. Exp. Biol. 210, 1491.

Fenical, W., Baden, D., Burg, M., De Goyet, C.D., Grimes, D.J., Katz, M., Marcus, N. Pomponi, S., Rhines, P., Tester, P., Vena, J., 1999. From Monsoons to Microbes: Understanding the Ocean's Role in Human Health. Washington, DC, National Academy Press.

Grosell, M., Walsh, P.J., 2006. Benefits from the sea: Sentinel species and animal models of human health. Oceanography 19, 126–133.

Hochachka, P.W., Somero, G.N., 2002. Biochemical Adaptation: Mechanism and Process in Physiological Evolution. New York, Oxford University Press.

Jørgenson, C.B., 2001. August Krogh and Claude Bernard on basic principles in experimental physiology. BioScience 51, 59–61.

Krogh, A., 1922. The anatomy and physiology of capillaries. New Haven, CT, Yale University Press.

Moyes, C., Schulte, P., 2005. Principles of Animal Physiology. New York, Benjamin Cummings.

Newby, N.C., Gamperl, A.K., Stevens, E.D., 2007. Cardiorespiratory effects and efficacy of morphine sulfate in winter flounder (*Pseudopleuronectes americanus*). Am. J. Vet. Res. 68, 592–597.

Russell, W.M.S., Burch, R.L., 1959. The Principles of Humane Experimental Technique, http://altweb.jhsph.edu/publications/humane_exp/het-toc.htm.

Schmale M.C., Nairn, R.S., Winn, R.N., 2007. Aquatic animal models of human disease. Comp. Biochem. Physiol. C. Toxicol. Pharmacol. 145, 1–4.

Sneddon, L.U., 2004. Evolution of nociception in vertebrates: Comparative analysis of lower vertebrates. Brain Res. Brain Res. Rev. 46, 123–130.

Somero, G.N., 2000. Unity in diversity: A perspective on the methods, contributions, and future of comparative physiology. Annu. Rev. Physiol. 62, 927–937.

Watabe, S., Johnston, I.A., Elgar, G., 2006. International symposium on functional genomics of pufferfish: Recent advances and perspective. Comparative Biochemistry and Physiology, Part D: Genomics and Proteomics 1, 4–5.

STUDY QUESTIONS

1. What is the August Krogh principle?
2. What are five generic traits of most aquatic organisms that make them inherently good models for biomedical research?
3. Go to the Nobel Prize Web site (http://nobelprize.org/nobel_prizes/medicine), and obtain information on at least five aquatic animal models that have been useful for biomedical research.
4. Based on some of the literature references from this chapter, do you think that fish feel pain?
5. What are the three R's of animal experimentation?
6. When are ESTs useful in the study of aquatic animal models of human health?
7. Make a current list of the aquatic animal models whose genomes have been sequenced or are in the sequencing pipeline? Are there models whose genomes have not been sequenced that you think should be sequenced? Why? What is a genome "white paper"? What would be the key elements of a white paper for an animal whose genome you believe should be sequenced?
8. What special feature of puffer fish led to them being one of the first genomes to be sequenced?
9. What aquatic animal resources are funded by the National Institutes of Health's National Center for Research Resources? Do you think there should be other resources?

CHAPTER

28

Aquatic Animal Neurophysiological Models

LYNNE A. FIEBER AND MICHAEL C. SCHMALE

INTRODUCTION

Aquatic animals are outstanding models of neuroscience because of their relatively simple nervous systems with few nerve cells that make pathways accessible, their well-defined and simple behaviors that permit analysis, and their faithful reproduction of conserved principles of function. This chapter gives an overview of how aquatic animals have contributed to our understanding of neurophysiological principles, rather than creating an exhaustively comprehensive list of the many aquatic animal species that have served as neurophysiological models. We hope the reader takes away an appreciation for the important roles aquatic species have served in enhancing our understanding of the workings of the nervous system.

NATURE OF EXCITABILITY

Our understanding about the basis of excitability, the mechanisms by which nerve impulses enable perception, maintenance, and movement, owes much to aquatic animal models. It was during the 1940s, using the giant axon of the squid, that Allan J. Hodgkin and Andrew F. Huxley discovered that the foundation of excitability lies in so-called gates in the cell membranes of nerves that pass charged ions. Before their work, the manner in which individual ions crossed the membrane to cause excitability was poorly understood. Hodgkin and Huxley were able to design and test the first kinetic model for excitability, but they did so without knowing anything about the morphology of the molecular gates that passed the ion currents. Through the work of these scientists we learned that gates that preferentially pass sodium ions into the cell and those selective for conducting potassium current out of the cell form the basis of the action potential (Fig. 28-1). This explosive electric signal is essential for initiating movement and thought, digestion and sensation, in short, every aspect of eukaryotic commerce. The squid giant axon was the preparation of choice in these studies because it is a nerve unique in the animal kingdom in size (typically 1 mm in diameter) and thus was conducive to the technology then available to study this phenomenon. Hodgkin and Huxley described with remarkable precision the intricate timing, on a millisecond timescale, with which these gates executed the action potential in their series of papers published in the *Journal of Physiology* between 1945 and 1952. This work was the seminal contribution to modern physiology of the second half of the 20th century. It won a Nobel Prize in physiology for Hodgkin and Huxley and colleague John C. Eccles in 1963, the first of three Nobels eventually awarded on this topic.

Eventually it was discovered that the membrane-bound gates are protein channels that have both a closed and an open configuration. When induced to open by an electrical or chemical trigger, they selectively conduct ions of various species down their electrochemical gradients across the cell membranes, temporarily dissipating the resting membrane potential, as shown for a potassium channel in Fig. 28-2a. Without these water-filled ion channels, such ions could not cross the membrane because charged ions cannot easily penetrate the lipid bilayer of the plasma membrane. The action potential itself is created when a sudden increase in the membrane's permeability to positively charged sodium ions via the opening of sodium channels causes an abrupt dissipation of the membrane voltage from near –60 mV to zero and then beyond in what it termed an overshoot. This depolarization of the membrane potential opens potassium channels, which opposes the depolarization of the membrane and restores the resting potential. The excitation caused by an

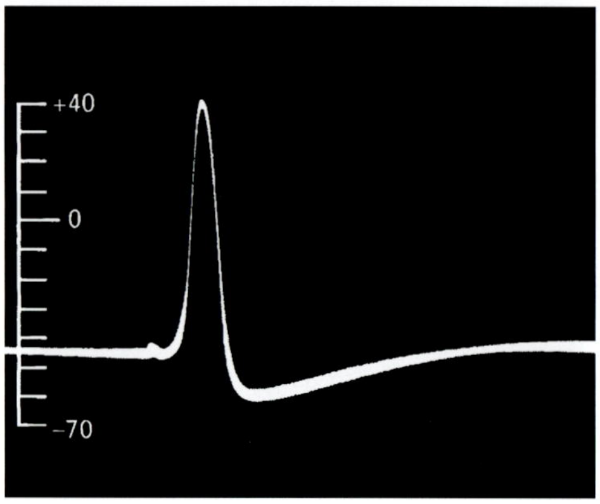

FIGURE 28-1. The first published action potential, recorded from the squid giant axon by Hodgkin and Huxley (1939).

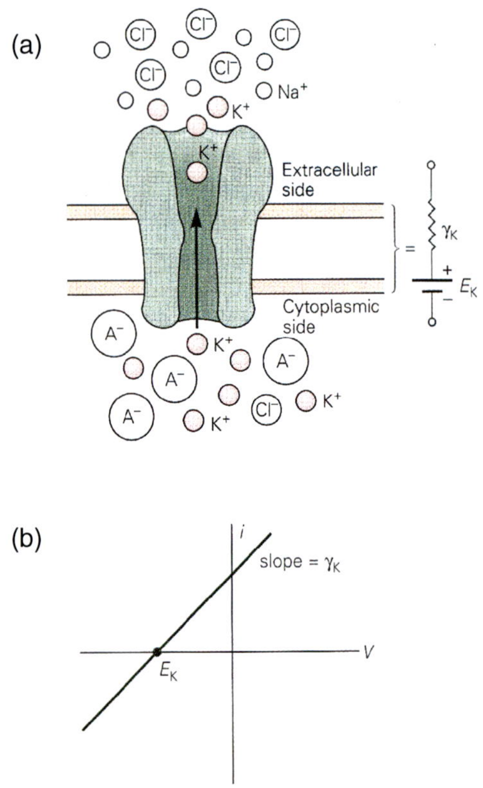

FIGURE 28-2. Chemical and electrical forces contribute to current flow. (a) Concentration gradient for K^+ gives rise to an electromotive force, represented by the battery E_k, the equilibrium potential for K. The battery is in series with a conductor, γ_K, representing the conductance of a channel that is selectively permeable to K^+ ions. (b) The current-voltage relation for a K^+ channel in the presence of both electrical and chemical driving forces. The potential at which the current is zero is equal to E_K. From Kandel *et al.* (2000).

action potential in a nerve spreads outward from the site of initiation in an all-or-none fashion, with action potentials initiated in new patches of membrane when depolarization reaches threshold for an action potential at that site. The channels, especially the sodium channels, become refractory to passing current once they have opened. In this way the nerve impulse spreads in one direction and without attenuation along the nerve.

Preparations from other aquatic species besides the squid giant axon have contributed to our understanding of excitability and basic ion channel properties before and after the work of Hodgkin and Huxley. The frog sciatic nerve, the crayfish stellate ganglion, and even the marine alga *Nitella* have contributed details to the chronicle of excitability. The freshwater snail *Lymnaea stagnalis* has contributed to our understanding of signal transduction mechanisms. The work in these and other simple nervous system preparations has expanded our knowledge to encompass hundreds of ion channels and many more hundreds of kinds of modulation of these channels by hormones and other signaling molecules both outside and inside cells.

NEURAL NETWORKS

Since the basic descriptions of the cellular basis of excitability, a major focus for behavioral neuroscience is to understand the organization of neural networks that initiate, maintain, and terminate specific behaviors. Research on neural networks has revealed which neurons contribute to a behavioral sequence, when they contribute, and how they can collectively control the behavior in question. Specific descriptions of neural networks are beyond the scope of this chapter, but a very short list of the contributing aquatic members would include squid (the giant fiber system, the chromatophore and statocyst networks), octopus and *Limulus*, the horseshoe crab (visual system), crustaceans (stomatogastric ganglion; Fig. 28-3), and the pteropod *Clione* and the nudibranch *Tritonia* (motor programs).

Morphological and electrophysiological data have been obtained showing how these networks of nerves and neurons are constructed, their response characteristics, and the network of interconnections that modulates and controls their operation. From such model systems we learn about the sequence of mechanical actions of nerves and muscles that constitute a behavior.

MEMORY AND LEARNING

In addition to the wiring diagrams of nerves that control behavior, it is critical to understand what learning is on a molecular level. The sea hare *Aplysia californica* is a marine opistobranch snail (Fig. 28-4) that has been a potent source of information about learning and memory. Many of the details of what we know from this animal was discovered by a trio of laboratories, those of Thomas Carew at Yale University and University of California-Irvine, John Byrne at University of Texas, and Eric Kandel at Columbia

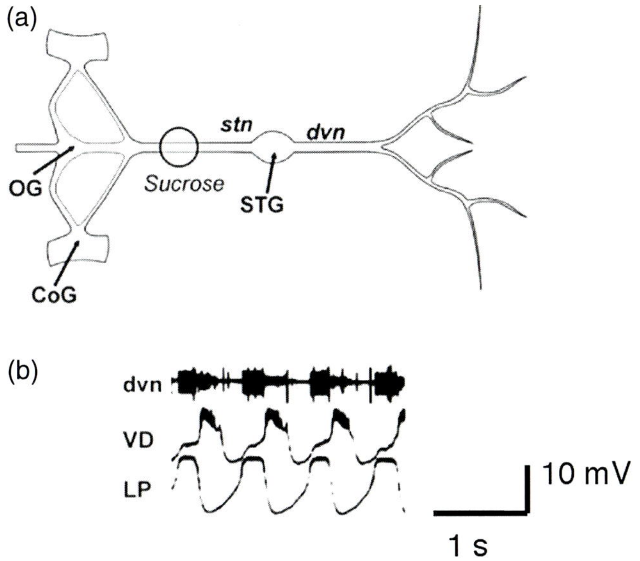

FIGURE 28-3. The neural network of the stomatogastric ganglion (STG). STG networks are embedded in a rich neuropilar innervation containing a large variety of neurotransmitters. Neuromodulatory neurons have been identified, and their effects on the operation of the network have been studied. (a) Schematic drawing of the dissected STG of the crayfish. The STG is connected to rostral ganglia (CoG and OG) via a single input nerve (stn) and a motor nerve (dvn) as an output. (b) Normal activity of the network is monitored by two intracellular recordings of VD and LP neurons and by an extracellular recording of the main motor nerve (dvn). Abbreviations: CoG, commissural ganglia; dvn, dorsal ventricular nerve; *dvn*, dorsal ventricular nerve; LP, lateral pyloric; OG, oesophageal ganglion; STG, stomatogastric ganglion; *stn*, stomatogastric nerve; VD, ventral dilator. Modified from Fenelon *et al.* (2003).

University. Three forms of simple learning in *Aplysia* lend themselves to study in the laboratory: habituation, sensitization, and classical conditioning. All these learning forms have adaptive value, which means they allow the animal to improve its fitness and be able to survive and to reproduce. All are associated with specific neural changes that constitute learning and memory on a cellular and molecular level. The most simple of the three forms of learning is habituation, which is a decrease in a response with repeated presentations of the stimulus. A straightforward way to elicit habituation in *Aplysia* is to cause the siphon, a muscular tube by which seawater is drawn over the gills, to withdraw by means of a light touch to the siphon skin (Fig. 28-5). The animal will withdraw its siphon and gill after the first touch, but subsequent touches to the siphon will elicit a progressively weaker response. We know that the siphon withdrawal reflex is effected by the action of a few sensory

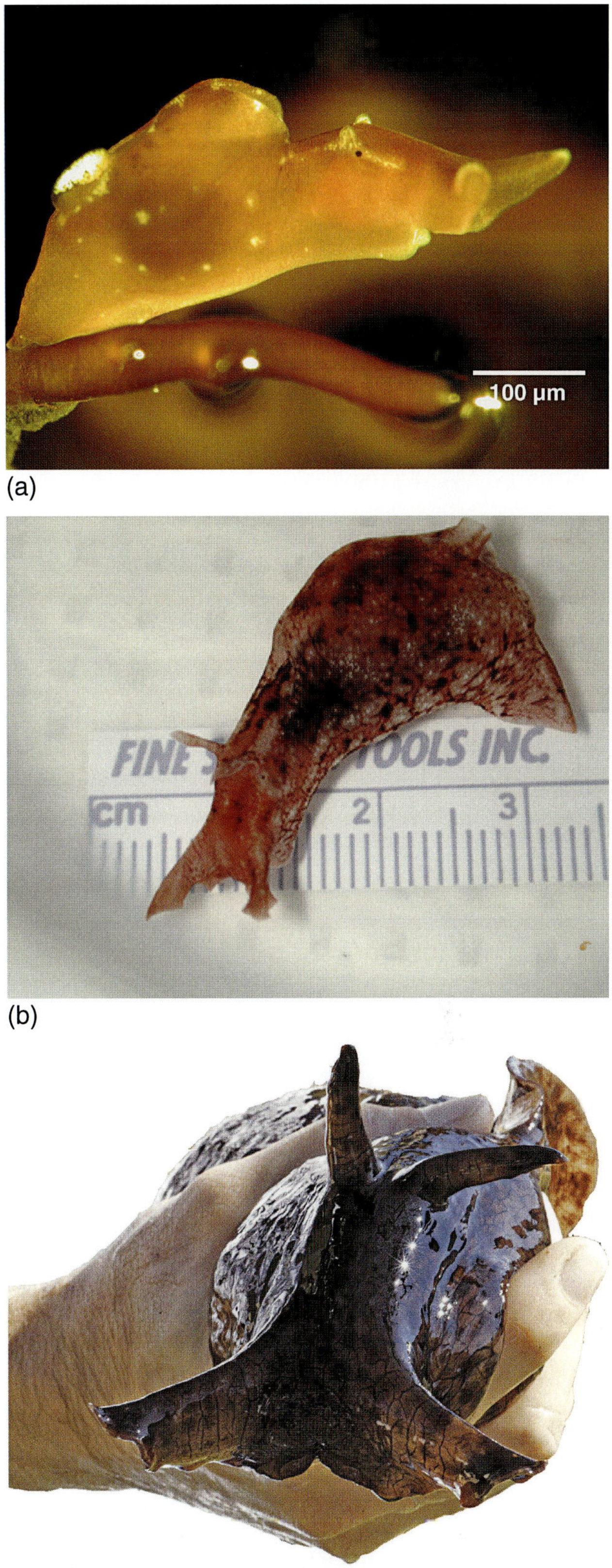

FIGURE 28-4. *Aplysia californica* in the University of Miami NIH-National Center for Research Resources facility. (a) Stage 11 animal. (b) Early stage 12. (c) Sexually mature adult stage 13. Photos (a) and (b) by L.A. Fieber, (c) courtesy of Mr. T. Capo, University of Miami.

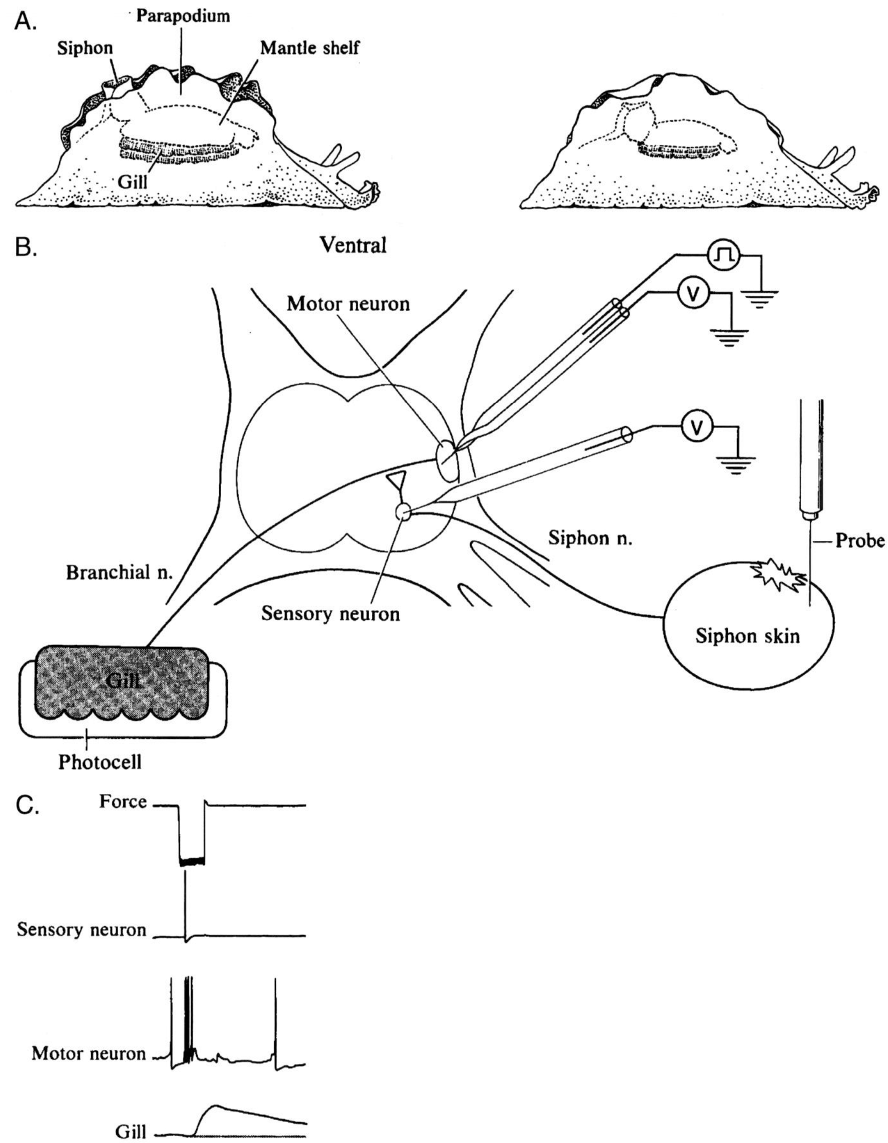

FIGURE 28-5. The gill-withdrawal reflex of *Aplysia californica.* (A) The gill-withdrawal reflex is produced by a tactile stimulus to the siphon. (B) Reduced preparation for elucidation of the sensory and motor components of this behavior. (C) Reconstruction of the reflex: simulation of the siphon skin produces an action potential in the sensory neuron that elicits repetitive firing in the motoneuron and contraction of the gill muscle. From Kandel (1976).

neurons that detect the touch and motoneurons that cause the movement of the siphon. With habituation, the synapse between these neurons changes such that less calcium is admitted into the presynaptic terminal of the sensory neuron every time it is stimulated, causing less release of the neurotransmitter, glutamate, that excites the motoneuron. The learning constituted by habituation is short term (hours) or long term (weeks), depending on the number of training sessions.

Sensitization is a form of learning in which a response is enhanced by a single, noxious stimulus. In our siphon withdrawal example, sensitization occurs when the animal withdraws the gill and siphon in response to siphon touch after experiencing an electrical shock to the tail. Sensitization occurs at the cellular level because of the involvement of interneurons recruited by the tail shock whose specific purpose is to facilitate excitation of sensory neurons innervating the siphon. The interneurons release the hormone serotonin, which causes elevation in concentration of the second messenger cyclic AMP in the sensory neurons, activating protein kinase A (PKA), and phosphorylating proteins that increase the activity of the sensory neurons.

Like habituation, sensitization can be short term or long term. Short-term sensitization requires nothing more than the serotonin-induced activation of siphon sensory neurons just described. Long-term sensitization occurs after repetition of training trials and requires morphological changes, such as new synapses on the siphon sensory neurons. The protein synthesis necessary for new synapse formation is caused by PKA translocation to the nucleus, where it activates transcription factors that induce transcription of specific genes.

Classical conditioning is the third type of learning easily studied in *Aplysia*. If a tail shock (the unconditioned stimulus) is preceded very shortly, on the order of 1 second, by a touch to the siphon (the conditioned stimulus), the animal learns to associate the shock with the siphon touch and retracts its gill, siphon, and tail when the siphon is touched. If the siphon touch occurs very shortly *after* the tail shock, in contrast, this association does not occur. On the cellular/molecular level, classical conditioning utilizes some of the same mechanisms as sensitization. That is, tail shock activates the serotonin-releasing interneurons, but it does this in the presence of elevated calcium in the sensory neuron activated by the siphon touch. Serotonin plus high local calcium in the sensory neuron's presynaptic element causes elevation of a calcium-sensitive, serotonin-sensitive adenylyl cyclase with a high capacity to generate cAMP. The subsequent phosphorylation of proteins creates a heightened release of neurotransmitter by the sensory neuron.

Other marine mollusks have contributed details to the molecular process by which classical conditioning occurs, for example, *Hermissenda crassicornis*. Using this animal, we learned that classical conditioning is mediated by cAMP-induced activation of PKA and calcium/calmodulin-associated protein kinases. *Hermissenda* also illustrates variation in how learning in classical conditioning occurs, because in this animal, depolarization, rather than neurotransmitter release, mediates the rise in presynaptic calcium. Depolarization-mediated learning is the mechanism also operating in vertebrates.

SENSORY SYSTEMS

The sensory systems of aquatic animals and particularly aquatic vertebrates such as fishes have responded to life in the extremely high density of water (relative to air) with some remarkable modifications in sensory systems compared to those observed in terrestrial environments. Fishes and invertebrates have evolved many types of eyes to adapt to unique aspects of the aquatic environment and particular visual needs. However, a key issue for vision in the aquatic realm is that visual stimuli are limited relative to what can be perceived in air. A combination of rapid light attenuation, particularly of very short and long wavelengths, even in "pure" water, combined with the high particulate loads found in most natural waters, puts severe limits on the useful range of visual stimuli in water. In contrast, the density and incompressibility of water facilitates the propagation of pressure waves and particle motion. As a result, sound travels faster (4.8 times faster than in air), and the acoustical near-field, where particle motion is important, extends much farther from a sound source of a given wavelength than in air. Indeed, the line between hearing and sensing vibration or movement through similar sensory organs in fishes and invertebrates is often obscured, with the senses widely overlapping and interrelated. In addition, many fishes have evolved the ability to sense weak electric fields, some with great precision, and several groups of fish have independently developed the capability to generate either weak or strong electric fields.

Vision

Although the major branches of the Bilateria, or animals with three germ layers, separated 100s of millions of years before the Cambrian explosion, it was during the Cambrian (543 mya) that the need for good vision accompanied the development of the first highly mobile animals. Thus, a major theory in the evolution of vision holds that eyes were derived many times in all animals, including those with only two germ layers, namely the sponges and cnidarians, as well as in the various phyla of the Bilateria, and converged into similarity following several different eye structural themes. An alternate theory, that eyes derived from a common ancestry, has developed from the discovery that the same master control gene, *Pax6*, codes highly conserved transcription factors that regulate the development of eyes in all organisms, no matter from what tissue the eyes form. In support of the common ancestry idea is the observation that a cnidarian, the box jellyfish, which possesses a possible precursor of Pax6 in the Pax B gene, has both complex eyes containing lens, retina, cornea, pigment layer, and iris, and more primitive eyes consisting only of photoactive pigments. Meanwhile, some Bilateria have eyes that are much more primitive than the complex eye of the simple box jellyfish. Numerous important aquatic animal models from jellyfish to horseshoe crabs and barnacles, squid and octopus, goldfish and sharks have made important contributions to our understanding of the evolution, anatomy, and function of diverse kinds of eyes.

The prototypic visual unit might consist of an optic nerve associated with pigment cells, or of two kinds of cells, photopigment and photoreceptor, such as found in planarians and often referred to as pinhole eyes. Pinhole eyes function without a lens by using optical light diffraction to form an image on a retina and pupils of variable size to focus (Fig. 28-6A). Three major classes of eyes derive from this basic design: camera type, compound (or apposition eyes), and mirror eyes (Figs. 28-6B–D). The camera-type, such as in octopus and box jellyfish, is an eye in which an inverted image is formed on a retina and focused by a lens. The compound eye, such as in many crustaceans and in

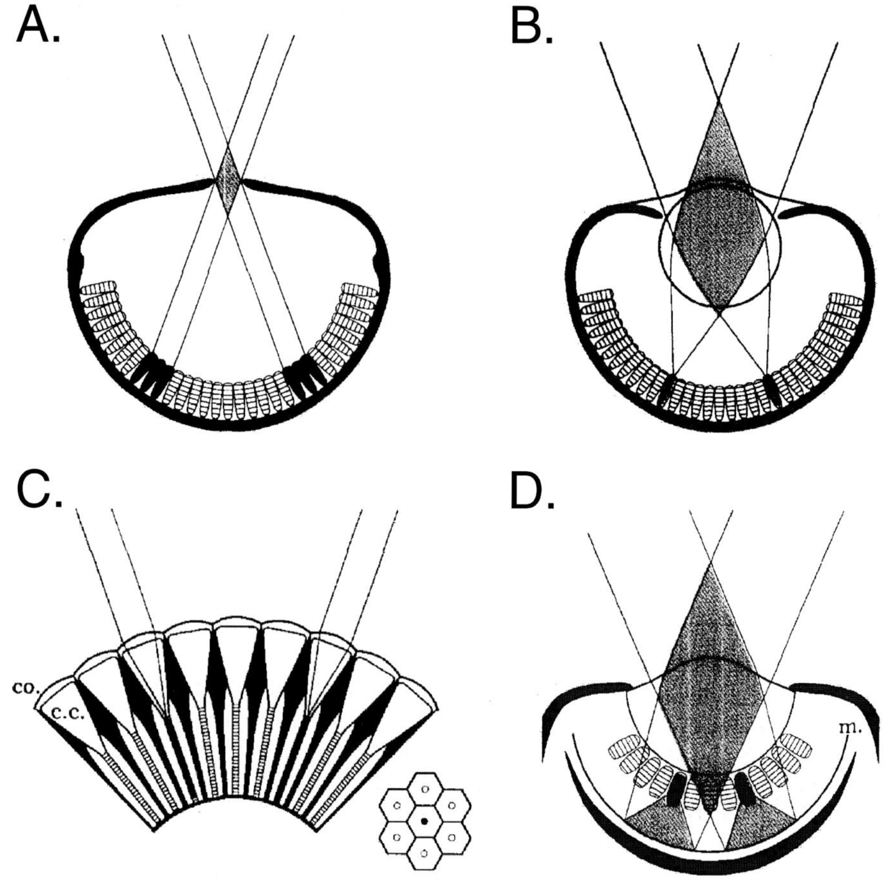

FIGURE 28-6. Eyes. (A) Pigment cup pinhole eye. (B) Camera eye. (C) Compound or appositional eye. (D) Concave mirror eye. From Nilsson (1990).

Drosophila, is composed of repeated units of eye units, the ommatidia, each with its own photoreceptor and fixed lens. The mirror eye, such as in giant clams, uses internal concave mirrors to form images on small retinas at the back of the eye. Interestingly, all three kinds of eyes are found in bivalve mollusks, lending credence to the theory that a common eye ancestor gave rise to all eyes, rather than eyes having evolved polyphyletically.

Two types of photopigments have been assembled into visual structures: flavoproteins and retinal-binding opsins. The flavoproteins include cryptochromes, which control the biological clock in many organisms, phototropins, and photo-activated adenylyl cyclase. Opsins are membrane-bound proteins with seven trans-membrane spanning regions. They belong to the family of proteins termed G proteins, and like other members of this family, they initiate transduction cascades that result in the generation of a receptor potential.

Two types of opsin-containing photoreceptor cells form eyes in the major invertebrate phyla and in vertebrates: rhabdomeres, characterized by microvilli and containing the photopigment r-opsin, and ciliary cells, which, as the name suggests, are ciliated, containing c-opsin. It is possible that the common ancestor to animals that see by means of photoreceptors possessed both rhabdomeric and ciliary cell types, because the opsins of both operate G proteins precipitating parallel transduction cascades. In rhabdomeric photoreceptors, the photon absorbed by r-opsin activates a G protein termed G_q, which activates both phospholipase C (PLC) and protein kinase C (PKC), which in turn play a role in activation of currents that depolarize the membrane potential. Ciliary receptors use G_i to activate phosphodiesterase, which generates a membrane hyperpolarization via its conversion of cyclic GMP to GMP [flame scallop (*Lima scabra*) and the bay scallop (*Pecten irradians*)].

We have already seen that the bivalves, such as the scallops and *Tridacna* (giant clam) have contributed much to our debate on the evolution of eyes, since species in this class of mollusks possess many different types of eyes. Bivalves in addition have revealed important information about the transduction cascades of rhabdomeric and ciliary photoreceptors because these animals contain both kinds of photoreceptors in their retinas: rhabdomeres in the proximal part and ciliated cells in the distal layer of the retina. As a result, these animals have aided understanding of the kinship of these receptors. *Hermissendra* and *Limulus* have contributed information about the function of rhabdomeric receptors. Cephalopods have contributed insights on the use of polarization that are an inherent property of visual pigments, but that our eyes cannot use. Cephalopods also have granted us an appreciation for how sophisticated eyes such as theirs can originate from different embryonic tissue precursors than in other Bilateria.

Hearing and Mechanoreception

Aquatic animals are the same density as the surrounding water and thus are relatively insensitive to pressure waves, the major component of sound heard by terrestrial vertebrates. Hearing in fishes has been studied extensively beginning in the early part of the 20th century when investigators were able to show response of fish to sounds played into the water. However, early studies were confused by several factors: a lack of understanding of the importance of particle motion in near-field hearing and the role of the swim bladder and associated bony structures in facilitating hearing responses. In contrast, relatively few studies have been conducted on hearing in aquatic invertebrates, and there remains considerable uncertainty concerning the hearing ability of most such animals.

The fundamental sensor for detection of particle motion at any frequency, from ultralow frequencies associated with water movements to frequencies above 100 Hz (often considered to encompass the range of "hearing"), is the hair cell. Hair cells feature a single kinocilia (usually true cilia) with multiple stereocilia (actually microvilli) (Figs. 28-7 and 28-8). Hair cells typically exhibit a steady tonic discharge, which is either excited (increased depolarization frequency) by bending of the kinocilia in one direction or inhibited (decreased firing) by deflection of the kinocilia in the opposite direction. The directionality is usually provided by the orientation of the adjacent stereocilia; movement of the kinocilia toward the stereocilia is excitatory while the reverse is inhibitory. Orthogonal movements, perpendicular to the two-dimensional axis of the cilia, generally do not elicit a response. The cells are contacted by paired afferent and efferent neurons. These cells often exist with the cilia attached to an accessory structure, such as cupula or otolith, which amplifies the sensation of vibration. The higher densities of these structures relative to the surrounding tissues and the water yield a differential response to displacement motion, bending the cilia. The neurophysiology of responses of such receptors to a wide variety of stimuli has been studied in many groups of fishes and in a few aquatic invertebrates.

Many aquatic invertebrates and essentially all fishes (and many aquatic amphibians) have external hair cells, either completely exposed or sheltered in lateral line type organs, which detect near-field vibration and water movements. Many, including crustaceans, cephalopods, and fishes have internal hair cell containing organs with calcified masses attached to the cilia to form statocysts. These organs seem to have evolved primary to detect gravity and linear acceleration, thereby providing balance. Although statocysts in

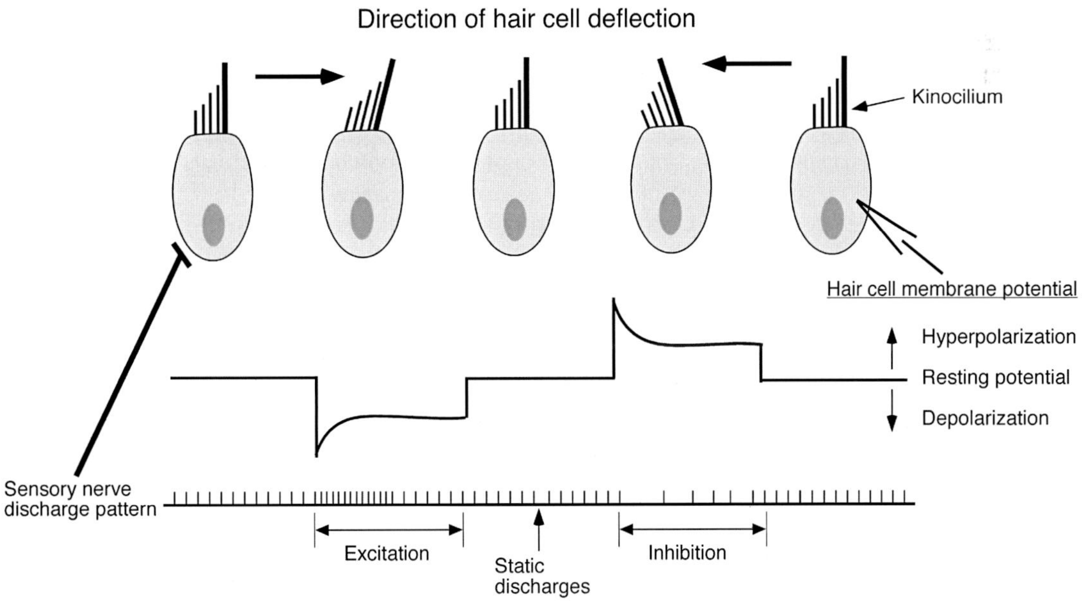

FIGURE 28-7. Hair cell structure and function. Each hair cell has one kinocilium and multiple shorter stereocilia. Deflection of the cilia either depolarizes or hyperpolarizes the receptor cell membrane depending on the direction of deflection. These changes in membrane potential produce either excitation or inhibition of tonic firing patterns in afferent neurons as shown. From Flock (1967).

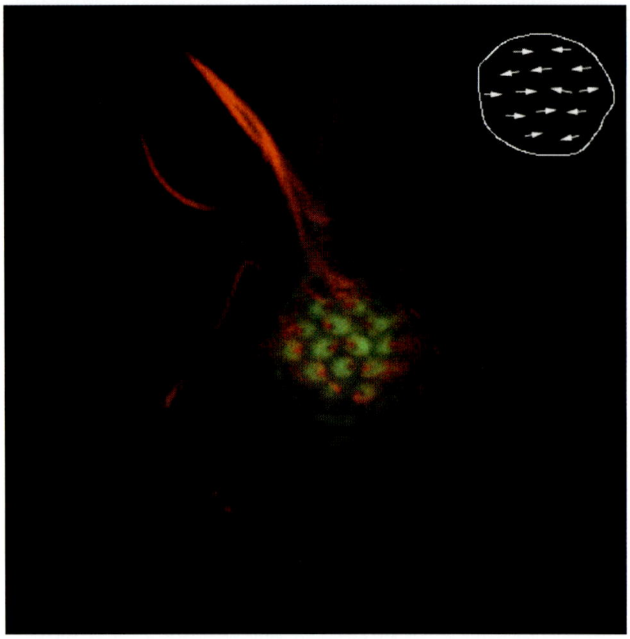

FIGURE 28-8. Polarity patterns in a lateral line neuromast of a 5-day-old zebra fish. View of a neuromast end on; fluorescein-phalloidin labels the hair cell stereocilia, seen here as green crescents, and antiacetylated tubulin stains the kinocilium as a red dot on the concave side of each crescent. *(Inset)* The polarities of these hair cells, which lie in two opposing directions. In the trunk, hair cells are aligned predominantly anteroposteriorly. In the head, neuromasts may have anteroposterior or dorsoventral polarities. From Tanya Whitfield (unpublished; from the zfin Web site, http://zfin.org/zf_info/anatomy/dict/lat_line/lat_line.html).

aquatic invertebrates can apparently sense vibration to varying extents, it is unlikely that they allow a true "hearing" sense to detect higher frequencies or pressure components of sounds in the water. As yet, no convincing studies have revealed any aquatic invertebrates with an ability to sense sounds.

In contrast, the semicircular canals of most fishes contain elaborate bony otolith structures (usually three pairs in two sets of three of canals) attached to numerous hair cells, which can clearly sense displacement components of sound in addition to providing balance cues in three dimensions. This arrangement is fundamentally different from that found in terrestrial vertebrates where a separate structure, the cochlea (in mammals), provides a precise response to sound pressure and the semicircular canals are only involved in sensing balance. The degree to which fish are able to detect and discriminate sounds, particularly the pressure (or far-field) components of sound, is determined primarily by the presence of accessory hearing structures such as air-filled spaces and bony attachments between such spaces and the semicircular canals. Most bony fish have swim bladders that, in addition to providing buoyancy control, act as transducing organs for converting pressure waves into near-field vibrations. Fish with such air-filled spaces clearly have better hearing abilities than those lacking swim bladders (such as elasmobranches, which respond primarily to low frequency sounds). Greater hearing abilities are found in the so-called hearing specialist groups, which have a bony connection between the swim bladder and the inner ear. Examples of these include the ostariophyisan fishes (including goldfish and carp), which have Webberian ossicles connecting these structures and exhibit far greater sensitivity and at most frequencies of sound important to hearing.

The best-studied aquatic systems for understanding the neurophysiological correlates of sound detection are several fish species: the goldfish (a hearing specialist) and the non-specialists the cod, the toadfish (see Chapter 29), the midshipman, and the sleeper goby. Using the sleeper goby, Lu *et al.* (2004) have studied the relative roles of all three otoliths, the large saccule and the smaller lagena and utricle, in sensing sounds. This goby has clearly separated afferent neurons for each otolith organ, allowing recording of electrical responses from each otolith individually using single-cell patch clamp techniques. By recording from all three otoliths in different studies, Lu and coworkers were able to determine that each otolith has a strong directional vector in its sensitivity to particle motion. This directionality is made possible by the structure of the particular otolith, its orientation in the head of the fish (parallel to the sagittal or cross-sectional plane, etc.), and the polarity inherent in the design of the hair cells (firing determined by direction of ciliary deflection). Thus, the arrangement of the three otolith organs in each set of semicircular canals can provide a sophisticated three-dimensional system for localization of sound sources (Fig. 28-9).

Another group of fishes in which has provided excellent models for study of the neurophysiology of hearing in aquatic environments is the Batrachoididae, the toadfish (see Chapter 29), and midshipmen. These fishes are soniferous, or sound producing, and use sound for communication among conspecifics. Andrew Bass and coworkers have investigated many aspects of physiological basis of both sound production and reception in the plainfin midshipman (*Porichthys notatus*) (Bass and Zakon, 2005). They have discovered that steroid hormones control the development of sonic muscles as well as receptor sensitivity in these fish. Cortisol, estradiol, and 11-ketosterone were all found to affect the firing patterns of nerves controlling sound production as well as the response spectra of saccular afferent neurons (Fig. 28-10).

The anatomy and innervation patterns of superficial hair cells have been well characterized in aquatic invertebrates and fishes. Similarly, many studies have examined how the orientation of these sensors, such as in the lateral line canals of fishes, reflects the responses of various species to water flow fields and nearby vibrations. However, few

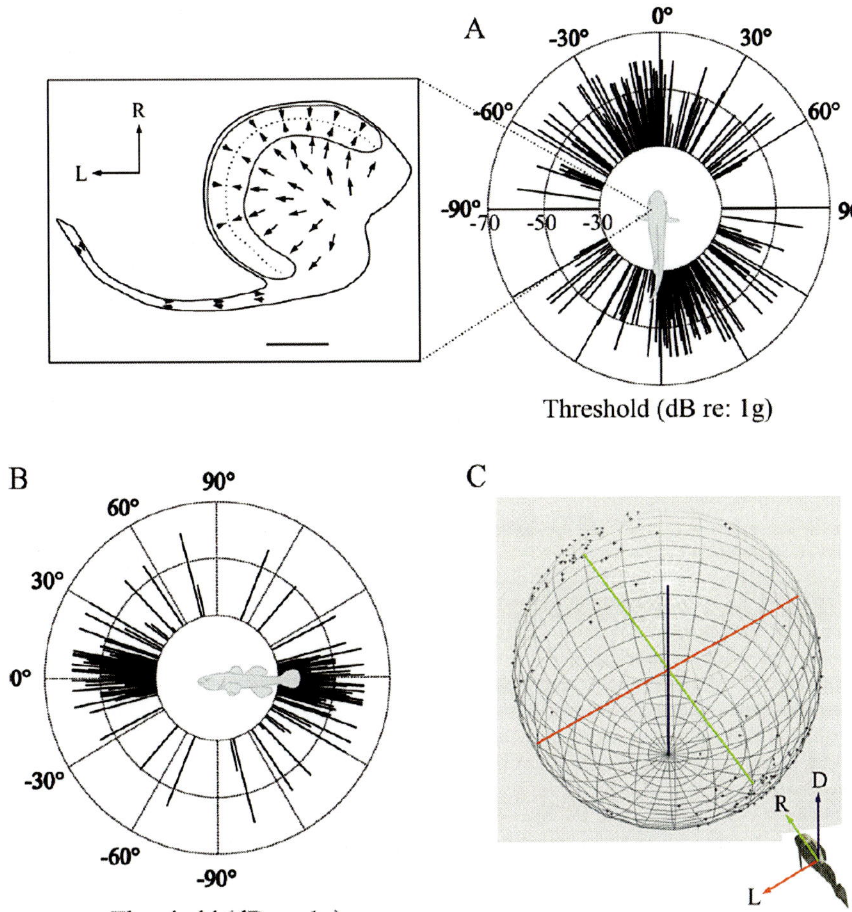

FIGURE 28-9. (A) Plots of directional threshold versus horizontal best response axis for 105 utricular fibers. The inset shows a left utricular macula with hair cell morphological polarizations indicated by arrows. The outlined area within the macula is the striolar region. The dotted curve in the striolar region divides two groups of hair cells with opposing polarity: R, rostral, L, lateral. Scale bar = 200 lm. (B) Plots of directional threshold versus midsagittal best-response axis for the 105 utricular fibers. The length of each line in (A) and (B) represents a threshold for a utricular fiber, whereas the angle of each line is the horizontal/sagittal best-response axis of that fiber. (C) Response directionality in three-dimensional space for the utricular fibers, illustrating a sphere with longitude and latitude lines in 15 steps on the surface of the sphere from a dorso-caudolateral view. Locations where each three-dimensional best-response axis penetrates the surface of the sphere are marked with two corresponding plus signs. The lit pole is the "northern" pole, and the green, red, and blue lines represent the longitudinal, side-to-side, and dorsoventral axes of the fish. Note that the fish was removed from the center of the sphere to avoid confusion: D, dorsal; L, left; R, rostral. From Lu *et al.* (2004).

experiments have used these receptors as model systems to study the neurophysiological correlates of stimuli and responses. One such investigation has demonstrated that, at least in some fishes, the sensitivity of these external hair cell receptors may overlap with what is usually considered the range of hearing by the otolith organs. Weeg and Bass (2002) reported that the superficial neuromast organs of the trunk lateral line of the plainfin midshipman fish could be classified into four distinct response groups based on spike rate and vector strength recorded from lateral line nerve primary afferents. These groups were termed low-pass, band-pass, broadly tuned, or complex. These findings of heterogeneous frequency response properties suggest multiple functions for these lateral line receptors. In addition to sensing currents and nearby movement of objects (at very low frequencies), these receptors were active at least up to 100 Hz. Vocalizations of the midshipman extend below 100 Hz, suggesting that these vocalizations may be detectable by the lateral line as well as the inner ear. This level of overlap may be common in many teleosts fishes; however, detailed neurophysiological studies of neuromast function have not been conducted on most species.

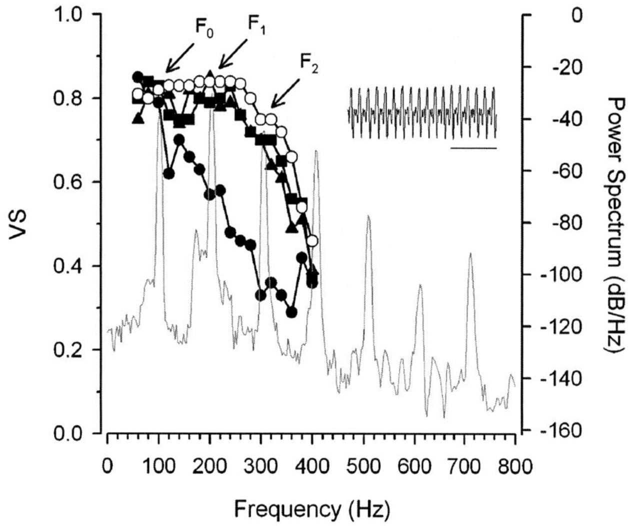

FIGURE 28-10. Vocal-auditory coupling in midshipman fish. This figure summarizes both seasonal- and steroid-dependent shifts in the temporal encoding of tone stimuli by individual afferents from the sacculus of adult females, the main auditory end organ within the inner ear of midshipman fish and many other teleosts. Among teleosts including midshipman, phase locking is a robust indicator of the frequency encoding properties of auditory neurons. For the plots shown here, the y-axis to the left indicates the vector strength of synchronization (VS), a measure of phase locking that can be used for the responses of individual saccular afferents to tone stimuli (a VS value of 1.0 would indicate perfect synchronization). The y-axis to the right shows a relative amplitude (dB/Hz) scale for the power spectrum of the sinusoidal-like hum advertisement call of a type I male midshipman (insert, upper right). The x-axis indicates frequency (Hz) for both the afferent recordings and the hum's power spectrum. Median VS values are plotted here for a population of saccular afferents recorded from nonreproductive females that are untreated (closed circles) or treated with either testosterone (triangles) or 17h-estradiol (squares). The highest VS values among nonreproductive females are close to the hum's fundamental frequency (F0), whereas steroid-treated females also show robust encoding of the higher harmonics of the hum (F1 and F2 indicate the second and third harmonics, respectively). Steroid induces a VS profile that closely resembles that found for wild-caught females in reproductive condition (open circles). From Bass and Zakon (2005).

Electroreception and Electric Field Generation

In contrast to air, which does not conduct electricity, natural waters are fair to good conductors of electricity, depending on ion concentrations. In apparent response to this, the vast majority of fishes have developed receptors sensitive to ambient electric fields. Several forms of these receptors arose independently in fishes apparently as modifications of lateral line hair cells. All fish lineages that evolved before the neopterygians have electroreceptor cells characterized by a prominent kinocilium, and these cells fire when the ambient electric field is negative relative to cell. The best-known examples of these receptors are the ampullae of Lorenzini found in elasmobranchs. These were first studied by A. J. Kalmijn in the early 1970s. He demonstrated that the ampullae were electroreceptors capable of detecting the weak electric fields emitted by live fish buried in the sand. This system became a widely used model for understanding how an animal cell can detect an electric field and allow the animal to orient to that field. Electrosensitivity appears to have been lost in neopterygians and then evolved in the form of two novel receptors in teleosts. These receptors in teleosts are unique in lacking a kinocilium and firing when the ambient environment is positive relative to the cell. The two distinct forms of these receptors are the tuberous and ampullary cells, which have distinctly different response criteria. The ampullary receptors respond to tonic stimuli, whereas the tuberous organs are more sensitive to rapidly changing or phase discharges. These receptors have been intensively studied in several weakly electric fish, described later. Electroreceptors have been identified in a small subset of aquatic tetrapods including many salamanders, some caecilians, and montotremes such as the platypus and echidna.

In addition to the widespread sense of electroreception, some fishes are also able to generate electric fields using specially modified myocytes arranged in parallel arrays and innervated to elicit synchronized depolarization of many cells. Electric fish represent diverse phylogenic groups in which electric field generation arose independently. Among elasmobranches several skates and one ray genus include weakly electric species and the Torpediniform or torpedo rays include many strongly electric species. These species and the teleost stargazers are the only electric species found in marine environments where the high conductivity of seawater results in rapid dissipation of electric fields. In freshwaters, three different groups have developed electric generation: a family of strongly electric catfish (Malapteruridae) found in Africa, the African Mormyriformes, and the South American Gymnotiformes. The later two groups have been extensively studied as neurophysiological models of orientation and communication based on weak electric fields (the one strongly electric member of these groups is the electric eel, which is a Gymnotiform fish and not generally used as a neurophysiological model). In freshwater, these weakly electric fish are able to set up complex electric fields that allow detection of nearby objects of varying conductivity, establishment of territorial boundaries and dominance hierarchies, and attraction of mates.

The Mormyriformes and Gymnotiformes are not closely related but exhibit remarkable parallel evolution of many aspects of both electric field generation and detection (Hopkins, 1995). Both groups include wave and pulse-discharging species as well as species producing complex multiphasic waveform pulse-discharges. The features of these electric organ discharges (EODs) are determined both by the morphology of the electrocytes and modifiable physiological properties. The polarity and number of phases of electric fields produced are determined by the orientation and number of excitable faces of the electrocytes. The simplest, monophasic EOD is generated by an electrocyte with only one excitable face. Because this face will become the pathway for inflow of positive current, its orientation in the fish (as summed over all the parallel electrocytes) will produce either head or tail positive polarity for the monophasic field. Fish having electrocytes with two excitable faces will first depolarize the innervated face creating a current flow in one direction along the body, the first action potential causes the second face to depolarize, creating a current flow in the opposite direction and producing a biphasic pulse. Some species fire a portion of the electrocytes or an accessory electric organ slightly out of phase from the primary pulse in order to produce a triphasic wave. The firing rates and timing of EODs are controlled by the pacemaker neurons in the midline medullary region of the brain. Some species produce sinusoidal wave EODs by matching the pulse duration to the frequency to yield a continuous current flux (Fig. 28-11). When changing EOD frequency, wave EOD–producing fish must vary pulse duration inversely to maintain a sinusoidal wave. Species that produce pulse-type discharges maintain short pulse durations relative to the pulse frequency. For both types of EOD generating species, frequencies and amplitudes typically vary during development, between mature males and females, in a circadian rhythm and during complex social interactions (Fig. 28-12).

To sense fluctuations in the electric field produced by the EOD but without interference from their own EOD, all electric fish utilize the efferent neurons synapsing on the receptor cells to inhibit signals from these cells during the precise time that EODs are being emitted. Thus, these receptors are monitored in perfect phase-shifted synchrony with EODs, which are occurring at hundreds of cycles per second and which may be rapidly varying in frequency and duration of pulses.

Numerous species of weakly electric fish from these two orders have been intensively studied as neurophysiological models of electric field detection and generation, capabilities that seem alien in a terrestrial environment. In addition to providing a unique window on an unusual sensory modality, these fish have been investigated as models of how excitability of membranes capable of firing action potentials can be modified by an animal. At least two distinct classes of mechanisms have been associated with such modifications. Long-term (ontogenic, etc.) changes in EOD have been shown to be induced by hormonal control of ion channel subtypes and numbers on membranes. One mechanism identified for short-term changes in EODs is the activation of membrane-bound receptors, which alter the function of working ion channels. The mechanisms by which electrocytes and associated pacemaker neurons in these fish are able to rapidly and reversibly alter firing patterns of excitable cells may provide insights relevant to understanding neural pathologies such as epilepsy and some cardiomyopathies as well as neuroendocrine disorders and abnormal circadian rhythms.

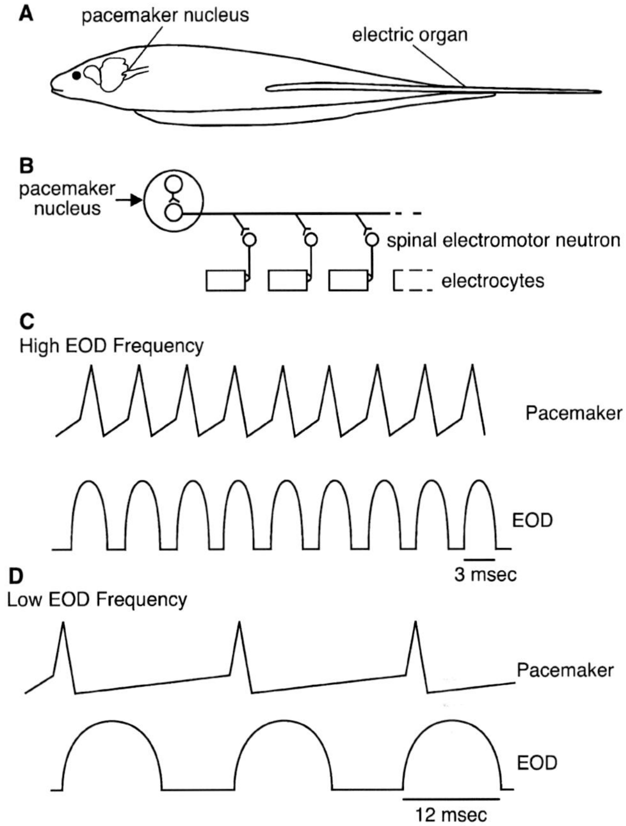

FIGURE 28-11. EOD-generating circuitry in Sternopygus (a Gymnotiform fish). (A) The EOD is triggered by neurons of the pacemaker nucleus in the hindbrain, which activate the electric organ via (B) descending projections to spinal electromotor neurons. Each EOD pulse is the summation of the action potentials of the cells of the electric organ, the electrocytes. (C) Each action potential in the pacemaker nucleus is followed by a pulse from the electric organ. In a wave fish, such as Sternopygus, the pacemaker neurons of females discharge at a high rate, and their electrocytes produce a brief action potential to generate a sinusoidal EOD at a high frequency. (D) The pacemaker neurons in males, on the other hand, fire at lower rates, and the EOD pulse is of longer duration to preserve the sinusoidal nature of the EOD. From Stoddard *et al.* (2006).

Spatial Perception

Aquatic vertebrates serve as excellent models for understanding how spatial perception, the sensing and integration of environmental cues from the three dimensions of space, is accomplished. Spatial perception is achieved by the central nervous system interpreting information collected by visual, vestibular, and proprioceptive sensing systems. The vestibulo-ocular reflexes, which are activated when looking at scenery outside a moving car and which are suppressed when tracking a moving object, integrate this information. Aquatic animals provide experimental power in informing about spatial perception because of the simple construction

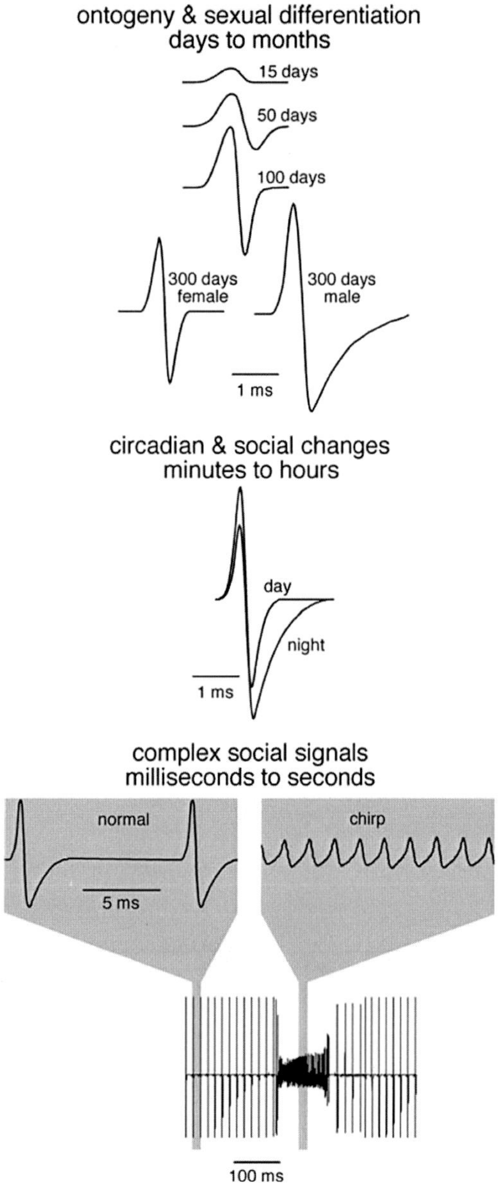

FIGURE 28-12. EOD waveforms change over timescales spanning 10 orders of magnitude. Shown here are three general classes of waveform plasticity in the EOD of Brachyhypopomus pinnicaudatus (a Gymnotiform fish). Developmental changes and sexual differentiation are the slowest, driven by growth factors and sex steroid hormones. Circadian and rapid social changes are intermediate, driven by monoamines and melanocortin peptides. Complex social signals are the fastest, driven by direct neural control. From Stoddard *et al.* (2006).

of their perceptive machinery and because stimuli can be provided to them easily in their natural water habitat. The goldfish, for example, possesses the simplest neural integrator of spatial information, composed of only about 50 neurons in the brain stem but capable of processing information that helps the organism maintain a stable eye position and also maintain its body position during rotations in any dimension as would be required when keeping track of a resting object outside its aquarium during rotation of the aquarium. This neural integrator, like all sensing systems of this type, produces action potentials continuously, but changes its firing rate proportionally with a change in input (such as variable speed rotations of its habitat). The time integral of the new firing rate is compared to that of the old firing rate, and constitutes the signal for activation of motor systems that help the animal maintain position.

Taste

Fish species respond to water-soluble chemical compounds with high sensitivity, possibly because in fishes, the taste buds are distributed in the lips, gill rakers, pharynx, oral cavity, and also on the body surface. Fish are a good model for studying taste discrimination, because several types of taste receptor cells are shared in fishes as well as in mammals suggesting that vertebrates may show common mechanisms of taste information processing. Aquatic mollusks, because of their history as models for learning and memory, have been very useful for understanding the relationship between taste and its memory at the cellular and molecular level. The freshwater snail *Lymnaea* has been used to understand conditioned taste aversion. Studies on this animal have led to the conclusion that taste discrimination on the cellular level is a process closely tied to consolidation of memory. Some studies on *Aplysia* have also aided this understanding.

Olfaction

Animals use molecular receptors on their cell membranes called G-protein coupled receptors to detect odors. These receptors activate complex protein signaling cascades when an odorant molecule binds to them. Vertebrates have large gene families of odorant receptors for unique odors, all derived from a common ancestral lineage, whereas invertebrates have a separate lineage for the odorant gene families coding their receptors. The number of odorant molecules recognized by a species can range from a few hundred (humans) to many thousands, with gene numbers for the receptors associated with them to match. This large variability in odorant receptor gene family size makes odorant receptor genes a fruitful avenue for studying how the genomes of species evolve. Gene duplication often leads to new functions for the duplicate, resulting in either sensation of a new odorant or a related new function based on G protein coupled receptors. Several aquatic animals are used as olfactory models, partly because of the ease with which the dissolved odorant can be delivered to the subject. It is possible, for example, to precisely regulate the concentration of odorant the animal is exposed to by dissolving it in water. Odorant plumes can be made clearly visible in water with dyes and then used to investigate the relationship between odorant exposure and the mode of sensation by the nervous system. Catfish, which can discriminate each of the amino acids individually or in mixtures, and lobsters, with a long history of scientific inquiry on this topic, are classical aquatic animal models of olfaction.

Breathing

The main utility of studying breathing in fishes and also in primitive cyclostomes such as the lamprey and the jawfish lies in the use of these models to understand the evolution of breathing mechanisms. Such studies are another example of the power of comparative physiology to engender understanding of the evolution of fundamental, conserved mechanisms. Rhythmic pattern generators in the reticular system of the brainstem appear to drive respiration in all vertebrates, with the muscles for breathing innervated by the hypobranchial nerve. Using aquatic models, researchers have followed the transition in the sensing trigger for respiratory movements being O_2 in fishes, but with progressive dependence on CO_2 in terrestrial vertebrates and mammals. The vagus nerve model for respiration-controlling innervation of the heart can trace its origins from the noninnervated hearts of cyclostomes, to fishes, in which the vagus nerve plays a dominant role.

CONCLUSION

In summary, aquatic animal models have provided simplified systems that have been essential for understanding some of the most fundamental aspects of neurophysiology as well as complex models that have permitted experiments on intricate neural systems and their interaction with environmental cues. Aquatic animal models have both blazed and illuminated the trail toward an understanding of many physiological processes.

References

Bass, A.H., Zakon, H.H., 2005. Sonic and electric fish: at the crossroads of neuroethology and behavioral neuroendocrinology. Horm. Behav. 48, 360–372.

Fenelon, V., LeFeuvre, Y., Bem, T., Meyrand, P.L., 2003. Maturation of rhythmic neural network: Role of central modulatory inputs. J Physiol. Paris 97, 59–68.

Flock, A., 1967. Ultrastructure and function of the lateral line organs. In P. Cohn (ed.), Lateral Line Detectors, pp. 163–197. Bloomington and Indianapolis, Indiana University Press.

Hodgkin, A.L., Huxley, A.F., 1939. Action potentials recorded from inside a nerve fiber. Nature 144, 710–711.

Hopkins, C.D., 1995. Convergent designs for electrogenesis and electroreception. Curr. Opin. Neurobio. 5, 769–777.

Kandel, E.R., 1976. Cellular Basis of Behavior: An Introduction to Behavioral Neurobiology. San Francisco, WH Freeman.

Kandel, E.R., Schwartz, J.H., Jessell, T.M., 2000. Principals of Neural Science, 4th ed. New York, McGraw-Hill.

Lu, Z., Xu, Z., Buchser, W.J., 2004. Coding of acoustic particle motion by utricular fibers in the sleeper goby, *Dormitator latifrons*. J. Comp. Physiol. A 190, 923–938.

Nilsson, D.-E., 1990. From cornea to retinal image in invertebrate eyes. TINS 13, 55–64.

Nilsson, D.-E., 2004. Eye evolution: A question of genetic promiscuity. Curr. Op. Neurobiol. 14, 407–414.

Stoddard, P.K., Zakon, P.H., Markham, M.R., McAnelly, L., 2006. Regulation and modulation of electric waveforms in gymnotiform electric fish. J. Comp. Physiol. A 192, 613–624.

Weeg, M.S., Bass, A.H., 2002. Frequency response properties of lateral line superficial neuromasts in a vocal fish, with evidence for acoustic sensitivity. J. Neurophysiol. 88, 1252–1262.

STUDY QUESTIONS

1. How does the action potential travel down the nerve fiber during nervous transmission? Why doesn't the strength of the impulse change during travel of the action potential? It is believed that eukaryotic ion channels originated from a voltage-sensitive model A potassium channel. What common principles dictate the function of voltage-gated ion channels?
2. Is sensitization the opposite of habituation? What critical local environmental factor enables classical conditioning to occur?
3. Nilsson (2004) thought that the controversy over whether eyes evolved one time or multiple times in animals is exaggerated because the difference is partly semantic: just as a sensor consisting of photoreceptor cell and pigment cell is not an eye, an arrangement of muscle cell and osteocyte is not a leg. Yet these examples have the potential to become eyes and legs with selection over time, perhaps uniquely so. Do you think this argument better supports single or multiple evolutions of eyes, and why?
4. Do aquatic environments favor vision or hearing for the long-distance propagation of information? Why? How does this differ from terrestrial environments?
5. How are displacement and pressure waves intercepted on land versus underwater?
6. How do hair cells detect sound?
7. Did electroreception in fishes evolve independently or from a common ancestor?
8. How do electric fields differ in fresh and marine waters, and how do these differences affect the generation of weak and strong electric fields?

C H A P T E R

29

Toadfish as Biomedical Models

PATRICK J. WALSH, ALLEN F. MENSINGER, AND STEPHEN M. HIGHSTEIN

INTRODUCTION

Fishes of the order Batrachoidiformes and its single family Batrachoididae (toadfish and midshipmen) are distributed worldwide in temperate and tropical waters with more than 70 described species (Nelson, 1994). Although largely marine, a number of species inhabit estuarine and true freshwater environs (e.g., the Amazonian River basin). Their exact phylogenetic position within teleost (bony) fishes has been challenged by sequences of their full mitochondrial DNA genome, with these newer molecular data suggesting that they are, in fact, among the most derived or advanced teleost fishes and should be included in the series Percomorpha. According to these new data, toadfish and midshipmen are most closely related to Synbranchiform fishes (e.g., swamp eels and spiny eels) and in the same series as scorpion fish, perches, flatfish, and puffer fish (Miya *et al.*, 2003)

Experimental biologists working on toadfish behavior and physiology have long appreciated their "advanced" functional traits (elaborate courtship calling/behavior and the associated sound production and reception systems; retained ability to synthesize and excrete urea as adults; aglomerulate kidney; ability to tolerate adverse environmental conditions, etc.). In an ecological setting, toadfish and midshipmen often occupy a rather important trophic level in that they prey on invertebrates and smaller fish. As these midlevel carnivores, they occur in rather high abundance (e.g., the gulf toadfish, *Opsanus beta*, is the second most abundant benthic fish species in Biscayne Bay, Florida) and thus form an important ecological link as prey to the highest marine carnivores (sharks, dolphins, birds, bonefish, etc.). Because they are a rather "noisy" group, their populations can be subject to intense predation by these top carnivores (Remage-Healey *et al.*, 2006), and the members of the family appear to have evolved several important devices to at least reduce predation. For example, members of one subfamily possess venomous dorsal spines. In another of the three subfamilies, bioluminescence in midshipmen (their name coming from the bioluminescent spots on their belly resembling the gold buttons of a sailor's coat) is believed to contribute to countershading such that they are less visible to predators below against the lighter waters above. Furthermore, the use of dinoflagellate bioluminescence in support of their visual system during fast strike predation is believed to minimize their vulnerability. In other members of the family, urea synthesis and co-excretion with ammonia has been suggested to act as a chemosensory "cloak" (discussed later). No doubt this advanced status and their general abundance and ease of capture, along with their aquarium hardiness, have contributed to these fish being important animal models in the understanding of basic physiological mechanisms and human diseases. This chapter explores several of the experimental areas in which toadfish have contributed to knowledge underpinning the general area of human health.

THE SONIC MUSCLE MACHINERY

Like many fish, toadfish and midshipmen possess an air-filled swim bladder that serves as a buoyancy organ, making the fish slightly less dense overall so that it can swim and position hold more easily. However, in this family the swim bladder is elaborated and doubles as a sound production organ. Indeed, the large heart-shaped swim bladder is the most obvious anatomical feature in the peritoneal cavity and is flanked by large sheaths of muscle on either side (Fig. 29-1). Both sexes can use these muscles to create grunts when the fish are upset or aggressive. However, during mating season, and particularly at dawn and dusk in many

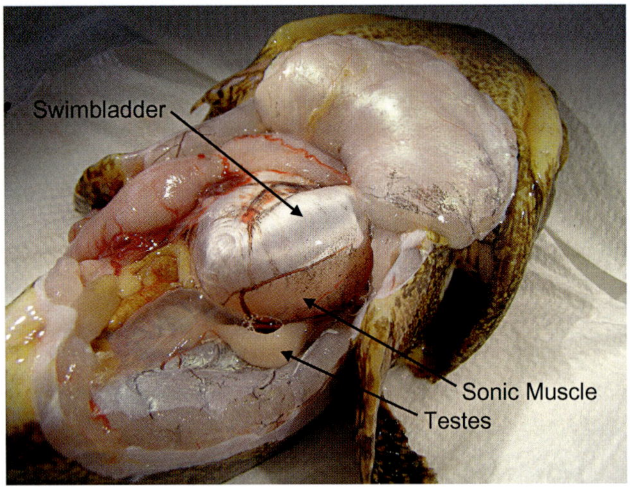

FIGURE 29-1. Peritoneal cavity of mature male gulf toadfish, *Opsanus beta*, dissected to reveal the swim bladder with attached sonic muscle. Photo by T. Rodela, University of Ottawa.

species, the males (only) are capable of producing a rather unique "boat whistle" courtship call for hours on end from the confines of their "nest," a conch shell, a hollowed out area under a rock, or in some cases refuse from humans such as cans. The sound propagates readily through water (and, in fact, can be heard easily by boaters). Gravid females are attracted to the nest and deposit eggs on the bottom of an overhanging surface, which the male then fertilizes. Typically, the male will then guard and fan the developing embryos with water for up to three weeks. At least in the case of *O. beta*, it is not unusual for a given male to attend to several clutches of eggs, presumably from multiple females. The attached embryos absorb a large yolk sac and eventually swim away as fully developed juveniles approximately 1 cm in length. In some species (e.g., *Porichthys notatus*), there is a "third" gender, the so-called sneaker male that has fully developed testes but is typically smaller with no nest of its own nor the ability to boat whistle; its reproductive strategy is to loiter near the nest of the larger male and inject sperm at the appropriate time when a female spawns (Bass, 1996).

Male toadfish and midshipmen have substantially more sonic muscle mass than females or sneaker males, and their muscles also contain higher activities of the enzymes of aerobic metabolism that produce ATP to fuel the contractions of the sonic muscle (Walsh *et al.*, 1989, 1995). Batrachoidid sonic muscle is among the fastest contracting and cycling muscles in the animal kingdom, literally the fastest vertebrate muscle known (with a contraction-relaxation cycle of 200 Hz), and has served as an interesting study system to examine muscle function (Rome and Lindstedt, 1998). In addition to the adaptations of the metabolic machinery, there are clear anatomical and ion transport specializations of the sonic muscle compared to swimming muscle: it has the ability to rapidly cycle the transient spike in intracellular Ca^{2+} concentration needed to initiate contraction and the resequestration of Ca^{2+} to permit relaxation. The development of the male version of the sonic muscle appears to be under the control of circulating testosterone and 11-ketotestosterone levels. Furthermore, it is not only the muscle that must be specialized but also the neural pathways from the brain to the muscle causing it to contract. (For a sample batrachoidid mating call, visit www.cbc.ca/quirks/archives/06–07/nov04.html and scroll down to the third story on "Toadfish Pee.")

THE ACOUSTICOLATERALIS SYSTEM

Study of the acousticolateralis system as a model system provides the opportunity to span systems level, cellular level, and molecular approaches to the nervous system. The acousticolateralis system is composed of the organ of equilibrium, the vestibular labyrinth, the organ of hearing in fish the saccule, and the lateral line organs used by fish in "schooling," feeding, and orientation. The labyrinth is composed of angular and linear motion detectors. The semicircular canals are fluid-filled tubes that detect angular acceleration, whereas the otolithic organs are composed of a calcified mass, the otolith atop the sensory hairs, and detect linear acceleration. The linear acceleration vector is composed of impulsive and static components. The impulsive component is akin to the acceleration of an automobile when it begins to move, whereas the static component is caused by gravity. The brain thus receives relevant information concerning angular and linear head motion and position from the vestibular labyrinth.

Lateral Line

Fish and aquatic amphibians have evolved a mechanosensory lateral line system to detect water movement and vibration. The lateral line is most apparent to fishermen or casual observers as a distinct line traveling along the body from behind the gill cover (operculum) to the tail. This posterior lateral line or trunk canal consists of neuromasts, which are the functional unit of the lateral line. Each neuromast consists of bundles of mechanoreceptive hair cells that respond to movement and vibration. Fish also posses an anterior lateral line that consists of numerous, species-specific lines or canals arrayed around the head. In most species, the majority of the anterior lateral line and trunk canal consists of canal neuromasts that are contained within subdermal canals and detect water acceleration. In contrast, superficial neuromasts are located on the skin surface and can be scattered anywhere along the head and body and determine water velocity (Kroese and Schellart, 1992).

Consistent with its other unique attributes, the toadfish has relatively few canal neuromasts. However, many of the superficial neuromasts are surrounded by two large finger-like projections (papillae) that may act to channel water past the neuromast in the same manner as the subdermal canals. As toadfish prefer soft-bottomed habitats, typical canal neuromasts could become clogged by sediment, which may explain their low abundance. It has been hypothesized that the papillae protects the neuromast from sediment and allow these superficial neuromasts to function as canal neuromasts.

Behavioral and electrophysiological experiments have demonstrated that the lateral line functions in schooling behavior (Partridge and Pitcher, 1980), rheotaxis (Montgomery *et al.*, 1997), and localization of underwater objects (Weissert and von Campenhausen, 1981). Numerous experiments have documented the ability of the lateral line to mediate predator prey interactions (Bleckmann and Topp, 2003; Montgomery and Macdonald, 1987). The sensitivity of this system is sufficient to allow fish to capture moving prey in complete darkness. Whereas many of the previous studies focused on short-range predator/prey encounters, technology has allowed the monitoring of the complex hydrodynamic trails generated by fish movement. As these trails can persist in the water column for several minutes, fish could use the lateral line to track prey long distances following these wakes (Hanke and Bleckmann, 2004).

Many of the previous studies on the lateral line have used vibrating spheres to provide the stimulus (Coombs, 1999; Kanter and Coombs, 2003). The forces produced by these dipole stimuli are easier to characterize and model compared to the complex waves produced by free-swimming prey. Though instrumental in determining neuromast characteristics (frequency and directional sensitivity), pure stationary dipole-like stimuli are rarely encountered in nature. The general consensus of these studies was that the lateral line was a short-range sensory system with a detection range of one to two fish lengths (Braun and Coombs, 2000). However, the response characteristics of the lateral line to naturally relevant stimuli remained unresolved.

Behavioral experiments with the toadfish suggested a shorter detection range. Fish were placed in the dark to restrict visual input (olfactory cues were not considered a factor because of experimental design), and the predator/prey interactions were viewed under infrared light (toadfish and prey were not sensitive to the far red wavelengths). Small toadfish (6 cm standard length = sl) did not strike at prey greater than 3 cm away, suggesting a shorter range for the lateral line. Unfortunately, behavior experiments are limited to determining when the predator reacts to the prey and not when the prey is detected. As ambush predators, toadfish may be able to detect prey at greater distances than those at which they attack.

To truly understand when animals detect external stimuli, it has long been the goal of neuroethologists to record from free-ranging animals in their natural environment. To determine when toadfish senses prey, it is necessary to record neural activity from the lateral line nerve. Ideally, this should be accomplished in a natural setting to avoid small tank artifacts and allow prey and predator to swim freely. An implantable electrode with a neural telemetry tag was developed to allow recording of neural activity from free-ranging and normally behaving toadfish. The electrode was implanted into the anterior lateral line nerve of a toadfish (30 cm sl), and following recovery, a small minnow was added to the tank. The position of the minnow was correlated with its distance from the toadfish and the neural activity of the nerve. The neuromasts on the anterior lateral line only responded to the minnow's movements if the prey was within 12 cm of the toadfish. These experiments supported the behavioral observations that the range of the lateral line to naturally relevant prey in the toadfish is significantly less than a body length. The toadfish experiments represent the first time that neural activity was recorded from the lateral line of a free-ranging fish in the presence of a natural prey and suggests that previous studies may have overestimated the range of the lateral line (Palmer *et al.*, 2005).

Vestibular System

Many vestibular studies have been pursued in fish. The oyster toadfish, *Opsanus tau*, has become an effective model system because of experimental convenience and availability of this species. The general view (e.g., Sarnat *et al.*, 1974) is that the vestibular endorgans in fish evolved early and relatively completely and therefore compare favorably to those of other vertebrates, including mammals in both form and function. For example, the semicircular canals of the toadfish are similar in size and morphology to those in humans.

The function of these vestibular semicircular canals is to detect and quantify angular head motion. Graphs that describe the responses of the canals to head rotation are remarkably similar across vertebrates. Any interspecies differences in morphology and dynamics are most likely related to the lifestyle of the particular animal, reflecting the range of angular accelerative forces experienced in different environments (e.g., birds versus lizards). Comparison of the physiology of toadfish canals to their mammalian counterparts reveals that fish canals encode angular accelerative stimuli in a similar fashion to mammals (Highstein *et al.*, 2005). Namely, fish canal afferents demonstrate the same three rough classes of afferent responses as mammals, i.e., low and high gain velocity sensitive afferents, and additionally demonstrate a third class of afferent that reports head acceleration across the entire bandwidth of motional frequencies experienced by the animal. Comparison of fish and mammalian canal microanatomy indicates that fish have only type II hair cells, instead of types I and II, and that they have fewer of these type II cells and fewer primary afferents than mammals. Note that type II hair cells are cylindrical and innervate vestibular nerves via typical bouton-type endings, whereas type I hair cells are flask shaped and are surrounded by a flask-shaped terminal or a so-called calyx terminal.

In toadfish, hair cells are arranged in a flat sheet atop the crista rather than in the typical three-dimensional mammalian array. Thus, the fish is able to produce a similar set of responses as mammals, but with a smaller set of hair cells and afferents when compared to mammals. These experimental facts provided an advantage in determining the structure versus function of the canal sensory apparatus. In toadfish, response dynamics were shown to be resident within a map in the crista (Boyle *et al.*, 1991). Hair cell-afferent complexes closer to the center of the crista have higher gain and those located more peripherally have lower gain to angular accelerative stimulation.

As well as receiving incoming afferent information from the ear, the central nervous system sends a set of efferent terminals *to* the vestibular labyrinth, ostensibly to modify incoming responses before they reach the brain. In toadfish, efferent vestibular action is mapped onto the crista with more profound action occurring in central hair cell-afferent complexes than in their peripheral counterparts. These maps of efferent action proved useful in discovering regional cellular attributes that might contribute to their construction.

Toadfish have also been employed aboard the space shuttle in studies of microgravity (Boyle *et al.*, 2001). In space, the angular accelerative and the impulsive component of linear acceleration are unchanged because they are independent of gravity, whereas the static component of linear acceleration caused by gravity is obviously decreased. Fish otolithic organs function in similarity to their mammalian counterparts housed in the ears of the astronauts who accompanied the fish into space. Because of the unique experimental amenability of toadfish, otolithic afferents could be recorded before, during, and following space flight to determine how the otolithic system reacted to a decrease in gravity. Generally, otolithic afferent nerves increased their sensitivity to what little gravitational force was present. This increased sensitivity lasted for more than 30 hours following return to Earth. This result seems to illustrate a general feature of biological systems, namely when the stimulus is decreased, the system turns its sensitivity up in search of the stimulus. For example, in Parkinson's disease when dopamine production is decreased, the d1 and d2 dopamine receptors are genetically up-regulated in search of dopamine.

The experimental approach to understanding labyrinthine function (Fig. 29-2) has entered into the modern era as new tools and techniques became available. The quantification

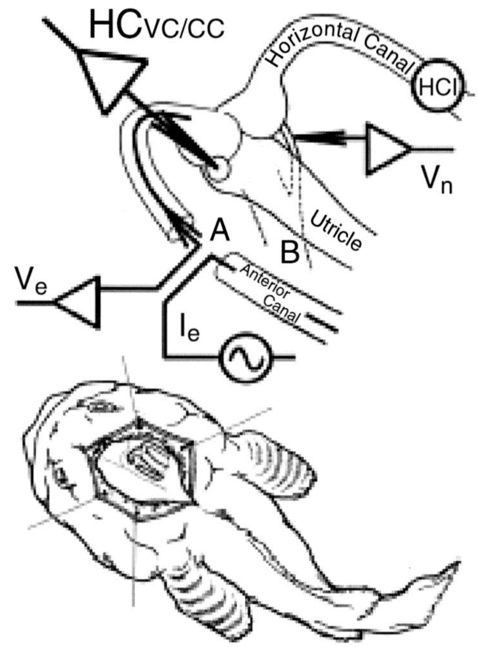

FIGURE 29-2. General experimental setup for vestibulary measurements. *(Top)* Expanded view of the labyrinth. *(Bottom)* Overview of the fish. A current-passing electrode (Ie) is placed in the posterior limb of the anterior canal and a second voltage-measuring electrode (Ve) is placed in the anterior canal ampulla for applying and recording electrical polarization stimuli, respectively. Mechanical stimulation of the horizontal semicircular canal is provided by a piezoelectrically driven indenter placed on the long and slender limb of the canal duct (HCI); this mode of canal activation can be tailored to closely mimic head rotation but can be applied independent of head movement in space. Extracellular afferent records (Vn) were obtained simultaneously with the applied stimuli. A small hole was made in the utricular side of the horizontal canal ampulla for access to the sensory epithelium by sharp electrodes to record hair cell receptor potentials or currents during canal indentation or polarization or a combination of the two stimuli. From Highstein *et al.* (2005).

and mathematical modeling of canal stimulation by a piezoelectrically controlled indenter placed on the long and slender limb of the canal (Rabbitt *et al.*, 1995) enabled many new experimental approaches, because the preparation no longer had to be rotated to activate canal receptors. It is experimentally cumbersome to rotate the animal and all of the attendant visualization and recording devices to evoke a response to adequate stimulation. However, rotation and indentation are mathematically related. Namely, rotation of the animal causes a pressure differential across the cupula or vane that closes the space above the crista to endolymphatic flow. This pressure differential causes endolymph movement relative to the canal. The fluid movement bends the vane that in turn deviates the sensory hairs atop the hair cells and thus begins the transduction cascade that encodes the parameters of angular motion. Indentation of the canal limb also causes fluid flow that results in sensory hair bundle motion and the subsequent encoding mentioned previously. Rotation and indentation have been experimentally shown to be related by a transfer function, namely, one micron of indentation causes the same neural responses as a 4-degree/second angular velocity (but see Rabbitt *et al.*, 1995, for details).

Hair cell potentials in response to rotation or canal indentation are called receptor potentials and can be recorded *in vivo* by placing a small hole in the canal ampullary wall and visually guiding a sharp microelectrode through the hole to penetrate hair cells (Highstein *et al.*, 1996). Utilizing a voltage clamp, we recorded hair cell receptor potentials and currents to sinusoidal stimuli applied via the indenter. Hair cell responses reflect the first half of the stimulation (transduction) cascade including (1) biomechanics (cupular motion and hair cell bundle displacement across the neuroepithelium or crista ampullaris), (2) mechanical transduction (hair cell transduction currents), and (3) basolateral currents (voltage-sensitive hair cell channel dynamics as they influence receptor potential generation). A further technical development has been the redeployment of endolymphatic polarization (Highstein and Politoff, 1978; Rusch and Thurm, 1990) to bypass canal mechanics and to directly activate hair cells. For this experiment, a current passing wire was placed into the endolymph and a voltage-sensing pipette employed to measure the translabyrinthine voltage drops cause by current passage. The effective stimulus for neurotransmitter release from the base of hair cells is a transcellular voltage change that opens or closes voltage-sensitive calcium channels, leading to the modulation of transmitter release. Constant current polarization and sinusoidally varying current waveforms were applied via endolymphatic electrodes to modulate hair cell transduction currents. Lumen-positive current injection increases the extracellular voltage acting on the apical faces of hair cells relative to both the intracellular and extracellular basal voltages. A certain number of hair cell stereocilliary transduction channels are open at rest and the pathway/mechanism through which endolymphatic polarization acts is directly through these open transduction channels. Thus, the system can be stimulated in the absence of fluid flow and bundle motion. This enables the study of responses that occurs during the second half of the transduction cascade, namely the posttransduction half.

Recordings taken from single afferents in response to lumen positive steps of voltage indeed demonstrate a maintained increase in spontaneous activity. The delay of the first recruited action potential from the onset of a strong lumen positive voltage pulse was always about 1 ms, indicating that the action potential was recruited monosynaptically, validating the presynaptic nature of the stimulus. In other words, the voltage drops were shown to act on hair cells and not directly on primary afferent nerves. The response dynamics of individual afferents to sinusoidal endolymphatic polarization were determined and compared to results for the same afferents generated by canal indentation. Because endolymphatic polarization bypasses the mechanics of the cupula-endolymph-stereociliary system, differences between the responses to the two types of stimuli can be used to isolate mechanical contributions to the overall response. Almost all of the low-frequency phase lead observed in primary afferent responses is the result of canal mechanics (<0.2 Hz sinusoidal stimuli). The high-frequency gain and phase enhancement reflected in the afferent population average is primarily a result of posttransduction current processing by hair cell basolateral currents (see item 3), (4) neurotransmitter release, or (5) by dynamics of the postsynaptic membrane and primary afferents and not by the canal mechanics (see items 1 and 2). Results also show that there are systematic deviations in the canal mechanics activating individual afferent classes, and that these differences contribute significantly to interafferent variability in response dynamics. Thus, canal mechanics accounts for approximately half of the complex diversity of afferent responses. The remaining processing is posttransduction current. Hair-cell receptor potentials recorded *in vivo* also fall short of the range represented by the afferent population and exhibit dynamics consistent with the mechanical activation of transduction.

We have also discovered heterogeneity in the neurotransmitters used by hair cells in their synapses that contact primary afferents (Holstein *et al.*, 2004a). Although most hair cells use an excitatory amino acid (glutamate), a subset of centrally located hair cells have been shown to be intensely immunoreactive for gamma-amino butyric acid (GABA). In fish, these GABAergic hair cells are present only in the central regions of the crista, exactly where the acceleration- and gain-sensitive afferents terminate. During mechanical labyrinthine stimulation, we employed the CGP class of compounds that specifically block GABAb receptors to study the effects of this GABAb receptor action on primary afferent dynamics. We discovered that GABA, in concert with glutamate had a major action in shaping the responses of acceleration and high gain afferent responses. For details, please refer to Holstein *et al.* (2004a, 2004b) and Highstein *et al.* (2005).

Studies of cupular structure have shown that the outer leaflets of both the canilicular and utricular faces of the cupula are composed of arrays of linear horizontally and vertically oriented fibers while the cupular internal structure seems to be a fine fibrous matrix or gel (Silver *et al.*, 1998). Stereocilliary bundles appear to be inserted into this gel. Visualization of the cupulary fibers and the stereocilliary bundles with the video microscope utilizing recently developed algorithms for the detection, tracking, and quantification of image motion promises to provide these data. Fluorescent beads tagged to a lectin were infused into the labyrinth and adhered to sugar residues on the surface of the cupula or vane. Macroscopic cupulary motion was imaged utilizing epi-illumination of these beads and video microscopic methods to visualize cupular motion during mechanical stimulation (indentation). The relative motion of individual stereocilliary bundles relative to the motion of the cupula could be determined. Each bead was visualized and its motion tracked during mechanical stimulation via a computer algorithm. Thus, a picture of the regional variations in cupular motion across the frequency bandwidth of stimulation can be constructed.

There is a long-term interest in the mechanisms and sequels of efferent action on the vestibular end organs. Hair cell receptor potentials were recorded (see the earlier discussion) during efferent stimulation *in vivo*, and profound efferent action on some hair cells could be demonstrated. The effects of efferent stimulation on the amplitude distribution of miniature (m) synaptic potentials recorded in primary afferents was also shown (Locke *et al.*, 1999). Efferent modification of synaptic potentials presumably occurs via efferent modulation of hair cell membrane potential, thus underlying efferent effects demonstrated in alert fish. Efferent stimulation seems to desynchronize transmitter release from hair cells, because a class of large, rapidly rising mEPSPs was removed following efferent action. Available pharmacological and physiological manipulations were employed to evaluate just how this happens. Previously, Boyle and Highstein (1990) demonstrated efferent action at the level of the primary afferent. The hair cell data cited earlier can now be related to that of the entire system previously demonstrated at the afferent level. We need to understand how the diminution of the receptor potential relates to efferent action on the entire spectrum of activity present on individual afferents. Considering the sophisticated signal detection in afferents, and the order in brainstem projections, the complex actions of the efferent system is not surprising. Efferent action seems to decrease the sensitivity of certain classes of afferent, raise the resting rate of others, and leave others unaffected. We envision that efferent action modifies the physiology of the vestibular periphery to emphasize certain aspects of incoming sensory signals in a context dependent way and may help the animal to emphasize, or pay attention to, certain aspects of head movement that have current behavioral relevance (Tricas and Highstein, 1990, 1991). Even the labyrinths of the earliest living vertebrates (e.g., Lamprey) are endowed with an efferent innervation similar to that of mammals (Lowenstein, 1955; Lowenstein and Sand, 1940; Wersäll, 1956; Wersäll and Bagger-Sjöbäck, 1974). It is our working tenet that information gleaned from the study of a lower animal, such as a fish, will provide clues to the operation of the efferent vestibular system in higher animals such as monkeys and humans.

In toadfish, goldfish, and monkeys, the efferent innervation of the labyrinth is both pre- and postsynaptic (Nakajima and Wang, 1974; Sans and Highstein, 1984; Wersäll and Bagger-Sjöbäck, 1974) with efferent synapses on both hair cells and primary afferent fibers. In toadfish, efferent neurons are spontaneously active and fire with a maximum rate of 100 Hz when activated via a variety of stimuli. The efferent system can be activated with electric pulse stimulation by placing electrodes within the efferent nuclei in the brain stem. This electric pulse stimulation produces monosynaptic excitatory postsynaptic potentials (EPSPs) in primary afferents (Highstein and Baker, 1985) and monosynaptic inhibitory postsynaptic potentials (IPSPs) in canal hair cells. The differential effect of efferent stimulation on different classes of semicircular canal primary afferent (Boyle and Highstein, 1990; Highstein and Baker, 1985) and different classes of lateral line afferent (Tricas and Highstein, 1991, 1990) has been illustrated. In a behavioral context, visual stimuli were employed to induce fish to activate their efferent system. When fish were shown smaller prey fish prey behind a glass barrier, they activated their efferents. A behavioral marker of this activation, namely the raising of the dorsal fin, was demonstrated. Raising of the dorsal fin probably results from the activation of phasic primary afferents that project directly to extensor motor neurons responsible for fin action.

We believe that the efferent system is in place to enhance the function of the peripheral sensory apparatus rather than to degrade it. We hypothesize efferent peripheral sculpturing of the signals carried by certain classes of afferent and suggest that the organism utilizes the efferent system to tune the octavolateralis periphery to emphasize certain behaviorally relevant aspects of incoming sensory information. It is not surprising that an organism can turn its attention to certain important aspects of its sensory environment, but it seems noteworthy that the brain is employed to tune peripheral receptors to accomplish this goal.

UREA PRODUCTION AND HYPERAMMONEMIA TOLERANCE AS A MODEL FOR HEPATIC ENCEPHALOPATHY

Most fish are sensitive to the toxic effects of ammonia, although generally the effects are not toxic until somewhat higher concentrations than in mammals. The general understanding of the mechanisms of ammonia toxicity in animals is important for many reasons but for purposes of this discussion largely because humans experience the disease hepatic encephalopathy (HE), whose pathology is associated with a rise in circulating plasma ammonia levels (Häusinger *et al.*, 2006). The liver/gallbladder (hepatobiliary) system is multifunctional, but a prime function is the detoxification of internal wastes and xenobiotics (foreign substances; see Chapter 6). Notably, in the human liver ammonia is detoxified metabolically through the production of urea in the ornithine-urea cycle (O-UC) (Fig. 29-3), which is then excreted via the kidneys or, to a lesser extent, through the production of glutamine (from the addition of ammonia to

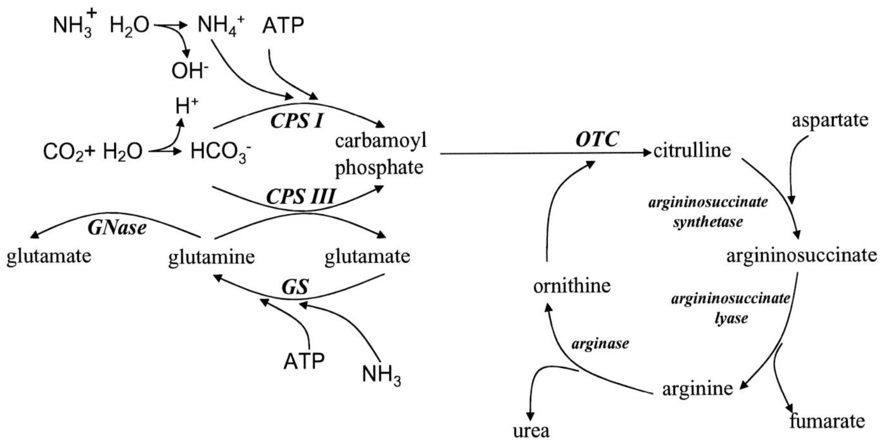

FIGURE 29-3. Schematic of the ornithine-urea cycle (O-UC) and accessory enzymes. Direct entry of ammonia into urea synthesis via CPS I is prevalent in mammals and other higher vertebrates, whereas entry via glutamine synthetase and CPS III is prevalent in many species of fish.

glutamate by the enzyme glutamine synthetase in a specialized subset of liver cells, the perivenous hepatocytes), which temporarily traps any overflow ammonia for later processing.

Liver failure in humans can occur in many ways. For example, on the acute time scale, ingestion of drugs or chemicals can immediately damage the organ (e.g., deliberate or accidental acetaminophen overdose appears to be more commonplace in emergency rooms as of late). In another short-term detrimental effect, portal hypertension (a constriction of the hepatic portal vein) can divert blood away from the liver, causing damage and less than optimal ammonia scavenging. On a longer time scale, liver cirrhosis (the impairment of blood flow caused by scarring of liver tissue from conditions such as chronic alcoholism), or simply inborn errors in the genes coding for the metabolic enzymes of the O-UC, can cause plasma ammonia concentrations to rise. Whether acutely or chronically, increased circulating ammonia levels causes severe toxicity to the main target tissue, the central nervous system. When circulating ammonia (normally at about 50 μM) even as much as doubles in humans, a toxic cascade in the brain results in a progression of neuropsychological deficits that depend on the length and severity of ammonia toxicosis (e.g., sleep disorder, cognitive impairment, coma, and eventually death). Thus, the phrase "when the liver ails, the brain fails" is commonplace among hepatologists and neurologists concerned with the treatment of HE.

Remarkably, after decades of study, although great progress has been made, the precise mechanism(s) of ammonia toxicity in the brain during HE and the causes of neuropathy are still not completely understood. However, we can discuss two of the most prominent features and hypothesized pathological pathways in HE, namely astrocytic swelling and N-methyl D-aspartate (NMDA) receptor/glutamate excitotoxicity. Astrocytic swelling appears to be caused by the direct action of the brain itself in attempting to detoxify the ammonia that the liver has failed to act on. Elevated plasma ammonia[1] freely crosses the blood-brain barrier (BBB), which is an anatomical/physiological complex that protects and nourishes the brain and consists of blood vessels, transport epithelia, and infiltrations of processes (feet) from astrocytes, or glial cells. The astrocytes ("white matter"), which in humans actually outnumber neurons ("gray matter") by approximately 10:1, support the neurons by supplying nutrition, removing waste and reprocessing a number of neurotransmitter molecules, including glutamate, from the synaptic cleft. Importantly, they also contain the sole pathway in the brain for detoxification of ammonia—that is, they are the exclusive brain location of the enzyme glutamine synthetase (GS). Under normal ammonia loads, astrocytic GS takes up glutamate (GLU) and processes it to glutamine (GLN) (by adding an ammonia molecule), which is then transported back to the neurons for reprocessing to GLU (Fig. 29-4). However, when exogenous ammonia swamps the brain, GS detoxifies this additional ammonia and creates more GLN than astrocytic export transport pro-

[1]"Ammonia" exists in biological solutions as the dissolved *ammonia* gas (NH_3) and as the *ammonium* ion (NH_4^+) with the proportions of the two compounds governed by the chemical equilibrium $NH_4^+ \leftrightarrow NH_3 + H^+$ and thus can be calculated by a reformulation of the Henderson-Hasselbalch equation such that $NH_4^+ = \text{Tamm}/\{1 + \text{antilog}(\text{pH-pK})\} = \text{Tamm} - NH_3$. Therefore, total Ammonia (Tamm) refers to the sum of ammonia (NH_3) and ammonium (NH_4^+). Since the pK for the above equilibrium is approximately 9 under biological conditions (although it varies with temperature, salinity, etc), at physiological pH (i.e., 7) ammonium exceeds ammonia in concentration by approximately 2 orders of magnitude. However, since generally biological membranes are far more permeable to ammonia than ammonium, and ammonia is also the more reactive form, it is believed that ammonia is the more toxic form of the two. Thus, in determining LC_{50} values and any other aspect of ammonia toxicity, it is important to compare studies based on *ammonia* concentrations.

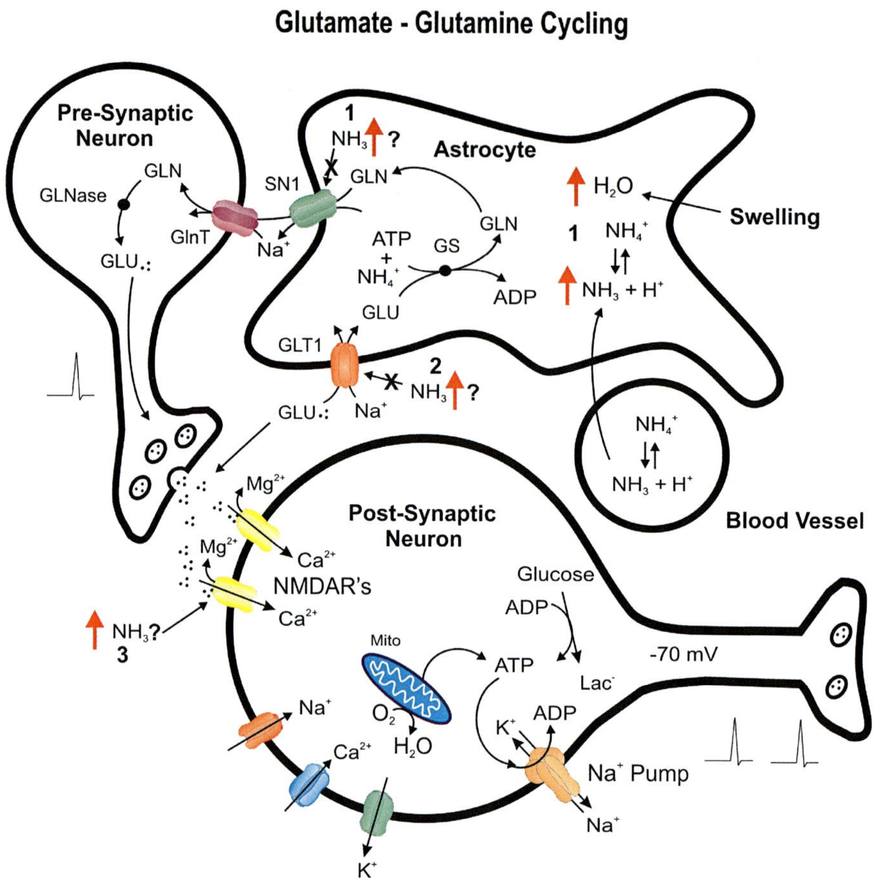

FIGURE 29-4. Glutamine/glutamate (GLN/GLU) cycle in neurons and astrocytes. Glutamate released into the synaptic cleft is taken up by Na^+ driven cotransport (GLT1) into astrocytes. Within the astrocyte GLU is enzymatically transformed to GLN by an ATP and NH_4^+ coupled reaction. GLN is transported by a glutamine/Na^+ cotransporter (SN1) to the extracellular space and taken up by neurons by another Na^+ coupled transporter (GlnT) and converted back to GLU by GLN synthetase and repackaged into presynaptic vesicles. There is also a Na^+ independent glutamine transport pathway, but it is not shown. Two major effects of ammonia toxicity are shown. (1) Astrocytic swelling from intracellular accumulation of GLN. This is the result of putative inhibition of GLN export from the cell and osmotic swelling. (2) Excess synaptic GLU is thought to be caused by NH_3 inhibition of astrocytic GLU uptake; the net effect is hyperstimulation of postsynaptic NMDAr. Acute NH_3 exposure may excite NMDAr directly. Legend: GS represents glutamine synthetase; GLNase, glutaminase; GLT1, glutamate transporter; GlnT, neuronal glutamine tranporter; SN1, astrocytic glutamine transporter. From Walsh *et al.* (2007).

cesses can handle. GLN accumulation leads to simple osmotic swelling of astrocytes, and this swelling (cellular edema) creates intracranial pressure that limits brain blood flow, causing brain damage. In rodent experimental models, the administration of a GS inhibitor, methionine sulfoximine (MSOX), before experimental ammonia loading effectively prevents astrocytic swelling and the subsequent neuropathology. Unfortunately, MSOX administration inhibits GS throughout the body and has other side effects such that it cannot be used as a treatment for HE.

Although the action of GS detailed here leads to a global depletion of brain GLU (and, in fact, an energy drain as the production of more GLU from Krebs' cycle intermediates is required), there is actually a localized synaptic *excess* of GLU in HE because the transporter that imports GLU into the astrocyte from the synaptic cleft is inhibited by ammonia. The failure of the astrocytes to clear GLU from the synaptic space leads to overstimulation of postsynaptic GLU receptors, most notably the NMDA receptors, so named because pharmacologically their high affinity for the glutamate analog N-methyl D-aspartate distinguishes them from other classes of GLU receptors. NMDA receptors are in essence glutamate activated Ca^{2+} channels that normally conduct a brief Ca^{2+} influx to the postsynaptic neurons allowing an action potential to be triggered. Their action is normally limited by the rapid removal of GLU from the synaptic cleft by the astrocytic GLU transporters. However, when GLU accumulates, too much Ca^{2+} enters the postsynaptic neuron, resulting in a toxic cascade of membrane disruption and oxidative stress, and ultimately to neuronal death. As in astrocytic swelling, glutamate excitotoxicity can be prevented pharmacologically: inhibition of NMDA receptors with the compound MK-801 prevents ammonia-induced neuropathy, but again nontissue specific action and side effects of MK-801 prevent its therapeutic use.

Given this background for HE, several groups of investigators working with fish models took note of the generally higher ammonia tolerances of fish, in particular the rather high (by even piscine standards) ammonia tolerances of toadfish and midshipmen (Walsh *et al.*, 2007). Although human and rodent LD_{50} values are on the order of 1 mM, similar LC_{50} values for plainfin midshipmen (*Porichthys notatus*) are 6 mM, for gulf toadfish (*Opsanus beta*) 10 mM, and for oyster toadfish (*Opsanus tau*) 20 mM. Researchers reasoned that if we can understand the mechanisms of these high batrachoidid ammonia tolerances, they might serve as useful animal models in the understanding and treatment of HE. Although research is still at an early stage with these species, several interesting insights are beginning to emerge.

The potential for brain swelling during ammonia exposure in fish has only been examined in one study of the gulf toadfish (*Opsanus beta*), where it was recently demonstrated that despite substantial chronic exposure to sublethal concentrations of ammonia (one third of the 96h LC_{50}) or acute exposure to lethal concentrations of ammonia (one to three times the 96h LC_{50}), toadfish showed no signs of brain swelling as assessed by water status using nuclear magnetic resonance imaging (Veauvy *et al.*, 2005). Fish, including toadfish, certainly have high levels of brain glutamine synthetase, and ammonia exposure did lead to substantial increases in toadfish brain [GLN]. Also as expected, preinjection of toadfish with MSOX prevented this increase in brain [GLN]. However, unlike mammalian models, MSOX *did not* change brain water status nor did it ameliorate symptoms and lethality from ammonia, but actually *accelerated* lethality, strongly suggesting that GS is a requirement for ammonia tolerance in the toadfish. Thus, it can be speculated that ammonia does not induce brain swelling in the toadfish model, for one of two reasons: (1) although brain GLN levels do rise, the *turnover* of GLN may be such that it can be rapidly exported from the brain to be processed or stored by other tissues; (2) if the accumulation of GLN in toadfish brain is localized to a very small area or number of cells, global brain swelling and its lethal effects may not occur. In this regard, recall that one of the reasons why astrocytic swelling in mammals can be so disruptive when ammonia is elevated is that brains have a high proportion of astrocytes to neurons. To date, data are scant in fish systems assessing the localization of brain GS, and the astrocyte to neuron ratio, but they suggest that GS may be localized to only a thin layer of brain cells (the ependymoglial cells) and that generally astrocyte:neuron ratios in brains of lower vertebrates may not be as high as in mammals. If further studies bear out the apparent lack of effect of MSOX in protecting fish from ammonia insult, this would suggest a fundamental difference in GLN handling and brain swelling during ammonia toxicity in fish versus mammals. (3) A third possibility for lack of brain swelling in fish is that the BBB may be structured differently, making fish less susceptible to vascular edema (differs from the cell edema mentioned previously and is caused by direct effects of ammonia on water transport by the BBB). Whereas the general functional permeability of the BBB in fish is similar to mammals (i.e., it is "tight"), virtually no information is available on the structure of the BBB in fish as it relates to water transport and ammonia transport and metabolism. Clearly, all of the preceding observations are fertile areas for research in understanding the mechanisms of how fish like the toadfish can resist brain swelling and perhaps discovering an explanation for why fish in general seem to be more tolerant of ammonia insults than do mammals.

If, as the limited dataset presented here suggests, ammonia-induced disruptions of brain swelling are *not* key causes of toxicity in fish, does glutamate excitotoxicity play the key role? Some evidence to date, in fact, suggests an involvement of NMDAr-mediated events for *Porichthys notatus*, as well as the nonbatrachoidid goldfish and weather loach, and the amelioration of ammonia toxicity by MK-801, although the results for all three species are largely unpublished and preliminary.

It has been known since the 1970s that, unlike the vast majority of fish, toadfish (at least *O. beta and O. tau*) have the capability of synthesizing urea in their liver (and muscle as well). Unlike mammals that directly consume ammonia in the first enzymatic step of their O-UC (namely carbamoyl phosphate synthetase, the "I" isozyme, or CPS I), fish that do express an O-UC have added a step to the pathway because their CPS prefers GLN over ammonia as the nitrogen donor, In fact, the same GS that detoxifies ammonia in the brains of both mammals and fish, and that is relegated to a "mopup" role in mammalian perivenous heptocytes, is the first step in urea production in fish. Thus, the presence of a pathway in fish that can directly consume the GLN produced (and exported?) in the brain could certainly account for an inherently higher turnover of GLN and ability to rid the brain of excess GLN as suggested earlier.

However, all of this laboratory experimentation begs the question, "Is there something inherent to the biology of toadfish that makes them excrete urea or makes them so resistant to ammonia?" An answer to this question recently began to emerge when it was discovered that the toadfish co-excretes approximately equal quantities of ammonia and urea across its gills, despite the fact that the fish is ostensibly capable of excreting only ammonia to an aqueous environment like other fish. Why bother to produce metabolically expensive urea? Researchers have discovered that whereas ammonia can be detected by the predators or toadfish, urea appears to blunt the response of predators to ammonia and effectively serves as a "chemical cloak" (Barimo and Walsh, 2006). Indeed, production of urea as cloaking molecule appears to be yet another antipredation strategy in the arsenal of this noisy family. However, there is still a great deal to understand about this interesting family of fishes relative to their use as a model for HE. For example, although *P. notatus* is just as sonically oriented as the *Opsanus* species, it does *not* produce urea. Recall that *P. notatus* may rely on other means (bioluminescence) to reduce the impact of predators. However, although it has a lower ammonia tolerance than the urea producing *O. beta* and *O. tau*, it still has a much higher ammonia tolerance than many other bony fishes. Thus, the family as a whole may have a component of their ammonia resistance that is not explained by GLN turnover and the ability to produce urea. Given these unknowns, undoubtedly, this family will be a useful model in the study of HE and ammonia toxicity in general for decades to come.

EPILOGUE: AQUACULTURE OF BIOMEDICAL SPECIES

Our discussion of toadfish and midshipmen has come full circle, indicating that many of the adaptations this species has to being so sonically oriented indeed make it such an interesting and important study species for biomedical research, indeed one that is highly sought after by researchers. In some geographical areas, fishes of the Family Batrachoididae are plentiful and caught in sufficient numbers to support biomedical research and not unduly harm the ecosystems from which they are taken. However, some areas have begun to experience shortages of these valuable species. With an eye toward maintaining an adequate supply of animals for research, to provide animals with consistently high health, *and* to protect environmental concerns, aquaculture studies of batrachoidids has been initiated.

The oyster toadfish has been a staple of scientific research since Ryder first reported on its development in 1887 with much of the research transpiring at the Marine Biological Laboratory (MBL) in Woods Hole, Massachusetts. In the 1960s, the toadfish was the focus of insulin research as its islet cells are segregated into a discrete mass rather than being scattered throughout the mammalian pancreas. More recently, the toadfish has served as a subject for auditory, vestibular, and muscle physiology studies, and ammonia toxicity studies, many of which have been reviewed in this chapter.

Since the 1960s, thousands of toadfish routinely were acquired by the MBL as a bycatch of the commercial eel fishery and used in research. However, since the fishery's collapse, the number of toadfish available for research has slowed to a trickle. Increased collection efforts by the MBL and investigators have proven unsuccessful indicating that the local population is in decline. Although the exact cause of the decline remains unknown, habitat degradation, the targeting of the fish by the Asian restaurant industry, and increases in the cormorant population all have been suggested as causes. In the past several years, toadfish have been imported from the Mid-Atlantic states to meet the need of researchers. Therefore, to relieve pressure on wild stocks, a mariculture program was initiated in 1998 to breed and culture toadfish for research. The goal of the project was to provide researchers with 500 fish (~25 cm; 400 g) per year.

Despite more than a century of investigation on the toadfish, there were little historical data to guide the mariculture efforts. The toadfish proved easy to breed and to be hardy in captivity with survival exceeding 90% after 2 years. However, raising the fish on locally available food (squid and butterfish) and at elevated temps (~20°C) resulted in disappointing growth rates of less than 25 g per year, which rendered the project scientifically and economically infeasible (Mensinger *et al.*, 2003). More recent experiments, which reared the fish on commercial fish pellets and at significantly higher year round temps (23 to 32°C), showed more promising results. Under these conditions, fish routinely reached 200 to 300 g within 2 years, indicating that it will take 3 years to culture research-sized animals (Mensinger and Tubbs, 2006). The scientific importance of the toadfish more than justifies the continuation of what is arguably the world's smallest mariculture project.

References

Barimo, J.F., Walsh, P.J., 2006. Use of urea as a chemosensory cloaking molecule by a bony fish. J. Exp. Biol. 209, 4254–4261.

Bass, A.H., 1996. Shaping brain sexuality. Am. Sci. 84, 352–363.

Bleckmann, H., Topp, G., 2003. Surface wave sensitivity of the lateral line organs of the topminnow *Aplocheilus lineatus*. Naturwissenschaften 68, 624–625.

Boyle R., Carey J.P., Highstein S.M., 1991. Morphological correlates of response dynamics and efferent stimulation in horizontal semicircular canal afferents of the toadfish, *Opsanus tau*. J. Neurophysiol. 66, 1504–1521.

Boyle, R., Highstein, S.M., 1990. Efferent vestibular system in the toadfish: Action upon horizontal semicircular canal afferents. J. Neurosci. 10, 1570–1582.

Boyle, R., Mensinger, A.F., Yoshida, K., Usui, S., Intravaia, A., Tricas, T., Highstein, S.M., 2001. Neural readaptation to Earth's gravity following return from space. J. Neurophysiol. 86, 2118–2122.

Braun, C.B., Coombs, S., 2000. The overlapping roles of the inner ear and lateral line: The active space of dipole source detection. Philos.Trans. R. Soc. Lond. B Biol.Sci. 355, 1115–1119.

Coombs, S., 1999. Signal detection theory, lateral-line excitation patterns and prey capture behaviour of mottled sculpin. Anim. Behav. 58, 421–430.

Hanke, W., Bleckmann, H., 2004. The hydrodynamic trails of *Lepomis gibbosus* (Centrarchidae), *Colomesus psittacus* (Tetraodontidae) and *Thysochromis ansorgii* (Cichlidae) investigated with scanning particle image velocimetry. J. Exp. Biol. 207, 1585–1596.

Häusinger, D., Kircheis, G., Schliess, F. (eds.), 2006. Hepatic Encephalopathy and Nitrogen Metabolism. Dordrecht, The Netherlands, Springer.

Highstein, S., Politoff, A., 1978. Relation of interspike baseline activity to the spontaneous discharges of primary afferents from the labyrinth of the toadfish, *Opsanus tau*. Brain Res. 150, 182–187.

Highstein, S.M., Baker, R., 1985. Action of the efferent vestibular system on primary afferents in the toadfish, *Opsanus tau*. J. Neurophysiol. 54, 370–384.

Highstein, S.M., Rabbitt, R.D., Boyle, R., 1996. Determinants of semicircular canal afferent response dynamics in the toadfish, *Opsanus tau*. J. Neurophysiol. 75, 575–596.

Highstein, S.M., Rabbitt, R.D., Holstein, G.R., Boyle, R.D., 2005. Determinants of spatial and temporal coding by semicircular canal afferents. J. Neurophysiol. 93, 2359–2370.

Holstein, G.R., Martinelli, G.P., Henderson, S.C., Friedrich, V.L., Jr., Rabbitt, R.D., Highstein, S.M., 2004a. gamma-Aminobutyric acid is present in a spatially discrete subpopulation of hair cells in the crista ampullaris of the toadfish *Opsanus tau*. J. Comp. Neurol. 471, 1–10.

Holstein, G.R., Rabbitt, R.D., Martinelli, G.P., Friedrich, V.L., Jr., Boyle, R.D., Highstein, S.M., 2004b. Convergence of excitatory and inhibitory hair cell transmitters shapes vestibular afferent responses. Proc. Natl. Acad. Sci. USA 101, 15766–15771.

Kanter, M.J., Coombs, S., 2003. Rheotaxis and prey detection in uniform currents by Lake Michigan mottled sculpin (*Cottus bairdi*). J. Exp. Biol. 206, 59–70.

Kroese, A.B., Schellart, N.A., 1992. Velocity- and acceleration-sensitive units in the trunk lateral line of the trout. J. Neurophysiol. 68, 2212–2221.

Locke, R., Vautrin, J., Highstein, S., 1999. Miniature EPSPs and sensory encoding in the primary afferents of the vestibular lagena of the toadfish, *Opsanus tau*. Ann. NY Acad. Sci. 871, 35–50.

Lowenstein, O., 1955. The effect of galvanic polarization on the impulse discharge from sense endings in the isolated labyrinth of the thornback ray (*Raja clavata*). J. Physiol. 127, 104–117.

Lowenstein, O., Sand, A., 1940. The mechanism of semi-circular canals: A study of single fiber preparations to angular acceleration and to rotation at constant speed. Proc. R Soc. Lond. B Biol. Sci. 129, 256–275.

Mensinger, A.F., Price, N.N., Richmond, H.E., Forsythe, J.W., Hanlon, R. T., 2003. Mariculture of the oyster toadfish: Juvenile growth and survival. N. Am. J. Aquaculture 65, 289–299.

Mensinger, A.F., Tubbs, M.E., 2006. Effects of temperature and diet on the growth rate of year 0 oyster toadfish, *Opsanus tau*. Biol. Bull. 210, 64–71.

Miya, M., Takeshima, H., Endo, H., Ishiguro, N.B., Inoue, J.G., Mukai, T., Satoh, T.P., Yamaguchi, M., Kawaguchi, A., Mabuchi, K., Shirai, S.M., Nishida, M., 2003. Major patterns of higher teleostean phylogenies: A new perspective based on 100 complete mitochondrial DNA sequences. Mol. Phylogenet. Evol. 26, 121–138.

Montgomery, J.C., Baker, C., Carton, A., 1997. The lateral line can mediate rheotaxis in fish. Nature 389, 960–963.

Montgomery, J.C., Macdonald, J.A., 1987. Sensory tuning of lateral line receptors in Antarctic fish to the movements of planktonic prey. Science 235, 195–196.

Nakajima, Y., Wang, D.W., 1974. Morphology of afferent and efferent synapses in hearing organ of the goldfish. J. Comp. Neurol. 156, 403–416.

Nelson, J.S., 1994. Fishes of the World, 3rd ed. New York, John Wiley & Sons.

Palmer, L.M., Deffenbaugh, M., Mensinger, A.F., 2005. Sensitivity of the anterior lateral line to natural stimuli in the oyster toadfish, *Opsanus tau* (Linnaeus). J. Exp. Biol. 208, 3441–3450.

Partridge, B.L., Pitcher, T.J., 1980. The sensory basis of fish schools: Relative roles of lateral line and vision. J. Comp. Physiol. 135, 315–325.

Rabbitt, R.D., Boyle, R., Highstein S.M., 1995. Mechanical indentation of the vestibular labyrinth and its relationship to head rotation in the toadfish, *Opsanus tau*. J. Neurophysiol. 73, 2237–2260.

Remage-Healey, L., Nowacek, D., Bass, A.H., 2006. Dolphin foraging sounds suppress calling and elevate stress hormone levels in a prey species, the gulf toadfish. J. Exp. Biol. 209, 4444–4451.

Rome, L.C., Lindstedt, S.L., 1998. The quest for speed: Muscles built for high-frequency contractions. News Physiol. Sci. 13, 261–268.

Rusch, A., Thurm, U., 1990. Spontaneous and electrically induced movements of ampullary kinocilia and stereovilli. Hear Res 48, 247–263.

Sans, A., Highstein, S.M., 1984. New ultrastructural features in the vestibular labyrinth of the toadfish, *Opsanus tau*. Brain Res. 308, 191–195.

Sarnat, H.B., Netsky, M.G., Martin G., 1974. Evolution of the Nervous System. New York, Oxford University Press.

Silver, R.B., Reeves, A.P., Steinacker, A., Highstein S.M., 1998. Examination of the cupula and stereocilia of the horizontal semicircular canal in the toadfish *Opsanus tau*. J. Comp. Neurol. 402, 48–61.

Tricas, T.C., Highstein, S.M., 1990. Action of the octavolateralis efferent system upon the lateral line of free-swimming toadfish, *Opsanus tau*. J. Comp. Physiol. A 169, 25–37.

Tricas, T.C., Highstein, S.M., 1991. Visually mediated inhibition of lateral line primary afferent activity by the octavolateralis efferent system during predation in the free-swimming toadfish, *Opsanus tau*. Exp. Brain Res. 83, 233–236.

Veauvy, C.M., McDonald, M.D., Van Audekerke, J., Vanhoutte, G., Van Camp, N., Van der Linden, A., Walsh, P.J., 2005. Ammonia affects brain nitrogen metabolism but not hydration status in the gulf toadfish (*Opsanus beta*). Aq. Toxicol. 74, 32–46.

Walsh, P.J., Bedolla, C., Mommsen, T.P., 1989. Scaling and sex-related differences in toadfish (*Opsanus beta*) sonic muscle enzyme activities. Bull. Mar. Sci. 45, 68–75.

Walsh, P.J., Mommsen, T.P., Bass, A.H., 1995. Biochemical and molecular aspects of singing in Batrachoidid fishes. In Hochachka, P.W., and Mommsen, T.P. (eds.), Biochemistry and Molecular Biology of Fishes, Vol. IV, Metabolic and Adaptational Biochemistry, pp. 279–289. New York, Elsevier.

Walsh, P.J., Veauvy, C.M., McDonald, M.D., Buck, L.T., Wilkie, M.P., 2007. Piscine insights into comparisons of anoxia tolerance, ammonia toxicity, stroke and hepatic encephalopathy. Comp. Biochem. Physiol. 147A, 332–343.

Weissert, R., von Campenhausen, C., 1981. Discrimination between stationary objects by the blind cave fish *Anoptichthys jordani* (Characidae). J. Comp. Physiol. 143, 375–381.

Wersäll, J., 1956. Studies on the structure and innervation of the sensory epithelium of the cristae ampullares in the guinea pig. Acta Otolaryngology Supplement (Stockh) 126, 7–85.

Wersäll, J., Bagger-Sjöbäck, D., 1974. Morphology of vestibular sense organ. In Kornhuber, H.H. (ed.), Handbook of Sensory Physiology, Vestibular System, Basic Mechanisms, 6: 123–170. Berlin, Springer-Verlag.

STUDY QUESTIONS

1. What factors have led to the use of batrachoidid fish as biomedical models?
2. What is a sneaker male, and what are the evolutionary advantages/disadvantages of this reproductive strategy?
3. What are the uses of the lateral line system in fish? What specific uses of the lateral line has the toadfish

emphasized? Is there an equivalent system in terrestrial animals?
4. What experimental advantage was gained by sending toadfish into space? Are there traits of the toadfish that made them especially amenable to space flight?
5. Why is urea production in fish so unusual?
6. What are the causes of hepatic encephalopathy? What is the mechanistic basis of this disease?
7. Think of mechanisms for how urea might act as a cloaking molecule? That is, at the molecular level, how might it interfere with ammonia sensing by a predator?
8. At normal seawater pH (~8), what is the proportion of total ammonia that is present as NH_3? How about at pH 7 or 9?

C H A P T E R

30

Lower Deuterostomes as Models of the Developmental Process

ROBERT W. ZELLER AND R. ANDREW CAMERON

INTRODUCTION

The plants and animals that inhabit the oceans constitute an astounding pool of morphological and physiological variation. It is from this diversity that individual model systems are adopted to address particular problems in cell and developmental biology. One such area is the mechanisms of development where morphologically simpler organisms that are closely related to humans have made a major contribution. This contribution is enhanced in this era of genomics, for it is the genome that dictates the form of an organism through the instructions hard-wired in the genome.

One of the advantages of studying the embryos of marine invertebrates is that the embryos develop outside of the mother within the seawater. For many species, it is easy to culture the embryos in seawater and thus have access to various developmental stages for study. In general, the embryos of marine invertebrates tend to be small and are composed of few tissue types and relatively small numbers of cells. Because the embryos are free living, it is possible to manipulate the embryos by a number of techniques and observe alterations in development. In many cases, similar experiments cannot be conducted in vertebrate models such as the mouse. Early developmental biologists recognized the advantages of working with externally developing embryos more than one hundred years ago, and an extensive body of knowledge has been assembled based on their early experiments. With advances in optical microscopy and labeling techniques, and molecular and cell biological approaches, many of the questions first posed by our scientific forefathers are being readdressed. With these modern approaches, we hope to one day be able to mechanistically explain many of the developmental processes that were first described more than a century ago.

PHYLOGENY OF SEA URCHINS AND ASCIDIANS

Deuterostomes

Sea urchins and ascidians (Fig. 30-1) belong to the same major branch of the bilaterians as the vertebrates, the deuterostomes. In the original classification (Grobben, 1908), the deuterostomes are distinguished from the other branch, the protostomes, by a series of developmental features: the blastopore (the first opening) that arises at gastrulation becomes the anus and the second opening becomes the mouth; the mesoderm arises by invagination from the endoderm, the patterns of cleavage are radial. Judging by the amount of controversy that has continued through the years following Grobben's classification, these characters were insufficient to unequivocally classify some less well-studied groups. However, modern molecular analyses have mostly set to rest the controversy and an accepted phylogeny based on ribosomal DNA sequences is available (Fig. 30-2). In the newest phylogenies combining molecular and morphological features, a contested group such as lophophorates or arrow worms is squarely situated in the appropriate bilaterian branch. In this scheme, the placozoans are placed in a position basal to the bilaterians based on extensive similarities in gene sequences. Outside this grouping lie the cnidarians, the group generally considered to be the best outgroup for comparison to bilaterians.

The bilaterians are now divided into three superclades: the deuterostomes, the lophotrochozoans, and the ecdysozoans (Aguinaldo *et al.*, 1997; Halanych, 2004). Included in the ecdysozoans are the nematodes and arthropods and related phyla, whereas the lophotrochozoans include groups such as annelids, mollusks, and their allies. The placement of the lophotrochozoans as more closely related to the deu-

FIGURE 30-1. The adult forms of the embryonic model systems from the nonvertebrate deuterostomes, the ascidian and the sea urchin. (A) Solitary ascidians: *Ciona savigny* on the left and *Ciona intestinalis* on the right. The larger incurrent siphon is vertical, whereas the smaller excurrent one is lateral. (B) Three purple sea urchins, *Strongylocentrotus purpuratus*, attached by their locomotory structures called tube feet to their preferred food the kelp, *Macrocystis pyrifera*.

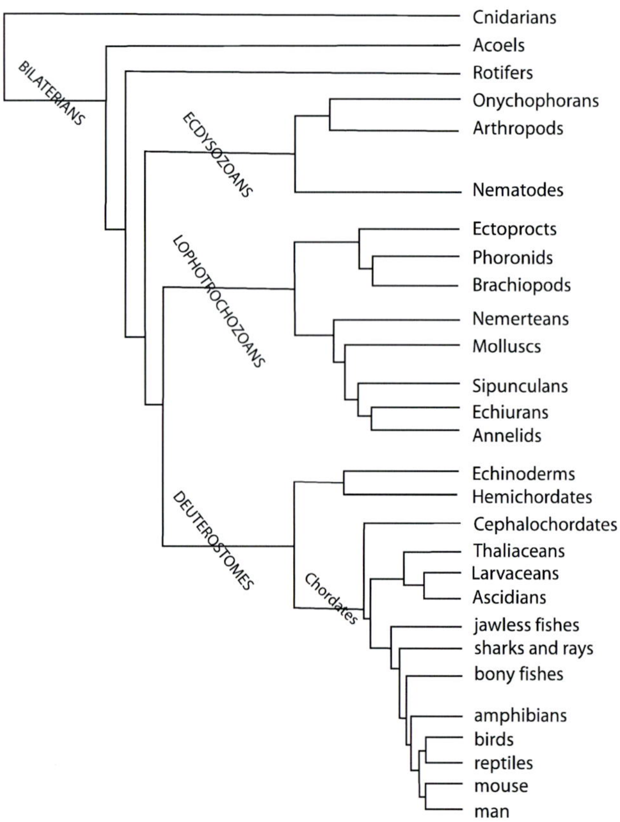

FIGURE 30-2. A phylogenetic tree of selected metazoan animals. The bilaterians are divided into three super-clades: deuterostomes, lophotrochozoans and ecdysozoans. The closest relative to the bilaterians is thought to be the placozoans but the cnidarians are a better bilaterian outgroup because they have a more extensive gene catalog.

terostomes is a conclusion based on molecular and morphological data (Eernisse and Peterson, 2004).

The explosion in genome sequencing among the deuterostomes has led to a new set of phylogenetic relationships there as well. Echinoderms and hemichordates are sister groups, the ambulacrarians, and arose at the base of the deuterostome lineage. The lineage of the urochordates (ascidians, thalacians, and larvaceans) was long considered to emerge above the ambulacrarians but positioned basally to the cephalochordates and vertebrates. Based on several analyses of large gene sets from multiple deuterostome species, the urochordates are now considered to be the sister group of the vertebrates, whereas the cephalochordates are positioned as basal chordates (Delsuc *et al.*, 2006; Vienne and Pontarotti, 2006). Expressed sequence tags (EST) from the enigmatic echinoderm-like Xenoturbella were used in a similar analysis to place this animal basal to hemichordates and echinoderms (Bourlat *et al.*, 2006).

Echinoderm Classes

Sea urchins belong to the echinoderm class, Echinoidea, which is an animal group with an exceptionally complete fossil record. The echinoderms are made up of five extant classes of echinoderms: crinoids or feather stars, holothuroids or sea cucumbers, asteroids or sea stars, ophiuroids or brittle stars, and echinoids or sea urchins. The first identifiable echinoderms possessing the characteristic skeleton were a stylophoran that appeared in the fossil record about 510 mya (Smith, 2005). Other stem group ancestors that possessed the diagnostic features of the phylum are found in the fossil record starting at about 520 mya. The earliest ancestor of the modern echinoderms occurred in the early Ordivician at about 485 mya (Sprinkle and Wilbur, 2005). However, modern sea urchins did not appear until slightly later in the Late Ordivician about 450 mya (Smith and Savill, 2001). The modern forms like the purple sea urchin trace their ancestry back to the Permian when the cidaroid or "pencil-spined" sea urchins appeared (Smith and Hollingworth, 1990). Many other lineages disappeared as a result a mass extinction at that time. This excellent fossil history offers the opportunity to correlate ancient structures with the modern genetic apparatus that underlie them in modern animals (Bottjer *et al.*, 2006).

Tunicates

There are two major groups of invertebrate chordates: the tunicates (urochordates) and the cephalochordates. As mentioned previously, the cephalochordates were once considered to be the sister group to the vertebrates, but they are now placed basal to the chordates. There are approximately 20 species of cephalochordates, and there are a small but important number of laboratories that actively study the development of these organisms (Holland *et al.*, 2004; Schubert *et al.*, 2006). The second major group of invertebrate chordates is the tunicates. All tunicates are marine organisms and are composed of several thousand species. Although several molecular phylogenies have been proposed for the tunicates, these are primarily based on 18S ribosomal RNA sequences. Given the evidence for the rapid evolution of the ascidian genome (Delsuc *et al.*, 2006), a more complete phylogenetic analysis must await larger-scale sequencing efforts that, given the rapid development of large-scale sequencing technologies, should come to fruition in the near future. These animals possess cellulose and form a distinctive outer covering called the "tunic" (hence the name *tunicates*). Tunicates are divided into three different classes, the largest of which are the ascidians, comprising about 2500 species. Ascidians are sessile marine organisms that inhabit all of the major oceans. Although the adult ascidians seemingly lack any chordate features, the ascidian tadpole larvae possess the key chordate characteristics of a dorsal hollow nerve cord and a notochord flanked by tail muscles. These features were first described by the famous Russian embryologist Alexander Kowalevsky (1866, 1871) and cemented their position within the chordates (until this time they were considered to be mollusks). There are several minor tunicate classes, and these are primarily pelagic animals. Some of them, such as the larvaceans, play major roles with carbon cycling in the world's oceans (Robison *et al.*, 2005). Superficially, larvaceans look like ascidian tadpole larvae except that they secrete a mucous "house," which serves as a feeding structure. When a larvacean house becomes occluded, the animal jettisons the house and builds a new one; the discarded house then sinks to the bottom of the ocean.

Ascidians are the best-studied tunicates and are divided into three orders: the phlebobranchs, the stolidobranchs, and the aplousobranchs. Ascidians are generally self-sterile hermaphrodites, so gametes from two individuals must mix for fertilization to occur. However, the degree of self-sterility varies, so that some individuals may successfully self-fertilize, whereas others, even from the same vicinity, may not. Many ascidians exist as solitary individuals in which reproduction is sexual. Interestingly, many ascidian species are colonial and both sexual as well as asexual reproduction occurs (Satoh, 1994). Extensive study of colonial ascidians has led to a detailed knowledge of histocompatability in colonial ascidians (De Tomaso *et al.*, 2005); little is known about this process in solitary ascidians. The adult ascidian is a sessile filter feeder and at least two chordate characteristics are found in the adults: gill slits in the brachial basket and an endostyle that is part of the filter-feeding structure (Whittaker, 1997).

TUNICATE MODEL SYSTEMS

The Ascidian Larva and Embryological History

The ascidian larva is bilaterally symmetrical and combines a basic chordate body plan typical of vertebrates together with a simple body organization typical of most invertebrate larvae. One of the most studied species, *Ciona intestinalis*, produces larvae about 18 hours after fertilization (Fig. 30-3). The larvae, composed of about 2500 cells, are a little more than 1 mm in length and will swim for several hours until they undergo metamorphosis (Satoh, 1994). Adhesive structures called palps, located at the anterior end of the larva, attach to a substrate and metamorphosis then ensues. It takes 6 to 8 weeks until gametes (sperm) are first visible in the gonoducts; egg production typically takes several weeks longer. The cell lineage of the embryo was first described by Edwin Conklin (1905) with subsequent minor corrections (Nishida, 1987; Nishida and Satoh, 1983, 1985). The larval tissue types are quite simple: 40 notochord cells, 36 muscle cells, and a central nervous system composed of about 330 cells. The bulk of the remaining cells of the larvae include about 800 epidermal cells that cover the outside of the embryo, 500 endoderm cells, and about 900 mesenchyme cells that will give rise to adult mesodermal structures after metamorphosis.

A rich history of experimental ascidian embryology exists and extends back more than 100 years. Early experiments by the French embryologist Laurent Chabry (1887) on ascidian embryos inspired the development of one of

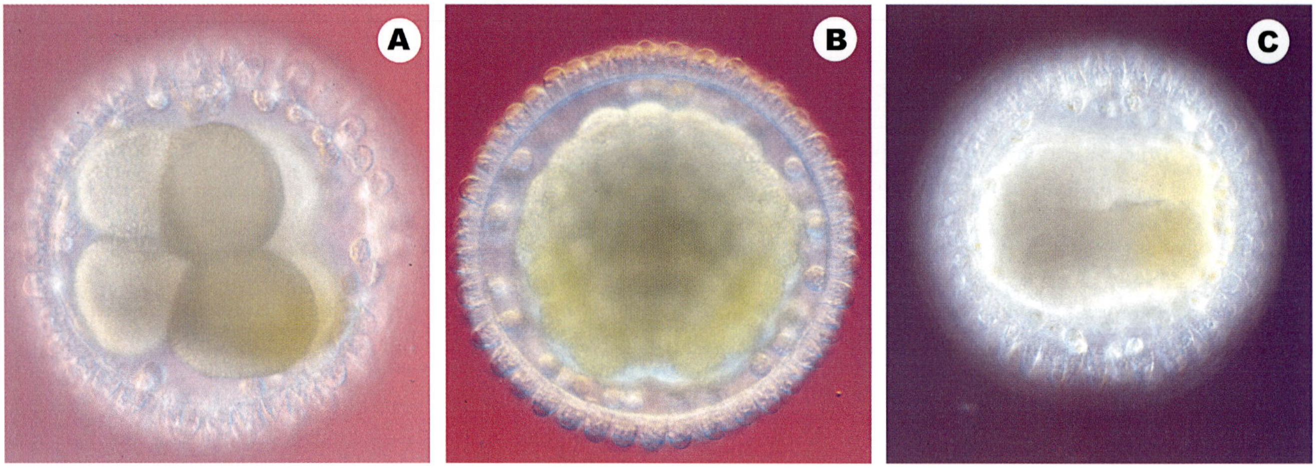

FIGURE 30-3. Ascidian development. Three representative stages of embryogenesis in the ascidian, *Styela plicata*. At these stages, the embryo is surrounded by a chorion, test cells, and follicle cells. (A) The 8-cell stage embryo viewed from the side. The yellow cytoplasm and the distinctive asymmetry identify the B4.1 blastomere pair (lower right pair of cells). (B) The 64-cell stage embryo viewed from the vegetal pole (dorsal side). The endoderm cells (dark gray color) are in the center and will begin to invaginate at the next cell division. The muscle cell precursors are identified by the yellow cytoplasm below and lateral to the endoderm. (C) The early tail-bud stage embryo viewed from the vegetal pole (dorsal side). The yellow myoplasm is visible in the muscle cells located in the emerging tail on the right side of the embryo. The neural tube is closing and can be seen running along the anterior-posterior axis of the embryo (left to right in the figure).

the first micromanipulators. Using this device, Chabry killed specific cells (early blastomeres) and observed the effect on subsequent embryonic development. Much of the early research on ascidian development relied on the histospecific differentiation of embryonic tissues, as molecular markers were not available until the late 1980s. In early experiments, S. Meryl Rose (1939) at the Marine Biological Laboratory (MBL), Woods Hole, relied on the development of two pigmented cells—the otolith (a primitive ear), and the ocellus (a primitive eye)—to indicate brain differentiation. He performed a series of experiments in which blastomeres were deleted or recombined and brain development was then assayed by pigment cell formation. Later investigators, including Reverberi and Mingati (1946, 1947), succeeded in linking cell lineage and cell fate in ascidian embryos. The key finding of their experiments was that partial embryos developed only those cell types predicted by the cell lineage map (Whittaker, 1997). In the 1970s, J. Richard Whittaker, also at the MBL, employed a series of histochemical stains to assay ascidian tissue differentiation (1973) and used these assays to investigate early ascidian cell fate specification in a series of elegant experiments (see Whittaker, 1987). These early experiments demonstrated that ascidian embryos contained numerous maternal determinants that were responsible for regulating early cell fate decisions.

Once established that maternal determinants played a major role in ascidian cell fate specification, researchers then asked (1) to what degree do autonomous (maternal determinants) and conditional (cell-cell signaling) specification play in the differentiation of embryonic tissues and (2) what is the identity or what are the identities of the determinant(s)? A series of blastomere separation and recombination experiments by Hiroki Nishida (reviewed by Nishida, 2002, 2005) further refined the role of maternal determinants and conditional specification in ascidian development. Nishida's major findings were that epidermis, muscle, and endoderm specification were highly dependent on maternal determinants, whereas notochord and nervous system development instead relied more on conditional specification mechanisms. Nishida and others have since identified several maternal determinants that are essential for the proper specification of different embryonic tissues. One key determinant encodes an mRNA for the zinc-finger transcription factor called Macho-1. The embryonic muscle cells normally inherit the maternally derived Macho-1 mRNA, and this step is crucial for the proper specification of embryonic muscle cells. If Macho-1 function is reduced, the number of muscle cells formed is reduced. Alternatively, if Macho-1 mRNA is expressed in cells that do not normally make muscle cells, those cells will start to express muscle-specific genes (Sawada *et al.*, 2005). Macho-1 also plays a key role in regulating the functional outcome of FGF signaling in the dorsal territories of the embryo. In the posterior-dorsal cells of the early embryo, the presence of Macho-1 activity causes cells to adopt a mesenchyme fate when induced by FGF, whereas in the anterior-dorsal cells, the absence of Macho-1 activity results in cells adopting a notochord fate when they receive the FGF signal (Kobayashi *et al.*, 2003).

Molecular Mechanisms and Tools in the Tunicate Toolbox

Beginning in the 1980s and 1990s, several groups, including Noriyuki Satoh and colleagues, embarked on efforts to identify the molecules responsible for regulating early development. Initial experiments identified genes involved in muscle cell specification (e.g., Meedel *et al.*, 1997), as well as notochord specification (e.g., Corbo *et al.*, 1997). These gene identification efforts were aided by several key developments in the acquisition of tools for studying ascidian development. The first of these key innovations was the discovery that exogenous DNA could be easily introduced by a simple electroporation procedure (Corbo *et al.*, 1997). Either commercial or custom-built electroporation devices (Zeller *et al.*, 2006a) may be used to generate thousands of transgenic embryos in less than 1 hour. Transgenic embryos may be generated that express reporter transgenes that express optimized fluorescent proteins (Zeller *et al.*, 2006b) or express genes encoding regulatory molecules that alter normal embryonic development (Di Gregorio and Levine 2003). Lastly, the availability of genomic sequences from two closely related species, *C. intestinalis* and *C. savignyi*, has provided the means for comparative genomics methods to identify putative cis-regulatory domains from numerous genes of interest (Johnson *et al.*, 2005; Shi *et al.*, 2005).

Efforts by two groups have led to the development of protocols for conducting traditional forward genetic screens in ascidians (Moody *et al.*, 1999; Nakatani *et al.*, 1999; Sordino *et al.*, 2001). Several mutants have been identified from either more traditional chemically induced mutagenesis or from spontaneously occurring mutations isolated from wild populations (Jiang *et al.*, 2005). The Satoh group has developed a transposon-based system, which may be used to generate stable transgenic embryos or to introduce insertional mutations into the ascidian genome (Sasakura *et al.*, 2003, 2005). The Satoh group has also sequenced nearly 700,000 ESTs from various stages of *Ciona* development and adult tissues (reviewed in Shi *et al.*, 2005). Nearly a third of the predicted genes in the *Ciona* genome (discussed later) have been analyzed by *in situ* hybridization, thus there is an extensive body of knowledge about both temporal and spatial gene expression during embryonic development. Lastly, an initial functional analysis of 76 regulatory genes has provided the first working "glimpse" of the gene regulatory network that is operating in the early ascidian embryo (Imai *et al.*, 2006). Although this network analysis is not nearly as extensive as that described for endomesoderm specification in the sea urchin embryo (described later), it has nonetheless provided insight into the mechanisms of early ascidian, and hence chordate, development.

Tunicate Genomes

Three different tunicate genomes have been sequenced to date. There are two ascidian genomes, for *Ciona intestinalis* (Dehal *et al.*, 2002) and *C. savignyi* (unpublished), and one genome for the larvacean *Oikopleura dioica* (Seo *et al.*, 2001). Because tunicate genomes are polymorphic, it was difficult to assemble the raw reads of genomic sequence (as discussed later in the Sea Urchin Model System section). Although three tunicate genomes have been sequenced, they are not equally assembled, analyzed, and annotated. The Oikopluera genome is estimated to be between 65 and 70 MB in size, the smallest chordate genome known to date (Seo *et al.*, 2001). The Oikopleura genome has not been assembled, and only about half of the genome has been shotgun sequenced and assembled into short contigs. Interestingly, although the genome contains nine Hox genes, these genes are dispersed throughout the genome and are not organized in a cluster (Seo *et al.*, 2004), but they are expressed in a predominately collinear anterior-posterior pattern within several tissues within the tail. The *C. savignyi* genome was sequenced and assembled by the Broad Institute at MIT; however, a paper describing the sequencing of this genome has not yet been published. Details of the sequencing and assembly strategy can be found at www.broad.mit.edu/annotation/ciona/assembly.html. Shotgun plasmid libraries were constructed from a single individual animal, and raw sequencing was obtained at 13× coverage. The level of polymorphisms between the two haplotype genomes is about 5%, but overall the level is closer to 19% if one includes large indels. The *C. savignyi* genome size is estimated to be 180 MB; 164 MB has been assembled in the first release.

The most complete genome, from *C. intestlinalis*, was sequenced and assembled at the Joint Genome Institute (JGI) (Dehal, *et al.*, 2002). The genome is 155 MB, 117 MB appears in the assembly, and it contains an estimated ~16,000 predicted genes. A size-selected plasmid genomic library was constructed from genomic DNA from a single individual. More than 2 million raw reads provided more than 8× coverage. Additional sequence was obtained from the end sequences of both bacterial artificial chromosome (BAC) and cosmid libraries. Several different gene prediction algorithms were then employed to identify the ~16,000 genes in the genome. Many of these genes have been annotated, but the assembly and annotation are incomplete. Because the *C. intestinalis* genome was the first invertebrate chordate genome sequenced, it provided an excellent reference point from which to determine which genes are deuterostome or chordate specific. The JGI determined that the gene content for *C. intestinalis* was intermediate between the known protostome and vertebrate genomes. About 60% of the genes were found to be in common with protostomes and presumably represent a core set of bilaterian genes (Dehal *et al.*, 2002). Nearly one sixth of the *Ciona* genes lack a protostome counterpart yet have clear orthologs in vertebrate genomes. This indicates that these genes are either chordate specific or deuterostome specific. Because the sea urchin genome was not yet available, this distinction is unclear. A new analysis incorporating the new sea urchin genome data has not yet been conducted. Lastly, about 20% of the genes appear to be tunicate or ascidian specific genes with no clear homologs in either protostomes or vertebrates.

The *C. intestinalis* genome does not contain extensive gene duplications, as is the case in vertebrates, supporting the hypothesis that genome duplications occurred during the evolution of vertebrates (Dehal *et al.*, 2002). However, there are some instances of duplicated genes within the genome and are most likely examples of ascidian- or tunicate-specific gene duplication events. The *C. intestinalis* genome contains examples of all of the major families of regulatory genes that encode either transcription factors or signaling molecules. *Ciona* lacks steroid hormones, yet the genome encodes several examples of nuclear receptors (Dehal *et al.*, 2002). The genome also lacks any of the genes that enable adaptive immunity as found in vertebrates, although many other genes that enable innate immunity are present. There are nine Hox genes, but they are not present in a single cluster as is commonly found in other species—the cluster is split between two different chromosomes (Ikuta and Saiga, 2005). Although the *Ciona* genome contains many genes in common with both protostomes and vertebrates, the genome has lost many genes and thus may not represent an ancestral chordate genome but may instead reflect the derived nature of tunicates (Hughes and Friedman, 2005).

Why sequence the genomes of tunicates? Ascidians are the only invertebrate chordates with extensive experimental embryology and molecular tools available to dissect the molecular mechanisms of early development. As previously mentioned, tunicates are now considered to be the sister group of the vertebrates; thus, they occupy a unique position within the tree of life. The other major invertebrate model organisms employed in biomedical research to date, *Drosophila*, *C. elegans* and the sea urchin, have yielded tremendous insight into human diseases and cellular and developmental processes such as apoptosis, microRNAs, and RNAi; however, they are not chordates and lack key features of the chordate body plan. Ascidians in particular therefore offer an excellent system in which to study the development of chordate-specific tissues and organs, such as the dorsal neural tube and the notochord—structures that are lacking in the other invertebrate models. In addition, the ease of embryonic manipulation and the wealth of tools and techniques for the functional analysis of ascidian development make them an excellent model organism. The combination of these assets is perhaps best illustrated by the work of Brad Davidson and Mike Levine on ascidian heart development. They showed that ascidians employ a core regulatory gene network for heart specification that is shared by many organisms (Davidson and Levine 2003, Davidson *et al.*, 2005). Recently, they showed that by manipulating key gene function in this network, they could convert the normal single-chambered heart to a two-chambered heart (Davidson *et al.*, 2006). Amazingly, this heart beats well, and the chambers appear to be coupled. This exciting result is thus providing insight into the development and evolution of the vertebrate/human heart, where an understanding of how heart chamber number changes in evolution has been sought for decades. Countless other examples are waiting to be discovered.

Because of several phylogenetic uncertainties among the various invertebrate deuterostomes, it would be desirable to sequence a large number of species to firmly establish their phylogenetic positions. Given the rapid evolution of large-scale sequencing technologies, this should be feasible in the near future. Apart from establishing phylogentic positions, large-scale genome sequencing may shed insight into the evolution of the deuterostome and chordate genomes. For example, an analysis of a large number of deuterostome genomes could determine which genes are deuterostome specific, which are chordate specific, and which are likely common to all bilaterians.

SEA URCHIN MODEL SYSTEM

The sea urchin is the best-used biomedical research model of the Phylum Echinodermata. Indeed, it is an iconic example of that phylum that is characterized by a water vascular system and a calcium carbonate endoskeleton (Brusca and Brusca, 2003). The water vascular system is a network of tubular vessels that acts as a hydraulic skeleton for the terminal branches, tube feet. The tube feet protrude through the body wall in five paired rows. Scans of the gene models predicted from the genomic sequence of the purple sea urchin show that many products of sensory genes are expressed in the tube feet, suggesting that they are sensory organs in addition to having locomotory function (Raible *et al.*, 2006). The skeleton is a complex of aligned crystals of calcium carbonate with about 5% calcium phosphate embedded in an organic matrix that is organized into an open meshwork, the spaces of which are occupied by dermal cells and fibers. The organic matrix proteins are a set of structural proteins unique to echinoids and probably echinoderms since this structure is representative of all five classes of echinoderms and is evident in fossils back as far as the beginning of the Cambrian, ~540 million years ago. These proteins are characterized by the possession of a c-type lectin domain and a unique glycine-proline rich amino acid repeat similar to the pericardin protein motif described in the protein motif database called PFAM. PFAM is a large collection of multiple sequence alignments and hidden Markov models that define many common protein domains and families (Finn *et al.*, 2006).

The regular sea urchins are free-living exclusively marine animals that have a bilaterally organized embryo and larva but a pentameral adult body plan. Embryos develop from broadcast spawnings of mature gametes and develop into a feeding larval stage in a few days in temperate species like the purple sea urchin (reviewed in Pearse and Cameron, 1991). The pattern of cleavage is radial and morphogenesis begins after the establishment of a hollow blastula and continues into the gastrula (Fig. 30-4). After a larval period of about 6 weeks, the purple sea urchin larva (Fig. 30-4) undergoes a drastic morphological metamorphosis and takes up the adult habit in the nearshore subtidal or intertidal zone. The purple sea urchin is a pivotal member of the kelp forest ecosystem where their grazing balances kelp productivity, and they are preyed upon by a variety of predators including sea otters and humans (Pearse, 2006).

History of the Sea Urchin Model

One way to measure the utility of the sea urchin as a model system is to consider its history in scientific research over the past 150 years for the discoveries made using the sea urchin egg or embryo paralleled the scientific advancements of the 19th century and beyond (reviewed in Ernst, 1997; Pederson, 2006). About the time that Ernst Abbe began to study optical problems and improved the microcopes built at the Carl Zeiss factory in Jena, Germany, Hertwig (1876) wrote of his observations on the fusion of sperm and egg pronuclei, and Fol (1879) showed that only one sperm is necessary to fertilize an egg. Before this work, the exact role of gametes in fertilization was debated vigorously (Baltzer, 1967; Briggs and Wessel, 2006). At the turn of the past century, Boveri reported a series of experiments showing unequivocally that a complete set of chromosomes (the nuclear material) is required to sustain development and heredity (Boveri, 1902). A particularly telling set of studies revealed that single blastomeres separated from eggs fertilized with multiple sperm usually did not develop normally as it was a matter of chance how many chromosomes each blastomere received. Thus, he concluded that all of the chromosomes are required for normal development. In this same time period, Driesch (1892) showed that separated blastomeres from four-cell stage sea urchin embryos could develop into normal plutei. He concluded that each cell has all the information necessary to develop an organism. These experiments joined the body of knowledge upon which later work leading to the identification of DNA as the nuclear material necessary for development and heredity was based.

In the early part of the 20th century, the focus on mechanisms of development using the sea urchin shifted to studies on the potency of individual blastomeres as investigators tried to understand how individual cells in an embryo came to know their fate. Horstadius and his coworkers (reviewed in Horstadius, 1939) separated and recombined blastomeres of sea urchin embryos from 8-cell, 16-cell, and late stages showing that the developmental potential was unequally distributed beginning with the micromeres at the 16-cell stage. By transplanting micromeres, they showed that these cells are the source of an intercellular signal leading to the speci-

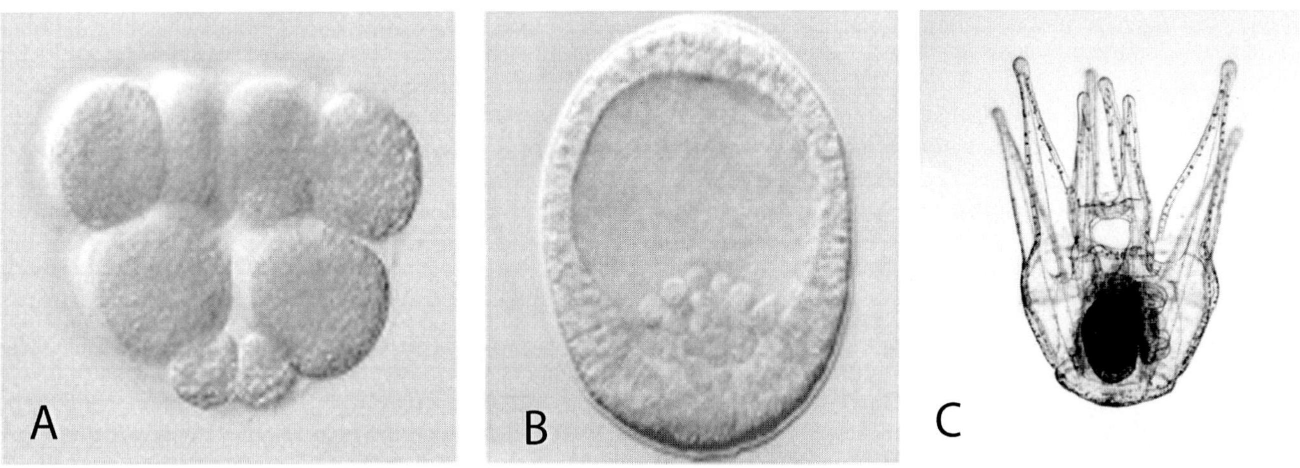

FIGURE 30-4. Sea urchin development. Two embryonic stages and a larval stage of the sea urchin, *Strongylocentrotus purpuratus*. (A) The 16-cell stage from a side view. The micromeres that derive from the asymmetrical cleavage are at the vegetal pole. (B) The blastula stage about 24 hours after fertilization. The cells destined to form the skeleton have ingressed into the hollow central blastocoel. (C) A larva at about 4 weeks after fertilization. The eight larval arms serve as locomotory and food-gathering organs, and the adult rudiment grows in the left side of the body.

fication of the endomesoderm (reviewed in Horstadius, 1939). Furthermore, descendants of these cells secrete the skeleton and they will do so even in isolated culture thus establishing the role of autonomous specification in development (Okazaki, 1975). These kinds of experiments were brought into a modern context when it was shown that transplanted micromeres not only led to a second gut but that the induced gut expressed a gene specific to that organ (Ransick and Davidson, 1995).

In the middle of the 20th century, the utility of the sea urchin model was again justified by a series of experiments that revealed the role of RNA in gene expression and development (reviewed in Davidson, 1986). First, it was shown that protein synthesis required nucleic acids (Brachet, 1947). The required RNA is cytoplasmic as shown by the measurement of protein synthesis in anucleate egg-halves (Denny and Tyler, 1964). Because RNA isolated from unfertilized eggs supports protein synthesis, there is a maternal RNA component (Monroy, 1965). Over the 20 years following this work, the application of radioactive labeling to biological molecules, advances in electron microsocopy, and the use of nucleic acid hybridization produced the knowledge base from which molecular biology and recombinant DNA technology could be applied to sea urchin development (Britten and Kohne, 1968, reviewed in Pederson, 2006). At this time, Britten and his coworkers determined the interspersed organization of repetitive sequences in the sea urchin genome and speculated on its function (Graham *et al.*, 1974; Britten and Davidson, 1969; reviewed in Pederson, 2006). Kedes and coworkers (1975) used the newly developed tools of recombinant DNA technology to first clone genes, in this case the sea urchin histone genes.

The discovery of the protein on sea urchin sperm, bindin, which is responsible for binding to the egg (Vacquier and Moy, 1977), was followed by a variety of studies ranging from the molecular biology of gamete recognition to the role of gamete recognition proteins in evolution. The sea urchin had long been a model for fertilization biology before this period, but the application of biochemical and molecular biological techniques expanded from here. The molecular activities leading up to the acrosome reaction that presents the bindin protein on the sperm tip are becoming clear (Su *et al.*, 2005). The receptor molecule for bindin on the egg has been characterized (Kamei and Glabe, 2003). The steps of egg activation that follow sperm-egg fusion are being revealed in great detail (Roux *et al.*, 2006). In the course of these studies, the sea urchin model has contributed to our understanding of a broad range of processes from the molecular basis of vesicle exocytosis (Bi *et al.*, 1995) to the role of positive selection in gamete recognition (Metz and Palumbi, 1996).

Prompted by the blastomere manipulation experiments performed by Horstadius and his coworkers demonstrating the importance of cellular interaction in early sea urchin development, studies have identified the function of all the major signal transduction pathways in the sea urchin (Sea Urchin Genome Sequencing Consortium, 2006). Components of the Wnt pathway play a role in the specification of micromeres at the vegetal pole of the embryo (Croce *et al.*, 2006; Logan *et al.*, 1999). The Notch system provides signals leading to the specification of mesoderm (Croce and McClay, 2006; Sherwood and McClay, 1999). Nodal, lefty, and BMP2/4 are ligands of the TGFβ signaling pathway and are expressed in the oral ectoderm where they provide positional information along the oral-aboral axis of the sea urchin embryo (Lapraz *et al.*, 2006). The full extent of the complement of signaling molecules that play a role in development emerged from the annotation of the sea urchin genome (Sea Urchin Genome Sequencing Consortium, 2006).

From this history developed the paradigm that development is realized through differential gene expression and the architecture of this process resides in hard-wired features of the genome. The tools to work in this paradigm were rapidly developed in the sea urchin model. The technique of *in situ* hybridization was used to show the localization of embryonic transcripts to particular territories of the developing embryo (Angerer and Angerer, 1981). Cell type monoclonal antibodies were developed and used in studies of regional gene expression (McClay and Chambers, 1978). The studies of spatial expression led to the discovery of lineage-specific gene expression in the early embryo (Angerer and Davidson, 1984). The development of easy gene transfer methods for sea urchin zygotes foreshadowed the first cis-regulatory studies of gene regulation (Flytzanis *et al.*, 1985; McMahon *et al.*, 1985).

Studies of the mechanisms of early development in sea urchins have lead to the erection of an intellectual model for understanding these processes at a fundamental level. Early cell specification events proceed as a result of the operation of gene regulatory networks (GRN). These networks are hard-wired in the genome as clusters of transcription factor binding sites that are organized into cis-regulatory modules that control the transcription of developmentally important genes (reviewed in Oliveri and Davidson, 2004). Typically cis-regulatory modules are a few hundred base pairs long, and they integrate incoming information through the interaction of the proteins that bind to the sites they contain (reviewed in Davidson, 2006). The endomesoderm specification pathway in the purple sea urchin has been described in great detail (Davidson, 2006). It is clearly the best understood GRN in development, and its operation can account for all of the experimental manipulations made with these embryos (Fig. 30-5). Even a quick glance at this burgeoning body of work on sea urchin GRNs illustrates its intimate connection with genomic sequencing efforts. Even before the whole genomic sequence became available, BAC inserts on the order of 100 Kb in length were sequenced for com-

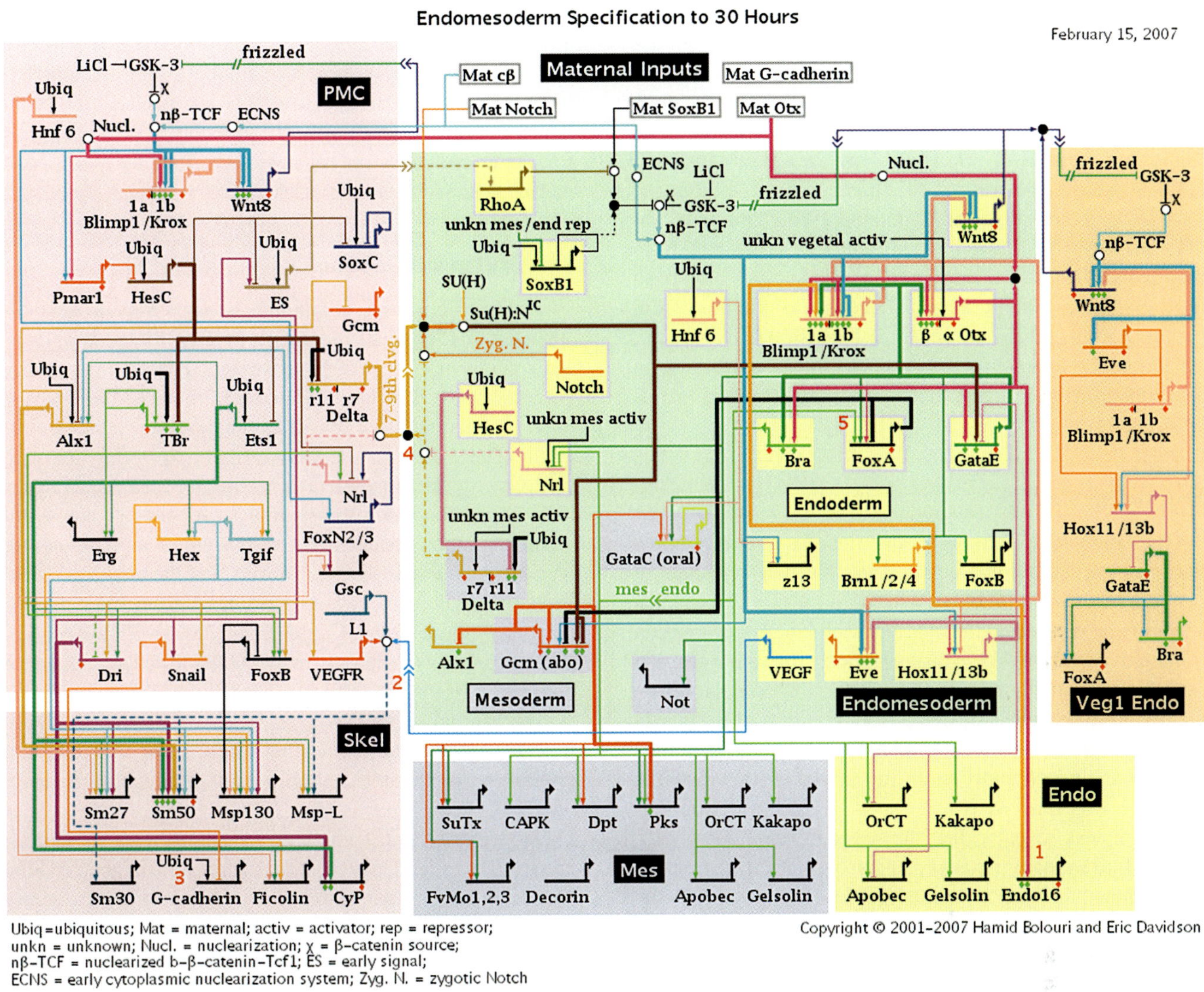

FIGURE 30-5. The gene regulatory network for endomesoderm specification in the sea urchin embryo viewed from the genome. The individual transcription units (genomic regulatory sequence and coding sequence) are represented by a horizontal line and a bent arrow. The lines between transcription units represent interactions between the units. An arrow indicates an activation, and a flat line indicates an inhibition. Filled and open circles represent cytoplasmic biochemical interactions at the protein level such as secretion of a signaling molecule or interaction between a ligand and its receptor. (See Davidson, 2006, for further explanation.)

parison between species to infer cis-regulatory modules (Brown *et al.*, 2002). It is clear that there is broad application of this kind of model, developed in sea urchins, to the solution of developmental mechanisms in many other developmental model systems (Davidson, 2006; Levine and Davidson, 2005).

The Sea Urchin Genome Sequence

Rationale

The sea urchin genome sequence was determined primarily because sea urchin eggs and embryos have proved to be a very useful model for modern studies in cell, developmental, and evolutionary biology as described earlier. There are also studies that have relevance to human biology. For example, the membrane bound receptor guanylate cyclase implicated in the important human disease heatstable enterotoxin dysentery was first isolated from sea urchin sperm. The ubiquitous Ca^{2+}-releasing second messenger, cyclic ADP ribose, was discovered in sea urchin eggs and subsequently found to be important in calcium release in the mammalian pancreas. The sea urchin sperm cell receptor for egg jelly (REJ) is the only known protein in GenBank with homology to human polycystin. REJ controls ion channel activity and by analogy polycystin may be an ion channel regulatory protein whose misregulation could be the basis for this human disease.

A particularly clever use of sea urchin material is the employment of sea urchin blastula basement membranes to analyze the role of extracellular matrix in metastatic cell invasion (Livant *et al.*, 1995). Because the sea urchin extracellular matrix is a relatively simple defined mixture of molecules, it gives more interpretable events than the variable mixtures of molecules in matrices extracted from mammals. This sea urchin material has successfully been used to demonstrate the peptide sequence of the domain in plasma fibronectin that enhances cell invasion (Livant *et al.*, 2000) and the interaction through this domain of fibronectin with integrins that is required to accomplish invasion (Jia *et al.*, 2004).

Sequencing Strategy

The rationale for the sequencing of the sea urchin genome became clear, but the strategy was not. The purple sea urchin genome was shown to have 4% to 5% intraspecific sequence variation by measuring the hybridization kinetics in solution (Graham *et al.*, 1974). This is almost 10× the variation between individual humans where about 0.5% is the norm. One effort to sequence the entire HOX complex of the purple sea urchin yielded a 50-Kb genomic region that overlapped between the two haplotypes in the library from which the sequence was derived (Cameron *et al.*, 2006). From this overlap it was revealed that the sequence variation in sea urchin, like in many other organisms, resides in regions of insertion and deletion as well as single nucleotide polymorphisms (SNP). There are almost three times as many base differences resulting from the insertions and deletions as from SNPs (Britten *et al.*, 2003).

This amount of variation presents a new problem for the assembly of the reads from the first part of the sequencing effort, the whole genome shotgun reads. It was difficult to merge the sequences and to determine if reads were from duplicated but diverged regions of the genome (Sea Urchin Genome Sequencing Consortium, 2006). A second group of sequence reads were made from 8248 BAC inserts that represented a minimal tiling path of the genome (Sodergen, 2006). The BACs were sequenced using an innovative clone-array shotgun strategy (CAPPS; Cai *et al.*, 2001). Software components that could handle local region assembly were added to the Atlas assembler to accommodate the polymorphism (Sea Urchin Genome Sequencing Consortium, 2006). The assembly statistics for the combined strategy showed a significant improvement over the whole genome shotgun (WGS) reads (Sodergen *et al.*, 2006). The whole assembled sequence was reduced from 980 MB to 847 MB, and the redundancy was reduced from 15% to 5%. Between the WGS assembly and the combined assembly, the number of scaffolds was reduced by a third to ~55,000 and the N50 length doubled to 142 Kb. (The N50 length is a measure of the distribution of scaffold sizes. If scaffolds are accumulated from largest to smallest, the N50 length is the smallest size when the accumulation reaches 50% of the total.) This sort of approach will probably become the norm for organisms with polymorphic genomes as many other marine organisms are expected to be.

The Sea Urchin Genome

The result of the combined strategy was a genomic sequence that covered more than 90% of the genome from sequencing to a level of 8× coverage. The final assembled genome size is 814 Mb, in good agreement with the size of 800 +/−5% measured by chemical means (Hinegardner, 1974). The sequences are necessarily organized by the scaffold structure because there is no linkage map available for the sea urchin. The average 36.9% GC content across the whole genome is uniformly low. Individual domains scored on average nearly the same (36.8%), and the predicted genes show no bias with respect to GC content (Sea Urchin Genome Sequencing Consortium, 2006). The genome assembly v2.1 is available on the NCBI Web site (accession number; AAGJ00000000).

The official set of predicted proteins or gene set (OGS) was derived by combining the output from four different prediction programs based on the WGS assembly (Elsik *et al.*, 2006; Sodergren *et al.*, 2006). The exercise yielded 28,945 gene models, which is an overestimate as there is about 15% redundancy in the assembly. The gene models were annotated in a web-based environment that presented automatically computed gene comparisons and received input from individual manual annotators. About 10,000 genes had been annotated by the time of this writing (http://sugp.caltech.edu).

Computational comparisons of the sea urchin gene predictions to other sequenced genome gene sets provide a rough overview of relationships between model systems and a first glimpse of the shared and unique features of these gene catalogs. To a first approximation, a classification of gene models as to conserved domain reveals homology and possible function for the encoded proteins. Using common databases of conserved domains like PFAM (Sonnhammer *et al.*, 1998) and Interproscan (InterPro Consortium, 2001), about three fourths of the Glean protein models could be assigned to at least one protein domain or motif. From similar searches with the nonredundant protein sets for mouse, fruitfly, and nematode, the classes of proteins that can be assigned to common shared gene families or are unique to sea urchin are revealed (Materna *et al.*, 2006a). The most prominent setoff expansions are proteins with homology to innate immunity proteins in vertebrates. These include the toll-interleukin1-receptors, the speract/scavenger receptors, and their intracellular signaling partners

(Hibino *et al.*, 2006). The expansion of these gene families points to a remarkable new way that long-lived sea urchins solve the problem of defense against pathogens (Rast *et al.*, 2006). Proteins that by homology can be assigned function in cell death processes are also greatly expanded (Robertson *et al.*, 2006). There are 49 predicted proteins that match histone motifs in the sea urchin compared to 75 in human and only 8 in the fly. A preliminary search of the star anemone assembly only shows three possible histone hits (E. Morin and R.A. Cameron, unpublished data). This distribution suggests that the expansion of histone genes is likely to be an invention of the deuterostomes. Proteins identified by the hyaline motif (HYR; PF02494) make up a large class in the sea urchin. They are involved in cell surface changes at fertilization (Whitaker *et al.*, 2006). The HYR motif was originally identified in the sea urchin as an extracellular matrix protein involved in fertilization (Wessell *et al.*, 1998). More recently it has been found in bilaterians generally associated with cell adhesion motifs (Callebaut *et al.*, 2000). Three classes of zinc finger motifs identify almost 900 predicted proteins. These motifs are usually present in multiple forms in a protein; therefore, this number may be an overestimate. Zinc finger structures are known to function in DNA binding and protein interaction. This may be an expanded group of proteins in sea urchins (Materna *et al.*, 2006b).

There are 1375 domains that identify proteins in other model organisms but are not found in sea urchins. Prominent among these are a class of proteins involved in oocyte maturation called spindlins, a unique class of C_2H_2 zinc finger transcription factors that bear a transcriptional repression domain called a Krab domain, and the immunoglobulins of adaptive immunity (Materna *et al.*, 2006a). The distinctive olfactory domains individually conserved in the genomes of sequenced model organisms are not present in the sea urchin, even though the general class of rhodopsin-type G-protein coupled receptors (GPCR) with a seven-transmembrane structure is the most abundant one. However, genomic signatures have been found for a sea urchin specific group of these receptors that suggest they may be olfactory in function (Raible *et al.*, 2006).

CONCLUSIONS

Sea urchins and tunicates are both unique systems that reveal in their body plans and genomes the survival of sets of traits crafted through a half billion years of evolution. From the base of the deuterostome clade, they share a common ancestor with vertebrates and as such provide useful comparisons to the complement of genes and the process of development. Because the gene regulatory networks of the developmental process are hard-wired in the genome, glimpses of the evolution of the deuterostome body plans are within reach through comparative studies of these model systems. These animals were selected as research models well before the modern era of molecular biology and genome sequencing. However, their value continues not in small part because of the historical body of information available. In both of these systems, the utility of gene regulatory networks in development has been realized and the complex interactions of regulatory molecules have been described. The synoptic rules of these networks (Ben-Tabou de-Leon and Davidson, 2006) and in some cases actual evolutionarily conserved network elements are being found in vertebrates (reviewed in Davidson, 2006). Though the utility of sea urchins and tunicates as developmental models is confirmed in the modern genomic era, there is yet much to learn.

References

Aguinaldo, A.M.A., Turbeville, J.M., Linford, L.S., Rivera, J.R., Garey, J.R., Raff, R.A., Lake, J.A., 1997. Evidence for a clade of nematodes, arthropods and other molting animals. Nature 387, 489–493.

Angerer, L., Angerer, R., 1981. Detection of polyA+RNA in sea urchin eggs and embryos by quantitative *in situ* hybridization. Nucleic Acid Res. 9, 2819–2840.

Angerer, R.C., Davidson, E.H., 1984. Molecular indices of cell lineage specification in the sea urchin embryo. Science 226, 1153–1160.

Baltzer, F., 1967. Theodor Boveri: The Life of a Great Biologist 1862–1915. Translated from the German by Dorothea Rudnick. Berkeley, University of California Press.

Ben-Tabou de-Leon, S., Davidson, E.H., 2006. Deciphering the underlying mechanism of specification and differentiation: The sea urchin gene regulatory network. Sci. STKE (361), pe47. [DOI: 10.1126/stke.3612006pe47], 2006.

Bi, G.Q., Alderton, J.M., Steinhardt, R.A., 1995. Calcium-regulated exocytosis is required for cell membrane resealing. J. Cell Biol. 131, 1747–1758.

Bottjer, D.J., Davidson, E.H., Peterson, K.J., Cameron, R.A., 2006. Paleogenomics of echinoderms. Science 314, 956–960.

Bourlat, S.J., Juliusdottir, T., Lowe, C.J., Freeman, R., Aronowicz, J., Kirschner, M., Lander, E.S., Thorndyke, M., Nakano, H., Kohn, A.B., Heyland, A., Moroz, L.L., Copley, R.R., Telford, M.J., 2006. Deuterostome phylogeny reveals monophyletic chordates and the new phylum Xenoturbellida. Nature 444, 85–88.

Boveri, T., 1902. Uber mehrpolige Mitosen als Mittel zur Analyse des Zellkerns. Verh. Phys. Med. Ges. Wurzburg 35, 67–90.

Brachet, J., 1947. The metabolism of nucleic acids during embryonic development. Cold Spring Harbor Symposia on Quantitative Biology 12, 18–27.

Briggs, E., Wessel, G., 2006. In the beginning . . . animal fertilization and sea urchin development. Dev. Biol. 300, 15–27.

Britten, R.J., Davidson, E.H., 1969. Gene regulation for higher cells: A theory. Science 165, 349–358.

Britten, R.J., Kohne, D.E., 1968. Repeated sequences in DNA. Science 161, 529–540.

Britten, R.J., Rowen, L., Williams, J., Cameron, R.A., 2003. Majority of divergence between closely related DNA samples is due to indels. Proc. Natl. Acad. Sci. USA 100, 4661–4665.

Brown, C.T., Rust, A.G., Clarke, P.J.C., Pan, Z., Schilstra, M.J., De Buysscher, T., Griffin, G., Wold, B.J., Cameron, R.A., Davidson, E.H., Bolouri, H., 2002. New computational approaches for analysis of cis-regulatory networks. Dev. Biol. 246, 86–102.

Brusca, R.C., Brusca, G.J., with illustrations by Nancy Haver., 2003. Invertebrates. Sunderland, MA, Sinauer Associates.

Cai, W.-W., Chen, R., Gibbs, R.A., and Bradley, A., 2001. A clone-array pooled shotgun strategy for sequencing large genomes. Genome Res. 11, 1619–1623.

Callebaut, I., Gilges, D., Vigon, I., Mornon, J-P., 2000. HYR, an extracellular module involved in cellular adhesion and related to the immunoglobulin-like fold. Protein Sci. 9, 1382–1390.

Cameron, R.A., Rowen, L., Nesbitt, R., Bloom, S., Rast, J.P., Berney, K., Arenas-Mena, C., Martinez, P., Lucas, S., Richardson, P.M., Davidson, E.H., Peterson, K.J., Hood, L., 2006. Unusual gene order and organization of the sea urchin Hox cluster. J. Exp. Zool. 306B, 45–58.

Chabry, L., 1887. Contribution A l'embryologie normale et teratologique des Ascidies simples. J. Anat. Physiol. (Paris) 23, 167–319.

Conklin, E.G., 1905. The organization and cell-lineage of the ascidian egg. J. Acad. Nat. Sci. 13, 1–119.

Corbo, J.C., Levine, M., Zeller, R.W., 1997. Characterization of a notochord-specific enhancer from the Brachyury promoter region of the ascidian, *Ciona intestinalis*. Development 124, 589–602.

Croce, J.C., McClay, D.R., 2006. The canonical Wnt pathway in embryonic axis polarity. Semin. Cell Dev. Biol. 17, 168–174.

Croce, J.C., Wu, S.-Y., Byrum, C., Xu, R., Duloquin, L., Wikramanayake, A.H., Gache, C., McClay, D.R., 2006. A genome-wide survey of the evolutionarily conserved Wnt pathways in the sea urchin *Strongylocentrotus purpuratus*. Dev. Biol. 300, 121–131.

Davidson, B., Levine, M., 2003. Evolutionary origins of the vertebrate heart: Specification of the cardiac lineage in *Ciona intestinalis*. Proc. Natl. Acad. Sci. USA 100, 11469–11473.

Davidson, B., Shi, W., Beh, J., Christiaen, L., Levine, M., 2006. FGF signaling delineates the cardiac progenitor field in the simple chordate, *Ciona intestinalis*. Genes Dev. 20, 2728–2738.

Davidson, B., Shi, W., Levine, M., 2005. Uncoupling heart cell specification and migration in the simple chordate *Ciona intestinalis*. Development 132, 4811–4818.

Davidson, E.H., 1986. Gene Activity in Early Development, 3rd ed. Orlando, FL, Academic Press.

Davidson, E.H., 2006. The Regulatory Genome. Gene Regulatory Networks in Development and Evolution. San Diego, CA, Academic Press/Elsevier.

De Tomaso, A.W., Nyholm, S.V., Palmeri, K.J., Ishizuka, K.J., Ludington, W.B., Mitchel, K., Weissman, I.L., 2005. Isolation and characterization of a protochordate histocompatibility locus. Nature. 438, 454–459.

Dehal, P., Satou, Y., Campbell, R.K., Chapman, J., Degnan, B., De Tomaso, A., Davidson, B., Di Gregorio, A., Gelpke, M., Goodstein, D.M., Harafuji, N., Hastings, K.E., Ho, I., Hotta, K., Huang, W., Kawashima, T., Lemaire, P., Martinez, D., Meinertzhagen, I.A., Necula, S., Nonaka, M., Putnam, N., Rash, S., Saiga, H., Satake, M., Terry, A., Yamada, L., Wang, H.G., Awazu, S., Azumi, K., Boore, J., Branno, M., Chin-Bow, S., DeSantis, R., Doyle, S., Francino, P., Keys, D.N., Haga, S., Hayashi, H., Hino, K., Imai, K.S., Inaba, K., Kano, S., Kobayashi, K., Kobayashi, M., Lee, B.I., Makabe, K.W., Manohar C., Matassi, G., Medina, M., Mochizuki, Y., Mount, S., Morishita, T., Miura, S., Nakayama, A., Nishizaka, S., Nomoto, H., Ohta, F., Oishi, K., Rigoutsos, I., Sano, M., Sasaki, A., Sasakura, Y., Shoguchi, E., Shin-I, T., Spagnuolo, A., Stainier, D., Suzuki, M.M., Tassy, O., Takatori, N, Tokuoka, M., Yagi, K., Yoshizaki, F., Wada, S., Zhang, C., Hyatt, P.D., Larimer, F., Detter, C., Doggett N, Glavina T., Hawkins T., Richardson P., Lucas S., Kohara, Y,. Levine, M., Satoh, N., Rokhsar, D.S., 2002. The draft genome of *Ciona intestinalis*: Insights into chordate and vertebrate origins. Science 298, 2157–2167.

Delsuc, F., Brinkmann, H., Chourrout, D., Philippe, H., 2006. Tunicates and not cephalochordates are the closest living relatives of vertebrates. Nature 439, 965–968.

Denny, P., Tyler, A., 1964. Activation of protein bio-synthesis in non-nucleate fragments of sea urchin eggs. Biochem Biophys. Res. Commun. 14, 245–249.

Di Gregorio, A., Levine, M., 2002. Analyzing gene regulation in ascidian embryos: New tools for new perspectives. Differentiation 70, 132–139.

Driesch, H., 1892. Entwicklungsmechanische Studien. I. Der Werth der beiden ersten Furschungzellen in der Echinodermenentwicklung. Experimentelle Erzeugung von Theil-und Doppelbildung. II. Uber die Beziehungen des Lichtes zur ersten Etappe der thierischen Formbildung. Z. Wiss. Zool. 53, 160–178, 178–184.

Eernisse, D., Peterson, K.J., 2004. The history of animals. In Cracraft, J., and Donoghue, M.J. (eds.), Assembling the Tree of Life. Oxford and New York, Oxford University Press.

Elsik, C.G., Worley, K.C., Zhang, L., Milshina, N.V., Jiang, H., Reese, J.T., Childs, K.L., Venkatraman, A., Dickens, C.M., Weinstock, G.M., Gibbs. R.A., 2006. Community annotation: Procedures, protocols, and supporting tools. Genome Res. 16, 1329–1333.

Ernst, S.G., 1997. A century of sea urchin development. Am. Zool. 37, 250–259.

Finn, R.D., Mistry, J., Schuster-Böckler, B., Griffiths-Jones, S., Hollich, V., Lassmann, T., Moxon, S., Marshall, M., Khanna, A., Durbin, R., Eddy, S.R., Sonnhammer, E.L.L., Bateman, A., 2006. Nucleic Acids Res., Database Issue 34, D247–D251.

Flytzanis, C.N., McMahon, A.P., Hough-Evans, B.R., Katula, K.S., Britten, R.J., Davidson, E.H., 1985. Persistence and integration of cloned DNA in postembryonic sea urchins. Dev. Biol. 108, 431–442.

Fol, H., 1879. Recherches sur la fécondacion etle commencement de l'hénogenie chez divers animaux. Mem. Soc. Phys. Hist. Nat. Genève 26, 92–397.

Graham, D., Neufeld, B., Davidson, E., Britten, R., 1974. Interspersion of repetitive and non-repetitive DNA sequences in the sea urchin genome. Cell 1, 128–137.

Grobben, K., 1908. Die systematische Einteilung des Teirreichs. Verh. Zool.-bot. Ges. Wien 58, 491–511.

Halanych, K.M., 2004. The new view of animal phylogeny. Annu. Rev. Ecol. Evol. Syst. 35, 229–256.

Hertwig, O., 1876. Beitrage der Kenntniss der Bildung, Befruchtung und Theilung des thierischen Eies. Morphol. Jahr. 1, 347–452.

Hibino, T., Loza-Coll, M.A., Messier, C., Majeske, A., Cohen, A., Terwilliger, D., Buckley, K., Brockton, V., Nair, S., Berney, K., Fugmann, S. D., Anderson, M.K., Pancer, Z., Cameron, R.A., Smith, L.C., Rast, J., 2006. The immune gene repertoire encoded in the purple sea urchin genome. Dev. Biol. 300, 349–365.

Hinegardner, R., 1974.Cellular DNA content of the Echinodermata. Comp. Biochem. Physiol. 49B, 219–226.

Holland, L.Z., Laudet, V., Schubert, M., 2004. The chordate amphioxus: An emerging model organism for developmental biology. Cell Mol. Life Sci. 61, 2290–2308.

Horstadius, S., 1939. The mechanics of sea urchin development, studied by operative methods. Biol. Rev. 14, 132–179.

Hughes, A.L., Friedman, R., 2005. Loss of ancestral genes in the genomic evolution of Ciona intestinalis. Evol. Dev. 7, 196–200.

Ikuta, T., Saiga, H., 2005. Organization of Hox genes in ascidians: present, past, and future. Dev. Dyn. 233, 382–389.

Imai, K.S., Levine, M., Satoh, N., Satou, Y., 2006. Regulatory blueprint for a chordate embryo. Science 312, 1183–1187.

InterPro Consortium, 2001. The InterPro database, an integrated documentation resource for protein families, domains and functional sites. Nucleic Acids Res. 29, 37–40.

Jia, Y.F., Zeng, Z.Z., Markwart, S.M., Rockwood, K.F., Ignatoski, K.M.W., Ethier, S.P., Livant, D.L., 2004. Integrin fibronectin receptors in matrix metalloproteinase-1-dependent invasion by breast cancer mammary epithelial cells. Cancer Res. 64, 8674–8681.

Jiang, D., Tresser, J.W., Horie, T., Tsuda, M., Smith, W.C., 2005. Pigmentation in the sensory organs of the ascidian larva is essential for normal behavior. J. Exp. Biol., 208, 433–438.

Johnson, D.S., Zhou, Q., Yagi,, K., Satoh N., Wong, W., Sidow, A., 2005. De novo discovery of a tissue-specific gene regulatory module in a chordate. Genome Res. 15, 1315–1324.

Kamei, N., Glabe, C.G., 2003. The species-specific egg receptor for sea urchin sperm adhesion is EBR1, a novel ADAMTS protein. Genes Dev. 17, 2502–2507.

Kedes, L.H., Chang, A.C.Y., Houseman, D., Cohen, S.N., 1975. Isolation of histone genes from unfractionated sea urchin DNA by subculture cloning in E. coli. Nature 255, 533–538.

Kobayashi, K., Sawada, K., Yamamoto, H., Wada, S., Saiga, H., Nishida, H., 2003. Maternal macho-1 is an intrinsic factor that makes cell response to the same FGF signal differ between mesenchyme and notochord induction in ascidian embryos. Development 130, 5179–5190.

Kowalevsky, A., 1866. Entwickelungsgeschichte der einfachen Ascidien. Mem. Acad. Imp. Sci. St. Petersbourg 10 (Ser. 7), 1–19.

Kowalevsky, A., 1871. Weitere Studien uber die Entwicklung der einfachen Ascidien. Arch. Mikr. Anat. 7, 101–130.

Lapraz, F., Rottinger, E., Duboc, V., Range, R., Duloquin, L., Walton, K., Wu, S,-Y., Bradham, C., Loza, M.A., Hibino, T., Wilson, K., Poustka, A., McClay, D., Angerer, L., Gache, C., Lepage, T., 2006. RTK and TGF-beta signaling pathways genes in the sea urchin genome. Dev. Biol. 300, 132–152.

Levine, M., Davidson, E.H., 2005. Gene regulatory networks for development. Proc. Natl. Acad. Sci. USA 102, 4936–4942.

Livant, D.L., Brabec, R.K., Pienta, K.J., Allen, D.L., Kurachi, K., Markwart, S., Upadhyaya, A., 2000. Anti-invasive, antitumorigenic, and antimetastatic activities of the PHSCN sequence in prostate carcinoma. Cancer Res. 60, 309–320.

Livant, D.L., Linn, S., Markwart, S., Shuster, J., 1995. Invasion of selectively permeable sea-urchin embryo basement-membranes by metastatic tumor-cells, but not by their normal counterparts. Cancer Res. 55, 5085–5093.

Logan, C.Y., Miller, J.R., Ferkowicz, M.J., McClay, D.R., 1999. Nuclear beta-catenin is required to specify vegetal cell fates in the sea urchin embryo. Development 126, 345–357.

Materna, S.C., Berney, K., Cameron. R.A., 2006a. The *Strongylocentrotus purpuratus* genome: A comparative perspective. Dev. Biol. 300, 485–495.

Materna, S.C., Howard-Ashby, M., Gray, R.F., Davidson E.H., 2006b. The C_2H_2 zinc finger genes of *Strongylocentrotus purpuratus* and their expression in embryonic development. Dev. Biol, 300, 108–120.

McClay, D.R., Chambers, A.F., 1978. Identification of four classes of cell surface antigens appearing at gastrulation in sea urchin embryos. Dev. Biol. 63, 179–186.

McMahon, A.P., Flytzanis, C.N., Hough-Evans, B.R., Katula, K.S., Britten, R.J., Davidson, E.H., 1985. Introduction of cloned DNA into sea urchin egg cytoplasm: Replication and persistence during embryogenesis. Dev. Biol. 108, 420–430.

Meedel, T.H., Farmer, S.C., Lee, J.J., 1997. The single MyoD family gene of Ciona intestinalis encodes two differentially expressed proteins: Implications for the evolution of chordate muscle gene regulation. Development 124, 1711–1721.

Metz, E.C., Palumbi, S.R., 1996. Positive selection and sequence rearrangements generate extensive polymorphism in the gamete recognition protein bindin Mol. Biol. Evol. 13, 397–406.

Monroy, A., 1965. An analysis of the process of activation of protein synthesis upon fertilization. Archives de Biologie 76, 511–522.

Moody, R., Davis, S.W., Cubas, F., Smith, W.C., 1999. Isolation of developmental mutants of the ascidian Ciona savignyi. Mol. Gen. Genet. 262, 199–206.

Nakatani, Y., Moody, R., Smith, W.C., 1999. Mutations affecting tail and notochord development in the ascidian *Ciona savignyi*. Development 126, 3293–3301.

Nishida, H., 1987. Cell lineage analysis in ascidian embryos by intracellular injection of a tracer enzyme. III. Up to the tissue restricted stage. Dev. Biol. 121, 526–541.

Nishida, H., 2002. Specification of developmental fates in ascidian embryos: Molecular approach to maternal determinants and signaling molecules. Int. Rev. Cytol. 217, 227–276.

Nishida, H., 2005. Specification of embryonic axis and mosaic development in ascidians. Dev. Dyn. 233, 1177–1193.

Nishida, H., Satoh, N., 1983. Cell lineage analysis in ascidian embryos by intracellular injection of a tracer enzyme. I. Up to the eight-cell stage. Dev. Biol 99, 382–394.

Nishida, H., Satoh, N., 1985. Cell lineage analysis in ascidian embryos by intracellular injection of a tracer enzyme. II. The 32-cell and 64-cell stage. Dev. Biol. 110, 440–454.

Okazaki, K., 1975. Spicule formation by isolated micromeres of sea-urchin embryo. Am. Zool. 15, 567–581.

Oliveri, P., Davidson, E.H., 2004. Gene regulatory network controlling embryonic specification in the sea urchin. Curr. Opin. Genet. Dev. 14, 351–360.

Pearse, J.S., 2006. The ecological role of purple sea urchins. Science 314, 940–941.

Pearse, J.S., Cameron, R.A., 1991. Echinodermata:Echinoidea. In Giese, A.C., Pearse, J.S., and Pearse, V.B. (eds.), Reproduction of Marine Invertebrates, pp. 514–662. Pacific Grove, CA, Boxwood Press.

Pederson, T., 2006. The sea urchin's siren. Dev. Biol. 300, 9–14.

Raible, F., Tessmar-Raible, K., Arboleda, E., Kaller, T., Bork, P., Arendt, D., Arnone. M.I., 2006. Opsins and clusters of sensory G-protein coupled receptors in the sea urchin genome. Dev. Biol. 300, 461–475.

Ransick, A., Davidson, E.H., 1995. Micromeres are required for normal vegetal plate specification in sea urchin embryos. Development 121, 3215–3222.

Rast, J.P., Smith, L.C., Litman, G., Coll, M.L., 2006. The sea urchin genome sequence provides a new perspective on animal immunity. Science 314, 952–956.

Reverberi, G., Mingati, A., 1946. Fenomeni di evocazione nello sviluppo dell'uovo di Ascidie. Risultati dell'indagine sperimentale sull'uovo di *Ascidiella aspersa* e di *Ascidia malaca* allo stadio di 8 blastomeri. Pubbl. Staz. Zool. Napoli 20, 199–252.

Reverberi, G., Minganti, A., 1947. La distribuzione delle potenze nel germe di Ascidie allo stadio di otto blastomeri, analizzata mediante le combinazioni e i trapianti di blastomeri. Pubbl. Staz. Zool. Napoli 21, 1–35.

Robertson, A.J., Croce, J., Carbonneau, S., Voronina, E., Miranda, E., McClay, D.R., Coffman, J.A., 2006. The genomic underpinnings of apoptosis in *Strongylocentrotus purpuratus*. Dev. Biol. 300, 321–334.

Robison, B.H., Reisenbichler, K.R., Sherlock, R.E., 2005. Giant larvacean houses: Rapid carbon transport to the deep sea floor. Science 308, 1609–1611.

Rose, S.M., 1939. Embryonic induction in the ascidian larva. Biol. Bull. 77, 216–232.

Roux, M.M., Townley, I.K., Raisch, M., Reade, A., Bradham, C., Humphreys, G., Gunaratne, H. J., Killian, C.E., Moy, G., Su, Y.H., Ettensohn, C.A., Wilt, F., Vacquier, V.D., Burke, R.D., Wessel, G., Foltz K.R., 2006. A functional genomic and proteomic perspective of sea urchin calcium signaling and egg activation. Dev. Biol. 300, 416–433.

Sasakura, Y., Awazu, S., Chiba, S., Satoh, N., 2003. Germ-line transgenesis of the Tc1/mariner superfamily transposon Minos in *Ciona intestinalis*. Proc. Natl. Acad. Sci. USA 100, 7726–7730.

Sasakura, Y., Nakashima, K., Awazu, S., Matsuoka, T., Nakayama, A., Azuma, J., Satoh, N., 2005. Transposon-mediated insertional mutagenesis revealed the functions of animal cellulose synthase in the ascidian *Ciona intestinalis*. Proc. Natl. Acad. Sci. USA 102, 15134–15139.

Satoh, N., 1994. Developmental Biology of Ascidians. Cambridge, Cambridge University Press.

Sawada, K., Fukushima, Y., Nishida, H., 2005. Macho-1 functions as transcriptional activator for muscle formation in embryos of the ascidian *Halocynthia roretzi*. Gene Expr. Patterns 5, 429–437.

Schubert, M., Escriva, H., Xavier-Neto, J., Laudet, V., 2006. Amphioxus and tunicates as evolutionary model systems. Trends. Ecol. Evol. 21, 269–277.

Sea Urchin Genome Sequencing Consortium, 2006. The Genome of the Sea Urchin *Strongylocentrotus purpuratus*. Science 314, 941–952.

Seo, H.C., Edvardsen, R.B., Maeland, A.D., Bjordal, M., Jensen, M.F., Hansen, A., Flaat, M., Weissenbach, J., Lehrach, H., Wincker, P., Reinhardt, R., Chourrout, D., 2004. Hox cluster disintegration with persistent anteroposterior order of expression in *Oikopleura dioica*. Nature 431, 67–71.

Seo, H.C., Kube, M., Edvardsen, R.B., Jensen, M.F., Beck, A., Spriet, E., Gorsky, G., Thompson, E.M., Lehrach, H., Reinhardt, R., Chourrout, D., 2001. Miniature genome in the marine chordate *Oikopleura dioica*. Science 294, 2506.

Sherwood, D.R., McClay, D.R., 1999. LvNotch signaling mediates secondary mesenchyme specification in the sea urchin embryo. Development 126, 1703–1713.

Shi, W., Levine, M., Davidson, B., 2005. Unraveling genomic regulatory networks in the simple chordate, *Ciona intestinalis*. Genome Res. 15, 1668–1674.

Smith, A., 2005. The pre-radial history of echinoderms. Geol. J. 40, 255–280.

Smith, A.B., Hollingworth, N.T.J., 1990. Tooth structure and phylogeny of the Upper Permian echinoid *Miocidaris keyserlingi*. Proceedings of the Yorkshire Geological Society 48, 47–60.

Smith, A.B., Savill, 2001. *Bromidechinus*, a new Ordovician echinozoan (Echinodermata), and its bearing on the early history of echinoids. Trans. R. Soc Edinb. Earth Sci. 92, 137.

Sodergren, E., Shen, Y., Song, X., Zhang, L., Gibbs, R., Weinstock, G., 2006. Shedding genetic light on Aristotle's lantern. Dev. Biol. 300, 2–8.

Sonnhammer, E.L.L., Eddy, S.R., Birney, E., Bateman, A., Durbin. R., 1998. Pfam: Multiple sequence alignments and HMM-profiles of protein domains. Nucleic Acids Res. 26, 320–322.

Sordino, P., Belluzzi, L., De Santis, R., Smith, W.C., 2001. Developmental genetics in primitive chordates. Philos. Trans. R. Soc. Lond. B Biol. Sci. 356, 1573–1582.

Sprinkle, J., Wilbur, B.C., 2005. Deconstructing helicoplacoids: Reinterpreting the most enigmatic Cambrian echinoderms. Geol. J. 40, 281–293.

Su, Y.H., Chen, S.H., Zhou, H.L., Vacquier, V.D., 2005. Tandem mass spectrometry identifies proteins phosphorylated by cyclic AMP-dependent protein kinase when sea urchin sperm undergo the acrosome reaction. Dev. Biol. 285, 116–125.

Vacquier, V.D., Moy, G.W., 1977. Isolation of binding: Protein responsible for adhesion of sperm to sea-urchin eggs. PNAS 74, 2456–2460.

Vienne, A., Pontarotti, P., 2006. Metaphylogeny of 82 gene families sheds a new light on chordate evolution. Int. J. Biol. Sci. 2, 32–37.

Wessel, G.M., Berg, L., Adelson, D.L., Cannon, G., McClay, D.R., 1998. A molecular analysis of hyalin: A substrate for cell adhesion in the hyaline layer of the sea urchin embryo. Dev. Biol. 193, 115–126.

Whitaker, C.A., Bergeron, K.-F., Whittle, J., Brandhorst, B.P., Burke, R.D., Hynes, R.O., 2006. The echinoderm adhesome. Dev. Biol. 300, 252–265.

Whittaker, J.R., 1973. Segregation during ascidian embryogenesis of egg cytoplasmic information for tissue-specific enzyme development. Proc. Natl. Acad. Sci. USA 70, 2096–2100.

Whittaker, J.R., 1987. Cell lineages and determinants of cell fate in development. Am. Zool. 27, 607–622.

Whittaker, J.R., 1997. Chordate evolution and autonomous specification of cell fate: The ascidian embryo model. Am. Zool. 37, 237–249.

Zeller, R.W., Virata, M.J., Cone, A.C., 2006a. Predictable mosaic transgene expression in ascidian embryos produced with a simple electroporation device. Dev. Dyn. 235, 1921–32.

Zeller, R.W., Weldon, D.S., Pellatiro, M.A., Cone, A.C., 2006b. Optimized green fluorescent protein variants provide improved single cell resolution of transgene expression in ascidian embryos. Dev. Dyn. 235, 456–467.

STUDY QUESTIONS

1. List five features of sea urchins that contribute to their use as a research model in cell and developmental biology.
2. Compare and contrast the modes of development used by the sea urchin *Strongylocentriotus purpuratus* and the ascidian *Ciona intestinalis* in terms of their use as a developmental model.
3. How do the gene transfer methods commonly used for sea urchins or ascidians differ from each other? What are the pros and cons of each method?
4. Briefly detail what the gene catalog of sea urchins tells us about vertebrate evolution.
5. What advantage does a combined whole genome shotgun and BAC-based sequencing strategy offer for a highly polymorphic genome?
6. What is a gene regulatory network, and how does it differ from a signal transduction pathway?
7. What are some advantages of studying marine embryos over other types of embryos?
8. Why is it important to understand developmental mechanisms in different kinds of embryos?
9. How can genomic information help us understand developmental mechanisms at the molecular level? How can this information help us understand the relationships between different animal species?
10. Why do ascidians make a good model for studying chordate development?
11. What are the advantages and disadvantages of the two major invertebrate chordate models?
12. Why has the ascidian genome lost many examples of genes that are present on both the sea urchin and vertebrate genomes?

C H A P T E R

31

The Zebrafish, *Danio Rerio*, as a Model Organism for Biomedical Research

JOCELYN J. LEBLANC AND LEONARD I. ZON

INTRODUCTION

The zebrafish, *Danio rerio*, is an invaluable model for the study of development, genetics, and disease. This tiny black and white striped fish originates from the Ganges River in India and is often a popular pet in home aquariums. In the scientific world, it has been selected for its many advantages in biological studies. At first glance, it may seem incredible that these small fish can teach us a tremendous amount about ourselves.

The obvious advantage of using mammalian models is that they closely resemble the human system in many biological aspects. The drawback is that the more sophisticated the organism is, the more limited the experiments must be because of time constraints and financial and ethical restrictions. The zebrafish is an ideal model organism because of its quick development, small size, low cost, and developmental, physiological, and genetic conservation with humans. Its increased use has provided insights into previously perplexing biological phenomena, potentially accelerating the process of drug discovery.

THE SYSTEM

The ease of maintaining zebrafish in a laboratory setting makes them an excellent model system for biomedical research. They are small fish; the adult is about 3 to 4 centimeters long, so as many as 600,000 animals can be housed together in one room (A. McCollum, Children's Hospital Boston, personal communication). One adult female can lay an average of 200 eggs per week, which are easy to collect and grow up. The large number of progeny means that large-scale genetic studies can be completed more easily than with most other vertebrate models. About 3 months after fertilization, the progeny reach sexual maturation and can generate their own offspring for up to 2 years. This short developmental period allows quick assessment of phenotypes and rapid production of multiple generations of fish.

Another benefit is that the embryo matures outside the mother and is transparent, allowing for easy visual monitoring of development. In a 48-hour period we can observe the morphological transformation of a one-cell embryo to a hatched and fully motile larval fish (Fig. 31-1).

The heart begins beating at 22 hours postfertilization (hpf), and the blood begins circulating at 24 hpf, so normal blood circulation can easily be seen under a dissecting microscope soon after fertilization. All of the major organs, such as the eye, brain, swim bladder, heart, liver, and spinal cord, are developed by 5 to 6 days postfertilization (dpf) and remain visible in the transparent larva (Rubinstein, 2003).

Zebrafish are especially useful to geneticists because they are a powerful link between invertebrate and mammalian systems. Zebrafish are vertebrates, which means that they exhibit extensive anatomic conservation during development and sequence homology with mammals. However, they are much easier to maintain and manipulate than any mammalian model. In addition, genetic resources continue to improve as the zebrafish emerges as a useful model. Three-fourths of the zebrafish genome is currently sequenced, and the updated status is available on the Sanger Institute's Web page (www.sanger.ac.uk/Projects/D_rerio). Sequencing the genome has been expedited by the excellent EST information available for zebrafish. An EST, or expressed sequence tag, is a piece of a gene whose partial sequence is known because part of the cDNA made from *expressed* mRNA was sequenced. The EST acts as a marker for a gene and can be used as a tag to locate this gene on a chromosome, making searching for particular genes much simpler.

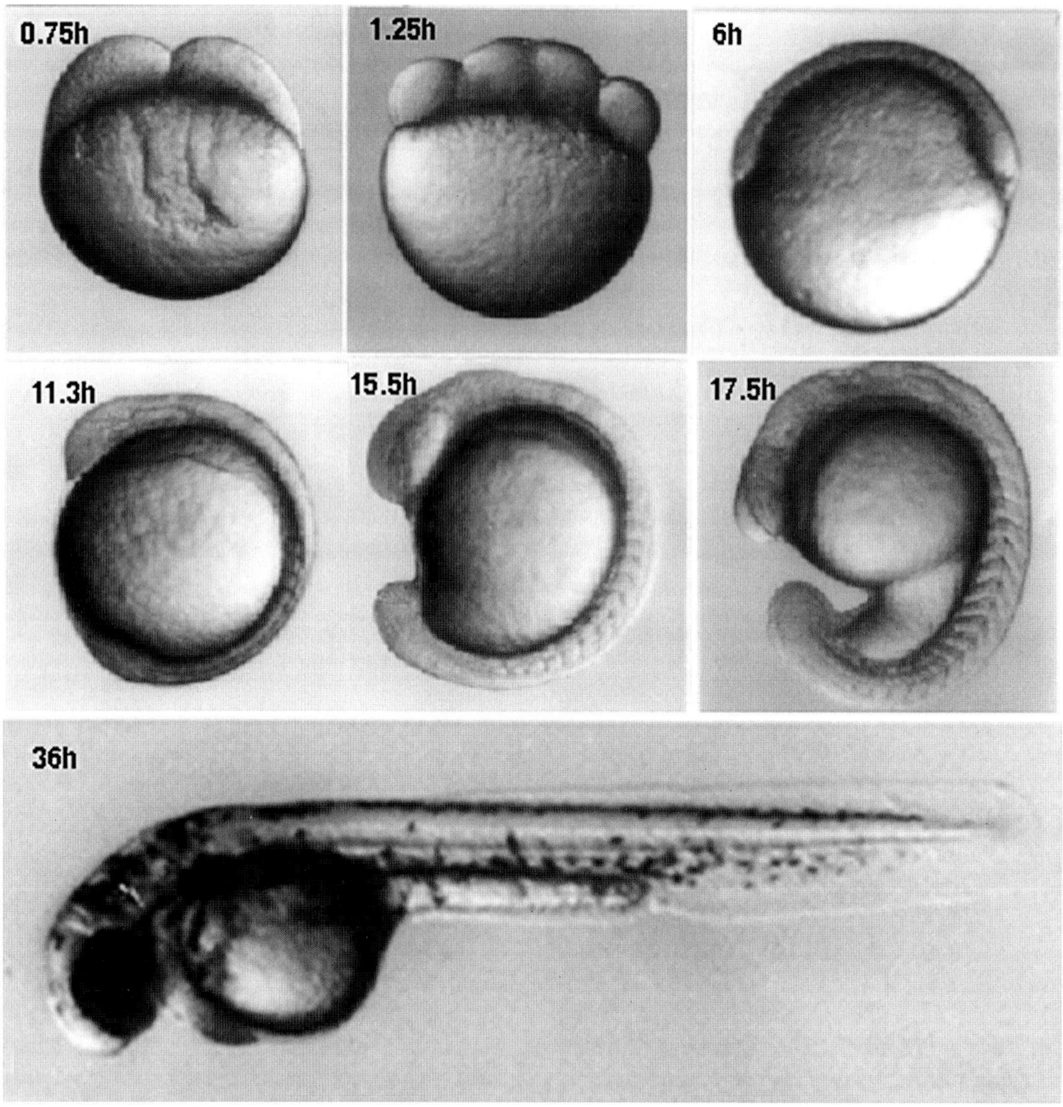

FIGURE 31-1. Photographs of the zebrafish at different stages of development. Hours postfertilization is indicated in the upper left corner of each photograph. Reprinted from Kimmel *et al.* (1995; images taken from Figures 4, 11, 15, and 29) with permission from Wiley-Liss, Inc.

More information on ESTs can be found on a Web site by the National Center for Biotechnology Information (www.ncbi.nlm.nih.gov/About/primer/est.html). Another incredibly useful resource for zebrafish researchers is http://zfin.org, an online database that allows access to genetic information, available mutant lines, job opportunities, programs, and an online book that details zebrafish development and laboratory techniques.

Many human diseases, such as cancer, leukemia, and neurological disorders, are also displayed by the zebrafish and have been shown to share a conserved genetic origin. Zebrafish have been notably valuable in assigning function to genes and positioning these genes in the correct genetic pathways. The power of the zebrafish lies in the short generation time and small size of the fish. The following sections discuss many useful genetic techniques that are specific to the zebrafish model.

GENETIC TECHNIQUES

Several signature techniques used in zebrafish research are especially effective because they exploit the exclusive advantages of the zebrafish system. The first step in searching for a cure to a certain disease is to find an animal model of that disease. This model could be identified simply based on phenotypic similarity to the human version of the disease, or it could be induced genetically or chemically, if the general mechanism of the disease is already known. We can use either approach, respectively called *forward* and *reverse genetics*, when developing a zebrafish model of a certain disease. For example, if we are trying to identify the gene for a certain disease, we could mutagenize fish and look for phenotypes that resemble the human symptoms of that disease, and later we could map the gene on the genome. This would be called forward genetics because we are looking for the genetic cause of a specific phenotype. If we are wondering about the function of a specific gene, we can knock down expression of the gene and see what phenotype results. This would be called reverse genetics because, in contrast to forward genetics, we already know the genetic cause but we do not know what phenotype to expect.

Forward Genetics: Genetic Screens

In forward genetics, a chemical mutagen is used to cause random mutations in the genome of fish, and the progeny of these mutagenized fish are then screened for a phenotype of interest (Patton and Zon, 2001). Once this phenotype is found, sequencing the genome of the mutant can identify the mutated gene responsible for that phenotype. This entire process is called a forward genetic screen (Fig. 31-2).

The traditional approach, known as the classic F2 screen, uses the following protocol. A male zebrafish is treated with a mutagenic chemical called ethylnitrosourea (ENU) to create hundreds of random point mutations in his genome. By mating this mutagenized male to a wild-type female, mutations should be passed on to their progeny, resulting in a generation of fish that is heterozygous for random mutations. This generation is called the F1 generation—the progeny of the first mating. The F1 fish are then outcrossed (mated to a wild-type female) to create an F2 generation that is 50% wild-type and 50% heterozygous. The heterozygous F2 siblings are then incrossed (mated to each other) to create an F3 generation that is theoretically 25% wild-type, 50% heterozygous, and 25% homozygous for the mutation. These F3 embryos can be scored for an unusual phenotype in a fourth of the clutch, which would indicate a homozygous mutation.

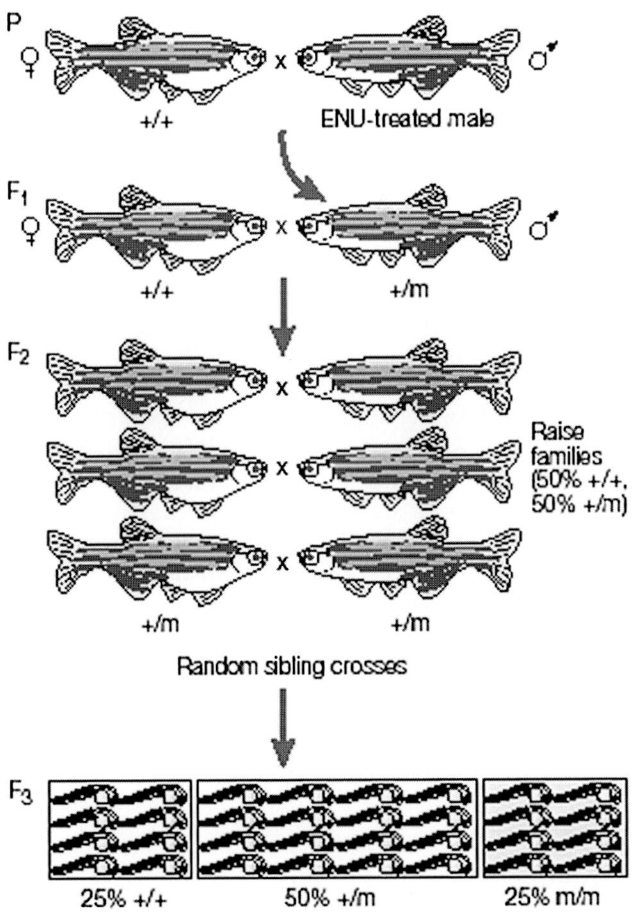

FIGURE 31-2. Classic F2 forward genetic screen in zebrafish. ENU-treated male is crossed to wild-type female. Male F1 fish are crossed with wild-type female to create F2 generation that is 50% wild-type and 50% heterozygous for a mutation. F2 siblings are then incrossed, and F3 embryos are examined for unusual phenotypes. Twenty-five percent of F3 embryos will be homozygous, 50% will be heterozygous for a particular mutation, and 25% will be wild-type. Reprinted from Patton and Zon (2001; Figure 1) with permission from Nature Publishing Group.

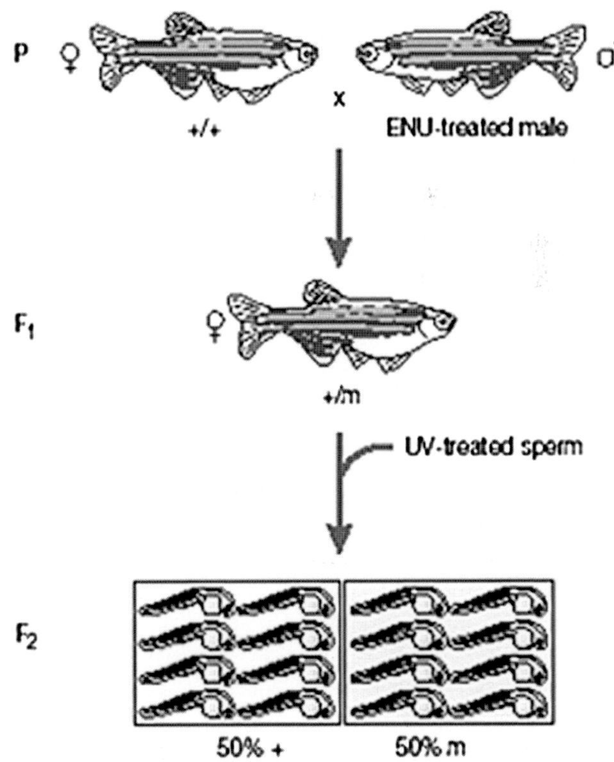

FIGURE 31-3. Haploid genetic screen. ENU-treated male is crossed to wild-type female creating F1 generation that is heterozygous for mutation. Female F1 fish is squeezed, and eggs are treated with UV-treated sperm to create haploid F2 generation that is 50% mutant and 50% wild-type. Reprinted from Patton and Zon (2001; Box 1a) with permission from Nature Publishing Group.

There are also several variations of this classic F2 screen that are designed to save time and space (Patton and Zon 2001). To bypass the time-consuming step of growing up and incrossing all of the fish from the F2 generation, one variation creates haploid embryos. A recessive mutation can be recovered in the F2 generation itself, because the embryo will contain only one copy of each allele (Fig. 31-3).

In this screen, an F1 female is gently squeezed to collect her eggs without harming her. The eggs are fertilized with UV-treated sperm so that the sperm stimulates the eggs to become embryos. But, because the sperm is UV-treated, its genetic content is destroyed and can no longer contribute to the genome of the embryos. The resulting F2 generation is therefore haploid, containing only DNA from the mother, and will contain 50% mutant and 50% wild-type embryos. This conveniently eliminates the need to grow up and mate the large number of F2 fish. Instead, only the mother of the clutch with an interesting phenotype needs to be mated in order to recover the mutant line. Haploid zebrafish embryos can survive up to 3 days with some developmental defects, but they are very useful to investigate embryogenesis and early genetic patterns.

Because of the defects inherent in haploid animals, sometimes false hits may be identified that are not due to heritable mutations. Considering this, another approach compensates for this potential problem (Patton and Zon, 2001). In this approach, the embryos that result from the squeezed eggs and inactivated sperm are subjected to pressure in the first few minutes after fertilization. The pressure is delivered by a French press, which exerts equalized pressure on the embryo. The purpose of the pressure is to break the meiotic spindle, preventing the segregation of sister chromatids in the second division of meiosis. The resulting gynogenetic diploid embryos contain two sets of chromosomes from the mother. The extra generation needed for the classic screen is still eliminated, but the developmental defects exhibited by the haploid embryos are ingeniously avoided.

Reverse Genetics: TILLING, Morpholinos, Transgenesis

Reverse genetics is a procedure by which a specific gene is targeted and then the resulting phenotype is determined. In the zebrafish, there are several ways to perform reverse genetics: TILLING, morpholino knock-down, and transgenesis. Each of these methods targets a particular gene of interest and disrupts it in some way, creating a phenotype that can be correlated to a change in the expression of this gene (either by overexpression or loss of function).

TILLING

One way to approach this goal is by a screening method called targeted-induced local lesions in genomes (TILLING) (Berghmans *et al.*, 2005a). A zebrafish male is mutagenized with ENU and outcrossed (mated with a wild-type female) to create heterozygous F1 embryos, as in the first steps of a classic forward genetic screen. These embryos are grown up and clips from their fins are used for genomic DNA analysis to screen for mutations in the specific gene of interest. This screening is done by either sequencing the genome or by using an enzyme that cuts the DNA where single base pair differences exist between wild-type and mutant alleles. The sperm of the F1 fish is frozen to facilitate reconstitution of a mutant line by *in vitro* fertilization once an interesting mutation is identified. The advantage of this method is that the DNA and the sperm are preserved, so that in the future the DNA can be screened multiple times and the mutant lines can still be created.

Morpholinos

Another way to investigate the role of a specific gene is to selectively knock down its expression and see what phenotype results. This ultimately phenocopies a real mutant, meaning that it is possible to mimic the phenotype of a mutant by manually knocking down the expression of that gene. In mammals, this is done through RNA interference, or RNAi. This process involves the introduction of fragments of double-stranded RNA that share the same sequence as a particular gene. The presence of this extra RNA interferes with the gene's normal expression. In zebrafish, this process can be imitated through the use of synthetic antisense oligonucleotides called morpholinos (Nasevicius and Ekker, 2000). A morpholino is injected into an embryo at the one-cell stage and transiently interrupts gene expression by blocking translational initiation or interfering with mRNA splicing. A morpholino will prevent the translation of a protein necessary for normal functioning, leading to a defective phenotype. The disadvantage of morpholinos is that they last less than 5 days before they are degraded, so they are mostly useful only for embryogenesis and very early developmental studies. *Gata1*, an erythroid-specific transcription factor involved in blood development, has been successfully knocked down by morpholino (Galloway *et al.*, 2005). The *morphants* (morpholino-injected animals) and the mutants are indistinguishable by characterizing with staining techniques.

Transgenesis

The zebrafish is also a good model for transgenesis, a technique used to create transgenic animals. Using this technique, an external DNA fragment is integrated into the genome of an animal. In zebrafish, creating a transgenic fish requires the injection of linearized plasmid DNA containing the gene of interest into one-cell-stage embryos, resulting in some embryos that will integrate the injected DNA into their genome. If the DNA integration occurs in the germ cells, some fish will even pass the integrated DNA onto subsequent generations in a phenomenon called stable integration.

Transgenic fish are useful for studying gene function for several reasons. For example, the gene of interest can be linked to a particular promoter driving the expression of the gene in a specific way. The promoter can be ubiquitous (expressed everywhere), tissue specific if only one cell type is of interest, or inducible if control of overexpression is desired. Target genes can have functions such as creating fluorescence or causing cancer. An example of a transgenic fish is one that ubiquitously expresses green fluorescence protein (GFP) (see Chapter 24) in all its tissues. An example of a ubiquitous promoter is ß-actin, so this transgenic would be called *β-actin* : *GFP*. This type of fish is useful when its cells are transplanted into a recipient that does not have fluorescent cells, because then the donor cells can be tracked in the recipient by their fluorescence.

On the other hand, a transgenic fish can be created to express a fluorescent protein linked to a tissue-specific gene of interest. For example, *fli1-GFP* is a transgenic line that expresses GFP only in its blood vessels, so that any defect in the vascular system becomes obvious under a fluorescent microscope (North and Zon, 2003) (Fig. 31-4).

Finally, a transgenic fish can have a gene under the control of an inducible promoter. An example of this is a heat-shock transgenic line, where heating the fish causes the expression of a particular gene. To continue using the example of fluorescence, a *Hs* : *GFP* line will only express GFP when the fish is heated, allowing control of the timing of fluorescence. The use of transgenic fish in cancer studies will be discussed later.

MODELING DISEASE

Now that some of the key techniques have been reviewed, we can discuss how researchers have used these techniques to create models for diseases that challenge the biomedical community. Zebrafish models are being developed for many types of diseases, including muscular dystrophy (Rubinstein, 2003), Alzheimer's disease (Tomasiewicz *et al.*, 2002), Huntington's disease (Keller and Murtha, 2004), and fetal alcohol syndrome (Barrett, 2005). For illustration, this chapter focuses on cancer and blood diseases.

Cancer

The zebrafish is a powerful model to study the causes, manifestations, and possible treatments of cancer. Cancer is a disease characterized by uncontrolled cell division, which results in the creation of tumors that can invade neighboring tissues. An essential feature of cancer is that several sequential mutations are needed to change a benign cell into a cancerous cell. Oncogenes that promote tumor growth must

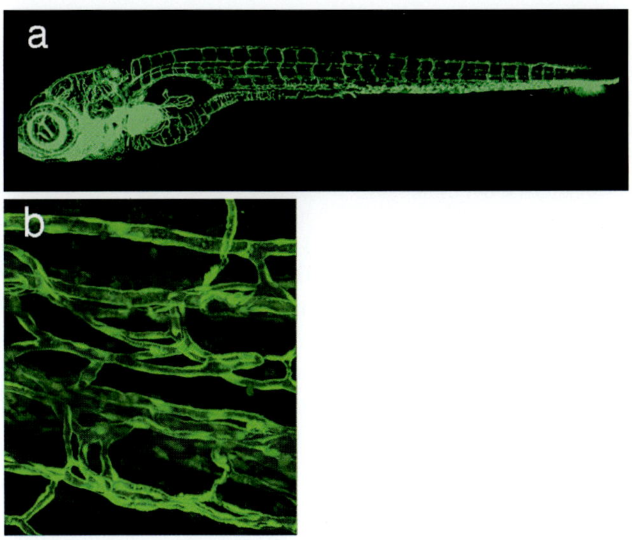

FIGURE 31-4. (a) Lateral view of a 7 dpf Fli1-GFP transgenic larval fish. Only the vasculature is GFP positive and therefore visible under a fluorescent microscope. (b) Tail fin microvasculature of Fli1-GFP adult transgenic fish. Part (a) is reprinted from Lawson and Weinstein (2002; Figure 4A) with permission from Elsevier. Part (b) is reprinted from Lawson and Weinstein (2002; Figure 4F) with permission from Elsevier.

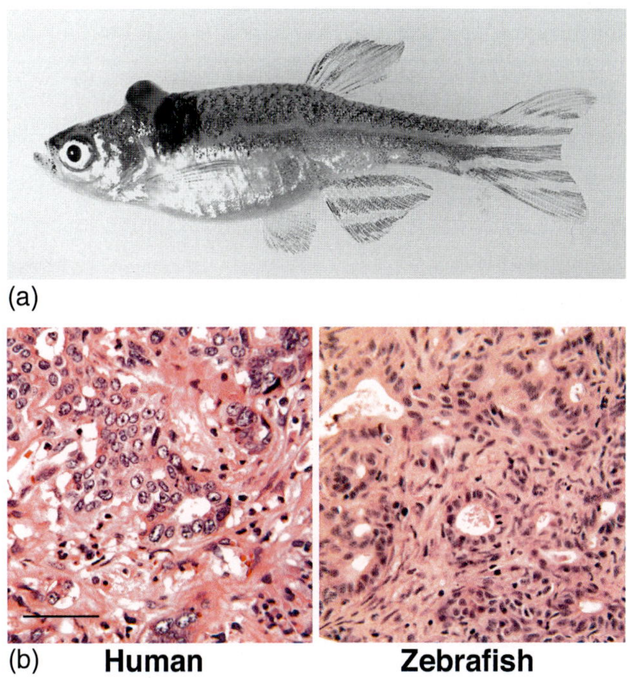

FIGURE 31-5. (a) Adult zebrafish with melanoma. (b) Histology of bile duct tumors, cholangiocarcinoma, in human *(left)* and adult zebrafish *(right)*. There are similarities in the histology of the tumors, including irregularly shaped glands and unusual nuclei. Part (a) is courtesy of Craig Coel, Children's Hospital Boston. Part (b) is reprinted from Amatruda *et al.* (2002; Figure 1) with permission from Elsevier.

be turned on, and tumor suppressor genes that prevent excessive cell division must be inhibited. Because it requires multiple mutations, cancer is usually developed later in life, when the body has been exposed to enough carcinogens that cause DNA mutations. This process can be accelerated if some mutations are inherited, so that fewer acquired mutations are necessary to cause tumor growth.

In the wild, zebrafish have been observed to develop cancer, bearing tumors that look strikingly similar in morphology to human tumors (Amatruda *et al.*, 2002). When exposed to carcinogens in the water, zebrafish also develop a wide variety of tumors that are histologically analogous to human tumors (Fig. 31-5). Not only is the physical manifestation of zebrafish cancer similar to that in humans, but zebrafish also share orthologs of human tumor suppressor genes, cell cycle genes, and oncogenes. Both the histological and genetic conservation of cancer make a convincing argument in support of using fish in general (see Chapter 32) and zebrafish in particular as cancer models. A great advantage of using zebrafish in cancer studies is that tumor progression can be monitored in real time by daily inspection of individual fish.

Zebrafish models for specific cancers can be developed by performing forward genetic screens. In this approach, cancer-related pathways are identified by sequencing the genomes of mutants that exhibit a phenotype that correlates with a cancer of interest. These mutants will not likely exhibit tumors because it takes several subsequent mutations for tumors to develop, which is why it is useful to look for phenotypes that reveal a predisposition to cancer. Shepard *et al.* (2005), for example, looked for cell cycle abnormalities in embryos containing random mutations, as an unusual cell cycle profile is characteristic of cancer because these cells are dividing without control. They performed a haploid forward genetic screen and stained the resulting embryos with anti-phosphohistone H3 (phospho-H3), which marks proliferating cells.

This screen uncovered several cell proliferation mutants, including one of particular interest called *crash&burn* (*crb*). *Crb* is caused by the loss of *bmyb*, a transcription factor with a role in cell proliferation and cancer. Mutation of *bmyb* results in the reduction of *cyclin B1* expression, a necessity for progression from the G2 phase to the completion of mitosis. To quickly review, the cell cycle is the process through which a single cell divides into two identical cells. The phases of the cell cycle include G1 (growth), S (duplication of DNA), G2 (preparation for mitosis), and mitosis (division into two identical daughter cells). Without *cyclin B1*, the cell cannot leave G2 and therefore cannot complete mitosis. *Crb* mutants exhibit mitotic arrest, genome instability, and increased *apoptosis* (a type of programmed cell death). This predisposes the mutants to cancer because subsequent mutations cause cells that already have genome instability and cell cycle abnormalities to ignore the normal checkpoints of the cell cycle and proliferate without control.

Further analysis of *Crb* mutants showed that they did indeed have a high rate of tumor formation after carcinogen treatment. Carcinogen treatment, incubating the fish in mutagenic chemicals, can cause a variety of tumors that are specific to the type of mutagen used and can create a large range of mutations, some of which will be in oncogenes, tumor suppressor genes, or cell cycle genes. In humans, exposure to UV light while sunbathing is an analogous process, predisposing an individual to skin cancer. Carcinogen treatment of a mutant zebrafish is an effective way to prove that the particular mutation predisposes the fish to cancer.

Reverse genetics is a useful tool to target a specific gene that has a suspected role in cancer development. Transgenesis can be performed to create a transgenic line of fish that overexpresses an oncogene. The specific gene can be manipulated and the resulting tumors can be labeled to distinguish them from naturally occurring tumors. In 2003, a group of scientists created a transgenic line of zebrafish that develops a leukemia characterized by GFP-labeled tumors that are specific to lymphoid cells (Langenau *et al.*, 2003). They injected wild-type embryos at the one-cell stage with a transgene that contains a promoter (*Rag2*) specific to lymphoid cells that drives the expression of an oncogene gene (mouse *c-myc*) linked to GFP. The transgenic fish developed GFP-positive tumors exclusively in lymphoid cells. This transgenic line of fish is then especially useful in screens to find drugs that prevent or treat leukemia.

In contrast to overexpressing an oncogene, reverse genetics using TILLING can be employed to target a putative cancer gene by inactivating a tumor suppressor gene. A particular TILLING screen targeted the gene *p*53, a tumor suppressor gene that normally induces apoptosis when the cell is damaged (Berghmans *et al.*, 2005b). When *p53* is mutated, it becomes inactive and allows a cell to avoid apoptosis and continue proliferating, making it the most common mutated gene in human cancer. By directly sequencing the genomes of about 2500 progeny of mutagenized fish, five point mutations in *p53* were uncovered, two of which were similar to human mutations. Homozygous *p53* mutants can live but develop malignant peripheral nerve sheath tumors before they reach 1 year old. These mutant fish lines can be used to investigate prevention, treatment, or cures for cancers caused by this type of mutation.

Blood Development

Because zebrafish embryos are transparent, develop quickly, and can be manipulated genetically, the zebrafish is an excellent system to study hematopoiesis (blood development). Blood development is similar in zebrafish and humans in terms of cellular structures, blood composition, and genetic pathways. Many zebrafish blood mutants that have been identified reveal the intricacies of certain aspects of hematopoiesis, model human blood diseases, and are therefore valuable for testing treatments. Research tools that are especially useful to investigate blood development in zebrafish include forward genetic screens, staining methods, and transplantation experiments.

First it is necessary to understand the hierarchy of blood development in the zebrafish. Cellular components of the blood are suspended in plasma and include myeloid and lymphoid cells. Myeloid cells include macrophages and red blood cells, also called erythrocytes. Erythrocytes contain hemoglobin, which binds to oxygen and carries it through the bloodstream. Lymphoid cells are the white blood cells, such as T cells, B cells, and natural killer cells, and are tools of the immune system. All of these differentiated cells begin as a hematopoietic stem cell (HSC), which is the most primitive blood cell. The HSC resides in the zebrafish kidney marrow, which is analogous to mammalian bone marrow. The HSC is able to create more HSCs, or it can begin to differentiate, forming cells called progenitors (Fig. 31-6). Progenitors cannot self-renew but go on to form the differentiated cells of the blood system. Progenitors are of specialized types: lymphoid progenitors form differentiated lymphoid cells, and myeloid progenitors form differentiated myeloid cells. The HSC can produce all of its progeny in the kidney marrow, and the progeny then enter into circulation or further mature in other organs like the spleen or thymus. Blood circulation in the zebrafish begins around 24 hpf.

Many blood mutants have been characterized, each exhibiting a specific deficit in some stage of blood development. One example is the many mutants with deficiency in red blood cell production, which were discovered during a large scale forward genetic screen performed simultaneously by two labs in 1996, one in Tubingen, Germany, and the other at Massachusetts General Hospital in Boston (Ransom *et al.*, 1996; Weinstein *et al.*, 1996). Thirty-three mutants were recovered from the screen and were classified into four groups based on blood cell count and concentration of hemoglobin in the blood cells. These groups include mutants that have no blood at all, have decreased blood counts, have decreased hemoglobin and decreased blood counts, or have photosensitive blood (Table 31-1).

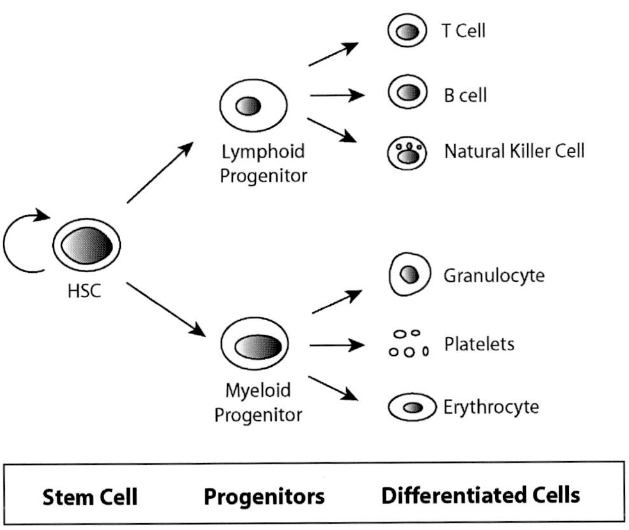

FIGURE 31-6. Hierarchy of hematopoiesis. A hematopoietic stem cell (HSC) has the ability to either self-renew or produce more specialized cells called progenitors. Lymphoid progenitors can only form lymphoid cells, such as T cells, B cells, or natural killer cells. Myeloid progenitors can only form myeloid cells, such as granulocytes, platelets, and erythrocytes.

The bloodless mutant group has two members: *moonshine* and *vlads tepes*. *Moonshine* is named for its increased iridophores (reflective skin cells), though its most interesting phenotype, as for *vlads tepes*, is that it lacks blood completely. Few survive until adulthood, and those that do remain severely anemic and sick. *Merlot* and *chablis* are mutants that have decreased blood cell counts. These mutations are recessive and embryonic lethal; these fish model a human disorder called hereditary elliptocytosis, in which patients exhibit anemia resulting from elliptical erythrocytes (North and Zon, 2003). Another mutant group suffers from decreased hemoglobin in addition to decreased blood cell counts. An example from this group is *sauternes*, the first animal model of the human disorder sideroblastic anemia, in which iron cannot be incorporated into hemoglobin in erythrocytes (Brownlie *et al.*, 1998). The final group has photosensitive blood, meaning that their red blood cells lyse upon exposure to light. One of these mutants is called *yquem*, and it models the human disorder hepatoerythropoietic porphyria, in which humans have defects in forming one of the components of hemoglobin (Ransom *et al.*, 1996).

These mutants were characterized by using staining procedures to mark erythrocytes, hemoglobin, and transcription. A simple way to visualize the presence or absence of blood in different parts of the embryo is by using benzidine stain to locate blood cells. Benzidine staining detects a com-

TABLE 31-1. Summary of some blood diseases that are modeled by mutant zebrafish lines.

Mutant Name	Phenotype	Related Human Disorder
Vlad tepes	Bloodless	Familial dyserythropoietic anemia
Merlot	Decreased blood	Hereditary elliptocytosis
Chablis	Decreased blood	Hereditary elliptocytosis
Weissherbst	Decreased blood	Type 4 hemochromatosis
Sauternes	Decreased blood and hemoglobin	Sideroblastic anemia
Yquem	Photosensitive blood	Hepatoerythropoietic porphyria

See North and Zon (2003) and Fraenkel *et al.* (2005).

ponent of hemoglobin, thereby identifying red blood cells and their precursors. A decrease in benzidine staining would reflect a defect in hemoglobin or in red blood cells. *In situ* hybridization can also be performed to probe for specific gene expression patterns.

In situ hybridization is a technique used to stain tissues that express a particular gene of interest. In this procedure, embryos are collected and fixed with paraformaldehyde at different stages of development. An antisense RNA probe is then applied to the embryos. The probe is designed to hybridize with the mRNA of a gene of interest and the tissues where the probe hybridizes will be stained. This gives a detailed visualization of blood development, because each gene is expressed in a specific pattern, both temporally and spatially, during development. This also allows for the possibility of identifying mutants by *in situ* staining, because a different gene expression pattern in the stained tissues indicates a deviation from normal development. For example, to investigate the genetic defects involved in the *moonshine* mutant, *in situ* hybridization can be performed for different hematopoietic genes to see what genes are lost. An *in situ* probing for *gata1*, a gene involved in hematopoiesis, would reveal that there is a loss of *gata1* expression in *moonshine* mutants (Fig. 31-7).

One standard way to investigate blood development in zebrafish is by using kidney marrow transplantation assays in the adult. This technique is analogous to bone marrow transplants in human patients. Bone marrow transplants are a medical procedure to help repopulate the blood of patients who have blood diseases. In this procedure, human bone marrow from a donor is harvested and transplanted into the recipient, with the intent that HSCs from the donor will go to the bone marrow of the recipient and form new differentiated blood cells, saving the patient's life. In zebrafish, the kidney marrow is analogous to bone marrow in humans, so kidney marrow transplants can be performed to effect blood repopulation in the recipient fish (Fig. 31-8).

Kidney marrow transplants can reveal a great deal about hematopoiesis because the whole process of a HSC differentiating into mature blood cells can be observed. Instead of

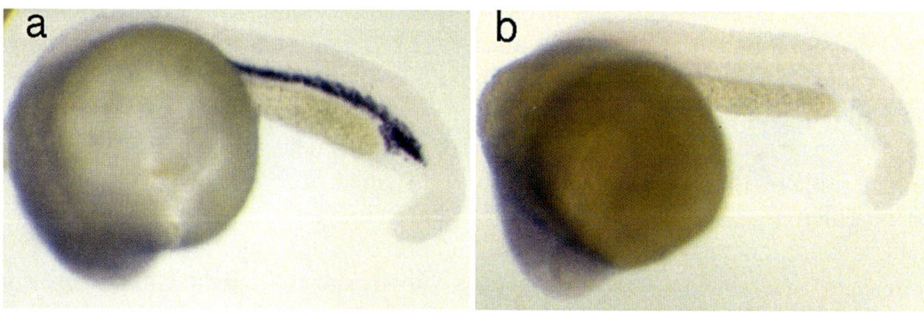

FIGURE 31-7. (a) *In situ* hybridization performed on 24 hpf wild-type embryo to stain *gata1* expression. Embryo's head is on the left and tail curves to the right. Dark stripe is *gata1* staining in a region involved in hematopoiesis. (b) *In situ* hybridization performed on 24 hpf *moonshine* mutant embryo to stain *gata1* expression. Notice the absence of *gata1* staining because of hematopoietic defects. Courtesy of Xiaoying Bai, Children's Hospital Boston.

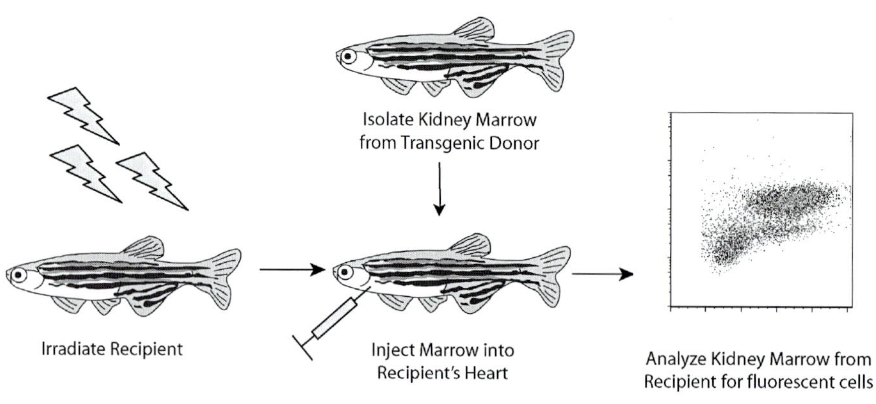

FIGURE 31-8. Kidney marrow transplantation in the zebrafish first involves irradiation of the recipient to kill resident blood cells. Then kidney marrow from a fluorescent transgenic donor is isolated, ground up, and injected into the heart of the recipient fish. This ensures that the HSCs will enter into circulation and find their way to the kidney marrow to begin production of new blood cells. Several months later, the recipient kidney marrow is isolated and analyzed by flow cytometry to sort out fluorescent cells. This plot shows fluorescent cells separated into clusters based on their size and granularity, indicating cell type. If the transplant is successful, the fluorescent cells will be from all blood populations.

using fish with blood diseases, we can irradiate a fish to mimic the same effects. Irradiation kills proliferating cells. When the adult is irradiated, theoretically all of its resident blood cells will die because blood cells are highly proliferative. If kidney marrow from a different fish is then transplanted into the recipient fish, the donor HSCs can engraft in the kidney marrow and repopulate the entire blood system indefinitely.

To ensure the donor cells are distinguishable from the host's cells, the donor kidney marrow is isolated from *β-actin : GFP* transgenic fish, so that the cells are labeled with a fluorescent protein and can be traced. In this way, the success of the transplant can be determined several months after the transplant by sacrificing the fish, isolating its kidney marrow, and using flow cytometry to see and quantify fluorescent cells. Flow cytometry is performed on a FACS machine (fluorescence activated cell sorter), which can identify the cell type and level of fluorescence. If the transplantation is successful, meaning that one or more HSCs were able to make it to the kidney marrow and produce newly differentiated cells, then there will be fluorescent cells that belong to all of the blood lineages (myeloid and lymphoid cells). The transplantation assay can be used to measure HSC activity and identify specific hematopoietic defects functionally in mutants. For example, a mutant that has HSC defects would not be a successful donor, and a mutant that has defects in the HSC niche would not be a successful transplant recipient.

The zebrafish has also been pivotal in helping to research blood disorders. In a developmental defect screen, a mutant fish was identified that exhibits anemia caused by abnormal hemoglobin production. This mutant, called *sauternes*, has a mutation that causes the lack of an enzyme, *ALAS2*, which is essential in hemoglobin production (Brownlie *et al.*, 1998). Its phenotype closely correlates with the symptoms of the human disease x-linked sideroblastic anaemia. This makes *sauternes* the first animal model of this disease. In a similar example, *weissherbst* is a mutant that exhibits anemia caused by impaired iron export (Fraenkel *et al.*, 2005). *Weissherbst* has been shown to have a mutation in *ferroportin1*, an essential iron exporter. This zebrafish mutant predicted the human disease type 4 hemochromatosis, which is also caused by a *Ferroportin1* deficiency. These mutant zebrafish, which share the same genetic etiology and symptoms as human patients, allow zebrafish to be used for research on treatments and cures of these diseases.

THE SEARCH FOR TREATMENTS AND CURES

The desired outcome of disease research is to develop treatments to relieve the symptoms and discover cures for these diseases. Zebrafish, because of their size and the environment in which they live, are excellent models to screen drugs that can treat a disease of interest. Embryos can be placed in different wells of a 96-well plate, and specific chemicals can be added to the liquid of each well. This way, genetically identical embryos that carry the gene for a disease can be exposed to different drugs, and any improvement in symptoms can be assessed. Using zebrafish for a chemical screen is both cost-effective, because a lower amount of a drug is needed than for larger animal models, and efficient, because a large number of age-synchronized embryos can be treated at once and raised to maturity after treatment. An additional advantage of the zebrafish is that the whole organism can be treated at once instead of only treating cells *in vitro*, which is valuable when investigating the effect of a certain drug on an entire system.

A significant conservation of drug response exists between zebrafish and humans (Rubinstein, 2003). With zebrafish it is easy to screen thousands of chemicals at once and discard those that elicit no effect. Those that appear to have therapeutic potential in zebrafish can then be tested in animal models closer to humans. Chemicals approved by the Food and Drug Administration (FDA) can also be tested efficiently to identify new therapeutic uses for drugs that are already being used on humans. The power of the zebrafish model lies in the large number of chemicals that can be investigated at once on thousands of embryos and the ease of seeing a change in phenotype, as there is no need to sacrifice or dissect the animal to see an effect. Different types of readouts from chemical screens include changes in morphology, tumor development, and differences confirmed by staining methods.

One type of chemical screen is a suppressor screen, which is used when a desired hit is suppression of a phenotype caused by a particular mutation. One such mutation in a gene called *hey2* inspired a group to perform a screen to suppress the characteristic phenotype of *hey2* mutants (Peterson *et al.*, 2004). The *hey2* mutants are called *gridlock* because the embryos exhibit a malformed aorta, causing lack of blood flow to the trunk and tail of the animal. The goal of this screen was to find a compound that corrected the disrupted aortic circulation. The embryos were exposed to a variety of chemicals for 48 hours and then blood circulation was examined. Of the 5000 chemicals that were screened, two caused circulation to be restored to the trunk and tail region of the *gridlock* embryos. This suppression of the *gridlock* phenotype was maintained after the chemical had been removed, and the rescued mutants were able to survive until maturity (Fig. 31-9).

The two similar compounds rescued the phenotype by up-regulating vascular endothelial growth factor (VEGF), which in turn allowed for correct development of the aorta. VEGF is a signaling protein involved in forming vasculature during development and repairing and modifying vascula-

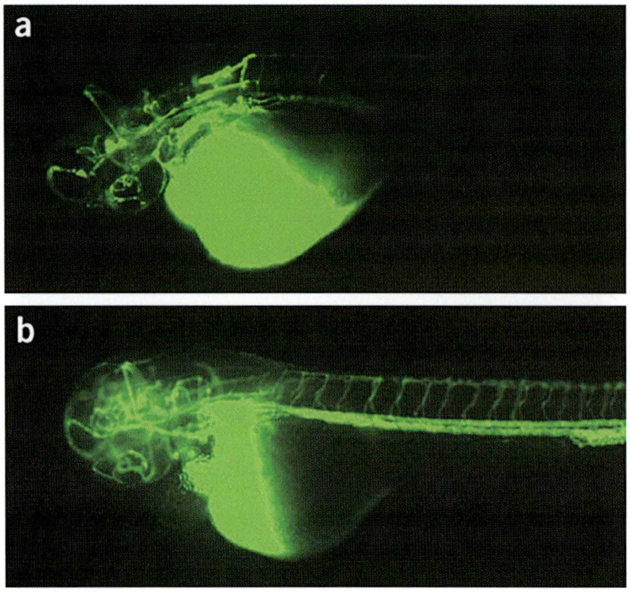

FIGURE 31-9. (a) Microangiogram of untreated 48 hpf *gridlock* embryo. Note the absence of blood circulation in tail because of the malformed aorta. (b) Microangiogram of 48 hpf *gridlock* embryo treated with a chemical that up-regulates VEGF. Note the restoration of blood circulation in the tail because of the rescued formation of the aorta. Part (a) is reprinted from Peterson *et al.* (2004; Figure 1a) with permission from Nature Publishing Group. Part (b) is reprinted from Peterson *et al.* (2004; Figure 1b) with permission from Nature Publishing Group.

ture during maturity. This example demonstrates that drug studies in zebrafish can be relevant to humans. VEGF inhibitors are already an FDA-approved treatment for colon cancer because they prevent vasculogenesis in tumors. The suppression of the *gridlock* phenotype represents another potential clinical function of VEGF.

Another suppressor screen was designed to rescue the *crash&burn* (*crb*) mutant, aptly named for the inability of its cells to complete the cell cycle (Shepard *et al.*, 2005). As discussed earlier, *crb* mutants arrest in the G2 phase of the cell cycle because they lack *bmyb*, a transcription factor required for the expression of *cyclin B1*, a gene that is necessary for the completion of mitosis. In this screen, 16,000 compounds were tested, and one compound called persynthamide (*psy*) was found to suppress the *crash&burn* phenotype (Stern *et al.*, 2005). *Crb* embryos treated with *psy* were examined by various staining methods that further characterized the extent of their rescue. The embryos exhibited more dividing cells, fewer chromosomal abnormalities (increased genomic stability), increased *cyclin B1* expression, and decreased apoptosis. Upon further investigation, it was determined that *psy* treatment caused the cells to delay in S phase, allowing *cyclin B* to accumulate to an amount high enough to drive the cells through mitosis.

SUMMARY

In biomedical research, it is important to find a model organism that can be efficiently used in a laboratory setting and that exhibits enough conservation with humans for discoveries to have enough immediate relevance. The zebrafish has proved to be an excellent model for biomedical research. It is very small, has a short developmental period, and can produce many offspring at once, making zebrafish amenable to large genetic and chemical screens. Early development is easy to follow because the embryo and larva are transparent. Zebrafish are vertebrates, and most genetic and biological pathways are conserved with humans. Mutant zebrafish lines now model many human diseases, and scientists are in the process of discovering new treatments and possible cures for these diseases using screens with the zebrafish. This little striped fish is emerging as a leading model organism for disease research.

References

Amatruda, J.F., Shepard, J.L., Stern, H.M., Zon, L.I., 2002. Zebrafish as a cancer model system. Cancer Cell 1, 229–231.

Barrett, J.R., 2005. A center of a different stripe. Environ. Health Perspect. 113, 160–163.

Berghmans, S., Jette, C., Langenau, D., Hsu, K., Stewart, R., Look, T., Kanki, J.P., 2005a. Making waves in cancer research: New models in the zebrafish. BioTechniques 39, 227–237.

Berghmans, S., Murphey, R.D., Wienholds, E., Neuberg, D., Kutok, J.L., Fletcher, C.D.M., Morris, J.P., Liu, T.X., Schulte-Merker, S., Kanki, J. P., Plasterk, R., Zon, L.I., Look, A.T., 2005b. *tp53* mutant zebrafish develop malignant peripheral nerve sheath tumors. PNAS 102, 407–412.

Brownlie, A., Donovan, A., Pratt, S.J., Paw, B.H., Oates, A.C., Brugnara, C., Witkowska, H.E., Sassa, S., Zon, L.I., 1998. Positional cloning of the zebrafish *sauternes* gene: A model for congenital sideroblastic anaemia. Nat. Genet. 20, 244–250.

Fraenkel, P.G., Traver, D., Donovan, A., Zahrieh, D., Zon, L.I., 2005. Ferroportin1 is required for normal iron cycling in zebrafish. J.Clin. Invest. 115, 1532–1541.

Galloway, J.L., Wingert, R.A., Thisse, C., Thisse, B., Zon, L.I., 2005. Loss of gata1 but not gata2 converts erythropoiesis to myelopoiesis in zebrafish embryos. Dev. Cell 8, 109–116.

Keller, E.T., Murtha, J.M., 2004. The use of mature zebrafish (*Danio rerio*) as a model for human aging and disease. Comp. Biochem. Physiol. C: Pharmacol. Toxicol.138, 335–341.

Kimmel, C.B., Ballard, W.W., Kimmel, S.R., Ullmann, B., Schilling, T.F., 1995. Stages of embryonic development of the zebrafish. Developmental Dynamics 203, 253–310.

Langenau, D.M., Traver, D.F., Ferrando, A.A., Kutok, J.L., Aster, J.C., Kaitki, J.P., Lin, S., Prochownik, E., Trede, N.S., Zon, L.I., Look, A.T., 2003. Myc-induced T cell leukemia in transgenic zebrafish. Science 299, 887–890.

Lawson, N.D., Weinstein, B.M., 2002. *In vivo* imaging of embryonic vascular development using transgenic zebrafish. Dev. Biol. 248, 307–318.

Nasevicius, A., Ekker, S.C., 2000. Effective targeted gene "knockdown" in zebrafish. Nat. Genet. 26, 216–220.

North, T.E., Zon, L.I., 2003. Modeling human hematopoietic and cardiovascular diseases in zebrafish, Dev. Dynam. 228, 568–583.

Patton, E.E., Zon, L.I., 2001. The art and design of genetic screens: Zebrafish. Nat. Rev. Genet. 2, 956–966.

Peterson, R.T., Shaw, S.Y., Peterson, T.A., Milan, D.J., Zhong, T.P., Schreiber, S.L., MacRae, C.A., Fishman, M.C., 2004. Chemical suppression of a genetic mutation in a zebrafish model of aortic coarctation. Nat. Biotechnol. 22, 595–599.

Ransom, D.G., Hafter, P., Odenthal, J., Brownlie, A., Vogelsang, E., Kelsh, R.N., Brand, M., van Eeden, F.J.M., Furutani-Seiki, M., Granato, M., Hammerschmidt, M., Heisenberg, C., Jiang, Y., Kane, D.A., Mullins, M.C., Nusslein-Volhard, C., 1996. Characterization of zebrafish mutants with defects in embryonic hematopoiesis. Development 123, 311–319.

Rubinstein, A.L., 2003. Zebrafish: from disease modeling to drug discovery. Curr. Opin. Drug Discov. Dev. 6, 218–223.

Shepard, J.L., Amatruda, J.F., Stern, H.M., Subramanian, A., Finkelstein, D., Ziai, J., Finley, K.R., Pfaff, K.L., Hersey, C., Zhou, Y., Barut, B., Freedman, M., Lee, C., Spitsbergen, J., Neuberg, D., Weber, G., Golub, T.R., Glickman, J.N., Kutok, J.L., Aster, J.C., Zon, L.I., 2005. A zebrafish bmyb mutation causes genome instability and increased cancer susceptibility. PNAS 102, 13194–13199.

Stern, H.M., Murphey, R.D., Shepard, J.L., Amatruda, J.F., Straub, C.T., Pfaff, K.L., Weber, G., Tallarico, J.A., King, R.W., Zon, L.I., 2005. Small molecules that delay S phase suppress a zebrafish *bmyb* mutant. Nat. Chem. Biol. 1, 366–370.

Tomasiewicz, H.G., Flaherty, D.B., Soria, J.P., Wood, J.G., 2002. Transgenic zebrafish model of neurodegeneration. J. Neurosci. Res. 70, 734–745.

Weinstein, B.M., Schier, A.F., Abdelilah, S., Malicki, J., Solinica-Krezel, L., Stemple, D.L., Stainier, D.Y.R., Zwartkruis, F., Driever, W., Fishman, M.C., 1996. Hematopoietic mutations in the zebrafish. Development 123, 303–309.

STUDY QUESTIONS

1. What are the advantages and disadvantages of using zebra fish as model organisms for biomedical research?
2. What is the general purpose of both reverse and forward genetics, and what is an example of a technique used for each method?
3. What could you use as a phenotypic readout in a screen where you were looking for rescue of a vascular defect? In other words, once you have embryos to examine, what technique would you use to visualize vascular integrity?
4. What do you think could be some rate-limiting factors during the execution of a forward genetic screen?
5. In researching cancer, why do you think scientists might prefer to work with zebra fish? What are some specific advantages of using zebra fish as model organisms in this type of research?

CHAPTER

32

Carcinogenesis Models
Focus on Xiphophorus and Rainbow Trout

RONALD B. WALTER, GRAHAM S. TIMMINS, SUSAN C. TILTON, GAYLE A. ORNER, ABBY D. BENNINGHOFF, GEORGE S. BAILEY, AND DAVID E. WILLIAMS

INTRODUCTION

Fish models have proven valuable in the study of a number of aspects of carcinogenesis (see Ostradner and Rotchell, 2005, for an outstanding review). These nonmammalian vertebrates are more economical (especially in the conduct of large, statistically challenging tumor studies discussed later), and their time to tumor development is often shorter, as are the generation times. Having access to an external embryo also presents advantages in experimental design (e.g., dosing), especially when the embryo is transparent (e.g., zebra fish, medaka). Fish respond to many of the same carcinogens as rodents. We will make an argument in this paper that, in some instances, fish represent a superior model compared to rodents, for assessing human cancer risk. In addition, feral populations of fish can be considered as the "canary in the coal mine" when it comes to early indicators of potential harm to humans, although we will not cover this topic in detail here. Fish bioaccumulate lipophilic chemicals from the environment (see Chapter 6). There are a number of reports in the literature of adverse health impacts in feral fish leading to the unveiling of environmental problems in surface water that may be used by humans for drinking or recreation. Examples include the high incidence of liver cancer in English sole in the Puget Sound area of Washington State [high polycyclic aromatic hydrocarbon (PAH) and polychlorinated biphenyl (PCB) sediment levels] (Malins *et al.*, 1987) and tomcod in the Hudson River (Cormier *et al.*, 1989). Adverse health effects observed in fish have served as red flags for other human health risks including endocrine disruptors (Sumpter and Jobling, 1995) and outbreaks in aquatic populations of microbial agents or dinoflagellates that are potentially infectious in humans or produce toxicants. As with all models, fish have limitations in the study of human cancer. Site-specific cancers that are important in humans, such as lung, breast, and prostate, obviously cannot be studied directly in fish. In the sections that follow, we demonstrate how fish, raised in the laboratory, have advanced the field of carcinogenesis. Because of the interests and expertise of the authors, this review focuses on *Xiphophorus* and rainbow trout.

XIPHOPHORUS: A MODEL FOR STUDY OF THE GENETICS OF MELANOMA

Background

Xiphophorus fishes represent a new-world live-bearing order with species derived from diverse habitats ranging from relatively stagnant waters to fast-flowing streams. *Xiphophorus* (commonly called platyfish and swordtails) are made up of 26 known species with a range that includes the eastern regions of northern Mexico, extending south through Guatemala, Belize, and ending in Honduras (Kallman and Kazianis, 2006). The currently described *Xiphophorus* species possess morphological phenotypes, behavior, and biochemical mechanisms as varied as the source habitats from which they are derived. The principal asset of using *Xiphophorus* in experimental science is this extreme natural variability among species (Fig. 32-1) coupled with the capability of producing fertile *interspecies* hybrid progeny. The ability to generate fertile interspecies *Xiphophorus* hybrids provides a powerful tool for investigations into the heritable bases of multigenic disease and the varied responses to experimental stimuli or environmental stressors.

Because of geographic distribution in their native habitat, most *Xiphophorus* species are reproductively isolated; however, there is sympatry for pairs of wide ranging species

FIGURE 32-1. Examples of a few *Xiphophorus* species to demonstrate the extreme morphological variations that exist among representatives of this order.

such as *X. maculatus* and *X. helleri*. In native sympatric regions, interspecies hybridization either does not occur or occurs only rarely (Rosenthal, 2003). In a laboratory setting, most *Xiphophorus* species may be mated to produce fertile offspring by forced pairing in aquaria or through artificial insemination. Upon interspecies mating, the resulting interspecies hybrid progeny (i.e., F_1 hybrid) are nearly always fertile.

The capability to produce fertile interspecies F_1 hybrids and the establishment of a *Xiphophorus* Genetic Stock Center (Walter *et al.*, 2005, 2006; www.xiphophorus.org), providing pedigreed *Xiphophorus* fish stocks to scientists worldwide, has driven the development of the *Xiphophorus* experimental system. *Xiphophorus* fishes and hybrids have been used in many diverse areas of research, including evolution (Meierjohann *et al.*, 2004; Morizot *et al.*, 2001; Wilkins, 2004), behavior (Basolo and Alcaraz, 2003; Wong and Rosenthal, 2006), basic physiology (Pinisetty *et al.*, 2005; Yang *et al.*, 2006), comparative biochemistry and genomics (Heater *et al.*, 2007, Oehlers *et al.*, 2004), sex determination (Luo *et al.*, 2005), development (Royle *et al.*, 2005), endocrinology (Flynn *et al.*, 1999), ethology and behavioral ecology (Rosenthal, 2003), toxicology (Kwak *et al.*, 2001), parasitology (Dove, 2000), and immunology (McConnell *et al.*, 1998; Roney *et al.*, 2004).

Because of their high degree of heterozygosity, *Xiphophorus* interspecies hybrids are extremely valuable models for molecular genetic study of gene regulation in physiology and behavior. Extensive use of these hybrids for gene mapping (Kazianis *et al.*, 2004; Morizot *et al.*, 2001; Walter *et al.*, 2004) has resulted in a robust *Xiphophorus* linkage map (resolution of 7.4 cM) that further expands the utility of this model to address complex genetic questions.

In addition to their many other uses, *Xiphophorus* fish have made major contributions to our understanding the genetics underlying tumorigenesis. Genetic control of tumor susceptibility in *Xiphophorus* has been determined both in pure strains and in select interspecies hybrids for a variety tumor types, including malignant melanomas, neuroblastomas, renal tumors, thyroid tumors, neurofibrosarcomas, and

Schwannomas (Kazianis *et al.*, 2001a, 2001b; Setlow *et al.*, 1989; Walter and Kazianis, 2001). With the accumulation of several fish whole genome databases, the use of the genetic power of the *Xiphophorus* model system in comparative genomics to approach complex scientific problems is becoming increasingly attractive.

Development of the *Xiphophorus* genetic system can be traced to pioneering studies in the 1920s detailing the production of select interspecies hybrids between *X. maculatus* and *X. helleri* that resulted in Mendelian segregation of malignant melanoma in backcross hybrid (BC_1) progeny (Gordon, 1927, 1931; Kosswig, 1927). With hindsight, results from these early *Xiphophorus* tumor crosses established the presence of what we now term "oncogenes" and were the initial indications that loss of gene function may be associated with tumorigenesis, thus suggesting the existence of "tumor suppressor" genes.

The Gordon-Kosswig Melanoma Model

In the late 1920s, Drs. Gordon, Kosswig, and Haussler independently found that crossing the platyfish *Xiphophorus maculatus* Jp 163 A with the swordtail *Xiphophorus helleri* produced hybrids that developed cancers that appeared similar to malignant melanomas in humans (for reviews, see Schartl, 1995; Walter and Kazianis, 2001). These investigators traced the origin of these tumors to pigmented cells normally localized to the dorsal fin area, a trait that has since been termed *Sd* (spotted dorsal). These platyfish cells, termed "macromelanophores," are observed as black spots in the dorsal fin region. In subsequent studies, it was demonstrated that melanomas only developed in backcross interspecies hybrid individuals in which both copies of a platyfish regulatory gene had been replaced with swordtail forms that apparently could not control pigment cell proliferation (Fig. 32-2).

While performing his studies, Dr. Gordon wished to precisely identify the genes responsible for development of melanoma and thus began an inbreeding program of stocks to produce genetically identical platyfish and swordtails. In so doing, he established the *Xiphophorus* Genetic Stock Center at the American Museum of Natural History in 1939. Several of the original genetic strains of platyfish and swordtails developed by Dr. Gordon in the 1930s are still available and remain in use today, long after his pioneering discoveries.

Since its initial description in 1927, a considerable amount of research has centered on what is commonly referred to as the Gordon-Kosswig (G-K) melanoma model (for reviews, see Walter and Kazianis, 2001; Schartl, 1995; Vielkind *et al.*, 1989). In this model, one observes development of malignant melanomas in BC_1 hybrid progeny produced by the backcross mating scheme: *X. helleri* × (*X. maculatus* Jp 163 A × *X. helleri*) (Fig. 32-2). The *X. maculatus* parental stock carries the macromelanophore pigment pattern designated *Sd* as an X-linked trait tightly linked (i.e., under 1% recombination) to a tumor inducing locus, often termed *Tu* (herein termed *Sd-Tu*). Macromelanocytes regulated by *Sd-Tu* may become hyperplastic and exhibit enhanced expression (melanization) when crossed into an *X. helleri* genetic background. Once the F_1 interspecies hybrid, which is 50% *X. maculatus* and 50% *X. helleri* in chromosomal constitution, is backcrossed to the *X. helleri* parent, approximately half of the offspring inherit the *Sd-Tu* pigment pattern while the other half are unpigmented (do not inherit an *Sd-Tu*). Of the pigmented BC_1 generation fish, half (25% of the total progeny) exhibit enhanced pigment cell expression like that observed in the F_1 hybrid, and half develop malignant melanoma characterized by invasive lesions initiating at the dorsal fin region (Meierjohann *et al.*, 2006; Schartl, 1995; Walter and Kazianis, 2001; Zunker *et al.*, 2006; Fig. 32-3).

Comparative biochemical, histological, and cytological studies of Gordon-Kosswig BC_1 melanoma cells show they exhibit a decline in differentiation. Pigment cells in amelanotic and more lightly pigmented BC_1 hybrids (e.g., phenotypically as F_1's) are able to undergo limited cell division. In contrast, malignant melanoma cells from heavily pigmented BC_1 fish retain rapid and unlimited cell division (Vielkind *et al.*, 1989). In addition, malignant melanoma cells from these fish are similar to mammalian melanoma cells at the histopathological, ultrastructural, and biochemical levels (Schartl and Peter, 1988). Like human melanoma cells, Gordon-Kosswig BC_1 hybrid melanoma cells have been shown to be capable of serial passage when transplanted into athymic mice (Manning *et al.*, 1973). These early studies served to stimulate scientific interest in the genetic etiology of melanoma using the *Xiphophorus* Gordon-Kosswig (G-K) model.

Genetics of the Gordon-Kosswig Melanoma Model

A two-hit genetic model has been developed to explain the Mendelian segregation of the tumor phenotype in the G-K melanoma model. The first event involves inheritance of oncogenic potential (*Tu*) that is tightly linked to the *Sd* pigment pattern on the platyfish X chromosome. A second genetic event must also concomitantly occur, whereby an individual fails to inherit an *X. maculatus* Jp 163 A autosomal gene that normally regulates *Sd-Tu* oncogenicity and thus differentiation of macromelanophore cells. Consistent with this model, parental *X. maculatus* carry two copies of the autosomal differentiation regulator that serves to maintain *Sd-Tu* expression balance, and thus parental fish do not develop melanoma at appreciable frequencies. The hypothetical regulator of *Sd-Tu* that behaves as a tumor suppressor locus in the Gordon-Kosswig cross has been variously

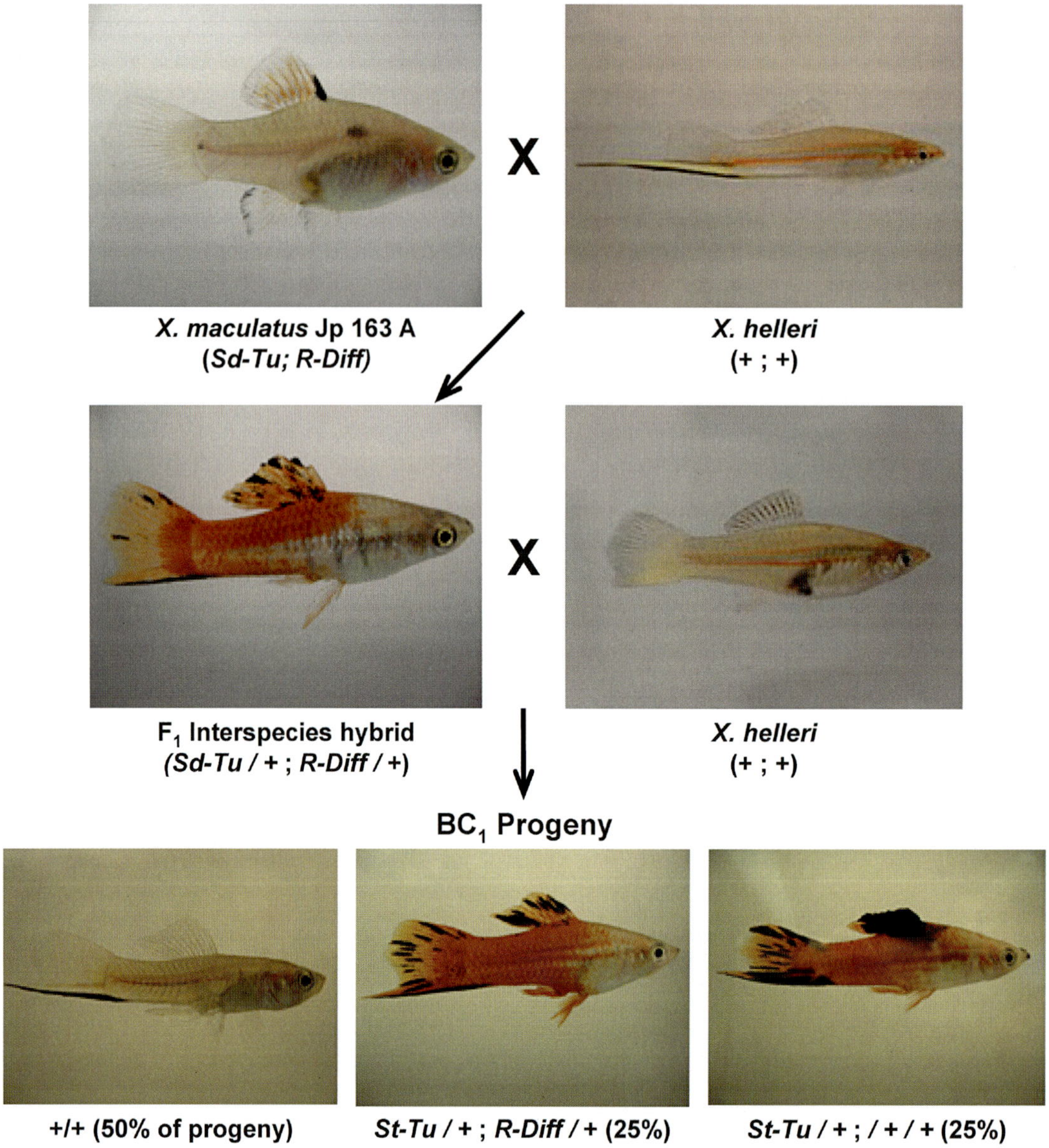

FIGURE 32-2. The Gordon-Kosswig hybrid melanoma cross. The *X. maculatus* Jp 163 A parent carries the sex-linked spotted dorsal (*Sd-Tu*) pigment pattern and the autosomal tumor regulator (*R-Diff*). Once crossed with the *X. helleri* parent, the F_1 hybrid progeny carry one allele from each of the parents. If F_1 hybrids are backcrossed the to *X. helleri* parent, about 25% of the BC_1 progeny will inherit the *Sd-Tu* oncogene but not the autosomal regulator of *Tu (R-Diff)*. In this scenario, invasive melanoma develops in these BC_1 progeny.

termed *R* (for regulator) or *Diff* (for differentiation) to account for its influence on the degree of differentiation of melanin-producing macromelanophore cells within BC_1 progeny. Herein we will designate this as *R-Diff*.

Molecular Genetics of Sd in the Gordon-Kosswig Melanoma Model

Studies have demonstrated that the *X. maculatus* X chromosome carrying the *Sd-Tu* pigment pattern and melanoma oncogenicity harbors two tightly linked copies of an epidermal growth factor-related (*EGFR, HER-1*) sequence (Wittbrodt *et al.*, 1989). These two gene copies have been designated *Xmrk-1* and *Xmrk-2* (Woolcock *et al.*, 1994). The two *Xmrk* (i.e., *Xiphophorus* melanoma receptor kinase) sequences exhibit structural similarities, suggesting they were derived through a duplication of an *EGFRα* gene homolog, and this duplication was followed by nonhomologous recombination events that involved at least one other unrelated gene (Schartl *et al.*, 1995). It has been suggested the postduplication nonhomologous recombination events served to bring the *Xmrk-2* gene copy under new regulatory

FIGURE 32-3. Examples of the melanoma expression in Gordon-Kosswig BC_1 hybrids. The initiation of melanoma invasion of the dorsal area *(top)*, continuing along the dorsal axis to the tail area *(middle)*, and finally becoming necrotic and affecting the overall health of the animals *(bottom)*.

control from the *R-Diff* regulator. Consistent with this, transcriptional expression patterns of the two *EGFR*-related gene copies have shown that *Xmrk-2* is highly expressed in BC_1 hybrid-derived melanoma tissues relative to the *Xmrk-1*, which appears to be the "normal" gene copy and is expressed at low levels in most tissues (Woolcock *et al.*, 1994). Experiments also indicate overexpression of *Xmrk-2* in melanotic tissues of interspecies hybrids is coincident with hypomethylation of the gene promoter regions (Altschmied *et al.*, 1997). These and many other results strongly implicate *Xmrk-2* as the primary *Sd* linked *Tu* oncogene in the G-K model.

Biochemistry of Signal Transduction in the Gordon-Kosswig Melanoma Model

The manner in which *Xmrk-2* expression affects gene expression of other cell cycle pathways resulting in unregulated pigment cell growth has been studied in some detail. Mutations identified in the *Xmrk-2* gene product would be expected to result in an EGFR-like protein that has gained new 5′ upstream controlling factors resulting in up-regulated expression with alterations in the external cellular domains that would induce dimerization of membrane bound *Xmrk-2* proteins in the absence of ligand binding (Gomez *et al.*, 2004). This autodimerization mimics the normal receptor bound state and leads to intercellular signaling via a tyrosine kinase mechanism, which in turn initiates a cascade of events signaling cell growth (Gomez *et al.*, 2001; Meierjohann *et al.*, 2006; Winnrmoeller *et al.*, 2005). Modulation of several varied cellular signaling pathways has been demonstrated to result from *Xmrk* expression. For example, *Xmrk* has been shown to induce expression of the Ras/Raf and MAP kinase pathways, thereby affecting cell differentiation and cell proliferation. The tyrosine kinase oncogene, Fyn, is activated by *Xmrk* expression, which may enhance MAP kinase expression and induce effects on cell migration factors (Geissinger *et al.*, 2002; Wellbrock and Schartl, 2000). In addition, in G-K melanoma *Xmrk* plays a role in up-regulation of the STAT5 pathway leading to expression of proteins that function to suppress apoptosis (Morcinek *et al.*, 2002). The cumulative effect of all of these events and perhaps others yet to be determined may explain how aberrant expression of a single protein factor, *Xmrk-2*, can lead to melanoma when crossed into in the proper genetic background.

Molecular Genetics of R-Diff in the Gordon-Kosswig Melanoma Model

Genetic analyses of G-K BC_1 hybrids (Fig. 32-2) allowed localization of the hypothetical melanoma tumor suppressor function controlling *Sd-Tu* oncogenicity (i.e., *R-Diff*) to an autosomal trait on *Xiphophorus* linkage group (LG) V (Kazianis *et al.*, 1998). Subsequently, a *Xiphophorus* gene homolog of the human cyclin-dependent kinase inhibitor gene family (*CDKN2*, AKA *p16, INK4*) was cloned and also assigned to the *Xiphophorus* gene map very close to the functional location of *R-Diff* on LG V (Kazianis *et al.*, 1999). The sequence analyses of this *R-Diff* candidate indicated the gene was equally similar to human *CDKN2A* (*p16*) and *CDKN2B* (*p15*), which are tumor suppressor genes already shown to be associated with familial and sporadic melanoma in humans (Hayward, 2003; Ortega *et al.*, 2002). Because of equal similarity to the human *CDKN2A* and -B genes, the *Xiphophorus* gene was initially designated *CDKN2X*. However, strong similarity between a *Takifugu* gene and the *Xiphophorus CDKN2X* sequence suggests that fish forms of this gene may predate the duplication producing human p15 and p16 gene copies (Gilley and Fried, 2001). Thus, this *Xiphophorus* gene has been renamed *CDKN2AB* (Kazianis *et al.*, 2004).

As expected for *R-Diff*, several studies have shown that inheritance of the *X. maculatus CDKN2AB* alleles in G-K BC_1 hybrids is strongly correlated with both the degree of melanin pigmentation expressed and the propensity for development of melanoma. Also, transcriptional expression of the *CDKN2AB* alleles appears to coincide with the degree of melanization in G-K backcross hybrid individuals (Kazianis *et al.*, 1999, 2000). For example, expression of the *Xiphophorus CDKN2AB* gene is relatively low in normal (i.e. nontumor) fin tissues but exhibits higher expression in F_1 interspecies hybrid fin tissues (i.e., enhanced melanotic expression but nonmalignant) and further, exhibits very high expression in the malignant melanoma tissue derived from BC_1 hybrids (≈sevenfold over parental expression levels). It has also been shown that BC_1 hybrids express *CDKN2AB* alleles derived from *X. maculatus* or *X. helleri* differentially in several tissues including melanomas (Kazianis *et al.*, 1999), and in virtually all cases, the *X. maculatus* alleles are very highly expressed relative to the *X. helleri* ones. Thus, *CDKN2AB* has been forwarded as a candidate gene for the hypothetical *R-Diff* tumor regulator (Kazianis *et al.*, 1999, 2000). Central to the idea that *R-Diff* and *CDKN2AB* are one and the same entity is the strong association that has been documented between human *CDKN2A* gene expression and heritable melanoma in humans where greater than 25% of familial melanoma cases have mutant copies of the *CDKN2A* gene that are thought to impart loss of activity of the protein (Hayward, 2003). However, in contrast to the human situation for *CDKN2A*, *Xiphophorus CDNK2AB* gene expression is found elevated in melanoma susceptible BC_1 hybrids. Thus, alternative hypotheses for the role of *CDKN2AB* in *Xiphophorus* melanoma have been forwarded (Kazianis *et al.*, 2000; Nairn *et al.*, 2001; Walter and Kazianis 2001). In contrast to the classical and hypothetical functions expected for *R-Diff*, there are no data indicating a mechanism whereby *CDKN2AB* might control the oncogenicity of *Xmrk*-2 (i.e., as a transcription factor). Thus, despite great strides in characterization of the genetics underlying the G-K melanoma model, our understanding of this tumor model is clearly not yet complete.

Induced *Xiphophorus* Tumor Models

Development of *Xiphophorus* interspecies hybrid experimental models that require exposure to ultraviolet (UV) light or treatment with mutagenic chemicals (i.e., *N*-methyl-*N*-nitrosourea; MNU) to induce tumorigenesis has opened new avenues for detailed study of the genetics underlying tumor susceptibility (Table 32-1). Setlow *et al.* (1989, 1993) first used *Xiphophorus* hybrids to study the role of UV light in inducing melanomas. In these studies, several new backcross hybrid models were developed that had varying response to UV-induced tumor development. For example, an *X. maculatus* stock (Jp 163 B) that carries the spotted side (*Sp*) trait, producing melanin spotting on the flank of the animal (instead of the dorsal fin as for *Sd*), was employed in crosses with recurrent parental species *X. couchianus* and *X. helleri*. In these and other studies, backcross hybrid progeny were generally exposed to UV irradiation at 5 days postbirth and melanoma or other tumors scored after 6 to 8 months growth.

TABLE 32-1. Example of *Xiphophorus* tumor models.

Species Utilized in Cross[a]	Tumor Type(s)	Induction
X. helleri × (*X. maculatus* Jp 163 A × *X. helleri*)	Melanoma	Spontnaeous
X. helleri × (*X. maculatus* Jp 163 B × *X. helleri*)	Melanoma	MNU and UV
(*X. maculatus* Jp 163 A × *X. andersi*) × *X. andersi*	Melanoma	Spontaneous
(*X. maculatus* Jp 163 B × *X. andersi*) × *X. andersi*	Melanoma Renal Adenocarcinoma	MNU
(*X. maculatus* Jp 163 A × *X. couchianus*) × *X. couchianus*	Retiinoblastoma Nuerofibrosarcoma Schwannoma	MNU
(*X. maculatus* Jp 163 B × *X. couchianus*) × *X. couchianus*	Melanoma	MNU & UV
X. helleri × (*X. variatus*^{Li} × *X. helleri*)	Fibrosarcoma Neuroblastoma	MNU
X. cortezi^{Sc}	Melanoma	Aging
X. variatus^{Pu2}	Melanoma	Aging

[a]The female parent is listed first, i.e., the first cross listed above would be a male F_1 hybrid derived from mating a female *X. maculatus* Jp 163 A with a male *X. helleri* backcrossed to a female *X. helleri*.

Concurrently with development of tumor models based on UV induction, other investigators were establishing models that required chemical treatment to induce tumor development. For example, MNU was used to induce a variety of tumor types in select *Xiphophorus* backcross hybrid crosses that included *X. maculatus* Jp 163 A and B crossed with several recurrent parental species, including *X. couchinaus, X. helleri*, and *X. andersi* (Nairn *et al.*, 2001; Kazianis *et al.*, 2001a, 2001b). MNU is a monofunctional alkylating agent that methylates DNA bases at nucleophilic sites (principally N^7 and N^3 alkylpurines). MNU is known to be a complete carcinogen and the primary mutagenic lesion is believed to be O^6-methylguanine and to a lesser extent O^4-alkylthymine (Hall and Monsanto, 1990). In the 1970s, Schwab *et al.* (1978a, 1978b, 1979) had established the propensity of MNU to induce a wide array of tumors in *Xiphophorus* hybrids. In particular, the development of neuroblastomas was strongly associated with a single hybrid cross involving *X. variatus* and *X. helleri* and not observed in 64 other nonhybrid species/strains or derived hybrid genotypes (Schwab *et al.*, 1978a, 1978b). These studies showed the potential of *Xiphophorus* interspecies crosses to serve as genetic models of rare tumor types.

In the 1990s, studies using the same *Xiphophorus* interspecies backcross hybrid models for both UV and MNU treatment showed that MNU-induced and UVB-induced melanoma were governed by different underlying genetics, an observation that has yet to be scientifically exploited or understood. Also, initial studies were performed using two new *Xiphophorus* interspecies hybrid models that (1) exhibited new MNU-induced tumor types in the absence of melanin pigment pattern enhancement and (2) showed pigment pattern enhancement but failed to exhibit melanoma induction after UVB exposure. In the latter case the BC_1 hybrids did show MNU-induced melanoma and retinoblastoma tumor induction but failed to develop tumors after UV irradiation. Next we discuss some of these models and observations.

UV-Induced Models: A Guide to UV Sources Used in *Xiphophorus* Studies

Much of the interest in the *Xiphophorus*-induced melanoma models has involved studying the effects of different wavelengths of UV light on tumor induction. In these experiments FS-40 fluorescent lamps have frequently been employed as broadband UV sources. A typical emission spectrum from an unfiltered FS-40 lamp is about equal amounts of UVB (280 to 320 nm) and UVA (320 to 400 nm) light, with a broad continuum of wavelengths with maximum around 310 nm and overlaid with visible Hg lines at 313 and 365 nm (Fig. 32-4a). Because solar light does not contain photons below 280 nm, various kinds of filtration are often used to experimentally remove the output from lower wavelengths, such cutoffs at 290 nm for acetate sheets, or at 304 nm for Mylar sheets placed between the light source and the fish to be exposed. The main other source of light used in studies has been monochromatic light, generated by use of a monochromator or interference filtration of Hg emission peaks from Hg arc lamps. These provide a narrow range of light spread about a central maximum (as shown for interference filtered 365-nm light in Fig. 32-4b) and so are suitable for studying the effects of these defined wavelengths. Rigorous characterization of light sources is essential, especially for studies of longer wavelengths such as UVA; even small amounts of shorter wavelengths that may be present in supposedly clean sources of longer wavelengths can dominate the effects of these light sources and be a source of considerable error.

Xiphophorus as a Model for UV-Induced Melanoma and Determination of Its Action Spectrum

Setlow and coworkers recognized the potential for using *Xiphophorus* models to answer important questions regarding the role and mechanisms of UV in melanoma causation (Setlow *et al.*, 1989). Two backcrossed hybrid models were developed that exhibited up to 40% melanoma prevalence

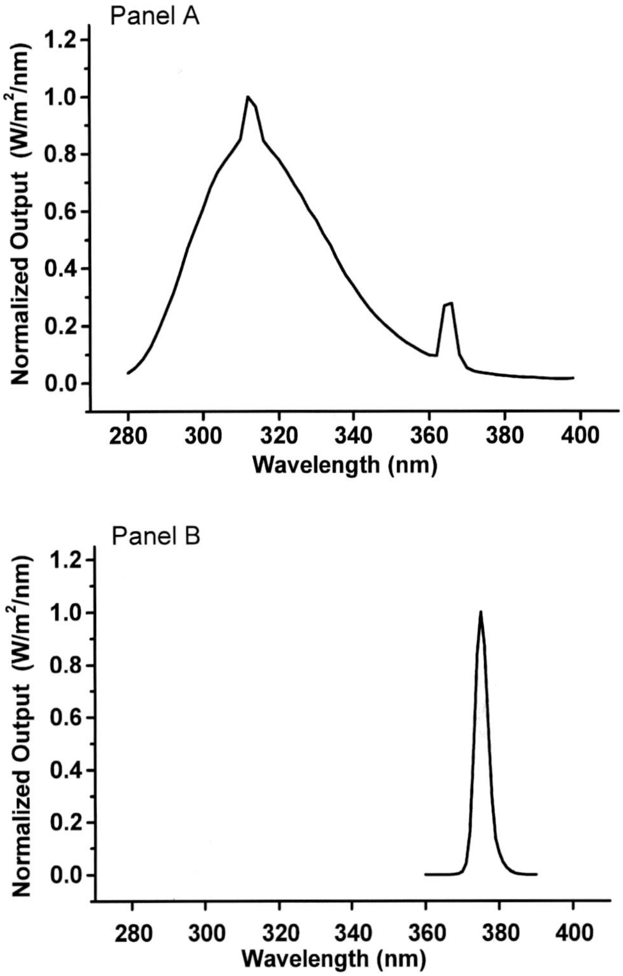

FIGURE 32-4. Outputs from commonly used UV sources for *Xiphophorus* melanoma induction experiments. (A) Normalized output of unfiltered FS40 sunlamps, measured using an Optronic 742 spectroradiometer. (B) Normalized output of a 365-nm interference filtered XeHg arc lamp, measured using an Optronic 742 spectroradiometer.

only 4 months after receiving UV light from filtered FS-40 lamps. When young fish (i.e., 6 days postbirth) were exposed to relatively low UV doses of only 150 to 1700 Jm^{-2}, repeated 5 to 15 times, this resulted in maximal melanoma yield, although single exposures were also surprisingly effective. The high tumor yield and speed of development, together with the small size of the fry during irradiation, made determination of the wavelength dependence, or action spectrum, for melanoma causation feasible. The action spectrum for melanoma induction in these two *Xiphophorus* backcross hybrid models was first reported by Setlow and coworkers for wavelengths from 302 to 436 nm (Setlow *et al.*, 1993) and later at 547 nm (Setlow, 1999). These specific wavelengths (302, 313, 365, 405, 436, and 547 nm) were isolated from the Hg lines of an arc lamp using a monochromator, because their relatively high intensity made it feasible to generate monochromatic output to achieve the observed biological effect. The crossing scheme studied employed *X. maculatus* strain Jp 163 B (i.e., Sp, spotted side) and showed effective melanoma causation with a single filtered FS-40 dose.

In Setlow's studies, melanoma exhibited significant induction by all wavelengths; however, the most significant wavelengths (in terms of the amounts of each present in sunlight) were found to be quite different from that expected should melanoma induction be a result of direct absorption of UVB by DNA. This was the first study to suggest the photochemical process driving melanoma causation may be distinct from the UV wavelengths known to cause DNA damage or mutagenesis in other experimental systems. Rather, it was hypothesized that the melanomagenic action spectrum could be best explained if melanin photosensitized processes played a major role in inducing melanoma, although direct DNA damage might also contribute (Setlow, 1999; Setlow *et al.*, 1993) as some photoreactivation of Mylar-filtered FS-40 sunlamp irradiation–induced melanoma was observed (although this not statistically significant). Furthermore, it was noted that should the human action spectrum follow the same form as *Xiphophorus*, the UVB component of sunlight would only be responsible for 5% to 10% of human melanoma, whereas longer wavelength components in sunlight would dominate melanoma causation from Sun exposures. In this scenario, ozone layer thinning would be expected to cause relatively little increase in melanoma rates, whereas the use of sunscreens widely utilized by humans, that absorbed mainly UVB, would not be predicted to protect well against melanoma. Adding to this "sunscreen abuse controversy" is the fact that such predictions are consistent with epidemiological analysis of melanoma, although the presence of many other confounding factors prevents definitive proof. To date, the *Xiphophorus* action spectrum has remained the only one available for melanoma, although fractionation of UVB and UVA wavelengths has been used in other animals such as transgenic mice (DeFabo *et al.*, 2004; Noonan *et al.*, 2001) and *Monodelphis domestica* (Ley, 1997, 2001). Work using *Xiphophorus* UV-induced melanoma has shown the action spectrum of melanoma causation was statistically identical to that of melanin photosensitization *in situ* within pigmented cells of *Xiphophorus* hybrid fry (Wood, *et al.*, 2006), supporting a central role for this process in this model.

In addition to the preceding, the overall importance of UVA and oxidative DNA damage in causing *Xiphophorus* melanoma remains as controversial as it does in other animal models, and humans. Thus, a different camp of researchers has emphasized the importance of UVB and direct DNA damage, and much of the DNA damage work (discussed later) continues to be performed with predominantly UVB light sources. However, even long-standing paradigms in this area are becoming challenged with reports that UVA may cause cyclobutane pyrimidine dimer (CPD) formation in nucleic acids through a sensitized process (Douki *et al.*, 2003; Mouret *et al.*, 2006) and that melanin can act as a UVB photosensitizer (Takeuchi *et al.*, 2004). Further studies are required to resolve these differences.

UV-INDUCED DNA DAMAGE AND REPAIR IN *XIPHOPHORUS*

With the utility and importance of the induced *Xiphophorus* hybrid melanoma models to UV photobiology clearly established, a range of studies have been undertaken to study UV-induced DNA damage and repair. Several of these studies used the strengths of backcross hybrid genetic analysis to determine the association and roles of specific genes in UV-induced melanoma. Filtered FS-40 lamps and also isolated UVB wavelengths (302 and 313 nm) were shown to produce cyclobutane pyrimidine dimers (CPDs) in a dose-dependant manner in skin cells of both adult and fry fish, whereas 365 nm light did not (Ahmed and Setlow, 1993). Significant photoreactivation was observed, as was photoprotection against CPD production by melanin pigmentation. Thus, it would appear that melanomas caused by 365 nm light (UVA) are not caused by CPD formation, although photoreactivation after UVB treatment reduced UVB-induced tumor incidence, showing direct DNA damage is important for UVB wavelengths. DNA damage and repair processes were studied in greater detail by Mitchell and coworkers (Meador *et al.*, 2000; Mitchell *et al.*, 1993) in *X. signum* using unfiltered FS-40 lamps to show CPD formation was much greater than induction of pyrimidine(6-4)pyrimidone photoproducts ([6-4]PD) in fish DNA. These studies also showed photoreactivation of CPDs occurred at roughly twice the rate of [6-4]PD removal. Nucleotide excision repair (NER, or dark repair) of CPDs and [6-4]PDs was also observed upon UV treatment, and the relative contributions of NER and photoreactivation were shown to exhibit complex tissue-specific patterns. Interestingly, Mitchell *et al.* (1993) had shown that photoreactivation could be induced by preexposure of *X. variatus* to photoreactivating light (i.e., before UV treatment), suggesting that compensatory adaptation to UV exposure may occur (Mitchell *et al.*, 1993). This work and other important data were comprehensively reviewed in 2001 (Mitchell *et al.*, 2001; Nairn *et al.*, 2001; Setlow and Woodhead, 2001).

Because susceptibility to UV-induced melanoma is only observed in interspecies hybrids, the rates of NER repair of CPD and [6-4]PD in a variety of relevant hybrids was compared to the repair capability in parents, using unfiltered FS-20 lamps as the UV source. In particular, the same cross that Setlow *et al.* (1993) had used to establish the melanoma action spectrum showed profoundly decreased NER of [6-4]PDs compared to parental strains, although repair of CPDs in these fish did not appear decreased in the hybrids

TABLE 32-2. Tumor development in various BC_1 hybrid models as a result of the hybrid cross (G-K model) or induced upon after UV or MNU treatment.

Cross Type	Tumor Type	Inducing Treatment[a]	Incidence (%)	CDKN2AB Associated[b]	Reference
X. helleri × (*X. maculatus* Jp 163 A × *X. helleri*)	Melanoma	None (G-K)	~25	Yes	Kazianis *et al.*, 1999
X. helleri × (*X. maculatus* Jp 163 B × *X. helleri*)	Melanoma	None	7.2	Yes	Nairn *et al.*, 1996a
	Melanoma	UVB	18.4	Yes	"
	Melanoma	MNU	36.8	No	Kazianis *et al.*, 2001b
X. maculatus Jp 163 B × *X. couchianus*) × *X. couchianus*	Melanoma	UV or MNU			Nairn *et al.*, 2001
(*X. maculatus* Jp 163 A × *X. couchianus*) × *X. couchianus*	All tumors below	None	<0.5	nd	Kazianis *et al.*, 2001a
	Retinoblastoma	MNU	3.8	nd	"
	Neurofibrosarcoma	MNU	6.6	nd	"
	Schwannoma	MNU	3.8	nd	"
	Af Unresticed Melanosis	MNU	10.4	nd	"
(*X. maculatus* Jp 163 B × *X. andersi*) × *X. andersi*	Melanoma	None	2.7	nd	Nairn *et al.*, 2001
		UVB	<2.5	nd	"
		MNU	29.7	No	Walter and Kazianis, 2001

[a]UVB, ultraviolet light B; MNU, *N*-methyl-*N*-nitrosourea.

[b]For association with *CDKN2AB*, individual BC_1 fish that developed melanoma were genotyped. The genotype and tumor development were subjected to linkage analyses, and a LOD (log of odds) score of 3.0 or greater was considered significant evidence for linkage/association. In this case, association was for the inheritance of *CDKN2AB* and latent tumor development. LOD score >3.0 is considered significant evidence for linkage/association.

(Mitchell *et al.*, 2004). Thus, in addition to the role that inheritance of different oncogenes, cell cycle regulators, and tumor suppressors may play in determining which particular hybrid fish develop melanoma, the hybrid genetic background also determines which of the many parental DNA repair proteins will interact in oligomeric complexes that may result in partially dysfunctional DNA repair leading to melanoma susceptibility.

Although there is increasing evidence that UVA may cause CPD formation via a photosensitization mechanism in some amelanotic cell types (e.g., keratinocytes; Mouret *et al.*, 2006), CPDs were not detected in the epidermis of either melanotic or amelanotic *Xiphophorus* hybrids treated with more than 31,000 J m^{-2} of monochromatic 365 nm UVA (Ahmed and Setlow, 1993). Thus, the effectiveness of different wavelengths to induce CPD formation, and any particular endpoint such as melanoma, is likely dependant upon the overall genetic background of the fish, the particular cell type involved, and tissue context of the cells. In contrast to studies that have detailed UV-induced CPDs and [6-4]PDs, other types of DNA damage, such as oxidative damage, that would be caused by melanin photosensitization have not received attention. Thus, full appreciation of how different solar wavelengths may cause different types of DNA damage and which damages are associated with latent melanoma remains to be determined. However, base excision repair (BER), a pathway principally responsible for repair of base alterations such as alkylated or oxidative case products, has begun to be studied (David *et al.*, 2004; Walter *et al.*, 2001a, 2001b). Although the specific probes used to study BER were designed with lesions that would be expected from MNU exposure rather than those from UV-induced oxidative stress, it is clear that BER occurs, shows modulation in select tissues of hybrid fish (compared to the parents), and may play major roles in cancer susceptibility among the various *Xiphophorus* tumor models (see Table 32-2 and the following discussion).

Genetic Analysis of UV Causation of Melanoma

A major advantage of the *Xiphophorus* model is its well-developed genetics. The first reports identifying specific genes important in UV-induced melanoma were by Nairn *et al.* (1996a, 1996b). In these studies, several interspecies cross models were used to identify the *R-Diff* tumor suppressor candidate, *CDKN2AB* (see Gordon-Kosswig model, discussed earlier). In a parallel with the fish, some germline mutations in *CDKN2A* are linked to an increased risk of melanoma in humans (Orlow *et al.*, 2007), although most germline mutations are not associated with increased risk. In one report, lack of inheritance of *X. maculatus CDKN2AB* genotypes was strongly linked to UV-B-induced (acetate filtered FS-40 lamps) melanoma development in hybrids (Nairn *et al.*, 1996b). Another showed segregation of *R-Diff* (i.e., LG V) was linked to melanoma induction in hybrids by monochromatic 405-nm light but was not associated with tumors induced by 313- or 365-nm light (Nairn *et al.*, 1996a,

1996b) using a light source developed by Setlow *et al.* (1993). However, the latter study was less rigorous, involving genetic analyses of LG V isozyme markers, and so it remains unclear how these reports relate to each other with regard to the different wavelengths studied. Variation in tumor response at different wavelengths could be entirely feasible, since it is becoming clear that in humans, different patterns of genes are activated and different mutations may result in melanomas when exposed to varying types of UV exposure (Rivers, 2004).

MNU-Induced Melanomagenesis in *Xiphophorus*

To provide comparison with the UVB tumor induction, *X. helleri* (x) [*X. maculatus* Jp 193 B (x) *X. helleri*] and [*X.maculatus* Jp 163 B (x) *X. couchianus*] (x) *X. couchianus* BC_1 hybrids (Fig. 32-5) were assessed for tumorigenesis after exposure to the monofunctional alkylating agent, MNU (Tables 32-1 and 32-2). Five-week-old BC_1 hybrid fish were subjected to four MNU treatments (1 mM MNU, 2 h.) every

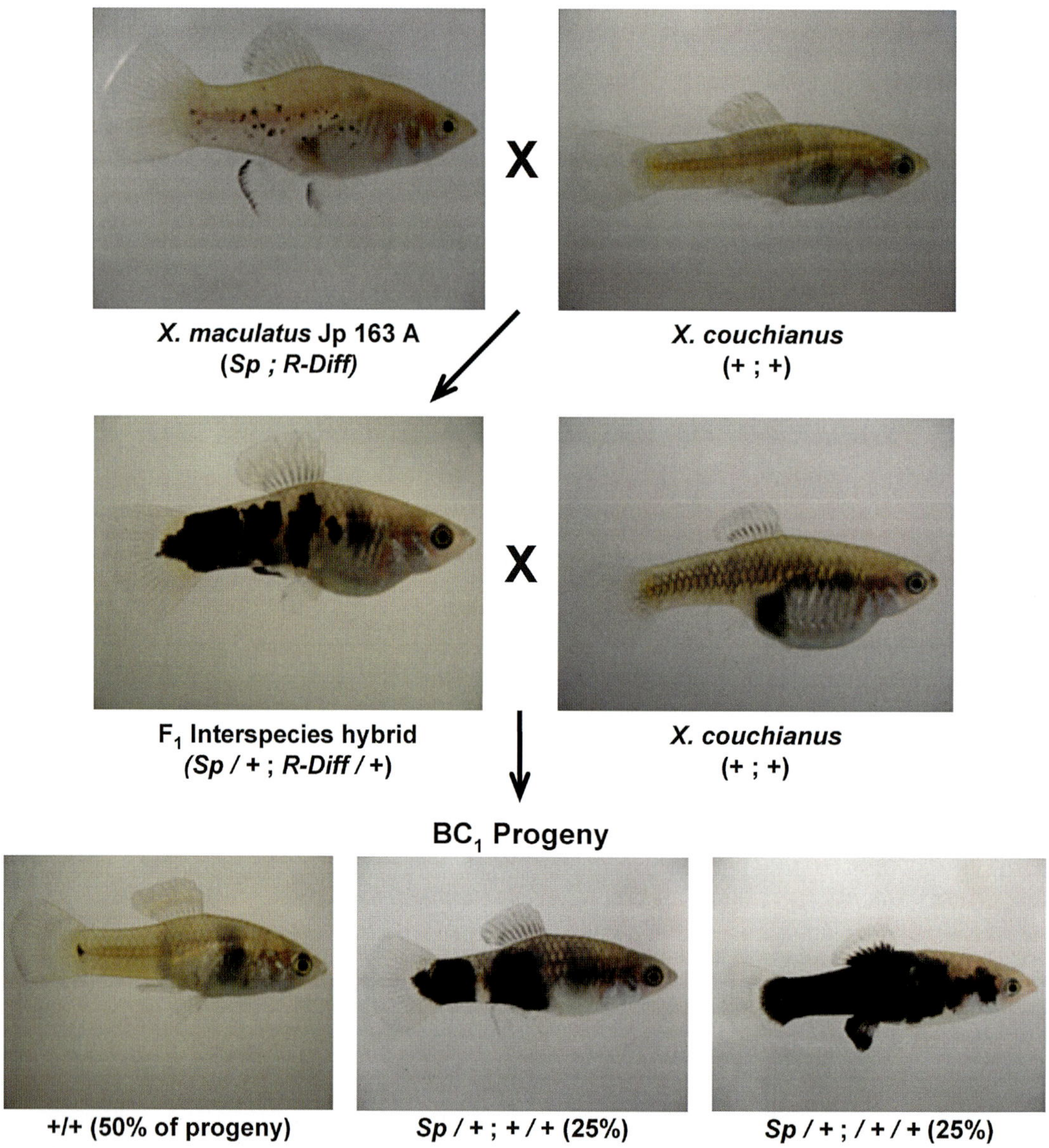

FIGURE 32-5. The *X. couchianus* x [*X. maculatus* Jp 163 B x *X. couchianus*] hybrid inducible tumor model. The *X. maculatus* Jp 163 B parent carries the sex-linked spotted side (*Sp*) pigment pattern and the autosomal tumor regulator (*R-Diff*). Once crossed with the *X. couchianus* parent, the F_1 hybrid progeny carry one allele from each of the parents. Once F_1 hybrids are backcrossed the to *X. couchianus* parent, about 25% of the BC_1 progeny will inherit the *Sp* and become highly pigmented. Tumor development in these BC_1 progeny does not normally occur unless the animals are treated soon after birth with UVB or MNU.

other day. Fish treated with MNU expressed large exophytic melanoma tumors anywhere from 4 to 8 months posttreatment (Fig. 32-6). As a comparative example, the data in Table 32-2 illustrate the level of MNU induction of melanoma is higher than induction by UVB, even in the same BC_1 hybrids. Surprisingly, genetic analyses of tumor-bearing MNU-exposed *X. maculatus* Jp 163 B-*X. helleri* interspecies hybrid cohorts (149 genetic markers per animal) did not produce significant association between inheritance of *CDKN2AB* (LG 5; Table 32-2) and tumor development. This absence of association of MNU-induced melanoma with *CDKN2AB* zygosity suggests that inheritance of an entirely different set of gene targets is needed to predispose BC_1 animals to UV or MNU tumor induction. Among the MNU-treated melanoma-bearing animals, approximately 20% exhibited multiple neoplastic lesions, a situation that is not observed among hundreds of UVB-induced tumor-bearing fish (Fig. 32-6). Comparatively, the UVB and MNU results indicate the existence of multiple genetic routes leading to melanoma susceptibility, even within BC_1 hybrids produced from pedigreed and highly inbred parental lines such as the *Xiphophorus* lines used in these studies.

FIGURE 32-6. Examples of MNU-induced tumors in *X. couchianus* x [*X. maculatus* Jp 163 B x *X. couchianus*] BC_1 hybrids. Top two panels, an MNU-induced nonpigmented melanoma. The highly pigmented BC_1 hybrids may become virtually black when treated with MNU (third panel from top). When induced with MNU one often observes multiple tumors developing in different body regions, such as a melanoma and retinoblastoma (bottom panel).

Development of a Nonmelanoma-Induced Tumor Model

It is known that F_1 and BC_1 fish from the *X. couchianus* (x) [*X. maculatus* Jp 163 A (x) *X. couchianus*] cross appear to have suppressed the spotted dorsal phenotype in their dorsal fins. This is contrary to the large enhancement in *Sd-Tu* pigmentation observed when the cross is performed with *X. maculatus* Jp 163 B (see Fig. 32-5). This is considered curious since the *Sd* and *Sp* pigment patterns are both very closely linked on the X chromosome. Because enhancement of pigmentation caused by interspecies hybridization generally results in fish that are susceptible to UVB- or MNU-induced melanoma, this suppression of melanin spots in these *X. maculatus* Jp 163 A-*X. couchianus* interspecies hybrids allowed determination of tumor types induced in the absence of enhanced melanin pigment pattern phenotypes. Figure 32-7 depicts the crossing scheme used to generate F_1 interspecies hybrids between *X. maculatus* Jp 163 A and *X. couchianus*. In this cross inheritance of the sex chromosomes harboring *Sd-Tu*, *Dr* pterinophore locus, and a pigment pattern locus referred to as Anal fin spot (*Af*; Kazianis *et al.*, 2001b) is shown. Within the F_1 and BC_1 hybrids to *X. couchianus*, the *Dr* pigment pattern is overexpressed, such that the fish are quite orange in color (see Fig. 32-7). In contrast, the *Sd-Tu* melanocyte pigment pattern normally expressed in the dorsal fin is phenotypically suppressed, where only about 14% *Dr*-bearing backcross hybrid animals exhibit any black spot pigmentation whatsoever. Treatment of these BC_1 fish with MNU produced three types of tumors in low incidence including; schwannomas, neurofibrosarcomas, and retinoblastomas (Table 32-2; Kazianis *et al.*, 2001a, 2001b). Sham-treated control fish did not develop any of these tumor types suggesting that a genetic component may underlie predisposition to tumor development in select BC_1 individuals of this cross. Interestingly, all three of these tumor types arise in cells that have their embryonic origin in the neural crest. Thus, a hypothesis has been forwarded that low incidence tumors were induced that had the same embryonic origin as the melanocytes previously induced to tumorigenesis at high incidence after MNU treatment in other *Xiphophorus* BC_1 models.

FIGURE 32-7. *X. maculatus* Jp 1634 A (x) *X. couchianus* interspecies hybrid. The F_1 interspecies hybrid from this cross exhibits suppression of the *Sd* (spotted dorsal) pigment pattern in contrast to the *Sp* (spotted side) pigment pattern that becomes enhanced when crossed into the *X. couchianus* genetic background (see Fig. 32- 5). In the cross pictured, the *Dr* (dorsal red) pigment pattern is enhanced coincident with suppression of *Sd*. BC_1 hybrids produced from crossing the F_1 pictured with *X. couchianus* do not exhibit melanization, and upon MNU treatment they may develop tumors in tissues derived from the neural crest including retinoblastoma, neurofibrosarcoma, and Schwannoma.

Results from this cross are important for two principal reasons: (1) they imply that *Xiphophorus* interspecies hybrids where *X. maculatus* is one of the parental species will produce neural crest cell derived lineages that are susceptible to MNU-induced tumor development; and (2) they indicate that if we were to expose appropriate numbers of such hybrids to MNU, we may be able to identify genetic regions associated with predisposition to schwannomas, or neurofibrosarcomas, or retinoblastoma. Genetic models for these three tumor types do not exist in other natural animals models.

Summary and Xiphophorus Resources

Because of their high degree of heterozygosity, *Xiphophorus* interspecies hybrids are extremely valuable models for molecular genetic study of gene regulation in physiology and behavior. The fact that interspecies hybridization leads to tumor bearing or tumor susceptible progeny provides valuable experimental models to uncover the genetics underlying various pathways to tumor development. Historically, *Xiphophorus* fish have made a contribution to understanding the genetics underlying melanoma, but the role this experimental model system may play in contemporary comparative biology has only just begun to be realized. In spite of the many advantages and importance of the model, a genome-level *Xiphophorus* sequencing project has not yet been initiated. To set a stage for a genome-level initiative, many *Xiphophorus* resources have been or are currently being developed. Completion of the *Xiphophorus* 24 chromosome gene linkage map, construction of microsatellite and EST databases, and the accessibility of BAC library resources all hallmark progress in future development of genomic capabilities of the *Xiphophorus* genetic system.

RAINBOW TROUT: MODEL FOR CANCER CHEMOPREVENTION, MECHANISMS OF PROMOTION, AND TUMOR RESPONSE AT ULTRALOW DOSES

Introduction to the Rainbow Trout Model of Hepatocarcinogenesis

As with many developments in science, discovery of the sensitivity of the rainbow trout to human carcinogens was serendipitous. In investigating the cause of an outbreak of liver cancer in hatchery rainbow trout in the Pacific Northwest, it was discovered that these fish are extremely sensitive to the human hepatocarcinogen, aflatoxin B_1 (AFB_1) (Sinnhuber *et al.*, 1968). AFB_1 is one of the most potent hepatocarcinogens characterized to date and was one of the first chemicals found in the environment (food) to be labeled as a class A carcinogen. Aflatoxins are produced by the mold *Aspergillus flavus* and *Aspergillus parasiticus* (reviewed in Eaton and Gallagher, 1994) found in hot, humid climates and contaminate primarily corn and peanuts, subject to such conditions. Primary liver cancer is relatively rare in the United States, and levels of aflatoxin on corn and peanuts is low (the average daily intake in the Southeastern United States is 110 ng/Kg (Eaton and Gallaher, 1994), in part because of postharvest storage and shipping conditions that are not conducive to the production of aflatoxins. Unfortunately, this is not the case in many parts of the world. For example, in certain regions of China, primary liver cancer is the number 1 cause of death in males (Yu, 1995). In addition to high dietary intakes of AFB_1, the high incidence of hepatitis B and C virus (HBV and HCV) contributes to these grim statistics. It is thought that AFB_1 functions as an efficient initiator and the chronic inflammation that comes with hepatitis infection is a very effective promoter (Ming *et al.*, 2002). Exposure to AFB_1 alone in the diet enhances the risk of liver cancer four- to sevenfold; this risk increases to 60-fold if the individual is HBV or HCV positive.

Studies employing rainbow trout as a model for AFB_1-initiated hepatocarcinogenesis were pioneered by individuals at Oregon State University, including Drs. Sinnhuber, Nixon, Wales, Hendricks and Bailey (reviewed in Bailey *et al.*, 1996). The high sensitivity of the rainbow trout is due, in part, to cytochromes P450 (CYP) with high activity toward bioactivation of AFB_1 to AFB_1-8,9-*exo*-epoxide (Williams and Buhler, 1983). This trout CYP is in the 3A subfamily; human CYP3A4 (and to some extent 1A2) likewise has high activity in production of the *exo*-epoxide (Gallagher *et al.*, 1996; Ueng *et al.*, 1995). Unlike mice, trout do not have a constitutive glutathione-S-transferase (GST) in the alpha class and so are not efficient in conjugation and detoxication of this electrophilic epoxide (Eaton *et al.*, 2001; McGlynn *et al.*, 2003; Raney *et al.*, 1992). Trout and rats can be made more resistant to AFB_1, if agents capable of induction of GST alpha are administered before or concurrently with the carcinogen (Stresser *et al.*, 1994b; Takahashi *et al.*, 1995a). In addition to induction of GSTs, protection could be caused by induction of CYPs that increase the yield of less carcinogenic metabolites or function as competitive inhibitors of CYPs that produce the epoxide (Stresser *et al.*, 1994a, 1995; Takahashi *et al.*, 1995a, 1995b). Another explanation for the sensitivity of the rainbow trout to AFB_1 may lie in the observation that the repair of bulky carcinogen-DNA adducts, such as *trans*-8,9-dihydro-(N^7-guanyl)-9-hydroxy-AFB_1, is not as efficient in trout as in more resistant species (Bailey, 1994).

The discovery of the high sensitivity of rainbow trout to AFB_1 was responsible for the initial development of this species as a model for human hepatocarcinogenesis. Since that time, a number of other chemical carcinogens have proven positive in this model, including nitrosamines, direct-acting methylating agents (methylazoxy-methanol acetate (MMA), ethylnitrosourea (ENU), N-methyl-N'-nitro-N-nitrosoguanidine (MNNG), and polycyclic aromatic hydrocarbons (PAHs) (reviewed in Bailey *et al.*, 1996). The model has also proven valuable in studies on dietary chemoprevention/promotion and assessing tumor response at ultralow carcinogen doses (Breinholt *et al.*, 1999; Dashwood *et al.*, 1989, 1991; Oganesian *et al.*, 1999; Orner *et al.*, 1995; Pratt *et al.*, 2007; Reddy *et al.*, 1999; Williams *et al.*, 2003), made possible because of our capability to design and carry out studies requiring the use of large numbers of animals (10,000 to 40,000) to address statistically challenging questions that would be too expensive to address in a rodent model. The sections that follow give specific examples of such studies. The low spontaneous background of 0.1% liver tumors (at 1 year of age) also contributes to the feasibility of using trout for such studies.

First, to conduct cancer studies of such magnitude, one must have access to a facility capable of breeding, dosing, housing, necropsy, and histopathological analysis of test animals on such a scale. Toward that goal, the Sinnhuber Aquatic Research Laboratory was constructed.

A Description of the Sinnhuber Aquatic Research Laboratory (SARL): A Unique National Resource for Utilization of Rainbow Trout in Cancer Studies

A schematic of the SARL facility is shown in Figure 32-8. This is a unique 15,270 ft^2 facility capable of spawning, maintenance, treatment, necropsy, and histopathological analysis of 40,000 rainbow trout (as well as thousands of zebrafish). The rainbow trout are housed in 440 3- or 4-foot diameter tanks, each with a working capacity of about 320 liters. Normally these trout spawn only in winter, but as the cues are entirely from the photoperiod, we constructed spawning rooms within our facility that are on a reverse photoperiod ("Southern Hemisphere" rooms). In this way, we are able to spawn fish twice a year, increasing the number of fish obtainable and, more important, the frequency with which fry are available.

A critical component of any such facility is water quality. When at maximum capacity, our facility runs 8.6 million liters a day through the lab. This high-quality, constant temperature (13°C) water is groundwater obtained from wells

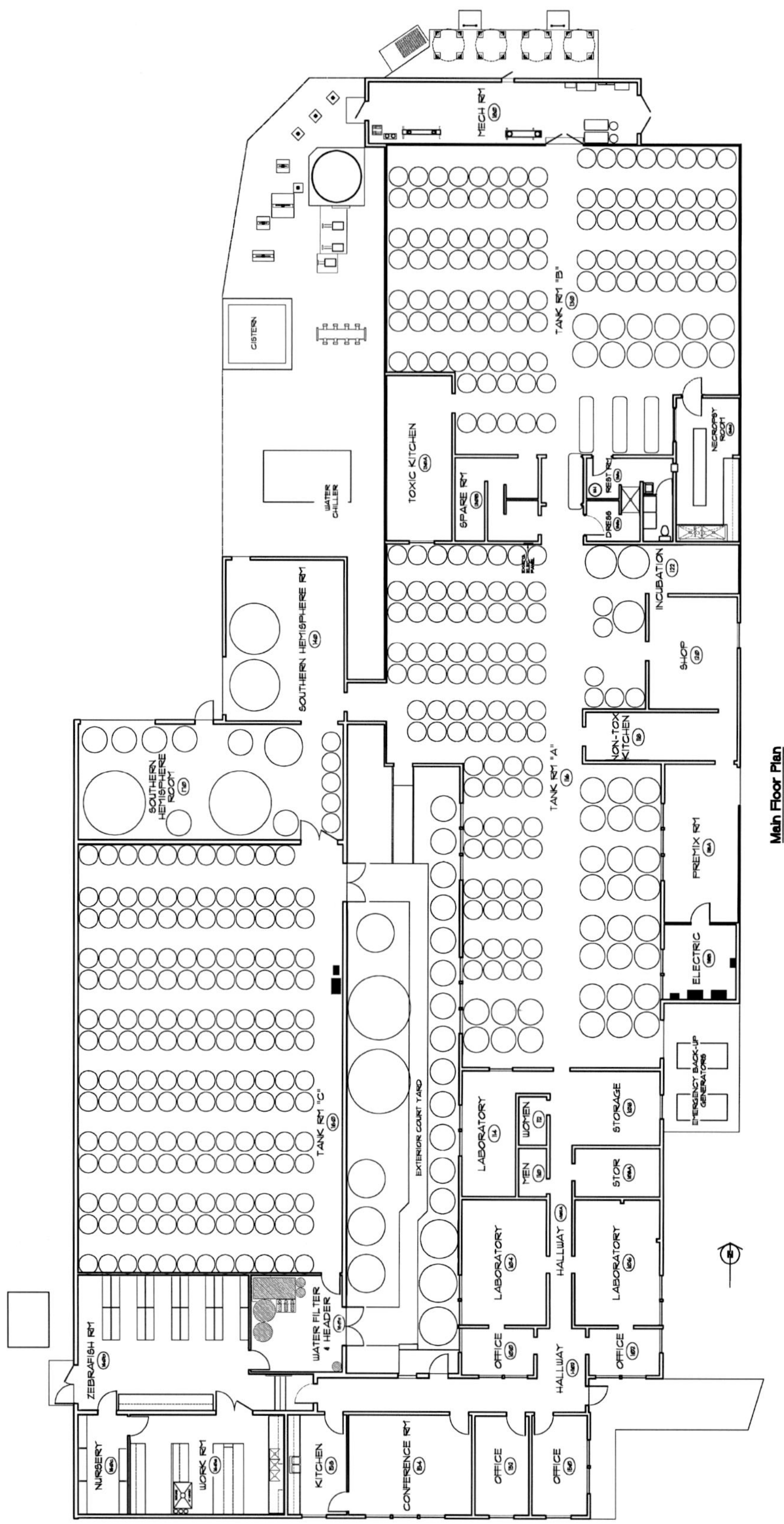

FIGURE 32-8. Schematics of the Sinnhuber Aquatic Research Laboratory at Oregon State University.

less than 1 mile from the Willamette River. Water from up to four operating wells is mixed and run through columns of plastic shards to drive off nitrogen and saturate the water with oxygen. The water then goes into a holding facility tank from which water is pumped through sand filters, then activated charcoal (more than 7000 Kg), passed over an industrial-size UV sterilizing lamp and then pumped into the laboratory. To reduce the flow required and to increase the survival time, if some major reduction or cessation of water flow occurred, the tanks are fitted with air stones and blowers that send 0.75 ft^3/min of air into the tanks. In the event of an interruption in electric power, a 55-kilowatt, propane-powered generator and a 100-kilowatt, natural gas generator, both automatically come online. The facility is equipped with sensors that constantly monitor individual well depth and pump output, holding tank capacity and, air, and water pressure. These parameters are displayed in real time on a computer in the laboratory. If any parameter falls outside of set limits, or if the power is interrupted, the computer sends an alarm and personnel are paged 24 hours per day, 7 days per week. These safeguards are critical in order for us to conduct yearlong tumor studies with confidence. The trout are fed a defined diet developed a number of years ago by researchers at Oregon State University. The complete composition of this diet, termed Oregon Test Diet (OTD), can be found in the publication of Lee *et al.* (1991). Trout can be exposed to the carcinogen or tumor modulator by introduction into OTD, fry bath exposure or microinjection of embryos or young fry (Bailey *et al.*, 1996).

Contributions of the Rainbow Trout Model in Carcinogenesis

Over the previous 40 years, the rainbow trout model has made a number of significant contributions to the field of carcinogenesis, especially with respect to mechanisms of dietary modulation. We will review a few of those studies in the sections that follow. Dietary inhibitors of hepatocarcinogenesis in the trout include indole-3-carbinol (I3C, a major constituent of cruciferous vegetables [broccoli, cauliflower, etc.] and sold as a human dietary supplement) (Dashwood *et al.*, 1988, 1989, 1991; Fong *et al.*, 1990), chlorophyllin (CHL), which is the water-soluble derivative of chlorophyll (Chl), used for many years in geriatric patients (Bailey, 1995; Breinholt *et al.*, 1999; Hayashi *et al.*, 1999; Pratt *et al.*, 2007; Reddy *et al.*, 1999) and more recently the natural chlorophylls extracted from spinach (Harttig and Bailey, 1998). Other agents that have proven positive in chemoprevention studies in mammalian models, such as D-limonene, green and white teas, oltipraz, menthol, vitamin E, freeze-dried onion or garlic, and mint oil, have given equivocal or negative results in the trout model (Bailey *et al.*, 1996).

Chlorophyllin (CHL) and the Natural Chlorophylls (Chl): An Example of Discovery and Elucidation of the Mechanism of Action with Subsequent Clinical Trials in Humans

Chlorophyllin (CHL) is the water-soluble derivative of the natural plant pigment chlorophyll, in which the phytol side chains have been removed and the magnesium replaced with copper. As stated previously, it has been used safely for many years in geriatric patients to control body odor. One of the mechanisms of action by which CHL is thought to act is by complexation with the carcinogen, especially those with aromatic, planar structures. This complexation is hypothesized to reduce carcinogen GI bioavailability. We have measured the binding constants for CHL with a number of trout carcinogens. For example, CHL forms a 1 : 1 complex with AFB_1, with an estimated K_d of 1.4 M (Breinholt *et al.*, 1995) and with DBP a 2 : 1 complex with a K_d of 1.6 M (Reddy *et al.*, 1999). A reduced bioavailability has been demonstrated when CHL and either carcinogen are administered at the same time (Hayashi *et al.*, 1999).

Studies conducted in the trout model were the first to demonstrate the anticarcinogenic potency and efficacy of CHL. These studies exploited the advantage of the model discussed earlier—that is, the ability to conduct studies with large numbers of individuals so that it is possible to test both multiple doses of initiator and multiple doses of chemopreventive agent. The study by Pratt *et al.* (2007) utilized 12,000 trout to test the tumor response with dietary levels of DBP from 0 to 371.5 ppm and 0 to 6000 ppm dietary levels of CHL. If the hypothesis is correct that the mechanism of CHL chemoprevention is due to reduced bioavailability through molecular complexation, then such a mechanism should apply across species (with the caveat that slight changes in conditions in the stomach [e.g., pH] do not play a major role in complexation). Specifically, this means that administration of CHL to individuals at high risk for AFB_1-initiated hepatocellular carcinoma should reduce bioavailability and thereby the risk for development of cancer. Researchers at Johns Hopkins and Oregon State Universities conducted a pilot study in Qidong province in China in which volunteers were given 100 mg of CHL (as Derifil) three times daily (instructed to take 1 tablet 20 minutes before each meal). After 4 months, urinary levels of aflatoxin-N^7-guanine, a biomarker of AFB_1 exposure, was reduced by 55% (Egner *et al.*, 2001). As CHL is inexpensive and has few side effects (leading to excellent compliance), the potential for CHL to impact the epidemic of liver cancer in Asia and other parts of the world is very exciting (Kensler *et al.*, 2004). What is important to note here is that this is an example of how *studies in an aquatic model, such as rainbow trout, can essentially go directly to human clinical trials*, providing the mechanism of action is known. More recently, Dr. Bailey's laboratory has demonstrated that natural chlorophylls could also act as chemopreventive agents (Simonich *et al.*, 2007) at levels that are possible to obtain by consuming spinach.

Dr. Bailey's laboratory has partnered with researchers at Lawrence Livermore National Laboratory to conduct the first study on the pharmacokinetics of AFB_1 in humans and alteration by Chl or CHL. By employing accelerator mass spectrometry, it is possible to safely dose human volunteers with AFB_1. Volunteers were dosed with 30 ng (5 nCi) of AFB_1 (equivalent to 1.5 g of peanut butter with the FDA allowable limit of 20 ppb). Following dosing, blood was collected periodically out to 72 hours. This protocol was done thrice for each volunteer. Subsequently, each volunteer repeated the study with the same AFB_1 dose plus 150 mg of either Chl or CHL. As predicted from the "complexation" hypothesis, co-administration with CHL or Chl delayed absorption and reduced the C_{max} and AUC (Bailey *et al.*, unpublished). Again, this is an excellent demonstration of how the trout model contributed to the development of a promising cancer chemoprevention strategy. The chemoprotection properties of both CHL and Chl were first demonstrated in trout, and the hypothesis of complexation, leading to reduced bioavailability, was first tested in this model.

I3C and 3,3′-diindolylmethane (DIM) from Cruciferous Vegetables: Utilization of the Trout Model in Studies of Mechanism(s) of Action and Risk Versus Benefit

I3C is found in cruciferous vegetables as glucobrassicin. Upon maceration of the cell, the enzyme myrosinase cleaves glucobrassicin to produce I3C. The average daily intake of I3C from cruciferous vegetables is 20 mg/day. Both I3C and DIM are available as dietary supplements with suggested doses of 200 to 800 mg/day.

In the acid environment of the stomach, I3C undergoes a series of condensation reactions and DIM is typically the most prevalent compound found *in vivo* after oral ingestion of I3C (McDanell and McLean, 1988). One of the pioneers of cancer chemoprevention, Dr. Lee Wattenberg, first demonstrated the chemopreventive properties of I3C in a rodent model (Wattenberg and Loub, 1978). Since that original observation, I3C has been demonstrated to inhibit a number of cancers in diverse models (reviewed in Kim and Milner, 2005). I3C was originally proposed as a blocking agent through aryl hydrocarbon receptor (AhR)-mediated induction of phase I and phase II enzymes, but a number of additional mechanisms of action have since been delineated. I3C is effective in estrogen-dependent cancers, such as mammary cancer (Grubbs *et al.*, 1995; Meng *et al.*, 2000). Induction of CYP1A-dependent metabolism of 17β-estradiol (E2) to 2-hydroxy-E2 is thought to lower E2 levels in women (Michnovicz and Bradlow, 1990) and produce more of the "good" E2 metabolite (as opposed to the "bad" 4-hydroxy and 16(-E2 metabolites (Michnovicz *et al.*, 1997). I3C has been evaluated in clinical trials for the chemoprevention of breast cancer (Lawrence *et al.*, 2000) and has also been safely and effectively used for the treatment of cervical cancers (Bell *et al.*, 2000).

In the trout model of AFB_1-initiated hepatocarcinogenesis, I3C has long been known as an effective chemopreventive agent if administered before or concurrent with AFB_1 (Bailey *et al.*, 1991). In this experimental protocol, I3C appears to be acting primarily as a blocking agent. Our laboratory raised concerns about the long-term usage of I3C in healthy individuals (I3C and DIM are available as dietary supplements) upon discovery that long-term, postinitiation ingestion of I3C or DIM led to promotion of hepatocellular carcinoma (HCC) (Dashwood *et al.*, 1991; Oganesian *et al.*, 1999). Again, the fact that we could conduct tumor studies with large numbers (9000 trout) allowed us to compare the efficacy and potency of I3C both as an inhibitor and promoter of AFB_1-initiated liver cancer (Oganesian *et al.*, 1999). Such studies are necessary to get an accurate picture of the risk/benefit of prolonged administration.

Dehydroepiandrosterone (DHEA) and Perfluorooctanoic Acid (PFOA)

In rodents, chemicals that induce peroxisome proliferation are hepatocarcinogens. Peroxisome proliferators (PPs), such as DHEA and PFOA, activate the PPARα receptor. The downstream events following PPARα activation include induction of palmityl CoA oxidation, catalase, and increases in the number of peroxisomes and oxidative stress. Chronic oxidative stress is thought to contribute to PP-dependent hepatocarcinogenesis (Reddy and Rao, 1989). Humans differ from rodents in that they do not respond to PPs (e.g., clofibrate, used for many years in patients to lower triglyceride levels in blood) with any of the parameters described above, including hepatocarcinogenesis (Ashby *et al.*, 1994). To determine if trout are a better model for human risk assessment, we conducted long-term feeding studies with classic PPs, including WY14,643 and clofibrate, in both tumor initiation and tumor promotion protocols. The response in trout to long-term PP treatment resembled the human response, with respect to biochemical, morphological, and pathological alterations in liver. However, two PPs tested (DHEA and PFOA) did promote hepatocarcinogenesis in trout (Benninghoff *et al.*, unpublished; Orner *et al.*, 1995, 1998).

The adrenal steroid DHEA and its sulfated ester are the second highest steroid (in terms of plasma concentration) in humans, but levels decrease rapidly with age. Individuals in their 70s typically have just 20% of the DHEA present in younger adults. This age-related decline in DHEA has led to speculation that the loss of DHEA contributes to the myriad of physical and mental ailments associated with aging, and supplementation with DHEA may assist in slowing the aging process. In laboratory studies and limited observational clinical trials, DHEA has shown promise with aging, Alzheimer disease, cardiovascular disease, certain cancers, loss of cognitive function, diabetes, HIV, multiple sclerosis, obesity, and systemic lupus erythematosus (Olech and Merrill, 2005). However, several randomized, placebo-controlled clinical trials have failed to confirm beneficial effects of DHEA toward age-related disorders (Baulieu *et al.*, 2000; Nair *et al.*, 2006). Nevertheless, DHEA continues to be aggressively marketed to slow aging, fight cancer, build muscle, burn fat, and improve sexual stamina.

Initial studies in our laboratory were driven by the apparent contradiction that although DHEA is a PP (Frenkel *et al.*, 1990), and therefore would be expected to be a rodent hepatocarcinogen (Rao *et al.*, 1992), most published reports show DHEA to have beneficial effects toward cancer (Schwartz *et al.*, 1988). Because trout, like humans, have a limited "peroxisome proliferation" response, they provide a unique opportunity to examine the effects of DHEA toward cancer in the absence of significant peroxisome proliferation.

DHEA is a complete carcinogen in rainbow trout. Twenty percent of trout fed 1800 ppm DHEA developed liver cancer (Orner *et al.*, 1995). DHEA also strongly enhanced the carcinogenic response of AFB_1 (Fig. 32-9A). Treatment with DHEA produced dose-dependent increases in tumor incidence, tumor multiplicity, and tumor size (Orner *et al.*, 1995). Subsequent research found that DHEA also reduced latency between initiation and tumor development in AFB_1-initiated trout (Orner *et al.*, 1998), and tumor promotion was not initiator-specific because DHEA also enhanced MNNG-initiated hepatocarcinogenesis (Orner *et al.*, 1996b).

The carcinogenic and tumor modulating effects of DHEA are independent of peroxisome proliferation. Hepatic palmityl CoA oxidation, lauric acid ω-hydroxylation, catalase activity, and electron microscopy did not provide any significant evidence of DHEA-induced peroxisome proliferation (Orner *et al.*, 1995). Instead, ultrastructural examination of livers from DHEA-treated trout suggested that hepatocytes were producing a secretory protein (Fig. 32-10). Serum levels of vitellogenin (Vtg), a classic biomarker for estrogen response in fish, were dramatically induced in DHEA-treated animals (Fig. 32-9B). This finding supported the hypothesis that modulation of hepatocarcinogenesis by DHEA may be due to its hormonal properties. Additional evidence in support of this hypothesis was provided by a comparison of DHEA with a fluorinated analog that is less readily metabolized to

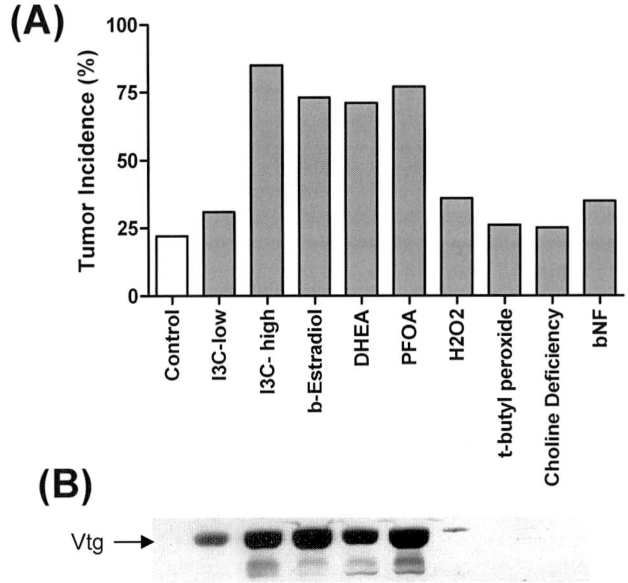

FIGURE 32-9. (A) Promotion of AFB_1-initiated hepatocellular carcinogenesis and (B) induction of plasma vitellogenin (Vtg) protein by dietary exposure to 17®-estradiol, I3C, DHEA, and PFOA in rainbow trout.

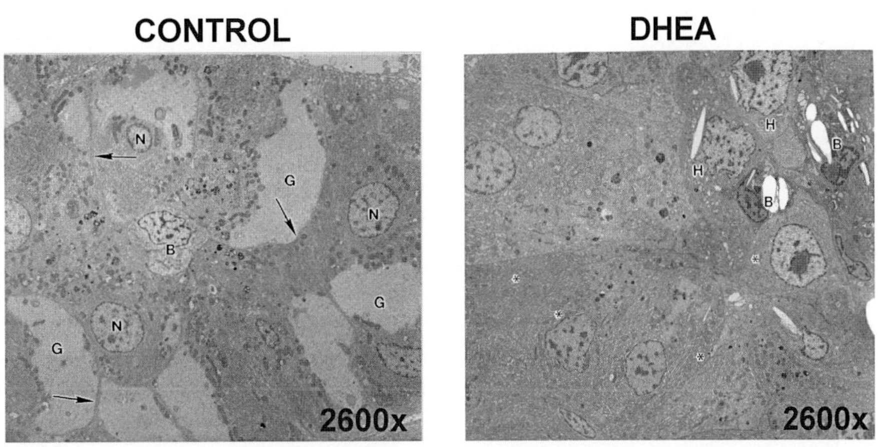

FIGURE 32-10. EM (2600 × magnification) of liver from trout treated with 1800 ppm DHEA in the diet for 14 days. H, hepatocellular hypertrophy; G, decreased glycogen storage; I, cytoplasm with increased rough endoplasmic reticulum.

androgens or estrogens (Orner *et al.*, 1996a). The analog was not carcinogenic in rainbow trout and did not significantly enhance AFB_1-induced hepatocarcinogenesis.

Perfluorooctanoic acid (PFOA) is a member of a broader class of perfluorinated chemicals (PFCs) that include shorter and longer chain perfluoroalkyl carboxylic acids and sulfonates that are by-products or intermediates in the manufacture of Scotchguard (3M Corporation), Teflon (DuPont) and other nonstick and stain-resistant products. These chemicals are notoriously refractory to degradation in the environment and have been detected in numerous wildlife species throughout the world at concentrations in the parts per billion (reviewed in Houde *et al.*, 2006). A number of epidemiological and risk assessment studies have demonstrated that many PFCs, such as PFOA and perfluorooctane sulfonate (PFOS), are present in humans worldwide at concentrations in the low to mid parts per billion. This widespread contamination suggests that environmental sources of PFOS and PFOA are likely related to the use and disposal of materials containing volatile PFCs; these compounds then degrade in the environment to the persistent PFCs that are found in humans and wildlife.

PFOA is a hepatocarcinogen in rodents acting via a peroxisome proliferation-dependent mechanism mediated by the peroxisome proliferator-activated receptor ⟨ (PPARα) (reviewed in Kennedy *et al.*, 2004). Humans are insensitive to PPs, in part because of the low expression of hepatic PPARα. Therefore, current data from rodent cancer studies are of limited use for extrapolating the risk of PFOA exposure to humans. Recent data obtained in our laboratory suggest that PFOA acts via alternate mechanisms to promote hepatocellular carcinogenesis. PFOA markedly enhanced hepatocellular carcinoma (HCC) in AFB_1-initiated rainbow trout, a species that is also resistant to peroxisome proliferation (Fig. 32-9A). PFOA was a strong inducer of vitellogenin (Vtg), a classic estrogen-responsive hepatic gene in fish that is often evaluated as a biomarker of xenoestrogen exposure (Fig. 32-9B). Exogenous estrogens have long been known to be associated with cancer in estrogen-responsive tissues including breast, ovary, and uterus, as complete carcinogens, promoters, or possible chemoprotective agents. Moreover, estrogens also influence carcinogenesis in other tissues not usually considered to be estrogen responsive, including liver (Nunez et al., 1989; Yager *et al.*, 1991).

The finding that PFOA acts as a xenoestrogen led to additional *in vivo* and *in vitro* studies in rainbow trout to evaluate the potential estrogenicity of other perfluoroalkyl carboxylic acids (Benninghoff *et al.*, 2006). Results from an *in vivo* Vtg induction screen of 10 perfluoroalkyl carboxylic acids (5 to 14 carbons in length) in juvenile rainbow trout suggest that three additional PFCs with structures similar to that of PFOA may also be xenoestrogens and potential promoters of hepatic tumorigenesis, including perfluorononanoic acid (PFNA), perfluorodecanoic acid (PFDA), and perfluoroundecanoic acid (PFUnDA). Furthermore, these four chemicals have weak binding affinity for the rainbow trout hepatic estrogen receptor, similar to that of other xenoestrogens such as I3C and nonylphenol (Benninghoff, unpublished observations).

Because humans are insensitive to peroxisome proliferation, PFOA exposure is not generally considered to be a major cancer threat. However, this is a potentially risky assumption given the possibility that PFOA and other PFCs could act via alternative mechanisms to promote liver cancer. Evidence obtained to date suggests that PFOA and other perfluorinated chemicals may act via an alternative estrogenic mechanism to influence carcinogenesis. Thus, greater focus on elucidating the role of PFC exposure and human cancer is needed, especially in light of the prevalence of these chemicals in humans worldwide.

Genomic Approaches in the Study of Carcinogenesis in the Trout Model

Rainbow trout have been widely utilized for research of carcinogenesis, comparative immunology, toxicology, physiology, reproduction, and nutrition. The economic importance of trout, combined with their use as a surrogate research model for human-related disease, has led to a significant investment in the development of genetic tools for trout (Thorgaard *et al.*, 2002). In particular, the Rainbow Trout Gene Index currently contains more than 240,000 ESTs representing more than 80,000 unique sequences (RTGI, 2006) as a result of sequencing efforts to increase the rainbow trout EST database. This influx of genetic information has allowed for the development and publication of several microarray platforms for use with rainbow trout (Table 32-3).

The available platforms differ in their type (cDNA v. oligonuncloetide), size, and overall design properties. Some arrays were developed using a large number of unique ESTs representing a wide variety of different gene classes, whereas other arrays were enriched for genes associated with specific mechanisms. For example, the rainbow trout 70-mer oligonucleotide array (OSUrbt v. 2.0; www.science.oregonstate.edu/mfbsc/facility/micro.htm) contains 1672 elements representing approximately 1400 genes known to be involved in the physiological processes of carcinogenesis, immunology, endocrinology, toxicology, and other stress responses. This array has been successfully utilized to examine mechanisms of carcinogenesis and immune function in rainbow trout (Gerwick *et al.*, 2007; Tilton *et al.*, 2005, 2006, 2007). Alternatively, the GRASP 16k v. 2.0 array contains some genes that were selected for relevance to immune function and reproduction; however, most ESTs were randomly chosen based on quality and uniqueness (http://web.uvic.ca/cbr/grasp/array.html). This cDNA array shows similar hybridization performance for all salmonids and has been applied to a broad range of research questions (Hook *et al.*,

TABLE 32-3. Example microarray platforms published for use with trout.[a]

Array Name	Type	Date[b]	GEO ID[c]	Institution	Citation
GRASP 3.7k	cDNA	Feb. 2003	GPL966	University of Victoria, Canada	Rise *et al.*, 2004
Salmonid fish v. 1.0	cDNA	May 2004	GPL1212	University of Kupio, Finland	Krasnov *et al.*, 2005a
GRASP 16k v. 2.0	cDNA	Feb. 2005	GPL2716	University of Victoria, Canada	Von Schalburg *et al.*, 2005
OSUrbt v. 2.0	Oligo	Sept. 2004	GPL2096	Oregon State University, USA	Tilton *et al.*, 2005 Gerwick *et al.*, 2007
AGENAE_Trout Generic2_9216	cDNA	Apr. 2006	GPL3650	INRA, France	Bobe *et al.*, 2006

[a]Based on information provided in Gene Expression Omnibus (GEO); www.ncbi.nlm.nih.gov/geo.
[b]Date microarray developed from manufacturer information or GEO submission data.
[c]GEO platform accession number.

2006; von Schalburg *et al.*, 2005). The overall increase in availability of DNA microarrays has allowed for molecular characterization of many physiological and toxicological mechanisms providing a powerful genetic tool for research in rainbow trout and other salmonid species.

Since the advent of DNA microarray technology, researchers have been collecting gene expression profiles, or transcriptional "fingerprints," that can be used as tools for classifying chemical exposures and predicting the potential mechanism of action. One such mechanism of considerable interest in cancer research is the estrogen pathway. Studies have identified transcriptional profiles resulting from natural, pharmaceutical, or environmental estrogens (Naciff and Daston, 2004; Terasaka *et al.*, 2004; Wang *et al.*, 2004). In our laboratory, we have utilized the custom oligonucleotide (70 mer) trout and salmon cDNA arrays (described earlier) to predict the mechanism of action for several promoters of HCC, including I3C, DHEA, and PFOA. Preliminary observations suggested these chemicals act as xenoestrogens, as *in vivo* exposure of rainbow trout results in elevated plasma Vtg. Therefore, we hypothesized that I3C, DHEA, and PFOA act in a manner similar to estradiol in promotion of liver carcinogenesis.

As discussed previously, I3C has been shown to suppress or enhance tumors in several animal models. The cancer-modulating effects of I3C (or its acid condensation products) are mediated, in part, by interaction with several nuclear transcription factors, including the AhR, the estrogen receptor (ER), Sp1, and nuclear factor IB (NFIB) to alter processes associated with carcinogen detoxification, cell cycle progression, apoptosis, and DNA repair (reviewed in Kim and Milner, 2005). In the trout liver, I3C differentially induces markers of ER and AhR activation (Vtg and CYP1A, respectively) depending on dose, whereas DIM is a potent estrogen as demonstrated by strong induction of Vtg (Oganesian *et al.*, 1999; Shilling *et al.*, 2001). A toxicogenomic approach was employed utilizing the salmon cDNA microarray (GRASP) to evaluate the relative importance of ER or AhR-mediated pathways in tumor promotion by I3C in trout (Tilton *et al.*, 2006). Hepatic gene expression profiles were examined in trout after dietary exposure to I3C and DIM; these profiles were compared to those of E2 and β-naphthoflavone, representative of ER and AhR-dependent, respectively, mechanisms of action. Clustering of gene responses revealed that both I3C and DIM acted similarly to E2 at the transcriptional level in liver following dietary treatment (Fig. 32-11A). Furthermore, *in vitro* studies with high precision cut liver slices demonstrated that, compared to other I3C acid condensation products, DIM was a potent activator of ER (as judged by Vtg induction), but had a relatively poor ability to activate the AhR (as judged by CYP1A1 induction) (Figs. 32-11B and C). These results provide further confirmation that I3C was acting through the ER signaling pathway in promotion of HCC in trout.

Based on the transcriptional similarities between I3C and DIM in trout liver, we determined in subsequent studies that DIM could also promote AFB_1-induced hepatocarcinogensis through estrogenic mechanisms (Tilton *et al.*, 2007). Strong transcriptional correlations were observed between DIM and E2 treatments in livers samples during promotion ($R = 0.84$ after 3 weeks; $R = 0.76$ after 15 weeks) and also in the resulting HCC tumors ($R = 0.87$). Trout are a sensitive model for estrogen-mediated promotion of hepatocarcinogenesis, which may be due to higher ligand promiscuity reported for trout ER compared to other species (Petit *et al.*, 1995). Interestingly, although DIM and E2 increased tumor incidence in initiated animals, transcriptional profiles in HCC tumors from these animals indicated lower invasive and metastatic potential compared to HCCs from control animals. It is possible that DIM is acting similar to E2 in trout liver and having a dual effect on tumorigenesis in which it may increase tumor incidence through ER-mediated mitogenic signaling while at the same time decrease the potential for metastasis. Overall, these data in trout may help explain the dichotomy of chemoprotective and enhancing effects of dietary indoles on cancer development.

A similar experimental approach was utilized to ascertain whether the tumor promoter PFOA enhanced tumorigenesis

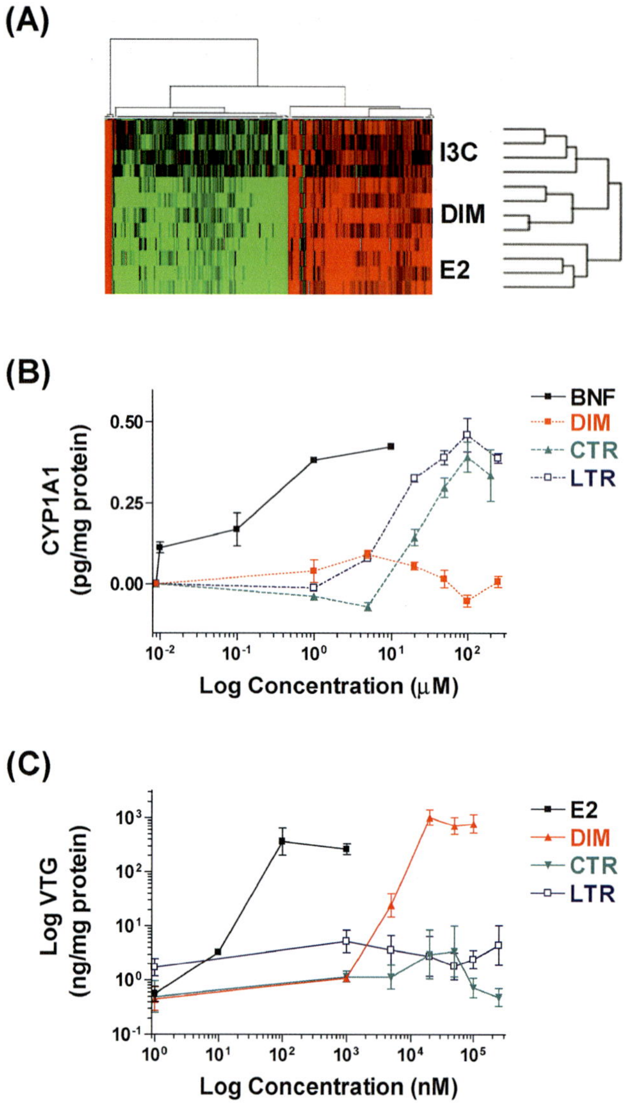

FIGURE 32-11. (A) Heatmap representation showing similarity in gene expression from liver of animals fed 24 mg/kg/day I3C and DIM compared to E2. Red represents genes up-regulated and green genes down-regulated at least twofold from control animals. (B) CYP1A and (C) VTG induction in trout liver slices after exposure to I3C acid condensation products, DIM, LTR, and CTR, compared to E2 and NF for 96 hours. Values are means with standard deviations (n = 6). Part (A) adapted from Tilton *et al.* (2006).

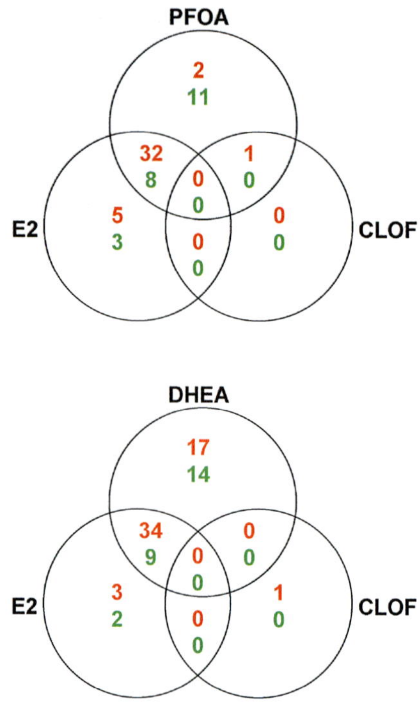

FIGURE 32-12. DHEA and PFOA promotion of hepatocarcinogenesis—role of ER and correlation between genes regulated by E2 and PFOA. Differential gene expression in trout liver after dietary treatment with 5 ppm E2, 1800 ppm CLOF, 1800 ppm PFOA, or 750 ppm DHEA. Venn diagrams show genes regulated ≥1.8-fold ($p \leq 0.05$) compared to controls in all replicates ($n = 3$). Red indicates genes up-regulated, and green indicates genes down-regulated.

in rainbow trout *via* an ER-mediated pathway rather than *via* peroxisome proliferation, the presumed mechanism of action for this chemical in mammalian animal models. DHEA, an aromatizable androgen, was also tested in this study, because it was suspected to act *via* the estrogen pathway by means of its metabolite, E2. For this study, the rtOSU microarray was utilized to compare the hepatic gene expression profile of PFOA and DHEA to other known tumor promoters, including E2 and clofibrate, a classic PP in mammals. Results of this toxicogenomics study revealed that the transcriptional profile of DHEA and PFOA was strongly correlated to that of E2 and poorly correlated to clofibrate (Fig. 32-12). Pairwise analysis of the 1672 features of this array by Pearson correlation demonstrates strong correlations in gene patterns between DHEA or PFOA and E2 ($p < 0.0001$), but not between DHEA and CLOF or CLOF and E2. Altogether, these observations suggest PFOA and DHEA increase incidence and multiplicity of hepatocellular carcinogenesis through ER- rather than PPARα-dependent signaling pathways.

The general pattern of E2-induced gene expression was similar across the array platforms utilized in these two studies, particularly when examining the list of major biological processes affected by the estrogenic tumor promoters (Tilton *et al.*, 2005, 2006, 2007). Aside from fish ER-responsive liver proteins (e.g., Vtg, vitelline envelope protein), genes of interest for carcinogenesis included those associated with cell growth and proliferation (e.g., serine/threonine protein kinase) and protein folding, stability, and transport (e.g., cathepsin D). Genes repressed by E2 exposure include the acute-phase immune response (e.g., chemotaxin), angiogenesis (e.g., angiogenin precursor), and metabolism (e.g., CYP3A\5). It should be noted, however, that unique gene targets for some tumor promoters were also identified, such as a set of complement component genes involved in the immune response that were uniquely repressed by DHEA.

The results of these studies may have direct human health implications. Humans are highly sensitive to the influence of hormones on certain cancers, particularly those of the breast and prostate. DHEA is clearly functioning as a steroid, regardless of whether it acts directly or as a precursor in the formation of androgens and estrogens, and should be regulated accordingly. PFOA and related compounds are widespread environmental contaminants. The resistance to metabolism results in a long half-life (estimated at 4 years) in humans. Little information is available on the risk of long-term exposure to PFOA, and our recent findings, that at least some of these compounds can act as xenoestrogens, highlight the usefulness of the trout model for discovery and the conduct of mechanistic studies.

We have utilized a toxicogenomic approach applied to gene expression profiles measured in trout AFB_1-induced hepatocellular carcinoma (HCC) utilizing the custom rainbow trout oligonucleotide microarray targeted to mechanisms of carcinogenesis, immunology, environmental toxicology, stress physiology, and endocrinology (Tilton *et al.*, 2005). Most genes or gene classes differentially expressed in the trout tumor samples were typical of those observed in HCC in humans or other mammalian models (Graveel *et al.*, 2001). Global analysis of molecular signatures revealed changes in a number of distinct gene functional classes and disease processes important for carcinogenesis in trout that can be extrapolated to human disease. For example, genes involved in normal liver function, including metabolism and homeostasis of drugs, lipids, glucose, and retinol, were down-regulated in HCC compared to the noncancerous adjacent liver indicating a role for dedifferentiation in neoplastic development in trout. Some histopathological evidence for this step-wise progression from foci to benign to malignant tumors exists in trout (Bailey *et al.*, 1996; Okihiro and Hinton, 1999) and is well documented in rodent tumors (Pitot *et al.*, 1996). Also of interest were gene profiles indicating rainbow trout may provide a novel and relevant model for the study of anemia and inflammation in chronic liver disease, including alteration of inflammatory cytokines and chemokines and genes important for iron absorption. Other genes altered in trout HCC were involved in matrix-membrane associations, cell migration, and metastasis and were indicative of a potentially invasive and aggressive tumor phenotype. These findings are consistent with the idea that HCC pathogenesis has been conserved during vertebrate evolution, particularly considering the different etiologies (many viral) reported for hepatocellular carcinogenesis, and enhance the merits of trout as a model for human hepatocarcinogenesis. A number of other studies have also been published recently utilizing microarray technology in trout and other salmonid models to examine stress responses, chemical toxicology, and immune function, further supporting the use of this model in biomedical research (Krasnov *et al.*, 2005a, 2005b; Rise *et al.*, 2004).

Utilization of the Trout Model to Test Dose-Response to Carcinogens at Ultralow Doses (the ED001 Trout Model)

Accurate risk assessment for carcinogens is a major challenge. The regulatory target is the dose of a chemical that would not produce more than 1 additional cancer in 1 million individuals. The challenge comes from having to extrapolate (five orders of magnitude) from rodent data that measures typically 1 cancer in 10. The largest mouse study ever done utilized 24,000 mice, dosed with the bladder and liver carcinogen, 2-acetylaminofluorene. The data from this study supported a linear model for liver, but a sublinear (with a potential threshold) model for bladder. The interpretation of the data from this hugely expensive rodent study is still being debated.

Dr. Bailey's group exploited the advantages of the trout model (low cost, low spontaneous liver tumor incidence, etc.) to conduct the largest cancer study in any animal model. This 42,000 trout study was designed to determine the statistically significant dose of DBP (dibenzo [a.1]pyrene) that resulted in 1 additional cancer in 1000 animals (an order of magnitude better than the mouse ED01 study) (Bailey *et al.*, in preparation). The target was actually exceeded and the dose of DBP resulting in 1 additional cancer (liver) in 5000 animals was determined. At lower doses, the tumor incidence deviated markedly from the conservative linear extrapolation model. In addition to tumor incidence, other endpoints were evaluated, including DNA adducts, cell proliferation, and Ki-ras mutagenesis) to examine biomarkers of DBP exposure. Interestingly, the biomarker of DBP-DNA adducts in liver remained linear. Utilizing this sublinear dose-response model, the predicted dose of DBP producing 1 liver cancer in 10^6 was actually 1000-fold higher than predicted by the conservative linear extrapolation model. We are currently conducting a second "ED001" study, this time with the hepatocarcinogen AFB_1. It is too early to assess what role this model may play in cancer risk assessment by regulatory agencies, but the initial studies clearly demonstrate the ability to conduct studies that would be too expensive in a rodent model. We believe that the trout model has great potential to contribute to risk assessment with some carcinogens and with some target organs.

CONCLUSIONS

The *Xiphophorus* melanoma and rainbow trout hepatocellular carcinoma models have made significant contributions to the study of genetic and environmental factors controlling cancers relevant to humans. These nonmammalian, vertebrate animals have a number of distinct advantages. Often they are more economical to maintain, have a relatively short time to maturation and breeding, can be exposed to carcinogens in a variety of ways (often using small doses at early life stages, which is an advantage for rare or expensive test chemicals), and have a relatively short time to tumor development. In addition, there is now a rich historical database for both of these models that stretches back to the 1950s. These fish models can also address hypotheses requiring many animals that would be economically impractical in a rodent model (e.g., an ED001 study in rainbow trout utilizing 40,000 mice or rats would cost $4 million to $8 million just in per diem fees to house and care for the animals). Of course, fish models have obvious limitations in that cancers in organs that the fish does not possess cannot be directly studied. Although none of the fish species currently being utilized are likely to replace rodent life-type studies as the "gold standard" for cancer studies, we believe that they provide important comparative data regarding mechanisms of action and, as stated earlier, may provide critical information not possible to obtain with rodent models. These are just a few of the many reasons why extramural agencies should nurture and encourage further development of these models.

References

Ahmed, F.E., Setlow, R.B., 1993. Ultraviolet radiation-induced DNA damage and its photorepair in the skin of the platyfish *Xiphophorus*. Cancer Res. 53, 2249–2255.

Altschmied, J., Ditzel, L., Schartl, M., 1997. Hypomethylation of the *Xmrk* oncogene promoter in melanoma cells of *Xiphophorus*. Biol. Chem. 378, 1457–1466.

Ashby, J., Brady, A., Elcombe, C.R., Elliott, B.M., Ishmael, J., Odum, J., Tugwood, J.D., Kettle, S., Purchase, I.F.H., 1994. Mechanistically-based human hazard assessment of peroxisome proliferator-induced hepatocarcinogenesis. Human Exp. Toxicol. 13, S1–S117.

Bailey, G.S., 1994. Role of aflatoxin-DNA adducts in the cancer process. In Eaton, D.L., and Groopman, J.D. (eds). The Toxicology of Aflatoxins: Human Health, Veterinary, and Agricultural Significance, pp. 137–147. New York, NY Academic Press.

Bailey, G.S., 1995. Chlorophylls as anticarcinogens and tumor modulators in fish and mammalian models. Environm. Mut. Res. Commun. 17, 1–5.

Bailey, G.S., Dashwood, R.H., Fong, A.T., Williams, D.E., Scanlan, R.A., Hendricks, J.D., 1991. Modulation of mycotoxin and nitrosamine carcinogenesis by indole-3-carbinol; quantitative analysis of inhibition versus promotion. In O'Neill, Chen, J., Bartsch, H. (Eds.), Relevance to Human Cancer of N-Nitroso Compounds, Tobacco Smoke and Mycotoxins. IARC, Lyon, France, pp. 275–280.

Bailey, G.S., Williams, D.E., Hendricks, J.D., 1996. Fish models for environmental carcinogenesis: The rainbow trout. Environ. Health Perspect. 104, 5–21.

Basolo, A.L., Alcaraz, G., 2003. The turn of the sword: Length increases male swimming costs in swordtails. Proc. Biol. Sci. 270, 1631–1636.

Baulieu, E.E., Thomas, G., Legrain, S., Lahlou, N., Roger, M., Debuire, B., Faucounau, V., Girard, L., Hervy, M.P., Latour, F., Leaud, M.C., Mokrane, A., Pitti-Ferrandi, H., Trivalle, C., de Lacharriere, O., Nouveau, S., Rakoto-Arison, B., Souberbielle, J.C., Raison, J., Le Bouc, Y., Raynaud, A., Girerd, X., Forette, F., 2000. Dehydroepiandrosterone (DHEA), DHEA sulfate, and aging: Contribution of the DHEAge Study to a sociobiomedical issue. Proc. Natl. Acad. Sci. USA 97, 4279–4284.

Bell, M.C., Crowley-Nowick, P., Bradlow, H.L., Sepkovic, D., Schmidt-Grimminger, D., Howell, P., Mayeaux, E.J., Tucker, A., Turbat-Herrera, E.A., Mathis, J.M., 2000. Placebo-controlled trial of indole-3-carbinol in the treatment of CIN. Gynecol. Oncol. 78, 123–129.

Benninghoff, A.D., Huset, C.A., Field, J.A., Williams, D.E., 2006. Estrogenicity of perfluoroalkyl carboxylic acids in rainbow trout: Results from a screen of 36 structurally diverse perfluorinated chemicals. Society of Environmental Toxicology and Chemistry (SETAC) North America 27th Annual Meeting, Montréal, Québec, Canada, November 5–9.

Bobe, J., Monfort, J., Nguyen, T., Fostier, A., 2006. Identification of new participants in rainbow trout (*Oncorhynchus mykiss*) oocyte maturation and ovulation processes using cDNA microarrays. Reprod. Biol. Endocrinol. 4, 39.

Breinholt, V., Arbogast, D., Loveland, P., Pereira, C., Dashwood, R., Hendricks, J., Bailey, G.S., 1999. Chlorophyllin chemoprevention in trout initiated by aflatoxin B_1 bath treatment: An evaluation of reduced bioavailability vs. target organ protective mechanisms. Toxicol. Appl. Pharmcol. 158, 141–151.

Breinholt, V., Schimerlik, M., Dashwood, R., Bailey, G., 1995. Mechanisms of chlorophyllin anticarcinogenesis against aflatoxin B1: Complex formation with the carcinogen. Chem. Res. Toxicol. 8, 506–514.

Cormier, S.M., Racine, R.N., Smith, C.E., Dey, W.P., Peck, T.H., 1989. Hepatocellular carcinoma and fatty infiltration in the Atlantic tomcod. J. Fish Dis. 12, 105–116.

Dashwood, R.H., Arbogast, D.N., Fong, A.T., Hendricks, J.D., Bailey, G. S., 1988. Mechanisms of anti-carcinogenesis by indole-3-carbinol: Detailed *in vivo* DNA binding dose-response studies after dietary administration with aflatoxin B_1. Carcinogenesis 9, 427–432.

Dashwood, R.H., Arbogast, D.N., Fong, A.T., Pereira, C., Hendricks, J.D., Bailey, G.S., 1989. Quantitative inter-relationships between aflatoxin B1 carcinogen dose, indole-3-carbinol anti-carcinogen dose, target organ DNA adduction and final tumor response. Carcinogenesis 10, 175–181.

Dashwood, R.H., Fong, A.T., Williams, D.E., Hendricks, J.D., Bailey, G. S., 1991. Promotion of aflatoxin B1 carcinogenesis by the natural tumor modulator indole-3-carbinol: Influence of dose, duration and intermittent exposure on indole-3-carbinol promotional potency. Cancer Res. 51, 2362–2365.

David, W.M., Mitchell, D.L., Walter, R.B., 2004. DNA Repair in Hybrid Fish of the Genus *Xiphophorus*. Comparative Physiology & Biochemistry, 138C, 300–310.

DeFabo, E., Noonan, F. P., Fears, T., Merlino, G., 2004. Ultraviolet B but not ultraviolet A radiation initiates melanoma. Cancer Res. 64, 6372–6376.

Douki, T., Reynaud-Angelin, A., Cadet, J., Sage, E., 2003. Bipyrimidine photoproducts rather than oxidative lesions are the main type of DNA damage involved in the genotoxic effect of solar UVA radiation. Biochemistry 42, 9221–9226.

Dove, A.D., 2000. Richness patterns in the parasite communities of exotic poeciliid fishes Parasitology. 120, 609–623.

Eaton, D.L., Bammler, T.K., Kelly, E.J., 2001. Interindividual differences in response to chemoprotection against aflatoxin-induced hepatocarcinogenesis: Implications for human biotransformation enzyme polymorphisms. Adv. Exptl. Med. Biol. 500, 559–576.

Eaton, D.L., Gallagher, E.P., 1994. Mechanisms of aflatoxin carcinogenesis. Annu. Rev. Pharmacol. Toxicol. 34, 135–172.

Egner, P., Wang, J., Zhu, Y., Zhang, B., Wu, Y., Zhang, Q., Qian, G.S., Kuang, Y.S., Gange, S.J., Jacobson, L.P., Helzlsouer, K.J., Bailey, G.S., Groopman, J.D., Kensler, T.W., 2001. Chlorophyllin intervention reduces aflatoxin-DNA adducts in individuals at high risk for liver cancer. Proc. Natl. Acad. Sci. USA 98, 14601–14606.

Egner, P.A., Munoz, A., Kensler, T.W., 2003. Chemoprevention with chlorophyllin in individuals exposed to dietary aflatoxin. Mutat. Res. 523–524, 209–216.

Flynn, K.M., Schreibman, M.P., Yablonsky-Alter, E., Banerjee, S.P., 1999. Sexually dimorphic development and binding characteristics of nmDA receptors in the brain of the platyfish. Gen Comp Endocrinol. 115, 282–291.

Fong, A.T., Swanson, H.I., Dashwood, R.H., Williams, D.E., Hendricks, J.D., Bailey, G.S., 1990. Mechanisms of anti-carcinogenesis by indole-3-carbinol. Studies of enzyme induction, electrophile-scavenging, and inhibition of aflatoxin B1 activation. Biochem. Pharmacol. 39, 19–26.

Frenkel, R.A., Slaughter, C.A., Orth, K., Moomaw, C.R., Hicks, S.H., Snyder, J.M., Bennett, M., Prough, R.A., Putman, R.S., Milewich, L., 1990. Peroxisome proliferation and induction of peroxisomal enzymes in mouse and rat liver by dehydroepiandrosterone feeding. J. Steroid Biochem. 35, 333–342.

Gallagher, E.P., Kunze, K.L., Stapelton, P.L., Eaton, D.L., 1996. The kinetics of aflatoxin B_1 oxidation by human cDNA-expressed and human liver microsomal cytochromes P450 1A2 and 3A4. Toxicol. Appl. Pharmacol. 141, 595–606.

Geissinger, E., Weisser, C., Fischer, P., Schartl, M., Wellbrock, C., 2002. Autocrine stimulation by osteopontin contributes to antiapoptotic signalling of melanocytes in dermal collagen. Cancer Res. 62, 4820–4828.

Gerwick, L., Corley-Smith, G., Bayne, C.J., 2007. Gene transcript changes in individual rainbow trout livers following an inflammatory stimulus. Fish Shellfish Immunol. 22, 157–171.

Gilley, J., Fried, M., 2001. One INK4 gene and no ARF at the *Fugu* equivalent of the human INK4A/ARF/INK4B tumour suppressor locus. Oncogene 20, 7447–7452.

Gomez, A., Volff, J. N., Hornung, U., Schartl, M., Wellbrock, C., 2004. Identification of a second EGFR gene in *Xiphophorus* uncovers and expansion of the epidermal growth factor receptor family in fish. Mol. Biol. Evol. 21, 266–275.

Gomez, A., Wellbrock, C., Gutbrod, H., Dimitrijevic, N., Schartl, M., 2001. Ligand independant dimerization and activation of the oncogenic Xmrk receptor by two mutations in the extracellular domain. J. Biol. Chem. 276, 3333–3340.

Gordon, M., 1927. The genetics of viviparous top-minnow Platypoecilus: The inheritance of two kinds of melanophores. Genetics 12, 253–283.

Gordon, M., 1931. Hereditary basis of melanosis in hybrid fishes. Amer. J. Cancer 15, 1495–1523.

Graveel, C.R., Jatkoe, T., Madore, S.J., Holt, A.L., Farnham, P.J., 2001. Expression profiling of novel genes in hepatocellular carcinomas. Oncogene 20, 2704–2712.

Grubbs, C.J., Steele, V,E., Casebolt, T., Juliana, M.M., Eto, I., Whitaker, L.M., Dragnev, K.H., Kelloff, G.J., Lubet, R.L., 1995. Chemoprevention of chemically-induced mammary carcinogenesis by indole-3-carbinol. Anticancer Res. 15, 709–716.

Hall, J., Monsanto, R., 1990. DNA alkylation damage: Consequences and relevance to tumour production. Mutat. res. 233, 247–252.

Harttig, U., Bailey, G.S., 1998. Chemoprotection by natural chlorophylls *in vivo*: inhibition of dibenzo[*a,l*]pyrene-DNA adducts in rainbow trout liver. Carcinogenesis 19, 1323–1326.

Hayashi, T., Schimerlik, M., Bailey, G.S., 1999. Mechanisms of chlorophyllin anticarcinogenesis: Dose-responsive inhibition of aflatoxin uptake and biodistribution following oral co-administration in rainbow trout. Toxicol. Appl. Pharmacol. 158, 132–140.

Hayward, N.K., 2003. Genetics of melanoma predisposition. Oncogene 22, 3053–3062.

Heater, S.J., Rains, J.D., Wells, M.C., Guerrero, P.A., Walter, R.B., 2007. Perturbation of DNA repair gene expression due to interspecies hybridization, Comparative Physiology & Biochemistry, Part C -Toxicology & Pharmacolgy 145, 156–163.

Hook, S.E., Skillman, A.D., Small, J.A., Shultz, I.R., 2006. Gene expression patterns in rainbow trout, *Oncorhynchus mykiss*, exposed to a suite of model toxicants. Aquat. Toxicol. 77, 372–385.

Houde, M., Martin, J.W., Letcher, R.J., Solomon, K.R., Muir, D.C., 2006. Biological monitoring of polyfluoroalkyl substances: A review. Environ. Sci. Technol. 40, 3463–3473.

Huang, C.Y., Wilson, M.W., Lay, L.T., Chow, C.K., Robertson, L.W., Glauert, H.P., 1994. Increased 8-hydroxydeoxyguanosine in hepatic DNA of rats treated with the peroxisome proliferators ciprofibrate and perfluorodecanoic acid. Cancer Lett. 87, 223–228.

Kallman, K.D., Kazianis, S., 2006. The genus *Xiphophorus* in Mexico, and Central America. Zebrafish 3, 267–281.

Kazianis, S., Coletta, L.D., Morizot, D.C., Johnston, D.A., Osterndorff, E.A., Nairn, R.S., 2000. Overexpression of a fish *CDKN2* gene in a hereditary melanoma model. Carcinogenesis 21, 599–605.

Kazianis, S, Gimenez-Conti, I., Setlow, R.B., Woodhead, A.D., Harshbarger, J.C., Trono, D., Ledesma, M., Nairn, R.S., Walter, R.B., 2001b. MNU induction of neoplasia in a novel fish model. Lab. Invest. 81, 1191–1198.

Kazianis, S., Gimenez-Conti, I., Trono, D., Pedroza, A., Chovanec, L., Morizot, D.C., Nairn, R.S., Walter, R.B., 2001a. Genetic analysis of MNU-induced neoplasia within *Xiphophorus* hybrid fish. Mar. Biotechnol. 3(S1), 37–44.

Kazianis, S., Morizot, D.C., Coletta, L.D., Johnston, D.A., Woolcock, B., Vielkind, J.R., Nairn, R.S., 1999. Comparative structure and characterization of a *CDKN2* gene in a *Xiphophorus* fish melanoma model. Oncogene 18, 5088–5099.

Kazianis, S., Nairn, R.S., Walter, R.B., Johnston, D.A., Kumar, J., Trono, D., Della-Coletta, L., Gimenez-Conti, I., Rains, J.D., Williams, E.L., Mamerow, M.M., Kochan, K.J., Schartl, M., Vielkind, J.R., Volff, J-N., Woolcock, B., Morizot, D.C., 2004. The genetic map of *Xiphophorus* fishes represented by 24 multipoint linkage groups. Zebrafish 1, 286–297.

Kennedy, G.L., Jr., Butenhoff, J.L., Olsen, G.W., O'Connor, J.C., Seacat, A.M., Perkins, R.G., Biegel, L.B., Murphy, S.R., Farrar, D.G., 2004. The toxicology of perfluorooctanoate. Crit. Rev. Toxicol. 34, 351–384.

Kensler, T.W., Egner, P.A., Wang, J.B., Zhu, Y.R., Zhang, B.C., Lu, P.X., Chen, J.G., Qian, G.S., Kuang, S.Y., Jackson, P.E., Gange, S.J., Jacobson, L.P., Munoz, A., Groopman, J.D., 2004. Chemoprevention of hepatocellular carcinoma in aflatoxin endemic areas. Gastroenterology 127(5 suppl. 1), S310–S318.

Kim, Y.S., Milner, J.A., 2005. Targets for indole-3-carbinol in cancer prevention. J. Nutr. Biochem. 16, 65–73.

Kosswig, C., 1927. Uber bastarde der teleostier *Platypoecilus* und *Xiphophorus*. Zeitschrift fur induktive Abstammungs- und Vererbungslehre 44, 253.

Krasnov, A., Koskinen, H., Pehkonen, P., Rexroad III, C.E., Afanasyev, S., Molsa, H., 2005a. Gene expression in the brain and kidney of rainbow trout in response to handling stress. BMC Genomics. 6, 3.

Krasnov, A., Koskinen, H., Rexroad, C., Afanasyev, S., Molsa, H., Oikari, A., 2005b. Transcriptome responses to carbon tetrachloride and pyrene in the kidney and liver of juvenile rainbow trout (*Oncorhynchus mykiss*). Aquatic Toxicol. 74, 70–81.

Kwak, H.I., Bae, M.O., Lee, M.H., Lee, Y.S., Lee, B.J., Kang, K.S., Chae, C.H., Sung, H.J., Shin, J.S., Kim, J.H., Mar, W.C., Sheen, Y.Y., Cho, M.H., 2001. Effects of nonylphenol, bisphenol A, and their mixture on the viviparous swordtail fish (*Xiphophorus helleri*). Environ. Toxicol. Chem., 20, 787–795.

Lawrence, J.A., Malpas, P.B., Sigman, C.C., Kelloff, G.J., 2000. Clinical development of estrogen modulators for breast cancer chemoprevention in premenopausal vs. postmenopausal women. J. Cell. Biochem. (suppl. 34), 103–114.

Lee, B.C., Hendricks, J.D., Bailey, G.S., 1991. Toxicity of mycotoxins in the feed of fish. In Smith. J.E. (ed.), Mycotoxins and Animal Feedstuff: Natural Occurrence, Toxicity and Control, pp. 607–626. Boca Raton, FL: CRC Press.

Ley, R.D., 1997. Ultraviolet radiation A-induced precursors of cutaneous melanoma in *Monodelphis domestica* Cancer Res. 57, 3682–3684.

Ley, R.D., 2001. Dose response for ultraviolet radiation A-induced focal melanocytic hyperplasia and nonmelanoma skin tumors in *Monodelphis domestica*. Photochem. Photobiol. 73, 20–23.

Luo, J., Sanetra, M., Schartl, M., Meyer, A., 2005. Strong reproductive skew among males in the multiply mated swordtail *Xiphophorus multilineatus* (Teleostei). J. Hered. 96, 346–355.

Malins, D.C., McCain, B.B., Myers, M.S., Brown, D.W., Krahn, M.M., Roubal, W.T., Schiewe, M.H., Landahl, J.T., Chan, S.L., 1987. Field and laboratory studies of the etiology of liver neoplasms in marine fish from Puget Sound. Environ. Health Perspect. 71, 5–16.

Manning, D.D., Reed, N.D., Schafer, C.F., 1973. Maintenance of skin xenografts of widely divergent phylogenetic origin of congenitally athymic (nude) mice. J. Exp. Med. 138, 488–494.

McConnell, T.J., Godwin, U.B., Norton, S.F., Nairn, R.S., Kazianis, S., Morizot, D.C., 1998. Identification and mapping of two divergent, unlinked major histocompatibility complex class II B genes in *Xiphophorus* fishes. Genetics. 149, 1921–1934.

McDanell, R., McLean, A.E.M., 1988. Chemical and biological properties of indole glucosinolates (glucobrassicins): A review. Food Chem. Toxicol. 26, 59–70.

McGlynn, K.A., Hunter, K., LeVoyer, T., Roush, J., Wise, P., Michielli, R.A., Shen, F.M., Evans, A.A., London, W.T., Buetow, K.H., 2003. Susceptibility to aflatoxin B1-related primary hepatocellular carcinoma in mice and humans. Cancer Res. 63, 4594–4601.

Meador, J.A., Walter, R.B., Mitchell, D.L., 2000. Induction, distribution, and repair of UV photodamage in the swordtail, *Xiphophorus signum*. Photochem. Photobiol. 72, 260–266.

Meierjohann, S., Schartl, M., Volff, J-N., 2004. Genetic, biochemical and evolutionary facets of Xmrk-induced melanoma formation in the fish *Xiphophorus*. Comp. Biochem. Physiol. C Toxicol. Pharmacol. 138, 281–289.

Meierjohann, S., Wende, I., Kariss, A., Wellbrock, C., Schartl, M., 2006. The oncongenic epidermal growth factor receptor variant *Xiphophorus* melanoma receptor kinase induces motility in melanocytes by modulation of focal adhesions. Cancer Res. 66, 3145–3152.

Meng, Q., Goldberg, I.D., Rosen, E.M., Fan, S., 2000. Inhibitory effects of indole-3-carbinol on invasion and migration in human breast cancer cells. Breast Cancer Res. Treat. 63, 147–152.

Michnovicz, J.J., Bradlow, H.L., 1990. Induction of estradiol metabolism by dietary indole-3-carbinol in humans. J. Natl. Cancer Inst. 82, 947–949.

Michnovicz, J.J., Adlercreutz, H., Bradlow, H.L., 1997. Changes in levels of urinary estrogen metabolites after oral indole-3-carbinol treatment in humans. J. Natl. Cancer Inst. 89, 718–723.

Ming, L., Thorgeirsson, S.S., Gail, M.H., Lu, P., Harris, C.C., Wang, N., Shao, Y., Wu, Z., Liu, G., Sun, Z., 2002. Dominant role of hepatitis B virus and cofactor role of aflatoxin in hepatocarcinogenesis in Qidong, China. Hepatology 36, 1214–1220.

Mitchell, D.L., Maedor, J.A., Byrom, M., Walter, R.B., 2001. The resolution of DNA damage in *Xiphophorus* fishes. Mar. Biotechnol. 3(S1), 61–72.

Mitchell, D.L., Nairn, R.S., Johnston, D.A., Byrom, M., Kazianis, S., Walter, R.B., 2004. Decreased levels of DNA excision repair in hybrid fish of the genus *Xiphophorus*. Photochem. Photobiol. 79, 447–452.

Mitchell, D.L., Scoggins, J.T., Morizot. D.C., 1993. DNA repair in the variable platyfish (*Xiphophorus variatus*) irradiated *in vivo* with ultraviolet B light. Photchem. & Photobiol. 58, 455–459.

Morcinek, J.C., Weisser, C., Geissinger, E., Schartl, M., Wellbrock, C., 2002. Activation of STAT5 triggers proliferation and contributes to anti-apoptotic signalling mediated by the oncogenic Xmrk kinase. Oncogene 21, 1668–1678.

Morizot, D.C., Nairn, R.S., Simhambhatla, P., Dell Coletta, L., Trono, D., Chovanec, L., Walter, R.B., Kazianis, S.., 2001. *Xiphophorus* genetic linkage map: Beginnings of comparative gene mapping in fishes. Mar. Biotechnol. 3, S153–S161.

Mouret, S., Baudouin, C., Charveron, M., Favier, A., Cadet, J., Douki, T., 2006. Cyclobutane pyrimidine dimers are predominant DNA lesions in whole human skin exposed to UVA radiation. Proc. Natl. Acad. Sci. USA 103, 13765–13770.

Naciff, J.M., Daston, G.P., 2004. Toxicogenomic approach to endocrine disrupters: Identification of a transcript profile characteristic of chemicals with estrogenic activity. Toxicol. Pathol. 32 (suppl. 2), 59–70.

Nair, K.S., Rizza, R.A., O'Brien, P., Dhatariya, K., Short, K.R., Nehra, A., Vittone, J.L., Klee, G.G., Basu, A., Basu, R., Cobelli, C., Toffolo, G., Dalla Man, C., Tindall, D.J., Melton 3rd, L.J., Smith, G.E., Khosla, S., Jensen, M.D., 2006. DHEA in elderly women and DHEA or testosterone in elderly men. N. Engl. J. Med. 355, 1647–1659.

Nairn, R.S., Kazianis, S., Della Coletta, L., Trono, D., Butler, A.P., Walter, R.B., Morizot, D.C., 2001. Genetic analysis of susceptibility to spontaneous and UV carcinogenesis in *Xiphophorus* hybrid fish. Mar. Biotechnol. 3(S1), 24–37.

Nairn R.S., Kazianis S., McEntire. B.B., Della Coletta, L., Walter, R.B., Morizot, D.C., 1996a. A *CDKN2*-like polymorphism in *Xiphophorus* LG V is associated with UV-B-induced melanoma formation in platyfish-swordtail hybrids. Proc. Natl. Acad. Sci. USA 93, 13042–13047.

Nairn, R.S., Morizot, D.C., Kazianis, S., Woodhead, A.D., Setlow R.B., 1996b. Nonmammalian models for sunlight carcinogenesis: Genetic analysis of melanoma formation in *Xiphophorus* hybrid fish. Photochemi. Photobiol. 64, 440–448.

Noonan, F.P., Recio, J.A., Takayama, H., Duray, P., Anver, M.R., Rush, W.L., De Fabo, E.C., Merlino, G., 2001. Neonatal sunburn and melanoma in mice. Nature 413, 271–272.

Nunez, O., Hendricks, J.D., Arbogast, D.N., Fong, A.T., Lee, B.C., Bailey, G.S., 1989. Promotion of aflatoxin B1 hepatocarcinogenesis in rainbow trout by 17®-estradiol. Aquat. Toxicol. 15, 289–302.

Oehlers, L.P., Heater, S.H., Rains, J.D., Wells, M.C., David, W.M., Walter, R.B., 2004. Gene structure, purification, and characterization of DNA polymerase β from *Xiphophorus maculatus*. Comp. Physiol. & Biochem. 138, 311–324.

Oganesian, A., Orner, G.A., Hendricks, J.D., Pereira, C., Bailey, G.S., Williams, D.E., 1999. The potency of indole-3-carbinol as a promoter of aflatoxin B_1-initiated hepatocarcinogenesis: Results from a 9000 animal tumor study. Carcinogenesis 20, 453–458.

Okihiro, M.S., Hinton, D.E., 1999. Progression of hepatic neoplasia in medaka (Oryzias latipes) exposed to diethylnitrosamine. Carcinogenesis 20, 933–940.

Olech, E., Merrill, J.T., 2005. DHEA supplementation: The claims in perspective. Cleve. Clin. J. Med. 72, 965–966.

Orlow, I., Begg, C.B., Cotignola, J., Roy, P., Hummer, A.J., Clas, B.A., Mujumdar, U., Canchola, R., Armstrong, B.K., Kricker, A., Marrett, L.D., Millikan, R.C., Gruber, S.B., Anton-Culver, H., Zanetti, R., Gallagher, R.P., Dwyer, T., Rebbeck, T.R., Kanetsky, P.A., Wilcox, H., Busam, K., From, L., Berwick, M., 2007. CDKN2A germline mutations in individuals with cutaneous malignant melanoma. J. Invest. Dermatol., in press.

Orner, G.A., Donohoe, R.M., Hendricks, J.D., Curtis, L.R., Williams, D.E., 1996a. Comparison of the enhancing effects of dehydroepiandrosterone with its structural analog on aflatoxin B_1 hepatocarcinogenesis in rainbow trout. Fund. Appl. Toxicol. 34, 132–140.

Orner, G.A., Hendricks, J.D., Arbogast, D., Williams, D.E., 1996b. Modulation of N-methyl-N'-nitro-nitrosoguanidine carcinogenesis by dehydroepiandrosterone in rainbow trout. Toxicol. Appl. Pharmacol. 141, 548–554.

Orner, G.A., Hendricks, J.D., Arbogast, D., Williams, D.E., 1998. Modulation of aflatoxin B_1 hepatocarcinogenesis in trout by dehydroepiandrosterone: Initiation/postinitiation and latency effects. Carcinogenesis 19, 161–167.

Orner, G.A., Mathews, K., Hendricks, J.D., Carpenter, H.M., Bailey, G.S., Williams, D.E., 1995. Dehydroepiandrosterone is a complete hepatocarcinogen and potent tumor promoter in the absence of peroxisome proliferation in rainbow trout. Carcinogenesis 16, 2893–2898.

Ortega, S., Malumbres, M., Barbacid, M., 2002. Cyclin D-dependent kinases, INK4 inhibitors and cancer. Biochim. Biophys. Acta 1602, 73–87.

Ostrander, G.K., Rotchell, J.M., 2005. Fish models of carcinogenesis. In Mommsen, T.P., and Moon, T.W. (eds.), Biochemistry and Molecular Biology of Fishes, vol. 6. Environmental Toxicology, pp. 255–288. Amsterdam, Elsevier.

Petit, F., Valotaire, Y., Pakdel, F., 1995. Differential functional activities of rainbow trout and human estrogen receptors expressed in the yeast *Saccharomyces cerevisiae*. Eur. J. Biochem. 233, 584–592.

Pinisetty, D., Huang, C., Dong, Q., Tiersch, T.R., Devireddy, R.V., 2005. Subzero water permeability parameters and optimal freezing rates for sperm cells of the southern platyfish, *Xiphophorus maculatus*. Cryobiology 50, 250–263.

Pitot, H.C., Dragan, Y.P., Teeguarden, J., Hsia, S., Campbell, H., 1996. Quantitation of multistage carcinogenesis in rat liver. Toxicol. Pathol. 24, 119–128.

Poirier, M.C., Beland, F.A., 1992. DNA adduct measurements and tumor incidence during chronic carcinogen exposure in animal models: Implications for DNA adduct-based human cancer risk assessment. Chem. Res. Toxicol. 5, 749–755.

Pratt, M.M., Reddy, A.P., Hendricks, J.D., Pereira, C., Kensler, T.W., Bailey, G.S., 2007. The importance of carcinogen dose in chemoprevention studies: Quantitative interrelationships between, dibenzo[*a,l*]pyrene dose, chlorophyllin dose, target organ DNA adduct biomarkers, and final tumor outcome. Carcinogenesis 28, 611–624.

Rainbow Trout Gene Index (RTGI), 1996. The DFCI Rainbow Trout Gene Index. [http://compbio.dfci.harvard.edu/tgi/cgi-bin/tgi/gimain.pl?gudb=r_trout].

Raney, K.D., Meyer, D.J., Ketterer, B., Harris, T.M., Guengerich, F.P., 1992. Glutathione conjugation of aflatoxin B_1 *exo* and *endo*-epoxides by rat and human glutathione *S*-transferases. Chem. Res. Toxicol. 5, 470–478.

Rao, M.S., Subbarao, V., Yeldandi, A.V., Reddy, J.K., 1992. Hepatocarcinogenicity of dehydroepiandrosterone in the rat. Cancer Res. 52, 2977–2979.

Reddy, A.P., Harttig, U., Barth, M.C., Baird, W.M., Schimerlik, M., Hendricks, J.D., Bailey, G.S., 1999. Inhibition of dibenzo[*a,l*]pyrene-induced multi-organ carcinogenesis by dietary chlorophyllin in rainbow trout. Carcinogenesis 20, 1919–1926.

Reddy, J.K., Rao, M.S., 1989. Oxidative DNA damage caused by persistent peroxisome proliferation: its role in hepatocarcinogenesis. Mutat. Res. 214, 63–68.

Rise, M.L., von Schalburg, K.R., Brown, G.D., Mawer, M.A., Devlin, R.H., Kuipers, N., Busby, M., Beetz-Sargent, M., Alberto, R., Gibbs, A.R., Hunt, P., Shukin, R., Zeznik, J.A., Nelson, C., Jones, S.R.M., Smailus, D.E., Jones, S.J.M., Schein, J.E., Marra, M.A., Butterfield, Y.S.N, Stott, J.M., Ng, SHS, Davidson, W.S., Koop, B.F., 2004. Development and application of a salmonid EST database and cDNA microarray: Data mining and interspecific hybridization characteristics. Genome Res. 14, 478–490.

Rivers, J.K., 2004. Is there more than one road to melanoma? Lancet 363, 728–730.

Roney, K.E., Cuthbertson, B.J., Godwin, U.B., Kazianis, S., Coletta, L.D., Rosenthal, G.G., Ryan, J.M., Schmidt, M., McConnell, T.J., 2004. Alternative splicing of major histocompatibility complex class II DXB transcripts in *Xiphophorus* fishes. Immunogenetics 56, 462–466.

Rosenthal, G.G., 2003. Dissolution of sexual signal complexes in a hybrid zone between the swordtails *Xiphophorus birchmanni* and *Xiphophorus malinche* (Poeciliidae). Copeia 2003, 299–307.

Royle, J.N., Lindstrom, J., Metcalfe, N.B., 2005. A poor start in life negatively affects dominance status in adulthood independent of body size in green swordtails *Xiphophorus helleri*. Proc. Biol. Sci. 272, 1917–1922.

Schartl, A., Malitschek B., Kazianis S., Borowsky R., Schartl M., 1995. Spontaneous melanoma formation in nonhybrid *Xiphophorus*. Cancer Res. 55.

Schartl, M., 1995. Platyfish and swordtails: A genetic system for the analysis of molecular mechanisms in tumor formation. Trends Genet. 11, 185–189.

Schartl, M., Peter, R.U., 1988. Progressive growth of fish tumors after transplantation into thymus- aplastic (nu/nu) mice. Cancer Res. 48, 741–744.

Schwab, M., Abdo, S., Ahuja, M.R., Kollinger, G., Anders, A., Anders, F., Frese, K., 1978a. Genetics of susceptibility in the platyfish/swordtail tumor system to develop fibrosarcoma and rhabdomyosarcoma following treatment with *N*-methyl-*N*-nitrosourea (MNU). Cancer Res. Clin. Oncol. 91, 301–315.

Schwab, M., Haas, S., Ahuja, M.R., Kollinger, G., Anders, A., Anders, F., 1978b. Genetic basis of susceptibility for development of neoplasms following treatment with *N*-methyl-*N*-nitrosourea (MNU) or x-rays in the platyfish/swordtail system. Experientia. 34, 780–782.

Schwab, M., Kollinger, G., Haas, J., Ahuja, M.R., Abdo, S., Anders, A. Anders, F., 1979. Genetic basis of susceptibility for neuroblastoma following treatment with *N*-methyl-*N*-nitrosourea and X-rays in *Xiphophorus*. Cancer Res. 39, 519–526.

Schwartz, A.G., Whitcomb, J.M., Nyce, J.W., Lewbart, M.L., Pashko, L.L., 1988. Dehydroepiandrosterone and structural analogs: A new class of cancer chemopreventive agents. Adv. Cancer Res. 51, 391–423.

Setlow, R.B., 1999. Spectral regions contributing to melanoma: A personal view. J. Invest. Derm. Symp. *P* 4, 46–49.

Setlow, R.B., Grist, E., Thompson, K., Woodhead, A.D., 1993. Wavelengths effective in induction of malignant melanoma. Proc. Natl. Acad. Sci. USA 90, 6666–6670.

Setlow, R.B., Woodhead, A.D., 2001 Three unique experimental fish stories: *Poecilia* (the Past), *Xiphophorus* (the Present), and Medaka (the Future) Mar. Biotechnol. 3, S17–S23.

Setlow, R.B., Woodhead, A.D., Grist, E., 1989. Animal model for ultraviolet radiation-induced melanoma: Platyfish-swordtail hybrid. Proc. Natl. Acad. Sci. USA 86, 8922–8926.

Shilling, A.D., Carlson, D.B., Katchamart, S., Williams, D.E., 2001. 3,3′-Diindolylmethane, a major condensation product of indole-3-carbinol, is a potent estrogen in the rainbow trout. Toxicol. Appl. Pharmacol. 170, 191–200.

Simonich, M.T., Roebuck, B.D., Orner, O., Jubert, C., Dashwood, R.H., Williams, D.E., Kensler, T., Egner, P., Groopman, J., Bailey, G.S., 2007. Natural chlorophyll inhibits aflatoxin B_1 induced multi-organ carcinogenesis in the rat. Carcinogenesis 28, 1294–1302.

Sinnhuber, R.O., Wales, J.H, Ayres, J.L., Engebrecht, R.H., Amend, D.L., 1968. Dietary factors and hepatoma in rainbow trout (*Salmo gairdneri*). 1. Aflatoxins in vegetable protein feedstuffs. J. Natl. Cancer Inst. 41, 711–718.

Stresser, D.M., Bailey, G.S., Williams, D.E., 1994a. Indole-3-carbinol and β-naphthoflavone induction of cytochromes P450 associated with bioactivation and detoxication of aflatoxin B_1 in the rat. Drug Metabol. Dispos. 22, 383–391.

Stresser, D.M., Bjeldanes, L.F., Bailey, G.S., Williams, D.E., 1995. The anticarcinogen 3,3′-diindolylmethane is an inhibitor of cytochrome P450. J. Biochem. Toxicol. 10, 191–201.

Stresser, D.M., Williams, D.E., McLellan, L.I., Harris, T.M., Bailey, G.S., 1994b. Indole-3-carbinol induces a rat liver glutathione transferase subunit (Yc2) with high activity towards aflatoxin B_1 *exo*-epoxide: Association with reduced levels of hepatic AFB_1-DNA adduction *in vivo*. Drug Metabol. Dispos. 22, 392–399.

Sumpter, K.P., Lobling, S., 1995. Vitellogenesis as a biomarker for estrogenic contamination of the aquatic environment. Environ. Health Perspect. 103 (suppl. 7), 173–178.

Takahashi, N., Dashwood, R.H., Bjeldanes, L.F., Williams, D.E., Bailey, G.S., 1995b. Mechanisms of indole-3-carbinol (I3C) anti-carcinogenesis: Inhibition of aflatoxin B_1-DNA adduction and mutagenesis by I3C acid condensation products. Food Chem. Toxicol. 33, 851–857.

Takahashi, N., Stresser, D.M., Williams, D.E., Bailey, G.S., 1995a. Induction of hepatic CYP1A by indole-3-carbinol in protection against aflatoxin B_1 hepatocarcinogenesis in rainbow trout. Food Chem. Toxicol. 33, 841–850.

Takeuchi, S., Zhang, W., Wakamatsu, K., Ito, S., Hearing, V., Kraemer, K., Brash, D., 2004. Melanin acts as a potent UVB photosensitizer to cause an atypical mode of cell death in murine skin. Proc. Natl. Acad. Sci. USA 101, 15076–15081.

Terasaka, S., Aita, Y., Inoue, A., Hayashi, S., Nishigaki, M., Aoyagi, K., Sasaki, H., Wada-Kiyama, Y., Sakuma, Y., Akaba, S., Tanaka, J., Sone, H., Yonemoto, J., Tanji, M., Kiyama, R., 2004. Using a customized DNA microarray for expression profiling of the estrogen-responsive genes to evaluate estrogen activity among natural estrogens and industrial chemicals. Environ. Health Perspect. 112, 773–781.

Thorgaard, G.H., Bailey, G.S., Williams, D., Buhler, D.R., Kaattari, S.L., Ristow, S.S., Hansen, J.D., Winton, J.R., Bartholomew, J.L., Nagler, J.J., Walsh, P.J., Vijayan, M.M., Devlin, R.H., Hardy, R.W., Overturf, K.E., Young, W.P., Robison, B.D., Rexroad, C., Palti, Y., 2002. Status and opportunities for genomics research with rainbow trout. Comp. Biochem. Physiol. B. 133, 609–646.

Tilton, S.C., Gerwick, L.G., Hendricks, J.D., Rosato, C., Corley-Smith, G., Givan, S.A., Bailey, G.S., Bayne, C.J., Williams, D.E., 2005. Use of a rainbow trout oligonucleotide microarray to examine transcriptional patterns in aflatoxin B_1-induced hepatocellular carcinoma compared to adjacent liver. Toxicol. Sci. 88, 319–330.

Tilton, S.C., Givan, S.A., Pereira, C., Bailey, G.S., Williams, D.E., 2006. Toxicogenomic profiling of the hepatic tumor promoters indole-3-carbinol, 17β-estradiol and β-naphthoflavone in rainbow trout. Toxicol. Sci. 90, 61–72.

Tilton, S.C., Hendricks, J.D., Orner, G.A., Pereira, C.B., Bailey, G.S., Williams, D.E., 2007. Gene expression analysis during tumor enhancement by the dietary phytochemical, 3,3′-diindolylmethane, in rainbow trout. Carcinogenesis 28, 1589–1598.

Ueng, Y.-F., Shimada, T., Yamazaki, H., Guengerich, F.P., 1995. Oxidation of aflatoxin B1 by bacterial recombinant human cytochrome P450 enzymes. Chem. Res. Toxicol. 8, 218–225.

Vielkind, J.R., Kallman, K.D., Morizot, D.C., 1989. Genetics of melanomas in *Xiphophorus* fishes. J. Aquat. Animal Health. 1, 69–77.

Von Schalburg, K.R., Rise, M.L., Cooper, G.A., Brown, G.D., Gibbs, A.R., Nelson, C.C., Davidson, W.S., Koop, B.F., 2005. Fish and chips: Various methodologies demonstrate utility of a 16,006-gene salmonid microarray. BMC Genomics. 6, 126.

Walter, R.B., Ju, Z., Martinez, A., Amemiya, C.T., Samollow, P., 2006. Genomic resources for *Xiphophorus* research. Zebrafish, 3, 53–63.

Walter, R.B., Kazianis, S., 2001. *Xiphophorus* interspecies hybrids as genetic models of induced neoplasia. ILAR J. 42, 299–321.

Walter, R.B., Kazianis, S., Hazlewood, L., Johnston, D., Kumar, J., 2005. The *Xiphophorus* genetic Stock center. In Uribe, M.C., and Grier, H. (eds.), Viviparous Fishes, pp. 343–350. Homestead, FL, New Life.

Walter, R.B., Rains, J.D., Russell, J.E., Guerra, T.M., Daniels, C., Johnston, D.A., Kumar, J., Wheeler, A., Kelnar, K., Khanolkar, V.A., Williams, E.L., Hornecker, J.L., Hollek, L., Mamerow, M.M., Pedroza, A., Kazianis, S., 2004. A microsatellite genetic linkage map for *Xiphophorus*. Genetics 168, 363–372.

Walter, R.B., Sung, H-M., Intano, G.W., Walter, C.A., 2001a. Characterization of O^6-methylguanine-DNA-methyltransferase (O^6-MGMT) activity in *Xiphophorus* fishes. Mutation Res. 493, 11–22.

Walter, R.B., Sung, H-M., Mitchell, D.L., Intano, G.W., Walter, C.W., 2001b. Relative base excision repair activity in *Xiphophorus* fish tissue extracts. Mar. Biotechnol. 3(S1), 50–61.

Wang, D.Y., McKague, B., Liss, S.N., Edwards, E.A., 2004. Gene expression profiles for detecting and distinguishing potential endocrine-disrupting compounds in environmental samples. Environ. Sci. Technol. 38, 6396–6406.

Wattenberg, W., Loub, W.D., 1978. Inhibition of polycyclic aromatic hydrocarbon-induced neoplasia by naturally occurring indoles. Cancer Res. 38, 1410–1413.

Wellbrock, C., Schartl, M., 2000. Activation of phosphatidylinositol 3-kinase by a complex of p59fyn and the receptor tyrosine kinase Xmrk is involved in malignant transformation of pigment cells. Eur. J. Biochem. 267, 3513–3522.

Wilkins A.S., 2004. Deciphering the swordtail's tale: A molecular and evolutionary quest. Bioessays 26, 116–119.

Williams, D.E., Bailey, G.S., Reddy, A., Hendrick, J.D., Oganesian, A., Orner, G.A., Pereira, C.B., Swenberg, J.A., 2003. The rainbow trout (*Oncorhynchus mykiss*) tumor model: Recent applications in low-dose exposures to tumor initiators and promoters. Toxicol. Pathol. 31 (suppl.), 58–61.

Williams, D.E., Buhler, D.R., 1983. Purified form of cytochrome P-450 from rainbow trout with high activity toward conversion of aflatoxin B_1 to aflatoxin B_1-2,3-epoxide. Cancer Res. 43, 4752–4756.

Winnemoeller, D., Wellbrock, C., Schartl, M., 2005. Activating mutations in the extracellular domain of the melanoma inducing receptor Xmrk are tumorigenic *in vivo*. Int. J. Cancer 117, 723–729.

Wittbrodt, J., Adam, D., Malitschek, B., Maueler, W., Raulf, F., Telling, A., Robertson, S.M., Schartl, M., 1989. Novel putative receptor tyrosine kinase encoded by the melanoma-inducing *Tu* locus in *Xiphophorus*. Nature 341, 415–421.

Wong, B.B., Rosenthal, G.G., 2006. Female disdain for swords in a swordtail fish. Am. Nat. 167, 136–140.

Wood, S.R., Berwick, M., Ley, R.D., Walter, R.B., Setlow, R.B., Timmins, G.S., 2006. UV causation of melanoma in *Xiphophorus* is dominated by melanin photosensitized oxidant production. Proc. Natl. Acad. Sci. USA 103, 4111–4115.

Woolcock, B.W., Schmidt, B.M., Kallman, K.D., Vielkind, J.R., 1994. Differences in transcription and promoters of *Xmrk-1* and *Xmrk-2* genes suggest a role for *Xmrk-2* in pigment pattern development in the platyfish, *Xiphophorus maculatus*. Cell Growth Differ. 5, 575–583.

Yager, J.D., Zurlo, J., Ni, N., 1991. Sex hormones and tumor promotion in liver. Proc. Soc. Exp. Biol. Med., 198, 667–674.

Yang, H., Hazlewood, L., Walter, R.B., Tiersch, T.R., 2006. Effect of osmotic immobilization on refrigerated storage and cryopreservation of sperm from a viviparous fish, the green swordtail *Xiphophorus helleri*. Cryobiology 52, 209–218.

Yu, S.-Z., 1995. Primary prevention of hepatocellular carcinoma. J. Gastroenterol. Hepatol. 10, 674–682.

Zunker, K., Epplen, J.T., Schartl, M., 2006. Genomic stability in malignant melanoma of *Xiphophorus*. Melanoma Res. 16, 105–113.

STUDY QUESTIONS

1. How many genes are thought to be involved in the Gordon-Kosswig melanoma model? Explain your answer.
2. Why don't parental X. maculatus Jp 163 A develop appreciable numbers of tumors, whereas interspecies hybrids using them as nonrecurrent parents develop tumors according to Mendelian expectations?
3. What evidence from these fish models support a hypothesis that exposure to UV light and MNU have distinct genetic targets to induced melanoma?
4. How valid do you think extrapolating the wavelength dependence (action spectrum) of melanoma causation from *Xiphophorus* to humans is?
5. What other evidence of conserved processes from "lower" life forms to humans, or the lack of, can you use to support your argument?
6. You learn that one of your parents has recently started taking DHEA in order to slow the aging process. How might you persuade him or her that supplementation with DHEA could be a bad idea?
7. Imagine that you are the president of a pharmaceutical company. Your company has developed a new drug that may dramatically reduce the risk of heart attacks. Unfortunately, this drug causes peroxisome proliferation and liver cancer in rodents. How might you convince the FDA that this drug is safe for use in humans?
8. Rainbow trout have a number of attributes that make them useful models for conducting carcinogen risk assessment studies. List these. In your opinion, which single attribute is the most essential for the successful completion of ultralow-dose carcinogen studies, and why?

C H A P T E R

33

New Approaches for Cell and Animal Preservation

Lessons from Aquatic Organisms

STEVEN C. HAND AND MARY HAGEDORN

INTRODUCTION

The vast majority of our marine environment has been stable for many millions of years, changing relatively slowly over time. However, for the first time in the history of our planet, a single species, *Homo sapiens*, has had the ability to influence and change our global ecosystems. Primarily this capability includes changing the Earth's climate, but concomitantly it includes global changes in ecosystem health. As with well-studied terrestrial ecosystems, wholesale degradation of the complexity of our marine ecosystems is being observed as weather patterns induce species extinctions. We are observing increasing temperatures, sedimentation, and stress on marine communities that fall victim to new types of diseases. Elevated water temperatures in the Indian Ocean have increased the incidence of dust storms over the past 40 years over the Sahel in Africa, and now millions of tons of African dust are deposited in the Caribbean each year, the contents of which are being modified in harmful ways by humans (Garrison *et al.*, 2003; Griffin *et al.*, 2002). In one study, sea fan mortality was directly related to the fungal spores, *Aspergillus* in the dust storms blowing over the Caribbean from Africa (Geiser *et al.*, 1998). Fortunately, conservation techniques that include genome resource banks for preservation of germplasm hold promise for preserving species within ecosystems. Successful biostabilization of germplasm is essential to the long-term future of stock centers because of inherent problems with breeding failure and unpredictable accidents and natural disasters.

This chapter evaluates current developments in the field of biostabilization of cells. One of the recurring themes in our treatment is that identifying and applying mechanisms from animals whose evolutionary history has provided the capacity for natural tolerance to freezing and drying has, and will continue, to inform us in our attempts to improve methods for stabilizing cells and embryos. Within our ocean ecosystems, there are many examples of animals that have adapted to severe environmental conditions, such as tidal and seasonal desiccation, hypersalinity, high temperatures, and extreme cold. We expand on these lessons from nature in this chapter.

Although numerous advancements in the field of germplasm cryopreservation can be traced to sound understandings of cryodamage and cryoprotection, our treatment here will not be limited to cryopreservation. We will evaluate advances and opportunities afforded for preservation that include metabolic depression, use of novel osmolytes and new classes macromolecules with stabilizing properties, and cell dehydration across a range of temperatures. Not surprisingly, the activity of water is a central issue in cell stabilization. Reducing the molecular mobility of water and of the macromolecules contained therein is critical to arrest cell processes and structural stabilization (Bruni and Leopold, 1990, 1992; Buitink *et al.*, 1998a, 1998b; Williams and Leopold, 1989). Understanding how aquatic animals deal with water stress has provided enormous insights into how, and in what directions, animal and cell preservation research should move. To succeed in preparing dried cells that retain high viability upon rehydration would offer tremendous economic and practical advantages over traditional cryopreservation protocols (Hand and Menze, 2007).

For example, lessons learned from organisms that are naturally desiccation tolerant are being applied to cell stabilization problems in the biomedical field with the goal of desiccating human cells for storage at ambient temperature. Storage of dried cells would increase the availability of blood to soldiers on the battlefield. One might envision combat soldiers carrying units of their own dehydrated blood that could be rapidly reconstituted with water when needed. This technological advancement would avoid problems with blood typing and immunological rejection. Similarly, in many regions of underdeveloped countries where refrigeration is not routinely available, the use of dried blood products would revolutionize medical care in clinical and emergency settings. Such long-term goals for the field of biostabilization are not unreasonable; it is appropriate to note that human blood platelets already have been successfully freeze-dried by preloading with trehalose, a naturally occurring sugar found in many dehydration-tolerant animals (Wolkers *et al.*, 2001). These freeze-dried platelets are now advancing toward clinical trials.

BASIC BIOLOGICAL PROCESSES: MAINTAINING OSMOTIC EQUILIBRIUM

Water is the basis for all life on Earth. Without it, virtually no cell process, such as anabolism, catabolism, or reproduction, could occur. Indeed, as we will emphasize at various points during this chapter, it is the restriction of water mobility in cells by either dehydration or low temperature (subfreezing temperatures) that is an essential aspect to successful biostabilization. First, we briefly describe concepts of water stress for cells and of intracellular osmolyte systems. Second, although the issue of how water moves across cell membranes is an extraordinarily complex process and cannot be fully reviewed here, we illustrate some key principles by focusing on water channels or aquaporins. Third, we highlight a second class of membrane channels, the so-called P2X receptors that are receiving interest in the context of transferring membrane-impermeant solutes into cells. Both of these families of membrane channels are finding utility in new protocols for cell preservation.

Water Stress, Cellular Osmolytes, and Late Embryogenesis Abundant (LEA) Proteins

Fluctuation in cellular water content is a universal problem confronting both aquatic and terrestrial organisms. Subfreezing temperatures, desiccating conditions prevalent in xeric climates, and the osmotic variation seen in aqueous habitats are common environmental occurrences that can pose severe problems of water stress (Yancey, 2005; Yancey *et al.*, 1982). The threat of desiccation for organisms inhabiting the intertidal zone occurs during emersion at low tides or when organisms are positioned in the high intertidal zone where wetting occurs primarily by spring tides, storm waves, and spray (Hand and Menze, 2007). Drying because of evaporative water loss is the most common mechanism for dehydration, although during winter in northern temperate regions freezing can also occur, which reduces the liquid water in extracellular fluids and can lead to intracellular dehydration in multicellular organisms. Freezing tolerance has been reported and characterized for a number of intertidal invertebrates, including gastropods like *Melampus* and *Littorina* (Aunaas, 1982; Loomis, 1985) and bivalve genera including *Mytilus* and *Geukensia* (Aunaas, 1982; Loomis and Zinser, 2001).

Physiological and biochemical features important for resisting or tolerating water loss are numerous in organisms of the intertidal zone. However, an instructive body of work with the great application to cell preservation deals with the maintenance of intracellular osmolytes for water retention and macromolecular protection at low water activities and the expression of protective proteins associated with desiccation tolerance. It is well established that many marine organisms frequently facing water stress possess systems of "compatible" solutes (Brown and Simpson, 1972), a term initially applied to small carbohydrates (i.e., polyhydric alcohols, or "polyols"). When accumulated at high levels, these compounds have no detectable effects on macromolecular structure and function, yet they provide osmotic balance with the external environment (Somero, 1986; Yancey *et al.*, 1982). Generally, these organic solutes are either uncharged or zwitterionic, often structurally resembling the stabilizing ions of the Hofmeister or lyotropic series (Clark and Zounes, 1977). The Hofmeister series ranks neutral salts based on their capacity to stabilize (salt out) macromolecules versus denature them (salt in). Other classes of organic osmolytes include amino acids and their derivatives and the combination of methylamines and urea. When cellular water is lost as a result of desiccation or some other form of water stress (freezing, fluctuating salinity), intracellular osmolytes become concentrated. The colligative effect (related to the total number of dissolved solute particles) of intracellular osmolytes is to retard water loss from the cell. Whereas some contribution to the intracellular osmotic pressure of cells comes from inorganic ions, a substantial fraction of the osmotically active solutes are organic osmolytes (Hand and Menze, 2007).

An evaluation of the dominant types of organic osmolyte systems and their evolution (Yancey *et al.*, 1982) suggested that a common feature of these osmolytes was that they exerted influences on macromolecular stability by acting on the solvent properties of water. In addition to organic osmolytes that can be considered nonperturbing or compatible, other organic osmolytes can actually stabilize macromolecules during water stress. Protective effects of osmolytes have been documented for proteins and for phospholipids that form membrane bilayers. It is appropriate to note that in the case of anhydrobiosis (life without water), disaccharides like trehalose are technically not serving the role of an osmolyte, because virtually all cellular water is lost in this state. However, trehalose is considered an osmolyte for other nonanhydrobiotic species. As discussed in the Dehydration of Cells section, trehalose stabilizes membrane structure through physical intercalation with phospholipids. In combination with transition metal ions, trehalose also protects proteins against denaturation incurred during freezing (Carpenter *et al.*, 1986) and drying (Carpenter *et al.*, 1987). Application of compatible and stabilizing osmolytes in cell preservation has substantially advanced the field.

It has become apparent that compatible or protective solutes of low molecular weight are not the only biochemical components that contribute to an organism's desiccation tolerance. For example, water loss during emersion can be fast among species of intertidal seaweeds, ranging from 10% to over 90% loss in water content after only a few hours under moderately desiccating conditions (Ji and Tanaka, 2002). Tolerance to water loss seems key to survival in the intertidal zone of seaweeds like brown algae of the genus *Fucus*, as opposed to possessing efficient mechanisms for the prevention of water loss. Fucoid algae constitutively express proteins that are immunologically related to dehydrins, some of which are specific for certain embryonic stages and certain species (Li *et al.*, 1998). Dehydrins are one family of proteins belonging to a larger group called late embryogenesis abundant (LEA) proteins. LEA proteins were first identified in land plants (Close *et al.*, 1989), and their expression is associated with desiccation tolerance in seeds and anhydrobiotic plants. The expression of LEA proteins is not restricted to plants, having been documented in bacteria, fungi, nematodes (Hoekstra, 2005 and references therein; Wise and Tunnacliffe, 2004), and most recently in desiccation-tolerant embryos of the brine shrimp *Artemia franciscana* (Hand *et al.*, 2007) and a chironomid insect larva (Kikawada *et al.*, 2006). Much higher in the intertidal zone than *Fucus* (supralittoral fringe), one can find cyanobacteria (*Calothrix, Lyngbya*) that are quite tolerant to the xeric conditions that exist, where seawater spray can often be the only mechanism for hydration. *Calothrix* is embedded in a gelatinous mass that helps retard water loss, and a dehydrin-like protein (40 kD) immunologically similar to ones in plants is inducible in this cyanobacterium upon osmotic challenge (Close and Lammers, 1993).

Among various physiological roles for LEA proteins, stabilization of sugar glasses (vitrified, noncrystalline structure in cells promoted by sugars like trehalose) is often suggested (Hoekstra, 2005), along with protein stabilization via protein-protein interaction (Grelet *et al.*, 2005), ion sequestration (Grelet *et al.*, 2005), and formation of structural networks (Wise and Tunnacliffe, 2004). Such networks have been hypothesized to increase cellular resistance to physical stresses imposed by desiccation (compare Goyal *et al.*, 2003). Two different LEA proteins were reported in *A. franciscana* embryos (Hand et al, 2007), raising an interesting issue relating to localization; targeting to different cellular locations might explain the functional significance of two LEA proteins. Evidence from plants indicates that targeting of LEA proteins to specific organelles like the mitochondria does occur (Grelet *et al.*, 2005). It is becoming clear that an ensemble of micromolecules and macromolecules are important for establishing the physical conditions required for cellular stabilization during drying in nature, and such approaches are being applied to cell stabilization (see the Dehydration of Cells section). Much of the data leading to these insights originated from studies of marine organisms.

Aquaporins

When a cell is placed in a hypo- or hyperosmotic environment, water can diffuse passively across the membrane or more rapidly through water channels, called aquaporins. Robert Macey unambiguously discovered water channels in 1970 while studying the water permeability of red blood cells. He identified these new water channels and discovered that mercuric chloride reversibly inhibited their action (Macey, 1970, 1984). Previously, no substance had been found that could alter a cell's water permeability. So Macey's studies provided strong evidence for the existence of water channels because they suggested that the mercury bound to and blocked a pore. However, at the time, most scientists did not agree with these observations.

Years later, Peter Agre and his colleagues at Johns Hopkins University found a large protein, CHIP28, in the red blood cells that they did not expect and thought it might be Macey's water channel, but they had no evidence to support this. They took a risk, cloned the gene, and then injected the cRNA into frog oocytes to test how it might change the water permeability of the oocytes. Frog oocytes typically have very low water permeability, so they can be placed in dilute frog ringers (~80 mOsm/kg) and they will not be damaged (Figs. 33-1A, B, left). However, after the injection of the CHIP28 cRNA, the modified oocytes swelled and burst from the increased water permeability after 3 minutes in the dilute frog ringers (Figs. 33-1A, B, right; Preston *et al.*, 1992). Preston and colleague's elegant experiments proved the identity and function of the water channel, renaming CHIP28 to AQP1 for aquaporin-1. To date, there are now more than 10 aquaporins found in the human body, and their impaired function is related to a few human diseases, such as congenital cataracts and nephrogenic *diabetes insipidus*.

All the aquaporins are members of the major intrinsic proteins (MIPs) family that is distributed in a wide variety of biological membranes throughout the plant and animal kingdoms (Agre *et al.*, 1993; Chrispeels and Agre, 1994; Heymann *et al.*, 1998; King and Agre, 1996; Lee *et al.*, 1997; Verkman *et al.*, 1996; Wintour, 1997). Generally, MIPs are transmembrane proteins that pass water, glycerol, or other small uncharged molecules. The aquaporin protein is extremely complex in its form and function. Its simplest elements consist of six membrane-spanning segments with two hemipores. Each of these membrane-spanning elements

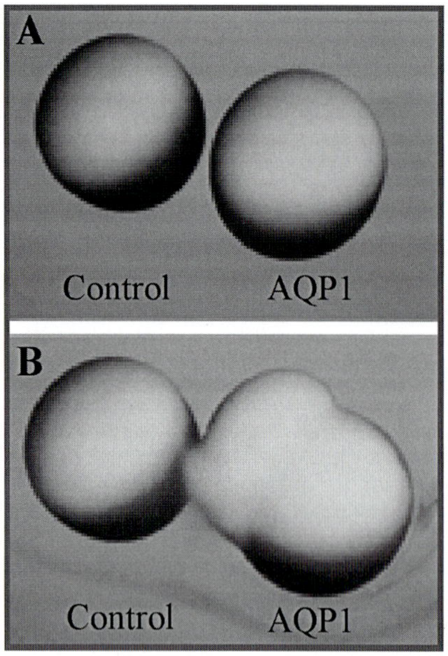

FIGURE 33-1. The discovery of water channels. (A) Both a control (water-injected) and AQP1-injected oocyte were placed in dilute frog Ringer's. (B) After 3 minutes, the AQP1-modified oocyte burst from the increased water permeability, whereas the volume of the control oocyte remained the same. Reprinted with permission from Preston *et al.* (1992).

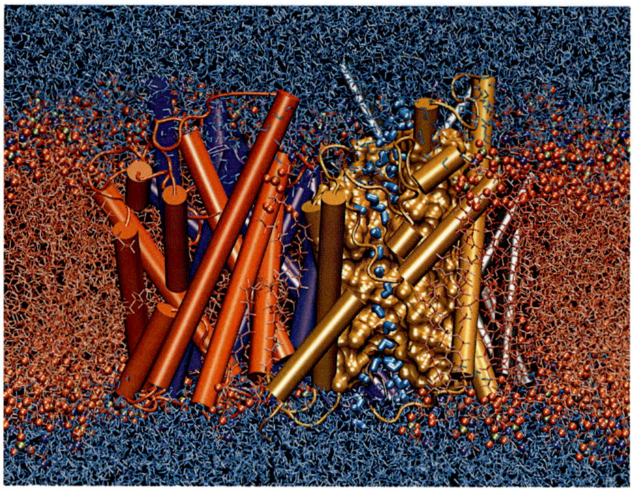

FIGURE 33-2. The structure of aquaporin depicts the tetrameric structure of the aquaporin molecule (each monomere is a different color) with the blue water molecules moving in single file through the gold pore. Reprinted with permission from D. Grimm (2004).

is called a monomer, and a functional water channel usually has four of these combined, but each monomer is a functional water-transporting unit (Fig. 33-2).

The most important feature of aquaporins is that they are completely selective, they transport water (aquaporins) and some uncharged polyols, such as glycerol (aquaglyceroporins). But they are completely impermeable to charged species, such as protons. This allows the cells to move water and still maintain the electrochemical potential of the cell. Although water usually transfers protons efficiently, it does not in the water channel, because of a unique type of somersault binding and unbinding of the water molecules as they pass through the channel in single file. Additionally, the narrow opening in the channel maintains this unique and biologically important selectivity (Kozono at al., 2002; Tajkhorshid *et al.*, 2002).

Why are aquaporins necessary? Although scientists debate the early origins of life, they do agree that life began in the oceans approximately 3.5 billion years ago. Because it took another 2 billion years for the protective ozone layer to form, all life continued to evolve in water. However, when organisms ventured out onto land, they took their watery past with them, but many modified it somewhat. Although most sessile organism within the oceans have an internal osmolality that matches that of the oceans (~1000 mOsm/kg), most vertebrate animals have an internal osmolality that is significantly lower (~300 mOsm/kg), presumably because of the long period of time many ancestral fishes spent in estuarine environs. Bony fishes are thought to have reinvaded marine environs with this lower body osmolarity. Regardless of the exact causes, most vertebrate animals that live in freshwater, in seawater, or on land face challenges of maintaining their internal osmolality. This is particularly true for cells that must adjust rapidly to changing conditions either inside the body, such as kidney and blood cells, the kidney processes hundreds of gallons of water a day as it cleanses the wastes from blood cells, or cells expelled into hostile environments, such as sperm cells, which must survive in a variety of new environments and still maintain their viability. Water must be able to move quickly across the cell membrane to allow these cells to adapt rapidly, and aquaporins permit this rapid water transit. So, the evolution and diversification of water channels enhanced the ability of organisms to exploit new environments. Because MIPs can be found in all organisms (Chrispeels and Agre, 1994), their molecular clock charts the evolution from our origins in the oceans (Zardoya *et al.*, 2002).

Aquaporins are now being incorporated into a number of cells to improve their preservation properties. Edashige *et al.* (2000) have shown that heterologous expression of AQP3 into mouse oocytes improves their survival after cryopreservation, as does heterologous AQP1 expression into yeast (Tanghe *et al.*, 2002). In addition, Valdez *et al.* (2006) found that aquaporin-3 expression increased the permeability of water and cryoprotectants in medaka oocytes. All of these cells were relatively permeable to water and cryoprotectant before aquaporin modification, but the "foreign" aquaporins increased their permeabilities.

An important area for marine conservation and aquaculture is the cryopreservation of fish embryos. Unlike fish sperm, fish embryos have never been successfully cryopreserved. One of the major obstacles to cryopreservation is the fish embryo's low membrane permeability. Specifically, in zebra fish embryos, the yolk syncytial layer prevents movement of water and appropriate cryoprotectants into and out of the yolk compartment (Hagedorn *et al.*, 1996, 1997, 1998). Hagedorn *et al.* (2002) demonstrated that introducing cRNA for aquaporin-3 increased membrane permeability to both water and propylene glycol. However, subsequent attempts to use this increased membrane permeability to cryopreserve the zebra fish embryos have been unsuccessful to date (Hagedorn unpublished data). Similarly, aquaporins were not found to be useful practically for freezing in overexpressing yeast strains. Tanghe *et al.* (2004) found that although aquaporins helped yeast in small laboratory batches, they had no potential for use in industrial-sized batches of frozen dough. In summary, aquaporins improve the permeability and freezing characteristics of single cells, but they have more complex characteristics when expressed in actively dividing multicellular organisms, like fish embryos.

$P2X_7$ Receptor Channel

To stabilize cells, one has to be able to exchange not only water but also solutes between the outside and inside of cells in order to adjust the osmotic and chemical properties of the intracellular milieu so that freezing and drying can be tolerated without damage to macromolecules and supramolecular structures. For example, the traditional methodology for cryopreservation of cells and tissues is to use high concentrations of solutes (cryoprotectants) that can penetrate the plasma membrane of cells—glycerol, dimethyl sulfoxide, ethylene glycol (Mazur, 1984). As introduced in the preceding sections and further expanded upon later, small carbohydrates like trehalose, sucrose, and maltose have utility for stabilization of cells during freezing and drying. Yet the impermeability to the cell membrane of these sugar-based preservation agents has to be first overcome.

An increasing number of artificial and natural mechanisms are being used to introduce solutes into cellular compartments enclosed by impermeable membranes. Included among these are transfection of cells with trehalose synthesis genes (Guo *et al.*, 2000), microinjection of solute solutions (Eroglu *et al.*, 2002; Janik *et al.*, 2000), fluid phase endocytosis (Elliott *et al.*, 2006; Hubel and Darr, 2002; Oliver *et al.*, 2004), temperature-induced phase transitions in lipid membranes that render cells permeable (Beattie *et al.*, 1997), and introduction of artificial/foreign protein pores into cell membranes that can be toggled open and closed for transfer of solutes (Eroglu *et al.*, 2000). Another approach is to exploit naturally occurring membrane channels, perhaps present in cells for unrelated functions, by coopting their properties for the transfer of protective solutes into cells.

One such channel that has been used effectively for such purposes is the $P2X_7$ receptor channel, which is an ATP-gated ion channel expressed naturally in a number of mammalian cell types (Dubyak and El-Moatassim, 1993; North, 2002; Steinberg and Silverstein, 1989). The P2X family of purinergic receptors opens in response to extracellular ATP. All P2X receptors are permeable to monovalent cations, and some have significant permeability to calcium and anions. P2X receptors are found in neurons, glia, various epithelia, endothelia, bone, muscle, and hematopoetic tissues (North, 2002). The normal physiological roles for this receptor family depends on the cell type as well as the subtype of the receptor, and in many instances, they are still a matter of debate (for review, see North, 2002). For example, ATP operates as a synaptic transmitter from sympathetic nerves to smooth muscle, and P2X receptors may be involved in mechanosensing and sensing of inflammatory pain. Extracellular ATP also serves as an autocrine and paracrine transmitter, and in this context opening of P2X receptor channels is likely involved in such signaling in ducted glands and epithelia. The $P2X_7$ receptor channel appears to be involved in signaling the release of cytokines in immune cells (macrophages) and inflamed tissues (Solle *et al.*, 2001).

Importantly for our discussion here, the $P2X_7$ receptor channel, once opened, exhibits a time-dependent dilation so that organic compounds larger than inorganic ions can pass (Cockcroft and Gomperts, 1979). Poration is thought to be mediated by binding of unliganded ATP^{4-} (i.e., not chelated with Mg^{2+}) to the $P2X_7$ receptor, which opens in milliseconds a membrane channel for selective passage of cations (Dubyka and el-Moatassim, 1993; Rassendren *et al.*, 1997; Steinberg and Silverstein, 1989; Surprenant *et al.*, 1996). After several seconds of exposure to ATP^{4-} concentrations above 100 micromolar, the pore dilates and becomes permeable to larger compounds (<900 Da, Fig. 33-3). The physiological relevance of this enhanced permeability to larger molecules is still a mystery (North, 2002).

Working with mouse J774 macrophages (a cell line that expresses substantial quantities of the $P2X_7$ receptor channel), we were able to optimize the ATP^{4-}-dependent poration technique and load these cells with metabolic effectors and sugars (Elliott *et al.*, 2006; Menze *et al.*, 2005a). Keys to the success of this method were optimizing the conditions under which the $P2X_7$ receptor channel was opened, as well as the duration of this event, to permit cells to recover without appreciable mortality. A poration medium was developed that mimicked the composition of the intracellular milieu and contained trehalose, a low sodium concentration, and a phosphate buffer system. It is appropriate to note that some reports indicate that the opening of the $P2X_7$ receptor channel may lead to cell death under certain conditions (Coutinho-Silva *et al.*, 1999; Humphreys *et al.*, 2000; Virginio *et al.*, 1999). Similar to these findings, we also have observed high mortality when ATP dependent

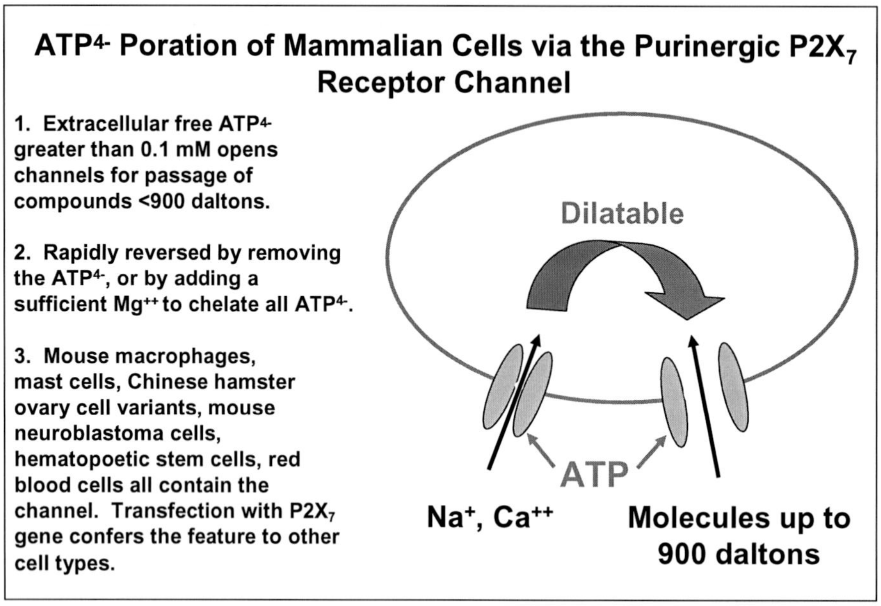

FIGURE 33-3. Addition of extracellular ATP to cells expressing the $P2X_7$ receptor channel promotes cation conductance within milliseconds. Continued exposure to ATP (seconds) causes permeability to larger molecules. One explanation for the time-dependent permeation by larger molecules is that the pore undergoes a conformational dilation, although alternative explanations exist.

poration was conducted in typical cell culture medium. Also, leaving the pore open longer than 15 min at 37°C significantly decreased viability. Extending the duration of pore opening to 1 hour or more without a substantially increased mortality can be accomplished by activating the $P2X_7$ channel at 37°C and then promptly lowering the temperature to 0° (Elliott *et al.*, 2006). Such a protocol is useful when loading large quantities of sugar (see the Dehydration of Cells section). The pore is closed simply by removal of ATP^{4-} from the medium, either by chelating with Mg^{2+} or diluting to an ATP^{4-} concentration <100 μM. Thus, by carefully optimizing the permeabilization conditions and the duration of pore opening, one can minimize cytotoxic effects of P2X7-mediated poration. The poration procedure also has been applied to the human TF-1 cell line, which expresses the $P2X_7$ receptor channel (Buchanan *et al.*, 2005). TF-1 cells display many features in common with hematopoietic stem and progenitor cells.

SURVIVAL UNDER EXTREME CONDITIONS: NATURAL AND PRESERVED STATES

Metabolic Preconditioning: Principles. Governing Biological Stasis

Tolerance of phylogenetically diverse organisms to reversible drying is well documented (Crowe and Clegg, 1978), and the ability of an animal to arrest development and metabolism before challenges like desiccation, anoxia, and freezing improves the survival of many species (Hand, 1999; Hand *et al.*, 2001). By extending this lesson from nature, one key prediction that emerges is that the placement of mammalian cells into stasis (defined here as cellular and developmental arrest under hydrated conditions at normal temperatures) will foster greater survival of cells when exposed to stabilization methods like dehydration, lyophilization, or cryopreservation. The more complex the cells and tissues, the more likely it is that induction of cell stasis will be required first in order to achieve acceptable survival during storage. For certain mammalian cells that prove to be intolerant of drying under any regime, the induction of the stasis phenotype, in and of itself, may be adequate for long-term storage.

These predictions are based on our understanding of the biological principles and criteria that underlie naturally occurring states of metabolic depression and dormancy, particularly diapause (Podrabsky and Hand, 1999; Reynolds and Hand, 2004). Diapause is a state of developmental or metabolic arrest controlled by endogenous cellular factors (as opposed to environmental cues), such that entry into diapause in nature precedes the onset of stressful environmental conditions. Diapausing organisms remain hypometabolic even under conditions that would normally promote active metabolism and development (for extended treatments, see Clegg *et al.*, 1996; Drinkwater and Crowe, 1987; Hand, 1991; Lavens and Sorgeloos, 1987; Lees, 1955). What occurs during entry into diapause is that animals interrupt their normal developmental program. Complex metabolic processes are carefully down-regulated in a coordinated fashion, minimizing imbalances in cellular processes that can cause pathological conditions to develop. It is critical to realize that diapause is a *regulated and natural process* for these organisms, *not a pathological one*. In this regulated state of stasis, diapausing animals are more tolerant of environmental stresses (Hand *et al.*, 2001; Podrabsky and Hand, 1999). For example, if sufficient time is not provided for metabolic restructuring before drying, the desiccation event is lethal to animals that are normally dehydration tolerant; if preconditioning occurs (e.g., converting glycogen into trehalose), survival of desiccation is dramatically improved (Crowe *et al.*, 1977). Consequently, it is reasonable to predict that mammalian cells, which normally do not experience drying, will also benefit from stasis.

Populations of the annual killifish, *Austrofundulus limnaeus*, survive seasonal drying of their ephemeral pond habitats by producing drought-tolerant, diapausing embryos (Wourms, 1972). Diapausing embryos of *A. limnaeus* may remain dormant for extended periods of time, spanning periods from months to years (Podrabsky and Hand, 1999; Wourms, 1972). A 90% depression of metabolism accompanies this state (Podrabsky and Hand, 1999), during which the fish embryo is remarkably resistant to environmental stresses such as anoxia and dehydration. A simultaneous down-regulation of ATP producing and ATP consuming processes is indicated by the maintenance of stable levels of ATP during the transition in metabolic rate (Podrabsky and Hand, 1999). Biosynthetic events are down-regulated during fish diapause. Protein synthesis decreases by over 93% in diapause II embryos compared to active embryos, resulting in a 36% reduction in energy demand because of the depression of protein synthesis during diapause.

Stable ATP/ADP ratios are maintained during diapause in these embryos (Podrabsky and Hand, 1999). However, the AMP levels increase, which causes a substantial rise in the AMP/ATP ratio. Importantly, there is a strong correlation between the increase in the AMP/ATP ratio and the decrease in metabolism during diapause (Podrabsky and Hand, 1999). Further, the elevated AMP/ATP ratios are tightly correlated with the decrease in rates of protein synthesis, suggesting that AMP/ATP ratios may be important in the regulation of protein synthesis and other anabolic processes in diapause. The effect of AMP could be mediated by the AMP-activated protein kinase (AMPK; for reviews see Hardie *et al.*, 1998; Hardie and Hawley, 2001), perhaps through phosphorylation of key biosynthetic enzymes and initiation/elongation factors of translation.

What emerges from the annual killifish example is that several requirements must be met if an organism is to enter, survive, and exit hypometabolic states like diapause. First, studies from several laboratories have underscored the importance of suppressing both energy production (e.g., oxidative pathways) and energy consumption (e.g., transcription, translation, ion pumping) in a coordinated fashion (Guppy and Withers, 1999; Hand, 1991; Hand and Hardewig, 1996; Hochachka and Guppy, 1987; Storey and Storey, 1990). Otherwise, cellular energy reserves will be depleted and organisms will reach an energetic state from which recovery is not possible (Hofmann and Hand, 1990). Second, up-regulation and down-regulation of specific gene products may be useful in promoting the diapause state (e.g., Denlinger, 2002). Third, in the face of an overall depression of biosynthesis, the existing macromolecules in cells must be preserved through reduced degradation rates (Anchordoguy and Hand, 1994; van Breukelen and Hand, 2000; Warner *et al.*, 1997) and the actions of protective intracellular solutes (Crowe *et al.*, 1998; Hoekstra *et al.*, 1997) and molecular chaperones (Clegg, 2001). Finally, initiation of apoptosis must be avoided during cell stasis (Menze *et al.*, 2005b; Menze and Hand, 2007).

To our knowledge, the most profound metabolic depression ever reported during a diapause state is that seen for embryos of the brine shrimp, *Artemia franciscana*. Over a period of about 4 days after release of diapause embryos from the female, the respiration rate of field-collected embryos dropped more than 90% (Fig. 33-4); the measured depression is even greater (97%) when embryos are synchronized for time of diapause entry (Clegg *et al.*, 1996). The metabolic arrest that accompanies diapause occurs under fully normoxic and hydrated conditions. Clegg *et al.* (1996) also reported a major depression in the rate of protein synthesis during diapause compared to active embryos. Similar to the killifish embryos discussed earlier, AMP/ATP increases under diapause in *A. franciscana* embryos (J Covi, J. Reynolds, S. Hand, unpublished observations).

Metabolic depression is also well known in many marine and aquatic species that display tolerance to oxygen limita-

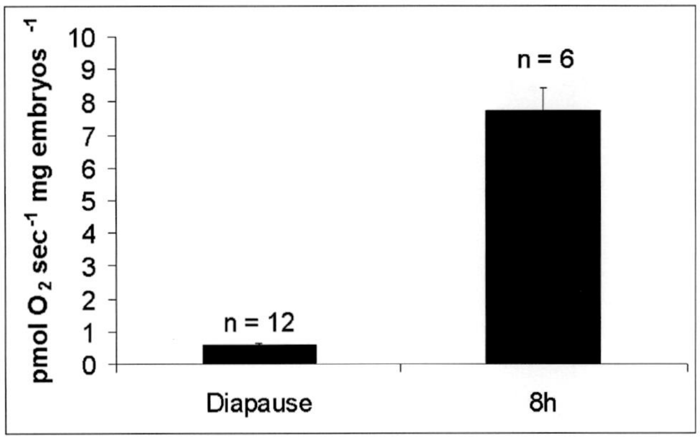

FIGURE 33-4. Oxygen consumption by embryos of *A. franciscana*. Respiration rates of embryos in diapause are significantly depressed compared to active postdiapause embryos that have developed at room temperature for 8 hours. Bars represent mean ±SEM. Modified from Reynolds and Hand (2004).

tion. In fact, the length of survival under anoxia is directly proportional to the degree of metabolic depression (Hand, 1998). When metabolism, expressed as a percentage of the normoxic rate, is depressed below 10%, the survival of a given species increases exponentially (Fig. 33-5). An extreme state of metabolic quiescence is induced in *A. franciscana* embryos in response to anoxia (Busa and Crowe, 1983), during which heat flow (which can be converted to metabolic rate with additional data) is depressed to very low levels. After 48 hours of anoxia, heat dissipation is below 0.2% of the aerobic value and is still declining (Hand, 1995). In keeping with the forgoing pattern, the tolerance of these embryos to anoxia is an amazing 3 to 4 *years* (Clegg, 1997)!

Entry into a state of estivation is a common occurrence in some species of prosobranch snails (*Bembicium vittatum*, *Nodilittoria unifasciata*) that live above the high tide mark in the intertidal zone (e.g., McMahon, 1990). The condition involves withdrawing the body deeply into the shell, occluding the shell aperture, and even cementing the edge of the aperture to the substratum with mucus. Estivation at least in land snails is commonly associated with major metabolic depression (Rees and Hand, 1990). In nonestivating snails in the upper intertidal, desiccation is normally associated with varying degrees of metabolic depression, either active or passive. Anaerobic metabolism and accumulation of end products (alanine, succinate) have been documented in at least one gastropod during emersion (*Littorina saxatilis*). However, the contribution anaerobiosis to overall energy production is less than 2% (Sokolova and Poertner, 2001). Specimens of *L. saxatilis* with greater tolerance to desiccating conditions also displayed greater metabolic depression.

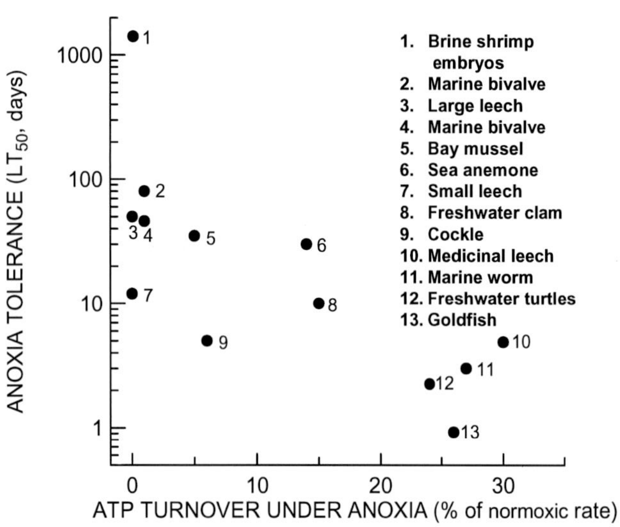

FIGURE 33-5. The dependence of anoxia tolerance for selected aquatic species as a function of the capacity for metabolic depression under anoxia, expressed as a percentage of the normoxic value. Modified from Hand (1998)

These examples are not intended to be a comprehensive treatment of the phenomenon of metabolic depression and its linkage to survival during environmental stress, but they do emphasize the basis for the premise that biostabilization of cells during storage may be improved by metabolic preconditioning.

Preservation of Cells

Genome Resource Banks

Even in the most remote, pristine marine bioreserves, such as the northwestern Hawaiian Islands (Maragos *et al.*, 2004), human activities are damaging fragile coral ecosystems (Bellwood *et al.*, 2004). Habitat preservation is one of the best ways to conserve ecosystems, but threatening global patterns show no signs of relief that will permit our ecosystems to recover. Fortunately, *ex situ* conservation techniques, such as genome resource banks using frozen samples, hold strong promise for rapid and profound improvements in preserving species within ecosystems. These banks have important passive and active functions. First, genetic material can remain frozen but alive for hundreds of years in liquid nitrogen (Fig. 33-6), allowing the time necessary to mitigate and restore habitats. Second, large samples of gene pools can be maintained, preventing species extinction. Third, the banks can be used actively to increase genetic diversity within an ecosystem through the use of thawed

FIGURE 33-6. Cells properly prepared for cryopreservation can remain frozen but alive in liquid nitrogen for hundreds of years. The cryostorage vessels must be constantly maintained and refilled with liquid nitrogen to ensure that all the stored tissues maintains a temperature of −196°C.

samples to "seed" shrinking populations. These resource banks provide (1) easy, inexpensive movement of genetic material among living populations; (2) an extended generational interval for individuals with valuable genetics; (3) reduced space constraints; and (4) biomaterials for scholarly research (Wildt *et al.*, 1997). However, it is crucial that these resource banks be developed in a coordinated fashion that maximizes diversity and ensures biosecurity of the material. The International Union for the Conservation of Nature and Natural Resources' Species Survival Commission has initiated guidelines for wildlife resource banks.

There are three main types of genome resource banks that encompass the genetic material for wildlife, agriculture, and biomedicine. They do not have to be mutually exclusive, but they have slightly different goals and needs. Presently, there are 551 Culture Collections in 66 countries of the world (http://wdcm.nig.ac.jp/hpcc.html). Certainly there are marine species within these extensive collections, such as algae, fungi, bacteria, fish sperm, and so on, but to date, there are no genome resource banks that are focused solely on the needs of the biodiversity of the oceans. This is especially alarming given the drastic depletion of our fisheries throughout the world. The availability of viable germplasm after cryopreservation could have a profound influence on the conservation of rare or threatened species. For conservation, the development of frozen or "insurance" populations would preserve genetic diversity and assist efforts to prevent extinction of wild species in natural aquatic ecosystems (Ballou, 1992; Wildt, 1992; Wildt *et al.*, 1993, 1997). There are some repositories in the United States and Europe whose focus is a single endangered marine species or a few species that are important to aquaculture. The Directorate for Nature Management in Norway contains frozen sperm from more than 6000 individuals from 155 salmon stocks (www.nordgen.org). The University of Idaho in partnership with Washington State University and the Nez Perce Tribe has a repository for Chinook salmon from the Snake River that contains sperm from more than 500 males from 12 tributaries (Faurot *et al.*, 1997). Many of these stocks are on the endangered species list. Finally, the U.S. Department of Agriculture's National Animal Germplasm Program contains the sperm from 11 taxa of freshwater fish, five species of marine fish, and the frozen spat from Pacific oysters (www.ars-grin.gov).

In medical science, fish have become one of the more important vertebrate models for the study of development, genetics, and human disease, especially *Danio rerio* (zebrafish), *Oryzias latipes* (medaka), and *Xiphophorus* sp. (swordtails and platyfish) (see Chapters 31 and 32). Transgenic and mutagenic studies of zebra fish are playing an important role in human health and disease, Therefore, systematic germplasm cryopreservation of these important lines is having a major impact on the management of medical resource dollars by (a) reducing the size and production costs of facilities; (b) allowing the maintenance of large gene pools and reducing inbreeding, while minimizing the amount of space required to hold living animals; (c) reducing pressure on wild populations from collection activities; and (d) facilitating global, regional, and institution-to-institution transport of genetic material. Successful cryopreservation of germplasm will have a significant impact on the future of stock centers because the productivity of these centers is continually threatened by accidents, natural disasters, and breeding failure, and many important mutant and transgenic lines are now maintained *only* in live culture.

There are several possible approaches to preserving genomes, but researchers working at stock centers (e.g., the Jackson Laboratory and the National Institutes of Health) emphasized the importance of including a variety of genetic material, such as embryos, sperm, oocytes, ovarian tissue, blood, and DNA in genome banks (Biomedical Model Report, 1998; www.nap.edu/catalog/6066.html).

In general, fish sperm cryopreservation is successful and practiced regularly in both field and laboratory settings (Harvey *et al.*, 1982; Leung and Jamieson, 1991; Rana *et al.*, 1995; Stoss, 1983), but for the rapidly increasing number of mutant lines now being produced, we do not have the high throughput techniques to freeze sufficient numbers of sperm samples to maintain these stocks. More important, there are not sufficient financial resources or trained individuals to staff these important resource centers. To meet the demands of maintaining our important disease models for the future, there must be a significant increase in the training of new cryobiologists.

Frozen Cells

Most of the technological innovations that have advanced the field of germplasm cryopreservation arose from a sound understanding of the mechanisms of cryodamage and cryoprotection (Mazur, 1970, 1984). Successful cryopreservation of germplasm must address intrinsic biophysical properties (e.g., osmotic tolerance limits), and cryopreservation procedures based on these biophysical properties are necessary to minimize cryodamage and maximize survival (Rall, 1993). Conventional cryopreservation of cells involves the use of cryoprotectants and slow freezing to produce cellular dehydration and shrinkage. To successfully cryopreserve a cell, water must exit the cells, and an appropriate cryoprotectant must enter. Cells can tolerate a certain amount of extracellular ice, but *not* intracellular ice formation—this results in 100% mortality of the cells. Figure 33-7 demonstrates some of the idealized routes to the freezing of a cell. Because cryopreservation can be damaging to cells, every aspect of the cell's reaction to the process must be understood, such as the type of cryoprotectant and concentration used, how quickly the cell is frozen and thawed, and how the cryoprotectant is removed from the cell.

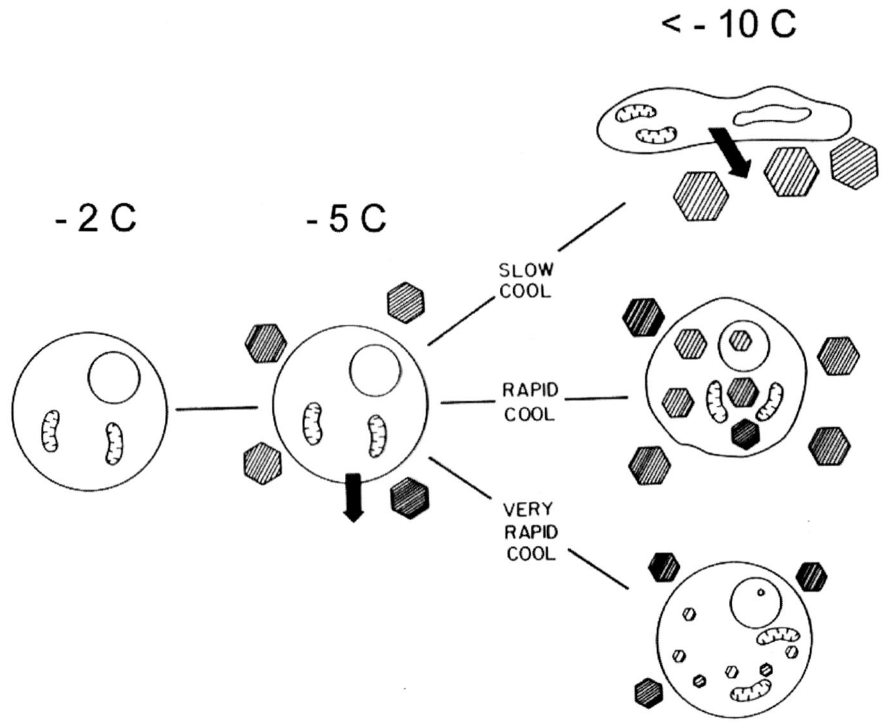

FIGURE 33-7. Idealized cooling in cells. At around –5°C, external ice crystals form, concentrating the external solution. During "slow cooling," cells have sufficient time to dehydrate so that lethal intracellular ice will not form. However, if the cooling is too rapid ("rapid cooling"), then lethal intracellular ice crystals will form. The very rapid cooling demonstrates vitrification where the internal cell is preloaded with a cryoprotectant then very rapidly cooled, so that no ice crystal form in the cell. Reprinted with permission from Mazur (1977).

The properties of cryoprotectants, such as dimethyl sulfoxide (DMSO) or propylene glycol (PG), are crucial for successful cryopreservation. Although their mechanism of action is not well understood, it appears that cryoprotectants depress the freezing point of solutions in and around the cells and may directly alter membrane bilayers or involve stearic interactions with bound proteins on the external surface (Hammerstedt *et al.*, 1990). Too little cryoprotectant entering the cell before cooling reduces effectiveness (Taylor *et al.*, 1974), whereas too much causes osmotic swelling and rupture during thawing and dilution (Levin and Miller, 1981). Therefore, cryopreservation procedures must be tailored for each type of cell, based on a thorough understanding of cellular properties.

Most cells have membranes permeable to water and certain solutes, and understanding how rapidly water and cryoprotectant move across the membrane is paramount to a successful cryopreservation process. When a cell is placed in water or solute, gradients produce a flux of these constituents across the membrane. The water flux across a cell membrane is given by

$$dVw/dt = LpART(Mi - Me) \quad (1)$$

where Vw is the cell's water volume, A the membrane area (assumed to be constant in this discussion), Lp the water permeability, R the gas constant, T the absolute temperature, and *Mi* and *Me* (the intracellular and extracellular solution osmolalities, respectively). Removal of most of the osmotically active water within the cell is crucial for successful cryopreservation. Dehydration times are dictated by the initial amount of water within the cell and how quickly it can move out (water permeability).

Typically, cryoprotectant concentrations of 1 to 2 molar are required for slow freezing. The conventional technique for assessing cryoprotectant permeation into embryos involves light microscopy. This method monitors the volume changes caused by the movement of cellular water and cryoprotectants into and out of the cell or tissue. If the cell is very small, like sperm, instruments such as a Coulter counter can measure the change volume in thousands of cells in a few minutes. Then, the water and cryoprotectant permeability can be deduced from experiments that examine the rate of volume change.

Cryobiology combines membrane physiology with the physics of ice. Therefore, a detailed understanding of the parameters in Equation (1) is essential for success. In addition to the more traditional slow-freezing cryopreservation process, described earlier, a more rapid process called vitrification uses high concentrations of cryoprotectants (5 to 6 M) that turn into a glass upon cooling. Vitrification avoids the problem of intracellular ice freezing, but it increases the likelihood of toxicity problems because of the high concentration of cryoprotectant needed for the process. If cells could be frozen at millions of degrees per second, then no cryoprotectant would be necessary and the cellular cytoplasm would vitrify on its own. This can be done with a new method called laser annealing. Mehmet Toner and colleagues are freezing some cells without the cryoprotectants (http://mghra.partners.org/narratives/TonerM.htm). These methods hold great promise in the future for safe and efficient freezing of whole organs, which is currently unavailable.

Room Temperature: Sperm Storage

Cryopreservation is not the only means of being able to preserve cells. Although it is the most commonly used method, there are many examples of marine organisms that have adaptations that have allowed them to solve preservation problems. If we can identify and understand some of these amazing processes, then perhaps the need for cryopreservation will become a technique of the past.

Many marine Crustaceans, sharks, and teleost fish can store sperm in the oviduct at body temperature for months to years after mating (Bauer, 1986; Muñoz *et al.*, 1999, 2000, 2002; Pratt, 1993; Takahashi *et al.*, 1991). Although little is know about how they mange this, nomadic sharks such as *Prionace glauca* and *Carcharhinus obscurus* can store sperm in specialized tubules for months to years (Pratt, 1993). The most complete understanding of sperm storage comes from the work of Muñoz and colleagues (Muñoz *et al.*, 1999, 2000, 2002) on the bluemouth rockfish, *Helicolenus dactylopterus dactylopterus*. Male bluemouth rockfish have mature sperm from July to February and the male's urogenital papilla is adapted as an exterior copulating organ, with spawning taking place only in January and February in the Western Mediterranean. However, sperm was observed in females for 10 months in highly vascularized storage crypts attached to the ovaries by a duct for 10 months. It was found that the ovarian fluid was acidic and that the ovarian wall secreted polysaccharides, protein, and lipid compounds throughout the storage and spawning period. Specifically glycoaminoglycans were found on the surface of the sperm in the ovarian crypts, but not in the testicular ducts. The combination of the glycoaminoglycans and the acidic environment may be responsible for maintaining the sperm for so many months.

Highly vascularized areas in the ovaries seem to be the most common morphological trait in sperm storage. The sperm remain immotile there, but the infusion of blood may feed them and keep them active while dormant. This system would be difficult to replicate in the laboratory because of the needs of the sperm metabolism.

Cooling: Fish Antifreeze Proteins

Arctic and Antarctic fishes live in some of the most extreme environments in the world. In 1969, Art DeVries and Donald Wohlschlag were working in the Antarctic and found notothenioid fish living between sea ice (Fig. 33-8, left) or immersed in icy slush, behaving normally at approximately −1.9°C (DeVries, 1983). A fish's body temperature conforms to their external environment and the freezing point of pure seawater is −1.3°. They wondered why these fish weren't frozen solid? Fish are hyposmotic to seawater, and normally their blood should freeze at −1.0°C, but DeVries and Wohlschlag (1969) discovered that the blood of these notothenioid Antarctic fish (Fig. 33-8, right) contained antifreeze glycoproteins that adhere to ice crystals and prevent their growth in solution. Since their initial work, four classes of antifreeze proteins (AFP I, II, III, and IV) as well as the original antifreeze glycoprotein have been identified in a wide variety of fish from northern and southern climates around the globe.

AFPs show a high affinity for ice. As an aqueous solution cools below the equilibrium freezing point, ice crystals begin to form. AFPs bind to the ice and limit the growth of crystals in the areas between the AFP molecules. The ice crystals then curve, making them thermodynamically unfavorable for additional growth (Davies *et al.*, 2002). Depending on the concentration of the AFPs in the blood, they can depress the freezing point about from 1 to 6°C (DeVries, 1983). Therefore, fish with a sufficient concentration of

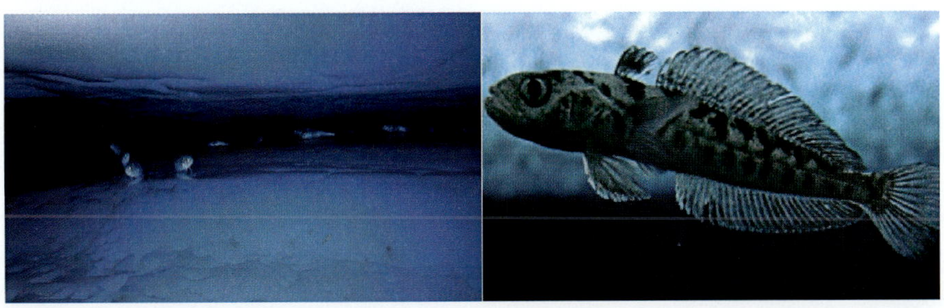

FIGURE 33-8. Antarctic fish living in super-cooled seawater between ice *(left)*. Closeup image *(right)* of the notothenioid Antarctic fish from which Devries and colleagues first discovered antifreeze glycoproteins.

AFPs in their blood and tissue can easily survive at −1.9°C. Jin and DeVries (2006) found that notothenioid fish living in deeper (<100 m), thermally more stable water (±0.1°C) have less antifreeze glycoproteins in their blood than those living closer to the surface with greater temperature swings.

Action of how APFs work is fairly well defined at low temperatures, such as maintaining fluids in a supercooled state, inhibiting growth of ice, and protecting membranes (Wang, 2000). Unfortunately, how they work at ultralow temperatures is contradictory. In some experiments, AFPs protected cells that do not ordinarily survive during cryopreservation (Rubinsky *et al.*, 1991, 1992), whereas in other experiments, AFPs damaged the cells (Larese *et al.*, 1996; Wang *et al.*, 1994). Wang (2000) summarized the contradictory nature of AFPs by correlating the affinity of the AFP-ice complexes with the nature of the cell membranes. He stated that in some systems, the APF-ice interact negatively with polar cell membranes, resulting in lethal ice formation, whereas in nonpolar cell membranes there is an inefficient interaction between AFP-ice. In the nonpolar situation, ice is less likely to form internally, thus providing a benefit.

Antifreeze proteins have already had some important practical applications. For example, Crowe and colleagues have used antifreeze glycoproteins to increase the storage life of blood platelets from 5 days at room temperature to 21 days at 4°C (Talbin *et al.*, 1996). Other areas where AFPs may be applicable is in extending the time of cold storage of organs for transplantation, reducing frostbite on fruits and vegetables, and improving the ice crystal stability of ice cream (Feeney and Yeh, 1998). Although AFPs are not the magic elixir to improve every cryopreservation protocol, with judicious use, they may help maintain many important cells lines, especially those of fish.

Dehydration of Cells

As emphasized in the Water Stress, Cellular Osmolytes, and Late Embryogenesis Abundant (LEA) Proteins section and the Metabolic Preconditioning: Principles Governing Biological Stasis section, the premise that mammalian cells under appropriate conditions can be reversibly dried when retention of viability is firmly grounded in examples from nature, where tolerance to extreme desiccation is widespread. It is our view that the better the operative mechanisms are understood for animals whose evolutionary history has provided the capacity for surviving dehydration, the more likely it is that investigators will formulate workable protocols to confer such phenotypes to cells initially lacking that ability.

The ultimate example of desiccation tolerance in animals is the state of anhydrobiosis, where certain nematodes and tardigrades typically survive dehydration down to 2% tissue water (Crowe and Madin, 1974). The residual water content tolerated by the anhydrobiotic brine shrimp embryo (Fig. 33-9) can be even lower when it is dehydrated under extreme

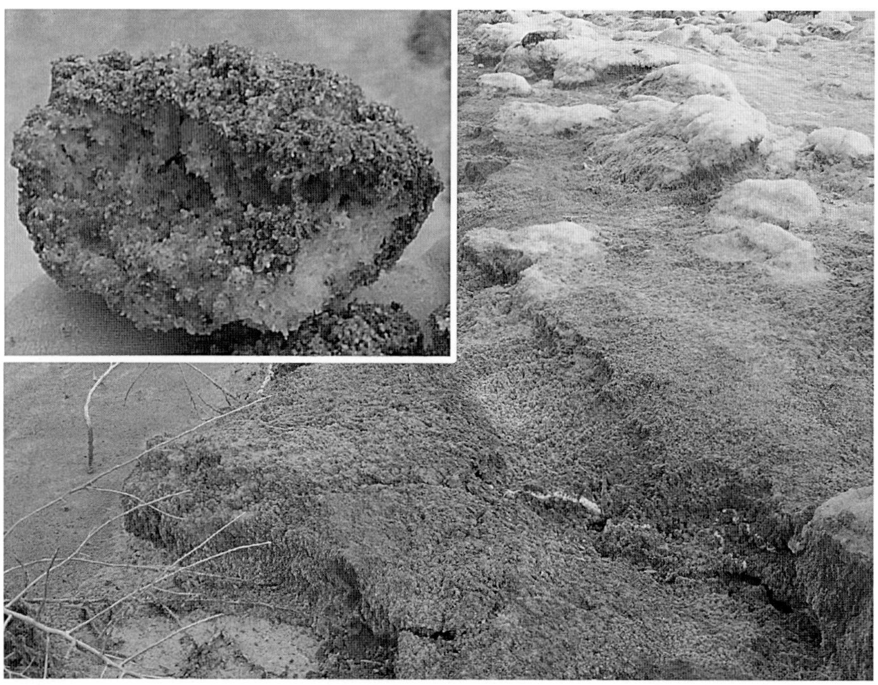

FIGURE 33-9. Salt-encrusted bank of anhydrobiotic embryos of the brine shrimp *Artemia franciscana*, located on the shore of the north arm of the Great Salt Lake (Utah, United States). *(Inset)* A clump of embryos encased in salt crystals.

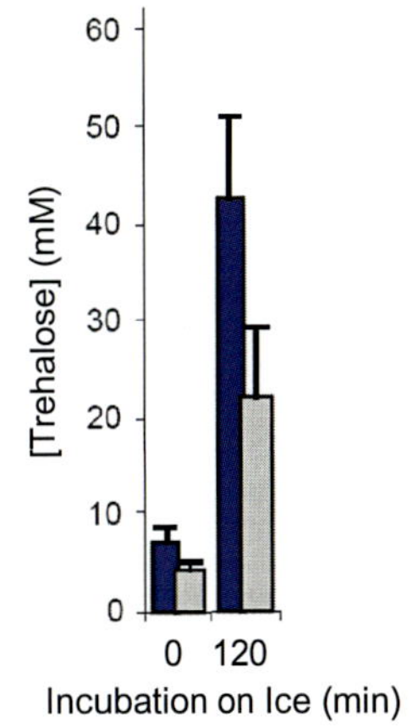

FIGURE 33-10. Estimated intracellular concentrations of trehalose loaded into J774 mouse macrophages by opening the endogenous $P2X_7$ receptor channel with exogenously added ATP^{4-}. Cells were first placed at 37°C for several minutes to open the channel with addition of 5 mM ATP (dark bars), and then cells were held on ice for the indicated times to load trehalose. An intracellular-like buffer containing 225 mM trehalose was used for incubation. Lighter bars are controls cells without ATP addition. Control cells show significant trehalose uptake because of the high rate of fluid phase endocytosis. Modified from Elliott *et al.* (2006)

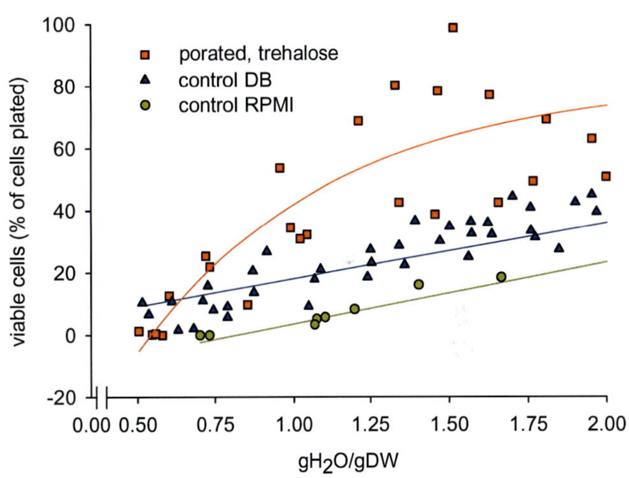

FIGURE 33-11. Viability of J774 mouse macrophages after drying and rehydration. Cells were air-dried to a range of final water contents in an intracellular-like drying buffer containing 250 mM trehalose. Cells that had been loaded with trehalose via the $P2X_7$ receptor channel before drying showed enhanced viability upon rehydration, compared to untreated control cells in drying buffer (DB) or in HPMI. Modified from Elliott *et al.* (2006).

conditions in the laboratory (Clegg *et al.*, 1978). Disaccharides are commonly accumulated to high levels intracellularly in animals possessing natural desiccation tolerance (Clegg, 1965, 2005; Crowe *et al.*, 1987, 2005; Crowe and Crowe, 2000; Hand, 1991). One of these sugars, trehalose, has received an enormous degree of attention because of its noteworthy chemical and physical features (reviewed by Crowe *et al.*, 2005). Earlier in the chapter, we mentioned that trehalose serves as an intracellular osmolyte in a number of species during water stress. It is appropriate to note that high trehalose levels are common in anhydrobiotic animals, yet some such species contain none or very little (e.g., Lapinski and Tunnacliffe, 2003; Womersley, 1990). The point here is that although trehalose is an excellent stabilizer of macromolecules, it is not magical. Other sugars also form glasses when dried, although most do not exhibit the high glass transition temperature (Tg) that trehalose possesses. Low mobility associated with glass formation is considered important for macromolecular stability, because for dry materials, the glassy state prevents surface interactions that can be detrimental (Crowe *et al.*, 2005).

Like many sugars, trehalose is not permeable to biological membranes. This is a practical problem, because much evidence clearly shows that trehalose is most effective as a stabilizer during drying when it is present on both sides of a membrane (Chen *et al.*, 2001; Leslie, 1995; Sun *et al.*, 1996). As water is removed, hydroxyl groups of trehalose form hydrogen bonds with phospholipid head groups of the lipid bilayer and prevent lipid phase transitions during rehydration. Similarly during the drying of proteins, trehalose may replace water molecules normally important in the hydrated state for the stability of proteins. This proposed action of trehalose during water removal has been referred to as the "water replacement hypothesis" (Crowe *et al.*, 1987; Webb, 1965). Thus, in the dried state, direct contact of trehalose is required with internal membrane surfaces and other macromolecules for optimal preservation. With J774 macrophages, we used the $P2X_7$ receptor channel to successfully load 40 to 50 mM trehalose into these cells with a two-step loading procedure, where the majority of loading occurred at 0°C for improved cell survival (Fig. 33-10). Note that as a result of the high rate of fluid phase endocytosis in macrophages, there is substantial uptake of sugar even when the pore is unopened. Cells were given an 18-hour recovery period after trehalose loading, and then they were subjected to air-drying to a range of water contents. The presence of trehalose improved cell viability (Fig. 33-11) when assessed after rehydration (Elliott *et al.*, 2006). Human blood platelets have been successfully freeze-dried with trehalose (Wolkers *et al.*, 2001), but stabilization of nucleated mammalian cells in the fully desiccated state has not been achieved using trehalose.

The approach of combining trehalose with protective macromolecules like stress proteins appears to hold considerable promise for improved desiccation tolerance. In the presence of trehalose loaded by fluid phase endocytosis, 293 cells that were genetically engineered to express a small α-crystallin stress protein (p26) showed marked improvement in survival during moderate dehydration (Ma *et al.*, 2005). The p26 protein was not effective alone, but it appeared to interact synergistically with trehalose in conferring tolerance. This protein was discovered in desiccation-tolerant stages of *A. franciscana* (Clegg *et al.*, 1994). The mechanisms of action of p26 during desiccation are not understood at present. We are currently evaluating the degree of desiccation resistance that LEA proteins may confer to mammalian cells when combined with stabilizing sugars. We have transiently transfected J774 macrophages with one of the LEA proteins recently discovered in *A. franciscana* (AfrLea1, Hand *et al.*, 2007). Preliminary data suggest that expression of AfrLea1 indeed increases desiccation tolerance in the presence of trehalose (M. Menze and S. Hand, unpublished). Simultaneously expressing in mammalian cells both LEA proteins from *A. franciscana* (AfrLea1, AfrLea2) may be beneficial, particularly if these proteins are naturally targeted to different regions/compartments of the cell.

The issue of subcellular compartments in nucleated cells is one that has received virtually no attention in the development of strategies for biostabilization. It may be insufficient to only deliver protective solutes across the plasma membrane for dispersal in the cytoplasmic compartment. A major reason for the lack of success in stabilizing eukaryotic cells against dehydration damage could be that organelles like mitochondria are not being protected. The inner membrane of the mitochondrion is impermeable to many metabolites and solutes. Trehalose loaded into the cytoplasm, for example, is unlikely to penetrate into the mitochondrial matrix. Species that are naturally dehydration tolerant seem to have mechanisms in place to provision the mitochondrial matrix with trehalose. Our preliminary data (G. Elliot, M. Menze, J. Reynolds, M. Toner, and S. Hand, unpublished) indicate that trehalose is present in the mitochondrial matrix of embryos of *A. franciscana*, and, conversely, very low levels are found in larval stages that are not resistant to drying.

As a proof of principle that trehalose in the mitochondrial matrix may be necessary for desiccation tolerance, we used isolated rat liver mitochondria to test the possibility that loading trehalose into the matrix would improve membrane stability during drying (Liu *et al.*, 2005). Trehalose was delivered to the matrix by transiently permeabilizing the inner membrane by opening the mitochondrial permeability transition pore (MPTP). Measurement of trehalose inside mitochondria using HPLC showed that the sugar permeated rapidly into the matrix upon opening the MPTP and reached approximately 190 mM in 5 minutes. Diffusive drying of mitochondria revealed that membrane integrity of the trehalose-loaded organelles was higher than for mitochondria without trehalose. Integrity was quantified as the ability of mitochondria to reestablish a membrane potential after rehydration. The capacity for oxidative phosphorylation, however, was compromised after drying and rehydration (i.e., the respiratory control ratio was substantially lower) (Liu *et al.*, 2005). Because a marked capacity to generate a membrane potential after drying/rehydration was retained in the presence of trehalose, it is likely that components of the electron transport chain remained functional and that proton leak across the inner membrane was not excessive. Rather, it was state 3 respiration that appeared compromised, which suggests the adenine nucleotide transporter or the ATP synthase may be particularly sensitive to the drying process (Liu *et al.*, 2005).

Future strategies need to more prominently address the likelihood that damage to the mitochondrion is a focal point for cell death during drying and that innovative ways to stabilize this organelle *in vivo* are required. Finally, lessons from nature indicate that adjustments made to the intracellular milieu designed to improve desiccation tolerance should be combined with strategies for metabolic preconditioning.

SUMMARY

Physiological and biochemical features important for resisting and tolerating water loss are numerous in marine, estuarine, and freshwater species. It is becoming clear that an ensemble of micromolecules (both compatible and protective solutes) and macromolecules (stress proteins, LEA proteins, antifreeze proteins) are important for establishing the physical conditions required for cellular stabilization during drying in nature. By extending the concept that metabolic depression in animals is a prerequisite for surviving many environmental stresses, the placement of mammalian cells into stasis, before preservation procedures like dehydration, lyophilization, or cryopreservation, may foster greater cell survival. Many of these concepts are currently being applied to cell stabilization in the biomedical field.

Aquaporins are being incorporated into a number of cells in attempts to improve the permeability and freezing characteristics of single cells, but they have more complex characteristics when expressed in actively dividing multicellular organisms, like fish embryos. Another approach co-opts naturally occurring membrane channels like the $P2X_7$ receptor for the transfer of protective solutes into cells. Human blood platelets have been successfully freeze-dried with trehalose, but stabilization of nucleated mammalian cells in the fully desiccated state has not been achieved using sugar alone. The approach of combining trehalose with protective proteins appears to hold considerable promise for improving the desiccation tolerance of cells. Future strategies need to more prominently address the likelihood that damage to the mitochondrion is a focal point for cell death during drying. Innovative ways to stabilize this organelle *in vivo* are required.

We are only beginning to understand how some of these important biological processes work *in situ*, and much more research is needed to understand how to apply these processes in nonnative cells. Regardless, we have standard freezing and drying practices being used in a number of genome resource banks around the world to help save biodiversity. Unfortunately, there are inadequate financial resources or trained individuals to staff these important resource centers in order to meet the demands of maintaining our important disease models for the future. There are no genome resource banks that focus solely on the needs of the biodiversity of our oceans, and this is an oversight that must be redressed soon. Finally, lessons learned from organisms that are naturally desiccation tolerant are improving our biostabilization procedures. Successes with the drying of human blood platelets are encouraging, and efforts are continuing to desiccate nucleated cells that retain high viability for storage at ambient temperature.

References

Agre, P., Preston, G.M., Smith, B.L., Jung, J.S., Raina, S., Moon, C., Guggino, W.B., Nielsen, S., 1993. Aquaporin CHIP: The archetypal molecular water channel. Am. J. Physiol. 265, F463–F476.

Anchordoguy, T.J., Hand, S.C., 1994. Acute blockage of the ubiquitin-mediated proteolytic pathway during invertebrate quiescence. Am. J. Physiol. 267, R895–R900.

Aunaas, T., 1982. Nucleating agents in the hemolymph of intertidal invertebrates tolerant to freezing. Experientia 38, 1456–1457.

Ballou, J.D., 1992. Potential contribution of cryopreserved germ plasm to the preservation of genetic diversity and conservation of endangered species in captivity. Cryobiology 29, 19–25.

Bauer, R.T., 1986. Phylogenetic trends in sperm transfer and storage complexity in decapod crustaceans. J. Crustacean Biol. 6, 313–325.

Beattie, H., Crowe, J.H., Lopez, A.D., Cirulli, V., Ricordi, C., Hayek, A., 1997. Trehalose: A cryoprotectant that enhances recover and preserves function of human pancreatic islets after long-term storage. Diabetes 46, 519–523.

Bellwood, D.R., Hughes, T.P., Folke, C., Nystrom, M., 2004. Confronting the coral reef crisis. Nature 429, 827–833.

Brown, A.D., Simpson, J.R., 1972. Water relations of sugar-tolerant yeasts: The role of intracellular polyols. J. Gen. Microbiol. 72, 589–591.

Bruni, F., Leopold, A.C., 1990. Glass transitions in soybean seed. Plant Physiol. 96, 660–663.

Bruni, F., Leopold, A.C., 1992. Pools of water in anhydrobiotic organisms: A thermally stimulated depolarization current study. Biophys. J. 63, 663–672.

Buchanan, S.S., Menze, M.A., Hand, S.C., Pyatt, D.W., Carpenter, J.F., 2005. Cryopreservation of human hematopoietic stem and progenitor cells loaded with trehalose: Transient permeabilization via the adenosine triphosphate-dependent P2Z receptor channel. Cell Preserv. Technol. 3(4), 212–222.

Buitink, J., Claessens, M.A.E., Hemminga, M.A., Hoekstra, F.A., 1998a. Influence of water content and temperature on molecular mobility and intracellular glasses in seeds and pollen. Plant Physiol. 118, 531–541.

Buitink, J., Walters, C.F.A., Hoekstra, F.A., Crane, J., 1998b. Storage behavior of *Typha latifolia* pollen at low water contents: Interpretation on the basis of water activity and glass concepts. Physiol. Plant. 103, 145–153.

Busa, W.B., Crowe, J.H., 1983. Intracellular pH regulates transitions between dormancy and development of brine shrimp (*Artemia salina*) embryos. Science 221, 366–368.

Carpenter, J., Hand, S.C., Crowe, L., Crowe, J., 1986. Cryoprotection of phosphofructokinase with organic solutes: Characterization of enhanced protection in the presence of divalent ions. Arch. Biochem. Biophys. 250, 505–512.

Carpenter, J.F., Martin, B., Crowe, L.M., Crowe, J.H., 1987. Stabilization of phosphofructokinase during air-drying with sugars and sugar/transition metal mixtures. Cryobiology 24, 455–464.

Chen, T., Acker, J.P., Eroglu, A., Cheley, S., Bayley, H., Fowler, A., Toner, M., 2001. Beneficial effect of intracellular trehalose on the membrane integrity of dried mammalian cells. Cryobiology 43, 168–181.

Chrispeels, M.J., Agre, P., 1994. Aquaporins: Water channel proteins of plant and animal cells. Trends Biochem. Sci. 19, 421–425.

Clarke, M.E., Zounes, M., 1977. The effects of selected osmolytes on the activity of lactate dehydrogenase from the euryhaline polychaete, *Nereis succinea*. Biol. Bull. 153, 468–484.

Clegg, J.S., 1965. The origin of trehalose and its significance during the formation of encysted embryos of *Artemia salina*. Comp. Biochem. Physiol. 14, 135–143.

Clegg, J.S., 1997. Embryos of *Artemia franciscana* survive four years of continuous anoxia: The case for complete metabolic rate depression. J. Exp. Biol. 200, 467–475.

Clegg, J.S., 2001. Cryptobiosis: A peculiar state of biological organization. Comp. Biochem. Physiol. B 128, 613–624.

Clegg, J.S., 2005. Desiccation tolerance in encysted embryos of the animal extremophile, Artemia. Integr. Comp. Biol. 45(5), 715–724.

Clegg. J.S., Drinkwater, L.E., Sorgeloos, P., 1996. The metabolic status of diapause embryos of *Artemia franciscana* (SFB). Physiol. Zool. 69, 49–66.

Clegg, J.S., Jackson, S.A., Warner, A.H., 1994. Extensive intracellular translocations of a major protein accompany anoxia in embryos of *Artemia franciscana*. Exp. Cell Res. 212, 77–83.

Clegg, J.S., Zettlemoyer, A.C., Hsing, H.H., 1978. On the residual water content of dried but viable cells. Experientia 34, 734–735.

Close, T.J., Kortt, A.A., Chandler, P.M., 1989. A cDNA-based comparison of dehydration-induced proteins (dehydrins) in barley and corn. Plant Mol. Biol. 13, 95–108.

Close, T.J., Lammers, P.L., 1993. An osmotic stress protein of cyanobacteria is immunologically related to plant dehydrins. Plant Physiol. 101, 773–779.

Cockcroft, S., Gomperts, B., 1979. ATP induced nucleotide permeability in rat mast cells. Nature 279, 541–542.

Coutinho-Silva, R., Persechini, P.M., Bisaggio, R.D., Perfettini, J.L., Neto, A.C., Kanellopoulos, J.M., Motta-Ly, I., Crowe, J.H., Carpenter, J.F., Crowe, L.M., 1998. The role of vitrification in anhydrobiosis. Annu. Rev. Physiol. 60, 73–103.

Crowe, J.H., Clegg, J.S., (eds.), 1978. Dry biological systems. New York, London, Academic Press.

Crowe, J.H., Crowe, L.M., 2000. Preservation of mammalian cells: Learning nature's tricks. Nat. Biotechnol. 18, 145–146.

Crowe, J.H., Crowe, L.M., Carpenter, J.F., Aurell Wistrom, C., 1987. Stabilization of dry phospholipid bilayers and proteins by sugars. Biochem. J. 242, 1–10.

Crowe, J.H., Crowe, L.M., Wolkers, W.F., Oliver, A.E., Ma, X., Auh, J.-H., Tang, M., Zhu, S., Norris, J., Tablin, F., 2005. Stabilization of dry mammalian cells: Lessons from nature. Integr. Comp. Biol. 45(5), 810–820.

Crowe, J.H., Madin K., 1974. Anhydrobiosis in tardigrades and nematodes. Trans. Am. Microsc. Soc. 95, 513–523.

Crowe, J.H., Carpenter, J.F., Crow, L.M., 1998. The role of vitrification in anhydrobiosis. Annu. Rev. Physiol. 6, 73–103.

Crowe, J.H., Madin, K., Loomis, S.H., 1977. Anhydrosis in nematodes: Metabolism during resumption of activity. J. Exp. Zool. 201, 57–63.

Davies, P.L., Baardsnes, J., Kuiper, M.J., Walker, V.K., 2002. Structure and function of antifreeze proteins. Phil. Trans. R. Soc. Lond. B 357, 927–935.

Denlinger, D.L., 2002. Regulation of diapause. Annu. Rev. Entomol. 47, 93–122.

DeVries, A.L., 1983. Antifreeze peptides and glycopeptides in cold-water fishes. Annu. Rev. Physiol. 45, 245–260.

DeVries, A.L., Wohlschlag, D.E., 1969. Freezing resistance in some Antarctic fishes. Science 163, 1073–1075.

Drinkwater, L.E., Crowe, J.H., 1987. Regulation of embryonic diapause in *Artemia*: Environmental and physiological signals. J. Exp. Zool. 241, 297–307.

Dubyak, G., El-Moatassim, C., 1993. Signal transduction via P2-purinergic receptors for extracellular ATP and other nucleotides, Am. J. Physiol. 265, C577–C606.

Edashige, K., Sakamoto, M., Kasai, M., 2000. Expression of mRNAs of the aquaporin family in mouse oocytes and embryos. Cryobiology 40, 171–175.

Elliott, G., Liu, X.-H., Cusick, J.L., Menze, M., Vincent, J., Witt, T., Hand, S., Toner, M., 2006. Trehalose uptake through $P2X_7$ purinergic channels provides dehydration protection. Cryobiology 52, 114–127.

Eroglu, A., Russo, M.J., Bieganski, R., Fowler, A., Cheley, S., Bayley, H., Toner, M., 2000. Intracellular trehalose improves the survival of cryopreserved mammalian cells. Nat. Biotechnol. 18, 163–167.

Eroglu, A., Toner, M., Toth, T.L., 2002. Beneficial effect of microinjected trehalose on the cryosurvival of human oocytes. Fertil. Steril. 77, 152–158.

Faurot, D., Kucera P., Armstrong, R., Blenden, M., 1997. Cryopreservation of adult male spring and summer Chinook salmon gametes in the Snake River Basin Project No. 1997–03800, (BPA Report DOE/BP-30423–2).

Feeney, R.F., Yeh, Y., 1998. Antifreeze proteins: Current status and possible food uses: Trends in Food Sci. Technol. 9, 1–5.

Garrison, V.B., Shinn, E.A., Foreman, W.T., Griffin, D.W., Holmes, C.W., 2003. African and Asian dust: From desert soils to coral reefs. BioScience 53, s469–s480.

Geiser, D.W., Taylor, J.W., Ritchie, K.B., Smith, G.W., 1998. Cause of sea fan death in the West Indies. Nature 394, 137–138.

Goyal, K., Tisi, L., Basran, A., Browne, J., Burnell, A., Zurdo, J., Tunnacliffe, A., 2003. Transition from natively unfolded to folded state induced by desiccation in an anhydrobiotic nematode protein. J. Biol. Chem. 278, 12977–12984.

Grelet, J., Benamar, A., Teyssier, E., Avelange-Macherel, M.-H., Grunwald, D., Macherel, D., 2005. Identification in pea seed mitochondria of a late-embryogenesis abundant protein able to protect enzymes from drying. Plant Physiol. 137, 157–167.

Griffin, D.W., Kellogg, C.A., Garrison, V.H., Shinn, E.A., 2002. The global transport of dust. American Scientist 90, 228–235.

Grimm, D., 2004. Visualization challenge: Illustration. Science 5692, 1905.

Guo, N., Puhlev, I., Brown, D.R., Mansbridge, J., Levine, F., 2000. Trehalose expression confers desiccation tolerance on human cells. Nat. Biotechnol. 18, 168–171.

Guppy, M., Withers P., 1999. Metabolic depression in animals: Physiological perspectives and biochemical generalizations. Biol. Rev. Camb. Philos. Soc. 74, 1–40.

Hagedorn, M., Hsu, E.W., Pilatus, U., Wildt, D.E., Rall, W.F., Blackband, S.J., 1996. Magnetic resonance microscopy and spectroscopy reveal kinetics of cryoprotectant permeation in a multicompartmental biological system. Proc. Nat. Acad. Sci. USA 93, 7454–7459.

Hagedorn, M., Kleinhans, F.W., Artemov, D., Pilatus, U., 1998. Characterization of a major permeability barrier in the zebrafish embryo. Biol. Reprod. 59, 1240–1250.

Hagedorn, M., Kleinhans, F.W., Freitas, R., Liu, J., Hsu, E., Wildt, D.E., Rall, W.F., 1997. Water distribution and permeability of zebrafish embryos, Brachydanio rerio. J. Exp. Zool. 278, 356–371.

Hagedorn, M., Lance, S.L., Fonseca, D.M., Kleinhans, F.W., Artemov, D., Fleischer, R., Hoque, A.T.M.S., Pukanzhenthi, B.S., 2002. Altering the membranes of fish embryos with aquaporin-3: An essential step for cryopreservation. Biol. Reprod. 67, 961–966.

Hammerstedt, R.H., Graham, J.K., Nolan, J.P., 1990. Cryopreservation of mammalian sperm: What we ask them to survive. J. Andrology 11, 73–88.

Hand, S.C., 1991. Metabolic dormancy in aquatic invertebrates. Adv. Comp. Environ. Physiol. 8, 1–50.

Hand, S.C., 1995. Heat flow is measurable from *Artemia franciscana* embryos under anoxia. J. Exp. Zool. 273, 445–449.

Hand, S.C., 1999. Calorimetric approaches to animal physiology and bioenergetics. In Kemp, R.B. (ed.), Handbook of Thermal Analysis and Calorimetry, From Molecules to Man, pp. 469–510. Amsterdam, Elsevier.

Hand, S.C., 1998. Quiescence in *Artemia franciscana* embryos: Reversible arrest of metabolism and gene expression at low oxygen levels. J. Exp. Biol. 201, 1233–1242.

Hand, S.C., Hardewig, I., 1996. Downregulation of cellular metabolism during environmental stress: Mechanisms and implications. Annu. Rev. Physiol. 58, 539–563.

Hand, S.C., Jones, D., Menze, M.A., Witt, T.L., 2007. Life without water: Expression of plant LEA genes by an anhydrobiotic arthropod. J. Exp. Zool. 305A, 62–66.

Hand, S.C., Menze, M.A., 2007. Desiccation stress. In Denny, M.W., and Gaines, S.D. (eds.), Encyclopedia of Tidepools and Rocky Shores, pp. 173–177. Berkeley, University of California Press.

Hand, S.C., Podrabsky, J.E., Eads, B.D., van Breukelen, F., 2001. Interrupted development in aquatic organisms: Ecological context and physiological mechanisms. In Atkinson, D., and Thorndyke, M. (eds.), Environment and Animal Development: Genes, Life Histories and Plasticity, pp. 219–234. Oxford, BIOS Scientific.

Hardie, D.G., Carling, D., Carlson, M., 1998. The AMP-activated/SNF1 protein kinase subfamily: Metabolic sensors of the eukaryotic cell? Annu. Rev. Biochem. 67, 821–855.

Hardie, D.G., Hawley, S.A., 2001. AMP-activated protein kinase: The energy charge hypothesis revisited. Bioessays 23, 1112–1119.

Harvey, B., Kelley, R.N., Ashwood-Smith, M.J., 1982. Cryopreservation of zebrafish spermatozoa using methanol. Can. J. Zoology 60, 1867–1870.

Heymann, J.B., Agre, P., Engel, A., 1998. Progress on the structure and function of aquaporin 1. J. Struct. Biol. 121, 191–206.

Hochachka, P.W., Guppy, M., 1987. Metabolic arrest and the control of biological time. Cambridge, MA, Harvard University Press.

Hoekstra, F.A., 2005. Differential longevities in desiccated anhydrobiotic plant systems. Integr. Comp. Biol. 45, 725–733.

Hoekstra, F.A., Wolkers, W.F., Buitink, J., Golovina, E.A., Crowe, J.H., Crowe, L.M., 1997. Membrane stabilization in the dry state. Comp. Biochem. Physiol. A 117(3), 335–341.

Hofmann, G.E., Hand, S.C., 1990. Arrest of cytochrome-*c* oxidase synthesis coordinated with catabolic arrest in dormant Artemia embryos. Am. J. Physiol. 258, R1184–R1191.

Hubel, A., Darr, T.B., 2002. Passive loading of trehalose into cells. Cryobiology 45(3), 227.

Humphreys, B.D., Rice, J., Kertesy, S.B., Dubyak, G.R., 2000. Stress-activated protein kinase/JNK activation and apoptotic induction by the macrophage $P2X_7$ nucleotide receptor. J. Biol. Chem. 275, 26792–26798.

Janik, M., Kleinhans, F.W., Hagedom, M., 2000. Overcoming a permeability barrier by microinjecting cryoprotectants into zebrafish embryos *(Brachydanio rerio)*. Cryobiol. 41, 25–34.

Ji, Y., Tanaka, J., 2002. Effect of desiccation on the photosynthesis of seaweeds from the intertidal zone in Honshu, Japan. Physiol. Res. 50, 145–153.

Jin, Y., DeVries, A.L., 2006. Antifreeze glycoprotein levels in Antarctic notothenioid fishes inhabiting different thermal environments and the effect of warm acclimation. Comp. Biochem. Physiol. B 144, s290–s300.

Kikawada, T., Nakahara, Y., Kanamori, Y., Iwata, K.-I., Watanabe, M., Brian McGee, B., Alan Tunnacliffe, A., Okuda, T., 2006. Dehydration-induced expression of LEA proteins in an anhydrobiotic chironomid. Biochem. Biophys. Res. Comm. 348, 56–61.

King, L.S., Agre, P., 1996. Pathophysiology of the aquaporin water channels. Annu. Rev. Physiol. 58, 619–648.

Kozono, D., Yasui, M., King, L.S., Agre, P., 2002. Aquaporin water channels: Atomic structure molecular dynamics meet clinical medicine. J. Clin. Invest. 109, 1395–1399.

Lapinski, J., Tunnacliffe, A., 2003. Anhydrobiosis without trehalose in bdelloid rotifers. FEBS Lett. 553: 387–390.

Larese, A., Acker, J., Muldrew, K., Yang, H. Y., McGann, L., 1996. Antifreeze proteins induce intracellular nucleation. Cryo-Letters 17, 175–182.

Lavens, P., Sorgeloos, P., 1987. The cryptobiotic state of *Artemia* cysts, its diapause deactivation and hatching: A review. In Sorgeloos, P., Browne, D.A., Decleir, W., and Jaspers, E. (eds.), Artemia Research and Its Applications, vol. 3. Ecology, Culturing, Use in Aquaculture, pp. 27–63. Wetteren, Belgium, Universa Press.

Lees, A.D., 1955. Physiology of diapause in arthropods. Cambridge, Cambridge University Press.

Lee, M.D., King, L.S., Agre, P., 1997. The aquaporin family of water channel proteins in clinical medicine. Medicine 76, 141–156.

Leslie, S.B., Israeli, E., Lighthart, B., Crowe, J.H., Crowe, L.M., 1995. Trehalose and sucrose protect both membranes and proteins in intact bacteria during drying. Appl. Environ. Microbiol. 61, 3592–3597.

Leung, L.K.P., Jamieson, B.G.M., 1992. Live preservation of fish gametes. In Jamieson, B.G.M. (ed.), Fish Evolution and Systematics: Evidence from Spermatozoa, pp. 245–269. Cambridge, Cambridge University Press.

Levin, R.L., Miller, T.W., 1981. An optimum method for the introduction and removal of permeable cryoprotectants. Cryobiology 18, 32–48.

Li, R., Brawley, S.H., Close, T.J., 1998. Proteins immunologically related to dehydrins in fucoid algae. J. Phycol. 34, 642–650.

Liu, X.-H., Aksan, A., Menze, M.A., Hand, S.C., Toner, M., 2005. Trehalose loading through the mitochondrial permeability transition pore enhances desiccation tolerance in rat liver mitochondria. Biochim. Biophys. Acta. 1717, 21–26.

Loomis, S.H., 1985. Seasonal changes in the freezing tolerance of the intertidal pulmonate gastropod, *Melampus bidentatus*. Can. J. Zool. 63, 2021–2025.

Loomis, S.H., Zinser, M., 2001. Isolation and identification of an ice-nucleating bacterium from the gills of the intertidal bivalve mollusc *Geukensia demissa*. J. Exp. Mar. Biol. Ecol. 261, 225–235.

Ma, X., Jamil, K., MacRae, T.H., Clegg, J.S., Russell, J.M., Villaneuve, T.S., Euloch, M., Sun, Y., Crowe, J.H., Tablin, F., Oliver, A.E., 2005. A small stress protein acts synergistically with trehalose to confer desiccation tolerance on mammalian cells. Cryobiology 51(1), 15–28.

Macey, R.I., 1984. Transport of water and urea in red blood cells. Am. J. Physiol. 246, C195–C203.

Macey, R.I., Farmer, R.E., 1970. Inhibition of water and solute permeability in human red cells. Biochim. Biophys. Acta. 1211, 104–106.

Maragos, J.E., Potts, D.C., Aeby, G.A., Gulko, D., Kenyon, J., Siciliano, D., VanRavenswaay, D., 2004. 2000–2002 Rapid ecological assessment of corals (Anthozoa) on shallow reefs of the northwestern Hawaiian Islands. Part 1: Species and distribution. Pacific Sci. 58, 211–230.

Mazur, P., 1970. Cryobiology: The freezing of biological systems. Science 168, 939–949.

Mazur, P., 1977. The role of intracellular freezing in the death of cells cooled at supraoptimal rates. Cryobiology 14, 251–272.

Mazur, P., 1984. Freezing of living cells: Mechanisms and implications. Am. J. Physiol. 247, C125–C142.

McMahon, R.F., 1990. Thermal tolerance, evaporative water loss, air-water oxygen consumption and zonation of intertidal prosobranchs: A new synthesis. Hydrobiologia 193, 241–260.

Menze, M.A., Clavenna, M., Hand, S.C., 2005a. Depression of cell metabolism and proliferation by membrane permeable and impermeable modulators: Role for AMP : ATP ratio. Am. J. Physiol. 288(2), R501–R510.

Menze, M.A., Hand, S.C., 2007. Caspase activity during cell stasis: Avoidance of apoptosis in an invertebrate extremophile, *Artemia franciscana*. Am. J. Physiol. (in press).

Menze, M.A., Hutchinson, K., Laborde, S.M., Hand, S.C., 2005b. Mitochondrial permeability transition in the crustacean *Artemia franciscana*: Absence of a Ca^{2+}-regulated pore in the face of profound calcium storage. Am. J. Physiol. 289: R68–R76.

Muñoz, M., Casadevall, M., Bonet, S., 1999. Annual reproductive cycle of *Helicolenus dactylopterus dactylopterus* (Teleostei: Scorpaeniformes) with special reference to the ovaries sperm storage. J. Mar. Biol. Assoc. UK 79, 521–529.

Muñoz, M., Casadevall, M., Bonet, S., Quagio-Grassiotto, I., 2000. Sperm storage structures in the ovary of *Helicolenus dactylopterus dactylopterus* (teleostei: scorpaenidae): an ultrastructural study. J. Environ. Biol. Fishes 58, 53–59.

Muñoz, M., Koya, Y., Casadevall, M., 2002. Histochemical analysis of sperm storage in *Helicolenus dactylopterus dactylopterus* (Teleostei: Scorpaenidae). J. Exp. Zool. 292, 156–164.

North, R.A., 2002. Molecular physiology of P2X receptors. Physiol. Rev. 82, 1013–1067.

Oliver, A.E., Jamil, K., Crowe, J.H., Tablin, F., 2004. Loading human mesenchymal stem cells with trehalose by fluid phase endocytosis. Cell Preserv. Technol. 2(1), 35–50.

Podrabsky, J.E., Hand, S.C., 1999. The bioenergetics of embryonic diapause in an annual killifish, *Austrofundulus limnaeus*. J. Exp. Biol. 202, 2567–2580.

Pratt, H.L., 1993. The storage of spermatozoa in the oviducal glands of western North Atlantic sharks. J. Environ. Biol. Fishes 38, 139–149.

Preston, G.M., Carroll, T.P., Guggino, W.B., Agre, P., 1992. Appearance of water channels in *Xenopus* oocytes expressing red cell CHIP28 protein. Science 256, 385–387.

Rall, W.F., 1993. Advances in the cryopreservation of embryos and prospects for the application to the conservation of salmonid fishes. In Cloud, J.G., Thorgaard, G.H. (eds.), Genetic Conservation of Salmonid Fishes, pp. 137–158. New York, Plenum.

Rana, K.J., Adam, M., McAndrew, B.J., 1995. Osmotic response of inseminated fish embryos exposed to various sucrose concentrations. Cryo-Letters 16, 66.

Rassendren F., Buell, G.N., Virginio, C., Collo, G., North, R.A., Surprenant, A., 1997. The permeabilizing ATP receptor, P2X7: Cloning and expression of a human cDNA. J. Biol. Chem. 272, 5482–5486.

Rees, B.B., Hand, S.C., 1990. Heat dissipation, gas exchange and acid-base status in the land snail *Oreohelix strigosa* during short-term estivation. J. Exp. Biol.152, 77–92.

Reynolds, J.A., Hand, S.C., 2004. Differences in isolated mitochondria are insufficient to account for respiratory depression during diapause in *Artemia franciscana* embryos. Physiol. Biochem. Zool. 77, 366–377.

Rubinsky, B., Arav, A., DeVries, A. L., 1991. Cryopreservation of oocytes using directional cooling and antifreeze glycoproteins. Cryo-Letters 12, 93–106.

Rubinsky, B., Arav, A., DeVries, A. L., 1992. The cryoprotective effect of antifreeze glycopeptides from Antarctic fishes. Cryobiology 29, 69–79.

Sokolova, I.M., Poertner, H.O., 2001. Physiological adaptations to high intertidal life involve water conservation abilities and metabolic rate depression in *Littorina saxatilis*. Mar. Ecol. Prog. Ser. 224, 171–186.

Solle, M., Labasi, J., Perregaux D.G., Stam., E., Petrushova, N., Koller, B.H., Griffiths, R.J., Gabel, C.A., 2001. Altered cytokine production in mice lacking $P2X_7$ receptors. J. Biol. Chem. 276, 125–132.

Somero, G.N., 1986. Protons, osmolytes, and fitness of internal milieu for protein function. Am. J. Physiol. 251, R197–R213.

Steinberg, T.H., Silverstein, S.C., 1989. ATP permeabilization of the plasma membrane. Methods Cell Biol. 31, 45–61.

Storey, K.B., Storey, J.M., 1990. Metabolic rate depression and biochemical adaptation in anaerobiosis, hibernation and estivation. Q. Rev. Biol. 65, 145–174.

Stoss, J., 1983. Fish gamete preservation and spermatozoan physiology. In Hoar, W.S., Randall, D.J., and Donalson, E.M. (eds.), Fish Physiology, vol. IXB, pp. 305–351. New York, Academic Press.

Sun, W.Q., Leopold, A.C., Crowe, L.M., Crowe, J.H., 1996. Stability of dry liposomes in sugar glasses, Biophys. J. 70, 1769–1776.

Surprenant, A., Rassendren, F., Kawashima, E., North, R.A., Buell, G., 1996. The cytolytic P2Z receptor for extracellular ATP identified as a P2X receptor ($P2X_7$). Science 272, 735–738.

Tablin, F., Oliver, A.E., Walker, N.J., Crowe, L.M., Crowe. J.H., 1996. Membrane phase transition of intact human platelets: Correlation with cold-induced activation. J. Cell. Physiol, 168, 305–312.

Tajkhorshid, E., Nollert, P., Jensen, M.O., Miercke, L.J.W., O'Connell, J., Stroud, R.M., Schulten, K., 2002. Control of the selectivity of the aquaporin water channel family by global orientational tuning. Science 296, 525–530.

Takahashi, H., Takano, K., Takemura, A., 1991. Reproductive cycles of *Sebastes taczanowskii*, compared with those of other rockfishes of the genus Sebastes. Environ. Biol. Fishes 30, 23–30.

Tanghe, A., Van Dijck, P., Colavizza, D., Thevelein, J.M., 2004. Aquaporin-mediated improvement of freeze tolerance of *Saccharomyces cerevisiae* is restricted to rapid freezing conditions. Appl. Environ. Microbiol. 70, 3377–3382.

Tanghe, A., Van Dijck, P., Dumortier, F., Teunissen, A., Hohmann, S., Thevelein, J.M., 2002. Aquaporin expression correlates with freeze tolerance in baker's yeast, and overexpression improves freeze tolerance in industrial strains. Appl. Environ. Microbiol. 68, 5981–5989.

Taylor, R., Adams, G.D.J., Boardman, C.F.B., Wallis, R.G., 1974. Cryoprotective- permeant vs nonpermeant additives. Cryobiology 11, 430–438.

Valdez, D.M., Jr., Hara, T., Miyamoto, A., Seki, S., Jin, B., Kasai, M., Edashige, K., 2006. Expression of aquaporin-3 improves the permeability to water and cryoprotectants of immature oocytes in the medaka (*Oryzias latipes*). Cryobiology 53, 160–168.

Van Breukelen, F., Hand, S.C., 2000. Characterization of ATP-dependent proteolysis in embryos of the brine shrimp, *Artemia franciscana*. J. Comp. Physiol. B 170, 125–133.

Verkman, A.S., Van Hoek, A.N., Ma, T., Frigeri, A., Skach, W.R., Mitra, A., Tamarappoo, B.K., Farinas, J., 1996. Water transport across mammalian cell membranes. Am. J. Physiol. 270, C12–C30.

Virginio, C., MacKenzie, A., North, R.A., Surprenant, A., 1999. Kinetics of cell lysis, dye uptake and permeability changes in cells expressing the rat $P2X_7$ receptor. J. Physiol. 519, 335–346.

Wang, J.H., 2000. A comprehensive evaluation of the effects and mechanisms of antifreeze proteins during low-temperature preservation. Cryobiology 41, 1–9.

Wang, T., Zhu, Q., Yang, X., Layne, J. R., DeVries, A. L., 1994. Antifreeze glycoproteins from Antarctic notothenioid fishes fail to protect the rat cardiac explant during hypothermic and freezing preservation. Cryobiology 31, 185–192.

Warner, A., Jackson, S., Clegg, J., 1997. Effect of anaerobiosis on cysteine protease regulation during the embryonic-larval transition in *Artemia franciscana*. J. Exp. Biol. 200, 897–908.

Webb, S.J., 1965. Bound water in biological integrity. Springfield, IL, Thomas.

Wildt, D.E., 1992. Genetic resource banking for conserving wildlife species: Justification, examples and becoming organized on a global basis. Animal Reprod. Sci. 28, 247–257.

Wildt, D.E., Rall, W.F., Critser, J.K., Monfort, S.L., Seal, U.S., 1997. Genome resource banks Bioscience 47, 689–698.

Wildt, D.E., Seal, U.S., Rall, W.F., 1993. Genetic resource banks and reproductive technology for wildlife conservation. In Cloud, J.G., and Thorgaard, G.H. (eds.), Genetic Conservation of Salmonid Fishes, pp. 159–173. New York, Plenum.

Williams, R.J., Leopold, A.C., 1989. The glassy state in corn embryos. Plant Physiol. 89, 977–981.

Wintour, E.M., 1997. Water channels and urea transporters. Clin. Exp. Pharmacol. Physiol. 24, 1–9.

Wise, M.J., Tunnacliffe, A., 2004. POPP the question: what do LEA proteins do? Trends Plant Sci. 9, 13–17.

Wolkers, W.F., Walker, N.J., Tablin, F., Crowe, J.H., 2001. Human platelets loaded with trehalose survive freeze-drying. Cryobiology 42, 79–87.

Womersley, C., 1990. Dehydration survival and anhydrobiotic potential of entomopathogenic nematodes. In Gaugler, R., and Kaya, H.K. (eds.), Entomopathogenic Nematodes in Biological Control, pp. 117–130. Boca Raton, FL, CRC Press.

Wourms, J.P., 1972. The developmental biology of annual fishes. 3. Pre-embryonic and embryonic diapause of variable duration in the eggs of annual fishes. J. Exp. Zool. 182, 389–414.

Yancey, P.H., 2005. Organic osmolytes as compatible, metabolic and counteracting cytoprotectants in high osmolarity and other stresses. J. Exp. Biol. 208, 2819–2830.

Yancey, P.H., Clark, M.E., Hand, S.C., Bowlus, R., Somero, G.N., 1982. Living with water stress: The evolution of osmolyte systems. Science 217, 1214–1222.

Zardoya, R., Ding, X., Kitagawa, Y., Chrispeels, M.J., 2002. Origin of plant glycerol transporters by horizontal gene transfer and functional recruitment. Proc. Natl. Acad. Sci. USA 99, 14893–14896.

STUDY QUESTIONS

1. What are the roles of compatible osmolytes in improving the tolerance of animals to desiccation stress? How have these solutes been applied to mammalian cell stabilization during drying? Organisms often express late embryogenesis abundant (LEA) proteins and stress proteins during desiccation. What evidence suggests that these proteins are required for cell drying and may act synergistically with protective solutes to confer dehydration tolerance?
2. Discuss the biological phenomena supporting the concept that metabolic depression is a prerequisite for surviving environmental stresses like desiccation, anoxia, and freezing. Why might the metabolic preconditioning of mammalian cells (i.e., conferring a phenotype of cell stasis) improve survivorship in the dried and frozen states?
3. To be maximally effective, solutes used for biostabilization of cells must be present on both sides of biological membranes. Describe the $P2X_7$ receptor channel, and explain how it has been utilized to load cells with protective solutes. What evidence suggests that protecting only the plasma membrane during

drying and freezing may be inadequate and that internal cell membranes also may require stabilization in order for cells to survive drying?

4. From an applied standpoint, why would the successful storage of dried cells at room temperature be of such high utility in biomedicine?
5. Please explain the benefits to conservation and medical research of maintaining genetic banks with frozen gametes, cells, and DNA samples.
6. One major issue with genetic banks is biosecurity, in that this resource may transmit diseases not only from place to place but through time. Provide specific examples that illustrate the problem, and suggest safeguards that might be used to minimize the concern.
7. Why do cryobiologists attempt to avoid intracellular ice formation during the cryopreservation of biological materials?
8. Arctic and Antarctic fishes often live in slushy ice, often even ingesting it. How do they avoid freezing solid?

Index

A

B

C

D

E

F

G

H

I

J

K

L

M

N

O

P

Q

R

T

U

V

W

Edwards Brothers Malloy
Ann Arbor MI. USA
September 6, 2012